T0186664

SAFETY AND RELIABILITY

PROCEEDINGS OF ESREL 2003, EUROPEAN SAFETY AND RELIABILITY
CONFERENCE 2003, 15–18 JUNE 2003, MAASTRICHT, THE NETHERLANDS

Safety and Reliability

Edited by

T. Bedford
Management Science Department, University of Strathclyde, Glasgow, Scotland

P.H.A.J.M. van Gelder
Faculty of Civil Engineering and Geosciences,
Delft University of Technology, Delft, The Netherlands

Volume 1

A.A. BALKEMA PUBLISHERS   LISSE / ABINGDON / EXTON (PA) / TOKYO

NVRB

Published by: A.A. Balkema, a member of Swets & Zeitlinger Publishers
www.balkema.nl and www.szp.swets.nl

For the complete set of two volumes: ISBN 90 5809 551 7
Volume 1: ISBN 90 5809 595 9
Volume 2: ISBN 90 5809 596 7

Cover photograph by: Meetkundige dienst, Rijkswaterstaat, Delft, The Netherlands

Printed in the Netherlands

Safety and Reliability – Bedford & van Gelder (eds)
© 2003 Swets & Zeitlinger, Lisse, ISBN 90 5809 551 7

Table of Contents

VII

IX

X

Volume II

XIV

XV

Safety and Reliability – Bedford & van Gelder (eds)
© 2003 Swets & Zeitlinger, Lisse, ISBN 90 5809 551 7

Preface

Scope of the conference

The breadth of safety and reliability applications has grown dramatically in recent years. In doing so they have led to the more general application of risk based models and risk management in which traditional safety and reliability modelling play an integral role.

Today's complex and sophisticated modelling techniques give us the potential to create models which express risk in a variety of different metrics. The potential for communication to stakeholders is great, but at the cost of a lack of understanding due to increased complexity.

Improved and more widely applicable models can be developed through the integration of existing approaches to risk. By a better understanding of the interaction between technical, project, financial and environmental risks we can hope to improve both communication and understanding, leading ultimately to better decision making.

The conference is of interest to industry, regulators, managers, investors, decision makers, universities, research organisations and government bodies and everyone interested by risk management, assessment, communication and stakeholder understanding.

The application of safety, reliability and risk management assessment techniques occurs across many different domains.

The principle application domains of interest in this meeting are:

- Aviation and Aerospace
- Civil Engineering
- Chemical Engineering and Processes
- Electronics
- Energy Production and Distribution
- Environmental Protection
- Food
- Information and Communication
- Insurance
- Manufacturing
- Maritime Engineering
- Medical and Health Systems
- Offshore Engineering
- Security and Protection
- Transport

Conference Chairs

- Conference Chairman: Paul Waarts (TNO)
- Conference Co-Chairman: Carlos Guedes Soares (IST/ESRA)

Honorary Conference Chairs

- Ben Ale (RIVM)
- Roger Cooke (Delft University of Technology)
- Louis Goossens (Delft University of Technology)

Technical Programme Board

- Tim Bedford (Strathclyde University), Chair
- Ben Ale (RIVM/Delft University of Technology)
- Aarnout Brombacher (Eindhoven University of Technology)
- Roger Cooke (Delft University of Technology)
- Pieter van Gelder (Delft University of Technology/ESRA)
- Rommert Dekker (Erasmus University, Rotterdam)
- Louis Goossens (Delft University of Technology)
- Carlos Guedes Soares (IST/ESRA)
- Andrew Hale (Delft University of Technology)
- Peter Kafka (Relconsult/ESRA)
- Theo Logtenberg (TNO)
- Lesley Walls (Strathclyde University)
- Enrico Zio (University of Milan/ESRA)

Local Organising Committee

- Paul Waarts (TNO), Chair
- Tim Bedford (Strathclyde University)
- Pieter van Gelder (Delft University of Technology)
- Ben van den Horn (Directorate-General of Public Works and Water Management)
- Theo Logtenberg (TNO)
- Ioannis Papazoglou (Demokritos//ESRA)
- Peter Sonnemans (Eindhoven University of Technology)

Technical Programme Committee

- Per Anders Akersten, Sweden
- Yulian (Julian) Argirov, Bulgaria
- Terje Aven, Norway
- Ioan Bacivarov, Rumania
- Guenter Becker, Germany
- Andreas Behr, Germany
- Joerg Blombach, Germany
- Rein Bolt, Netherlands
- Marco Carcassi, Italy
- Andrea Carpignano, Italy
- Marko Cepin, Slovenia
- Eric Chatalet, France
- Palle Christensen, Denmark
- David Coit, United States
- Giacomo Cojazzi, Italy
- Martin Cottam, United Kingdom
- Vinh Dang, Switzerland
- Claude Degrave, France
- Richard Denning, United Kingdom
- Yves Dutuit, France
- Mohamed Eid, France
- Elie Fadier, France
- Joseph R. Fragola, United States
- Peter van Gestel, Netherlands
- Andrew Hale, Netherlands
- Lars Harms-Ringdahl, Sweden
- Eduard Hofer, Germany
- Per Hokstad, Norway
- Bernt Leira, Norway
- Bo Lindqvist, Norway
- Sebastian Martorell, Spain
- Marino Mazzini, Italy
- Menso Molag, Netherlands
- Martin Newby, United Kingdom
- Jan van Noortwijk, Netherlands
- Richard van Otterloo, Netherlands
- Ioannis Papazoglou, Greece
- Lars Pettersson, Sweden
- Norberto Piccinini, Italy
- Florin Popentiu-Vladicescu, United Kingdom
- Christian Preyssl, Netherlands
- Urho Pulkkinen, Finland
- John Quigley, United Kingdom
- Maria Fernanda Ramalhoto, Portugal
- Antoine Rauzy, France
- Bill Robinson, United Kingdom
- Jan Rouvroye, Netherlands
- Fabrizio Ruggeri, Italy
- Peter Sander, Netherlands
- Tjerk van der Schaaf, Netherlands
- Gerhart Schueller, Austria
- Jean-Pierre Signoret, France
- David Smith, United Kingdom
- Peter Sonnemans, Netherlands
- Gigliola Spadoni, Italy
- Inge Svedung, Sweden

- Harry Hopkins, United Kingdom
- Peter Kafka, Germany
- Krzysztof Kolowrocki, Poland
- Henrik Kortner, Norway
- Dorota Kurowicka, Netherlands
- Pierre-Etienne Labeau, Belgium
- Gerald Laheij, Netherlands
- André Lannoy, France
- Vladimir Trobjevic, United Kingdom
- Giovanni Uguccioni, Italy
- Paul Uijt de Haag, Netherlands
- Jussi Vaurio, Finland
- Jan Erik Vinnem, Norway
- Han Vrijling, Netherlands
- Dik de Weger, Netherlands
- Sten de Wit, Netherlands

Message from the Chairman of the Technical Programme Board

ESREL 2003 covers all aspects of risk and reliability applications. This highly important area is undergoing continual development as a result of technological changes and theoretical progress. These factors justify the unique position of the ESREL conference in Europe as the principal conference that brings together practitioners and academics working in risk and reliability to contribute to the generation of new theory and practice.

In putting together the programme we have sought to illustrate the breadth of risk applications and to stress the multidisciplinary nature of the subject. We intend to stimulate the debate about the application of new quantitative and qualitative methods, and to facilitate the spread of methods for risk and reliability management across the countries of Europe.

The papers presented in these proceedings represent the current developments in Europe and outside, in the areas of risk and reliability. In putting together this volume, we should thank the authors for contributing their ideas, but also the referees and reviewers from the Technical Programme Board and the Technical Programme Committee. They have together reviewed more than 350 submissions, giving advice to authors and recommendations to the conference organisers.

ESREL2003 will take place in the city of Maastricht, one of the most charming and friendly of all Dutch cities. Famous for its food, culture, and history the city became even more prominent after the signing of the *Treaty of Maastricht*. We hope to welcome you to Maastricht, either as speaker or participant.

Tim Bedford

Message from the Chairman of NVRB

Risk and reliability has been an issue in the Netherlands for a long time. In the last decades it has been given increasing attention. Our awareness of risks is not only because our country lies partly below sea level, with threats from the sea and rivers, but also because we are a highly industrialised country with the highest population density in Europe. It requires a continuous search for the best methodologies to assess hazards and prevention techniques, and to disseminate experience and knowledge. The ESREL 2003 conference enables us to meet and to discuss with each other, and to improve our understanding in these fields. We are therefore proud to be your hosts in Maastricht and wish all participants a pleasant and fruitful stay in our country.

Theo Logtenberg
Chairman NVRB (Netherlands Society for Risk Analysis and Reliability)

Message from the Chairman of ESRA

The ESREL Conferences have always been an important part of the activity of ESRA, the European Safety and Reliability Association. They started in 1989 with the objective of providing a European Conference that could be a central meeting point for people interested in Safety and Reliability, across technical disciplines as well as bringing together industry and academia.

Despite the various specialized Conferences dealing with the specifics of one given industry, including safety and reliability aspects, there was still the need for one conference that would cover different industrial sectors and technical disciplines allowing for the transfer of technology to be achieved more easily. The same applies with the nature of the Conferences, which is not only for academics nor only for industrial specialists. Despite being sometimes difficult to find an adequate balance between these conflicting requirements, it is exactly that aspect which is one of the main interests of this series of Conferences.

Since its start, it was clear that this series of Conferences should not conflict with the national conferences, but instead should use the synergies that they offer. Therefore, efforts have been made to use the cycle of the national conferences in different countries to make the ESREL Conferences coincide with them and to have the national associations as the local organisers of the Conference.

The first two Conferences were in 1990 and 1991 in France and UK, respectively coinciding with lambda-miu 7 and Reliability 91, which represented two series of established National Conferences in those countries where there were many specialists in Safety and Reliability.

In 1972, the Conference was organised by SRE in Copenhagen and this was indeed the first time that the name ESREL was used. In 1993, it was in Germany and in 1994 and 1995 the cycle of France and UK was again repeated in La Baule and Bournemouth respectively.

New countries hosted the Conference afterwards: it was in Crete, Greece in 1996 in association with the PSAM Conference, in Lisbon, Portugal in 1997 and in 1998 it was again organised by SRE in Scandinavia, this time in Trondheim. In 1999, the location was again Germany, now in Munich, in 2000 it was in Edinburgh, UK, in 2001 it was in Turin, Italy and ESREL2002 was in Lyon, France.

The ESREL Conferences have already created expectations about the contents, the format and the quality of papers. Many authors and participants appear every year and many others every two years, creating a core group of participants in addition to the ones that are more attracted locally in the country of the Conference.

This year the Conference has the Dutch Society for Risk Analysis and Reliability as the local organiser who has provided the main infrastructure to support the organisation of the Conference. The preparation of the Conference was achieved by a dedicated Local Organising Committee and a Technical Programme Board, who's Chairs have to be especially acknowledged for taking initiative and mobilizing activity. The Technical Programme Committee is a group of volunteers, who have reviewed the papers and ensured the usual high quality of contributions. The authors are also a very important asset of the Conference as they contribute with their knowledge and experience to the quality of the technical aspects of the event.

On behalf of ESRA, I would like to thank all the contributors to this initiative for their dedicated effort in making this series of Conferences an event of reference, by ensuring that the various aspects of this Conference were fittingly organised.

I look forward to a successful Conference that can represent a useful technical meeting for all participants and that is at the same time an important social and cultural event making it one more Conference to remember with pride and satisfaction.

Carlos Guedes Soares,
Instituto Superior Técnico, Portugal,
Chairman ESRA

Safety and Reliability – Bedford & van Gelder (eds)
© 2003 Swets & Zeitlinger, Lisse, ISBN 90 5809 551 7

Main sponsors

TNO Public Safety

Directorate-General of Public Works
and Water Management

Sub Sponsors

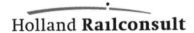

The Pre-ESREL workshop is sponsored by Pro-ENBIS. The Pro-ENBIS
project is supported by funding under the European Commission's
Fifth Framework 'Growth' programme via The Thematic Network "Pro-ENBIS"
contract reference: G6RT-CT-2001-05059.

Small Sponsors

Arteca Consultancy
Delta PI
NRG
Simtech Engineering

Safety and Reliability – Bedford & van Gelder (eds)
© 2003 Swets & Zeitlinger, Lisse, ISBN 90 5809 551 7

Keynote lecture: Living with risk: a management question

Ben J.M. Ale

Faculty of Technology, Policy and Society, Delft University, Delft, The Netherlands

ABSTRACT: Public authorities started to be really involved in risk management of hazardous materials some 30 years ago. Recent developments have led to fresh attention for this matter and many further developments are underway. The history of risk management and safety regulation is one of strongly variable interest, forgotten lessons and rude awakenings. The impetus exerted by accidents is short lived. Safety cases become documents to satisfy regulation rather than instruments to reduce risk. Deregulation, privatisation, outsourcing pose new challenges to safety and risk management. Some of the unfortunate side effects already become apparent. This invariably leads to the next disaster, which will have a striking resemblance to the previous one when abstracted from the immediate technological context. Lessons can be learned if we really want. The question remains: "Do we?".

1 INTRODUCTION

In January 2003 an atmospheric tank was being filled with ortho-cresol, when it split at the bottom seam, spilled its contents and collapsed (Figure 1).

The cause of the accident was strikingly familiar. The normal vents were did not have sufficient capacity to cope with the normal fill rate. Therefore at each loading operation an operator would climb to the top of the tank to open a few covers. Apparently he did not open enough covers. There had been a provision to have the tank split at the top seam, should the tank ever be over-pressurised. This would keep the contents of the tank in the tank even in case of a failure. When additional insulation was fitted this provision was put out of order, which was the reason for the tank to split at the bottom.

This accident did not happen at a badly run small of medium size enterprise. This accident happened to one of the largest storage facilities in the Rijnmond Harbour area, which operates large tank farms with hundreds of these tanks. The company has a safety management system and a quality system. Nevertheless a situation persisted, known to many, which was a sure way to get an accident: A vent not designed accord-ing to the loading rate and a top seam that was made stronger than the bottom seam.

It should be noted that a parent of the present company had been under study in the early 80s. It was part of the COVO study, a study into the risks of six hazardous installations in the Rijnmond area [1]. At the time some serious doubts were raised about the safety

Figure 1.   Collapsed tank (Rijnmond 2003).

culture of the company. As a result considerable improvements were announced.

One could draw a variety of conclusions form this accident: the company forgot the lessons from the past; the companies safety management system did not function properly; the inspections by the authorities were not adequate; the safety culture in the company was insufficient. In any case the system as a whole, consisting of the company inside the regulatory framework in which it operates, proved not to be capable of learning lessons form past accidents sufficiently to prevent another "normal accident" to happen.

This raises the question whether the risk management systems set up in industry and government really led to risks being managed, or just contained.

The recent history of dealing with industrial risk in the Netherlands is an example of how volatile public and political interest in risk management is, and how difficult it is to keep the message alive that risks are not just virtual social construct. The materialisation of a risk as an accident leads to real loss and real human suffering.

2 POLICY DEVELOPMENT

Industrial risk is not a new phenomenon. Indeed large scale explosions of munitions storages have been known all over Europe for centuries. Also the storage and handling of ammonium nitrate have been accompanied by series of devastating explosions. Nevertheless there was not much of nation wide safety policy until the mid-seventies. A concentrated series of large scale releases of flammable and toxic clouds and a number of large scale explosions formed indications that the rapid development of the chemical process industry and the associated transport may conflict with increasing demands on space for housing and other community building.

In the Netherlands space is extremely limited. At present the mean density of population is 475 persons per square km. Space limitations preclude policies based on precautionary principles, if this means creating safety zones based on effect distances. It always has been necessary to take probability into account. In the early eighties the foundations were laid for a policy based on quantitative risk analysis.

On a European level the Decree on Major Hazards, which commonly is known as the Seveso directive, was issued, binding all members of the European Union to set up a reporting system [2]. After the accident in Bhopal more emphasis was put on decision-making especially with regards to sitting of and spatial planning around chemical facilities.

All these efforts did not lead to a reduction in the number of major hazards in the european countries. However it was only after explosions in a fireworks warehouse in the Netherlands and a fertiliser plant in France and the attack on the world trade centre in New York that the slowly decaying interest in safety was rekindled.

In the Netherlands some large scale accidents with explosives materials occurred as well. In 1654 the centre of Delft was demolished by the explosion of a powder tower. This explosion, which could be heard 80 km away, created the "horse market", which still exists as an open space. (Figure 2)

In 1807 a similar explosion took place. Now a barge laden with black powder exploded in the centre of Leiden. The van der Werf park today is still witness of this event. 150 people were killed among who 50 children, whose school was demolished by the blast.

This explosion led to an imperial decree by Napoleon. The emperor stated that from then on a permit was needed for having an industrial facility. Three classes of industry were designated:

1. Industries that were considered too dangerous to be inside a city. The authorities would indicate a location.
2. Industries for which location inside a city could be considered if it could be demonstrated that there was no danger for the community.
3. Industries that always could be located inside city limits.

Figure 2.   The big thunder of Delft in 1654.

In addition Napoleon stated that objections of future neighbours should be noted and addressed by the authority, which made a decision.

Interestingly the safety regulations in France can be traced back to the same imperial decree.

The origin of modern risk management lies in the industrial accidents after World War II. In 1966 a fire in a storage facility for LPG in Feyzin, France killed 18 and wounded 81. This accident led to re-emphasis on design rules for bottom valves on pressure vessels. In the realm of physical planning no actions from the French or the European authorities seemed to have resulted from that accident.

Ten years later a number of similar accidents occurred: Flixborough (1974, 28 dead), Beek (1975, 14 dead) and Los Alfaques (1978, 216 dead). These accidents showed that the Feyzin accident was not a unique freak accident. Apparently LPG and other flammable substances could pose a serious threat to the workforce and to the surroundings.

3 GOVERNMENT IN ACTION

In 1979 Prime-Minister van Agt, just as his predecessors, wrote a letter to parliament about the development of environmental policies as integral part of the nations policies. In this letter he introduced "External Safety" as separate from occupational safety. The Prime-Minister introduced and announced three elements of a new policy:

1. Appointment of the minister of environment as co-ordinator for hazardous materials;
2. Founding of a new separate policy body dealing with external safety and announcement of new legislation covering external safety.

At the same time a major change in the energy market appeared imminent. This among other lead to a major market push for LPG as motor fuel. In 1978 a tank car exploded in a tank station. Although nobody was hurt in this accident, it made the point that such accidents could happen in the Netherlands as well and that the population around the stations should be limited. The chief inspector for the environment decided not to wait for legislation. He issued an instruction for his inspectors to not approve a permit unless the conditions for distances and population densities as indicated in the Table 1 were satisfied[3]. This was the first explicit zoning measure around a hazardous activity.

A further potential increase in the transport of LPG through the Netherlands resulted from the desire to use LPG as feedstock for the production of ethylene. A committee was charged with developing a policy. A study was commissioned into the safety of the whole chain from import to final use. It became apparent that a policy aimed at insuring that no accident ever

Table 1. Zoning arond LPG stations.

Distance to tank and/or fillingpoint (m)	Allowed building	
	Houses	Offices
0–25	None	None
25–50	Max 2	Max 10 people
50–100	Max 8	Max 30 people
100–150	Max 15	Max 60 people
>150	No limit	No limit

would harm the population would not be compatible with the limited space in the Netherlands. The committee decided that there should be a level of risk below which it is neither desirable nor economical to strive for further reduction. This statement implied that the level of risk should be established and that acceptability limits should be set.

At the same time authorities in the Rijnmond area started to be worried about the safety of the population around the large petro-chemical complexes in the area. Taking the Canvey Island study as an example [4, 5], the Rijnmond authority embarked on a study to establish whether quantification of risk was feasible and would give results that would be useful in decision-making. The results [6] were promising with regards to the usefulness of the results. The quantification of risk as a routine exercise was judged not to be feasible unless information technology could be used to take away the burden of the many complicated calculations and reduce the time needed.

The Rijnmond Authority together with the ministry of environment embarked on the venture towards an automated method for quantification of risk. Now, twenty plus years later the process still is not fully automated. Such a level of automation no longer is desired either. But the techniques developed since together with the rapid development of computational capability has lead to workable systems with reasonable return times.

4 CRITERIA

Having decided that risk quantification is the way to go the inseparable counterpart had to be developed as well. What to do with the results, and how to make sure the analyses would actually be made and used in decision making.

Regional and local authorities as well as industry asked for guidance regarding the acceptability of risk. The bases for this guidance was found in documents and decisions taken earlier.

An important base line was found in decisions made regarding the sea defences of the Netherlands. In 1953 a large part of the south west of the Netherlands was flooded as a result of a combination of heavy storms, high tides and insufficient strength and maintenance of the diking system. Almost two thousand people lost their lives and the material damage was enormous especially because the Netherlands was still recovering from World War II. The Netherlands embarked on a project to strengthen the sea defences, including a drastic shortening of the coastline by damming off all but one of the major estuaries of the Rhine/Maas delta. The design criteria were determined on the basis of a proposal of the so-called "Delta Committee" who proposed that the dikes should be so high that the sea would only reach the top once every 10,000 years. [7]. The probability of the dike collapsing is a factor of 10 lower. The probability of drowning is another factor of 10 lower, so that the recommendation of the Delta Committee implies an individual risk of drowning in the areas at risk of 1 in a million per year. This recommendation was subsequently converted into law.

This value of risk was reaffirmed when a decision had to be taken about the construction of the closure of the Oosterschelde estuary. For reasons of preserving the ecosystems the design was changed from a closed solid dam, to a movable barrier. This barrier should give the same protection as the dams. In this manner Dutch parliament had a history of debating safety in terms of probabilistic expectations, which came in handy when industrial risk had to be discussed.

The value of 1 in a million per year corresponds to about 1% of the probability of being killed on the road in the mid 80s. This became the maximum acceptable addition to the risk of death for any individual resulting from industrial accidents.

For societal risk the anchor point was found in the "interim viewpoint" regarding LPG points of sale mentioned above. When combined with value already chosen for individual risk this led to the point 10 people killed at a frequency of 1 in 1,00,000 per year. As societal risk usually is depicted as an FN curve having the frequency of exceeding N victims as a function of N, the limit had to be given the same form. Thus the slope of the limit line had to be determined.

A first attempt dates back to 1976: In a since forgotten document about environmental standards, limits on the acceptability were presented. In region where the consequences included people being killed, the slope was -2 as a representation of the aversion of people against large-scale accidents. It had a four orders of magnitude grey area between the unacceptable and the negligible level.

It was decided to incorporate the apparent aversion against large disasters in de the national limit by having the slope steeper than -1 in there as well. An additional argument for this choice is that at a slope of -1 the expectation-value of the limit would be infinite. The only way to have a finite expectation value at a -1 slope is to set an absolute maximum to

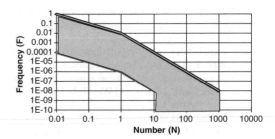

Figure 3. Risk criteria according to the Groningen policy document.

the scale of an accident, which practically not feasible, although admittedly the maximum number of people killed in any one accident is the population of the earth (presently some seven billion).

Several values circulated in literature at the time, ranging from -1.2 to -2 [8–15]. In the end it was decided to adopt a slope of -2 for the limit line.

In order to bind the decision space at the lower end of the risk spectrum limits of negligibility were set for individual risk and societal risk alike at 1% of value of the acceptability limit.

The resulting complex of limit values was laid down in a policy document called "premises for risk management" [16]. It should be noted that although the document laying out the policy has the title "Premises For Risk Management" in the English version, the original title probably better translates as "coping with risks". The document sets out acceptability criteria. The document does not say how compliance with these criteria should be achieved. The document describes the boundaries of the playing field and leaves it to the players: industry, and local and provincial authorities.

The accident in Bhopal, where some 3000 people were killed as a result of a release of methyl iso-cyanate, helped to promote the adoption of European legislation. The SEVESO directive, named after a small village in Italy where dioxine was released in an accident, became the vehicle to implement these policies into law in the Netherlands just as in many other members of the EU. The "Hazards of Major Accidents Decree" [17] demanded that top tier establishments would submit a safety report, in which a quantified risk analysis performed according to the set standards, would be presented. This information then subsequently could be used by local planners for zoning decision and by the emergency services for disaster abatement planning.

5   METHODOLOGY

As stated above the ministry of environment and the Rijnmond authority were involved in a long term effort to make risk quantification techniques more readily available, using ICT tools. As the national policy demanded that risk were limited to the values given in the policy document various stakeholders now were started to demand that the government would give guidance of how to do the calculations. In response to this demand the methods to be used in calculating risks were described in a series of handbooks [18–20].

The handbooks did not prevent that different results could be produced for the risk analysis of the same establishment. In a study performed in 1986 the band-with of risk calculations were estimated to be two orders of magnitude [21] Even today the band-with between different computer implementations of the methodology is an order or magnitude [22]. International studies into the uncertainties of quantified risk analysis lead to similar results: the band-with in the estimates has decreased over the years but is still considerable, when in decision making on safety distances the desired accuracy is measured in centimetres rather than meters [23, 24]. Let alone orders of magnitude. Several situations arose where parties with certain interests tried to use the uncertainty to their advantage. Especially local authorities in their role of issuer of licenses put increasing pressure on the ministry to harmonise the methodology to a further extent.

In their role as exploiter of limited space these authorities on the other hand used conflicting results as an argument to not having to actually implement the policy. This was in particular the case around a number of railroad shunting yards, where large numbers of railcars with hazardous materials such as Chlorine, Ammonia and LPG were processed. These shunting yards often are situated in the inner cities, behind the central station, where land is almost priceless.

Industries were part of this game as well. Sometimes together with authorities risk was reduced by calculation rather than safety measures. These practices slowly eroded the value of the safety reports.

6 PERCEPTION

A major factor influencing the people's reaction to potentially hazardous activities is what generally is described as risk perception. There are many factors shaping the perception of risky activities [25–27]. The top 10 of the most listed are:

1. Extent and probability of damage
2. Catastrophic potential
3. Involuntariness
4. Non-equity
5. Uncontrollability
6. Lack of confidence
7. New technology
8. Non-clarity about advantages
9. Familiarity with the victims
10. Harmful intent

As these factors are different from differing activities it cannot be expected that a single set of risk criteria is applicable to all activities. Nevertheless a policy may look more organized as the set of applicable criteria in small.

On the other hand it is argued that these factors make it impossible to set general standards, as every situation and every activity is different. In a more extreme stance it is argued that risk is a social construct rather than something that in principle can be determined scientifically. In this view there are so many subjective choices made in risk analyses that they cannot be called objective science at all. [28]. Scientists are just other lay-people. There judgement is influenced by the same factor, but in addition they let their science influence by their political judgements. It is no surprise that the more objectivist risk analysts argue that scientific judgements and political judgements are not the same thing and that objective quantification of risk is a scientific exercise. Indeed such objectivity is necessary make cost benefit based decisions. In such argumentation the value of the risk should be as objective as the – monetary – value of potential risk reducing measures [29].

Any policy should conform to general principles of justice and democracy, be it setting a speed limit or a limit on risk. The results should be predictable for the stakeholders and for the public and execution should measurable against objective standards. This holds even when arguments are formulated in more qualitative terms such as "As Low As Reasonably Achievable" or "gross disproportionality". It should always be borne in mind that any stakeholder in any regulatory system can resort to getting a dispute settled in court.

How valid the arguments may be, they nevertheless are of great help to stakeholders that have no interest in having risks limited by a government policy in the short run. And as the last accident disappears in past history the pressure to be firm on risk dissolves.

7 DECLINE

In the early nineties the joint cities under the leadership of the mayor of Dordrecht/Zwijndrecht put pressure through parliament to reduce the impact of especially the societal risk criteria in favour of large scale development of offices near railway stations, where large scale transport of hazardous materials takes place as well. The case was raised that central government had issued two conflicting demands: the promotion of the development of areas close to railway stations to reduce the use of cars by commuters and the limitation of these developments close to marshalling yards, which in a number of cases are just behind these central stations.

The situation was worsened by the development around Schiphol Airport. Here it became apparent that

Figure 4.   Explosion in Enschede.

a decision about societal risk could not be made at all. On the one hand the accident in the Bijlmermeer, where a Boeing 747 crashed into an apartment building made it difficult to defend the acceptance of a risk in excess of 10 times the total risk of the combined SEVESO sites. On the other hand the expansion of the airport which produced such a risk was deemed essential for the Dutch economy. This lead to a decision-making stalemate, which could be broken only, if the binding of societal risk policy was broken.

Thus in 1994 the lower – negligibility – limits were abandoned and the remaining acceptability limit for societal risk was declared advisory rather than binding.

As in many other European nations the 90s were a period where privatisation of services and the preference for market driven developments dominated the socio-political processes. Many governmental services were sourced out and sometimes largely discontinued.

The inspection services were hit very hard by these developments. They lost a significant part of their staff and most of their specialised experts because the general line of thought was that no specialist knowledge were needed for assessing whether the correct procedures were in place. As a result checking the paperwork against reality on site became a rare activity.

The whole of the nineties went by without major accident, if one discounts the crash of the airliner in an apartment building in Amsterdam mentioned above, an accident that public authorities and the air traffic sector were eager to forget.

8   REBIRTH

In the beginning of the current century a series of accidents revealed the deficiencies that had developed in the regulatory system and in the supervision and inspection.

On 13 may 2000 an explosion occurred in a fireworks storage and trading facility in Enschede, the Netherlands. Twenty-two people were killed and some

900 injured. The material damage was approximately 400 million Euro.

Immediately after the accident an investigation was started into the causes of the accident. Special attention was given to the unexpected violence of the explosion. The investigative committee installed by the Government used results and advice of domestic and international institutes to obtain results.

It appeared that the firm had a long history of violating permits, that the city had legalised these violations and that inspectorates and state institutions were not aware of the hazards thus created. Especially the importance of the correct classification of the fireworks and of the storage of the correct types and quantities went unnoticed. As a result prior to May 13, 2000 most of the fireworks stored at the premises were more powerful than the labels indicated and in fact a significant part of the storage was mass-explosive contrary to the current permit.

After the disaster inspections were held in the other storage facilities for fireworks. The results were dramatic. In fact none of the facilities operated within the legal boundaries and sometimes for many years.

This lead the inspectorate for the environment to check on a number of other widely spread activities. Of the LPG stations about a third had houses inside the prescribed exclusions zone. A few hundred houses were built after the legislation had come into effect in 1984. The maintenance of ammonia-cooling facilities proved to be consistently below standard to the extent that most facilities would have to be closed should the law be upheld strictly.

On New Years eve of 2000 a fire broke out in a bar in Volendam. Thirteen young people were killed and more than a hundred seriously injured. The bar had no valid license, there were too many people in the building, contrary to legal requirements the Christmas decorations had been fireproofed, firework items (so-called "cold fire") were lighted in the building and emergency escape routes were either blocked or the stairs were removed. Again a nationwide inspection had dramatic results. There were in fact only a few public buildings that actually had a license to operate, even though the legislation existed for over a decade. This was not only true for bars, restaurants and disco's. Most governmental offices had no license to operate either. In fact it looked as if the legislation was forgotten from the day it was adopted.

Both incidents were investigated, each by its own committee. Both committees came to similar conclusions:

- legislation does not implement itself after it is adopted
- trust is fine, direct inspection is better
- the existing policies are fine, but they should be implemented.

On March 7, 2001 Prime-Minister Kok wrote a letter to parliament about the further development of safety policies as integral part of the nation's policies. The Prime-Minister introduced and announced three elements of a new policy:

affirmation of the minister of environment as co-ordinator for hazardous materials;

founding of a new separate policy body dealing with external safety and announcement of new legislation covering external safety.

This letter was issued almost 22 years after the letter of Prime Minister Van Agt, which in essence had the same content.

On August 20, 2002, a small leak occurred in an ACN railroad tank car at the station in the centre of Amersfoort. A circle with a radius of 500 m had to be evacuated. As a result train traffic between the west and the east of the Netherlands came to a halt, as most of this traffic normally passes through the now closed station. Also the command centre of the local fire-brigade was in the zone and had to be evacuated.

After the accident the union of cities raised the question whether it made sense to transport hazardous materials close to large scale urban developments, thereby ignoring that the same union pushed the relaxation of the policy in favour of building large scale real estate close to the transport of hazardous materials 10 years earlier.

9 SLOWDOWN

Soon thereafter a quick scan into the consequences of a strict implementation of existing policies was published [30]. Several of the measures figuring in that report were measures that should have been taken in the past if the set policy would have been fully implemented, but were postponed or discarded because of the cost.

One example is the points of sale of LPG. When the LPG policy was set in the late 70s, early 80s 200 LPG stations were closed or relocated. However another 600 should have been relocated due to the vicinity of houses. The cost of this operation was deemed excessive in the mid 80s and looks even more excessive now: 22 Billion Euro. A more recent analysis shows that it is much cheaper to actually stop using LPG as motor fuel [31]. Because space becomes available for housing development which currently is inside safety zones, to stop using LPG would bring at least 400 million euro's, a conclusion the LPG industry does not like at all.

Another example regards the railroad marshalling yards mentioned earlier. One of these is Venlo. This yard has not had a valid license to operate since it started over 80 years ago. However the yard is essential for the operation of the railways. The railways in the Netherlands and in Germany use different voltages for their electric locomotives. It is therefore necessary

to change locomotives at the border, the main function of the Venlo yard. A side effect of this is that large quantities of tankcars pass through Venlo and have to wait on the yard for further transport for considerable time. In fact the yard can be seen as a permanent storage facility for every conceivable hazardous substance around. It is impossible to change operations such that Venlo may meet safety criteria, nor noise criteria for that matter. Therefore a plan was made already in the mid 70s to relocate the yard. Estimated cost at the time 73 Million guilders (some 35 Million Euro). At the time this was considered too expensive. The problem however did not disappear, so Venlo figures again as a problem in the quick scan. Now the cost estimate is 173 Million Euro.

These costs led to a renewed discussion about to what extent the societal risk constraints should be made a legal requirement, a discussion still going on. In the wake of this discussion also the problem resurfaced of having differing results of analyses performed on the same system or establishment.

Finally the question was raised in how far emergency response and self-help of citizens were beneficial in reducing the consequences of potential disasters. If indeed the effect of disaster abatement and escape could be quantified, this potentially could lead to the conclusion that the consequences of disasters in chemical establishments and especially with transport of hazardous materials was overestimated and thus that more real estate could be developed near transportation routes with retaining the current limits than previously was assumed.

In the mean time the city of Dordrecht/Zwijndrecht pushes for more city development in the railway-zone, which incidentally is one of the most dangerous of the country. Also a further growth of Schiphol airport was announced. This growth will lead to a violation of the risk limiting criterion which now is expressed of the mean weight of aircraft falling to the ground each year. The development of this "Risk Weight" is depicted in Figure 5, together with the present criterion. Again it

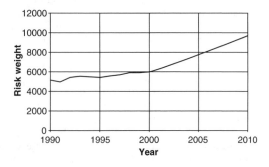

Figure 5. Development of the "risk weight" at Schiphol Airport.

7

did not take long for accidents to loose their importance in the political debate. The question therefore rises whether society really wants to reduce risks or whether public outcry is just a phase in a mourning process that is necessary but passes.

10   MORE RESEARCH

In order to solve these questions a number of new investigations were launched.

The first is aimed at establishing the risks of chains activities related to a single hazardous material. The materials chosen are those that potentially create the most problems: LPG, ammonia and chlorine. To a certain extent the study into LPG is a repeat of the study performed 20 years ago, and thus there should not be any surprises in the answers.

The study should lead to the identification of measures that could be taken to reduce the risk of these activities and remove problems for the developments in city centres. A first measure already has been taken: the transport of chlorine by rail will be stopped by 2007. The production units will be relocated such that chlorine is produced were it is used.

It should be expected that for ammonia and LPG this will be a lot more difficult. The transport of chlorine essentially occurs only along a single route between three facilities. The transport of LPG and Ammonia occurs along many routes and between many locations. In addition these materials have a wide spread and large user-base.

The second investigation is aimed at establishing whether the unification of results of risk analysis can be promoted beyond the present level. As mentioned before the current handbooks in the colors series have not led to the desired uniformity, not even after the release of the purple book, which was meant to take away ambiguities and freeze parameters to agreed values [32]. The obvious problem with driving unification further is the loss of flexibility in the analyses. In this sense authorities and industry have conflicting demands. They want uniform results on the one hand, but on the other hand they want to be able to cater for local differences in the way installations are constructed and managed. Especially the latter factor creates ample room for interpretation by the individual analyst. There is abundant theory about the management influence but very little data [33, 34]. Also data on failure frequencies are seldom suitable for taking account of particular design solutions.

Finally a series of investigations have been started into quantification of the success of emergency response operations. Among other, it is noteworthy how many people survived the 9/11 attack on the Twin towers. Most if not all people below the points of impact survived, indicating that when people know what to do and do it survival rate can be high. This is in contrast with findings in experiments in tunnels, where it was noted that most people stay and wait in their cars until help arrives [35], ignoring the signs pointing to emergency exits.

In any case the emphasis already is shifting again, from a strong desire to address the problems of risk and eliminate particularly nasty hazards to authorities trying to find ways to cope with these risks without really reducing them.

11   CONCLUSION

History indicates that lessons of the past are not remembered very long. The growth of the population leads to ever denser use of available space. This makes it very tempting to forget accidents. Thereby the perception of risks slowly but certainly changes form serious to remote as the last major accident shifts into the past. These developments fit well to human nature, which does not seem to be adapted to manage risks. Certainly not when the advantages of an activity are immediate and the chance of negative effects are remote or the occurrence of these effects remote. Our minds are shaped such that we accept information that confirms our existing conceptions much more readily than information that would let us change out mind or opinion [36, 37].

Humans are mortal, but do not like to be reminded of the fact. Everyday that we wake up we take as a confirmation of our immortality. As our life expectance increases, we are less and less confronted with the reality of the opposite: that indeed we are mortals. This way the human mind works is the basis for progress. Only if man faces hazards and perils can man develop. Therefore courage is an acclaimed human trait. To go on in the face of danger is built into human behaviour. Columbus would not have sailed west and discovered America, if not for his courage nor would anybody wantonly make a trip in a space shuttle. Religions preach that man has to go through death to reach eternal bliss, be it heaven, nirwana or whatever this state is called in a particular belief system. Thus to face up to peril, and steadfastly go on in the face of doom is considered a good thing.

Unfortunately the line between courage and recklessness is difficult to draw. Often the difference is determined by the result rather than the deed itself. Only because they conquered Troj is the handful of Greek soldiers who hid in the Trojan Horse remembered for their courage. Would they have lost the deed would have been described as foolish. Similary David facing up to Goliath only lives on as a courageous act because the latter was defeated [38].

To operate a chemical plant at the boundary of its operating envelope may be interpreted as courageous way of increasing production to the maximum. And there still are many companies who reward this type of behaviour. May be not consciously. But a bonus based on sales or throughput has the effect of rewarding taking risk. After the accident the investigation will undoubtedly interpret the same behaviour as reckless playing with the lives of innocent bystanders

Risk managers and safety people have a function to point out the evidence that conflicts with the dominating perception. Things can go wrong and if they can, they will. And given the overwhelming domination of the optimistic view that "it cannot happen", or at least "not to us" it looks almost advisory to make it an obligation for every company, or other entity on the verge of performing a courages act to consult a morbid pessimist like a risk analyst.

Safety people therefore are trying to balance the drive for courage, insisting on caution, thinking before acting, not taking risks if it can be avoided and not taking unnecessary risks at all.

In this role risk managers are like the court jesters of yesteryear. They say what hardly anybody wants to know but the King better listen.

REFERENCES

1. *Report on the COVO study to the Rijnmond authority*, Reidel, 1979.
2. EU Directive 82/501EEC, 1982.
3. *Het interim standpunt LPGstations*, HIMH, The Netherlands, 1978.
4. HSE , Canvey: *An investigation of Potential Hazards from Operations in the Canvey Island/Thurrock Area*, Londen (HMSO) 1978.
5. HSE, *Canvey: Second Report, A Review of the Potential Hazards from Operations in the Canvey Island Thurrock Area Three Years after Publication of the Canvey Report*, Londen (HMSO) 1981.
6. Cremer and Warner, *Risk Analysis of Six Potentially Hazardous Objects in the Rijnmond Area*, Londen 1981.
7. *Rapport van de Delta Commissie*, 1960, Delta wet 1957.
8. R. Wilson, *The Cost of Safety*, New Scientist, 68 (1975) 274–275.
9. J. Okrent, *Industrial Risk*, Proc. R. Soc. 372 (1981) 133–149, Londen.
10. Ph Hubert, M.H. Barni, J.P. Moatti, *Elicitation of criteria for management of major hazards*, 2nd SRA conference, April 2–3 1990, Laxenburg, Austria.
11. F.R. Farmer, *Reactor Safety and Siting, a proposed risk-criterium*, Nuclear Safety, 8 (1967) 539.
12. W.C. Turkenburg, *Reactorveiligheid en risico-analyse*, De Ingenieur, vol 86 nr 10 (1974) 189–192.
13. M. Meleis and R.C. Erdman, *The development of reactor siting criteria based upon risk probability.* Nuclear Safety, 13 (1972) 22.

14. D.J. Rasbash, *Criteria for Acceptability for Use with Quantitative Approaches to Fire Safety*, Fire Safety Journal, 8 (1984/85) 141–158.
15. H. Smets, *Compensation for Exceptional Environmental Damage Caused by Industrial Activities*, Conference on Transportation, Storage and Disposal of Hazardous Materials, IIASA, Laxenburg, 1985.
16. *Omgaan met Risico's*, Tweede Kamer, vergaderjaar 1988–1989, 21137, nr 5. Under the same number the translation in English is titled: Premises for Risk Management.
17. *Besluit Risico's Zware Ongevallen,* Staatsblad 1988, 432, Staatsdrukkerij, The Hague, The Netherlands.
18. *Methods for the calculation of physical effects*, Committee for the Prevention of Disasters, CPR 14E, The Hague, The Netherlands, 1997.
19. *Methods for determining and processing probabilities*, Committee for the prevention of Disasters, CPR 12E, The Hague, The Netherlands, 1997.
20. *Methods for the determination of possible damage*, Committee for the Prevention of Disasters, CPR 16E, The Hague, The Netherlands, 1990.
21. R. Geerts et al, *De onzekerheid in effectberekeningen in Risico analyses*, AVIV, Enschede, 1986.
22. B.J.M. Ale, G.A.M. Golbach, D. Goos, K. Ham, L.A.M. Janssen, S.R. Shield, Benchmark Risk Analysis models, RIVM report 6100066015, Bithoven, The Netherlands, 2001.
23. A. Amendola, S. Contine, I. Ziomas, Uncertainties in Chemical Risk Assessment, Joint Research Centre EU, ISPRA, 1992.
24. F. Markert, I. Kozine, K. Lauridsen, A. Amendola, M. Christou, Sources of Uncertainties in Risk Analysis of Chemical Establishments, first insights from a European Benchmark Exercise, EFCHE Loss Prevention Symposium, Stockholm, 2001.
25. P. Slovic, *Emotion, sex, politics and science: surveying the risk assessment battlefield*, Risk Anal, vol 19 nr 4 (1999) pp 689–701.
26. L. Sjoberg, *Factors in Risk Perception*, Risk Anal, vol 20 nr 1 (2000) pp 1–11.
27. C. Vlek, *A multi–stage, multi-level and multi-attribute perspective on risk assessment decision making and risk control*, Risk Decision Policy vol 1 (1996) pp 9–31.
28. M.B.A. van Asselt, *Perspectives on uncertainty and risk, The PRIMA approach to decision support, Kluwer, 2000.*
29. T.O. Tengs, M.E. Adams, J.S. Pliskin, D.G. Safran, J.E. Siegel, M. Weinstein, J.D. Graham, *Five hundred life saving interventions and their cost effectiveness*, Risk Anal, 15 (1995) 369–390.
30. *Quick Scan Gevolgen Beleidsvernieuwing Externe Veiligheid*, ARCADIS TNO, 140221/BA1/0X5/539/kg dd 9 January 2002.
31. *Kosten en Ruimtelijke baten van mogelijk uitfasering van LPG en grootschalig vervoer van propaan en butaan*, Ecorys-NEI, RVW-EVH/128 Rotterdam, the Netherlands, January 21, 2003.
32. B.J.M. Ale and P.A.M. Uitdehaag, (1999) Guidelines for Quantitative Risk Analysis, (CPR18) RIVM, 1999.
33. L. Bellamy, J.I.H. Oh, A.R. Hale, I.A. Papazoglou, B.J.M. Ale, M. Morris, O. Aneziris, J.G. Post, H. Walker, W.G.J. Brouwer, A.J. Muys. elaar, IRISK, *Development of an integrated technical and management*

risk control and monitoring methodology for managing on-site and off-site risks, EC contract report ENV4-CT96-0243 (D612-D), 1999.

34. I.A. Papazoglou, O.N. Aneziris, J.G. Post, B.J.M. Ale, *The Technical Modeling in Integrated Risk Assessment of chemical installations, Journal of Loss Prevention*, 15 (2002) 545–554.

35. *Safety proef, Rapportage Brandproeven*, Ministerie van Verkeer en Waterstaat, Den Haag, Augustus 2002.

36. James Reason, *Human Error*, p 39, Cambridge University Press, 1990.

37. J. Groeneweg, *Controlling the Controllable*, p 184, DWSO press, Leiden, 1998.

38. I Sam 17 45:51.

Safety and Reliability – Bedford & van Gelder (eds)
© 2003 Swets & Zeitlinger, Lisse, ISBN 90 5809 551 7

Keynote lecture: Uncertainty analysis for flood defence systems in the Netherlands

A.C.W.M. Vrouwenvelder
TU-Delft / TNO-Bouw, Delft, The Netherlands

ABSTRACT: Reliability and risk assessment becomes more and more important as a background for decision making in the field of flood defence design in the Netherlands. In this development reliability experts have to cooperate with a large group of hydraulic, geotechnical, structural and mechanical engineers. This cooperation leads to many discussions related to the exact meaning of both input and output of the failure probability calculations. Additionally, there are from time to time also discussions with civil servants, politicians and the public in general. In those discussions confusion often seems to be a major key word. This holds in particular for the debates on so called epistemological or knowledge uncertainties. This paper summarises a number of the discussions that have taken place in the past few years and hopefully contributes to a gradual better communication on this difficult but important issue.

1 INTRODUCTION

Already since 1958 the design and assessment of flood protection systems in The Netherlands is based on probabilistic and risk based principles. In their elaboration, however, the present operational methods are still rather primitive, but intensive research is going on to bring them on a higher level of sophistication. Practical application of these more advanced risk and reliability procedures for the design and assessment of flood protection measures are to be foreseen within the near future. A number of papers in this conference deal these research efforts and results reached so far (Lassing, 2003, Buijs, 2003, Kok, 2003).

The introduction of full probabilistic and risk analysis into practice brings, among other things, along the question of the interpretation of probability for a broad class of engineers. And not only engineers. Also politicians, civil servants and citizens will be confronted with safety statements in terms of probability. To some extent, of course, they already do. In the present design process the starting point is a safety level chosen by the politicians. The Dutch Water Act, approved by the parliament, specifies return periods for water levels and river discharges of for instance 3000 or 10000 years. This already is a matter of confusion for many people and the message that these flood levels or discharges still may occur next year (although with small probability) has to be repeated every time.

The design of the corresponding structures, however, is a task that is completely left to the engineers and scientist. It should be noted that the steps to arrive at for instance the dike crest levels corresponding to the design conditions, is a complex and often difficult traceable process. It involves statistical extrapolations, estimation of load effects for all kinds of mechanisms, estimation of resistance parameters, sometimes on the safe side, sometimes best guesses, and so on. As a result the safety against flooding may deviate in a rather uncontrolled way in a positive or negative sense from the politically formulated starting points.

In the future, this situation will change in so far that the engineering world will make explicit statements about the probabilities of inundation and the corresponding risks. They will tell the public that the probability to have an inundation in a certain city of say at least 1.0 meter will be, for instance, once every 100 or 1000 years. It seems to be the only fair way to present the safety of the system to society and to underpin the necessity of protection measures by economic considerations and individual human safety and group risk criteria. The problem, however, is that these estimates will include all "dark parts" of the design process. The first problem is to find a complete and adequate probabilistic modelling for all aspects and parameters that show scatter or uncertainty. The second point is to communicate the resulting failure probabilities and flood risks with the non engineering world. Questions will be asked as "how good do we know this probability" and this may be difficult to answer. As engineers, we may simply state that the whole procedure is now more consistent and more logical compared to the approach used before, and that maybe we should not give too much attention to the numbers. This may be true to some extent, but the outside world will simply not accept such an answer.

This paper will discuss a number of the typical issues raised above. The intention is not to convince the layman directly, but to open the discussion between experts and make sure that the engineering community will try to develop a common view on it.

2 TYPES OF UNCERTAINTY

The starting point of the discussion is that not all uncertainties seem to be of the same type. The main distinction usually made is the one between so called inherent or aleatory uncertainties on the one side and knowledge or epistemological uncertainties on the other. The first category can be associated with experiments like throwing dices or unpredictable phenomena in nature like river discharges, sea levels, wind speeds and so on. It is the scatter that one can observe in practice, that can be measured and can be described in objective terms. The second category is related to our lack of knowledge, for instance the limited amount of data in statistical data bases or the fact that our models in general are incomplete or inaccurate. These uncertainties may give rise to great difficulties in interpretation. Usually it is for instance impossible to find exact methods to describe these uncertainties and one has to fall back on concepts like "intuition", "expert opinion" and "engineering judgment".

The central question is whether the two types of uncertainty can be treated in the same way or that different concepts and procedures should be followed. There is in this respect a direct relationship with the interpretation of probability. The Joint Committee on Structural Safety [JCSS, 2001] makes a distinction into three possible interpretations:

1. the frequentist's interpretation;
2. the formal interpretation;
3. the Bayesian interpretation.

The *frequentist's* interpretation is quite straight forward and allows only "observable and countable" events to enter the domain of probability theory. Probabilities should be based on a sufficient number of data or on unambiguous theoretical arguments only (as in coin flipping or die throwing games). Such an interpretation, however, can only be justified in a stationary world where the amount of statistical or theoretical evidence is very large. It should be clear that such an interpretation is out of the question in the field of civil

engineering applications. In almost all cases the data is too scarce and often only of a very generic nature.

The *formal* interpretation gives full credit to the fact that the numbers used in the analysis are based on ideas and judgement rather than statistical evidence. Probabilistic design is considered as a strictly formal procedure without any physical interpretation. Such a procedure, nevertheless, is believed to be a more rich and consistent design procedure compared to classical and deterministic design methods.

However, in many cases it is convenient if the values in the probabilistic calculations have some meaning and interpretation in the real world. One example is that one should be able to improve (update) the probabilities in the light of new statistical evidence. This leads into the direction of a *Bayesian* probability interpretation, where probabilities are considered as the best possible expression of the degree of belief in the occurrence of a certain event. The Bayesian interpretation does not claim that probabilities are direct and unbiased predictors of occurrence frequencies that can be observed in practice. The only claim is that the probabilities will be more or less correct if averaged over a large number of decision situations. The requirement to fulfil that claim, is that the purely intuitive part is neither systematically too optimistic nor systematically too pessimistic. Calibration to common practice on the average may be considered as an adequate means to achieve that goal.

It should be clear that a frequentist's approach does not allow a subjective interpretation of probability and, within this framework knowledge, uncertainties cannot be treated in the same way as inherent and measurable uncertainties. This does not imply that somebody who favours this view on probabilities fully neglects the other types of uncertainty. He will only use "other ways" to deal with them, as for instance classical tools like confidence bounds, safety factors, safety margins and conservative estimates or additional formalisms like fuzzy set theory [Zadeh, 1975], belief mass [Lair, 1999] and so on.

In the Bayesian approach, on the other hand, the two types of uncertainty are treated in the same way. Shortly speaking, the inherent uncertainties are treated in the frequentistic way and the knowledge uncertainties in a degree of belief way. The basic advantage of the Bayesian approach above the other approaches is that the "degree of belief" becomes exactly equal to a "frequentistic probability" in the limiting case of strong evidence like huge statistics or closed theoretical arguments. This property ensures a clear interpretation of the calculations and enables the combination of several sources of evidence. Another advantage is that one has the fully developed and strong theory of probability at ones disposition for both types of uncertainty.

The conclusion of the foregoing is that distributions and probabilities in most civil engineering applications should be given a Bayesian degree of belief type of interpretation. For further discussion on this point the reader is referred to for instance Benjamin and Cornell (1970), Savage (1972), Lindley (1991), Paté-Cornell (1991, 1996), ISO2394 (1998), Ditlevsen (1988, 1996) and Vick (2002).

3. DOES THE "REAL FAILURE PROBABILITY" EXIST?

The subjectivity in the Bayesian approach leads to the problem that the failure probability of some system (in our case the flood protection system for the Netherlands) may depend on the person or the group that is making the design. If another group of experts is asked, they may come up with different estimates. Also when research is going on and new information becomes available, the safety estimate of a dike may change. For many people this is a difficult point to swallow. People like to see the reliability of a dike as an unambiguous property of the dike. The subjectivity is destroying this ideal picture. However, it is worth to realise that the element of unambiguity is also present in the world of the objective probabilities. Also in an objective analysis without any intuition or expert opinion, one may arrive at different probabilities for the same event. In our project we try to "sell this point" by presenting examples of the following type.

3.1 Car example

As an example from the daily practice, consider a car with two types of braking systems, 10 percent of the cars have type A and 90 percent have type B installed. Unfortunately the brakes of type A prove to be unreliable, but it is not known in which car which type of brake is present. Some accidents already have happened. So the car industry invites all cars to come to the garage for a repair.

An arbitrary owner of the car type has driven without any concern until the message comes that something might be wrong. After that he will drive only very carefully and preferably to the closest garage. If one finds there that brakes of type A are present, the drivers' carefulness was justified: his car was unsafe indeed. It will be repaired and he will drive safe again. However, if one finds type B brakes, the car was considered as unreliable without justification (from the posterior point of view) and it can continue its journey safely without anything done. Now notice that the car of type B never changed intrinsically: neither when the message came, nor when it was inspected. The reliability of the car as "a car property" never changed. But nevertheless its reliability to the owner changed. It is very important to understand this distinction. The point

is that usually we cannot base our decisions on the "real reliability" of the car, because it often is unknown. We have to base our decisions on our knowledge of the car. The lack of knowledge is sometimes the only thing that matters and cannot be neglected.

It may also be illustrative to give some numbers for this example: Let the failure probability of the brakes type A be 10^{-6} per km en type B 10^{-9} per km, then the failure probability for the driver of an arbitrary car is given by:

$$p = 0.1 \times 10^{-6} + 0.9 \times 10^{-9}$$
$$= 0.1009 \ 10^{-6} \approx 10^{-7} \text{ per km}$$

The intellectually difficult point for most people will be that there simply is no car with this probability: there exists only cars of 10^{-6} and 10^{-9}, but none of 10^{-7}.

3.2 Dike example

Consider an ideal ground structure (e.g. a foundation or embankment) where all calculation models are without dispute and all statistical models are well known. Let, in this model, the friction angle φ be a spatially varying stochastic field with the following properties.

- the field is described by the mean value $\mu(\varphi)$, the standard deviation $\sigma(\varphi) = 4°$ and the correlation distance $d(\varphi) = 100$ m;
- for the mean $\mu(\varphi)$ one has found in comparable layers a value of 30° on the average with a scatter 10 percent.

We may now proceed in three different directions:

(1) Consider $\mu(\varphi)$ as a random variable, having a mean of 30° and a standard deviation of 3°.
(2) Do a local soil mechanics survey resulting in a deterministically known value for $\mu(\varphi)$.
(3) Do a very detailed local soil mechanics survey leading to a known deterministically known value for φ at every point of the layer; the friction angle φ then in fact stops to be a random variable.

All three situations then lead to a certain value of the failure probability. In most cases we may expect $P_{f1} > P_{f2} > P_{f3}$, but not always. Nevertheless all three failure probabilities may be considered as "objective and real". It emerges that the possibility of reducing the uncertainty is also present for intrinsic uncertainty and not a property of epistemologic uncertainty, as often believed. Note that even without explicit research, the failure probability may change. For instance if the dike has survived a number of load situations without failure, its reliability has increased. How much depends on the variability of the load and the resistance.

13

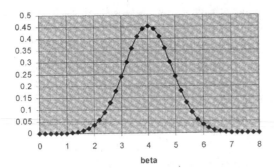

Figure 1. The reliability index β presented as a random variable; the randomness is due to the epistemological uncertainties.

4   PRESENTATION OF THE UNCERTAINTIES

Consider the following simplified limit state function:

$$Z = R - S - E \qquad (1)$$

where R is the resistance part with "real inherent uncertainties", S is the load part with "real inherent uncertainties" and E is the part that represents the "epistemological uncertainties", for instance the statistical uncertainties in the loads or the model uncertainty in the resistance. The values of means and standard deviations have been summarised in the following table:

Var	Type	μ	σ	Description
R	Normal	14	2.0	Resistance
S	Normal	5	1.0	Load
E	Normal	0.0	2.0	Lack of information

The reliability index may then be calculated as $\beta = 3$ corresponding to a failure probability of about 10^{-3}; however, in this analysis the epistemological uncertainties have been treated exactly in the same way as the inherent uncertainties in R and S. A possible indication for the importance of the epistemological uncertainties, of course, is the influence coefficient α as calculated in standard FORM. As an alternative and visually attractive alternative one might present β as a distribution indicated in Figure 1. The curve can be found by calculating the reliability index, conditional upon the variable E:

$$\beta = \Phi^{-1} P(Z < 0 \mid E) \qquad (2)$$

and using this relation for the transformation from E to β. In (2) Φ is the standard normal distribution function. Figure 1 shows that the reliability index would be 4.02 if all epistemological uncertainties would be absent (this is easy to verify). The scatter indicates "how well we know" this index. A narrow band indicates a small influence of the subjective influences, a wide band the contrary. The standard deviation of the curve may be proven to be equal to $\alpha_E / \sqrt{(1 - \alpha_E^2)}$, where α_E is the FORM influence coefficient for the variable E.

The reason that the two types of uncertainty are not considered as fully equivalent is, as said before, that one type is "real" and the other type seems to be present "in the mind of the engineer" only. The "taste" is different and so people want to know what is the one and what is the other. Uncertainty in the mind of the engineer looks less harmless. Maybe the situation is not "really dangerous" after all. Some people would even tend to forget it at all. This of course is not acceptable. It was already mentioned that no designer should ever forget about these uncertainties. If not allowed or accustomed to presenting them by way of probabilities, they will use other methods as indicated in section 2.

Separate presentation gives a feeling of the sensitivity to assumptions and estimates. It is interesting to observe that if one combines the both types of uncertainty, the question to the sensitivity of the assumptions and estimates pops up again. It is true that one will get other answers if different input values are used. However, this may never happen in such a way that a change of say a mean value into another "equally likely value" results in a major difference in final outcome. If that is the case, the uncertainty was not quantified in a proper way. As a rule of thumb one could say that if a shift of about a standard deviation is considered as a shift to an equally possible value, the standard deviation in the analysis was too small.

Another reason to make a distinction between the different types of uncertainty is that uncertainties that have to do with a lack of knowledge, at least at first sight, might be removed by doing research (model uncertainty) or gathering data (statistical uncertainty) and "inherent uncertainties" cannot. So if there is a large failure probability and the inherent uncertainties dominate, we go for redesign or reconstruction, while in case the epistemological uncertainties dominate we will start investigations. After further reflection, however, this matter is more complex. Some statistical uncertainties cannot be reduced because it is physically impossible to gather them within the time available to take the decision. We cannot gather the wind speeds for the next 1000 years just tomorrow. And on the other hand, as we have seen in the dike example of section 3, time independent inherent uncertainties like soil properties can be reduced. Faber (2003) even proposes to use the word epistemological for all uncertainties that can be reduced. Structural properties

Table 1. Types of uncertainty (example).

Uncertainty	Type	Can be reduced	Examples
In time	Inherent	No	Sea level, wind speed
In space	Inherent	Yes	Ground parameters
Statistical uncertainty (time)	Epistemo-logical	No	Wind and sea level statistics
Statistical uncertainty (space)	Epistemo-logical	Yes	Kriging-procedure
Model uncertainty	Epistemo-logical	Yes	Slip circle model

then change then from inherent before building to epistemological after building.

In the ONIN project (Slijkhuis, 1998) initiated by Rijkswaterstaat in the Netherlands to discuss this issue, the subdivision of uncertainties was proposed (See table 1).

This way already two different subdivisions of the uncertainties are visible. In the project even 5 different options were considered.

5. LACK OF DATA FOR INHERENT RANDOMNESS

The whole system of flood protection structures in the Netherlands is some thousands of kilometres long and for a large part of it data is not complete. This holds in particular for the many geometrical and material properties of the inside parts of the dike. For many parts of a dike ring there may not be any data at all. This is an extreme case of statistical uncertainty. Physically it is possible to gather all the data by extensive soil mechanics surveys. This however is not feasible from the economical point of view. The pragmatic strategy that has been formulated is the following:

(1) Start by using conservative estimates for the unknown parameters, using expert opinions, supported by data of other related places.
(2) If the resulting failure probability is uncomfortably high, calculate the failure probability also for more likely values of the parameters.
(3) If the difference of (2) and (3) is large the knowledge uncertainty is substantial, so perform a full probabilistic analysis including all statistical and model uncertainties.
(4) If the result is still unsatisfactory: reduce the uncertainty by doing measurements and observations.
(5) If the difference between (2) and (3) is small or if the resulting failure probability is still too high

after step (4), the dike has to be strengthened in one of the many possible ways.

As an alternative one might start with (3) and have a look at the FORM α-coefficients of the relevant parameters in the end result. The above approach, however, seems to more understandable to the non trained engineer. It appeals a bit on the popular worst case scenario thinking. The basic message is that large knowledge uncertainties are acceptable as long as they do not have a disturbing influence on the final result.

The strategy formulated above for the parameters may hold equally well for model scenarios. For instance, one does not know whether a certain thin clay layer is present in the sub soil. One may have to model the subsoil including and excluding the clay layer, and make two calculations, giving each one a certain amount of belief. Finally they can be combined using the Total Probability Theorem (step3).

Another issue in this respect is that one does not want to incorporate all dike sections of a dike ring in the analysis for all the mechanisms. The total number may be up to 50 or 100. The procedure is to select, again based on engineering judgement, a limited set of sections that may be expected to be the most dominant ones. If, for a certain mechanism, the contribution of say the 5 most dangerous believed sections is very limited one may conclude that the other dike sections may be neglected. If not, some more sections will be included. Even if the most dominant sections were present, the inclusion of more sections is necessary because of possible series system effects on the reliability.

6 MODEL UNCERTAINTIES AND EXPERT OPINIONS

Within the Bayesian framework lack of data may be replaced by engineering judgement and expert opinions. We may ask an expert to estimate the probability that a certain event will happen or we may ask his best guess for a certain variable. Additionally we may ask him to express his uncertainty with respect to the number given. In the case of a random variable we may also ask directly for the whole distribution. Essentially, however, this comes down to ask for a mean, a standard deviation and/or a set of other parameters. So it is not necessary to discuss this separately.

In some cases we have only a single expert and in other cases a group. In the last case, which by the way is to be preferred, we have to combine the results. Let us start the discussion with the simple case that we ask two experts for the probability of event A occurring. Suppose we get the answers:

Expert 1: P(A) = 0.8
Expert 2: P(A) = 0.6

The most obvious thing now to do now is to conclude is that we will use 0.7 in our analysis. But is this always correct? It turns out that the details of the background are important. Consider first the following case: expert 1 looks for typical characteristics in the problem at hand that may lead to the occurrence of A or not. Let it be his experience that he is correct in 80 percent of the cases. A similar mechanism may be present behind the reasoning of expert 2. If their indicators are independent and if there is no prior preference regarding A or $\overline{A}$ (that is $P(A) = P(\overline{A}) = 0.5$ if no expert is consulted), one may derive on the basis of Bayes' Theorem:

$$P(A \,|A_1\, A_2) = c\, P(A_1 \cap A_2 \,|A)\, P(A)$$
$$= c\, 0.6\, 0.8\, 0.5 = 0.86$$

$$P(\overline{A} \,|A_1\, A_2) = c\, P(A_1 \cap A_2 \,|\overline{A})\, P(\overline{A})$$
$$= c\, 0.4\, 0.2\, 0.5 = 0.14$$

Here A_1 indicates that expert 1 "expects A" and A_2 that expert 2 "expects A"; c is a normalising constant. The result is that adding expert 2 increases the probability that A will occur, rather then that it decreases. This makes sense: we have already the opinion of expert 1 that A most probably will occur and we find another expert who more or less supports his opinion. This should increase the probability that A will occur. It is only the non-expert, who can only predict A with a probability of 0.5, that will not increase the estimate by expert 1.

However, this is not the only way the experts may have arrived at their opinion. In another situation, maybe, both experts had just observed a small number of observations, say expert 1 had seen A occurring four times out of five similar situations and expert B had seen it happen only 3 times. If those observations were not overlapping one might in that case indeed better of with the first approximation of simply averaging the estimates.

Of course one might say that these are not purely expert opinions as in both cases some elementary observations and data are being used. On the other hand, there is always some kind of reasoning behind the estimate of the expert. It is absolutely essential to get this reasoning on the table as it may help to construct a combined judgement.

In a similar way we may ask a number of experts the value of X representing some deterministic physical entity. Let us aks them for their best guess and let us also ask them to quantify their uncertainty. Assuming normal distributions we them might have:

$$F(X) = \frac{1}{n} \Sigma\, \varphi \left(\frac{X - \mu_i}{\sigma_i} \right) \tag{3}$$

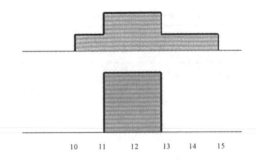

Figure 2. Two possible options to combine expert opinion 1 (X is uniform in [10,13]) and expert opinion 2 (X is uniform in 11,15]).

This seems a very straightforward approach, but again we have other options. First let us consider a slightly different case where we use a block distributions instead of normal distributions. So then the result for instance of two experts could be:

Expert 1: X is in the interval [10, 13]
Expert 2: X is in the interval [11, 15]

Following the procedure from above we would get the distribution indicated in Figure 2, upper diagram. However, following elementary rules from formal logic we should conclude that X must lie with 100 percent probability in the interval [11,13], resulting in Figure 2, lower diagram. Of course, this requires that we trust the experts for a full hundred percent, and this usually is exactly the bottleneck of experts opinions.

The equivalent of this second type of reasoning for normal distributions could be reached by considering the experts opinions to be data of X_i, where every measurement has some uncertainty. We use Bayesian updating formula. There are two approaches:

(1) We consider the uncertainties defined by the experts as "calibrated measurement uncertainties". We then have:

$$f'(X) = C \prod \varphi \left(\frac{X_i - X}{\sigma_i} \right) f(X) \tag{4}$$

Where $f(X)$ is a prior estimate, X_i and σ_i are the best estimate and uncertainty from expert i (i = 1 ... n) and $f''(X)$ is the updated distribution for the unknown quantity X. This approach is proposed Mosleh and Apostolakis (1982, 1986).

(2) We treat the best guesses of the experts just as data points in an experiment to find X. The measurement error equals the scatter in the estimates, enlarged by the uncertainty expressed by each

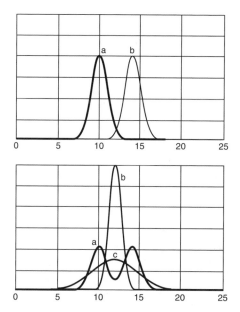

Figure 3 above. Two expert opinions: (a) expert 1 X = 10 with uncertainty σ = 1; expert 2 X = 14 with σ = 1;
Figure 3 below. Combination of expert opinions according to equation (3) (line A), equation (4) (line B) and equation (5) (line C).

individual expert. In that case, using Bayesian analysis, we find:

$$f(X) = \iint \varphi \left(\frac{X - \mu_X}{\sigma_X} \right) f''(\mu_X \, \sigma_X) \, d\mu_X \, d\sigma_X \qquad (5)$$

where

$$f''(\mu_X \, \sigma_X) = C \prod \varphi \left(\frac{X_i - \mu_X}{\sqrt{\sigma_X^2 + \sigma_i^2}} \right) f(\mu_X \, \sigma_X)$$

For this estimate one also needs a prior. We may refine the procedure by giving every expert not just one data point, but by letting his estimate count for a number of say 3 or 10 data points. Especially in the case of one expert this is necessary in order to avoid extremely large uncertainties.

Figure 3 shows a number of results for the case of two experts. We consider the situation that the two experts have different views. One expert thinks that X = 10 and the other one that X = 14. Both give an uncertainty of σ = 1. Which is the alternative to be chosen? Equation (4) is attractive and is the equivalent from our "formal logic approach" in Figure 2, lower diagram. But which engineer is inclined to put so

Table 2. Examples of model uncertainties found by expert elicitation in The Netherlands.

	Mean	Scatter
Local water levels along rivers	0.0	$\sigma = 0.15\,m$
Wave development due to wind	1.0	V = 0.15
Wave run up on the dike slope	1.0	V = 0.50
Critical crest discharge	1.0	V = 0.50

much faith in his experts that he really dares to make the choice. Equations (3) and even better (5) seem to be more acceptable from this point of view. Different views by experts is a clear indication that things have not been settled and one should not be too optimistic. In addition one should keep in mind that experts often tend to be overconfident in expressing the accuracy of their own estimates (e.g. De Wit, 2001).

Until now we have treated all experts to be equally skilful. An interesting option is to give the experts a number of questions for which they do not know the answer, but the interrogaters do. We then may adjust the weights of the experts on the basis of their score on these "seed questions". We will not go into this matter in this paper. The reader is referred to Cooke (1991).

In order to get some impression of the reliability of the various models used in the Dutch safety project for the flood protection systems, a number of experts have been interrogated in order to get a grip on the accuracy of the various models used (Cooke et al., 1998). In this particular project the experts were asked to give a 0.50 fractile as best estimate and express their uncertainty by stating the 0.05 and 0.95 fractiles.

The distributions obtained this way have been added by giving a weight to every expert. One procedure was to give equal weights and the second one was to give weights based on how the experts scored on a number of questions for which the interviewers knew the expert answer but not the experts. There were two criteria: the answers should be within the margins of the expert and the margins should not be too narrow, otherwise the expert would be of no help. Details of the analysis can be found in Cooke and Cooke. Some results are, for the case of illustration, presented in Table 2.

7 CONCLUSIVE DISCUSSION

The future design and assessment of the sea and river flood protection systems in the Netherlands will be explicitly based on risk analysis principles. The failure probabilities to be calculated in such a procedure will be influenced by a large number of knowledge uncertainties. This raises questions on how these uncertainties should be dealt with, both in preparing input for the calculations as well as in presentations and drawing conclusions with respect to possible flood protection measures.

This difficult and highly philosophical issues are now brought under the direct attention of engineers not trained in this type of reasoning and even civil servants and politicians. It is a challenge for the reliability community to convince all those groups that this is still the most rational way to proceed and that all other methods suffer from exactly the same difficulties, although they may be less obvious on the surface.

Both in the technical and the political society however, there is a tendency to distinguish as long as possible between inherent and epistemological uncertainties. There is sometimes even a reluctance to quantify knowledge uncertainties. Making safe side assumptions seems to be more popular. By careful communication, however, it is possible to arrive at procedures that appeal on the one side to the intuition of the average engineer who is not trained in reliability theory and on the other side to proper and theoretically correct Bayesian reliability concepts.

REFERENCES

Benjamin, J.R. and Cornell, C.A. (1970), *Probability, Statistics and Decision for Civil Engineers*, McGraw-Hill, New York.

Buijs, F.A., et al. (2003), *Application of Dutch reliabilty based flood defence design in the UK*, European Safety and Reliability Conference ESREL 2003 Maastricht, The Netherlands.

Cooke, R.M. (1991), *Experts in Uncertainty. Opinion and Subjective Probability in Science,* Oxford University Press, U.K. 1991.

Cooke, R.M. (1998), *Expertmeningen Onzekerheidsanalyse,* Rapport WB 1083, Rijkswaterstaat, The Netherlands (in Dutch).

Ditlevsen, O. (1988), *Uncertainty and Structural Reliability. Hocus Pocus or Objective Modelling?* Danmarks Tekniske Højskole, Lyngby.

Ditlevsen, O. and Madsen, H.O. (1996), *Structural Reliability Methods,* John Wiley & Sons, Chichester, U.K. 1996.

Faber, M.H. (2003), *Uncertainty Modelling and Probabilities in Engineering Decision Analysis*. Proceedings to the International Conference on Offshore Mechanics and Arctic Engineering in Cancun, Mexico (in press).

Ferry-Borges, J. and Castanheta, M. (1971), *Structural safety*, 2nd Edition, National Civil Eng. Lab., Lisbon, Portugal.

International Standard ISO 2394, second edition (1998), *General principles on reliability for structures,* Switzerland.

JCSS. (2001), *Probabilistic Model Code for Reliability Based Desig*n. The Joint Committee on Structural Safety.

Kok, M., Stijnen, J.W. and Silva, W. (2003), *Uncertainty analysis of river flood mangement in the Netherlands,* European Safety and Reliability Conference ESREL 2003, Maastricht, The Netherlands.

Lair, J. (2000), *Evaluation de la durabilité des systèmes constructifs du bâtiment*. Université Blaise Pascal de Clermont-Ferrand, France.

Lassing, B.L., et al. (2003), *Reliability analysis of flood defence systems in the Netherlands*, European Safety and Reliability Conference ESREL 2003, Maastricht, The Netherlands.

Lindley, D.V. (1976), *"Introduction to Probability and Statistics from a Bayesian Viewpoint"*, Cambridge, Cambridge University Press.

Mosleh, A. and Apostolakis, G. (1982), *Models for the Use of Expert Opinions*, Workshop on low-probability high-consequence risk analysis, Society for Risk Analysis, Arlington, Va.

Mosleh, A. and Apostolakis, G. (1986), *The Assessment of Probability Distributions from Expert Opinions with an Application to Seismic Fragility Curves*, Risk Analysis, vol. 6, no. 4, pp. 447–461.

Paté-Cornell, M.E. and Fischbeck, P.S. (1991), *Aversion to epistemic uncertainties in rational decision making: effects on engineering risk management,* Engineering Foundation Conference on Risk-based decision making Santa Barbara, November 1991.

Paté-Cornell, M.E. (1996), *Uncertainties in Risk Analysis: Six Levels of Treatmen*t, Reliability Engineering and Systems Safety 54, Elsevier Science Limited, Northern Ireland. pp. 95–111.

Savage, L. (1954), *The foundation of statistics*, Wiley, republished Dover, New York.

Slijkhuis, K.A.H., et al. (1998), *Probability of flooding, an uncertainty Analysis*, ESREL 1998.

Vick, S.G. (2002), *Degrees of Belief. Subjective Probability and Engineering Judgment,* ASCE Press, Reston, U.S.A.

Wit, de, M.S. (2001), *Uncertainty in predictions of thermal comfort in buildings,* Delft University The Netherlands.

Zadeh L. *Fuzzy sets and their applications to cognitive and decision process*, Academic Press, New York, 1975.

Safety and Reliability – Bedford & van Gelder (eds)
© 2003 Swets & Zeitlinger, Lisse, ISBN 90 5809 551 7

Reliability modeling and optimization for distributed fault-tolerant computing systems

G. Albeanu
University of Oradea, Oradea, RO

Fl. Popentiu-Vladicescu
City University, London, UK

L. Serbanescu
Astronomic Institute of the Romanian Academy

ABSTRACT: The optimal design of the fault-tolerant distributed systems, from the point of view of the reliability requirements, is considered. A Monte Carlo controlled search method is used to estimate the failure rates appearing in a constraint optimization model.

1 INTRODUCTION

Reliability is an important feature of the quality of any large information system. Recently, distributed data bases, distributed objects and remote hardware resources have been used in order to provide on-line information processing. Designing and managing distributed systems with predictable reliability and availability are difficult tasks. A good architecture based on optimal reliability allocation is needed.

Different reliability models are available for distributed computing systems. Reliability block diagrams and Markov models are used by Wattanapongsakorn & Levitan (2001), for the reliability modeling of a large three-tiered e-commerce site. Graph theory and combinatorial methods are used to analyse deterministic and probabilistic networks (Lin et al. 1999; Shier 1991). When imperfect nodes are considered, efficient algorithms to evaluate the distributed software reliability are proposed by Lin et al. (1999). Also Loman & Wang (2002) and Nguyen (2001), report some modeling applications. Such approaches and the techniques described by Albeanu & Popentiu (2002), Popentiu et al. (2001) and Wattanapongsakorn Levitan (2001) can be integrated to optimize the architecture of fault-tolerant distributed computing systems.

Previous results in the fault-tolerant distributed systems (Andres et al. 1997; Sens & Folliot 1998; Folliot & Sens 1999; Popentiu et al. 2001; Popentiu &

Sens 1999) and in the optimal reliability allocation for hardware and software projects (Albeanu & Popentiu 2002) already obtained, encourage us to start a project which integrates models and fast algorithms to provide an optimal architecture.

This paper describes both models and algorithms concerning the optimal design of the distributed systems from the viewpoint of the reliability requirements.

2 FAULT-TOLERANT ARCHITECTURES AND RELIABILITY REQUIREMENTS

Designing and managing complex distributed computing systems with predictable reliability and availability is generally difficult. The development of reliable systems can be characterized by two different approaches: fault-prevention and fault-tolerance. The prevention approach deals with methodologies, testing and other software engineering techniques that help the developer to avoid the introduction of the faults into the system. Fault-tolerance acknowledges the existence of faults and provides techniques for dealing with failures. Fault-tolerance is mainly characterized by the following metrics: reliability, availability, mean-time-to-failure (MTTF), and expected number of failures.

The reliability of a distributed fault-tolerant computing system can be expressed by analyzing both the distributed software reliability (DSR) and distributed hardware reliability (DHR). One of the most suited approaches to formulate these reliability performance indexes is to generate all disjoint file spanning trees in the distributed computing system graph such that the DSR and DHR can be expressed by the probability that at least one of the file spanning trees is working for the period of time under the operation conditions encountered.

The availability can be characterized as the probability that a service will be available when clients attempt to use it. It is important to mention that the availability of the service can be high even if it has relatively frequent failures.

The value MTTF is the expected time until the system fails, and the expected number of failures is the number of failures expected during a certain period of time. Another important metrics is the time to repair a system: mean-time-to-repair (MTTR). When optimal aspects are addressed, such metrics will be incorporated into the mathematical models formulated as to capture important information about failure allocation with minimum cost.

Avoiding the introducing of errors both in hardware and software or making hardware and software less sensitive to failures is possible by hardware and/or software replication strategies.

The architecture of a software system supporting fault-tolerating computing is based on modules providing the following functions: failure detection, recovery and coordination. The failure detection module detects when a server has failed. For complex systems, the failure detection modules have to detect if one component in a set of replicated components provides incorrect answers. The recovery module helps the system to recover from a failure state. For distributed data base management applications, the recovery process is based on logs (files registering transaction states) which are processed to update previous state and to continue the processing. The coordination module coordinates the answers received from different replicated modules.

In transaction systems, a series of actions is treated as one unit, called transaction. If all the actions are successful, the transaction is committed. The requested updates are guaranteed only if the commit process is successful. When transaction fails, the actions within the transaction are not visible outside the transaction. That is, the fault tolerant attribute is applied in the sense that the system can expose failures without going to inconsistent states.

The most reliability and availability metrics, for transactions based systems, are application dependent.

Another class of fault-tolerant systems is dedicated to message delivering. To increase the system's reliability, a local message queue handler is used. Once the client module delivers the message to the message queue handler, the client is relieved from any

additional concerns of delivering the message. The message queue handler delivers the message to the server. If the server is down when the client sends the message, the message queue handler waits until the server restarts. If the message queue handler is down, the message remains in the storage. Only when the message queue is recovered, the message will be delivered to the next destination.

The N-version programming approach requires voting and coordination among several servers and, usually asks for high development and maintenance costs. It is not clear how such an approach will produce reliable software, in practice, because different programmers can introduce errors, for the same functions of a module specification. Due to the human nature, the economical results are not as good as expected. However, using this approach, it is possible to improve the availability of replicated systems.

Reliability modeling is an important task in software reliability engineering. The software designer can set and interpret reliability requirements, predict the reliability of different system configuration, evaluate alternatives etc. This is why different types of reliability models are used: parts-count models, combinatorial models, and state-space models.

For real distributed computing systems, an important amount of time is used to obtain the expression of the reliability function. In the following, a simple model is used to illustrate the optimization issue.

3 RELIABILITY ALLOCATION MODELS

3.1 *Reliability evaluation functions*

Let n be the number of software components, σ the reliability system target ($0 < \sigma < 1$) and T be the length of the mission interval ($T > 0$). In order to simplify the presentation, the time interval $[0, T]$ will be considered. Let m be the number of operations, and p_i the probability of executing operation i ($0 < p_i < 1$, $i = 1, 2, \ldots, m$) according to the operational profile investigation. A natural assumption considers that

$$\sum_{i=1}^{m} p_i = 1.$$

Let φ_{ij} be the expected proportion of the execution time i spent using component (node) j ($0 \leq \varphi_{ij} \leq 1$, $i = 1, 2, \ldots, m; j = 1, 2, \ldots, n$).

Let Φ denotes the matrix $(\varphi_{ij})_{i=1, \ldots, m; j=1, \ldots, n}$. The following equality holds:

$$\sum_{i=1}^{n} \varphi_{ij} = 1, i = 1, 2, \ldots, m.$$

When the distributed software reliability is under computation, the most natural approach is based on the fact that the reliability characteristics are independent of the age of the software product. That is why the exponential model is the most suited choice in modeling. However, the performance of the software would decrease in time due to the inappropriate temporary file management (less free space for virtual memory). Therefore, the software without periodical maintenance would loose fractions of initial running speed, with important drawbacks for the world of safety-critical applications. This will affect the quality of the software from the customer's point of view.

Let us consider the exponential model. For all $j = 1, 2, ..., n$, let λ_j be the failure intensity of the jth component. Hence, the probability density function of the random variable giving the time to failure of the jth module is:

$$f_j(t) = \lambda_j \exp(-\lambda_j t),$$

and, according to Catuneanu & Mihalache (1989), the corresponding reliability function is:

$$R(t) = \exp(-\lambda_j t).$$

The purpose of the investigation is to determine the intensity failures when the time t is given. Let $R(\lambda_j; t)$ be the reliability of the jth component: $R(\lambda_j; t) = \exp(-\lambda_j t)$. For a serial composition, the reliability of the system can be computed by the product:

$$R(\lambda_1, \lambda_2, ..., \lambda_n; t) = \prod_{j=1}^{n} R(\lambda_j; t).$$

When T is the time mission (according to the reliability contract), p_i is the probability of executing the operation i, and φ_{ij} is the time allocated to the component j, then

$$\tau_j = \sum_{i=1}^{m} p_i \varphi_{ij}$$

is the expected proportion of the total mission time that the software spends executing in component j. Therefore, the reliability of the jth component with respect to the proportion of time it runs is:

$$R(\lambda_j; T) = \exp\left(-\sum_{i=1}^{m} p_i \varphi_{ij} \lambda_j T\right) = \exp(-\tau_j \lambda_j T).$$

Finally, the entire system reliability with respect to time mission T is given by (Albeanu & Popentiu 2002):

$$R(\lambda_1, \lambda_2, ..., \lambda_n; T) = \exp\left(-\sum_{j=1}^{n} \sum_{i=1}^{m} p_i \varphi_{ij} \lambda_j T\right).$$

When a fault-tolerant architecture is built, the reliability function is defined adequately. Wattanapongsakorn & Levitan (2001) consider two fault-tolerant architectures: N-version Programming (NvP) and Recovery Block (RB) approaches. The NvP model assumes a voter and N (an odd number) independent developed software versions functionally equivalent. All N software versions are executed for the same operation at the same time, their outputs are received and analyzed by the voter which decides the correct answer (in general, using the majority criterion). The RB approach considers a special module running an acceptance test and at least two alternate software components. When the system is started, the first alternate component executes the operation. If its output is not accepted by the special module, the process will roll back to the initial state and the second alternate executes the operation and the output will be tested for acceptance. This game continues until the output of an alternate is accepted or all outputs have been rejected.

For reliability allocation reasons, some parameters have to be evaluated (example: the probability that an unacceptable result occurs during an operation) or considered for estimation after an optimization process.

For a general model, intensity failure rates, and finally the reliability function, are functions depending on some explanatory variables, like: component complexity, execution time etc. Let $\theta = (\theta_1, \theta_2, ..., \theta_S)$ be the parameters describing the intensity failure function, in a linear or nonlinear model based on the explanatory variables. For this case, the analytical expression of the reliability function is a complicated one, and will depend on the parameter θ. In the context of this paper, applying optimization techniques it is possible to obtain an estimation of the unknown θ, and then to estimate the intensity failure rates for prescribed levels of reliability.

3.2 The cost modeling

Boehm (1981) classifies the mathematical models for software cost estimation, which are definable in algorithmic manner, in the following branches: (1) linear models; (2) multiplicative models; (3) analytic models that are neither linear nor multiplicative; (4) tabular representation: cost driver – development effort, and (5) combinative models. Linear models try and fit a simple line to the observed data. A multiplicative model describes the cost effort as a product of constants with various cost drivers as their exponents. In a tabular model, a matrix describing the relationship between cost drivers and development resources is given. Composite models are general enough. The most used models in cost estimation are the COCOMO model and the Putnam's SLIM model (Albeanu & Popentiu 2002; Fenton, N.M., Pfleeger 1996).

The cost driver attributes, in the COCOMO model, are grouped in four classes: (a) product attributes, including required software reliability; (b) hardware attributes; (c) personnel attributes, and (d) project attributes. Each attribute is rated on a η-point scale (at least 6 levels), and an effort adjustment factor is obtained by multiplication.

The Putnam's algorithm, based on the Norden/Rayleigh function, it is known as a macro estimation model. SLIM (Software Lifecycle Management) model enables a software cost estimator to perform the following functions (Fenton & Pfleeger 1996):

(a) the calibration of the model to represent the local software development environment by interpreting a historical database of past projects;
(b) the econometric modeling of the software system, collecting software characteristics, personal attributes, computer attributes, and
(c) the estimation of the software size. The formula proposed is

$$E = (NLOC/(C * t^{4/3})) * 3,$$

where E is the total life cycle effort in working years, t is the development time in years, C is the technology constant (varying from 610 to 57314), combining the effect of using tools, languages, methodology and quality assurance, and NLOC is the number of lines of code.

Considering the failure intensity as a parameter, the cost function must decrease as the failure intensity gets larger. Also the function cost must be a convex one relative to the failure intensity. Such models (a linear model, a logarithmic-exponential model and the inverse power model) are used by Albeanu & Popentiu (2002).

The difficult aspects of the cost modeling process are related to the reduction of data and the parameter estimation. However, working with approximations obtained by different deterministic or stochastic methods could be a good solution. A stochastic approach will be applied when a model based on software complexity and cost drivers will be fully considered.

Let $T_C (\lambda_1, \lambda_2, ..., \lambda_n)$ be the total cost of achieving the failure intensities $\lambda_1, \lambda_2, ..., \lambda_n$. A first approach is based on a separable form of the total cost function:

$$T_C(\lambda_1, \lambda_2, ..., \lambda_n) = \sum_{i=1}^{n} C(\lambda_i),$$

with $C(\lambda_i)$ the cost of developing the component i.

If a non-separable model is considered or a non-convex one is given, then the numerical processing by deterministic methods could not provide a global optimum. In such a case, stochastic algorithms have to be applied.

3.3 Optimal reliability allocation

Let $\Lambda = \{(\lambda_1, \lambda_2, ..., \lambda_n) \in \mathbf{R}^n \,|\, \lambda_i \geqslant 0, i = 1, 2, ..., n\}$. The minimization of the total cost of achieving the target reliability σ it is modeled by (Albeanu & Popentiu 2002):

$$\begin{cases} \min \ T_C(\lambda) \\ R(\lambda;T) \geq \sigma \\ \lambda \in \Lambda. \end{cases}$$

Denoting $g_0(\lambda) = \sigma \ln(\sigma) - \sigma \ln R(\lambda;T)$, and $g_j(\lambda) = -\lambda l_j, j = 1, 2, ..., n$, the model can be written as:

$$\begin{cases} \min \ T_C(\lambda) \\ g_j(\lambda) \leq 0, j = 1,2,..., n, \end{cases}$$

which is a standard form in the nonlinear optimization.

Another allocation model is based on the maximization of the system reliability subject to a budget constraint:

$$\begin{cases} \max \ R(\lambda;T) \\ T_C(\lambda) \leq B \\ \lambda \in \Lambda, \end{cases}$$

where B is the maximum cost effort allowed. Also, a standard form can be obtained:

$$\begin{cases} \max \ R(\lambda;T) \\ T_C(\lambda) - B \leq 0 \\ -\lambda_j \leq 0, j = 1, 2, ..., n. \end{cases}$$

It is obvious now, that any optimization method (deterministic or stochastic) can be used depending on the properties of the functions which appear in the model. A standard Kuhn-Tucker approach is described in Helander at al. (1998). Other methods are used by Wattanapongsakorn & Levitan (2001). Our approach is based on stochastic search. A Monte-Carlo controlled search algorithm is used in a 'divide and conquer' approach.

4 OPTIMUM SEARCH METHODS

Let consider the following optimization problem:

$$\begin{cases} \mathrm{opt} \ f(\lambda) \\ g_j(\lambda) \leq 0, j = 0,1,2,...,n, \end{cases}$$

where "opt" is *min* for the first problem, respectively *max* for the second. The function f is the goal function and g are the appropriate constraint.

Let $D = \{\lambda \mid g_j(\lambda) \leq 0, j = 0, 1, ..., n\}$. Then the optimization problem consist of finding f^* and λ^* so that $f^* = f(\lambda^*) = \text{opt } \{f(\lambda) \mid \lambda \in D\}$. For practical implementation D is generated as a subset of $X := [u_1, v_1] \times [u_2, v_2] \times \cdots \times [u_n, v_n]$, $u_i < v_i$, $i = 1, 2, ..., n$.

Some major classes of stochastic search methods can be used depending on the desired accuracy or the functions f and g_j properties. Any stochastic algorithm consists of three distinct major parts: a sampling part, an optimization procedure and a stopping test. For the sampling part, the researcher provides the cardinality of the sample (fixed or adaptively updated) and a sampling procedure. For the above mentioned situation, points are drawn in X as uniform random numbers and rejecting all points outside D. New points are generated in the same manner until a stopping rule is valid or the search domain is splitting in subsets according to Albeanu (1997). For the general case, no local search is performed, that means an algorithm to identify the optimum point (for maximum or minimum) is applied. However, a local search of single-start or multi-start type can be used. For the subject under discussion a controlled search is chosen. The method considers a decreasing convergent sequence $(h_k)_{k>0}$ with zero limit and generates a sequence $F_0, F_1, ..., $ of sets which contains pairs (λ, y), with λ in D and $y = f(\lambda)$. The procedure is a computer intensive one, but works for general situation. The procedure will be described for $n = 2$. However, it can be implemented for the general case. Let $k \geq 1$,

$$S_{st}^{(k)} = [u_1+sh_k, u_1+(s+1)h_k] \times [u_2+th_k, u_2+(t+1)h_k] \cap D,$$

$$\Delta^{(k)} := \bigcup_{s=0}^{p_{(k),1}} \bigcup_{t=0}^{p_{(k),2}} S_{st}^{(k)},$$

where $p_{(k),i} = [(v_i - u_i)/h_k]$. Let q be the maximum number of iterations, ε a threshold used to select an optimum for f, ε_1 a threshold used to choose between two function values and ε_2 a threshold used to select between two candidates. The sequence $(F_k)_{k \geq 0}$ can be generated according to the following steps:

1. /Initial/ $F_0 = D \cap ([u_1, v_1] \times [u_2, v_2])$, $k = 1$;
2. /Iteration/ Repeat steps 2.1 to 2.5 until $k > q$
2.1. Generate the set $\Delta^{(k)}$ and establish the sample size matrix $N_{ls}^{(k)}$
2.2. Generate $N_{st}^{(k)}$ uniform random vectors in every set $S_{st}^{(k)}$, $s = 0, 1, ..., p_{(k),1}$; $t = 0, 1, ..., p_{(k),2}$.
2.3. Evaluate the corresponding function values for every set $S_{st}^{(k)}$ and store in f_{st}^* the optimum value and in $\lambda_{st}^{(*)}$ the optimization point from $S_{st}^{(k)}$ only if $|f_{st}^*| < \varepsilon$. This step selects the optimization point needed in the step 2.4.
2.4. /Reduce/ Let G_k be the set of all pairs (λ^*, f^*) such that $\lambda^* \in \Delta^{(k)}$ is an optimization point

selected in the previous step, and $f^* := f(\lambda^*)$. Obtain F_k from G_k by a reducing rule:
For any pair $((\lambda_1^*, f_1^*), (\lambda_2^*, f_2^*))$ such that $|f_2^* - f_1^*| < \varepsilon_1$ and $\|\lambda_2^* - \lambda_1^*\| < \varepsilon_2$, insert into the set F_k the closer point according to the optimum criterion.
If a minimization problem is considered then add the corresponding pair according to min $\{f_2^*, f_1^*\}$.
2.5. /Divide/ For every point in F_k a small neighborhood is considered and a grid size $h_{k+1} < h_k$ is built. Set $k := k + 1$ and iterate from step 2.1.
3. Stop, with F_q the set of all pairs candidate to the optimum solution.

The described method will find the global optimum even for non-convex models.

5 CONCLUSIONS

Designing and managing complex distributed computing systems with predictable reliability and availability is generally difficult. The paper presents a general framework to obtain different parameters in order to optimize an objective function. However, the reliability function has to be obtained for any computing system (difficult enough when fault-tolerant systems are considered). This explains why the researchers use particular software tools for reliability allocation.

Numerical experiments, when simple architectures are considered, show a good behavior, but for complex architectures are expensive enough from the resources involved.

REFERENCES

Albeanu, G. 1997. A Monte Carlo approach for controlled search. *Mathematics and Computers in Simulation* 43: 223–228.
Albeanu G. & Popentiu, Fl. 2002. Optimal Reliability Allocation for Large Software Projects. In *Optim 2002 – The 8th International Conference on Optimization of Electrical and Electronic Equipment*: 602–606. Brasov: Transilvania University.
Andres, F., Folliot, B., Cadinot, P., Kaneko, K., Makinouchi, A., Ono K. & Sens P. 1997. The TOSDHIM System Management of Distributed Heterogeneous Multimedia Information. *Proc. of 8th Int. Workshop on Database and Expert Systems Applications (DEXA'97)*, Toulouse: IEEE Computer Press.
Boehm, B.W. 1981. *Software Engineering Economics*. Englewood Cliffs, N.J: Prentice Hall.
Catuneanu, V.M. & Mihalache, A.N. 1989. *Reliability fundamentals*. Amsterdam: Elsevier.
Fenton, N.M. & Pfleeger, S.L. 1996. *Software metrics: A rigorous and practrical approach*. London: International Thomson Computer Press, PWS Publishing Company.

Folliot, B. & Sens, P. 1999. Load Sharing and Fault Tolerance Manager. In Rajkumar Buyya (ed.) *High Performance Cluster Computing: Architectures and Systems*: Chapter 22, Vol 1, NJ: Prentice Hall PTR.

Hecht, M. 2001. Reliability/Availability Modeling and Prediction for E-Commerce and other Internet Information Systems. *Proc. Ann. Reliability & Maintainability Symp.*: 176-183. Philadelphia: IEEE.

Helander, M.E., Zhao, M. & Ohlsson, N. 1998. Planning models for software reliability and cost. *IEEE Trans. Soft. Eng.* 24: 420–434.

Lin, M.S., Chen, D.J. & Horng, M.S. 1999. The Reliability Analysis of Distributed Computing Systems with Imperfect Nodes. *The Computer Journal* 42(2): 129–141.

Loman, J. & Wang, W. 2002. On Reliability Modeling and Analysis of Highly-Reliable Large Systems. *Proc. Ann. Reliability & Maintainability Symp.*: 456–460. Seatle: IEEE.

Nguyen, D. 2001. Reliability Modeling and Evaluation in Real-time Distributed Multimedia Systems. *Proc. Ann. Reliability & Maintainability Symp.*: 200-206. Philadelphia: IEEE.

Popentiu, Fl., Albeanu, G., Sens, P. & Thyregod P. 2001. Software Architecture for distributed Systems: NN and EV approaches. In E. Zio, M. Demichela, N. Piccinini (eds.) *ESREL 2001. Proc. of the European Conference On Safety and Reliability*: 1031–1039. Torino: Politechnico di Torino.

Popentiu, Fl. & Sens, P. 1999. A Software Architecture for Monitoring the Reliability in Distributed Systems. *European Safety and Reliability Conference (ESREL '99)*. Munchen.

Sens, P. & Folliot, B. 1998. The STAR Fault Tolerant manager for Distributed Operating Environments. *Software Practice and Experience.* 28(10): 1079–1099.

Shier, D.R. 1991. *Network Reliability and Algebraic Structures*. New York: Oxford University Press.

Taghelit, M., Haddad, S. & Sens, P. 1991. An Algorithm Providing Fault Tolerance for Layered Distributed Systems. *Proc. of the IMACS-IFAP International Symposium on Parallel and Distributed Computing in Engineering Systems*. Corfou-Grèce.

Wattanapongsakorn, N. & Levitan, S.P. 2001. Reliability Optimization Models for Fault-Tolerant Distributed Systems. *Proc. Ann. Reliability & Maintainability Symp.*: 193–199. Philadelphia: IEEE, 2001.

Safety and Reliability – Bedford & van Gelder (eds)
© 2003 Swets & Zeitlinger, Lisse, ISBN 90 5809 551 7

An analysis of barriers in train traffic using risk influencing factors

E. Albrechtsen & P. Hokstad

SINTEF Industrial Management, Dept. of Safety and Reliability, Trondheim, Norway

ABSTRACT: A method to evaluate/rank the effectiveness of various risk reducing measures, based on barriers and factors influencing these barriers are presented. As an example a specific railway scenario is investigated, considering the barriers to prevent a train from leaving a station when there already is another train on the anterior block section. The various barriers are first investigated by a fault tree analysis. The basic events are ranked according to their importance. By linking risk influencing factors, RIFs, (rather stable conditions related to the equipment and personnel of the operating railway company) to the basic events risk influencing models are established. By linking these RIFs to the basic events, it is possible to give a more in depth overall analysis of the relevant safety measurers and their effect. The analysis provides a ranking of the most critical RIFs, and thus offers decision support regarding the prioritising of risk reducing measures.

1 INTRODUCTION

A large number of technical and human/organisational barriers may be implemented to secure a safe operation of train traffic. This paper presents an overall approach for overall evaluation of barriers, in order to arrive at recommendations regarding the cost effectiveness of various risk reducing measures. The present paper is built on work by Albrechtsen (2002) presented in a SINTEF memo.

The barriers are analysed by a combination of a fault tree analysis (FTA) and a risk influence analysis (RIA). Throughout the paper the definitions of barriers in the ISO standard 17776 is used. A barrier is defined as a measure which reduces the probability of realizing a hazard's potential for harm and which reduces its consequence. Barriers may be physical (materials, protective devices, shields, segregation, etc.) or non-physical (procedures, inspection, training, drills, etc.) The general approach for the barrier analysis is shown in figure 1. A FTA, where the top event is barrier failures, is carried out. The final step in the FTA is a ranking of the basic events according to their importance for the top event. The basic events in the FTA are connected to the related risk influencing factors (RIFs). RIFs are relatively stable conditions affecting risk of an activity (Rosness, 1998, Hokstad et al., 2001). Some examples of RIFs related to helicopter transport taken from Hokstad et al. (2001) are: operators' maintenance, operations working conditions, human behaviour and environment. The RIFs contributing most to the top

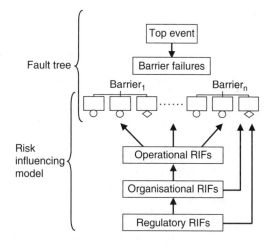

Figure 1. Overall approach of barrier analysis.

event are identified by combining the ranked basic events and their importance measure with the linked RIFs. Thus, it is possible to identify the RIFs for which it will be most effective to implement countermeasures.

The risk influencing model (the lower part of figure 1) used in this paper is based on an approach applied in previous studies at SINTEF related to transport (helicopter, high-speed craft and ferries) (Hokstad et al., 1999; 2000; 2001; Klingenberg Hansen et al., 1999). Other examples of modelling

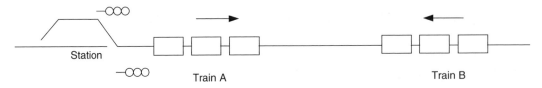

Figure 2 Case scenario.

influence diagrams are described in e.g. Wu et al. (1991), Embrey (1992), Moosung and Apostolakis (1992) and Paté-Cornell and Murphy (1996).

This paper first describes the case scenario used throughout the paper. In section 3a FTA is carried out, where the basic events are ranked by the help of Birnbaum's reliability measure. Section 4 shows how the fault tree is combined with a risk influencing model. Furthermore, the section gives a ranking of the RIFs by using the ranked basic events from the FTA. Finally some conclusions are given.

2   CASE DESCRIPTION

A scenario where a train A leaves a station when the approaching train B already is on the anterior block section is used to illustrate the approach of the barrier analysis. It is assumed that train B is driving legally, i.e. in accordance with rules and orders. There are two barriers that should prevent train A from leaving the station; the signalling system and the automatic train control (ATC). A third barrier become operative when the other two barriers fail; verbal communication by the use of mobile cell phone or train radio from the rail traffic control centre (RTCC) to the locomotive drivers. Figure 2 shows the situation where train A has left the station and entered the block section, where the approaching train B is located. Train A has passed the exit signal as well as the ATC. The two trains are supposed to cross at the station. In this situation, the only barrier that cans stop the trains from meeting is that the traffic controllers at the RTCC warn the locomotive drivers by verbal communication. This verbal communication can either be telephonic contact with the locomotive driver's cell phones or by communication by train radio. The phone/radio call depends on (1) that the traffic controllers at the RTCC discovers the situation and (2) that it is possible to get in contact with the drivers by calling their mobile cell phones or using train radio.

The sequence of the three barriers should be noted. The first barrier, the signalling system should prevent the train from passing an illegal signal. If train A passes the exit signal, the ATC should stop the train from leaving the station. The ATC is an automatic system that should prevent the train from passing an illegal exit signal. The ATC consist of equipment on the track

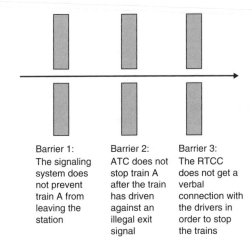

Barrier 1: The signalling system does not prevent train A from leaving the station

Barrier 2: ATC does not stop train A after the train has driven against an illegal exit signal

Barrier 3: The RTCC does not get a verbal connection with the drivers in order to stop the trains

Figure 3   Barrier failures creating the case scenario.

as well as on the train. Should train A in some way pass the ATC as well, the only barrier left to stop the train is verbal communication to the trains from the RTCC. The railway is not electrified, so it is not possible to break the power in order to stop the trains. All three barriers must fail in this order if the two trains actually shall be heading towards each other. Figure 3 shows the three barriers and their order.

The departure procedure at the station regarding signalling system includes the locomotive driver as well as the conductor. The conductor is in this setting a crew member who is responsible for, among other things, the safety regarding departure from a station. The locomotive driver must get a go-ahead signal from the conductor before he can drive on a legal signal. Thus, the conductor must interpret the exit signal correctly, the locomotive driver must interpret the go-ahead signal from the conductor correctly, the locomotive driver must interpret the exit signal correctly and the signalling system must function correctly if the train shall pass the exit signal legally.

3   FAULT TREE ANALYSIS

The barriers are first investigated in a traditional way using a standard fault tree analysis (FTA). The top

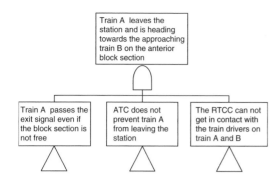

Figure 4.   The two first levels of the FTA.

event is a situation where all three barriers have failed; "Train A leaves the station and is heading towards the approaching train B on the anterior block section". The occurrence of this top event imply that all three barriers have failed; the exit signal does not prevent train A from leaving the station, the ATC does not prevent train A from passing the exit signal and the RTCC does not manage to warn the train crews at the two trains. Thus, the two first levels of a fault tree regarding the case description in section 2 will be as illustrated in figure 4.

It would have been possible to create even more detailed levels in our FTA. However, for our risk influencing model it was suitable to stop at a level not being too detailed.

In the following the basic events of the FTA is presented. The probabilities of failures per year of the basic events were categorised into four groups:

LL : $\leqslant 10^{-7}$
L : $10^{-5} - 10^{-6}$
M : $10^{-4} - 10^{-3}$
H : $\geqslant 10^{-2}$

Some basic events are either present (probability 1) or absent (probability 0). These basic events are indicated by 1/0.

The basic events were assigned to these groups of probabilities. In the presentation below the related groups are given in parenthesis after the basic events.

The executed FTA gave the following basic events:

- Technical failure in the signalling system gives a legal exit signal even if conditions is not safe (LL)
- The locomotive driver deliberately drives against an illegal exit signal (LL)
- The locomotive driver believes the exit signal is legal when is illegal (L)
- The conductor believes the exit signal is legal when is illegal (L)
- The locomotive driver interprets that the conductor gives go-ahead signal when the conductor does not give this signal (L)

- The locomotive driver deliberately drives against an exit signal that is out of order (LL)
- The locomotive driver believes the exit signal is legal, when it is out of order (L)
- The conductor believes the exit signal is legal, when it is out of order (L)
- Technical failure gives signalling system out of order (LL)
- The traffic controller is distracted by other conditions in the RTCC (L)
- The traffic controller focuses on another section (M)
- The traffic controller does not observe the acoustic alarm (L)
- The acoustic alarm in the RTCC is out of function (LL)
- Acoustic alarm in the RTCC is not installed (1/0)
- Instructions for use of mobile cell phone as a communication media between RTCC and locomotive driver/conductor not implemented (1/0)
- The traffic controller calls the wrong phone number (M)
- The correct phone numbers are not available at the RTCC (M)
- The locomotive driver does not answer the call from the RTCC (L)
- The traffic controller's phone is out of function (LL)
- Train radio not installed (1/0)
- Train radio out of function (L)
- The locomotive driver does not answer the train radio (L)
- Train without ATC (1/0)
- Failure in ATC (L)
- ATC deliberately disabled (M)
- It is forgotten to enable ATC (M)
- ATC not installed on track (1/0)
- Failure in ATC on train (LL)

The basic event can be a technical failure, a human/organisational error and lack of implementation of certain equipment or procedures. Figure 5 illustrates the fault tree for the barrier failure of the ATC.

A qualitative analysis of the fault trees is carried out to provide cut sets related to the various barriers. For our case, this qualitative analysis showed the importance of implementing the basic events with probability 1/0, since several of the cut sets of the lowest order included these basic events.

The basic events are ranked by the use of Birmbaum's reliability measure. The main objective of this part of the barrier analysis is to rank the basic events with regard to their importance for the system. Thus, it should be noted that other measures for component importance could be used for this ranking as well. In our analysis the basic events related to the ATC and the basic events "Technical failure in signalling

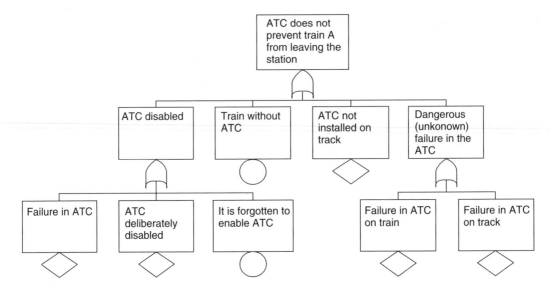

Figure 5.   Fault tree for the barrier failure of the ATC.

system gives a legal exit signal even if conditions are not safe" and "The locomotive driver deliberately drives against an illegal exit signal" are the most critical.

By the help of this ranking, the basic events were classified into five groups on the basis of their importance for the top event. The groups were weighted according to their importance, the most important/most critical group was given the weight 5 and the least important group was given the weight 1. In next section this weighting will be used to decide which RIFs are most important for the barriers.

4   RISK INFLUENCE ANALYSIS (RIA)

The main objective of the approach is to identify which conditions that have the highest effect on the safety level of the various barriers. Out of these conditions it is possible to identify which measures that will be most effective to implement in order to reduce the overall risk level. The conditions are evaluated by a risk influence analysis (RIA). The Risk influence analysis (RIA) provides decision support within a conceptual framework that integrates technical, individual and organizational factors (Rosness, 1998). The objectives of the RIA is according to Rosness (1998): 1) identifying important RIFs, 2) identifying and describing risk reduction strategies defined in terms of actions to change the RIFs and 3) assessing the effects on the total risk level of implementing each risk reduction strategy.

Next risk influencing models are established based on the FTA presented in section 3. These models are constructed by linking relevant RIFs to the relevant basic events found in the fault tree analysis. In that way the RIFs are connected to the fault tree. In addition influences between the various RIFs are constructed. In figure 6 such a construction is illustrated by linking RIFs to the basic events in the fault tree of the event "ATC does not prevent the train from leaving the station". Similar diagrams related to failure of the other two barriers are presented in the memo by Albrechtsen (2002).

The first step in our RIA is to define which RIFs should be used in the analysis. The RIFs are classified into three levels; operational RIFs, organizational RIFs and regulatory related RIFs. Furthermore, the operational RIFs can be divided into technical, human and external factors. The RIFs used in this analysis is found in the lower part of figure 6. These RIFs are defined as relatively stable conditions (Rosness, 1998, Hokstad et al., 2001). For example is the RIF "Maintenance of signalling system" defined as the quality of the maintenance work on the signalling system in order to keep the signalling system on a defined/wanted safety level. The organisational RIF "Train crew" is defined to be the way the train crew plans and carries out their work that directly and indirectly influences the safety level. The RIF "Regulators" is defined to be the quality of the national and international legal framework related to train safety.

The operational RIFs used in our analysis are:

– Design of trains
– Maintenance of trains
– Design of signalling system

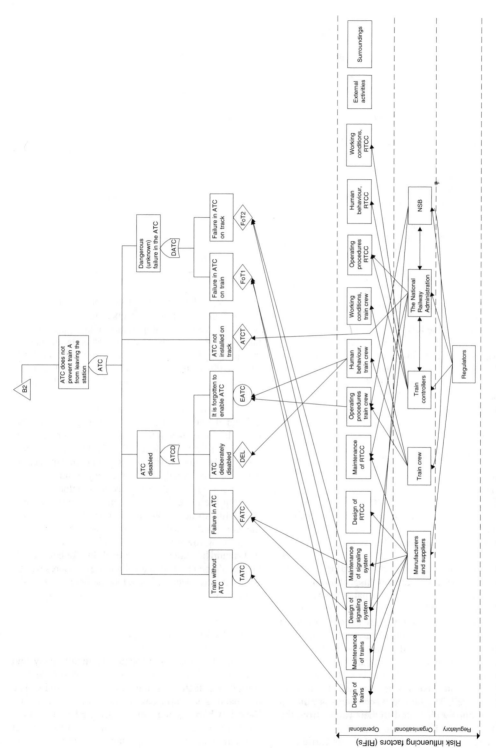

Figure 6. Risk influencing model for event ATC does not prevent the train form leaving the station.

Table 1. Weighted sums for operational RIFs.

RIF	Barrier 1					Barrier 2 & 3					W. sum
	I	II	III	IV	V	I	II	III	IV	V	
Design of trains						2	1	1			17
Maintenance of trains						1	1	1			12
Design of sign. system	1		1			2					24
Maintenance, sign syst	1		1			2					24
Design of RTCC							3	1			15
Maintenance of RTCC							3	1			15
Op. proc. train crew		2			2	1		2			16
Human behavior, t. crew	1	3	1		3	2	1	3			61
Working cond., t. crew		3			3		1	2			34
Op. proc. RTCC								3			9
H. behavior, RTCC								3			9
Working cond., RTCC								3			9
External activities		3			3			2			30
Surrounding		3			3						24

– Maintenance of signalling system
– Design of the RTCC
– Maintenance of the RTCC
– Operation procedures for train crew
– Human behaviour, train crew
– Working conditions, train crew
– Operation procedures in the RTCC
– Human behaviour, RTCC
– Working conditions, RTCC
– External activities
– Surroundings

The organisational RIFs used in our analysis are:

– Manufacturers and suppliers
– Train crew
– Traffic controllers
– The National Railway Administration ("Jernban-everket")
– NSB, the Norwegian State Railway

Finally we have used the RIF "Regulators" at regulatory level (Norwegian Railway Inspectorate).

In the analysis presented in this paper we have only taken the operational RIFs into consideration. A more comprehensive analysis should take the organisational and regulatory RIFs into consideration as well.

By linking the RIFs to the basic events it is possible to give a more in depth overall analysis of the relevant safety measures and their effect. By the help of the ranking of the basic events it is possible to rank the RIFs as well. The ranked basic events are placed into five groups with regard to their values from the calculations in our FTA (section 3).

The ranking of the operational RIFs are carried out by counting the number of basic events each RIF is influencing. The objective of this counting is to arrive at a weighted sum for the operational RIFs. When weighting the sum the importance of the basic events is taken into account. The sum is weighted by the help of the groups of ranked basic events found in our FTA. This weighted sum will provide a ranking of the operational RIFs. In table 1 the weighted sums for the ranked operational RIFs are presented. The number of times an operational RIF influences a basic event is counted and combined with the associated ranked group of basic event. It is also taken into consideration that barrier 1 is the most critical one, since failure in this barrier is necessary if the other two barriers should come into action. Thus the operational RIFs influencing basic events for barrier 1 is additionally weighted by a factor 2.

From the table it can be seen that the operational RIF "Human behaviour, train crew" is the condition that contributes most to the situation where two trains are heading towards each other. Thus, this is the RIF that should be reduced if the probability for barrier failure should be reduced. Risk reducing measures will be most effective when directed at the behaviour of the train crew.

5 CONCLUSIONS

A general method is presented to qualitatively and quantitatively assess the importance of various risk influencing factors on the safety of an activity. The approach is scenario based, and starts by analysing the safety barriers of a scenario using FTA. By linking the RIFs to the basic events of the fault tree, direct measurement of the importance of the RIFs are obtained (gives a rather sound basis for evaluation).

Such a method provides a good tool for evaluation and priority of safety reducing measures related to the RIFs, which not necessarily are directly linked to the barriers.

Any measure can be related to the improvement of a RIF, thus the resulting risk reduction can be assessed. The approach has been successfully applied to a railway scenario, providing specific results regarding the importance of the RIFs relevant for this scenario.

In an actual situation, a number of scenarios should be analysed, thus giving safety management a good tool for evaluation of the effectiveness of risk reducing measures.

The method is simple (based on well-known techniques) and transparent, and can be supported by illustrative diagrams showing relative importance of the relevant factors.

ABBREVIATIONS

ATC	Automatic train control
FTA	Fault tree analysis
NSB	the Norwegian State Railway
RIA	Risk influence analysis
RIF	Risk influencing factor
RTCC	Rail traffic control center

REFERENCES

Albrechtsen, E. 2002. *Barriere analyse ved hjelp av risikopåvirkende faktorer*. SINTEF Industrial Management, memo, August 2002. (in Norwegian).

Embrey, D.E. 1992. Incorporating management and organisational factors into probalistic safety assessment. *Reliability Engineering and System Safety* 38: 199–208.

Hokstad, P., Jersin, E., Klingenberg Hansen, G., Sneltvedt, J. & Sten, T. 1999. Helicopter safety study 2, SINTEF Report STF38 A99423.

Hokstad, P., Jersin, E. & Sten, T. 2000. Helicopter safety study. 12th Annual European Aviation Safety Seminar (EASS), pp. 179–90.

Hokstad, P., Jersin, E. & Sten, T. 2001. A risk influence model applied to North Sea helicopter transport. *Reliability Engineering and System Safety* 74: 311–322.

Klingenberg Hansen, G., Hokstad, P. & Sneltvedt, J. A risk influence model for helicopter traffic. IN: Schueller, G.I. & Kafka, P. (eds). Safety and reliability, Rotterdam: Balkema, 1999, pp. 1267–72 (ESREL'99).

Moosung, J.A. & Apostolakis, G.E. 1992. The use of influence diagrams for evaluating severe accident strategies *Nucl Tech* 99: 142–57.

Paté-Cornell, E. & Murphy, D.M. 1996. Human and management factors in probalistic risk analysis: the SAM approach and observations from recent applications. *Reliability Engineering and System Safety* 53: 115–26.

Rosness, R. 1998. Risk Influence Analysis. A methodology for identification and assessment of risk reduction strategies. *Reliability Engineering and System Safety* 60: 153–164.

Standard ISO 17776. 2000. *Petroleum and natural gas industries. Offshore production installations. Guidance on tools and techniques for hazard identification and risk assessment.*

Wu J.S., Apostolakis, G.E. & Okrent, D. 1991. On the inclusion of organisational and management factors into probabilistic safety assessments of nuclear power plants. In Garrick & Gekler (eds). *The analysis, communication and perception of risk*. New York; Plenum Press, 1991, pp. 619–24.

Safety and Reliability – Bedford & van Gelder (eds)
© 2003 Swets & Zeitlinger, Lisse, ISBN 90 5809 551 7

Exploring an optimal maintenance strategy for repairable assets

Jake Ansell & Tom Archibald
The School of Management, The University of Edinburgh, UK

Lyn Thomas
Department of Accounting and Management Science, University of Southampton, UK

ABSTRACT: Equipment in the process industry is often subject to decay and requires maintenance, repair and eventual replacement. The challenge of competition and the accompanying regulatory regime requires that actions be integrated and cost effective. In Ansell, Archibald and Thomas (2001) an approach to the assessment of asset life of maintained equipment in the process industry was explored using a semi-parametric approach. Using stochastic dynamic programming techniques an approach to find the optimal strategy was developed in Ansell, Archibald and Thomas (2001a).

A major aspect to the development of optimal strategy for repair and replacement is the costs of these activities. Often the costs can only be roughly ascertained in terms of hours expended on the activity. Detailed costings are rarely available. The discount factor will depend on the interest rate in place. Generally a conservatively high level has been taken but with current low rates one needs to explore the sensitivity of solution to the discount factor. Also there is a need to explore the sensitivity of the solution to changes in the costs of the maintenance activities involved. In this paper we explore the stability of the results to changes in the relative costs. It is seen that two costs seem to be more important than the others, hence the accounting effort appears to be best directed towards these costs.

1   INTRODUCTION

Billions of euros are spent on maintenance within the European Community. Solely on Capital Maintenance the UK water industry spent £814 million in 1998/99 on maintenance (Competition Commission 2000). This is therefore a crucial area of expenditure which should be, where possible, managed in an optimal way. In the literature there are a number of solutions which have been offered to the optimal planning of maintenance, see Christer (1999) and Dekker et al (1997). Sadly these approaches have not been frequently implemented. The reasons are many fold; assumption that the process is well understood, that it is documented and the relevant data is available to the decision-maker, see Zheng and Love (2000). Often the models over-simplify the situation, though, a number of authors have attempted to address this, see Christer (1976) and Ansell and Ansell (1987).

In Ansell et al (2001) an approach was introduced which was data driven. The model based on Cox's Proportional Hazard Model, Cox (1972, 1975) used smoothing approaches to obtain the underlying rate of failure of the system. When this approach was applied to a number of processes in the water industry it highlighted the cyclic behaviour of the system. A new system would slowly deteriorate and at some stage an overhaul would take place which rejuvenated the system. This meant that the equipment's life could be extended. This led to the introduction of the concept of refurbishment to the other maintenance strategies of repair and replacement.

The optimal repair strategy for equipment with these features was developed using a discrete time stochastic dynamic programming formulation in Ansell, Archibald and Thomas (2002a). This approach was applied to equipment used in the water industry, specifically the rapid gravity filter. The optimal strategy was based on assumptions about the discount factor and the costs of repair, refurbishment and replacement. Unfortunately the accounting systems are rarely adequate to capture this information accurately. The discount rate applicable over time will vary, and hence the choice of a single value will either be optimistic or pessimistic. In the past the authors have assumed a high rate but this may have an impact on the maintenance policy. In this paper a range of discount rates are explored to evaluate the discount rates effect. From the study it would appear that the effect of the choice of discount rate is limited in terms of the maintenance policy having a slight impact on the time of preventative maintenance.

Finding the average cost of repair, refurbishment and replacement can prove difficult. Whilst in the previous paper the figures used were the "best" available reflecting the relative costing of these activities as perceived by engineers, there must be some concern about the accuracy of such figures. Hence establishing the sensitivity of the optimal policy in relation to the relative cost is important. It also will indicate the level of effort that should be used to obtain such costs. An aim of this paper, therefore, is to consider the sensitivity of the optimal strategy to changes in the cost structure. Changes in the relative costs of maintenance activity have a comprehensible effect on the optimal strategy. Two of the costs seem to have the greatest impact on the optimal policy and its cost. This suggests that the accounting effort should be centred on these activities since it seems important to accurately reflect these costs in the model.

The first two sections of the paper provide the background to the model and the optimal maintenance strategy is developed. The next section explores the sensitivity to the discount factor and costs of the repair, refurbishment and replacement.

2 BACKGROUND

The work arose out of a consultancy for the water industry in which an estimate of the cost of maintenance of equipment over time was required. It was assumed that the cost would be related directly to the frequency of unplanned maintenance events. The estimation suggested was detailed in Ansell et al (2001). It was based on Cox regression (1972, 1975) following Lawless and Nadeau (1995) and kernal density smoothing, see Bowman and Azzalini (1997). The key assumptions was that the unplanned maintenance events followed a Non-homogeneous Poisson Process and that differences between process was proportional and dependent on set of covariates. The model was fitted to data from rapid gravity filter, details again are given in Ansell et al (2001). For the rapid gravity filter the baseline rates is shown in Figure 1. It should be noted that the curve is not monotonic and there does appear to be the impact of cyclic behaviour. The cyclic behaviour was believed to be related to maintenance activity, refurbishment of the equipment.

The system itself will eventually be replaced. This usually occurs when the failure rate of events passes through some economic threshold. The threshold is often represented as a horizontal line, though, in

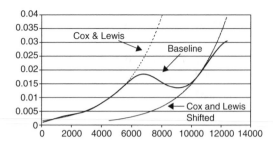

Figure 1. Estimate of underlying "failure rate" curve for rapid gravity filter, rate per day over 12000 days.

practice it is unlikely to be horizontal, due to the discount rate.

In order to estimate the impact of refurbishment Ansell et al (2002b) fitted Cox and Lewis's formulation of the NHPP, Cox and Lewis (1966). The fit illustrates the benefit from refurbishment, see Figure 1.

The process seems to follow the model proposed by Kijima (1989). The underlying failure rate for a system can be described by three parameters $(\mathbf{x}, t_1, t_2)$ where $\mathbf{x}$ are the covariates of the system, t_1 is the actual age of the system and t_2 is the operating age of the system. The difference between t_1 and t_2 is that a maintenance intervention may have occurred which may have led to the system being rejuvenated. The effect of the rejuvenation is to produce a virtual age of the system, which is represented by the "operating" age t_2, $t_2 < t_1$. This is equivalent to reducing the age of the system by δ. Before the first maintenance intervention the operating time is equal to the actual age, $t_2 = t_1$, and after this intervention, the operating age is equal to the actual age minus the time equivalent to the rejuvenation, δ, $t_2 = t_1 - \delta$.

3 OPTIMAL MAINTENANCE POLICY

The optimal maintenance policy may be obtained by using a stochastic dynamic programming formulation on the discrete time version of the model developed in the previous section. The state of the system is given by three parameters $(\mathbf{x}, t_1, t_2)$ where $\mathbf{x}$ are the covariates of the system, t_1 is the actual age of the system and t_2 is the operating age of the system.

If no action is performed on an operational system in state $(\mathbf{x}, t_1, t_2)$ then at the next time period the system will be in state $(\mathbf{x}, t_1 + 1, t_2 + 1)$ and will either still be operational or will have failed. The discrete-time rate of unplanned maintenance events for a system in state $(\mathbf{x}, t_1, t_2)$ is $\exp(\mathbf{a'x}) \, \lambda_0(t_2)$, and this gives the probability that system will fail during the next period. We assume that $\lambda_0(t)$ is non-decreasing in t, so that the probability of a system failure is non-decreasing in the operating age of the system.

Alternative actions available for an operating system are to perform preventive maintenance or preventive replacement. The cost of preventive maintenance is C. The maintenance is assumed to be instantaneous and its effect is to transform the system into state $(x, t_1, m(t_1, t_2))$. Thus $t_2 - m(t_1, t_2)$ is the "rejuvenation" in age that the maintenance induces. It is assumed that the effect of this rejuvenation wears off with both the actual age and the operating age of the system, so $m(t_1, t_2)$ is assumed to be non-decreasing in t_1 and t_2 separately. We also assume that $m(t_1, t_2) \geqslant 0$, so that the operating age of the system will always be non-negative. The cost of preventive replacement is P. The new system will be in state $(y,0,0)$ with probability $p(y)$, where y can be a new set of covariates. This models the possibility that the new system will not have the same characteristics as the old one, but otherwise will be aged 0.

If the system is in state (x, t_1, t_2) and has failed, then two actions are possible. The system can be repaired at a cost F. Repair is instantaneous and after repair the system is in state $(x, t_1, n(t_1, t_2))$. We assume that the effect of a repair on the system is no greater than the effect of preventive maintenance, so that $n(t_1, t_2) \geqslant m(t_1, t_2)$. One would expect the repair to be less effective as the actual and operating ages of the system increase, so we assume that $n(t_1, t_2)$ is non-decreasing in t_1 and t_2 separately. The failed system can also be replaced at a cost R. Failure replacement has the same effect on the system as preventive replacement, a move to state $(y,0,0)$ with probability $p(y)$.

We assume $R > P > F > C > 0$, since a failure replacement costs more than preventive replacement which is more costly than a failure repair which in turn costs more than preventive maintenance. Let $V(x, t_1, t_2)$ be the expected infinite horizon discounted cost for running the system under the optimal maintenance, repair and replacement policy if the system is currently in state (x, t_1, t_2). If we assume the discount factor each period β, standard dynamic programming approaches (Puterman 1992) show that V satisfies the following optimality equation:

$V(x, t_1, t_2) =$
$\min[C + V(x, t_1, m(t_1, t_2)),$
$\quad \beta h(x, t_2) \min\{F + V(x, t_1+1, n(t_1+1, t_2+1)),$
$\quad\quad R + \Sigma p(y)V(y,0,0)\} + (1-h(x,t_2))V(x,t_1+1, t_2+1)),$
$\quad P + p(y)V(y,0,0)],$
$\quad$ where $h(x,t_2) = e^{a.x} h_0(t_2)$.

The minimisation in the middle row describes the repair/replace decision for a failed system. In Ansell, Archibald and Thomas (2002a) the theoretical properties where established.

For application the model is simplified in Ansell, Archibald and Thomas (2002a) to explore the optimal maintenance policy for RGF's baseline hazard function. The baseline hazard rate for a system is taken to be $h_0(t) = 1-\exp\{10^{-2.80783 + 0.000176t}\}$. Given the use of the baseline function then $V(x, t_1, t_2)$ will not depend on x and so it can be expressed as $V(t_1, t_2)$. A repair is assumed to restore the system to an operational state, but to have no effect on the operating age of the system $(n(t_1, t_2) = t_2)$. It is assumed that preventive maintenance can reduce the operating age of the system by up to 4,500 days. Hence

$$m(t_1,t_2) = \begin{cases} 0 & \text{if } t_2 \leq 4500 \\ t_2 - 4500 & \text{otherwise} \end{cases}$$

The relative costs were assumed to be C = 1 are F = 2, P = 4 and R = 5 and discount factor taken to be 0.995.

The optimal preventive maintenance/repair/replacement policy achieves a minimum cost of 1.319. The policy requires that preventive maintenance should be instigated whenever the operating age of the system is 3,280 days and to repair the system on failure.

4 SENSITIVITY TO COSTS

The choice of 0.995 as discount factor is very conservative since it represents a high yearly interest rate of approximately 20%. Given the current low inflation rates and hence low interest rates this seems rather extreme and so the model was explored with a range of discount factors equivalent to interest values from approximately 20% to approximately 3%. The results are presented in Table 1. From Table 1 it is clear that as the discount rate increases (the interest decreases) then the cost rises as expected since more cycles become included. This also predictably has the effect of decreasing the time to preventive maintenance. It does not appear though to have an effect on the type of policy developed. The marginal change in the time to preventive maintenance changes relatively little for a large change in the interest rate. Hence one can argue that the discount rate is not having a significant effect.

In terms of cost of maintenance activity again the desire was to investigate changes in the relative costs.

Table 1. The effect of discount factor on optimal strategy.

Discount rate	Approximate interest rate	Minimum cost	Time to first PM
0.995	20	1.32	3280
0.9962	15	1.88	3100
0.9969	12	2.43	3060
0.9974	10	3.00	3010
0.9979	8	3.87	2960
0.9984	6	5.31	2920
0.9987	5	6.73	2910
0.9989	4	8.14	2870

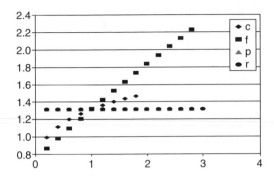

Figure 2. Graph of minimum cost as C, F, P and R as increased by 0.2.

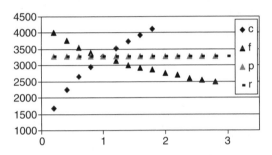

Figure 3. The times to first optimal preventive maintenance are C, F, P and R increased by 0.2.

There are, of course, constraints on costs by dint of their ordering. Within the constraints a range of values are considered rising incrementally by 0.2. Obviously one could argue that the range of values explored for the smaller costs are relatively wider than for the larger values, but this seems marginal. Again the focus is on minimum cost and optimal time for preventive maintenance. The results are presented in Figures 2 and 3. It is clear that changes in C and F have the more major impact. While apparently changes in P and R have little effect. Unsurprisingly the costs minimum cost rises for increases in c and f, also the time of PM rises if c increases and falls as f increases.

An explanation of the above results is that preventive maintenance is so effective that the number of unplanned maintenance events never passes through the threshold for replacement or that it does so after such a long time that the discount factor effectively reduces the cost to zero. One reason for this occurring is that the effect of maintenance activity is too good.

5 CONCLUSION

The aim of the paper has been to investigate sensitivity to some of the assumption of the model. The

results seem to indicate that the model is relatively stable with respect to costs, with predictable changes in minimum cost and time to preventive maintenance. The discount rate does not appear to give any concern. The results on costs suggest that accounting effort should be focused on the costs of preventive maintenance and maintenance on failure rather than preventive replacement and replacement on failure.

REFERENCES

Ansell, RO, and Ansell, JI, (1987), "Modelling the reliability of Sodium Sulphur batteries", *Reliability Eng*, **17**, 127–137.

Ansell, JI, Archibald, TW, Dagpunar, J, Thomas, LC, Abell, P, and Duncalf, D, (2001). "Assessing the maintenance in process industry using semi-parametric approach", Quality and Reliability Engineering International 17, 163–167.

Ansell, JI, Archibald, TW, and Thomas, LC, (2002a). "The Elixir of Life: Using a maintenance, repair and replacement model with virtual and operating age in the water industry", (Revised for IMA J Man Math).

Ansell, JI, Archibald, TW, Dagpunar, J, Thomas, LC, Abell, P, and Duncalf, D, (2002b), "Analysing maintenance data to gain insight into systems performance", (forthcoming JORS).

Bowman, AW, and Azzalini, A, (1997), Applied Smoothing Techniques for Data Analysis, Oxford University Press, Oxford.

Christer, AH, (1976), "Economic cycle period for painting", *Opl Res Q*, **27**, 1–14.

Christer, AH, (1999), "Developments in delay time analysis for modelling plant maintenance", JORS, **50**, 1120–1137.

Competition Commission (2000). Mid Kent Water Plc: A report on the references under sections 12 and 14 of the Water Industry Act 1991 ISBN 0-11-702527-5.

Cox, DR, and Lewis, PAW, (1966), The Statistical Analysis of Series of Events, Methuen: London.

Cox, DR, (1972), "Regression models and life tables", Journal of Royal Statistical Society Series B, **74**, 187–220.

Cox, DR, (1975), "Partial Likelihoods", Biometrika, **62**, 269–76.

Cox, DR, and Lewis, PAW, (1966), *The Statistical Analysis of Series of Events*, Methuen: London.

Dekker, R, Van der Duyn Schouten, FA, and Wildeman, RE, (1997), "A review of multicomponent maintenance models with economic dependence", *Mathematical Methods of Operational Research*, **45**, 411–435.

Kijima, M, (1989), "Some results for repairable systems with general repair", J Appl Prob, 26, 89–102.

Lawless, JF, and Nadeau, C, (1995), "Some simple robust methods for the analysis of recurrent events", Techometrics, 37,158–168.

Puterman, M, (1992), Markov Decision Processes, John Wiley: New York.

Zheng, ZG, and Love, CE, (2000), "A simple recursive Markov chain model to determine the optimal replacement policies under general repair", *Computers and Operational Research*, **27**, 321–333.

Safety and Reliability – Bedford & van Gelder (eds)
© 2003 Swets & Zeitlinger, Lisse, ISBN 90 5809 551 7

Analysis of high speed craft accidents

P. Antão & C. Guedes Soares

Unit of Marine Technology and Engineering, Technical University of Lisbon Instituto Superior Técnico, Av. Rovisco Pais, Lisboa, Portugal

ABSTRACT: High Speed Crafts (HSC) represent, at the present day, one important component of intermodal transport due to their flexibility, cost, time efficiency and high speed. This high speed leads to risks on ship operations and therefore to potential accidents. The analysis of such accidents is of great importance as it provides information that can be used to improve safety in design and operation. In this paper an analysis of HSC accidents is performed with a methodology proposed recently. The objective of this study is to understand the causes of these accidents and to compare with the findings of accidents of traditional commercial vessels.

1 INTRODUCTION

The constant demand for high-speed transportation at sea has led to the development of High Speed Craft (HSC). This fast waterborne transport, and in particular fast transport of passengers, is rapidly developing all over the European waters in which fast ferries carrying up to 1,500 passengers and 100 cars are already in operation at a speed exceeding 35 knots in the North Sea and the Mediterranean basin. The possibility of carrying passengers and goods as fast as possible between two ports has become a competitive factor and therefore HSC constitute a rapidly developing sector of the shipping industry. The consequence of increased speed was an increase of fuel costs, and in order to reduce these costs new low weight materials such as aluminium and composites were required. This has resulted in the development and adoption of the International Code of Safety for High Speed Craft (HSC-Code) (IMO, 1994) by the International Maritime Organisation (IMO).

The basic concept of the HSC code was the recognition that High Speed Craft represent new and technologically advanced features and hence are normally operated under more stringent operational procedures. These stringent operating conditions can originate the relaxation of the demand for safety equipment on the craft itself. HSC was also expected to receive support from shore-based infrastructure related to a particular route (Papanikolaou, 2002). The revision of the HSC code, in July 2002, did not change the basic concepts of the code but upgraded the regulations for damage extend, fire regulations and modifications in terms of the Safety of Life at Sea (SOLAS).

Although maritime transport and travel has a relatively low death and injury rate when compared with other transport modes, the consequences of one accident are significant and sometimes of far reaching consequences (Guedes Soares and Teixeira, 2001). Therefore is not surprising the existing concern towards HSC. This is due to the fact that as speed at sea increases, risks of ship operation are also increasing and safety is consequently being challenged. HSC owners can maintain and even increase safety by placing higher demands on the operators (the crew) and the organisation. HSC operation places new and higher demands on safety and prevention of accidents. Increased speed must, by definition, increase both risks and possible consequences, thereby decreasing safety at sea compared to conventional crafts (Kallstrom, 2000). This statement is, however, valid only under the condition that all other variables are constant, i.e., when no changes are made in other safety-related areas. As technology changes make ships more vulnerable and exposed to greater risks, it is necessary to make improvements in other significant areas to maintain at least the necessary level of safety (Schager, 1998).

Many changes in maritime safety have resulted from changes following large-scale loss of life. Design and operation of high-speed vessels can benefit from advances in safety engineering to minimise risk of learning through new or old mistakes. So, accidents and incidents have to be carefully studied in order to establish their causes and possible measures to reduce their numbers and minimize their consequences.

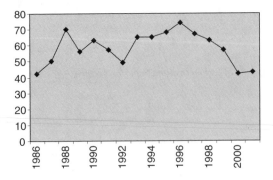

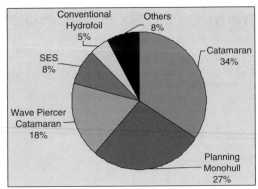

Figure 1.   Number of new constructions of HSC worldwide (Incat Design Statistics).

Figure 2.   Distribution of types of HSC (Zaraphonitis, 2002).

2   EVALUATION OF THE HSC INDUSTRY

High-Speed Crafts it is not a complete new concept since it already exists since the middle 60's. A good example is the crossing to Cowes by the Red Funnel in U.K. Nevertheless, it was in the last decade that one could see a large growth of passenger vessels with design speeds greater than 25 knots. In fact in 2001, more 1,200 passenger ferries and 100 car/passenger ferries of different types were in operation worldwide (Phillips, 2002). This trend can be seen in figure 1 where a saturation of the annual increase of ship until 1995 originated a substantial drop of new buildings 1996.

This early high trend of the HSC market was due to a sudden increase of the ship sizes, power and speed. This has given rise to new technical challenges. It has encouraged a shift of production from small boat-builders to larger shipyards. An increase in the range of different concepts (e.g. crafts with 3 or more hulls) has also presented new challenges. This includes Monohulls, Catamarans, SWATHs, Foil-catamaran or Pentamarans.

Today the development of prototypes designs of hybrid air lubricated catamarans with very high speeds of up to 70 knots are under research in some EU projects. In figure 2 a distribution of types of HSC operating around the world is presented. This constitutes a significant sample of the existing high-speed crafts. One may see that catamarans and planning mono-hulls represent more than 50% of the 353 high-speed craft database created by Zaraphonitis et al. (2002).

Another aspect to consider is the traffic and routes worldwide. The traffic of HSC in North-European waters and in Mediterranean is particularly intensive in South France and Irish Sea with attractive schedules and quality are factors that can influence the choices of passengers. There were more than 9 regular routes circulating in velocities higher than 35 knots during 1996 in UK waters, with a regularity of 14 vessels daily during summer. Others are in service in Scandinavia, Middle East and South-America. However, safety is more a visible concern of passengers and a demand for the professionalism of the ship owners. In this time passengers have a choice between conventional ferries and high-speed crafts with the advantage of speed. Nevertheless, it is important to keep in mind that a single serious accident can change the overall picture not only for the company involved in the accident but to the whole industry of HSC.

HSC within the concept of short sea shipping represents the most environmental friendly mode of transport, in particular, because of its comparatively low external costs and high-energy efficiency (Nilson et al., 2002). This is also supported by the economical fact that the cost of construction and exploration of such vessels are 2/3 to 3/4 less (Hynds, 1997). It is then not surprising the extreme interest of the EC on the development of short sea shipping in Europe for goods and also as a point of environmental issues i.e. active investigation on transport of goods. There is always an economical imperative associated with speed. This leads to high-standards services, speed and efficiency in the movement of people and goods as elements of increased competitiveness. The fast ferries, with their impressive lines, speed and highly attractive express the existence of a valid alternative to conventional vessel.

Therefore it should not be surprising with the growth within the Mediterranean Sea which is the area destined to the greatest development in the future, with already active various traffic lines between E.U. Countries (Italy/Italian Islands, Italy/Corsica, France/Corsica only, Spain/Balearic Islands, Greece/its Archipelagos either in the Aegean or Ionic Seas). On the contrary, the connections between these E.U. Countries (Italy, France, Spain and Greece) are still very poor. Other areas are concentrated in the connections to Cyprus and Malta.

3 FORMAL SAFETY ASSESSMENT

Formal Safety Assessment is a systematic process to assess the risks associated with an activity and evaluating the costs and benefits of the options for reducing these risks. Within IMO several studies have already been performed using FSA to support decisions about implementation of international regulations. These have included studies on the application of FSA to High-Speed Craft, disabled oil tankers, and in particular to emergency propulsion and steering devices, and helicopter landing areas. All of these studies have been reported to the IMO and are briefly described below.

An application of FSA was made in UK to study the safety of high-speed passenger catamaran vessels, concentrating on the safety of passengers and crew (IMO, 1997b). This study addressed the quantified risk levels for high-speed craft operating under the present regulatory regime. A selection of risk control options was developed to address these risks and an evaluation of the costs and benefits of those options were also presented. Recommendations based on these results are also included. Within the range of accident categories studied, collision was identified as the main contributor to the overall risk level. Two risk control options, specific to the collision accident category, and an option common to all accident categories were considered in detail. Assuming effective implementation, all three options would significantly reduce risk levels, typically by at least 40%, and are economical.

The experience gained from this trial application of the FSA process to high-speed catamaran ferries has been reported by the United Kingdom (IMO, 1998a). This document discusses the application of both established and new techniques, and assesses their usefulness. In addition, areas where further research and development are considered desirable to improve the robustness, accuracy and audit ability of the FSA process were identified.

Another study on the safety assessment of high-speed craft operations was performed among the Scandinavian Countries (IMO, 1998e). They addressed not only the craft itself with the crew, but also the fairway and environmental conditions. The study includes risk analysis, risk acceptance and safety measures, which together form the platform for the outline of a uniform Safety Assessment methodology. Comparative considerations, in particular with Ro-Ro passenger traffic, are used as references for high-speed craft risk predictions and discussion on acceptance criteria.

In general terms the methodology was based on the IMO FSA Interim Guidelines, but designed to meet the need of each Flag State to evaluate individual high-speed craft on various routes. The preliminary results of the Scandinavian high-speed craft safety study (IMO, 1998d) indicate that either head-on collision with a large ship or uncontrollable fire in the accommodation

area will be the most severe type of accidents. Judgement from a group of experts indicated that up to 10% of persons might get killed in both types of "worst-case" accidents. It should be noted that in collision accidents the number might be reduced to less than 5% by avoiding accommodation areas in the assumed damaged zone. A range of possible risk control options was considered in the study and the relative risk reducing effects analysed. In particular, the considerable importance of route specific conditions on high-speed craft safety is incorporated.

The complete FSA application should comprise of the following five steps:

1. Hazard identification
2. Risk assessment
3. Risk control options
4. Cost benefit assessment
5. Decision making

3.1 Hazard identification

As stated in the FSA process description: the purpose of hazard identification is to help identify the causes of accidents and to generate a prioritised list of accident categories specific to the problem under review. Is then of major importance the analysis of accidents.

Then as input to the risk analysis it is necessary to identify the hazards on board. By reviewing statistics and experiences, different main scenarios can be determined. These main scenarios are thereafter used in the safety evaluation process.

3.1.1 Statistics

When evaluating an existing safety level it is of certain interest to monitor recent statistics. The statistics can, if they are interpreted in a proper way, answer the main questions namely if the safety level is satisfactory. With a deeper analysis one can also ascertain the most common hazardous associated to each casualty and from this adopt effective measures in order to improve the safety level. Unfortunately, when examining existing statistics and safety records for HSC one will immediately run into problems. Since the HSC code was published in 1994, the operational hours of large HSC have been too few to find reliable data and thereby recognise trends. However, some general conclusions can be drawn regarding accident frequencies and causes. Historical data 1981–1996, mainly for craft built according to the DSC-Code (Dynamical Supported Craft) is presented in Table 1 (IMO, 1997).

The statistics presented in Table 1 derived from the application of Formal Safety Assessment to HSC (IMO, 1997b), which includes all category A and B passenger HSC. The risk levels were obtained by the combination of historical data and expert judgement from a group of experts familiar with HSC operation.

Table 1.   Data for frequent accident categories.

Accident Category	Number of Accidents	Number of Fatalities	Historical Accident Frequency (per vessel operating year)	Predicted Fatality Risk (per vessel operating year)
Collision	104	14	0.067	0.065
Contact	84	0	0.047	0.006
Grounding	36	–	–	–
Fire	16	0	0.007	0.004
Loss of hull integrity	9	0	0.004	0.002
Other	45	–	–	–
Total	301	14		0.077

For example analysing this table in terms of historical fire frequency, one can conclude that is one fire on every craft every 125 years and the predicted future equivalent fire fatality frequency is one fatality on every craft every 250 years. According to this historical statistics fire represents 5% of all accidents on board HSC. For comparison, on board traditional vessels 20% of the major and total losses are caused by fire (Institute of London, 1997). Finally, the existing total risk is only one sixth of the predicted future risk, i.e. the total risk level and thereby also the fire risk is expected to increase in the future (IMO, 1997b).

3.1.2 *Hazardous scenarios*
Throughout the survey performed for this paper it became clear that many were the hazardous associated with this type of vessels and some of them are specific from this type of vessels.

3.1.2.1 Environmental factors
The wash of high-speed ferries has became a well recognised issue. It not only affects the environment but also the safety of other waterway users world wide, which caused concern in researchers and operators to look closely at the wash generated by fast vessels. This concern is more visible in coastal areas. In this environment, the generated disturbance is substantially different from waves generated by conventional displacements ships as a consequence mainly of the higher speed and the increasing size of these modern vessels (Dand et al., 1999).

Complaints due to wave wash are appearing in Denmark, Sweden, Ireland (Whittaker et al., 1999), Australia, New Zealand, UK, Portugal and the USA. Based on measurements in Denmark it could be establish that waves generated by a fast vessel having a speed of 35 knots have periods of 9–10 sec whereas from conventional ferries are producing waves with periods between 4 and 5 sec.

Also problems with the erosion of coastal areas due to the HSC operations as well as overall environmental impact due to large power required were identified.

3.1.2.2 HSC dynamics
The dynamic behaviour of high-speed craft in a seaway is directly related to its safety. The large kinetic energy associated and the greater dynamic rather than static behaviour since it is the combination of power, displacement and speed that have given rise to new potential behaviour can induce in the possibility of large accelerations and in the possibility of dynamic instabilities.

3.1.2.3 Hazard associated with navigation
Many were the problems identified related to navigation. Most of them are external to the vessel himself but nevertheless represent issues that should be pinpointed: First related with the vessel traffic control and traffic separation schemes and the lack of different rules for this type of vessels; Second the external navigational aids – many existing maritime navigational environmental designed for traditional crafts. Visibility issues, which increases risk for HSC due to relative high speeds, especially during fog conditions. Also identified were issues related with the radar navigation, particularly in the danger of complacency, improper training and installation of equipment at board.

Also navigation restrictions of Fast ferries that operate under the High Speed Craft Code, which limits their operations to times when significant wave heights are below 4 meters. In fact, there has been reported that some European operators self-imposed a lower restriction of 3 meters, due to the structural damage being caused to their vessels and also due to passengers discomfort and consequent affects on their image as a suitable transport mode.

3.1.2.4 Human factors
Today modern high-speed crafts also involve increased operating complexities and demands, with systems capable of solving increasingly complex tasks. At the same time, reliance on human ability has diminished. The human factor is now the most common explanation for accidents and the operator is often regarded as the weakest link in the system (Schager, 1998).

Then is not surprising that the number of safety issues be quite extensive like the level of type-training for the crew of high speed craft, the quality of route assessment; adequacy of route information provided for the master, the interpretation of the "worst expected conditions" that the vessel can encounter, the short reaction times due to high speeds and less reaction time for avoidance, the requirements for night-time high-speed navigation and finally the direct influence on the quantity and quality of the information available to the officers in charge of the vessel for ship handling.

4 ANALYSIS OF HSC ACCIDENTS

Any analysis of safety related matters is frequently done by the systematic collection of casualty data followed by its detailed analysis. This provides information about number of fatalities, types of casualties, human and technical causes for accidents, etc. However the fact that HSC are comparatively new on the maritime scene leads to little historical casualty data being available. This fact can have two main explanations: one, the fact that there does not appear to be many organisations collecting, collating and publishing such data on a regular basis and second the fact that the HSC have to date in general a good safety record since their casualties, both in ports and harbours are of less consequences than those of conventional vessels.

Other factor is the fact that even the organisations that analyse, collect and maintain databases of accident data, often do not make a clear distinction between HSC vessel and other conventional ones in their coding scheme.

Accidents appear to be the result of highly complex coincidences and a chain of failures, which could rarely be foreseen. The unpredictability is caused by the large number of causes associated and with the role of the persons involved. It is in this perspective that it is necessary to analyse both the technician and operator's roles in the chain of the events as well as the compatibility between operators, technology and nature. The question is thus no longer whether a human factor is involved in an accident, but where in the chain of events the human factor is to be found.

For the purpose of this paper forty-one maritime accidents and incidents involving High Speed Crafts were analysed with an approach similar to the one proposed by Kristiansen et al. (1999) and Guedes Soares et al. (2000). The maritime accidents analysed complemented some of the major accidents that involved HSC. The sources of the descriptions of the accidents were the Maritime Agencies (MAIB, U.S. Coast Guard, Australian M.A., New Zealand and Canada M.A.).

4.1 The accident modelling approach

The analysis of the accidents used is based on finding the chain of events that led to the accident and their associated contributory factors and causes. In this methodology the casualty events are numbered chronologically in relation to the other identified accidental events. The Casualty Events are qualified with two different attributes, namely Class and State. These attributes express certain aspects of the situation when the casualty took place. The accidental events are divided in five categories, namely: human error, equipment failure, hazardous materials, environmental effects, or other vessel or agent. It is important to keep in mind that accidental events are strictly related to the casualty as it is observed from initiation to final outcome.

These events are then categorized according to specific attributes depending on the context that they occur. This means if a particular event is coded as human error certain attributes (codes) will be given to it, namely: position (master, pilot, engineer...), task affected by the error (mooring, close door, trip planning...), behaviour (detection, activation, analysis...) and finally the type of error (delayed, ignored, underestimated...). The same procedure will be used for the other types of accidental events. The analysis process has so far concentrated on identifying the accidental events, or in other words give an account of *what* happened and *who/what* were involved. The next step is to give the diagnoses or answers *why* the casualty took place by identifying the conditions or actions external to the accidental event sequence itself that have arise prior to the casualty. These causal factors may be given for either or both decision levels: Daily Operations and Management and resources.

These two levels correspond to relevant actions taken on-board or ashore in the management company or by the owning company. As with the accidental events, the causal factors are attributed codes. For the Daily Operations the codes can include: Supervision (inadequate work methods, inadequate work preparation...), Personnel (Lack of skills, Lack of knowledge...), Social environment, Manning, etc. The same way for the Management & Resources, one may have: Operations Management (inadequate procedures and checklists, pressure to keep schedule...), Personnel Management (inadequate training program, selection/training of officers), Organisation & General management, etc.

5 DISCUSSION OF THE RESULTS

Despite the fact that only forty-one cases were analysed and therefore the reliability of the conclusions is limited, they can give some indications where to look for in terms of identification of safety issues. The analysis has consisted on the identification of the accidental events as human errors, equipment failures or environmental factors and their correspondent categorization depending on the context. The sample used in this

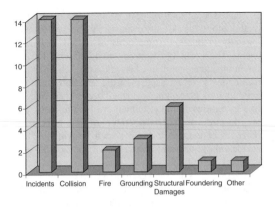

Figure 3. Sample of the incidents and distribution of accidents analysed in terms of terminal casualty.

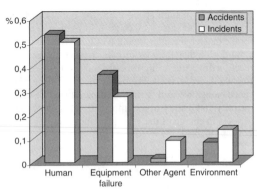

Figure 5. Distribution of accidental events for the sample – accidents and incidents.

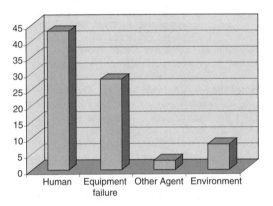

Figure 4. Distribution of accidental events for the sample.

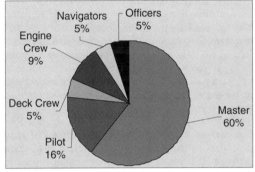

Figure 6. Distribution of human accidental events.

analysis resulted from 27 cases of accidents and 14 incidents as one can see in figure 3. Between the accidents, collisions and structural damages were the most frequent. These damages on the structure are highly associated with bad weather conditions, fog or high waves, and to wrong berthing procedures. Figure 4 presents the distribution of the accidental events where mainly are related to human errors. In fact 55% of the accidental events are human error related which goes toward the figures usually associated with marine casualties. This high percentage of human errors on-board ships are the ones that caused the entire industry to be concerned about the quality of the people who run the ships. These human errors are mainly related with decisions taken by the bridge crew in terms of navigational procedures. Therefore, it is not surprising that the master of the crafts involved in the accidents performed more than half of the human error as shown in figure 6.

From the analysis of figure 5, one may see that the distribution that characterises the accidents is also similar in the cases of incidents. In fact the variance between the two is less that 2%, which means that the same factors that lead to incident trigger as well accidents. It is also relevant to notice that is the navigational system the one more involved in accidental events (figure 7). One may then conclude that these systems were not very reliable but this would be a false conclusion. This distribution of accidental events only emphasises that the navigation systems are very "sensitive", which mean their failure conducts usually to incidents or accidents, as expected.

The distribution of the Human Task affected presented in figure 8 and one may see that the majority of these tasks are related with navigation and with bridge operations where Set speed is the most frequent. In fact eleven accidents had in high speed of the vessel one of the main causes. Figure 9 and figure 10 present the distributions of the Daily Operation causal factors and Management & Resources causal factors respectively. On the first one may see that the problems with supervision and personal causal factors are the more frequent with 71%. On the second, personnel management and

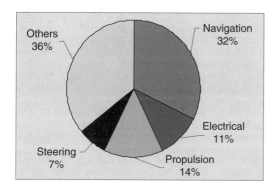

Figure 7.   Distribution of equipment accidental events.

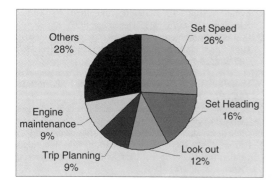

Figure 8.   Distribution of affected human task.

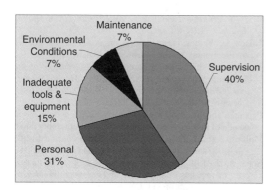

Figure 9.   Distribution of daily operations casual factors.

design causal factors have a frequency of 67%. From the results of both causal factors one may conclude that problems with adequate training, adequate knowledge, work preparedness and adequate procedures are the points to be concentrated on if your objective is to reduce the frequency of HSC incidents and accidents.

These are important findings that can help in decision making for a proactive approach to HSC

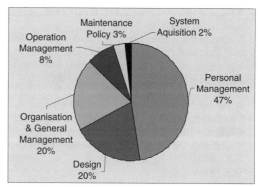

Figure 10.   Distribution of management & resources casual factors.

operations. It is also important to compare these finding with conventional vessel in order to see if there is any specificities. Antão et al. (2002) presented a similar study for forty maritime accidents. In both studies human factors are the most frequent accidental events (60% for conventional vessel and 55% for HSC). However environmental effects present the double of the frequency in HSC accidents. This is due to frequent presence of *fog*, *significant wave heights* and the formation of *wave wash*. In terms of the human task affected there is also some significant differences. In the commercial vessel the main task affected are the *Set heading* (16%) and *Trip Planning* (19%) in comparison with the *Set Speed* (26%) in the HSC. Also an important contrast can be found on the equipment failure events. In the commercial vessels they are concentrated in the structure of the vessels (25%) however in the HSC vessels *Navigation systems* present a frequency 5 times higher. Then one can easily conclude that the main hazardous that can lead to an accident are concentrated mainly in the bridge and bridge operations.

6   RECOMMENDATIONS

After the analysis of the accidents and incidents it became clear that there are important improvements to be made towards safe HSC operations. Therefore there are some recommendations based on the findings of the accidental events and causes:

– Adoption of control systems in busy regions similar to the existing at board of airplanes for the reduction of the probabilities of collisions on traffic areas.
– Strict routing policies.
– Redundant system such as charts to double check radar.

- Navigational lights onboard HSC easily recognised, even in fog situations.
- Upgrading equipment on other traditional vessels and increased knowledge of small craft operators.
- Review of rules for assigning limiting speed/wave height near coastal areas.
- Ensure that all pilots employed on HSC have an understanding of the problems of the wave wash.
- The International Code of Safety of High-Speed Craft requires that two officers be on duty in the vessel when operating in international water, but is our belief that should be extended to national waters as well.
- Provide an operator with the best possible information and give him the opportunity to choose it himself. Furthermore, since people are different we should choose for critical tasks those people who have good perceptiveness, high capacity and judgement.
- Ensure different routes and quays for HSC in port areas in order to reduce the probability of collisions.
- Transverse Watertight Bulkheads in the lower deck of the vessels in order to reduce the consequences of an accident.

7 CONCLUSIONS

The problem of safety of High Speed Craft was addressed. Forty-one maritime accidents involving this type of vessels were analysed and their findings presented. They are mainly concentrated in bridge and bridge operations. The human factors represented the majority of the accidental events found. A comparison with commercial vessels was presented and differences discussed. Also some recommendations were presented having as base the findings of the accident analyses.

ACKNOWLEDGMENTS

The work presented was performed within the EU-project Tools to Optimise High Speed Craft to Port Interface Concept (TOHPIC) (C.N. G3RD-CT-2000-00491) funded partially by the European Commission.

REFERENCES

Antão, P. and Guedes Soares, C., 2002, Organisation of data-bases of accidental data, *3rd International Conference on Computer Simulation in Risk Analysis and Hazard Mitigation (Risk Analysis III)*, Witpress (Ed.), Southampton, pp. 394–403.

Dand, I.W., Dinham-Peren, T.A. and King, L., 1999, "Hydro-dynamic aspects of a fast catamaran operating in shallow water", RINA, *International Conference: Hydrodynamics of High Speed Craft*, 24–25 November 1999, London, UK.

Guedes Soares, C. and Teixeira, A. P., 2001, "Risk assessment in maritime transportation", *Reliability Engineering and System Safety*, 74, pp. 299–309.

Guedes Soares, C., Teixeira, A.P. and Antão P., 2000, "Accounting for Human Factors in the Analysis of Maritime accidents", *Foresight and Precaution*, M.P. Cottam, D.W. Harvey, R.P. Pape & J. Tait (Eds), A.A. Balkema, Vol. I, pp. 521–528, Pergamon.

Hynds, P., 1997, Safety culture is still parochial, *Safety at Sea*, Vol. 3, Issue 2 May 1997.

IMO, 1998a, Aspects of FSA methodology: Experienced gained from the trial application undertaken by the United Kingdom, submitted by United Kingdom, *MSC 69/INF.14*, 12 February.

IMO, 1997, The code of Safety for Dynamical Supported Craft.

IMO, 1997b, Trial application to high-speed passenger catamaran vessels, final report, submitted by United Kingdom, *DE 41/INF.7*, 12 December 1997.

IMO, 1998d, Joint Nordic Project on Safety Assessment of High-Speed craft operations, submitted by Sweden, *MSC 70/INF.15*, 9 January 1998.

IMO, 1998e, Joint Nordic Project on Safety Assessment of High-Speed craft operations, submitted by Sweden, *MSC 70/INF.15*, 9 October 1998.

Incat Design Statistics, www.incatdesigns.com.au/pages/salesstats.asp

Institute of London, 1997, Hull Casualty Statistics.

Kallstrom, C.G., 2000, "Autopilot and track-keeping algorithms for High Speed Craft", *Control Engineering Practice*, 8, pp. 185–190.

Kristiansen, S., Koster, E., Schmidt, W.F., Olofsson, M., Guedes Soares, C. and Caridis, P., 1999, "A new Methodology for Marine Casualty Analysis Accounting for Human and Organisational Factors", *Proceedings of the International Conference on Learning from Marine Accidents*, RINA, London.

Nilson, I., Winnes, H. and Ulfvarson, A., 2002, Integrating environmental performance in a logistic approach to short sea shipping – a case study. *Proceedings of the International Conference ENSUS 2002: Marine Science and Technology for Environmental Sustainability*, 16–18 December, New-castle, UK, 2002.

Papanikolaou, A. and Papakirillou, A., 2002, Safety based design of fast ships, *Safety at sea*, 18–21 September 2002.

Phillips, S (Ed.), 2002, "Jane's High Speed Marine Craft", Published annually by Jane's Information Group Ltd., UK.

Schager, B., 1998, Increased safety for high-speed marine craft by focusing on operations and organization. *Marine profile UK Ltd.*, London, UK.

Whittaker, T., Bell, A., Shaw, M. and Patterson, K., 1999, "An Investigation of Fast Ferry Wash in Confined Waters", RINA, *International Conference: Hydrodynamics of High Speed Craft*, 24–25 November 1999, London, UK.

Zaraphonitis, G., Papanikolaou, A., Oestvik, I., Eliopoulou, E., Georgantzi, N. and Karayannis, T., 2002, Review of technoeconomic characteristics of fast marine vehicles. *Proceeding of International Conference: HSCV 2002, Naples, September 2002.*

Safety and Reliability – Bedford & van Gelder (eds)
© 2003 Swets & Zeitlinger, Lisse, ISBN 90 5809 551 7

CFD simulations to better evaluate technical improvements to reduce fire damages

P.G. Aprili & D. Franciotti
LNGS-INFN (Safety Plants Services) Assergi (AQ)

ABSTRACT: The Gran Sasso National Laboratory is an underground facility unique in the world for the dimensions of the halls, for the ease of access to experimental apparatuses of great dimensions and for the high technology of the supporting external infrastructures. Leading experiments in the field of nuclear and sub-nuclear physics are hosted in the Laboratory. The analysis made has considered three simulation of fires and smoke transport, both involving the Galleria Auto – a tunnel in which service cars are parked – and the other connected tunnels. Version 2 and 3 of NIST's FDS was used. A fire originated from a car parked in the gallery and a fire of the electrical cables running across the gallery were considered.
The aims of the analysis were:

- to describe the diffusion of smoke and the temperature trend in the Galleria Auto tunnel, predicting visibility and species concentrations (such as CO and O_2) during the fire;
- to investigate two proposed positioning schemes for heads of the Water Mist system that will be installed in the area. Two different spray heads configurations were considered in order to estimate the time of intervention of the system. From the analysis of the simulation predictions it was decided to change the first planned configuration of the system;
- to determine the improvements about smoke spread and visibility in the tunnels during the fire produced by a division into compartments of the tunnels and by an exhausting plant.

1 INTRODUCTION

The Gran Sasso Laboratory and the local Fire Brigade have been cooperating since long time to always improve the security in the underground facilities. Inside this collaboration, the Safety Plant Service and the L'Aquila Fire Brigade commander, Ing. Basti, decided to conduct a set of CFD simulations and real scale tests to better understand the dynamics of a fire in the underground facilities and in the highway tunnel. The goal was to have a real comprehension of the situation in case of a fire, such as the spread of the smokes, the possible direction of smokes in the Laboratory, the temperature trend and the concentration of O_2, CO, HCL and other pollutant, the time available for people to escape, and the conditions in which the fire brigades would have to operate.

2 DESCRIPTION OF THE FACILITY

The Gran Sasso Laboratories are unique in the world for the dimensions of their underground halls and for the ease of access to experimental apparatuses of great dimensions.

INFN's Gran Sasso Facility houses leading experiments in the world of nuclear and sub-nuclear physics.

The Gran Sasso Laboratories are located beside the Gran Sasso Tunnel (10.4 km long) which is on the highway connecting Teramo to Rome, Italy (Fig. 1). They are about 6 km from the west entrance of the tunnel. They consist of three experimental halls, simply designated as Laboratories A, B, and C, and a series of connecting tunnels and service areas. The three experimental halls are more than 100 m long and about 18 m high and wide, enclosing a total volume that exceeds 180,000 m³. The natural temperature inside the laboratories is 6°C and the relative humidity is near 100% but an air conditioning system keeps them at more comfortable values of 18°C and 50%, respectively. Ventilation introduces fresh air at a normal rate of 35,000 m³/h. The Gran Sasso rock has a density of 2.71 g/cm³ and a mean nuclear charge number (Z) equal to 9.4. The laboratory is located at 963 m above sea level and the maximum thickness of the rock overburden is 1400 m.

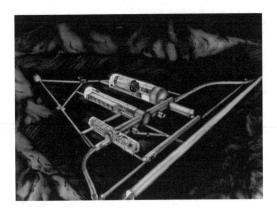

Figure 1. INFN underground laboratories of Gran Sasso.

3 COMPUTATIONAL MODELS

Several tools are used to evaluate the hazards from fires to the tunnel complex. This section describes the models used in this investigation.

3.1 CFAST

CFAST (Jones et al. 2000) is a zone model capable of predicting conditions in a multi-compartment structure subjected to a fire. It calculates the time-dependant distribution of smoke, gases and temperature throughout a building during a user-specified fire. Although it may not be all-inclusive, CFAST has demonstrated the ability to make reasonably good predictions. Also, it has been subject to close scrutiny to insure its correctness. Thus it forms a prototype for what constitutes a reasonable approach to modeling fire growth and the spread of smoke and toxic gases.

3.2 Fluent

Fluent (Fluent Inc), as other CFD codes like StarCD and CFX, is a computer program for modeling fluid flow and heat transfer in complex geometries. Fluent solves flow problems with unstructured meshes that can be generated about involved geometries. These meshing technologies in Fluent have been proven over time and multiple validations. Fluent provides a wide array of advanced physical models for turbulence, combustion, and multiphase applications. All functions required to compute a solution and display the results are accessible within Fluent through an interactive menu driven interface.

3.3 FDS (versions 1, 2 and 3)

NIST's Fire Dynamics Simulator (McGrattan et al. 2001a, b) predicts smoke and/or air flow movement caused by fire, wind, ventilation systems and other sources of momentum. A post-processor called Smokeview can be used to visualize the predictions generated by NIST FDS (McGrattan & Forney 2001c).

FDS solves a form of the Navier-Stokes equations appropriate for low-speed, thermally-driven flows of smoke and hot gases generated in a fire. FDS was developed to specifically address fire-hazard scenarios.

Smokeview visualizes FDS computed data by animating time dependent particle flow, 2D slice contours and surface boundary contours. Data at a particular time may also be visualized using 2D or 3D contour plots or vector plots.

4 CASE STUDIES

In this section, the scope of this safety analysis project is presented. The titles of the subsections identify which model is used.

4.1 Electronic rack fire in the MACRO experiment (Fluent)

The Fluent model is utilized to analyze the effects of a fire on an external section of the Macro Experiment. The exterior of the experiment is primarily composed of PVC and the aim of the study is to determine the smoke concentration, temperature, and O_2 level in the Laboratory B, where the experiment is installed. The simulation considered different scenarios with normal and emergency air flow.

4.2 Smoke spread within the service gallery from fire originating in MACRO experiment (CFAST)

This simulation investigates the smoke accumulation within the "Tir Gallery" originating from electric cable fire in the nearby "Service Gallery". The Fluent model was used to analyze the concentration and movement of smoke in "Service Gallery" to obtain a more realistic and accurate results. The Fluent output data were used as input data for CFAST which calculates the height layer of the smoke in the connected "Tir Gallery".

4.3 Truck fire in the highway tunnel (FDS)

This simulation investigates the consequences of a truck fire located just in front of the laboratory's entrance. Two main aspects are studied:

– the smoke movement and the temperature levels near the laboratory exit
– the direct effects of heat release on thae main entry door and the smoke propagation into the secured area 1.

A more in depth discussion of this case is provided in (Aprili et al. 2001).

4.4 Car fire in the Galleria Auto (FDS)

The *Galleria Auto* is a service gallery that is also used as a car parking area. Therefore, the fire of a typical service car is modeled. A detailed discussion of this case is provided in Section 5.

4.5 Electric cable fires in the Galleria Auto (FDS)

The *Galleria Auto* is also the facility area where the highest concentration of electrical cables can be found. These lines provide electric power to the halls and to the experiments installed in the service galleries. The oldest cables (1986) in the facility are PVC insulated and can be found in the trays in the *Galleria Auto*.

Although FDS does have a built-in fire-spread model, the need to simulate a large section of tunnel limits the possible resolution within the cable tray arrays. However, the results presented in (Sumitra 1982; Lee 1985) clearly establish the sensitivity of flame-spread to thermal communication between cables. Hence, it was decided to model the spread based on the data in (Sumitra 1982; Lee 1985). Since PVC cables in trays have a high smoke yield – 11.5% (Tewarson 1995) – and a high enough heat flux – 283 kW (Lee 1985) –, it was decided to consider fire only amongst the PVC cables. The procedure was to determine the exposed cable area (top and bottom) within each tray, determine the total heat output based on this area, and then scale the flux to the resolution of the model. The spread rate of $\dot{S} = 0.24$ m/s was determined from the data in (Lee 1985). Sumitra (1982) establishes that a 2.4384 m long tray with PVC cables can burn for up to 15 min. This was taken as a worst-case value upon which to scale the duration of each burning cell. In the computational domain, a long obstruction was used to represent the group of trays. Heat will emanate from the top cells of this obstacle. The algorithm is as follows. The cell in the center "ignites" at $t = 0$. At a time of $\Delta x/S$, its neighboring cells "ignite." The process repeats itself until the end of the run. Each cell will "extinguish" itself after the duration time is passed. The goal is to analyze the concentrations of different species and the cooling effect of the water mist.

4.6 Electrical cable fire in Luna 2 experiment area (FDS)

The cables fire algorithm described in Section 4.5 is also applied to the Luna 2 experiment.

The hazard posed by the quantity of cables located in this area may be considered to be not very significant. The main is to assess the safety of the experiment operators working within Luna 2 buildings. A follow-up analysis will determine appropriate procedures for emergency actions.

4.7 Fire from combustible liquid spreading from rupture during tank refilling in Laboratory C (FDS)

The Borexino experiment is located in Laboratory C where a large quantity of a combustible liquid (pseudo-cumene) is housed for its operation. The liquid is stored in 4 tanks (*storage area*). The filling procedure may pose a spilling or rupture risk.

Supposing the failure of a refilling hose and reasonable operator intervention time, the pool fire arising from a realistic amount of liquid spread is modeled. The main goal of the simulation is to verify the safety conditions for people involved in the area. Different scenarios are studied. The oxygen and smoke concentrations are tracked in order to verify the adequacy of standard and emergency ventilation.

4.8 Electronic rack fire in the Borexino experiment control room in Laboratory C (FDS)

The same domain described in the previous Section (Laboratory C) is now studied assuming the scenario resulting from an electrical rack fire inside the Borexino control room. The aim of this study is to estimate the possible damage to equipment from smoke and elevated temperatures. A water mist system is slated for installation in the room. The simulation will analyze the capability of the system to control the temperature.

5 FIRE SIMULATION AND SMOKE SPREAD ANALYSIS IN GALLERIA AUTO

5.1 General

The fire scenario chosen is the one resulting from a fire of car parked in the *Galleria Auto* parking tunnel. The fire is located at the beginning of the allowed parking area, just behind the entry of the Service tunnel. The floor of the tunnels was considered adiabatic. On the other hand, the walls and the ceiling of the tunnels were each considered as a 30 cm concrete layer, with its characteristics of thermal conductivity and thermal diffusivity.

The doors at both ends of the tunnel were considered inert, because they are classified as fire resistant for 120 min and so they can be considered inert in the time scale of the simulation.

The car fire is simulated by a pool fire under an obstruction of the dimensions of the car in order to comply with the thermodynamics of the hot fumes. For these simulations were used the results of car fire tests performed in a closed car park area, carried out by:

- ProfilARBED, Luxemburg
- University of Liègi, Belgium
- CTICM, France
- TNO, The Netherlands
- LABEIN and ENSIDESA, Spain

The test considered here is test number 3, where a Renault 5 (1996), equipped as in practice with oil, 4 tyres and a spare tyre and the fuel tank 2/3 full, was burned. The car was ignited with 1.5 l of petrol in an open tray under the left front seat.

Version 2 of NIST's FDS was used. For the FDS simulations an input file was created that makes use of the results of the experimental data for the duration of the fire, the total released energy, the mass loss and the heat released. The fire simulated lasted for 32 minutes, had a peak heat release rate of 3.5 MW, a mass loss of 138 kg and an energy release of 2.1 GJ. It was considered that the car is the only thing burning in the tunnel.

The analysis made has considered three main simulation scenarios of fires and smoke transport, involving the *Galleria Auto* and the other connected tunnels.

The three scenarios were:

- study of the diffusion of smoke and the temperature trend in the *Galleria Auto* tunnel, predicting visibility and species concentrations (such as CO and O_2) during the fire;
- examination of two proposed positioning schemes for heads of the Water Mist system that will be installed in the *Galleria Auto*;
- verification of efficiency in smoke spread and visibility in the tunnels during the fire brought by a division into compartments of the tunnels and by an exhausting plant.

5.2 Smoke spread study

5.2.1 Geometry
FDS allows describing the simulation domain as a three-dimensional grid of hexahedral cells. All geometrical details are represented by means of free or obstructed cells.

The simulation domain (Fig. 2) is $416 \times 77 \times 8.6$ m and is divided in 384,000 ($400 \times 80 \times 12$) cells, with side approximately $1 \times 1 \times 0.7$ m.

The domain includes the *Galleria Auto* tunnel and the areas directly linked to it, the so called Service tunnel, ISPESL tunnel and Tir Tunnel.

5.2.2 Hypothesis and assumptions
The actual vault ceiling of tunnels wasn't simulated because FDS is limited to hexahedral cells and the geometry of this case was very complex, also

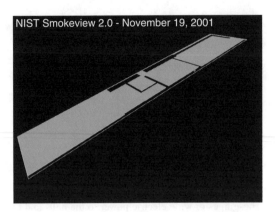

Figure 2. Overview of the computational domain used in the smoke spread simulation

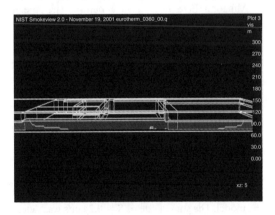

Figure 3. Visibility values at time $t = 360$ s, shown in a vertical plane cut along the *Galleria Auto* tunnel.

considering that the geometrical resolution, due to the cell size, is roughly 1 m. The various tunnels rectangular cross section areas are equivalent to the real ones by means of a proper equivalent height.

It is also assumed that no forced ventilation is active in the area of the simulation.

5.2.3 Goal of the study
The main goal of the analysis is to describe the diffusion of smoke and the temperature trend in the *Galleria Auto* tunnel and in its connected tunnels, estimating species concentrations (such as CO, and O_2) during the fire.

5.2.4 Results
As expected, high temperatures are present mostly in the area near to the fire.

On the other hand, smoke spread is very fast in the *Galleria Auto* tunnel, but also towards the most remote part of the domain. Figure 3 shows the visibility values at the time $t = 360$ s in a vertical plane cut along

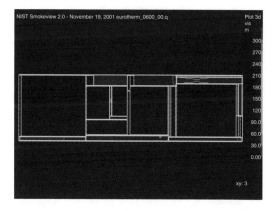

Figure 4. Visibility values at time $t = 600$ s, shown in a horizontal plane about 2 m high from ground on the whole domain.

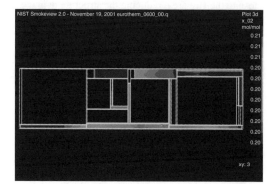

Figure 5. Oxygen concentration values at time $t = 600$ s, shown in a horizontal plane about 2 m high from ground on the whole domain.

the *Galleria Auto* tunnel. It is clear that at $t = 360$ s the visibility in most part of the *Galleria Auto* drops rapidly towards 0 m. Figure 4 shows visibility on a horizontal plane about 2 m high from ground on the whole domain at the time $t = 600$ s, the time at which the fire reaches its peak.

Oxygen and CO concentrations the same time $t = 600$ s on a horizontal plane cut at about 2 m level are illustrated on Figure 5 and 6.

The simulation predicts that the first aspect becoming dangerous for the occupants of the area can be assumed to be the lack of visibility due to smoke spread.

At the same time, CO concentration begins to become dangerous in the area near the fire, considering also that due to the lack of visibility, it can be assumed that people still in the area could have difficulties heading for the escape exits.

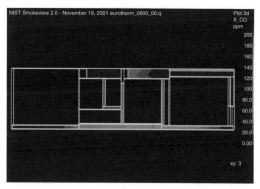

Figure 6. CO concentration values at time $t = 600$ s, shown in a horizontal plane about 2 m high from ground on the whole domain.

5.3 *Positioning of water mist heads examination*

The results are not shown here. Only the conclusions are reported.

5.4 *Smoke spread study with a division into compartments of the tunnel and an exhausting plant*

The smoke spread simulation predictions were used in order to evaluate if current arrangement of the area could have been improved to better contain the smoke spread in *Galleria Auto* and annexed tunnels. Apart from the water mist system which is installed for the area and that marks a point towards the improved safety of the *Galleria Auto*, it could have been evaluated the feasibility of the introduction of new smoke/fire resistant partitions and doors in order to limit the extension of the compartment.

Ventilation strategies were also investigated in order to evaluate and improve emergency ventilation procedures.

5.4.1 *Geometry*

A new simulation domain was set up (Fig. 7) but with the same fire scenarios of previous simulation. The only *Galleria Auto* tunnel was included ($416 \times 6.5 \times 5.5$ m and was divided in 720,000 cells, with side approximately 0.27 m).

All the openings to the annexed tunnels were considered closed with doors. And the doors were considered inert, because they would be classified as fire resistant for 120 min and so they can be considered inert in the time scale of the simulation.

Two velocity boundaries conditions were defined: one at the end of the tunnel, extracting the air, and another one, in which the air enter in the domain, placed close to the Hall A entrance.

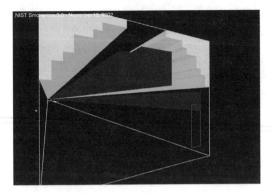

Figure 7. Overview of the computational domain used in the smoke spread simulation.

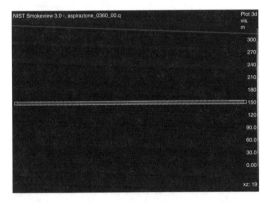

Figure 8. Visibility values at time $t = 360$ s, shown in a vertical plane cut along the Galleria Auto tunnel.

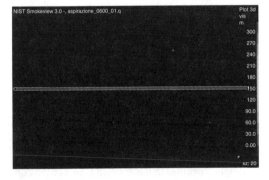

Figure 9. Visibility values at time $t = 600$ s.

5.4.2 Goal of the study

The main goal of the analysis is to describe the diffusion of smoke in the tunnel during the fire and to check if with this configuration of the tunnel the safety for the personnel of laboratory is increased.

5.4.3 Results

It was calculated that the time needed for the detection of the fire by the smoke detector installed in the tunnel and the following switching on of the exhausting plant, are 6 minutes. At 360 s. the simulation shows that the smoke spread is worst, but this time confined to the only Galleria Auto tunnel. It is evident the turbulence effect and the destruction of the stratification of the smoke layer in the area close to the fan jet.

After 10 minutes, it's evident that the part of the tunnel before the fan jet maintains an high visibility and no smoke is present. This situation permit to escape in/out the south entrance of Hall A, and a safer way for the fire brigade to reach the fire. On the other hand the exhausting plant create a turbulence that completely destroy the smoke layer. But, also in the previous simulation, it was clear that the smoke layer after 10 minutes was to low for the human safety. Furthermore this turbulence keeps the air temperature more uniform ($\sim 50°C$ in the upper part of the vault ceiling) and lower close to the fire area, reducing the structure damages. Due to the big dimensions of the tunnel and the short time of intervention of the LNGS fire brigade, the fresh air injected doesn't furnish more power to the fire.

6. CONCLUSIONS

6.1 Smoke spread study

The simulation predictions can be used in order to evaluate if current arrangement of the area can be improved to better contain the smoke spread in *Galleria Auto* and annexed tunnels.

Apart from the water mist system which is planned for the area and that would mark a point towards the improved safety of the *Galleria Auto*, it can be evaluated the feasibility of the introduction of new smoke/fire resistant partitions and doors in order to limit the extension of the compartment.

Ventilation strategies are also being investigated in order to evaluate and improve emergency ventilation procedures.

6.2 Water mist positioning study

The simulations carried out show that to obtain a good time response of the water mist system installed in the parking tunnel, the activation must be placed on the top of the vaulted ceiling. The first plant layout (spray heads at a 2.9 m height only), that seemed efficient for the water mist distribution and for the time of activation because of the presence of numerous obstructions in the vaulted ceiling, in reality takes too much time to be activated. The best plant layout results the one with the spray heads at a height of 2.9 m and the thermal heads on the top of the ceiling. In this way we combine a more efficient water mist distribution (for the fire loads present in the tunnel) with a faster activation of the system. Thanks to this configuration of the water mist system, it is possible to control the fire since the beginning, to keep the temperature low and to reduce considerably the smoke production.

6.3 Smoke spread study with a division into compartments of the tunnel and an exhausting plant

The simulation shows that the safety for the personnel of LNGS and for the intervention on the fire by the fire brigade result increased with these technical interventions. In fact the exhausting plant permit to keep an high visibility and no smoke spread in 100 m of the tunnel. This situation permit to escape in/out the south entrance of Hall A, and a safer way for the fire brigade to reach the fire. On the other hand the exhausting plant create a turbulence that completely destroy the smoke layer. But, also in the previous configuration of the tunnel, it was clear that the smoke layer after 10 minutes was to low for the human safety. Furthermore this turbulence keeps the air temperature more uniform ($\sim$50°C in the upper part of the vault ceiling) and lower close to the fire area, reducing the structure damages. Due to the big dimensions of the tunnel and the short time of intervention of the LNGS fire brigade, the fresh air injected doesn't furnish more power to the fire.

REFERENCES

Aprili, P.G., Daquino, G., Franciotti, D., Zappellini, G. & Ferrari, A. 2001. CFD Simulations of a Truck Fire in the Underground Gran Sasso National Laboratory. In E. Zio, M. Demichela & N. Piccinini (eds), *ESREL 2001 – Towards a safer world – Proceedings of the European Conference on Safety and Reliability – Torino, 16–20 settembre 2001.* Torino: Politecnico di Torino.

Babrauskas, V. 1995. Burning Rates. In P.J. DiNenno (ed.) *SFPE Handbook of Fire Protection Engineering, Second Edition.* Bethesda: Society of Fire Protection Engineers.

Development of design rules for steel structures subjected to natural fires in closed car parks – ProfilARBED, Universite' de Liege, CTICM, TNO, LABEIN.

Drysdale, D. 1998. *An Introduction to Fire Dynamics – Second Edition.* Chichester: John Wiley & Sons Ltd.

Fluent Inc. *Fluent 5 User's Guide.* Centerra Resource Park: Fluent Inc.

Jones, W.W., Forney, G. P., Peacock, R.D. & Reneke, P.A. 2000. *A Technical Reference for Cfast.* Gaithersburg: National Institute of Standards and Technology.

Lee, B.T. 1985. *Heat Release Rate Characterisitics of Some Combustible Fuel Sources in Nuclear Power Plants, Technical Report NBSIR 85–3195*, Gaithersburg: National Institute of Standards and Technology.

McGrattan, K.B., Baum, H.R., Rehm, G.P., Forney, G.P., Floyd, J.E. & Hostikka, S. 2001a. *Fire Dynamics Simulator (Version 2), Technical Reference Guide, – Technical Report NISTIR 6783.* Gaithersburg: National Institute of Standards and Technology.

McGrattan, K.B., Forney, G.P., Floyd, J.E. & Hostikka, S. 2001b. *Fire Dynamics Simulator (Version 2) User's Guide, – Technical Report NISTIR 6784.* Gaithersburg: National Institute of Standards and Technology.

McGrattan, K.B. & Forney, G.P. 2001c. *User's Guide for Smokeview Version 2.0 – A Tool for Visualizing Fire Dynamics Simulation Data, – Technical Report NISTIR 6761.* Gaithersburg: National Institute of Standards and Technology.

Quintiere, J.G. 1998. *Principles of Fire Behavior.* Albany: Delmar Publishers.

Rosenbaum, E.R. (SFPE chairman) 2000. *SFPE Engineering Guide to Performance-Based Fire Protection Analysis and Design of Buildings.* Bethesda: Society of Fire Protection Engineers.

Sumitra, P.S. 1982. *Categorization of Cable Flammability: Intermediate-Scale Fire Tests of Cable Tray Installations, Interim Technical Report NP-1881.* Palo Alto: Electric Power Research Institute.

Tewarson, A. 1995. Generation of Heat and Chemical Compounds in Fires. In P.J. DiNenno (ed.) *SFPE Handbook of Fire Protection Engineering, Second Edition.* Bethesda: Society of Fire Protection Engineers.

Safety and Reliability – Bedford & van Gelder (eds)
© 2003 Swets & Zeitlinger, Lisse, ISBN 90 5809 551 7

Software reliability and safety: a real case study in the transportation area

Mrs. Jacques ARDON & Benoît GUILLON
FAIVELEY TRANSPORT, Z.I. des Yvaudières, St-Pierre-des-Corps Cedex – France

Mrs. Olivier LECOMTE & Emmanuel ARBARETIER
SOFRETEN, Parc Saint Christophe, CERGY PONTOISE CEDEX

ABSTRACT: The aim of this article is to present original software reliability method concerning transport domain with odometer system of safety application. When a new software is created, the phase of software validation is essential to development cycle. That's why some qualitative and/or quantitative analysis of software potential errors, and of their effect, are achieved to estimate the software criticality.

FAIVELEY TRANSPORT, as a software designer, performs this kind of analysis for the software validation and in particular has applied it for a speed measure station ERTMS (European Rail Traffic Management System). The study presented in this project is in particular supported by software of ERTMS odometer system.

The SEEA (Software Effect and Error Analysis) is a qualitative method applied to software, for safety, in order to estimate it and to improve it, through the adoption of some recommendation emerging study. In fact, the fundamental principle of SEEA is to consider some software intrinsic hypothesis of design simple mistake, to deduce the effect from mistakes and to analyse the sturdiness of software compared with mistakes.

The setting up of a software fault tree, with a top event representing the feared event, enables to extract the first order minimum cut set and to submit it to the SEEA method. In this case, the weak point emphasized by this method are improved with recommendations.

In the first time, the program functional comprehension allows to set up the skeleton main part of the software fault tree, with the top event representing the top event and with top down analysis in the program code lines. In fact, some functions are not studied systematically because they are totally beyond the limits of the study. Indeed, the generation of software fault tree has to assume analysis hypotheses such as relevant life phases of the product or its operational process.

As a result from this analysis, recommendations to introduce means of protection within the software have allowed to remove the most serious software errors and to assure a high level of software safety. This method has allowed to validate an original reliability process of real time data-processing application, through a joint approach with the SEEA and the modelling by fault tree, based on the rereading of source code.

1 INTRODUCTION

The more and more systematic use of software in system operated in the transportation area necessitates generalization of RAMS studies for these applications. Following article presents an original RAMS methodology based on a software safety study applied to the odometer system of a safety application used in the railway transportation area.

In the design process of a software, the analysis process is a very important phase of its development cycle. Therefore qualitative or quantitative analyses about potential software errors and their effects must be performed to assess its criticality. As a software designer, FAIVELEY TRANSPORT performs this type of analysis during the validation phase, and did apply it on a speed measurement device ERTMS (European Rail Traffic Management System). The analysis presented in this publication particularly deals with the software of the odometer system which is defined as a component of the "on board" equipment. Its function consists in providing in a real time process cinetic parameters of the train (speed, position, acceleration) to the main "on board" equipment, which is called EVC (European Vital Computer).

In the particular case of this analysis, validity of the analysis is based on the exact knowledge about hardware and functional architecture of the system.

Thus it is necessary to determine precisely on what manner and through what component information transmitted by captors to odometer system are processed and validated, and according to what parameters. The hardware architecture of the odometer system is based on a set of two redundant captors providing to three independent processing cards, speed values which are calculated by three different mathematic methods based on different algorithms.

2 ARCHITECTURE AND VOTING SYSTEM

The voting system which is integrated in the odometer system is made of two selection levels (cf. Figure 1). The first is based on a choice of redundant speed values, provided by two wheel sensors (Wiegand) and two external radar boxes. This way, the system is able to manage a set of six input data owing to the selection of at most three signals issued from different acquisition processes, and these measures which are selected on this manner are compliant with validity criteria such as availability and coherence, and if not, they are unvalidated and eliminated. The role of the second selection level is to provide a theoretical estimation of the dynamic magnitudes of the train through Kalman filtering process.

Wheel sensors are composed of six active sensors for speed measurement based on Wiegand effect, plus two redundant ones. The sensors are divided into two groups of four, and the difference between the phasing of these two groups is of the magnitude of 90°. This way, owing to separate cable links, these sensors can provide totally independent information to the three odometers detecting cards which constitute the system (each of these cards has to choose the specific wheel sensor which will provide to it the measure which will be necessary for the next computations).

Radars' mission is to provide in a real time process information about train speed owing to two different computation methods (speed assessment based on Doppler effect and INTERCORRELATION). Differently from wheel sensors, radars do provide respectively the same measures, Doppler and inter correlation, to the three odometer detecting cards through the same CAN device. The architecture which is based on this principle enables to increase the availability of Doppler and inter correlation measures and the safety of not using a correct value. Indeed, the voting system is not limited to a simple choice of radar A or B, but in the contrary to a selection which permits to reduce the risk of processing wrong measures.

To justify the architecture of the voting system which is used for radars, logical truth tables have been built. After an analysis of these tables, it appears that only 108 scenarii among 256 are able to generate a bad selection of radar, that is to say somewhat more than 42%. Therefore selection of speed value is achieved according to a mark which is associated to each radar, and which is based on the availability and consistency

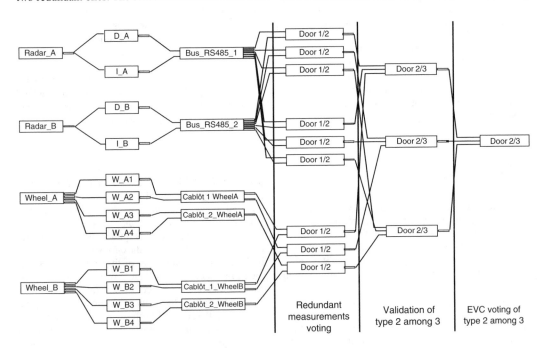

Figure 1. Architecture and voting system.

54

of Doppler and inter correlation measures. Thus the radar which obtains the highest mark is selected and provides Doppler and inter correlation measures which are associated to it.

The three speed measures which are selected this way are now going to be submitted to a second level of selection based on the comparison of the two parameters standard deviation and validity related with the values of these measures. However the computation achieved in the framework of this second level integrates a theoretical computation of the train's dynamics, and because this computation has to be performed with a high level of security and safety, it is necessary to take into account at least two of the three measures taken previously. When it's not possible, for example in case of degraded or transient modes, it is necessary to produce a fatal error of the system after a certain time of KALMAN filtering. The value of this time, which depends on a trade off between safety and availability, is related with the number of remaining measures.

The evolution of the train dynamics estimation is computed on the base of the updated estimation which has been calculated in the last update performed from the last valid sensor measures. The evolution duration corresponds to the difference between this update time and the date of the new measures. Moreover the estimation of the train's dynamics is updated only if at least one measure is valid. In this case, updated estimation and date of update are kept for future computation of modifications of estimations. As output of this processing, information generated by Kalman filter are available and sent back to train's dynamics. As soon as theoretical computation of the train's dynamics is performed, theoretical value of the speed of the train is used as a reference for the whole voting system, to determine coherency and therefore validity of the new measures of the sensors.

3 QUALITATIVE ANALYSIS FOR SOFTWARE SAFETY

From the knowledge about hardware and functional complexity of this odometer system, the approach adopted for software analysis is going to be applied on the basis of two famous RAMS methods. First, a failure analysis is performed from the building of a software fault tree, taking as the top event the dreaded event issued from the Preliminary Hazard Analysis (PHA): this top event is called "Computed distance out of the confidence interval".

As soon as the fault tree is built, a Software Error and Effect Analysis (SEEA) leads to the evaluation of the criticality of the software from the value of each component (procedures, modules, …), through the identification of the different Single Point Failures appearing in previous Fault Tree. SEEA is a qualitative analysis of the Single Point Failures (1st order) which is performed on the software for RAMS analysis purpose, for their evaluation and their improvement owing to some recommendations issued from the analysis. Indeed the fundamental principle of SEEA is to consider simple hypotheses of software inherent design errors, and to determine the consequences of these errors and analysis integrity of the software in respect to them.

The starting point of software analysis of ERTMS odometer system will be based on an object oriented approach, from HOOD methodology and program shortcuts written in ADA. HOOD method's principle consists in the description of the system as an object hierarchy, where each of them is decomposed from different abstract breakdown levels. Leading principle of an abstract breakdown level means that objects belonging to the same breakdown level share the same abstract level from the real world. Therefore, objects belonging to a same level exchange data which correspond to a common point of view of the real world.

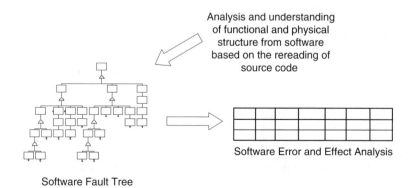

Analysis and understanding of functional and physical structure from software based on the rereading of source code

Software Fault Tree

Software Error and Effect Analysis

Figure 2.   Qualitative analysis for software safety.

4   FAULT TREE ANALYSIS

From this approach, the objective is to build the fault tree associated to the dreaded event which is called "Computed distance out of the confidence interval". For this purpose, the analyst applies STOOD software to develop the understanding and analyse the functional and hardware architecture of the odometer. Moreover, in order to build a fault tree in best conditions, he also uses SIMFIA software.

As soon as the fault tree is built, results obtained from this analysis have to enable a progressive improvement of the software, owing to specified corrective actions. These actions are applied to prove and justify the coherency and consistency of the software. The performance of this analysis will have to be supported in different steps, to be completed exhaustively in the framework of considered life cycle.

Indeed, at first, it is necessary to understand properly architectural and functional logics of the program, to be able to satisfy analysis completeness through the verification of the whole set of the functions of program. For this purpose, the building of the fault tree must be performed from different hypotheses concerning the program, such as: life phase of the product, operational mode used, objects or operations to be analysed. For this analysis, the program will be studied in the life phase "Exploitation" and in the functional mode "Operational" or "Normal". This phase is the one where the odometer performs periodic measure of the odometer data and transmits them to the EVC. However, the functional mode "Operational" of the software includes two phases, intimately integrated, but however different, which are respectively "starting phase" and "normal exploitation". The Fault Tree is purposely built, taking into account these two phases, and in the framework of SEEA, errors and failure modes concerning these two phases will be duplicated, to underline the two possible effects.

However, functions related with functional modes other than "operational" like "test mode", "remote loading" or "maintenance mode" must not be taken into account in this tree. Moreover, the set of functions related with operators $(+, -, *, /, <, >)$ and also "check assertions" has not been processed.

When performing this study, it appears that functional understanding of the program enables to build in a first step, the major part of the tree from the top event, and then to carry on the analysis in a more precise way through a top down analysis in the program. However the size of the tree becoming very important, it is necessary to use a detailed list of the already existing failure states, which is exported owing to the dictionary command of SIMFIA software, to check constantly that a new failure mode to create does not already exist in another branch of the tree. Indeed, there are different failure modes located in different branches of the tree and marked with reference symbols, and being able to generate effects on the software which are of very different nature.

During this very step, the building of the tree will be performed, using three different softwares (STOOD, SIMFIA, and EXCEL). This way, double occurrence of the failure modes will not be possible, and owing to the reference symbols, leaves events of the tree will only appear one time, corresponding to cut sets of the first order. After this first step, one has to check that all functions of the software have been processed, because some functions have to be left apart, as they are not in the scope of the study. Moreover, with HOOD software, it is possible to explore the activation graph of the functions of the program, on a hierarchised way. This way, the check of the processing of every function can be performed, comparing the activation graph to the sub Fault Tree related with considered function. When this is not possible, the analysis can be performed. Besides, owing to this activation graph, it is possible to control the progression through the whole program, and this way, to warranty exhaustivity of the study.

5   SEEA

Once the fault tree is achieved, extraction of the cut sets of the first order will lead to the definition of the failure modes of the software. Indeed, single point failures which can generate dreaded event are identified throughout this Fault Tree and thus will lead to protection measures. For this purpose, every single point failure is listed in a worksheet and submitted to a SEEA. This way, the weakest points of the software which are emphasized this way are processed through recommendations, to make the software more reliable as to its potential defaults. Moreover, the introduction of protection barriers inside the software aims at removing most serious software errors and this way maintain a high level of software safety.

6   PROGRAMMING PROTECTION BARRIERS

Besides, during the design phase of a software, designers try as far as possible, either to put in the program "traps" ensuring that called function is supported in a correct way, either to apply secure programming methods. Therefore, protection barriers introduced in the software can have different types and origins. Indeed, a software can be controlled, either through internal devices either through external devices. The latter ones can be protections like "watch dogs" controlling the good operating of the software considered, or protections through hardware logical input ("key" or "lockers") which can impose manually operational modes of the odometer (beware human errors …).

The software also includes "Check Assertions" which constitute a first level of protection within difficult parts of the program, and also uses buffers in the software which are adapted to the situation.

The "checks" object is useful to manage exceptions of the software detected during the execution without using ADA exceptions. It only contains one operation "assertion" called by the objects needing controls able to detect an exception. If dreaded condition exists, the parameter of the "assertion" operation is equivalent to "false", and the "assertion" operation executes an infinite loop, waiting for new initialisation and producing a debug message on the series backup link.

Buffers are finite information aggregates able to keep data for reading functions. Its operating mode is very simple, as it only consists in writing and then reading information contained in these tables. Indeed, one buffer has to prepare and keep data ready to be read when they have been written physically and completely. However, potential failure modes do exist, and therefore depending on the type of buffer and on the context where it is used, considered failure modes are no longer realistic.

The use of "FIFO" buffers in a sequential environment only imposes to look after buffer management, only when it is full. Obviously, when the environment is not sequential, one has to protect oneself from competitive use of the buffers, i.e. prevent from the inversion of read/write and reversely. For this purpose, another type of buffer is used. The set of the buffers of the software of the odometer system is positioned in a sequential environment, except for two of them which are put under interruption (IT) in the processing of CAN process. An interruption enables to interrupt processing of the program at any time, after reception of predefined messages. A priority order is affected to these messages, in the case where two different messages would arrive simultaneously. At the level of the microchip and the hardware, possible inputs for this type of action are located. The two buffers, positioned under interruption, are of a circular type and enable, on one side, to check constantly that a buffer is never full, and on the other side to be sure, that one information which is ready to be read has been physically and totally written at previous step.

About CAN system, specific protections used for different buffers used, and preventing the occurrence of their failure modes, would be sufficient to justify security of the program. Indeed, other failure modes of CAN system produce loss of messages, and in this type of situation nothing can happen. However, these internal protections are not sufficient to warranty a nominal sequencing in the execution of the rest of the programming of the odometer system, and again, one problem consists in identifying if such or such function has really been called.

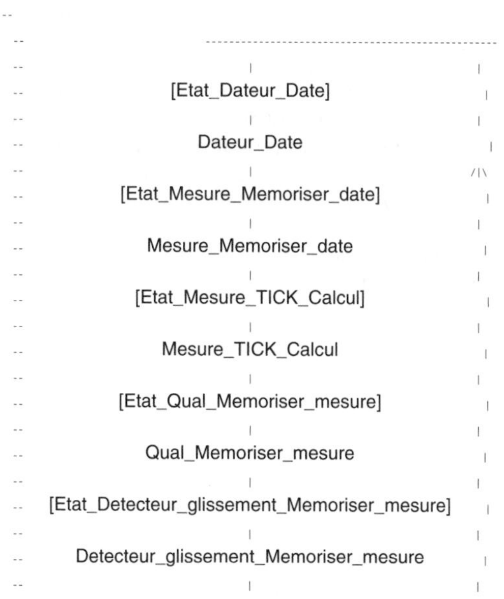

7 RECOMMENDED PROTECTION BARRIERS

For this purpose, the introduction of traces generated when functions are called, among the critical ones, enables to warranty their proper sequencing. A trace generator is useful to check correct sequencing of critical functions in the operation of an object of a software. This trace generator is called every time one of these critical functions is called. If the calling sequence of watched critical functions is not compliant with the possible activation graph, which is stored in the trace generator, the latter immediately causes the interruption of the program.

At last, failure modes which take benefit from a protection device inherent to the software, will be associated to a justification notice explaining why they have not been taken into account in the fault tree or the SEEA. Moreover, failing states associated to a sub tree containing multiple causes (AND gate), have not been developed in the fault tree, because minimal cut set of an order higher than one have not been taken into account in the SEEA.

To introduce trace generators only on critical functions, from the safety point of view, SEEA is applied

Figure 3. Trace generator of critical function based on sequencing of called functions.

57

on the fault tree realized as previously, and the structure of which reveals the structure of software programming (activation tree). It emphasizes failure modes concerning loss and failures of functions the loss of which can produce the dreaded event. Therefore operation sequences which present these weak points are processed through recommendations of trace generators to make the software more reliable, as to its potential defaults. This study is a complete analysis of functions of odometer system, apart from environmental objects, the level of analysis has been limited to functions present in the activation tree.

8   RESULTS

The SEEA which is performed this way, includes 599 analysed scenarios, among them 96 have an unacceptable effect on the system: appearance of the dreaded event. Validation of these 96 functions selected for the introduction of trace generators, has been performed in the framework of working groups. It must be noticed that some of these functions are called several times in different contexts, and therefore such functions can have their activation traced in all cases or partially. Following the recommendation to introduce these trace generators, reliability of the software is ensured, according to design assumptions envisaged about the software, as well as error consequences.

9   QUANTITATIVE ANALYSIS FOR SOFTWARE SAFETY

Qualitative analysis which is performed this way is not sufficient to prove the satisfaction level of the software according to all safety levels. Indeed, quantification of the occurrence rate of the dreaded event is necessary to this justification. Therefore, owing to information about system architecture used in SIMFIA, and to the integration of failure rates provided by FAIVELEY TRANSPORT, for every system or subsystem used, computation of occurrence rate of dreaded event called "computed distance out of the confidence interval" from all hardware failure modes is the proof of an odometer system which is more reliable and more available.

About computation conditions, occurrence rate of the top event is computed according to a statistical law named "Periodic Test" as well as parameters which are associated to it. This type of law applies very well to component which fail according to an exponential law, and for which the failure is detected owing to a periodic test. Repair action is then performed on an instantaneous way. Therefore assuming the safety target for odometer system with a voting architecture of type 2 among 3, and that failure rate of a non safety situation is less than 10^{-11}, than obtained occurrence rate for dreaded event satisfies initial reliability threshold.

10   CONCLUSIONS

Through this analysis, architecture of odometer system has been analysed deeply, so that it might be modelled on a proper way. Indeed, the qualitative or quantitative analyses have enabled to identify on one side measure acquisition networks related with the sensors, and on the other side the two levels of selection of the voting system which are respectively used for selection and validation of acquired measures. This way, computation of dreaded event "Computed distance out of the confidence interval" proves odometer system to satisfy reliability requirement corresponding to a maximal threshold of 10^{-11}, which means it's completely safe from this point of view.

Software safety analyses of a system performed in the framework of RAMS studies determine the impact design and realisation process may have on the safety of a system. Indeed whatever may be used, programming method or language implemented, requirements are satisfied for every scale of software safety integrity, as well as for level 0 which is not related with safety.

Safety and Reliability – Bedford & van Gelder (eds)
© 2003 Swets & Zeitlinger, Lisse, ISBN 90 5809 551 7

On the economic value of safety

F. Asche & T. Aven
Stavanger University College

ABSTRACT: In this paper we discuss the hypothesis that safety and accident risk is in general not adequately incorporated into the economic decision processes. Our starting point is firms' incentives for investing in safety. A simple economic model is used as a basis for the discussion, that points at and reveals a way of thinking that strongly influence decision-makers. In particular, the official language used when decision-makers communicate safety is not necessarily the same as the underlying driving forces. It is an aim of the paper to demonstrate, in particular to safety people, analysts and managers, that there is a need to show that safety measures have a value in an economic sense. If such values cannot be demonstrated, businesses cannot be expected to invest in higher safety. Regulators need to develop a framework that give the "correct" incentives for safety, and such incentives need to be linked to the key economic performance measures.

1 INTRODUCTION

The purpose of this paper is to discuss the hypothesis that safety and accident risk is in general not adequately incorporated into the planning and decision processes of a business. Safety is often perceived to be in conflict with the process of adding value (or maximizing profit). As a consequence greater efforts seem to be made to reduce investments in safety than to add value by investing in safety.

A causal look at a financial newspaper quickly reveals that safety does not hold a prominent place with respect to what is relevant information for decision makers. To satisfy investors, one often sees firms promise a given level of return on their investments. Whether this is EBIT, ROACE or some other measure of capital return do not matter much. What matters is that this clearly put the focus of the firm on its economic performance, and there seems to be little room for other considerations, including safety issues. Does this mean that the firm does not take safety issues into account? We will in this paper look closer at this question from an economic perspective.

Our starting point is a discussion of the business incentives for investing into safety. What are the fundamental principles that rule the decision-making process from an economic perspective? To what extent and how is other factors than the economic performance measures given weight? Are factors not directly related to the firms' core activity such as potential loss of goodwill in the society taken into account? And to what extent are risk reducing measures external effects for the firm, in the sense that they are beneficial for the society but not necessarily for the firm? Moreover, how are these issues influenced by public regulations and actions?

We will try to answer these questions by looking at firm behavior from an economic perspective. We will not be concerned with investment decision as such, whether in safety or other factors but rather how the firm as an economic unit under different market structures finds it in its own interest to operate. This is important as it can be regarded as the step before an investment analysis, as it reveals the incentives of the firm. From this perspective an investment in safety measures can be compared with any other investment or purchase by the firm. As such, this is the first step if one is to show that safety has an intrinsic value to the firm. This can then in the next stage be fed into traditional investment or benefit/cost models. To the extent that safety is evaluated on its own basis and not in relation to the firms' economic performance, it will not be perceived to have any economic value. There is accordingly a need for showing that safety has a value in an economic sense. If such values cannot be demonstrated, businesses cannot be expected to invest in higher safety.

The economic value of safety has also been addressed by many other researchers, such as Jones-Lee (1989), see also Calow, P. (ed.) (1998) and Fischoff et al. (1981). Much of the existing literature in this field has a focus on public safety, and the problem of balancing safety and costs. Our concerns, however, relates to firms and the relationship between firms and the society. The aim has been to obtain new insights by using a simple, standard economic model as a basis for the discussions. The primary target group for our work is the safety profession, including regulators, and the main contribution of the paper is the communication and discussion of fundamental principles that drive businesses and their attitude to safety. We refer to Gibson (1978), Kotz and Schafer (1993), Pape (1997), UKOOA (1999) and Viscusi (1986, 1989) for other works related to these issues.

2   THE FIRM'S ECONOMIC INCENTIVES

We will here start with the basic economic model of a competitive firm, and highlight how safety issues are taken into account. A more general treatment of the competitive firm can be found e.g. in Varian (1992). When markets are competitive, a firm can sell as much as it would like at an price p, and buy as much as it would like of any input factor at a price w_i, and can accordingly not influence the prices by its actions. We disregard the possibility of price uncertainty, which can be introduced by using expected prices. Furthermore, we look at a static problem. This can then be regarded either as spanning the total time of a project or the profit maximisation problem in each period (which is then the cash flow variable in a net present value formula). We are then not able to capture dynamic aspects in the problem. However, this does not take much away from our primary focus, the firm's incentives to invest in safety. Formulating the firm's maximisation problem as a dynamic problem means increased mathematical complexity, but the additional insights gained is marginal.

Let the function $y = f(x)$ be a firm's production function, where x is a vector of inputs and y is expected output based on the levels of the elements in x. The function f can be regarded as a black box that transforms the levels of the input factors in the x vector to a quantity of the output y. Examples of elements of the x vector are man hours of labour (or of a specific type of labour), units of capital used (i.e. machine hours), units of materials used (e.g. tons of steel). The elements of x will differ depending on what the product y is, and also on the technology used. The output y is measured as an expected quantity, e.g. bbl/day, tonnes etc. We assume standard regulatory conditions apply to $f(x)$. The function is assumed to be increasing in all elements of x, that is, the marginal product for each factor is positive, i.e. $\partial f(x)/\partial x_i > 0$, $\forall i$, but decreasing i.e. $\partial^2 f(x)/\partial x_i^2 < 0$. Finally, one can in general at least partly compensate lower utilisation of one input factor by using more of the others, but there can also be factors that are essential in the

sense that one cannot produce a given level of y without utilization of a given level of that factor.

When we let y denote expected output, the firm is assumed to be risk neutral, i.e. the firm's expected utility value of the possible outcomes/consequences is measured by the expected value. No outcomes/consequences are given increased or decreased utility value to reflect aversion or attraction. This assumption makes decisions with respect to the tradeoffs between production technology and risk (uncertainty in outcomes/consequences not reflected in the expected value) irrelevant. This is to keep the analysis simple, as we otherwise would need to express it in an expected utility framework where we would have to take the decision makers attitude to risk into account. Just and Pope (1978) and Asche and Tveteras (1999) provide applications.

Based on the production function, the firm will maximize expected profits.

$$\underset{x}{Max} \quad \Pi = pf(x) - w'x \qquad (1)$$

Assuming that the second order conditions indicates a maximum, the first order condition for factor i is

$$p\frac{\partial f(x)}{\partial x_i} = w_i \qquad (2)$$

This indicates that a firm will demand input factors until the value of their marginal product is equal to the price of the factor. Please note that this is the user price (rental price) and not the investment cost. For a firm that invests in a capital good, one can regard it as paying a rental to itself in terms of depreciating the good and taking into account the alternative value of the capital locked into this good.

Equation (2) gives us the first insights with respect to investments in safety. If one input factor is a risk reducing measure, the firm should invest in this factor up to the point where the value of its marginal product is equal to its cost. This is optimal for the firm since higher as well as lower utilisation of a factor would reduce profits. Hence, firms have economic incentives to invest in risk reducing measures. However, if risk reducing measures has a value of their marginal productivity that is lower then their cost they should not be undertaken. This is of course also true if the marginal product of a risk reducing measure is zero. An economist will therefore say that a (profit maximising) firm that is not investing in a possible risk reducing measure does not do it because it does not contribute to profits in a positive way. If it had, the investment should have been undertaken as this would have increased profits. Assuming that the model is fully implemented, the firm will then never under invest in risk reducing measures. Of course, in practice, there are factors beyond those incorporated in the model that need to be addressed, but for the time being let us think within the framework of the model.

It can be useful to discuss the tradeoffs involved with an example. Assume that we are looking at a production process, where a machine is used that has some positive probability of a breakdown. Each breakdown reduces the production as it takes time to fix and restart the machine. Hence, any measure that reduces the number of breakdowns will increase the expected production. The firm can influence the probability of a breakdown by the use of a risk reducing measure, maintenance. How much will the firm then spend on maintenance? It will spend up to the point where the cost of the maintenance is equal to the value of its marginal benefit. However, note that as long as maintenance is not free, there will always be a positive probability for breakdowns. Hence, there will in general always be breakdowns, and the fact that it is costly to reduce the probability of breakdowns gives a limit for how much maintenance it is profitable for the firm to undertake. The firm will allow some breakdowns since this is less costly then more maintenance.

In general the firm will be able to substitute between the different inputs. For instance, better quality on the capital may reduce labour and maintenance expenditure etc. What type of technology the firm will use depends on the (expected) price levels for the different inputs. A firm that is not risk neutral will also look at the effect of different input compositions on total risk.

From the assumption of competitive markets it follows that the firm can buy as much as it would like of any input factor at given prices. In itself this implies that for the firm no factor has value in itself, and no effort will be given its condition or well being with exception of its influence on the production. Note that with respect to capital, the firm is interested in the flow of services from the capital equipment, not the capital in itself. Hence, it is immaterial whether the firm rents the capital or purchase it. This changes if the factors are not easily replaceable, that is, there are substantial adjustment costs involved in training or in other ways make a new factor fully productive. This gives the firm incentives to further protect some factors, but again one can measure the effect in terms of expected output loss of output, as the output are reduced in the period it takes for new factors to be trained/made operational.

These points hold in general, also when the basic model is made more complex and realistic in different dimensions. The firm's incentives to invest in risk reducing measures are directly linked to how these measures influence the expected output given their cost. However, before we close this section, we will briefly discuss two more issues that can change this picture somewhat.

Many safety issues are characterized by a low probability for them occurring together with substantial consequences if they occur. Does this change the picture above? No, not in theory, but it may in practice if it is difficult to assess their marginal benefit. To some extent this is likely to be mitigated by substitution as e.g. better capital and labour quality reduce the need for maintenance. However, if one cannot show the benefit of an investment, the investment is not likely to be undertaken. This can then lead to under investment in risk reducing measures, relative to conditions set by the model. However, many economists will still argue that this is rational, as it is not prudent to invest in a factor for which the benefit has not been shown. One can also argue that the one of the advantages with right organisational culture or structure is that one has the right level of many of the factors for which it is difficult to demonstrate the benefit. Moreover, often industry structure will in itself provide some insurance for the society. In particular, if there are several firms, an unlikely event can force one or some firms to bankruptcy, but it will not destroy the industry. Hence, it is often hard to justify additional costs to prevent unlikely events for any single firm. One may also note that relatively large firms will often have greater incentives then relatively smaller firms to prevent such accidents, as such unlikely events could have larger repercussions for these firms. Indeed, in very risky industries, one often observes an industry structure with many small firms which individually can be hit by negative events, but which still will not influence the industry.

If firms are not risk neutral this will influence their choice of production technology or input mix may be different from a risk neutral firm. Hence, risk neutral firms can choose riskier production technologies than risk averse firms. For investors this is not a problem as they can use the market to diversify away unwanted risk. This also implies that firms that do not act as if they were risk neutral will not be good investments, as they will give lower returns then similar firms that act as if they were risk neutral. However, if the society at large is risk averse, the activity of the firm may cause an additional cost to the society.

3 EXTERNALITIES

An externality is an economic significant effect due to the activities of an agent/firm that is not influencing the agent's/firm's production y, but which influence other agents decisions. Externalities can be positive and negative, and is dependent on the model and framework used. An example of a negative externality is pollution from a firm that reduces productivity or well-being for other firms or individuals. An example of a positive externality is the effect of a bee farmer's activity on surrounding fruit farmers. If the firm is in some way made to take the externality into account, we say that the externality is internalised and. This can be accomplished e.g. by government regulations or by a merger (Varian, 1992).

If there are no externalities, a competitive equilibrium will give an efficient allocation of the societies resources. This also means that the private cost and the social cost of any activity is equal. If there are externalities this changes this outcome as there will be a difference in the private and the social cost of an activity, as the firm will under or over invest from society's point of view, depending on the direction of the externality. This gives the society a reason to regulate firms behaviour.

To illustrate the effect of an externality, we expand our model from Section 2 by including a new factor, z, the externality, and we operate in an economy with two firms. Firm 1, that produces the externality has the following optimisation problem

$$\underset{x,z}{Max} \; \Pi^1 = pf^1(x,z) - w'x \qquad (3)$$

The problem of firm 2 has a similar structure

$$\underset{x}{Max} \; \Pi^2 = pf^2(x,z) - w'x \qquad (4)$$

What is important here is that for firm 1, z is an uninteresting output that it does not take into account in its problem definition and $\partial f^1/\partial z = 0$ However, for firm 2, $\partial f^1/\partial z < 0$ if z is a negative externality and $\partial f^1/\partial z > 0$ if z is a positive externality. The externality therefore influences the productivity of firm 2, but since it is due to firm 1's activity, firm 2 cannot influence it, without a say in the decision process of firm 1 that it does not have.

The society will take both firms' problems into account, so that the society's problem is

$$\underset{x,z}{Max} \; \Pi^1 = pf^1(x,z) + pf^2(x,z) - w'x \qquad (5)$$

In addition to the standard first order conditions, this also gives the condition

$$p\frac{\partial f^1(x)}{\partial z} = -p\frac{\partial f^2(x)}{\partial z} \qquad (6)$$

That is, the value of the marginal effect of the externality from firm 1 is equal to the negative of the value of the marginal effect for firm 2. This means that firm 1 should be compensated by firm 2 to produce more of the externality if it is positive, or it should compensate firm 2 if it is negative. Indeed, this is what will happen in a system with clear enforceable property right as firm 2 then has a legal right to be compensated by firm 1 which if necessary can be enforced by a court (Coase, 1960). However, it will

not happen if these rights are not clear, as will often be the case e.g. if a polluting firm influences a number of other agents only a little and if there are uncertainties related to the severity of the pollution. This will then give a scope for the society to do something, for instance by imposing a tax or quota on a firm producing a negative externality, or subsidising a firm producing a positive externality.

In a safety context we can interpret firm 1 in the model as the firm (or industry) of interest and firm 2 as the rest of the society. If the externality is the number of accidents, suitably defined, and accidents give a negative externality on the society, the firm's activity should be restricted. However, in general it should not be stopped (this should only happen if the value of the externality exceeds the positive effect of the firms activity). It should only be reduced to the point where the marginal benefit of the externality to the firm is equal to the marginal cost for the society. This also means that if one is to use economic arguments when looking at risk that may cause death, one implicitly also put a value on statistical lives if one does not close down the activity. By the value of a statistical life means the expected cost per expected saved life.

An example may here be in order. Let the firm of interest be a steel press that makes spoons. This is often a risky process in particular with respect to the employees fingers, as they will be lost if they are caught in the steel press. Hence, we regard z as the expected loss of fingers. Most employees are able to function just as well even if they lose a finger, and as such, it is not in the direct economic interest to be concerned whether the employees keep their fingers (with exception of the time the employee spends in hospital). However, for the society, the loss will often be a cost as these individuals will be less productive in other settings, and one will also regard the loss of limbs as bad in itself. This will then give the right hand side of (6) a positive value as it is the negative of the marginal cost of the lost limbs. It is then optimal for society to force the firm to cover that cost. This can take different forms, from forcing the firm to invest in risk reducing measures to directly compensate the employees. A problem with regulated measures, is that it is hard to specify the right level for any external agency. Some kind of fee often gives better incentives for the firm, as it will invest in risk reducing measures as long as these are less costly then the expected compensation. The example of the steel press is from an actual firm, where the z variable became a five dimensional vector depending on finger, and the employees got an automatic compensation whenever they lost a finger with the highest value for the thumb.

A somewhat uncomfortable implication of the results above is that one can explain the lack of safety measures in poor countries by stating that human life there has less value. This can further be explained along at least two lines. People in poor countries tend to receive less education and therefore possess less human capital. Alternatively, the wealth creation in rich countries are so big that one can here afford to correct an externality that has little value in a poor country because of the lower wealth creation.

4 WHEN REPUTATION MATTER

In the simple competitive model presented above, the firm can sell and by as much as it would like of any input or output as long as it is competitive on price. Hence, whether it is prone to accident is not a problem for the firm in any other dimension than the extent to which it reduces output, as it will not matter for its economic performance. This is likely to be the case when one is selling a homogenous product like crude oil. Hence, in such settings, the firm's reputation does not matter at all. However, when the identity of the firm matters in its output or factor markets, as is associated for instance with a brand name, this changes, as the firm then face (at least partly) firm specific demand or supply conditions. In economics terms this is the case for firms that have market power, as they can influence the price they receive in their product market or have to pay in their factor markets. Hence, for instance when selling retail gasoline, the firms record with respect to accidents, environment or whatever may matter, as experienced e.g. by Shell in the Brent Spar episode. Similarly, a firm can find it harder to recruit labour and other factors if it has a poor reputation in the relevant factor market, but again only if the providers of the service cares about the firms specific identity.

To model this, we need to specify the firm specific demand or supply schedules. Let us start with the case when the firm is a monopolist for the product it produce, and face a competitive supply of labour and other input factors. The firm then face a firm specific demand schedule as it supplies the whole market, but has no identity in the factor markets. Economists will denote a demand function as ordinary if quantity is the dependent variable, while it is denoted as inverse if price is taken to be the dependent variable. In most cases a one to one mapping is assumed to exist between price and quantity (Diewert, 1982). A competitive firm then perceives that the demand schedule for its product is flat since it receives the same price independently of how much it produces, even though aggregate market demand is downward sloping. A firm that has market power is able to influence the price and therefore face downward sloping demand. Similarly, a competitive firm face a flat supply schedule for its inputs, since it cannot influence the price it pays.

In our model reputation or market power is recognized by making the price the firm receives a function of the quantity it supplies, $p = p(y)$. This is basically to put total (inverse) market demand into the firm's optimisation problem. To account for reputation, we also need to include a variable in the demand equation that measures reputation. Let this variable be denoted a, where a higher value of a is an indication of a better reputation. The demand equation can then be denoted as $p = p(y, a)$ where we assume that this function is strictly concave in a to ensure the existence of a maximum, and where $\partial p(y,a)/\partial y < 0$ and $\partial p(y,a)/\partial a > 0$ The firm then has incentives to maintain a good reputation as this enables it to charge a higher price for its product.

Assuming that there is a cost involved in maintaining the reputation, the firm's problem is

$$\underset{x,a}{Max}\ \Pi = p(y,a)f(x) - w'x - w_a a \qquad (7)$$

The first order condition for each factor is

$$p\frac{\partial f(x)}{\partial x_i} + \frac{\partial p(y,a)}{\partial x_i} f(x) = w_i \qquad (8)$$

where the second term on the left hand side is the monopolists mark-up based on this factor and

$$\frac{\partial p(y,a)}{\partial a} f(x) = w_a \qquad (9)$$

Two things are immediately to be noted by from (9). First, reputation does not influence the production function at all and therefore not the productivity of the firm. It only influences the price the firm receives for its product. Second, there is an optimal expenditure on reputation as long as it is costly to build reputation, and this puts a limit on how much the firm is willing to spend on its reputation. Although reputation is not directly an input factor for the firm, it can still in many ways to be regarded as an input factor that is purchased until the relevant marginal condition is satisfied. It is of interest to note that this problem is similar to the firm's expenditure on advertising, and this has indeed the same structure and can be modelled simultaneously with reputation building by making a a vector. Several models investigating the effect of advertising can be found in Tirole (1988).

A firm that is the only source of demand for an input factor is known as a monopsonist. For such a firm, the price that the firm has to pay for a factor becomes a function of the quantity it demands. Letting b denote the impact of reputation the firm's problem is similar to that of a monopolist, but with factor prices of the form $w_i = w_i(x_i, b)$.

When there are more firms present in a market but where a firm has a degree of market power, the firm is known as an oligopolist or an oligopsonist depending on whether it has market power on the supply or demand side. This problem can be treated in two different ways, and for brevity we only look at the oligopoly case. The demand schedule facing the firm then becomes $p = p(y + y', a)$, where y is the quantity produced by the firm of interest and y' is the quantity produced by its competitors. The smaller the firm's share of total supply the less influence changes in y will have on price, and the less market power the firm has, and the competitive case is the limit as the firm then cannot influence the price at all. This also limits the importance of reputation, and again in the competitive case, it does not matter at all.

To conclude this section we can say that the firm's reputation matters only when the identity of the firm is of importance in some market, that is, the firm has some degree of market power in this segment. Moreover, as the degree of market power increases, the value of the firm's reputation will also in general increase. For firms that purchase and sells in different markets, there can be different reputations that need to be maintained (Nilsen, 2001). Hence, a firm can be perceived differently by e.g. consumers, the government, labour and other providers of factors. For instance, a firm can have a very poor environmental record giving it problems with respect to consumers, but have a good financial and accident record giving it a good reputation with its labour and suppliers.

5   REGULATIONS AND FIRM INCENTIVES

We saw in Section 3 that the society may want the firm to invest more in safety then the firm finds it optimal in its private optimisation problem. Governments have a wide menu of tools they can use to enforce its preferences. These can broadly be grouped into two main categories, quantity measures like minimum standards and quotas and value measures like taxes and fees. In addition, the government can provide services itself like a health system. However, a problem with all regulation is that the regulated party in general will regard it as a constraint, that it will do what it can to avoid the effect of. Hence, regulations often has unintended consequences (Laffont and Tirole, 1993).

The problem is that whenever there is a requirement, say that a minimum level of a factor is used, a firm will recognise this and account for it when making its decisions. However, the goal for the firm is still to maximize profits, and if the firm can obtain higher profits by (partly) avoiding the measure, it will do so. For instance, assume that a competitive firm has to spend a given quantity of a factor z, and that this quantity is larger than what the firm would find it worthwhile to use without a regulation. The firm's

problem then becomes

$$Max_x \ \Pi = pf(x;z) - w'x - w_z z \qquad (9a)$$

This is more restrictive then (1) since the input factors that are choice variables for the firm is the subset that excludes z. The first order conditions then become

$$p \frac{\partial f(x;z)}{\partial x_i} = w_i \qquad (10)$$

which in general differs from (2) since the level of z differs from what it would be in (2). It then follows that also the levels that the firm use of the other factors as well as output will then in general change with the regulation. Moreover, different input factors can in general partly substitute each other, as one for instance can compensate e.g. lower use of capital by higher labour use. The degree of substitutability will vary between any pair of inputs. The only case when the regulation will not lead to a change in the input mix is if no other factors can substitute z. A likely effect of a regulation imposing a safety measure is therefore that expenditure on other measures will be reduced. Moreover, it will to some extent change the firm's focus from finding the best safety measures to fulfil the required measures. If the restrictions are sufficiently strong, they can also lead to changes in behaviour and the production technology that in many cases will not be risk reducing. An interesting example is that one tends to drive more recklessly with safer cars (Risa, 1994)

Also public provision of safety will in most cases influence firm behaviour. When the government is providing services this can act as a positive externality for the firm. One can then see from Section 3 if the agent causing the externality is taken to be the government and the externality is positive that in general this will increase the firm's activity, and reduce the firm's use of factors that covers similar areas. Hence, publicly provided goods will partly or completely crowd out similar private services. For instance, a public emergency service is likely to reduce the firm's internal capacity to handle emergencies. Public measures that are intended to increase the general safety level will accordingly in most cases have a less than anticipated effect as firms adjust to their new working environment. With a sufficiently high level of public safety services, one can well end up in a situation where there will be no incentives for the firm to provide safety.

How firms respond to public regulations is a large field in itself, and are mostly treated in the subject called principal agent theory (Laffont and Tirole, 1993). Also regulations that focus on safety are likely to fall under this general topic, and certainly warrants more research with greater depth. What we hope to do

64

in this brief section is mainly to raise the issue as it is important in relation to how firms respond to regulations. In particular, it is not necessarily bad that public provision of services crowd out or lead to a reduction in the firms own actions, as the public may well be a more efficient provider as it can exploit economies of scale. However, one needs to be aware of these issues when making decisions to assess their full impact.

6   ECONOMICS AND SAFETY

So far we have primarily looked at safety from a strictly economic point of view. We think this is interesting as economic factors are important for firms' decisions. However, we would also like to raise some points that are important for many people concerned with safety that are not economists who would like to take other factors into account.

Economists prefer to consider the expected value of a cash flow when analysing projects and to think this gives the most efficient use of the firm's or the society's resources. One is then reluctant to use other criteria as this is perceived to give a welfare loss. However, people from other disciplines often think that this gives little weight for instance, avoidance of low probability but high consequence events. This may well be the case as the economists' models certainly are limited in scope and although one pays lip-service to the notion that a model is a simplification of reality that leaves a number of issues out, one tends to operate on the basis that only what is in the model is relevant.

We also saw in the section on externalities that society can have a different value on subjects than the firm, but even economic models for the society tend to focus on the use value of factors, including labour, and do not give them a value in itself (existence value). Researchers from other disciplines will still often assign a higher existence value to a number of subjects than economists (and pay less attention to where one takes resources from to achieve these goals). If these are to be taken into account, the decision process must treat the economic analysis as only one of several inputs. This opens up for the sort of managerial review as in Hertz and Thomas (1983), Aven (2003) and Aven and Kørte (2003), to also take other information into account.

However, one needs to be careful with terminology when discussing these issues. In a strict economic interpretation, all deviations from the economic model reduce value added and therefore welfare in the society. If one recognize that economic models do not capture all relevant aspects, as most people would agree on, this opens the possibility that one can increase value added and social welfare by deviating from economic models in the "right" way. More likely though is it that deviations will reduce value added, as seen from a business point of view, but it may increase social welfare given that a number of things can have a societal value that economic models do not pick up.

7   CONCLUDING REMARKS

In this paper we have provided an overview of safety's role in firm behaviour from an economic perspective. The analysis provides two main insights. First, safety can be regarded as an input factor on an equal basis as any other in the firm's optimisation problem. Hence, the firm will invest in safety as long as it contributes to the profits. In an uncertain world, most firms will have incentives to invest in safety as this will increase profits, but there is also a maximal level over which the firm will not provide additional coverage. Although one at times can get a different impression when reading the safety literature, this should be as expected as firms normally spend substantial funds either at directly reduce risk e.g. through maintenance or to sell their risk exposure to agents that can bear the burden with less cost through e.g. insurance and derivative markets. Second, the firm is often likely to under invest in safety from society's perspective of a number of reasons. One important reason is that firms utilise factors because of their use value, and factors do not have any intrinsic value. Hence, a building is appreciated for the services it provides but not for, say, its historical value. Another reason is that it is difficult to justify the cost for factors for which it is difficult to explicitly show their value, or where this is very uncertain. To some extent this is mitigated as firms with the "right" organisational culture do best, as right organisation culture often will provide reasonable levels of efforts that are difficult to explicitly show the value for.

In a modern economy many firms are concerned about their reputation as their brand gives them an advantage in the market place. This amounts to say that the firms have a degree of market power (Tirole, 1988). From a safety perspective this is positive as this make the firm susceptible to the society's values as the firm do better by investing in practices that win public approval or at least avoid public disapproval. However, the firm is still likely to under invest in safety from society's perspective.

If one accepts that the notion that firms' behaviour is primarily guided by its search for profits this gives a number of challenges for safety research. The most straightforward issue to address is the extent to which firms under invest in safety because the value is not clear. One would preferably like to be able to demonstrate the effects, and accordingly better techniques of measurement are called for. Short of being able to demonstrate the benefits in every specific case, solid empirical evidence could influence rules of thumb that often are present in business decisions. For instance, firms routinely adjusts discount rates to account for risk of doing business in politically instable countries, and similar empirical evidence could boost investment in safety. This is also true if one can demonstrate that successful organisational cultures are related to the cultures view on safety. As this involves factors that are hard to measure in monetary terms, it lends support to the notion that managerial reviews are essential, as in Hertz and Thomas (1983) and Aven (2003), to see beyond the calculations. Future research is, however, required to provide more detailed understanding of the need for and extent of such managerial review. From a traditional safety perspective, safety measures and improvements are to large extent justified by the existence of uncertainty related to future performance, which is not covered by the expected economic analysis supporting the business decisions.

From the society's point of view the issues are more difficult as the firm will invest in safety based on its private value of safety. This seems to call for some set of regulations. Whether these are of a European flavour with e.g. minimum standards or a more American flavour where the firm is liable if something happens is to a large extent a matter of taste. However, regulations must be carefully designed as firms will in general try to avoid their intention if that is profitable. Moreover, as all regulations also have a cost in terms of lower value added in the society, one should also be careful in not over regulate firms and thereby stifle innovation. A further discussion of the nature of regulation and the dilemmas introduced in actual applications can be found in Hale (2000).

ACKNOWLEDGEMENTS

The authors are indebted to the Norwegian Research Council for financial support, and Jan Erik Vinnem, Per Hokstad, Morten Sørum and Anders Toft for valuable comments and suggestions.

REFERENCES

Asche, F. and R. Tveterås (1999) Modeling Production Risk with a Two-Step Procedure. *Journal of Agricultural and Resource Economics*, 24, 424–439.

Aven, T. (2003) Foundations of Risk Analysis. Wiley, N.Y. To appear.

Aven, T. and Kørte, J. (2003) On the use of cost/benefit analyses and expected utility theory to support decision-making. Forthcoming in *Reliability Engineering and System Safety*.

Calow, P. (ed.) (1998) Handbook of Environmental risk assessment and management. Blackwell Sciences, Oxford.

Coase, R (1946) The problem of social cost, *Journal of Law and Economics*, 3, 1–44.

Diewert, W.E. (1982) Duality Approaches to Micro-economic Theory. Handbook of Matematical Economics. K.J. Arrow and M.D. Intriligator. Amsterdam, North-Holland.

Fischhoff, B. et al. (1981) Acceptable Risk, Cambridge University Press, Cambridge.

Gibson, S.B. (1978) Major hazards – should they be prevented at all costs? *Engineering and Process Economics*, 3, 25–34.

Hale, A. (2000) Issues in the regulations of safety: setting the Scene. In Kirwan, Hale & Hopkins (eds) (2002) Changing Regulations. Controlling risks in society. Pergamon.

Hertz, D.B. and H. Thomas (1983) Risk Analysis and its Applications, Wiley, New York.

Jones-Lee, M.W. (1989) *The Economics of Safety and Physical Risk*, First Blackwell, Oxford

Just, R.E., and Pope, R.D. (1978) Stochastic Specification of Production Functions and Economic Implications. *Journal of Econometrics*, 7, 67–86.

Kotz, H. and Schafer, H-B (1993) Economic incentives to accident prevention. *International Review of Law and Economics*, 13, 19–33.

Laffont, J.-J. and J. Tirole (1993) *A Theory of Incentives in Procurement and Regulation*, Cambridge, MIT Press.

Nilsen, E.F. (2001) Economic Accidental Risk Analysis, Ph.D. thesis, NTNU, Trondheim.

Pape, R.P. (1997) Developments in the tolerability of risk and the application of ALARP. *Nuclear Energy*, 36, 457–463.

Risa, A.E. (1994) Adverse incentives from improved technology: Traffic safety regulation in Norway, *Southern Economic Journal*, 60, 844–857.

Tirole, J. (1988). *The Theory of Industrial Organization*, Cambridge, MA, MIT Press.

UKOOA (1991) A framework for risk related decision support – Industry guidelines. UK Offshore Association.

Varian, H.R. (1992). *Microeconomic Analysis*, New York, Norton.

Viscusi W.K. (1989) Safety through markets. *Society/ Transaction, social Sceince and modern*, 27, 9–10.

Viscusi W.K. (1986) Market incentives for safety. *Harward Business Review*, 63, 133–138.

Safety and Reliability – Bedford & van Gelder (eds)
© 2003 Swets & Zeitlinger, Lisse, ISBN 90 5809 551 7

Electronic systems manufacturing: some aspects of plastic encapsulated microcircuits (PEM) and the evaluation of their reliability

T.I. Băjenescu
Company for Consulting C.F.C., La Conversion, Switzerland

M.I. Bâzu
National Research Institute for Microtechnology I.M.T., Bucharest, Romania

ABSTRACT: The paper gives an overview of the present situation in the domain. After a short introduction, the main problems arisen in reliability predicition and evaluation of monolithic integrated circuits (ICs), the evaluation itself and some new points of view concerning the dynamic life testing, screening and burn-in, humidity environment, accelerated thermal tests, physics of failure of plastic encapsulated microcircuits (PEM) and reliability building are presented. Characterization techniques for $Hg_{1-x}Cd_xTe$ epilayers are mentioned.

1 INTRODUCTION

The intrinsic reliability of a transistor from an IC improved with two orders of magnitude (the failure rate decreases from $10^{-6}h^{-1}$ in 1970, to $10^{-8}h^{-1}$ in 1997). But also, in the same period of time, the number of transistors per device increased with 9 orders of magnitude! Therefore, the IC reliability increased even faster than the prediction given by Moore's Law. The model for reliability growth was called "Less's Law", taking into account the known philosophy from the architectural design: "Less is More" [1]. Actually, Less's Law means a tremendous increase of the requirements for the IC's failure rate: from 1000 failures in 10^9 devices × hours (or 1000 Fits), now only some Fits in a single digit are required.

Basically, two types of integrated circuits were developed since now: bipolar and MOS, taking into account the basic cell: bipolar transistors or MOS ones, respectively. In the beginning, the MOS ICs were n-channel MOS ICs or p-channel MOS ICs, but sooner complementary MOS ICs (or CMOS ICs) including both types were developed. The main characteristic of CMOS circuits is the small supply voltage. As the portable–electronics market increases, low-power and low-voltage technologies, such as CMOS, became the most used. Also, the technological improvements leading to the remove of sodium contamination in the Si–SiO$_2$ system encourage the use of CMOS ICs, because a high reliability level becomes possible to

get. Recently reduced standard digital CMOS power supply voltage of 1.8 V was obtained, reducing the power consumption by 70%. These ULP (Ultra Low Power) CMOS ICs were deeply investigated [2] and proved to have a large potential.

The last challenge in the IC family is the *microsystem*. Arisen in the early 90's, the micro-system represents a superior integration step compared with common ICs: the "intelligent" element (the signal processing part) is integrated with micro sensors and with micro actuators, in a single component, basically still an IC [3]. In fact, the microsystem is a "smart" sensor, able also to actuate. This device determines the development of some new microtechnologies. Many disciplines being involved, hybrid terms, such as: mechatronics, chemtronics, bioinformatics were used [4], but the term *microtechnology* (technology of microfabrication) seems to be the most adequate. In Europe, the term microtechnology includes both microelectronics (the "classical" devices) and micro system technology (MST) [5]. Other related terms are MEMS (Micro-Electro-Mechanical Systems), system technology (MST) [6]. BIO-MEMS (BIOlogical MEMS) and MEOMS (Micro-Electro-Opto-Mechanical-Systems). Silicon is still the basic material and the CMOS technology can be used for the manufacturing.

Recently, a new term, *nanotechnology*, was proposed, because the structures have now characteristic features of a few nanometers. Accordingly, the tools used for manufacturing these new technologies (micro- and nano-) are called micro-machines and nanomachines, respectively.

1.1 Modeling IC reliability

First, only simulators for one or two subsytems or failure mechanisms were arise, such as: RELIANT [7], only for predicting electromigration of the interconnects and BERT EM [8]. Both use SPICE for the prediction of electromigration by derivating the current. Other electromigration simulators were CREST [8], using switch-level combined with Monte-Carlo simulation, adequated for the simulation of VLSI circuits and SPIDER [9].

Other models were built for hot-carrier effects: CAS [10] and RELY [11], based also on SPICE. An important improvement was RELIC, built for three failure mechanisms: electromigration, hot-carrier effects and time-dependent dielectric breakdown [12].

A high-level reliability simulator for electromigration failures, named GRACE [13], assured a higher speed simulation for very large ICs. Compared with the previously developed simulators, GRACE has some advantages [13]:

- an orders-of-magnitude speedup allows the simulation of VLSI many input vectors;
- the generalised Eyring model [14] allows to simulate the ageing and eventually the failure of physical elements due to electrical stress;
- the simulator learns how to simulate more accurately as the design progresses.

If the typical failure mechanisms are known, by taking into account the degradation and failure phenomena, models for the operational life of the devices can be elaborated. Such models, in contrast with the regular CAD tools determining only wearout phenomena, predicts also the failures linked to the early-failure zone.

2 RELIABILITY EVALUATION

2.1 Some reliability problems

From a theoretical viewpoint, a higher reliability is expected for integrated circuits than for discrete components. In practice, these expectations were surpassed, because some basic conditions were fulfilled:

- high level and automated industrial fabrication,
- constant quality materials, screening tests for finished products.

Before stating which type of integrated circuit is adequate for a specific use, one must carefully analyse the design, the most important parameters, the dimensions, the costs and the limits imposed by the reliability.

Such studies must take into account:

- a comparison of the total costs (development, manufacturing and testing costs),
- a comparison of some important parameters (resistor tolerances, temperature coefficient, speed, voltage level) with limitations specific to each type,
- an evaluation of the dissipated power of the circuit and of the thermal resistance of the encapsulated circuit, because an acceptable temperature of the substrate must be assured.

2.2 Evaluation of integrated circuits reliability

Generally, three main problems arise at the evaluation of integrated circuits reliability.

1. For modern devices, the failure rates decrease under a certain limit and the conventional methods become less usable. To overcome these difficulties, two solutions may be discussed:
   a) To perform on a very high number of integrated circuits reliability tests in normal operational conditions with the duration of a couple of years. Obviously, this solution is unacceptable. As an example, if one has to test a failure rate of $10^{-9}h^{-1}$ (called also 1 FIT), 1000 devices must be tested for 114 years and only one device to be found defect.
   b) To perform on some integrated circuits reliability tests in higher than normal conditions, the so-called accelerated tests This method may be applied only if at the accelerated tests the failure mechanisms are the same as for normal operational conditions. And this fact must be indubitably demonstrated.

The accelerated tests are used in the purpose to obtain quickly and with a minimum of expense information about the reliability of the product. The used stresses are higher than for normal operational conditions, the results are extrapolated and the failure rate for normal conditions is obtained. Usually, the accelerated tests contain combinations of stresses such as: temperature bias, pressure, vibrations, etc. [15]. If the temperature is the only variable of the accelerated tests, the Arrhenius model may be used. To obtain reliable results, relatively short testing times must be used.[1] So, using various levels of the same stress factor one may follow the real behaviour.[2] The analysis of the physico–chemical process leading to failure allows

---

[1] Even if the purpose is to minimise the testing time, a too stronger stress level must not be used, because new failure mechanisms may be induced.
[2] If the time is the accelerated variable, this means that an hour of tests at high stress level produces the same effect on the component reliability as n hours at normal operation time.

68

obtaining the correlation between the speed of these phenomena and the stress and, as a result, the real dependence of time to failure on stress levels.

2. The rapid development of the manufacturing technology for integrated circuits needed by the aim to improve the control and to reduce the costs, makes difficult the reliability evaluation. Usually, any modification in the technology or used materials is followed by the appearance of a new failure mechanism. Consequently, any manufacturing modification must include a new reliability evaluation.
3. The last problem is linked to the increasing complexity and costs of integrated circuits vs. discrete devices. Although the cost of a certain electronic function decreases substantially by integration, the basic costs are always higher for an integrated circuit than for a discrete component fulfilling the same function.

The definition of the failure criterias, unavoidably, very difficult because the complexity of ICs is increasingly higher. Even for a simple device, like a transistor, it is hard to define the failure limits. For an integrated circuit, the basic parameters are more complex and hard to be measured and the degradation of these parameters differs from an utilisation to another.

For evaluating the various stresses able to be used in reliability accelerated tests, the following aspects must be taken into account [16][17]:

• The stress must be encountered in the operational environment. In principle, one must note that the failure rate of integrated circuits is influenced by the thermal, electrical and mechanical conditions of the operational environment. But for common industrial use, mechanical shock and vibrations have a little influence on the integrated circuits encapsulated in epoxy packages, able to assure the necessary mechanical stability and a good protection. For instance, the acceleration measured at a sudden stop of a running car reaches 40 g, for airplanes take-off and landing – up to 5 g and for missiles – up to 50 g. Compare these values with the acceleration level used for periodic tests: 30,000 g. Consequently, mechanical factors will be used only rarely for accelerated tests. On the contrary, the temperature is the most used stress for this kind of tests. The experimentally observed correlation between failure rate and temperature is based on the fact that the speed of chemical reactions arising in the device is thermally increased.
• The failure mechanisms must be always those arising in the operational environment.
• All samples of integrated circuits used in accelerated tests must behave in the same way at a stress modification: the same circuits should be the first to fail at any stress level.

3 DYNAMIC LIFE TESTING

The failure of an IC in operational life is an unpleasant event not only because the owner of the equipment must replace it, but also because this failure may induce serious damages to the equipment, loss of important information or even of human life. Therefore, it is desirable to replace a IC before failure. From economical reasons, this replacement must take place shortly before the anticipated failure. This implies that the lifetime of the IC be accurately estimated. This operation may be done only if laboratory tests simulating as closed as possible the real operational life are performed. In this respect, in the laboratory, not only static, but also dynamic testing must be done. The purpose is to quantify the performance degradation during IC operation. An example of such testing is given by Son and Soma [18]. First, the IC parameters which will be monitored during dynamic life testing are chosen by two criteria: (i) to be measurable at existing pin-outs, and (ii) to predict progressively IC degradation. Then, the typical failure and degradation mechanisms must be studied. In fact, there are two major types of degradation mechanisms: electrical ones (such as: latchup, hot-carrier effect, dielectric break-down electromigration, etc.) and environmental ones (produced by thermal and mechanical stress, humidity, etc.). By means of appropriately chosen electrical parameters (such as: static/transient current level change, noise level in current, cut-off frequency, input offset voltage of CMOS differential amplifier, etc.), these mechanisms are monitored during dynamic life testing.

Eventually, aging models for various failure mechanisms must be elaborated. In [18] a model for hot-carrier effect is given. Starting from a widely accepted empirical relationship between parameter deviation and the elapsed stress time for the hot-carrier degradation mechanism, given in [19], an aging curve due to hot-carrier effect under the static or periodically repeated AC stress was obtained [9], defined by the equations:

$$\Delta V/V_o = k \cdot t^a \qquad (1)$$

$$k = C \cdot [(I_{ds}/w) \cdot exp(-\Phi\sqrt{(q \cdot p \cdot E_{ch})})]^a \qquad (2)$$

where C is a constant, a and k are coefficients of the aging curve (depending on the technology and on IC structure), I_{ds} is the drain current, w is the channel width of a MOS transistor, Φ_i – the minimum energy required to cause impact ionization, q – the electron charge (1.6×10^{-19} C), p – the hot electron mean free path, E_{ch} – the channel electric field, $\Delta V/V_o$ – circuit aging and t – elapsed stress-time. In a log($\Delta V/V_o$) vs. log t plot, a straight line with slope a and y-intersection log k is obtained (see Fig. 1).

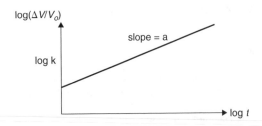

Figure. 1   A log($\Delta V/V_o$) vs. log t plot for hot-carrier degradation mechanism.

From the case study given in [18], one may understand the procedure for IC replacing before to occur a failure by hot-carrier effect. For a 31-stage inverter chain, designed according to MOSIS 0.8 μm HP technology rules, the device operation was simulated, the ageing being modeled by randomly changing device parameters. Based on this model, the probability of survival until the next inspection may be quantified at each inspection of dynamic-life testing. Then, the optimal moment for replacement may be calculated with respect to maintenance cost (the recovery cost of an unanticipated failure and the wasted cost of replacing an IC too early).

4   RELIABILITY BUILDING

The reliability is built at the design phase and during the manufacturing. This means that reliability concerns must be taken into account both at the design of the process/product (the so-called *design for reliability*) and also at the manufacturing (*process reliability*). A special attention must be given to the last step of the manufacturing process, *the screening (or burn-in)*.

The component reliability is influenced by the materials, the concept and the manufacture process, but strongly depends on the taking over input control conditions, so not only the component manufacturer, but the equipment manufacturer too must contribute greatly to the reliability growth of the equipment. If the failure rate is constant during the operation period, this is a consequence of a good component selection during the manufacturing process. But there are, also, components that frequently fail, without a previous observation of a wearout effect.

The early failures – usually produced as a consequence of an inadequate manufacturing process – must be avoided from the beginning, in the interest as much of the manufacturer, as of the user. Unfortunately, this wish is not always feasible; before all because physical and chemical phenomena with unknown action can produce hidden errors which appear as early failures.

Generally, the selection is a 100% test (or a combination of 100% tests), the stress factors being the temperature, the voltage, etc. followed by a parametric electrical or functional control (performed 100%), with the aim to eliminate the defect items, the marginal items or the items that will probably have early failures (potentially unreliable items). To overcome these problems, recently, a method was proposed [6]. The method was called MOVES, an acronym for Monitoring and Verifying a Screening sequence. MOVES contains five procedures, namely VERDECT, LODRIFT, DISCRIM, POSE and INDRIFT.

5   ACCELERATED THERMAL TESTS

By definition, an accelerate test is a trial during which the stress levels applied to the components are superior to these foreseen for operational level; this stress is applied with the aim to shorten the necessary time for the observation of the component behavior at stress.

The use of accelerated tests starts from the presumption that the possible failure mechanisms are well known. Received intially with hesitation and scepticism, the technique became useful for component producers and users. The experience in the utilisation of electronic components shows that the life duration is not an infinite one and the operational conditions are important. For initially «good» integrated circuits, failures before the end of the normal life duration were found, such as:

- catastrophic failures, breaking the normal operation,
- drift failures, producing defective operation by an important time variation of the electrical characteristics.

One must understand that the appearance of a failure is not a proof that the life duration is smaller. The drift failure is hard to define, depending of the drift threshold stated as a failure criterium. In practice, the accelerated thermal tests are not sufficient for estimating the reliability of a product. Step stress tests must also be performed [14]. In this case, the initial hypotheses are that the stress has no memory and that the wearout does not arise. It is important to note that without knowing the value of the activation energy, no correct analysis of the data obtained from laboratory tests is possible. It is worth to be mentioned that the international standardisation body do not take always into account the importance of the activation energy for data processing.

6   PHYSICS OF FAILURE

Until now, no main wearout mechanism for a mature technology of manufacturing semiconductors components is known. The discussed defects are reportable to the structure anomalies of the component. The knowledge of these structural weaknesses of some constructive or technological insufficiencies and their causes is a useful premise for the uninterrupted ageing of the semiconductor component reliability.

There are several problems that must be solved in the next years. First, to identify the acceleration laws with different stress factors. Then, the idea to take into account, at the design of the accelerated tests, the synergies between the stress factors encountered in the operational environment [21].

7 HUMIDITY ENVIRONMENT

Compared with high temperature testing, the experience acquired in high humidity testing is very small. However, the law for failure of Ics due to high humidity seems to be a log-normal one [21] and the average life time to depend directly proportionally to the vapour pressure of the humid environment. In the early days, to perform tests in such an environment, a temperature of 25°C and a relative humidity of 75% was recommended. A test performed in these conditions and with a duration of 20 days, with bias, simulated an ageing of 20 years [5]. Lately, a more efficient test was used, the so-called "85/85 test" (85°C and 85% relative humidity).

8 RELIABILITY PREDICTION

In the concurrent engineering era, Reliability Prediction Procedures (RPP) are valuable tools for designers and users of any product. The designer needs RPP to avoid the lag in feedback occurring when the predictions are made by a reliability team. The user, which in turn is a designer of a complex system, wants to have correct information about the part reliability. All the usual RPP (the most known being MIL-HDBK-217) have some common features which diminish the prediction accuracy:

(i) a constant failure rate model is used;
(ii) the failure mechanisms (FM) are not analysed. Lately, improved RPP – with failure rate models other than the exponential and starting from the physics of failure – were considered desirable and as "a change in direction for reliability engineering" [8]. In 1995, an improved methodology, called SYRP, for predicting the failure rate of a lot was proposed, with the following characteristics: (*a*) a log-normal distribution for each FM used involved, and (*b*) the interaction (synergy) between the technological factors, depending on the manufacturing and control techniques, is considered.

9 CHARACTERIZATION TECHNIQUES FOR MERCURY CADMIUM TELLURIDE EPILAYERS

The approach of p-on-n junctions in a heterostructure configuration with extrinsic p-type and n-type doping for photovoltaic devices [22] has taken a critical role in the long-wavelength IR (LWIR) technology based on a $Hg_{1-x}Cd_xTe$. The effect of the acceptor species diffusion commonly observed is the difficulty in achieving precise control of the electrical junction which in turn translates into nonuniform and nonreproducible device performance. Lacking suitable characterization techniques, the heterojunction cannot be evaluated, and the effects of As implant/diffusion on the composition of the material at various depths cannot be studied. In addition, the electrical activation of the p-type species in a p-on-n structure cannot be assessed because of the high carrier mobility ratio (~200) in HgCdTe.

The dissertation [22] reports on the study of As diffusion mechanism in HgCdTe grown by metalorganic chemical vapor deposition (MOCVD-IMP) technique from an ion-implanted source and on application of the resultant findings to the As diffusion from a grown source. A diffusion model was developed from a chemical analysis (secondary ion mass spectroscopy – SIMS) and a Gaussian theoretical modeling. The nature of p-type species diffusion mechanism and associated components was investigated. The most desired mechanism found in the controlled atomic vacancy-based component in which As starts on the Te-sublattice. The dislocation density and distribution as the main cause for the uncontrollable enhancement of As diffusion in the tailing components was demonstrated. It was found that the proper choice of ion implantation and anneal conditions helped reduce the tails of the implantation and anneal conditions helped reduce the tails of the diffusion profiles for a given dislocation density in the material. High performance of p-on-n photodiodes, at 77 K and at 40 K, limited by thermal processes, was demonstrated. Also reported is a very sensitive novel technique which was developed for analyzing the relative variations of material composition, and for quantitatively measuring the absolute value of composition at various depth. This technique is based on matrix effects on Te-electronegative ion yield which occurs in SIMS when a reactive primary ion beam of Cs is used. Another novel technique was developed to measure the p-type carrier concentration in a p-on-n structure, which was differential Hall effect analysis of the p-type region after the entire structure is converted into p-type by a Hg-vacancy anneal at low temperatures.

A reliability study is in preparation.

10 CONCLUSIONS

This overview showed the dynamic in the field of the evaluation and prediction of monolithic ICs reliability. Starting from some reliability problems, the main aspects arisen in dynamic life testing, reliability building, accelerated thermal tests, humidity environment, physics of failure of plastic encapsulated microcircuits (PEM) and reliability evaluation were presented.

REFERENCES

[1] Spicer England, J., England, R. W.: *The reliability challenge: new materials in the new millenium Moore's Law drives a discontinuity.* International Reliability Physics Symposium, Reno, Nevada, March 31–April 2, 1998, pp. 1–8
Băjenescu, T. I., Bâzu, M. I.: *Reliability of Electronic Components. A Practical Guide to Electronic Systems Manufacturing.* Springer-Verlag, Berlin, Heidelberg, New York, 1999

[2] Schrom, G., Selberherr, S.: *Ultra-low-power CMOS technologies.* International Semiconductor Conference, Oct. 9–12, 1996, Sinaia/Romania, pp. 237–246

[3] Coppola, A.: *The Status of Reliability Engineering Technology.* Rel. Soc. News, April, 1997, p. 7

[4] Fluitman, J. H. (1994): *Micro systems technology: the new challenge.* International Semiconductor Conference, Oct. 11–16, Sinaia / Romania, pp. 37–46

[5] Preston, P. F.: *An industrial atmosphere corrosion test.* Trans. Ind. Metal finish (Printed Circuit Suppl.), vol. 50 (1972), pp. 125–129
Dascălu, D.: *From micro- to nano-technologies.* Proceedings of the International Semiconductor Conference, October 6–10, 1998, Sinaia/Romania, pp. 3–12

[6] Bâzu, M, Dragan, M.: *MOVES – a method for monitoring and verifying the reliability screening.* Proceedings of the International Semiconductor Conference CAS'97, Oct. 7–11, 1997, Sinaia/Romania, pp. 345–348

[7] Liew, B. J., Fang, B., Cheng, N. W., Hu., C.: *Reliability simulator for interconnect an intermetallic contact electro-migration.* Proceedings of the International Reliability Physics Symposium, March 1990, pp. 111–118

[8] Najm, F., Burch, R., Yang, P., Hajj, I.: *Probabilistic simulation for reliability analysis of CMOS VLSI circuits.* IEEE Trans. Computer-Aided Design, vol. 9, April 1990, pp. 439–450

[9] Hall, J. E., Hocevar, D. E., Yang, P., McGraw, M. J.: *SPIDER – a CAD system for modeling VLSI metallisation patterns.* IEEE Transactions on Computer-Aided Design, vol. 6, November 1987, pp. 1023–1030

[10] Lee; Kuo; Sek; Ko; Hu : *Circuit aging simulator (CAS).* IEDM Technical Digest, December 1988, pp. 76–78

[11] Shew, B. J., Hsu, W. J., Lee, B. W.: *An integrated circuit reliability simulator.* IEEE J. Solid State Circuits, vol. 24, April 1989, pp. 517–520

[12] Hohol, T. S., Glasser, L. A.: *RELIC – a reliability simulator for IC.* Proceedings of the International Conference Computer-Aided Design, November 1986, pp. 517–520

[13] Kubiak, K., Kent Fuchs, W.: *Rapid integrated-circuit reliablity-simulation and its application to testing.* IEEE Transactions on Reliability, vol. 41, n° 3, Sept. 1992, pp. 458–465

[14] Băjenescu, T. I.: *Look for cost/reliability optimisation of Ics by incoming inspection.* Proceedings of EURO-CON'82, pp. 893–895
Băjenescu, T. I.: *Pourquoi les tests de déverminage des composants.* Électronique, no. 4 (1983), pp. 8–11

[15] Adams, J., Workman, W.: *Semiconductor network reliability assessment.* Procceddings of IEEE, vol. 52, no. 12 (1964), pp. 1624–1635
Bâzu, M.; Tăzlăuanu, M.: *Reliability testing of semicon-ductor devices in humid environment.* Proceedings of the Annual Reliability and Maintainability Symposium, January 29–31, 1991, Orlando, Florida (USA), pp. 237–240
Băjenesco, T. I.: *Quelques aspects de la fiabilité des microcircuits avec enrobage plastique.* Bulletin SEV, vol. 66, no. 16 (1975), pp. 880–884

[16] Peck, D. S., Zierdt Jr., C. H.: *The reliability of semiconductor devices in the Bell System.* Proceedings of the IEEE, vol. 62, no. 2 (1974), pp. 185–211

[17] Colbourne, E. D.: *Reliability of MOS LSI circuits.* Proceedings of the IEEE, vol. 62, No. 2, 1974, pp. 244–258

[18] Peck D. S.: *The analysis of data from accelerated stress tests.* Proceedings of the International Reliability Physics Symposium, March 1971, pp. 69–78
Son, K. I., Soma, M.: *Dynamic life-estimation of CMOS ICs in real operating environment: precise electrical method and MLE.* IEEE Transactions on Reliability, vol. 46, no. 1, March 1977, pp. 31–37

[19] Băjenescu, T. I., Bâzu, M. I.: *Reliability of Electronic Components.* Springer-Verlag, Berlin, Heidelberg, New York, 1999
Băjenescu, T. I.: *Pourquoi les tests de déverminage des composants.* Électronique, no. 4 (1983), pp. 8–11
Hu, C., Tam, S. C., Hsu, F. C.: *Hot-carrier induced MOSFET degradation: model, monitor and improvement.* IEEE Trans. on Electron Devices, vol. 32, Feb. 1985, pp. 375–385

[20] Birolini, A. : *Reliability of technical systems.* Springer, Berlin and New York, 1997

[21] Bâzu, M., Tăzlăuanu, M.: *Reliability testing of semiconductor devices in a humid environment.* Proceedings of the Annual Reliability and Maintainability Symposium, January 29–31, 1991, Orlando, Florida (USA), pp. 237–240
Băjenescu, T. I. : *A Particular View of Some Reliability Merits, Strengths and Limitations of Plastic Encapsulated Microcircuits (PEMs) vs Hermetically Sealed Microcircuits (HSMs) Utilised in High-Reliability Systems.* Proceedings of the 8th European Symp. on Reliab. of Electron Devices, Failure Physics and Analysis (ESREF 97), Bordeaux (France), October 7–10, 1997
Băjenescu, T. I., Bâzu, M. I.: *Semiconductor devices reliability: An overview.* Proceedings of ESREL '99 – The Tenth European Conference on Safety and Reliability, Munich-Garching, Germany, 13–17 September 1999, pp. 283–288

[22] Bubulac, L. O.: *Diffusion and activation of p-type species for p-on-n junction formation and novel characterization techniques for mercury cadmium telluride epilayers.* Ph. D. Thesis, UCLA, Los Angeles, 1991

Safety and Reliability – Bedford & van Gelder (eds)
© 2003 Swets & Zeitlinger, Lisse, ISBN 90 5809 551 7

Application of Bayesian techniques for corrosion state assessment of reinforced concrete bridge girders

K. Balaji Rao, M.B. Anoop, S. Gopalakrishnan & T.V.S.R. Appa Rao
Structural Engineering Research Centre, Chennai, India

B. Satish
Formerly B.E. student, Birla Institute of Technology and Science, Pilani, India

ABSTRACT: A methodology for the corrosion state assessment of reinforced concrete bridge girders using Bayesian techniques is proposed in this paper. The methodology will be useful for realistic service life assessment of these members based on data from field inspection. An attempt has been made to show how the engineering judgment can be used in formulating the likelihood function used in Bayesian decision making. Likelihood functions are formulated for two different cases, which will arise in practice. The usefulness of the proposed methodology is demonstrated by applying it to the chloride concentration data obtained from field investigations on Gimsøystraumen bridge girder (Fluge 2001). The predictive probability obtained using the proposed methodology is more realistic than the point estimate of probability (computed using relative frequency approach), thus helping in making better decisions.

1   INTRODUCTION

The emerging design philosophy for reinforced concrete (RC) structures is to design the structures for a specified service life. Durability is one of the major parameters affecting the service life of RC structures. For RC structures located in marine and other aggressive environments, chloride-induced corrosion of reinforcement is identified as a major mechanism of resistance degradation. In general, the service life of RC structural members, with respect to chloride-induced corrosion of reinforcement, can be divided into two phases, namely, corrosion initiation and corrosion propagation. Since the corrosion propagation phase is small compared to the corrosion initiation phase, initiation time is considered as the service life in the design. The workmanship, which determines the quality of concrete in the cover region, is one of the important factors governing the time to corrosion initiation. Due to the variations in the workmanship, there will be uncertainty associated with the estimated corrosion initiation time. The variations in the workmanship should be considered at least as random while quantifying the risks arising from workmanship and for determining the service life of the structure. The estimation of service life at the design stage will be useful for scheduling the inspection for maintenance. However, the estimations made at the

design stage on service life should be updated based on the performance of the structure in actual service conditions. Bayesian techniques provide a scientific and rational framework for updating the estimations based on data from field inspection.

In this paper, a methodology has been proposed for the condition assessment of RC bridge girders against corrosion initiation using Bayesian techniques. Bayesian techniques help in incorporating additional information from field investigations during maintenance period, in updating the estimated service life. In the estimation of service life, the uncertainties associated with the quality of concrete, cover to reinforcement and exposure conditions are treated as random.

Development of reliability based service life models require that the models can incorporate the information generated during in-service inspection; that is, the models/model parameters can be updated based on in-service inspection data. The importance of development of such models has been clearly brought out by Mori & Ellingwood (1994a, b). Use of Bayesian methods for incorporation of information obtained during in-service inspection in condition assessment and thus in realistic service life estimation of existing structures is well established (viz., Mori & Ellingwood 1994a, b). Condition assessment and service life estimation would help in making decisions regarding repair/retrofit to be carried out to realise the expected service life of the structural component/structure. However, in most of the above investigations, conjugate distributions are used in decision making. While the use conjugate distribution helps in making the problem more mathematically tractable, it may not be possible to include the greater degree of engineering judgment in decision making regarding expected service life (Ang & Tang 1975, Benjamin & Cornell 1968). In this paper, an attempt has been made to show how the engineering judgment can be used in formulating the likelihood function used in Bayesian decision making. An example problem, in which a real RC bridge girder, which was inspected at an age of 11 years and found to have undergone corrosion of reinforcement requiring extensive repair, is considered to demonstrate the usefulness of the proposed methodology. The predictive probability of corrosion initiation, computed using Bayesian techniques is found to be in corroboration with the engineering decision taken to repair the bridge at the end of 11 years.

2   CORROSION OF STEEL IN CONCRETE

The reinforcement in concrete is normally protected against corrosion by a microscopically thin oxide layer formed on the surface of the reinforcement due to the high alkalinity of the surrounding concrete (Kropp & Hilsdorf 1995). Chloride ions diffuse from the surface of the concrete through the cover concrete. Chlorides can diffuse into concrete as a result of (i) sea salt spray and direct sea wetting, and (ii) use of deicing salts. The diffusion of chlorides through the cover concrete is generally modeled using Fick's second law of diffusion. Using Fick's second law of diffusion, chloride concentration, $C(x, t)$, at a particular depth, x, from the exposed surface, at any time "t", can be obtained from

$$C(x, t) = C_s \left[1 - erf\left(\frac{x}{2\sqrt{Dt}} \right) \right] \qquad (1)$$

where C_s is the surface chloride concentration, D is the diffusion coefficient for chlorides in concrete and erf is the error function. When the chloride concentration around the reinforcement exceeds a critical value (critical chloride concentration, c_{cr}), the protective oxide layer dissolves and corrosion initiates. Using Eq. 1, time to corrosion initiation can be computed from

$$t_i = \frac{d^2}{4D} \left[erf^{-1}\left(1 - \frac{C_s}{C_{cr}} \right) \right]^{-2} \qquad (2)$$

where d is the clear cover to the reinforcement.

3   CORROSION INITIATION LIMIT STATE FOR BRIDGE GIRDERS

To account for variations in workmanship and exposure conditions, the diffusion coefficient at any given time, surface chloride concentration, critical chloride concentration and the clear cover to reinforcement should be considered as random variables. Thus, the time to corrosion initiation obtained using Eq. 2 will be a random variable. By properly defining a limit state function, it is possible to compute the probability of corrosion initiation due to chloride ingress into concrete (viz. Geiker et al. 1999, Vrounwenvelder & Schiessel 1999). The probability of corrosion initiation at any time "t" is given by

P{$T_i \le t$}= P{chloride concentration at the level of
reinforcement at time 't' ≥ critical
chloride concentration}
= P{$C(d, t) \ge c_{cr}$}   (3)

Thus, corrosion initiation time depends on the probability density function (pdf) of chloride concentration at any given time. Thus, to compute the probability of corrosion initiation, the form of the pdf should be known. Let the chloride concentration at any time $\tilde{t}$ at the level of reinforcement be described by the random variable $X(\tilde{t})$. During the design stage, by carrying out Monte Carlo simulation studies on chloride concentration distribution at different times (Balaji Rao & Appa Rao 1999), the time at which probability of corrosion initiation is going to be high (or equal to a pre-fixed value) can be determined. This time corresponds to the service life of the structure. It is noted that the evaluation of pdf of chloride concentration at $\tilde{t} = \tilde{t}_i$ does take into account the random variations in the basic variables corresponding to a specified quality of construction. Let the pdf of chloride concentration be given by $f_x (x; \tilde{t}_i)$. At $\tilde{t} = \tilde{t}_i$ the following equation is valid.

$$F_x \left(x^*; \tilde{t}_i \right) = P\left\{ x \le x^* \right\} = \int_{-\infty}^{x^*} f_x \left(x; \tilde{t}_i \right) dx \qquad (4)$$

Based on the initial data available on variables required to compute expected corrosion initiation time (either assumed during the design or data obtained just after completion of the project), the designer would infer that at $\tilde{t} = \tilde{t}_i$ the reinforcement in the structure would start corroding and hence would like to plan an inspection for determining the actual chloride concentration at the level of reinforcement. The chloride concentrations are determined by obtaining samples by either incremental percussive drilling, or from cored samples which are subsequently cut into sections and ground up for laboratory chemical analysis (fib 2002). There are a number of commercially available kits, such as the Hach, RCT and Quantab, for determining chloride concentration from drilled samples. Depending on quality control exercised in the field (in the project construction), the results of field investigations may indicate that: (i) there is no enough evidence to assume that corrosion has initiated, or (ii) in many samples, the actual chloride concentration $\geq$ critical chloride concentration, and hence corrosion might start or has started. In the first case, the departure from the estimated pdf may occur due to deviation in construction quality from the specified quality. The field observations suggest that the pdf of chloride concentration has to be updated in view of field investigations. In the second case, a decision has to be taken regarding repair since corrosion initiation itself is assumed as service life.

4 PROPOSED METHODOLOGY FOR CORROSION STATE ASSESSMENT

In this paper, a methodology based on Bayesian theory is proposed for corrosion state assessment of RC bridge girders. Within the Bayesian framework, the parameters of the distribution of $X(\tilde{t})$ are treated as random variables. In this study, a parameter associated with mean value and variance of the distribution is considered as random variable.

Using the Bayes theory, the predictive distributions before and after field investigations are respectively given by

$$F_X\left(x^*;\tilde{t}_i\right)= \int_{\substack{\text{overall possible}\\ \text{value staken by }\lambda}} F_X\left(x^*;\tilde{t}_i \mid \lambda\right) f'_\Lambda(\lambda)\, d\lambda \qquad (5)$$

$$F_X\left(x^*;\tilde{t}_i\right)= \int_{\substack{\text{overall possible}\\ \text{value staken by }\lambda}} F_X\left(x^*;\tilde{t}_i \mid \lambda\right) f''_\Lambda(\lambda)\, d\lambda \qquad (6)$$

where Λ is the parameter of distribution of random variable X (which affects the mean and variance of X); $f'_\Lambda(\lambda)$ and $f''_\Lambda(\lambda)$ are prior and posterior distributions of Λ. The form of prior distribution is generally fixed based on engineering judgment. The posterior pdf depends on engineering judgment and results of field investigation. The posterior pdf is obtained from

$$f''_\Lambda(\lambda) = \kappa L(\lambda)\, f'_\Lambda(\lambda) \qquad (7)$$

where κ is a normalisation constant, $L(\lambda)$ is the likelihood function, $f'_\Lambda(\lambda)$ is the prior distribution of Λ.

From a statistical analysis of chloride concentration obtained from field studies (Zemajtis 1998), it has been reported that pdf of chloride concentration follows Gamma distribution. A similar trend was noted for the pdf obtained from Monte Carlo simulation studies carried out at SERC (Balaji Rao & Appa Rao 1999). Hence, in this investigation, it is assumed that the chloride concentration at the surface of steel, $X(\tilde{t}_i)$, follows a Gamma distribution. Thus,

$$f_X(x) = \frac{\lambda\, e^{-\lambda x}(\lambda x)^{n-1}}{(n-1)!} ; \quad n \geq 1;\ \lambda > 0 \qquad (8)$$

with mean $= n/\lambda$ and variance $= n/\lambda^2$. The coefficient of variation of $X(\tilde{t}_i)$ is $1/\sqrt{n}$ and hence, the shape of distribution depends on value of "n". For a given value of "n", the mean and variance of chloride concentration depends on λ. As the value of λ increases, the mean and variance of chloride concentration would reduce. Within the Bayesian framework, n or λ or both can be treated as random variables. In this paper, λ is treated as a random variable and the random variable is represented as Λ.

In general, for engineering applications, conjugate distributions are used in Bayesian inferencing. For example, if the basic variable is lognormal, and if the parameter related to the mean of distribution is considered random, the conjugate distribution of the random variable is normal. However, there is neither any mathematical justification nor engineering justification for the use of conjugate distribution except that it would ease the computations. Hence, in this paper, an attempt has been made to determine the appropriate distribution for Λ. Using this distribution and knowing that $X(\tilde{t}_i)$ follows Gamma, posterior distribution, $f''_\Lambda(\lambda)$ is obtained using Eq. 7. The form of likelihood function is generally not known. Determination of the form requires engineering and statistical judgment or background. The form of likelihood function should be so chosen that it will increase the likelihood of observations made based on data obtained from field investigations. For example, during an inspection, engineer takes powder samples from the cover region at four or five different locations along the reinforcing bar to determine the corrosion state of reinforcement. Based on the chloride concentration at these locations, he may find that the actual mean chloride concentration is less than critical value. For this case, the form of likelihood function in Eq. 7 should be so chosen that the predictive probabilities estimated using Eq. 6 should be less than that estimated using Eq. 5. It is also noted that the likelihood function should also take into account the quantity of information available (i.e., more the information, more the confidence in predictive probabilities).

4.1 Prior probability density function

In this investigation, the parameter λ of $X(\tilde{t}_i)$ is assumed to follow an exponential distribution, and hence

$$f'_\Lambda(\lambda) = \alpha e^{-\alpha\lambda} ; \quad \alpha > 0,\ \lambda > 0 \qquad (9)$$

75

where $\alpha = 1/\lambda$, λ is the value of the para-meter obtained from the Gamma distribution of $X(\tilde{t}_i)$ fitted by the designer at the design stage or based on information obtained after the construction, but at $\tilde{t} \ll \tilde{t}_i$.

4.2   Likelihood function

The likelihood function is formulated based on the additional information available from field investigation. In order to formulate the likelihood function $L(\lambda)$ where λ is the location parameter, two cases need to be considered; (i) in more number of cases the chloride concentration obtained from field investigation is less than the mean chloride concentration estimated earlier by the designer, and (ii) in more number of cases the chloride concentration obtained from field investigation is more than the mean chloride concentration estimated earlier by the designer.

4.2.1   Case 1: In more number of cases the chloride concentration obtained from field investigation is less than the mean chloride concentration estimated earlier by the designer

This case suggests that there is a need to update the predictive probabilities estimated using Eq. 6 and thus the service life of the structure.

Since more number of observed chloride concentrations are lesser than the mean, to account for the actual condition, the mean of $X(\tilde{t}_i)$ has to shift to the left (Fig. 1). Hence the likelihood function is an increasing function of λ. The suitable likelihood function was found to be as given below.

$$L(\lambda) = e^{-\left(r/\lambda + 1/r\lambda\right)} \qquad (10)$$

where r is the number of cases in which actual chloride concentration at the level of steel is less than the mean value estimated by the designer. A plot of $L(\lambda)$ vs λ is shown in Figure 2.

4.2.2   Case 2: In more number of cases, the chloride concentration obtained from field investigation is more than the mean chloride concentration estimated earlier by the designer

In this case, a decision has to be taken regarding repair since corrosion initiation itself is assumed as service life.

Since more number of observed chloride concentrations are greater than the mean, to account for the actual condition, the mean of $X(\tilde{t}_i)$ has to shift to the right (Fig. 3). As the mean increases, λ has to decrease. Hence the likelihood function is a decreasing function

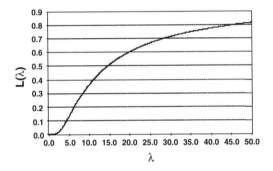

Figure 2.   Likelihood function when more often the chloride concentration obtained from field investigation is less than the mean chloride concentration estimated by the designer.

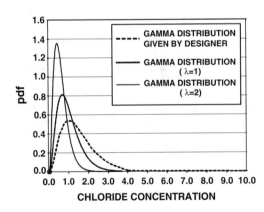

Figure 1.   Schematic representation of shifting of mean to the left when more often the chloride concentration obtained from field investigation is less than the mean chloride concentration estimated by the designer.

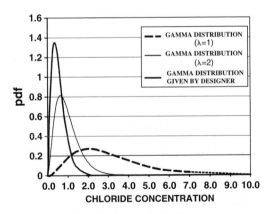

Figure 3.   Schematic representation of shifting of mean to the right when more often the chloride concentration obtained from field investigation is more than the mean chloride concentration estimated by the designer.

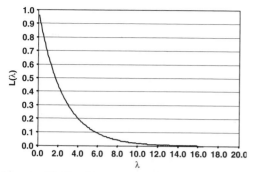

Figure 4. Likelihood function when more often the chloride concentration obtained from field investigation is more than the mean chloride concentration estimated by the designer.

Table 1. Predictive probabilities for Case 1 – Less data/information available*.

n	10	10	10	10
r	1	2	3	4
P [X ⩽ 0.08]	0.653	0.635	0.623	0.615

* Note: n – number of samples obtained (at the level of reinforcement).
r – number of cases in which actual chloride concentration is greater than the mean chloride concentration.

of λ. The suitable likelihood function was found to be as given below.

$$L(\lambda) = e^{-r^2\lambda/n} \tag{11}$$

where r is the number of cases in which actual chloride concentration at the level of steel is more than the mean value estimated by the designer. A plot of $L(\lambda)$ vs λ is shown in Figure 4.

As already pointed out, the decision criteria regarding service life of the RC structure with respect to chloride-induced corrosion of reinforcement involves calculation of probability of chloride concentration less than the critical chloride concentration (Eq. 4).

A computer program has been developed implementing the proposed methodology. The posterior pdf required for estimating the predictive probabilities are obtained numerically (using Eq. 7). The predictive probabilities are computed numerically. A sensitivity study is carried out to determine the effect of field information and the number of cases in which the observed chloride concentration at the level of reinforcement is less (or more) than the mean estimated by the designer, on the predictive probabilities for Case 1 (or Case 2). The results of the sensitivity analysis are presented in Tables 1–4.

Table 2. Predictive probabilities for Case 1 – More data/information available.

n	100	100	100	100
r	10	20	30	40
P [X ⩽ 0.08]	0.818	0.802	0.791	0.773

Table 3. Predictive probabilities for Case 2 – Less data/information available.

n	10	10	10	10
r	6	7	8	9
P [X ⩽ 0.08]	0.405	0.398	0.390	0.380

Table 4. Predictive probabilities for Case 2 – More data/information available.

n	100	100	100	100
r	60	70	80	90
P [X ⩽ 0.08]	0.262	0.223	0.188	0.158

From these tables, it can be noted that the $L(\lambda)$ and hence the predictive probabilities are sensitive to both size of data and also the number of cases in which the actual chloride concentration exceeds or equals the mean chloride concentration estimated by the designer.

5 APPLICATION OF BAYESIAN DECISION MAKING – AN EXAMPLE

The purpose of this example is to demonstrate the usefulness of the proposed methodology. The super-structure of the Gimsøystraumen bridge, Norway, constructed during 1979–81 in a marine environment, is considered in this study (Fluge 2001). Inspection data on Gimsøystraumen bridge girder is one of the well-documented information available in literature for carrying out Bayesian decision making. The concrete used is of Grade 40 MPa with a water–cement ratio ≈0.52. The specified minimum concrete cover to the reinforcement is 30 mm. The bridge was thoroughly inspected and repaired after 11 years.

From a database of D values created at SERC, Chennai, by collecting the values of D reported by various researchers based on different laboratory tests and field exposure tests (Anoop et al. 1999), the value of D for concrete with grade of 40 MPa is taken as $5.5 \times 10^{-12} m^2/s$. Based on the values of C_s reported by Bamforth (1999) for concrete exposed to marine environment, the value of C_s is taken as 0.30% by mass of concrete. The value of C_{cr} is taken as 0.13% by mass of concrete as reported by Fluge (2001). Using these values, the concentration of chlorides at the level of reinforcement at the end of 11 years is obtained using Eq. 1 as

C = C(30 mm, 11 years) = 0.2344% by mass of concrete.

Using the information that chloride concentration follows Gamma distribution and that prior distribution of λ is exponential, the predictive probability estimated by the designer at an age of 11 years is

$P[C \leqslant C_{cr}] = 0.195$

This predictive probability needs to be updated based on the information from the field investigations carried out at the end of 11 years. From the measured chloride profiles at the end of 11 years, C_s and D values were determined and were reported in Fluge (2001). Based on the C_s values reported for the leeward side of the superstructure (more critical due to large amount of chlorides deposited compared to the windward side), the average C_s value is taken as 0.55% by mass of concrete. The values of D for the leeward side, as determined from the chloride profiles, along with its frequency are given in Table 5 (Fluge 2001). The values of C(30 mm, 11 years), determined using Eq. 1 using these values of Cs and D are also given in Table 5.

It can be noted from Table 5 that out of the 236 observations, in 163 cases, the chloride concentration determined based on field investigation exceeds the mean predicted by the designer (i.e., 0.2344% by mass of concrete). Since in more number of cases, the chloride concentration from field investigations is more than the mean chloride concentration estimated by the designer (Case 2), the likelihood function given by Eq. 11 is used. After incorporating the additional information from field investigation (with n =236 and r =163) and updating the probability predicted earlier,

Table 5.   D values determined from the chloride profiles at the end of 11 years along the leeward side of superstructure and C(30 mm, 11 years) computed.

D (m²/s)	Frequency	C(30 mm, 11 years)* (% by mass of concrete)
0.2×10^{-12}	21	0.1096
0.4×10^{-12}	52	0.1987
0.6×10^{-12}	65	0.2496
0.8×10^{-12}	38	0.2831
1.0×10^{-12}	28	0.3073
1.2×10^{-12}	17	0.3257
1.4×10^{-12}	9	0.3404
1.6×10^{-12}	1	0.3524
1.8×10^{-12}	4	0.3625
2.6×10^{-12}	1	0.3909
Total	236	

Note: *computed using Cs 5 0.55% by mass of concrete.

$P[C \leqslant C_{cr}] = 0.0407$

From the values of C(30 mm, 11 years) given in Table 5, the point estimate of $P[C \leqslant C_{cr}]$ is 21/236 = 0.088. The predictive probability obtained using Bayesian theory does give importance to the quantity of information/data available. From the field investigations it was reported that that the reinforcing bars have already started corroding with severe corrosion on several locations, and the bridge girder had to be repaired at the end of 11 years. From the three values of probability of corrosion initiation obtained, namely, 0.805 (based on the prediction at the design stage), 0.912 (based on the point estimate – computed using relative frequency approach – information obtained from field investigations), 0.960 (based on updated chloride concentration using the proposed methodology), it is noted that the value obtained using the proposed methodology corroborates with the engineering decision taken to repair the bridge girder at the end of 11 years (Fluge 2001). This also suggests that the forms of the prior distribution and the likelihood function used in this investigation are appropriate. Thus, the prediction made using the Bayes techniques is more realistic, and the use of proposed methodology helps in making better decisions.

6   CONCLUSIONS

For RC structures located in marine and other aggressive environments, chloride-induced corrosion of reinforcement is identified as a major mechanism of resistance degradation. In this paper, a methodology which applies Bayesian techniques for corrosion state assessment of RC bridge girders is presented. The paper brings out the need to use engineering judgment and the background of statistics in formulating the likelihood functions. Likelihood functions are formulated for two different cases, which will arise in practice. Gamma distribution has been used for representing the pdf of chloride concentration at the level of reinforcement based on the information from field studies reported by Zemajtis (1998) and from the Monte Carlo simulation studies carried out at SERC (Balaji Rao & Appa Rao 1999). The posterior pdf required for estimating the predictive probabilities are obtained numerically. The usefulness of the proposed methodology is demonstrated in updating the chloride concentration at the level of reinforcement in a real RC bridge girder, which was inspected at an age of 11 years and found to have undergone corrosion of reinforcement requiring extensive repair. The predictive probability of corrosion initiation, computed using Bayesian techniques is found to be in corroboration with the engineering decision taken to repair the bridge girder at the end of 11 years.

ACKNOWLEDGEMENT

This paper is being published with the kind permission of the Director, Structural Engineering Research Centre, Chennai. The authors are thankful to Dr F. Fluge for providing very useful, detailed information on Gimsøystraumen bridge girder used in the example presented in this paper, which is taken from the proceedings of 3rd Duranet Workshop, Tromsø, Norway.

REFERENCES

Ang, A.H. & Tang, W.H. 1975. *Probability concepts in engineering planning and design – Volume I*. John Wiley & Sons.

Anoop, M.B., Balaji Rao, K. & Appa Rao, T.V.S.R. 1999. *Durability of reinforced concrete with respect to chloride ingress – a critical review*. Project Report No. SS-OLP 07741-RR-99-2. Structural Engineering Research Centre, Chennai.

Balaji Rao, K. & Appa Rao, T.V.S.R. 1999. *Chloride induced corrosion of reinforcement in reinforced concrete members – Brief review of literature and results of preliminary studies on reliability analyses*. Project Report No. SS-OLP 07741-RR-99-1. Structural Engineering Research Centre, Chennai.

Bamforth, P.B. 1999. The derivation of input data for modeling chloride ingress from eight-year UK coastal exposure trials. *Magazine of Concrete Research*. 51(2): 87–96.

Benjamin, J.R & Cornell, C.A. 1968. *Probability, statistics and decision making for civil engineers*. McGraw-Hill Book Company.

fib. 2002. *fib bulletin 117 – Management, maintenance and strengthening of concrete structures*. Technical Report, fib.

Fluge, F. 2001. Marine chlorides: A probabilistic approach to derive provisions for EN 206-1. In *Service life design of concrete structures – from theory to standardization*: 63–83. 3rd Duranet Workshop, Tromsø, Norway.

Geiker, M., Edvardsen, C. & Rostam, S. 1999. Building from first to last. *Concrete Engineering International* 3(6): 33–37.

Kropp, J. & Hilsdorf, H.K. 1995. *Criteria for concrete durability*. E & FN Spon. London.

Mori, Y. & Ellingwood, R. 1994a. Maintaining reliability of concrete structures. I: role of inspection/repair. *Journal of Structural Engineering (ASCE)*. 120(3): 824–845.

Mori, Y. & Ellingwood, R. 1994b. Maintaining reliability of concrete structures. II: optimum inspection/repair. *Journal of Structural Engineering (ASCE)*. 120(3): 846–862.

Vrounwenvelder, A. & Schiessel, P. 1999. Durability aspects of probabilistic ULS design. *HERON*. 44: 19–30.

Zemajtis, J. 1998. *Modeling the time to corrosion initiation for concretes with mineral admixtures and/or with corrosion inhibitors in chloride-laden environments*. Ph.D. Dissertation. Virginia Polytechnic Institute and State University. Blacksburg. Virginia.

Safety and Reliability – Bedford & van Gelder (eds)
© 2003 Swets & Zeitlinger, Lisse, ISBN 90 5809 551 7

Merging cut sets methods and reliability indexes for reliability and availability analysis of highly meshed networks

G. Ballocco, A. Carpignano & M. Gargiulo
Politecnico of Torino, Department of Energy, Torino, Italy

M. Piccini
RAMS&E pscarl, Asti, Italy

ABSTRACT: The paper addresses the problem of reliability and availability assessment for meshed network systems, as gas or heat distribution systems. An introduction highlights the specific associated problems, related to the structural complexity and to the dynamic character of these systems. An approach merging cut sets method based on the graph theory and reliability indexes is described and applied to an hypothetical network. A critical analysis follows, where the results of the study are compared to those achieved by previous works. The activity is based on the experience acquired by the authors about reliability and availability analyses of district heating systems.

1 INTRODUCTION

The reliability, availability and safety assessment of meshed network systems, as gas or heat distribution systems, can today be considered an open issue. Few studies have been performed so far about this domain, in spite of the wide proliferation of such systems, in particular in the urban context and then in highly complex and vulnerable zones. In addition, the studies already done concern the application of classical techniques, more suitable for small and radial systems: a consequence of that is an high uncertainty in reliability and availability estimations.

The main features of this type of systems leading to a difficult assessment, can be summarised as follows.

- Heating and gas distribution networks are complex systems made of heat production plants, transportation and distribution pipelines, pumping stations, regulation and shut-off valves: they are then constituted by a large number of components and are characterised by a wide territorial extension.
- The final users of this type of systems are highly distributed in space and characterised by different expectances. Therefore it is difficult to clearly characterise a precise goal for the system. It is then very difficult to define the event "unavailability of the system", as the concept of unavailability must take into account for user's distribution, coupling the out of service probability to each user and the related loss of service.
- There is an high interaction between probabilistic and thermo-fluidodynamic aspects: in network systems, used for fluid distribution, there is an high interdependence between component failures and the physics behaviour of the system, particularly regarding the redistribution of pressure at the nodes of the network (thermo-hydrodynamic aspects); this means that networks cannot be treated only from a logical-probabilistic point of view but require the use of hybrid techniques able to study the problem in an integrated way.
- These systems rarely have a radial structure: the number of meshes in the composition of the network is a factor that can highly increase the complexity of the reliability/availability analysis, as in this way the possibility for different supply ways for some portions of the network is amplified. This situation intuitively increases the reliability and availability of the network, but imposes the need of considering multiple failures in the assessment approach and makes these systems very difficult to be modelled, because of the large number of possible working configurations.
- Heat and gas distribution networks are usually structures characterised by a high rate of changing and upgrading design activities: this need a frequent updating of the analysis and the evaluation of an

overall parameter able to qualify the reliability and availability of the network at a whole, and leading to highlight the changes in network performances.

On the basis of these considerations, the work aims to find an efficient methodologies for reliability and availability analysis of meshed networks. A technique based on the combination between a cut sets method an reliability indexes will be proposed and applied to an hypothetical network. Results will be discussed in comparison with those achieved with the Fault Tree Analysis, applied to the same case.

2   USE OF RELIABILITY INDEXES FOR THE ASSESSMENT OF OVERALL NETWORK CONDITIONS

When approaching the analysis of a network system, the first step concerns the calculations of reliability and unavailability parameters of each users connected to the system, in order to assess the *local* behavior of the network; then a technique, able to gather and con-sider all these local results, is needed in order to define few parameters expressing the *overall* condi-tions of the system.

Previous works presented by the authors (Carpigano et al. 2001, Carpignano et al. 2002) demonstrated as an approach to the problem with modified reliability indexes, could be considered an excellent solution for the assessment of overall conditions of the network, when the local parameters are known.

Reliability Indexes have been developed and vali-dated mainly for the electro-technical domain, for electrical distribution systems (Billinton 1984). They allow to evaluate, under some restraints and boundary conditions, overall and synthetic parameters repre-senting the reliability and availability of network dis-tribution systems, eventually exportable to domains others than the electro-technical one.

New indexes have been developed, based on two categories of results:

- Unavailability and Expected Number of Failures (Q_i and W_i);
- losses of supplied volume of heated buildings in case of each section failure (V_i), where different boundary conditions have been applied for the def-inition of consequences of failures.

The two main indexes are described in Table 1.

The choice of reliability indexes has been done to identify parameters able to give synthetic and overall information about reliability and availability char-acteristics of the systems: they allow to calculate average values of unavailability (Q) and Expected Number of Failures (W), weighted on all connected users. Indexes represent very significant parameters

Table 1.   Reliability indexes.

System Average Interruption Frequency Index (SAIFI), where : W_i represents the Expected Number of Failures related to each network section, V_i the supplied volume of heating buildings related to section i and T the mission time; it represents the Expected Number of Failures per unit of time, weighted on the not-supplied volume.

$$SAIFI = \frac{\sum W_i V_i}{\sum V_i} \cdot \frac{1}{T}$$

System Average Interruption Duration Index (SAIDI), where Q_i represents the Unavailabilities related to each network section and V_i the supplied volume of heating buildings related to section i; it then represents the Unavailability of the system, versus the not-supplied volume; as Unavailability can be interpreted as the ratio between the duration of the failure and the theoretical duration of all mission, this index represents the percentage duration of the failure, weighted on the not-supplied volume.

$$SAIDI = \frac{\sum Q_i V_i}{\sum V_i}$$

in case of evaluation of different design and technical configurations and allow to take into account for dif-ferent expectancies by the connected users.

Once the reliability indexes have been chosen for the assessment of the overall availability and reliabil-ity, a method is needed in order to calculate the local conditions, i.e. the reliability and availability values for each user connected to the network.

At first, the work by the authors concerned the use of Fault Tree Analysis, which furnishes the values of Unavailability (Q) and Expected Number of Failure (W), taking into account multiple failures and thermo-hydrodynamic aspects (Carpignano et al. 2001). Nevertheless this technique is characterised by an high associated workload since a fault tree for each user has to be constructed; then it is very inefficient for wide networks and in case of hydraulical/structural modifications of the systems.

3   USE OF CUT SET METHODS FOR THE ASSESSMENT OF LOCAL NETWORK CONDITIONS

Literature offers several examples of techniques to face the problem of reliability and availability assessment of networks. A critical review of methodologies has been done (Carpignano et al. 2002), considering several alternative techniques such has *Monte Carlo methods* (Marseguerra 2001), *Neural networks* (Pasquet et al. 1997) and *Markov chains* (Buchsbaum 1995), and the attention has been finally focused on the *cuts sets* method based on the *graph theory*.

According to this approach, on the basis of a graphical representation of the network, all the possible cut sets can be identified and evaluated (Yan et al. 1994), giving substantially the same results of the fault tree approach but through a different way. The graph of the network can be translated in a matricial representation and algorithms exist that produce all minimal cut sets in directed network: the time per cut set varies linearly with the number of nodes but decreases exponentially with the density of the graph.

One of the more common assumptions is that nodes do not fail and hence all network failures are consequence of link failures: new approaches are now under development, that introduce in the analysis also the failure of nodes (Fard 1999) and try to associate to different failures also the impact on thermo-hydraulic characteristics of the network (Yeh 2001).

This approach presents several advantages:

- the most part of the workload can be transferred to software algorithms: the analyst is only in charge of the graphical representation of the network;
- thermo-fluidodynamic aspects and multiple failures can be taken into account;
- high flexibility in case of structural changes of the system;
- capability of considering the dynamics of the system, defining adequately the different nodes;

The main possible lacks and limitations of the approach seems to be:

- the system must be constituted by only one source and only one sink; to characterise more sources or sinks the introduction of "virtual" nodes seems to be the only way;
- high amount and times of calculations: the matrixes associated to meshed networks can assume very large dimensions; anyway, there exist factoring methods (Page 1989, Theologou 1991) that could reduce calculation problems.

4   APPLICATION EXAMPLE

In order to evaluate its effectiveness, the cut sets method has been applied to an hypothetical network for district heating, and compared with the Fault Tree Analysis. The example network is reported in Figure 1; in spite of its plainness, it concerns the same problems connected to a real case.

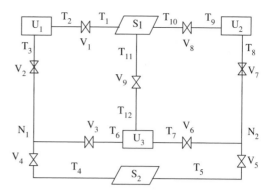

Figure 1.   Hypothetical networks for district heating.

Table 2.   Value of failure rates.

Component	Failure modes	Failure rate λ (h^{-1})	MTTR (h)
Pipeline sections	Leakage	1.541E-05 ($h^{-1}Km^{-1}$)	24
Valves	Stuck closed	3.000E-06	1
	Leakage	5.000E-08	24
Users	Stuck closed	3.000E-06	1
	Leakage	5.000E-08	24
Sources	Out of order	5.337E-04	3.5

T_x: main transportation sections made of a couple of pipelines (one for the delivery of the hot water, and one for the return of the cold water); the considered failure mode is leakage.

V_x: motor operated valves whit two failure modes: stuck closed and leakage.

S_x: sources (heat production and pumping plants) with a single failure mode: out of order

U_x: user with two failure modes: stuck closed and leakage.

The presence of two sources makes the analysis more realistic, involving the possibility of considering several solution to feed the users. On the other hand the presence of three users allows the distinction between *local* and *global* results, as defined in 2.

The considered network is of the indirect type, i.e. each pipeline section can be fed in both direction. When a user is selected as target of the analysis, in order to evaluate its (*local*) availability, it has to be considered as a point of arrival of the flow. When the user is not selected as target, it behaves as valve, allowing the flow in both directions. A source is a starting point of the flow, i.e. each pipeline section straightway connected to a source has to be considered as direct.

The Table 2 reports the values for Failure Rates (λ) and the Mean Times To Repair (MTTR) used in this example.

Table 3. FTA results.

Name	Top event 1	Top event 2	Top event 3
Unavailability	9.617E-006	9.617E-006	7.679E-006
ENF	4.863E-002	4.863E-002	4.416E-002
MCSs	425	425	261
Max Order	5	5	4
Top Gate	Top Event 1	Top Event 2	Top Event 3
MissionTime	8760.00	8760.00	8760.00
Use BC	0	0	0
Use CutOff	0	0	0

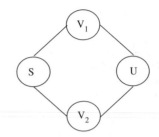

Figure 3. Associated graph of a network.

Table 4. Value of reliability indexes.

Index	Value
SAIFI	5.492E-06
SAIDI	9.391E-06

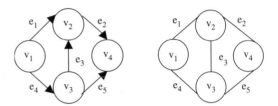

Figure 4. Direct and indirect graph.

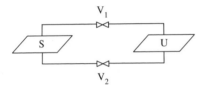

Figure 2. A simple network.

	e_1	e_2	e_3	e_4	e_5			e_1	e_2	e_3	e_4	e_5
v_1	1	0	0	1	0		v_1	1	0	0	1	0
v_2	−1	1	−1	0	0		v_2	1	1	1	0	0
v_3	0	0	1	−1	1		v_3	0	0	1	1	1
v_4	0	−1	0	0	−1		v_4	0	1	0	0	1
a.							b.					

Figure 5. Incidence matrix.

5 FAULT TREE ANALYSIS

The FTA allows to evaluate the local reliability and availability of the network, by calculating the values of expected number of failure (ENF) and unavailability for each user of the network.

In the case referred to Figure 1, the presence of three users involves the construction of three fault trees. The Top Event is the loss of heated volume for the examined user.

The Table 3 reports the FTA results, achieved with the software STARS Studio.

Once the local parameters are found, reliability indexes can be calculated as defined in 3. The values are reported in Table 4.

6 CUT SETS METHOD

6.1 Graph theory

In order to explain the technique based on cut sets method, the graph theory must be briefly introduced.

A graph is a relational structure made of an finite set of object, named nodes, and a finite set of connections, named arcs or edges. The Figure 2 represents a simple networks for district heating, composed of a source (heating production plant), two valves, a user and four pipelines sections; the associated graph, is shown in Figure 3.

This duality allows to consider indifferently the network or the associated graph, since both these terms refer to the same set of relationships.

A graph is defined oriented or direct, when its edges refer to relationship between orderly couples of nodes while it is indirect if edges refer to relationship between not orderly couples of nodes.

Each graph can be expressed as an incidence matrix. The incidence matrix D of a graph G, is formed in the following way: a column of the matrix is allocated to each edge and a row to each node. An entry d_{ij} is 1, −1 if node i is an end point of edge j and is zero otherwise.

The incidence matrices for graphs of Figure 4 are represented in Figure 5.

Obviously when G has A edges and N nodes, the dimension of the matrix is E×N. The incidence matrix stores all the information of a graph, in a format that computers can easily receive and manipulate.

6.2 The recursive algorithms of enumerating minimal cut set

The first step in order to analyse a network with the cut sets method, concerns the construction of the incidence matrix. The incidence matrix for the network of Figure 1 is given below:

	Links													
Nodes	T_1	T_2	T_3	T_4	T_5	T_6	T_7	T_8	T_9	T_{10}	T_{11}	T_{12}	N_1	N_2
V_1	1	1	0	0	0	0	0	0	0	0	0	0	0	0
V_2	0	0	1	0	0	0	0	0	0	0	0	0	1	0
V_3	0	0	0	0	0	1	0	0	0	0	0	0	1	0
V_4	0	0	0	1	0	0	0	0	0	0	0	0	1	0
V_5	0	0	0	0	1	0	0	0	0	0	0	0	0	1
V_6	0	0	0	0	0	0	1	0	0	0	0	0	0	1
V_7	0	0	0	0	0	0	0	1	0	0	0	0	0	1
V_8	0	0	0	0	0	0	0	0	1	1	0	0	0	0
V_9	0	0	0	0	0	0	0	0	0	0	1	1	0	0
U_1	0	1	1	0	0	0	0	0	0	0	0	0	0	0
U_2	0	0	0	0	0	0	0	1	1	0	0	0	0	0
U_3	0	0	0	0	0	1	1	0	0	0	0	1	0	0
S_1	1	0	0	0	0	0	0	0	0	1	1	0	0	0
S_2	0	0	0	1	1	0	0	0	0	0	0	0	0	0

Given the incidence matrix, the minimal cut sets can be enumerated through two successive step:

- Enumerating the minimal cut sets (mcs_i) of a graph subject to link failures and perfect nodes;
- Obtaining the deduced cut sets (DCS_i) of a graph subject to both link and node failures from the mcs_i of the network, for which perfect nodes are assumed; at the end of the procedure it is possible to obtain the minimal cut sets for the system (MCS_i) by using a systematic process of elimination.

In order to analyse a network with more users, it is necessary to consider in every recursive process a single target; in other words, if the graph presents three users, it will be necessary to repeat three times the enumeration of the cut sets, one for each user. The way to proceed is the same used for Fault Tree Analysis, but in this case the minimal cut sets are gained directly, without having to construct the fault tree.

6.3 Network with perfect nodes

The approach of the first algorithm presented in this paper is based on the partitioning principle (Jensen 1969, Yan 1994). The partitioning principle, as defined for networks with only one source, concerns the definition of two mutually exclusive sets; the first (X) containing only the source node, the second (Y) including all the other nodes. The general procedure of the algorithm consists in removing one node from set Y and add it to set X after it has been analysed.

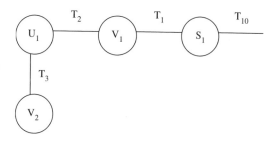

Figure 6. Example.

In order to face networks with several sources, the algorithm here proposed consists in defining three mutually exclusive sets of nodes X, Y & S; the solution starts by assigning X the target node (nodes already analysed), S the source nodes and Y the remaining nodes (nodes to analyse). This algorithm enumerates the minimal cut sets by investigating each node, starting from the target node (a user) to end with sources nodes. When a source has been reached, the algorithm stops and passes to the successive node.

Once a target node has been chosen, the algorithm captures from the incidence matrix the incident edges for that node, which are related to the columns showing 1 or −1 at the crossing with the row of the examined target node. These edges represent the first minimal cut sets of the network since their failure does not allow the operation of the user. The following cut sets are then generated by using the recursive equation:

$$CS_i = [\, CSR \cup C_i \,] - [\, CSR \cap C_i \,] \tag{1}$$

where CSR, represents the route followed until this moment;

C_i, represents edges of the currently examined node;
CS_i, represents the new evaluated cut set;

The recursive process terminates when the group Y is empty, i.e. all the nodes have been analysed and therefore the cut sets of the local target have been listed. Minimal cut sets are then found from the cut sets with a simple procedure of reduction.

With regard to Figure 6, where a part of the graph of Figure 1 is shown, choosing as target the node (user) Ut_1, it is obtained:

$$X = \{U_1\} \quad S = \{S_1, S_2\}$$
$$Y = \{U_2, U_3, V_1, V_2, V_3, V_4, V_5, V_6, V_7, V_8, V_9\}$$

Given the target node U_1, the algorithm captures from the incidence matrix the incident edges for that node (T_2, T_3).

These edges represent the first minimal cut sets of the network; in fact their failure does not allow the operation of the user U_1.

$CS_1 = \{T_2, T_3\}$

For this example the edge T_3 is neglected and the attention will be focused on T_2.

The edge T_2 characterises two nodes: U_1 and V_1. U_1 belongs to the set X (nodes analysed), therefore it comes discarded. The node V_1 belongs to the set Y (nodes to be analysed) it comes therefore conserved.

From the node V_1 two incidents edge, T_2 and T_1, are obtained; they have to be inserted in the carrier edges of the currently analysed node:

$C_1 = \{T_2, T_3\}$

Now the edge V1 can be removed from set Y and added to set X:

$X = \{U_1, V_1\}$ $Y = \{U_2, U_3, V_2, V_3, V_4, V_5, V_6, V_7, V_8, V_9\}$

By using the recursive equation (1), the results is the following:

$CSR = CS_1$
$CS_2 = [\{T_2, T_3\} \cup \{T_1, T_2\}] - [\{T_2, T_3\} \cap \{T_1, T_2\}]$
$\quad = \{T_1, T_3\}$

The associated nodes of the links (T_3 & T_1) can be simply determined from the incidence matrix, by identifying the corresponding nodes in the rows whose elements in the columns of the corresponding links are one.

The associated nodes of the link T_1 are V_1 & S_1. The node V_1 belongs to the set X, therefore it comes discarded. The node S_1 belongs to the set S, therefore it comes discarded. In order to continue the enumeration of the cut sets it is necessary to return at T_3 and resume with the search. Once all the cut sets have been listed, a simple procedure provides for their reduction to minimal cut sets.

6.4 Real network with unreliable edges and nodes

Until this point, the procedure has provided the minimal cut sets for the network with perfect nodes. A second algorithm is needed for deducing minimal cut sets for the network with unreliable nodes.

This method is fundamentally based on the following concept: the failure of a node inhibits the working of all edges associated to it.

This algorithm starts choosing a target node from which is possible to recover the mcs_i obtained with the first algorithm. In the next step first mcs_i is selected, and the incidence matrix is determined from the nodes associated to its edges. From those associated nodes the algorithm generate the set $\{VS_i\}$ of all

possible combinations of the associated nodes of the links stored in $\{mcs_i\}$, except for the target node; since a failure of the target node results in a network failure, that node itself constitutes a Deduced Cut Set (DCS_i).

The following DCS_i can be deduced from the mcs_i by the following recursive equation:

$$DCS_i = \{VS_i\} \cup [\{mcs_i\} - \{L_i\}] \qquad (2)$$

where VS_i, represents a set of all possible combinations of the associated nodes of the links in $\{mcs_i\}$;

L_i, links connected to $\{VS_i\}$.

Considering again the example of Figure 6 and choosing as target the node U_1, the first mcs_i is:

$mcs_1 = \{T_2, T_3\}$

The associated nodes of links T_2 e T_3 in mcs_1 are first determined from the incidence matrix, by identifying the nodes corresponding to the rows, which have 1 or -1 in the columns related to T_2 and T3. Excluding the target node U_1, the associated nodes are determined to be $\{V_1, V_2\}$, and hence the VS_i of this node set is generated as:

$VS_1 = \{V_1\}$ $VS_2 = \{V_2\}$ $VS_3 = \{V_1, V_2\}$

Then the first node combination $\{V_1\}$ s selected and the incident edges for this combination $\{L_1\}$, are obtained from the incidence matrix, by identifying the edges corresponding to the columns which have 1 or -1 in the row related to V_1. This step has to be repeated until all the VS_i are examined. The sets L_i are:

$L_1 = \{T_1, T_2\}$ $L_2 = \{T_3, N_1\}$ $L_3 = \{T_1, T_2, T_3, N_1\}$

Then by applying the equation (2), the DCS_i are deduced.

$DCS_1 = \{V_1\} \cup [\{T_2, T_3\} - \{T_1, T_2\}] = \{V_1, T_3\}$
$DCS_2 = \{V_2\} \cup [\{T_2, T_3\} - \{T_3, N_1\}] = \{V_2, T_2\}$
$DCS_3 = \{V_1, V_2\} \cup [\{T_2, T_3\} - \{T_1, T_2, T_3, N_1\}] = \{V_1, V_2\}$

The above procedure must be repeated until all mcs_i, associated at the target node selected, are encountered.

Simplification and reduction of the DCS_i are the final step to obtain the minimal cut sets MCS_i.

6.5 Results

Once the minimal cut sets are listed, the values for Expected Number of Failure and Unavailability can

Table 5. Results of FTA.

Name	Top event 1	Top event 2	Top event 3
Unavailability	9.617E-006	9.617E-006	7.679E-006
ENF	4.863E-002	4.863E-002	4.416E-002
MCSs	425	425	261
Max Order	5	5	4
Mission Time	8760.00	8760.00	8760.00

Figure 7. Selection menu.

be calculated straightway. With regards to the example of Figure 1, results are the same founded with Fault Tree Analysis and reported in Table 5. In order to assess the overall performance of the network indexes SAIDI an SAIFI can be determined, as previously done.

7 SOFTWARE TOOL

On the basis of the algorithms presented in the previous chapters, the authors have developed a software tool, named Nettuno, as a support for availability and reliability analyses of meshed networks.

Given the incidence matrix, the tool directly provides the values of Unavailability and Expected Number of Failures for each user of the network, and the criticality indexes (Barlow-Proschan and Fussell-Vesely indexes) for the systems.

The software has been implemented in Matlab 6 with multiple windows, in which buttons and field exist with which the users can converse effectively with the program, as shown in Figure 7.

The incidence matrix can be simply created, as shown in Figure 8. In Figure 9 an incidence matrix is shown as it appears within the software.

The panels for configurations of arcs and nodes are shown in Figure 10 and Figure 11.

The simulations results are displayed within the software as shown in Figure 12 and they can be saved in html format and exported towards the Office environment (Figure 13).

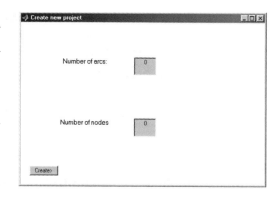

Figure 8. Creation of incidence matrix.

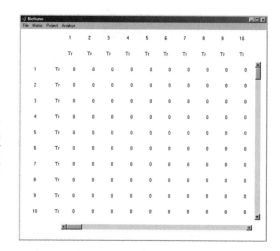

Figure 9. Incidence matrix.

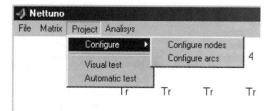

Figure 10. Configurations panel.

In a first phase the tool has been tested on some simple networks, furnishing very satisfying results; the minimal cut sets are the same obtained with the FTA while differences in values of Unavailability and Expected Number of Failures are in the order of 0,1%.

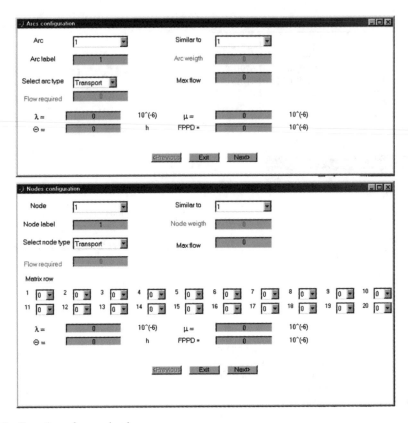

Figure 11.   Configurations of arcs and nodes.

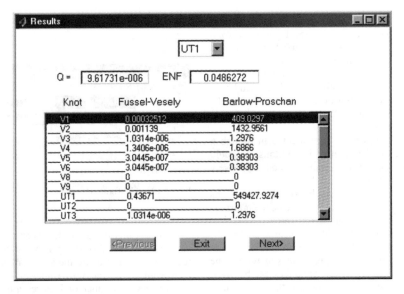

Figure 12.   Results visualisation.

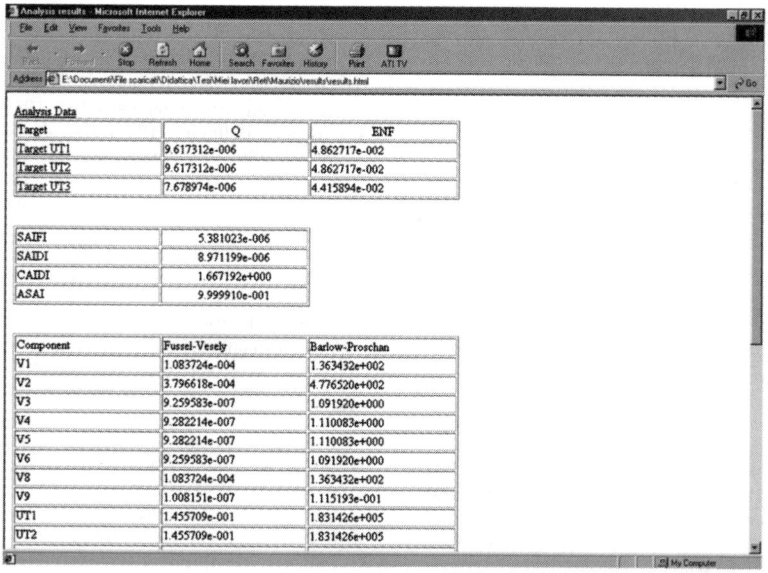

Analysis Data

Target	Q	ENF
Target UT1	9.617312e-006	4.862717e-002
Target UT2	9.617312e-006	4.862717e-002
Target UT3	7.678974e-006	4.415894e-002

SAIFI	5.381023e-006
SAIDI	8.971199e-006
CAIDI	1.667192e+000
ASAI	9.999910e-001

Component	Fussel-Vesely	Barlow-Proschan
V1	1.083724e-004	1.363432e+002
V2	3.796618e-004	4.776520e+002
V3	9.259583e-007	1.091920e+000
V4	9.282214e-007	1.110083e+000
V5	9.282214e-007	1.110083e+000
V6	9.259583e-007	1.091920e+000
V8	1.083724e-004	1.363432e+002
V9	1.008151e-007	1.115193e-001
UT1	1.455709e-001	1.831426e+005
UT2	1.455709e-001	1.831426e+005

Figure 13. Exporting of results.

8 CRITICAL REVIEW OF RESULTS

As previously pointed out, FTA allows the evaluation of the local reliability and availability of a network. Once an users is selected as object, the FTA provides the MCSs, the unavailability and the ENF, taking into account multiple failure. It also allows to appreciate the contribution of each components to the Top Event ENF and unavailability by calculating the criticality index. Nevertheless this technique requires an heavy overall workload in order to find out the results since a fault tree for each user is needed; moreover the number of fault trees grows with the increasing in the meshing of the network. Finally, this technique is very inefficient in case of hydraulical/structural modifications of the systems, which require the reconstruction of the fault trees.

The same results found out with FTA, are provided by the analysis with recursive algorithms which revealed to be faster and more flexible. The analyst's work is reduced to the definition of the incidence matrix, the major workload being transferred to the software. In case of modifications of the network, it is sufficient to update the matrix and re-run the algorithm.

Really the time of calculation is not negligible. With regard the application example, the algorithm, implemented within MATLAB 6 on a 1000 MHz processor, took about few seconds, but for the network presented in previous works by the authors the time calculation took about 18 minutes. Considering the overall time of calculation the method is then comparable with the Fault Tree Analysis, but in the case of algorithm, the presence of the analyst is not required for the most time. Moreover, the constant increasing in processors speed retrenches the problem.

With regard to the evaluation of the global parameters of the network, the application of Modified Indexes furnished very satisfying results. In fact this approach provide synthetic and overall information about reliability and availability characteristic of the whole network, also in case of very extended systems; it allows to evaluate different design and technical configurations and to take into account different expectancies by the connected users.

9 CONCLUSION

With the aim to assess the problem of reliability and availability analysis of complex networks systems, this work proposed a methodology, based on the combination between the cut sets method and the reliability indexes approach, which revealed itself very efficient. This technique allows to assess the reliability and availability of a network, both from a local and global point of view, taking into account multiple failure, and requiring reasonable workload and times of calculation, also in case of highly meshed systems and/or hydraulical/structural modifications of the systems.

The future developments of the project concern the set up of the method in order to consider in detail thermo/fluidodynamic aspects of the problem, the testing of alternative methodologies and the optimisation of the software Nettuno.

ACKNOWLEDGMENTS

We would like to thank AEM (Azienda Energetica Metropolitana) and in particular eng. Andrea Ponta for the useful co-operation.

REFERENCES

R. Billinton, R. Allan, (1984). "Reliability evaluation of power systems", New York: Plenum.

A.L. Buchsbaum, M. Mhail, (1995). "Monte Carlo and Markov chain techniques for network reliability and sampling". Networks, vol.25, 1995, pp.117–130.

A. Carpignano, A. Mosso, M. Piccini & A. Ponta, (2001). "Merging FT/ET approach with the index approach to assess the reliability and availability of heating distribution network", Proc. of European Safety and Reliability Conference ESREL 2001, 2001 September 16–20, pp.365–372.

A. Carpignano, M. Piccini, M. Gargiulo, (2002). "Reliability and availability evaluation for highly meshed network systems : Status of the art and new perspectives". In: proceedings of Annual Reliability and Maintainability Symposium, 2002, Seattle, WA USA.

N.S. Fard, T.H. Lee, (1999). "Cutset enumeration of network systems with link and node failures". Reliability Engineering and System Safety, vol.35, 1999, pp.141–146.

P. Jensen, M. Bellmore, (1969). " An algorithm to determine the reliability of a complex system", IEEE Trans.Reliab, vol.R-18, November 1969, pp.169–174.

M. Marseguerra, E. Zio, (2001). "Principles of Monte Carlo simulation for application to reliability and availability analysis". Tutorials notes of ESREL 2001 European Safety and Reliability Conference, September, 2001, Torino, Italy, pp.37–61.

L.B. Page, J.E. Perry, (1989). "Reliability of directed networks using the factoring theorem", IEEE Trans. Reliab., vol.38, December 1989, pp.556–562.

S. Pasquet, E. Chatelet, P. Thomas, Y. Dutuit, (1997). Analysis of a sequential non coherent and looped system with two approaches: Petri nets and Neural networks. In: proceedings of European Safety and Reliability Conference ESREL' 97, June 17–20, 1997, Lisbon, Portugal, pp.2257–2264.

O.R. Theologou, J.G. Carlier, (1991). " Factoring & reductions for networks with imperfect vertices", IEEE Trans. Reliab., vol.40, June 1991, pp.210–217.

L. Yan, H. Taha, T.L. Landers, (1994). "A recursive approach for enumerating minimal cutsets in a network". IEEE Trans.Reliab., vol.43, September, pp.383–387.

W.-C. Yeh, (2001). "A simple approach to search for all d-MCs of a limited-flow network". Reliability engineering & system safety, vol.44, June 2001, pp.15–19.

Safety and Reliability – Bedford & van Gelder (eds)
© *2003 Swets & Zeitlinger, Lisse, ISBN 90 5809 551 7*

A prospective application of ATHEANA for a research reactor PSA

J.H. Barón, J.E. Núñez Mc Leod & S.S. Rivera
CEDIAC Institute, Cuyo National University, Mendoza, Argentina

ABSTRACT: The human errors possible to occur in a research reactor constitute a significant part of the associated risk of such installations. To include the human errors in a PSA for such a facility is a complicated issue, as there are serious limitations.

In the present work the ATHEANA (A Technique for Human Error ANAlysis) technique has been adapted and used to understand and quantify typical human errors in the PSA of a research reactor, during its design phase. The adaptations were required to include not only the characteristics of these reactors (that make them quite different from NPPs) but also the fact that the case study was performed on a research reactor during its design phase.

For four safety issues, an ATHEANA analysis has been performed, starting from generic procedures, identifying the unsafe actions that can lead to human error events, and establishing the Error Forcing Contexts. The safety issues have been studied and several conclusions drawn, in order to reduce the most significant human error contributors.

1 INTRODUCTION

The new generation of research reactors over the world, provided their public acceptance and the advanced regulatory requirements in different countries, require the performance of a Probabilistic Safety Assessment to demonstrate their safety level. The techniques used to perform these PSAs are those initially developed for Nuclear Power Plants (NPPs) PSAs. These techniques provide tools and models to treat internal events, external events, human errors, common cause failures, and associated studies related to uncertainty and importance calculations.

Despite the fact that the existing techniques are mature in their development, there remain areas where the application is not straightforward for a research reactor, provided that substantial differences exist between the last and the NPPs. Furthermore, the data usually used for the quantitative evaluation of research reactor's PSAs is taken from data bases for NPPs. As many of the individual components are similar in both applications, and they are subjected to similar quality assurance procedures, test and maintenance requirements, this is assumed valid. However, when treating failures related to human errors or to common causes, the extrapolation from NPPs to research reactors is not clearly applicable.

The case of analyzing and including common cause failures was treated for test, maintenance and design common cause failures, and presented in Barón et al. (2002). The case of treating human errors for its inclusion in the research reactor's PSA is the object of the work presented herein.

To include the human errors in a PSA for such a facility is also a complicated issue, as there are serious limitations, namely: (a) most of the Human Reliability Analysis (HRA) models have been developed for Nuclear Power Plants (NPPs); (b) the modern research reactors include many inherently safe design concepts, and therefore the errors of omission are much less important than those errors of commission; (c) there is a virtual lack of any specific data with the sufficient degree of detail to apply the classic HRA approaches; and (d) the operational, emergency, test and maintenance procedures are usually less specific than those existing for NPPs.

The most widely used approach for the inclusion of human errors in the PSAs is that of the Human Reliability Handbook (Swain & Gutmann 1983). The work contained in this handbook is a very good piece of work, and it has proved as a valuable tool for a large number of studies. It has, however, some limitations that arise from the philosophy behind it, where the human errors are considered random failures that depend on a series of performance shaping factors and other influences.

In particular, the treatment of errors of commission is very difficult to treat with Swain's techniques, and these are very important contributors to the overall human-error-related risk of a research reactor. This is so for several peculiarities of research reactors that allow for committing errors:

- the research reactors operate in a wider set of conditions than those of NPPs (i.e. change of "operation mode" is possible)
- many routine operations can be performed inside or near the core of a research reactor during full power operation (i.e. movement of irradiation rigs)
- test and maintenance procedures for research reactors are usually more flexible than those for NPPs, and in some cases written updated procedures do not exist
- as the research reactors are used precisely for research, several innovative actions are usually taken on the operation of the reactor and its related facilities (i.e. irradiation beams, irradiation specimens).

These peculiarities call for a new, different approach. In the present work, the ATHEANA (NUREG 2000) technique was explored, expanded, and applied to various safety issues related to the treatment of human errors for the PSA of a research reactor.

The next section briefly indicates the capacity of ATHEANA and its possible application. Then a section on the analyses performed on four relevant safety issues is included, with the proper discussion of the steps carried out for each case. Finally, some conclusions on the followed approach are indicated.

2 ATHEANA TECHNIQUE

ATHEANA is the result of development efforts sponsored by the US NRC, and it is said to be "a second-generation of human reliability analysis method". Its most recent version was published in May 2000.

The premise of the ATHEANA HRA method is that significant human errors occur as a result of "error-forcing contexts" (EFCs), defined as combinations of plant conditions and other influences that make operator error very likely. ATHEANA is distinctly different from other HRA methods in that it provides structured search schemes for finding such EFCs.

ATHEANA can be applied in two different approaches:

- Retrospective approach (i.e., when analyzing already occurred events)
- Prospective approach (i.e., for HRA purposes)

When used in its prospective approach, ATHEANA pursues the search for EFCs. This search goes well beyond the simple identification of Performance Shaping Factors (PSFs) of previous methods. ATHEANA identifies the (unexpected) plant conditions that, coupled with relevant PSFs, can have significant impact

on human information processing, enabling a wide range of error mechanisms and error types. The result of this change is that quantification becomes more an issue of calculating the likelihood of specific plant conditions, for which unsafe actions are much more likely than would be true under anticipated (expected) conditions.

The ATHEANA prospective search process is indicated in Figure 9.1 from (NUREG 2000). Each step is briefly described herein and discussed according to the applicability of it to the present work.

2.1 Step 1: Define and interpret issue

The objective of the specific analysis should be clearly stated here. The expected output of this step is a succinct definition of the issue to be analyzed, indicating the boundaries for the analysis, its overall goal, and the relationship of the issue to the PSA.

For the research reactor case, the application of this step is straightforward. Several issues related to either maintenance, operating or emergency procedures are independently analyzed. For each issue, a specific definition is performed.

2.2 Step 2: Define scope of analysis

This step limits the scope of the analysis, based either on inherent limitations of the issue, or by setting priorities.

For the research reactor case, the scope of the analysis is indicated for each safety issue.

2.3 Step 3: Describe base case scenarios

The base case scenario is summarized and defined for a chosen initiator. This scenario represents the most realistic description of expected plant and operator behavior for the selected issue, and provides the basis from which to identify and define deviations from such expectations in Step 6.

For the research reactor case, for each procedure (issue) to be analyzed, the information existing in the PSAR and other supporting documents, is used to define the "reference case". This reference case is a detailed description of the several steps to be performed, assuming a proper plant and operators response (i.e., with neither errors nor deviations from expected behavior).

The result is a detailed description of the procedure, with a phenomenological analysis of the "plant" behavior expected during this reference case.

2.4 Step 4: Define HFEs and UAs of concern

This is a very important step in the ATHEANA prospective analysis. The Human Failure Events

(HFEs) and Unsafe Actions (UAs) are identified at this step. This step may be revisited in an iterative process, when HFEs or UAs are identified during the evaluation of Complicating Factors or Recovery Actions. A HFE is a basic event already modeled in the PSA that represents a failure of a function, system, or component that is the result of one or more UAs. An UA is an action inappropriately taken or not taken when needed, by plant personnel that results in a degraded plant safety condition.

For the research reactor case, the definition of the HFEs comes from the proper PSA if they exist, and the UAs are identified for each specific issue. A HFE can be described as either an error of omission (operator fails to actuate system X) or an error of commission (operator inappropriately actuates system X) and, in a complex system, several UAs can be identified as possible causes for a HFE.

In the research reactor case, as the issues under analysis are specific (simple) procedures, a straightforward relationship between HFEs and specific UAs is usually present.

2.5 Step 5: Identify potential vulnerabilities

The search for vulnerabilities in the operator's knowledge base is oriented mainly to identify which biases may influence the operator's understanding of the situation under consideration. This identification should mainly focus on the expected cause-effects response of the system or procedure.

For the research reactor case, this step focuses on the basic information and training that is needed to perform each task.

2.6 Step 6: Search for deviations from base cases

Experience has shown that no serious accidents have occurred for a base case (expected) scenario. On the contrary, only significant deviations from the base case scenario are troublesome for operators.

This step has, therefore, the objective of identifying the deviations from the base case that may result in risk-significant unsafe actions. This is the key step in the whole ATHEANA process.

2.7 Step 7: Identify and evaluate complicating factors

This step expands and further refines the EFC definition that begun in step 6. This step may or may not be needed in a specific issue, depending of the development of the previous step.

The completion of step 7 results in the addition of PSFs and other physical conditions to the descriptions of the deviation scenarios so that the EFC is now considered sufficiently strong to make the likelihood of the HFEs or UAs worth of concern.

2.8 Step 8: Evaluate recovery potential

This step is devoted to the consideration of opportunities to recover from initial errors. This is iterative with the previous steps 6 and 7, and it may direct to issue resolution if the recovery can be assured.

If a recovery action is found to be relevant, it is clearly defined and its requirements (both in personnel and in the necessary time) clearly stated.

2.9 Step 9: Resolve issue (including quantification)

The issue resolution may be qualitative or quantitative, and has to be performed in agreement with the objectives of the PSA. The quantitative evaluation may require the quantification of the probabilities of EFCs, UAs, and of recovery actions.

For the quantitative evaluation of EFCs two components are needed: the probability of certain plant conditions (using classic PSA tools and data) and the likelihood of certain PSFs (obtained from steps 6 and 7). Then the quantitative evaluation of UAs is needed, which can be performed by expert opinion, engineering judgement, or modelled by several methods. The recovery actions can also be quantified with several models.

For the research reactor case, a qualitative approach is performed in the first term. Then, a quantification of the most significant errors is performed, by using classic quantification methods like human error fault trees and standard PSA generic data.

2.10 Step 10: Incorporation into PRA

This step indicates the incorporation of the HFE in the PSA, which can be performed either at fault-tree or at event-tree level.

For the research reactor case, the incorporation of the relevant HFEs is discussed for each safety issue.

3 SAFETY ISSUES ANALYSES

Four safety issues are analyzed, the detailed steps performed according to the ATHEANA methodology are not described, but the most relevant concerns that have been found are indicated, as well as a simplified description of the whole process, in the following sections.

3.1 Primary pump mechanical maintenance

3.1.1 Definition, interpretation, scope and base case scenario of the issue

The objective is the analysis of the mechanical maintenance procedure of a primary pump, focusing on pump disassembly and assembly tasks.

The analysis focuses on the tasks that could produce an initiating event (loss of flow in the primary cooling system due to primary pumps trip). The tasks may imply errors of omission and errors of commission. Several assumptions have been made about success criteria, possible errors with pump identification; other procedures related to the pumps maintenance and test activities were not considered. Other aspects like administrative procedures, cleaning, use of proper solvents and lubricants, and avoidance of disposable materials able to remain unnoticeable in the pump, list of tools needed, list of spare parts, previous and post actions were taken into account.

The base case scenario is the mechanical maintenance of a primary pump during reactor shutdown. In this status the three pumps are offline.

3.1.2 *Definition of important human failure events and unsafe actions*

The HFE is defined as the unsuccessful performance of the primary pump maintenance that could produce a pump trip during normal operation of the reactor. This trip is the initiating event on the primary cooling system loss of flow.

The most relevant UA that can contribute to the identified HFE are:

1. The miss of one step during the assembly procedure. The missed step may be due to an Error of Omission (EOO). For example, a specific step was not made implying that a part removed during the disassembly is not reinstalled.
2. One step is wrongly made during the assembly procedure. This may be due to an Error of Commission (EOC). For example, a worn bearing is not replaced or the lubricant is different to the specifications.
3. A deviation occurs during a verification step, the maintenance personnel notice it but they decide to take no action on it.
4. One verification step is skipped.

3.1.3 *Potential vulnerabilities in the plant personnel knowledge base*

The repairman expectations and training are related to the normal and routine procedure, that has well defined steps. No unpredictable event is expected from this point of view.

The repairman tendency is not to exceed the time frame scheduled for the maintenance tasks. If the repairman has in a high workload within the specified timeframe, he could feel himself forced to work "against the clock". Under these circumstances, he may commit some errors of commission, for example, decide not to replace some part of the internal pump components, judging that its wearing is still acceptable.

3.1.4 *Deviations from the base case scenario*

Three deviations from the base case scenario were identified:

3.1.4.1 Maintenance of standby primary pump when the reactor is in normal operation: The main deviation from the base case scenario is the maintenance of standby primary pump when the reactor is in normal operation. This case is not forbidden, and its realization allows diminishing the workload during reactor outage.

In this new scenario two pumps are running and one is in standby. The main problem arises from the context. In this situation the third pump can not start due to electrical limitations, and a specific interlock prevents this action. Under these conditions, the verification with the pump running is not possible (EFC).

The procedure is not adequate for the new scenario. The procedure does not give any guidance for this scenario (PSF). In this scenario, no indication of failure is available. The repairman guided by another verifications is leading to complacency and the pump verification is not done.

On the other hand, the situation assessment could lead to the repairman to schedule the pump test adequately on the next reactor outage. The response plan will be to do the pump test. However, the complacency is the main error mechanism in this case. Thus the workload and the lack of discipline or trained practice are relevant (PSF).

To evaluate the potential for recovery must be taken into account that the HFE/UA has been performed. In this stage there is time available to situation assessment. If the situation assessment is correct the repairman will schedule the pump verification on the next reactor outage.

3.1.4.2 A deviation on the expected results arises during the verification step: In this scenario the routine maintenance is performed. The reactor is shut down and the three primary pumps are off. During verification steps a deviation arises. This deviation could rise from any verification point (motor/pump elastic coupling alignment, bearing lubrication oil level, rotor assembly turn free and so on).

In this situation (EFC) the repairman could take an UA. The repairman decides that the deviation is within expected conditions and no action is needed.

In this scenario an indication of failure is available. The repairman confirms the deviation but he does not respond properly.

On the other hand the situation assessment could lead the repairman to disassemble the pump and fix the deviation. The workload and the lack of discipline or trained practice are relevant (PSF).

The procedure is not adequate for the new scenario. The procedure does not give any guidance for this scenario (PSF).

To evaluate the potential for recovery must be taken into account that the task supervision becomes the most important issue. The verification on the repairman tasks is considered as a recovery action.

3.1.4.3 The maintenance is performed with a weak administrative control: The routine maintenance is performed without using the written procedures. This behaviour is allowed due to a weak administrative control (EFC). In this situation the repairman could be following a customised procedure and one step could be wrongly made, omitted or modified.

The administrative control is the most relevant issue (PSF).

No recovery action is identified in this scenario.

3.1.5 *Recommendations for these activities*

The most important results that can be derived from the analysis performed by using ATHEANA methodology, is that the most relevant contributors to the human error contribution have been identified, and very simple corrective actions can be taken.

The recommendations from this analysis can be ranked as follows:

- Consistency: the procedures must be specific with the operational modes,
- Completeness: the procedures must include all the related tasks, regular tasks as well as tasks expected when a deviation arises,
- Specificity: the procedures must be specific and not for generic assemblies,
- Training: the training must include deviations and their solutions and the procedures must be written in a checklist style.

Furthermore, the strict administrative control should guarantee that the maintenance tasks are properly performed, and properly supervised.

The implementation of a proper maintenance scheme under strict administrative control is needed, and its impact on the values presented may reduce the likelihood of human errors during maintenance by one or two orders of magnitude.

3.2 *Second shutdown system pneumatic actuators maintenance*

This safety issue is quite similar to the previous one from the analysis point of view. Therefore, a simplified description is indicated.

3.2.1 *Definition, interpretation, scope and base case scenario of the issue*

The objective is the analysis of the maintenance procedure of the Second Shutdown System, focusing on ball valves pneumatic actuators.

The analysis focuses on the tasks that could produce the unavailability of the Second Shutdown System on demand. These tasks may imply errors of omission and errors of commission. The basic scenario is the maintenance of ball valve pneumatic actuators during reactor shutdown. The valve room is accessible only under shutdown conditions.

The base case scenario is, therefore, the performance of the maintenance procedure at the right moment during the operational scheme, by the right personnel, under the right administrative control, and in the prescribed time period.

3.2.2 *Deviations from the base case scenario*

Two deviations from the base case scenario were identified:

3.2.2.1 A deviation on the expected results arises during the verification step: In this scenario the routine maintenance is performed. The reactor is in shut down state. During verification steps a deviation arises. This deviation could rise from any verification point (air tightness, water tightness in dump ball valves or in the test of the dump ball valves with their actuators). Moreover, several actions related with seals, cylinder, pinion greased and so on, are not clearly defined on how they will be tested.

In this situation (EFC) the repairman could take an UA. The repairman decides that the deviation is within expected conditions and no action is needed. On the other hand, the repairman could not detect the deviation and in a similar way than before, no action is taken.

The procedure is not adequate. The procedure does not give any guidance for this scenario (PSF).

In this scenario an indication of failure is available in some cases. The repairman may not notice this indication (Error of Omission) or otherwise he confirms the deviation but he does not respond properly (Error of Commission).

On the other hand the situation assessment could lead the repairman to disassemble the pneumatic actuator and fix the deviation. The workload and the lack of discipline or trained practice are relevant (PSF).

The task supervision becomes the more important issue. The check on the repairman tasks is considered as a recovery action.

3.2.2.2 The maintenance is performed with a weak administrative control: The routine maintenance is performed without using the written procedures. This behavior is allowed due to a weak administrative control. In this situation the repairman could be doing a customized procedure and one step could be wrongly made.

The administrative control is the most relevant issue (PSF).

No recovery action is identified in this scenario.

3.2.3 Recommendations for this activities

The recommendations from this analysis can be ordered as follows:

- Completeness. The procedures must include all the tasks related as regular tasks as well as those to be performed when a deviation arises. This should significantly reduce the conditional probability that a detected deviation is not corrected.
- Unambiguity. The procedures must be specific rather than for generic assemblies. The figures included in the procedures must be specific to the device configuration.
- Training. The training must include deviations and their solutions. This is a cultural approach, complementary of the previous procedural approach.
- Traceability. The procedures must be written with a checklist style. This fact should significantly reduce the likelihood of omission errors.

Furthermore, the need of a strict administrative control should guarantee that the maintenance tasks are properly performed, and properly supervised.

The implementation of a proper maintenance scheme under strict administrative control is needed, and its impact on the values presented may reduce the likelihood of human errors during maintenance by one or two orders of magnitude.

3.3 Rig removal from an irradiation position

3.3.1 Definition, interpretation, scope and base case scenario of the issue

The objective is the analysis of the task sequence for a Rig Removal from its irradiation position, its transfer to the Service Pool and its subsequent transfer to the Transfer Hot Cell.

The analysis focuses on the task that could produce incidents or accidents with the rigs handling. These tasks may imply errors of omission and errors of commission.

The overall procedure and its tools are in a conceptual stage of development.

The basic scenario is the successful transfer in three stages of one irradiated rig from its irradiation position in its irradiation position to the Transfer Hot Cell. This procedure is performed with the reactor at full power. The transfer tool is assumed to work properly.

In this base case scenario, the operators follow exactly the procedure, without any deviation, and the plant does not present any deviation from the expected behavior. This means, for example, that all the verification steps give the right indications and no further actions are needed.

3.3.2 Deviations from the base case scenario

The base case scenario is highly relevant in this analysis, where PSFs related to familiarity and complacency becomes important. Since the tasks related to the transfer of irradiated rigs will be performed with a high frequency and are very repetitive, the likelihood of developing a customized (simplified) procedure is assumed to be high.

Two deviations from the base case scenario have been identified:

3.3.2.1 The procedure is performed with a loss of attention level:
In this scenario the routine rig removal is performed. The reactor is operating at full power.

The operator attention level is diminished by its familiarity with the procedure. When anxiety arises complacency is the error mechanism to right assessment. In this scenario the operator could remove the incorrect rig from the irradiation tubes.

The operator has no point of control to verify that he has removed the desired rig.

After putting the rig in the reactor pool rack he confirms the end of the operation.

From this point on, the rig is handled as other rig (the right rig). This error will be noticed in the Transfer Hot Cell. No accident is expected as a consequence of this error.

Workload is the main PSF.

The error may be noticed in the Transfer Hot Cells or when another rig is removed. In both cases no accidental consequences are expected

One additional control point is needed. This control point must be related to the right identification of the rig. The operator must be confirming (visually or by other means) the rig identification to take an action that corrects the error.

3.3.2.2 The procedure is performed with a weak administrative control:
The routine procedure is performed without following the written procedures. This behavior is allowed due to a weak administrative control. In this situation the operator could be doing a customized procedure and one step could be wrongly made or skipped.

In this case it is conservatively assumed that the operator selects the erroneous rig from the irradiation tubes. After the operator selects erroneously a rig to remove, he may transfer the rig immediately from the reactor pool to the service pool rack. With this action none radiological incident is expected due to the water shield.

In this situation the operator works without written procedures. The operator could customize the procedure.

The workload is a PSF to consider. But the administrative control is the most relevant issue (PSF).

One additional control point is needed. This control point must be related with the right identification of the rig. The operator must be confirming (visually or by other means) the rig identification to take an action that corrects the error.

It is assumed that the radioactivity is not adequate, gamma sensors detect this and the rig is returned to Service Pool automatically.

No radiological consequences are expected.

3.3.3 *Recommendations for this activities*

The recommendations from this analysis can be ordered as follows:

– Unambiguity. The rigs should be identified without ambiguity for their removal process. This may be made of several forms.
– Training. The training must highlight the procedure control points. More efforts must be done in this sense, incorporating additional control points.

Furthermore, the need of a strict administrative control should guarantee that the transfer tasks are properly performed, and properly supervised.

The implementation of a proper transfer scheme under strict administrative control is needed, and its impact on the values presented may reduce the likelihood of human errors during transfer tasks by one or two orders of magnitude.

3.4 *EOP, presence of water in an irradiation beam*

3.4.1 *Definition, interpretation, scope and base case scenario of the issue*

The objective is the analysis of the Emergency Operating Procedure (EOP) for the presence of water in an Irradiation Beam, focusing on the tasks to be performed after an indication of water presence exists.

The analysis focuses on the tasks that could be done erroneously during the EOP application (in response to the presence of water in an irradiation beam). The tasks may imply errors of omission and errors of commission.

The base case scenario starts with the presence of one or more leakage alarms with the reactor in normal operation. The base case scenario implies the performance of all the steps of the EOP. The EOP instructions must be followed and executed with priority, and requires some actions from the operators to be taken immediately.

The EOP requires the shutdown of the reactor within of determined time from the moment the water is detected in the irradiation beam. This means that the operators must act in a time-constrained scheme. The EOP tasks are assumed to be performed by a trained operator and supervised by the Operation Manager.

The base case scenario assumes that the EOP is successfully performed without any deviation, error or mistake. This means, for example, that during the performance of the verification steps, no wrong indications are found that may require further action.

In this base case scenario, the operating personnel follows exactly the procedure, without any deviation, and the plant does not present any deviation also (other than the presence of water in the irradiation beam).

Control points of the successful EOP were identified.

3.4.2 *Deviations from the base case scenario*

The analysis is based on the normal reactor operation, when one or two leakage alarms are activated. For the particular situation of an emergency occurring, all cases are deviations from the base case scenario in this analysis.

3.4.2.1 A high level alarm in one of the liquid leakage collector vessels in the neutron guide bellows: In this scenario the abnormal procedure is performed. The reactor is operating. The source and type of failure has not been identified. Multiple lines of reasoning are present related to the consequences. The main error mechanisms expected are simplification, polarization of thinking, anxiety and stress. The error types associated with these error mechanisms are lack of, or reduced, attention paid to other parameters and their changes, application of incorrect procedure step or no response and taking an inappropriate action or failing to take a needed action in time.

Two possible errors were considered: the operator closes the isolation manual valves of Helium Cooling System when this is not needed and the operator performs the erroneous procedure. The second error embraces both.

Training is the main PSF. The operator lack of training or practice for off-normal emergency conditions.

Additional consequences are not expected.

3.4.2.2 A high level alarm in one of the liquid leakage collector vessels in the neutron guide: In this scenario the abnormal procedure is performed. The reactor is operating. The source and type of failure has not been identified. Multiple lines of reasoning are present related to the consequences. The main error mechanisms expected are simplification, polarization of thinking, anxiety and stress. The error types associated with these error mechanisms are lack of, or reduced, attention paid to other parameters and their changes, application of incorrect procedure step or no response and taking an inappropriate action or failing to take a needed action in time.

Two possible errors were considered: the operator does not close the shutter and the operator closes the erroneous isolation manual valves of Helium Cooling System.

Training is the main PSF. The operator lack of training or practice for off-normal emergency conditions. The error related to the shutter may not be noticed. The shutter may be operable after a long incident time.

The error with isolation manual valves may be noticed with dew point alarm in another neutron beam tube.

Additional consequences are not expected.

3.4.2.3 Two high level alarms of a neutron guide: This case is considered similar to the previous case.

3.4.3 *Recommendations for this activities*

The recommendations from this analysis can be ordered as follows:

- Training. The training must highlight the procedure control points. More efforts must be done in this sense, incorporating additional control points.
- Correctness. The EOP must include check statements about previous actions. The EOP must clarify the context for their use, considering that similar procedures exist.

Furthermore, the need of a strict administrative control should guarantee that the EOPs are properly performed, and properly supervised.

The implementation of a proper training scheme under strict administrative control is needed, and its impact on the values presented may reduce the likelihood of human errors during maintenance by one or two orders of magnitude.

4 CONCLUSIONS

The problem of human errors associated with the PSA of a research reactor during its design phase, has been addressed using ATHEANA technique.

The ATHEANA technique has been adapted in a prospective approach, in order to understand the nature and implications of the possible human errors of commission and omission to four representative "safety issues". These include a contributor to an initiating event, a contributor to a safety system unavailability, an event during in-pool operations at full power, and a contributor to errors during the performance of an Emergency Operating Procedure.

The technique has proved to be useful in understanding the safety issues, and in identifying the most relevant basic human errors and conditions that may lead to unsafe actions.

Several specific conclusions on the needs for procedures adequacy, supervision, training, administrative control, etc, have been identified.

It is concluded that the ATHEANA approach is very useful, and it is possible to be applied to a research reactor PSA, even during its design stage. This application has the advantage that the lessons learned can be retrofitted to the designers of the plant, and to the designers of the maintenance, test, and emergency procedures.

More work is needed to fully exercise the potential of the technique, mainly in the quantitative aspect of it.

REFERENCES

Barón, J., Núñez Mc Leod, J. & Rivera, S. 2002. Human Reliability Analysis to consider Common Cause Failures for a Research Reactor during its Design Phase. In Bonano et al (eds), *Probabilistic Safety Assesment & Management PSAM6. Proc. intern. conf., San Juan, Puerto Rico, 23–28 June 2002.* Amsterdam: Elsevier.

NUREG-1624, Rev.1, Technical Basis and Implementation Guidelines for A Technique for Human Event Analysis (ATHEANA), NRC, USA, 2000.

Swain, A. & Guttmann, H. 1983. Handbook on Human Realiability Analysis with Emphasis on Nuclear Power Plant Applications, NUREG/CR-1278, *Nuclear Regulatory Commission*, USA.

Safety and Reliability – Bedford & van Gelder (eds)
© 2003 Swets & Zeitlinger, Lisse, ISBN 90 5809 551 7

On the hazard rate process for imperfectly monitored multi-unit systems

A. Barros, C. Bérenguer & A. Grall
University of Technology of Troyes, Troyes, France

ABSTRACT: The aim of this communication is to present a stochastic model to characterize the failure distribution of multi-unit systems when the current units state is imperfectly monitored. The definition of the hazard rate process given by (Arjas 1981a) with perfect monitoring is extended to the case where the units failure time are not always detected (non-detection events). The so defined observed hazard rate process is calculated on a practical industrial system (two parallel units). Its monotony property is studied when the monitoring quality deteriorates (increase of the non-detection probabilities).

1 INTRODUCTION

The classical failure rate (or hazard rate) is often used in maintenance and reliability studies to characterize the failure distribution of a single-unit system. It corresponds to the probability per unit of time that a system fails at a time t knowing that it has not failed until t. Many reliability measures and maintenance optimization schemes are based on the evolution of this probability (Barlow and Proschan 1996). For multiunit systems, the classical failure rate is also often used to calculate reliability quantities but its connection with the current stochastic behavior of the system can be discussed: the current state of each unit is not considered and the classical failure rate is an approximation of the actual system failure probability per unit of time. That is why a hazard rate process is defined in (Arjas 1981a) as a stochastic extension of the failure rate notion in order to take into account the current units states. For each combination of the units states (which are supposed to be perfectly known) the hazard rate process gives the actual system failure probability per unit of time without any expectations. Contrary to the classical failure rate which is a deterministic function of the time, the hazard process is a stochastic process whose paths depend on the units states. It is closely linked to the stochastic behavior of the multi-unit system and as a result it can be shown that maintenance optimization schemes based on this hazard process allow important savings in comparison with the ones based on the classical system failure rate (Heinrich and Jensen 1992).

In this paper we propose to extend the hazard rate process notion for multi-unit systems to the case where only imperfect information about the units state is available. From a practical point of view the assumption of perfect units information used in (Arjas 1981a) to define the hazard process is unrealistic. In many situations, some units are often impossible or two expensive to monitor continuously and perfectly. Here, the usual case of non detection events is considered (Simola and Pulkkinen 1998): the units are not directly observable and their state must be inferred from a monitoring system which can fail itself. A failure of the unit i monitoring system (probability p_i) implies a non detection of the unit i failure (with the same probability p_i). An observed hazard rate process is defined taking into account these non detection events.

The second part of this paper is devoted to the description of a particular system S (two dependent parallel units) for which the classical failure rate, the hazard rate process and the observed hazard rate process will be successively calculated. It gives a realistic illustration of the three rates and allows comparisons. The perfect and imperfect monitoring cases are also defined in this part. In the third part, the definitions of classical failure rate and hazard rate process are given with applications on system S. In the forth part, the extension of hazard rate process to observed hazard rate process is presented. In the last part the evolution of the observed hazard rate process is studied as a function of the monitoring quality (i.e the non-detection probabilities p_i) in the case of system S. Conditions on the non-detection probabilities are identified to specify when the monotony property of the hazard rate process calculated with perfect information still holds as the monitoring deteriorates.

2 SYSTEM AND MONITORING

2.1 System S

Results presented below are applied on a particular system (called hereafter system S) to give relevant illustrations of the so called classical failure rate, hazard rate process, and observed hazard rate process. We consider the high voltage system presented in (Pham 1992) consisting of a power supply and two pieces of rectifier equipment for converting AC power to DC power. The power plant should provide constant DC, current throughout the mission. When both rectifiers are operating each of them provides one part of the desired current. If one rectifiers fails, the other can tune the entire range with a resultant change in the expected time to failure.

This system is modelled by a two-units parallel system (Figure 1) with random lifetimes T_1 for unit 1 (rectifier 1) and T_2 for unit 2 (rectifier 2). When both units are running, their respective failure rate are supposed to be constant and are noted λ_1 and λ_2 (exponential lifetime laws). When unit i fails, the extra stress is placed on the surviving one whose failure rate is switched to a superior one $\bar{\lambda}_j$ ($\lambda_j < \bar{\lambda}_j$). To model common cause failure (external traumatic events), external s-independent shocks are added which occur at rate λ_{12} (random occurrence time Z) and make the whole system fail. Consequently, if the natural lifetimes of units 1 and 2 are respectively T_1 and T_2, their respective real lifetimes are $\min\{T_1, Z\}$ and $\min\{T_2, Z\}$. The total system lifetime is noted T.

2.2 Perfect monitoring

In the following, a n-unit system is perfectly monitored when the failure times of all the units $T_{i,1 \leq i \leq n}$ are exactly detected at their occurrence time, without any delay or any measurement error. Hence no degradation level is measured, and the state of each unit is

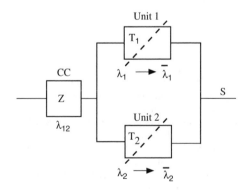

Figure 1. System S model – CC: common cause failure – S: system output.

reduced to failed or running. A failure of the whole system is also detected perfectly at its occurrence time.

2.3 Imperfect monitoring

The imperfect monitoring studied here corresponds to non-detection events on the units failure: a unit i failure is detected with a probability $1 - p_i$ and never detected with a probability p_i. The actual units failure times $T_{i,1 \leq i \leq n}$ are not available and must be inferred from the observed ones $T^o_{i,1 \leq i \leq n}$ with the following equations:

$$P(T^o_i = T_i) = 1 - p_i \quad \text{and} \quad P(T^o_i = \infty) = p_i. \quad (1)$$

The failure times of the whole system T and the common cause failure Z are always perfectly detected. This case of imperfect monitoring corresponds to many realistic situations: as soon as a unit i (a rectifier for system S) is not directly observable, a monitoring device must be used (detectors i) and the possible failures of this device must be also taken into account (failure probability p_i). In the same time, the failure of the whole system can remain obviously detectable (for example power shortage).

3 HAZARD RATE PROCESS

3.1 Classical failure rate

The classical failure rate (or classical hazard rate) Cocozza-Thivent 1997) is defined as follows:

$$\lambda(t) = \lim_{h \to 0} \frac{1}{h} P(T \in [t, t + h] | T > t) \quad (2)$$

where T is the lifetime of the whole system. The function $\lambda(t)$ is deterministic. This quantity is relevant to study single-unit systems: the current information on the system lifetime T, which is taken into account in the conditional part, corresponds to the current state of the single unit. Hence the function $\lambda(t)$ is closely linked to the stochastic behavior of the system and is exactly equal to the system failure probability per unit of time. On the other hand, when considering multi-unit systems, the relevance of such a hazard rate is less obvious: current information on the whole system lifetime T does not reveal the current state of each unit. Taking for example system S, the information $T > t$ can correspond to ($T_1 > t$, $T_2 > t$, $Z > t$), ($T_1 < t$, $T_2 > t$, $Z > t$) or ($T_1 > t$, $T_2 < t$, $Z > t$). Consequently, the classical failure rate is calculated by expectations on all the unknown states, and $\lambda(t)$ is an approximation of the actual system failure probability per unit of time $\lambda_a(t)$.

Let us illustrate the gap between the classical failure rate $\lambda(t)$ and the actual failure rate $\lambda_a(t)$ in case of system S. $\lambda_a(t)$ can be of three different forms:

- When $T_1 > t$, $T_2 > t$, $Z > t$, the system can only fail because of shocks and the actual system failure rate corresponds to the occurrence rate of shocks. Hence $\lambda_a(t) = \lambda_{12}$.
- When $T_1 < t$, $T_2 > t$, $Z > t$, the system fails when unit 2 fails or when a shock occurs. Since these events are independent, the actual failure rate is equal to $\bar{\lambda}_2 + \lambda_{12}$.
- When $T_1 < t$, $T_2 > t$, $Z > t$, the system fails when unit 1 fails or when a shock occurs. Since these events are independent $\lambda_a(t) = \bar{\lambda}_1 + \lambda_{12}$.
- In other cases the system has failed before t.

By comparison, we get for the classical failure rate $\lambda(t)$ calculated with information on $\min\{T, Z\}$ only:

$$\lambda(t) = \frac{N}{D} \tag{3}$$

with

$$N = \lambda_2 L_2(\bar{\lambda}_1 + \lambda_{12}) + \lambda_1 L_1(\bar{\lambda}_2 + \lambda_{12})e^{-(\bar{\lambda}_2 - \bar{\lambda}_1)t}$$

$$- \lambda(\lambda_1\bar{\lambda}_2 + \lambda_2\bar{\lambda}_1 - \bar{\lambda}_1\bar{\lambda}_2)e^{-L_1 t},$$

$$D = \lambda_2 L_2 + \lambda_1 L_1 e^{-(\bar{\lambda}_2 - \bar{\lambda}_1)t} - (\lambda_1\bar{\lambda}_2 + \lambda_2\bar{\lambda}_1 - \bar{\lambda}_1\bar{\lambda}_2)e^{-L_1 t},$$

$$L_i = \lambda_1 + \lambda_2 - \bar{\lambda}_i, i = 1, 2 \quad \text{and} \quad \lambda = \lambda_1 + \lambda_2 + \lambda_{12}.$$

Figure 2 sketches an example of $\lambda_a(t)$ and $\lambda(t)$ for system S.

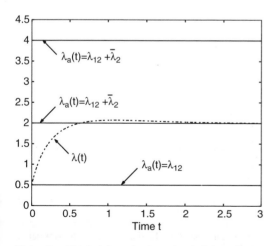

Figure 2. Classical hazard rate and actual hazard rate: $\lambda_{12} = 0.5$, $\lambda_1 = 1$, $\lambda_2 = 3$, $\bar{\lambda}_1 = 1.5$, $\bar{\lambda}_2 = 3.5$.

3.2 Hazard rate process

The hazard rate process $\Lambda(t)$ is defined by Arjas in (Arjas 1981a) to give a mathematical framework for the so called actual failure rate $\lambda_a(t)$ described in Paragraph 3.1. The monitoring is considered to be perfect (see Paragraph 2.2), so all the units failure times are always perfectly known until the current time t. For a n-unit system, this perfect monitoring information is modelled by the filtration $\mathcal{F}_t$ (Barros et al. 2002) generated by the stochastic processes $I_{(T \leq t)}$, $I_{(Z \leq t)}$, $I_{(T_i \leq t, 1 \leq i \leq n)}$. It corresponds to the exact history of the system until time t, i.e. it indicates at each time which units have failed beforehand (with the exact failure dates) and if the whole system is running. $\mathcal{F}_t$ is added in the conditional part of the hazard rate given by Equation 2. Hence a new function $\Lambda(t)$ is defined which is no more a deterministic function but a stochastic process and which is called here hazard rate process:

$$\Lambda(t) = \lim_{h \to 0} \frac{1}{h} P(T \in [t, t+h]|T > t \cap \mathcal{F}_t). \tag{4}$$

In the case of system S, $\mathcal{F}_t$ is a filtration generated by $I_{(T_1 \leq t)}$, $I_{(T_2 \leq t)}$, $I_{(Z \leq t)}$ and $I_{(T \leq t)}$. It indicates on $[0, t]$ if one of the unit has failed ($I_{(T_1 \leq t)} = 1$, or $I_{(T_2 \leq t)} = 1$) or if the whole system has failed ($I_{(Z \leq t)} = 1$, or $I_{(T \leq t)} = 1$). Hence the hazard rate process can be rewritten as follows:

$$\Lambda(t) = \Lambda_1(t)I_{(X>t,T_2>t,Z>t,T>t)}$$

$$+ \Lambda_2(t)I_{(X<t,T_2>t,Z>t,T>t)} \tag{5}$$

$$+ \Lambda_3(t)I_{(X>t,T_2<t,Z>t,T>t)}$$

with

$$\Lambda_1(t) = \lim_{h \to 0} \frac{P\{T \in [t, t+h]|T > t, X > t, T_2 > t, Z > t\}}{h},$$

$$\Lambda_2(t) = \lim_{h \to 0} \frac{P\{T \in [t, t+h]|T > t, X < t, T_2 > t, Z > t\}}{h},$$

$$\Lambda_3(t) = \lim_{h \to 0} \frac{P\{T \in [t, t+h]|T > t, X > t, T_2 < t, Z > t\}}{h}.$$

The cases $I_{(T_1 \leq t, T_2 \leq t, Z > t)} = 1$ and $I_{(Z \leq t)} = 1$ are not considered because there are mutually exclusive with $T > t$. Calculating the limits, we obtain for $\Lambda(t)$:

$$\Lambda(t) = \lambda_{12}I_{(X>t,T_2>t,Z>t,T>t)}$$

$$+ (\bar{\lambda}_2 + \lambda_{12})I_{(X<t,T_2>t,Z>t,T>t)} \tag{6}$$

$$+ (\bar{\lambda}_1 + \lambda_{12})I_{(X>t,T_2<t,Z>t,T>t)}.$$

101

Contrary to the classical failure rate, $\Lambda(t)$ is a stochastic process whose paths depend on the realizations of the units lifetime. For each of these paths we obtain the actual failure rate $\lambda_a(t)$ described in Paragraph 3.1. Hence the hazard rate process is exactly connected to the stochastic behavior of the system and can be very efficient in the framework of maintenance optimization (Barros et al. 2003). However the assumption of perfect monitoring can be unrealistic from a practical point of view: some units are often impossible or two expensive to monitor perfectly. Here we want to take into account imperfect monitoring in the hazard rate process calculations. The aim is to define, in case of imperfect monitoring, an "observed hazard rate process" $\Lambda^o(t)$ which is obviously different from $\Lambda(t)$, but which is more connected to the stochastic behavior of the system than the classical failure rate.

4 OBSERVED HAZARD RATE PROCESS $\Lambda^o(t)$

4.1 Definition

The definition of the hazard rate process given by Equation 4 is extended to the imperfect monitoring case by replacing the perfect information on all the units $\mathcal{F}_t$ by the imperfect information $\mathcal{G}_t^o$. For any n-unit system, $\mathcal{G}_t^o$ is a filtration generated by the stochastic processes $I_{(T^o \leqslant t)}, I_{(Z^o \leqslant t)}, I_{(T_i^o \leqslant t, 1 \leqslant i \leqslant n)}$. The variable T^o, Z^o, $T_{i,1 \leqslant i \leqslant n}^o$ are the observed system lifetime, the observed common cause failure time, the observed units lifetime, given by the monitoring device. $\mathcal{G}_t^o$ corresponds to the observed history of the system until time t and indicates on $[0, t]$ if a unit i failure has been observed ($I_{(T_i^o \leqslant t, 1 \leqslant i \leqslant n)} = 1$)or if a system failure has been observed ($I_{(Z^o \leqslant t)} = 1$, or $I_{(T^o \leqslant t)} = 1$). Hence an "observed hazard rate" can be defined as follows:

$$\Lambda^o(t) = \lim_{h \to 0} \frac{1}{h} P(T \in [t, t+h]|T > t \cap \mathcal{G}_t^o). \tag{7}$$

4.2 $\Lambda^o(t)$ for system S

In the case of system S with imperfect monitoring described in Paragraph 2.3, we have observed unit lifetimes T_1^o and T_2^o with $P(T_1^o = T_1) = 1 - p_1$ and $P(T_2^o = T_2) = 1 - p_2$. The variable T and Z are perfectly known. The observed hazard rate process is calculated with this imperfect information, i.e. the filtration added in its conditional part is generated by ($I_{(T_1^o \leqslant t)}, I_{(T_2^o \leqslant t)}, I_{(Z \leqslant t)}$ and $I_{(T \leqslant t)}$. Hence $\Lambda^o(t)$ can be rewritten as follows:

$$\Lambda^o(t) = \Lambda_1^o(t) I_{(T>t, T_1^o>t, T_2^o>t, Z>t)}$$
$$+ \Lambda_2^o(t) I_{(T>t, T_1^o<t, T_2^o>t, Z>t)} \tag{8}$$
$$+ \Lambda_3^o(t) I_{(T>t, T_1^o>t, T_2^o<t, Z>t)}$$

with

$$\Lambda_1^o(t) = \lim_{h \to 0} \frac{P(T \in [t, t+h]|T_1^o>t, T_2^o>t, Z>t, T>t)}{h},$$

$$\Lambda_2^o(t) = \lim_{h \to 0} \frac{P(T \in [t, t+h]|T_1^o<t, T_2^o>t, Z>t, T>t)}{h},$$

$$\Lambda_3^o(t) = \lim_{h \to 0} \frac{P(T \in [t, t+h]|T_1^o>t, T_2^o<t, Z>t, T>t)}{h}.$$

The $\Lambda_i^o(t)$ are the observed hazard rates corresponding to the three possible running states of the system at t (not mutually exclusive with $T > t$). They are calculated as follows:

- When the monitoring indicates ($T_1^o > t$, $T_2^o > t$, $Z > t$, $T > t$), because of non-detection events, the actual system state can be either ($T_1 > t$, $T_2 > t$, $Z > t$), ($T_1 \leqslant t$, $T_2 > t$, $Z > t$) or ($T_1 > t$, $T_2 \leqslant t$, $Z > t$). All the possible paths leading to a system failure on $[t, t + h]$ must be considered by expectation. The result is presented here for $L_1 \neq 0$ and $L_2 \neq 0$ where $L_i = \lambda_1 + \lambda_2 - \bar{\lambda}_i$ for $i = 1, 2$. Similar calculations can be led for the complementary cases. We get:

$$\Lambda_1^o(t) = \frac{N_1}{D_1} \tag{9}$$

with

$$N_1 = e^{-(\lambda_1+\lambda_2)t}(\lambda_{12} - p_2\lambda_2 \frac{\lambda_{12} + \bar{\lambda}_1}{L_1} - p_1\lambda_1 \frac{\lambda_{12} + \bar{\lambda}_2}{L_2}),$$

$$+ \frac{p_2\lambda_2}{L_1}e^{-\bar{\lambda}_1 t(\lambda_{12} + \bar{\lambda}_1)} + \frac{p_1\lambda_1}{L_2}e^{-\bar{\lambda}_2 t(\lambda_{12} + \bar{\lambda}_2)},$$

$$D_1 = e^{-(\lambda_1+\lambda_2)t}(1 - \frac{p_1\lambda_1}{L_2} - \frac{p_2\lambda_2}{L_1}) + \sum_{i \neq j \in (1,2)} \frac{p_i\lambda_i}{L_j}e^{-\bar{\lambda}_j t}.$$

- When the monitoring indicates ($T_1^o > t$, $T_2^o \leqslant t$, $Z > t$, $T > t$), the failure of unit 2 has been detected ($T_2^o \leqslant t$) and the system is still running since $T > t$. The "true" information on both units can then be obviously inferred to be $T_1 > t$, $T_2 \leqslant t$, and we get:

$$\Lambda_2^o(t) = \lambda_{12} + \bar{\lambda}_1. \tag{10}$$

- When the monitoring indicates ($T_1^o \leqslant t$, $T_2^o > t$, $Z > t$, $T > t$), $\Lambda_3^o(t)$ can be calculated like $\Lambda_2^o(t)$ and we get:

$$\Lambda_3^o(t) = \lambda_{12} + \bar{\lambda}_2. \tag{11}$$

Let us note that the classical failure rate $\lambda(t)$ calculated in Section 3.1 with only information on Z and T is a particular case of $\Lambda^o(t)$ when $p_1 = p_2 = 1$, and the

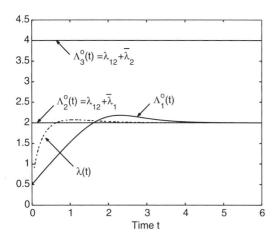

Figure 3. $\lambda(t)$ compared to $\Lambda^o(t)$ with $\lambda_{12} = 0.5$, $\lambda_1 = 1$, $\lambda_2 = 3$, $\bar{\lambda}_1 = 1.5$, $\bar{\lambda}_2 = 3.5$, $p_1 = 0.3$, $p_2 = 0.01$.

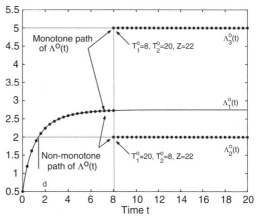

Figure 4. $\Lambda^o(t)$ paths with $\lambda_{12} = 0.5$, $\lambda_1 = 1$, $\lambda_2 = 3$, $\bar{\lambda}_1 = 1.5$, $\bar{\lambda}_2 = 4.5$, $p_1 = 0.5$, $p_2 = 0$.

hazard rate process $\Lambda(t)$ is a particular case of $\Lambda^o(t)$ when $p_1 = p_2 = 0$.

Figure 3 sketches an example of the functions $\Lambda_1^o(t)$, $\Lambda_2^o(t)$, $\Lambda_3^o(t)$ for system S. At the beginning of a cycle, the system is new and the on-line information indicates: $T_1^o > t$, $T_2^o > t$, $Z > t$, $T > t$. Hence $\Lambda^o(t)$ is equal to $\Lambda_1^o(t)$. Then if unit 1 (resp. unit 2) failure is detected, $\Lambda^o(t)$ "jumps" from $\Lambda_1^o(t)$ to $\Lambda_3^o(t)$ (resp. $\Lambda_2^o(t)$). If no unit failure is detected e $\Lambda^o(t)$ is equal to $\Lambda_1^o(t)$ until a shock or the system failure occurs. Hence, as soon as one unit failure is detected, the paths of $\Lambda^o(t)$ and $\Lambda(t)$ merge together. If it is not, $\Lambda^o(t)$ is an approximation of the actual failure rate.

In the following section, we further investigate the impact of the monitoring quality on the observed hazard rate process. We look particularly at the monotony property of $\Lambda^o(t)$. This property is actually of great interest to characterize the failure distribution of the system. An important result should be to know for any system structure, if the monotony of the hazard rate process with perfect monitoring still holds with imperfect monitoring. Hence, even if $\Lambda^o(t)$ does not correspond at any time with $\Lambda(t)$, it can be a correct representation of the system failure distribution. At this moment, we do not have general results for any system structure. The study presented here concerns only system S, but it gives a good illustration of monotony problems and allows to present future work.

5 MONOTONY OF $\Lambda^o(t)$

The paths of $\Lambda(t)$ can be described as random jumps from $\Lambda_1(t) = \lambda_{12}$ to $\Lambda_2(t) = \lambda_{12} + \bar{\lambda}_1$ or to $\Lambda_3(t) = \lambda_{12} + \bar{\lambda}_2$. Hence $\Lambda(t)$ for system S has only monotone

increasing paths. So we look for the same property of the observed hazard rate process paths.

5.1 Increasing paths of $\Lambda^o(t)$

The identification of monotone increasing paths is led through the study of $\Lambda^o(t)$ paths. From Equation (8) these paths depend on $\Lambda_1^o(t)$, $\Lambda_2^o(t)$, $\Lambda_3^o(t)$. The paths of $\Lambda^o(t)$ have been described in Paragraph 4.2 as random jumps from $\Lambda_1^o(t)$ to $\Lambda_2^o(t)$ or $\Lambda_3^o(t)$.

Examples of such paths for $\Lambda^o(t)$ are sketched on Figure 4. As a consequence the observed hazard rate process has always monotone increasing paths if the functions $\Lambda_1^o(t)$, $\Lambda_2^o(t)$ and $\Lambda_3^o(t)$ are non-decreasing in t and if the values $\Lambda_2^o(t)$ and $\Lambda_3^o(t)$ are greater than $\Lambda_1^o(t)$ at any possible jump time i.e. at any $t > 0$. Hereafter conditions are first identified for which $\Lambda_1^o(t)$, $\Lambda_2^o(t)$ and $\Lambda_3^o(t)$ are non-decreasing. Only the monotony of $\Lambda_1^o(t)$ is studied since the other functions are constant. Then the value of $\Lambda_1^o(t)$ is compared to $\Lambda_2^o(t)$ and $\Lambda_3^o(t)$ for $t > 0$.

5.1.1 Variations of the function $\Lambda_1^o(t)$

Table 1 sketches the conditions for the deterministic function $\Lambda_1^o(t)$ to be increasing in time. When one of the p_i is zero, $\Lambda_1^o(t)$ is increasing for any value (i.e. unconditionally) of the parameters p_i, λ_i, $\bar{\lambda}_i$ and λ_{12}. When none of the p_i is zero, the monotony of $\Lambda_1^o(t)$ can be conditioned by the units failure rate only or by the units failure rate and the monitoring quality ($p_i \leqslant C_i$). In this last case, when the monitoring quality of one unit i deteriorates (i.e. p_i overpasses $C_i = (\lambda_1 + \lambda_2 - \bar{\lambda}_j)\bar{\lambda}_i/\lambda_i(\bar{\lambda}_i - \bar{\lambda}_j)$, $i \neq j = 1, 2$), the function $\Lambda_1^o(t)$ is no more increasing in time, if unit i is the unit with the lower failure rate under stress condition ($\bar{\lambda}_i < \bar{\lambda}_j$).

Table 1. Conditions for $\Lambda_1^o(t)$ monotone increasing.

$\Lambda_1^o(t)$	$p_1 = 0$	$p_1 \neq 0$
$p_2 = 0$	No condition	No condition
$p_2 \neq 0$	No condition	If $\quad \bullet \lambda_1 + \lambda_2 < \bar{\lambda}_1$ and $\lambda_1 + \lambda_2 < \bar{\lambda}_2$ Or if $\quad \bullet \bar{\lambda}_i < \lambda_1 + \lambda_2 < \bar{\lambda}_j$ and $p_i < C_i$

Table 2. Conditions for $\Lambda_1^o(t) < \Lambda_{k,k=2,3}^o(t)$.

$\Lambda_1^o(t)$	$p_1 = 0$	$p_1 \neq 0$
$p_2 = 0$	No condition	If $\quad \bullet \bar{\lambda}_2 < \bar{\lambda}_1$ Or if $\quad \bullet \bar{\lambda}_2 < \bar{\lambda}_1$ and $p_1 \leqslant C_1$
$p_2 \neq 0$	If $\quad \bullet \bar{\lambda}_1 < \bar{\lambda}_2$ Or if $\quad \bullet \bar{\lambda}_1 < \bar{\lambda}_2$ and $p_2 \leqslant C_2$	Same conditions as for $\Lambda_1^o(t)$ monotone increasing

Table 3. Conditions for $\Lambda^o(t)$ increasing paths.

$\Lambda^o(t)$	$p_1 = 0$	$p_1 \neq 0$
$p_2 = 0$	No condition	$\Lambda_1^o(t) < \Lambda_2^o(t)$ and $\Lambda_1^o(t) < \Lambda_3^o(t)$
$p_2 \neq 0$	$\Lambda_1^o(t) < \Lambda_2^o(t)$ and $\Lambda_1^o(t) < \Lambda_3^o(t)$	$\Lambda_1^o(t)$ monotone increasing

5.1.2 Comparison of $\Lambda_1^o(t)$ with $\Lambda_2^o(t)$ and $\Lambda_3^o(t)$

Table 2 sketches the conditions for the deterministic function $\Lambda_1^o(t)$ to remain below $\Lambda_2^o(t)$ and $\Lambda_3^o(t)$. When both p_i are zero, $\Lambda_1^o(t) < \Lambda_{k,k=2,3}^o(t)$ for any value (i.e. unconditionally) of the parameters. When one of the p_i is zero, conditions on the units failure rates $\bar{\lambda}_2$ and $\bar{\lambda}_1$ are identified. The probability p_j which is non-zero has an impact only if $\bar{\lambda}_i > \bar{\lambda}_j$. When none of the p_i is zero, we have $\Lambda_1^o(t) < \Lambda_{k,k=2,3}^o(t)$ and $\Lambda_1^o(t)$ increasing under the same conditions.

5.1.3 Increasing paths of $\Lambda^o(t)$

Table 3 sketches the conditions for the stochastic process $\Lambda^o(t)$ to always have increasing paths. It is obtained by taking the intersection of Table 1 and Table 2. Figure 4 sketches an example of a monotone increasing path and a non-monotone path for $\Lambda^o(t)$ when one of the p_i is zero. In this case $p_2 = 0, p_1 > C_1$ and $\bar{\lambda}_2 > \bar{\lambda}_1$. Hence the condition $\Lambda_1^o(t) < \Lambda_2^o(t)$ is not verified and non-monotone increasing paths can be observed when $T_2^o > d$ and $T_2^o = \min\{T_1^o, T_2^o, Z, T\}$.

Figure 5 sketches examples of non-monotone paths when none of the p_i are zero. It corresponds to the case $\bar{\lambda}_2 < \lambda_1 + \lambda_2 < \bar{\lambda}_1$ and $p_2 > C_2$. Hence nonmonotone

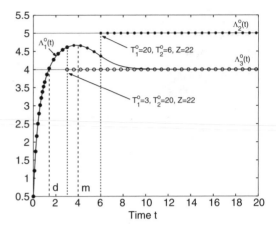

Figure 5. $\Lambda^o(t)$ paths with $\lambda_{12} = 0.5$, $\lambda_1 = 2$, $\lambda_2 = 3$, $\bar{\lambda}_1 = 4.5$, $\bar{\lambda}_2 = 3.5$, $p_1 = 0.01$, $p_2 = 0.6$.

paths occur for example when $T_1^o > d$ and $T_1^o = \min\{T_1^o, T_2^o, Z, T\}$ (downwards jump) or when $T_2^o > m$ and $T_2^o = \min\{T_1^o, T_2^o, Z, T\}$ (non-monotone path of $\Lambda_1^o(t)$).

5.2 Frequency of non-monotone paths

The study of system S shows that the observed hazard rate process $\Lambda^o(t)$ can have non-monotone paths whereas the hazard process rate $\Lambda(t)$ is monotone increasing. Hence with imperfect monitoring, the monotony property of $\Lambda(t)$ does not always hold. It depends on the system parameters (failure rates) and the monitoring quality (non-detection probabilities). When such non-monotone paths are possible it can be interesting to evaluate their frequency: for example on Figure 4, we want to know if the non-monotone paths obtained when $T_2^o > d$ and $T_2^o = \min\{T_1^o, T_2^o, Z, T\}$ are frequent. Such evaluations are obtained by simulations (Monte Carlo simulation) on 10^5 histories. Figure 6 sketches the probability P of non-monotone paths for any non-detection probabilities p_1 and p_2 in the following case: $\bar{\lambda}_1 < \lambda_1 + \lambda_2 < \bar{\lambda}_2$. When $p_1 \leqslant C_1$, the paths of $\Lambda^o(t)$ are always monotone increasing (see Table 3) and $P = 0$. When $p_1 > C_1$, $\Lambda^o(t)$ can have non-monotone paths. In this case P increases with p_1 and p_2 from 0 to 0.14. This estimation of P must be combined with the conditions given in Table 3 to evaluate the real impact of monitoring quality on the observed hazard rate process. Hence, even if the conditions are verified to have nonmonotone paths, the probability P that such paths occur can be very low. In this case, $\Lambda^o(t)$ can still be a good approximation of $\Lambda(t)$ in reliability studies or maintenance optimization. Analytical evaluation of P is for further work.

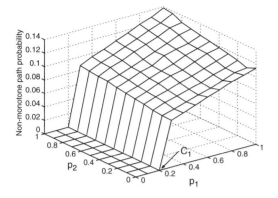

Figure 6. Probability of non-monotone paths of $\Lambda^o(t)$ with $\lambda_{12} = 0.5, \lambda_1 = 1, \lambda_2 = 3, \bar{\lambda}_1 = 1.5, \bar{\lambda}_2 = 3.5$.

6 CONCLUSIONS

A new observed hazard rate process $\Lambda^o(t)$ has been defined to approximate the actual failure rate $\lambda_a(t)$ of an imperfectly monitored (non-detection events) multi-unit systems. The application on a practical industrial example shows that there exists some configurations (depending on the system parameters and the monitoring quality) for which the monotony property of $\lambda_a(t)$ is not always kept by $\Lambda^o(t)$. The frequency of non-monotone paths of $\Lambda^o(t)$ (whereas $\lambda_a(t)$ is monotone increasing) has been evaluated by simulations. The study of these configurations is for further work in order to specify for any system structure, if non-monotone paths for $\Lambda^o(t)$ are possible and to evaluate their occurrence probability.We look for particular classes of multi-unit systems (*MIFR*, (Arjas 1981b))

whose observed hazard process would be always monotone (whatever the monitoring quality is).

REFERENCES

Simola, K. and U. Pulkkinen (1998). Models for Non-destructive Data. *Reliability Engineering and System Safety 60*, 1–12.

Barlow, R. and F. Proschan (1996). *Mathematical Theory of Reliability*. Classics in Applied Mathematics. Society for Industrial and Applied Mathematics.

Cocozza-Thivent, C. (1997). *Processus Stochastiques et Fiabilité des Systémes*. Mathématiques et Applications. Springer. In French.

Arjas, E. (1981a). The Failure and Hazard Process in Multivariate Reliability Systems. *Mathematics of Operations Research 6*(4), 551–562.

Heinrich, G. and U. Jensen (1992). Optimal Replacement Rules Based on Different Information Levels. *Naval Research Logistics 39*, 937–955.

Pham, H. (1992). Reliability Analysis of a High Voltage System with Dependent Failures and Imperfect Coverage. *Reliability Engineering and System Safety 37*, 25–28.

Barros, A., C. Bérenguer, and A. Grall (2002). Maintenance Policy for a Two-components System with Stochastic Dependences and Imperfect Monitoring. In *Proc. of the 3th. Int. Conf. on Mathematical Methods in Reliability – MMR'02 – 17–20 june 2002, Trondheim, Norway*, pp. 81–84.

Barros, A., A. Grall, and C. Bérenguer (2003). A Maintenance Policy Optimized with Imperfect and/or Partial Monitoring. In *Proc. Annual Reliability and Maintenability Symposium, IEEE-RAMS 03*. To appear.

Arjas, E. (1981b). A Stochastic Process Approach to Multivariate Reliability Systems: Notions Based on Conditional Stochastic Order. *Mathematics of Operations Research 6*(2), 263–276.

Safety and Reliability – Bedford & van Gelder (eds)
© 2003 Swets & Zeitlinger, Lisse, ISBN 90 5809 551 7

High contrast consumer test: a case study

G. Baskoro, J.L. Rouvroye & A.C. Brombacher
Eindhoven University of Technology, Netherlands

N. Radford
Philips Electronics, Hasselt, Belgium

ABSTRACT: This paper addresses the application of high contrast consumer tests for product reliability improvement. The purpose of these tests is to define already during product development levels- and effects- of extreme use to a product and in doing so determining product weaknesses already early during product development. It has been designed, especially for innovative products, to prevent possible gaps between (often implicit) customer requirements and the actual specification. This is done by provoking interactions between early prototype products and a limited number of well defined (extreme) users during the early phases of product development.

1 INTRODUCTION

For strongly innovative products manufacturers often do not know in detail how customers will perceive the product. One of the ways to overcome this problem is to bring such a product as fast as possible to the market. This will reveal with a high degree of certainty unanticipated problems or opportunities (Trott 1998). From a quality or reliability perspective this approach will be very effective but, due to costs involved in launching immature products, probably not very efficient. In order to prevent these costly iterations, companies apply numerous quality assurance techniques during design, development, and production. However, for "first of a kind" product development processes, there is a high likelihood that the product (or the way the customer will use it) contains some novel elements that are not (yet) captured by the available quality assurance techniques. This will require a different approach for ensuring product reliability as well as capturing its limitations.

The aim of the HCCT, as described in this paper, is to determine whether an alternative test strategy is able to capture potential product deficiencies already during product development (Lu et al. 1999, Sander et al. 1999).

2 CONCEPT OF HIGH CONTRAST CONSUMER TEST (HCCT)

HCCT is based on the assumption that a well-structured product development process it is not very likely that "normal" products fail at "normal" customers. Normal product validation tests usually take care of these issues. The aim of HCCT is to create an efficient method to capture more unlikely and/or unanticipated events such as may occur when a new product is subjected to new customers. As a consequence this type of test is different in a number of ways from existing tests (Trott 1998).

In order to create added value from a manufacturers point of view such a test should not only be able to capture events; it should also contain adequate details to allow root-cause analysis and subsequent product improvement.

2.1 What is HCCT?

The high contrast consumer test is a method for product testing that is designed not to test for compliance with specifications but to maximize the variability in the interaction between product and customer in order to provoke, early in the product development process, failures that would normally occur in the field of quality or (early) reliability problems. These failures, especially in high volume consumer product, identified as phase 1 and phase 2 failure on the bath-tub curve (Fig. 1). According to the model presented by Brombacher these events are caused by (combinations of-) extreme product and extreme users (Brombacher et al. 1992, Lu et al. 2000).

Waiting for field failure information from the end consumer feedback is considered too slow especially in high volume consumer product with short

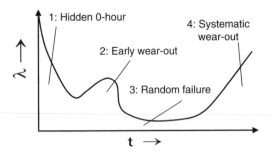

Figure 1. Bath-tub curve.

time-to-market and product cycle time. Therefore HCCT is introduced as a method of testing in product development process to provoke failures that may lead to phase 1 and phase 2 failure in the field.

2.2 Test scheme

Since in HCCT the focus is on provoking failure phase 1 and phase 2 the goal is to get, in a short amount of time, a large number of realistic combinations of (extreme) users and/or (extreme) products.

This results in two ingredients required (either independently or in combination) and one additional requirement for the test:

- *Getting extreme users*: able to generate to generate extreme but realistic user profiles.
- *Getting extreme products*: fully functional but close to the specification limits.
- *Analysing the resulting interactions*: since the test may result in events with a relatively low probability of occurence it will be important to analyse in detail whether the events generated are realistic.

In conventional test the focus is to comply with product specification and/or standards. Therefore the path of testing is predetermined. From the requirements mentioned above it is obvious that users in an HCCT can not use predetermined test patterns. Therefore the user is free to define a test path. The user can stop, extend, restart, continue, redefine, or repeat whenever necessary.

2.3 Test covered by HCCT

Products are usually designed to fulfill certain needs of (potential) customers. Satisfying the always increasing customer expectation is a challenge for a company. In order to ensure that the product satisfies customer needs and expectations, companies struggle to capture customer needs and deploy this information into the pipeline of new product development. Part of these expectations are captured in explicit product specifications and most test methods, used during product

development, ensure compliance with these specifications. Especially for new classes of products it may not be possible to fully specify the product since it is not yet known how users will interact with new functionality. HCCT focuses especially on these unknown and often implicit specifications. Therefore, by definition, HCCT will never use standard test protocols. The test will vary for every product type and it is very much dependent on the purpose of the products and their ability in satisfying customer expectation. In broad lines an HCCT can follow the phases during the operational life of a product: 1. (Un)packaging Test, 2. Installation Test, 3. Software Test, 4. Usability Test, 5. Torture (or endurance) Test. During each of these phases mis-match between the actual product and customer use can be found. Failures in each of these steps can contribute to the failure rate of a product. Since the HCCT concentrates especially on extreme users and/or extreme products the faiures found will more likely relate to phase 1 and 2 of the roller coaster model than to phases 3 and 4. The sections below outline some of the activities that can take place in the context of an HCCT:

1. The *Packaging Test* addresses the weaknesses on the packaging such as miss-information, clarity of the language, and easy-does-it guideline.
2. The *Installation Test* explores the possible weaknesses during product installation. Some users may read the instruction manual but some others may directly carry out an installation without carefully reading the manual. This test will explore any weaknesses of the installation.
3. The *Software Test* is necessary for electronic products with many functionalities achieved via (embedded) software. These functionalities generate mostly problems for unfamiliar users, therefore on this type of situations this test is required.
4. The *Usability Test* is designed to explore the use of the products by the user. It addresses to test the extend that users are satisfied by the product to fulfill their needs.

As a 5th phase it is possible to extend the tests into the early use of a product. In this, so called, torture tests, extreme users, derived from the results in the previous phases, will use the product for a prolonged time to determine the potential for mainly phase 2 reliability problems. Experiences with this 5th step, however, are currently limited.

2.4 HCCT in the product development process (PDP)

Since the HCCT bases itself on the interaction of a limited number of (prototype) products with a limited number of customers it is possible to carry out this test before the production phase or earlier. The advantage

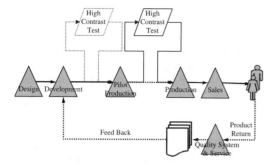

Figure 2.   HCCT in new product development.

of early application of this test is to provide more time for any product improvement. Therefore having this test earlier will reduce any risk of being late on the market. Figure 2 shows two possible choices of conducting high contrast consumer test i.e. after development or after pilot production. It is also possible that the test is carried out after development and pilot production consecutively, if necessary. Since the test is carried out with a limited number of prototype or early production products care needs to be taken for the extrapolation of the data found to (potential) field events for the real product.

2.5   HCCT success factors

The important success factors of the HCCT are Timing (position in the product development process), Speed (time required to obtain information) and the Quality (level of detail) of the information gathered. Timing and Speed of the test can be very important factors when the products are developed with a strong focus on time-to-market. Obtaining high quality information fast provides manufacturers in a position to improve their product in earlier phases of product development. To be able to do an early and fast HCCT a very clear scope of the test should be defined. The test scheme can than be designed to satisfy this need e.g. for products with a conceptual new functionality the focus can be on the instalation process and the first hours of use.

   The quality of the information gathered during the test will have to have sufficient level of detail to give people in product development root-cause insight in the product weaknesses and the ability to introduce product improvement. To ensure that the test generates high quality information without affecting the user too much the think-aloud-protocol is applied (Ching Yang 2003, Spronken 2002). Audio-visual aides (photo/video cameras) are used to store relevant information and the participant is encouraged to speak-up on any findings that he or she considers as relevant. As an enhancement it can be usefull to include people from the product improvement group as a (passive!) observer

during the test, in order to build up additional knowledge by observing during the test. These people will observe and experience the difficulties faced by the users when using the products. This knowledge is not only important for the understanding the feedback given by participants but also during future product development processes.

3   CASE STUDY

An HCCT, fulfilling the requirements mentioned above, has been carried out on a newly developed consumer product; an internal DVD writer for use in a PC. This test was designed with the aim of capturing the product strengths and weaknesses and observing the user perception of the product. Although, conceptually, HCCT considers both extreme users and extreme products this HCCT concentrated especially on the user aspects. The test participants consisted of 5 groups with 5 people in each group. In order to define the contrast, the participants were selected in such away that significant differences exist. As an example: the backgrounds of participants was chosen as engineering and non-engineering, also the participant's, gender, PC knowledge, and age were varied.

3.1   The classification of the participants

Before conducting the test it is necessary to classify the participants based on the following criteria[1]:

1. *Gender*
   Male and female participant.
2. *Profession*
   The profession was chosen to cover a wide range of professions; from blue-collar worker to administrative staff and management.
3. *Age*
   The range of participant age was designed from 27 to 55 years, emphasising on the extremes.
4. *Engineering background*
   Engineering/technical and non engineering/technical background.
5. *PC usage*
   The method of application (professional or normal end user).

3.2   Operational test setup

The participant task is to unpack the product, do-it-yourself setup installation, and make a video DVD on their own. The result should be demonstrated on a

---

[1] The following section presents a simplified structure of the used contrast groups. Further details are available at the authors but are not disclosed for reasons of confidentiality.

common DVD player and television. They were limited only two hour for doing this task, otherwise the result was considered a fail. The result were recorded as OK, Fail, writing Data, making different video system (NTSC). The participants were not allowed to ask question(s) to their peers, however they were allowed to call a helpdesk when necessary.

3.3   Participant background on using PC

Participant knowledge in using PC which was categorized based on their ability on setup and hardware installation, programming, internet, and application software. It is shown that most participants were familiar with using software applications but less with the programming.

The think-a-loud method was applied to encourage participant to speak up on their findings. Besides that, the participants were also encouraged to write down their observations regarding the product. When participants were having major difficulty they were allowed to call the helpdesk for assistance. These calls were also registered.

The test was then executed in the sequence as what the participant preferred based on their preference. When the test was finished the participant filled in the final questionnaire regarding their assessment of the whole process, as well as the product.

3.4   Case study result

The result of the high contrast test was measured based on the success rate of performing the required task. Figure 3 shows that some portion of the participants fail to perform the test while others were successful. The rest, marked as Data and NTSC, manage to write a DVD although not a playable video DVD.

The result showed that the problems generally were caused by the software, that disabled participants to perform their required task. The problems have been

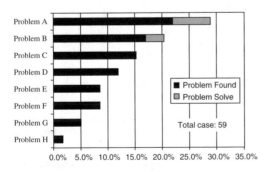

Figure 3.   Pareto of identified problem.

identified, prioritized from the think-a-loud as shown on Figure 3.

Later, in depth, analysis showed that the failures identified during the HCCT had a close relation to similar faults identified as customer complaints in earlier products. The conceptual understanding of the root-causes of the failures was, however, much better than with the earlier analysis of field failures. Using the data generated with HCCT it was possible to confirm earlier field failures but also, thanks to the much higher level of detail of the information, to look into improvements of the product. For some of the issues identified it was already possible to introduce product improvements for the existing product without affecting the time-to-market too much.

4   DISCUSSION

The above case study demonstrated that, in order to introduce a sucessful HCCT, the following points will have to be clearly defined:

– *Goal of the test* (phases of the product lifecycle that are relevant, classes of failures that are relevant, main contrast groups, etc.)
– *Method* (test structure (lab-conditions/at home), duration, relevant data, data gathering process, etc.)
– *Selection of participants* in accordance with the profiles defined earlier
– *Execution*
– *Analysis* of the obtained data and relating it to (potential) reliability issues

It is also necessary to note when designing the method and model for a high contrast consumer test, the analysis methods must be defined in advance. The question of what kind of problems can be observed during the test has to be clear so that the test can be executed in an efficient manner.

Defining the analysis method early-on may lead to a better design of the questionnaire and participant groups. For example, what the ages of participant and the test should be, and how many people represent these ages. Another important issue to be considered is the function, occupation, gender, and level of education. It is difficult to have the right combination of these backgrounds when the number of participants will be kept low, therefore finding a correct compromise between test expectation and the requirement should be a major consideration.

Having these questions leads to further research on the effectiveness of the high contrast consumer test itself and the way to speed up the test. A further topic of research is to find out for which type of products and related PDPs this type of test is most beneficial.

Due to its (uncommon) focus on phase 1 and 2 reliability problems it is beneficial to introduce the high contrast consumer test early in the PDP. The test result, which has not only the ability to capture the product weakness and customer perception but also to increase the awareness of those who are involved on product development.

REFERENCES

Brombacher, A.C., Van Geest, E., Arendsen, R., Van Steenwijk, A. & Herrmann, O. 1992 Simulation, a Tool for Designing-in-Reliability, *ESREF'92 Conference, 92.*

Ching Yang, S. 2003 Reconceptualizing Think-aloud Methodology, *Computer in Human Behavior* 19: 95–115.

Kuczmarski, D.T. Managing New Products, 3rd Edition, *The Innovation Press, Book Ends Publishing, Chicago Illinois.*

Lu, Y., Loh, H.T., Ibrahim, Y. & Brombacher, A.C. 1999 Reliability in a Time-Driven Product Development Process, *Quality Reliability Engineering International* 15: 427–430.

Lu, Y., Loh, H.T. & Brombacher, A.C. 2000 Accelerated Stress Testing in a Time-Driven Product Development Process, *Int. J. Production Economics* 67: 17–26.

Sander, P.C. & Brombacher, A.C. 1999 MIR: The Use of Reliability Information Flows As Maturity Index For Quality Management, *Quality And Reliability Engineering International* 15: 439–447.

Spronken, R.M.L. 2002 A New Drive: A Guideline to Better Reliability, *Master Thesis, Technische Universiteit Eindhoven.*

Trott, P. 1998 Innovation Management & New Product Development, Financial Times Pitman Publishing.

Safety and Reliability – Bedford & van Gelder (eds)
© 2003 Swets & Zeitlinger, Lisse, ISBN 90 5809 551 7

Reviewing the safety management system by incident investigation and performance indicators

B. Basso, C. Dibitonto, G. Gaido, A. Robotto & C. Zonato
ARPA Piemonte – Unità di Coordinamento Rischio Tecnologico (UCRT), Turin, Italy

ABSTRACT: Monitoring performance is one of the issues that must be addressed by the Safety Management System (SMS), but operators usually disregard it. By the experience gained during inspections on the SMS in establishments covered by Seveso II Directive, the Regional Agency for Environmental Protection (ARPA Piemonte) has developed a methodology to improve this monitoring. It combines incident investigation and analysis of performance indicators, allowing operators to measure the achievement of the objectives stated in the major accident prevention policy and to evaluate their adequacy. On the basis of data collected during the inspections on the SMS about incidental events (major accidents, accidents and near misses), a database was organized, showing the most critical elements of the SMS that operators should improve.

1   INTRODUCTION

On 17 August 1999 Council Directive 96/82/EC on the control of major accidents hazard involving dangerous substances, the so-called Seveso II Directive, has been adopted in Italy through a national act: D.Lgs. n. 334/99. Seveso II Directive replaces Directive 82/501/EEC and contains some innovative elements, like the introduction of the Safety Management System (SMS).

D.Lgs. n. 334/99 has required that since April 2000 operators drew up a document setting out their major accident prevention policy and demonstrated that this policy was properly implemented, putting into effect a Safety Management System.

In compliance with Annex III of the Directive, the following issues must be addressed:

1   organization and personnel
2   identification and evaluation of major hazards
3   operational control
4   management of change
5   planning for emergencies
6   monitoring performance
7   audit and review

Concerning the monitoring performance, operators must adopt and implement procedures for the ongoing assessment of compliance with the objectives set by the major-accident prevention policy and the Safety Management System, and the mechanisms for investigation and taking corrective actions in case of non-compliance.

In Regione Piemonte inspections on the SMS began on April 2001 and are carried out by the unit of the Regional Agency for Environmental Protection (ARPA Piemonte) specialized in industrial risk control (UCRT).

Till now the inspections showed a generalised lack of the SMS about the monitoring performance, even when the other elements of the system had been well implemented. Therefore UCRT developed a methodology based on the combination of incident investigation and the analysis of performance indicators, that allows to measure the achievement of the objectives set in the major accident prevention policy and to evaluate their adequacy.

2   INCIDENT INVESTIGATION

2.1   *Criteria for incident investigation*

In compliance with Annex III of the Directive, operators must adopt procedures for reporting accidents and near misses, particularly those involving failure of protective measures, and their investigation and follow-up on the basis of lessons learnt.

During the inspections, UCRT found out that these activities had often been disregarded. Therefore, operators should adopt effective means to increase the personnel participation, explaining that reporting is not used to judge persons for their errors, but to analyse events, learn from them, avoiding their repetition.

Furthermore, they should illustrate the risk analysis, with particular attention to the accidents and near

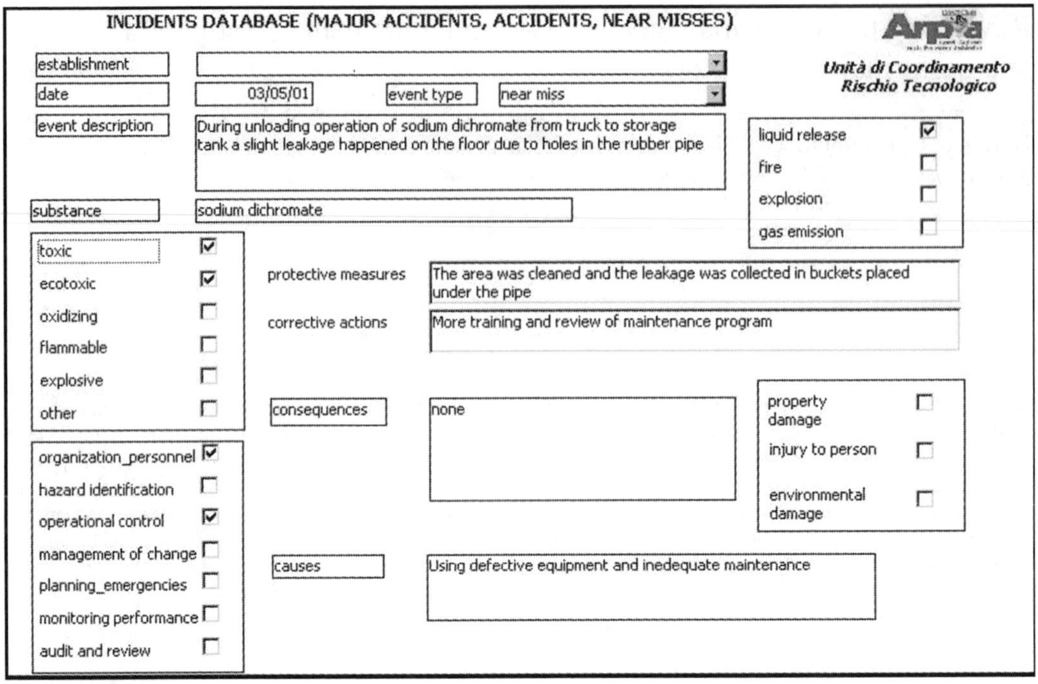

Figure 1. UCRT incidents database.

misses that could happen in the establishment, give the personnel a specific form to fill up with information about all incidents, regardless of their seriousness, and eventually introduce a sort of reward.

After the reporting, operators should guarantee an effective activity of investigation and analysis: incidents should be discussed among technicians, in order to find out the root causes and the corrective actions, and managers, in order to plan and define the appropriations.

2.2 UCRT incidents database

During 2001–2002 about 50 establishments covered by Seveso II Directive have been inspected; in the course of the inspections UCRT required operators to fill up a form with the following information:

– incidents description (date, substances and plant involved, causes and consequences);
– aspects of the SMS involved.

Up to now 38 operators produced data about more than 250 incidents happened in the last ten years; collecting these data, UCRT organized a MS Access database, in order to perform statistical analysis.

The database is born for public authority, but it can also be useful for operators as an example of organization of data that should be collected during the incident investigation.

The main window of the database (Figure 1), contains the following information:

– name or trade name of the operator and address of the establishment;
– date of the event;
– type of event: major accident, accident, near miss;
– incidental scenario (liquid release, fire, explosion and gas emission);
– dangerous substance involved and indication of its category (toxic, dangerous for the environment, oxidizing, flammable, explosive or other);
– Immediate and delayed measures of protection;
– circumstances of the event, in term of causes and consequences;
– critical elements of the SMS which caused the event.

With reference to the severity, different types of incidents has been defined, according to the following definitions:

– major accident: an occurrence such as a major emission, fire or explosion resulting from uncontrolled developments in the course of the operation, involving one or more dangerous substances and leading to serious danger to human health and/or the environment, immediate or delayed, inside or outside the establishment (criteria are the same contained in Annex VI of the Directive);

- accident: an occurrence such as an emission, release, fire, or explosion resulting from uncontrolled developments, involving one or more dangerous substances and leading to consequences for man, the environment, real estate and/or property less serious than the consequences of a major accident;
- near miss: an incident that has the potential for injury and/or property damage, involving dangerous substances or not (in this case it is called anomaly).

The example shown in Figure 1 is a near miss because it involved a release of a dangerous substance, sodium dichromate, without consequences. The event was caused by a defective equipment and human error; therefore, the critical elements of the SMS were the operational control (for the inadequate maintenance of the pipe) and the organization and personnel (for the inadequate training). As corrective actions, the operator planned to review the maintenance program and improve the safety training.

The database contains other information about the establishments, such as the number of employees and the type of activity, because the experience teaches that the SMS structure depends on these two parameters. With reference to the type of activity, it is possible to divide the establishments into two groups, according to their level of standardization : high for warehouses and low for chemical plants.

The SMS structure is generally simpler in establishments with a high level of standardization and few employees.

In Piemonte, on the 115 establishments covered by Council Directive 96/82/EC, the competent authorities organized a system of inspections that gave higher priority to chemical plants. Among the establishments inspected, more than 70% has a low level of standardization (Figure 2); the establishments not yet inspected are scheduled for the end of 2003.

2.3 Statistical analysis of collected data

In Tables 1 and 2 the distribution of the events collected in the database, in terms of severity and scenario is shown.

The database allows to select more than one scenario for each incident, as the frame of Figure 1 shows: for example, both the major accidents happened in Piemonte were caused by a LPG emission, that caught fire but only in one case had an explosive evolution.

By statistical analysis about the collected events, it has been found that the most critical element of the SMS is operational control, as shown in Table 3 and Figure 3. In the course of many inspections it has been found out that the outcomes of the risk analysis had not been used to draw up suitable procedures addressing each operating phase. Morever, the risk analysis had not identified the critical equipment, that requires a preventive maintenance program; so the safety-related

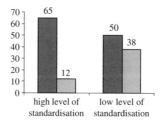

■ Establishments in Piemonte covered by Seveso II Directive
□ Inspected establishments in Piemonte (2001–2002)

Figure 2. Distribution of establishments in terms of level of standardization.

Table 1. Distribution of events in terms of severity.

Type of event	Number
Major accidents	2
Accidents	33
Near misses	242
Total	277

Table 2. Distribution of events in terms of severity and scenario.

Type of event	Emission	Fire	Explosion	Release
Major accidents	2	2	1	0
Accidents	8	10	6	15
Near misses	73	20	11	67
Total	83	32	18	82

Table 3. Distribution of critical elements of the SMS.

Elements of SMS	Number
Organization and personnel	76
Identification of major hazards	77
Operational control	129
Management of change	3
Planning for emergencies	15

Number of establishments considered: 38
Number of events collected: 277

work had not been given higher priority than routine maintenance.

As the percent incidence is upper than 30%, also the organization and personnel and the identification and evaluation of major hazards should be considered critical. The reason of those incidences is that in many establishments the roles and responsibilities of personnel involved in the management of major hazards

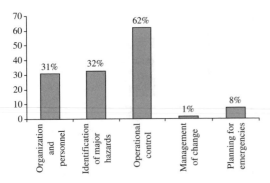

Figure 3. Distribution of critical elements of the SMS.

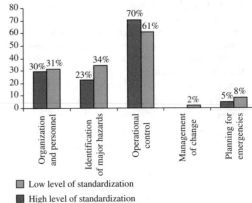

□ Low level of standardization
■ High level of standardization

Figure 4. Percent distribution of critical elements of the SMS in establishments with high and low level of standardization.

Table 4. Distribution of critical elements of the SMS in establishments with high level of standardization.

Elements of SMS	Number
Organization and personnel	13
Identification of major hazards	10
Operational control	31
Management of change	0
Planning for emergencies	2

Number of establishments considered: 12
Number of events collected: 44

Table 5. Distribution of critical elements of the SMS in establishments with low level of standardization.

Elements of SMS	Number
Organization and personnel	73
Identification of major hazards	80
Operational control	141
Management of change	4
Planning for emergencies	19

Number of establishments considered: 26
Number of events collected: 233

had not been well identified, so had their training needs. Moreover, the risk analysis had often turned out incomplete.

In Tables 4 and 5 and in Figure 4 a comparison between establishments with high and low level of standardisation is shown.

As the number of inspections in establishments with high level of standardisation is low (Figure 2), the collected events are only 44 out of 277.

However, although some of the differences shown in Figure 4 are yet statistically not significant, they confirm widely the outcomes of the inspections carried out up to now. In fact, in the establishments with high level of standardisation the repetition of the same operations often induces operators to neglect the adoption and implementation of procedures and instructions for safe operation.

On the contrary, the identification and evaluation of major hazards and the management of change are more critical in chemical plants, because the risk analysis is in general more complex and changes more frequent.

3 MONITORING PERFORMANCE

3.1 Performance indicators

In many inspected establishments the monitoring performance has been neglected. Even when the operator introduced some indicators to monitor the SMS performance, they regarded only some issues of the system and were not important for the assessment of compliance with the objectives set by the operator's major accident prevention policy and the SMS.

Therefore, the first step of an effective methodology should consist in the choice of one or more indicators for each objective stated in the policy, that in its turn should cover all the issues of the SMS or at least the most critical ones.

Both the competent authorities and other technical organizations have found out some examples of significant indicators that can help the operator in his choice.

The evaluation of their adequacy in reference to the activity of the establishment is due to the operator himself: an indicator is adequate if it helps the operator in the expression of the appraisal about the system and if the calculation of its value is simple in terms of collecting data sources.

On the basis of these last considerations, some indicators for each issue of the SMS are listed in the following; besides, between brackets, it is mentioned the possible source of data to calculate such indicators.

1 Organization and personnel
   – Investments in preventive or protective measures to limit the consequences of major accidents (from budget)
   – Hours for safety training per person (from training minutes)
   – Percentage of right answers for person (from tests to evaluate the effectiveness of training)
2 Identification and evaluation of major hazards
   – Number of incidents happened in the establishment not previewed by the risk analysis (from risk analysis and incident reports)
3 Operational control and management of change
   – Non-compliance about procedures, instructions and all documents necessary to describe dangerous substances, processes, plants and equipments (check list of audits)
   – Technical inspections for control and maintenance of critical plants and equipments (from records)
4 Planning for emergencies
   – Number of incidents happened in the establishment not considered or badly planned by the emergency plan (from emergency plan and incident reports)
5 Audit and review
   – Number of audits
   – Number of reviews of the major accident prevention policy and the SMS.

The operator should review the SMS at regular time intervals. In this occasion he should calculate the values of the chosen indicators and use them to assess the system performance.

The real value of each indicator must not be considered as an absolute number; on the contrary, it should be calculated in the reference period (e.g. one year) and compared with an expected value, called threshold. For example, considering the indicator number of audits, if at the beginning of the year two audits have been planned and at the end of the year only one audit has been made, the degree of satisfaction is not high because the threshold has not been reached.

For positive indicators (e.g. hours for safety training per person, number of audits) the more the real value approximates its threshold, the more the satisfaction degree is high. For negative indicators (e.g. number of incidents, non-compliances) the satisfaction degree is higher when the real value is far below its threshold.

As shown in Figure 5, in the first case, there is a direct proportion between the ratio real value / threshold and the satisfaction degree; in the second case, there is a direct proportion between the ratio threshold/ real value and the satisfaction degree.

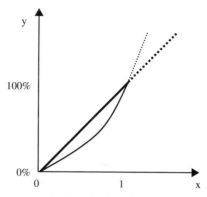

x = real value / threshold for positive indicator
x = threshold / real value for negative indicator
y = satisfaction degree of the indicator

Figure 5.   Correlations for satisfaction degree of the indicator.

If the operator wants to adopt more severe criteria, he could use an hyperbola instead of a straight line to determine the degree of satisfaction.

3.2   Reviewing the SMS

By the definition of a threshold for each chosen indicator, it is possible to express an appraisal about the system on the basis of real data.

As said in section 3.1, the chosen indicators should be the correct ones: they must be related with all the issues of the system and they must be simple to calculate. Moreover, the choice of the thresholds should be adequate.

In Table 6 a methodology to evaluate the adequacy of the threshold of each indicator is proposed, based on the correlation between performance indicators analysis and incident investigation.

The severity of the issue of the SMS connected with the indicator is evaluated by the incident analysis and the following data aggregation, as shown in Figure 3.

The satisfaction degree of the indicator is calculated by the ratio real value / threshold or its inverse, as shown in Figure 5.

If the satisfaction degree of the indicator is equal or higher than 80%, we could state that the objective correlated with the indicator has been reached, but this statement is premature, because it does not consider the outcomes of incident investigation. In fact, if the issue of the SMS connected with the indicator is not critical (lower left box), the threshold has been chosen adequately. On the other hand, if the issue of the SMS connected with the indicator is critical (upper left box), the threshold is not adequate, because it has been underestimated (overestimated in case of negative indicator).

Table 6. Correlation between the satisfaction of the indicator and the severity of the issue of the SMS.

| Severity of the issue of the SMS | Satisfaction degree of the indicator | |
	High (≥80%)	Low (0–79%)
High (30–100%)	The threshold is not adequate: increase it for positive indicators and lower it for negative ones.	The threshold has not been reached: define proper actions aiming to it.
Low (0–29%)	The threshold is adequate: if the satisfaction degree is not 100%, try to reach it.	The threshold has been overestimated for positive indicators or underestimated for negative ones: it may be lowered or increased.

If the satisfaction degree of the indicator is lower than 80%, we could state that the objective correlated with the indicator has not been reached, but this statement is not exhaustive. In fact, if the issue of the SMS connected with the indicator is critical (upper right box), the operator should try to reach the threshold. On the other hand, if the issue of the SMS connected with the indicator is not critical (lower right box), it is likely that the threshold has been overestimated (underestimated in case of negative indicator).

To evaluate the severity of the issues of the SMS, the percentage of 30% has been selected on the basis of the experience gained in Regione Piemonte (Figure 3). About the satisfaction degree of the indicators the percentage of 80% is conservative.

The operator should use the criteria shown in Table 6 for each chosen indicator and then define the priority for the corrective actions necessary for the follow-up of the SMS.

First of all the operator should plan the actions for the achievement of the objectives connected with the most critical elements of the SMS (upper boxes of Table 6), redefining the inadequate thresholds of the indicators when necessary.

Then, with lower priority, he should plan the actions to guarantee the satisfaction of all the other unsatisfied objectives (lower right box).

To express a more significant appraisal about the system, the operator might choose two or more indicators for each issue of the SMS. For example, considering the organization and personnel, he can choose hours of safety training per person per year and average percentage of right answers. If the satisfaction degree for the hours of training is 100%, but the average percentage of right answers for person in the tests is only 60%, while the threshold is 90%, we could say that the operator should increase the training effectiveness.

The monitoring performance should be a process of continuous improvement; so, when the objectives stated in the safety prevention policy have been reached, the operator might define more severe thresholds and/or increase the percentage for a high satisfaction degree.

4 CONCLUSIONS

During the inspections on the Safety Management System, carried out in Piemonte between 2001 and 2002, a general lack about monitoring performance has been found. In fact, even when operators introduced some indicators to monitor the SMS performance, they regarded only some issues of the system and were not important for the assessment of compliance with the objectives set by the operator's major accident prevention policy and the SMS.

The proposed methodology allows the expression of an appraisal about the SMS performance and the identification of the corrective actions for its follow-up, by the following steps:

- incident investigation to find out the critical issues of the SMS;
- choice and analysis of performance indicators and their thresholds;
- correlation between the satisfaction of the indicator and the severity of the issue of the SMS, in order to evaluate the adequacy of the objectives stated in the safety policy.

For the application of the methodology the operator must define an effective system for incident reporting and investigation. So the methodology can be an incentive to implement these important activities, often disregarded.

Moreover, the collection of data deriving from the incident investigation is very important also for the competent authorities responsible for the controls on the SMS. They can schedule the next inspections giving particular attention to the most critical issues of the system.

Safety and Reliability – Bedford & van Gelder (eds)
© 2003 Swets & Zeitlinger, Lisse, ISBN 90 5809 551 7

Estimating the maximum allowable outage time for redundant safety systems of a nuclear power plant – a RAMS application

G. Becker
RISA Sicherheitsanalysen GmbH, Germany

L. Camarinopoulos
University of Piraeus, Greece

D. Kabranis
ERRA, Greece

B. Schubert
Hamburgische Electricitätswerke AG, Germany

ABSTRACT: A mixture of Boolean and Markovian state models has been developed and applied to determine the influence on core melt frequency of maintenance under full power of the residual heat removal system (RHRS) in a nuclear power plant. The results of the analysis show a positive influence on development of the unavailability of the RHRS system and indicate that further applications in optimisation of component maintenance in nuclear power plants are thinkable.

1 INTRODUCTION

The annual refueling period for a nuclear power plant is a period of time, which is not immediately productive, as no electricity is produced. Hence, it is attempted to make this period as short as possible. A lot of maintenance work is scheduled during this period. So, annual refueling is not only unproductive, but also a stressful challenge to the personnel of the utility.

Although the technical state of the reactor during refueling may be considered safer than during power production, the sheer workload produces a tendency towards the opposite phenomenon, yielding a potential unsafe state. For this reason, there is a strong tendency to schedule maintenance tasks for the safety systems outside the refueling phase, where this is possible.

Preventive maintenance under full power operation requires proper adjustment of Allowed Outage Times (AOTs) and appropriate Surveillance Test Intervals (STIs). With the proper choice of these parameters it is ensured that the operation meets the safety specifications. Much work has been done in recent years in evaluating and optimizing the AOTs (Cepin & Martorell 2002) , STIs (Cepin & Mavko 1997) as well as their interaction (Martorell et al. 1995), (Becker et al.

2001). The proposed methods use probabilistic measures such as component and system unavailability or core damage probability, in order to evaluate and optimize AOTs and STIs.

A common setback with PSA calculations for complex systems is their approximating nature. Due to the underlying complexity, certain assumptions and simplifications are made in order to obtain quantitative results. A consequence of this practice is the conservative character of the results as well as the low sensitivity to individual changes down to component level.

In order to investigate the influence of different choices of STIs and AOTs, it is necessary to adopt a more detailed model of component behavior as well as to improve the accuracy of the calculations at the cut set level in order to include the influence of the existing dependencies or restrictions.

Markovian state models are an adequate tool for the analysis of small systems, given that a state graph can be given, where the system is memoryless (Papazoglou 2000). Unfortunately this method becomes prohibitive when systems are getting larger, because the number of system states is rapidly increasing. In this work we propose a method to quantify fault trees using markovian graph modeling for single components or small groups of components. This method is able to treat local dependencies among components, resulting in more accurate quantitative results. An additional advantage of the methodology is that the effort to model a system stays in manageable levels. A "real world" example of a boiling water reactor in Germany will also be presented. The intension in the presented case is to schedule maintenance to the residual heat removal system (RHRS) during full power operation rather than during refueling.

2 FAULT TREES AND QUANTITATIVE MEASURES

Let $\Phi(\underline{x})$ be a Boolean function corresponding to a fault tree. Formally a coherent Boolean function can be reduced to the canonical form:

$$\Phi(\underline{x}) = \bigcup_{i=1}^{p} C_i = C_1 \cup C_2 ... \cup C_p \tag{1}$$

where:

$$C_i = \bigcap_{j=1}^{p_i} x_i^j = x_i^1 \cap x_i^2 \cap ... x_i^{p_i} \tag{2}$$

and x_i^j the j event of minimal cut set C_i.

Unavailability: For safety systems where the occurrence probabilities of basic events are low, a sufficient conservative approximation for system unavailability is:

$$U(t) = P\{\Phi(\underline{x})\} = \sum_{i=1}^{p} P\{C_i\} = \sum_{i=1}^{p} U_{cut}(i, t) \tag{3}$$

Under the assumption of independency between basic events, cut set i unavailability $U_{cut}(i, t)$ can be estimated by:

$$U_{cut}(i, t) = \prod_{j=1}^{p_i} P\{x_i^j\} = \prod_{j=1}^{p_i} P_i^j(t) \tag{4}$$

where $P_i^j(t)$ is the probability of occurrence of basic event X_i^j.

Failure probability: It is estimated using failure frequency $H_s(t)$ according to relation (1):

$$F_s(t) \approx H_s(t) = \int h_s(s)ds = \sum_{i=1}^{p} \int h_{cut}(i, s)ds \tag{5}$$

where we refer to the whole system using the subscript s, p is the number of cut sets and $h_{cut}(i, t)$ is the failure frequency density of cutest i.

3 MARKOVIAN STATE MODELS WITH DISCONTINUOUS TRANSITIONS

When the markovian property is possessed by a state model, then the state probabilities over time are readily calculated solving the Kolmogorov equations:

$$dP_i(t)/dt = -P_i(t)\sum_{j \neq i} \lambda_{ij}(t) + \sum_{j \neq i} P_j(t)\lambda_{ji}(t) \tag{6}$$

Another useful measure for state graph analysis is frequency density of reaching and leaving a state. Let $H_r(i, t)$ be the expected value of the number of times state i is reached in the interval [0, t], and $H_l(i, t)$ the expected value of the number of times it is left. Then, $dH_r(i, t) = h_r(i, t)dt$ and $dH_l(i, t) = h_l(i, t)dt$ define the corresponding frequency densities functions. It is proved in [2] that a finite stochastic process with independent transitions and without loops obeys for all of its states (indexed with i)

$$dP_i(t)/dt = h_r(i, t) - h_l(i, t) \tag{7}$$

In order to include periodical inspection or maintenance, the Markovian state graph model has been expanded to contain discontinuous transitions from one state to another. If the period of analysis is [0, t] and $\{t_1, t_2, ...\}$ are the points of discontinuous transitions, then, the state probabilities at period $[t_k, t_{k+1}]$ are estimated by solving the Kolmogorov equations:

$$\frac{\partial P_j(t)}{\partial t} = -P_j(t)\sum_{\substack{i=1 \\ i \neq j}}^{n} \lambda_{ji} + \sum_{\substack{i=1 \\ i \neq j}}^{n} [\lambda_{ij}P_i(t)] \tag{8}$$

$$j = 1,...n \quad , t \in [t_k, t_{k+1}]$$

where:
n: Number of states
$P_j(t)$: Probability of state j
λ_{ij}: Continuous transition rates (failure, repair...) from i to j state.

The initial conditions will be the state probabilities immediately following the transition at t_k, which shall be represented by $p_j(t_k + dt)$. The state probabilities $p_j(t_k)$ are already known from the solution of the previous time period $[t_{k-1}, t_k]$. So, initial condition are calculated by:

$$p_j(t_k+dt)=p_j(t_k)\cdot\left[1-\sum_{\substack{i=1\\i\neq j}}^{n}q_{ji}(t_k)\right]+\sum_{\substack{i=1\\i\neq j}}^{n}q_{ij}(t_k)p_i(t_k)\right] \quad (9)$$

$j=1,..n$

where:

q_{ij}: The probability of a discontinuous transition from i to j state at t_k.

An in depth analysis of markovian models with discontinuities along with applications can be found in (Becker et al. 1994b).

4 A JOINT MODELING APPROACH

In many realistic cases the assumption of independency between basic events is not accepted. Such cases, where specific dependencies between components exist, are treated introducing markovian state graph modeling in the fault tree analysis. The obvious approach is to create a markovian state graph model for each cut set and introduce in the state graph all the existing dependencies between basic events.

While this option is able to provide a precise solution for the probabilistic measures of the cut set, it has the following disadvantage: The number of states is rapidly increasing when the number of the cut set components is increasing or when more detailed models of individual components are adopted. Suppose for example a cut set consisting of three components, each modeled by five states. Then the state graph of this cut set will have 125 states, all possible combinations of the states of the three components.

The user of such methodology, from the engineer who works with his own preferred tools, to the user of a PSA tool with special facilities for state graphs editing, will find the scenario of creating or updating such complicated state graphs, very discouraging. A new methodology is proposed against this drawback. It introduces the notion of state cut set.

A state cut set is a combination of failure states of the components included in a cut set, leading obviously to system failure. A state cut set list is a list of combinations of failure states for the components of a specific cut set, from which are excluded all those combinations that are not possible due to an existing policy or regulations, or are of no interest for the purposes of the analysis (the risk assessment). Such a list can be generated according to a simple and convenient set of rules. A possible implementation of such a rules schema, adopted in the VardaGen risk assessment tool, can be found in appendix A.

A state cut set can be represented as:

$SC = \{S_1, S_2, ..., S_{pi}\}$

where:

p_i is the number of components of cut set i

S_k is a failure state of component k.

4.1 System unavailability estimation

The unavailability of state cut set j of cut set i is given by:

$$U^i_{sc}(j,t) = P\left\{\bigcap_{k=1}^{p_i} S^i_j(k)\right\} \quad (10)$$

where:

$S^i_j(k)$ is the failure state of component k of state cut set j of cut set i

p_i is the number of components of cut set i.

Given the state cut set list the unavailability of cut set i is given by:

$$U_{cut}(i,t) = P\left\{\bigcup_{j=1}^{r_i}\left[\bigcap_{k=1}^{p_i}S^i_j(k)\right]\right\} = \sum_{j=1}^{r_i}P\left\{\bigcap_{k=1}^{p_i}S^i_j(k)\right\} \quad (11)$$

where:

r_i is the number of state cut sets of cut set i.

Assuming that the states of each state cut set occur independently from one another, then system unavailability becomes:

$$U_s(t) = \sum_{i=1}^{p}U_{cut}(i,t) = \sum_{i=1}^{p}\left\{\sum_{j=1}^{r_i}\left[\prod_{k=1}^{p_i}P\{S^i_j(k)\}\right]\right\} \quad (12)$$

where:

p is the number of cut sets.

4.2 System failure probability estimation

Failure probability of the system is approximated (for reasonable small times where $H_s(t)$ is less than 1) by:

$$F_s(t) \approx H_s(t) = \int_0^t h_s(s)ds = \int_0^t\left\{\sum_{i=1}^{p}h_{cut}(i,s)\right\}ds \quad (13)$$

where:

$h_{cut}(i, t)$: is the failure frequency density of cutest i

p: is the number of cut sets.

4.3 Cut set failure frequency density

Cut set failure frequency density is estimated using the following equation:

$$h_{cut}(i, t) = \sum_{j=1}^{r_i} h_{sc}^i(j, t) = \sum_{j=1}^{r_i}\sum_{k=1}^{p_i} I_j^i(k, t) h_r\left\{S_j^i(k), t\right\} \quad (14)$$

where:

$h_{sc}^i(j, t)$: is the failure frequency density function of state cut set j of cut set i

$I_j^i(k, t)$: is the criticality of component k of state cut set j of cut set i

$S_j^i(k)$: is the failure state of component k of state cut set j of cut set i

$h_r\{S_j^i(k), t\}$: is the frequency density function reaching failure state $S_j^i(k)$.

The criticality $I_j^i(k, t)$ of a component k is the probability that all components of cut set i besides k are in failure states of state cut set j.

$$I_j^i(k, t) = \prod_{\substack{e=1 \\ e \neq k}}^{p_i} P\left\{S_j^i(e)\right\} \quad (15)$$

Let $S_j^i(k) = s$.

Assume also that discontinuous transitions from state x to state y occur periodically starting at TO_{xy} and repeated every T_{xy} time units. Then frequency density function $h_r\{S_j^i(k), t\} = h_r(s, t)$ reaching state s is given by:

$$h_r(s, t) = \sum_{m=1}^{N_I(k)}\left(P\{i_m, t\}\left(\lambda_{i_m s} + \sum_{n=0}^{\infty} q_{i_m s}\delta(t - TO_{i_m s} - nT_{i_m s})\right)\right) \quad (16)$$

where:

$N_I(k)$: number of intact states of component k

i_m: is an intact state of component k

$\lambda_{i_m s}$: transition rate from intact state i_m to defect state s

$q_{i_m s}$: discontinuous transition probability from intact state i_m to defect state s.

$TO_{i_m s}$: time of first discontinuous transition from i_m to s.

$T_{i_m s}$: Discontinuous transition period from i_m to s.

Equation 16 can be further expressed as:

$$h_r(s, t) = \sum_{m=1}^{N_I(k)} P\{i_m, t\}\lambda_{i_m s} + \sum_{m=1}^{N_I(k)} P\{i_m, t\}\sum_{n=0}^{\infty} q_{i_m s}\delta(t - TO_{i_m s} - nT_{i_m s}) \quad (17)$$

$$= h_r^c(s, t) + h_r^d(s, t)$$

where $h_r^c(s, t)$ and $h_r^d(s, t)$ are the continuous and discontinuous parts of frequency density function reaching state s.

Finally cut set failure frequency density is estimated as

$$h_{cut}(i, t) = \sum_{j=1}^{r_i}\sum_{k=1}^{p_i} I_j^i(k, t)\left[h_r^c(s, t) + h_r^d(s, t)\right] \quad (18)$$

$$h_{cut}(i, t) = h_{cut}^c(i, t) + h_{cut}^d(i, t)$$

where:

$$h_{cut}^c(i, t) = \sum_{j=1}^{r_i}\sum_{k=1}^{p_i} I_j^i(k, t)h_r^c(s, t) \quad (19)$$

$$h_{cut}^d(i, t) = \sum_{j=1}^{r_i}\sum_{k=1}^{p_i} I_j^i(k, t)h_r^d(s, t) \quad (20)$$

the continuous part and discontinuous parts respectively, of cut set failure frequency density.

It must be noted that the preceding analysis stands generally for non periodical discontinue transitions if we replace the term $\sum_{n=0}^{\infty} q_{i_m s}\delta(t - TO_{i_m s} - nT_{i_m s})$ with $\sum_{n=0}^{n_D} q_{i_m s}(t_n)\delta(t - t_n)$:

where:

n_D is the number of process discontinue transitions

t_n are the time points of discontinuous transitions

$q_{jk}(t_n)$ is 0 if at t_n does not occur a discontinue transition from state j to state k.

5 COMPONENT MODELS

In this section are presented the state graph models of some used component types. These are state models with periodical discontinue transitions. With a dashed line are presented the discontinuous transitions and with a solid line continuous.

5.1 Inspected component with fixed yearly periods of maintenance

This is an inspected and repairable component. It includes also periodical maintenance of fixed duration.

5.2 Fixed probabilities

This is a special model describing the behaviour of components which unavailability U(t) is given by a step function.

Such behaviour is modelled by a two states markovian graph where $p_1(t) = U(t)$. If the unavailability from $t = 0$ to $t = T$ is U1 and at $t = T$ changes to U2 then: The initial states probabilities (at $t = 0$) are given by: $p_0(t) = 1 - U1$ and $p_1(t) = U2$.

If U1 < U2 a discontinue non periodic transition is added from node 0 to node 1 at $T_i = T$. The transition probability is given by $P = (U_2 - U_1)/(1 - U_1)$.

If U2 < U1 a discontinue non periodic transition is added from node 1 to node 0 at $T_i = T$. The transition probability is given by $P = 1 - (U_2/U_1)$.

For each new change of unavailability a new discontinue non periodic edge is added according to the above procedure.

6 NUCLEAR POWER PLANT CASE

In a BWR in Germany, the intention has been developed to perform scheduled maintenance on the residual heat removal system (RHRS) during full power operation rather than during refueling. Specifically, the following operating conditions appeared desirable:

- The trains can be taken out of service for scheduled maintenance for at most 14 days per year, one at a time.
- If by periodic test, a train is found to be defect, it may be repaired for 28 days. If the unavailability period exceeds 28 days, the plant is shut down.
- If during either scheduled maintenance or repair, failure of an additional train is revealed, the plant is shut down after 24 hours, unless at least one of the trains can be put in service again before 24 hours have elapsed.

In order to investigate the influence of these changes to the safety measures of the relevant systems, a new PSA analysis was performed using the proposed methodology.

First a new fault tree for the RHRS and the corresponding power supplies was constructed and the corresponding cut sets were calculated. In a next step each component was modeled with the proper markovian state model.

In Figure 1, the state model of an inspected-repairable component with fixed maintenance periods is depicted, while in Figure 2, a basic event with fixed probabilities changing over time is modeled, used to model common cause failures.

In the case of maintenance under operation, the common cause failure probabilities are step functions, with steps at the time points where the redundancy level is changed because the train is put in maintenance.

After the repair and inspection policy was determined, the maintenance schedule should be decided. In order to have conservative results, the most unfavorable period for preventive maintenance is at the end of an operational year. It is also favorable during a preventive maintenance to proof the functionality of the two trains sharing the same pumping facility, which is different from the corresponding of the train under maintenance. Additionally is reasonable to follow right after the end of a maintenance an inspection of the corresponding train.

Figure 2. State graph model for components with fixed probabilities.

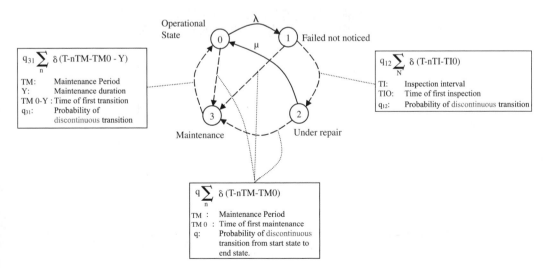

Figure 1. Observable component state graph model.

A maintenance schedule which satisfies the preceding demands is shown if Table 1. Note that a surveillance test cycle lasts 4 weeks.

During an operational year, dependencies exist between maintainable components. For example when a train or any other relevant component is under repair, it is obvious that a scheduled maintenance will be shifted or postponed. Also, all the combinations of failures which are not possible due to a preceding transition to a safer state should be excluded from unavailability assessment. A set of rules which takes into account these dependencies was developed and is given in Appendix A.

The result of the analysis is shown in Figure 3 where the RHRS system unavailability is drawn.

The developed method will be used to underline and quantify the possiblity of preventive maintenance in npp with four train redundancies during full power operation. Further applications in optimisation of the plant's techspecs are thinkable.

REFERENCES

Becker, G., Camarinopoulos, L. 1993, Failure frequencies of non-coherent structures, Reliability Engineering and System Safety 41, pp 209–215.

Becker, G., Camarinopoulos, L., Zioutas, G. 1994a, A Markov type model for systems with tolerable down times, Journal of the Operational Research Society 45, No. 10, pp 1168–1178.

Becker, G., Camarinopoulos, L., Ohlmeyer, W. 1994b, Discontinuities in homogeneous markov processes and their use in modeling technical systems under inspection, Microelectronics Reliability, Vol. 34, No. 5, pp 771–788.

Becker, G. Camarinopoulos, L., Zioutas, G. 1995, An inhomogenous state graph model and application for a phased mission and tolerable downtime problem, Reliability Engineering and System Safety, 49, 51–57.

Becker, G., Camarinopoulos, L., Zioutas, G. 2000, A semi-Markovian model allowing for inhomogenities with respect to process time, Reliability Engineering and System Safety 70, pp 41–48.

Becker, G., Camarinopoulos, L., Nagel, J., Zioutas, G. 2001, Simultaneous optimization of test and maintenance intervals of given components using non-homogenous semi-Markovian models, Journal of the Operations Research society 52, 559–566.

Camarinopoulos, L., Obrowski, W. 1981, Construction of tolerable down times in the analysis of technical systems. Nuclear Engineering and Design 64, 185–194.

Cepin, M., Martorell, S. 2002, Evaluation of allowed outage time considering a set of plant configurations, Reliability Engineering and System Safety, 78, 259–266.

Table 1. Schedule of RHRS preventive maintenance.

Maintenance	Tested train	week
Train 3	Tr. 2	40
	Tr. 1	41
Train 2	Tr. 3	42
	Tr. 4	43
	Tr. 2	44
Train 4	Tr. 1	45
	Tr. 3	46
Train 1	Tr. 4	47
	Tr. 2	48
	Tr. 1	49

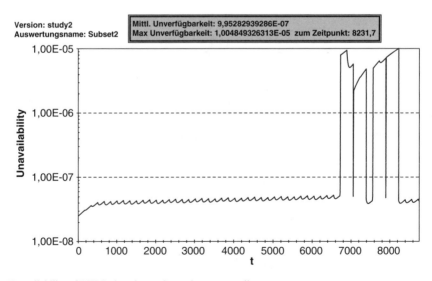

Figure 3. Unavailability of RHRS given in-service maintenance policy.

Cepin, M., Mavko, B. 1997, Probabilistic safety assessment improves surveillance requirements in technical specifications, Reliability Engineering and System Safety, 56, 69–77.

Martorell S. A., Serradell V. G., Samanta P. K. 1995, Improving allowed outage time and surveillance test interval requirements: a study of their interactions using probabilistic methods, Reliability Engineering and System Safety, 47, 119–129.

Papazoglou, I.A. 2000, Semi-Markovian reliability models for systems with testable components and general test/outage times, Reliability Engineering and System Safety, 68, 121–133.

Vilemeur, A. 1992, Reliability, availability, maintainability and safety assessment, John Wiley & Sons.

APPENDIX A: STATE CUT SETS FILTERING

In order to generate a state cut set list, must be established a state cut set rejection scheme. This scheme is a set of rules where each state cut set is tested upon and rejected or not.

In VardaGen application each state of a state graph model is assigned a set of attributes. To the inspected-repairable component of Figure 1 the following attributes are assigned:

Table 2. Inspectable-maintenable component attributes.

State number	Attributes	Description
0	I	Intact
1	F, FU	Failed, Failed unnoticed
2	F, URL	Failed, under repair Less than 14 days
3	F, URM	Failed, under repair more than 14 days
4	F, M	Failed, in maintenance

A rejection rule is represented by the following form:

(A, y), (B, z),…

The action of the rule is

IF (N(A) $\geqslant$ y AND N(B) $\geqslant$ z AND …) Reject State Cut Set

where N(A) represents the number of states that posses the attribute A.

The state cut sets filtering mechanism consists from any desired number of rules combined with an **OR** operation:

IF (Rule1 OR Rule2 OR Rule3 OR …) Reject State Cut Set

For filtering the state cut sets of the RHRS system of Krümmel nuclear power plant in Hamburg, the following set of rules was used:

(M, 1), (UR, 1)
(URL, 2)
(M, 2)
(URM, 1)
(I,1)

Safety and Reliability – Bedford & van Gelder (eds)
© 2003 Swets & Zeitlinger, Lisse, ISBN 90 5809 551 7

Safe transmission of the end position of a railway switch using a cable with four wires

G. Becker, U. Hussels
RISA Sicherheitsanalysen GmbH, Germany

A. Behr
Siemens AG Transportation Systems, Germany

ABSTRACT: An old and well known technical solution to transmit the status of a railway switch using a cable with four wires has been investigated using fault tree analysis and modern indicators for common cause failures. Results show, that the old technique is acceptable even in the light of much younger methods of investigation.

1   INTRODUCTION

During the first decades of last century, when copper was extremely expensive in Germany, a supervision for switches has been developed, which requires only four wires to notify about the position of the switch and to operate the motor of the switch. As a switch has two end positions and one unsafe intermediate position, the absolute minimum to code this information using only electrical switches inside the mechanical switch is three wires. If a fourth wire is used, this means, that there can be some redundancy in the system to be used for safety purposes. Although this four wire supervision is old and proven technique, which requires no new safety case, it is interesting to measure it using modern probabilistic approaches for fault tree analysis and common cause failure analysis.

2   SYSTEM DESCRIPTION

2.1   *Description of 4-wire supervision*

Between the four wires, there are four contacts realized by electrical switches, which can connect or disconnect

one pair of wires each. The following coding is used to denote final positions a, and b, and intermediate position M.

Only the wires given must be connected, all other ones are disconnected to define a position. All other combinations are identified by the signal box as illegal. Figure 1 denotes the four wires and contacts in final position a.

For the purpose of this analysis, the system is defined to include the external four wire cable, the electrical switches, and contactors between cable, and between cable and electrical switches.

Table 1.   Position coding.

Connected wires	Position
1–4, 2–3	Final position a
1–3, 2–4	Final position b
1–2–4	Intermediate position M

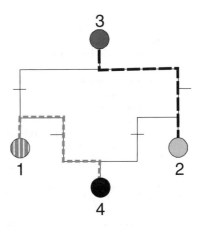

Figure 1.   Four wires and contacts in final position a.

Table 2. Failure modes of electrical components.

Failure mode	Description
UL	Interruption of conductivity
LL	The concerning conductor is loosing its fixed position on the neighbouring component
II	Decrease of insulation resistance or short-circuit inside the component
FI	(at least) One conductor of the component is loosing its fixed position and contacts another conductor inside the component
FA	(at least) One conductor of the component is loosing its fixed position and contacts a conductor outside the component
BI	Bridge circuit over foreign substance or third components inside the component
BA	Bridge circuit over foreign substance or third components outside the component

Table 3. Failure modes of switches.

Failure mode	Description
ON	The switch does not open on demand
SN	The switch does not close on demand
OF	The switch opens erroneously
SF	The switch closes erroneously

2.2 Safety target

The 4-wire supervision has to notify the final positions of the switch rail correctly. That means an end position of the switch rail must not be notified erroneously. The intermediate position may be given erroneously from safety viewpoint.

2.3 Failure modes

The components which are part of the 4-wire supervision can be divided in electrical components like the 4-wire cable (external and internal), and various connectors and in electro-mechanical components like the switches.

There are different ways how these components can lose their operational functions.

To obtain a regular view of possible failure modes a survey in accordance with Failure Mode Effect Analysis (FMEA, IEC 812 1985) has been conducted. The results for the electrical components and the switch are shown in Table 2 and Table 3.

3 OPERATING CONDITIONS

Before providing a fault tree the operating conditions for 4-wire supervision must be identified. Proceeding from the safety target the corresponding event in the fault tree is "notifying a switch rail end position erroneously".

This erroneous feedback to the signal box can occur under different operational conditions:

– Failure on changing position/subsequent to the change of position
– Failure without changing position (without supply of energy)

In addition, it has to be distinguished, if the change of position starts at an end position or at immediate position respectively if the switch rail is located in end position or immediate position.

4 PROBABILISTIC ANALYSIS

4.1 Fault tree analysis

The fault tree of the 4-wire supervision is divided in one main tree and several sub trees. These fault trees are given in form of tables, which present the information in a more condensed manner than the normal graphical representation does. Each line denotes an event, which can be a gate (OR; AND) or basic event. Basic events are representing single failure of 4-wire supervision and are defined by the name of the component and a failure mode (see Table 4).

Events are indented in columns; the inputs of a gate are idented one position to the right of the gate itself. These columns are called fault tree levels.

Sub tree "2–3 remain connected" shall serve to demonstrate the construction off all sub trees of this level. In its first column ("Level 8") the operational feedback signals of the 4-wire cable are specified, which determine a hazard (see Table 5).

4.2 Hazardous single failures and independent multiple failures

The cut sets of the safety model have been determined. The top event results in cut sets of order one and two. Cut sets of order one are summarized in Table 6. Basic events are coded using three letter abbreviations for the component, two letters for the failure mode, and some additional letters to distinguish different locations of the component, where failure might occur. Cut sets of higher orders will be considered in chapter 5.

On "Level 9" the sub tree is divided into an electrical and mechanical part. This partition will be continued down to the component level, where the appropriate basic events are specified.

The results shown in Table 6 disclose, that

– the switch failures "does not open on demand" (ON) and "closes spuriously" (SF) and

Table 4. Main fault tree 4-wire supervision.

Event type	Level 1	Level 2	Level 3	Level 4	Level 5	Level 6	Level 7	Level 8
OR	Safety target failure							
OR		Supervision failure						
OR			Failure on changing position / subsequent to the change of position					
OR				Failure on changing position on demand				
OR					Failure on changing position going from end position to immediate position			
OR						Supervision erroneously notifies an end position (further on)		
AND							Supervision erroneously notifies end position a (further on)	
OR								2 – 3 remain connected (see **Table 5**)
OR								2 – 4 remain disconnected
AND							Supervision erroneously notifies end position b (further on)	
OR								1 – 3 remain connected
OR								1 – 4 remain disconnected
OR						Supervision erroneously notifies the opposite end position		
AND						...		
OR					Failure on changing position starting from immediate position and reaching an end position			
OR						Supervision erroneously notifies an (opposite) end position		
AND						...		
OR				Failure without changing position (without supply of energy)				
OR					Failure without changing position			
OR						Failure in immediate position		
OR							Supervision erroneously notifies end position	
AND							...	
OR						Failure in end position		
OR							Supervision erroneously notifies (opposite) end position	
AND							...	

Table 5. Sub tree "2–3 remain connected".

Event mode	Level 8	Level 9	Level 10	Level 11	Level 12
OR	2 – 3 remain connected				
OR		Electrical connection by failure in the electrical part			
OR			Failure inside of components of the electrical part		
OR				4-wire cable failure	
BE					decrease of insulation resistance or short-circuit inside the component (2_3)
BE					Loss of fixed position and contacting (2_3)
BE					bridge circuit over foreign substance or third components (2_3)
OR				terminal block failure	
BE					...
OR				internal cable failure	
BE					...
OR			Faulty electrical connection between components of the electrical part (or third components)		
OR				4-wire cable failure	
BE					Loss of fixed position and contacting (2_3)
BE					bridge circuit over foreign substance or third components (2_3)
OR				terminal block failure	
BE					...
OR				internal cable failure	
BE					...
OR			Electrical connection because of electrical failures inside the switch (connector)		
BE				Loss of fixed position and contacting (2_3)	
BE				bridge circuit over foreign substance or third components (2_3)	
OR			Electrical connection because of mechanical inside the switch		
BE				Closed switch does not open (2_3 remain connected)	

– the loss of fixed position and contact inside (FI) or outside (FA) of the component

lead to an incorrect feedback signal concerning the switch position.

5 COMMON CAUSE FAILURE ANALYSIS

5.1 *Disclosure mechanisms and prevention*

In each relevant life-cycle phase of components, that means from the production to the operation, hazardous

Table 6. Cut sets at first order (single failures).

Nr.	Basic event	Failure mode
1	4LZ_FA_1_3/2_3	Loss of fixed position and contact outside the component
2	4LZ_FI_1_3/2_3	Loss of fixed position and contact inside the component
3	KAB_FA_1_3/2_3	Loss of fixed position and contact outside the component
4	KAB_FI_1_3/2_3	Loss of fixed position and contact inside the component
5	KLL_FA_1_3/2_3	Loss of fixed position and contact outside the component
6	KLL_FI_1_3/2_3	Loss of fixed position and contact inside the component
7	KPL_FI_1_3/2_3	Loss of fixed position and contact inside the component
8	SCH_ON_1_3/2_3_BV	Does not open on demand
9	SCH_SF_1_3/2_3_WV	Closes spuriously

Table 7. ECF-causes in the different life-cycle phases.

Life-cycle phase	ECF-causes
Production	Operations in production instructions
Storage	Environmental influences EN60721-3-1 (EN60721-3-1 1998)
Transportation	Environmental influences EN60721-3-2 (EN60721-3-2 1998)
Installation	Operations in installation instructions
Operation	Environmental influences EN60721-3-3 (EN 60721-3-3 1995); electromagnetic compatibility EN50124-2 (EN50124-2 1996)
Inspection/repair	Operations in instructions for maintenance

Table 8. Assessment criteria of ECF-causes.

Assessment	Description
Low	In the same or subsequent life-cycle phase a disclosure mechanism exists; the disclosure mechanism can be a cyclic element or self-signalling
Middle	In the same or subsequent life-cycle phase a disclosure mechanism does not exist, but an uninterrupted prevention between the last disclosure mechanism and the life-cycle phase "operation"
High	In the same or subsequent life-cycle phase a disclosure mechanism and an uninterrupted prevention does not exist or the specified conditions in instructions and standards are too restrictive or there are components which are existing in several copies in the system or

failures can occur. Only failures during operation or persisting until operation are hazardous in terms of the safety target.

For each component considered it must be ensured, that extern adverse conditions in all life-cycle phases can be controlled by prevention mechanisms (ambience conditions and handling conditions) and disclosure mechanisms. It has to be verified for each single failure, in which life-cycle phase a prevention or disclosure mechanism exists.

5.2 Analysis of failures caused by external reasons (ECF)

First, externally caused failures (ECF), that means failures which were caused by non conventional factors acting on the components, have to be identified. The main reasons of external caused failures (ECF) are:

- environmental influences
- operator errors of omission
- operator errors of commission

These reasons for the different life-cycle phases arise from the documents (see Table 7).

After identification, external reasons will be assigned to the failure events (basic events). Thereby an assessment of the external reasons will be accomplished. The assessment distinguishes between high (H), middle (M) and low (L) based on the cognitions of chapter 5.1 (see Table 8).

5.3 Analysis of common cause failures (CCF)

The analysis of common cause failures considers the conditional dependence between failures, that means the independent components A and B will fail if the condition C occurs.

All single failures which have an assessment of "H", "M" or "L" relating to the same ECF cause can occur if that specific ECF cause occurs. Dependencies between signal failures influence the safety of 4-wire supervision, when these single failures are part of the same hazardous failure combination.

The evaluation concerning the ECF causes and life cycle phases is realized in two steps:

1. determination of ECF causes which influence many cut sets (chapter 5.4) and
2. determination of components especially susceptible to many common ECF causes (chapter 5.5).

5.4 Frequently appearing ECF causes

Frequently appearing ECF causes have been determined using the following indicator:

$$\text{part } q = \frac{\sum \text{cut sets caused by ECF cause } z}{\sum \text{cut sets (higher order than 1)}} \times 100$$

Table 9. Frequent appearing GVA-reasons.

Nr.	Percentage	Description of ECF-cause
1	2	Contactor (consistency of design)
2	100	Wiring inspection (maintenance)
3	100	Material; non-iron-metal (production)
4	80	Wiring replacement (repair)
5	80	electrical passive (production)
6	80	Fauna (operation)
7	72	Electrical connection (installation)
8	42	visual control wiring (maintenance)
9	42	Fauna (transportation)
10	42	Fauna (storage)

The results shown in Table 9, are sorted on decreasing percentages, with the exception of line 1. The ECF cause "consistency of design" of the connector, is relevant only in 2% of the cut-sets but with the assessment "H" (see 5.2).

The consideration of the percentages shows that GVA-reasons appear frequently in the life-cycle phases "repair", "production", "maintenance", "opertion" and "installation".

Conditional on the multiplicity of instructions in the life-cycle phases "repair", "maintenance" and "installation", the percentages of ECF causes in these life cycle phases are high.

5.5 Components especially susceptible to ECF causes

To determine components especially susceptible to ECF causes (and thus to CCF) the cut-sets detected have been reduced to component types. E.g. the cut set "4LZ_BI_1_3 AND 4LZ_UL_2_3" has been transformed to "4LZ", consisting of two failures of the component type 4LZ, which means four wire cable (external).

The results of the analysis are represented in Table 10. In the first three columns the ECF causes are sorted from "high" to "low". The percentage and the absolute number of concurrent ECF causes are listed in the columns 4 and 5. In the last two columns the involved component types are specified. If only one component type is listed, the cut-set is a combination of failures in that component type (see above).

The contactor (KPL) is the only component type which exists in several copies in the 4-wire supervision (consistency of design). Therefore, influence of this ECF cause has been assessed "high". Underneath combinations of two component types with a conformity of hundred percent in their ECF cause are listed. The absolute number of ECF reasons relevant

Table 10. Component types susceptible to CCF.

Strength of ECF cause

High	Middle	Low	Relative (%)	Absolute	Component involved	Types
1	0	49	100	50	contactor	
0	0	49	100	49	terminator block	
0	0	49	100	49	internal cable	
0	0	49	100	49	contactor	terminator block
0	0	49	100	49	internal cable	contactor
0	0	49	100	49	internal cable	terminator block
0	0	24	100	24	external cable	
0	0	24	100	24	external cable	contactor
0	0	24	100	24	external cable	terminator block
0	0	24	100	24	external cable	internal cable
0	0	18	62	18	switch	contactor
0	0	18	62	18	switch	terminal block
0	0	18	62	18	switch	internal cable
0	0	10	41,6	10	switch	external cable

to the wole combination of component type, decreases from top to bottom.

In the last four lines the switch in combination with other component types is shown. A cut-set only consisting of switch failures does not exist, because the switch failures "does not open on demand" and "closes erroneously" are already single failures (see also 4.2).

6 SUMMARY AND CONCLUSIONS

By means of modern probabilistic methods like fault tree analysis and common cause failure analysis the proven technique of 4-wire supervision has been analysed. The results of the evaluation of the 4-wire supervision model are the hazardous single and multiple failures.

By using the ECF and CCF analysis, external influences could be identified, which can cause the concurrent arising of independent multiple failures.

The main result of the analysis is that the 4-wire supervision is a safe method for observing switch rail positions.

REFERENCES

EN 50124-2: Bahnanwendungen – Isolationskoordination, Teil 2: Überspannungen und geeignete Schutzmaßnahmen; prEN 50124-2; Oktober 1996

EN 60721-3-1: Klassifizierung von Umweltbedingungen, Teil 3; Hauptabschnitt 1: Langzeitlagerung; März 1998

EN 60721-3-2: Klassifizierung von Umweltbedingungen, Teil 3; Hauptabschnitt 2: Transport; März 1998

EN 60721-3-3: Klassifizierung von Umweltbedingungen, Teil 3; Hauptabschnitt 3: Ortsfester Einsatz, wettergeschützt; September 1995

IEC 812: Analysis techniques for system reliability – Procedure for failure mode and effects analysis (FMEA); 1985

Safety and Reliability – Bedford & van Gelder (eds)
© 2003 Swets & Zeitlinger, Lisse, ISBN 90 5809 551 7

Risk reduction prioritization using decision analysis

Tim Bedford & John Quigley
University of Strathclyde, Glasgow, Scotland

ABSTRACT: The ALARP principle is applied in many areas to regulate the tolerable level of risk. Usually the principle is operationalized by assigning a value per fatality. A cost-benefit analysis is used to trade the expected value of lives saved with the costs of technical measures required to reduce risks.

In sectors in which risks have been reduced over a period of years, it is difficult to pinpoint those areas in which further risk reduction might be sought. In this paper we show how many different risk reduction mechanisms can be considered simultaneously in a decision analysis framework. Using influence diagrams it is straightforward to build mini decision analysis models in which competing alternatives addressing the same risk can be compared. The mini-model decision alternatives are assembled into decision strategies representing the best possible combination of alternatives at different cost/benefit ratios. Disynergies between the different alternatives are highlighted through the model.

1 INTRODUCTION

The Railway Group is legally bound to use the ALARP framework as the basis for decision making to reduce risk on the Railways. Although many decisions taken with the objective of reducing risks will be made without a formal Cost-Benefit Analysis (for example, because all stakeholders are in agreement that the changes should be made or because of compatibility with European regulations [1]), the larger scale investment decisions necessary to reduce risks substantially will be subject to such an analysis.

The Railway Group adopts a risk reduction alternative, under the ALARP-CBA framework [2], when the alternative is "reasonably practicable". That is, when the net expected costs of implementing an alternative (discounted, currently by 6% per year, if necessary to be measured in present costs) divided by the expected net number of fatalities saved (discounted, currently by 2% per year, if necessary to be measured in present numbers of lives saved) is less than the prescribed "value per fatality" (VPF) currently (2003–4) set to £M 1.30. Where the alternatives would reduce catastrophic risks involving multiple fatalities a prescribed VPF of around £M 3.65 is used (see [7]). For any risk reduction measure one can, in principle, determine the savings in the number of statistical lives and the expenditure in cost of implementation. The ratio of these is what we call the *implicit VPF* of the measure. If that implicit VPF is above the prescribed VPF then the measure is considered not to be

reasonably practicable and is not implemented. An exception has been made with the introduction of the Train Protection and Warning System (TPWS), which costs approximately £M 10 per life saved, but which has been mandated. The prescribed VPF (of £M 1.30 or £M 3.65) is determined using a formula of the UK Department of Transport, based on the cost to society of a road fatality [7].

It should be noted that there is a large literature on the valuation of (statistical) human lives. The Treasury document [5] discusses the economic context of determining a VPF. Since there is no market valuation of a human life (as there might be for commodities in simple cost-benefit analysis), it is necessary to assess implicit valuations through questioning "willingness to pay" for risk reduction and "willingness to accept" (compensation) for risk exposure. The difficulties with this approach are nicely discussed in [6].

The ALARP-CBA framework for risk reduction is one of the ways of determining which risk reduction alternatives are mandatory. Other alternatives (such as TPWS) are mandatory if the government (*in casu* the Railway Inspectorate of HSE) decides to mandate them. The Railway Group might also voluntarily decide to adopt risk reduction alternatives even if they were not mandatory.

The Railway Group has, however, defined a set of Safety Objectives for 2009 [3]. Broadly speaking these state that a wide range of risks should be halved from their 1999 level. These safety objectives were identified and agreed independently of the ALARP framework. An important question therefore is whether it is possible to link those objectives to the ALARP framework and determine what levels of risk reduction might correspond to what levels of VPF.

This paper addresses the question of whether a model can be built to consider the impact of different prescribed VPF levels on risk reduction. Such a model should address issues such as:

- Are the objectives feasible? If so, what is the required prescribed VPF to meet the objectives?
- If objectives are feasible, can we support the process of stakeholder involvement in prioritizing measures, for example by determining alternative measures that would increase stakeholder insights into the problem?
- Can we use the model to optioneer, that is, to creatively generate new decision alternatives that can be used to reduce risks?
- Can the model be used to illustrate the changes in different risk measures?
- Can the model be used to illustrate diverse impacts of decision alternatives?

The model would also have to build on existing risk assessment work carried out by Railway Safety, primarily the Safety Risk Model, and would output information in terms of:

- Reductions in Equivalent Fatalities per category (Staff-major, staff-minor, staff-fatalities, passenger-major, passenger-minor, passenger-fatalities, public-major, public-minor, public-fatalities).
- Expected operational costs per alternative
- Expected total costs per alternative.
- The risk measures mentioned in the Railway Group Safety Plan.

This paper arises from work carried out by the University of Strathclyde for Railway Safety. The contents of this paper do not represent policy of Railway Safety or of the Railway Group. In Section 2, we provide some background information to Railway Safety and describe the risk assessment currently conducted. We introduce decision analysis models in Section 3 and propose a five-step procedure for identifying appropriate risk reduction measures. Illustrative examples are provided in Section 4 and in Section 5 we reflect on the challenges associated with implementation.

2 RISK ASSESMENT IN THE RAILWAY SECTOR

The UK Railways are engaged in an extensive program of risk reduction. This is partly driven by external regulation (both at a European level and by UK Government policy) which lead to mandated changes to infrastructure and/or working practices. It is also driven by internal policy in two ways. Firstly, there is the general introduction of improved working practices and improved infrastructure where a benefit can clearly be gained. Secondly for improvements that have a less clear cost-benefit the ALARP framework is applied to decide on whether a particular risk measure should be put in place. The Railway Group is in a somewhat strange position, for despite it being a privatized industry, large infrastructure projects are still dependent on the injection of government money. The Department of Transport VPF is around £M 1. However, the government is not entirely consistent as is shown by the decision to mandate the use of TPWS, which has been estimated to cost around £M 10 per human life saved. This raises the question of whether other risk reduction measures, which would cost between £M 1 and £M 10 per human life saved should be implemented.

Railway Safety (currently a subsidiary of Network Rail) is responsible for assessing risk and advising on Railway Group policy for risk reduction. Amongst many other initiatives taken by Railway Safety, a large scale risk model has been built called the Safety Risk Model (SRM). The SRM is a combined fault tree/ event tree model that estimates risks across a variety of different categories (third party, passengers and staff involved in minor, major and fatal incidents). It can be used to gain an understanding of how the current risk is built up, and therefore to indicate what the effect of different risk reduction measures might be.

Despite some (known) conservatism in the SRM, it has the potential to play an important part in supporting decision making on risk reduction.

3 DECISION ANALYSIS MODELS

Figure 1 illustrates conceptually how decisions about risk reduction alternatives can be viewed in a decision analysis framework. The rectangle represents the

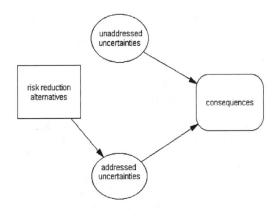

Figure 1. Decision analysis framework for risk reduction.

decision alternatives (risk reduction alternatives) available to the decision maker. The ovals represent uncertainties, only some of which are impacted upon by the risk reduction alternatives. The rectangle with rounded corners represents the consequences as measured by the performance measures. Arrows on the diagram illustrate the influence of one object on another, and which can be probabilistic (changing the likelihood of an event) or deterministic (one node being a function of the input nodes).

The set of all risk reduction mechanisms and alternatives is rather large, indeed, one might usefully think of these as portfolios of possible risk reduction mechanisms rather than individual mechanisms since individual risk reduction alternatives may be combined to produce a risk reduction strategy.

Decision analysis, with its emphasis on considering different decision alternatives takes a slightly different perspective to that commonly used in considering safety policy. Consider for example measures to reduce the number of signals passed at danger (SPAD's). Crudely, there are three main options: Status quo; TWPS, and ATP (Automatic Train Protection). These measures could potentially be implemented together. However as nearly all the benefit of TPWS would disappear if ATP was implemented there would be little point. (Clearly there are all sorts of "in between" options in which regional versions of these systems might be implemented, but we are ignoring these for simplicity of the discussion). Hence it is useful to consider the options as being real alternatives. In fact, ALARP cases usually present a number of different alternatives and in this sense are structured as a decision problem.

The main difficulty in approaching the problem of analysing the effect of different measures is the sheer number of potential measures, and the number of different possible impacts they can have on the system. The SRM is a collection of more than a hundred integrated fault trees and event trees relating relevant hazards to risks.

The philosophy of the approach used here is that different risk reduction mechanisms are considered individually in so-called mini-models, where a substantial proportion of the overall risk associated with various key hazards are modeled as a linear function of a few root causes that can be influenced through various measures. These mini-models just consider the impact of individual mechanisms, so that suboptimal alternatives can be excluded. Then the remaining mini-models are assembled into a global model in which alternatives achieving a similar level of risk reduction cost-effectiveness are put together in a so-called strategy table. A strategy table is used in DPL [4] to reduce the number of *combinations* of decision alternatives considered (this is important in reducing the complexity of the overall problem, for example it

is unlikely that a combination of an alternative which gives a marginal risk reduction at high cost with an alternative giving a high risk reduction at low cost would give an overall optimal combination. Using the strategy table these combinations can be ignored.)

Step 1
Review the SRM to determine parameters that potentially can be influenced by Railway Group initiatives. This is a considerably reduced set of parameters compared to the full set of parameters in the SRM, e.g. we discovered fault trees where a three quarters of the risk was explained with a quarter of the root causes. To speed up calculations a simplified model of the SRM was made in Excel that only contained the parameters of interest. In general the set of precursors that are worth modeling is a relatively small set of the full collection of parameters. This is because we not only restrict to "main effect" parameters but also to those that would be influenced by plausible risk reduction measures.

Step 2
Identify potential risk reduction mechanisms, each with its own set of alternatives. These are mechanisms which may impact on the SRM input parameters. However, these mechanisms may also impact on the SRM output parameters in a way which differs from that explicitly modeled in the SRM.

This means that extra relationships have to be built explicitly into the decision model.

Step 3
For mechanism (set of alternatives) build a mini model. This enables the model builder to concentrate on the factors relevant to those particular alternatives. Often it is not clear which sorts of trade-offs (for example between operations impacts and investment) would be optimal. The mini model enables the exploration of such issues and the selection of a better defined alternative (which then takes account of such trade-offs) for the overall model.

Step 4
Determine implicit VPFs for alternatives in the mini-models. When two alternatives have a similar implicit VPF, but one has a lower risk reduction then the alternative with the lower risk reduction is discarded.

Step 5
The different alternatives of a similar implicit VPF across different risk reduction mechanisms are assembled into a *strategy*. A particular strategy therefore represents the set of alternatives that would be adopted at a particular level of prescribed VPF. The reason for assembling the alternatives is to look at the way in which the alternatives interact. Such interaction

(when two different alternatives address the same underlying root cause) can reduce the effectiveness of the individual alternatives. Because of interactions between the different alternatives, the implicit VPF of the strategy may be higher than that of each of the alternatives in the strategy.

The implicit VPF and equivalent fatalities reduction is determined using the global model for each strategy. The results are plotted to show how increasing the prescribed VPF reduces the equivalent fatalities.

There are many states of knowledge uncertainties in the high level modelling that is inevitably used in the construction of the mini-models. These impact on the frequency of certain kinds of events appearing in the SRM.

4 EXAMPLE MINI MODELS

To illustrate the way a mini-model works we give a couple of examples. In each case the impact of the

measures on costs, operations, and on precursor events in the SRM are shown. Those precursor events are shown in a dashed box. Figure 2 illustrates the potential effects of different measures taken to improve fencing along the track. The rectangular decision node represents various alternatives (e.g. only at high risk locations; only at high or medium risk locations; everywhere; and the status quo). The total cost and operations impact nodes are used to represent the total costs and the costs to train operators. The two DR nodes represent derailment events due to incursion on the line. The fencing measures will directly reduce the frequency of these precursor events in the mini-model. Expert meetings are used to assess the likely reduction in frequency due to each fencing measure.

Figure 3 illustrates a mini model used to assess the impact of object sensors for detecting unwanted objects on the rail. The alternatives represent different locations in which the sensors will be placed (Urban bridges only, Urban only; Urban and Civil Structures; and the status quo). In this case the expert group believe there to be a high chance of a negative impact on operations through false positive detection of objects.

A third model (not shown) shows the impact of changes to working practices for track workers. These are combined with the impact of different strategies about prioritizing preventive maintenance, which of itself has more operational implications than direct safety implications for trackside workers.

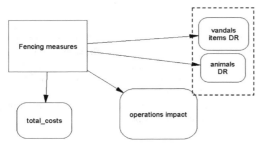

Figure 2. Fencing decision.

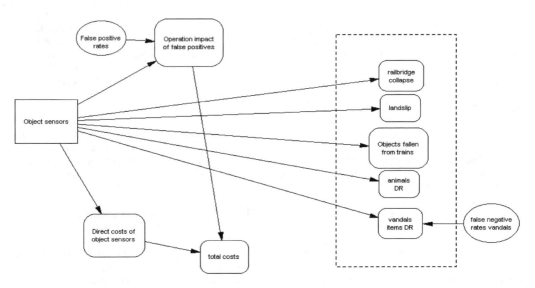

Figure 3. Object sensors decision.

4.1 Strategies of measures

A *strategy* in the global model consists of the combination of alternatives that would be chosen at a specific prescribed VPF. Within DPL this is modeled using a strategy table (see Figure 4). The strategies are named after the corresponding level of implicit VPF achieved by individual alternatives. The strategy table shows the strategies in the left hand column. The remaining columns correspond to the different risk reduction mechanisms and show which strategy each alternative belongs to.

For example, none of the options considered had an implicit VPF of under 10, so the strategy "Under 10" consists of just carrying out the current baseline alternative in each class of risk reduction mechanisms.

For the "Under 20" strategy, only the *Urban bridges* alternative within the set of *Object sensor* alternatives is allowed. For the other mechanisms we would still adopt the current (baseline) alternatives. Hence the results for the "Under 20" strategy will be identical to those of the *Urban bridges* alternative in the mini-model for Object sensors.

The "Under 40" and "Under 60" strategies are actually identical here, as they use the same combination of alternatives – this can be seen because the associated boxes lie next to each other in each column.

The main point of the global model is to understand how the different alternatives interact where they address the same precursor event. Because the two mechanisms impact on the same precursors their combined effect will be non-linear. In other words the effect on the overall risk reduction will be less than the sum of the risk reductions of the individual alternatives. This actually occurs in the examples used, where vandal and animal derailment occur in more than one model. Because we are using the set of precursors already defined within the SRM, we are able to ensure that we will be able to identify those

precursors which are addressed by more than one risk reduction measure. It is necessary to think about the joint impact on the precursor, although in most cases the impact will be multiplicative (as with fencing and object sensor reduction on animal and vandal derailment frequencies).

The overall risk reduction is given by the different prescribed VPF strategies shown in Figure 5. Each point on the graph corresponds to a strategy in the strategy table. The strategies are ordered (left to right) in the same order as they appear in the table. (Note that because the two strategies "Under 40" and "Under 60" are identical there is one less point on the graph then there are strategies in the strategy table.)

This shows that there is a strong tail-off in the extra benefit engendered by a higher prescribed VPF. An adjusted plot (Figure 6) in which the equivalent fatalities are broken down into the sub-categories tells a similar story:

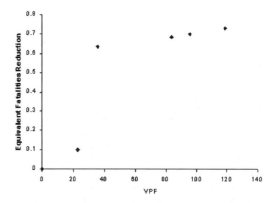

Figure 5.   Overall risk reductions as a function of VPF.

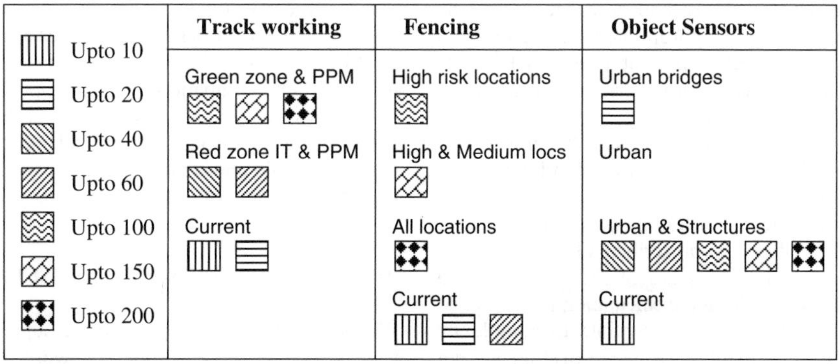

	Track working	**Fencing**	**Object Sensors**
▥ Upto 10			
▤ Upto 20	Green zone & PPM ▨ ▧ ▦	High risk locations ▨	Urban bridges ▤
▧ Upto 40			
▨ Upto 60	Red zone IT & PPM ▧ ▨	High & Medium locs ▧	Urban ▧
▨ Upto 100	Current ▥ ▤	All locations ▦	Urban & Structures ▧ ▨ ▨ ▧ ▦
▨ Upto 150			
▦ Upto 200		Current ▥ ▤ ▨	Current ▥

Figure 4.   Strategy table.

137

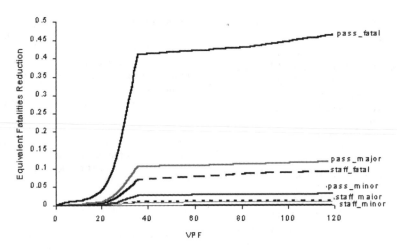

Figure 6.   Risk reduction as function of prescribed VPF, per category.

5   CONCLUSIONS

In this example we have shown how the dependence of risk model outputs on the prescribed can be mapped. Any of the consequences determined by the SRM can be studied in this way, as well as direct costs and operational costs.

The centre of our proposed decision support model contains the SRM or some simplification thereof. The SRM captures relationships between precursors and specified performance measures. Using the SRM is not enough however, as it does not capture the impact of risk reduction alternatives on SRM parameters, as such it is necessary to involve stakeholders in judging the impacts of risk reduction alternatives on SRM parameters (such as precursor frequencies), in discussing the prioritization of performance measures, and in providing a requisite level of modeling.

Stakeholder involvement in assessing the qualitative structure of the model is an important step in building up understanding and confidence that the model is addressing the appropriate issues. The structure of the mini-models can be at a level that is relatively easy to understand during stakeholder meetings. The mini-models can be built up to a global model in a relatively straightforward way because the interactions can be easily flagged. Thus, this framework will facilitate stakeholder involvement.

The main difficulty in building the model of risk reduction dependence on prescribed VPF is to capture the complexity of the myriad of different potential measures without making the calculation too complex. Decision trees are highly susceptible to complexity problems. However in this case the use of strategies not only corresponds nicely to the way management thinks

about the problem, but also enables a big reduction in complexity.

Complexity is also an issue when building the mini-models. This is because there has to be a consistency of approach in order to ensure that the process of "patching together" mini models into a global model using the strategy tables can be done consistently. The major problem here is the potential for inconsistent definitions of events and risks. Fortunately, the SRM acts as a check because it provides a list of pre-defined precursor events. It is unfortunate that no such model exists for analyzing the effects of financial consequences. This suggests that an important area for future work is the definition of a standardized cost breakdown structure that could be used to ensure consistency of approach between different mini-models on the cost side.

Finally, we mention one complex aspect of this problem that has to be modeled: the impact of the decisions on different stakeholders. This is largely a result of the privatization of the rail network. Different cost elements of risk reduction measures are borne by different stakeholders (track owners, maintainers, train operating companies, etc). An important issue in introducing risk reduction measures is therefore the equity of cost distribution. A model such as we have described could play an important role in aiding stakeholder assessment of cost distribution.

ACKNOWLEDGEMENTS AND DISCLAIMER

We would like to acknowledge the support of Railway Safety in carrying out this work. The results published do not represent policy of the Railway

Group, nor do any figures have any other than a purely illustrative meaning.

REFERENCES

1. Blacker, C. and Kirwin, A. "Safety Decision Criteria for Railway Group Standards."
2. Bedford, T. and Cook, R. (2001) *Probabilistic Risk Analysis: Foundations and Methods*. Cambridge University Press.
3. http://www.railwaysafety.org.uk/railplan0102.asp
4. http://www.adainc.com/software/
5. HM Treasury (1996) *The Setting of Safety Standards*. HMSO, London.
6. Dorman, P. (1996) *Markets and Mortality: Economics, Dangerous Work, and the Value of Human Life*. Cambridge University Press.
7. http://www.railwaysafety.org.uk/download/PDF00133.PDF

Safety and Reliability – Bedford & van Gelder (eds)
© 2003 Swets & Zeitlinger, Lisse, ISBN 90 5809 551 7

Time consuming commodity flows in transport networks with vertex buffering

A. Behr

Oranienburger Straße, Berlin, Germany

ABSTRACT: Transport networks can be often modeled by graphs in which commodity units are to be transported between two excellent vertices, a source vertex and a terminal vertex. If the links between the vertices in the transport network are subject to failure then the probability to transport the commodity units successfully depends on the operating probabilities of the links. For networks, in which transport times of the commodity units can be neglected, the problem to compute this probability can be solved in most cases using well-known and broadly discussed results from network reliability (see e.g. Colbourn (1987) for connectivity problems or Ford & Fulkerson (1962) for flow problems in graphs). This does not hold for networks, in which commodity unit transport is time consuming. Here, the probability to transport commodity units from the source to the terminal vertex within a given time depends not only on the (time-dependent) operating probabilities of the links but also on the decision, along which (of the possible) operating links the commodity units are to be transported. Different decisions lead to different probabilities for a successful transport. The present analysis examines this peculiarity. In the end, an efficient algorithm for computing the probability of a successful transport of a single commodity unit is rendered as well as an efficient algorithm for computing this probability by means of an optimum decision strategy.

1 PRELIMINARIES

The topological structure of a transport network can often be modeled by a graph $G = (V, E)$ having a single source vertex $v_s \in V$ and a single terminal vertex $v_t \in V$, ($v_s \neq v_t$). V denotes the set the vertices of G and $E \subseteq (V \times V)$ the set of edges of G with $e_{ab} = (v_a, v_b)$ denoting an edge from vertex v_a to vertex v_b. W.r.t. (v_a, v_b), v_a is called the starting vertex and v_b is called the end vertex of e_{ab}. Let $V_o (v_a) = \{v_b \mid (v_a, v_b) \in E\}$ denote the set of end vertices of edges in G having starting vertex v_a and $V_i (v_b) = \{v_a \mid (v_a, v_b) \in E\}$ the set of starting vertices of edges in G having end vertex V_b. Moreover, (v_a, v_b) is called a leaving edge of vertex v_a and a reaching edge of vertex v_b. Although the notation (v_a, v_b) implies that edges are directed, the results to be presented hold for graphs with directed and/or undirected edges. However, for simplicity, let the graph G be originally loopless, i.e. without edges of the form (v_a, v_a).

We observe the transport network over a given time period T, $T > 0$, modelled in discrete time steps. Over T the vertices of G are assumed to be perfectly reliable, whereas the edges of G may be either failed or operating in a given time step. Referring to edge e_{ab} at time step τ, let p_{ab}^τ be the probability that edge e_{ab} is operating in time step τ. The objective of this analysis is to determine the probability that a set of commodity units, which is located at vertex v_s at time step 1, is transported through G to vertex v_t within time T. For that, travel time of a commodity unit along a single edge should equal one time step – if the edge is operating in this time step. If not, the commodity unit cannot be transported along this edge, which means that the commodity unit has either to be transported over another – operating – edge or it has to be buffered at the starting vertex of the edge as long as the edge remains failed. The travel time through a vertex itself should be neglectable. Finally, the maximum flow along an edge in a given time step should be restricted to a single commodity unit. This restriction is not essential w.r.t to practical applications, since one can include multiple edges between two vertices in order to model larger edge flow capacities. However, it makes explanations easier.

2 THE SINGLE COMMODITY UNIT CASE

A simple case to start with is the transport of a single commodity unit. For illustration, let the well-known

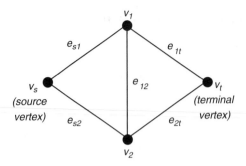

Figure 1. Bridge graph G^B.

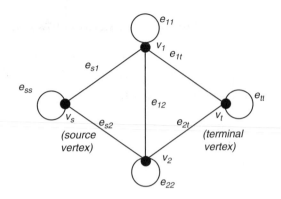

Figure 2. Graph G' for the bridge graph.

bridge structure depicted in Figure 1 be a transport network, through which a commodity unit is to be transported from vertex v_s to v_t.

Since the shortest paths from v_s to v_t in the bridge graph consists of two edges, namely either of e_{s1} and e_{1t} or of e_{s2} and e_{2t}, the minimal time to transport a commodity unit from v_s to v_t may take two time steps. Thus, the probability that the commodity unit is transported from v_s to v_t in G^B in time $T \leq 1$ is zero. For $T = 2$, two observations are worth to be realized: First, travelling along edge e_{12} imposes a travel time of at least three time steps such e_{12} does not play any role in the case of $T = 2$. Secondly, the probability to transport the commodity unit from v_s to v_t in two time steps depends on which edge, e_{s1} or e_{s2}, is tried to be travelled first in the first of the two time steps. If preferably e_{s1} is travelled in the first time step, which would cause that only e_{1t} can be travelled in the second time step in order to reach v_t in two time steps, then the probability to transport the commodity unit from v_s to v_t in G^B in $T = 2$ is given by

$$p_{es1}^1 p_{e1t}^2 + (1 - p_{es1}^1) p_{es2}^1 p_{e2t}^2 =$$
$$p_{es1}^1 p_{e1t}^2 + p_{es2}^1 p_{e2t}^2 - p_{es1}^1 p_{es2}^1 p_{e2t}^2 \quad (1)$$

Here, the second summand of the left-hand side of the equation covers the fact that edge e_{s1}, although it is preferred over e_{s2}, is not operating in the first time step and therefore cannot be travelled. The operating edges e_{s2} and e_{2t} have to be travelled instead.

On the other hand, if preferably e_{s2} is travelled in the first time step, which would cause that only e_{2t} can be travelled in the second time step in order to reach V_t in two time steps, then the probability to transport the commodity unit from v_s to v_t in G^B in $T = 2$ is given by

$$p_{es2}^1 p_{e2t}^2 + (1 - p_{es2}^1) p_{es1}^1 p_{e1t}^2 =$$
$$p_{es2}^1 p_{e2t}^2 + p_{es1}^1 p_{e1t}^2 - p_{es2}^1 p_{es1}^1 p_{e1t}^2 \quad (2)$$

Since the decision to travel along e_{s1} or along e_{s2} in the first time step predetermines which edge, e_{1t} or e_{2t}, may only be travelled in the second time step, eqns. 1 and 2 obviously need to be different. Note that this effect is inherent and essential for transport networks with time consuming commodity flows. Widely discussed counter examples in literature are communication networks: if information flow time is – as it is often done – neglected, then the probability that information is transmitted from v_s to v_t over the edges e_{s1}, e_{s2}, e_{1t} and e_{2t} in G^B is expressed by

$$p_{es1} p_{e1t} + (1 - p_{es1} p_{e1t}) p_{es2} p_{e2t} =$$
$$p_{es1} p_{e1t} + p_{es2} p_{e2t} - p_{es1} p_{e1t} p_{es2} p_{e2t}, \quad (3)$$

i.e. the well-known reliability formula of a simple series-parallel network. It should be clear that due to the lack of flow times eqn. 3 reflects inherently a more reliable structure than eqn. 1 or eqn. 2. In formal terms, it is $p_{es1} p_{e1t} p_{es2} p_{e2t} < p_{es1}^1 p_{es2}^1 p_{e2t}^2$ and $p_{es1} p_{e1t} p_{es2} p_{e2t} < p_{es2}^1 p_{es1}^1 p_{e1t}^2$ in case of identical, non-zero operating probabilities for e_{s1}, e_{s2}, e_{1t} and e_{s2}.

The peculiarity, that the decision which edges are tried to be travelled first determine the probability that a commodity unit is successfully transported from v_s to v_t in G^B, exists obviously also for $T > 2$. We skip therefore the discussion of this case here and proceed to a more general model. In order to account for a possible decision to buffer commodity units at vertices over time, the graph G modelling some transport network in question is extended by loops at its vertices. In formal terms, instead of G we consider the graph $G' = (V, E')$ with $E' = E \cup (v_a, v_a)$ for all $v_a \in V$. The probability that edge (v_a, v_a) is operating at time step τ should be one for all $v_a \in V$. In Figure 2, G' is depicted for the bridge graph.

If D is a fixed decision strategy, by means of which an ordering $(e_{a1}, \ldots, e_{ak})$ for all k, $k \geq 1$, edges leaving an arbitrary vertex v_a in a given graph G' for all time steps τ, $1 \leq \tau \leq T$, is obtained, then

$R_{G',D}^{\tau...T}(v_a, v_t)$ – denoting the probability that the commodity unit is transported from vertex v_a to vertex v_t in G within time $T - \tau + 1$ is expressable by

$$R_{G',D}^{\tau...T}(v_a, v_t) =$$
$$\sum_{i=1}^{k} [\prod_{j=1}^{i-1}(1-p_{aj}^{\tau})] \cdot p_{ai}^{\tau} \cdot R_{G',D}^{\tau+1...T}(v_i, v_t) \qquad (4)$$

This formula is a simple recursive application of pivotal decomposition on all edges leaving vertex v_a (Barlow & Proschan 1975). In using G' instead of G, buffering the commodity unit at vertex v_a may be preferred over transporting it in time step τ. For proving eqn. 4 note that time step τ is different from the time steps $\tau +1...T$, such that the terms on the right-hand side of eqn. 4 are statistically independent. Finally, eqn. 4 is recursively applied for computing $R_{G',D}^{1...T}(v_s, v_t)$.

3 THE MULTI-COMMODITY UNIT CASE

The considerations which led the analysis of the single unit commodity case, are also valid for the multi-commodity unit case. Here, G is considered. Assume $V_b = \{v_{b1}, ..., v_{bn}\}$ to be the set of vertices, where those n, $n \geq 1$, commodity units are located, which should be transported to vertex v_t in graph G within time $T - \tau +1$. Since in principle more than one commodity unit may be located at a vertex in time step τ, V_b needs to be a multiset (Anderson 1989), that is vertices may appear more than once in V_b (The set V_b and the set V_e defined below are the only multisets in this analysis). Let $E_o(V_b) = \{e \in E | e = (v_b, v_e) \cap v_b \in V_b\}$ be the union set of leaving edges of the vertices in V_b. From the assumption, that edge flow capacity is restricted to a single commodity unit in one time step, it follows that $E_o(V_b)$ is an ordinary set and not a multiset. Then, in time step τ, a decision concerning the ordering of all subsets $S \subseteq E_o(V_b)$ with $|S| \leq n$ is required for determining which $S \subseteq E_o(V_b)$ is tried to be travelled first, which secondly, which thirdly and so forth. In contrast to the single commodity unit case, the decision strategy D, by which this ordering is obtained, yields in general an ordered tupel $(S_1, ..., S_k)$ of all subsets of $E_o(V_b)$ with $|S| \leq n$, i.e. it is ordering on the members of the set family $S = \{S \subseteq E_o(V_b)| |S| \leq n\}$ having cardinality $k = |S|$. Clearly, the empty set $\emptyset$ is contained in S. It represents the possible decision to leave over time step τ all n commodity units at the vertices where they are. Now, in analogy to the term

$$[\prod_{j=1}^{i-1}(1-p_{aj}^{\tau})] \cdot p_{ai}^{\tau} \qquad (5)$$

in eqn. 4, which expresses the decision to travel edge e_{ai} in time step τ due to the fact that all preferred edges e_{ai} are failed,

$$\overline{Pr\{\text{all edges in } S_1 \text{ are operating at } \tau \cap ...}$$
$$...\cap\text{all edges in } S_{i-1} \text{ are operating at } \tau$$
$$\cap\text{all edges in } S_i \text{ are operating at } \tau\}$$
$$= Pr\{\bigcap_{j=1}^{i-1}\text{some edges in } S_j \text{ are failed at } \tau\ |$$
$$\text{all edges in } S_i \text{ are operating at } \tau\} \qquad (6)$$

denotes the probability that exactly the edges in set S_i are travelled due to the fact that all sets S_j preferred over S_i contain at least one failed edge and cannot be travelled therefore.

With eqn. 6, the probability that n, $n \geq 1$, commodity units, which are located at the vertices $v_{b1}, ..., v_{b-1}$ and v_{bn}, are all transported to vertex V_t in some graph G within time $T - \tau +1$ can be given as

$$R_{G,D}^{\tau...T}(\{v_{b1}, ..., v_{bn}\}, v_t)) = \sum_{i=1}^{k}[\prod_{e \in S_i}p_e^{\tau}] \cdot$$

$$Pr\{\bigcap_{j=1}^{i-1}\text{some edges in } S_j \text{ are failed at } \tau\ |$$

$$\text{all edges in } S_i \text{ are operating at } \tau\} \cdot$$
$$R_{G,D}^{\tau+1...T}(\{v_{e1}, ..., v_{en}\}, v_t) \qquad (7)$$

with

v_{e1} being the end vertex of (v_{b1}, v_{e1}) if $(v_{b1}, v_{e1}) \in S_i$, otherwise $v_{e1} = v_{b1}$,
v_{e2} being the end vertex of (v_{b2}, v_{e2}) if $(v_{b2}, v_{e2}) \in S_i - \{(v_{b1}, v_{e1})\}$, otherwise $v_{e2} = v_{b2},...,$ and v_{en} being the end vertex of (v_{bn}, v_{en}) if $(v_{bn}, v_{en}) \in S_i - \{(v_{b1}, v_{e1}),..., (v_{bn-1}, v_{en-1})\}$, otherwise $v_{en} = v_{bn}$.

By the forestanding rules the multiset $V_e = \{v_{e1}, ..., v_{en}\}$ is constructed in a way that all edges in S_i are travelled exactly once.

Eqn. 7 is proven in analogy to eqn. 4.

4 OPTIMUM STRATEGY AND COMPLEXITY

From the view of computational complexity it is instantly seen that an exact computation of eqn. 7 is NP-hard. The reason is the #P-completeness of the right-hand side of eqn. 6 (cf. Valiant 1979). However, in practical applications, cases exist, where the decision strategy D leads for all time steps to an ordered tupel $(S_1, ..., S_k)$ which is linearizable, for instance in the way that

143

$$\frac{\Pr\{\text{all edges in } S_1 \text{ are operating at } \tau \cap \dots}{\dots \cap \text{all edges in } S_{i-1} \text{ are operating at } \tau}}{\cap \text{ all edges in } S_i \text{ are operating at } \tau\}}$$

$$= \Pr\{ \text{ all edges in } S_{i-1} \text{ are operating at } \tau$$
$$\cap \text{ all edges in } S_i \text{ are operating at } \tau \}. \qquad (8)$$

Example transport networks are analysed in (Behr & Lüttgert 2001). But, as already stated, for the general case of eqn. 7 there is no efficient algorithm.

It is therefore the more interesting, that the single commodity unit case can be generally solved in $O(|E'| \cdot T)$ time. Moreover, it is solved in $O(|E'| \cdot \log_2 (|E'|) \cdot T)$ time when searching and following an optimum strategy D, i.e. a strategy that maximizes the probability that the commodity unit is transported from vertex v_s to vertex v_t in G within time $T - \tau + 1$. The Algorithm 1 given below performs this computation. It consists of a main procedure CALCR and a recursive procedure CALCR1. The inputs of the main procedure CALCR are the graph, $G' = (V, E')$ which is obtained in $O(|V|)$ time from the original graph G by breadth- or depth-first search, the operating probabilities of all edges of G' over time T – referred to by $\mathbf{p}$ -, time T and, finally, the minimal (topological) distances of all vertices of G to the source vertex v_s – referred to by $\mathbf{mindist}()$. The minimal topological distance between some vertex $v \in V$ and the source vertex v_s equals to the minimal number of edges to be travelled in G in order to reach v from v_s. For all $v \in V$, these distances are obtained by a breadth-first search starting at v_s in $O(|E|)$ time (if broader understanding is needed, analyse Dijkstra's algorithm in the unweighted case (Dijkstra 1959)). The program flow of procedure CALCR is as follows: First, representing $R_{G',D}^{\tau..T}(v_a, v_t)$, an array $\mathbf{R}()$ is initialised for all vertices $v \in V$ and all time steps τ. It is the exact computation of all entries in $\mathbf{R}()$ in procedure CALCR1, that renders $R_{G',D}^{1..T}(v_s, v_t)$ in accordance to eqn. 4.

The computation of the entries in $\mathbf{R}()$ is performed through a breadth-first search on G' starting at vertex v_t at time step T. These are the arguments with which procedure CALCR1 is called by procedure CALCR. In procedure CALCR1, V_{act} denotes the set of vertices from which v_t can be reached in $T - \tau + 1$ time and V_{new} denotes the set of vertices from which v_t can be reached in $T - (\tau - 1) + 1$ time, i.e. generating V_{new} from V_{act} constitutes one step in the breadth-first search. For each $v_a \in V_{new}$, the vertices $v_i \in V_o(v_a)$, that is those vertices that can be reached in a single time step are examined. If v_t can also be reached from v_i in $T - \tau + 1$ time, then

$$R_{G,D}^{\tau+1..T}(v_i, v_t) > 0, \qquad (9)$$

since it was computed in the preceding step of the breadth-first search. Otherwise,

$$R_{G,D}^{\tau+1..T}(v_i, v_t) = 0. \qquad (10)$$

This distinction causes consequently only those terms $q \cdot p_{ai}^{\tau} \cdot R_{G',D}^{\tau+1..T}(v_i, v_t)$ to be summed up to, for which inequality (9) holds. Herein, q is defined as

$$q = [\prod_{j=1}^{i-1}(1 - p_{aj}^{\tau})] \qquad (11)$$

w.r.t eqn. 4.

Since the breadth-first search guarantees that all contributing terms $R_{G',D}^{\tau+1..T}(v_i, v_t)$ are computed prior to $R_{G',D}^{\tau..T}(v_a, v_t)$, each entry of $\mathbf{R}()$ is computed exactly once. This may suggest a computational complexity of $O(|V| \cdot T)$ for Algorithm 1. But, since $\mathbf{mindist}()$ is obtained only in $O(|E|)$ time and since in each step of the breadth-first search in the worst case all edges (v_a, v_i) are needed to determine q and p_{ai}^{τ}, respectively, Algorithm 1 has a computational complexity of $O(|E'| \cdot T)$, which is nevertheless linear in the number of the entries of $\mathbf{R}()$ and therefore efficient.

A slight enhancement of procedure CALCR1 leads to an algorithm that renders $R_{G',D}^{1..T}(v_s, v_t)$ using a strategy that maximises $R_{G',D}^{\tau..T}(v_a, v_t)$ in each step τ of the breadth-first search. By sorting the terms $p_{ai}^{\tau} \cdot R_{G',D}^{\tau+1..T}(v_i, v_t)$ for every $v_a \in V_{new}$ and all $v_i \in V_o(v_a)$ such that those terms $p_{ai}^{\tau} \cdot R_{G',D}^{\tau+1..T}(v_i, v_t)$ with the higher values are preferred over those values with lower values, in each time step τ the next edge to travelled is chosen according to the maximal probability of a successful transport of the commodity unit. The resulting strategy D is thus optimal. In procedure CALCR1, sorting is performed by the function $\mathbf{sortgt}()$. It is assumed that this function sorts the members v_i of $V_o(v_a)$ according to the values of $p_{ai}^{\tau} \cdot R_{G',D}^{\tau+1..T}(v_i, v_t)$ in descending order. For transparency of the program flow, the v_i's are sorted in an array $\mathbf{varray}()$. Sorting the vertices of G' is typically done in $O(|V| \cdot \log_2(|V|))$ time if $\mathbf{varray}()$ is initially unsorted. If a single v_i is sorted into a so far sorted array $\mathbf{varray}()$ – as it is done in procedure CALCR1, this takes typically $O(\log_2(|V|))$ time. By the same reason, by which $O(|E'| \cdot T)$ holds instead of $O(|V| \cdot T)$ in the unsorted case, the required add-on time for a single call of $\mathbf{sortgt}()$ is $O(\log_2(|E'|))$. The overall computational complexity of Algorithm 1 is hence $O(|E'| \cdot \log_2(|E'|) \cdot T)$.

As a final observation, it should be remembered that, if commodity flow times are neglected, the probability that v_t can instantly be reached from v_s through a set of operating edges is known to be an NP-hard

1   **procedure** CALCR($G' = (V, E')$, **p**, T, **mindist**())
/* *CALCR computes* $R_{G',D}^{1...T}(v_s, v_t)$ *for a single commodity unit. The strategy D is chosen either randomly or optimally – see line 23 in procedure CALCR1. This main procedure is used for initialisation.* */
2   **begin**
3   **for all** $v \in V$ **do**
4   **begin**
/* *Here the R's to be computed are initialised. Entries* $R_{G',D}^{T+1...T}(v, v_t)$ *are used for technical reasons only.* */
5   **for** $\tau = 1$ **to** T + 1 **do** $R_{G',D}^{\tau...T}(v, v_t) := 0$;
6   **end**
/* $R_{G',D}^{T+1...T}(v_t, v_t)$ *represents the starting point in the subsequent computation of* **R**() *beginning at vertex* v_t *at time step T.* */
7     $R_{G',D}^{T+1...T}(v_t, v_t) := 1$;
8   CALCR1(G', **p**, $\{v_t\}$, T, T, **mindist**(), **R**());
9   **end**

10 **procedure** CALCR1(G', **p**, V_{act}, τ, T, **mindist**(), **R**())
/* V_{act} *is the set of vertices from which* v_t *can be reached in* $T - \tau + 1$ *time. First in this procedure, for all* v_a, *which can be reached from some vertex* $v_b \in V_{act}$ *in a single time step,* $R_{G',D}^{\tau...T}(v_a, v_t)$ *is computed. Secondly, the procedure calls itself recursively with the set of vertices* V_{new}, *i.e. those vertices from which* v_t *can be reached in* $T - (\tau - 1) + 1$ *time* */
11 **begin**
12    $V_{new} := \emptyset$;
13   **for all** $v_b \in V_{act}$ **do**
14 **begin**
15   **for all** $v_a \in V_i(v_b)$ **do**
16   **begin**
/* *If vertex* v_a *cannot be reached from* v_s *within time* τ *then* $R_{G',D}^{\tau...T}(v_a, v_t)$ *does not contribute to* $R_{G',D}^{1...T}(v_s, v_t)$, *i.e. vertex* v_a *can be skipped.* */
17   **if** $mindist(v_a) \leq \tau$ **then**
18   **begin**
19         $V_{new} := V_{new} \cup \{v_a\}$;
/* **varray**(i) *is an array for all vertices* $v_i \in V_o(v_a)$, *i.e. all vertices that can be reached from* v_a *in one time step* */
20   **for** i := 1 **to** $|V_o(v_a)|$ **do**
21   **begin**
22   **varray**(i) := v_i;
/* *The following line 23 is only to be considered, if an optimum strategy D is to be followed. Otherwise it can be skipped. See the explanations in the text above.* */
[23   **sortgt**(**varray**(i), $p_{a\,varray}^{\tau}(i) \cdot R_{G',D}^{\tau+1...T}(\textbf{varray}(i), v_t)$);]
24   **end**
/* *In the following loop* $R_{G',D}^{\tau...T}(v_a, v_t)$ *is computed according to formula (E1)* */
25   q := 1;
26   **for** i := 1 **to** $|V_o(v_a)|$ **do**
27   **begin**
28           $R_{G',D}^{\tau...T}(v_a, v_t) := R_{G',D}^{\tau...T}(v_a, v_t) = q \cdot p_{a\,varray}^{\tau}(i) \cdot R_{G',D}^{\tau+1...T}(\textbf{varray}(i), v_t)$;
29           $q := q \cdot (1 - p_{a\,varray(i)}^{\tau})$;
30   **end**
31   **end**
32   **end**
33   **end**
34   **if** $V_{new} \neq \emptyset$ **then** CALCR1 (G', **p**, V_{new}, $\tau - 1$, T, **mindist**(), **R**());
35 **end**

Algorithm 1.   Computation of $R_{G',D}^{1...T}(v_s, v_t)$ in $O(|E'| \cdot T)$ time or, respectively, in $O(|E'| \cdot \log_2(|E'|) \cdot T)$ time if an optimum strategy D is searched and followed.

problem (Provan & Ball 1984). In other words, including time dependency into transport network modelling may lead to a computationally solvable problem whereas avoiding time dependencies – perhaps due to pretended complexity considerations – does not.

REFERENCES

Anderson, I. 1989. *Combinatorics of Finite Sets*. Oxford University Press.

Barlow, R. E. & Proschan, F. 1975. Statistical Theory of Reliability and Life Testing. New York: Holt Rinehart, and Wilson.

Behr, A. & Lüttgert, M. (2001). Minimum Delivery Time Transportation with Guaranteed Supply Times – A Discrete Commodity Flow/Discrete Time Reliability Model for Series Digraphs with Buffers. In: E. Zio, M. Demichela, N.Piccinini (eds), Towards a safer World; Proceedings of the European Conference on Safety and Reliability 2001. Torino: Politecnico di Torino.

Colbourn C. J. 1987. The Combinatorics of Network Reliability. Oxford University Press.

Ford, Jr. L. R. & Fulkerson D. R. 1962. Flows in Networks. Princeton, N.J.: Princeton University Press.

Provan, J. S. & Ball, M. O. 1984. Computing Network Reliability in Time Polynomial in the Number of Cuts. Operations Research 32: 516–526.

Valiant, L. G. 1979. The Complexity of Enumeration and Reliability Problems. SIAM Journal on Computing 8: 410–421.

Dijkstra, E. W. 1959. A Note on Two Problems in Connection with Graphs. Numerische Mathematik 1: 269–271.

Safety and Reliability – Bedford & van Gelder (eds)
© 2003 Swets & Zeitlinger, Lisse, ISBN 90 5809 551 7

On the consideration of long-range consequences of decisions

S. Benedikt
Computer and Automation Institute of the Hungarian Academy of Sciences, Budapest, Hungary

I. Kun & G. Szász
Dennis Gabor College, Budapest, Hungary

ABSTRACT: Factors influencing laymen's feeling of danger are not results of any sound reasoning, characteristic of experts. Cognitive psychology has discovered heuristics typical of the way of thinking of everyday people. It can be stated in general that laymen's feeling of danger is usually realistic. It often proves to be a better assessment of risk than expert opinion is on the same subject. Moreover, in a democratic society laymen's opinion cannot be neglected: their taxes form the main financial resource of public investments and they must bear the negative consequences of wrong decisions. In the present paper we show how one can determine a socially acceptable assessment of long-range dangers, appearing possibly more than a human generation later.

1 INTRODUCTION

In an earlier paper (Benedikt et al. 1999) we gave an account on our research concerning application of entropy for the characterisation of danger. The extent of danger of a decision under risk can be appropriately described numerically by thermodynamic entropy while subjective feeling of danger by information theoretic entropy. This latter depends on the occurrence probabilities of individual events.

It is a fact justified by psychological experiments that people look at future risks in a "perspective" way. This means that the farther is risk in time, the less it worries people. The concave character of the time dependence of risk, detailed in the paper, also supports the soundness of the approach based on entropy.

While judging a risk situation laymen recall the set of unfavourable events stored in their memories rather than taking the set of objective facts. This heuristics is known in cognitive psychology as "accessibility", see Simon (1983). Despite its evident lack of perfection this heuristics delivers often a better estimation than expert opinion.

Under the term "socially acceptable" we mean a mathematical consideration of points of view of different characters. That is, in addition to the natural drive for direct profit increase, environmental and other not production oriented factors are given attention which are normally hard to assess but involve laymen's anxiety.

For this end we have to determine probabilities belonging to different time horizons, which leads to a problem that had been solved earlier.

This problem has classical solution methods, belonging to the scope of decision theory. In all such methods however points of view of the decision-maker (risk awareness, empathy, profit) have an almost exclusive role. The solution proposed by us has the advantage that it makes possible for the decision-maker to consider public opinion even in cases when a public opinion poll is not possible.

In order to see the motivation of these ideas, as an example, we have to remember that floods have become regular in Central Europe in the past years, leading to catastrophic situations in several countries of this region in 2002. We know from climatology that, as a consequence of glasshouse effect, atmosphere has warmed up, therefore water circulation moves a larger mass of water, leading on our rivers to floods breaking water height records of several centuries. Henceforth, there is a growing interest in society for a long-range forecast of effects, especially that of dangers, of human actions (Gayer 2000).

2 INDIVIDUAL AND COLLECTIVE RISK

In this section we build up a model describing how the change of risk depends on the change of occurrence probability of the risk event.

We simplify time scale by discretising it, i.e. the length of a time interval can only be the nonnegative integer multiple of a given length.

Following a partly similar reasoning in Chow et al. (1988) Ch. 13, we suppose to have a sequence of time intervals. In each interval a risk event can occur with probability p, independently of what has happened in previous intervals.

Then the number of intervals elapsing until the first risk event and the number of intervals elapsing between two consecutive risk events has the same geometric distribution. All have the same distribution parameter p. The expected value of this distribution is $1/p$. So the expected number $\tau = \tau(p)$ of intervals elapsing before the first or between two consecutive risk events is:

$$\tau(p) = \frac{1}{p} - 1 = \frac{1-p}{p} \tag{1}$$

This is an objective degree of individual safety for a person who faces the given risk event.

Now suppose that we have a group of people and each of them assesses risk on the basis of the above probability structure, according to his/her own probability parameter formed by personal experience. This means that each person has his/her own conditional probability space identical in structure and different only in the parameter value of the geometric distribution, denoted hereafter by z. Now z and $\tau(z)$ are random variables. That is, people know – instead of the exact value of p – only an interval $[s, q] \subseteq [0, 1]$ in which the risk probability parameter is supposed to be distributed uniformly. (This is actually a common *a priori* hypothesis, used in statistics until a better fitting *a posteriori* distribution is found.) The expectation of $\tau(z)$ is

$$E\tau(z) = \frac{1}{q-s} \int_s^q \frac{1-u}{u} du$$

$$= \frac{1}{q-s} \left[\ln q - \ln s - (q-s) \right] \tag{2}$$

If s decreases to 0, then $E\tau(z)$ increases to infinity. This means roughly that when the probability parameter z is very low for some members of the given group, that is the risk event seems to occur practically never for them, then the average safety level of the group tends to increase to infinity.

3 PERSPECTIVE VIEW OF TIME

As it was discussed in Benedikt et al. (1999), time elapsed since the last occurrence of the risk event, denoted by t, as mental stimulus can be handled by the following formula where the feeling of safety, denoted by b, is the response.

$$b \approx \ln t \tag{3}$$

Actually, this is the Fechner-Weber (F-W) physiological law. Its application is justified for mental situations.

The same formula holds if t denotes the time elapsing until the first occurrence of the risk event in the future. (3) reflects the fact justified by psychological experiments that people look at past and future risks in a "perspective" way. This means that the farther is the risk event in time, the less it worries people. "Farther" can be understood for both the past and the future, depending on whether we have factual data relating to past or theoretical data relating to future risk events.

Application of the F-W law for mental responses is supported by descriptions of human intellectual processes, well known in cognitive psychology. These are the heuristics of "fixation and adjustment" (we adjust our own subjective perception of subsequent numbers to the number noticed first), see Tversky & Kahneman (1988).

As an illustration, let stimulus be the average length of the time interval elapsing between two consecutive risk events. E.g. a type of a nuclear power plant accident for a given nuclear plant is supposed to occur once in 10^4 years, another type once in 10^5 years, i.e. ten times more rarely. For the sake of simplicity let $k = 1/\ln 10$. Response is the feeling of safety (in the above situation the change is from 4 to 5, i.e. only one unit). Here we have theoretical information on future occurrence of risk events, computed by means of probabilistic safety analysis, well known in engineering practice.

4 SUBJECTIVE RISK ASSESSMENT

We apply the F-W principle for the risk perception situation. Starting from (1), we observe that $\tau(p)$ expresses actually the length of a time interval, therefore we may apply (3).

$$\ln \tau(p) = \ln \frac{1-p}{p} \tag{4}$$

We estimate the community perception level of the riskless time horizon – analogously to Section 2 – by computing the integral mean (i.e. the expectation of conditional expectations, where the condition is a given value of z) of τ on this interval:

$$E \ln \tau(z) = \frac{1}{q-s} \int_s^q \ln \frac{1-u}{u} du$$

$$= \frac{1}{q-s} \left[u \ln \frac{1}{u} + (1-u) \ln \frac{1}{1-u} \right]_s^q$$

$$= \frac{1}{q-s} \left[H(q) - H(s) \right] \tag{5}$$

where $H(u)$ is the entropy of the simple alternative distribution with probability parameter $0 < u < 1$.

This means that entropy is a relevant measure of the feeling of risk.

In case that we consider the interval $(0, p]$ (i.e. members of the observed community assess risk probability to be between 0 and p, 0 excluded for the sake of mathematical precision), equation (5) reduces to

$$E \ln \tau(z) = \frac{H(p)}{p} \tag{6}$$

In order to illustrate the above reasoning, we can consider a situation, not unknown in present day Eastern Europe. A river flows across two neighbouring countries. Flood control dams for all rivers were designed some decades ago to resist in one country a 50 years maximum flood level (a risk event with $p = 0.02$), in the other country a 5 years maximum flood level (a risk event with $p = 0.2$). Environment devastation accompanying the change of political system in the past decade has led to a change of the discharge curve of the river, with an essential increase of the maximum flood level. In the second country, considering all rivers, bursting of a dam had earlier been an event occurring several times each year. Public in that country did not worry about the increase of p. The same increase of p however shocked public in the first country, where, considering all rivers, bursting of a dam had earlier been quite unusual. This is in good agreement with the shape of the graph of $H(p)/p$: convex decreasing with $\lim_{p \to 0} H(p)/p = \infty$.

5 ESTIMATION OF LONG-RANGE RISK

Suppose that we have an estimation for the perceived length of the riskless time period in a social group. We get this in (4) for the case when there is a unique value p accepted by the whole group while in (6) for the case when possible values are uniformly distributed between 0 and p.

We denote by $S(p)$ the number of time intervals up to the first occurrence of the risk event, including the latter, as perceived in the group according to the F-W law. Evidently S is greater by 1 than the perceived length of the riskless period since we add now the first period when the risk event occurs.

Now if we denote by L the loss due to the risk event and by $R(L)$ the average loss for a perceived time unit then

$$R(L) = \frac{L}{1+S} \tag{7}$$

In case of an individual value of the risk probability, we get from (4):

$$R(L) = \frac{L}{1 + \ln \frac{1-p}{p}} \tag{8}$$

while in case that the values of the risk probability are dispersed, we get from (6):

$$R(L) = \frac{L}{1 + \frac{H(p)}{p}} \tag{9}$$

In both cases, since for a risk event $p \ll 1$ therefore $\ln(1-p) \approx 0$, and we get the following simpler approximate formula:

$$R(L) = \frac{L}{1 - \ln p} \tag{10}$$

where, according to information theoretic concepts, $-\ln p$ is actually the information gained by the occurrence of the risk event.

It is easy to check that for the function

$$f(p) = \frac{1}{1 - \ln p} \tag{11}$$

the following relations hold:

$$\lim_{p \to 0} \frac{f(ap)}{f(p)} = 1 \quad \text{for } 0<a<1 \tag{12}$$

$$\frac{df(p)}{dp} > 0 \quad \text{for } 0<p<1 \tag{13}$$

$$\lim_{p \to 0} \frac{df(p)}{dp} = \infty \tag{14}$$

We note that $f(p)$ is the special case of a more general weighting function $f(p, B)$ for $B = 1$, discussed in more detail in Benedikt et al. (1999).

(12) means that the smaller risk probabilities actually are, the less we can distinguish among their effects on frequencies of occurrence. This is absolutely in accordance with the fact that an average person with no mathematical expertise hardly feels the difference between two different orders of magnitude.

6 EXAMPLE

We give an illustration for the accessibility heuristics and perspective view of time, based on real life data. The following data were collected in Hungary in 1999 during a poll about the danger of bursting of the flood control dam on the river Tisza (Vári 2000, p. 41), based on past events.

Data in column of zone 2.55 are somewhat ambiguous. Elderly people answer 1948, younger people 1970 or later, everybody according to either hearsay or personal experience. The flood in 1970 was actually

149

Table 1. Numerical values of a poll – year of the last dangerous flood before 1998 according to the memory of people living on the territory concerned, given as percent values of relative frequencies.

Year	Lower Tisza zone 2.54	Upper Tisza zone 2.55	Upper Tisza zone 2.57	Altogether
1948	–	54.5	1.3	22.2
1970	71.4	37.9	97.4	69.3
Other year	28.6	7.6	1.3	8.5

Table 2. Numerical values of poll on danger of bursting of the flood control dam, given as percent values of relative frequencies.

Probability of bursting of dam (%)	Lower Tisza zone 2.54	Upper Tisza zone 2.55	Upper Tisza zone 2.57	Altogether
Under 25	84.4	38.1	69.1	62.5
26–50	13.0	27.8	23.0	21.7
Above 51	2.6	34.1	7.9	15.8

much less dangerous in zone 2.55, therefore we can speak in this zone about a 50 years quiet period. For younger people however the quiet period is much shorter.

Data are quoted exactly in the form appearing in the reference. Nevertheless, feeling of danger is complementary to feeling of safety, therefore data can easily be interpreted in the above context.

As Table 1 shows, the last extremely dangerous flood on river zones 2.54 and 2.57 was in 1970. As the cited paper points out, it was on river zone 2.55 in 1998. We can easily see the differences in the feeling of danger. Formula (3) – more exactly, the perspective view of time – is supported by Tables 1 and 2, in so far as where the last flood was in 1970 there is much less feeling of danger (i.e. much more feeling of safety) in population than where it was in 1998.

The dramatic difference between zone 2.55 and the other two zones, demonstrated in Table 2, is due partly to the fact that the flood in zone 2.55 followed a 50 years long quiet period.

7 CONCLUSION

Opinion of potentially endangered people should be considered at the installation of any equipment involving risk. Laymen measure the size of danger not through expert reasoning. Cognitive psychology has discovered heuristics characteristic of the way of thinking of everyday people. This paper presents a model for an individual with unique judgement of risk as well as for a group with dispersed judgements. Using information theoretic concepts, the former depends on individual information while the latter on entropy.

REFERENCES

Benedikt, S., Kun, I. & Szász, G. 1999. Determination of Safety Minimum for a Risk of Very Small Probability. In G. I. Schuëller and P. Kafka (eds), *Safety and Reliability*: 1355–1358. Amsterdam: Balkema.

Chow, V.T., Maidment, D.R. & Mays, L.W. 1988. *Applied Hydrology*. New York: McGraw Hill.

Gayer, J. (ed.) 2000. *Participatory Processes in Water Management. Proc. of the Satellite Conf. to World Conf. on Sciences*. Paris: UNESCO.

Simon, H.A. 1983. *Reason in Human Affairs*. Stanford University Press.

Tversky, A. & Kahneman, D. 1988. *Risk and Rationality: Can Normative and Descriptive Analysis be Reconciled?* Institute of Philosophy and Public Policy.

Vári, A. 2000. Risk Evaluation and Willingness to Make Sacrifices in Three Zones of the River Tisza. In: Rozgonyi, T., Tamás, P., Tamási, P., Vári, A. (eds), *The Flood on the River Tisza. Opinions, Risks, Strategies* (in Hungarian): 33–64. Budapest: Sociological Research Institute of the Hungarian Academy of Sciences.

Safety and Reliability – Bedford & van Gelder (eds)
© 2003 Swets & Zeitlinger, Lisse, ISBN 90 5809 551 7

Bayesian modelling of failure prediction using maintenance information

M. Bhattacharjee & E. Arjas
Rolf Nevanlinna Institute, University of Helsinki, Finland

ABSTRACT: In this paper we illustrated that even for large, complex systems with evident heterogeneity and sparse failure data, incomplete maintenance information, simplistic models with nominal but intuitive assumptions can bring out interesting and useful features of the system. Moreover, inference for both population and individual level inferences can be carried out. A hierarchical Bayesian model based on a latent variable structure is proposed to take into account such heterogeneity in reliability/survival data. This is illustrated by an analysis of failure data from several motor operated closing valves at two nuclear power plants.

1 INTRODUCTION

A characteristic feature of certain types of complex repairable systems is the small number of observed failures. Often, many of the systems that were monitored in the data are totally failure free, and the failures are observed in some systems only. Although incidents of such heterogeneity are quite common with large and complex engineering systems, this issue is not well addressed in the literature. Bhattacharjee et al. (2002) demonstrated that even quite simplistic models could give meaningful results for such a sparse database and in presence of considerable heterogeneity.

We propose that, in addition to spontaneous failure records, scheduled maintenance information could be utilised to provide a better understanding of the system condition and consequently improve failure prediction. Unfortunately, often maintenance information is only partly available. Also, for complex systems, it may not be feasible to analytically express the effect of different maintenance actions on the failure process.

We illustrate the use of hierarchical Bayesian models with a latent structure to address some of these problems by analysing a real data set.

2 DATA

9 years follow up data on 104 motor operated closing valves in different safety systems at two nuclear reactor power plants in Finland are considered. These valves are electromechanical systems consisting of many parts and have quite a complicated structure. The valves are either continuously monitored or regularly tested for malfunctioning.

A valve can experience a failure of the type "External Leakage" when there is a leakage from one of its subcomponents, such as a "Bonnet" or "Packing". Valves are continuously monitored for such failures and are rectified/repaired without delay. 37 such leakages pertaining to only 16 of the 104 systems were recorded altogether. Details of the failure time and their distribution according to different safety systems could be found in Bhattacharjee et al. (2002).

Valves undergo several tests for preventive maintenance purposes. These are categorised typically as one of the following four types, (1) Internal leakage (2) Valve not closing (3) Valve not opening, and (4) Other/Miscellaneous. A valve may or may not fail during such a test.

The systems undergo repairs typically under two types of circumstances. Firstly, if an external leakage has been observed then it is immediately repaired, and secondly, if a system fails during one of its preventive maintenance test. We assume that all repair times are negligible.

Apart from the 37 repairs following external leakages, the numbers of repairs following failures during different maintenance tests were, respectively, 53, 23, 23 and 37 for the four types of tests mentioned above. The maximum number of repairs recorded for any valve was 15, which for this particular valve included 8 external leakages. Altogether repairs of any type were carried out on 67 of the 104 valves, so that remaining 37 valves neither failed spontaneously nor failed in any of the scheduled maintenance tests during the entire study period of 9 years.

3 MISSING DATA AND CONSEQUENCES

Unfortunately the data contain only partial information on the preventive maintenance tests. That is, the failure time and repair action are recorded if and only if a system fails when such a test is carried out.

These tests are carried out periodically, but their periodicities are not same and are mostly not available from the records. Therefore the failure rate for such failures can't be related directly to real time. To emphasize this point, consider the following (fictitious) example. Suppose that for a certain type of preventive testing, say T, the records show that two consecutive failures of the system during the test T were on dates D_1 and D_2. The gap $D_2 - D_1$ between these successive failures can have very different implications regarding the susceptibility of the system to this test depending on the periodicity P of the test T. $D_2 - D_1$ will be approximately a multiple of P, say $K \cdot P$, which will also be the number of times the system was tested during that time. Then an obvious estimate of failure probability would be $1/K = P/(D_2 - D_1) \propto P$. Therefore it would be difficult to infer about the systems failure probabilities in the preventive maintenance test T without having any knowledge about its frequencies/periodicities P.

It is unfortunate that even though the necessity of recording data as complete as possible has been stressed by many and for a long time, it is still not implemented for many systems. This appears to be the case for as critical systems as parts of safety systems in nuclear power plants. Even when the system passes a test and doesn't have to undergo a repair, the test schedule followed, carries significant information regarding the state of the system.

4 PREVIOUS ANALYSIS

The data exhibited remarkable variation with respect to the number of external leakages from any single valve during the period of study, with the majority experiencing no failures (88 of the 104 valves) whereas some recording as many as 8.

Also the inter-failure times of the valves were observed to be highly varying, with very small as well as very large values being observed from the same valve. No specific pattern like systematic deterioration was observed.

Bhattacharjee et al. (2002) modeled the probabilistic behaviour of these systems using a hierarchical Bayesian model with a latent structure. Of the several models that were attempted, evident heterogeneity rejected models treating the valves in an i.i.d. manner. Also models without valve specific control on the models for the inter-failure times performed poorly from a prediction point of view. From these we noted

that even if heterogeneity was allowed between valves, within valve heterogeneity, arising possibly due to repairs and other environmental factors, should be modelled carefully. An appropriate model should be able to capture valve specific behaviour also.

5 PROPOSED ANALYSIS

It is apparent that in order to be able to infer about the state or health of the system as a whole one would require complete information regarding the preventive maintenance tests also. As at any given time point whether the system will fail, either spontaneously or during maintenance tests, is generally affected by whether or not the given time point is one of scheduled maintenance times.

On the other hand, the systems are continuously monitored for spontaneous failures, which implies that complete information on this type of failures is available. However, the inter-relationship of the different failure types, and consequent repairs, is not known. Each of these repairs may have an effect on the health of the system, which in turn can affect the process of spontaneous failures. Therefore, as an indirect assessment of the state of the system, we attempt to identify the effect of these repairs to the spontaneous failures.

5.1 *Model*

For the i-th valve consider a counting process $N_i^0(t)$, $t \in (0, T)$, which counts the number of spontaneous failure epochs (i.e. external leakages) from this valve. For simplicity, we assume that all repairs following any failure affect the failure intensity of the external leakages. The effects of these 5 types of repair need not be the same, although we assume individually that they do not change over time.

Let us denote the four counting processes counting the number of failures in the i-th valve during the four types of scheduled tests by $N_i^j(t), t \in (0;T), j = 1, \ldots, 4$.

Earlier analysis of these data has strongly supported the hypothesis of heterogeneity amongst the valves. Following an earlier suggestion, we assume that there are two types of valves namely "good" and "bad". For valve i a latent variable c_i indicates the membership of that valve in one of these two classes.

Assume the failure intensity of $N_i^0(t)$, $t \in (0;T)$ to be $\lambda_i^0(\theta, t \mid F_{t-})$, where F_t is the observed history of the process up to time t. Let θ_1^*, θ_2^* be the initial failure intensities for the two categories of valve, with which they enter the study, where $\theta_1^* \leqslant \theta_2^*$. We further assume that the same type of repair affects the two classes of valves in different ways. Let θ_k^j, $k = 1, 2$ and $j = 0, 1, \ldots, 4$ be the failure intensities (for external leakage) of the two categories of valves immediately after the system has undergone j-th type of repair. No further ordering was assumed on θ_k^j's.

Then for the i-th valve λ_i^0 is assumed to be given by the following;

$$\lambda_i^0(t) = \theta_{c_i}^* + \sum_{s \le t} \sum_{j=0}^{4} \left((\theta_{c_i}^j - \lambda_i^0(s-)) \cdot \Delta N_s^j \right)$$

The latent variable (c_i) for two valves were decided to be fixed in advance to represent the two categories of the valves. These two valves were chosen for this purpose based on the data. One of the 37 valves which did not fail either due to external leakage or during any of the scheduled tests during the entire 9 years period of study was chosen at random to represent the "good" group, whereas the valve experiencing the largest number of external leakages was chosen to represent the "bad" group.

Of the remaining 102 valves the latent variables (c_i) were assumed to be have Multinomial (p) distribution with p having Dirichlet (I), where I is a vector of appropriate dimension of all ones. Although we have assumed only two possible latent classes for each valve, this is not essential and a larger number can be assumed.

θ_1^* was assumed to have a priori Gamma distribution. The θ_2^j's were assumed to be $\eta_1^j + \theta_1^j$, where the η_1^j's are Gamma distributed, $j = 0, 1, \ldots, 4$. The hyperparameters were chosen so as to give rise to vague enough priors.

5.2 Analysis

The model was implemented by using WINBUGS software and computations were carried out using Markov Cain Monte Carlo methods. After ignoring the initial (burn-in) iterations of the chain, model parameters were estimated based on 75 000 samples from the posterior.

The proportions of "good" and "bad" valves and estimates of the initial hazard parameters are given.

Surprisingly, although the repair-specific hazards θ_1^j and θ_2^j, $j = 0, \ldots, 4$, were not assumed to be ordered over the two categories of valves, their estimates indicated them to be so.

The message here is that not only there is considerable heterogeneity in the quality of valves when they enter the study, but they continue to be so throughout the study period. Also the different repair types seem to be affecting the valves in widely different ways. Which is clearly visible for the "bad" valves.

These estimates of hazards are also quite comparable with those indicated by Bhattacharjee et al. (2002), where the hazards were assumed to be generated by two exponential distributions with parameters β. From the estimates of β the posterior expected hazards would be 0.00007 for the "good" group and 0.002 for the "bad" group, which are of similar order as obtained here.

The estimated distributions of the very first failure-type experienced by the valves in the two categories were clearly distinct from each other.

Note that by superposing $N_i^j(t)$ $j = 0, 1, \ldots, 4$, we can obtain the repair process of the i-th valve. Combining the repair processes from all the 104 valves, transitions between different repair types can also be estimated. The estimated transition probability matrices for the two groups of valves showed distinct characteristics with respect to their susceptibility to different types of failure.

Clearly valves belonging to the "bad" group are typically susceptible to external failures, irrespective of the scheduled tests and consequent repairs. On the other hand the overall chance of failing during an

Table 2. Estimates of repair specific hazards for "good" and "bad" categories of valves.

Failure intensity followed by repair type	Posterior expectation	
	"good"	"bad"
External leakage (0)	0.00001	0.00117
Internal leakage (1)	0.00006	0.00124
Valve not closing (2)	0.00000	0.00842
Valve not opening (3)	0.00000	0.00071
Others (4)	0.00006	0.00039

Table 3. Estimated proportions of first repair type for "good" and "bad" categories of valves.

Repair type	Posterior proportion for valve type	
	"good"	"bad"
(0)	0.06	0.63
(1)	0.34	0.10
(2)	0.21	0.00
(3)	0.17	0.11
(4)	0.22	0.15
All	1.00	1.00

Table 1. Estimated proportions and initial hazards for "good" and "bad" categories of valves.

Characteristic	Posterior expectation
Proportion of "good" valves	0.88
Proportion of "bad" valves	0.12
Initial hazard for "good" valves	0.00002
Initial hazard for "bad" valves	0.0005

Table 4a. Posterior estimates of transitions from one repair to different repair types for "good" categories of valves.

Repair type	Repair type					
	(0)	(1)	(2)	(3)	(4)	All
(0)	0.00	0.65	0.00	0.24	0.11	1.00
(1)	0.06	0.71	0.00	0.07	0.16	1.00
(2)	0.00	0.25	0.17	0.17	0.42	1.00
(3)	0.00	0.11	0.33	0.44	0.11	1.00
(4)	0.15	0.04	0.30	0.15	0.36	1.00
All	0.05	0.40	0.14	0.17	0.24	1.00

Table 4b. Posterior estimates of transitions from one repair to different repair types for "bad" categories of valves.

Repair type	Repair type					
	(0)	(1)	(2)	(3)	(4)	All
(0)	0.61	0.10	0.09	0.04	0.15	1.00
(1)	0.44	0.26	0.00	0.04	0.26	1.00
(2)	1.00	0.00	0.00	0.00	0.00	1.00
(3)	1.00	0.00	0.00	0.00	0.00	1.00
(4)	0.26	0.44	0.00	0.01	0.29	1.00
All	0.55	0.18	0.05	0.03	0.18	1.00

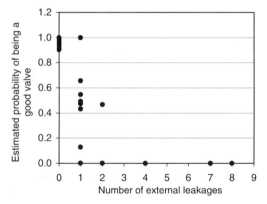

internal leakage test was more prominent for the "good" group of valves.

From the posterior distribution of c_i's for the i-th valve the chance of being a "good" (or "bad") can be estimated. The following estimated probabilities of being "good" were plotted for all valves against the observed numbers of external leakages.

5.3 An additional analysis

Noting that although in the model only the initial hazards for the valves (at the time of entering the study) were constrained by ordering them, the two groups of valves were estimated to have ordered intensities

throughout. In an additional model the parameters were constrained pairwise by $\theta_1^j \leq \theta_2^j$, for $j = 0, 1, ..., 4$.

The priors were assigned accordingly as follows. θ_1 was assumed *a priori* to have Gamma distribution. The conditional distributions $\theta_2^j | \theta_1^j$, $j = 0, 1, ..., 4$, were assumed to be truncated Gamma distributed, truncated from below at θ_1^j.

The latent variables (c_i) for all 104 valves were assumed to have *a priori* a Multinomial distribution. The prior specifications for other parameters continued to be the same.

The resulting estimates were quite comparable and so were the specific structures in the estimates found in the proposed analysis.

6 CONCLUSIONS

From the estimates of the repair-specific hazards for external leakage, we observe that the estimates differ widely across the different types of repair. Recall that these are preventive maintenance tests, that is, at regular intervals each system is put through a specific kind of stress and if it fails during such a test it is repaired. The amount of stress may vary from test to test. For example, the annual leakage tests (internal leakage) is known to be quite rigorous. Hence, if a system fails such a test even though it is then repaired, the excessive stress put on the system leaves its mark on the system and this is reflected in the hazard of the external leakages.

Moreover, the effects of repair were found to differ across quality of the valves. This makes it essential to account for heterogeneity, if any, while inferring about the systems. Non-optimal or erroneous system reliability assessment of systems as crucial as parts of nuclear power plants could result in misleading safety related decisions.

The added information from the repairs carried out at failures during scheduled tests certainly provided a better understanding of the system condition and also an understanding of the failure process of the spontaneous failures.

In this paper we illustrated that even for large, complex systems with evident heterogeneity and sparse failure data, incomplete maintenance information, simplistic models with intuitive but nominal assumptions can also bring out interesting and useful features of the system. Moreover inference for both population and individual level can be carried out.

ACKNOWLEDGEMENTS

We are thankful to Prof. Urho Pulkkinen for providing us with the data. Thanks are also due to Dr. Dario Gasbarra for helpful discussions.

REFERENCES

Arjas, E. and Bhattacharjee, M. (2000). Modelling dependence by using latent variables: a simple example from repairable systems. In *Proceedings of the Joint Statistical Meeting, Dec. 00–Jan. 01, New Delhi, India.*

Bhattacharjee, M. and Arjas, E. (2002). Modelling heterogeneity in nuclear power plant valve failure data. In *Proceedings of the third International Conference on Mathematical Methods in Reliability, Jun. 17–20, Trondheim, Norway.*

Pulkkinen, U. and Simola, K. (2000). Bayesian Models and Ageing Indicators for Analysing Random Changes in Failure Occurrence. *Reliability Engineering & System Safety* 68, 255–268.

Simola, K. and Laakso, K. (1992). Analysis of Failure and Maintenance Experiences of Motor Operated Valves in a Finnish Nuclear Power Plant. *VTT Research Notes 1322.*

Safety and Reliability – Bedford & van Gelder (eds)
© 2003 Swets & Zeitlinger, Lisse, ISBN 90 5809 551 7

Establishing a safety culture in a distributed offshore logistics activity

R. Bye & T. Kongsvik
Norwegian University of Science and Technology

L. Hansson
Norwegian Marine Technology Research Institute

ABSTRACT: Most of the research published about safety culture focuses on how safety culture can be described theoretically or measured. This paper suggests concrete actions to improve safety on offshore service vessels. The paper sketches out a method for developing actions in order to improve safety by conceptualising safety culture as an abstract analytical concept for interpretations of complex connections between safety and other conditions. This conceptualization of safety culture is based on a discussion of the use of the term culture in the fields of safety management, organizational theory and cultural anthropology. The use of Action Research as a strategy in developing a safety culture is described. The empirical foundation of this paper is an ongoing project that aims to improve personnel safety on the vessels in the supply services in the Norwegian oil company Statoil. The number of injuries has decreased after the implementation of the project.

1 INTRODUCTION

The service vessels in the North Sea transport goods to the oil and gas installations by supply vessels. Special equipped vessels do anchor-handling activities and the emergency preparedness is taken care of by stand-by vessels. The service vessels are working places with high personnel risk. The risk is mainly related to occupational accidents and to collisions between vessels and installations. The number of personnel injuries has increased in later years as shown in figure 1.

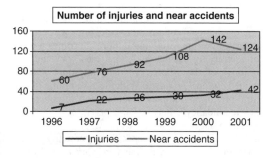

Number of injuries and near accidents

Figure 1. Number of injuries and near accidents on Statoil's service vessels in the period of 1996–2001.

The aim of the study reported in this paper is to improve personnel safety on the service vessels.

The supply vessels transport food, chemicals, piping, equipment and spare parts from shore bases to the installations offshore and empty cargo back to the shore base. Shipment is done from five different bases along the west cost of Norway. The anchor handlers are involved when the mobile drilling rigs shall move from one location to another. The stand-by vessels are located in a position close to the installations. Their main task is to assist the installation in an emergency situation.

The logistic chain of the supply vessels is complex. Different departments in the oil company are involved in addition to the seamen and the ship-owners office. The vessels are contracted from several ship-owners and are operated by the oil company. The crew on the vessels represents a small and partly isolated organisation with defined roles in a hierarchic structure with the captain in charge. The vessel crew is partly a sub organisation in the oil company and partly a sub organisation in the ship-owner organisation. The installation crew is a partly isolated, but larger organisation. The control room operator, the deck manager and the crane operator also play important roles in the supply service activities.

The safety management process is challenging in this distributed organisation with many different units in a close co-operation. This makes it a challenge to create a common safety culture in this organisation.

Several papers have been written on the subject safety culture in later years. The approaches presented in these papers are mainly either theoretical discussions on safety culture or descriptions of how to measure safety culture. Only a few of these papers describe safety culture in relation to the safety management process. This paper describes a new and different approach. Based on theory on safety culture we have proposed a practical approach to how the concept of safety culture can be used to improve the safety level. The safety level has increased in the reported study after application of this approach, and a description is given for how the work is actually done and how this approach relates to the presented theory.

The problem definition of this paper is: How can the concept of safety culture be applied in a practical way to improve the safety level?

2 SAFETY CULTURE

2.1 The use of the concept safety culture

The use of safety culture may be seen as a reaction of the lack of explanatory power of conventional theories in the field of risk and safety studies. By this the use of the concept safety culture has some parallels to the use of the concept of *corporate culture* in the field of organizational studies during the late 70s and 80s[i]. The concept of *safety culture* was coined in the middle of the 80s (Cox & Flin 1998, Pidgeon 1991). It was used as a term to explain several major accidents, such as the King's Cross Underground fire in London, the Exxon Valdes accident and the Chernobyl meltdown. In the investigations the accidents were explained as a result of *complex recursive chains of human actions, inadequate procedures, and insufficient technology that over time generated the disastrous accident*. This generating system was abstracted as poor safety culture (Cox & Flin op.cit.) With the

---

[i] As *cultural differences* was an appropriate explanation for the Chernobyl accident, cultural differences could also serve as an appropriate explanation for the *shocking competitive problems with Japanese corporations*. Consultants, authors of management literature (Peters & Waterman 1982, Deal & Kennedy 1982) and practitioners started the quest to build "winning cultures", and academic researchers tried to construct theories of corporate culture. The attention towards organizational/corporate culture in the fields of academic organizational studies may also be seen as a search for an alternative approach and methods amongst members of the community of organizational researchers which, especially in North America, was dominated by a strong positivistic conventions (Alvesson & Berg 1992:22).

concept of poor safety culture, is it possible to picture the concept of the opposite, i.e. a *good safety culture*, i.e. cultures where the probability of accidents is minimal.

2.2 "De-abstractionalization" and "factorization"

Attempts to define culture in the field of risk and safety and safety management have led to what we may call "de-abstractionalization" and "factorization" of the term. Kennedy & Kirwan (1995), Pidgeon (1991), Ek, Olsson & Akselsson (2000) claim that the study of safety culture has been unsystematic, fragmented and underspecified in theoretical terms. Culture has been defined by using concepts such as *values, symbols, artefacts, beliefs, norms or behaviour*. By defining culture as "shared values", "shared models" etc., and by this considering the concept as different from e.g. "technology" (equipment, architecture, safety systems etc.), and "social structure" (organization of people, procedures, reward systems etc.) the term culture becomes no longer a signifier of the "complex whole", as it was used in the aftermath of the Chernobyl accident. By referring to Russel & Whiteheads theory of Logical Types (1913) culture is treated as belonging to the same "logical type" as "structure" and "technology". A consequence of the "de-abstractionalization" may be that culture becomes a "new brand" of conventional concepts such as "values", "attitude", "preferences" etc.

If the concept of culture should have any "excess value" in the work of understanding and handling safety issues, the term must be used as a "high logical level" analytical concept.

2.3 Culture as a "high logical level" analytical concept

As an academic term, the concept of culture has been the main tool of social and cultural anthropologists. Despite of different definitions, the term has been treated as a *holistic concept* focusing on what *people learn and develop as members of a society*, contrasted to *pregiven capabilities by nature*. Culture may be seen as a "quantum field" that explains "human actions", or any other categories we define as a part of culture. Studying cultures becomes a quest of identifying "structures", "patterns" or "configurations" of the whole, or what Bateson (1958:24) named *cultural premises*. With the assumption of cultural premises, one may assume a sort of coherence, or "redundancy", between the parts, as in a "holographic image". For instance may what we may define as *work practices* in a specific community reflect concepts that we may define as *organizational structure, technology, formal procedures, beliefs, moral statements etc.*

2.4 Interaction, time, and existing cultural premises

Culture may be seen as a consequence, and as a condition of human *interaction*. The maintenance and modifications of culture may be seen as a consequence of (a) *human interaction*, (b) *time* and (c) *existing cultural premises and belonging conventions*. People who take part in relatively durable interactions may be considered as forming and sharing *communities* of space and time. As Hannerz (1992:15) points out: *"people shape social structures and meanings in their contact with one another, it proposes; and societies and cultures emerge and cohere as results of the accumulation an aggregation of these activities"*. A minimum condition for cultural production is *human contact*. Formation of different *communities* may also mean forming differentiated cultures[ii].

In distributed organizations there are special challenges regarding interaction. At the outset, the possibilities for interaction are limited. To develop a safety culture in a distributed organization should therefore imply establishing better conditions for interaction.

2.5 Safety culture and cultural relativism

Instead for a search for "general rules" of "safe cultures", the quest is to use safety culture as a concept for *understanding* the cultural configurations in a specific situated community.

The term *safety culture* is addressing a specific aspect of culture in general, i.e. how *cultural premises* may explain the existence of safe work practices in a particular community of practitioners. The analytical quest is to *contextualize* the work practices by the attempt of *interpret* the actions in relation other defined parts of culture with a focus on coherence (or lack of coherence). For instance may the use of *precise written procedure* be a partial explanation of safe practice in one community, but this *doesn't mean that precise written procedures is a general condition for safety in any community*. In some cases an introduction of procedures may actually be a disturbance to existing safe practices. The significance of precise written procedures in relation to work practices must be interpretated in relation to the total configurations of "parts" in a specific community. A defined part in one community *may* have a total other significance in another

community. The implication of this view is that there may not exist generic knowledge of what constitute safe working practices in *any* community, i.e. the efforts in one community may not work in another. Based on *understanding,* normative appropriate initiatives (e.g. initiatives aiming to reduce the numbers of personal injuries) may be carried out.

2.6 Understanding culture as an interpretative act

Using culture as an analytical concept for understanding complex relations in a specific community implies an attempt to understand the relation between defined parts and the "complex whole". This analytical process may be seen as loops of *abductions* and *inductions*[iii]. Sampling a n d using categories may be considered inductive acts. These categories may be further abstracted by new acts of abduction, resulting in the meta-category of e.g. "safety culture". Using the meta-categories, i.e. high logical levels, new dimensions on "lower logical levels" may be conceptualised. By performing these acts of abductions and inductions, the *meaning* of the phenomenon studied may be constructed[iv]. This analytical activity is illustrated by Geertz (1973:69) as: *" Hopping back and forth between the whole conceived through the parts that actualize it and the parts conceived through the whole...."* However, sampling qualitative or quantitative data (or both) must be seen as an approach to *understand* the cultural configurations without any *researcher made* normative standards of the "best culture". This statement refer to the assumptions that; (1) culture may not be seen as a "set of stable factorized parameters", and (2) the state of a part (e.g. safe working practices) is not dependant on one specific configuration of other parts. The ambition is not to evaluate or measure if a community of practitioners correspond (or fails to correspond) with a generalized conceptualization of a

---

[ii] The borderlines between different communities are often relatively blurred. Members of a company may for instance be conceived as a community, but at the same time be regarded as contenting different communities. The formation of differentiated or corresponding cultural traits are anyhow dependent on the specific *social networks*, i.e. the actual communication- or interactional network between people involved.

[iii] The acts of abduction, or what neurophysiologist McCulloch (1965) calls *homomorphic modelling*, is done by constructing a relation between a phenomenon A and B, named as C. The *homomorphic modelling* gives categories for *inductional analyses.*

[iv] These interpretative activities correspond with Scheins (1985) description of cultural analyses. By defining 5 dimensions (the organizations relation to its surroundings, ontological assumptions about the world, assumptions about humans, assumptions about human activities, assumptions about inter-human relations) sampling data (qualitative data) belonging to these dimensions, he constructs the metacategory culture, which unites the selected dimensions. Cultural analyses may not be restricted to qualitative data. The work of Hofstede (1994) is an evident example of such an approach. Hofstede are constructing dimensions of culture, sampling *quantitative data.*

general safe community. The aim is rather to figure out *what can be done to increase safety in the community studied*, and *then carry out the actions*.

2.7 Subjective orientation and involvement

This interpretative challenge calls for a *subjective orientation*, i.e. an attempt to grasp the "*native point of view*" Geertz (ibid.:58). The subjective orientations implies an interpretation of individuals interpretation of their environment, due to a assumption that subjective meaning has consequences for individual's behaviour, and further: how these interpretations may be understood in relation to the concept of culture. As an observer, the conscious quest is to interpret what might be the meaning produced by the individuals situated in a specific cultural context. This ambition calls for an intensive involvement of members of the particular community studied. There has been an ongoing controversy among social scientists if organizational culture is something that can be managed, or whether it is a global property that somehow lives it own life and springs out of the total values, beliefs and ideologies of the organization. In short the debate has been about whether culture is something an organization "has" or something it "is" (Reason 1997). Our discussion of the safety culture concept underlines our argument, that culture is something an organization "is". This does not imply however, that it is impossible to change a culture. *Collective practices* can be regarded as an important part of organizational culture (Hofstede 1994), and that such practices can be changed by collective *reflections* on such practices. The most efficient way for such changes is not top-down, but bottom-up.

Besides better opportunities to grasp the "*native point of view*", the involvement in reflection about safety and culture may in itself contribute to changes (Schein 1985, Argyris & Schön 1996, Lewin & Grabbe 1945, Greenwood & Levin 1998). By changing significant "parts", cultural configurations *may* be converted, and through this changes of human practice.

Employee participation (EP) is a principle that is used in a number of ways in organizations, and is assumed to have a number of advantages. Some of them are increased motivation for work, increased democracy, increased quality of solutions and decisions and increased learning. This line of strategy is more *bottom up* and involves the work force in general to a greater extent.

In Scandinavia, in the 60s and 70s, employee participation was considered as a goal and especially important for ideological reasons – as a way of ensuring the democratic influence of the workers. During the 90s, EP was increasingly considered a *tool* for better quality and more effective production. The view of EP as a tool is also evident in some efforts to improve

safety in industry. One example is from a five-year project in the oil company Statoil, where safety improvements in crane and lifting operations were accomplished through development of "*Best practices*" (Hepsø & Botnevik 2002). Supplies and cargo are transported by supply ships to the installations in the North Sea. The loading and unloading of these ships involves fixed cranes on the installations. These operations are high-risk activities, partly because of the harsh weather conditions. How the crane operators, banks men and seamen on the supply boat deck coordinate their work is of great importance to the safety of the operation. Sometimes there is also time pressure involved, for example when spare parts critical for the oil production is needed. The crane operators saw the need to develop joint practices across installations to improve safety, and a project group of involved personnel was established. An espoused practice was written down as a product of several activities, all involving a participatory approach. An intranet based virtual bulletin board was established, where discussions and reflections took place. Work practices were also discussed in search conference seminars. The development of the written formulations describing the best practices could also be followed on a day-to-day basis and were available for comments for those interested. The project was characterized as a grass root movement, and is a good example of a bottom-up strategy of improving safety by means of stakeholder involvement and dialog. The safety results have also improved in terms of decreasing numbers of unwanted occurrences in crane and lifting operations in 2000–2001, although one is cautious to link this directly to this particular project.

A strategy that has a very long tradition in behaviour modification is to reward desired practices and punish or apply sanctions against unwanted acts. Early in the history of psychology, rewards and punishments were regarded as the core mechanisms of *learning* through the concepts of classical and operant conditioning. In the later and very influential school of thought called Behaviourism, this was developed further. Although there is very few that still call themselves behaviourists, parts of the principles and the ideology are still in use – also in the management of organizations.

With regard to safety, it is common to establish reward systems to encourage practices that presumably enhance the level of safety. Two examples could be bonuses for good safety results and to give safety awards to employees or departments that have done an especially good job regarding safety. Sanctioning or punishment of unsafe acts is also common in many organizations, although this is an area were there seldom are simple answers and were there is difficult to establish clear-cut systems and responses. Reason (1997) maintains though, that the engineering of a *just* culture is an important part of establishing a safety culture in general. Although situational and systemic factors are important in many instances, clearly individual responsibility must also be addressed when reckless, negligent or even malevolent behaviour is the case (ibid.:209). Reason outlines a decision tree that separates between acceptable and unacceptable individual actions. We would nevertheless argue that punishment is not a well-suited strategy for changing a safety culture – instead change requires reflection and interaction over time.

3 RESEARCH ACTIVITIES FOR MUTUAL LEARNING

3.1 Our participatory approach: action research

Our attempt at improving the safety culture and safety results builds on two basic assumptions: First, attempts at improving the safety culture should be situated and grounded in the community of practice in question. Second, the safety culture can be changed through changing the configuration of factors at a lower logical level.

Action research (Greenwood & Levin 1998) was considered an approach that met both these assumptions. As a tool for accomplishing democratic social change, action research (AR) involves three basic elements – research, action and participation. The AR-process is described in the following manner (ibid.:4):

> *"AR is social research carried out by a team encompassing a professional action researcher and members of an organization or community seeking to improve their situation. […] Together, the professional researcher and the stakeholders define the problems to be examined, cogenerate relevant knowledge about them, learn and execute social research techniques, take actions, and interpret the results of actions based on what they have learned."*

The captains on the vessels were considered key personnel and important stakeholders in safety questions, since they have the overall responsibility for the activity on board. We therefore have established an arena – "Captain's Forum" – where the AR-process takes place. We organize regular *search conferences*, where safety questions are addressed. A search conference is a methodology where planning, creative problem solving and concrete action are integrated activities (ibid.:155–156). The first search conferences where devoted to reflections upon why accidents happened, and what could be done to reach the vision of zero accidents. Later seminars have focused on how different actors in the logistics chain can cooperate in order to improve the safety results.

3.2 Research activities

Two important activities from the researchers have originated from the conferences. First, it was made an analysis of accidents involving injuries in the period from 1996 to 2001. The purpose of the analyses was to identify trends and accident-prone groups among the crews. The results were later presented in the conference for reflection upon how to tailor and prioritize measures. The captains report accidents to Statoil according to a form with fixed response alternatives. All reports are registered in a database system. Due to dubious quality of the earlier registration of accidents, it was decided to limit the analyses to accidents that had happened after 1995. It was registered 153 accidents involving injuries and 579 near-accidents in this six-year period. The number of accidents increased in the period, both in absolute terms and adjusted for an increase in the level of activity. Furthermore, there were more lethal and severe accidents in 2000 and 2001 than in the previous years. Most of the lethal and severe accidents happened during anchor handling operations. When injuries where distributed according to work experience on board, the group with least experience had most injuries.

Second, a survey was made, where organisational safety factors and work environment was measured. A questionnaire was distributed to all members of the crews on vessels that had at least a 2-month contract with Statoil. The population was 630 people, and 487 questionnaires were returned, giving a 77% response rate. In general, there was a positive evaluation of the organizational safety factors and the work environment. The youngest among the crews on vessels that switched between anchor handling and stand by operations were less positive than other groups. In an open question, the respondents were also given the opportunity to comment on how the safety could be improved. There were 43% of the respondents who gave such comments, and this material represented a rich source of information on what could be done.

3.3 Search conferences

The results were discussed in the search conferences, where actors that had no regular face-to-face interaction on a regular basis were brought together. The results were used as a starting point for mutual reflection on how the present safety results could be explained. Based on mutual interpretation of the present state, the participants went on discussing what measures could be taken to improve the safety results.

The discussions in the conferences were structured by using a "high level" logical concept of culture. Three dimensions were selected for the task to describe the "parts" of culture. The selected dimensions were *structure*, *interaction*, and *attitude* (fig. 2).

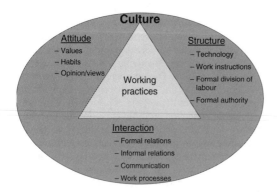

Figure 2. "Model of culture" used in structuring the discussions at the search conferences.

Table 1. Seriousness and number of accidents involving injuries from 1997 to 2002.

	1997	1998	1999	2000	2001	2002
Death	–	–	–	1	1	–
Serious injury	3	–	–	1	4	–
Other injuries	19	26	30	30	39	25

The conditions that the researchers and the participants assumed had an impact on working practices and safety level, were categorized according to these three selected dimensions. Using an assumption of coherence, or "redundancy" between the dimensions, specific configurations of conditions were interpreted as having an influence on the safety level. Conditions along one dimension (i.e. attitude), could be seen as a reflection of conditions along the other dimensions (structure and interaction). Based on this i.e. technical conditions (categorized as part of the *structural dimension*) that might have no *direct impact* on the safety level, could be seen as crucial important as a part of a configuration of other conditions. For instance configurations of conditions such as "*spill water from the platforms*" (categorized as part of the *structural dimension*), the sailors "*feeling of being treated as underdogs*" (categorized as part of the *attitudinal dimension*), and the "*lack reciprocity in the communication between platform crew and sailors*" (categorized as part of the interactional dimension), were interpretated as influential on the working practices of the sailors, resulting in "*lack of motivation*" and a focus on "*finishing the job without any commitment*".

Based on such interpretations alterations where suggested. A large number of measures have been proposed as a result of the conferences, some of them are implemented, including technological, organizational and communicational changes.

3.4 Safety results

The safety results have improved since the program started in the autumn of 2001. The number of accidents involving injuries has declined, and no lethal accidents or serious injuries involving sick leave have taken place (table 1).

Of course, this positive development could have a number of reasons, and it is difficult to say whether it

is a result of the program and a better safety culture in these services. An alternative explanation could be that the improvement is caused by a "Hawthorne-effect", an effect first described by Mayo (1933): It could be that the attention of outside parties is the main reason for the improvement. If this is the case, it is likely that the effect will vanish when the program period ends. If the program has resulted in a change in the safety culture, the effect should be more enduring.

4 CONCLUSIONS

The problem definition of this paper was: How can the concept of safety culture be applied in a practical way to improve the safety level?

We have argued that safety culture should be treated as a high logical, holistic concept. The study of safety cultures should be a quest of studying configurations of the whole. Interaction and involvement from the stakeholders are prerequisites for developing a safety culture – or changing configurations of the whole. Action research as a strategy seems well suited because participation is regarded as a basic element in change processes.

Search conferences were used as a tool, where captains on service vessels have reflected upon causes of accidents and accident prevention, and where research activities were used in the reflection process. In these reflections, safety culture was treated as a holistic concept. Bottom-up processes that an action research approach supports, generates many specific ideas on how to improve safety in a particular context. This is in line with a situated use of the safety culture concept.

The number of injuries decreased in the period after the implementation of this project. A future research activity will be aiming at deciding to what degree this is the result of the project itself.

REFERENCES

Alvesson, M. & Berg, P.O. (1992): *Corporate Culture and Organizational Symbolism*. NY: Walter de Gruyter.

Argyris, C. & Schön, D.A. (1996): *Organizational Learning II*. NY: Addison-Wesley Publishing Company.

Bateson, G. (1958 [1936]): *Naven*. California: Stanford University Press.

Cox, S. & Flin, R. (1998): Safety culture: philospher's stone or man of straw. *Work & Stress* vol. 12, no.3, 189–201.

Deal, T. & Kennedy, A. (1982): *Corporate Cultures: The Rites and Rituals of Corporate Life*. Ma: Addison Wesley.

Ek, Å., Olsson, U. & Akselsson, K.R. (2000): Safety culture onboard ships. *Conference proceedings of the International Ergonomics Association/Human Factors and Ergonomics Society*, San Diego, vol. 4, pp 320–322.

Geertz, C. (1973): *The Interpretation of Culture*. NY: Basic Books.

Greenwood, D.J. & Levin, M. (1998): *Action research*. London: SAGE Publications.

Hannerz, U. (1992): *Cultural Complexity*. NY: Columbia University Press.

Hepsø, V. & Botnevik, R. (2002): Improved crane operations and competence development in a community of practice. Paper presented at the Participatory Design Conference, Malmø University, 23–24 June 2002.

Hofstede, G. (1994): *Cultures and organization: Intercultural cooperation and its importance for survival*. London: Harper Collins.

Kennedy, R. & Kirwan, B. (1995): The failure mechanisms of safety culture. In Carnio & Weimann (eds), *Proceedings of the International Topical Meeting on Safety Culture in Nuclear Installations*. Wien: American Nuclear Society of Austria. 281–290.

Lewin, K. & Grabbe, P. (1945): Conduct, Knowledge and acceptance of New Values. *The Journal of Social Issues*, vol. 1, no. 3, August.

Mayo, E. (1933): *The human problems of an industrial civilization*. New York: Macmillan.

McCulloch, W.S. (1965): *Embodiments of Mind*. Cambridge: MIT Press.

Peters, T.J. & Waterman, R.H. (1982): *In search of excellence*. NY: Harper & Row.

Pidgeon, N.F. (1991): Safety culture and risk management in organizations. *Journal of Cross-Cultural Psychology*, 22, 129–140.

Reason, J. (1997): *Managing the risks of organizational accidents*.

Russel, B. & Whitehead, A.N. (1913): *Principia Mathematica*. Cambridge: Cambridge University Press.

Schein, E. (1985): *Organizational Culture and Leadership*. San Francisco: Jossey-Bass Publisher.

Safety and Reliability – Bedford & van Gelder (eds)
© 2003 Swets & Zeitlinger, Lisse, ISBN 90 5809 551 7

Using total process costing to measure self-insurance safety costs

Huguette Blanco

School of Commerce and Administration, Laurentian University, Sudbury, Ontario, Canada

ABSTRACT: Safety is one of the strategic cost drivers that affect the long term cost competitiveness of a firm, but the information available from conventional cost and information systems is insufficient for managing an effective safety strategy. This paper proposes a model designed to capture and display the cost of all the activities that a firm incurs as a result of a safety failure, across units and in time. The list of activities, drawn from investigation reports of actual injuries and vetted by plant managers, uses mind-map format for displays and Microsoft Excel for data handling. In several parallel tests on real life events, the model identified safety costs much larger than conventionally recognized. The difference represents large preventable self-insurance costs. Generalizing the observations requires more testing, but access to data is restricted by secrecy reasons and doubt. The scale of safety self-insurance cost should help persuade doubtful management to adopt safety as a strategic cost driver.

1 INTRODUCTION

In the introduction to a series of papers on management control systems, (Oyon and Mooraj, 1999) argue that management systems can be used to add real value to an organization if they are used as a framework enabling alignment of the organization strategic and operational goals and objectives. This paper presents safety management as one of the fundamental or executional cost drivers that determine for the long term the cost position of a firm compared to its competitors. As such, good safety performance, essential to create long-term value, should be a strategic goal. Total process costing and the model presented in the paper provide management with the information necessary to drive this strategic goal and align it with operational goals. Because costs are collected from across the various cost or profit centres, management has the information to think globally and consequently to manage on the basis of the firm overall skills at executing, of its ability to perform well.

2 STRATEGIC OR EXECUTIONAL COST DRIVERS

2.1 Definition

They are defined as "those determinants of a firm's cost position that hinge on its ability to "execute" successfully" (Shank and Govindarajan, 1992). Quality is a prime example of an executional cost driver. Companies have recognized that quality drives their ability to compete, and that has driven them to invest in quality improvement activities. A firm will gain a sustainable competitive advantage by controlling the quality driver better than its competitors (Ittner, 1999).

2.2 Safety is a strategic cost driver.

The consequences on the business environment of poor safety performance can be obvious. Some are drastic: Westray and Swiss Air, for example, went out of business; some are long lasting: Ford and Bridgestone, British Rail, for example. Most, such as the higher costs of financing or the higher costs of human resources are insidious. Although the consequences of poor safety performance on operating costs are pervasive, they are often ignored. Some companies have identified safety performance as a strategic issue, recognizing that "good safety performance correlates directly with increased productivity and quality" (Weyerhaeuser, 2000), but the practice is not widespread.

3 MANAGEMENT INFORMATION SYSTEMS

Although, when asked, company executives are all in favour of improving safety performance, statistics show that improvement actions are often taken in reaction to incidents. Executives do not have the information necessary to identify preventative actions with expected economic return competitive with that of other investment projects. The model presented here begins to fill this gap.

3.1 Management accounting shortfalls

The primary reason for this gap is the lack of adequate method for determining the financial consequences of poor safety performance. In contrast with costs in the core areas that are reasonably well defined and tracked, costs related to safety have not received much attention beyond the direct costs. Safety related costs such as safety department expenditures, insurance premiums and fines may be readily available from existing accounting records. However, many costs are hidden in other cost pools, and many are not even measured. There is not much data about the injury-driven costs that occur beyond the immediate areas because accounting systems are not designed to bring together costs incurred as a result of a safety failure but spread in the operations. Costs that continue in time, and those that may appear at a later date are not attributed back to the failure itself.

Existing costing methods give only crude estimates of the real cost of incidents and accidents. They are usually based on costs of damaged equipment and lost wages, resulting in estimates that may be as much as 20 times below the real losses (Hammer, 1989).

It is reasonable to expect that improving injury cost knowledge would improve event recognition and response, thus leading to the double effect of fewer events, and lower costs. Until they are identified, such costs represent a hidden self-insurance whose magnitude and scope cannot be managed.

In addition to not providing the pertinent information for sound investments and operating decisions that would prevent injuries and generate long-term value, conventional management accounting may also prevent management from acting (Reinhardt, 2000). Operating incentives often motivate managers to make short-term savings and to avoid cost increases that would result in longer-term benefits to the whole organization. Budgeting can result in an entrenched level of failures and inefficiencies. Prevention expenditures may be flagged as negative variances that have to be explained, but built-in inefficiencies will not attract scrutiny. In the absence of information on the true full cost of incidents and accidents, the widely accepted insurance system of workers compensation, limiting company liability and lessening litigation, may lead to a higher threshold of injury than what would otherwise be possible.

3.2 Total process costing

It is not a new concept. It has been discussed for example in the context of procurement and of ownership.

The total cost of procurement allows a firm to make better-informed decisions resupplier choices and design. The total cost of ownership incorporates not only the acquisition cost of a resource but also all the costs associated with integrating it with other needs, using and replacing it later on. Total process costing applied to safety management is aimed at developing an understanding of the "true cost" directly related to an incident or an accident.

3.3 Activity based costing (ABC) and Activity based management

Many companies are applying various forms of activity-based costing to assess hidden costs of environmental and quality drivers. This paper reports on an adaptation of ABC to measure the total cost of safety failures. However, since the objective is to understand causality and to provide management with information to manage costs at the root, and not to calculate product cost, ABC here stands for Accounting Based on Causality (Lebas, 1999). The total cost of an incident includes the cost of all the activities carried out as a result of the incident. Known activities can be analyzed and managed by eliminating the non-value-adding ones and streamlining others for increased efficiency and productivity.

4 THE MODEL

This paper advances a cost gathering protocol and a cost handling technique to explore the extent of hidden safety self-insurance costs throughout the organization. The model also helps identify activities that may be required to stem both the sources of costs and their spread.

The difference between the costs that are known and the total cost resulting from a safety failure is essentially a hidden form of self-insurance. This cost is potentially large, and because it is hidden it cannot be eliminated. The model can also reveal activities that may shelter avoidable costs if left unattended.

To facilitate the management objective embedded in total process costing, the model was developed on a mind-map frame for increased comprehension and visual display, so that total process costing teaches as it works.

4.1 Developing the list of activities

The total costing of safety incidents calls for assigning costs to all the activities that are being carried out as a result of the incident. These include the obvious activities such as all the incident investigation activities, as well as the hidden activities such as hiring and training temporary workers to replace injured ones or revising the training material, or having idled employees because of equipment failure or investigation. Other activities include, for example, caring for injured workers, or modifying the environment or attending meetings, or if appropriate activities related to added regulations or to legal intervention.

The model was developed using incident investigation reports from several firms in the mining industry. The objective was to determine whether from these reports it was possible to identify lists of recurring activities and to list all that were considered relevant. Approximately 200 reports were initially screened for an adequate level of written detail. Fifty-three accident and incident reports were selected for in-depth analysis.

The resulting exhaustive list of activities was very large, suggesting that total costing would not be manageable without proper classification and automation. The final classification and list of activities was achieved through interactive discussions with managers with responsibility for safety. Figure 1 shows the six broad categories of activities agreed upon.

Microsoft Excel, a spreadsheet program widely used in the workplace, was selected to provide for the needed automation feature.

4.2 Mind-mapping

Earlier satisfactory experience with mind-maps in the context of accident investigation made them an

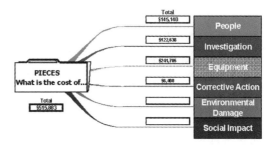

Figure 1. The cost pools: the first four include activities that to differing degrees would happen in any incident. The last category is included for the firms that want to have a true total cost of an incident, not limiting the measurement to their own loss.

attractive solution to the challenge of displaying long lists of activities. Mind-mapping builds on evidence that shows that the human mind is far more multidimensional and pattern making and far less linear than previously thought (Busan, 1989). The activities in the six broad categories were organized and displayed into mind maps with several layers. Figure 2 shows the map for one of the six categories of activities. The mind-map layout also facilitates the assigning of costs to the pertinent activities in ways that are accessible to the non-specialist.

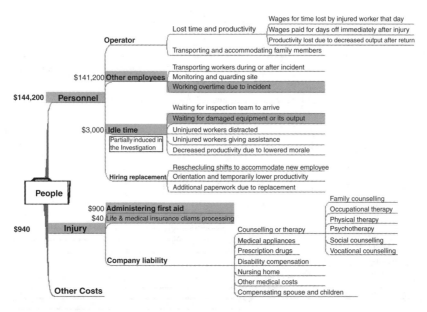

Figure 2. The list of activities: the activities in each cost pool are laid out in a visible non-linear manner to facilitate discussion within the organization. Only activities pertinent to the specific incident need to be considered. Upper management may only be interested in the summary numbers whereas operators may want the analysis at a detailed level.

4.3   Using the model

Activities are selected from the mind-maps as they occur, and their cost is entered in a corresponding Excel spreadsheet, as shown in Figure 3. Figure 4 shows an aggregate cost worksheet. The activity analysis level can be adapted to the firm requirements, time frame and resources.

The steps required to arrive at a total cost are simple. As the incident investigation proceeds, the cost incurred on each related activity should be determined. This includes identifying people involved, keeping track of time, looking at work orders, and interviewing personnel remotely connected with the activities. Employees can estimate the amount or percent of time spent on a given task. Hourly wages can be accurate or estimated

if not readily available. For best results one employee should be assigned the task of collecting and entering the information.

Total costing of safety failures can be done concurrently with the failure investigation, retroactively for past failures, and prospectively in "as if" scenarios. It is therefore useful for forensic as well as planning exercises.

4.4   Case studies

Following an enthusiastic reception of the model at a safety conference, several cases were made available by mining companies not part of the development group. Two of the cases are discussed below.

Working Overtime due to Incident				Report Number

Description	Time (hours)	Number of Employees	Wages Per Hour ($)	Total Wages Paid
Production delayed due to closing the incident site				
4 cycles (each 12 hours), 17 employees				
Overtime to catch up with the production schedule	48	17	$75.00	$61,200
Cost of regaining the 4 cycles lost due to stack fire	1	1	$80,000.00	$80,000
			Total Cost	$141,200

Figure 3.   Costing the activities: costs for each activity are entered in a worksheet and the total is automatically computed and carried to several summary schedules to facilitate analysis.

Incident Costing - Summary Table

Incident Costs are divided into following six cost-pools

Pool		Total Costs of the Incident	Revenue to recover the cost
1	People	$145,140	
2	Incident Investigation	$122,638	
3	Equipment and Property	$241,705	
4	Corrective Action & Follow-UP	$8,400	
5	Environmental Damage		
6	Social Impact		
7	Lump sum add-on		
		$515,883	

Figure 4.   The total cost: in this final schedule the total cost of the cost pools are aggregated to give the final total cost of the incident under scrutiny.

4.4.1 The scoop tram fire case

This was a retroactive investigation of an incident. In this case a fire started in the engine compartment of a scoop tram – a large articulated heavy-duty loader designed for mining – in service at a mine. This was the second fire in less than a week for this scoop tram, and there were no injuries. Traditional estimation methods placed the loss in the low end of the $10 000 to $100 000 standard classification range. Total costing arrived at $251 573, including $122 250 in production losses.

4.4.2 The stack fire and line rupture case

This case involved a furnace stack fire that resulted a few days later in the rupture of a boiler feed water line. One employee was exposed to heat, dust and gas. Costing was done concurrently with the investigation, but it stopped before implementation of all the corrective actions. The total cost summary is shown in Figures 1 to 4 below. Discussions of the results with the firm employees and management showed the immediate value of the costing exercise. Neither the employees nor management expected that the total cost would be so high. But more importantly, the discussions generated by the activity maps were unprecedented. People were questioning all aspects of the incident, its effects and the actions that would possibly have prevented the incident. They were actively learning from the incident and internalizing the lesson. This case confirmed without a doubt the value of the methodology and the mind-maps to promote discussions across unit boundaries concerning costs, operations and incident prevention.

4.4.3 Total costing by WSIB account managers

The Ontario Workplace Safety Insurance Board saw in the model an opportunity to help them overcome the reluctance of plant managers and small business owners to invest in prevention measures that they see as producing no return. Several account managers were instructed to approach their clients and calculate with them the total cost of a safety failure. Although the conditions for introduction of the model were somewhat strained, the majority of clients exposed to the model understood and supported the reasoning behind the model and recognized its value. Comment from clients included "brings reality of how much it costs", "a platform for debate", "interesting to see how many people it affects (ripple effect), the downtime other than the person who is hurt", "surprised the president of the company". Account managers also found the model valuable to make the business case for injury prevention to their client. One account manager commented after a case presentation: "I feel confident the upper management left the meeting with a commitment to H&S".

5 DISCUSSION

The test cases reported above and others not reported confirm that the costs of safety failures are very high. In all cases the calculated costs were much larger than the estimated costs. Given that competitive and secrecy reasons restricted the access to cost data, it is not unreasonable to suspect that total costs in the cases studied could be five to ten times greater than the originally estimated costs. These unrecognized costs, called here "the self-insurance safety costs", are large and un-managed.

Included in the model are the costs of all the disruptions to the operations, such as production lost or production delayed, because of the incident. Company employees and management showed great reluctance in including these costs, arguing that since the production targets had been met there were no additional costs. In all cases these costs were very high, suggesting great potential for efficiency and productivity improvement with improved safety performance.

Field-testing of the model on isolated incidents showed its value. However, the model value to a company comes also from costing several incidents, over a lengthy period of time and building a database. Cost and activity analysis and management would show the true potential for improvement in operation efficiency and consequently in competitiveness and profitability.

The teaching and learning capability of the model was established by the discussions with the initial group of managers involved in developing the model. It was confirmed by the people involved in the case studies and by the responses from the WSIB account managers and their clients.

Beyond the human costs of injury and their impact on process flow and quality, the steady rise in injury treatment, hospitalization, rehabilitation and insurance costs contribute to making safety an urgent focus of management attention. As good safety performance becomes essential to create long-term value, it becomes a strategic goal. This places safety management among the strategic cost drivers that define a firm long-term cost competitiveness and its position with respect to its markets. However, few companies have recognized safety management as a strategic goal, and this may explain why cost and management information systems are not yet providing the needed information for improvement in safety performance. Without the broad perspective that total process costing provides, management will continue to maintain a one incident perspective and to motivate its employees to act at the local level.

6 BARRIERS TO IMPLEMENTATION

Based on the case studies, there are several barriers that make total costing of safety incidents difficult to implement. The major problem is the lack of integration with the management information systems in place. The person compiling the information cannot readily get the data from the firm current systems. Information has to be gathered across departments with sometimes reluctant cooperation. Obtaining valid and unbiased information may be a challenge, especially in retroactive costing.

On a longer-term basis, maintaining a high level of cooperation and interest in applying the results throughout the firm is crucial. This can only occur if the fit between short or local and long or strategic interests is understood and accepted, which may require an overhaul of the incentive systems. Sustained interest depends on the perceived vision of safety that upper management conveys to the employees.

This perceived vision of safety is the greatest barrier and consequently the greatest opportunity. The regulators and insurance executives to whom the model was presented were enthusiastic about the potential of the model for injury prevention. Employees of Health and Safety departments also saw the potential, but many lacked the authority they would need to obtain the information and to use the results in a meaningful way. By encouraging testing of the proposed model in their own firms, upper management could validate the model and clarify their doubts about the savings and efficiency gains that could result from making safety a strategic goal of their firm.

ACKNOWLEDGEMENTS

The author wishes to acknowledge the financial and contact support from the Mines and Aggregates Safety and Health Association of Ontario and from the Ontario Workplace Safety Insurance Board, and the contributions to the project of Dr. J. Lewko, Jane Djivre and Tiina Kopinnen.

REFERENCES

Oyon, D., Mooraj, S. 1999. Overview: Management Control Systems as a Source for Creating Value. *European Management Journal* 17(5): 478–480. Great Britain: Elsevier Science Ltd.

Shank, J.K., Govindarajan, V., Winter. 1992. Strategic Cost Management and the Value Chain. *Journal of Cost Management*: 5–21.

Ittner, C.D. 1999. Activity-Based Costing Concepts for Quality Improvement. *European Management Journal* 7(5): 492–500.

Weyerhaeuser Company Annual Report 2000.

Hammer, W. *Occupational Safety Management and Engineering*. 4th Edition. New Jersey: Prentice Hall.

Reinhardt, F.L. 2000. *Down to Earth*. Boston: Harvard Business School Press.

Lebas, M. 1999. Which ABC? Accounting Based on Causality Rather than Activity-Based Costing. *European Management Journal* 17(5): 501–511. Great Britain: Elsevier Science Ltd.

Busan, T. 1989. *Use your Head*. London: BBC Books.

Safety and Reliability – Bedford & van Gelder (eds)
© 2003 Swets & Zeitlinger, Lisse, ISBN 90 5809 551 7

Sorting out performance patterns from noise in safety data

J.A. Blanco & P. Chaddock

Health, Safety, and Productivity Group, Laurentian University, Sudbury, Ontario, Canada

ABSTRACT: Aggregate failure data show that we learn to prevent failures as we accumulate experience. This implies that some recent newsworthy failures may have been preventable. Managers need a system that milks incidents for hints of prevention, accelerates learning, and leads to permanent correction. Managers need technical aids because there are fewer opportunities for prevention and the data become less stable as we improve. Simple noise filters are needed to sort transient conditions from trends, to identify when management must initiate change, and to spot types of work with exceptional exposure to risk. This paper discusses noise filters based on techniques proven in production environments. Fact-based noise filters are more effective than "seat of the pants", "specifications", "period comparisons", or "benchmarking", which may foster system failures such as complacency, faulty attribution, and superstition. Practical examples from industry illustrate how Poisson distribution and control charts provide an effective focus for prevention work.

1 INTRODUCTION

1.1 *Learning curves*

This paper discusses why and how statistical process control methods can help focus the search for factors that contribute to failure so that we can learn from them and accelerate prevention.

Aggregate failure data for a number of industries show that we learned to prevent failures as we accumulated experience. The normalized data fall into what is called a "learning" or "decay" curve.

For example, the near-miss data for 11 years of air carriers in the US prepared by Duffey & Saull (2001).

Note that Air Carriers report and act on "near misses" as accident precursors which are far more numerous than the collisions that they seek to prevent.

The data shows a clear trend to rapid initial decrease when the number of near misses was high, reaching an asymptote as the rate approached one near miss per 100,000 flight hours.

Duffey & Saull (2002) propose a universal learning curve and advance the importance of learning from errors and accidents for safety improvement.

As another example, here is the curve representing about 50 years of fatality data for mining in the province of Ontario, Canada.

The decay curves tell us that the number of accidents decreased as we gained experience. However, they also suggest that we learned how to prevent accidents after

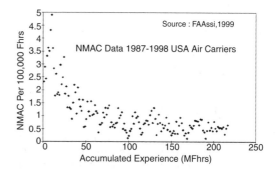

Figure 1.   12 Years of North American aviation.

we were aroused to action by the casualties. Once we recognize the damage, we search the records of these failures and find information that we then use for prevention.

Just as improvements would become more difficult, the stimulus of undesirable events would decrease and we would be tempted to slacken our efforts.

1.2 *Pursuing improvement*

Recent events such as train derailments, air crashes, and automobile rollovers all reveal failures that could have been detected and prevented.

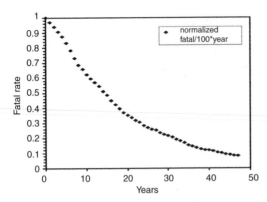

Figure 2.   Normalized mining fatals per 100 worker years, 1952–1999, in Ontario, Canada.

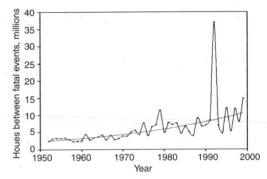

Figure 3.   Hours between fatal accidents in Ontario mining, 1952–1999, in Millions.

If we can sensitize ourselves to the numerous precursor events and conditions which lead to incidents, we should be able to accelerate improvement.

The global, long-term, decay curves reflect the contributions made in individual workplaces. Obviously, quicker improvement at the global level can not happen without earlier recognition and faster response at the individual workplaces. It is reasonable to assume that managers are already doing for safety and prevention what they can with the tools available. If the individual managers are to improve prevention in the workplaces they manage, they need better systems that help track incidents, separate the wheat from the chaff, and milk them for hints of prevetion.

2   NOISE IN THE DATA

Managers need technical aids because as the number of injuries decreases the number of opportunities for prevention also decreases and, moreover, the data tend to become less stable and therefore harder to interpret.

Figure 3 covers 52 years of fatality data in Ontario mining and shows a five-fold improvement in hours worked between consecutive fatalities but it also shows greater instability.

The instability reflects the deterioration of the "signal/noise ratio" as the number of undesirable events decreases.

If the noise is apparent in the data for aggregate industry performance, what do we see in data for one company or facility?

2.1   Noise impairs analysis

Undesirable incidents occur randomly in the sense that the rare combinations of circumstance leading to damaging release of energy are essentially unpredictable.

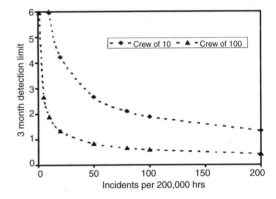

Figure 4.   Three month detection limit for incident rate change as a multiple of the current average rate for two work group sizes.

As Ladislaus Bortkiewicz discovered while studying cavalry deaths from horse kicks over 100 years ago, rare events tend to follow the Poisson distribution. A count of the number of incidents within a work group follows the Poisson process and the variability in the data (noise) is strictly dependent on the mean rate of incident occurrence.

Figure 4 illustrates that detection becomes more difficult as the incident rate declines, and more difficult again for small crew sizes. The graph shows detection limit for data accumulated over a 3-month period; obviously, detection over shorter periods is still more difficult.

For example, if a group of 100 workers has a historical rate of 50 incidents per 200,000 hours worked, then in a given 3 month period the rate must jump by almost 100% (at a detection limit of 1) to be deemed a significant change (be detected).

A group of 10 workers with the same historical rate must generate incidents over 3 months sufficient to raise the rate by almost 300% (detection limit of 3) in order to signal a significant deterioration.

A manager armed with this knowledge in a plant where lost time accidents, medical aid visits, and first aid visits average 2, 12, and 60 respectively would understand that realistic shifts in safety performance (of 100% for example) would be essentially undetectable based on LTA or medical visit data. A deterioration might be observable in first aid visit data, but three months would pass before the higher rate could be called significant.

A logical progression along this line of thinking would have the manager focusing on "precursor incidents" instead of "incidents that produce injury", since they are more numerous. Another step might be to track and chart incident precursors extracted from the historical injury report records.

Note that designing our measurement systems to account for the noise in the data naturally redirects us to detecting and correcting precursors before the failures occur.

The wisdom of the air carriers in charting near misses is multiple: there are many more near misses than collisions, and forestalling them prevents major loss of life and equipment. The airline industry has also adopted a "human factors" strategy that greatly increases the rates monitored, the observed crew size, and the number of observations. These moves are the equivalent of moving one or more orders of magnitude to the right in Figure 4, thus increasing the resolution and speed of the detection step, and speeding up prevention.

2.2 Perceptual factors

People (managers and workers) tend to be unaware of the level of noise in the injury data, and have little sense of the impact of random factors in small data sets. Kahneman and Tversky (1984) have studied misperceptions of probability extensively. Similar work on Prospect Theory in the field of economics resulted in Kahneman receiving the Nobel prize.

Injury data may tempt managers and workers alike to make invalid attributions. A cluster of accidents may be interpreted as a sudden rise in carelessness or poor morale, whereas an absence of accidents may lead to complacency. Randomness in data can trigger "superstitious thinking".

For example a manager may schedule special safety meetings with the workforce after a worrisome uptick in injury rate. In the next month the injury rate drops, so the manager concludes that safety meetings have a beneficial impact on safe work behaviour. The fact that the injury rates may fluctuate up and down well within the noise band is not generally recognized.

Managers trying to improve injury prevention need not overreact to apparent clusters, nor wait for damage

to occur. As discussed earlier, they could adopt potentially useful strategies such as increasing the number of observations by switching from "incidents that cause LTA" towards "precursor incidents", and enlarging the cohort under observation by searching the industry record for incident precursors. These are some of the drivers for the very successful "human factors" approach in civil aviation (Blanco & Lewko 2002).

Whether focusing on failure or precursor data managers need simple, objective, noise filters that will prevent attribution error.

Are there transient conditions that require attention? Are there trends which portend a long-term change? Do the existing technologies or management systems need upgrading? Are there work groups or individuals whose safety records suggest exceptional exposure to risk? What tasks, equipment or types of work are most in need of change to improve safety?

Fortunately, there are simple statistical techniques to milk information out of the data.

2.3 Operational definitions

Organizations risk reducing the apparent number of undesirable events when they tamper with the operational definitions for categorization of data. A "lost time" or "disabling" injury could reclassify as "medical aid" when the employee performs modified work, but in doing so it would make problem recognition more difficult and likely promote a "no-reporting" environment.

Throughout this paper we have assumed that the operational definitions of incident or injury are unchanged.

3 USING THE POISSON DISTRIBUTION

We will demonstrate the use of Poisson calculations by testing the oft held assumption that safety awards are useful in motivating workers to work safely. Poor understanding of the effect of random variation on injury occurrence leads managers and workers alike to assume a priori that a person who has been injured has done something wrong. The observation that the majority of workers function in the same environment without incident often leads people to conclude that the injured party has chosen to work in an unsafe manner.

In the examples that follow we will show that randomness alone is a sufficient explanation; it is also the simplest. We base the calculations on the assumption that all workers behave identically (no possibility of "accident proneness" or "unusual risk exposure") in an environment with a stable empirical hazard rate of 1.5 lost time accidents (LTA) per 200,000 hours worked.

Table 1. Poisson estimates at 1.5 lost time accidents per 100 worker-years (200,000 hours).

Years LTA-free	5	10	30
Estimated probability	0.928	0.861	0.638
People LTA-free in group of 1000	928	861	638

Table 2. Poisson estimates for a subgroup of 30 within a population with a global average of 1.5 lost time accidents per 100 worker-years.

LTA per year	0	1	2	3	4
Estimated probability	0.985	0.015	0.0	0.0	0.0
Workers uninjured	29.6	0.4	0.0	0.0	0.0

Table 3. Poisson estimates for a subgroup of 30 within a population with a global average of 1.5 lost time accidents per 100 worker-years.

Years LTA-free	1	2	3	4	5
Estimated probability	0.985	0.97	0.956	0.942	0.928
Uninjured workers	29.6	29.1	28.7	28.3	27.8

3.1 Individual safety awards

Awards are commonly given to workers who have been injury free for long periods. Managers interpret absence of LTA as a reliable indicator of safe work behaviour. Many believe that a safety award can reinforce the desirable behaviour thus inferred.

Table 1 shows that the count of accidents will fail to correctly classify all the employees as consistent in behaviour. Even after 30 years over 60% of workers will have avoided significant injury. This likens the process of rewarding those who remain LTA-free for 5, 10 or even 30 years to a lottery in which all enter and the majority that did not get injured win regardless of their contribution.

3.2 Annual group safety awards

The Poisson distribution can also help a manager supervising a group of 30 workers: would a year worked without an LTA in the group indicate behaviour worthy of award?

These estimates show that the supervisor or manager may reasonably expect the group to be LTA free in any given year.

Most such groups will enjoy their awards blissfully unaware that their environment and behaviour harbour an ongoing hazard of 1.5 LTA per 200,000 hours worked. There is no motivation for them to seek improvement.

The groups which experience one LTA will be regarded as somewhat inferior. Management may examine the group work behaviour in a vain attempt to discover a significant difference.

We conjecture that even though a true "outlier" group that harvests 2 lost time accidents is perhaps demonstrating a lower standard of behaviour, management may miss the opportunity for learning if they aggregate the "outlier" group together with those with a single injury. The noise will thus obscure an otherwise clear signal.

3.3 Long term group safety awards

The manager could also use the Poisson calculation to estimate the group's chances of working several years without LTA. Would 5 years worked without LTA indicate safety behaviour better than that assumed to underly our global average of 1.5?

Clearly if the group is LTA free for a number of years, improved safety behaviour cannot be inferred. There is over a 90% chance that a group of 30 can get to 5 years LTA free by chance alone. The "typical" group will reap about 2 LTA. Even after 5 years the chance of the group escaping an LTA is appreciable.

At best, awards show no promise as motivators; in typical award systems the majority of people are rewarded for their good luck, and the few that are injured suffer award denial in addition to injury. At worst award systems can lead to suppression of injury reporting behaviour and thus further hamper measurement. Safety awards clearly are not an appropriate way to encourage performance.

3.4 Implications for incident analysis

Even though one injury or incident does not necessarily indicate a negative trend, each deserves close inspection because as the long term evidence shows, it is likely to contain information that can lead to prevention, and such information would otherwise be lost. The findings of incident investigations tend to be conditioned by our expectations. If we expect to find the worker behaviour deficient we will risk selecting the evidence that supports this belief. When we are focussed on individual attribution we will be blind to the constant aspects of the system that facilitated the incident, thereby ensuring that the incident will recur.

When awareness of randomness compels us to look beyond individual contribution we are poised to detect the system constants and factors that are the precursors to failure and we may refocus our preventive activities. Any systematic approach, for example the five point

174

method for investigation shown below (Blanco et al. 1996), can be used to accelerate learning:

– Reconstruct the event
– Analyze the situation and the relationships
– Identify significant and insignificant information
– Formulate and test tentative cause-effect links
– Develop solutions to fundamental causes

It is critical that the process of learning from incidents must include and impact all affected members of the organization. Learning restricted to the "executive suite" will accomplish little.

4 USING CONTROL CHARTS

We can use the u-chart, a time series, to make the noise level explicit and thus identify performance that is truly exceptional (good or bad) and which merits exceptional attention. U-charts allow supervisors and managers to detect real trends early on with a view to investigation, prevention and improvement. U-charts plot the averages and bracket them with upper and lower control limits; these are calculated from the sample data:

$$CL = \bar{u} \pm 3 \sqrt{\frac{\bar{u}}{N}} \qquad (1)$$

where CL = upper or lower Control Limit; u = incident rate; and N = subgroup size

4.1 U-chart for plant data

To illustrate some uses for the u-chart, consider a retrospective look at the 17-year injury history of a plant with about 1000 employees.

U-charts were not being used for managing safety at that time; rather, management relied on "specifications", "period comparisons" and "benchmarking".

Figure 5 shows the mean, upper control limit and lower control limits for the first seven years:

Since management did not have the benefit of the u-chart analysis they were unaware of the magnitude of variation due to noise in the system. As the disabling injury rate fell from year 1 to 4 management believed their safety efforts were effective. People tended to become complacent about work behaviour and the implicit level of hazard in the workplace.

When the disabling rate rose in year 5 and continued upward in an apparent trend in years 6 and 7 a futile search for causes ensued. Since no change in working environment or process had occurred the tendency was to attribute blame to those injured. Efforts were focused on exhorting workers to be more careful.

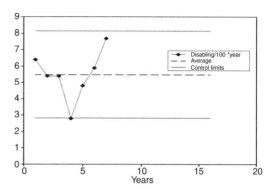

Figure 5.   Seven year history of lost time accidents per 100 worker years.

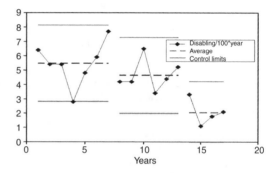

Figure 6.   Seventeen year history of lost time accidents per 100 worker years.

At that point, the manager experienced a rare personal epiphany and concluded that the "status quo" approach to safety was not effective. A detailed examination of the injury history revealed simple patterns that stood out clearly once attribution to individual character was abandoned.

In one pattern electricians were suffering injury while rolling large spools of cable over uneven surfaces. Simple transport wagons were bought and that type of injury ceased to occur.

In another pattern workers using 2 meter long wrenches for adjustment of large equipment were suffering back injuries. Powered wrenches were substituted and another cluster of injury type was eliminated.

Figure 6 plots the data for the next 10 years. The points have not been connected between years 7 and 8 or years 13 and 14 as an aid in visualizing the periods under discussion.

The concerted efforts of management produced a small reduction in average rate from year 8 to 13. Through year 13 the improvement was meager. Technically there is insufficient evidence in these data to justify a recalculation of control limits. Usually nine

175

consecutive points to one side of the original mean or 4 of 5 points one sigma or more to one side of the original mean would be required. We have given the manager "the benefit of the doubt" and have recalculated the limits based on our knowledge that real system change was beginning to take place during this period as the management safety culture evolved.

The data for the remaining 4 years show significant change (points outside the limits seen in years 8 to 13), confirming that the prevention initiatives were producing a measurable improvement. The on-going benefits of sticking to a program of searching the injury data for patterns and setting up a "safety-work-order-system" became visible.

Although the eventual trend is satisfying, the noise on the signal had forestalled improvement for years. If the manager had remained a prisoner of traditional thinking, the improvement may never have been achieved.

With u-charts in use from year one and injury data plotted every month, the manager would have seen early on that the system of work was stable despite the noise and that change was required. The temptation to attribute injury cause to personal character would have been removed, and the manager would have directed the finite energies of the organization toward changing the "status quo" of about 5 disabling injuries per 200,000 hours worked. The analysis and plotting of precursor data would naturally have followed. Learning and improvement would have been accelerated.

Fact-based noise filters such as Poisson distribution and u-charts are rigorous and dependable, and unlikely to feed attribution or complacency.

4.2 Diligent analysis

Faulty attribution to character is typical of situations where noise or random variation obscures the true state of the system. Managers may be tempted to perceive the injured worker as deficient in behaviour. This is rooted in the observation that others working in the identical environment escaped injury. The most salient variable in an informal analysis is the identity of the injured worker.

Although safety professionals and researchers may have long ago refuted the myth of "accident proneness" (Haight 2001) our experience with managers in North American industry indicates that belief in the existence of the "unsafe worker" remains deeply rooted. The following model is suggested to promote education of management by first checking for the existence of anomalous behaviour in groups and individuals, then moving toward analysis of system effects:

1. U-chart overall plant incident data versus time (as in the example above, but monthly).
2. U-chart total incident data in categories by work group (shift or area or function).

3. Check individual incident history accumulated versus time. Update for the affected individual as incidents occur.
4. Focus charting on total incident data sorted by task group or individual task to identify activities which entail significantly more risk.

Steps 1 through 3 will establish whether or not there is significant change over time for either groups of workers or individual workers. Total incident data should be used rather than LTA or disabling data in order to have an average rate high enough to facilitate detection (as per Figure 4).

Step 2 is illustrated in Figure 7, which shows data for a crew of 15 workers plotted as a monthly u-chart.

The data points are all well below the upper control limit for all months, indicating that this crew harbours no safety behaviours worthy of special investigation.

Step 3 is illustrated by Figure 8, which shows the incident history of an individual worker plotted on a cumulative basis over a period of 4 years. This chart has a construction differing from the u-chart yet keeps the viewer "noise aware" by plotting the upper control limit associated with the cumulative level of hazard exposure experienced by the worker.

The accumulated incidents are well below the upper control limits until month 34, where they equal the limit. This is a borderline case where the performance of the individual is right at the 3 sigma limit of

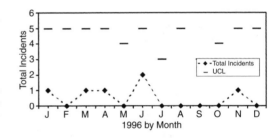

Figure 7   U-chart of monthly total incidents for crew A.

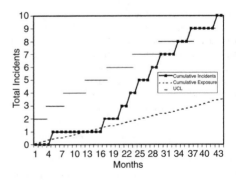

Figure 8.   Personal incident control chart (cumulative).

probability. A cautious investigation is indicated. This analysis is valid only if the incidents are independent, so if subsequent injury is conditioned by incomplete recovery from earlier trauma, then these control limits would not apply.

When steps 1 through 3 fail to reveal individuals or groups with abnormal safety behaviour a manager is forced to conclude that the system of work should get a closer look.

In step 4 the incident data is categorized by activity or task group.

Consider a group of 113 people distributed in four crews to provide 24-hour coverage. Group data covering 41 months had shown that the injury rates of the four crews (occupations pooled) were statistically similar, and all were within limits.

Figure 9 reveals significant differences in the incident rates for various occupational classifications.

Individuals holding job class E are more likely to suffer injury than their counterparts on job class J. Without knowing that that job may entail greater risk than the others, managers might blame workers in class E for their injuries. Figure 9 points to the job itself rather than personal contribution of the workers. Using this control chart, the manager could rationally conclude that job classes E, I, G entail greater risks than job classes P and J, and therefore the workers in those classifications need more help than their counterparts at P or J.

4.3 Control charts guide action

It is important to emphasize that control chart analysis never suggests that efforts to improve safety should slow or cease. Using "noise aware" tools like control charts guides us in selection of the course of action. If control limits or trend/run rules are violated we are guided to conduct intensive investigations of the anomalous situation.

If the system of work is stable and the charts indicate normal noise variation about the mean we have a

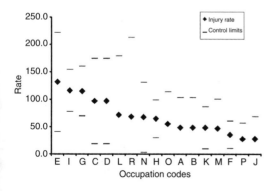

Figure 9. Plant incident rate by occupation.

duty of care to study the system of work to understand why our workers are subject to this constant level of hazard. The noise on the signal must never be allowed to deflect us from our target of continuous improvement of safe work.

5 ON DOWN THE LEARNING CURVE

Injuries and other failures are beacons that light the areas where there is malfunction; if the factors that contribute to the failure are not removed, the failure will recur, although it may have different effects.

When injury rates are high our attention is focussed on the problem and we study causation and implement prevention. We move down the learning curve of Duffey and Saull.

When injury rates are low the noise implicit on the signal masks the truth and can misdirect our energies via errors of attribution or lull us into complacency.

The use of Poisson calculations and control charts will render the manager "noise aware" and prevent both attribution error and complacency.

Realization that measurement of rare events entails high implicit noise levels will naturally lead the manager to seek events for measurement that occur at much higher rates. The precursors to incidents are much more numerous and will provide guidance to effective action to reduce hazard.

Removing contributing factors may require someone to change something they do, or it may require physical changes in equipment or process.

A systematic approach as is used in a maintenance work order system (defect detection, planning, scheduling and completion) is a good model to follow to ensure that prevention initiatives are handled efficiently and systematically. There is a great similarity between reactive/preventive/predictive measures in either maintenance or safety.

Whether in maintenance or safety, weak systems are a source of reaction, short cuts, recurring failure and injury.

Disciplined consistent use of Poisson calculations and control charts can help managers, supervisors and workers to improve the focus and effectiveness of their "safety-work-order" and injury prevention systems, and, in the process, improve all failure prevention systems. The results will follow in due time: a safer and more efficient operation.

REFERENCES

Blanco, J., Lewko, J.H. & Gillingham, D. 1996. Fallible Decisions In Management: Learning From Errors, *Disaster Prevention and Management* 5(2): 5–11.

Blanco, J.A. & Lewko, J. 2002. Human Factors in Aviation Maintenance. Conference Proceedings, *16th Symposium*, Federal Aviation Administration, Transport Canada, Civil Aviation Authority. San Francisco

Duffey, R.B. & Saull, J.W. 2001. Errors in Technological Systems. Paper for the World Congress *Safety of Modern Technical Systems*. Saarbruecken.

Duffey, R.B. & Saull, J.W. 2002. *Know the Risk: Learning from Errors & Accidents: Safety and Risk in Today's Technology*. Butterworth-Heinemann.

Haight, F.A. 2001. Accident Proneness: The History of an Idea. Institute of Transportation Studies. University of California. *http://www.its.uci.edu/its/publications/ papers/ WP-01–4.pdf.*

Kahneman, D. & Tversky, A.1984. Choices, Values, and Frames. *American Psychologist* 39(4): 341–50.

Safety and Reliability – Bedford & van Gelder (eds)
© 2003 Swets & Zeitlinger, Lisse, ISBN 90 5809 551 7

Full stochastic analysis of a cofferdam

A. de Boer
Civil Engineering Division, Ministry of Transport, Public Works and Water Management, Utrecht, Netherlands

P.H. Waarts
TNO Building and Construction Research, Delft, Netherlands

ABSTRACT: There are no design rules for cofferdam structures. Research has been carried out to close this gap by developing a design procedure for cofferdam structures in a semi-probabilistic way, close to sheet pile guidelines. During the past few years a complete probabilistic analysis tool has become available as part of an existing FE code. With this code, a full stochastic analysis of a cofferdam structure example may be demonstrated. The paper shows the results of the probabilistic code and compares the results of this stochastic analysis with the results of an analysis based on the design procedure.

1 THE COFFERDAM STRUCTURE

The cofferdam structure used in this contribution was published in Geotechniek by Weersink et al., 2002 (figure 1).

The structure consists of two sheet piles connected to each other with an anchor. The two sheet piles were drilled into an existing dike system of a two-layer soil system of sand. This basic soil system has a small slope at surface level (#1). After installing the sheet piles and the anchor, the water level on the right side can be raised. Additional filling sand will be put between both sheet piles. On the left and right side of the cofferdam structure the soil may then be excavated.

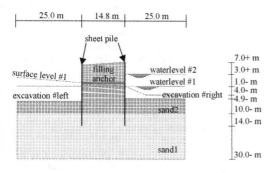

Figure 1. Overview cofferdam structure.

2 DESIGN PROCEDURE

The design procedure of the cofferdam structure in current use consists of two phases:

– predesign phase of the cofferdam structure to obtain the dimensionial properties.
– checking phase of the cofferdam structure for the serviceability limit state (SLS) and the ultimate limit state (ULS). Apart from this, a stability analysis has to be carried out using an FE code.

There are several (CUR 1997) methods to design a cofferdam structure, like Homberg, Jacoby or Terzaghi, as well as an FE code. However, at this stage in the design process an analytical method is preferable. Afterwards the result of this pre-design process is to be checked, using an FE code on SLS, ULS and stability aspects.

The SLS checks deformations. Normally, a maximum horizontal deformation of 100–150 mm at the top of the sheet pile is tolerated. Apart from this, the relation of $\delta_{max} < 0.01 * H$ is used, in which H is the retaining height of the structure. The soil properties used are the so-called representative values.

The ULS checks the strength of the sheet pile itself, the strength of the filling of the cofferdam structure and the overall stability of the structure. The values used for the properties of soil, water heights, retaining height and additional loads are extracted from the existing guidelines of retaining walls[..]. This retaining wall guideline consists of three safety classes. Each safety class is made up of partial safety factors for the soil properties, determined in a probabilistic way. However, the density of the soil and water level in the soil are not stochastic. For these additional soil properties any expectation value will be a deterministic value. The values of water heights, retaining height and additional load cases are also deterministic, and are derived from common guidelines of the usual structures.

3 THE PROBABILISTIC METHOD

At Delft University of Technology, in cooperation with TNO Building and Construction Research, as well as the Dutch Ministry of Transport, Public Works and Water Management, research has been carried out to compute the structural reliability using a combination of finite element analysis (FEA) and probabilistic methods.

The structural behaviour of a complex structure is often calculated using a Finite Element Analysis (FEA). Stresses and deformations of the structure may be computed given the (deterministic) parameters of loads, geometry and material behaviour.

Structural codes require a certain level of structural reliability. The Dutch Building code, for example, demands a maximum failure probability of 10^{-4} within a given reference period (service life of structure). This failure probability is ideally translated into partial safety factors by which variables like strength and load have to be divided or multiplied to arrive at the so-called design values. These design values are to be used as input for a Finite Element Analysis. The outcome of the calculations is compared with the limit states (for example, collapse or maximum deformation). The structure meets the reliability requirements if the limit states are not exceeded.

Reality is different. First: the method using partial safety factors renders it only plausible that the reliability requirements are met, without there being any certainty. A second aspect is that safety factors are often based on experience only. There is often no link with the required reliability on a theoretical basis. The third aspect is the system behaviour of structures. Safety factors are often derived from components of a structure; for instance, single sheet piles, anchors or single failure surfaces. A structure as a whole will behave like a system of these components. As a result, and depending on the system under consideration, a structure may be more or less reliable than its separate components. It would therefore be useful to have at one's disposal a method to calculate at once the accurate (system) failure probability of a total structure. Standard reliability methods compute the failure probability given a limit state and stochastic parameters. Limit states may

be, for instance, exceeding of yield stress in a structural member, exceeding of maximum deformation or global collapse. Well-known methods for computing the reliability are the Monte Carlo simulation (MC) (Rubinstein 1981) and the First Order Reliability Method (FORM) (Hasover & Lind 1974). In this paper an unusual method is applied: an adaptive method based on Directional Sampling (Bjerager 1988).

For large and complex structures it is almost impossible to provide an explicit limit state function. Points of the limit state function may, however, be calculated using the Finite Element Analysis (FEA). Combining reliability methods and Finite Element Analysis is often referred to as Finite Element Reliability Methods (FERM). Instead of computing structural behaviour (with FEA) in terms of deformations and stresses, behaviour is computed in terms of failure probability and uncertainty contributions. In this way the basic demands of the codes are met, i.e. meet the required failure probability.

The problem is that the above mentioned standard reliability methods are traditionally used for problems with only a few random variables, taking up little time to evaluate the limit state function. In a combination with FEA, the opposite occurs, as there are many random variables, so that evaluating the limit state function takes much computational effort. To speed up computations, research at Delft University of Technology has led to the introduction of the so-called "Directional Adaptive Response surface Sampling" (DARS) (Waarts 2000). In short: the improvement to the standard directional sampling lies in the use of FE for important directions and a response surface for those less important. In practice this means that after the response surface is constructed, only a few FE computations will have to be carried out.

In the DARS procedure for the construction of the response surface, all variables are varied individually and increased or decreased until failure. An FE model with n stochastic variables results in 2n (directional) samples in the principal directions. Consequently a quadratic response surface is fitted to these results. Following this starting procedure random directional sampling takes place. The response surface is used in case of a large distance from the origin to the response surface. FE computations are used to calculate the real distance in case of a small distance from the origin to the response surface. In that case the response surface is updated (adapted).

Influence factors give an insight into the importance of stochastic variables on the limit state. After finishing the directional sampling procedure, the influence factors α are computed using a FORM analysis on the response surface.

In this research project the probabilistic method is implemented in an existing FE code, namely Diana (Diana 1998), release 7 of 1998.

4 CALCULATION PROCESS

The calculation process follows the calculation steps of a single sheet pile. The checking codes describe a partial safety factor for the soil material properties, such as:

– Young's modulus ($\gamma = 1.3$),
– cohesion ($\gamma = 1.0$) and
– internal friction angle ($\gamma = 1.15$ for both sands and $\gamma = 1.39$ for the filling sand).

In this case the partial safety factor of cohesion is ignored. Also, no partial safety factors are used for dimensioning the thickness of the sheet pile.

The cross-section of the anchor originally has a partial safety factor ($\gamma = 1.25$), but this part of the structure is always overdimensioned by a factor. In this case the total partial factor will be almost 2, so both dimensions may be taken as a deterministic part of the calculation. In one of the variations the anchor cross-sections will be taken as a stochastic parameter. Apart from this extra partial safety factor an additional structure safety factor is calculated using the so-called φ-c reduction technique method (Brinkgreve and Bakker 1991).

Phased analysis means that the construction process of a sheet pile structure will be simulated. This simulation allows one to include the initial stresses of the construction stages into the analysis. Each construction stage has its own geometry, its own supports and its own initial stress stage. The construction process of this cofferdam structure may be divided into six stages. Table 1 gives an overview of these stages during construction of the sheet pile structure.

One result from this phased non-linear analysis is a diagram of the bending moment M_z of the sheet piles, shown in figure 2.

In this case the maximum or minimum values of the bending moments M_z of the sheet piles are not crucial. The shape of the bending moment M_z determines the subdivisions of the soil. Over the height of the left sheet pile the soil needs to be subdivided into three areas. For the right sheet piles two subdivisions will suffice. Apart from this, there are two different soil material properties (sand1 and sand2). As for height, during the first step of a probabilistic analysis a set of two areas for the sand1 – with at least three horizontally oriented subareas, left, middle and right of the cofferdam structure – should be created. At a later stage the middle section between the sheet piles may be split up into more subsections over the sand2 and filling area.

5 VARIATIONS OF STOCHASTICS

In any type of analysis the engineer will have to choose which properties are liable to change from

Table 1. Construction stages of the sheet pile during construction.

Stage	Construction stage
1	Initial stage
2	Drilling sheet pile into the soil
3	Initial stage of the filling soil layer between the sheet piles
4	Placing the anchor between the sheet piles + add the filling to the virgin state
5	Excavate the soil on both sides of the sheet piles from surface level to the indication of surface left and surface right
6	Increase the water level from #1 to #2

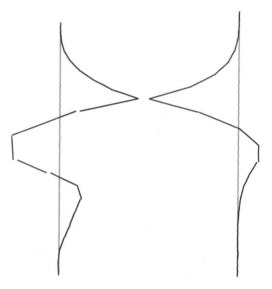

Figure 2. Bending moment M_z of the sheet piles.

determinist to stochastic. In this case the outcome is more or less neutral. The guidelines for sheet piles show partial safety factors for φ and for the Young's modulus E. The basic full probabilistic analysis includes the same amount of stochastic variables.

This will be discussed after the results of the first basic analysis.

Table 2 gives an overview of the relevant stochastics of the cofferdam structure.

Common to geotechnical calculations, sand types are split up into several sections in case of active or passive behaviour of the soil, depending on the load case.

In the horizontal plane the three subdivisions are: the area left of the cofferdam structure, the area in the middle and the area to the right of the cofferdam

Table 2. Distribution and values of the relevant stochastics.

Material/Property	Unit	Mean	St. dev.	Lower bound	Distribution type
Sand1					
1. $\sin(\phi)$	–	0.57	0.057	–	Lognormal
2. E	kN/m^2	46.e6	4.6e6	1.0e4	Sh-lognormal
Sand1 interface					
3. $\tan(\phi)$	–	0.7	0.07	0.1	Sh-lognormal
Sand2					
4. $\sin(\phi)$	–	0.53	0.053	–	Lognormaal
5. E	kN/m^2	30.e6	3.0e6	1.0e4	Sh-lognormal
Sand2 interface					
6. $\tan(\phi)$	–	0.62	0.062	0.1	Sh-lognormal
Filling					
7. $\sin(\phi)$	–	0.62	0.062	0.1	Sh-lognormal
8. E	kN/m^2	30.e6	3.0e6	1.0e4	Sh-lognormal
Filling interface					
9. $\tan(\phi)$	–	0.62	0.062	0.1	Sh-lognormal

structure. In the vertical plane the sand1 area will be subdivided into two parts:

- the lower fixed sand1 part and
- the sand1 part between the sheet piles

This results in a total of 24 stochastic variables for the soil and 3 stochastic variables for the interface between the soil and the sheet piles.

6   RESULTS OF THE BASIC CALCULATION

Within almost a day of calculation time (on a common workstation computer) the designer of the cofferdam structure will obtain the results of the basic probabilistic calculation. A reliability calculation yields a reliability index $\beta = 3.92$. This reliability index meets the requirements of the structural codes.

Standard design calculations do not give a reliability index. The partial safety factors used and an extra structural safety factor of 2.0 – resulting from the ϕ-c reduction technique method – accounts for an acceptable design.

An additional output result of this full probabilistic method is the overview of the influence factors of the various stochastic variables in relation to the reliability index. In table 3 only the relevant stochastic variables are shown. Here the relevancy limit of the stochastic parameters means higher than 5%.

The remaining 5% is accounted for by the ignored 18 stochastics, so the halfway conclusion is here that the Young's modulus of the soil is very important for the cofferdam structure. Attention must be given to the fact that not all of the material or physical properties in this calculation are stochastic variables. Special attention has to be given to the soil properties of the fixed sand1 layer and the left side of the sand2 layer.

Table 3. Influence factors stochastics to the index of β.

Number	Stochastic	Influence [%]
2. sand2 left	Young's modulus	25
6. sand2 mid	Young's modulus	8
12. sand1 fixed left bottom	Young's modulus	33
14. sand1 fixed left top	Young's modulus	5
16. sand1 fixed mid bottom	Young's modulus	14
20. sand1 fixed right bottom	Young's modulus	10

The middle section of the sand2 layer and the filling sand is not influenced to the degree that was expected of the active and passive behaviour of the soil.

At this stage the two main soil layers are to be subdivided into more horizontally oriented subsections in the calculation described above. It is preferable to subdivide the belonging interfaces between the soil and the sheet pile into more subsections. Thus the probabilistic procedure is more consistent with the soil layers, including the interface sections. This approach has been analyzed using a total of 35 stochastic material parameters, resulting in a reliability index $\beta = 3.89$. The influence factors in this calculation do not differ much from the basic calculation. The largest influence factor of the interface material property is less than 1%.

This procedure allows one to present an overall contour plot of the given partial safety factors as a mean value of the elements, which describe the soil layers. Each element has its own starting value of the Young's modulus. At the end of the calculation each element obtains its value within the belonging distribution of the Young's modulus. This means that the ratio of those two values may be calculated, which should be the partial safety factor of each FE element.

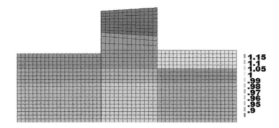

Figure 3. Contour plot partial safety factor Young's modulus.

Table 4. Density properties sand types.

Material/ Property	Unit	Mean	St. dev. bound	Lower	Distribution type
Sand 1					
ρ	kg/m³	18.	0.9	0.1	Sh-lognormal
Sand 2					
ρ	kg/m³	17.	0.85	0.1	Sh-lognormal
Filling					
ρ	kg/m³	17.	0.85	0.1	Sh-lognormal

Figure 3 shows these partial safety factors for each soil element.

In this ratio figure of the Young's modulus each set of soil elements obtains its own final Young's modulus. The maximum value of this plot is $\gamma_{max} = 1.2$ and the minimum $\gamma_{min} = 0.86$. As for the checking code, its partial factor was set to the uniform value of 1.3.

In this case the analysis shows a small underestimation of the soil property. The overestimation of the Young's modulus is, however, far more important. It demonstrates that in some regions the Young's modulus is a strength parameter and not a load parameter.

7 VARIATIONAL CALCULATIONS

Section 6 described a second calculation beside the basic calculation of the cofferdam structure. It utilized the additional aspect of the influence factor only, adding the interface material properties to the stochastic variables. In this paragraph other properties are added as well to the total set of stochastic parameters. The first variation is the density parameter of the soil layers. The basic set of stochastics without the subdivided stochastics for the interfaces forms the foundation of this variant calculation. This means that each of the 12 sublayers has its own density stochastic. There is of course a strong correlation between the porosity of these soil layers and the density, but this option is not yet available to this pilot version of the probabilistic module, coupled to the existing FE code.

The main property variants of this density addition are given in table 4.

The analysis, using the additional density material parameter beside the basic explained material parameters, leads to a reliability index $\beta = 4.08$.

The influence parameters do not change very much. The highest density influence factor falls within the range of 4%, so that the given influence factors of the Young's modulus in table 3 remain approximately the same. The basic calculation is still acceptable.

The second variation on the basic calculation is the addition of a load case as a stochastic variable. However, within the design of a cofferdam structure there

Table 5. Load case stochastic slib layer.

Load case	Unit	Mean	St. dev.	Lower bound	Distribution type
Phase5 and 6					
Left side	kN/m	7.2	0.72	–	Normal
Right side	kN/m	7.2	0.72	–	Normal

are only indirect load cases, such as the dead weight or the porosity of the groundwater (flow). The dead weight may be a stochastic, although the gravity value of 9.81 naturally is no stochastic variable. In this pilot version porosity is also not an option, so for the time being the only additional load case is the weight of the mud on the left and right side on the surface level of the soil in the water. This mud layer has a thickness of 0.6 meter, simulated by an edge load on the soil layer with a value of 7.2 kN/m.

Within this cofferdam structure design the edge load case differs from phase to phase, so it is preferable to keep the water and mud layers apart from each other. The water edge load case stays deterministic.

This results in the stochastic variables are shown in Table 5. Merging these stochastic variables with the basic configuration produces no significant additional results, which was to be expected. The reliability index β does not change drastically because of the low value of the standard deviation of both load cases. The ratio between the mud loadmaster of 7.2 kN/m and the water load case of 46 kN/m is too small. During the final stage the water load case increases to a value of 86 kN/m, so at this stage the mean value of the mud load case and the additional standard deviation is very low.

At the same time the influence factors for the additional stochastic variables do not change.

The final variation on the cofferdam structure, within this contribution, is the change of the deterministic parameters in the physical property field.

There are three relevant physical property parameters: the cross-section A_x of the anchor and the thickness of both sheet piles. The cross-section A_x is the most relevant of these three parameters, so this cross-section

Table 6. Cross-sectional area anchor.

Physical property	Unit	Mean	St. dev.	Lower bound	Distribution type
Anchor A_x	m^2	3.05e−3	0.152e−2	–	Lognormal

of the anchor A_x will be changed into a stochastic parameter. The value of this parameter was $A_x = 3.05 \; 10^{-3} \, m^2$ in the phased analysis. It is not our intention to change the reliability index dramatically, but it is only to emphasize the influence of using a stochastic physical property in the design of the cofferdam structure. The standard deviation will be small with a value of 5% of the mean value, namely $0.152 \; 10^{-2}$.

The additional stochastic parameters for the cross-section are shown in table 6.

The analysis results in a reliability index $\beta = 4.02$.

The influence factor of this new stochastic parameter is large, namely 98%. This means that special attention will have to be paid to the construction of the anchor in the overall cofferdam structure. The same holds true for the thickness of the sheet piles. These parameters are therefore very important, not only with a view to construction time, but also for the overall service life of the cofferdam structure. A monitoring program should be initiated to check this behaviour of parameters during the service life of the structure.

8 CONCLUSIONS

The following conclusions may be drawn:

1 The DARS method is fast enough to fulfill a full probabilistic analysis.
2 Linking the DARS method directly to an existing FE code allows the setting up of a new probabilistic calculation model without waste of time.
3 Structures without dedicated checking code rules may be checked in a modern way.
4 Influence factors provide designers with the opportunity to give more attention to the real issues.
5 Influence factors may become important to maintenance issues during the total service life of structures.
6 More research should be carried out in the field of soil and common material properties to arrive at a distribution type and standard deviation database.
7 In the future the human influence factor to design and construction may also be qualified.

REFERENCES

CUR, 1997, CUR publication 166, 3e edition, Retaining Walls (in Dutch), Gouda.

Rubinstein, "Simulation and the Monte Carlo method", John Wiley and Sons, New York, 1981.

Hasofer, A.M., N.C. Lind, "An exact and invariant first order reliability format", J. Eng. Mech. Div., ASCE, Vol. 100, 1974, p. 111–121.

Bjerager, P., "Probability integration by directional simulation", J. of Eng. Mech., Vol. 114, No. 8, 1988.

Waarts, P.H., "Structural reliability using Finite Element Methods", Delft University Press, Delft, The Netherlands, 2000.

Diana User's Manuals, Release 7, TNO Building and Construction Research, Rijswijk, The Netherlands, 1998.

Weersink, R.G.J. et al, 2002, Probabilistic design procedure cofferdam structures (in Dutch), Geotechniek, jaargang 6, number 3, july 2002, p. 32–41.

Bouwdienst RWS and DWW RWS, Retaining walls in water-retaining structures (in Dutch), WBA-N-900136, february 1991, Report 2e phase: Cofferdam and Retaining Walls, including appendices.

Bakker, H.L. et al, Cofferdams and retaining walls in water-retaining structures (in Dutch), DWW-1401, DWW, Delft, 6 december 1998.

Brinkgreve and Bakker, Non-linear finite element analysis of safety factor., Proc. 7th. Int. Conf. Computer Methods and Advances in Geomechanics., Vol 2, p. 1114–1122, 1991.

Safety and Reliability – Bedford & van Gelder (eds)
© *2003 Swets & Zeitlinger, Lisse, ISBN 90 5809 551 7*

Design based safety engineering applied to railway systems, part I

J. de Boer & B. van der Hoeven
CMG Public Sector B.V., Den Haag, The Netherlands

M. Uittenbogaard
ADSE B.V., Hoofddorp, The Netherlands

E.M. Dijkerman & W. Kruidhof
Holland Railconsult B.V., Utrecht, The Netherlands

ABSTRACT: The Dutch railinfra provider, ProRail, is developing a new system for railtraffic management and safety. This system, BB21 – Better Utilisation in the 21st century, implements the European ERTMS standard, takes into use a new communication system and improves the traction current supply.

For railway projects in Europe the Cenelec-standards for RAMS and safety are mandatory. Due to technical and organisational aspects the application of the Cenelec-standards is not straightforward.

In this paper we present a new method for hazard identification, based on the specification of the dynamic behaviour of this system. We also describe the organisation used for safety management and suggest some improvements on the organisation as described in the Cenelec standards.

In a second part of this paper the further analysis of these hazards is elaborated.

1 INTRODUCTION

The BB21 project aims to improve the utilisation of the Dutch rail track capacity by using the European railway standard ERTMS, improving the planning and the communication systems and introducing a higher traction voltage.

Furthermore, in the context of the European unification of transport, it was very desirable that the Dutch rail track systems would become highly interoperable with the Belgian, French and German systems, such that cross-border transportation will be much easier in the future. Another important objective was to improve the overall safety of the application of rail systems, for track-workers, railway personnel as well as passengers, the train and its environment.

The introduction of the European Railway Traffic Management System (ERTMS) on the Dutch rail is a major change in the way the safe train movement and improved utilisation of the capacity of rail transport is reached.

Therefore a thorough analysis is needed and prescribed by law to prove that the safety of the specific composition of systems is at least as good as the old systems. Besides this technical driver also the National

Governmental regulations prescribe that there is a Safety Approval of the Safety Authority required. In the Netherlands this is IVW Rail, on behalf of the Dutch Ministry of Transport.

We focus on the safety at integration level, where the different "safe" parts of the system are glued together. Because at this level the system consists of various different components like electronics, mechanicals, software, procedures, etcetera a multidisciplinary team reviewed all different kinds of hazards of the system.

2 APPROACH

2.1 *Organisation*

BB21 (Improved Utilisation 21st century) is a project of ProRail, an organisation that is responsible for:

– control and development of the Dutch railway network, and
– planning and control of the railway traffic.

The aim of BB21 is to improve the utilisation of the Dutch railway network in the 21st century and to create international interoperability. The organisation of BB21 consists of the following four projects (see figure 1), each responsible for the risk management and safety of its own product:

- The Bev21 project introduces the new railway safety system meeting the European Rail Traffic Management System requirements.
- The GSM-Rail project introduces a new voice and data radio communication system.
- The VPTI + project primarily integrates Bev21 in the existing traffic control system.
- The 25 kV project develops changes to increase the traction voltage from 1.5 kV DC to 25 kV AC.

Because BB21 has to supply the results of the four projects as one integrated product to railway operators, the Design and Integration project (D&I) was created. The D&I project is responsible for the structured system engineering, risk management, safety and validation of the interfaces and the integrated product. In order to analyse the integrated BB21 system the D&I project identifies the following hazards:

- Hazards contained in the functionality specified in the BB21 top-level specifications, the System/Subsystem Design Description (SSDD) and the System/Subsystem Specification (SSS) (Department of Defence 1989),
- Hazards identified in multidisciplinary hazard operability sessions,
- Hazards transferred by the four projects, Bev21, GSM-R, VPT+, 25 kV which are not their responsability, and hazards from parties outside BB21.

If a hazard is discovered in one of the four subsystems, or the interfaces, the problem will be analysed and the responsible project will be addressed by means of a Problem Report which could result in a Change Request.

The Design and Integration project presents the results of its analyses in the form of a Technical Safety Report, which is part of the generic application safety case for BB21.

2.2 Safety engineering activities and standards

The overall standard for our safety assessment process is the compulsory European standard Cenelec EN 50126 (CENELEC 1999). This standard describes the necessities for a new railway system before it will be accepted and taken into operation. At a high level these necessities are: proper quality management, change management, safety management and safety engineering. We will focus on the safety engineering activities.

The safety engineering activities basically consist of the following steps:

1. The set-up of a hazard log;
2. The identification of hazards;
3. Selection of the analysis method;
4. The actual analysis;
5. If necessary, initiation of corrections;
6. Reporting the results.

The results will be reported in a "safety case", for which the Cenelec EN 50129 (CENELEC 2000) standard describes the structure and mandatory elements.

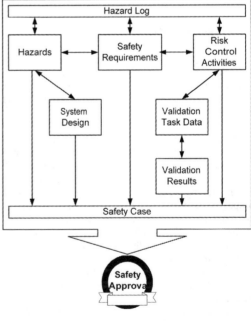

Figure 2. The relation and tracability between the system specification, the safety engineering activities and the safety approval.

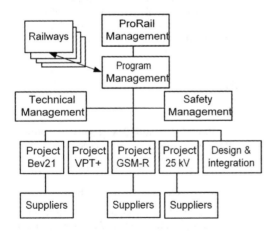

Figure 1. The BB21 project organisation.

In this paper we focus on part 4 of the safety case, the Technical Safety Report. In this report evidence is delivered of safe system behaviour for both nominal functioning and malfunctioning of the system.

The hazard log is a central entity in the safety engineering process. In this log all identified hazards are recorded and the analysis is traced. The log also provides tracebility between the system specification, the hazard identification and the hazard analyses. In figure 2 this is illustrated.

The gathering of the evidence has to comply to the Cenelec 50126 standard. However, this standard does not provide a detailed method for the risk/safety assessment. We adopted an advisory Railtrack document, the Yellow Book Iss. 3 (Railtrack 2000), containing a structured and detailed guideline for the risk/safety assessment.

3 HAZARD IDENTIFICATION

The Yellow Book describes a seven stage process for the identification, analysis and closeout of hazards. We adopted this guideline and added a sophisticated hazard identification method. In our hazard identification process, it proved very important to clearly define the scope of the system, and with that the scope of the analysis.

3.1 Identification methods

The difficulty with hazard identification at integration level was that hazards had to be identified on a system that did not yet exist, built together from systems that did not exist either, and for which were no stable specification available.

Another difficulty that arose during the analysis of the hazards was that there was little experience available on the use and reliability of the BB21 system and its subsystems. Only experience of the "oldfashioned" railway systems could be used. The analysis of the identified hazards is covered in part 2 of this paper.

The following hazard identification methods were used to investigate the safety of the proposed system:

- Design based structured hazard identification, a new design based hazard identification method. Faulty behaviour of the *specified*[1] system is analysed to reveal hazards;
- Traditional HazOp (Hazard-Operability) sessions, in which an expert panel is presented a topic on

---

[1] We only identify hazards on the already specified system. We don't try to analyse the completeness of the specification and whether the specification is internally sound. This is partially examined in the HazOp sessions and partially the responsibility of the designers of the system.

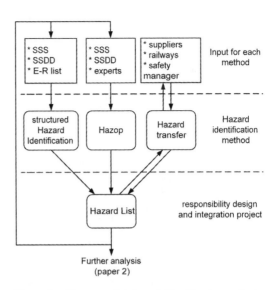

Figure 3. Three parallel hazard identification methods. The analysis of the identified hazards is covered in part II of this paper.

which a structured discussion is held. HazOps are used to identify hazards concerning human interaction with the system and to identify possible incompleteness the specification;

- Transfer of identified hazards from third parties. Typically hazards identified within the subprojects or by the Railways, but also the results of the creative minds of individuals can now contribute to the safety of the railway system.

These three methods are illustrated in figure 3. Also note that the process is iterative, after closing out a hazard by a change in the system design the analysis is reviewed again. This is to verify that the chosen solution reduces the risk to an acceptable level.

3.2 Scope

In order to be able to define the scope of the hazard identification, we have to agree on the definition of a hazard. In the literature several definitions can be found. The "official" Cenelec definition is:

DEFINITION 1: *A hazard is a physical situation with a potential for human injury (CENELEC 1999).*

As a result an almost unlimited number of hazards can be identified. From the safety engineering view of BB21, we only consider hazards caused by the BB21 system. For example, if we consider hazards caused by the train driver, we only consider those hazards when the train driver acted upon a false action of the BB21 system. For example, we do not analyse hazards that are the result from a train driver who intentionally breaks the rules. We do consider the hazards that result from a train driver who exceeds the speed limit when a wrong speed limit is displayed on the BB21 man-machine interface.

This limitation in the hazard identification scope-gives us the following definition for a hazard we use within the D&I project of BB21:

DEFINITION 2: *A hazard is a physical situation caused by an action of the BB21 system, which can potentially lead to a derailment, or a collision with another train, infra workers or road traffic.*

We now define a scope that limits the results of the hazard identification to a manageable data set, and we show that this scope is precisely right for confidently proving the safety of the *specified* system.

THEOREM: *In order to find all possible hazards on integration level, it suffices to enumerate the failures of every action on every subsystem interface for every system function.*

We "prove" this as follows. As pointed out before, the task of the Design & Integration project was to collect evidence that proves that integrating the four subsystems of BB21 does not cause intolerable risks on its environment. The safety of the subsystems in isolation is assumed proven, because safety cases are created for them.

Since we assume each subsystem "safe" enough when operating on its own, the only possibility for a hazard to occur is when this safe system receives an external stimulus. Stimuli are received on an interface, so we have to focus on these interfaces. Therefore an enumeration of failures on the interfaces will reveal all possible hazards.

3.3 *SSDD based structured hazard identification*

It is difficult to apply a structured hazard identification to a new system that is under development, definitely when considering this hazard identification process as a very creative process. Furthermore, such identification must have certain provable completeness. Another issue is the innovative aspect of the system under development; this implies that there is no experience based on reality.

Therefore the well-based deliverables as a result of the Systems Engineering process, applied conform the MIL-STD-498 (Department of Defence 1989), are used as starting point for the structures hazard identification. Two engineering areas are melting together.

The event-response list is used for the hazard identification. This list is a part of the SSDD and describes the expected system behaviour technically. By defining failure modes for the external events and responses within this system behaviour it is possible to define hazards at the boundary of the system. The failure modes are defined by using the Failure Modes Effects Analyses technique of guidewords.

Table 1. An example of the construction of BB21 Core Hazards from one or more "hazards" from the structured identification method. The BCH numbers indicate a BB21 Core Hazard, the E numbers are an internal identification number.

BCH001: A point switch is operated unjustified or unintentionally
• E211: When the "wrong" point switch is operated, and this point is part of a train route, the train could end up on the wrong track
• E367: When operating an infra element, BB21 sends the command to the wrong element

BCH002: A point switch is not blocked from operation without justification
• E205: BB21 should warn the signalman about unblocking a point switch, but doesn't do this
• E219: The signalman received a message of acceptance for a block command, but the command is not executed
• E220: A block command was sent to the wrong point switch
• E246: A block command is not executed at all
• E248: BB21 should unblock a point switch from operation, but unblocks the wrong switch

---

This approach makes it possible to define hazards systematically for the whole system within the defined scope. The considered hazards can occur by malfunctioning of the system. The reason for this is the identification of hazards based on the failure modes of the systems behaviour. Unfortunately this hazard identification results in an enormous amount of hazards. However, these hazards can be categorised in a couple of so-called "core-hazards".

Some of the hazards resulting from the structured identification method are identical, since some interface messages are used in more than one system function. Other hazards are very much alike, for example when they result in the same hazardous situation in the environment. This is illustrated in table 1, where two hazards from the structure failure mode identification are joined in one Core Hazard. As can be seen, the BB21 Core Hazard are at a fairly abstract level and describe actions of the system that lead to the hazard. In that way, the analysis of this hazard can focus on the system itself and does not have to include trains or track layout. This was very desirable, since those will not change, only the management and safety systems will.

3.4 *HazOp sessions*

In addition to the structured SSDD hazard identification the HazOp technique is used. This was needed to fill the gap round the nominal operation of the system and the areas where humans are highly interact with the system.

These kind of HazOp sessions are time-consuming and therefore it was needed that the list of subjects was as short as possible, without being incomplete. A list of criteria was set up and used to select subjects. Two main criteria were used, human interaction must be involved in the functionality and the functionality had to be different or additional to the current system. To apply these criteria a lot of interviews took place with technical experts of the current system.

This selection process ends up with about five subjects for HazOp-sessions with technical experts, users, designers and safety engineers. These sessions provided a list of hazards in the environment for which the base is the nominal operation of the system included human operational procedures. This together with the SSDD hazard identification, which namely provided hazards when the system is malfunctioning, gives a good insight in all kinds of hazards where the system is involved.

3.5 Third-party hazard identification

Of course, it is possible that one of the suppliers of subsystems identifies a hazard that cannot be resolved within the subsystem, but which has consequences to the other parts of the system.

The responsibility for analysis and closeout of these hazards was transferred to the Design & Integration project for further elaboration. This transfer of hazards, equal to transfer of responsibilities, is logged and signed in the Hazard Log. These transferred hazards are treated and analysed in the same manner as the other hazards.

Similarly, if the Design & Integration project identifies a hazard that can be resolved only by the suppliers of one of the systems or the Railways by for example operational procedures, then the responsibility for analysis and closeout is transferred to that party.

4 DIFFICULTIES ENCOUNTERED

4.1 Process

An initial difficulty with using a multi-disciplinary team were differences in vocabularies. This was soon solved by intensifying the communication.

Further difficulties were caused by the project structure prescribed by the Cenelec 50126 standard. The split between safety management as a staff function and safety engineering as a seperate project proved to make communication between the two more difficult.

For example, it was not always clear who was responsible for the safety analysis documents and who was responsible for the safety engineering method descriptions. Would the Safety Manager be responsible or the the project manager of the Design & Integration project? Formally, Safety Management should be responsible for method descriptions and safety engineering for the engineering activities. But in practice it proved necessary to evolve the methods during engineering activities.

On the other hand, the Safety Manager BB21 sometimes proposed changes in the safety engineering activities which would cause a significant delay to these activities. This is also true for changes to safety engineering methods described by safety management. In short, the goals of safety management sometimes conflicted with the goals of the Design & Integration project. Maybe the integration of the safety engineering and management teams would prevent this kind of problems.

As a result the BB21 Safety Plan was not always a guide to the safety engineering activities, but often rather described what activities had been carried out. The delays in communication between the two separate projects made it difficult to keep the safety plan updated. These communcation problems were solved by informal communication between the safety engineering and safety management teams and a more structural communication between safety manager and the project management of the Design & Integration project.

We recommend thinking about integrating the safety management and safety engineering teams, which may could prove useful in future projects. A more direct line of hierarchy between safety management (methodology) and safety engineering (activity) teams could be more productive and reduce miscommunication and delays.

4.2 Hazard identification

As pointed out before, safety engineering within BB21 has been approached as part of the Systems Engineering process. A difficulty with this approach was that the system design changed during or as a result of the safety engineering process. Some requirements of the system were loosened, while others were sharpened. There were changes in functionality, new functionality and some functions were removed.

Within the BB21 programma change management was used for changes on formal releases of documents. The period between two releases of the system/subsystem design descriptions (SSDD, (Department of Defence 1989)) proved too long for the safety engineering team to effectively identify hazards. We chose to identify hazards on the working versions of the system design description. A major problem with this proved to be the communication of changes, as changes on working versions are not recorded and the hazard identification could not always concentrate on the latest developments.

To overcome this, both the dynamic behaviour of the system and the results of the hazard identification based on this dynamic behaviour were integrated in one database. Some queries were designed to identify which identified hazards were no longer applicable and to identify new functionality. The problem remained to notice small changes in functionality after the hazard identification for that function was finished. This is solved by intensifying the communication between the safety engineering and system engineering teams.

However, even after finishing the hazard identification based on the final design description, change requests from other projects kept coming in. Although the change management procedures worked, it was hard to keep the identification and finished analyses up to date with all the changes. We strongly felt that following the V-model for systems engineering would simplify the work of the safety engineers, as that would give less changes and a more direct influence of the hazard analyses on the system design. Now every solution for an untolerable hazard has a great impact on the other projects.

Another challenge was to identify hazards on a system that was still under construction. No user experience was available with this new system, so we had to rely on a "desk-based" hazard identification. Applying the standard Yellow Book guidelines to all possible failure modes of the design, we would have ended up with hundreds of hazards for analysis. This was solved by narrowing the scope, divide-and-conquer techniques to spread the workload over several projects and by defining an abstraction layer (BB21 Core Hazards). This proved very useful, resulting in less than 40 Core Hazards for the complete design, consisting of about 1000 design requirements.

4.3 *Hazard analysis*

The hazard analysis is covered in part II of this paper.

5 CONCLUSIONS

The structured hazard identification method presented in this paper proved to be very successful. After solving problems with the enormous amount of identified hazards, this method proved to be an efficient method to check the complete dynamic design of a new railway safety and management system. Several problems showed up in an early stage of the engineering and could be solved even before the hazard analysis itself started.

The use of a structured hazard identification methods, based on the Cenelec and Yellow Book guidelines, failure modes and a structured iteration of the dynamic system behaviour proved very useful. This approach is probably useful in many more situation where safety risks are investigated. We think this method is more precise and complete than the traditional situation where only HazOp sessions are used.

The use of a multi-disciplinary team helped to inspect the system design from different points of view. A complex system like BB21 does not only contain traditional railway components, but computer hardware and software as well. Direct communication between the systems engineering and safety engineering teams (located in the same room) was very useful to improve the system design and to improve the understanding of the system with the safety engineers.

Between the systems engineering and safety engineering teams, the failure modes on the dynamic system description became a familiar means of communication and helped to improve the design as well as the understanding of the system in both teams.

The split between safety management and safety engineering as described in the Cenelec standards may be subject to improvement. The problem is that the communication is not straight forward as the teams are managed with different objectives, and work in a different physical location. It may be helpful to place the safety engineering team directly under safety management in hierarchy and then place one of two members of the safety engineering team in the same room as the systems engineers.

REFERENCES

CENELEC (1999). *NEN-EN 50126: The specification and demonstration of RAMS*. Delft: Nederlands Normalisatie instituut.
CENELEC (2000). *NEN-EN 50129: Safety related electronic systems for signalling*. Delft: Nederlands Normalisatie instituut.
Department of Defence, U. S. A. (1989). *MILSTD-498, Software development and documentation*. Arlington, VA: Department of Defence, United States of America.
Railtrack 2000. *Engineering Safety Management – Yellow Book 3*, Volume 1 and 2: Fundamentals and Guidance. London: Railtrack PLC.

Safety and Reliability – Bedford & van Gelder (eds)
© 2003 Swets & Zeitlinger, Lisse, ISBN 90 5809 551 7

Reducing product rejection via a high contrast consumer test

J. Boersma, G. Loke & H.T. Loh
Design Technology Institute, National University of Singapore, Singapore

Y. Lu & A.C. Brombacher
Eindhoven University of Technology, Eindhoven, The Netherlands

ABSTRACT: Product rejection of a new innovative product is one of the major uncertainties for high volume consumer companies. It is well known that some of the field failures are consequences of unexpected consumer behavior [Brombacher (1999)]. Conventional in house tests, during product development have not been able to fully prevent product rejections. This paper describes a new consumer testing concept called the high contrast consumer test, which tries to deal with unexpected consumer behavior. This new consumer testing concept was attempted in a company that developed an innovative domestic iron. The results showed that this concept for this particular company was able to highlight consumer issues and therefore design issues that the company's conventional tests could not show.

1 INTRODUCTION

The increasing need for products that are able to deliver reliable, complex functionality with a high degree of innovation presents a major challenge to modern day product development processes. At the same time, product quality & reliability needs to be maintained at a high standard because high reject rates ruins a company's image and profit. This is further compounded by short product life cycles.

Presently, product rejection by customers within the warranty period is a major problem for many companies. This is especially the case for innovative products. Therefore a lot of research is done to find the root causes of these product rejections. Service centres are one of the instruments companies use to analyze customer rejection. However, feedback from service centres is usually too slow to improve the product [Petkova (1999) & Petkova (2000)].

To find possible product rejections, another method is to make use of consumer tests. The advantage of consumer tests is that possible issues can be detected already during the product development process and subsequently changes can still be made to the product. This makes consumer testing a more pro-active consumer feedback method in comparison with a service centre feedback approach.

This paper introduces a new concept of consumer testing called a High Contrast Consumer Test (HCCT).

It tries to solve some of the current problems related to consumer testing. In addition, it describes a successful implementation of such a concept in a Multi-National Company (MNC) which manufactures domestic irons. The structure of the paper is as follows. Chapter 2 discusses consumer test methods. Chapter 3 introduces a new concept. Chapter 4 will describe the use of such a high contrast consumer test. A conclusion will be presented in Chapter 5.

2 PRODUCT REJECTION AND CONSUMER TESTS

Consumer reliability tests are commonly used by companies to observe how customers react to new innovative products and how the product will "react" when subjected to real customer stresses. However, such tests are often being conducted with ideal products in a controlled environment with well defined "target customers" and pre-determined test procedures. Therefore it is questionable how well these tests results reflect real customer use of the product.

Research by one of the authors [Boersma (2001)] showed that in the first weeks after purchase, the product rejection rate by customers is high. Figure 1 shows very clearly that in the first weeks after consumers purchased the product the rejection rate is high, while after those initial weeks the rejection rate becomes

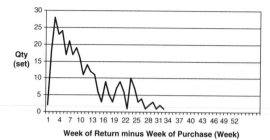

Figure 1. Amount of returns versus the week of purchase in a high volume consumer electronics industry [Internal company document].

lower. This could be caused by two reasons [Boersma (2001)]: first, there could be a high rate of so-called "hidden zero-hour" failures. These faults however would be found in the company's service centre [Petkova (1999)]. The second, and more likely reason, is that the customer didnot know how to operate the product or that the product didnot meet his/her expectations [Brombacher (1996)]. It makes sense that these kinds of misfit would occur in the first few weeks after purchase instead of after many months.

The findings from that research indicate that the initial use, right after the product is purchased, is very important. More specifically, this means that the "out of box experience", the product installation process and first usage are important experiences that will determine whether or not a consumer rejects the product.

A new concept that tries to focus more on the first phases after product purchase is the High Contrast Consumer Test. In the next section this concept will be described more in detail.

3 NEW CONSUMER TEST CONCEPT: THE HIGH CONTRAST CONSUMER TEST

In the attempt to make consumer testing more effective, an idea was initially developed by the authors to observe critical and extreme customers using a product in realistic operating conditions. The purpose is to accelerate failures and expose product-usage issues as soon as possible. The setup of a HCCT involves the following steps:

1. Identifying all new innovative product features to be tested.
2. Identifying extreme customers of the product and of these new features via brainstorming.
3. Initiating a test session which allows observation of these extreme customers utilizing the test product under near-realistic operating conditions. The focus is on the "out of the box experience", the installation process and the first use.

4. Prompting for the customers' thought-process via a unique "Think-aloud-Protocol".
5. Feeding-back information to the product development team; a consolidation of observations, thought-processes and customer interviews for follow-up actions.

Some specific benefits that are expected by doing a consumer test in this fashion are:

- Improved understanding of how the product would be used by customers.
- Identifying a relatively wider range of potential design issues about the product.
- Highlighting problem areas that should be addressed as early as possible in the product development process.

4 A HIGH CONTRAST CONSUMER TEST FOR AN INNOVATIVE IRON

The company in which the concept was used was in the business of developing a new iron with a new functionality. This functionality was not seen before in the ironing industry. During the development process, questions were raised about the amount of product rejection within the warranty period. Normally, this was predicted using reliability test data and historical data from previous products. However for this iron prediction was difficult, due to its new functionality.

In addition, the company faced difficulties obtaining enough information regarding consumer and design issues simply using their own tests. In previous innovative products, this resulted in a high amount of so-called "fault not found" product rejections. These are products which were rejected by consumers, yet were functioning well within its specifications.

In order to overcome these problems, the company was keen to try out the new HCCT concept. Below is a description of the steps that were taken to accomplish this test.

1. Identifying all new innovative product features to be tested.
   a. The new functions of the product needed to be defined. For the iron in question, there was basically one new function that differentiated it from other irons. However, this iron also functioned as a typical steam iron.
2. Identifying extreme customers of the product and its features via brainstorming. (Because of confidentiality reasons, full details were not disclosed in this paper, but are known to the authors)
   a. After brainstorming a number of characteristics were identified for a normal steam iron. Some of the characteristics were: frequency of use, type of iron (dry or steam), age and gender.

b. Another brainstorm session was conducted to determine what kind of users were considered "extreme" for such an iron.
3. Initiating a test session to observe customers utilizing the product under realistic operating conditions.
   a. In order to test the product under realistic operation condition it was necessary to test the new iron at the customers' homes. The advantage of this was that the customers can iron their own laundry. In addition, the customer was located in a familiar environment and would most likely show his/her own ironing habits.
   b. During the test, the customer did not receive any help. Similar to real product use, the only help available was the manual or a telephone helpline. The test was planned deliberately in this manner to simulate the stresses and problems that the customer and iron will face.
4. Prompting for the customers' thought-process via a unique "Think-aloud-Protocol".
   a. During the test, the customer was prompted to think aloud. This constituted the "Think-aloud" protocol. It provided insight into what the customer thinks. In addition, the customer can name all his/her likes, dislikes and other problems with the iron. All these were recorded to capture as much useful information as possible.
   b. Simultaneously, some motion pictures were taken of the iron installation process and of other problems faced. This information would be useful for subsequent analysis and presentation.
5. Feeding-back information to the product development team.
   a. The last phase of the test was to consolidate all the available information. Matrixes were developed to analyze the relations between the customers' background and specific problems.
   b. Subsequently, the results were presented to the product development team. The movies assist in showing what kind of problems customers were facing in the field.
   c. Recommendations were made for the problems observed during the test. These recommendations range from small changes to the manual to proposed product design changes.

5 CONCLUSION

While work is still on-going, it can be seen that the HCCT concept showed excellent potential for product development processes with innovative products. The HCCT concept was very useful for capturing initial consumer installation & usage problems, because of its realistic test conditions. In addition, the "think-aloud protocol" was very effective in capturing the expectations, opinions and line of thoughts of the test persons. HCCT helped to identify possible product rejection issues that the company was not able to find in their conventional consumer testing methods.

REFERENCES

Boersma, J. 2001. How to improve customer feedback in a product development process. *MSc. Thesis, Eindhoven University of Technology.*

Brombacher, A.C. 1996. Predicting reliability of high volume consumer products; some experiences 1986–1996. *Symposium "The Reliability Challenge".* London: Finn Jensen Consultancy.

Brombacher, A.C. 1999. The use of Reliability Information Flows as a maturity index for quality management. *Quality and Reliability Engineering International* 15.

Petkova, V.T. et al. 1999. The role of the Service Centre in Improvement Processes. *Quality and Reliability Engineering International* 15.

Petkova, V.T. et al. 2000. The Use of Quality Metrics in Service Centres. *International Journal for Production Economics* 67.

Safety and Reliability – Bedford & van Gelder (eds)
© 2003 Swets & Zeitlinger, Lisse, ISBN 90 5809 551 7

Expert judgement methodology for failure anticipation in nuclear power plants

L. Bouzaïène, J.C. Bocquet & F. Pérès
Laboratoire LGI, Ecole Centrale Paris, Chatenay Malabry, France

F. Billy, Ph. Haïk & A. Lannoy
EDF-R&D, Chatou Cedex, France

ABSTRACT: Risk analysis is a tool for investigating and reducing uncertainty related to outcomes of future activities. We are interested here in failure anticipation in nuclear power plants. This involves very specific systems with little or no existing historical failures. In such cases, both engineering judgement and historical data are used to quantify uncertainty related to the predictions, like probabilities and failure rates. This paper is focussed on this aspect. The purpose is to provide an expert judgement elicitation methodology for anticipating the failures of a component, up to the end of its design life cycle period, including eventually an extension period.

1 INTRODUCTION

Failures of a component are generally well known during the design process. However if some failures are effectively observed, others are never observed, the degradation speeds being very low, often lower than the previously expected ones.

Moreover, generally for economic reasons, when degradation mechanisms are considered well controlled, the question is to extend the lifetime of the component beyond its design lifetime. New problems not considered at the design stage by functional analysis or FMEA, can occur. These problems can occur when modifying for instance the operation procedures or when improving the performance of installations or when ageing has not been detected or is not correctly managed. Consequently failures not predicted can occur, maintenance programmes can be inadequate and it is indispensable to anticipate these potential failures which can occur during the end of life phase, or during the extension life phase, of the component. This anticipation problem and its consequences in terms of decreasing the performance of a component (availability, safety, costs) have to be determined.

This paper is focussed on this aspect. The purpose is to provide an expert judgement elicitation methodology for anticipating the failures of a component, up to the end of its design life cycle period, including eventually an extension period.

We define failure anticipation as "the identification of events which are potentially objectionable as concerns cost, safety or availability, before they occur to evaluate the risks which they represent and to prepare and implement the appropriate preventive or exceptional measures which may be required."

This paper is divided into three parts.

The first part deals with the use of expert judgement as an essential source of information in a decision-making context. As risk analysis typically deals with rare events, this makes relevant data scarce. For this reason, the use of expert judgement is strengthened. This is even truer when dealing with nuclear systems with a high quality design and a very demanding maintenance programme, where failures are very rare.

The second part, is a state of art on expert judgement methodologies. Some expert judgement methodologies already used in nuclear studies are presented, analysed and compared. Each methodology is described in a sheet including the characteristics, the phases, the strong and weak points and the references. Theses methodologies are then compared to our case study and classified according to the effort required for implementation and their appropriateness to anticipation. The objectives of the methodology, the creativity aspects, the expert team (multidisciplinary or not), and the existing applications are the main criteria to evaluate this appropriateness.

This leads us to identify recommendations aimed at building an expert judgement methodology well suited to failure anticipation.

2 FAILURE ANTICIPATION AND EXPERT JUDGEMENT

This study has been carried out within the framework of equipment life cycle considerations (Life Cycle Management). Replacements of certain equipment represent major investments for the company. In addition to the cost of design, manufacturing and installation, such equipment can often require significant maintenance. However, if an equipment has been designed with a high level of quality and is properly maintained, it is possible to envisage extension of its service life beyond the service life defined during the design process. This life cycle extension would make it possible to further amortise the initial investment.

Problems other than those identified during the defined process can appear. For this reason, it is useful to anticipate these potential failures which can occur during the end of life cycle period. To anticipate, it is necessary to take account of past feedback concerning the equipment and also of feedback relative to similar equipment installed in other units under the same environment, operating and maintenance conditions. It is also necessary to take account of modifications with respect to the design, current and forecast operating and maintenance conditions.

Two important aspects must be considered:

– management of physical ageing of component,
– cost management.

Due to the very special framework which the nuclear context represents, equipment used in this context presents several special characteristics:

– specific equipment,
– importance of safety,
– high quality design,
– stringent maintenance.

These characteristics result in limited feedback (low number of failures) which can make a statistical study difficult.

To compensate for this limited information, the classic solution consists in gathering expert survey information. The expert survey contributes to filling in the gaps of the feedback data. The expert is considered as a relevant source of information.

3 STATE OF THE ART ON EXPERT JUDGEMENT METHODOLOGIES

For this study, we have considered 10 expert judgement methodologies already used in nuclear studies. Six of

them were considered on a benchmark exercise for a PSA Study, experiment L-24 of the JRC-ISIS, FARO facility for fuel coolant interaction studies in a nuclear reactor accident [1]:

– NNC methodology,
– FEJ-GRS methodology,
– STUK-VTT methodology,
– NUREG-1150 methodology,
– KEEJAM methodology,
– CTN-UPM methodology.

Other methods covering different safety applications have also been studied:

– Procedure guide for structured expert judgement,
– LCM methodology developed by EPRI (Life Cycle Management),
– TRIZ-AFD methodology (failure anticipation),
– RIPBR, Risk-Informed, Performance-Based Regulation, developed by the Department of Nuclear Engineering, MIT (risk management and maintenance optimisation).

3.1 Presentation of the methodologies

3.1.1 NNC methodology[1]

This methodology was developed in 1996. It is based on the quality principles and procedures in the NNC Quality Procedures and Engineering Manual, U.K. NNC is a Quality based methodology: based on quality assurance methods of the sources of information and of the problem solving processes, this approach is based on individual estimates. It involves a multi-disciplinary team, defined as a set of individuals with different but complementary skills.

As there is no rigorous formal elicitation process, the NNC approach may be called informal expert judgement.

3.1.2 FEJ-GRS methodology[1]

This methodology was developed in 1985 by GRS, Germany. The methodology has been developed to quantify the state of knowledge in elements of a breakdown of the question and to propagate it through this breakdown to arrive at a quantitative uncertainty statement for the answer.

The methodology aggregates the judgements at lower levels and propagates them through the breakdown to arrive at a quantitative expression of the resulting state of knowledge at the model output level.

3.1.3 VTT-STUK methodology [1]

This methodology was developed in 1997 by VTT Automation, STUK, Finland. It is based on the NUREG-1150 method (next paragraph). The use of belief networks allows an adaptation of the elicitation efforts according to the available resources. This is a simplification of NUREG-1150. The methodology was originally intended for use in various kinds of quantitative risk and reliability assessments, and in engineering and economical analyses, where remarkable uncertainties are present.

The methodology is based on probabilistic representation of uncertainties. The predictions obtained from experts are expressed as probability distributions. The combination of these assessments is based on hierarchical Bayes models (belief networks). Due to this property, it is also possible to deal with experts who are not familiar with the concepts of probability. Although, there are no restrictions as to the applicability of the method, it is at its best when applied to generate predictions to physical parameters.

3.1.4 NUREG-1150 methodology[1, 2, 3]

This methodology was developed in 1987–1990 by US-NRC, USA.

Highly structured, this approach includes training of the experts, review of discussions, individual elicitations, composition and aggregation of the opinions and review by experts.

In the NUREG-1150 approach, the domain experts write reports on the issue and their final estimates are elicited individually after expert's discussions, then averaged on an equal weight basis.

3.1.5 KEEJAM methodology[1, 3]

This methodology was developed in 1997 at JRC-ISIS in collaboration with the University of Brescia and the University of Bologna, Italy. Knowledge based methodology: the method employs Knowledge Engineering techniques, and includes explicit modelling of the knowledge and problem solving procedure of the domain expert.

The approach provides structured and disciplined support to the knowledge engineer in eliciting the knowledge and reasoning strategies of the experts, building consistent knowledge models, and applying these models to the solution of the expert judgement task.

3.1.6 CTN-UPM methodology[1]

This methodology was developed in 1997 by the Department of nuclear engineering, University of Polytechnics of Madrid, Spain.

It was developed and adapted on the basis of the NUREG-1150 methodology, although there exists a very important difference between them regarding the way to aggregate experts evaluations. The CTN protocol has been developed to get estimates of subjective probabilities for unknown parameters and uncertain events. It consists of nine steps executed sequentially.

3.1.7 Procedure guide for structured expert judgement SEJ [4]

This methodology was developed in 2000 by Delft University of Technology, The Netherlands.

This is a European Guide for Expert Judgement in Uncertainty Analysis. It deals with procedures to perform an expert judgement study with the aim of achieving uncertainty distributions for an uncertainty analysis. In that field of application, the methods developed at the Delft University of Technology have benefited from experience gained with expert judgement in the US with the NUREG-1150 methodology. The procedure guide represents a mix of these developments.

3.1.8 LCM methodology[5]

This methodology was developed by EPRI, USA as part of the Life Cycle Management/Nuclear Asset Management studies.

In order to guarantee long-term equipment reliability risk in nuclear power plants, LCM helps managing ageing degradation and obsolescence of important systems, structures and components. It gives an optimal solution for life cycle management based on an economical comparison between the different possible solutions.

3.1.9 TRIZ-AFD methodology[6]

This methodology was developed in 1997 by KAPLAN, USA. Il allows identification and analyses of failures based on the TRIZ methodology. AFD (Anticipatory Failure Determination) was recently developed in the United States.

AFD consists of two tools: AFD 1 and AFD 2. AFD 1 is used to analyse failure causes. AFD 2 completes AFD1 with a number of steps for failure anticipation.

3.1.10 RIPBR, Risk-Informed, Performance-Based Regulation developed at the Department of Nuclear Engineering, MIT [7]

RIPBR is an evolving alternative to the current prescriptive method of nuclear safety regulation. RIPBR is goals oriented while the prescriptive method is means oriented.

RIPBR is capable of justifying simultaneous safety and economic nuclear power improvements. It includes the formulation of probabilities through expert elicitation and the review of risk-informed, performance-based engineering analyses used to evaluate proposed changes to existing technical specifications.

3.2 Description of the methodologies

For each of these methodologies, a method sheet has been prepared to provide a summarised description of each.

Each sheet contains:

– the date and country of development,
– the organisation which developed the methodology,
– the characteristics of the method (presented to underscore its originality),
– the input data available to the expert, and the output data are both described.

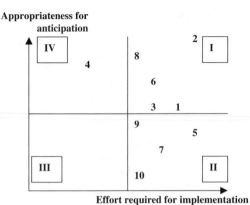

Appropriateness for anticipation

IV

2

I

8

4

6

3   1

9

5

7

III

II

10

Effort required for implementation

1 NNC	6 CTN-UPM
2 FEJ-GRC	7 Procedure Guide for SEJ
3 STUK-VTT	8 LCM
4 NUREG-1150	9 TRIZ-AFD
5 KEEJAM	10 RIPBR

Figure 1.   Classification of the expert judgement methodology. Note: The effort/anticipation diagram represents an initial look at the various methodologies. A more precise classification, based on the expertise of a few major experts will be issued in the near future.

The sheet then presents the various phases involved in the method and the existing tools. The main applications of the methodology are given, as well as the methodology's weak and strong points. Finally, the background references are given.

3.3   *Classification of the methodologies*

To compare the methods studied, we have classed them with respect to their appropriateness for anticipation and to the effort which they require, in an anticipation/effort diagram. To evaluate this appropriateness we considered for each methodology the objectives, the creativity aspects, the expert team (multidisciplinary or not), and the existing applications.

Part I at the top right shows those methods which are more appropriate to anticipation but which require high elicitation efforts.

Parts II and III at the bottom bring together the methods which are only moderately appropriate for anticipation purposes.

Part IV at the top left corresponds to those methods which are appropriate for anticipation and which do not require major efforts for implementation.

Method NUREG-1150 is located by expertise in this frame. In our context, methodology NUREG-1150 appears to provide the best basis and it would be useful to adapt it to our failure anticipation context by further developing the aspects specific to anticipation and reducing the elicitation efforts. In this respect, the

experts are not very available and the expertise time must therefore be reduced.

3.4   *Analysis of the methodologies*

The comparison of the various methodologies reveal a set of generic phases which have been developed to a greater or lesser extent in each depending on its objectives.

These generic phases are:

1. Definition of elicitation objectives.
2. Choice of experts to be elicited.
3. Training session in probabilities for experts.
4. Preparation of a questionnaire.
5. Elicitation.
6. Aggregation of expert replies.
7. Synthesis.

With respect to the generic phases described above, the phase concerning training of experts in probabilities has not been opted for at this time. Furthermore, this phase can be replaced by questions adapted to the experts interviewed and by work involving translation of the qualitative replies into probabilities. This would lighten the load of the expert and best responds to the expert's availability constraints.

4   SPECIFICATIONS OF AN EXPERT JUDGEMENT METHODOLOGY WELL SUITED TO FAILURE ANTICIPATION

The objective of the methodology is to allow the analyst to call on the expert to anticipate potential failures of a given equipment based on his own knowledge and on the data gathered by the analyst. The expert, here, is not only required to apply the knowledge which he has in tacit form, but also to provide imagination and creativity in anticipating an event.

4.1   *Constraints*

4.1.1   *Limited study time and experts which have only limited availability*
This constraint will limit the choice with respect to the type of elicitation to be chosen. The accent is placed on individual interviews. However, a return to the experts, as used in the Delphi method, should not be excluded.

4.1.2   *Reticence of experts with respect to elicitation*
The objective of the study is to stimulate the expert's creativity to anticipate failures which may never have yet occurred. It is important for the expert to be able to express himself free of any constraints or pressure which can be created by interactive groups. The Delphi approach therefore does not seem very well suited to

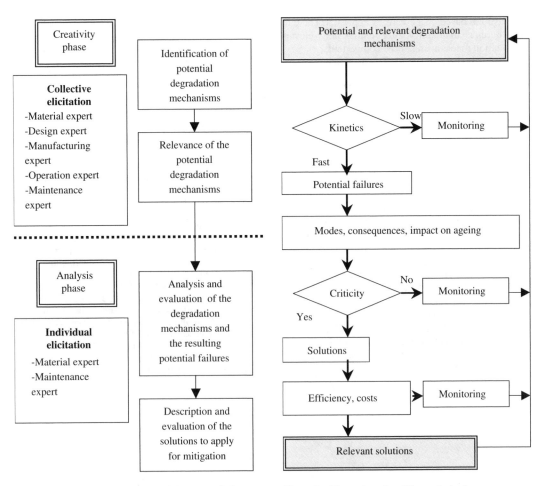

Figure 2. The two main phases of the expert judgement methodology for failure anticipation.

Figure 3. The main tasks of the analysis phase.

our study context as it results in systematically elimi-nating the most original replies [3]. This could be counter-productive in the anticipation context.

4.2 *Preliminary inputs for elicitation input data generally available before expertise*

Preliminary inputs for elicitation input data generally available before expertise:

- Objectives and context of the elicitation
- Data concerning the studied component: bound-aries, design, functions, materials, operating conditions, environment, procedures (safety, maintenance,...) ,...

These data are generally very heterogeneous. Oper-ating feedback, procedures (like maintenance proce-dures), knowledge reports (rules, reliability reports,...), physical data,... can be found.

4.3 *Outputs*

1. Identification of potential and relevant degradation mechanisms and failures of the component.
2. Assessment of degradation and failure evolution.
3. Evaluation of potential failure effects: safety, unavailability and maintenance costs, dosimetry.
4. Solutions to apply to avoid, postpone or mitigate failures (and their efficiency and costs).

4.4 *Expert judgement methodology for failure anticipation*

In our present state of advance, the two main elicita-tion phases of the NUREG-1150 approach have been retained: the first one is a collective elicitation phase and the other one is an individual elicitation phase. The contents of these two phases are defined accord-ing to failure anticipation requirements.

Phase 1 is based on a creativity approach. The purpose here, is to identify all potential and relevant degradation mechanisms of the component.

Phase 2 is the analysis phase. Identified degradation mechanisms and kinetics and the solutions to be applied are described and evaluated. The main tasks of the phase 2 are described in figure 3 above.

5 CONCLUSION

Through this state of the art on expert judgement methodologies used in nuclear studies, we have been able to compare the existing approaches. They have been classified according to their appropriateness to failure anticipation and to the effort required for their implementation. This has allowed us to identify the methodology that seems the most useful. This identified methodology, NUREG-1150, must, nevertheless, be better adapted to anticipation problems.

In our present state of advance, two main elicitation phases have been retained: the first one, the creativity phase based on collective elicitation and the second one, the analysis phase based on individual elicitation.

The question we have to answer now is "how to formulate the questions to be easily understood by experts according to their skills?"

In order to carry on and validate these results, this failure anticipation methodology, here presented, will be applied to a nuclear power plant component.

REFERENCES

[1] G. COJAZZI, D. FOGLI, Benchmark exercise on expert judgement techniques in PSA level 2, Extended final report, 2000.

[2] R.L. KEENEY et al., Use of expert judgement in NUREG-1150, Nuclear Engineering and Design, Vol. 126, pp. 313–331, (1991), North-Holland.

[3] A. LANNOY, H. PROCACCIA, L'utilisation du jugement d'experts en sûreté de fonctionnement, Editions TEC & DOC, Décembre 2001.

[4] R.M. COOKE, L.J.H. GOOSSENS, Procedure guide for structured expert judgement, Project report, Nuclear Science and Technology, European Commission, 2000.

[5] G. SLITER, Life cycle management planning source-books, EPRI, December 2001.

[6] F. GUARNIERI, P. HAIK, AFD, une nouvelle méthode pour l'identification et la maîtrise des défaillances: présentation, illustration et perspective. Proceedings of the conferences of ESREL 2002, Lambda Mu 13, 2002.

[7] M.W. GOLAY, Improved nuclear power plant operations and safety through performance-based safety regulation, Journal of hazardous materials, Vol.71, pp. 219–237, 2000.

Safety and Reliability – Bedford & van Gelder (eds)
© 2003 Swets & Zeitlinger, Lisse, ISBN 90 5809 551 7

CARE.CAME.CAFDE (C³) A new technology for designing high reliable systems at minimum life-cycle cost

Yizhak Bot

BQR Reliability Engineering Ltd

ABSTRACT: In today's insecure and uncertain climate in the high-tech industry, where money is less accessible, companies are constantly searching for ways to cut expenses. Two important financial factors to consider are maintenance and down time when the equipment/system/processes are not functioning/operating.

1   GENERAL

1.1   *The technology*

In the last decade, BQR has developed a new technology that combines traditional methods as well as innovative ideas and algorithms, to help increase the availability of high-tech systems, while reducing their maintenance cost. Many companies that attempt to reduce maintenance operating cost fail to address all aspects of the issue. For example, companies may optimize the required number of spares needed for a system, without optimizing repair, repair level or disposal of the part. Other companies try to perform conditioning based maintenance, without combining tasks despite the risk of delaying certain tasks or limiting other tasks which could lead to a money loss.

BQR's integrated tools provide solutions for reliability and maintenance issues and enable users to make decisions regarding the system as early as the design phase. BQR technology is also suitable once the system is produced and is operating out in the field. BQR's unique technology uses standard GUI (Graphic User Interface) and databases that can be interfaced and integrated with other IT systems.

BQR technology includes: CARE – Computer Aided Reliability Engineering, CAME – Computer Aided Maintenance Engineering and CAFDE – Computer Aided Field Data Engineering (C³).

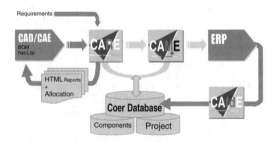

1.2   *CARE*

CARE is used for analyzing possible failures in electronic and mechanical parts and calculates all failures, causes and their effect on the system's operating ability. The database includes stress and redundancy, failure time distributions, failure detection and isolation degree with Built-In-Tests, effects of inspections and any critical parts that could affect the safety and critical operation of the system. CARE is used under the designer CAD/CAE and offers solutions in real time and enables any necessary modifications in the system more efficiently before manufacturing.

1.3 CAME

CAME is used for optimizing the maintenance concept to help reduce the total life cycle cost of the product. This is achieved by applying several optimization steps such as: Optimal Repair level Analysis, Preventive Maintenance Optimization (Conditioned Monitoring), Sparing to Availability, Reliability Centered Maintenance and MSG-3. Calculating LCC following each optimization step, demonstrates how the maintenance cost is reduced while availability is increased. CAME presents a full maintenance plan including applied tasks, required personnel and skills, support equipment, materials and facilities, types of resources used for each maintenance site and a field failure diagnostics database. In addition, CAME features the IETM (Interactive Electronic Technical Manual), with drawings, schemes, instructions, user written texts, as well as database. CAME is applicable under any ERP/CRM system and offers solutions in real time to enable any necessary modifications in the system more efficiently.

1.4 CAFDE

CAFDE is used for analyzing field data and for providing reliability and maintainability measures for the system, its assemblies and parts. The results are then applied to CARE and CAME and the data is used on operating systems to predict subsequent failure occurrences, to correct PM schedule times and provide health monitoring for each system. In addition, CAFDE provides recommendations for any necessary modifications in the design concept. For example: Is a total overhaul required for the system, or should it be replaced with new produced or developed parts?

1.5 Core Data Base

CARE, CAME and CAFDE projects may be linked to each other through one Core Data Base (CDB). The inputted data and calculation results of other projects from other program modules (or C^3) can be used by C^3. Any inputted data is automatically updated in all the linked projects and is included in all the calculations. Different parts of parameters from various projects taken from several modules can be gathered and used for any calculation purposes. A project in one module can be used for creating a new project in another module. All the links and parameters are automatically updated through the core database.

2 APPLICABILITY

2.1 For electronic systems

The MTBF module imports a net list from a CAD and CARE Spice module and calculates the maximal electrical stresses for each element (Power, Current, Voltage). The Stress De-rating and Thermal Analysis (SDTA) is then performed, identifying overstressed and under stressed components with recommendations for enhancing the scheme.

2.2 For mechanical systems

The MRS module predicts material degradation processes and geometric size changes, simulates failure occurrences and forms predicted distributions of the failure time. These results are imported into CAME and used for Preventive (scheduled) Maintenance Optimization using PMO (Preventive Maintenance Optimizer), in order to reduce the total maintenance cost by providing the required mission reliability and system availability.

2.3 Applicable organizations

Primes and integrators – The prime contractor is responsible for the system. Different subcontractors provide the assemblies for the system (allocation techniques of requirements are needed from the system level to the assemblies level).

Designers – The designers are responsible for designing the equipment (estimation techniques of reliability and maintenance figures are required).

Maintenance service provider – must be cost effective and provide low maintenance cost.

Governments and large corporations (banks) – that are looking to purchase effective and efficient systems.

2.4 Applicable industries

Chemical plants, train companies, aviation, trucks, automotive, telecom and defense.

The C^3 have already been used successfully by many organizations which reported a reduction of at least 35% in maintenance cost, while increasing the availability of their products.

3 CARE – IN DETAILS

3.1 Allocation

At the first stages of design the configuration of many blocks, their reliability models and parameters are unknown. There are reliability requirements for the system (or sub system) level. In order to satisfy these requirements during design, it is necessary to allocate them sequentially to sub systems, assemblies, blocks and components as the project evolves and develops in more detail. The allocation should be as realistic as possible for each stage, i.e. it should take into account the relative complexity of each block, as well as other factors that might affect its failure rates.

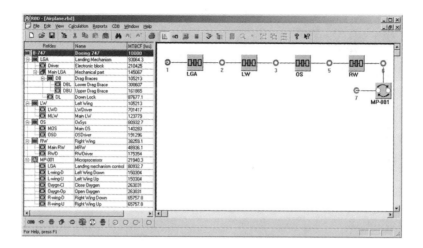

The Reliability Block Diagram (RBD) module is used for allocating reliability or availability requirements for the most current configurations available to the designer. The user selects one of 4 reliability requirements for the system from which he wants to allocate: Reliability, Availability, MTBF or Down time, and defines the required value for each one. The user assigns up to 5 failure rate factors and their relative weights to the failure rate of any block. The user then inputs the relative value of each factor reflecting the block's complexity, operation conditions, quality, etc.

RBD calculates the reliability parameters for all the blocks so that: a) the selected reliability requirement of the system is satisfied, b) all known reliability models of all the designed assemblies that are included in the allocation RBD are satisfied, c) the failure rates of all the lowest blocks are related to each other as the total sum of weights of their factor values.

3.2 Failure rates prediction for electronic parts

If the system design is detailed enough, the MTBF module can calculate the failure rates for all the components more accurately, as well as, the total number of all failures (critical and non critical) that occur in each assembly per time unit. The user can add a component manually, obtain it from one of the 3 existing CARE libraries (Master, Personnel or Design), import components from a CAD file (View Logic, Integraph, Mentor, Or-CAD, Pacific-Numeric, View Logic Plus, Zuken-Redac, Pspice) or import components from a net list with the nominal data of each part.

The operational conditions (environment type, temperature, cooling) of the system, as well as, the components' stress data (Voltage, Current, Power) are used for the MTBF prediction calculation. If the components are included in the net list, the Spice module can be used for calculating the stresses for each component.

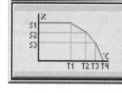

The following are the various types of prediction methods used for calculating failure rates:

CARE is also used for Stress De-rating and Thermal Analysis (SDTA). SDTA calculates the values of stress parameters for each component and each temperature. SDTA compares the applied stressed values and detects overstressed and under stressed components and recommends changing them in order to achieve their optimal stress load.

3.3 Failure rates prediction for mechanical parts

MRS analyzes mechanical systems with end elements and subsystems. Each end element is described with several inputted parameters (geometric sizes, material properties, loads, environment, etc.). These values are defined with a distribution type (normal, lognormal, etc.) and distribution parameters. The failure time for each end element and subsystem is simulated, taking into account degradation processes (failure modes): wear, fatigue, creep and stress corrosion cracking, as well as, the following factors: environment temperature, Hydrogen sulfide H_2S concentration, Oxygen concentration, Hydrogen ion concentration – pH, Total Pressure, Partial Pressure of CO_2, Glycol concentration and Inhibitor efficiency.

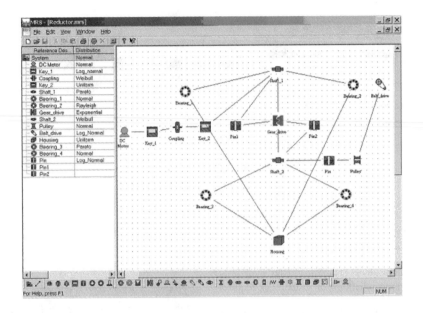

MRS includes a definition, description of one or more failure mechanism(s) and one or more mathematical model(s) for each failure mode. The mathematical models include the number of cycles for fatigue analysis and the rates of wear or corrosion, depending on physical and chemical conditions. Following the failure time simulation, MRS selects a suitable standard type of distribution and defines the Failure density, Reliability and MTBF for each block.

MRS considers the following groups of elements:

- Machine elements: Bar, Rotating disk, Thin wall cylinder, Shafts/Axle, Ring, Seal, Spring, Coupling, Rope, Pulley, Case/Housing, Plate, Customized.
- Drives/Transmissions: Spur Gear drive, Bevel Gear drive, Friction – roller drive, Rack-pinion drive, Worm drive, Chain drive, Belt drive, Screw drive.
- Bearings: Ball/roller bearing, Rotary Sliding bearing/bushing, Line Sliding bearing/bushing.
- Actuators: Hydraulic/Pneumatic line actuator, Electric Motor.

3.4 FMEA/FMECA

The FMECA (Failure Modes, Effects and Criticality Analysis) module defines all the possible failure modes, internal failure causes and failure ratios and effects for each block or function of a considered system. The system configuration and failure rates of all the lowest blocks can be imported from CDB (Core Data Base) allowing data exchange between various C^3 projects.

System failure modes are referred to as 'end effects'. FMECA allocates end effect severities (a degree of danger) to all failures of all blocks and calculates the probability and criticality for all failure modes of all blocks for each severity. These results are used to rank each failure mode using 2 criterions: the probability a failure will occur during a mission phase and the severity of the failure. The user can assign 5 levels of probabilities and severities. Each failure mode is assigned a color that corresponds to its probability and severity, resulting in a visual and clear presentation of all possible failures and their criticality level.

3.5 Failure detection and isolation – built-in test

The Testability Analysis (TA) module uses all data from FMECA in order to calculate the percentage of failure detection and isolation for the system and for each failure mode and block. Failure detection is the fraction of failures identified with one of the tests that are performed during a testing session (BIT).

The tests for each BIT can be defined separately or in a table format. Each test can detect one or more failure modes.

The isolation degree of the system, a mode or a block is the fraction of detected failures for which a single cause can be found, using all the test results. All the failure modes of any replaceable blocks are considered as failure causes. If there is more than one possible cause in the final test results, the isolation is incomplete. The isolation degree is important for reducing the elimination time of a failure and for increasing the system's availability. TA defines all levels of isolation for all failure modes and blocks and estimates the system's self-testing ability. TA also calculates the fraction of non-detected failures (BIT fail) and the frequency of False Alarms.

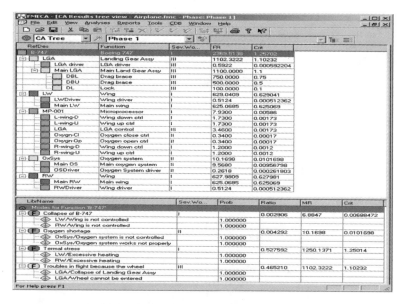

3.6 FTA

CDB can be used for exporting a FMECA project into the FTA (Fault Tree Analysis) module in the form of a Fault Tree. It is also possible to create a Fault Tree manually. FTA features: a) a visual presentation of all cause-effect relationships, b) reliability models other than just OR for the failure events, c) in addition to failure modes, FTA also takes into account restoration and d) FTA Performs cut set and dormant failures analysis.

A cut set is a collection of end causes that result together in a system failure. FTA can identify all minimal cut sets, calculate the probability and rate of the system failure caused by a cut set, show them in Fault Tree and list them in a special cut set report. The cut sets list can be used for preparing repair tools, personnel, materials, facility and instructions for failures restoration.

FTA takes into account "common causes", when one cause affects more than one failure event (modes). A "common cause" can be copied from one event and pasted under a different parent gate. The original and copied events are considered as one event.

3.7 RBD

RBD is usually applied to system models with redundancy, different types of repair, multiple states of a block and network configuration. Any RBD project can be imported from CDB, or be created manually with a diverse set of RBD design options. RBD presents a simultaneous view of the project as a functional tree (shown on the left hand side of the screen) and the

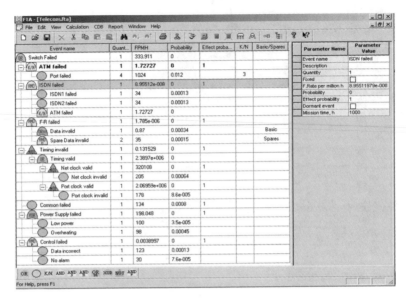

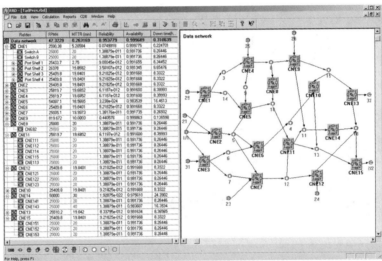

Reliability Block Diagram (shown on the right hand side of the screen) for each selected block in the tree.

The following models are available: Simple, Simple Markov, Serial, Parallel, K out of N identical sub blocks, K out of N different sub blocks, Standby, Composite Markov, Network. Each block can be non-repairable, replaced or repaired. Its sub blocks may be cold repairable (after the parent stops operating) hot repairable (during the redundant 'brothers' operation) or non-repairable. The palette is used for designing all block diagrams.

The inputted failure time distributions are used for the lowest blocks (the leaves of the tree). The advanced analytical and numerical methods are used

for calculating reliability parameters of all composite blocks bottom to top (system) level: Failure time distribution, Reliability, Availability, MTBCF, MTTR and down time per year.

4 CAME

CAME is a powerful tool for describing and optimizing a system's maintenance concept from all aspects: system configuration, maintenance capabilities and policy, preventive maintenance, failures and failure restoration, transportation, spares and other maintenance resources, as well as, documentation.

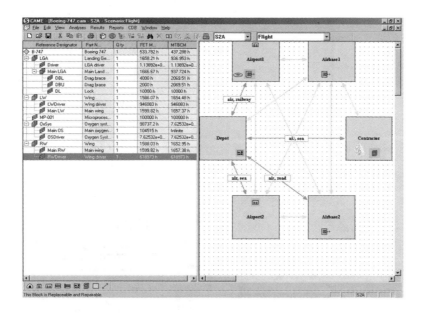

4.1 Maintenance concept modeler

The system is presented in a form of a maintenance tree showing the sequence of disassembly: In order to remove a block, it is first necessary to remove the "father" block. If the parent block is an entire system, it must be available for disassembly. The maintenance tree can include blocks of different maintenance types: repaired without replacement, replaced and then repaired and replaced and then discarded.

The maintenance facilities are presented on a maintenance map: geographic sites that include: operation locations, different levels (up to 5) of repair shops, spare stocks (central and forward) and new spare purchase sites (suppliers, contractors, stores). One or more types of transportation to be used between the maintenance sites can also be defined.

All possible block failures (failure modes) are described with mode ratios, probabilities (during a mission) and effects to failures of higher blocks (Failure Effects Tree). Certain maintenance tasks (corrective and preventive) can be applied for each failure mode with the required resources (Personnel, Support Equipment, Materials, Facility and Utility). These resources should be listed in the repair shops, where the failure mode is restored.

4.2 Failures and maintenance data

For each leaf the user must define its quantity per parent assembly, failure time distribution (6 types), MTTR and replacement time. These data may vary for different scenarios that correspond to different system configurations or maintenance concepts. The scenarios share most of the project data and allow for trade-off between different versions of the system and its maintenance concept.

In order to save valuable time for the user of having to manually input all the names, descriptions and parameters for the system, CAME supports several libraries for: part numbers, failure modes, tasks, resource categories, and resources.

4.3 ORLA

ORLA is used for optimizing the maintenance concept for a system: repair/discard policy, repair, stock and spare purchase sites, as well as, transport choice for failed and repaired items. ORLA provides the required system availability at a minimal cost for repair, replacement, storage and transportation. Following the ORLA calculation, all recommendations for locations of maintenance shops, stocks and spare sources and for transportation types are shown. ORLA recommendations are presented on the maintenance map. The ORLA Summary dialog box and graph show the achieved availability and the optimal maintenance cost for the system. In addition, ORLA also shows the optimal cost for all possible availability values higher than required, to enable the user to compare the results and make a reasonable choice.

4.4 PMO

PMO recommends preventive maintenance for all the blocks. For failures with an increasing hazard rate, as well as, down time damage (in cost units) or failure damage (in addition to restoration cost) due to a system

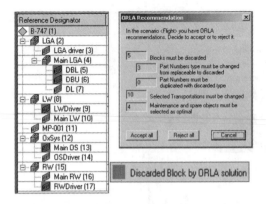

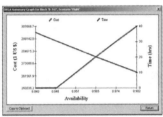

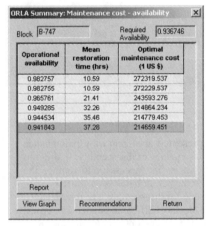

failure, a preventive replacement or repair might be proficient for the corresponding components. PMO analyzes all the possible PM actions and their frequency and selects a PM schedule providing the required availability, mission reliability and failure frequency at minimum cost and damage during operation, preventive and corrective maintenance.

PMO recommends an optimal time interval for each PM session and lists all the lowest blocks (leaves of the maintenance tree) that should be replaced or repaired during this session. The PMO summary and graph show the total maintenance cost, availability and reliability versus the number of PM sessions during the system's lifetime satisfying the availability and reliability requirements. The optimal choice for maintenance is highlighted in the PMO results dialog box.

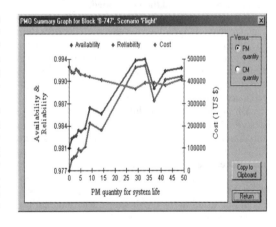

4.5 *S2A*

S2A (Sparing to Availability) analyzes all possible stock sizes (initial number of spares on different stocks) in order to provide the required system availability at minimal cost for spare parts, storage, transportation, as well as, down time damage due to system failure.

S2A allocates spares cost between blocks and sub blocks, and decides which stock is more suitable (central or forward stock). If there are blocks with the same part number and stock site, all the required spare parts use the same stock. CAME allows optimizing combined spare services for some systems with blocks that have the same part numbers.

208

4.6 RCM

RCM module presents a standard procedure for examining the safety of a system and determining the most suitable maintenance for it. The goal is to plan all inspections or preventive maintenance for each emergent failure during the system mission. RCM is composed of 7 leading questions. As the user follows the questions, he is able to analyze each failure and assign a preventive maintenance for it. The questions might also lead to a decision that the system needs to be redesigned. The basis of manual analysis and calculations is the Failure Effect Tree (FET) – the fault tree presents all possible failure events and their cause effect relationships. If each block has only one failure mode, the FET is built automatically. At the event of a block having more than just one failure mode, the FET is built manually by adding a new failure cause for each selected effect event.

Once the FET is completed, RCM calculates the failure rate and mission probability for each FET event, as well as the probability of a system failure during mission. RCM also shows the damage of each end failure cause and its and restoration cost. The user is then able to judge the severity of each failure and find ways of preventing it from happening. RCM also shows the end effects (failure modes) that are generated by each end cause.

RCM includes a flowchart to help the user to plan maintenance actions. The flowchart is composed of leading questions which the user follows sequentially in order to reach the appropriate conclusions regarding: redesign, preventive tasks, on-condition tasks, inspections, etc. Each question on the flowchart opens up a corresponding dialog box, where the user defines all the required tasks for the selected failure. All the decisions are stored in the CAME database and are available for review upon request.

4.7 MSG-3

MSG-3 is used for preparing electronic maintenance documents during the maintenance decision-making process. Most of the maintenance data and instructions are saved in the CAME database, including the documents' changes. All the illustrations, drawings and diagrams are stored as separate files, but their paths are also saved in the database and can be extracted with MSG-3 interface.

MSG-3 allows selecting Maintenance Significant Items (MSI) – the blocks for which the failures and restoration actions are analyzed. All drawings, schemes and supporting data are prepared, collected and stored for each MSI. Each MSI failure is then analyzed according to its severity, economic damage and the possibility of detecting such a failure. All the corresponding tasks are selected for preventing or restoring the failures. Each step is controlled with a two level flow chart, stored in the database and documented. More than 10 different reports may be formed, viewed and printed for each MSI, showing the process of all maintenance decisions. In addition the electronic documents contain detailed tasks and

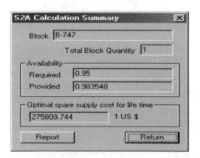

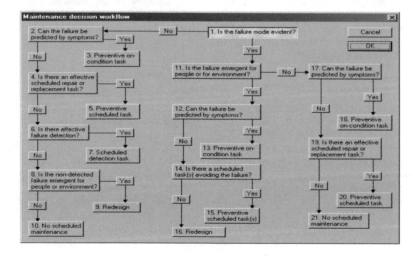

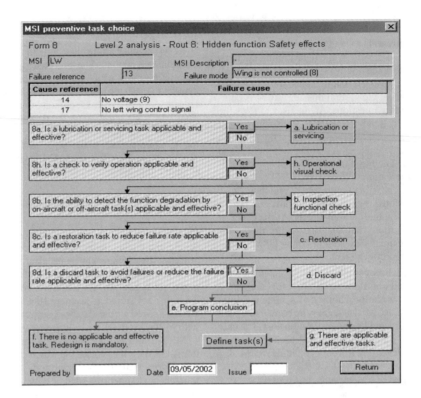

instructions for operators, inspectors and maintenance personnel that can also be edited and printed.

5 CAFDE

CAFDE (Computer Aided Field Data Engineering) is a convenient tool for test or field verification planning, data treatment and decision-making. CAFDE is linked through one core database into CARE and CAME. This unique feature enables the user to import any configuration from CARE and CAME into CAFDE, as well as, all test results pertaining to failure rates, MTBF and distributions.

5.1 Operation and maintenance data

CAFDE provides the means for defining tested systems with their locations and serial numbers. The user can input or import the system block tree containing LRU, SRU and components. CAFDE presents a log for collecting operation and maintenance data for all levels of the tree. Most of the records in the log relate to the system level. Each record corresponds to a time interval between a system's "turning on time" and "turning off time". The cause of turning off is inputted in the field "Off type". If the cause is a failure, the responsible LRU, SRU and components are selected with their serial numbers. If a stoppage of LRU, SRU or component did not cause the higher block to halt, the corresponding record is added only to a "stop block". If there is no turning off records for the sub block within the specified interval, the operation time interval for the block is considered as the operational time interval for each of its sub blocks. The log is used for calculating the operational time of each failure for all serial numbers that are necessary for all the test conclusions and results.

In addition, the log includes a record of all start and finish times for the maintenance of each serial number. This enables the user to calculate the maintenance time.

5.2 Failures distribution

Having calculated time between failures for all serial numbers, the program can recognize one of 7 standard distributions for each tested nominal block or system. The list of the distributions that is used in RBD and CAME, is also used in CAFDE, so the recognition results can be used in these modules.

The recognition is performed using Hi – square criteria. The operation time observations between failures are used for all serial numbers of the considered nominal block. The significance level is calculated for each distribution and the distribution with the maximal level is recommended. Significance level

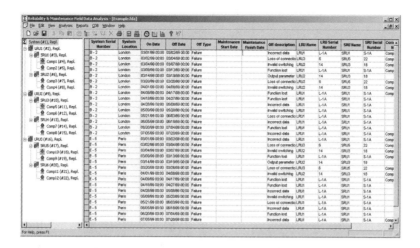

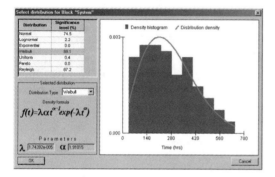

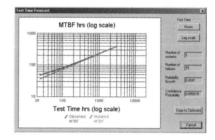

presents the probability that this distribution is not rejected. However, the user can select the distribution he wants. The parameters and significance levels for each selected distribution are presented in a density histogram and a theoretical density curve.

5.3 MTBF estimation

CAFDE calculates MTBF estimation, upper and lower two-sided confidence intervals and lower one-sided confidence intervals for MTBF, using the total observed operational time and the number of observed failures for all serial numbers of a nominal block or system, according to the confidence probability specified by the user. These estimations are presented in a reliability histogram approximating reliability curve.

The time to failure observations for each serial number is used for calculating the MTBF growth (or fall) rate for the Duane reliability growth model and Confidence level, estimating the degree of how the model matches the observed data. MTBF growth takes place, if the system is subjected to an enhancement of the design, operational conditions or maintenance.

The MTBF fall results from general aging of most of the system's components.

If the Confidence level is high enough, the growth rate can be used for planning the time necessary to achieve a target MTBF during the process of enhancement.

5.4 Reliability growth

A producer will most often use failure time observations for MTBF demonstrations. Both parties (producer and consumer) select a test with the appropriate consumer and producer risks, discrimination ratio and type. There are tests with fixed time and sequetial tests, when the test time is not known.

CAFDE presents some standard test plans (Mil-Std-782) for MTBF demonstration and the possibility of preparing a non-standard test plan for customized test conditions (risks and ratio). If a fixed duration test is chosen, CAFDE shows or calculates the required test durations for providing the desired risks, and maximal acceptable number of observed failures.

If a sequential test is chosen, CAFDE calculates and shows the decision chart for this test.

CAFDE enables the user to record and monitor failure times while the test is being performed, showing the current state and final test results.

Safety and Reliability – Bedford & van Gelder (eds)
© *2003 Swets & Zeitlinger, Lisse, ISBN 90 5809 551 7*

Application of chance constraint programming to water system control

A.A.J. Botterhuis
HKV Consultants, Netherlands

A. McCaffrey
University of Strathclyde, Department of Management Science, Glasgow, Scotland

ABSTRACT: In the Netherlands water system control limits the flooding of polders. Applying adaptive control can reduce the frequency of system failure. The estimation and predictions of the state of the polder system in on line application are uncertain. A Gaussian distribution represents the information about the uncertainties. The state of the system should be kept within predetermined boundaries. Exceedance probabilities of these boundaries are obtained using a linear model of the water system. Exceedance is accepted for small probabilities. The optimisation problem is called chance constraint. The objective is to minimise operational costs. A genetic algorithm is used to solve the optimisation problem. This method has been successfully applied to simulate control of a water reservoir with unknown inflow while taking account of the uncertainties involved.

1   INTRODUCTION

This study is concerned with the decrease in the frequency of water system failure by applying real-time control. The subject of the paper is the adaptive optimal control of a Dutch polder system. The Netherlands can be divided into a hilly region and a flat and low-lying region with each being of approximately the same size (figure 1). The low-lying region mainly

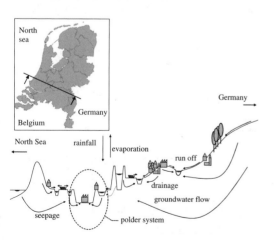

Figure 1.   Schematic crossection of the hydrological system of the Netherlands from the North Sea to Germany.

consists of polders, the majority of which are below mean sea level. Pumping stations are used to drain these polders. The capability to control the water level in a polder depends mainly on the *pumping* capacity and *storage* capacity (Schultz 1992). A couple of times per year an undesirable rise in the water level occurs, due to limited capacity. Flooding of polder sections occurs once every 10–50 years. These events will here be called *system failure*. The frequency of system failure or even better the design of pump and storage capacity depends on the social and economic acceptance of the effects. Research (Schilling 1991, Lobbrecht 1997) shows that usually not all the capacity available is used at the moment of failure and that unused capacity remains in the system. System failures can be reduced considerably by installing adaptive control systems, which reallocate available storage and discharge capacity before the moment of failure.

2   ADAPTIVE OPTIMAL CONTROL

In current practice, an operator in a pumping station controls the polder. Generally, he uses measurements and his experience to decide upon control actions. Lobbrecht (1997) suggested semi-automatic adaptive optimal control. An on-line computer application determines a control strategy for the near future. It runs on a centrally placed computer that receives monitoring data and sends the control actions supervised by the operator. The application:

- predicts the drainage discharge in the polder;
- determines the associated system behaviour and
- optimises the control strategy.

There are similarities in several optimal control methods (Martin-Garcia 1996). The strategy is determined by building and solving an optimisation problem, which incorporates a model of both the system and the objectives. In this article we follow the same approach. Within this approach, the evaluation of the state of the system is complicated by the uncertainty in the information. Estimates or predictions of the system behaviour incorporate uncertainties. Even measurements or not without uncertainties. Neglecting this aspect, Lobbrecht (1997) found a considerable decrease in the frequency of system failure by applying adaptive control.

3 UNCERTAINTY

3.1 Uncertainty of state estimation

One of the reasons why optimal control has not always been successful in practice is that the model of the system does not always represent the actual state of the polder system. When a discrepancy occurs between the estimated and actual state, it cannot be guaranteed that the solutions of the optimisation problem indeed reflect the optimal control. Filtering methods can help. Filtering methods compare the simulated state with real-time monitoring data. For example, based on this comparison the well-known Kalman filter (Kalman 1960, Anderson & Moore 1979) estimates the state of the system and yields a measure for the uncertainty in the information.

3.2 Uncertainty of state prediction

Predictions of the drainage discharge in a polder system can play an important role in the determination of the optimal control strategy. For polders that react slowly (e.g. a day or longer), it has been proven that only an indicative prediction of the drainage is needed to enhance system performance considerably (Lobbrecht 1997). For polders with rapid runoff characteristics (e.g. a couple of hours), the suitability of the control strategy depends much more on the accuracy of the prediction available (Botterhuis 1998). Weather forecasts are input for such drainage predictions. These forecasts are generally inaccurate and very dependent on the type of weather, the time of year, the intensity predicted and the period the prediction is made for (Jilderda 1995, van Meijgaard 1994, de Rooy 1992). The forecast accuracy is far from ideal to effectively use the anticipative control in fast-reacting polder systems (Lobbrecht 1997).

4 ADAPTIVE OPTIMAL CONTROL OF STOCHASTIC SYSTEMS

4.1 Decision under uncertainty

In semi-automatic adaptive optimal control the operator has to choose between different strategies for the near future. As previously stated, the accuracy of the state estimation should be known. The accuracy of the weather forecast should be supplied with the forecast itself. This permits the operator to use this assessment of the uncertainty in his decision. An optimisation method, which includes uncertainty, is described in this paper to advise the operator. In this approach, the amount of water stored in the polder and the drainage discharge are stochastic in nature. The solution of the optimisation gives the probability of the water level being at a certain level, as a result of control strategies and uncertain system behaviour.

4.2 State estimation

A polder system can be modelled as a system of two reservoirs: the open water reservoir and the boundary reservoir. These are connected as shown in figure 2. The open water reservoir represents a network of canals. The excess water is pumped to a boundary, which could represent the sea or a river. In this paper, the real system has been simplified in order to concentrate on the optimisation problem. The state estimation gives the initial values for the state prediction. This estimation is calculated with a Kalman filter. As a result the estimation of the current drainage discharge $\hat{X}_Q(k|k)$ and the water level in the open water $\hat{X}_h(k|k)$ have a Gaussian distribution.

$$\hat{X}_Q(k|k) \sim N\left(\mu_{X_Q}(k), \sigma_{X_Q}(k)^2\right), \quad k = 1,\ldots,n \qquad (1)$$

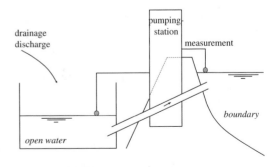

Figure 2. Simple model of a rural water system.

$$\hat{X}_h(k|k) \sim N\left(\mu_{X_h}(k), \sigma_{X_h}(k)^2\right), \ k = 1,\ldots,n \qquad (2)$$

In these equations k is the time index.

4.3 State prediction

The drainage discharge $\hat{X}_Q(k|k)$ is the outcome of the transformation of the rainfall into a discharge that drains from the agricultural fields to ditches, streams and canals. This discharge can be predicted with equation (3). In this equation ℓ is the prediction index.

$$\hat{X}_Q(k+l|k) = \hat{X}_Q(k+l-1|k) + Y_{\Delta Q}(l), \ l = 1,\ldots,m \qquad (3)$$

The predicted drainage $\hat{X}_Q(k+l|k)$ is considered to be dependent on the previous discharge. The operator assumes the discharge for the coming period is equal to the discharge of the previous period. From one period to another the discharge is changed by a small increment $Y_{\Delta Q}(l)$. This adjustment is related to the measured or predicted amount of rainfall and evaporation. $Y_{\Delta Q}(l)$ is considered to be a random variable, Gaussian distributed and independent of the water level or discharge.

$$Y_{\Delta Q}(l) \sim N\left(\mu_{Y_{\Delta Q}}, \sigma_{Y_{\Delta Q}}^2\right), \ l = 1,\ldots,m \qquad (4)$$

The water level in the open water $\hat{X}_h(k|k)$ can be predicted with equation (5). $\hat{X}_Q(k+l|k)$ is dependent on the previous value, the drainage discharge and the discharge of the pumping station $u(k)$.

$$\hat{X}_h(k+l|k) = $$
$$\hat{X}_h(k+l-1|k) + \frac{\Delta t}{a}\hat{X}_Q(k+l-1|k) - \frac{\Delta t}{a}u(k+l) \qquad (5)$$

In equation (5) a stands for the surface area of the open water and Δt stands for the time step of the prediction.

4.4 Control strategy optimisation

4.4.1 Chance constraint programming

Chance constraint programming (CCP) is a type of mathematical programming, with stochastic elements incorporated into the constraint functions (Birge 1997). Charnes and Cooper (1959) introduced CCP as a means of handling randomness or uncertainty in data. CCP models the decision by specifying a confidence level α for each constraint. The constraint can only be violated $(1 - \alpha)$ percent of the time. α is provided as an appropriate safety margin by the operator. The objective function in CCP may contain expected-value functions. However in our application the objective

function is deterministic. The method has been used in many stochastic programming problems, for example inventory control, portfolio selection and reservoir management. The application for offline reservoir management (Meter et al. 1972, Arunkumar & Yeh 1973) is most closely related to our application for on-line polder system control.

4.4.2 Objective function

The aim of polder system control is to minimise the difference between the predicted water level $\hat{X}(k+1|k)$ and the preferred water level in the open water reservoir against the lowest pumping costs. Therefore the objective function can be defined as:

$$\min \sum_{l=1}^{m} u(k+l) \qquad (6)$$

The restrictions concerning minimising the difference between the preferred and predicted water level will be included in the stochastic constraints.

4.4.3 Stochastic state constraints

For any given time-step, the water level should not be allowed to rise Δh above or below the preferred water level h_{pref}. In fact, the water level $\hat{X}_h(k+1|k)$ should be equal to h_{pref}. These constraints are not deterministic due to the stochastic behaviour of $\hat{X}_h(k+1|k)$; there is still a possibility that they will be broken. Therefore, incorporating the probabilities for which these constraints must hold, the constraints can be defined as stochastic constraints, as follows:

$$\Pr\left\{\hat{X}_h(k+l|k) > h_{pref} + \Delta h = h_{upper}\right\} \le 1 - \alpha_1$$
$$\Pr\left\{\hat{X}_h(k+l|k) \le h_{pref}\right\} \le \alpha_2$$
$$\Pr\left\{\hat{X}_h(k+l|k) > h_{pref}\right\} \le 1 - \alpha_3 \qquad (7)$$
$$\Pr\left\{\hat{X}_h(k+l|k) \le h_{pref} - \Delta h = h_{lower}\right\} \le \alpha_4, l = 1,\ldots,m$$

In other words, the first stochastic constraint states that the probability of the water level remaining below the upper boundary should be greater than α_1. The second states that the level should not fall below h_{pref} any more than α_2 percent of the time. The probability of the predicted level being less than the preferred level has to be greater than or equal to α_3. The last constraint states that the probability of falling below the lower bound has to be less than α_4. The probability distribution of $\hat{X}_h(k+l|k)$ can be drawn, as shown in figure 3.

It helps illustrate a clearer picture of how the values for probabilities are calculated. In order for the problem to have a feasible solution, the distribution has to cross the continuous lines between the arrows. In figure 3 it can be seen that the drawing of the probability distribution of $\hat{X}_h(k+l|k)$ satisfies the first three water level

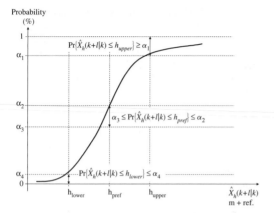

Figure 3. Probability distribution of the water level compared with the constraints.

constraints. It does not satisfy the fourth constraint; therefore this is not a feasible solution.

4.4.4 Deterministic equivalents of stochastic constraints

The stochastic constraints of equation 9 can be rewritten using equations 3, 4, 5 and state estimations 1 and 2.

$$\hat{X}_h(k+l|k) = \hat{X}_h(k+l-1|k) + \frac{\Delta t}{a}\hat{X}_Q(k+l-1|k)$$

$$-\frac{\Delta t}{a}u(k+l) \le h_{pref} \qquad (10)$$

With mathematical induction it can be proven (appendix 1) that this is equal to:

$$\hat{X}_h(k|k) + l\frac{\Delta t}{a}\hat{X}_Q(k|k) + \frac{\Delta t}{a}\sum_{i=1}^{l}(l-i)Y_{\Delta Q}(i)$$

$$-\frac{\Delta t}{a}\sum_{i=1}^{l}u(k+i) \le h_{pref} \qquad (11)$$

$$\Pr\left\{\frac{\Delta t}{a}\sum_{i=1}^{l}u(k+i) \ge h_l\left(\cdot,h_{pref}\right)\right\} \le \alpha_2 \qquad (12)$$

with

$$h_l\left(\cdot,h_{pref}\right) = h_l\left(k,\hat{X}_Q(k|k),\hat{X}_h(k|k),Y_{\Delta Q}(k),h_{pref}\right) =$$

$$\hat{X}_h(k|k) + l\frac{\Delta t}{a}\hat{X}_Q(k|k) + \frac{\Delta t}{a}\sum_{i=1}^{l}(l-i)Y_{\Delta Q}(i) - h_{pref}$$

The mean and the variance of $\hat{X}_Q(k|k)$, $\hat{X}_h(k|k)$ and $Y_{\Delta Q}(k)$ can be used to define the deterministic equivalents of the stochastic constraints. It is assumed that these three stochastic variables are independent. Since the system is assumed to be linear, it is proven that the stochastic behaviour of $h_l(\cdot,h_{pref})$, $h_l(\cdot,h_{upper})$ and

$h_l(\cdot,h_{lower})$ is Gaussian. The mean $\mu_l(\cdot,h_{pref})$ and standard deviation $\sigma_{h_l(\cdot)}$ of $h_k(\cdot,h_{pref})$ is derived in appendix 2 and written in equation (13). The mean of $h_k(\cdot,h_{upper})$ or $h_k(\cdot,h_{lower})$ can be derived when h_{pref} is substituted with h_{upper} or h_{lower}. The standard deviations of the functions $h_k(\cdot,h_{upper})$ and $h_k(\cdot,h_{lower})$ are equal to $\sigma_{h_l(\cdot)}$.

$$h_k\left(\cdot,h_{pref}\right) \sim N\left(\mu_{h_l(\cdot,h_{pref})}(k),\sigma_{h_l(\cdot)}^2(k)\right) \qquad (13)$$

with

$$\mu_{h_l(\cdot,h_{pref})} = \mu_{X_h}(k) + l\frac{\Delta t}{a}\mu_{X_Q}(k)$$

$$+ \tfrac{1}{2}l(l-1)\frac{\Delta t}{a}\mu_{Y_{\Delta Q}} - h_{pref}$$

$$\sigma_{h_l(\cdot)}^2 = \sigma_{X_h}(k)^2 + l^2\left(\frac{\Delta t}{a}\right)^2\sigma_{X_Q}(k)^2$$

$$+ \tfrac{1}{6}l(l-1)(2l-1)\left(\frac{\Delta t}{a}\right)^2\sigma_{Y_{\Delta Q}}^2$$

As the probability levels can be chosen using figure 2, the deterministic equivalents can then be computed and the constraint (12) can be rewritten as:

$$\Pr\left\{h_l\left(\cdot,h_{pref}\right) \le \frac{\Delta t}{a}\sum_{i=1}^{l}u(k+i)\right\} =$$

$$F_{h_l(\cdot,h_{pref})}\left(\frac{\Delta t}{a}\sum_{i=1}^{l}u(k+i)\right) \le \alpha_2$$

where $F_{h_l(\cdot,h_{pref})}(\cdot)$ is the distribution of $h_l(\cdot,h_{pref})$.

$$\frac{\Delta t}{a}\sum_{i=1}^{l}u(k+i) \le F_{h_l(\cdot,h_{pref})}^{-1}(\alpha_2)$$

$$\frac{\Delta t}{a}\sum_{i=1}^{l}u(k+i) \le \mu_{h_l(\cdot,h_{pref})} + \Phi^{-1}(\alpha_2)\sigma_{h_l(\cdot)}$$

with Φ is the cumulative distribution function of the standard normal distribution.

Deterministic equivalents of the chance constraint problem of equation 9 are written in equation 14.

$$\frac{\Delta t}{a}\sum_{i=1}^{l}u(k+i) - \mu_{h_l(\cdot,h_{upper})} - \Phi^{-1}(\alpha_1)\sigma_{h_k(\cdot)} \ge 0$$

$$-\frac{\Delta t}{a}\sum_{i=1}^{l}u(k+i) + \mu_{h_l(\cdot,h_{pref})} + \Phi^{-1}(\alpha_2)\sigma_{h_l(\cdot)} \ge 0$$

$$\frac{\Delta t}{a}\sum_{i=1}^{l}u(k+i) - \mu_{h_l(\cdot,h_{pref})} - \Phi^{-1}(\alpha_3)\sigma_{h_l(\cdot)} \ge 0 \qquad (14)$$

$$-\frac{\Delta t}{a}\sum_{i=1}^{l}u(k+i) + \mu_{h_l(\cdot,h_{lower})} + \Phi^{-1}(\alpha_4)\sigma_{h_l(\cdot)} \ge 0,$$

$$l = 1,\ldots,m$$

4.4.5 Deterministic pumping station constraints

The pumping station restrictions involved in this problem relate to the capacity of the pumps. The feasible pumping rate is not continuous between zero and the maximum rate of the station $Max_{station}$. The possible options are discrete between no pumps in operation and all pumps in operation. In polder system control the possible strategies are limited. The decision problem can be simplified to a selection of the optimal strategy from a set of known strategies. We use a *genetic algorithm* (GA) to search within the population of feasible strategies. For this article we have used the work of Goldberg (1989) to design the algorithm. For an introduction and description of GAs the interested reader is referred to the book of Goldberg.

Each possible strategy is coded in a string where each position can have a value of 0 or 1. For example, a string with two positions has four possible codes (00, 10, 01 and 11). The code 00 could represent no pumps in operation, and the code 11 could represent three pumps in operation. The number of string positions per time step is related to the number of possible control actions per time step. Due to the string having a fi-nite number of positions, the possible strategies are limited. The string can only represent strategies that do not violate the pumping station constraints. All possible codes are linked to a strategy.

4.4.6 Transformation into unconstrained maximisation problem

The constraint problem in section 4.4.4 is transformed into an unconstrained problem by associating a penalty with all constraint violations. Because of the lack of capacity (section 1), it is not possible to keep the state of the system between the upper and lower boundary. The formulation of an unconstrained problem makes constraint violation possible within the search for the optimal strategy. Even when the drainage discharge causes the water level to rise above the upper boundary, the algorithm still finds an optimal strategy. The penalty for violating a constraint is calculated as follows:

$$g_{h_l(\cdot,h_{upper}),\alpha_1}\big(u(k+1),\ldots,u(k+l)\big) =$$

$$\frac{\Delta t}{a}\sum_{i=1}^{l}u(k+i)-\mu_{h_l(\cdot,h_{upper})}-\Phi^{-1}(\alpha_1)\sigma_{h_l(\cdot)}\geq 0 \tag{15}$$

$$\Psi\big(g_{h_l(\cdot,h_{upper}),\alpha_1}(\cdot)\big)=\begin{cases}g_{h_l(\cdot,h_{upper}),\alpha_1}(\cdot)^2 & if\ g_{h_l(\cdot,h_{upper}),\alpha_1}(\cdot)<0\\0 & otherwise\end{cases}$$

$$\tag{16}$$

Likewise, the penalty function $\Psi\big(g_{h_l(\cdot,h_{upper})},\alpha_1(\cdot)\big)$ is used to transform the other constraints in (14). This

results in the following minimisation problem:

$$\min \sum_{l=1}^{m}u(k+l)+\lambda_1\sum_{l=1}^{m}\Psi\Big(g_{h_l(\cdot,h_{upper}),\alpha_1}(\cdot)\Big)$$

$$+\lambda_2\sum_{l=1}^{m}\Psi\Big(g_{h_l(\cdot,h_{pref}),\alpha_2}(\cdot)\Big)$$

$$+\lambda_3\sum_{l=1}^{m}\Psi\Big(g_{h_l(\cdot,h_{pref}),\alpha_3}(\cdot)\Big) \tag{17}$$

$$+\lambda_4\sum_{l=1}^{m}\Psi\Big(g_{h_l(\cdot,h_{lower}),\alpha_4}(\cdot)\Big)$$

The unconstrained problem is considerably more suited to being handled by a GA (Goldberg 1989). A GA maximises an objective function that is guaranteed to be nonnegative. To transform a minimisation problem to a maximisation problem (suitable for GAs) the following objective transformation is commonly used (Goldberg 1989):

$$\min\ C^*_{max}-\sum_{l=1}^{m}u(k+l)$$

$$+\lambda_1\sum_{l=1}^{m}\Psi\Big(C^1_{max}-g_{h_l(\cdot,h_{upper}),\alpha_1}(\cdot)\Big)$$

$$+\lambda_2\sum_{l=1}^{m}\Psi\Big(C^2_{max}-g_{h_l(\cdot,h_{pref}),\alpha_2}(\cdot)\Big) \tag{18}$$

$$+\lambda_3\sum_{l=1}^{m}\Psi\Big(C^3_{max}-g_{h_l(\cdot,h_{pref}),\alpha_3}(\cdot)\Big)$$

$$+\lambda_4\sum_{l=1}^{m}\Psi\Big(C^4_{max}-g_{h_l(\cdot,h_{lower}),\alpha_4}(\cdot)\Big)$$

$C^{(\cdot)}_{max}$ is the largest possible value of the summation following it in equation (18). When the minimum of the constraint function (15) is smaller then zero, the square of this minimum is equal to maximum of the penalty function. These maximums have to be summed to get the largest possible value:

$$C^1_{max}=\sum_{l=1}^{m}\Big(-\mu_{h_l(\cdot,h_{upper})}-\Phi^{-1}(\alpha_1)\sigma_{h_l(\cdot)}\Big)^2$$

$$C^2_{max}=\sum_{l=1}^{m}\Big(-l\frac{\Delta t}{a}Max_{station}+\mu_{h_l(\cdot,h_{pref})}+\Phi^{-1}(\alpha_2)\sigma_{h_l(\cdot)}\Big)^2$$

$$C^3_{max}=\sum_{l=1}^{m}\Big(-\mu_{h_l(\cdot,h_{pref})}-\Phi^{-1}(\alpha_3)\sigma_{h_l(\cdot)}\Big)^2$$

$$C^4_{max}=\sum_{l=1}^{m}\Big(-l\frac{\Delta t}{a}Max_{station}+\mu_{h_l(\cdot,h_{lower})}+\Phi^{-1}(\alpha_4)\sigma_{h_l(\cdot)}\Big)^2$$

$$l=1,\ldots,m$$

The largest possible pumping costs C^*_{max} is equal to maximum pumping rate of the station multiplied by the amount of prediction steps m.

$$C^*_{max} = m \frac{\Delta t}{a} Max_{station}$$

4.5 Tests of algorithm

To solve the unconstrained maximisation problem the GA code of Lindfield and Penny (1995) was used. For the simple problem in figure 2 it provides good results. In the test the mean and variances of $\hat{X}_Q(k|k)$, $\hat{X}_h(k|k)$ and $Y_{\Delta Q}(l)$ were chosen arbitrarily. Further testing of the algorithm is required. To compare the results with current polder system control, system identification methods have to be used to identify the expectations and variances. An Extended Kalman Filter can be used to increase the complexity and subsequently increase the accuracy of the state estimation and prediction. Such a filter will make it possible to translate uncertain rainfall forecasts into a stochastic prediction of system behaviour in the near future.

5 CONCLUSIONS

It is possible to model the control problem for a polder system as a chance constraint optimisation problem. The state prediction based on persistence of the drainage discharge can be translated to a stochastic dynamical system. An increase of the complexity of the model of a rural water system will make the prediction of the system state more accurate. Physically based modelling of the rainfall-runoff process can be translated with the Extended Kalman Filter to a stochastic dynamical system. Chance Constraint Optimisation of such estimations and/ or predictions will make the method applicable to adaptive control for a broad variety of rural water systems.

REFERENCES

Anderson, B.D. & Moore J.B. 1979. Optimal Filtering. New Jersey: Prentice Hall.

Arunkumar, S. & Yeh, W.W.G. 1973. Probabilistic models in the design and operation of a multi-purpose reservoir system. ..: Davis.

Birge, J.R. 1997. Stochastic Programming Computation and Applications. INFORMS Journal on Computing Vol. 9 No 2: 111–133.

Botterhuis, A.A.J. 1998. Optimal water system control during and after a period of heavy rainfall, the use of weather forecasts in the operational watermanagement of Flevoland (in Dutch: Optimalisatie van de peilhandhaving gedurende en kort na een zware bui; toepassing van neerslagvoorspelling in het operationele beheer van Oostelijk en Zuidelijk Flevoland), MSc Thesis. Delft University of Technology: Delft.

Charnes, A. & Cooper, W.W. 1959. Chance-constrained programming. Management Science, Vol. 6 No. 1: 73–79.

De Rooy, W.C & Engeldal, C.A. 1992. Rainfall verification LAM (in Dutch: Neerslagverificatie LAM), TR-143. KNMI: de Bilt.

Goldberg, D.E. 1989. Genetic Algorithms in Search, Optimization & Machine Learning. Addison Wesley Longman.

Jilderda, R.,Van Meijgaard, E. & De Rooy, W. 1995. Rainfall in the catchment area of the Meuse in January 1995 (in Dutch: Neerslag in het stroomgebied van de Maas in januari 1995), TR-178. KNMI: de Bilt.

Kalman, R.E. 1960. A new approach to linear filtering and prediction problems. Journal of Basic Engineering Vol. 82: 34–35.

Lindfield, G.R. & Penny, J.E.T. 1995. Numerical methods using Matlab, 2nd ed. Prentice Hall: New Jersey.

Lobbrecht, A.H. 1997. Dynamic Water-system control, Design and Operation of Regional Water-Resources Systems. STOWA: Utrecht.

Martin-Garcia, H.J. 1996. Combined logical-numerical enhancement of real-time control of urban drainage networks. Balkema: Rotterdam.

Meter, W., Helm, J. & Curry, G. 1972. Development of a Dynamic Water Management Policy for Texas. Texas A&M University.

Schultz, E. 1992. Water management of the drained lakes in The Netherlands (in Dutch: Waterbeheersing van de Nederlandse droogmakerijen), Van zee tot land 58. Directoraat-Generaal Rijkswaterstaat: Lelystad.

Schilling, W. 1991. Real-time Control of Urban Drainage Systems: From suspicious attention to wide-spread application. Advances in Water Resources Technology: 561–576. Balkema: Rotterdam.

APPENDIX 1

Let $\hat{X}_Q(k+l-1|k) = \hat{X}_Q(k|k)$,

$\hat{X}_h(k+l-1|k) = \hat{X}_h(k|k)$, $l = 1$, $k = 1,\ldots,n$

as well as

$$\hat{X}_Q(k+l|k) = \hat{X}_Q(k+l-1|k) + Y_{\Delta Q}(l),$$

$$\hat{X}_h(k+l|k) =$$

$$\hat{X}_h(k+l-1|k) + \frac{\Delta t}{a}\hat{X}_Q(k+l-1|k) - \frac{\Delta t}{a}u(k+l),$$

$l = 1,\ldots,n$

then

$$\hat{X}_Q(k+l|k) = \hat{X}_Q(k|k) + \sum_{i=1}^{l} Y_{\Delta Q}(i)$$

$$\hat{X}_h(k+l|k) = \hat{X}_h(k|k) + l\frac{\Delta t}{a}\hat{X}_Q(k|k)$$
$$+ \frac{\Delta t}{a}\sum_{i=1}^{l}(l-i)Y_{\Delta Q}(i) - \frac{\Delta t}{a}\sum_{i=1}^{l}u(k+i)\,,$$

The proof follows by mathematical induction. As the basis of the induction,

$$\hat{X}_Q(k+1|k) = \hat{X}_Q(k|k) + Y_{\Delta Q}(1)$$

$$\hat{X}_h(k+1|k) =$$
$$\hat{X}_h(k|k) + \frac{\Delta t}{a}\hat{X}_Q(k|k) - \frac{\Delta t}{a}u(k+1)$$

Using the induction hypothesis it follows that,

$$\hat{X}_Q(k+l|k) = \hat{X}_Q(k|k) + \sum_{i=1}^{l-1}Y_{\Delta Q}(i) + Y_{\Delta Q}(l)$$
$$= \hat{X}_Q(k|k) + \sum_{i=1}^{l}Y_{\Delta Q}(i)$$

$$\hat{X}_h(k+l|k)$$
$$= \hat{X}_h(k+l-1|k) + \frac{\Delta t}{a}\hat{X}_Q(k+l-1|k)$$
$$-\frac{\Delta t}{a}u(k+l)$$
$$= \hat{X}_h(k|k) + (l-1)\frac{\Delta t}{a}\hat{X}_Q(k|k)$$
$$+\frac{\Delta t}{a}\sum_{i=1}^{l-1}(l-1-i)Y_{\Delta Q}(i) - \frac{\Delta t}{a}\sum_{i=1}^{l-1}u(k+i)$$
$$+\frac{\Delta t}{a}\left(\hat{X}_Q(k|k) + \sum_{i=1}^{l-1}Y_{\Delta Q}(i)\right) - \frac{\Delta t}{a}u(k+l)$$
$$= \hat{X}_h(k|k) + l\frac{\Delta t}{a}\hat{X}_Q(k|k) - \frac{\Delta t}{a}\sum_{i=1}^{l}u(k+i)$$
$$+\frac{\Delta t}{a}\left(\sum_{i=1}^{l}(l-1-i)Y_{\Delta Q}(i) - (l-1-l)Y_{\Delta Q}(l)\right)$$
$$+\frac{\Delta t}{a}\left(\sum_{i=1}^{l}Y_{\Delta Q}(i) - Y_{\Delta Q}(l)\right)$$
$$= \hat{X}_h(k|k) + l\frac{\Delta t}{a}\hat{X}_Q(k|k) + \frac{\Delta t}{a}\sum_{i=1}^{l}(l-i)Y_{\Delta Q}(i)$$
$$-\frac{\Delta t}{a}\sum_{i=1}^{l}u(k+i)$$

The mean of $h_l(\cdot,h_{pref})$ is

$$\mu_{h_l(\cdot,h_{pref})}$$
$$= E\left(\hat{X}_h(k|k) + l\frac{\Delta t}{a}\hat{X}_Q(k|k) + \frac{\Delta t}{a}\sum_{i=1}^{l}(l-i)Y_{\Delta Q}(i) - h_{pref}\right)$$
$$= \mu_{X_h}(k) + l\frac{\Delta t}{a}\mu_{X_Q}(k) + \frac{\Delta t}{a}E\left(\sum_{i=1}^{l}(l-i)Y_{\Delta Q}(i)\right) - h_{pref}$$

The expectation of the summation of the increment of the drainage discharge $Y_{\Delta Q}(l)$ can be calculated as follows:

$$E\left(\sum_{i=1}^{l}(l-i)Y_{\Delta Q}(i)\right) = \sum_{i=1}^{l}(l-i)E\left(Y_{\Delta Q}(i)\right) = \sum_{i=1}^{l}(l-i)\mu_{Y_{\Delta Q}}$$
$$= \tfrac{1}{2}l(l-1)\mu_{Y_{\Delta Q}}$$

The last step in this denotation can be proven by mathematical induction. As the basis of the induction,

$$\sum_{i=1}^{1}(1-i)\mu_{Y_{\Delta Q}} = \tfrac{1}{2}1(1-1)\mu_{Y_{\Delta Q}} = 0$$

Using the induction hypotheses it follows that,

$$\sum_{i=1}^{l+1}(l+1-i) = \sum_{i=1}^{l+1}(l-i) + \sum_{i=1}^{l+1}1$$
$$= \sum_{i=1}^{l}(l-i) + (l-l-1) + (l+1) = \sum_{i=1}^{l}(l-i) + l$$
$$= \tfrac{1}{2}l(l-1) + l = l\left(\tfrac{1}{2}(l-1)+1\right) = l\left(\tfrac{1}{2}l + \tfrac{1}{2}\right)$$
$$= \tfrac{1}{2}l(l+1) = \tfrac{1}{2}(l+1)(l+1-1)$$

The variance of $h_l(\cdot,h_{pref})$ is

$$\sigma_{h_l(\cdot,h_{pref})}$$
$$= VAR\left(\hat{X}_h(k|k)\right) + VAR\left(l\frac{\Delta t}{a}\hat{X}_Q(k|k)\right)$$
$$+VAR\left(\frac{\Delta t}{a}\sum_{i=1}^{l}(l-i)Y_{\Delta Q}(i)\right) - VAR\left(h_{pref}\right)$$
$$= \sigma_{X_h}(k)^2 + l^2\left(\frac{\Delta t}{a}\right)^2\sigma_{X_Q}(k)^2$$
$$+\left(\frac{\Delta t}{a}\right)^2 VAR\left(\sum_{i=1}^{l}(l-i)Y_{\Delta Q}(i)\right)$$

The variation of the summation of the increment $Y_{\Delta Q}(l)$ can be calculated as follows:

$$VAR\left(\sum_{i=1}^{l}(l-i)Y_{\Delta Q}(i)\right)=\left(\sum_{i=1}^{l}(l-i)\right)^2\sigma_{Y_{\Delta Q}}{}^2$$

$$=\tfrac{1}{6}l(l-1)(2l-1)\sigma_{Y_{\Delta Q}}{}^2$$

The last step in this denotation can also be proven by mathematical induction. As the basis of the induction,

$$\left(\sum_{i=1}^{1}(l-1)\right)^2=\tfrac{1}{6}1(1-1)(2\cdot1-1)=0$$

Using the induction hypothesis it follows that,

$$\left(\sum_{i=1}^{l}(l-i)\right)^2=\tfrac{1}{6}(l+1)(l+1-1)(2(l+1)-1)$$

$$=\tfrac{1}{6}l(l-1)(2(l+1)-1)+\tfrac{1}{6}l(2(l+1)-1)$$

$$+\tfrac{1}{6}(l+1-1)(2(l+1)-1)$$

$$=\tfrac{1}{6}l(l-1)(2l-1)+\tfrac{2}{6}l(l-1)+\tfrac{2}{6}(2(l+1)-1)$$

$$=\tfrac{1}{6}l(l-1)(2l-1)+\tfrac{3}{6}l^2+\tfrac{4}{6}l^2+\tfrac{2}{6}l$$

$$=\left(\sum_{i=1}^{l}(l-i)\right)^2+l^2=\left(\sum_{i=1}^{l+1}(l+1-i)\right)^2$$

Safety and Reliability – Bedford & van Gelder (eds)
© 2003 Swets & Zeitlinger, Lisse, ISBN 90 5809 551 7

BHDL: Formalization of digital circuits by coupling VHDL and the B method

J.L. Boulanger
Univeristé de Technologie de Compiègne, HEUDIASYC UMR CNRS 6599, BP 20529, 60205 Compiègne cedex

G. Mariano
INRETS, ESTAS, 20 rue élisée Reclus, BP 317, 59666 Villeneuve d'ascq

ABSTRACT: The goal of this paper is to show how it is possible to combine the advantages of the VHDL programming language and of the B method in order to design secure digital circuit that may be easily developed and does not need a design test. Secure in this paper means more than correct, the design perform what the client wants, furthermore it guarantees to note achieve the unwanted cases.

1 INTRODUCTION

In safety critical applications it is mandatory to fabricate chips which are design error free. With the increasing complexity of designs this goal is hard to satisfy without methods specially dedicated to this task. The reliability level of a critical application may not be assessed quantitatively. Confidence on the acceptability of a system must be built all along the development process by combining stringent fault avoidance and fault tolerance techniques. In this context, *Formal methods* are good candidate to address fault avoidance: they can make significant contribution to both fault prevention and fault removal.

VHDL (VHSIC – Very High Speed Integrated Circuits – Hardware Description Language)(1076–1993) and (Airiau et al. 1998) is an IEEE Standard since 1987 and many commercials design tools are based on. It is "a formal notation intended for use in all phases of the creation of electronic systems [...] it supports the development, verification, synthesis, and testing of hardware designs, the communication of hardware design data ..."[1]. VHDL is a well known programming language used to express the hardware components with a high level of abstraction. It is a good utility to describe single integrated circuits, or complete system of hardware and software. Also it can be used to

declare the circuit behaviour. VHDL methodology is based on the conception of modules. The hierarchy enables the programmer to write the program as units. Some of these units are elementary expressions which can be directly compiled while others may be decomposed into several units and so on. This mechanism makes the work of team easier, especially when the complexity of the desired system increases. A special library is built for each programmer using his correct units which have been compiled. The programmer may use his libraries and the libraries of his colleagues and the general libraries.

On the other hand, the *B* method due to J.R Abrial (Abrial 1996) is a formal method for the incremental development of mathematical abstract specifications and their refinements down to an executable implementation.It is a model-based approach similar to Z (Spivey 1992) and VDM (Jones 1990). The software design in *B* starts from mathematical specifications. Little by little, through many refinement steps (Morgan 1990), the designer tries to obtain a complete and executable specification. This process must be monotonic, that is any refinement has to be proved coherent according to the previous steps of refinement. The abstract machine (Abrial 1992) is the central concept of a *B* development. It encapsulates some state data and offers some operations or "services" to the outside. The content of an abstract machine is composed of three parts:

• the *declarative part* which describes the states and their properties,

---

[1] Preface to the IEEE Standard VHDL Language Reference Manual.

- the *execution part* which introduces operations and
- the *composition clauses* which introduces architectural links between abstract machines.

In the *B* development process, the proofs accompany the construction of software. Each time an abstract machine is defined or modified, there are proof obligations related to its mathematical internal consistency are automatically generated. When the considered machine is a refinement or an implementation, additional proof obligations are generated to ensure its correctness with respect to the previous steps of the development chain. The *B* tool allows us to generate automatically the proof obligations (noted PO) for each abstract machine. Available commercial tools supporting the *B* method are used to automatically generate these proof obligations (also named PO) for each abstract machine. Then these proof obligations are discarded either automatically for the simple ones or by the mean of an interactive prover, manually by the designer for the complex ones. Finally, at the last refinement step called the implementation, we are able to automatically generate the code of a secure software. Commercial tools are provided with C, C++ or ADA automatic translators. Our main goal is, by designing and implementing the suited methodology, to extend this set of target languages with VHDL

2 MODELING OF DIGITAL CIRCUITS

In this section, our aim is to give a general method to model a circuit without knowing the internal details of the desired circuit.

In the main formal verification tools, based on the formal proof, the specification and the implementation are represented as the formulas in logic. The relationship between a specification and an implementation is specified as a theorem in logic which is proved. In our methodology, we define an equivalent model of hardware and proof the equivalence.

2.1 Modeling methodology

A synchronized circuit is view as a box, within which an (or more) input line is entering, and out of which an output line (or more) is emerging, see an example in figure 1. A synchronized circuit is supposed to be synchronized by a clock. Then an example of more complicated circuit is used, in section 2.2 to show who many components may be connected. Using VHDL terminology, a module is called entity and one entity is coded in one *B* abstract machine. All the inputs and outputs are called ports.

In the table 1, the analogy of development between the *B* method and the numeric circuits design is presented.

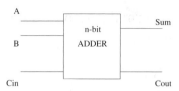

Figure 1. n-bit adder.

Table 1. Analogy between VHDL and the *B* method.

Circuits synthesis	B Method
Functional Specifications	Abstract machine
Architecture Specification & Behaviour Details Validation "functional to material"	Refinements Proof
Physical Description	Implementation Machine
Port	Global Variable
Connection	Invariant of relation between variables
Signal Propagation	Operation call (transmission of values)
Reusing	Importation + Renaming

In a first step, we define the properties (Safety, Liveness,...) of the new components. Actuators in safety critical systems must be driven by failsafe signals. Under a failure in the system, such a signal must be either on the correct or on the safe state (e.g. red colour in traffic control lights). The second step, is graphical, it introduce the components synthesis by composition of basic components with a VHDL graphic tool. In (Boulanger, Mariano, and Aljer 2001), we demonstrate how a VHDL packages can be translated as *B* circuit components for giving to the designer a high-level view. In fact one VHDL component (entity) is translated in one *B* circuit specifications which is one abstract machine. From VHDL component, the "entity" part provide a list of procedure and the "architecture" part provide a description of principal properties and introduce the implementation. We add some annotation (commentary) in the VHDL description. In fact, the subset of VHDL that we are using is that produced by the tools called VGUI[2]. VGUI is a free graphical tool for capturing, drawing, editing, and navigating hierarchical block-diagrams, and for producing corresponding structural VHDL or Verilog code. In third step, the refinement direction is determined by the basic elements which are used to construct the desired circuit. So the designer can orient the development to the needed level. This level can be

---

[2] http://www.atl.external.lmco.com/rassp/vgui/

found as a basic library in *B* (see 2.3). So, at the last refinement, the implementation, we obtain a secure software. In the implementation machine, the behavior of the circuit is deterministic and it may be easily converted to another programming language. We define a tool that provide the possibility of an automatic transformation of the *B* implementation to VHDL language. In this case, we obtain the possibility of co design between the *B* method and VHDL. Using this approach, one can develop a circuit of which each part of the specification has proved to be correct. Using this method, 80% of the needed proofs are proved automatically, and the others need cooperation with the designer. Sometimes little changes of the source machines are necessary to create the desired proofs. This enables the designer to correct the possible errors of his design.

2.2 Examples

This section describes and models some synchronized basic components which will be then reused. A 4-bit full adder circuit is used as an example to show how to reuse components previously modeled in order to obtain complex integrated circuits. Basing ourselves on the logical specification of a component, we can supply an assembly of many simple components to achieve the desired function. This is called the synthesis of a numeric circuit.

The n-bit full adder is a circuit with 3 inputs (2 n-bit vectors and 1 bit) and two outputs (1 n-bit vector and 1 bit). The n-bit full adder is described by a graphical specification (sees the figure 1), but we can complete these specifications with a boolean expression that described the behaviour:

$$Adder\ (a,\ b,\ cin,\ sum,\ cout) ==$$
$$2^{(n+1)}BV\ (cout) + V\ (sum) =$$
$$V(a) + V(b) + BV(cin)$$

with

$$V:\quad bitVector\ - - > INTEGER$$
$$BV:\ bit\ - - > INTEGER$$

We can use the last expression to write a *B* abstract machine. This abstract machine contains the abstract specifications of the desired circuit (or software):

MACHINE
  B_ADD_4bits_2_0
SEES
  B_type_0
DEFINITIONS
  COMPUTE_(AA,BB,CIN,SUM,COUT) ==
  SUM + 16*COUT = AA + BB + CIN

In our methodology we represented, in the abstract machine, each port of the circuit by a local variable (AA, BB, SUM, CIN, COUT). So that the connection between the ports is represented as a fixed relation

between the global variables. These relations are represented in the *INVARIANT* clause which contains all the relations that must be satisfied in a machine and in its refinements. The behaviour of the port is described using the *Compute_* definition[3], which gives us the possibility to express systematically the definition of the function. The *INVARIANT* clause states the static laws, in our case the properties, that the data must obey whatever the operation that is applied to it.

VARIABLES
  AA,BB,CIN,COUT,SS
INVARIANT
  IS_A_DECIMAL_16_(AA)
  & IS_A_DECIMAL_16_(BB)
  & IS_A_DECIMAL_16_(SUM)
  & IS_A_DECIMAL_1_(CIN)
  & IS_A_DECIMAL_1_(COUT)
  & COMPUTE_(AA,BB,CIN,SUM,COUT)
INITIALISATION
  AA, BB, CIN, SUM, COUT := 0,0,0,0,0

These variables can be assigned by all of the defined operations. Each output is attached to a read operation and each input is attached to a store one. The environment of the circuit will use these operations to know the circuit output or to change the input. So, we choose In and Out as names to these operations.

OPERATIONS
in_AA(yy) =
PRE
  IS_A_DECIMAL_16_(yy)
THEN
  AA, SUM, COUT
      :(IS_A_DECIMAL_16_(AA)
  &   AA = yy
  &   IS_A_DECIMAL_16_(SUM)
  &   IS_A_DECIMAL_1_(COUT)
  &   COMPUTE_(AA,BB,CIN,SUM,COUT))
END;

...

xx <− out_COUT =
  xx := COUT;
xx <− out_SS =
  xx := SS;
END

This abstract machine is not deterministic since we use the operator *list_var :(predicate)* in the *OPERATION* clauses. This operator indicates that the list of variable become such that the predicate is true. We can generalize this method for any combinatory logical circuit, for the complex ones as well as for the simple ones. The 4-bit adder can be split-up in bit slices.

---

[3] a definition may be seen as a function, but the associated code will be expanded.

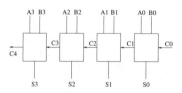

Figure 2.   n-bit carry-ripple adder.

We obtain the well-known carry ripple adder, also denoted as ripple through carry adder, consists of n cascaded full adders for a n bit adder. Each slice performs the addition of the bits Ai, Bi and the Carry-in bit Ci (= Carry-out bit of the previous slice).

Next piece of code show how we split the 4-bit adder in its implementation. Each slice consists of a 1-bit full adder, illustrated below. A 1-bit full adder (figure 2) is a combinational circuit that computes the arithmetic sum of three input bits of the same magnitude (i.e., 1-bit numbers). As expected in an object-oriented environment, you are able to reuse the single Full_Adder (B_FULL_ADD_1bit_0) component by instantiating (*IMPORTS* clause) four copies of it (FA0 through FA3). The *B* method gives us the possibility to reuse other machines. So many already defined machines may be reused as components of more complex ones. An already defined machine may also be renamed and reused by adding other operations to produce a more developed circuit.

IMPLEMENTATION
  B_ADD_4bits_2_n
REFINES
  B_ADD_4bits_2_1
SEES
  B_type_0
IMPORTS
  Fa0.B_FULL_ADD_1bit_0
, Fa1.B_FULL_ADD_1bit_0
, Fa2.B_FULL_ADD_1bit_0
, Fa3.B_FULL_ADD_1bit_0
...

Each copy is instantiated differently in the four port-mapping statements, with actual signal names being substituted for the desired input/output connections.

INVARIANT
  Fa0.AA = A0
& Fa0.BB = B0
& Fa0.CIN = CIN
& Fa0.COUT = CI1
& Fa0.SS = SS0
...

In the *B* method the OPERATIONS clause represents the dynamic part of a machine in which the values of the global variables may be changed. So signal propagation is done by an operation call.

OPERATIONS
in_AA(yy) =
BEGIN
A0, A1, A2, A3 <− Decimal16_To_Bit(yy) =
;   Fa0.in_AA(A0)
;   Fa1.in_AA(A1)
;   Fa2.in_AA(A2)
;   Fa3.in_AA(A3)
;   Fa0.in_CIN(CIN)
;   CI1 <− Fa0.out_COUT
;   Fa1.in_CIN(CI1)
;   CI2 <− Fa1.out_COUT
;   Fa2.in_CIN(CI2)
;   CI3 <− Fa2.out_COUT
;   Fa3.in_CIN(CI3)
;   COUT <− Fa3.out_COUT
;   SS0 <− Fa0.out_SS
;   SS1 <− Fa1.out_SS
;   SS2 <− Fa2.out_SS
;   SS3 <− Fa3.out_SS
END;

...
END

We declare the 1-bit adder with the inputs and outputs shown inside the port(). This will add two bits together(x,y), with a carry in(cin) and output the sum (sum) and a carry out(cout).The next equations describes a concise view of the functionality of the 1-BIT adder.

sum = a xor b xor cin

cout = ((a or b) and cin) or (a and b)

But, we use the next description:

```
sum = (not a and not b and cin) or
      (not a and b and not cin) or
      (a and not b and not cin) or
      (a and b and cin);
```

```
cout = (not a and b and cin) or
       (a and not b and cin) or
       (a and b and not cin) or
       (a and b and cin);
```

MACHINE
  B_FULL_ADD_1bit_0
SEES
  IEEE.B_STD_LOGIC_1164_0
  ,B_STD_ENTRIE
DEFINITIONS
... ; Sum_(inA, inB, Cin, Sum) = =
     (Sum = bool_to_bit(
     bool(((not(Val_(inA)))
     & ((not(Val_(inB)) & (Val_(Cin)))
     or ((Val_(inB)) & (not(Val_(Cin)))))))
     or ((Val_(inA))

& ((not(Val_(inB)) & not(Val_(Cin)))
    or ((Val_(inB)) & (Val_(Cin))))))))
; Carry_Expr_(inA, inB, Cin, Cout) = =
    (((Val_(inA)) & not(Val_(inB)) & Val_(Cin)) or
    (not(Val_(inA)) & (Val_(inB)) & Val_(Cin)) or
    ((Val_(inA)) & (Val_(inB)))
)
; Carry_(inA, inB, Cin, Cout) = =
(
Cout = bool_to_bit(bool(
    Carry_Expr_(inA, inB, Cin, Cout)))
)
; Compute_(inA, inB, Cin, Sum, Cout) = =
    (Sum_ (inA, inB, Cin, Sum)
    & Carry_(inA, inB, Cin, Cout))
...

OPERATIONS
in_AA(yy) =
PRE
    yy : BIT
THEN
    AA, COUT, SS
    :(
    AA : BIT & AA = yy
    & COUT : BIT
    & SS : BIT
    & Compute_(yy,BB,CIN,SS, COUT)
)
END;
...
END

It is assumed that the behavioral models of each component are provided elsewhere, that is, there are entity-architecture pairs describing a half-adder and a two-input or gate.

IMPLEMENTATION
    B_Full_ADD_1bit_n
REFINES
    B_Full_ADD_1bit_0
IMPORTS
    add1.B_Half_ADD_1bit_0
    ,add2.B_Half_ADD_1bit_0
, Or.B_Or_0
...

INVARIANT
    AA = add1.AA
&   BB = add1.BB
&   add1.SS = add2.AA
&   CIN = add2.BB
&   add1.COUT = Or.in1
&   add2.COUT = Or.in2
&   COUT = Or.out
&   SS = add2.SS
...

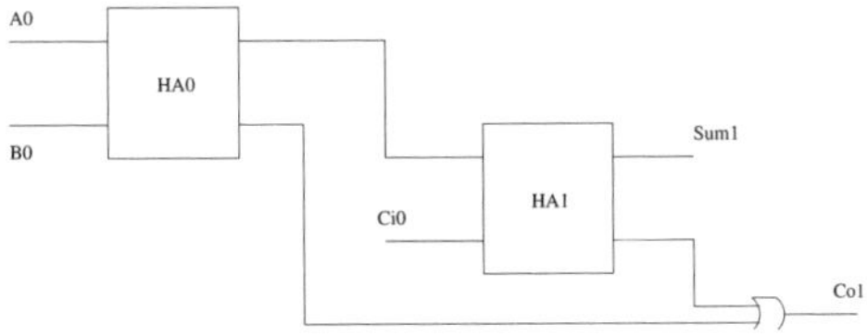

Figure 3.   The full adder implementation.

OPERATIONS
in_AA(yy) =
VAR xx IN
    AA := yy
;   add1.in_AA(AA)
;   xx <- add1.out_SS
;   add2.in_AA(xx)
;   xx <- add1.out_COUT
;   OU1.In_1(xx)
;   xx <- add2.out_COUT
;   OU1.In_2(xx)
;   COUT <- Or.Out
;   SS <- add2.out_SS
END;
...
END

In figure 3 you can see the internal structure of the full adde is shown. Two half adders (HA) and an OR gate are required to implement a full adder.

MACHINE
    B_Half_ADD_1bit_0
SEES
IEEE.B_STD_LOGIC_1164_0
, B_STD_ENTRIE
DEFINITIONS
Val_(xx) = = (bit_to_bool(xx) = TRUE)
; Compute_(inA, inB, Som, Cout) = =
    (Som = bool_to_bit(
    bool ((not(Val_(inA)) & (Val_(inB)))or
        ((Val_(inA)) & not(Val_(inB)))))
&Cout = bool_to_bit(bool (Val_(inA) & Val_(inB)))
)

CONCRETE_VARIABLES
    AA, BB, COUT, SS
INVARIANT
    AA : BIT
& BB : BIT
& COUT : BIT
& SS : BIT
& Compute_(AA, BB, SS, COUT)
...
OPERATIONS
...
in_BB(yy) =

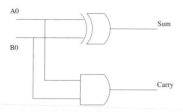

Figure 4. The half adder implementation.

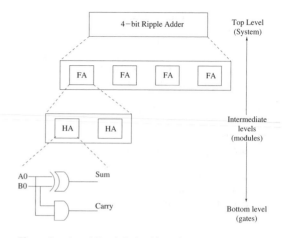

Figure 5. A multilevel design hierarchy.

```
PRE
  yy : BIT
THEN
  BB, COUT, SS
  :(
    BB : BIT
    &COUT : BIT
    &SS : BIT
    &BB = yy
    &Compute_(AA,BB,SS,COUT)
  )
END;
…
END
```

The last piece of the design is the Half_Adder entity (see figure 4), which is instantiated twice in each Full_Adder entity. A Half Adder is the simplest form of an adder circuit. The more significant sum bit is called carry-out (*carry*) because it carries an overflow to the next higher bit position. It has two operand bits $A0$ and $B0$ (for bit position 0) that are added to form a sum bit (*sum*) and a carry bit (*carry*), we can write $sum = A0\ XOR\ B0$ and $carry = A0\ AND\ B0$.

```
IMPLEMENTATION
  B_Half_ADD_1bit_n
REFINES
  B_Half_ADD_1bit_0
IMPORTS
  xor.B_XOR
  , and.B_And_0
SEES
  BASIC_IO
  ,IEEE.B_STD_LOGIC_1164_0
  ,B_STD_ENTRIE
INVARIANT
      AA = xor.in1
&     BB = xor.in2
&     SS = xor.out
&     AA = and.in1
&     BB = and.in2
&     COUT = and.out
…
```

```
OPERATIONS
…
in_BB(yy) =
BEGIN
  BB := yy
; xor.In_2(BB)
; SS <− xor.Out
; and.In_2(BB)
; COUT <− and.Out
END;
xx <− out_COUT =
  xx := COUT;
…
END
```

A multilevel design hierarchy detailing components of a full adder is used in VHDL designs. The structure of the *B* development is the same. The design method of starting with the topmost level and adding new levels of increasing detail is called top-down design. This method may be used for any logical circuit, for the complex ones as well as for the simple ones. We build a library that contains all standard gates (OR, AND, NOT,..), structure which is based on the method described below. Using the B tool (Atelier B 3.5) the coherence of theses B abstract machines is proved.

2.3 *VHDL in B*

The general method described in this section provides the capabilities to define some *B* abstract machines that modeling the behavior of some circuit. But we want modeling some realistic VHDL component and we need some *B* abstract machines that correspond to VHDL standard library such STD_LOGIC_1164. (Boulanger et al. 2002) introduces the STD_LOGIC_1164 library and its equivalent in *B*.

226

In (Boulanger and Mariano 2002) and (Boulanger et al. 2002), we present the *B* counterpart of the VHDL STD_LOGIC_1164 library mostly consists in three machines. The first machine, named B_STD_LOGIC_1164_0, contains all the definitions of types and the operations which concern the extended bit. The second machine, B_STD_LOGIC_1164_VECTOR_0, depends on the first to define the vectors and the corresponding operations. We also created a machine, B_Signal_0, to process the signals.

3 CONCLUSIONS

Our method cover the two approaches distinguished in hardware verification : verifying the equivalence of two implementations (by refinement) and property verification. The most important characteristic of the method is that it produces a secure circuit. In fact, introduction and verification of properties provides the capabilities to demonstrate the compliance between the informal specification and the formal description. This *internal* verification, based on proof, significantly contributes to fault avoidance.

When the circuit is designed in *B*, we can prove that the circuit which are obtained in the implementation satisfies 100% the specifications in the abstract machine. It is possible to add the required safety conditions under the *INVARIANT* clause in the abstract machine. So the circuit design is completely correct. The error may occur only when we describe the specifications or if the physical circuit does not correspond to the proposed model. With this coupling from VHDL to the *B* method, it is possible to represent the circuits as near as it is needed to the physical level. In (Boulanger and Mariano 1999) authors refined the abstract machine that modelise the NOT gate based themselves on abstracts machines of CMOS transistors.So the designer may use all the circuits which are designed before, he can make some changes and then prove a small part related to the new characteristics. It is not necessary to reprove all the old ones. The cost and the time of the test are won. A circuit may be easily improved and it may be integrated with the other elements in the environment to satisfy safety conditions.

However, there is still the problem of the *external* verification. External verification aims at tracking down faults originating from a misunderstanding of the requirements, or from the failure to express adequately an understood requirement.

REFERENCES

1076–1993, I. S. (1993). *Standard VHDL Reference Manual.* IEEE.

Abrial, J.-R. (1992, 7–11 Sept.). On constructing large software systems. In *Algorithms, Software, Architecture. Information Processing 92. IFIP 12th World Computer Congress*, Volume A-12, pp. 103–12.

Abrial, J.-R. (1996, August). *The B Book – Assigning Programs to Meanings.* Cambridge University Press.

Airiau, R., J.-M. Bergé, V. Olive, and J. Rouillard (1998). *VHDL – Langage, Modélisation, Synthèse.* Collection Technique et scientifique des télécommunications. Presses Polytechniques et universitaires romandes.

Boulanger, J.-L., A. Aljer, and G. Mariano (2002). Formalization of digital circuits using the *B* method. In *CompRail VIII, Eighth International Conference on Computer Aided Design, Manufacture and Operation in the Railway and Other Advanced Mass Transit Systems, Lemnos, Greece.*

Boulanger, J.-L. and G. Mariano (1999, Octobre). Modélisation formelle de circuits numériques par la méthode B. Technical Report TR 99–25, INRETS-ESTAS.

Boulanger, J.-L. and G. Mariano (2002, May 21st–24th). Formalization of digital circuits using the *B* method. In *AFIS, EUSEC 3rd European Systems Engineering Conference Systems Engineering: a focus of European expertise, session 10 Modelling & Tool, Toulouse (France).*, pp. 281–290.

Boulanger, J.-L., G. Mariano, and A. Aljer (2001, 4–6 décembre). Conception sûre de circuit basée sur la notion de propriété. In *14èmes Journées Internationales GÉNIE LOGICIEL & INGÉNIERIE DE SYSTÈMES et leurs APPLICATIONS. SESSION 8: MÉTHODES FORMELLES.* ICSSEA'01.

Jones, C.B. (1990). *Systematic Software Development Using VDM* (Second ed.). Englewood Cliffs, New Jersey: Prentice-Hall International. ISBN 0-13-880733-7.

Morgan, C. (1990). *Deriving programs from specifications.* Prentice Hall International.

Spivey, J.M. (1992). *The Z Notation: A Reference Manual* (2nd ed.). Prentice Hall International Series in Computer Science.

Safety and Reliability – Bedford & van Gelder (eds)
© 2003 Swets & Zeitlinger, Lisse, ISBN 90 5809 551 7

The simulation of emergency dynamic scenarios through HARIA-2, a tool for the planning and analysis of technological emergencies

R. Bovalini, A. De Varti & M. Mazzini
University of Pisa, Via Diotisalvi, Italy

ABSTRACT: The paper begins by introducing HARIA-2, a methodology setup for external emergency planning and analysis, which takes into account the physical phenomena and the sociological aspects of technological emergencies.

The paper then focuses on the module *Dynamic Scenarios*, which addresses emergency planning issues. The model includes road traffic and control of ingress and exit models, with the aim of minimizing population exposure. It integrates in simplified mode the data bases needed for simulating accident scenarios, including population behaviour and evacuation plan.

The evacuation of people at risk and the intervention on the accident scene of the rescue teams are the most intriguing countermeasure to be undertaken during an emergency. Given the road network, described in terms of arcs and nodes, data on the maximum allowed speed on one-way roads, etc., the traffic model calculates for each arc the time evolution of density and flux of vehicles. This model uses the continuous fluid-dynamic analogy with the Greenshield correlation between speed and density of vehicles.

To outline the main functions of the *Dynamic Scenarios* module, an example of application to an Italian industrial site is briefly described.

1   INTRODUCTION

HARIA-2 (Handling Algorithms for Risk evaluation in Industrial Activities) is a research project aiming at the development of integrated software for external emergency planning, analysis and response. The National Research Council (CNR), under the sponsorship of the Civil Protection Department, supports the research. In Italy, emergency planning and management is a responsibility of the public Administration in charge of civil protection, in collaboration with the companies owning the plants at risk. The intent of HARIA-2 project is primarily to assist the implementation of this important public responsibility. In particular, HARIA-2 organizes the information needed for the preparation of the emergency plan, including the simulation of the physical evolution of an accident scenario. The main concerns are the social dimensions of the emergency and the complexity arising from the interaction of physical, technical and organizational factors Mazzini, Contini, Volta (2001).

A technological emergency is a dynamic process that can last some hours or days, as a result of the interaction of various physical and social factors. The main issues addressed by the project are the modeling and the simulation of the three main dynamic systems interacting during a major accident at hazardous chemical and petrochemical sites Mazzini, Contini, Volta (2001), Mazzini (2001):

– Physical System: the plant, as source of the hazard and the accidents that could impact the surrounding environment.
– Territorial and Social System: population and other targets which can be affected by the accident.
– Civil Protection System: rescue services and resources available to mitigate the consequences of the accident.

The output of the project is a set of models and tools for the analysis and the optimization of emergency planning, including the evacuation and rescue phases. However the software can be useful also for training of rescue services, in such a way to test the efficiency and the effectiveness of various rescue strategies.

Research is in progress on three parallel lines strictly integrated:

– development of a Demonstrator system, a relatively simplified tool, which does not need particular software licenses;

- development of a Prototype system using Arc View® GIS platform, for applications to risk areas;
- technological transfer, in order to have a verified system that can be used by different Italian Bodies.

The project is carried out by a multi-disciplinary team, in such a way as to consider adequately not only the physical phenomena, but also the sociological aspects of the emergency, i.e. the management organization and the population behaviour in the surrounding area Mazzini, Contini, Volta (2001), Mazzini (2001), Bellezza, Christou, Contini, Kirchsteiger (2000), Barlettani, Mazzini, Scarselli (2000).

The HARIA-2 package is designed to be used in an emergency that covers an area of up to $20 \times 20\,\text{km}^2$, around the industrial complex. In many cases, the area affected by the physical consequences of an industrial accident is considerably smaller (a few km^2). However, the social impact can easily extend beyond these limits; thus, the boundaries for the accident model need to be chosen with adequate margins.

2 FUNCTIONAL MODULES OF HARIA-2 SYSTEM

Figure 1 shows the set of functional modules that represent the product of the overall HARIA-2 project,

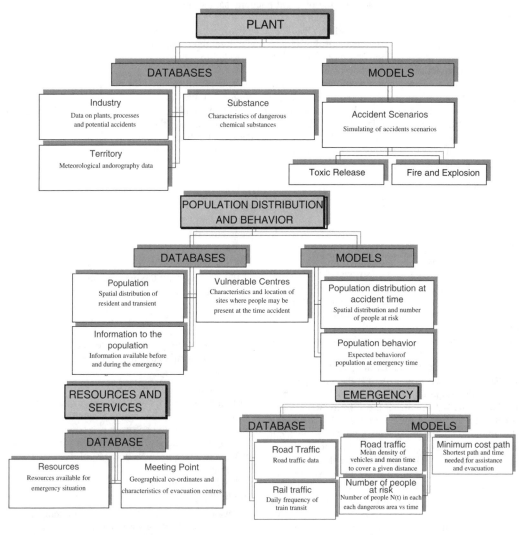

Figure 1. Functional Modules of HARIA-2 System.

grouped according to the three main systems interacting during an industrial accident

The first block of Figure 1 comprises the data bases and the chain of models needed to represent the accident sequence. The chain allows the simulation of set accidental sequences:

- physical explosions, pool and jet fires are considered as potential initial events of domino sequences or a source of toxic clouds;
- in case of toxic or flammable releases, the models allow the treatment of heavy gas dispersion in the atmosphere with subsequent neutral cloud dispersion taking into account terrain slope and obstacles in the diffusion process Wuertz (1998). This chain has been tested on a number of experimental cases and on the real case of the Mexico City accident Bovalini, Gabbrielli, Mazzini, Petea (2000).

The set of models is complemented by a "quick-and-dirty" method ("Metodo Speditivo"), recommended for preliminary analysis by the Italian Authorities.

The second block (Fig. 1) is devoted to the treatment of the social system. The area is subdivided into polygons with a characteristic population density, distributed over the area and concentrated in vulnerable centers. The corresponding values are recorded in the database as a function of day hour, week day, season, etc.

People are classified into four categories: residents, non-resident workers, commuters and tourists. For each category, the mean flow from one polygon to the others is defined in input by the user. The population behaviour model needs other data such as capability of autonomous movement, degree of education, accident time and status of information prior and during the emergency. The output provides an estimate of the percentage of people in each of the following four classes of behaviour: in compliance with the instructions (guided), independent not in compliance, panic/random, static (as though nothing has happened).

An updated inventory of the availability of rescue resources is an essential part of any emergency plan (third block in Fig. 1). In this context, HARIA-2 may be used like a simple control tool. It will be able to give valuable support to the systematic check-up of the completeness and coherency of existing emergency plans.

The fourth block in Figure 1 allows the simulation of various emergency measures and is discussed with some detail in the next chapter.

3 THE MODULE "DYNAMIC SCENARIOS"

The following discussion refers mainly to the case of a toxic release, the only one in which an emergency

plan, prepared in advance and accurately tested, can reduce significantly the consequences of major industrial accidents. In the other cases, the only measure to be planned is the rescue action. HARIA-2 can be used also to this aim, although it is not finalized yet.

The emergency measures which are possible to simulate by HARIA-2 are:

a. Control of access and exit from the area at risk;
b. Emergency shelter;
c. Evacuation.

To do this, other than the data briefly indicated in the previous chapter, the Territorial Data Base comprises the description of the road network, usually available under Arc View®.

The first step of the dynamic simulation of the emergency is the definition of the risk area (usually the area between the contours corresponding to the dose limits IDLH and DL50). This area is subdivided by the user in at least two zones, one upwind and the other downwind the release source. The user can divide the area in more zones, i.e. subdividing the downwind zone in two parts, separated by the semi-rect exiting from the release source position and with the direction of the wind.

At the second step, the user chooses in each zone the main escape road, with the safe meeting point as node 0 and node 1 at the end of this road inside or at the border of the considered risk zone (Fig. 2). Then it chooses the principal roads, connected to the main one by nodes of order 2, 3, 4 or more (where the order means the number of roads interconnected by the considered node), as well as the roads of second and third order (Fig. 2).

A node of order 2 means that two roads have this node in common, one as its end and the other (which can have equal or different characteristics, like allowed speed, jam density, etc.) as initial point. A node of order 3 means that in it two roads supply cars to the escaping road and so on.

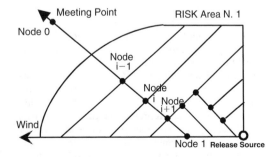

Figure 2. Scheme of a road network construction.

All the roads are assumed one way. Two parallel roads model a road with two ways, each one way (as a highway).

The construction of the simulation road network may be done by the user directly on ARC-VIEW®.

To have a road network that is not too complicated, secondary roads are simulated by generation rates which give traffic input to the road network specified by the user (number of cars, delay between two successive cars, etc.). The road network is initially assumed in conditions of equilibrium corresponding to normal traffic, with the usual traffic lights.

At the time assumed as time zero of the emergency declaration, the first measure put into operation is the control of accesses and exits from the risk zone: the rescue teams take control of all the nodes of the road network at the contour of risk zone, according to a time table specified by the emergency Authority. If this is the only emergency measure applied, no entry in the risk zone is allowed, while inside the zone the traffic continues as before.

Usually emergency planning foresees at least the actuation of the second measure: the emergency shelter. This mitigates the consequences to the population with guided behaviour, reducing the doses according to a simple exponential model.

Those with static behaviour take full doses, as though nothing were done. The other two groups of population (autonomous and random) try to escape and modify the traffic inside the risk zone.

Finally, if the third emergency measure (evacuation) is actuated, HARIA-2 foresees that rescue teams control the nodes of the road network inside the risk zone according to a pre-planned timetable. The road network becomes like a tree (Fig. 2) with only one arm open at each time. The normal traffic lights system no longer operates and rescue teams control the opening of the various routes, according to the situation in the field.

The model uses the continuous fluid-dynamic analogy, with Greenshields' correlation between speed and density of vehicles Barlettani, Mazzini, Scarselli (2000).

The above discussion refers to the population distributed in the territory and in uncontrolled vulnerable centers (supermarkets, churches, campaings, etc.). The population inside controlled vulnerable centers (prisons, hospitals, etc.) is assumed to have a guided behaviour. Therefore the emergency shelter is always applied to it. In case of evacuation, the model aims at minimising population exposure to dangerous substances by a "minimum cost" path algorithm, where the "cost" can represent the travel time or the dose likely absorbed by people. To this end an allocation model is implemented that establishes the optimal assignment of resources (Fig. 3), in relation to the paths from their location to the vulnerable centers and

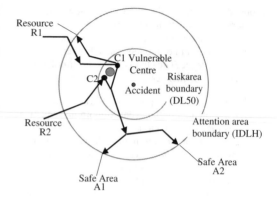

Figure 3. Minimum cost path model for the evacuation times calculation.

from these to destinations Bellezza, Binda, Contini, Giardina (2000).

4 SOFTWARE IMPLEMENTATION AND APPLICATIONS

To obtain the congruence of data and their control in the input phase, HARIA-2 uses a system of interfaces. The output interface, being devoted to operators and decision makers, needs to be particularly user friendly. In this connection, Arc View® commercial GIS software was chosen, because it is the most widely used by the Italian Administrations.

As already stated, HARIA-2 is divided in two packages: Demonstrator and Prototype.

HARIA-2 Demonstrator integrates the following main modules:

- a database on plants, components, hazardous substances, population, territory, etc.;
- a module Static Scenarios, which simulates the accident evolution and its consequences in absence of countermeasures;
- a module Dynamic Scenarios, which simulates the intervention of rescue teams and population behaviour;
- a simplified GIS, which allows the handling of the information related to a specific territory and the representation of the output of the modules Static and Dynamic Scenarios.

HARIA-2 Demonstrator has been partially validated through the simulation and the analysis of the Mexico City accident, Bovalini, Gabbrielli, Mazzini, Petea (2000), Bovalini, De Varti, Mazzini, Petea (2002).

HARIA-GIS Prototype Bellezza, Binda, Contini, Giardina (2000) implements an architecture which is

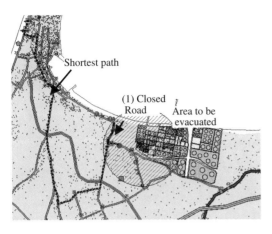

Figure 4. Map of Milazzo (Italy) area with paths for transport means and vulnerable centers.

open to various upgraded solutions, like connection to Internet and to additional off-line modules. The software has been tested on the site of Milazzo (Italy), which hosts an important refinery plant. Figure 4 is a screen shot of HARIA-GIS Prototype, showing the map of the area, the zone with vulnerable centers to be evacuated, the road network available to the rescue means and the alternative paths in case of interruption of the main road. The system has been applied to the assessment of the evacuation time of the vulnerable centers under different hypotheses about the accident time, the availability of resources and the road network. The simulation has been carried out following these steps:

– plot of the area of interest for evacuation;
– identification of populated vulnerable centers within the area and calculation of the total number of people to be evacuated;
– analysis of the transportation means available and of their readyness;
– calculation of the shortest path for the arrival of rescue services at the accident scene and of transportation means at the meeting points.

The last step was retraced after considering the impracticability of one road ((1) in Figure 4). This test provided an evacuation time ranging from 50 to 70 min, depending on readiness of transportation means. If the road (1) in Figure 4 is closed to traffic, the time increases by 15% Bellezza, Binda, Contini, Giardina (2000).

As second example, we applied HARIA-2 Demonstrator to the evacuation of the city of Milazzo. In this case the evacuation is purely hypothetical as no accident scenario implies the evacuation of the city. Figures 5 and 6 show the map of the city with the actual road network and the scheme of this network

Figure 5. View of Milazzo (Italy) area, with the network of main roads.

considered in HARIA-2 evacuation model. The population considered accounts for about fifteen thousands people. According to data available for a similar industrial site we assumed that about 60% of people have static behaviour, 22% guided, 17% random and 1% autonomous. Because all the road network nodes are controlled by rescue teams, the emergency plan foresees that all moving people assume a guided behaviour.

The model assumes that people evacuate with their own cars (3 persons for each car, as an average) and that the speed limits are 50 km/h for urban roads and 90 km/h for extra-urban roads. With these assumptions, the model calculates the position of the population at each time step. The territory is subdivided in cells of 100×100 m. In each cell the concentration of toxic substance is assumed to be uniform and equal to the one calculated at the cell center. The integral of the concentration (eventually elevated to an exponent n, in accordance with the vulnerability model assumed for the considered toxic substance) over the whole area and for all the time during which the release

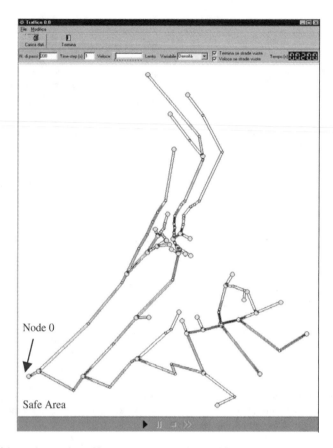

Figure 6.   Scheme of the road network used by HARIA-2 evacuation model.

occurs, allows to determine the consequences of the accident with such evacuation. The results can be compared with ones related to the other emergency measures (control of access and shelter). This allows the emergency Authority to judge the efficiency of the options available.

According to the assumptions made above, the time of evacuation calculated by HARIA-2 is about 45 minutes, starting from the time in which all nodes in Figure 6 are under the control of rescue teams.

5   CONCLUSIONS

Research on HARIA-2 is almost completed. Important results have already been achieved through significant applications, such as the Mexico City disaster analysis. The applications to the industrial areas of Rosignano Solvay and Milazzo are in progress but results so far obtained are already of interest. However, the evacuation of people at risk and the intervention on the accident scene of the rescue teams are the most

intriguing countermeasures during an emergency: in most of the accident scenarios the only measure to be applied is the assistance to people by the rescue teams. Infact the time available for emergency action is very small, much lower of that needed not only for evacuation, but also for access control. Besides, to perform the test indicated in previous chapter a lot of assumptions have to be made. In this respect, it should be also noted that HARIA-2 software is not fully validated because the lack of an exhaustive application to a real case. This can be done only with the active collaboration of the area Authorities in charge of the emergency planning. Only such a collaboration allows the collection of the whole set of data necessary for a good simulation (as an example, traffic conditions during the day and in the various season, people present on the area during the various hours of the day, and so on).

Another task to be completed is the functional linking of the module "Static and Dynamic Scenarios" to the HARIA-GIS software in development at Ispra JRC.

The final conclusion however is that because HARIA-2 takes into account both physical phenomena and sociological aspects, it constitutes a good advancement in emergency planning.

ACKNOWLEDGMENTS

The authors thank the various contributors to the realization of the project: M. Barlettani, F. Bellezza, M. Binda, S. Contini, G. Giardina, S. Gabbrielli, L. Pellizzoni, M. Petea, M. Scarselli, G. Sica, D. Ungaro, F. Verri, G. Volta, J. Wuertz, F. Zani.

The work was carried out with the financial support of CNR (National Group for Defense from Industrial and Ecological Risks), under the sponsorship of the Italian Civil Protection Department.

REFERENCES

Mazzini M., Contini S., Volta G. 2001. HARIA-2. a computer supported approach to emergency planning, analysis and response. *J Risk Decision and Policy*, Vol 7, p. 131.

Mazzini M. 2001. An informatic system for planning and analysis of technological emergencies. *Workshop: "Risk Management – Defining the Needs for the Future"*, Vol. 1, Ascona (CH).

Bellezza F., Christou D., Contini S., Kirchsteiger C. 2000. The Use of Geographic Information System in Major Accident Risk Assessment and Management. *Journal of Hazardous Materials 78, 223–245.*

Bovalini R., Mazzini M., Petea M., De Varti A. 2002. *Developments in HARIA-2 Research Project: a computer supported approach to emergency planning analysis and response.* Conference PSAM 6, San Juan di Portorico.

Wuertz J. 1998. *Modellistica HARIA-2 per la dispersione dei gas pesanti e relativa convalida.* Conference VGR' 98, Pisa, Italy.

Bovalini R., Gabbrielli S., Mazzini M., Petea M. 2000. *HARIA-2: una metodologia per l'analisi e la pianificazione delle emergenze tecnologiche. Validazione del sistema informatico in base all'analisi dell'incidente di Città del Mexico.* Conference VGR'2K, Pisa, Italy.

Barlettani M., Mazzini M., Scarselli M., 2000. *HARIA 2: una metodologia per la pianificazione e l'analisi di emergenze tecnologiche. Sviluppo di un modello matematico per la simulazione del piano di evacuazione.* Conference VGR'2K, Pisa, Italy.

Bovalini R., De Varti A., Mazzini M., Petea M. 2002. *HARIA-2: una meteodologia per l'analisi e la pianificazione delle emergenze tecnologiche. Sviluppo della modellistica per la simulazione degli scenari incidentali.* VGR2002, Pisa, Italy.

Bellezza F., Binda M., Contini S., Giardina G. 2000. *Il sistema proptipo HARIA-GIS per la pianificazione delle emergenze,* Conference VGR'2K, Pisa, Italy.

Safety and Reliability – Bedford & van Gelder (eds)
© 2003 Swets & Zeitlinger, Lisse, ISBN 90 5809 551 7

ESACS: an integrated methodology for design and safety analysis of complex systems

M. Bozzano[1], A. Villafiorita[1], O. Åkerlund[2], P. Bieber[3], C. Bougnol[4], E. Böde[5], M. Bretschneider[6],
A. Cavallo[7], C. Castel[3], M. Cifaldi[8], A. Cimatti[1], A. Griffault[9], C. Kehren[3], B. Lawrence[10], A. Lüdtke[11],
S. Metge[4], C. Papadopoulos[10], R. Passarello[8], T. Peikenkamp[5], P. Persson[12], C. Seguin[3], L. Trotta[7],
L. Valacca[8] & G. Zacco[1]

[1] *IRST, Trento, Italy*

[2] *Prover, Technology AB, Stockholm, Sweden*

[3] *ONERA, Centre de Toulouse, Toulouse, France*

[4] *AIRBUS, Blagnac, France*

[5] *OFFIS, Oldenburg, Germany*

[6] *Airbus, Deutschland GmbH, Hamburg, Germany*

[7] *Alenia Aeronautica S.p.A, Caselle (TO), Italy*

[8] *Società Italiana Avionica, Strada Antica di Collegno, Torino, Italy*

[9] *LaBri, Université de Bordeaux, Talence, France*

[10] *Airbus UK Ltd., Filton, Bristol, UK*

[11] *University of Oldenburg, Oldenburg, Germany*

[12] *Saab AB, Linköping, Sweden*

ABSTRACT: The continuous increase of system complexity – stimulated by the higher complexity of the functionality provided by software-based embedded controllers and by the huge improvement in the computational power of hardware – requires a corresponding increase in the capability of design and safety engineers to maintain adequate safety and reliability levels. Emerging techniques, like formal methods, have the potential of dealing with the growing complexity of such systems and are increasingly being used for the development of critical systems (e.g. aircraft systems, nuclear plants, railways systems), where at stake are not only delays in delivering products and economical losses, but also environmental hazards and public confidence. However, the use of formal methods during certain critical system development phases, e.g. safety analysis, is still at an early stage. In this paper we propose a new methodology, based on these novel techniques and supported by commercial and state-of-the-art tools, whose goal is to improve the safety analysis practices carried out during the development and certification of complex systems. The key ingredient of our methodology is the use of formal methods during both system development and safety analysis. This allows for a tighter integration of safety assessment and system development activities, fast system prototyping, automated safety assessment since the early stages of development, and tool-supported verification and validation.

1 INTRODUCTION

The dramatic improvement in the computational power of hardware brings about increasingly sophisticated software-based embedded controllers that take over complex functionality in an efficient, precise, and flexible way. These benefits allow the use of such systems in environmental conditions where delays in delivering products and/or economical losses are not the only things at stake, but also environmental hazards and public confidence.

This development, however, entails an unavoidable increase in the complexity of systems, which is expected to continue in the future. Therefore, in order to retain the benefits from more sophisticated controllers, a corresponding increase in the capability of the design and safety engineers to maintain adequate safety and reliability levels is required.

One of the most challenging issues in system development today is to take into consideration, during development, all possible failures modes of a system and to ensure safe operation of a system under all conditions. Current informal methodologies, like manual fault tree analysis (FTA) and failure mode and effect analysis (FMEA) (Vasely, 1981; Liggesmeyer & Rothfelder, 1998; Rae 2000), that rely on the ability of the safety engineer to understand and to foresee the system behavior are not ideal when dealing with highly complex systems, due to the difficulty in understanding the system under development and in anticipating all its possible behaviors.

Emerging techniques like formal methods (Wing, 1990) have the potential of dealing with such a complexity and are increasingly being used for the development of critical systems (see, for instance, (Hinchey & Bowen, 1999)). Formal methods allow a more thorough verification of the system's correctness with respect to the requirements, by using automated procedures. However, the use of formal methods for safety analysis purposes is still at an early stage. Moreover, even when formal methods are applied during system development, the information linking the design and the safety assessment phases is often carried out by means of informal specifications. The link between design and safety analysis may be seen as an "over the wall process" (Fenelon et al., 1994).

A solution to the issues mentioned above is to perform the safety assessment analysis in some automated way, directly from a formal system model originating from the design and safety engineer.

This approach is being developed and investigated within the ESACS project (Enhanced Safety Assessment for Complex Systems), an European Union sponsored project in the area of safety analysis, involving several research institutions and leading companies in the fields of avionics and aerospace. The methodology developed within the ESACS project is supported by state-of-the-art and commercial tools for system modeling and traditional safety analysis tools and is being trialed on a set of industrial case studies.

Outline of the paper. This paper is structured as follows. In the next section we present the ESACS approach and illustrate its use through a simple example. In section 3 we present the architecture of the ESACS platform. Finally, in section 4 we draw some conclusions and discuss related work.

2 ESACS METHODOLOGY

2.1 *Methodology*

The ESACS methodology aims to:

- Support system development and safety analyses processes.
- Provide a tight link between design and safety analysis.
- Support automated verification and validation of the design.
- Support automated safety assessment of the design.

The goals mentioned above are being achieved through two basic ingredients: firstly, the use of formal notations both for design and for safety assessment of systems and secondly, the extension of state-of-the-art tools, to provide users with a set of basic functionality that can be combined in different ways.

The use of formal methods allows for automation of analyses and for a tighter integration between system design and safety analysis (as the information exchanged between design and safety engineers is based on the same formal models, with a clear and unambiguous syntax and semantics). Tool support, achieved by implementing a set of basic functionality in state-of-the-art tools, allows for automated support of the methodology and for its flexible integration in various development and safety processes.

The ESACS methodology takes into account and supports two main scenarios. In the first scenario the process is initiated by the design engineer, who provides a formal model of the design (at a given level of abstraction) using formal modeling tools (System Model Definition). The model is then automatically enriched with failures in order to perform safety analyses (Failure Mode Definition and Failure Mode Injection). The second scenario is useful when safety assessment activities must be carried out when there is still no formal design model to start with, e.g. to evaluate a proposed system architecture. In this case, the process is initiated by the safety engineer, who builds a high-level model of the system using a library of components enriched with failure modes. Such model can then be used to perform safety assessment on the system (System Model Prototyping for Safety Assessment).

The scenarios sketched above entail for the definition of the following, tool-supported, basic functionality:

- **System Model Definition:** definition, using formal notations, of an executable specification (at a given level of abstraction) of the model of the system under development (what we call design model).
- **System Model Prototyping for Safety Assessment:** definition, using a library of pre-defined components, of a model suitable for safety assessment. This functionality allows performing safety analyses in the early phases of the development process.
- **Failure Mode Definition and Failure Mode Injection:** definition of the failure modes of the components constituting the design model. The failure modes can be automatically injected into the design model, in order to produce what we call an *extended system model* (ESM). The ESM is an executable specification of the design model in which components can fail according to the failure mode specification. Extended system models can be used by safety engineers to perform safety analyses.
- **Functional and Safety Requirements Definition:** definition of the functional requirements and of the

238

safety requirements of the system, using a formal language (e.g. linear temporal logic and computation tree logic (Emerson, 1990)).

– **Design Assessment:** automated assessment of the design model against the functional requirements, using standard formal methods techniques (e.g. simulation, theorem proving (Boyer & Moore, 1979), and model checking (Clarke et al., 1999)).

– **Safety Assessment:** automated safety assessment of the extended system model against the safety requirements, using automated techniques based on formal methods for automated fault tree analysis and automated failure mode and effect analysis.

Note that the basic functionality provided by the platform, which we have described, can be combined in different ways, in order to comply with any given development methodology one has in mind. For instance, it is possible to support an incremental approach, based on iterative releases of a given system model at different levels of detail (e.g. model refinement, addition of further failure modes and/or safety requirements). Furthermore, it is possible to have iterations in the execution of the different phases (design and safety assessment), e.g. it is possible to let the model refinement process be driven by the safety assessment phase outcome (e.g. disclosure of system flaws requires fixing the physical system and/or correcting the formal model).

2.2 A simple example

To illustrate the basic concepts of the methodology we will show its application to the design of a controller that regulates the level of a fluid in a tank.

The goal of the controller is computing the capacity of two pumps that regulate the inflow and the outflow of a liquid contained in a tank, so that the tank level remains between the two "Activation levels" in the middle of the tank. At the same time, the controller services a request, coming from the user of the system, for a minimum outflow capacity. If the tank level, obtained through a dedicated sensor, becomes either too low or too high, the controller must issue an alarm, and close the input or the output emergency valves. The user request for a minimum outflow is the only "external" input to the system. The controller monitors the tank level through a sensor providing the actual tank level. Figure 1 illustrates the components in the system.

In the following we focus on one of the two scenarios supported by the methodology, namely the scenario in which the model is built by the design engineer.

The first step of the methodology, therefore, is the definition of the design model of the system on which we will perform the automated analyses, using formal notations. Such step is performed using standard modeling tools: Figure 2 illustrates the high level

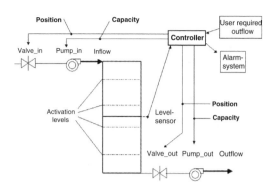

Figure1. A simplified tank controller.

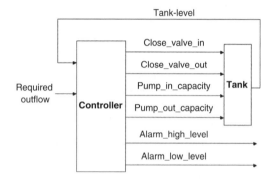

Figure 2. The model of the tank controller.

decomposition of the model highlighting the inputs/outputs of the system. The model is written in SCADE[1]. The model also includes a definition of the environment with which the controller interacts: this allows the verification of the whole system (i.e. controller + sensors + actuators + tank) to ensure that it behaves as expected.

The next steps of the methodology are the definition of the functional requirements of the system and the formal assessment, to verify and validate the system against the functional requirements, in its nominal working conditions. The goal of this activity is to acquire confidence on the behavior of the system when all the components work as expected. Thus, we assume that the environment is functioning correctly and verify that the controller satisfies certain requirements. For example, we assume that the pumps and valves are functioning according to controller demand and that the level indicator is sending correct information. We investigate properties like: (1) is it possible to have an overflow, (2) is it possible to have a drain, (3) can there be a false alarm or (4) can there be an overflow without

---
[1] See http://www.esteral-technologies.com

239

alarm, etc. The properties are formalized using standard logical formalisms and the verification of the nominal system behavior is performed automatically using theorem proving and model checking techniques. After this step we are confident that the design model behaves as expected in nominal situations.

This is, however, not sufficient to ensure that the system behaves as expected in all the possible situations. Having confirmed that all the requirements are fulfilled in the nominal case, it is then necessary to investigate if failures modes can make the system fail to meet the defined requirements. In order to do so, we must first identify and formalize the possible failure modes (FM) of the various components of the system. This is done in the failure mode definition phase, during which safety engineers identify and allocate all the possible failures of the components. This is performed, in the ESACS methodology, by retrieving, from a library of pre-defined failure modes, the failure modes of the different components of the system and by allocating such failures to the various components. The allocation of the failure modes to the components (Failure Mode Injection) automatically transforms the design model into a new model, that we call an *extended system model*. The extended system model enriches the possible behaviors of the system taking into account all the situations in which some of the components may fail according to the specifications provided by the safety engineers.

In the tank example, for instance, we can assume that the input pump (*Pump_in*, in Figure 1) can fail in two ways: (1) no flow can be produced or (2) the pump delivers full capacity all the time. In this case the model of the pump (whose output is modeled by the variable *Pump_in_capacity*) becomes, in the extended system model:

> if (Pump_in_status = stuck_zero) then
>   Pump_in_capacity = 0
> elseif (Pump_in_status = stuck_max_flow) then
>   Pump_in_capacity = max_capacity
> else
>   Pump_in_capacity = nominal, (that is, the flow
>   required by the controller)

Notice that the occurrence of failure is revealed by the variable *Pump_in_status* that is automatically generated during failure injection. So, if no failures occur, then the pump will react as required by the controller. However, if any of the failures occurs, then the output produced by the pump is determined by the failure. The model extension algorithm also takes care of defining the behaviour of the variables encoding the failures: it is for instance possible to express the condition that no two different failures can occur at the same time on the same component or that no more that four different failures can happen for a certain simulation, etc.

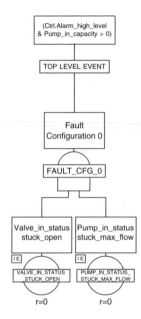

Figure 3. A simple fault tree, generated by the ESACS platform, of the safety requirement "it is not possible to have a high level alarm while the input valve is not closed".

Once we have an extended system model (either built using failure injection on the design model or by directly building a model with failures from the library of components), we can perform safety assessment on the model, either via failure mode and effect analysis (FMEA) or via fault tree analysis. This is done using algorithms based on model checking and theorem proving techniques for, e.g. automatically building fault trees. The algorithms supported by the ESACS methodology are similar to those described in (Liggesmeyer & Rothfelder, 1998; Coudert & Madre, 1993) and allow to analyze both monotonic and non-monotonic systems; the results produced by the analyses can be shown using commercial safety analysis tools. Figure 3, for instance, shows a simple fault tree automatically generated by ESACS algorithms and imported in FaulTree+.

3 ESACS PLATFORM

As illustrated in the previous section, an important aspect of the ESACS methodology is the support provided to the design and safety engineer by tools. This support is provided by the ESACS Platform, which has been defined and is being developed within the ESACS project.

240

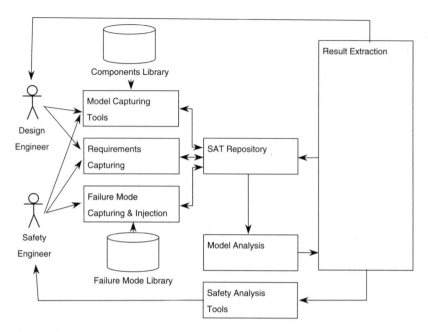

Figure 4.   The ESACS platform reference architecture.

The starting point for the development of the ESACS platform has been the tools used by the industrial partners for design and for performing safety assessment, that include (but are not limited to) Cecilia-OCAS, Statemate, SCADE, Simulink, FaultTree+. However, the task of providing a single platform that integrated all these tools and delivers to the user an interoperable environment, in which different tools and different modeling languages could be used interchangeably, was judged to be too risky for the time span and goals of the project. For instance, the problem of providing sound translators for the various input languages would have been a project in itself.

The approach taken in the ESACS project, thus, was to provide different configurations. Each configuration uses its own input language (tailored to the needs of an industrial partner) and its own set of support tools. All the configurations have been developed using the same underlying principles and the same architectural schema.

This has allowed the development and the delivery of the following four different configurations:

- **Altarica configuration**, based on the Cecilia-OCAS tool.
- **NuSMV configuration**, based on the NuSMV model checker.
- **SCADE configuration**, based on the on SCADE tool and on the PROVER plug-in.

- **Statemate configuration**, based on Statemate tool and on the VIS model checker.

3.1   ESACS architecture

The ESACS reference architecture is depicted in Figure 4. In particular, we distinguish the following blocks:

- **Model Capturing Tools,** the block used for defining the models for the analyses. It is used by the design engineer to define the design model and by the safety engineer to design models enriched with failures for safety assessment.
- **Components Library,** a library of components, extended with failures, which can be used for providing fast prototypes for safety assessment. It is used by the safety engineer to fast prototype systems on which to perform preliminary safety analysis.
- **Requirements Capturing,** the block for defining the functional and safety requirements. It is used both by design and safety engineers. The design engineer uses this block to define functional requirements against which to check the consistency of the design model. The safety engineer uses this block to define the safety requirements, against which to test the extended system model.
- **SAT Repository,** the central repository of the ESACS platform. The SAT (Safety Analysis Task)

241

repository stores all the information about a system, like, for instance, the analyses to be performed.

- **Failure Mode Capturing & Injection**, the block for defining failure modes and injecting them into the system model; the injection yields the extended system model. This block is used by the safety engineer to define what are the degraded behaviors of the system model and to generate the extended system model, on which to perform safety analyses.

- **Model Analysis**, the block for performing all the analyses both on the design model and on the extended system model. The ESACS Platform configurations support standard formal verification analyses (e.g. simulation, model checking, and theorem proving) and safety analyses (fault tree construction and FMEA table construction).

- **Result Extraction**, the block responsible for presenting the results produced during the analyses in formats understood by commercial tools (e.g. fault trees in the FaultTree+ format).

- **Safety Analysis Tools**, a block that comprehends various commercial safety analysis tools (e.g. FaultTree+), that are used to display results in a way that is familiar to safety engineers.

Some of the components of the ESACS platform (i.e. Model Capturing tools and Safety Analysis Tools) are based on standard commercial tools: therefore, we do not illustrate them here any further. The rest of this section, instead, focuses on the components that are particular to the ESACS Platform (namely, Components Library, Generic Failure Model Library, Failure Mode Capturing and Failure Injection, Model Analysis, and Result Extraction).

Components Library. The library of components gathers formal models of basic system components immediately suitable for safety analysis. Thus, each formal model describes both nominal and faulty behaviour of a specific system component. In the earliest phases of the system design, details of physical components are not fixed and safety engineers work with simple functional blocks and failure modes. Usually, a block offers a service (outputs) provided that some inputs and resources are available in the nominal case. Only permanent failures are considered and, after a failure, the block no longer provides an output. Some other blocks do not require inputs or resource to play their role (e.g. source of energy), or they do not provide any output (receptor). The library contains a first family of such generic blocks. It is responsibility of the safety engineer to select the appropriate failure modes to be assigned to the different components, among the available ones. During the "preliminary system safety assessment", the system components are better known and the safety engineers take into account more specific failure modes.

At this stage, the library contains two other families of components. One is dedicated to components of hydraulic systems (reservoir, pump, valve, pipe, ...) and deals with failure modes such as total loss, leak in a component, ... The second family of components is dedicated to components of electrical systems (generator, bus, switch, receptor, ...) and handles failures such as total loss, short-circuit, ... The library is currently written in the AltaRica language (Arnold et al., 2000). Basically, each AltaRica model consists of two parts. An automation describes which failure or nominal mode may be activated when a failure or a normal event occurs. Then, a set of logical assertions describes the relationship between the input/output of the components according to the current modes (see for instance (Bieber et al., 2002)). The library is implemented within the Cecilia-OCAS environment. The environment offers graphical facilities and a system model is built by drag and drop of icons of the components from the library panel to the graphical editor window. Then, the components are connected by drawing links between icons. A translator from Alta-Rica to Lustre has been developed [http://altarica.labri.fr/Tools/AltaLustre/] and will allow the ability to have similar libraries written in the Lustre language, usable in the SCADE environment.

Failure Mode Capturing and Failure Injection. In the design model of a system, the goal is to specify the nominal behaviour of the system fulfilling all the functionality. Suppose it is verified that this design, i.e. its nominal behaviour, fulfils all requirements. Also assume some of the inputs to and outputs from the design model represent signals coming from or going to "components", which can malfunction – i.e. there is a possibility that the nominal behaviour can be changed due to a component failure mode (FM), e.g. a valve may fail in a stuck position. In order to include FM in the analyses we extend all input and output signals, which can malfunction, with FM nodes. Having this extended system model, it is now possible to investigate if requirements are fulfilled when FMs are allowed.

FM are defined through the Failure mode capturing module that safety engineers use firstly to identify the components of the model that have to be enriched, secondly to specify the parameters of the failure, and thirdly to assign possible failures to the elements of the design. Failures are retrieved from a Generic Failure Mode Library. The Generic Failure Mode Library defines and specifies the kind of failures that can be injected into the system model: the library supports all the common failure modes, such as "stuck at", "inverted", "glitch" (the "glitch" failure allows for transient changes in the outputs delivered by a component).

Once defined, failure modes can be automatically injected into a design model via the Failure Injection module. Failure injection takes as input a design model coming from the design engineer, the failure modes defined by the safety engineers, and produces an extended system model, namely a model in which components may fail.

Only permanent failures are supported so far. The modeling of FM nodes is done such that the occurrence of a FM has priority over the nominal behaviour (see the example in Section 2).

Requirements Capturing. This module is used to define the functional requirements and the safety requirements of the extended system model. The requirements are written using standard logic formalisms (e.g. linear temporal logic and computation tree logic) using the facilities provided by the tools supporting the design of models.

SAT Management. The SAT is the central repository of all the information related to the (safety) assessment of a system. Within the SAT are stored references to design model, functional and safety requirements, references to the extended system model, and analyses task specifications, namely, the specification of the (safety) analyses that have been performed on the system.

In order to provide easy accessibility and portability, the SAT is stored in XML format (W3C, 2000). The XML tags provide the structure of the SAT and encode, through attributes, information on how the content of the tags shall be interpreted. In this way, all the different configurations share the same XML structure to encode the information and can share the same computing facilities (e.g. transformation into HTML).

The SAT is currently supported by the FSAP/NuSMV-SA configuration line, an extension of the NuSMV model checker (Cimatti et al., 2002). We conclude this section by discussing the model analysis and result extraction modules of the ESACS architecture.

3.2 ESACS verification engine

The model analysis and result extraction modules include the verification engines to perform verification and safety analyses on the design, and the necessary conversions algorithms to present the results of safety analyses using commercial safety analysis tools. At the moment, the following analyses are supported.

Bottom up analysis. As soon as a simulator of formal specification is available, it can be used to assist the failure mode and effect analysis. The safety engineer injects one or more failure event; the simulator computes the effects of the failure according to the propagation laws encoded in the formal texts; finally, the engineer inspects the reached states and analyses the effects. Thanks to the hierarchies of formal models, local effects can be propagated to higher views so that global effects can be identified. Let us consider an aircraft whose surfaces (spoilers, flap, ...) and some other devices are displaced thanks to

hydraulic power. The aircraft model is the top level of the hierarchy. It includes surfaces and hydraulics system models. In turn, the hydraulics system model consists of models of atomic components (pumps, pipes, etc.). Failures affect these atomic components whereas the global impact is perceptible at the aircraft level. It is worth noting that a good graphical simulator makes this kind of analysis easy and intuitive for safety engineers (see a snapshot of a Cecilia OCAS simulation for instance).

Dynamic failure behaviour. Traditional FTA is a static analysis – i.e. it is done for a given system configuration – investigating the influence of failure modes on unwanted system behaviour. In our approach, since the general design includes the dynamic system behaviour, we can also investigate the influence of FM in dynamic situations. This gives us the possibility of doing new types of dynamic analyses, e.g. to see if the order of occurrence of FM is important or if an intermittent FM has the same impact as a constant FM. Given that the analysis result shows the system can malfunction, then a so-called counter-model – showing a sequence of values for all system variables – is generated. In such a case it is important to analyze the whole sequence and not only the last time-step of the sequence. Another possibility is to define the top level event, i.e. the unwanted system behaviour, with regard to the system dynamics. For example we could regard a top event to occur not until the unwanted event has existed continuously for a certain time.

Traditional fault tree generation. The ESACS Platform can compute fault trees using algorithms based on formal methods techniques. The implemented algorithms support both monotonic and non-monotonic systems; the minimization of non-monotonic systems is based on algorithms presented, e.g. in (Coudert & Madre, 1993). The underlying principle for model-checking based fault tree construction is as follows. The algorithm starts from an extended system model and a top level event and generates, using standard symbolic model checking techniques, a formula representing all the possible ways in which the top level event is not satisfied by the extended system model. From such a formula it is possible to extract all the possible combinations of failures of components. If the design model behaves correctly with respect to the top level event (i.e. if the top level event is verified by the design model), such combinations of failures are exactly the reason for the top level event not being fulfilled anymore in the extended system model (failure modes being the only change between design model and extended system model). Standard minimization techniques are then run on the combination of failures identified thus far, in order to extract from them the combinations that are minimal. The algorithm produces outputs that are suitable for integration with commercial safety analysis tools (e.g. FaultTree+).

Fault tree with ordering information. When the failure mode behaviour of a system includes both primitive and derived failures, it is also possible to perform ordering analyses on the model. This is done by selecting a minimal cut set and by verifying whether the failures related to a particular top level event only occur in a particular order. We refer the reader to (Bozzano & Villafiorita, 2003) for a detailed discussion on this topic.

4 RELATED WORK AND CONCLUSIONS

In this paper we presented the methodology that we are investigating within the ESACS project. The ESACS methodology is based on a tight integration between system design and safety analysis and on the use of formal methods for performing both design and safety assessment. The methodology is supported by state-of-the-art tools that have been extended with innovative algorithms for the safety assessment of systems.

Currently, the ESACS methodology and the ESACS platform are being tested in the following case studies:

- Air inlet control system APU (Auxiliary Power Unit) JAS39 Gripen, related to a critical subsystem of an airplane;
- Wheel Steering System, related to a critical subsystem of a family of Airbus airplanes;
- A controller of the Airbus A340 High Lift System;
- Hydraulic System A320, related to the hydraulic system of the Airbus A320;
- Secondary Power System (SPS) related to power system of the Eurofighter Typhoon.

All the case studies have been chosen to show a reasonable degree of complexity. For instance, the SPS comprises two independent channels controlled by two independent computers. The SPS normal operation consists in transmitting the mechanical power from the engines to the relevant hydraulic and electrical generators. In case of an engine failure the SPS computers automatically initiates a "cross-bleed" procedure consisting in driving the hydraulic and electrical generators by means of an air turbine motor using bled air from the opposite engine. This is an example of one safety requirement of the system.

A set of formal models have been produced for the different case studies and a first series of tests have been run. The results are interesting. As common with formal methods techniques, the algorithms are sensitive to the models and to the properties provided as input and are subject to the state explosion problem. In particular, failure injection contributes to increasing the size of models, as it increases the possible behaviors.

Current work is focused on the direction of making the algorithms more efficient and in the direction of making interaction with the user easier.

ACKNOWLEDGMENTS

The work described in this paper has been and is being developed within the ESACS Project, an European sponsored project, G4RD-CT-2000-00361.

Several other people contributed to the work presented in this paper. We wish in particular to thank: André Arnold from LaBri-Université de Bordeaux, Jack Foisseau from ONERA, Jean Gauthier from Dassault Aviation, Jean-Pierre Heckmann from Airbus, Torgny Knutsson from SAAB, Sylvain Lajeunesse from GFI Consulting, Antoine Rauzy from IML-Université de Marseille, and Paolo Traverso from IRST.

REFERENCES

Arnold, A., Griffault, A., Point, G. & Rauzy, A. 2000. The AltaRica formalism for describing concurrent systems. *Fundamenta Informaticae*, 40:109–124.

Bieber, P., Castel, C. & Seguin, C. 2002. Combination of Fault Tree Analysis and Model Checking for Safety Assessment of Complex System. *In proceedings of 4th European Dependable Computing Conference, LNCS 2485*, pp. 19–31.

Boyer, R.S. & Moore, J.S. 1979. *A Computational Logic*. Academic Press, New York.

Bozzano, M. & Villafiorita, A. Integrating Fault Tree Analysis with Event Ordering Information. In *Proc. European Safety and Reliability Conference (ESREL 2003)*.

Cimatti A., Clarke, E.M., Giunchiglia, E., Giunchiglia, F., Pistore, M., Roveri, M., Sebastiani, R. & Tacchella, A. 2002. NuSMV2: An OpenSource Tool for Symbolic Model Checking, *International Conference on Computer-Aided Verification (CAV 2002)*. Copenhagen, Denmark.

Clarke, E., Grumberg, O. & Peled, D. 1999. *Model Checking*. MIT Press.

Coudert, O. & Madre, J. 1993. Fault Tree Analysis: 10^{20} Prime Implicants and Beyond. In *Proc. Annual Reliability and Maintainability Symposium*.

Emerson, E. 1990. Temporal and Modal Logic In *J. van Leeuwen (Ed.), Handbook of Theoretical Computer Science*, Volume B, pp. 995–1072. Elsevier Science.

Fenelon, P., McDermid J.A., Pumfrey D.J., & Nicholson M. 1994. Towards Integrated Safety Analysis and Design. In *ACM Applied Computing Review*.

Hinchey, M.G. & Bowen, J.P. 1999. *Industrial Strength Formal Methods in Practice*. Springer-Verlag, London.

Liggesmeyer, P. & Rothfelder, M. 1998. Improving System Reliability with Automatic Fault Tree Generation. *In Proc. 28th International Symposium on Fault Tolerant Computing (FTCS'98)*, Munich, Germany, pp. 90–99. IEEE Computer Society Press.

Rae, A. 2000. Automatic Fault Tree Generation – Missile Defence System Case Study. *Technical Report 00-36, Software Verification Research Centre, University of Queensland.*

Vesely, W., Goldberg, F., Roberts, N. & Haasl D. 1981. Fault Tree Handbook, *Technical Report NUREGF-0492, Systems and Reliability Research Office of Nuclear Regulatory Research U.S. Nuclear Regulatory Commission.*

W3C 2000. Extensible Markup Language (XML) 1.0 (Second Edition), *W3C Recommendation. Available on the internet* http://www.w3.org/TR/2000/REC-xml-20001006.

Wing, J.M. 1990. *A specifier's introduction to formal methods. IEEE Computer* 23(9):8–24.

Safety and Reliability – Bedford & van Gelder (eds)
© 2003 Swets & Zeitlinger, Lisse, ISBN 90 5809 551 7

Integrating fault tree analysis with event ordering information*

Marco Bozzano & Adolfo Villafiorita

ITC – IRST, Via Sommarive 18, Povo, Trento, Italy

ABSTRACT: Fault tree analysis is a traditional and well-established technique for analyzing system design and robustness. Its purpose is to identify sets of basic events, called *cut sets*, which can cause a given *top level event*, e.g., a system malfunction, to occur. In this paper we present an algorithm that extracts *ordering* information, i.e., finds out possible ordering constraints which are required to hold between basic events in a cut set. The algorithm is completely *automatic*, and has been incorporated into a more general framework, based on model checking techniques, for automatic fault tree generation and analysis.

1 INTRODUCTION

The development of safety critical systems requires to check that the system behaves as expected not only in nominal situations, but also under certain degraded situations. Thus, on the one hand, system models are developed by the design engineers in order to specify and to analyze the expected behaviour of the system under consideration. On the other hand, the envisaged system is analyzed by safety specialists with respect to malfunctions, i.e., unintended behaviour. The safety analysis, performed at each stage of the system development, is intended to identify all possible hazards with their relevant causes. Traditional safety analysis methods include, e.g., Functional Hazard Analysis, Failure Mode and Effect Analysis (FMEAs), and Fault Tree Analysis (FTA) (Vesely et al. 1981).

Fault tree analysis (Vesely et al. 1981), in particular, is a deductive and top-down method to analyze system design and robustness. Roughly speaking, the FTA process consists in picking a *top level event* (e.g., a system malfunction condition) and identifying all possible *sets* of basic events, called *cut sets*, which can cause the top event to occur. Among them, one would like to isolate *minimal* cut sets, that is, cut sets which do not include events that ultimately do not affect the occurrence of the top event. The information on cut sets is then collected in a *fault tree*, which consists of system and component events, connected by *gates* which

define the logical relations between events. The *cut set* representation provided by traditional fault tree analysis is not structured. A cut set is simply seen as a flat collection of basic events, and no information is provided about their mutual relationship. Although events are often allowed to happen in any order, in general there may be *timing constraints* which enforce a particular event to happen before or after another one. This can happen as a result of a causality relation, a functional dependency, or more subtle reasons related to dynamic scenarios where system behaviour can be affected by, e.g., automatic control systems or operator actions (Siu 1994).

In this paper, we are interested in *automatically* computing ordering information of basic events. Specifically, given a top level event and a minimal cut set computed via fault tree analysis, we want to find out whether there are *ordering constraints* which hold between pairs of basic events in the cut set. We call this *event ordering analysis*. We present an algorithm which integrates traditional fault tree analysis by providing event ordering information for basic events in a cut set. The algorithm is completely *automatic*, and has been incorporated into a more general framework for automatic fault tree generation and analysis. The core of our ordering analysis algorithm is based on known procedures for *minimization* (i.e., computation of *minimal cut sets*) of *boolean functions* (Coudert and Madre 1992; Coudert and Madre 1993; Manquinho et al. 1998) represented as Binary Decision Diagrams (BDDs) (Bryant 1992). The encoding of the problem and some adjustments necessary to deal with *inconsistency* are original. The encoding is based on ordering information variables, that is, variables which

---

* This work has been and is being developed within ESACS, an European sponsored project, contract no. G4RD-CT-2000-00361. See also URL http://www.esacs.org/

relate pairs of different basic events, tracking the information about the mutual order in which the two events may or may not occur.

Our framework is based on model checking (Clarke et al. 2000), a well-established method for formally verifying temporal properties of finite-state concurrent systems. Model checking has been applied for the formal verification of a number of significant safety-critical industrial systems (Holzmann 1997; Larsen et al. 1997; Cimatti et al. 2002). We have incorporated fault tree and ordering analysis functionalities into the model checking tool NuSMV (Cimatti et al. 2002), a BDD-based symbolic model-checker developed at ITC-IRST, originated from a re-engineering and re-implementation of SMV (McMillan 1993). NuSMV is a well-structured, open, flexible and well-documented platform for model checking, and it has been designed to be robust and close to industrial system standards (Cimatti et al. 2000).

This line of research has been carried on inside the ESACS project, an European-Union-sponsored project whose main goals are to define a methodology to improve the safety analysis practice for complex systems development, to set up a shared environment based on tools supporting the methodology, and to validate the methodology through its application to case studies. The fault tree and ordering analysis functionalities which we discuss in this paper have been included in a more general safety analysis platform which we are developing inside the ESACS project (Bozzano et al. 2003).

Structure of the paper. The rest of the paper is structured as follows. In Section 2 we give a brief overview of the basics of fault tree analysis and we introduce *event ordering analysis*, explaining its significance and its relationship with model checking. In Section 3 we introduce a simple example which we will use in Section 4, where we present our minimization algorithm for ordering analysis and we briefly discuss its integration with fault tree analysis based on model checking. Finally, in Section 5 we discuss related work and draw some conclusions.

2 EVENT ORDERING ANALYSIS

Fault Tree Analysis (FTA) (Vesely et al. 1981; Liggesmeyer and Rothfelder 1998; Rae 2000) is a deductive, top-down method to analyze system design and robustness. It usually involves specifying a *top level event* (TLE hereafter) to be analyzed (e.g., a *failure state*), and identifying all possible sets of basic events (e.g., basic *faults*) which may cause that TLE to occur. Benefits of FTA include, e.g.: identify possible system reliability or safety problems at design time; assess system reliability or safety during operation; identify root causes of equipment failures. *Fault*

trees provide a convenient symbolic representation of the combination of events resulting in the occurrence of the top event. Fault trees are usually represented in a graphical way, structured as a parallel or sequential combination of AND/OR gates.

In this paper we are interested in deductive methods which can be used to automatically generate fault trees starting from a given system model and top level event. In particular, we focus on analysis techniques based on model checking. Model checking (Clarke et al. 2000) is a well-established method for formally verifying temporal properties of finite-state concurrent systems. System specifications are written as temporal logic formulas, and efficient symbolic algorithms (based on data structures like BDDs (Bryant 1992)) are used to traverse the model defined by the system and check if the specification holds or not. The application of model checking to fault tree generation works in the following way. Given a system model and a top level event (TLE) to analyze, model checking techniques can be used to extract *automatically* all collections of basic events (called *minimal cut sets*) which can trigger the TLE. The generated cut sets are minimal in the sense that only events that are strictly necessary for the TLE to occur are retained.

In this paper, we discuss and propose an algorithm for extending FTA with *event ordering information*. In traditional FTA, cut sets are simply flat collections (i.e., conjunctions) of events which can trigger a given TLE. However, there might be timing constraints enforcing a particular event to happen before or after another one, in order for the TLE to be triggered (i.e., the TLE would not show if the order of the two events were swapped). Ordering constraints can be due, e.g., to a causality relation or a functional dependency between events, or caused by more complex interactions involving the dynamics of the system under consideration. Whatever the reason, event ordering analysis can provide useful information which can be used by the design and safety engineers to fully understand the ultimate causes of a given system malfunction, so that adequate countermeasures can be taken.

3 AN EXAMPLE

We present below an example which we will use in Section 4 to explain our methodology. The example is deliberately simple for illustration purposes and should not be regarded as modeling a realistic system. We refer to (Bozzano et al. 2003) for more meaningful examples to which the methodology and the algorithm have been applied. Let us consider the circuit drawn in Figure 1. The circuit is composed of two JK flip-flops and an OR gate, and it has three input bits and one output. In short, a JK flip flop is a (clock driven) logical component with two input bits ("J" and

"K") and two output bits ("Q" and "!Q", the latter simply being the negation of the former). The truth table of the JK flip flop is such that whenever "J" and "K" are low the output signal "Q" (which can be either low or high) remains unchanged, whenever either "J" or "K" is set to high the output "Q" is set to, respectively, one or zero, and, finally, a high signal on both "J" and "K" is used to *toggle* the current value of "Q".

In the circuit drawn in Figure 1, the three input bits are set to zero, but they can non-deterministically fail, at any time and in any order, in which case their value is inverted (i.e., it is set to one) forever (note

that we assume *persistent* failures). Initially, we assume all signals to be low, i.e., the input bits, the "Q" outputs of the flip-flops and, consequently, the output of the circuit, are all set to zero. A NuSMV model of the circuit is shown in Figure 2. It is composed of three modules, one for modeling an input bit, one for modeling a JK flip-flop (note that this module simply implements the truth-table of a JK flip-flop) and the main module, which puts all the components together and defines the output signal of the circuit. For simplicity, we have not modeled flip-flop clocks explicitly. We assume that the input values "J" and "Q" are transferred to the flip-flop outputs at each NuSMV transition (i.e., a NuSMV transition can be thought of as causing a triggering edge of the clock pulse).

Top level events to be used for fault tree analysis can be expressed in the temporal logic CTL (Emerson 1990). Arbitrary CTL formulas can be used to perform FTA in NuSMV. Some examples are:

$$\text{AG(out)} \qquad\qquad (T_1)$$
$$\text{AG((out} \rightarrow \text{AX(!out)) \& (!out} \rightarrow \text{AX out))} \qquad (T_2)$$
$$\text{EG((out} \rightarrow \text{AX(!out)) \& (!out} \rightarrow \text{AX out))} \qquad (T_3)$$

The top level event T_1 is a CTL formula specifying all the states of the system in which the output of the circuit is *forced* to be set to value one *forever*, i.e., for

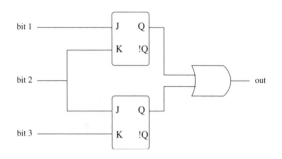

Figure 1.   A simple circuit with two JK flip-flops.

```
MODULE bit(input)
VAR
    out          : boolean;
    FailureMode : {no_failure,inverted};

ASSIGN
    init(FailureMode) := no_failure;
    next(FailureMode) := case
            FailureMode = no_failure : {no_failure,inverted};
            FailureMode = inverted    : inverted;
    esac;
    out := case
            FailureMode = no_failure : input;
            FailureMode = inverted    : !input;
    esac;
```

```
MODULE ff(J,K)                      MODULE main
VAR                                 VAR
    Q : boolean;                        bit1   : bit(0);
                                        bit2   : bit(0);
                                        bit3   : bit(0);
ASSIGN                                  ff1    : ff(bit1.out,bit2.out);
    init(Q) := 0;                       ff2    : ff(bit2.out,bit3.out);
    next(Q) := case                     out    : boolean;
            !J & !K :  Q;
            !J &  K :  0;           ASSIGN
             J & !K :  1;               out := ff1.Q | ff2.Q;
             J &  K : !Q;
    esac;
```

Figure 2.   A NuSMV model for the circuit in Figure 1.

every possible path (evolution of the system) the output is *globally* set to value one on that path. Similarly, the CTL formula T_2 is a specification of all the states such that the output of the circuit is *forced* to *oscillate* forever back and forth between the values zero and one. Finally, the CTL formula T_3 is a specification of all the states such that there exists *one* path on which the output is *globally forced* to oscillate forever.

As mentioned in Section 2, we have implemented a procedure for performing fault tree analysis in NuSMV. As an aside, we mention that the safety analysis platform we are developing inside the ESACS project provides additional features for managing failure modes. Specifically, fault tree computation starts with the user assigning a set of failures to the various components, which are then automatically inserted into the original model of the system. The result is an *extended system model* with failure variables (e.g., the variable FailureMode of the bit module in Figure 2). Model checking techniques can then be applied to the extended NuSMV model (e.g., the model in Figure 2) to extract *automatically* all collections of basic events, i.e., all *minimal cut sets*, which can cause any of the above TLEs. Cut sets are expressed in terms of *sets of failure events*, i.e., pairs consisting of a failure variable and a failure mode. The results of fault tree analysis for the model in Figure 2 and the CTL formulas T_1, T_2, and T_3 are shown below (hereafter, we shorten (bit*i*.FailureMode, inverted) with bit*i*_inv). In this particular case, exactly one minimal cut set M_i is computed for each formula T_i (note that in general more than one cut set can be computed for a TLE).

{bit1_inv, bit3_inv}	(M_1)
{bit1_inv, bit2_inv, bit3_inv}	(M_2)
{bit2_inv}	(M_3)

For M_1, we have that, in order for the output of the circuit to be *forced* to return value one forever (property T_1), it is necessary that both the first and the third bit fail. Notice that the output of the circuit can also get stuck at value one as a result of a failure of the first bit only. In this case, however, the output of the circuit is not *forced* to that value, i.e., as the reader can verify, there exist possible evolutions of the circuit such that the output can assume value zero. In Figure 3 we show a simple graphical representation for the fault tree corresponding to T_1. The cut set is minimal in the sense that only events that are strictly necessary for the TLE to occur are retained. Similarly, minimal cut set M_2 states that all bits must be failed in order for the output of the circuit to be forced to oscillate forever. Notice that failure of all input bits is not a *sufficient* condition for oscillation of the circuit. In fact, there are some *timing constraints* which must be satisfied in order for the circuit to show this oscillating behaviour. Extracting information about these timing constraints is exactly

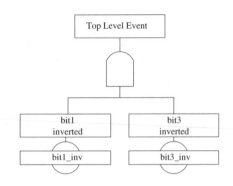

Figure 3. A fault tree for T_1.

the purpose of the *ordering analysis* which is described in the next section.

4 THE MINIMIZATION ALGORITHM

In this section we explain in detail our algorithm for event ordering analysis. Specifically, we describe a procedure which takes in input a system model (e.g., the NuSMV model in Figure 2) and a cut set, and is able to extract event ordering information. The core of the algorithm is based on procedures for computing *prime implicants* of boolean functions (Coudert and Madre 1993), and exploits the BDD-based representation for boolean functions, which is used extensively in NuSMV. The algorithm is made up of a number of different phases, which are detailed below.

(0) Pick a minimal cut set. A prerequisite of our procedure is having a system model SM and a top level event TLE at hand. Then, as explained in Section 3, we run NuSMV on SM and TLE, and we get a collection of minimal cut sets. Assuming the collection is not empty, we pick one MCS. The purpose of the ordering analysis algorithm is to extract ordering constraint information from MCS.

(1) Generate the ordering information model. Assuming MCS is composed of a set of n failure events, say $(fm_1, var\_name_1), \ldots, (fm_n, var\_name_n)$, for each pair of distinct failure events with indexes i and j in MCS ($i \neq j$), we introduce a new ordering variable order_var_name$_{ij}$ in the NuSMV model SM, to keep track of the mutual order in which the two failure events may happen. In order to give a complete encoding for ordering information, we thus need a total of $\frac{1}{2}n(n-1)$ ordering variables. We call the resulting model *ordering information model* (OIM hereafter). The NuSMV skeleton for defining an ordering variable is shown in Figure 4. The skeleton is instantiated each time with different actual parameters for failure variables and failure modes. Every ordering variable can assume one

```
VAR
    ORDER_VAR_NAME_AUX : {UNKNOWN, BEFORE, AFTER, SIMULT};
    ORDER_VAR_NAME : {BEFORE, AFTER, SIMULT};

ASSIGN
    init(ORDER_VAR_NAME_AUX) := UNKNOWN;
    next(ORDER_VAR_NAME_AUX) := case
        ORDER_VAR_NAME_AUX = UNKNOWN : case
                (VAR_NAME_1 = FM_1 & VAR_NAME_2 = NO_FAILURE) : BEFORE;
                (VAR_NAME_1 = NO_FAILURE & VAR_NAME_2 = FM_2) : AFTER;
                (VAR_NAME_1 = FM_1 & VAR_NAME_2 = FM_2)       : SIMULT;
                1                                             : UNKNOWN;
            esac;
        1                                    : ORDER_VAR_NAME_AUX;
    esac;
    ORDER_VAR_NAME := case
        ORDER_VAR_NAME_AUX = UNKNOWN : SIMULT;  -- does not matter
        1                            : ORDER_VAR_NAME_AUX;
    esac;
```

Figure 4.   NuSMV code skeleton for ordering variable definition.

among the three values {before, after, simult}, the intuition being that the first event happens before, after, or at the same time with the second one (note that the notion of simultaneousness is relative to the granularity of the NuSMV *step*, e.g., one clock pulse for the circuit in Figure 1). An auxiliary variable definition (which may assume the additional value unknown) is necessary in order to code the fact that the value of a given variable is still unknown during the computation (the unknown value will be eventually overwritten, because all failure events in MCS are forced to occur, see below).

(2) Re-run NuSMV on the ordering information model. In this phase NuSMV is re-run on the OIM, with the same TLE, in order to track the information captured by the ordering variables. The analysis is specialized to the given MCS, i.e., the formula provided to NuSMV for the analysis forces the failure events contained in MCS (*and only them*) to occur. The result is a BDD representing all the different configurations (including system variables and ordering variables) which can cause TLE in presence of the failure events in MCS. For instance, consider the NuSMV model, the top level event T2 and the relevant minimal cut set M_2 described in Section 3. The minimal cut set and the top level event are combined together, yielding the CTL formula

$$((AG\ (out) \rightarrow AX\ (!out))\ \&\ (!out \rightarrow AX\ out))\ \&$$
bit1_inv & bit2_inv &bit3_inv

NuSMV is fed with this formula in order to generate a BDD representing all states causing T_2 because of the failure events in M_2. This BDD also includes the information about ordering variables.

(3) Abstract away non-ordering variables. In this phase we simply abstract away variables other than ordering ones. The result is still a BDD representing all the possible failure event orderings.

(4) Extract ordering constraints. This phase contains the core of the minimization algorithm, and is composed of three interrelated sub-phases.

4.1. Add inconsistent configurations. The ordering variable encoding described in point 1 above is redundant in the following sense. Consider, e.g., three variables v_{ij}, v_{jk}, and v_{ik}, representing the order in which failure events i and j (j and k, i and k) occur. Clearly, if, say, v_{ij} and v_{jk} are both set to the value before, for *transitivity* also v_{ik} will be (necessarily) set to before. In other words, the encoding allows for *inconsistent* configurations which will never be the result of the model checking analysis (e.g., ⟨before, before, after⟩ in the previous example). During this phase we extend the BDD resulting from phase 3 with a BDD representing such inconsistent configurations (i.e., we consider their disjunction). This amounts to admitting inconsistent configurations as if they were perfectly legal. In this way (and with a little adjustment explained in phase 4.3 below) we can ensure that the minimization algorithm (see phase 4.2) works properly. Intuitively, the minimization algorithm works by picking variables whose value is *irrelevant* (in the sense that they can assume any among the possible *legal* values). Therefore, phase 4.1 is necessary to give the minimization algorithm enough information to correctly recognize which variables are irrelevant and which are not.

4.2. Compute prime implicants. We simply run the standard algorithm for computing *prime implicants* of a boolean function (Coudert and Madre 1993) on the BDD resulting after phase 4.1. For instance, for the top level event T_2 and cut set M_2 this phase computes 16 prime implicants. An example of prime implicant is the following (it represents the timing constraint enforcing

251

the first bit to fail before the second one and the second one before the third one):

```
p0) ------------------------
    bit1_inv  **before**  bit2_inv
    bit2_inv  **before**  bit3_inv
```

4.3. Run simplification subroutine. The purpose of this phase is to cut *inconsistent* results and *subsumed* (i.e., logically implied) results from the output of phase 4.2. Inconsistent results can arise as a side effect of phase 4.1, and must be discarded after the minimization phase, whereas the purpose of the simplification subroutine is to retain only *minimal* results. For the top level event T_2 and cut set M_2, the previously generated 16 implicants are reduced to 3 after the simplification phase. For instance, the prime implicant p0 is discarded because it is subsumed by prime implicant p2 (see below).

(5) Show results. The final output consists of the collection of prime implicants resulting after phase 4. For the top level event T_2 and cut set M_2, NuSMV outputs the following three prime implicants:

```
p1) ------------------------
    bit1_inv  **simult**  bit2_inv
p2) ------------------------
    bit2_inv  **before**  bit1_inv
    bit3_inv  **before**  bit1_inv
p3) ------------------------
    bit1_inv  **before**  bit3_inv
    bit2_inv  **before**  bit3_inv
```

Each prime implicant is a list of ordering constraints between failure events, and represents a different alternative (i.e., a different ordering possibly causing the TLE). Each list of failure events represents a *precedence graph*, showing which failure event must happen before which, for a given prime implicant. The output might be drawn in a more suggestive way, as shown in Figure 5. Each node in the graph contains one or more failure events (an index *i* denotes failure event bit*i*_inv), which are supposed to happen simultaneously, whereas each arrow represents a *before* relation between two (sets of) failure events.

The precedence graphs shown in Figure 5 represent the ordering information which has been obtained by running our algorithm on the model, top level event

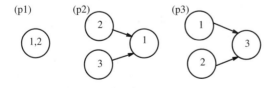

Figure 5.   Prime implicants as precedence graphs.

T_2 and minimal cut set M_2 described in Section 3. As discussed in Section 3, the CTL formula T_2 is a specification of all the states such that the output of the circuit is *forced* to oscillate forever between the values zero and one, whereas minimal cut set M_2 shows that all three input bits of the circuit must fail in order for such a behaviour to be observable. The ordering analysis results give us further information about this oscillating phenomenon, i.e, show that some further *timing constraints* between the failure events must hold as well. In particular, one of the orderings shown in Figure 5 must hold: either the first and the second bit must fail simultaneously (and the third one at any time, including simultaneously with the other two), or the second and third bit must fail before the first (the order between the former two bits is left unspecified), or the first and second must fail before the third (again, the order between the former two bits is left unspecified). A possible ordering which is *ruled out* by the results of our ordering analysis is the one in which the three bits fail in the following order: the first bit, then the third one, and finally the second one.

4.1   Ordering and fault tree analysis

We conclude this section with a brief explanation about how the minimization algorithm can be integrated with fault tree analysis. A system model is assumed to be given. The verification process consists of the following phases. First of all, a top level event to analyze is chosen (clearly, the analysis can be repeated for different top level events). Then, we run the minimization algorithm of (Coudert and Madre 1993) to compute the *minimal cut sets* of basic events causing TLE. Finally, for each cut set we generate an ordering information model and we perform the ordering analysis. The output is a fault tree for TLE, where each cut set in the tree is equipped with ordering information (that is, a precedence graph).

For the example of Section 3, the analysis can proceed in the following way. First, a top level event is chosen (e.g., T_1, T_2, or T_3) and NuSMV is run to perform fault tree analysis. The result is a collection of minimal cut sets of failure events. For each cut set, ordering analysis is performed on a suitable ordering information model. The results show that: for T_1 there are no timing constraints (all orderings between the two failure events are possible); for T_2, the output of NuSMV is shown in Figure 5; finally, T_3 only includes one basic event, hence it is useless to perform ordering information analysis on it.

5   CONCLUSIONS AND RELATED WORK

In this paper we have presented an algorithm which improves the results given by traditional fault tree analysis (Vesely et al. 1981). In particular, our algorithm performs what we call *event ordering analysis*. This analysis can be conveniently used to extract possible ordering constraints holding between basic events in a given cut set, thus providing a deeper insight into the causes of system malfunction and supporting the reliability and safety analysis process. Although very simple, the example in Section 3 shows the importance and significance of event ordering information. It also suggests that timing constraints can arise very naturally in industrial systems.

As explained in the paper, our algorithm for ordering analysis is based on classical procedures for *minimization* of *boolean functions*, specifically on the implicit-search procedure described in (Coudert and Madre 1992; Coudert and Madre 1993), which is based on Binary Decision Diagrams (BDDs) (Bryant 1992). This choice was quite natural, given that the NuSMV model checker makes a pervasive use of BDD data structures. For alternative explicit-search and SAT-based techniques for computation of prime implicants, see (Manquinho et al. 1998). The results computed by the algorithm may differ depending on the exact order in which ordering variables are chosen in the minimization step. This a consequence of the non-determinism which is inherent in the prime implicant computation (Coudert and Madre 1993). However, we conjecture that the minimization procedure enjoys some *optimality* properties which we are studying as part of our future work.

A large amount of work has been done in the area of probabilistic safety assessment (PSA) and in particular on *dynamic reliability* (Siu 1994). Dynamic reliability is concerned with extending the classical event or fault tree approaches to PSA by taking into consideration the mutual interactions between the hardware components of a plant and the physical evolution of its process variables (Marseguerra et al. 1998). Examples of scenarios which dynamic reliability tries to take into consideration are, e.g., human intervention, expert judgment, the role of control/protection systems, the so-called failures *on demand* (i.e., failure of a component to intervene), and also the ordering of events during accident propagation (Senni, Semenza, and Galvagni 1991; Cacciabue and Cojazzi 1994; Cacciabue and Cojazzi 1995). Different approaches to dynamic reliability include, e.g., state transitions or Markov models (Aldemir 1987; Papazoglou 1994), the dynamic event tree methodology (Cojazzi et al. 1992), and direct simulation via Monte Carlo analysis (Smidts and Devooght 1992; Marseguerra et al. 1998). The work which is probably closer to ours is (Cojazzi et al. 1992), which describes dynamic event trees as a convenient means to represent the timing and order of intervention of a plant sub-systems and their eventual failures. With respect to the classification the authors propose, our approach can support *simultaneous*

failures, whereas, at the moment, we are working under the hypothesis of *persistent* failures (i.e., no repair is possible).

The most notable difference between our approach and the works on dynamic reliability mentioned above is that we present *automatic* techniques, based on model checking, for both fault tree generation and ordering analysis, whereas traditional works on dynamic reliability rely on manual analysis (e.g., Markovian analysis (Papazoglou 1994)) or simulation (e.g., Monte Carlo simulation (Marseguerra et al. 1998), the TRETA package of (Cojazzi et al. 1992)). Automation is clearly a point in favour of our framework. Furthermore, we support automatic verification of arbitrary temporal CTL properties (in particular, both safety and liveness properties). Current work is focusing on a number of improvements and extensions in order to make the methodology competitive with existing approaches and usable in realistic scenarios. First of all, there are some improvements at the modeling level. The NuSMV models used so far are discrete, finite-state transition models. In order to allow for more realistic models, we are considering an extension of NuSMV with hybrid dynamics, along the lines of (Henzinger 1996; Henzinger et al. 1997). This would allow both to model more complex variable dynamics, and also a more realistic modeling of time (currently, time is modeled by an abstract transition step). Furthermore, we need to extend our framework in order to deal with *probabilistic* assessment. Although not illustrated in this paper, associating probabilistic estimates to basic events and evaluating the resulting fault trees is straightforward. However, more work needs to be done in order to support more complex probabilistic dynamics (see, e.g., (Devooght and Smidts 1994)). Also, we want to overcome the current limitation to permanent failures.

We also mention (Manian et al. 1998; Sullivan et al. 1999), which describe DIFTree, a methodology supporting (however, still at the manual level) fault tree construction and allowing for different kinds of analyses of sub-trees (e.g., Markovian or Monte Carlo simulation for dynamic ones, and BDD-based evaluation for static ones). The notation the authors use for non-logical (dynamic) gates of fault trees and the support for sample probabilistic distributions could be nice features to be integrated in our framework.

Finally, the line of research concerning ordering analysis has been carried out inside the ESACS project. As a contribution to the project, we are developing an integrated platform providing the safety engineers with tools for the specification, analysis and validation of complex systems. Formal verification functionalities of the platform are based on model checking, and in particular on NuSMV (Cimatti et al. 2002). In this paper we have focused only on the aspects related to the minimization procedure. We refer the reader to (Bozzano et al. 2003), where a more detailed description of the project goals, the ESACS methodology, and more realistic examples to which the methodology has been applied can be found.

REFERENCES

Aldemir, T. (1987). Computer-assisted Markov Failure Modeling of Process Control Systems. *IEEE Transactions on Reliability R-36*, 133–144.

Bozzano, M. et al. (2003). ESACS: An Integrated Methodology for Design and Safety Analysis of Complex Systems. In *Proc. of European Safety and Reliability Conference (ESREL'03)*.

Bryant, R. (1992). Symbolic Boolean Manipulation with Ordered Binary Decision Diagrams. *ACM Computing Surveys 24*(3), 293–318.

Cacciabue, P. and G. Cojazzi (1994). A human factor methodology for safety assessment based on the Dylam approach. *Reliability Engineering and System Safety 45*, 127–138.

Cacciabue, P. and G. Cojazzi (1995). An integrated simulation approach for the analysis of pilotaeroplane interaction. *Control Engineering Practice 3*(2), 257–266.

Cimatti, A., E. Clarke, E. Giunchiglia, F. Giunchiglia, M. Pistore, M. Roveri, R. Sebastiani, and A. Tacchella (2002). NuSMV2: An Open Source Tool for Symbolic Model Checking. In *Proc. 14th International Conference on Computer Aided Verification (CAV'02)*, LNCS. Springer-Verlag.

Cimatti, A., E. Clarke, F. Giunchiglia, and M. Roveri (2000). NUSMV: a new symbolic model checker. *International Journal on Software Tools for Technology Transfer 2*(4), 410–425.

Clarke, E., O. Grumberg, and D. Peled (2000). *Model Checking*. MIT Press.

Cojazzi, G., J.M. Izquierdo, E. Meléndez, and M.S. Perea (1992). The Reliability and Safety Assessment of Protection Systems by the Use of Dynamic Event Trees. The DYLAM-TRETA Package. In *Proc. XVIII Annual Meeting Spanish Nucl. Soc.*

Coudert, O. and J. Madre (1992). Implicit and Incremental Computation of Primes and Essential Primes of Boolean Functions. In *Proc. 29th Design Automation Conference (DAC'98)*, pp. 36–39. IEEE Computer Society Press.

Coudert, O. and J. Madre (1993). Fault Tree Analysis: 10^{20} Prime Implicants and Beyond. In *Proc. Annual Reliability and Maintainability Symposium*.

Devooght, J. and C. Smidts (1994). Probabilistic Dynamics; The Mathematical and Computing Problems Ahead. In T. Aldemir, N.O. Siu, A. Mosleh, P.C. Cacciabue, and B.G. Göktepe (Eds.), *Reliability and Safety Assessment of Dynamic Process Systems*, Volume 120 of *NATO ASI Series F*, pp. 85–100. Springer-Verlag.

Emerson, E. (1990). Temporal and modal logic. In J. van Leeuwen (Ed.), *Handbook of Theoretical Computer Science*, Volume B, pp. 995–1072. Elsevier Science.

Henzinger, T. A. (1996). The Theory of Hybrid Automata. In *Proc. 11th Annual International Symposium on Logic in Computer Science (LICS'96)*, pp. 278–292. IEEE Computer Society Press.

Henzinger, T.A., P.-H. Ho, and H. Wong-Toi (1997). HyTech: A Model Checker for Hybrid Systems. *Software Tools for Technology Transfer 1*, 110–122.

Holzmann, G. (1997). The Model Checker SPIN. *IEEE Transactions on Software Engineering 23*(5), 279–295.

Larsen, K., P. Pettersson, and W. Yi (1997). UPPAAL in a Nutshell. *International Journal on Software Tools for Technology Transfer 1*(1–2), 134–152.

Liggesmeyer, P. and M. Rothfelder (1998). Improving System Reliability with Automatic Fault Tree Generation. In *Proc. 28th International Symposium on Fault-Tolerant Computing (FTCS'98)*, Munich, Germany, pp. 90–99. IEEE Computer Society Press.

Manian, R., J. Dugan, D. Coppit, and K. Sullivan (1998). Combining Various Solution Techniques for Dynamic Fault Tree Analysis of Computer Systems. In *Proc. 3rd International High-Assurance Systems Engineering Symposium (HASE'98)*, pp. 21–28. IEEE Computer Society Press.

Manquinho, V., A. Oliveira, and J. Marques-Silva (1998). Models and Algorithms for Computing Minimum-Size Prime Implicants. In *Proc. International Workshop on Boolean Problems (IWBP'98)*.

Marseguerra, M., E. Zio, J. Devooght, and P.E. Labeau (1998). A concept paper on dynamic reliability via Monte Carlo simulation. *Mathematics and Computers in Simulation 47*, 371–382.

McMillan, K. (1993). *Symbolic Model Checking*. Kluwer Academic Publ.

Papazoglou, I.A. (1994). Markovian Reliability Analysis of Dynamic Systems. In T. Aldemir, N.O. Siu, A. Mosleh, P. C. Cacciabue, and B.G. Göktepe (Eds.), *Reliability and Safety Assessment of Dynamic Process Systems*, Volume 120 of *NATO ASI Series F*, pp. 24–43. Springer-Verlag.

Rae, A. (2000). Automatic Fault Tree Generation – Missile Defence System Case Study. Technical Report 00–36, Software Verification Research Centre, University of Queensland.

Senni, S., M. Semenza, and R. Galvagni (1991). ADMIRA: An Analytical Dynamic Methodology for Integrated Risk Assessment. In G. Apostolakis (Ed.), *Probabilistic Safety Assesment and Management*, pp. 407–412. Elsevier Science.

Siu, N.O. (1994). Risk Assessment for Dynamic Systems: An Overview. *Reliability Engineering and System Safety 43*, 43–74.

Smidts, C. and J. Devooght (1992). Probabilistic Reactor Dynamics II. A Monte-Carlo Study of a Fast Reactor Transient. *Nuclear Science and Engineering 111*(3), 241–256.

Sullivan, K., J. Dugan, and D. Coppit (1999). The Galileo Fault Tree Analysis Tool. In *Proc. 29th Annual International Symposium on Fault-Tolerant Computing (FTCS'99)*, pp. 232–235. IEEE Computer Society Press.

Vesely, W., F. Goldberg, N. Roberts, and D. Haasl (1981). Fault Tree Handbook. Technical Report NUREG-0492, Systems and Reliability Research Office of Nuclear Regulatory Research U.S. Nuclear Regulatory Commission.

Safety and Reliability – Bedford & van Gelder (eds)
© 2003 Swets & Zeitlinger, Lisse, ISBN 90 5809 551 7

Bayesian estimation of the parameters in safety and reliability models for the subjective priors

A. Brandowski

Faculty of Ocean Engineering and Ship Technology, Technical University of Gda'nsk, Poland
Mechanical Faculty, Gdynia Maritime University, Poland

F. Grabski

Chair of Mathematics, Naval University, Gdynia, Poland
Chair of Mathematics, Gdynia Maritime University, Poland

ABSTRACT: The problem of estimation of some unknown reliability characteristics using the nonparametric method of Bayesian estimation is considered in this paper. In many cases, the experts' opinion is the only way to obtain the priors. To construct the subjective priors, we use some methods of elicitation. The Bayes nonparametric estimators for uncensored and censored data are constructed based on Dirichlet process, which is a key notion in the Ferguson theory.

1 INTRODUCTION

We consider the problem of estimation of some unknown reliability characteristics using nonparametric method of Bayesian estimation. It is well known, that the most controversial and most criticized point of Bayesian estimation theory deals with the choice of the prior distribution. In many cases, the experts' opinion is the only way to obtain the priors. To construct the subjective priors, we use some methods of elicitation. Some elicitation techniques and practical guidelines for eliciting opinions from experts are presented by Cooke (1991). Construction of the nonparametric estimator for uncensored and censored data is based on Dirichlet process, which is a key notion in the Ferguson theory.

2 THE BASIC NOTIONS

To explain the main aim of our paper, we have to recall basic Ferguson's definition and results.

DEFINITION: [Ferguson (1973)]. Let $(\mathcal{X}, \mathcal{A})$ be a measurable space, and let $\alpha(\cdot)$ be a finite non-null measure on $(\mathcal{X}, A)$. Let $\{A_1, ..., A_k\}$ be a partition of $\mathcal{X}$, i.e. $A_1, ..., A_k \in \mathcal{A}, A_i \cap A_j$ for $i \neq j$ and $A_1 \cup ... \cup A_k = \mathcal{X}$. A *Dirichlet process with parameter* α, denoted by $\mathcal{D}(\alpha)$, is a random process (random measure) $P(\cdot)$, indexed by elements of $\mathcal{A}$ if for every measurable partition $\{A_1, ..., A_k\}$ $k = 1, 2, ...$ of $\mathcal{X}$ the random vector $(P(A_1), ..., P(A_k))$ has a Dirichlet distribution with parameter $(\alpha(A_1), ..., \alpha(A_k))$.

The basic result is:

THEOREM: [Ferguson (1973)]. Let P be a Dirichlet process on $(\mathcal{X}, \mathcal{A})$ with parameter $\alpha(\cdot)$ and let $X_1, ..., X_n$ be a sample of size n from P. Then the conditional distribution of P given $X_1, ..., X_n$ is a Dirichlet process with parameter $\alpha + \Sigma_{i=1}^{n} \delta_{X_i}$ where $\delta_{X_i}(A) = 1$ for $X_i \in A, \delta_{X_i}(A) = 0$ for $X_i \notin A$.

It is convenient to replace the random probability measure $P(\cdot)$ by its distribution function $F(\cdot)$: $F(t) = P((-\infty, t))$ or by function $R(\cdot)$: $R(t) = P((t, \infty))$.

3 ESTIMATION BASED ON UNCENSORED DATA

3.1 *Estimation of a reliability function*

Let $(\mathcal{X}, \mathcal{A}) = (\mathbb{R}, \mathcal{B}(\mathbb{R}))$, where $\mathcal{B}(\mathbb{R})$ is a Borel σ-field in $\mathbb{R}$. Consider the problem of estimating an unknown reliability function $R(\cdot)$ by function $\hat{R}(\cdot)$ with a loss function

$$L(R, \hat{R}) = \int_{\mathbb{R}} (R(x) - \hat{R}(x))^2 d\mu(x),$$

where μ is a known finite nonnegative measure on $\mathcal{B}(\mathbb{R})$ and let the space of action of a statistician be the space of probability distribution on $\mathbb{R}$ represented by the reliability function $R(t) = P((t, \infty))$.

If P is a Dirichlet process than the Bayes rule for no-sample problem is

$$\hat{R}(t) = E(R(t)) = R_0(t) = \frac{\alpha((t, \infty)}{\alpha(\mathbb{R})} \qquad (1)$$

We can treat this function as our prior guess at the shape of the unknown reliability function $R(t)$.

For a sample $X_1, \ldots, X_n$ the Bayesian nonparametric estimate of the reliability function has the form (see Ferguson (1973))

$$\hat{R}_n(t|X_1, \ldots, X_n) = p_n R_0(t) +$$
$$+ (1 - p_n) R_n(t|X_1, \ldots, X_n), \qquad (2)$$

where

$$p_n = \frac{\alpha(\mathbb{R})}{\alpha(\mathbb{R}) + n}, \quad R_0(t) = \frac{\alpha((t, \infty))}{\alpha(\mathbb{R})}$$

and

$$R_n(t|X_1, \ldots, X_n) = \frac{1}{n} \sum_{i=1}^{n} \delta_{X_i}((t, \infty)).$$

We see that the Bayes rule is a mixture of our guess at R and the empirical reliability function, with the weights p_n and $(1 - p_n)$ respectively. If $\alpha(\mathbb{R})$ is small in comparison with n, a small weight is given to the prior guess at R. If $\alpha(\mathbb{R})$ is large in comparison with n, a small weight is given to the observations.

3.2 Estimation of the mean time to failure

Suppose that we want to estimate with squared error loss $L(P, \hat{m}) = (m - \hat{m})^2$ the mean $m = \int x dP(x)$ of the random distribution P based on a sample $X_1, \ldots, X_n$ with prior which has a finite first moment. By the theorem of Ferguson the random variable m exists. The Bayes rule for no-sample problem is $\hat{m} = E(m) = m_0 = \int x d(\alpha(x))/\alpha(\mathbb{R})$. For a sample $X_1, \ldots, X_n$ the Bayes estimator of the mean is given by the formula

$$\hat{m}(X_1, \ldots, X_n) = E(m|X_1, \ldots, X_n)$$
$$= p_n m_0 + (1 - p_n)\overline{X}, \qquad (3)$$

where p_n is given above and

$$\overline{X} = \frac{X_1 + \ldots + X_n}{n}$$

4 ESTIMATION BASED ON CENSORED DATA

4.1 Estimation of the reliability function based on censored data

Up to now, the theory of Ferguson has been developed and many important results have been obtained. The treatment of censored data is an important extension of that theory. The nonparametric Bayes approach to this problem was made by Susarla & Van Ryzin (1976, 1979) and by Ferguson & Phadia (1979). They considered the problem of the nonparametric Bayes estimation of an unknown reliability (survival) function $R(t) = 1 - F(t)$ based on the right censored data. They used a Dirichlet process as a prior for $F(\cdot)$. That problem is formulated as follows. Let $X_1, \ldots, X_n$ be the instants of failure. The sequence $X_1, \ldots, X_n$ is assumed to be a simple sample from right sided function $R(\cdot)$. Let the censoring points $Y_1, \ldots, Y_n$ be the i.i.d. random variables independent of the X's. Set

$$Z_j = \min(X_j, Y_j), j = 1, \ldots, n$$

and

$$\Delta_j = 1 \quad \text{for } X_j \leq Y_j$$

and

$$\Delta_j = 0 \quad \text{for } X_j > Y_j.$$

The sufficient statistics can be written as a vector (Z, Δ), where $Z = (Z_1, \ldots, Z_n)$, $\Delta = (\Delta_1, \ldots, \Delta_n)$. Let λ_j be the number of observations equal to $Z_j, j = 1, \ldots, n$ and let $N^+(t)$ denote the number of observations greater than t, $(>t)$.

From the main result of Susarla & Van Ryzin (1976) it follows that if $F \in \mathcal{D}(\alpha)$, then the nonparametric Bayes estimator of the reliability function $R(t) = 1 - F(t)$ given observations is

$$\hat{R}(t|(Z, \Delta)) = E(R(t)|(Z, \Delta)) =$$
$$= \frac{\alpha(t) + N^+(t)}{\alpha(\mathbb{R}) + n} \times$$
$$\times \prod_{j=1}^{n} \left(\frac{\alpha(Z_j^-)) + N^+(Z_j) + \lambda_j}{\alpha(Z_j^-) + N(Z_j)} \right)^{I\{\Delta_j = 0, Z_j \leq t\}}, \qquad (4)$$

where $\alpha(t) = \alpha((t, \infty))$ and $I\{\Delta_j = 0, Z_j \leq t\}$ is an indicator of the event $\{\Delta_j = 0, Z_j \leq t\}$.

5 THE SUBJECTIVE PRIOR MEASURE

We construct a subjective prior measure $\alpha(\cdot)$ in the following way. Let $\alpha_k((t, \infty))$, $k = 1, \ldots, r$ be the subjective prior reliability function obtained from expert k. Let us notice that $\alpha_k(\mathbb{R}) = 1$. We define the measure $\alpha(\cdot)$ as

$$\alpha((\cdot)) = r \sum_{k=1}^{r} w_k \alpha_k(\cdot),\qquad (5)$$

where w_k, $k = 1, \ldots r$ are the experts' weights such that $\Sigma_{k=1}^{r} w_k = 1$. It is obvious that $\alpha(\mathbb{R}) = r$.

Sometimes the nature of the quantities whose distributions are estimated suggests to choose a particular class of probability measures indexed by some parameters. In such cases we may use the parametric elicitation procedures to obtain unknown prior distributions, (see R.M. Cooke (1991)).

The Weibull distribution is convenient to explain those procedures. Let's recall that the Weibull distribution of the random variable X describing the time to failure of an object is given by

$$F(t) = 1 - e^{-\lambda t^\gamma}, \quad t > 0,$$

where $\gamma > 0, \lambda > 0$. The Weibull reliability function is

$$R(t) = e^{-\lambda t^\gamma}, \quad t > 0.$$

The Weibull distribution is defined if we know two parameters γ and λ. The elicitation technique of the Weibull distribution parameters consists in eliciting from expert two quantiles t_{p1} and t_{p2} of distribution for given probabilities p_1 and p_2.

5.1 Elicitation procedure for the Weibull distribution $\mathcal{W}(\gamma, \lambda)$

1. The first question to expert dealing with time to failure X is: For what t_1 do you have $p_1\%$ confidence that the event $\{X \le t_1\}$ will occur. His answer is $t_1 = t_{p1}$.
2. The expert is asked to estimate the value t_2 of the random variable X such that he has $p_2\%$ confidence that the event $\{X \le t_2\}$ will occur. His answer is $t_2 = t_{p2}$.

Eliciting the numbers $t_1 = t_{p1}$ and $t_2 = t_{p2}$ gives us a possibility to get the parameters γ and λ of the Weibull distribution. To obtain those parameters we have to solve the system of equations.

$$1 - e^{-\lambda t_1^\gamma} = p_1 \quad \text{and} \quad 1 - e^{-\lambda t_2^\gamma} = p_2 \qquad (6)$$

where γ and λ are unknown.
Finally we obtain

$$\gamma = \frac{\ln\left[\frac{\ln(1-p_1)}{\ln(1-p_2)}\right]}{\ln\left[\frac{t_1}{t_2}\right]}, \quad \lambda = -\frac{\ln(1-p_1)}{t_1^\gamma}. \qquad (7)$$

Now, the subjective prior measure obtained from expert k is

$$\alpha_k((t,\infty)) = e^{-\lambda_k t^{\gamma_k}}, \quad k = 1,\ldots,m \qquad (8)$$

where λ_k and γ_k satisfy equations (7).

To obtain the weights w_k, $k = 1, \ldots m$ we may use one of the methods presented by Cooke (1991).

5.2 Example

The time to failure of a ship engine is a random variable X which has Weibull distribution. Our aim is to elicit opinions from experts which allow us to assess the unknown parameters γ and λ of that distribution.
Suppose $p_1 = 0, 5$ and $p_2 = 0, 99$.
We use the events $\{X \le t_1\}$ and $\{X \le t_2\}$.
Assume that four experts take part in the elicitation procedure, $(k = 1, 2, 3, 4)$. Suppose that their weights are: $w_1 = 0.32$, $w_2 = 0.24$, $w_3 = 0.28$, $w_4 = 0.16$ respectively.

Procedure of elicitation
1. You have a 50% confidence that the number of days to failure of the engine does not exceed t_1.

The answers of the experts are:
1. $t_1 = 400$ [days], 2. $t_1 = 500$ [days],
3. $t_1 = 420$ [days], 4. $t_1 = 600$ [days].

2. You have 99% confidence that the number of days to failure of the engine does not exceed t_2.

The answers of the experts are:
1. $t_2 = 1000$ [days], 2. $t_2 = 1200$ [days],
3. $t_2 = 1000$ [days], 4. $t_2 = 1400$ [days].

Applying (7) for the expert 1 we obtain

$$\gamma_1 = \frac{\ln\left[\frac{\ln(0.5)}{\ln(0.01)}\right]}{\ln\left[\frac{400}{1000}\right]} = 2,06669$$

and

$$\lambda_1 = -\frac{\ln(0.5)}{400^{2,0666}} = 2.90512 \times 10^{-6}.$$

In the same way we get

$$\gamma_2 = 2.16306, \quad \lambda_2 = 1.00644 \times 10^{-6},$$

$$\gamma_3 = 2.182393, \quad \lambda_3 = 1.30154 \times 10^{-6},$$

$$\gamma_4 = 2.23498, \quad \lambda_4 = 4.28269 \times 10^{-7}.$$

From (5) and (8) we obtain the subjective prior measure in this case

257

$$\alpha((t,\infty)) = 4\sum_{k=1}^{4}\exp[-\lambda_k t^{\gamma_k}] \qquad (9)$$

6 SUBJECTIVE PRIOR MEASURE IN ESTIMATION OF RELIABILITY CHARACTERISTICS

6.1 Estimation of a reliability function

If we have no observations, the Bayes estimator of the unknown reliability function in this case is

$$\hat{R}_0(t) = \sum_{k=1}^{r} w_k \alpha_k((t,\infty)). \qquad (10)$$

The Bayes nonparametric estimator of the reliability function for the sample $X_1, \ldots, X_n$ is

$$\hat{R}_n(t|X_1,\ldots,X_n) = \frac{r}{r+n}\sum_{k=1}^{r} w_k \alpha_k((t,\infty)) +$$

$$+\frac{n}{r+n}\frac{1}{n}\sum_{i=1}^{n}\delta_{X_i}((t,\infty))$$

We see that this estimator depends on a sample of size n, on the number of experts m, experts' weights w_k, $k = 1, \ldots, r$ and on the subjective prior reliability functions $\alpha_k((t,\infty))$, $k = 1, \ldots, r$. The opinion of one expert weighs the same as one observation.

For the Weibull subjective prior measure the nonparametric Bayes estimator of reliability function has the form:

$$\hat{R}_n(t|X_1,\ldots,X_n) = \frac{r}{r+n}\sum_{k=1}^{r} w_k e^{-\lambda_k t^{\gamma_k}}$$

$$+\frac{1}{r+n}\sum_{i=1}^{n}\delta_{X_i}((t,\infty)) \qquad (11)$$

6.2 Continuation of the example

If we have no information on the failure of the engine, the value of reliability function estimator in the above case is

$$\hat{R}_0(t) = \sum_{k=1}^{4}\exp[-\lambda_k t^{\gamma_k}]. \qquad (12)$$

This function is shown on Figure 1.

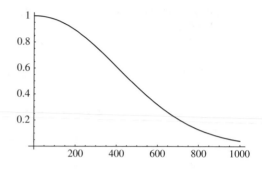

Figure 1. The function $\hat{R}_0(t)$, $t \geq 0$.

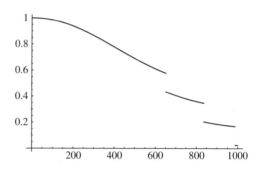

Figure 2. The function $\hat{R}_3(t|t_1, t_2, t_3)$, $t \geq 0$.

Now suppose that observations of the time to failure are:

$$t_1 = 836, \quad t_2 = 651, \quad t_3 = 987 \quad [\text{days}]$$

According to (11), we obtain the value of the reliability function nonparametric Bayes estimator

$$\hat{R}_3(t|t_1, t_2, t_3) = \frac{4}{7}\sum_{k=1}^{4} w_k e^{-\lambda_k t^{\gamma_k}} \qquad (13)$$

$$+\frac{\delta_{t_1}((t,\infty)) + \delta_{t_2}((t,\infty)) + \delta_{t_3}((t,\infty))}{7} \qquad (14)$$

which is shown in Figure 2.

6.3 Estimation of mean time to failure

Applying subjective prior measure in formula (3), for sample $X_1, \ldots, X_n$ we get the Bayes estimator of the mean.

$$\hat{m}(X_1,\ldots,X_n) = E(m|X_1,\ldots,X_n) =$$

$$= \frac{r}{r+n}m_0 + \frac{n}{r+n}\overline{X}, \qquad (15)$$

where

$$m_0 = \sum_{k=1}^{r} w_k \int_0^{\infty} \alpha_k((t,\infty))dt$$

and

$$m_0 = \sum_{k=1}^{r} w_k \int_0^{\infty} \alpha_k((t,\infty))dt.$$

For the Weibull subjective prior measure the nonparametric Bayes estimate of mean time to failure is

$$m_0 = \sum_{k=1}^{r} w_k \lambda^{-\frac{1}{\gamma_k}} \Gamma(\frac{1}{\gamma_k}+1) \qquad (16)$$

6.4 Continuation of the example

Using (16) we obtain the value of mean time to failure Bayes estimator in our case.

$$m_0 = 443.24 \text{ [days]}.$$

6.5 Estimation of the reliability function for censored data

According to (4), the Bayes nonparametric estimate of the reliability function for censored data and subjective prior is

$$\hat{R}(t|(Z,\Delta_j)) = E(R(t)|(Z,\Delta))$$

$$= \frac{r \sum_{k=1}^{r} w_k \alpha_k(t) + N^+(t)}{r+n} \times$$

$$\prod_{j=1}^{n} \left(\frac{r \sum_{k=1}^{r} w_k \alpha_k(Z_j^-) + N^+(Z_j) + \lambda_j}{r \sum_{k=1}^{r} w_k \alpha_k(Z_j^-) + N^+(Z_j)} \right)^{I\{\Delta_j=0,Z_j\le t\}} \qquad (17)$$

where $I\{\Delta_j = 0, Z_j \le t\}$ is an indicator of the event $\{\Delta_j = 0, Z_j \le t\}$, $\alpha_k(t) = \alpha_k((t,\infty))$, λ_j is the number of observations $= Z_j$, $j = 1, \ldots, n$ and $N^+(t)$ is the number of observations $> t$.

For the Weibull prior measure we have

$$\hat{R}(t|(Z,\Delta)) = \frac{r \sum_{k=1}^{r} w_k e^{-\lambda_k t^{\gamma_k}} + N^+(t)}{r+n} \times$$

$$\prod_{j=1}^{n} \left(\frac{r \sum_{k=1}^{r} w_k e^{-\lambda_k Z_j^{\gamma_k}} + N^+(Z_j) + \lambda_j}{r \sum_{k=1}^{r} w_k e^{-\lambda_k Z_j^{\gamma_k}} + N^+(Z_j)} \right)^{I\{\Delta_j=0,Z_j\le t\}} \qquad (18)$$

6.6 Continuation of the example

Assume that vector

$$z = (651, 836, 987, 1200, 1500, 1500)$$

is the value of statistics Z and the vector

$$\delta = (1,1,1,0,0,0)$$

is the value of statistics Δ. It means that numbers 651, 836, 987 are the instants of failure while the numbers 1200, 1500, 1500 are the censoring points. In our case we have $r = 4$, $n = 6$, $z_1 = 651$, $z_2 = 836$, $z_3 = 987$, $z_4 = 1200$, $z_5 = 1500$, $z_6 = 1500$, $\lambda_j = 1$ for $j = 1, 2$, 3, 4 and $\lambda_j = 2$ for $j = 5, 6$.

From (18) we obtain the value of nonparametric Bayes estimator for censored data.

As $I\{\Delta_j = 0, Z_j \le t\} = 1$ for $t < 1200$, then

$$\hat{R}(t|(z,\delta)) = \frac{4 \sum_{k=1}^{4} w_k \exp[-\lambda_k t^{\gamma_k}] + N^+(t)}{10}, \qquad (19)$$

where

$$N^+(t) = \begin{cases} 6 & \text{for } t \in [0,651) \\ 5 & \text{for } t \in [651,836) \\ 4 & \text{for } t \in [836,987) \\ 3 & \text{for } t \in [987,1200) \\ 2 & \text{for } t \in [1200,1500) \\ 0 & \text{for } t \in [1500,\infty) \end{cases}$$

For $t \in [1200, 1500)$ and $j = 4$, we obtain $I\{\Delta = 0, Z_j \le t\} = 1$. Hence

$$\hat{R}(t|(z,\delta)) = c_1 \frac{4 \sum_{k=1}^{4} w_k \exp[-\lambda_k t^{\gamma_k}] + 2}{10}, \qquad (20)$$

where

$$c_1 = \frac{4 \sum_{k=1}^{4} w_k \exp[-\lambda_k 1200^{\gamma_k}] + 3}{4 \sum_{k=1}^{4} w_k \exp[-\lambda_k 1200^{\gamma_k}] + 2} = 1.48969.$$

The realization of this estimate is shown in Figure 3.

7 CONCLUSIONS

The nonparametric Bayesian estimation is one of the strongly developing methods in mathematical statistics. This method may be used in reliability problems.

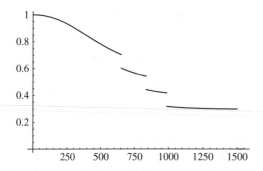

Figure 3. The function $\hat{R}(t|(z, \delta))t \geq 0$.

An important advantage of this method is the fact that it allows us to combine the subjective prior knowledge of experts with some data coming from observations. Bayes nonparametric estimator of the reliability characteristics depends on a subjective prior measure and a sample. The techniques of elicitation allow to construct a subjective prior measure based on experts' knowledge. Our subjective prior measure is such that for uncensored data opinion of one expert weights the same as one observation.

REFERENCES

Cooke, R. M. 1991. Experts in uncertainty. Opinion and Subjective Probability in Science. Oxford University Press. New York, Oxford.

Ferguson, T. S. 1973. A Bayesian analysis of some nonparametric problems. The Annals of Statistics, Vol.1, No. 2, 209–230.

Ferguson, T. S. & Phadia, E. G. 1979. Bayesian nonparametric estimation based on censored data. The Annals of Statistics, Vol. 7, No. 1.

Susarla, V. & Van Ryzin, J.1976. Nonparametric Bayesian estimation of survival curves from incomplete observations. J. Amer. Statist. Soc. 71, 897–902.

Susarla, V. & Van Ryzin, J. 1979. Large sample theory for survival curve estimators under variable censoring. *Optimizing Methods in Statistics*, New York, Acad.Press, 16–32.

Safety and Reliability – Bedford & van Gelder (eds)
© 2003 Swets & Zeitlinger, Lisse, ISBN 90 5809 551 7

Cost benefit analysis and flood damage mitigation in the Netherlands

M.Brinkhuis-Jak & S.R. Holterman
Road and Hydraulic Engineering Division, Ministry of Transport, Public Works and Water Management

M. Kok
Road and Hydraulic Engineering Division, Ministry of Transport, Public Works and Water Management,
Delft University of Technology, Faculty of Civil Engineering and Geosciences, Section of Hydraulic Engineering
HKV Consultants, Lelystad

S.N. Jonkman
Road and Hydraulic Engineering Division, Ministry of Transport, Public Works and Water Management,
Delft University of Technology, Faculty of Civil Engineering and Geosciences, Section of Hydraulic Engineering

ABSTRACT: Aim of this paper is to investigate the application of cost benefit analysis methods in decision-making on a desired flood protection strategy in the Netherlands. After a discussion of history and developments in flood protection in the Netherlands the method of cost benefit analysis is presented as a useful instrument in decision-making. In the second part of the paper the economic analysis of flood protection strategies is firstly approached from a theoretical point of view. Subsequently the economic analyses carried out in practice are described for two more practical cases, the study on "emergency retention areas" and the dike reinforcement program in the river system. The paper shows that an economic analysis, when correctly applied, can provide important rational information in the decision-making process.

1 INTRODUCTION AND OVERVIEW OF DEVELOPMENTS IN FLOOD PROTECTION

1.1 *Introduction*

Large parts of the Netherlands lie below sea level and are threatened by river floods. The flood depths in some areas can therefore become higher than 7 meters. Without the protection of dunes, dikes and hydraulic structures more than half of the country would be almost permanently flooded as is shown in figure 1. Therefore, flood protection has always received much attention. There is always the possibility of flooding. But how serious is this danger? It is difficult to say. On the one hand we sail on a feeling of safety and on the other hand, especially short after a disaster, on a feeling of unsafety. Dunes and water defences protect the country, yet never 100%. There is no such thing as absolute safety against flooding. The question is which risks are acceptable and which ones are not. This is an ever recurring socio-political consideration, one which is always fed by knowledge developments.

In the last decade of the 20th century methods have been developed to determine the probability of flooding

Figure 1. The Netherlands without flood protection (the dark area can be flooded due to influence from the sea).

and the consequences of a flood. The outcomes of this research offer new insights and moreover new possibilities to carry out a cost benefit analysis for various flood protection strategies. Aim of this paper is to investigate the application of cost benefit analysis methods in decision-making on a desired flood protection strategy in the Netherlands.

The paper is structured as follows. The remainder of section 1 will give a short overview of history, and new developments in flood protection. Section 2 will describe the method of cost benefit analysis and section 3 will give an overview of the available methodology for flood damage estimation. Section 4 shows how the decision on an economically acceptable level of flood protection can be approached from a more theoretical point of view. Section will 5 analyse the application of methods for cost benefit analysis for a practical case. The conclusions from this study are summarized in section 6.

1.2 History

Due to its location, the Netherlands is always threatened by floods. Life in the delta of the Rhine and Meuse involves risks, but has also enabled the Netherlands to develop into one of the main gates of Europe. In the past river floods provided fertile soil and clay for brick-works, but the opposite of this positive effect were the negative effects, such as the loss of goods and cattle and the danger of drowning. As welfare increased and population density grew, more and better protection systems were built to prevent flooding. Since the Middle Ages more and more dikes, quays and hydraulic structures have been constructed. Whether the protection against the water is sufficient, is an all-time question; one that is asked at this moment and that will be asked constantly in the future.

Since the danger of flooding is difficult to determine in advance, politics and society usually adopted a reactive position until recently. An "almost" flood should not repeat itself. Until 1953 dikes were constructed to withhold the highest known water level. In 1953 a flood from the North sea occurred in the south of the Netherlands, killing 1800 people and causing the disruption of a large part of the Netherlands. This flood disaster resulted in major investments to improve the water defences, based on a more pro-active base. After the 1953 flood, the Delta Committee was installed to investigate the possibilities for a new safety approach. Safety was not based on the highest occurred level anymore, but on a rough cost-benefit analysis. In an econometric analysis the optimal safety level was determined for the largest flood prone area, Central Holland. This work laid the foundations for the new safety approach, in which dikes are dimensioned based on a design water level with a certain probability of occurrence.

The Deltaworks, which were constructed to protect the Southwest part of the country against inundation from sea, were given priority over protection against river floods. After the completion of the Deltaworks, the strengthening of the river dikes began at full speed.

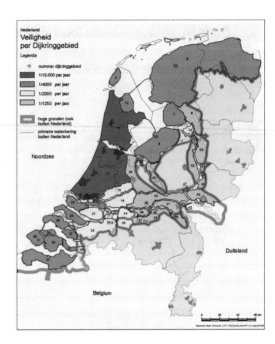

Figure 2. Overview of protection standards for dike rings given in the "water defence act".

As the last big river flood dated back to 1926, there was strong opposition from environmentalists who were against the strengthening programme. The strengthening of river dikes resulted in loss of nature areas, landscape and sites of cultural value. In 1993 the Government and Parliament agreed upon a new approach, saving the landscape, nature areas and places of cultural value (see the Boertien-case in section 4 of this paper). The river floods in 1993 and 1995 once again drew attention to the risks of life in a delta; afterwards the water defences were reinforced at an accelerated rate. In 2001 most water defences were at strength, and in accordance with the safety standards referred to in the Water Defence Act. For the coastal areas design water levels (see above) have been chosen with frequencies between 1/4000 [1/year] and 1/10.000 [1/year]. For the Dutch river area the safety standards were set between 1/1.250 [1/year] and 1/2.000 [1/year]. These safety standards are shown in figure 2.

1.3 Developments in coping with floods in the Netherlands

While damage protection in the Netherlands traditionally aimed at reduction through improved dike construction, nowadays new political movement can be seen that searches for measures to prevent flooding *without* raising the dikes along the rivers. For example by giving the river more space. This "Room for the Rivers" is a widespread and accepted idea in the Netherlands now. In the coming document of the national policy on spatial planning in the Netherlands it is expected that room for rivers will be described. An other "hot item" are the so-called "Emergency Retention Areas": areas that can be inundated in a controlled way to prevent uncontrolled flooding of other areas. They will be used when a discharge occurs that exceeds the design discharge. These Emergency Retention Areas are, though controversial, a serious item in today's Dutch political discussion. Also more attention can be found for evacuation planning, early warning systems, insurance of flood damage and for the link with spatial planning.

The expected developments of the rise of the sea level, higher river discharges and soil subsidence require a pro-active policy, in which the increase of the interests and investments to be protected will have to be taken into consideration. Knowledge of water and water defences is indispensable when considering the desired protection level against flooding. The Technical Advisory Committee for Water Defences has developed a method to determine the probabilities of flooding and has successfully tested its application for four dike ring areas (TAW, 2000). Based on this study the Ministry of Public Transport, Public Works and Water Management carries out a project in which the probabilities of flooding are calculated for all 53 dike ring areas (definitive results 2004). The method to determine probabilities of flooding can be distinguished from the current "design water level" approach at three levels:

- The transition from an individual dike section to a dike ring approach: the strength of a dike ring (consisting of dikes, engineering structures and dunes) can be calculated as a whole. The method will considerably increase understanding of possible weak links in the protection system.
- Taking equal account of various failure mechanisms of a dike (ring). This is different from the current approach, in which the safety analysis of the dike dominated by the mechanism of overtopping and overflow of water.
- Discounting in advance, in a systematic and verifiable way, all uncertainties when calculating the probabilities of flooding. In the current approach uncertainties are for the greater part discounted afterwards by building in an additional safety margin.

Knowing this probability of flooding gives the opportunity to use a risk-assessment (or cost benefit analysis) to determine if the current – or in future expected-flooding risks are acceptable.

When evaluating the acceptability of the probability of flooding in an area the potential damage caused by floods and danger for the population are key-information. In the next years the Ministry wants to portray the damage as a result of a flood together with other parties involved. At that moment it will also be possible to calculate the costs and benefits of the entire range of measures. These measures might include research (inspection and testing, study and research), reinforcement and elevation of the water defences, "room for the rivers", retention areas as well as restriction of flood consequences by means of spatial planning or technical and administrative measures.

2 COST BENEFIT ANALYSIS

The basic principle of cost benefit analysis requires that a project results in an increase of economic welfare, i.e. the benefits generated by the project should exceed the costs of it. In a general cost benefit analysis the benefits of an activity are compared with the costs of the activity. If the benefits are higher than the costs, the activity is attractive (it generates an increase in economic welfare). If the benefits are lower, the activity is not attractive. In flood management this means that the costs of measures for increasing the safety against flooding (for example dike strengthening of flood plain lowering) are compared with the decrease in expected flood damage. In the cost figure different type of costs have to be included: costs of investment (fixed and variable) and the costs of maintenance and management. The benefits include the reduction of damage costs, which are often subdivided in direct costs (repair of buildings and interior damage), costs of business interruption of companies in the flooded area, and indirect costs outside the flooded area (mainly due to business interruption. But outside the flooded area there may also be companies which have benefits of the flood). Also the potential economic growth due to improved flood defence should be taken into account in a full cost benefit analysis. Decisions with respect to the safety of flooding are in general not done in the private sector, but it is a "common good" and part of the societal preferences. It is therefore important that the value concept is included in the decision problem.

The cost benefit approach can be criticised because it necessitates quantification of all costs and benefits in monetary terms. However, in our opinion it is one of the essential parts of information, which is necessary for rational decision-making. Other elements can be added, in order to achieve a complete overview of all relevant aspects of the decision problem. Yet, there is no general accepted framework available where the relevant pieces of information are put together, and there are different ways of monetising the non-monetary impacts (for example contingent valuation by surveying the willingness to pay). Therefore it is investigated in the context of the flood management in the Netherlands how subjective attitude towards flood risks can be incorporated into a cost benefit analysis. In a utility framework, see for example (French, 1988) non-monetary measurable impacts can be expressed in the utility function, which describes the usefulness of the decision. The same can be done for the monetary impacts, so that all impacts can be summed up in one variable. An important notion is the attitude (of the people who might be a victim of a flood, or the decision maker) towards the costs and the flood damage reduction. There are three basic attitudes: the risk neutral (which is assumed in a cost benefit analysis), the risk-prone attitude (where costs are valued lower and damage reductions are valued higher than they are) and the risk-averse attitude (where costs are valued higher and flood damage is valued lower). It turns out in the context of flood management that a risk-averse decision maker will choose higher protection levels than a risk-neutral of risk prone one (Voortman, 2002). However, a reliable method to quantify the risk aversion factor is in the context of events with "high impacts, low probabilities" not yet available.

Summarizing this section it can be stated that despite the limitations, a cost benefit analysis still can provide significant rational information to the decision makers.

3   ESTIMATION OF FLOOD DAMAGE

Cost benefit analysis of flood protection alternatives requires insight in the magnitude of flood damage. This section will give a brief overview of the available methodology for flood damage estimation.

Firstly damage is analysed from an economical point of view. An inundation of one of the densely populated, highly economically developed areas in the Netherlands will undoubtedly cause enormous economic damage. The extent of this damage depends on the nature of the flood, for example from sea or river, and the properties of the area, for example terrain height and land use. A method has been developed for estimation of the economic damage due to flooding (Vrisou van Eck et al, 2001). The procedure for damage estimation is schematically shown in figure 3. Information on land use is combined with flood data (water depth). Stage-damage relations have been developed for different types of land use, which estimate (part of maximum) damage as a function of water depth. The result of the damage assessment is the total economic damage that can be expected, given that particular scenario.

While the described method focuses mainly on estimation of the direct economic damage, ongoing research is carried out to gain more insight in the indirect effects of floods for the national economy. It is expected that neglecting of the indirect effects of floods will lead to an under-estimation of damage numbers. In the Netherlands for example the loss of gas-supply

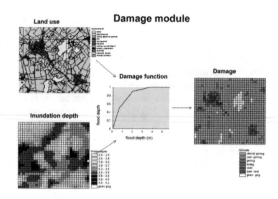

Figure 3.   Principle of the method for assessment of economic damage.

from the fields in the Northern part of the country due to flooding will result in an economic damage. Other examples could be the loss of the national airport of Schiphol or the Rotterdam harbour.

However, when carrying out a complete cost benefit analysis, also other types of damage have to be included in the analysis. For example loss of life caused by floods. It is likely that floods in the Netherlands have inundation depths of more than 4 meters and it may even be 7 metres. The big flood of 1953 in South-West Netherlands caused 1800 victims. In a international survey of more than 1000 floods all over the world it is included that on average 1% of the people at risk is a victim of the flood (Jonkman, 2002). However, the final number will depend on both the properties of the flood and evacuation.

Another potential problem is the environmental damage due to pollution of all kind of chemical elements. The number of potential objects, which might cause the pollution, is enormous (chemical factories, stocks, oil tanks, …). However, the conditional probability of releasing polluted substances (given that a flood has happened) is not known yet.

Damage of Nature, Landscape and Ecological values: A flood may damage unique values, which cannot be recovered (due to human interaction or by nature) after a flood has happened. Four aspects (woodland area, flora and vegetation, freshwater ecosystems and historic buildings) are charted and valued, depending on the properties of the flood, by expert judgement.

Economic valuation of these "intangible" damage types is a difficult subject. Although some methods have been developed, which attribute an economic (monetary) value to loss of life, ecological damage, these are generally not taken into account in a cost benefit analysis. Yet, there is no general accepted framework available where the relevant pieces of information are put together, and there are different ways of valuing the non-monetary impacts.

4 ECONOMIC OPTIMISATION AND COST BENEFIT ANALYSIS IN THEORY

Firstly, the method of economic optimisation is presented as a framework for the derivation of an economically optimal level of risk in section 4.1. This method is closely related to the cost benefit analysis, as is shown in section 4.2.

4.1 Economic optimisation

The derivation of the (economically) acceptable level of risk can be formulated as an economic decision problem. According to the method of economic optimisation, the total costs in a system (C_{tot}) are determined by the sum of the expenditure for a safer system (I) and the expected value of the economic damage ($E(D)$). In the optimal economic situation the total costs in the system are minimised:

$$\min(C_{tot}) = \min(I + E(D))$$

The method of economic optimisation was originally applied by van Danzig (1956) to determine the optimal level of flood protection (i.e. dike height) for Central Holland (this polder forms the economic centre of the Netherlands). An exponentially distributed flooding probability (P_f) was assumed, which depends on the flood level h and the parameters A and B of the exponential distribution:

$$P_f = e^{-\frac{h-A}{B}}$$

The total investments in raising the dikes (I_{tot}) are determined by the initial costs (I_{h0}) and the variable costs (I_h). The dike is raised X metres, the difference between the new dike height (h) and the current dike height (h_0).

$$I_{tot} = I_{h0} + I_h \cdot X \qquad and \qquad X = h - h_0$$

In this study a more general formulation has been chosen between investments and flood protection level (denoted by flooding probability). Based on van Danzig a linear relation between the two has been adopted, but for more practical applications another relation can be chosen. The investment function is reformulated by substitution as a linear function of the negative logarithm of the flooding probability with parameters constant I_0 and steepness I' (By substitution with the equations presented above it can be shown that: $I_0 = I_{h0} + I_h(A - h_0)$ and $I' = I_h \cdot B$):

$$I_{tot} = I_0 + I' \cdot (-\ln(P_f))$$

The expected value of the economic damage can be calculated from the probability of flooding (P_f), the

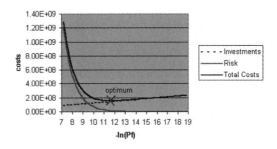

Figure 4. Relation between total costs and decreasing failure probability, for the example and corresponding variables analysed in (van Danzig, 1956):[$I_0 = 3.9.10^7$ (Dfl), $I' = 0.33 \cdot 10^6$ (Dfl), $D = 24.10^9$ (Dfl), $r' = 0.015$ (/yr)].

damage caused by the flood (D). The expected value has to be discounted with the so-called reduced interest rate (r'), which takes into account the interest rate (r) and the economic growth rate (g), for a very long time period considered, this can be written as:

$$E(D) = P_f \cdot D / r' \qquad r' = r - g$$

The total costs are the sum of investments and the expected value of the economic damage. The economic optimum is found by minimising the total costs. The derivative of the total costs and the flooding probability results in the economically optimal flooding probability ($P_{f,opt}$), from which the optimal dike height can be derived:

$$C_{tot} = I_0 + I' \cdot (-\ln(P_f)) + P_f \cdot D / r'$$

$$dC_{tot} / dP_f = 0 \quad \Rightarrow \quad P_{f,opt} = I' \cdot r' / D$$

The relation between (decreasing) flooding probability and investments, risk and total costs is shown in figure 4.

4.2 Cost benefit analysis

However, this economic optimisation merely takes into account the cost side of the flood protection problem, and does not consider the potential economic benefits in the area due to improved flood protection. It has been shown by Voortman (2002) how economic benefits can be taken into account in the framework presented above. This shows that the economic optimisation as presented above is a special (limited) case of this full cost benefit analysis. A cost benefit analysis can be carried out to assess the profitability of a project, as has been described in section 2. In a simplified approach it should be checked that the costs in the initial situation should exceed the total costs after completion of the project. After determination of an economically optimal level of system protection, using the economic optimisation as presented in section 4.1, this cost benefit criterion should also be applied. Following the formulations given above the criterion can be written as follows:

$$I_0 + I' \cdot (-\ln(P_f)) < (P_{f,0} - P_f) \cdot D / r'$$

Where: $P_{f,0}$ – flooding probability in the initial situation.

From the criterion shown above it can be seen that the profitability of a measure will depend on the ratio between investments and risk or damage reduction. Decision makers may choose the most cost effective strategy, i.e. the project that achieves the largest risk reduction with the smallest investments. This is the project for which the highest protection level is found (i.e. the smallest optimal failure probability) at lowest cost. However, it should be noted that based on other values, such as ecological, social and political considerations, an alternative could be chosen that would not be the most favourable when merely economic aspects are considered.

5 THE PRACTICE OF COST BENEFIT ANALYSIS

Although theoretical concepts are nice and attractive, it is interesting to investigate the application of the theory in practice. Therefore we studied two cases in the Dutch river-area. Both studies were carried out by a (different) advisory committee, which advised the Dutch government.

5.1 1992: "River dike reinforcement criteria testing commission"

Reason of this project was the finding of the 1977 Commission (called the Becht commission), which recommended that river dikes be designed tot resist water levels that would be exceeded with an expected frequency of 1/1.250 [1/year]. While dike improvements based on these standards were underway, protests grew against their harmful impacts on the landscape and natural and cultural values on and along the dikes. In response, the Minister of Transport, Public works and Water Management established the "River dike reinforcement criteria testing commission" (also called the Boertien Commission after its chairman) and contracted for the research. (Walker et al, 1993). Primary objective of the study was to identify policies that would provide a high level of safety, would not cost too much, and would preserve as much as possible of the existing landscape, natural, and cultural values (LNC values) along the rivers.

In the study it was stated that a flood protection policy is composed of two parts: a safety level, and a strategy for improving the dikes and/or reducing the water level of the rivers to provide the chosen level of safety. The minimum safety level considered was a level of 1/200 [1/year] and the maximum level was the level of 1/1250 according to the Becht commission. The diverse consequences (or impacts) of the policies examined were estimated and were displayed in a scorecard. The scorecard which summarises the arguments of the committee is given in table 1.

The results show that the benefits through reduction of projected flood damage greatly exceed the financial costs of improving the dikes. The table shows that the return on the M€l 75 ($= 375 - 300$) that it would cost tot build dikes to a safety level of 1/1250 instead of 1/200 is a present-value benefit of at least M€ 994. The commission recommended to maintain the safety-level in the river area on the level of 1/1250 [1/year] and the government followed this line.

The score table shows that, if monetary costs and benefits are the only desiderata, even the 1/1250 safety level is less than what a pure financial cost/benefit analysis would be recommend. In the study no higher safety-levels were regarded. Apparently the decision makers considered a higher safety-level not acceptable from other than economic (LNC values) point of view. Our opinion is that it would have been better to take this extra step to complete the economic analysis. In this way the cost benefit approach is not used in an optimal way to support explicit and rational decision-making.

Table 1. investment cost and estimation of risk reduction for the alternatives analysed by the "Boertien Commission" (derived from table III in (Walker et al., 1993).

Alternative	Investment costs (M€ $=10^6$€)	Present value of reduction in expected flood damage (M€)		
		(max)	(med)	(min)
safety level 1/200	300	0	0	0
safety level 1/500	331	2872	1997	726
safety level 1/1250	375	4089	2809	994

5.2 2002: "Committee Emergency Retention Areas"

Start of this project was the growing awareness in political regions that, even with the high safety levels in the Netherlands, absolute safety does not exist. The Minister of Transport, Public works and Water Management established the "*Committee Emergency Retention Areas*" (2002), which is also called the Luteyn Commission after its chairman. Objective of the committee was to advise about the attractiveness of a spatial reservation of certain areas which can then be used as a storage basin to store access of water along the big rivers Rhine and Meuse. The committee was asked to advise about the usefulness, effectiveness and necessity of these storage basins, and if the idea is attractive, to choose (select) certain areas. The committee advised that it is indeed attractive to have these basins, and they proposed three areas, see figure 5 (previous page) for an overview.

The basic argumentation of the committee is that controlled flooding is to be preferred above uncontrolled flooding, and therefore it is wise to invest more than an estimated one billion €. The scorecard which summarises the arguments of the committee is presented in table 2. This table can be found on page 32 in the report of the committee (Committee Emergency areas, 2002).

On the basis of table 2, the committee concluded that: "The committee has assessed that the total investment costs is about 1,25 10^9 €. But on the other hand, with a controlled flooding less people have to leave their homes, the societal disruption is smaller and the flood damage will be substantially lower. With other words, the benefits are far bigger than the costs" (page 33).

However, if we apply the concepts as described in the sections above, we may conclude that the cost benefit analysis in the report cannot pass the test. The committee compares the total flood damage with the investments of the retention areas. In such a comparison, the present value of the economic risk has to be calculated (flood damage multiplied with the probability divided by the discount rate). Note of authors of this paper: it is our opinion that the assumed flood damage in table 2 is unrealistically high: it assumes that all dike rings along the river will be flooded. In reality, however, if one of the dike rings is flooded, the expected damage of the other dike rings will be lower, because the water levels in the river will drop after failure of one of the water defences. Another observation is that in the calculations of the committee and in table 2 it is assumed that the emergency areas reduce the flooding probability downstream these areas completely. This, however, is not a valid argument, as noted by the committee on page 22 (the flooding probability of these areas will be reduced to 1/4000). This part of the criticisms is also been remarked by the committee of water defence experts (Technical Advisory committee on Water defences). Applying the method as described above and in previous sections, table 3 is obtained. From the table it can be concluded that the costs of the emergency areas are (much) higher than in the current situation under the assumption of maximal damage. We also remark that the reduction of the flooding probability due to the impact of retention areas may be lower than is assumed in table 3 (see Kok et al, 2003).

Table 3 indicates that for case of realistic damage the creation of the emergency areas will lead to a

Table 2. Scorecard of the two alternatives as presented in the report of the Committee Emergency Areas, 2002.

Alternative	Number of people to be evacuated	Flood damage (10^9 €)	Investment costs (10^9 €)
Current situation (without emergency areas)	500.000	55	0
New situation (with 3 emergency areas)	35.000	0.7	1.25

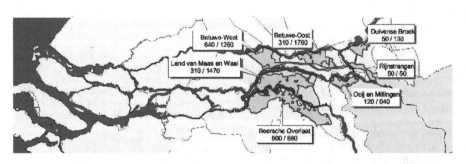

Figure 5. Proposed areas along the Rhine and Meuse for selection of an emergency retention area (Numbers indicate economic damage in (mln of Euros) with and without additional protective measures).

Table 3. Scorecard of the alternatives using the theoretical concepts of section 4 and using more realistic assumptions.

Alternative	Investment Costs (10^9 €)	Flood damage (10^9 €)	Present value Flood damage (10^9 €)	Total costs (10^9 €)
Current situation (maximal damage)	0	55	$55/(1250 * 0.04) = 1.10$	1.1
Current situation (realistic damage)	0	15	$15/(1250 * 0.04) = 1.10$	0.3
New situation (with 3 emergency areas)	1.25	0.7	$0.7/(1250 * 0.04) = 0.014$	1.264

reduction of present value of flood damage of 0.29 billion € a year, at a cost of 1,25 billion €. This difference between benefits (= risk reduction) and costs can be overcome by assuming non-monetary values of the three emergency areas. At the moment of writing this paper the government has not yet decided whether to adopt the commission's advise or not. If a decision would only be based on the cost-benefit analysis, the commission's advise would not be adopted. However, there are more values than the economic values, such as landscape, natural and cultural (LNC) values and social values. The final weighing of economic, cost-benefit aspects and other aspects is a political choice.

6 CONCLUSIONS

The aim of this paper is to investigate the application of cost benefit analysis methods in decision-making on a desired flood protection strategy in the Netherlands. The following conclusions and recommendations can be given:

1. The basic principle of cost benefit analysis indicates whether a project results in an increase of economic welfare, i.e. whether the benefits generated by the project exceeds the costs of it. An economic optimisation can be carried out to determine the optimal level of system. The information provided by the cost benefit analysis and /or the economic optimisation should be considered as a technical advice to the decision- and policy-makers. In the decision making process it should be combined with other types of relevant information.
2. An important issue in the economic analysis is the estimation of flood damage. Various types of damage will occur in case of a flood, such as material (direct) and indirect economic damage, cultural and ecological losses and loss of life.
3. Analysis of two recent case studies shows that the theoretical cost benefit concepts are not fully applied in practice. In the two case studies the government asked an advise to an independent committee with respect to the level of investments in river flood

management. In the first case study the committee compared the costs and benefits in a sound way, but the optimal level of protection is not determined. In the second case study the committee did not compare the costs and benefits correctly, and compared the investment costs directly with the flood damage. These shortcomings may have influenced the decision.

4. It is recommended to apply the concept of cost benefit analysis in decision problems with respect to flood damage mitigation more explicitly. Providing information to decision makers generated by these concept will increases the possibilities that the alternative is chosen which optimises the societal needs.

From an economic point of view decision makers may choose the flood protection strategy that achieves the largest risk reduction at lowest costs. The final decision on a desired flood protection level should not only consider economic aspects, but it should involve a comparison of all relevant alternatives. The economic optimisation and the cost benefit analysis can provide important rational information in this decision-making process.

DISCLAIMER

Any opinions expressed in this paper are those of the authors and do not necessarily reflect the position of the Dutch Ministry of Transport, Public Works and Water Management.

REFERENCES

Danzig D. van, 1956. Economic decision problems for flood prevention, *Econometrica 24*, p. 276–287.
S.N. Jonkman, P.H.A.J.M. van Gelder, J.K. Vrijling, 2002. Loss of life models for sea- and river floods, *in Proceedings Flood Defence '2002, Wu et al. (eds), Science Press*, New York Ltd., ISBN 7-03-008310-5.
Committee Emergency areas, 2002. Controlled flooding. Report of committee emergency areas. The Hague, May 2002.

French, S. 1988. Decision theory: an introduction to the mathematics of rationality, Chichester Horwood.

Kok, M., Stijnen, J.W. & Silva, W. 2003. Uncertainty analysis of river flood defense design in the Netherlands. *Proceedings of the ESREL conference*, Maastricht.

TAW, Technical Advisory Committee on Water Defences. 2000. Towards a new safety approach. A calculation method for probabilities of flooding.

Voortman H. 2003. Risk-based design of large-scale flood defence systems, Delft University Press.

Vrisou van Eck N., Kok M., Vrouwenvelder A.C.W.M., 2000. Standard method for Predicting Damage and Casualties as a Result of Floods, Lelystad: HKV Conultants and TNO Bouw.

Walker, W., Abrahamse, A., Bolten, J., Kahan, J.P., Van de Riet, O., Kok, M. & Den Braber, M., 1994. A Policy Analysis of Dutch River DIke Improvements: Trading off safety, cost and environmental impacts. *Operations Research* (42), 5, 823–836.

Safety and Reliability – Bedford & van Gelder (eds)
© 2003 Swets & Zeitlinger, Lisse, ISBN 90 5809 551 7

Optimization of maintenance cost under asymptotic reliability constraint

R. Briš
Technical University Of Ostrava, Faculty Of Electrical Engineering and Computer Science, Czech Republic

E. Châtelet
System Modeling and Dependability Laboratory, University of Technology of Troyes, France

F. Yalaoui
Industrial Systems Optimization Laboratory, University of Technology of Troyes, France

ABSTRACT: General preventive maintenance model for input components of a system, which improves the reliability to "as good as new", is used to optimize the maintenance cost. The cost function of a maintenance policy is minimized under given availability constraint. An algorithm for first inspection vector of times is described and used on selected system example. A special ratio-criterion, based on the time dependent Birnbaum importance factor, was used to generate the ordered sequence of first inspection times. Problem called as "reliability assurance" is theoretically solved and answered, i.e. finding the cost of maintenance when asymptotic availability value (*WRV*-Worst Reliability Value) conforms to a given availability constraint (asymptotic availability value is supposed as the limiting value for time going to infinity). System representation using acyclic graph is briefly introduced. Basic system availability calculations of the paper were done by using Matlab program (analytical) for computing of the *WRV* value of any coherent system under maintenance. A genetic algorithm optimization technique is used and briefly described to create the algorithm (in Matlab as well) to solve the problem of finding the best maintenance policy with a given restriction.

1 INTRODUCTION

The evolution of system reliability depends on its structure as well as on the evolution of the reliability of its elements. The latter is a function of the element age on a system's operating life. Element ageing is strongly affected by maintenance activities performed on the system. Preventive maintenance (PM) consists of actions, which improve the condition of system elements before they fail. PM actions such as the replacement of an element by a new one, cleaning, adjustment etc. either return the element to its initial condition (the element becomes "as good as new") or reduce the age of the element. In some cases the PM activity does not affect the state of the element but ensures that the element is in operating condition. In this case the element remains "as bad as old".

Optimizing the policy of preliminary planned PM actions is the subject of much researches. In the past, the economic aspects of preventive and corrective maintenance have been extensively studied for monitored components in which failures are immediately detected and subsequently repaired. Far less attention has been paid to the economics of systems in which failures are dormant and detected only by periodic testing or inspections. Such systems are especially common in industrial safety and protection systems. For this kind of systems, both the availability evaluation models and the cost factors assessment differ considerably from those of monitored components (Vaurio, 1999).

This paper develops availability and cost models for systems with periodically inspected and maintained components subjected to some maintenance strategy. The aim of our research is to optimize, for each component of a system, the maintenance policy minimizing the cost function, with respect to the availability constraint such as $A(t) \geq A_0$, for time t going to infinity.

A genetic algorithm (GA) is used as an optimization technique. GA is used to solve the above mentioned problem, i.e. to find the best maintenance policy under given asymptotic reliability constraint. The solution comprises both the availability and the cost evaluation.

WRV: worst reliability value
N: total number of components
$T_0 = (T_0(1), T_0(2)...T_0(N))$: first inspection time vector
$T_0^{ord} = (T_0^{(1)}, T_0^{(2)}, ...T_0^{(N)})$: ordered first inspection time vector; $T_0^{(1)} \leqslant T_0^{(2)}...\leqslant T_0^{(N)}$
$T_P = (T_P(1), T_P(2)...T_P(N))$: solution vector of system component inspection periods
T_M: mission time
$C(e(i,k))$: cost of one inspection of ith component in the kth parallel subsystem.
$A(t)$: system availability at the time t
A_0: availability constraint (lower limit)

2   PREVENTIVE MAINTENANCE MODEL FOR GENERAL SERIES-PARALLEL SYSTEMS

2.1   *Input component's model*

In the paper we will assume that the PM actions improve the reliability of input component to "as good as new". It means that the element's age is restored to zero (replacement). The model is demonstrated in Figure 1 (T_F: Time to Failure).

The problem to find the optimal vector T_P is closely connected with another problem, i.e. to find the optimal first inspection time vector T_0. Of course, it makes no sense to carry out inspections in the beginning of the life of system, when both the system and its input components are very reliable. Consequently the preliminary calculations must be realized to find the optimal T_0 for each of input components. The optimal vector T_0 must be constructed so that it takes into account both cost and reliability view.

2.2   *General series-parallel structure*

Optimal PM plan is found for a general series-parallel structure that is showed in the Figure 2.

2.3   *Cost model*

Cost of the above mentioned preventive maintenance policy of a given system is simply given by summarizing each of the PM inspections done on the components that are under maintenance:

$$C_{PM} = \sum_{k=1}^{K} \sum_{i=1}^{E_k} \sum_{j=1}^{n_{e(i,k)}} C_j(e(i,k))$$

- $n_{e(i,k)}$ represents the total number of inspections of the ith component in the kth parallel subsystem in the course of mission time

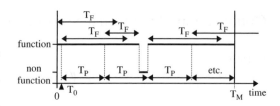

Figure 1.   PM model for periodically tested elements.

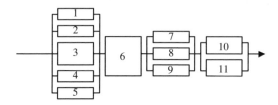

Figure 2.   General series-parallel structure.

- $C_j(e(i,k))$ is the cost of the jth inspection of the ith component in kth parallel subsystem
- E_k is the number of components in given kth parallel subsystem
- K is the number of parallel subsystems
- the total number of components is:

$$N = \sum_{k=1}^{K} \sum_{i=1}^{E_k} e(i,k)$$

In most cases, the cost of inspection of a given input component is constant in the course of mission time, i.e.

$$C_{PM}(e(i,k)) = \sum_{j=1}^{n_{e(i,k)}} C_j(e(i,k)) = n_{e(i,k)} \times C(e(i,k)),$$

where $C(e(i,k))$ is the cost of one inspection of the ith component in kth parallel subsystem.

$$n_{e(i,k)} = \left\lfloor \left| \frac{T_M(e(i,k)) - T_0(e(i,k))}{T_P(e(i,k))} \right| \right\rfloor$$

is the integer part of the fraction, and T_M, T_0, T_P, are respectively, mission time, first inspection time and inspection period of a given component.

2.4   *Problem formulation*

A system consisting of subsystems connected in series is considered. Each subsystem contains different elements connected in parallel (see, for example, Figure 2). Each component is characterized by its failure rate function $h_j(t)$ and PM cost of an inspection $C(e(i,k))$.

The time in which a component is not available due to PM activity, is negligible if compared to the time elapsed between consecutive activities.

Basic assumptions of this paper are as follows:

1. Testing actions (or inspections) are carried out periodically, for jth component with the period of $T_P(j)$. Inspections are ideal which means that the component is renewed (as good as new). The inspection of the jth component begins at the time $T_0(j)$.
2. A system consisting of subsystems connected in series is considered. Each subsystem contains different components connected in parallel. Each component is characterized by its failure rate function $h_j(t)$, and PM cost of one inspection; $C(e(i,k))$ is cost of one inspection of the ith component worked in kth parallel subsystem.

First aim of the research (was successfully solved in Bris et al. 2002) is to optimize, for each component of a system, the maintenance policy minimizing the cost function C_{PM}, and respecting the availability constraint $A(t) \geq A_0$, for all t, $0 < t \leq T_M$, and a given mission time T_M.

Second aim of the research and the main aim of the paper is to optimize, for each component of a system, the maintenance policy minimizing the cost function C_{PM}, during the mission time T_M, and respecting the availability constraint $A(t) \geq A_0$, for time going to infinity.

In both cases, it is necessary to find optimal vectors (cost minimizing) $T_P = (T_P(1), T_P(2)...T_P(N))$, and $T_0 = (T_0(1),T_0(2)...T_0(N))$, under given availability constraint.

3 ASYMPTOTIC AVAILABILITY COMPUTATION, SYSTEM REPRESENTATION AND FINDING OPTIMAL T_0

3.1 Basic concepts from the reliability theory

Consider structures capable of two states of performance, either complete success in accomplishing an assigned function or complete failure to function. Similarly, the components from which the structures are constructed are assumed capable of only the same two states of performance. The performance of the structure is represented by an indicator φ which is given the value 1 when the system functions, 0 when the system fails. The performance of each of the n components in the structure is similarly represented by an indicator x_i which takes the value 1 if the ith component is functioning, 0 if ith component is failed ($i = 1,2,...,N$).

It is assumed that the performance of a structure depends deterministically on the performance of the components which is characterized by the function φ of $x = (x_1, x_2, ..., x_N)$; $\varphi(x)$ is called the *structure function* of the structure (Misra 1992).

For structures in which each component if functioning contributes to the functioning of the structure, certain hypotheses appear intuitively acceptable:

(i) $\varphi(1) = 1$, where $1 = (1,1, ..., 1)$,
(ii) $\varphi(0) = 0$, where $0 = (0,0, ..., 0)$,
(iii) $\varphi(x) \geq \varphi(y)$ whenever $x_i \geq y_i$, $\forall\, i = 1,2, ..., N$.

Hypothesis (i) states that if all the components function, the structure functions. Hypothesis (ii) states that if all the components fail, the structure fails. Finally, hypothesis (iii) states that functioning components do not interfere with the functioning of the structure. Structures satisfying (i), (ii), and (iii) are called *coherent*. Sometimes also the *monotonic* term is used, since such structures are characterized by a monotonic structure function which is equal to 0 at 0, and 1 at 1.

Assuming a probability distribution for the performance of the components, we obtain the *availability of the ith component* as follows:

$p_i = P[X_i = 1] = E[X_i]$, where X_i is the binary random variable designating the state of component i (this expression is the reliability with the same X_i but conditioned to the component has not failed).

The *availability (reliability for non-repairable systems)* of the structure is:

$$h = P[\varphi(X) = 1] = E[\varphi(X)]$$

The structure function now becomes a binary random variable. When components perform independently, we may write $h = h(p)$, where $p = (p_1, p_2, ..., p_N)$, $h(p)$ is **called reliability function** of the structure.

3.2 Asymptotic availability computation

Let us suppose the system from Figure 2, which is, no doubt of coherent structure. The components of the system have exponentially distributed failures, and are periodically inspected. The inspections of jth component are carried out periodically at the times given by the period $T_P(j)$. They are ideal which means that each inspected component is renewed (as good as new). Worst case from reliability point of view is the moment when the inspection times of all components will meet in one time point. Let us call the point as a *worst point*, which assigns the system a **worst reliability value (WRV)**.

Theorem: Let us assume a coherent system with randomly generated periods of inspection $T_P(j)$. Apparently, the availability of the system $A_S(t)$ satisfies the condition: $A_S(t) \geq WRV$. Then the minimum value of system availability **min** $A_S(t)$, converges for time going to infinity, to the value of *WRV*, which is given as follows:

$$\lim_{t\to\infty}\left[\min A_S(t)\right]= \text{WRV}$$

$WRV = h(A);\quad A = (A_1, A_2, ..., A_N)$, where $A_j = \exp[-T_P(j)/\text{MTTF}(j)]$ is the availability of the jth component at the end of its inspection period $T_P(j)$, $j = 1, ..., N$.

Proof: Immediately follows from the assumptions of coherent structure.

Answering the main problem, we will try to find the optimal vectors $T_P = (T_P(1), T_P(2)...T_P(N))$, and $T_0 = (T_0(1), T_0(2)...T_0(N))$, minimizing the cost function C_{PM}, and respecting the availability constraint such as $WRV \geq A_0$. Cost is minimized within given mission time T_M.

Matlab program (analytical) for computing of the *WRV* value of any coherent system represented by acyclic graph was completed and successfully used within the research. The program is not limited by the structure of the system (not only the series-parallel system but general systems).

3.3 *Oriented Acyclic Graph – mathematical representation of a system used for availability calculation*

Newly developed computing code (Bris 2002) for availability calculation is based on the following system representation that is very effective for using in both analytical and simulation codes. Relations between subsystems and basic events of input components of a system must be given by an oriented Acyclic Graph. Example of the Acyclic Graph for the system from Figure 2 is demonstrated in Figure 3. Corresponding Success Tree is demonstrated in Figure 4, for better understanding of the basic conception of the Acyclic Graph.

System under analysis must be mathematically represented by the oriented Acyclic Graph that is constructed under following rules:

1) Graph is acyclic which means that two immediately bound nodes are connected just by one edge.
2) On the top of the graph is the only SS node.

3) Orientation of the graph is given by the relation: inferior – superior.

superior

inferior

4) An internal node (nonterminal) represents a subsystem of given system: The subsystem is functioning just in the case when at least **m** inferior nodes (either terminal or nonterminal nodes) are functioning. Consequently the integer number **m** must be an element of the interval: <1, number of input edges $>$.

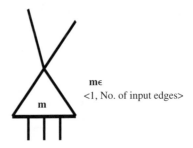

$m\epsilon$
$<$1, No. of input edges$>$

Examples (see Figures 3 & 4):
if **m** = 1 then the internal node represents OR gate
if **m** = "number of input edges" then the internal node represents AND gate

5) Terminal nodes are basic events of input components of the system. Each of them is characterized by a probability density function.

3.4 *Finding the optimal first inspection time vector T_0*

Naturally, the problem of finding of the optimal vector T_P is closely connected with another problem, namely the finding of vector T_0 which represents the beginning of inspections of each input component, i.e. the vector of first inspection times. We will not carry out inspections in the beginning of the life of a component, when the component is very reliable.

274

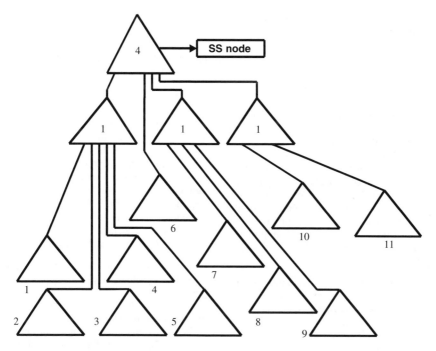

Figure 3.   Acyclic Graph for the system from Figure 2.

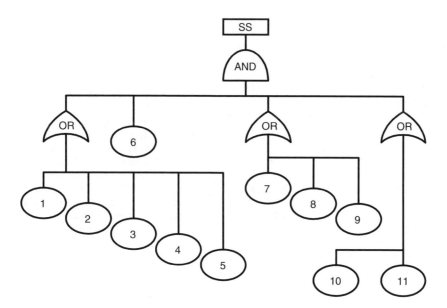

Figure 4.   Corresponding success tree for the system from Figure 2.

Consequently, the preliminary calculations must be realized to find the optimal vector T_0. The starting point for the finding of T_0 is based on the idea that only such interventions into the system must be made, that are maximally effective from both reliability and cost point of view. Measure of efficiency is more or less subjective question and in many situations in practice may be dependent on concrete reliability data files. For the research we decided to use the time dependent ratio-criterion of efficiency that is defined as follows:

$$\min\{R_j(t)|j=1,...,N\}, \quad R_j(t)=\frac{C(j)}{IF_j^B(t)}$$

where: $C(j)$ is cost of one inspection of jth component, $IF_j^B(t)$ is Birnbaum's measure of importance of jth component at time t (definition see in 0, for example).

Actually, the Birnbaum's importance measure provides the probability that the system is in a state in which the functioning of component j is critical to system failure. The system fails when jth component fails.

For given time point, we obtain number of the component, inspection of which is optimal, i.e. for which the ratio-criterion defined above is minimal.

The following procedure determines the vector $T_0 = (T_0(1), T_0(2)...T_0(N))$:

Step 1: Calculate dependence of reliability (availability) of given system on time for mission time T_M, supposing no maintenance; $i = 1$; $T_0 = (+\infty, +\infty, ..., +\infty)$.

Step 2: Obtain the time point t_i in which the system availability value A_0 is reached.

Step 3: If $t_i < T_M$, then t_i is ith component of ordered first inspection time vector T_0^{ord}; $T_0^{(i)} = t_i$; $T_0^{ord} = (T_0^{(1)}, T_0^{(2)}, ..., T_0^{(N)})$.

Step 4: Determine component N^o of j, using the above mentioned ratio-criterion applied in the time t_i; $1 \leq j \leq N$. Then $T_0(j) = T_0^{(i)} = t_i$.

Step 5: Recalculate dependence of availability of the system on time under the maintenance actions given by first inspection times of all relevant components in all time points $T_0^{(k)} = t_k$; $k = 1, ..., i$.

Step 6: $i = i + 1$, $i \leq N$, return to step 2.

Using the procedure we obtain full vector T_0. However, in some cases there is not necessary to use all components of the vector. That is just in the case when repeating inspections of one or more system elements brings more effective way (under the given criterion) to satisfy given availability constraint A_0. Consequently in such cases, it is necessary to select those of elements that will be maintained. Final decision about system interventions must be made in good accordance with given cost matrix.

4 COST OPTIMIZATION TECHNIQUE

The genetic algorithms were developed by John Holland in 1967 (Davis 1991, Goldberg 1994 & Elegbede et al. 2003) at the Michigan university. This method is based on the species reproduction principle, which consists in selecting the best adapted individuals among a population and in procreating by a crossing process. The implementation of the genetic algorithm consists to first create an initial population with given size (number of individuals). Then by a selection process similar to that of the natural selection, which is defined by an adaptation function, the second step is to select the individuals who will be crossed. These individuals are represented by a chromosome in the Genetic Algorithm. Then a current population is created by crossing of the individuals. The passage from a current population to another is called generation. For each generation, the algorithm keeps the individual with the best criterion value. The coding and the construction of the chromosome, representing the individual in the population, is the most important step of the algorithm. (Generally, a chromosome corresponds to a solution of the problem.)

The general structure of the genetic algorithm according to (Davis, 1991) is as follows:

Step 1: Initialization of the chromosomes population.

Step 2: Evaluation of each chromosome of the population.

Step 3: Creation of new chromosomes using crossing and mutation operators.

Step 4: Evaluation of the new chromosomes.

Step 5: Removing of the not selected chromosomes.

The last step is the final stop test (one considers for example the iteration count, or the no improvement of the solution value on a certain iteration count...). If the test is not verified, go to Step 3. Full details concerning the application of the technique to our optimization problem are in (Briš et al. 2002).

5 RESULTS AND ILLUSTRATIVE DATA

Consider a series-parallel system consisting of four parallel subsystems connected in series (Figure 2). The system contains 11 components with different reliability and PM cost data. The reliability of each component is defined by an exponential distribution with the failure rate $\lambda_0 = 1/MTTF$ presented in Table 1. This table also contains the PM cost $C(e(i,k))$ of each component. The basic data are exponential modification of those Weibull data presented in (Levitin 2000).

5.1 Calculations for the mission time $T_M = 25$ years

5.1.1 Availability constraint $WRV \geq A_0$; $A_0 = 0.9$
Obtained solution:

N^o	3	5	6	8	11
Tp	2.40	5.96	9.34	2.81	4.94
T_0	18	14	9.5	20	12
C_{PM}	133.1	133.1	133.1	133.1	133.1

The components N° 1,2,4,7,9,10 are not maintained.

Table 1. Parameters of system components.

Comp. No	Prob. Distr.	MTTF $=1/\lambda_0$ [years]	$C(e(i,k))$
1	EXP	12.059	4.1
2	EXP	12.059	4.1
3	EXP	12.2062	4.1
4	EXP	2.014	5.5
5	EXP	66.6667	14.2
6	EXP	191.5197	19.0
7	EXP	63.5146	6.5
8	EXP	438.5965	6.2
9	EXP	176.0426	5.4
10	EXP	13.9802	14
11	EXP	167.484	14

5.2 Calculations for the long term mission time $T_M = 50$ years

5.2.1 Availability constraint $WRV \geq A_0$; $A_0 = 0.9$
Obtained solution:

N°	3	5	6	8	11
Tp	2.67	7.35	8.62	4.33	4.75
T_0	18	14	9.5	20	12
C_{PM}	370.6	370.6	370.6	370.6	370.6

The components N° 1,2,4,7,9,10 are not maintained.

5.2.2 Availability constraint $WRV \geq A_0$; $A_0 = 0.8$
Obtained solution:

N°	1	2	3	5	6	8	11
Tp	17.86	12.18	10.14	18.84	12.74	8.84	10.97
T_0	26.5	26.5	21	32.5	15	32.5	18
C_{PM}	154.3	154.3	154.3	154.3	154.3	154.3	154.3

The components N° 4,7,9,10 are not maintained.

6 RESULT COMMENTS

We computed the results for two levels of reliability constraint, i.e. $A_0 = 0.9$ and 0.8 and, two levels of mission time, 25 and 50 years. We solved the problem of finding optimal policy respecting the availability constraint $WRV \geq A_0$, which in fact answers the following question: which policy is necessary to apply to prevent that the availability constraint will never be overstepped even if the worst case from reliability point of view has happened? Other words, the policy means something like "reliability assurance", i.e. if we do not care for instantaneous availability $A(t)$ and

even if the first inspection time vector T_0 is not respected, the policy assures that the given constraint A_0 will never be reached.

7 CONCLUSIONS

This paper shows the efficiency of an optimization method (based on Genetic Algorithms) to minimize the preventive maintenance cost of series-parallel systems based on the time dependent Birnbaum importance factor. A theoretical approach based on the asymptotic availability value is proposed. Starting from the results obtained for series-parallel systems, this approach can be extended to more complex systems (no exponential failure rates, complex structures different than series-parallel ones, etc.) according to the ability of the chosen methods. Another extensions seem possible: the improvement of the importance factor (other interesting importance factor should be studied), the study of other constraints than a minimal availability (minimal distance to an average availability), additional constraints (safety), or more realist characteristics of the maintenance (imperfect maintenance, logistic delays). Also, other optimization methods would be developed and compared (simulated annealing for example) to the GA (present work or modified improved forms).

REFERENCES

Briš R. 2002. Parallel Simulation Algorithm for Reliability and Availability Optimization, paper No. RESS 314 – under process of revision in *Reliability Engineering & System Safety*.

Briš R., Châtelet E. & Yalaoui F. 2002. An efficient method to minimise the preventive maintenance cost of series-parallel systems, Proceedings of RBO'02 Conference, Warsaw, Poland, http://www.ippt.gov.pl/amas/workshops/rbo02/.

Davis L. 1991. Handbook of genetic algorithms. Van Nostrand Reinhold, New York.

Elegbede, A.O.C., Chu, C., Adjallah, K.H. & Yalaoui, F. 2003. Reliability Allocation by Cost Minimization. To appear in *IEEE Transactions on Reliability*.

Goldberg D. 1994. Algorithmes génétiques. Edition Addison Wesley, France.

Levitin G., Lisnianski A. 2000. Optimization of imperfect preventive maintenance for multi-state systems; *RESS* 67, 193–203.

Misra K.B. 1992. Reliability Analysis and Prediction; A Methodology Oriented Treatment. Elsevier, ISBN 0-444-89606-6; P. 765–775.

Vaurio J.K. 1999. Availability and cost functions for periodically inspected preventively maintained units; RESS 63, 133–140.

Safety and Reliability – Bedford & van Gelder (eds)
© 2003 Swets & Zeitlinger, Lisse, ISBN 90 5809 551 7

GIS based emergency management tool for road and rail transport of hazardous materials

R. Bubbico, S. Di Cave, B. Mazzarotta

Dipartimento di Ingegneria Chimica, University of Rome «La Sapienza», Italy

ABSTRACT: An emergency occurring while transporting a hazardous substance may have a rather high-risk potential, depending on the amount of transported product, on its hazardous characteristics and on the features of the surrounding environment. This paper addresses road and rail transport of hazardous materials, with the objective of outlining some guidelines and providing a tool for the management of transport emergencies. An effective management of a transportation emergency derives from a careful balance of the opposite needs for the immediate availability of the information and its reliability. The procedure here proposed suggests to divide the data of concern into product-related information and environment-related information, estimating in advance the impact zones for some reference scenarios. By including all product related information into a database and all territorial information in a GIS (Geographic Information System) application, and combining the two, a fast and powerful tool for transport emergency management is obtained.

1 INTRODUCTION

The transportation of hazardous materials by road and rail is a very common industrial activity and represents a potential hazard both for the exposed population and for the environment. As a matter of fact, a number of accidents occurred in the past years have demonstrated that this activity can involve a large number of people, comparable or even larger than in accidents occurred at fixed installations (Haastrup & Brockhoff 1990, Egidi et al. 1995, Vilchez et al. 1995).

For example it is well known that in 1978 at San Carlos de la Rapita (Spain), the occurrence of a fireball of a tanker containing 22 t of propylene caused about 200 fatalities.

1.1 Regulations

The EC 96/82 Directive (European Community 1996), usually referred to as "Seveso 2" Directive, which is the most important European legislation addressing risk prevention in the handling of dangerous materials, does not cover the issue of their transportation.

However, some international regulations such as the ADR and RID codes, for road and rail transportation, respectively, and the UN Recommendations on the Transport of Dangerous Goods, cover some aspects of this topic.

In these regulations:

- the various substances are classified based on their chemical, physical and hazardous properties;
- some directions about transport modalities, type of carrier, loading and unloading phases are given;
- the requirements in terms of on-board documentation and on driver training are identified;
- finally, some generic instruction about action to be adopted during an emergency are suggested.

However, with the exception of some general information, all these procedures are mainly aimed at accident prevention. Above all, the possible outcomes of an accident are completely neglected.

For road transportation, the European Council of the Chemical Industries (CEFIC) issued its safety cards (CEFIC, 1997) to be kept on board, which contain information about the substance itself (name, UN and ADR number, hazard type) and about actions to be taken, or to be avoided, in case of release and/or fire.

1.2 State of the art

In recent years a number of studies have been devoted to the application of risk analysis techniques to the transportation of hazardous materials, mainly by road and rail (Purdy 1993, CCPS 1995, Goh et al. 1995, Bubbico et al. 1998, Gadd et al. 1998, Tiemessen & van Zweeden 1998, Bubbico et al. 2000).

Again, the suggested methodologies can be applied mainly for:

- the calculation of the risk associated to a given route;
- the comparison between different routes and modalities of transportation (amount per trip, number of trips, road or rail, etc.), and hence on the selection of the best one;
- the scheduling of different transportation activities;
- and so on.

In other words, they were mainly interested with the priori reduction of the risk. Less attention has been devoted to the emergency phase after an accident has occurred.

1.3 Features of the problem

Based on various past accidents, it is well known that, given a release of hazardous materials, a timely and effective action by the emergency team can reduce the number of people possibly involved or the extension of environment affected substantially. In order to accomplish that, the knowledge of a large number of factors which affect the evolution of the accident and its consequences is very important.

However, since the accident can happen at any point along the route, any operations in emergency conditions can be much more complex with respect to a similar situation occurring in a fixed installation, where a local emergency team can already be present at the site, the team components may probably already know all the relevant information about the substances involved in the accident and they can be provided with all the necessary equipment.

Conversely, in the case of transportation, the team in charge of the emergency can be located relatively far from the site of the accident. As a consequence, they first need to be alerted by some witness (hopefully the driver himself). Then they need to get as much information as possible about the accident (materials, location, dynamics of the accident, its evolution, characteristics of the surrounding environment, etc.); this is an important topic since the local conditions can be deeply variable along the route. They must reach the site as soon as possible, usually by means of the ordinary road network, and this implies choosing the best itinerary in terms of length, traffic conditions and so on. Finally, when arrived on the site and based on the collected information, they must decide about the actions to be taken.

Another important aspect of the problem is that in the case of a release of dangerous materials during transportation, the on-road and on-rail population and those living in the surroundings are very likely unaware of the material involved, of its possible consequences (toxic cloud, fire, explosion), of the emergency actions

to be taken and of the actions to be avoided (for example whether to stay indoor or to evacuate the area), or even of the accident occurrence itself. This very important aspect must be taken into account by the emergency team as well.

2   ACCIDENTAL SCENARIOS

In case of a loss of containment of a hazardous material, the possible damages are due to its toxicity and/or its flammability and the evolution of the accident depends on a number of parameters. For example, the physical properties of the substance, its physical conditions during transport (pressure, temperature, degree of filling), location and size of the release hole, determine whether the spill will be liquid, vapor or two-phase.

The amount of material released is a function of:

a) the total amount of transported material. It depends on the size of the tank, and therefore also on the transportation modalities (by road or rail), and on the filling degree;
b) the size of the release area (leak from the relief valve, from a pinhole, from a larger fracture in the vessel wall, etc.);
c) the release duration (whether it is stopped or not).

In case of immediate ignition, a flammable material will form a jet fire, whose consequences will be limited to a relatively small area near the release site. Even in this case, serious consequences can occur, either because of the presence of many people nearby, or because of secondary effects (domino effect), such as the heating effect on the tank itself or the impingement of the flame on other objects (collapsible structures, other vessels containing hazardous materials, buildings, cars and so on).

In the absence of immediate ignition or for non-flammable materials, also depending on the external (ambient) conditions, the liquid spill may or may not generate a liquid pool. In both cases a vapor cloud will be produced, either directly or from pool evaporation.

The dispersion modalities of the vapor cloud will strongly depend on both ambient conditions, such as ambient temperature, wind velocity, wind direction, Pasquill stability class and humidity and on the characteristics of the surrounding territory such as orography, presence of obstacles and so on.

In case of a delayed ignition of a flammable cloud, a flash fire or an Unconfined Vapor Cloud Explosion (UVCE) will occur, and the impacted area will be much larger than in all other outcomes, depending on the distance traveled by the vapor cloud during dispersion until the ignition time. In case of flammable clouds, the ignition probability of the cloud will be affected by the characteristics of local environment, while the flammability limits will determine the amount of material in the flammable fraction of the cloud, and chemical reactivity will determine the potential for explosion.

Finally, a delayed ignition of the pool, will produce a pool fire, which is a less hazardous event than flash fires or UVCEs, but it still has the potential of causing severe consequences. It is worth noting that in some cases deciding what are the best actions to be performed can be quite hard to do. For example, would the release of a flammable material not be stopped safely, extinguishing the fire might produce much worse effects, since the delayed ignition of the resulting vapor cloud would affect a much larger area. Based on the actual situation and on the surrounding environment, this possibility should be taken into account by the emergency team.

For toxic clouds, the extension of the impact zone will be determined by the toxicity limits (IDLH, LC_{50}, etc.) and the dispersion modalities.

In summary, the parameters which affect the possible outcomes of an accident can be divided into the following four groups:

a) properties of the material;
b) conditions during transportation;
c) release type;
d) characteristics of the environment.

Of course, given an accident, the number of people involved depends on the accident location (either a crowded urban area, or the country), and, when taking into account a non-uniform population density distribution, on the wind direction. The same holds true also for its impact on the environment.

From the above considerations, it is apparent that the hard task of an emergency responding team is that of collecting many news and taking a number of decisions in a very short time interval. As we will see later, some of such data will require a lot of time to be obtained, therefore, also due to the high variability of the parameters involved, preparing a detailed emergency plan in advance is practically impossible, even for a single transportation case.

3 PROPOSED GIS APPROACH

The proposed approach consists in combining an application developed on Geographic Information Systems (GIS) with a product database. The former makes available the relevant territorial information, and some useful tools, such as routing, while the latter contains the extension of the impact zones precalculated for the expected accidental scenarios. They will be described in the following paragraphs.

3.1 *GIS application*

As already demonstrated when dealing with Risk Analysis (Bubbico et al. 2002), the introduction of Geographical Information Systems (GIS) would allow a great improvement even in the emergency management. In fact, by the use of these systems, much information needed at the time of the emergency can be collected in advance and simply recalled when required with remarkable time saving.

As far as the territorial data are concerned, a number of commercial packages are available associating data layers containing different types of information to specific objects shown on maps with their geographical coordinates.

The information usually available in these packages concerns:

– land use;
– location and extension of populated areas;
– waterways, road and rail networks.

If not already present, the location of particular structures, such as hospitals, schools, emergency team stations, fire-fighter departments, police stations, industrial sites and so on, can be manually introduced.

However, other very useful data for emergency management are not immediately available, but they need to be calculated from data of different origin, modified in a format suitable for the GIS, and finally introduced in the system.

3.1.1 *GIS application MapRisk*
The GIS application MapRisk, developed based on the commercial GIS ArcView, was specifically designed for the analysis of the risks due the transportation of dangerous materials, and the details can be found elsewhere (Bubbico et al. 2002). At present it contains a map of the Italian territory provided with the whole rail and road networks (including highways, state and local roads) obtained by using the commercial TeleAtlas Roadnet database.

An example of the road and rail network in the surroundings of Gela, town in Sicily located close to an important industrial site, is given in Figure 1.

The road and rail networks are subdivided into segments with average length 400 and 700 m respectively, and to each of these segments a table containing data about local population, weather conditions, traffic data and so on, has been associated.

In particular, based on the extension of the populated areas and on the corresponding total number of inhabitants, which are the data obtained from census records (ISTAT 1992), a population density distribution has been calculated, which is more useful when combined with the impact area of an accident.

As an average, the dispersion of a toxic cloud (especially for highly toxic materials) will affect an area much larger than fires or explosions do. Therefore, the population density was calculated for two buffers from both sides of the road/rail, namely at a distance of 500 and 1500 m, respectively.

Weather data, in terms of monthly average high, medium, low temperature and wind velocity and direction, are available from the meteorological stations.

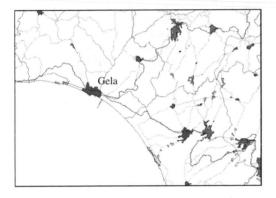

Figure 1. Road and rail network in the surroundings of Gela.

For any road/rail segment, the data corresponding to the weather stations falling on the same side of a mountain divide, have been correlated, based on their distance from the segment, in order to obtain average seasonal high, medium and low temperatures, as well as wind velocities and direction.

Traffic data were obtained from ANAS (National agency for roads), from AISCAT (National company for toll highways) and from Trenitalia (National railway company). On-road and on-rail population density was also calculated based on these data.

An example of the tables containing the data relevant to a given segment of rail route, running inside the town of Gela, is shown in Figure 2.

3.2 Product database

The impact area of any possible outcome of an accident is a very important information for setting up an effective emergency plan. In fact, this allows the emergency team to know exactly which types of outcome cases should be expected and what would be the surface extension (and the safety distance) of their respective consequences.

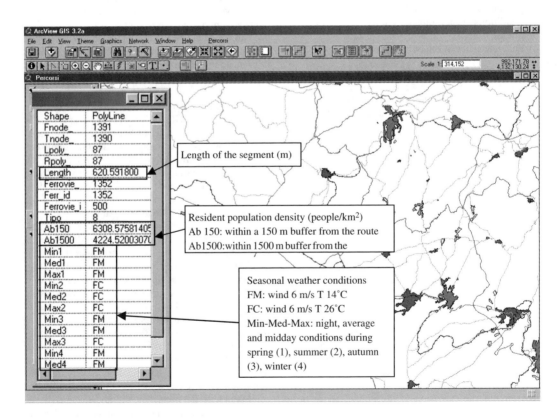

Figure 2. Example of data tables associated to a rail segment.

282

The impact areas due to thermal radiation, over-pressure or toxic concentration, deriving from the accidental scenario, can be calculated by means of any consequence calculation software code. However the use of these codes is not easy, and requires the presence of a specialist. Furthermore, the introduction of all the input data required by the modeling algorithm, repeated for all the outcomes which are going to be studied, will take a long time, which can be unacceptable during an emergency.

In order to speed up this step of the procedure, the impact areas of a number of events were previously calculated, and the results introduced into a specific database (Bubbico et al. 2002).

Taking into account that the impact areas depend on many of parameters, in order to keep the number of cases to be studied reasonably low, some "reference conditions" were selected.

Two standard volumes were assumed for the transportation vessel, i.e. 35 and $70\,m^3$, for road and rail transportation respectively, and the filling degree was set at 85% in both cases.

As far as the amount of released material is concerned, two cases were considered:

– a medium release, represented by a continuous release from a 15 mm hole located at the bottom of the tank (i.e. the release is conservatively assumed as liquid);
– a catastrophic release from a large hole, where the whole contents of the tank is discharged within a few minutes.

Six reference weather conditions were assumed, by varying ambient temperature (5, 14, and 26°C) and wind velocity (3, 6 m/s). These values were derived from a statistical analysis of the data obtained from about 60 meteorological stations in Italy (ISTAT 1994). The Pasquill atmospheric stability class was D in all cases.

For each combination of substance, release size and weather conditions, all the possible outcome cases (fireball, UVCE, flash fire, pool fire, jet fire) have been simulated by means of a commercial software code (TRACE 8.2 by Safer Systems LLC), and the resulting danger areas have been stored into the database.

The impact areas were calculated with reference to the following damage thresholds:

– 10 kW for thermal radiation;
– 5000 Pa for overpressure;
– IDLH for toxic concentration.

Presently, the database TrHazDat contains the impact areas, calculated for 2 road and rail release scenarios and 6 weather conditions, for all the outcome cases expected for 29 flammable and/or toxic substances, listed in Table 1.

TrHazDat also contains the probabilities, given an accident, of the two types of release, and, given a release size, those of the possible outcome cases. This information is required for Quantitative Risk Analysis; however, for emergency management, and in the absence of more information, the user may choose the worst conditions by assuming a catastrophic release.

4 EMERGENCY MANAGEMENT

In order to set up an effective emergency plan for a transportation accident, the competent authorities need to know a lot of information. For the sake of simplicity, two phases can be identified. The first one is composed of the emergency actions carried out by the travelling personnel, while the second one is carried out by the emergency team upon arrival on the accident site.

1. *First emergency actions*

Based on the substance properties, on the condition of the release (large/small, liquid/vapor) and on the external conditions (weather conditions, presence of

Table 1. Substances in TrHazDat database.

N.	Hazardous substance
1	Acethaldheyde
2	Acetic acid
3	Acetone
4	Acetonitrile
5	Acrylonitrile
6	Ammonia (anhydrous)
7	Benzene
8	Bromidric acid
9	Carbon tetrachloride
10	Chloridric acid
11	Chlorine
12	Chlorobenzene
13	Cyanidric acid
14	Cyclohexane
15	Dichloropropane
16	Ethanol
17	Ethylacetate
18	Ethylene oxide
19	Fluoridric acid
20	Isopropylammine
21	Methanol
22	Methylacrylate
23	Nitrobenzene
24	n-Octane
25	Propane
26	Sulfur dioxide
27	Toluene
28	Trichloroethane
29	Vynil chloride

ignition sources, presence of nearby population), the driver should be able to evaluate the severity of the situation and decide about the actions to be taken: whether to immediately alert the emergency teams, which could act more effectively, providing them with all the necessary information, or to intervene without delay.

All the present regulations and practices (safety cards, specific driver training, etc.) are mainly concerned with this aspect of the problem.

4.1 Emergency response

In the second phase, the emergency teams are involved. First of all, the exact location of the accident should be known. Based on that, the main operative center can select the emergency teams to be activated and the best itinerary to reach the site of the accident: it will be the shortest or the fastest one (they might be different), also taking into account possible interruptions or unavailability due to the accident itself. For example structural damages could result in a line interruption; traffic jam could make some of the roads near the site of the accident unavailable, and so on.

Also here the use of a GIS application would be of great help. In fact, the site of the accident might be easily identified on the map and, by selecting the surrounding area, other information such as the presence of important targets (hospitals, schools, offices, stations, etc.), the type and location of other roads, the location of emergency stations, and so on, would be immediately shown on the map itself. Furthermore, local data such as the population density and distribution, average weather conditions and traffic flows, might be obtained by recalling the corresponding data tables.

By this way, once the origin (emergency stations) and destination (accident site) have been identified on the map, it would be possible to rapidly identify the best itinerary to be used by means of some "routing" subroutine. In fact, in order to select the fastest route, an algorithm can be prepared, in which the traffic flows of all the possible roads and/or rails can be automatically loaded and processed, also taking into account interrupted segments or other bottlenecks.

Figure 3 shows, for example, the fastest route from an emergency management center in the town of Ragusa to the site of an accident in the town of Gela.

To properly organize an effective response, the knowledge of the actual weather conditions (above all the wind direction) would be of the greatest importance, since the areas possibly affected by the dispersing cloud, and therefore the corresponding population, may vary remarkably.

However, should this information not be available, it is still possible to recall the prevailing wind directions in the zone from the GIS database.

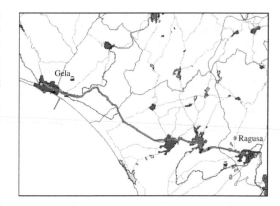

Figure 3.   Example of routing to the emergency site.

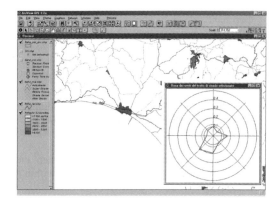

Figure 4.   Prevailing wind directions in the zone of Gela.

For example, Figure 4, shows the prevailing wind direction for the town of Gela, which is from South-West in more than 20% of the cases.

The use of the GIS tool in an emergency is as follows. First the location of the incident is identified on the road or rail networks displayed on the GIS map. Then, the product is selected and the expected size of the spill is also chosen: should this information not be available at that moment, the large release size is conservatively assumed. Finally the season, moment of the day (full day, night, or early morning/late afternoon) is selected, thus recalling the prevailing seasonal weather conditions from the database, and the wind direction is inserted. The impact areas of all the expected outcome cases are immediately displayed on the GIS map, in different colors, thus allowing to determine useful information, such as expected safety distances, populated areas possibly falling in the impact zone, etc.

For example, Figure 5 shows the impact areas for a large release of cyanidric acid in case of a rail accident

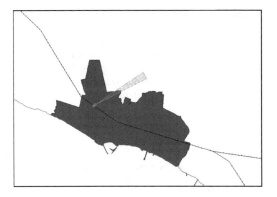

Figure 5. Impact area of the outcome cases for a large release of cyanidric acid during rail transport.

in the town of Gela, assuming wind blowing from West-South-West.

The knowledge of the impact areas first allows the emergency team to know which outcome cases have to be expected for the transported product and what will be the impact area of each of them. Second, it also allows to determine the total number of people possibly involved in the consequences of the release of the transported substance.

Based on the possible evolution of the accident, such as the dispersion of toxic products or the likelihood of further dangerous events (explosions, delayed fires or production of toxic chemicals, and so on), it can be decided whether to evacuate the area or force the population indoor. In the first case, as done previously, the routes to be used for evacuation and those to be interdicted can be easily determined on the map.

Besides its use during an actual emergency, the presented tool might also be used in emergency plan preparation. Once selected a location, a number of accidents could be simulated, with various dangerous materials and under different external conditions (wind directions, traffic conditions, period of the year, etc.), thus identifying possible critical situations: for example, it can be found that the area might be difficult to be reached by the emergency team, or to be evacuated by the population, and so on. If the probabilities of wind direction are considered, it is possible to identify the areas which are more likely prone to be affected by the consequences of the dispersing cloud, and, therefore, the roads to be used for evacuation.

4.1.1 Emergency management

When the type and the amount of the released material are known, the impact areas of all the possible outcomes which might occur can be recalled from the products database.

This can be done directly from the GIS program: once selected the substance, the operator should choose the size of the release and insert the season and the moment of the day.

Among all the impact areas calculated during the consequence calculation step under the different weather conditions, those closest to the actual local meteorological conditions will be automatically selected. In the absence of real weather data, the seasonal average ones contained in the same database will be used.

A further useful application of such a methodology would be in the investigation of past accidents. In fact, the simulation of an earlier emergency phase would allow, on one hand, to check the values of the many parameters stored in the database (incorrect prediction of the outcome of the accident, of the affected areas, of the population distribution), and, on the other hand, to examine whether wrong actions had been performed (wrong evacuation routes, etc.) and safer choices might have been taken.

4.2 Discussion

The proposed tool is mainly intended to be used by public authorities, such as Civil protection officers, firefighters, etc., who are the people who will be directly involved in an emergency occurring during the transport of hazardous substances. However, also road and rail transportation companies would benefit from it, since the driver(s) of the vehicle would usually start the first response.

The management of the system would be properly done from an emergency center room, for example at a firefighter station, or from the communication center of the transportation company. In fact the alarm would come from the driver, if able to communicate, or from passersby, who would probably provide less complete information, but may give useful data by answering questions of the operator.

Provided that the operators make use of the same data set they will obtain the same scenario: in practice, the main variable concerns the estimate of the release size, which may be conservatively assumed as catastrophic. Obviously, the actual accidental scenarios may be different from the simulated ones, due to release sizes and/or weather conditions different from those assumed as reference ones, when calculating the impact areas.

Just to give an idea of the accuracy of the tool, it was used to simulate a road accident occurred in Italy on 29 December 1982, at 10:25 a.m., at Capannori to a tanker transporting 10 t of LPG (Ministero dell'Interno, 1986). The scenario was an impact, due to misty weather, followed by a fire causing a BLEVE and a fireball, causing 4 casualties. The walls of the buildings in a 150 m radius were severely damaged. Assuming a catastrophic release of LPG and the seasonal weather conditions expected for that portion of the highway, the database of the GIS software estimates an impact distance of 190 m for the resulting fireball, which represents the worst outcome case for a catastrophic release of LPG. It is worth noting that the Fire Brigades empirically set a safety distance of 150 m, but this was not actually safe since heavy damages occurred even beyond that distance. The prediction of the software would therefore represent a "safer" and more realistic safety distance for the case under exam.

It should be taken into account that any consequence analysis software is based on mathematical modeling which may more or less accurately simulate the actual scenario. However, the main purpose of the tool is not that of determining exactly the impact area of a given scenario, but that of very rapidly providing the emergency team with a complete set of the possible outcome cases they may have to oppose and information about areas at risk and safety distances.

The system is user friendly: the spot where the accident occurs is selected with the mouse on a map, the selection of the scenario is started clicking on an icon and the following choices are made by selecting the proper items from a list. However, the use of some typical GIS options, such as determining the fastest route from an emergency center to the accident location, would require an operator who is familiar with such tools. This would not be a very stringent requirement, since after some hours training, any individual with basic computer education would be able to use the main functions of ArcView or other similar GIS softwares.

5 CONCLUSIONS

Emergency management of an accident occurred during the transportation of a hazardous product is a quite complex duty, since a lot of information is needed and decisions have to be taken timely and effectively. In this paper, a GIS application has been presented, where all the territorial information have been previously associated to the basic segments of the whole Italian road and rail networks. The use of other built-in options, such as the routing algorithm, would allow to carry out very useful steps (e.g. best route selection), otherwise impossible to be performed. Similarly, a

database has been prepared, containing the properties of 29 hazardous substances and the impact areas of the corresponding possible hazardous events.

The combination of the two, would provide the competent authorities with a fast and powerful tool which would allow a remarkably safer and more effective management of the emergency.

REFERENCES

Bubbico, R. Di Cave, S. Mazzarotta, B. 1998a. *Proc. SRA Annual Conference "Risk analysis: opening the process", Paris 1998* (2): 665–676

Bubbico, R. Dore, G. Mazzarotta, B. 1998b. *J. Loss Prev. in the Proc. Ind.* (11): 49–54

Bubbico, R. Ferrari, C. Mazzarotta, B. 2000. *J. Loss Prev. Proc. Ind.* (13): 27–31

Bubbico, R. Di Cave, S. Mazzarotta, B. 2002. A GIS supported transportation risk analysis approach. *CCPS Conference on Risk, Reliability and Security Jacksonville FL, USA, October 2002.* 361–374

CEFIC 1997. Schede per singole sostanze. Ministero della sanità. Roma

CCPS, 1995. Guidelines for Chemical Transportation Risk Analysis. AIChE. New York

European Community 1996. Council Directive 96/82/EC, On the control of major accident hazards involving dangerous substances. *Official Journal of the European Community*, L 10, 14.1.1997, 13–33

Egidi, D. Foraboschi, F. Spadoni, G. Amendola, A. 1995. *Rel. Eng. and System Safety.* 75–79

Gadd, S.A. Leeming, D.G. Riley, T.N.K. 1998. *Proc. 9-th Symp. on Loss Prevention and Safety Promotion, Barcelona, 4–7 May 1998.* (1): 308–317

Goh, C.B. Ching, C.B. Tan, R. 1995. *J. Loss Prev. in the Proc. Ind.*, (8): 35–39

Haastrup, P. Brockhoff, L.H. 1990. *J. Loss Prev. Proc. Ind.*, (3): 395

ISTAT 1992. 13° Censimento generale della popolazione e delle abitazioni, Roma

ISTAT 1994. Statistiche meteorologiche anni 1984–1991. Ann. n.25. Roma

Ministero dell'Interno 1986. Rassegna comparata di incidenti di notevole entità, Roma.

Purdy, G. 1993. *J. Haz. Mat.*, (33): 229–259

Tiemessen, G. van Zweeden, J.P. 1998. *Proc. 9-th Symp. on Loss Prevention and Safety Promotion, Barcelona, 4–7 may 1998.* (1): 299–307

Vilchez, J.A. Sevilla, S. Montiel, H. Casal, J. 1995. *J. Loss Prev. Proc. Ind.*, (8): 87

Safety and Reliability – Bedford & van Gelder (eds)
© 2003 Swets & Zeitlinger, Lisse, ISBN 90 5809 551 7

Methodology for post-incident investigation in process industry

R. Bubbico, I. Circelli, S. Di Cave & B. Mazzarotta
Dipartimento di Ingegneria Chimica, University of Rome "La Sapienza", Italy

F. Geri
Italian Civil Protection Department, Rome, Italy

ABSTRACT: This work presents a brief overview of the main steps required for post-incident analysis, outlines some procedures which enable to effectively investigate the causes of accidents in process plants and shows how they may be applied to real cases. In fact, a better understanding of past incidents can be obtained also applying only some steps of the procedure, based on the available information. The applications refer to three major accidents occurred in Italy, at an oil products storage facility in Naples (1985), at a maleic anhydride plant in Colleferro (1981) and at an oil refinery in Milazzo (1993), respectively. Different procedures are applied to these study cases according to the detail of the available information.

1 INTRODUCTION

In order to avoid the recurrence of incidents in process plants, it can be very useful to apply the lessons learnt from incidents or near-misses occurred in the past. As a matter of fact, a number of regulations, directives and codes of practice took their origin from the analysis of case histories, such as those of Flixborough (UK) in 1974 and of Seveso (Italy) in 1976. Therefore, performing a correct and complete post-incident investigation is of the greatest importance in order to extract all the valuable information from each case under study. However, such investigation would generally require extensive and in deep information, which may be not available in real cases.

This work presents an overview of the main steps required for post-incident analysis according to the literature and outlines a procedure which can be followed to effectively investigate the causes of such events in process plants to get a better understanding based on the available information.

The application cases concern some major accidents occurred in Italy: the procedure is applied to each case according to the different detail of the available information concerning these incidents.

2 LEARNING FROM INCIDENTS

Learning from design and operational errors is essential for process industry in order to decrease the probability of recurrence of incidents they may originate, as well as in order to mitigate their consequences. General lessons can be drawn from reports concerning historical cases, such as major incidents, but also from other events such as "near miss", losses of containment of toxic and/or flammable materials, operating errors, etc.

Accidental events may be classified according to the following 4 levels of severity (CCPS, 1992):

– level 1: minor incident and near miss without catastrophic potential;
– level 2: incident with limited impact;
– level 3: incident not catastrophic but with a certain impact;
– level 4: major incident and serious near miss.

For incidents of level 1 and 2 an investigation carried out by a local supervisor is generally sufficient, while for events of level 3 and 4 it would be preferable to activate an investigation team. The team composition may be largely variable depending on the size and the type of the incident: it would include a team leader, operation and/or manufacturer representatives, process engineers, safety department representatives, personnel from the plant and, in some cases, some specialists as consultants.

The team should proceed according to a specific plan including the following:

– investigation on the scene of the incident, site photographs, witnesses interviews;

- evidence investigation, conservation and recording;
- safety of the scene;
- documentation;
- site remediation, clean-up, rebuild and restart.

It is important to properly select the team leader, who will take care of the management and of the coordination and communication aspects.

2.1   Incident investigation

A number of investigation procedures exist, and some of them are reviewed in the guidelines published by the Center for Chemical Process Safety (CCPS, 1992). However, any complete investigation would generally require the steps listed below.

First, the investigator(s) should make an initial site visit. It allows to physically identify and secure the zone to be investigated, and should be mainly aimed at assessing potential hazards to ensure safety of the scene.

All the technical documents concerning the process and the plant should be examined. They include P&Is, operating procedures, training manuals, project data, accidental scenarios assumed for safety systems design, alarms and set points of trip systems, safety data sheets of the products, characteristics of the chemical reactions involved, logs on past incidents, control software, site plans, etc.

All the potential sources of information should be identified. They include control instrumentation records, shift logs, maintenance records, run histories, laboratory analyses, equipment process data sheets, etc. Some information is also obtained from the analysis of the physical proofs.

The interviews to potential witnesses also represent an important source of information. In addition to eye-witnesses, others may give their contribution: maintenance personnel, workers with previous experience with the plant, retired and transferred personnel, etc. It is important to interview eye-witnesses as soon as possible, avoiding the exchange of information between them.

The actual situation of the site is then surveyed, by taking photographs, identifying the locations and positions of victims (if any), finding and classifying the fragments, and collecting further information. It is particularly important to analyze damages associated to explosion and fire, such as permanent deformations, fire effects, etc., and to examine the damages to buildings and windows.

Physical proofs examination is one of the most critical steps: they may include tests on damaged equipment and experiments performed to verify some hypotheses.

Evidence should be preserved both immediately after the incident and for longer times (for example for legal considerations): all the proofs should be carefully identified and controlled.

2.2   First cause determination

Incidents generally derive from multiple first causes, whose combinations represent the fundamental reasons why the incident occurred. Determining the first causes requires evidence collection and the logical reconstruction of all the events according to their chronology to determine the most probable scenario. The logical procedure may be deductive (from the general to the particular) or inductive (viceversa). The most complete deductive method is probably the Causal Tree Method (Boissieras, 1983) which moves from the top event (incident) proceeding back to the possible scenarios causing the incident. Hazops represent an example of inductive methods, which starts from deviations from the normal operating conditions to determine the effects of such deviations.

In any case, the analysis of multiple first causes involves searching for all the credible modes which would make the incident possible.

A significant role can be played by modeling and simulation of the entire sequence from the loss of containment (for example, the release of a hazardous fluid from a pipe) to the incident (fire, explosion, toxic dispersion) and to its consequences (burns, overpressure, missiles, injuries), which allow to check hypotheses and assumptions concerning the cause and the evolution of the incident.

2.3   Report and recommendations

The results of the investigation should appear in a report including some recommendations, aimed at:

- reducing the frequency of occurrence;
- reducing/mitigating the consequences;
- limiting the exposition of persons.

The report should also clearly state the priorities of the listed recommendations. In some cases the plant is no longer working after the incident, and prerequisites to be satisfied or modifications to be made before restarting the activities should also be specified.

3   POST-INCIDENT INVESTIGATION

Generally, the proposed methods make use of logical trees, and, among them, the causal tree method is probably the most complete. However any thorough incident investigation would generally require collecting a great number of data, performing proofs and/or simulations. In most cases, it is difficult to apply this method to historical cases, where the information is limited to that available in the reports. However, a better understanding of past incidents can be obtained also applying only some steps of the procedure, based on the available information.

Some data about the 13 worst major accidents occurred in Italy from 1979 to 1987 are listed in Table 1 (Ministero dell'Interno, 1986), VVF, 1987): three of them, selected as those where more information was available, are examined as study cases.

From the data of Table 1 it results that in most cases a first incident (Incident 1) evolved into up to three subsequent incidents (Incidents 2–4) due to domino effects. The main accident was a fire in 61.5% of the listed cases and an explosion in the remaining 38.5%; operating and human errors were the cause of more than 53% of the reported incidents.

3.1 Oil products storage facility (Naples, 1985)

The oil storage facility covered an area of about 75,000 m^2 and included a number of tanks containing fuel and other combustible liquids. During the night of December 21, 1985, a gasoline spill occurred from a tank, due to overfilling. Following the release, the vapor cloud exploded, giving rise to an UVCE which involved damages and fire of 5 fixed roof and 9 floating roof tanks containing oil, fuel oil, gasoil and gasoline (Ministero dell'Interno, 1986).

As a matter of fact, some gasoline was being transferred from an oil-tanker to tank N.17: this transfer was scheduled to finish at 3:08 a.m., but, actually, at 5:10 a.m., when the UVCE took place, the oil-tanker was still pumping in tank N.17. Therefore, the gasoline overfilled the tank and started discharging in the bund, for a total volume of about 5,000 m^3. A vapor cloud was generated, with an estimated height of 7–8 m, which found an ignition in some area nearby.

The cause of the overfilling was a human error of the operators, who failed in switching the flow from tank N.17 to the next tank. It is likely that the 2 operators attempted some response during the emergency, since both of them died in the incident.

The effects of the UVCE were registered in the seismologic station as that of an earthquake of 6.5 on the Mercalli scale, and included the following:

- the roofs of 6 fixed roof tanks were found at a distance of 50 m from the tanks;
- the tanks partially filled were damaged and moved from their seats, and their content was released in the respective bunds;
- a highway connection nearby was heavily damaged;
- all the cooling and fire protection devices were put out of order, with the pipes thrown at some tens meters of distance;
- damages to 220 houses and total destruction of some of them, damages to 12 large buildings and 448 small industrial units.

3.1.1 Analysis of the damages due to the explosion
In this study case, available data are scarce, so hypotheses and assumptions about cause and evolution of the incident were simply checked by using a consequence analysis software (ChemPlus 2.0, 1991).

The average mass flux of gasoline released in the bund was estimated as 34 kg/s, which, due to the geometry of the area and the weather conditions, generated about 0.44 kg/s of gasoline vapors. The spill continued for about 2 hr, generating a flammable cloud covering an area of more than 300 m^2.

The following UVCE was simulated with the TNT model, obtaining the overpressure profile shown in Figure 1. The observed data were set by correlating the observed damages caused by the UVCE at various distances from the tank to the overpressure values typically associated to such damages. It can be noticed that the agreement is good only very close or very distant from the explosion center, while the used model underestimates the overpressure actually suffered at midrange distances.

It has to be remarked that in this study case the uncertainty concerning the release modality is extremely limited, so that the observed differences mainly depend on the characteristics of the used model. In fact, it is well known that, even if the TNT model is capable of describing the overpressure profile of UVCEs at some distance from the explosion center, the duration of the positive phase is longer for a UVCE than for a TNT explosion (Lees, 1996). This means that, at fixed overpressure, an UVCE is more destructive (due to its longer duration) than a TNT explosion. Therefore, at a fixed distance, the observed damages are greater for an UVCE than for a TNT explosion, as shown in Figure 1.

3.2 Maleic anhydride plant (Colleferro, 1981)

The production process of maleic anhydride included a reaction step of benzene with air, followed by condensation, separation, absorption and dehydration; a final distillation step removed the impurities. This step was a batch azeotropic distillation, where o-xylene was added: first, water and maleic acid were removed as distillate and separated from the organic product; then, pressure was reduced to 350 mm Hg to remove o-xylene; finally, pressure was further reduced to 50 mm Hg, to obtain maleic anhydride of the desired purity. After 5 to 10 cycles, the distillation system was completely washed by performing the distillation from a water batch for some hours to remove solid deposits.

At 20:30 of March 21, 1981, during the initial phase of the batch distillation, the temperature in the kettle increased rapidly and unexpectedly. In spite of the attempts of opposing to this abnormal situation, in some time the kettle and the column exploded (Ministero dell'Interno, 1986; Di Cave, 2002).

Table 1. Some major accidents occurred in Italy (Ministero dell'Interno, 1986).

Plant	Main accident	Incident 1	Incident 2	Incident 3	Incident 4
Priolo petrochemical plant (1979)	Fire and explosion	Release of gaseous propane from a drain Cause: human error	Fire of gaseous propane Cause: ignition at a furnace	Explosion of a propylene tank Cause: exposed to fire for 20 min	Fire of two benzene tank Cause: jet fire from a broken pipe
Colleferro maleic anidride plant (1981)	Explosion and fire	Explosion and fire of kettle and column Cause: exothermic reaction (human error)			
Falconara refinery (1981)	Fire	Cloud of flammable vapors Cause: unknown	Ignition of the cloud of flammable vapors Cause: pump	Fire of a tank Cause: ignition of the flammable vapors cloud	Collapse of the tank and of the floating roof Cause: exposed to fire
Porto Marghera petrochemical plant (1984)	Fire	Release of virgin naphta with fire and explosion Cause: human error	Explosion of a pipe containing a liquefied ethylene mixture Cause: exposed to fire for 10 min		
Porto Marghera refinery (1984)	Fire	Release and fire of gasoil Cause: human error	Explosion of a control valve Cause: exposed to fire		
Priolo plant of (1985)	Fire and explosion	Release and fire of LPG Cause: discharge valve vibration	Explosion of an LPG pipe Cause: exposed to fire	Explosion of an ethylene pipe Cause: exposed to fire	Explosion propylene tank Cause: exposed to fire

Milazzo refinery (1993)	Explosion and fire	Explosion and fire of gasoi surge tank Cause: operating errors		Collapse of a tank Cause: exposed to fire for 10 hr
Priolo petrochemical plant (1971)	Explosion and fire	Explosion and fire of acetic aldheyde Cause: ignition of an explosive mixture in a tank	Collapse and fire of an acetic aldheyde tank Cause: damage to the tank	Fire of acrylonitrile tanks Cause: fragments and product on fire from tank explosion
Storage farm in S. Dorlingo della Valle (1972)	Fire	Release and fire of crude oil Cause: sabotage	Collapse and fire of 3 tanks Cause: exposed to fire and boilover after 11 hr	Collapse of a tank Cause: exposed to fire for 24 hr
Naples refinery (1980)	Fire	Partial collapse of the floating roof Cause: no drainage of rain water (human error)	Tank fire Cause: sparks for mechanical attrition of the floating roof	Collapse of the tank Cause: exposed to fire for 13.5 hr
Massa Carrara plant (1980)	Fire	Fire of a fungicide with formation of SO_2 Cause: unknown		
Storage farm in Naples (1985)	Explosion	Release of gasoline followed by UVCE Cause: overfilling of the tank (human error)	Fire of a number of storage tanks Cause: domino effect of the explosion	
Storage farm in Genoa (1987)	Explosion	Explosion of a tank during degassification Cause: accidental sparks	Roof explosion and fire of 2 hexane tanks Cause: flammable vapor cloud	

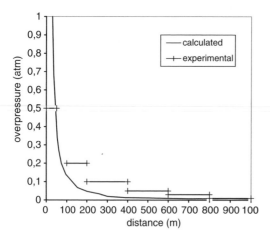

Figure 1. Calculated and experimental overpressure profiles.

Table 2. Characteristics and distance traveled by the main fragments.

Fragment	A	B	C
Part of the column	bottom	middle	head
Weight (t)	31.4	18.8	8.6
Diameter (m)	2.5	2.5	2.5
Length (m)	7.9	7.5	4.5
Actual distance (m)	100	280	439
Estimated maximum range of distance (m)	220	450	869

3.2.1 Analysis of the damages due to explosion

The effects of such explosion were as follows:

– heavy damages to one of the production lines and slight damages to the other one;
– noticeable damages to the control room;
– damages and partial collapses of concrete sheds nearby.

The kettle remained in place with the shell completely open longitudinally and the bundle substantially integer; the distillation column was broken into three main sections, which were found at distances ranging from 100 to 430 m.

The weight and length of the 3 main fragments of the column are not known exactly, but were estimated based on the geometry of the column, and the number of plates of each fragment: the obtained results are listed in Table 2.

The maximum distance that these fragments were expected to travel was estimated according to simplified models based on their initial velocity and drag coefficients (CCPS, 1989). These maximum values exceed the observed ones by a factor in the range 1.6 to 2.2, as shown in Table 2.

3.2.2 Causal sequence

In this case study available information allows to investigate the causal sequence which gave rise to the incident. It can be determined based on the logical tree shown in Figure 2, where the effective (i.e. the most probable) sequence is identified by bold lines.

Starting from the incident, the absence of fire signs on the fragments of the column and their distance from the plant demonstrated that the incident was an explosion and not a fire. Localized collapse was excluded to be the origin of such explosion, since the

rupture was extended to the entire kettle and column. So, the explosion derived from a pressurization of the equipment: due to the material (AISI 316 L) and the thickness of the walls, the burst pressure should be around 30 atm. Taking into account that the column is batch operated, the possible causes of the pressurization should be related to this part of the plant and are listed in the 3rd level of Figure 2. The inspection of the kettle tube bundle and the hydraulic test of the head condenser demonstrated that both did not present any problem. A combustion of the product in the column can be also excluded based on the absence of fire signs on the fragments launched away and taking into account the pressure in the equipment was increasing from the atmospheric to the burst value, thus preventing any air inlet. Therefore, the cause of the pressurization should be related to an undesired reaction of the product with some other substance. In particular, highly reactive alkaline substances and water are recognized as possible sources of problems for maleic anhydride plants. The former, which may residuate from chemical washing operated about 20 days before, react very rapidly with maleic anhydride giving rise to large amounts of carbon dioxide, and to a black friable polymer. The reaction with water, originating maleic acid, is highly exothermic, and its rate greatly increases with increasing the temperature.

No polymer generation was observed on the fragments, and, moreover, the plant worked regularly for several days after the chemical wash, thus allowing to exclude the hypothesis of a reaction with alkaline substances. On the contrary, the operation cycle includes the use of water, as demonstrated by the sequence of steps performed the same day of the incident:

– washing with water the distillation system;
– discharge of the washing liquid;
– addition of o-xylene to remove the traces of water by azeotropic distillation.

As temperature approaches 140°C, which is the atmospheric boiling temperature of o-xylene, maleic anhydride is added and purification starts. Instrumentation records showed that azeotropic distillation was slower

292

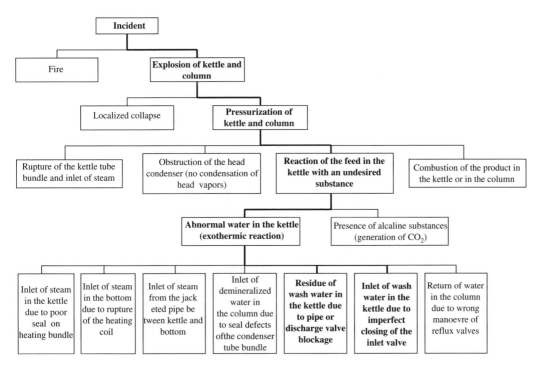

Figure 2. Logical tree for maleic anhydride incident.

than usual, possibly due to an incomplete discharge of water at the end of the washing phase. The introduction of maleic anhydride, contacting water at temperature higher than 100°C caused the release of heat, which increased the evaporation rate. Head condenser was unable to condense the large amount of vapors, and pressure in the column increased, thus causing also the temperature to increase. As a matter of fact, the temperature values recorded in the column were rapidly increasing, and the same did the temperature of the cooling water from the condenser, clearly indicating that the design duty of the equipment was exceeded.

Both the column and the kettle were provided with safety valves, set at 2.25 atm, which regularly opened. However, the vapor production rate exceeded the maximum flow rate from the valve, so that pressure increased further, up to the burst value.

The presence of water in the system may depend on a number of reasons, listed in the lowest level of Figure 2. Seal defects of the tube bundles of kettle and condenser were excluded by an hydraulic test; the rupture of the column heating coil was excluded from the visual exam of the large bottom fragment; the return of water in the column from the head condenser was also excluded, since all the switching valves were found in the correct position.

Due to the damages to the kettle and bottom, none of the remaining causes could be excluded. However,

the most likely cause appeared to be the presence of wash water in the kettle, due to either the blockage of the discharge pipe/valve or to the imperfect closing of the inlet valve.

3.3 Hot oil plant (Milazzo, 1993)

Hot oil plants are generally present in refineries, to heat process equipment by an oil. The hot oil is heated in a furnace and flows in a closed loop including a surge vessel. In fact, due to the difference between ambient and work temperature (about 300°C), the volume of the hot oil increases of about 20%. Hot oil is generally a specific high boiling mixture of diatermic substances and the surge vessel is generally pressurized with nitrogen. However, in the refinery of Milazzo, the original hot oil plant was modified using heavy gasoil as hot oil, and fuel gas to pressurize the surge vessel.

At 13:15 of June 3, 1993, the explosion of the surge vessel, followed by a fire caused the death of some nearby workers (Sebastiani, 2002).

The information available for this incident enables to use the Causal Tree Method (Boissieras, 1983), which includes the following steps:

- Data collection;
- Incident reconstruction;
- Facts formulation;

– Cause tree generation;
– Recommendations.

3.3.1 *Data collection*
The collected data included: P&Is and instruments data; operating procedures, plant data, analysis of product samples.

It was preliminarily assessed that the used gasoil presented an ignition temperature around 200°C, i.e. lower than the maximum work one (300°C). At the plant pressure of 0.5 atm, the boiling temperature of this gasoil was 305°C and the dew temperature 370°C. The fuel gas usually consists of methane and ethane, which are insoluble in gasoil, with traces of propane and butane, which, on the contrary, are rather soluble in gasoil.

Even though the physical properties of gasoil and fuel gas were rather different from those of a diathermic oil and nitrogen, the size of the venting line remained unchanged. During start-up, when gasoil temperature increased from the ambient value to about 300°C, this line was known to be insufficient to vent the vapors of the most volatile compounds, so that the relief valve by-passing the safety valve was maintained manually open for 5–6 hr.

It was also found that some operating procedures were not defined, most instruments were not working properly, and that data recording was poor, especially during start-up, when data were collected at 2 hr intervals. In particular, the level and temperature indicators of the surge vessel were out of order, and its automatic pressure controller worked in manual modality for most of the time. The controller of fuel gas flow rate to the furnace was also out of order and the controller of gasoil temperature at furnace exit worked only as indicator. Only the pressure controller on these circulation pump appeared to work properly.

3.3.2 *Incidental sequence*
The plant was started at 19:45 of June 2: however, at 1:45 of June 3, the pumps lost priming due to the presence of vapors in the line. At that time the hot oil temperature, increasing regularly at a rate of about 50°C/hr, had reached 225°C. That day, fuel gas was richer in propane and butane, due to a production surplus of LPG. Propane and butane were dissolved into the gasoil, and the boiling point of the obtained mixture was much lower than 300°C, so that the liquid started boiling in the surge vessel. Due to vaporization and foaming the gasoil level in the surge vessel decreased and, starting from 7:45 it was refilled up to 50% in 2 hr, introducing some heavy gasoil from the topping. Temperature decreased to 120°C: the furnace was started-up again, but the temperature increased at higher rate (100°C/hr) than before. It was estimated that, at 13.15, when the surge vessel exploded, the temperature was around 350°C.

3.3.3 *Facts formulation*
With reference to the incident, the following facts can be formulated:

– the plant was modified, by merging 2 distinct hot oil circuits into a single one, provided with only one surge vessel;
– the process was modified, by substituting the system diathermic oil-nitrogen, with that gasoil-fuel oil;
– some of the instruments were not working correctly;
– gasoil temperature increased rapidly as the furnace was started-up again;
– the temperature of gasoil returning into the surge vessel increased;
– the process heat exchangers did not require much hot oil;
– the blow-down line presented a vertical bend;
– the fuel gas, rich in propane and butane, was partly dissolved in the gasoil;
– no temperature or pressure control were present on the surge vessel;
– the manual interception valve of the control valve on the blow down vents was closed,
– the gasoil level in the surge vessel was decreasing;
– some water was probably present in the surge vessel.

3.3.4 *Causal tree generation*
The causal tree relevant to the case study is shown in Figure 3. In addition to the inefficient management of process modifications and to the poor state of the instrumentation other events are also present as co-causes of the incident:

– High solubilization of the fuel gas in the gasoil; rapid increase of hot oil temperature at the second start-up of the furnace; return temperature of the hot oil in the surge vessel much higher than the project one. This was due to process modification but also to the absence of proper operating procedures, coupled with the scarce heat request from the process.
– Inadequate pressure control in the surge vessel; valve of interception of the control valve on the blow down line manually closed; permanent connection with the blow down system by a by-pass of the safety valve. This was done to increase the vapor flow rate to the blow down system.
– Presence of a vertical bend in the blow down line. This was done "temporarily" but in this zone some liquid gasoil could accumulate, blocking the discharge of vapors and gases towards the flare.
– Nobody investigated about the cause of the decreased level in the surge vessel and about the exact level. The level controller, provided with high and low level alarms was out of service. Although the level was low, it was first attempted to start circulating the gasoil, with the pumps loosing their prime, instead of refilling immediately the surge vessel.

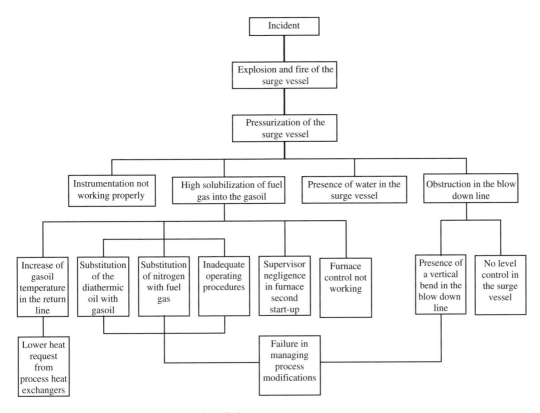

Figure 3. Causal tree for the incident in the hot oil plant.

– Possible presence of water in the surge vessel. All the events reported above could not result in the explosion of the vessel, unless some other event, such as the inlet of air or water, occurred. Due to the absence of ignition sources in the vessel, the presence of air could lead to an explosion only if the flammable vapors–air mixture concentration was within the flammability limits and the autoignition temperature was reached. On the other hand, the inlet of water may cause a physical explosion due to the sudden vaporization at the temperature inside the surge vessel. Water in this vessel could derive from steam purging of the blow down line: the water may have condensed on the walls of the surge vessel, remaining on the bottom, being immiscible with gasoil. Indeed some water was purged from the suction line of the pumps.

3.3.5 Recommendations

The lessons learned from this incident include the following:

- Operating procedures:
  – for furnace start-up;
  – for shut-down and start-up of the hot oil plant;

  – preheating gasoil to maintain sufficiently high its temperature in the surge vessel.
- Plant improvements:
  – gasoil should enter the surge vessel from the bottom.
- Management of modifications:
  – preventive check of the physical and chemical properties of the fluids used to substitute the original ones;
  – check of the venting capacity of the blow down line.
- Improvement of the maintenance procedures:
  – inspection and testing of control instrumentation;
  – inspection and testing of safety systems.

4 CONCLUSIONS

The causal tree method is probably the most effective technique for determining the first causes of an incident, but it requires to strictly adhere to the suggested procedure, in order to avoid omissions and delays in detecting evidence, which may result in the impossibility to determine the first causes. Some simplified procedure, such as the use of logical trees can be adopted when less complete information is available, as may happen when investigation starts some time after the incident. Very often the method would include a simulation of the incident, based on a consequence analysis approach. In all cases, the final report of the incident should include information about protection and prevention measures to be adopted to avoid the recurrence of the incident or mitigate its consequences.

REFERENCES

Boissieras, J. 1983. Causal Tree. Description of the Method. Princeton, N.J.

CCPS 1989. Guidelines for Chemical Process Quantitative Risk Analysis, AIChE, New York.
CCPS 1992. Guidelines for Investigating Chemical Process Incidents, AIChE, New York.
ChemPlus 2.0, A.D. Little, 1991.
Di Cave, S. 2002. Personal communication.
Lees, F.P. 1996. Loss Prevention in the Process Industries, 2nd Ed. Butterworths, London.
Ministero dell'Interno 1986. Rassegna comparata di incidenti di notevole entità, Roma.
Sebastiani, E. 2002. Personal communication.
VVF (Italian Fire Brigades), 1987. Personal communication.

Safety and Reliability – Bedford & van Gelder (eds)
© 2003 Swets & Zeitlinger, Lisse, ISBN 90 5809 551 7

Monte Carlo reliability analysis of a safety plant of the Italian Gran Sasso high energy physics national laboratory by means of the MARA code

Gabriele Bucciarelli & Dino Franciotti
INFN-LNGS, S.S., Assergi L'Aquila

Marzio Marseguerra & Enrico Zio
Politecnico di Milano, Dipartimento di Ingegneria Nucleare, Via Ponzio, Milano

Francesco Muzi & Marco Ventulini
Università degli Studi dell'Aquila – Dipartimento di Ingegneria Elettrica, L'Aquila

ABSTRACT: The Gran Sasso National Laboratory (LNGS) of the Istituto Nazionale di Fisica Nucleare (INFN), the Italian government agency for nuclear physics, is the biggest underground physics laboratory in the world. The Laboratory is located inside the left motorway tunnel crossing the Gran Sasso mountain in the motorway connecting Rome to Teramo in central Italy. The Laboratory consists of three big Halls, connected through a network of tunnels, where nuclear physics experiments are performed. The unique access of the Laboratory is through the left motorway tunnel. The physics experiments run inside the Laboratory involve the use of considerable quantity of flammable, toxic and other dangerous materials that in case of an accident could pose hazards to the personnel as well as to the highway users. This situation requires that detailed safety analyses be made in order to evaluate the risk associated to the experimental activities and that state-of the-art safety systems be installed for prevention of accident occurrences, early accident detection, warning and protection so as to minimize the consequences of an accident inside the Laboratory. Then, reliability and availability analyses of the safety systems are mandatory in order to keep the control of their overall performance. In this paper we present the results of a Monte Carlo reliability analysis activity which was performed on the high-expansion foam fire-fighting safety system installed in one of the main halls of the Laboratory. The analysis was carried out by means of the MARA (Montecarlo Availability Reliability Analysis) code.

1   INTRODUCTION

The Gran Sasso National Laboratory is one of the most important underground laboratories for nuclear physics research in the World. The Laboratory consists of three large Halls, A, B and C, and a network of service and emergency tunnels connecting the three Halls (Fig. 1). Each Hall is about 100 m long and 20 m wide, with a maximum height of 20 m and a volume of about 25,000 m³. Each Hall has three emergency exits (dimension 5 m by 5 m), two at the north and south ends of the Halls and one in mid position, to allow the access of big loads.

In this laboratory, international collaborations run large and complicate experiments, taking advantage of the screening action offered by the overlaying natural rock to study rare processes such as neutrino interactions and proton decays. Some of these experiments are of very big dimensions and use huge quantities

of potentially hazardous chemicals whose associated risk is enhanced by the confined environment. For this reason, several safety systems have been installed. For instance,

– fire-detecting systems
– fire-fighting systems
– gas-detecting systems

The reliability and proper operation of these systems are crucial for the safety of the Laboratory's personnel and for the protection of the surrounding environment. For this reason, an activity of reliability assessment of these systems is under way.

In this paper we present the steps of the analysis performed, and the results obtained, with respect to the high-expansion foam fire-fighting system installed in Hall C. The analysis, aimed at verifying the capability of the system of fulfilling its safety objectives, has been divided into two steps: 1) logic analysis, by the

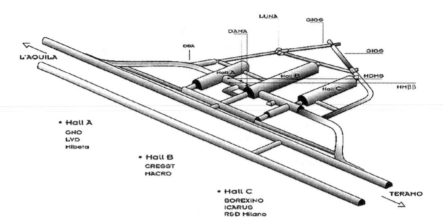

Figure 1.   3D view of LNGS.

fault tree method, of the system configuration and components' connections for systematically identifying the combination of faults (cut sets) that produce the top event, identified as insufficient water supply to the high expansion foam system protecting Hall C; 2) Monte Carlo quantitative evaluation of the reliability of the system. This second step was performed by means of the MARA (Montecarlo Availability Reliability Analysis) code, developed by some of the authors at the Department of Nuclear Engineering of the Polytechnic of Milan (www.cesnef.polimi.it/mara.htm). It is a user-friendly, powerful simulation tool which allows taking into account realistic aspects of system operation such as components' ageing, inspection and maintenance, stand-by and load-sharing configurations, etc.

1.1   The fire fighting water pumping system

In order to supply water to the high-expansion fire fighting system realized to protect the Hall C volume from fire, 4 electric pumps have been installed in a gallery in front of Hall C.

Each pump is capable of providing more than 1900 litres per minute of water. The high-expansion foam system consists of 14 foam producers, each one demanding an optimal flow of 400 litres/min of a mixture of water and foaming liquid: thus, the total flow requested by the foam producers installed is of 5600 litres/min. Actually, the foam producers can function also with a reduced flow of 3500 litres/min, although with a lower efficiency. At least two out of the four 1900 litres/min pumps must then operate successfully in order to provide the required flow to the high-expansion foam system. Since the available water reservoir is only of 130 m³ and safety regulations at LNGS establish that the fire-fighting water supply should last for at least one hour, a supporting water pumping system has been installed with the

capability of refilling the reservoir at a flow rate of 45 litres/min (see Fig. 2).

For the reliability analysis, the water pumping system can be divided into three sections: electric power supply, main water pumping system and supporting water pumping system.

1.2   Power supply

LNGS has its own electric power station which is fed at a medium level voltage (20,000 kV). Three transformers then reduce the voltage at low level (Fig. 3).

In normal condition, the main and supporting water pumping systems are fed via the transformer T2, but in case of needs it is possible to switch the power to the transformer T3. The commuting switch gear is located before the general switch gear of the low voltage section, so that a fault in the low voltage section does not produce a fault in the power supply of the fire-fighting water pumping systems.

1.3   Main water pumzping system

As explained earlier, the main water pumping system consists of 4 electric pumps, each one capable of providing a maximum of 1900 litres/min with a head of 100 meters of water (Fig. 4). There is also a compensation pump that maintains at 8 bars the pressure inside the water pipe collector. The electric power absorbed by each pump is 37 kW while the compensation pump absorbs only 2.2 kW. The system operates in an automatic way in the sense that a drop in the pressure inside the collector causes a first pump to start and then the other main pumps are started sequentially when the value of the pressure drops below three other preset levels. Once started, the pumps can be stopped only manually. Since the pumps are overhead pumps, an automatic priming system has been realized in such a way that via a tank of 1000 litres

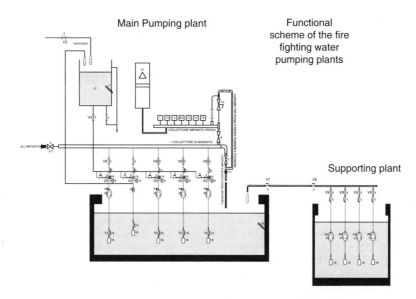

Figure 2. Main and supporting water pumping systems.

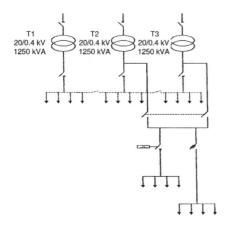

Figure 3. Power supply for main and supporting pumping systems.

Figure 4. Main pumps picture.

capacity the suction pipes of the pumps is guaranteed to remain full of water also in case of failure of the bottom valves of the suction pipes. If the level of water inside the priming system tank decreases below a certain level, the plant produces an alarm. A tester and a safety circuit are also present to allow testing the flow output of the pumps without loading and discharging possible overpressures.

1.4 *Supporting water pumping system*

The supporting pumping system has been installed to refill the water reservoir so as to comply with the rule of providing sufficient fire-fighting water for at least one hour to feed the high expansion foam system protecting Hall C, which requires 5600 litres/min. The supporting pumping system consists of 4 submerged electric pumps (2 automatically operating and 2 used for back up). The 2 automatic pumps are started by a signal coming from a level probe placed in the main water reservoir. The pumps are dimensioned to provide a refilling water flow rate of 45 litres/min with a head of 20 m of water column. The starting of the pumps is alerted also by a siren and they also can be stopped only manually.

299

The correct working of both the main and the supporting pumping systems is monthly tested by specialised technicians.

2 RELIABILITY ANALYSIS BY MONTE CARLO

The reliability analysis performed follows two successive steps; 1) logic analysis, by the fault tree method, of the system configuration and components' connections for systematically identifying the combination of faults (cut sets) that produce the top event; 2) Analog Monte Carlo quantitative evaluation of the reliability of the system.

The Monte Carlo method is a powerful tool that can be of great value in the reliability analysis of complex systems, due to the inherent capability of achieving a closer adherence to reality [1–7]. Indeed, Monte Carlo simulation allows to embrace a system transport formulation which represents the best model for assessing quantities of interest in system engineering, such as reliability and availability, under realistic conditions. The resulting integral transport equation, which still lacks of an explicit general analytical solution, takes into account explicitly the various phenomena (failure, repair, maintenance, etc.) which affect the system life. An estimate of the solution may be pursued by a Monte Carlo simulation approach which follows the individual fates of a large number of like-systems, one at the time, from the beginning of their operation to the end of their mission, i.e. to the mission time TM. The system state, i.e. the configuration of its components is represented by a point in a discrete NC-dimensional phase space, where Ci is an integer which codes the configuration of the i-th component: for example, in case of two-state components, the integer might be 1 for the nominal state and 2 for the failed state (e.g. for a system of two-state components, up and down, a generic configuration could be C1 = up, C2 = up, ... Cj =down, ... CNC = up). Each system stochastic transition from one state to another corresponds to a move of the system from one point to another so that each Monte Carlo trial, which simulates the system history during the mission time, generates a random walk of the system across several points P. The set of trials of a standard analog Monte Carlo calculation then provides an ensemble of realizations of the system stochastic life process from which ensemble averages of the quantities of interest, e.g. the system reliability and availability, are performed to obtain the relevant estimates.

In our case, the calculations of the reliability of the fire-fighting water pumping system were performed by means of the MARA code. The code is provided with a user friendly graphic interface for fault tree construction and a crude method for its solution in terms of cut sets. These, together with the failure and repair rates of the components, are the main input to the computational Monte Carlo engine for the quantification of the system reliability and availability measures.

Given the logic structure of the system under consideration, it has been possible to subdivide the system into three independent subsystems, linked with an OR gate with respect to the top event of overall system failure: electric power supply, main pumping system, supporting pumping system. For each subsystem we developed the fault tree, determined the cut sets and ran the quantitative reliability computations by MARA. Once the reliability of each subsystem is computed, the calculation of the reliability of the entire safety system is trivial.

In the next subsections we report the results of the fault tree analysis and Monte Carlo simulation for each subsystem and for the aggregate system. All the simulations were performed with 10E6 trials. The mission time has been set equal to 750 hours, taking into account that every month the safety system undergoes accurate maintenance by specialised technicians.

3 RESULTS

3.1 Power supply

The reliability study of the electric power supply system took into account failures of transformers, cables, switch gears and the absence of power supply at the primary section of the transformers. In Figure 5, the fault tree of the electric supply system is reported. The top event is the absence of power supply to the pumps. The minimal cut sets are 5: 3 cut sets of first order and 2 cut sets of second order. The components failure rates are listed in Table 1. The values have been taken from technical literature [8], except for the failure rate of presence of power supply at the primary section of the transformers that comes from historical data recorded by LNGS.

The failure rates are assumed to remain constant throughout the mission time, thus neglecting the effect of ageing and maintenance.

In Figure 6, we report the unreliability of the power supply system as a function of time. The Mean Time Between Failures (MTBF) turns out to be 29.15 10E2 h.

3.2 Main pumping system

The main pumping system is certainly the core of the entire fire fighting water pumping system. It includes the 4 main pumps and the priming system. In Figure 7 we report the corresponding fault tree.

The top event is the failure of starting at least three out of the 4 pumps installed. In this case, the number of cut sets is very high: 256 cut sets of order 3, 924 cut sets of order 4, 48 cut sets of order 5 and 4 cut sets

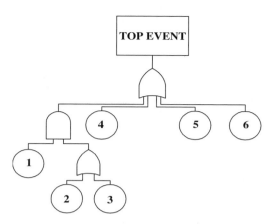

Figure 5. Electric power supply fault tree.

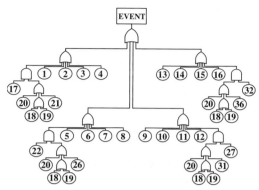

Figure 7. Main pumping system fault tree.

Table 1. Power supply – components failure rates.

Basic event	Description	Failure rate $(10^{-6}h^{-1})$: Mean
1	Transformer T_2 failure	6
2	Transformer T_3 failure	6
3	Commuting switch gear failure	0.51
4	Protection switch release	0.69
5	Power supply cable failure	0.01
6	Power supply absence before transformers	342

Table 2. Main pumping system – components failure.

Basic event	Description	Failure rate $(10^{-6}h^{-1})$: Mean
1	Pump n.1 mechanical failure	89.36
2	Pump n.1 starting manostat failure	20
3	Manual valve V6 of pump n.1 closed	0.27
4	Low voltage electric board failure or electric motor of pump n° 1 failure	8.3
5	Pump n.2 mechanical failure	89.36
6	Pump n.2 starting manostat failure	20
7	Manual valve V6 of pump n.2 closed	0.27
8	Low voltage electric board failure or electric motor of pump n° 2 failure	8.3
9	Pump n.3 mechanical failure	89.36
10	Pump n.3 starting manostat failure	20
11	Manual valve V6 of pump n.3 closed	0.27
12	Low voltage electric board failure or electric motor of pump n° 3 failure	8.3
13	Pump n.4 mechanical failure	89.36
14	Pump n.4 starting manostat failure	20
15	Manual valve V6 of pump n.4 closed	0.27
16	Low voltage electric board failure or electric motor of pump n° 4 failure	8.3
17	Bottom valve V2 of pump n° 1 broken	3.18
18	Level probe of priming plant broken	2.8
19	Priming plant valve V3 closed	0.27
20	General valve V4 of priming plant closed	0.27
21	Valve V5 for pump n.1 priming closed	0.27
22	Bottom valve V2 of pump n° 2 broken	3.18
26	Valve V5 for pump n.2 priming closed	0.27
27	Bottom valve V2 of pump n° 3 broken	3.18
31	Valve V5 for pump n.3 priming closed	0.27
32	Bottom valve V2 of pump n° 4 broken	3.18
36	Valve V5 for pump n.4 priming closed	0.27

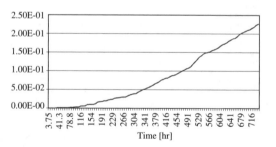

Figure 6. Power supply unreliability.

of order 6. The failure rates listed in Table 2. The values have been taken from technical literature [8]. Figure 8 shows the unreliability of the plant as a function of time. The MTBF is 30.86 10E4 h.

3.3 Supporting pumping system

Figure 9 shows the fault tree for the supporting pumping system. It takes into account faults that could occur to the pumps, to the electric power supply, to the water level signaling system in the water reservoir and to the water reservoir itself. The top event considered is the water reservoir dry out.

301

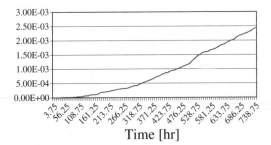

Figure 8. Main pumping plant unreliability.

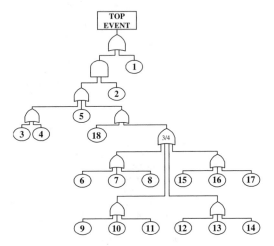

Figure 9. Supporting plant fault tree.

Table 3. Support pumping – components failure rates.

Basic event	Description	Failure rate $(10^{-6}h^{-1})$: Mean
1	Water reservoir empty for maintenance	0.21
2	Leak in the water reservoir	0.001
3	General valve V7 closed	0.27
4	General valve V8 closed	0.27
5	Water reservoir level probe broken	2.8
6	Mechanical failure of pump n° 1	190.73
7	Low voltage electric board failure or electric motor of pump n° 1 failure	8.3
8	Interception valve V9 of pump n.1 closed	0.27
9	Mechanical failure of pump n° 2	190.73
10	Low voltage electric board failure or electric motor of pump n° 2 failure	8.3
11	Interception valve V9 of pump n.2 closed	0.27
12	Mechanical failure of pump n° 3	190.73
13	Low voltage electric board failure or electric motor of pump n° 3 failure	8.3
14	Interception valve V9 of pump n.3 closed	0.27
15	Mechanical failure of pump n° 4	190.73
16	Low voltage electric board failure or electric motor of pump n° 4 failure	8.3
17	Interception valve V9 of pump n.4 closed	0.27
18	Absence of power supply	342

Also in this case the number of cut sets is large. We have 1 cut set of order 1, 4 cut sets of order 2, no cut sets of order 3 and 108 cut sets of order 4. The components failure rates are listed in Table 3. The values have been taken from technical literature [8]. In Figure 10 we report the unreliability of the system as a function of time. The MTBF is 47.62 10E5 h.

3.4 Fire fighting water pumping system

Once we have computed the reliability of the three subsystems, we can combine them to calculate the reliability of the entire fire fighting system. We recall that the system is logically composed by the main and supporting pumping systems and the power supply, linked with an OR gate with respect to the failure of the system. In addition, we considered also the event that the valve on the delivery collector could be accidentally closed. The failure rate of this latter event has been taken from the technical literature [8]. In Figure 11 we report the unreliability of the whole system as function of time. The corresponding MTBF is 28.84 10E2 h.

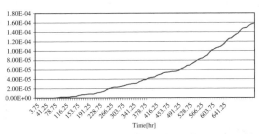

Figure 10. Supporting system unreliability.

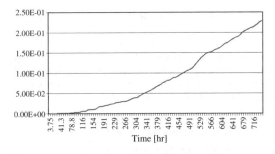

Figure 11. Unreliability of the whole plant.

302

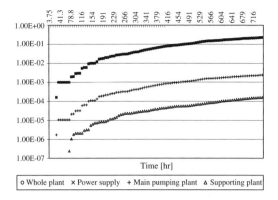

Figure 12. Unreliability breakdown.

Table 4. Power supply – failure rates.

Basic event	Description	Failure rate (10^{-6}h^{-1}): Mean
1	Failure of transformer T_2	6
2	Failure of transformer T_3	6
3	Commuting switch gear failure	0.51
4	Protection switch release	0.69
5	Failure in the cable	0.01
6	Fault in the power supply before transformers	342
7	Fault in the diesel generator	740
8	Fault in the automatic starting	8.33
9	Fault in the manual starting	9.51

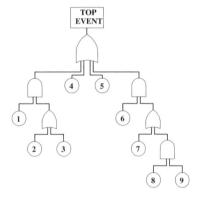

Figure 13. Power supply fault tree with diesel generator.

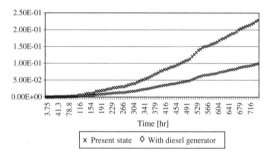

Figure 14. Unreliability of the whole system comparison.

The reliability of the system appears to be very low. This is due largely to the unreliability of the power supply system which is more than two orders of magnitude higher than the main and supporting pumping systems unreliability.

The unreliability of the power supply, on the other hand, is due to the unreliability of the power supply to the transformers, i.e. to the unreliability of the primary source of energy. Figure 12 shows in the same graph the previous results for the unreliability of the entire system and of the three subsystems separately, as a function of time. The unreliability of the system is basically coinciding with the unreliability of the power supply (the two curves are indistinguishable in the Figure). To improve the reliability of the system, one of the possible solutions is to install inside the underground laboratories a diesel generator with automatic start. In case of failure of the automatic start, the technical personnel, continuously present, could start the generator manually within a short time.

Figure 13 gives the power supply fault tree with the presence of the additional diesel generator and Table 4

lists the failure rates used [8]. The unreliability of the system decreases significantly as shown in Figure 14 and the corresponding MTBF increases to 72.08 10E2 h.

4 CONCLUSIONS

The reliability and proper operation of the installed safety systems are crucial for the safety of the LNGS's personnel and for the protection of the surrounding environment with respect to the hazards posed by the undergoing experimental activities. For this reason, an activity of reliability assessment of these systems is under way.

This work reports the results obtained with respect to the first system considered, the high-expansion foam fire-fighting plant installed in one of the main hall of the Laboratory, the so called Hall C. The reliability assessment was performed by Monte Carlo simulation, with the use of the MARA code. The work has shown the user friendliness of the code with respect to the construction of the fault trees, the cut sets determination and the actual Monte Carlo simulation running.

All results were obtained within few minutes of calculation on an Athlon 1400 MHz computer.

The results obtained have provided an idea of the capability of the system of fulfilling its safety objectives. In particular, a high value of system unreliability was found due to the unreliability of the electrical power supply system, whereas the unreliability contribution due to faults of the fire-fighting system components is negligible. For this reason, the potential improvement coming from the installation of an ad hoc motor generator has been evaluated: the results show the significant beneficial effect that such additional component could have on the reliability of the safety system.

This work represents the first step of a full-scope reliability analysis of the safety systems at LNGS. Further developments include accounting for the ageing of the components (failure rates variable in the time) and for physical dependences and common mode failures, as well as propagation of the uncertainties in the failure rates values.

REFERENCES

1. E. D. Cashwell and C. J. Everett, *A practical manual on the Monte Carlo method for random walk problems*, Pergamon Press, N. Y., 1959.
2. R. Y. Rubinstein, *Simulation and the Monte Carlo method*, Wiley, 1981.
3. M. H. Kalos and P. A. Whitlock, *Monte Carlo methods. Volume I: basics*, Wiley, 1986.
4. I. Lux and L. Koblinger, *Monte Carlo particle transport methods: neutron and photon calculations*, CRC Press, 1991.
5. E. J. Henley and H. Kumamoto, *Probabilistic Risk Assessment*, IEEE Press, 1991.
6. A. Dubi, Monte Carlo Applications in Systems Engineering, Wiley, 1999.
7. M. Marseguerra, E. Zio, *Basics of the Monte Carlo Method with Application to System Reliability.* LiLoLe-Verlag GmbH (Publ. Co. Ltd.), 2002.
8. *Reliability data bank.* ENI.

Safety and Reliability – Bedford & van Gelder (eds)
© 2003 Swets & Zeitlinger, Lisse, ISBN 90 5809 551 7

Safety analysis of the prototype plant of ICARUS experiment of INFN – National Laboratory of Gran Sasso

G. Bucciarelli
INFN – National Laboratory of Gran Sasso (Safety Plants Servise)

D. Franciotti & P. Aprili
INFN – National Laboratory of Gran Sasso (Safety Plants Servise)

ABSTRACT: The Gran Sasso National Laboratory (LNGS) of the Istituto Nazionale di Fisica Nucleare (INFN), the Italian governmental authority for nuclear physics, is the biggest underground physics laboratory in the world. The Laboratory is located inside the left motorway tunnel crossing the Gran Sasso mountain in the motorway connecting Rome to Teramo in central part of Italy. The physics experiments run inside the underground laboratory request the use of considerable quantity of flammable and other dangerous materials that in case of an accident could pose hazards also for the highway users. This situation requires that detailed safety analyses have to be carried out in order to evaluate the risk associated to the experimental activities and to new plants that will be installed in the Laboratory. In this paper we present the results of HAZOP analysis of the prototype plant of ICARUS experiment. The analysis was carry out to avoid the problems linked to a leak of cryogenic liquids in one of the main hall of the Laboratory.

1 GRAN SASSO NATIONAL LABORATORY GENERAL INFORMATION

The National Laboratory of Gran Sasso is an underground laboratory. It is located under the mountain of Gran Sasso (Italy), alongside the motorway tunnel crossing the Gran Sasso.

It consists of three experimental halls, simply designated as Laboratories A, B, and C, and a series of connecting tunnels and service areas. The three experimental halls are more than 100 m long and about 18 m high and wide, enclosing a total volume that is about 25,000 m³ each. Moreover, around the Laboratory there is an aqueduct that feeds a lot of towns. Then the risks of water and air pollution require that detailed safety analysis have to be carried out in order to associated to the experimental activities and to new plants that will be installed in the Laboratory.

The experiments inside the laboratory go from astrophysics of elementary particle to rare decays, and from dark matter research to solar neutrino research. The reason why the laboratory was built inside a mountain is that 1400 meters of rocks are a perfect filter to screen huge number of particles that hit the land surface; these particles are produced by the interaction of cosmic primary rays with the atmosphere's atoms. Muons

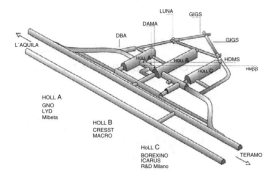

Figure 1.   View of laboratory.

(particles with positive or negative charge and mass 200 times inferior to it of electron) and neutrinos are the particles that are able to cross this filter. Solar and stellar neutrinos are studied in the Laboratory of Gran Sasso.

2 AIM OF THE ICARUS EXPERIMENT

The aim of the ICARUS experiment, that will be carried out in the Laboratory of Gran Sasso, is to study the rare process such as decay of protons and oscillation of neutrinos and to understand if these particles have a mass or not.

ICARUS is an liquid argon detector; it's able to find the transit of neutrinos through the ionization that particles produce when they cross the detector. Moreover ICARUS is able to make a three-dimensional reconstruction of events having the wire chambers submerged in the liquid argon.

Since studied interactions are very rare, it necessary to build very big detectors. The first detector will contain 600 tons of liquid argon and we expect to get to a final volume of 1200 tons.

3 PROTOTYPE PLANT OF ICARUS

Due to the complex technology and the demand for very pure liquid argon, has been decided to construct a prototype plant of 10 tons of liquid argon, before realizing a 600 ton detector. The aim is to resolve problems linked to three principal aspects of ICARUS 600 ton:

- test the cooling of the aluminium structure with liquid nitrogen. Nitrogen is necessary to take under control the temperature of the liquid argon in the cryostat.
- test insulation.
- resolve structural problems linked to mechanics and thermal resistance of materials.

It was possible to extract a lot of information from the construction of the prototype, such as:

- check seal of panels that form cryostat structure.
- find the optimal gaskets for flange and manhole.
- establish a standard procedure for the construction and assemblage of the 600 ton cryostat.
- test that the cooling system fulfil design specifications.

Therefore the cryostat is the heart of the plant; in fact on inside it there are liquid argon and electronics to detect particles, while inside the wall of the cryostat there is a cooling system made up of coils crossed by liquid nitrogen. Moreover argon has to be very pure ($O_2 < 0.1\,ppb$) to discover the passage of neutrino, therefore there are a set of equipments used to purify both liquid and gas argon.

3.1 Top events

The hazards of this experiment don't result from the nature of used substances, as they aren't either toxic or harmful, but the hazards are due to their physical state and the confined environment where the plant will be installed. In fact, if cryostat or pipes break liquid argon rapidly will go out increasing its specific volume 800 times. This causes a drastic drop in oxygen concentration and temperature inside the Hall.

Another hazard is linked to the fact that the plant is provided with safety magnetic disk (DMS); they preserve the cryostat from pressure increase. Therefore if the pressure goes over 1.45 bar the DMS will open. When a DMS opens, in the surrounding area, cold gas argon goes out that could strike people around the plant.

4 RISK EVALUATION

Before studying the eventual consequences that might originate from one of the as above described incidents, it's very important to find all the possible causes that can generate a top event.

There are a lot of methods to find the causes that can generate it, but here we have chosen the HAZOP analysis. The reason why we prefer HAZOP and not FMEA analysis, for example, it's that in this case we have to analyze a prototype plant for which it doesn't exit standards to apply. The realisation of ICARUS is a international collaboration, therefore not all the components of the plant have been made in the same country and under the same rules. So with the HAZOP analysis we are able to carry out a complete, systematic and documented evaluation of each component in the whole plant. HAZOP, in fact, provides an understanding of the causes and consequences of deviations from expected behaviour and facilitates decision-making on actions needed to eliminate or reduce the risk. It must remembered, however, that HAZOP is an identifying technique and is not intended as a means of solving problems.

Moreover, HAZOP analysis document is accepted by Gran Sasso National Laboratory safety authorities as a document in order to assess the acceptability of the proposed experiments.

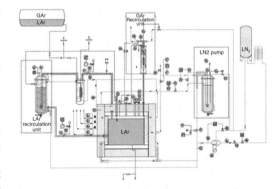

Figure 2.   ICARUS 10 ton.

5 HAZOP ANALYSIS

The objective of HAZOP is to stimulate the imagination of a review team, in a systematic way so that they can identify potential hazards in a design.

The P&I is the principal instrument to start the study (Fig. 3). In this document there are the components of the plant and also control systems to manage it. But to start HAZOP study it still lacks forms that in general are paper documents as in Fig. 4 where the team introduces its own evaluation.

In our case, we haven't used paper forms but we realised spreadsheets. These simplify a lot the carrying out of the work; in fact with a PC linked to a projector all the members of the team can see the same spreadsheet and listen what the team leader is saying. Moreover there are never problems due to different handwriting or erasures. Therefore in this way it's also very simple to consult different schedule at the same time.

The use of spreadsheets allows also to introduce a help to understand the items of the schedule.

The first thing to do is to find the functioning phases of the plant and then for each of them we will carry out a HAZOP analysis. In this paper we are considering only the running phase.

At this point we divide the plant in sections and we carry out a section analysis considering the credible deviations of physic parameter (flow rate, temperature, etc.) in the single junction, evaluating the causes, the consequences and the efficient precautions that are present.

If a consequence is no more developable it means that it is a degraded situation and we'll call it "Top event".

The considered sections are:

- Coil with liquid nitrogen into cryostat.
- Liquid nitrogen feeding the purification system of liquid argon.
- Liquid nitrogen distribution line to the purification system of gas argon.
- Purification circuit of liquid argon.
- Purification circuit of gas argon.
- Liquid argon line from the purification system to the cryostat.
- Feeding circuit of liquid nitrogen to the pumping system.

HAZOP				Company:			Factory:		chart /
Plant:									
Reference plan:									
Section:						Symbology:			
Junctions:									
Not esteemed junctions:									
Analysed junctions:									

1 2	3	4	5	6		7		8	9	10
N°	GUIDE WORD	DEVIAT- IONS	CAUSES	CONSE- QUENCES	SIGNALS		INTERVENTIONS		HINTS	NOTE	TOP
					OPTICAL	ACOUSTIC	AUTOMATIC	MANUAL			

Figure 4.   HAZOP schedule.

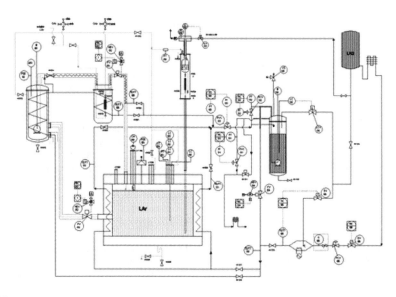

Figure 3.   P&I.

5.1 Coil with liquid nitrogen into cryostat

To be able to find the possible causes of cryostat cooling system failure we considered the following parameters: temperature and liquid nitrogen flow rate.

The first deviation is flow absence of cooling fluid for one of the following reasons:

- Lack of nitrogen.
- M510 pump broken.
- Closed line for erroneous closing of the HV511 valve or for closing of HV503 and HV508 valves.
- Pneumatic PVC513 valve closed for compressed air lacking.
- Interrupted line.

The most important consequences are an increase of pressure inside the cryostat and the discharge of gas argon to the outside trough the safety magnetic disk (indicated on the P&I PSE001). To prevent this problem it was installed a pressure gauge on the cryostat to allow the operator to know quickly what is happening and to operate before the safety system starts.

Another deviation of liquid nitrogen flow is that the flow rate is lower than the design value. The causes that can produce such a wrong functioning are:

- Wrong functioning of the M510 pump.
- Partial breaking of the line.
- Presence of narrow passages.

Also in this case the most important consequence is the pressure increasing inside the cryostat, even if the pressure increasing is slower because the cooling is partially guaranteed.

The last deviation linked to liquid nitrogen flow is due to the higher flow of the M510 pump.

In these case there isn't pressure increasing, but a higher heat transfer.

The deviation linked to the temperature produces biphasic fluid inside the pipes and can be due to the breaking of the pipes insulation or to an external source of heat. In both cases the consequences are similar.

5.2 Liquid nitrogen feeding to the purification system of liquid argon

Here we consider as parameter the temperature and liquid nitrogen flow. The aim is to find the possible consequences on the liquid nitrogen flow when wrong functioning is present.

The first deviation is the absence of liquid nitrogen caused by one of these reasons:

- Lack of Liquid nitrogen.
- M512 pump broken.
- Line interrupted for wrong closing of the HV512 valve or of the HV502 and HV501 valves.
- Pneumatic FV514 valve closed for lacking of compressed air.
- Interrupted line.

The only consequence is the heating of liquid argon present in the purification system and then the pressure increasing due to liquid evaporation. A DMS (PSE102 on P&I) is also installed on the tank of the liquid argon pump that at 1.45 bar opens. The instruments that shows this situation are the L013 level indicator and the FT018 orifice plate. The first measures the level in the argon liquid tank, it will block the pump when the level is too low, the second indicates the gas argon flow rate that arrives to the purification system of gas phase.

The second deviation is the lower flow rate of liquid nitrogen caused by one of the following reasons:

- Wrong functioning of the M510 pump.
- Presence of narrow passages.

In this case we report only a higher production of gas argon that will be condensed again by the purification system of gas argon.

The last deviation linked to the nitrogen flow is the higher flow rate due to a wrong regulation of the M510 pump. The only consequence is a higher cooling of the cryostat showed by TE/TT509 and TE/TT508 thermometers.

The deviation linked to the temperature is a higher temperature of liquid nitrogen due to the loss of insulation or heating coming from an external source of heat. The consequence is the presence of an insufficient heat transfer with an overproduction of gas argon in the cryostat.

5.3 Liquid nitrogen distribution line to purification system of gas argon

The only parameter considered is the liquid nitrogen flow rate. The first deviation is the absence of liquid nitrogen due the following reasons:

- Closing of the valves for error or failure.
- Interrupted line.

The fact that liquid nitrogen doesn't arrive to the purification system is very serious, because in this condition it's impossible to condense again the gas argon that arrives from the cryostat and from the purification system of liquid argon. In addition it's impossible to control the pressure that causing the opening of the DMS. The instruments that are able to show this kind of event are the LC518 (it measures the nitrogen level in the tank), the PT511 and PT011 (respectively the first measures the pressure of nitrogen and the second the pressure of argon in the cryostat).

The second deviation is the lower flow rate of liquid nitrogen due to a wrong functioning of the LCV518 system that provides to fill the nitrogen tank of the purification system when its level is too low. Therefore it is not possible to condense all the gas argon that arrives to the purification system causing an increasing of the pressure in the cryostat.

5.4 Purification circuit of liquid argon

The first deviation is the lack of flow of liquid argon due to the following causes:

- HV014 valve erroneously closed.
- FV011 or FV012 pneumatic valves closed for lack of compressed air.
- M001 argon pump broken.
- Pipes breaking.
- Cryostat empty because the opening of the drainage valve.

The consequence of the first three causes is the lack of liquid argon pureness, but it's only an operative problem and there are no problems for the safety.

The last two, on the contrary, cause an emission of liquid argon to the outside with serious consequences to the people that are in the proximity. The argon, in fact, coming out evaporates and increases its specific volume causing a heavy fall of temperature and a drastic drop of oxygen level. But the plant is provided of instruments to measure the level of liquid argon whether in the cryostat and in the tank of the pump, so the operator can understand what is happening.

The second and last deviation considered in this section is a lower liquid argon flow rate. This can be caused by:

- Wrong functioning of the pump.
- Presence of narrow passages.
- Wrong functioning of the FV012 valve.

Also in this case the consequence is only operative, because it needs more time to obtain the pureness level required.

5.5 Purification circuit of gas argon

The aim is to find the consequences of an anomalous rate of gas argon.

The first deviation is the absence of argon flow caused by:

- On-off valve closing.
- Pipe from cryostat to purification system of gas argon closed and purification system of the liquid argon switched off.

This deviation has a double effect on the plant: is not possible to purify the gas argon and to control the pressure in the cryostat. This is very important for safety.

The second deviation is the higher flow rate of gas argon due to the fact that the safety systems of the cryostat and of the liquid argon tank did not operate well. Also in this case the consequences are critic. In fact it's possible that the gas argon that arrives to the purification system doesn't condense completely. So the pressure in the cryostat increases causing the opening of the DMS.

5.6 Liquid argon line from the purification system to cryostat

The only parameter considered is the liquid argon flow and the first deviation concerns the flow absence the pipe. The causes of this event are the following:

- M001 pump breaking.
- HV014 valve erroneously closed.
- FV011 closed for absence of compressed air.
- Pipes breaking.
- System off.

The consequence of the first and last causes is that it's impossible to purify the liquid argon, whereas the other causes produce a failure with more serious consequences. In fact the interruption of the line increases the pressure in the tank of the argon pump and produces at the same time a depression in the cryostat. On the contrary in case of breaking of pipe there is an emission of liquid argon.

In the plant there are level indicators that are able to point out this wrong functioning.

A further deviation concerns a lower flow rate of liquid argon due to a bad functioning of the M001 pump or to the presence of narrow passages. Both these deviations haven't important effects on the safety.

The last case is connected to the presence of biphasic argon in the pipe due to a wrong functioning in the cooling nitrogen system. In this case it's enough to switch off the liquid argon purification plant.

5.7 Feeding circuit of liquid nitrogen to the pumping system

Trough the analysis of this section we want to find the consequences linked to wrong functioning of the nitrogen pipe network.

The only important deviation is the absence of liquid nitrogen for the following reasons:

- Liquid nitrogen tank empty.
- HV514 or LCV518 valves erroneously closed.
- Interrupted line.

For all the causes the consequence is unique: the tank of nitrogen pump doesn't fill causing the stop of the cooling system.

The only instrument to check the fill of the pump tank is the LC510 level indicator, that when the level is low, it stops the pump.

6 CONCLUSIONS

The success or failure of HAZOP analysis depends:

- The accuracy of drawings and other documents used in the study.
- The expertise and experience of the team.

- The ability of team to visualize deviations, causes and consequences.
- The ability of team to assess the seriousness of hazards.
- The skill of the chairman in keeping the study on track.

Of these the skill of the chairman is perhaps the most important. But in this analysis, we had problems with availability of designers, because a lot of them were French engineers, therefore we had lost time to understand the operation of the plant.

However all team members and ICARUS' team were satisfied of analysis, particularly the members who didn't ever do HAZOP analysis. In fact, it carried out a new HAZOP analysis for ICARUS 600 ton and its results are being considered by safety authority of National Laboratory of Gran Sasso.

As far as ICARUS 10 ton the most important event that may occur is the breaking of the cryostat.

From risk analysis we deduced that this event is not very probable, because it must occur many failures in different parts of the plant to determine it; but we found some weak points. One of these is the line that transports the nitrogen to the cryostat. On this line there are different equipment that allow the operator to check the process parameter, for example the FV514 valve, the pressure gauge, thermometers. The operator, in fact, reading the pressure value and the argon and nitrogen temperatures controls if the flow rate of liquid nitrogen is correct or not and eventually corrects it. But if there is a breaking down of the FV514 valve the operator doesn't notice it, also because the increasing of the pressure in the cryostat is not immediate. For this reason the operator could notice it too late. Then even if the breaking occurs up to the FV514, this one is not able to switch off the motor pump that is on the contrary controlled by the level indicator placed on the tank of the pump of liquid nitrogen.

Another improvement could be to introduce some acoustic alarms to call immediately the operator attention in the control room.

The experience showed that a lot of incidents occur exactly for the absence of operator action which doesn't realise the failure moment for a lot of different reasons.

For what concerns the pneumatic valves in the plant, these in case of lack of compressed air, closed and can cause serious consequences for the safety as seen with HAZOP analysis. So it could be better to provide the plant with a backup system for compressed air of the valves.

REFERENCES

1. *Process Safety Analysis,* Bob Skelton – Institution of chemical engineers, 1997
2. *Progetto ICARUS: studio di ingegneria di base,* Air Liquide
3. *Nota di funzionamento relativa all'impianto e al processo del prototipo ICARUS 10 m³,* Cervini Francesco, J. M. Disdier – Air Liquide, 1999
4. *Simulations of Argon accident scenarios in the ATLAS experimental cavern – a safety analysis,* Fabrizio Balda – Tesi di laurea Politecnico di Torino, Dicembre 1999
5. *Chemical Process Safety Fundamental with Applications,* Daniel A. Crowl, Joseph F. Louvar – Prentice Hall, 1990
6. *Handbook of Heat Transfer,* Warren M. Rohsenow, James P. Hartnett, Young I. Cho – McGraw Hill, 1998
7. *Heat Transfer a basic approach,* M. Necati Ozisik – McGraw Hill, 1998
8. *Physical Chemistry,* Ira N. Levine – McGraw Hill, 1988
9. *Gas data book,* William Braker and Allen L. Mossman – Matheson
10. *Encyclopédie des gaz,* Air Liquide – Elsevier, 1976

Safety and Reliability – Bedford & van Gelder (eds)
© 2003 Swets & Zeitlinger, Lisse, ISBN 90 5809 551 7

Application of Dutch reliability-based flood defence design in the UK

F.A. Buijs, P.H.A.J.M. van Gelder, J.K. Vrijling & A.C.W.M. Vrouwenvelder
Delft University of Technology, Delft, Netherlands

· J.W. Hall
University of Bristol, Bristol, UK

P.B. Sayers
HR Wallingford, Wallingford, UK

M.J. Wehrung
Road and Hydraulic Engineering Division of the Dutch Ministry of Transport, Public Works and Water Management, Delft, Netherlands

ABSTRACT: Increasing concern about recent and potential impacts of flooding in the UK are leading to the adoption of risk-based methods for planning, appraisal, design and operation of flood defences. Probabilistic methods for assessment and design of flood defences are relatively well developed in the Netherlands because of the potentially devastating impacts of flooding. However, for reasons described in this paper, reliability methods developed in the Netherlands are not universally applicable in the UK context. This paper describes a test application of a reliability method developed for ring dykes in the Netherlands to the flood defence system at the Caldicot Levels in South Wales. Although constrained by data limitations (even at this relatively data-rich site, by UK standards) the reliability method provides estimates of the probability of failure of the flood defence system, identifies weak system components and identifies which parameters contribute most to the probability of failure. The study has provided recommendations for the future development of reliability-based methods for flood management in the UK.

1 INTRODUCTION

In The Netherlands as well as in the UK flood protection policy is currently undergoing changes towards a risk-based approach of flood defences. A risk-based safety approach takes into account the strength and the loading conditions of the flood defence system as part of the probability of inundation as well as the consequences of inundation in case of failure of the flood defence system. TAW (2000) points out that a risk-based analysis of flood defence systems can result in the identification of the system's weak areas and can therefore enable the decision-maker to target improvement schemes and maintenance activities. Another advantage is that in case of large scale flood defence improvements the decision-maker can compare different design options in terms of the actual risk reduction and the costs which are associated with the improvement options. In the light of the shift to a risk-based safety approach the UK Environment

Agency and the Department of the Environment, Food and Rural Affairs, which together have responsibility for flood defence policy and implementation in the UK have launched a research and development project called RASP: Risk Assessment of flood and coastal defences for Strategic Planning. This project aims to develop tiered methodologies for risk assessment of flood defence systems: a high level methodology supporting national policy making, an intermediate level methodology supporting regional policy making and a detailed level approach supporting policy making at the scale of one flood defence system.

2 OBJECTIVE AND DETAILED WORKING METHOD

The objective of this research is to apply a reliability analysis to the Caldicot Levels' flood defence system in the UK using Dutch reliability methods for flood defences in order to support an evaluation of the appropriateness of these methods as part of the detailed level methodology in RASP. The main working-method is derived from CUR report 190 (1997) and involves carrying out the reliability analysis by taking the following steps (see Figure 1 details).

Definition of the Caldicot Levels' flood defence system and its components.

– Analysis of the failure modes connected to the components
– Modelling the Caldicot Levels' flood defencec system and expressing this model into data

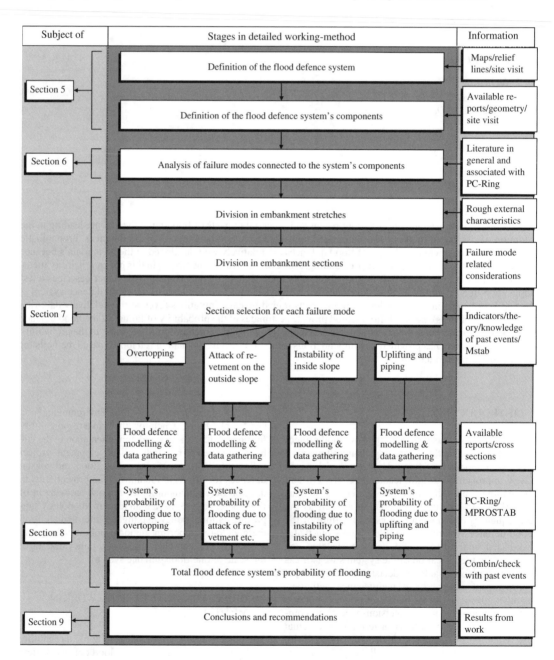

Figure 1. Detailed working-method.

– Calculation of the probability of flooding of the Caldicot Levels' flood defence system.

Two scenario's are subject of the reliability analysis:

1. The present Caldicot Levels' flood defence system
2. The same system after a number of planned improvements have been taken into account

Before these steps are discussed in more detail the boundary conditions and the Dutch reliability methods for flood defences which form the framework of the reliability analysis are addressed.

3 DESIGN LEVELS

The Caldicot Levels' flood defence system is located at the south coast of Wales in the UK. The system borders the Severn Estuary in the south, the river Usk in the west and a distinct line of hills in the north and east, see Figure 2. ABP Research (2000) provides information on the local water level, wind speed and wave conditions. In the Severn estuary one of the largest tidal ranges in the world occurs, varying between 9 m and 15 m. The largest fetches and the most severe wind speeds are related to the south westerly wind directions. The River Usk is a small river, however the water levels can reach relatively high values especially in case of high water levels at the Severn Estuary. The mean elevation of the Caldicot Levels is OD + 5.5 m which compares to mean tide high water levels of OD + 4.8 m, mean spring tide high water levels of OD + 6.5 m and a 200-year return period water level of OD + 8.55 m. The crest levels of the flood defence system vary between OD + 8 and OD + 10 m. According to WS Atkins (1999), the embankments and soil underneath consist mainly of clay. Finally, in WS Atkins (2000) information is available of areas which have been subjected to damage caused by a number of storms in the past and which can be considered as weak.

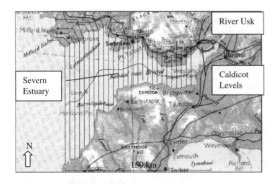

Figure 2. Location of the Caldicot Levels' flood defence system with respect to the Severn Estuary and the Usk.

4 DUTCH RELIABILITY METHODS FOR FLOOD DEFENCES

4.1 Calculation of annual probability of failure

4.1.1 Failure modes
The following failure modes are included in the calculation of the annual probability of flooding of the flood defence system:

– Overtopping or overflow discharges pass the embankment crest and consequently failure occurs either due to damage and erosion of the inside slope or due to saturation of the clay cover layer soil leading to instability of the inside slope.
– Instability of the inside slope of the embankment. The extreme outside water levels result in different water pressure distributions in the embankment body. The geotechnical equilibrium (according to Bishop) of the ground body is affected in such a way that instability of the inside slope occurs.
– Uplifting and piping. The hydraulic uplifting force exerted by the water head difference between the outside and inside water levels leads first to bursting of the impervious foundation layer of the embankment. After uplifting of the impervious layer, water flow as a result of the hydraulic head difference causes the development of pipe shaped erosion in the burst impervious layer.
– Damage of the revetment on the outside slope and consequently erosion of the embankment body. Attack of the revetment on the outside slope by the water and wave conditions causes damage to the revetment. The embankment body is exposed to the same hydraulic conditions after the revetment has been damaged. Erosion of the embankment body can lead to breach.

A more detailed description of the failure modes and the reliability functions which represent these failure modes is given in Vrouwenvelder et al. (2001) and Lassing & Vrouwenvelder (2003).

4.1.2 Statistics
The statistical character of the random variables in the reliability functions consists of statistical distribution functions and correlations in time and space. The correlation in time is modelled according to Borges Castanheta. This model assumes constant correlations during and between time intervals. The spatial correlation of random variables in a section is assumed to decrease to a constant value after a certain length along the flood defences. The statistical models of the water levels and wind speeds involves a set of basic random variables of water levels and wind speeds at the mouth of the Severn Estuary (Figure 2). These basic variables are transferred to local water levels and wave conditions by use of a numerical model like for instance

Mike11, Sobek, SWAN, HISWA, DELFT 3D, etc. The statistical distribution functions and the correlations between water level and wind speed are applied to the basic random variables at the mouth of the estuary. Detailed information can be found in Vrouwenvelder et al. 2001.

4.1.3 Calculation methods

Calculations at the level of one reliability function are made with: FORM, SORM, crude Monte Carlo, Directional Sampling. Detailed information about these methods can be found in Vrouwenvelder (2001) and Vrijling & Van Gelder (2002).

In Vrijling & Van Gelder (2002) a method is presented to calculate an annual probability of failure of a flood defence system involving a number of different reliability functions representing: one failure mode, cross section, tide, wind direction. These different reliability functions are combined to one remaining equivalent reliability function. During this process mutual correlations between the functions are taken into account.

In order to perform these calculations software called PC-Ring has been applied, see Figure 1. Besides the failure modes of dikes, PC-Ring contains the possibility to calculate the probability of failure of dunes and structures. The latter is calculated in connection to failure due to piping and failure to close of the structure in case of a storm.

4.1.4 Description of calculation results

The calculation of the annual probability of flooding of the Caldicot Levels' flood defence system results among others in:

– Annual probabilities of failure and reliability indices of β the sections which are included in the calculations.

– Coefficients of influence, α-values, that point out which random variables contribute most to the total uncertainty of the reliability function.
– The total annual system's probability of flooding due to one failure mode and the weakest link in the system.
– The total annual system's probability of flooding.

4.2 Selection of weak flood defence sections

The amount of work related to data gathering is reduced by a process which aims to select the sections that are most representative of the system's probability of failure. This process starts with dividing the flood defence system into stretches, and more detailed sections. The cross sectional and statistical properties are assumed to be constant along one section. By use of indicators, which are based on rough information, sections are selected which are considered as weak. These selected cross sections dominate the total probability of failure and are therefore included in the calculations with PC-Ring. This process is described in Calle et al. (2001).

5 DEFINITION OF THE FLOOD DEFENCE SYSTEM AND ITS COMPONENTS

The Caldicot Levels' flood defence system is defined as shown in Figure 3. The line south of the locations marked A and B represents the relevant defence length for the calculation of the probability of inundation. The OD + 10 m relief line defines the boundary formed by the high grounds. The area between the lines suffers consequences in the form of partial or complete flooding if the flood defence system fails at one or more locations. The components are also indicated in Figure 3.

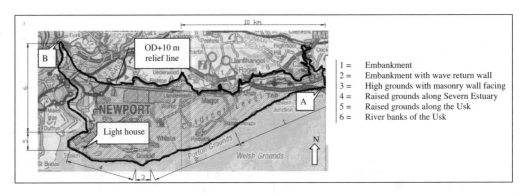

Figure 3. Definition of the Caldicot Levels' flood defence system's boundaries. The OD + 10 m line represents the high grounds which do not contribute to the system's probability of flooding. Additionally, the figure includes a rough indication of the position of the main flood defence components (from Chatterton (2001)).

6 ANALYSIS OF THE FAILURE MODES CONNECTED TO THE COMPONENTS

6.1 Failure modes of present flood defence system

All of the components except for the embankment with wave return wall, are calculated with the failure modes which are present in PC-Ring (see subsection 4.1.1). Below the main flood defence components which are also mentioned in Figure 3 are listed with the failure modes which have been considered in the calculation:

- Embankment: Overtopping, instability inside slope, piping, attack of the revetment on the outside slope (grass).
- Embankment with wave return wall: Overtopping, instability inside slope, piping, attack of the revetment on the outside slope (rock armour). In case of the present flood defence system the effect of the wave return wall on the overtopping is assumed to be negligible. The approach of the wave return wall in case of the improved system is described in subsection 6.2.
- High grounds with masonry wall facing: Only overtopping is taken into account and approached with a limit critical discharge value instead of the grass/erosion or saturation models. The high grounds are regarded as broad embankments with shallow slopes. Therefore, the contributions to the total probability of flooding by the failure modes instability inside slope, piping, attack of the revetment on the outside slope are assumed to be small.
- Raised grounds along the Severn Estuary are approached in a similar way as high grounds with masonry wall facing.
- Raised grounds along the River Usk: see raised grounds along the Severn Estuary.
- The river banks of the Usk: failure is represented by the return period of the river water level exceeding the highest elevation of the river bank.

6.2 Failure modes of improved flood defence system

The main flood defence components will be improved in future. These improvements concern mainly raising of the flood defences along a considerable length and the replacement of the present wave return wall by a higher and more effective one. For all the flood defence components except the embankment with wave return wall the selection of failure modes remains the same for the present and the improved system. For the improved form of the embankment with wave return wall an approach is developed which takes the wave return wall into account.

Theoretically speaking, the wave return wall reduces the amount of overtopping and failure of the wave return wall does not necessarily need to lead to failure of the complete embankment (upper part of Figure 4). However, this approach leads to practical complications of the implementation of the wave return wall in PC-Ring. Therefore, the practical approach as illustrated in the lower part of Figure 4 was applied. This approach is based on the assumption that when the wave return wall fails the entire embankment also fails.

Analysis of the results of the calculations will point out whether the probability of failure of the wave return wall is high or low relative to that of the embankment. If this is high, the probability of failure of the embankment dominates the probability of failure. If it is low, the practical approach applies, the probability of failure is dominated by the highest of the following two values: probability of failure of the wave return wall or the probability of failure of the embankment with wave return wall as a whole.

Two failure modes of the wave return wall are taken into account: horizontal sliding or tilting (see Table 1). The force is formed by the wave impact and the strength is mainly determined by the weight of the wave return wall. The model which is applied to determine the wave impact pressures is described in Martin (1999).

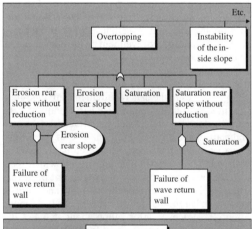
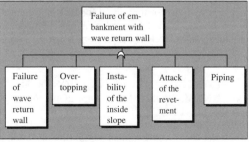

Figure 4. Fault tree of theoretical approach of wave return wall (top) and of practical approach as implemented in PC-Ring (bottom).

Table 1. Failure modes of the wave return wall which have been incorporated in PC-Ring.

Reliability function

Name	Function	Short description
Wave return wall Horizontal sliding	$Z = {}^2/_3\tan(\varphi)\,\Sigma V - \Sigma H$	Failure of the wave return wall due to horizontal sliding: the resulting wave impact horizontal forces ΣH exceed the friction force as a result of the weight of the wave return wall. ΣV = resulting vertical weight, friction coefficient = ${}^2/_3\tan(\varphi)$, in which φ = effective angle of internal friction of the soil.
Tilting	$Z = {}^1/_6 b_f - (\Sigma M/\Sigma V)$	Failure of the wave return wall due to tilting: the resulting force $\Sigma M/\Sigma V$ is not within the core of the foundation plane. ΣM = the resulting moment of the horizontal wave impact forces with respect to the centre of the foundation plane, ΣV see sliding, b_f = width of the foundation plane.

7 MODELLING THE CALDICOT LEVELS' FLOOD DEFENCE SYSTEM AND EXPRESSING THIS MODEL INTO DATA

As is mentioned in subsection 4.2, the Caldicot Levels' flood defence system is divided into stretches and more detailed sections. For each failure mode the relevancy in terms of contribution to the probability of system failure has been determined and the weak sections have been selected according to rough indicators. The failure modes which have been taken into account are: overtopping/wave return wall, instability of the inside slope and attack of the revetment on the outside slope. The contribution of failure due to piping is assumed to be negligible as the seepage length is large at all locations along the Caldicot Levels' flood defences. The main obstacles in the data requirements are the statistical models of the hydraulic boundary conditions, the numerical models of the local water levels and the general statistical data availability. In case of the Severn Estuary a model of the local water levels has been set up with Mike11 based on limited information with respect to: geometry of the estuary and the actual occurring water levels which can serve to calibrate and validate the model. Part of the applied network can be found in Figure 2. Moreover, information about statistics did not appear in the form required for PCRing required form. In case of the River Usk a numerical model of the local water levels is available in Mike11, though only limited discharge statistics were available.

8 RESULTS OF THE CALCULATION OF THE ANNUAL PROBABILITY OF FLOODING, PRESENT AND IMPROVED SYSTEM

The calculations of the annual probability of flooding of the Caldicot Levels' flood defence system result in:

– Annual probabilities of failure and reliability indices β of the selected sections (see Figure 5).

– Coefficients of influence, or α-values.
– The total annual system's probability of flooding due to one failure mode and the accompanying weakest link in the system (see Table 2).
– The total annual system's probability of flooding (see Table 2).

The dominating failure mode turns out to be overtopping (see Table 2). The weak areas which result from the calculations correspond with the more severely attacked areas in the past storms.

The coefficients of influence point out that the uncertainty associated with in the first place the water levels and in the second place the wind speed and wind direction, contribute most to the total probability of failure. From this information the main reasons causing the above mentioned areas to be weak are derived: a low crest level in combination with the orientation of the embankment with respect to the south westerly wind directions. These wind directions are related to high wind speeds and large fetches and are therefore associated with high levels of wind set up and more severe local wave conditions.

The annual probability of failure of the wave return wall is determined by failure due to tilting and is relatively high (see Table 2). Because of this high probability, the assumption that the complete embankment fails if the wave return wall fails is not justified in this case. Therefore, based on these results the actual probability of failure is expected to be a combination between failure of the embankment with the influence of the wave return wall on wave overtopping and failure of the embankment without a wave return wall on the crest (see Figure 4, top). In Figure 5 the former scenario is referred as "improved, no failure w.r.w.", whilst the latter scenario is referred as "improved, no w.r.w. present on crest".

As overtopping is the dominating failure mode, in Figure 5 the reliability indices in connection to failure due to overtopping of the selected sections are given for the present and improved flood defence system.

316

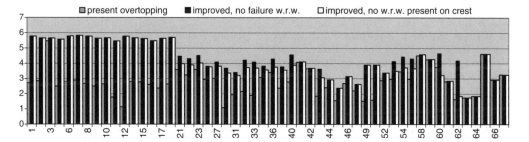

Figure 5. Reliability indices of failure due to overtopping of the present flood defence system and the improved flood defence system. For the improved flood defence system the reliability indices of embankments with and without the influence of the wave return wall on the overtopping discharges are included. Below the flood defence components and the corresponding section numbers are given.

Section no. 1 t/m 19 and no. 49 t/m 52   = embankment without additional structures
Section no. 20 t/m 48 and no. 53 t/m 57   = embankment with wave return wall
Section no. 41 and 42   = high grounds with masonry wall facing
Section no. 58 t/m 62   = raised grounds along the Severn Estuary
Section no. 62 t/m 67   = raised grounds along the river Usk
Section no. 68 t/m 78   = Usk river banks (not included in the plot)

Table 2.   Results from the calculations of the annual probability of flooding of the Caldicot Levels' flood defence system in its present and improved form.

	Overtopping			Instability of the inside slope			Attack of the revetment on the outside slope			Total flood defence system		
	Weakest link		Total	Weakest link		Total	Weakest link		Total	Weakest	Total	
	No.	β	β	No.	β	β	No.	β	β	link no.	β	Pf
Present system												
With Usk	68	0.643	0.51	79	1.19	1.18	14	3.108	3.106	68	0.293	0.385
Without Usk	28	1.09	0.90							28	0.59	0.278
Improved system												
With Usk	68	0.643	0.51							68	0.532	0.297
Without Usk												
Failure w.r.w.	31	1.076	1.01									
Failure embankment with effect w.r.w. on overtopping	63	1.75	1.67									
Failure embankment without w.r.w.	63	1.75	1.67	46	3.73	3.73	29	4.5	4.36	63	1.568	0.584 *10^{-1}

w.r.w. = wave return wall.

Figure 5 points out that the planned improvements are unbalanced:

- Sections 1 to 20 are much improved compared to the sections with wave return wall, 21 to 40, 43 to 57, 60 and 62.
- The latter mentioned sections are moderately improved, moreover their reliability indices are very irregular.
- Table 2 points out that the sections along the Usk provide the weakest link. These sections are not improved at all.

- Without considering the sections of the river Usk, section 63 turns out to be the weakest link: this is one of the sections for which no improvement is planned.

9   CONCLUSIONS AND PRACTICAL IMPLICATIONS

9.1   Conclusions

The study has demonstrated how reliability analysis of a flood defence system identifies defence sections

317

and system components that make the greatest contribution to flood risk. This information can then be used to target inspection, maintenance and upgrade activities. The combination of probabilistic analysis of the flood defence system with quantified analysis of potential impacts of flooding in the zone protected by the defences provides a quantified estimate of flood risk, which can be used to justify and optimize economic investment in flood defence improvements.

In order to apply the Dutch methodology to the UK, a new failure mechanism relating to failure of the wave return wall had to be introduced into the PC-Ring program. Moreover, as is often the case for the complex failure of flood defence dikes, this new mechanism interacts with other mechanisms during the failure process. Combining separate mechanisms with logical OR gates is a simplification. However, the mathematical relations in the program are suitable. Furthermore, even though the Caldicot site was relatively well provided with data but UK standards, there was still not all of the data that would ideally be necessary for application of PC-Ring.

The study has highlighted some of the difference between flood defence systems in the UK and the Netherlands (see also Hall et al. 2000) and the implications for quantified risk analysis. There are 35,000 km of flood defences in the UK ranging from low earth embankments protecting just a few fields to the Thames Barrier protecting central London. Because of this diversity, a range of appropriate methods are required for risk analysis that are suited to the potential severity of the consequences of flooding and the available information. All of these methods should be risk-based in some sense, the intention being that risk should form the basis of decision making at all levels in the UK, from national policy decisions, regional strategic plans, to project specific appraisal, design, operation and maintenance decisions. The detailed reliability methods being promoted in the Netherlands and implemented in PC-Ring are best suited to the design and appraisal of, by UK standards, relatively large and highly engineered flood defence systems. Even for these important systems, the study described in this paper has demonstrated that some adaptation of the Dutch reliability methods and additional data collection is inevitable.

9.2 Practical implications

Recommendations with respect to possible adaptations of the Dutch methods follow from this study:

- Whether a "tailor-made" set up, such as in PC-Ring, or a "one-size-fits-all" set up of reliability software is desired in the UK. The tailor-made set up is based on implementing all possible failure modes in the program code, whereas the one-size-fits-all set up is based on flexibly entering the reliability functions, statistical data and the desired mathematical relations between the reliability functions. A number of typical systems can be set up default in the program, but are easier to adjust.
- With respect to models which are for instance applied in order to calculate overtopping discharges, it is recommended to choose models with an as wide application range as possible so that the results are comparable across different types of flood defences and flood defence systems.
- It is recommended to investigate whether the model which is applied in PC-Ring to transfer the hydraulic boundary conditions from a set of basic random variables to local conditions by using models like Mike11, Sobek, etc., can be applied to other numerical models, which are used e.g. in calculations with respect to other flood defence types than dikes. Moreover, this model should also be compared with currently applied joint probability methods in the UK taking into account the quality and laboriousness of both models.
- Considering the often rather low data availability in the UK, an efficient time-saving method to select the weak areas may be beneficial. To this end an approach similar to the one which is described in 4.2 could serve after adaptation to the desired risk-based method in the UK.

REFERENCES

ABP Research, *Gwent Levels hydraulic study, Objective A: Joint probability surge and tide wave analysis*, Southampton 2000.

Calle, E., Jonkman, B., Lassing, B., Most, van der, H., *Data gathering and flood defence modelling to support the calculation of probabilities of flooding of ring dike systems (in Dutch), Manual (version 3.2)*, Delft 2001.

Chatterton, J.B., *Caldicot Sea Defence Strategy Plan: Benefit Assessment*, JCA, January 2001.

CUR, *Probabilities in Civil Engineering, part 1: Probabilistic designing in theory (in Dutch), CUR 190*, Gouda 1997.

Hall, J.W., Meadowcroft, I.C., Roos, A., Development of risk management in flood and coastal defences: perspectives from the UK and The Netherlands, *ICCE 2000, Proc. Intern. Conf., book of abstracts, Sydney Australia, July 17–21 2000*.

Lassing, B.L., Vrouwenvelder, A.C.W.M., Application of reliability analysis methods in The Netherlands, *Proceedings ESREL 2003*.

Martin, F.L. 1999, Experimental study of wave forces on rubble mound break water crown walls, *PIANC bulletin*, no. 102: pages 5–17.

TAW, *from probability of exceedance to probability of flooding*, June 2000.

Vrijling, J.K., Van Gelder, P.H.A.J.M., *Probabilistic Design in Hydraulic Engineering, Lecture notes TUDelft, Faculty of Civil Engineering and Geosciences, Delft 2002.*

Vrouwenvelder, A.C.W.M., Steenbergen, H.M.G.M., Slijkhuis, K.A.H., *Theoretical manual of PC-Ring, Part A: descriptions of failure modes (in Dutch),* Nr. 98-CON-R1430, Delft 2001.

Vrouwenvelder, A.C.W.M., Steenbergen, H.M.G.M., Slijkhuis, K.A.H., *Theoretical manual of PC-Ring, Part B: Statistical models (in Dutch),* Nr. 98-CON-R1431, Delft 2001.

Vrouwenvelder, A.C.W.M., *Theoretical manual of PC-Ring, Part C: Calculation methods (in Dutch),* 98-CON-R1204, Delft 1999.

WS Atkins, *Caldicot Levels sea defence improvements, Ground investigation, interpretative report and wave return wall condition survey,* Swansea 1999.

WS Atkins, *Caldicot Levels sea defence improvements, Design parameters report,* Swansea 2000.

Safety and Reliability – Bedford & van Gelder (eds)
© 2003 Swets & Zeitlinger, Lisse, ISBN 90 5809 551 7

Two-stage Bayesian models – application to ZEDB project

C. Bunea, T. Charitos & R.M. Cooke
Delft University of Technology, Delft, Netherlands

G. Becker
RISA, Berlin, Germany

ABSTRACT: A well-known mathematical tool to analyze data for nuclear power facilities is the two-stage Bayesian models. In this paper we review this mathematical model, its underlying assumptions and supporting arguments. Furthermore, we will verify the software implementations of the ZEDB database and compare the results. Lastly, an assessment of the relevance of new developments will take place, while the viability of the two-stage Bayesian approach will be discussed.

1 INTRODUCTION

ZEDB [Becker and Schubert, 1998] is the major German effort to collect data from nuclear facilities. The goal of the project is to create a reliability data base which contains all major plant events: failure events, operational experience, maintenance actions. As a mathematical tool to analyse ZEDB data, a two-stage Bayesian model was chosen. Firstly we identify the standard conditional independence assumptions and derive the general form of the posterior distribution for failure rate λ_0 at plant of interest 0, given failures and observation times at plants 0, 1, ... n. Any departure for the derived mathematical form necessarily entails a departure from the conditional independence assumptions. Vaurio's one stage empirical Bayes model is discussed as an alternative to the two-stage model [Vaurio, 1987]. Hofer [Hofer et al., 1997, Hofer and Peschke, 1999, Hofer, 1999] has criticized the standard two-stage model and proposed an alternative, which is also discussed. Finally, the methods of Pörn and Jeffrey for choosing a non-informative prior distribution are discussed.

2 BAYESIAN TWO-STAGE HIERARCHICAL MODELS

Bayesian two-stage or hierarchical models are widely employed in a number of areas. The common theme of these applications is the assimilation of data from different sources, as illustrated in Figure 1. The data from agent i is characterized by an exposure T_i and a number of events X_i. The exposure T_i is not considered

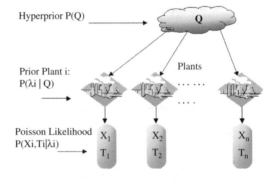

Figure 1.   Bayesian two-stage hierarchical model.

stochastic, as it can usually be observed with certainty. The number of events for a given exposure follows a fixed distribution type, in this case Poisson. The parameter(s) of this fixed distribution type are uncertain, and are drawn from a prior distribution. The prior distribution is also of a fixed type, yet with uncertain parameters. In other words, the prior distribution itself is uncertain. This uncertainty is characterized by a hyperprior distribution over the parameter(s) of the prior.

In Figure 1, the hyperprior is a distribution $P(Q)$ over the parameters Q of the prior distribution from which the Poisson intensities $\lambda_1, \ldots \lambda_n$ are drawn. In sum, our model is characterized by a joint distribution:

$$P(X_1, \ldots X_n, \lambda_1, \ldots \lambda_n, Q) \tag{1}$$

To yield tractable models, such models must make two types of assumptions. First, conditional independence assumptions [Pörn, 1990, Iman and Hora, 1990] are made to factor (1). Second, assumptions must be made regarding the fixed distribution types and the hyperprior distribution $P(Q)$. The conditional independence assumptions may be read from Figure 1, by treating this figure as a "belief net". In particular, this figure says:

CI.1 Given Q, λ_i is independent of $\{X_j, \lambda_j\}_{j \neq i}$

CI.2 Given λ_i, X_i is independent of $\{Q, \lambda_j, X_j\}_{j \neq i}$

The expression "X_i is independent of $\{Q, \lambda_j, X_j\}_{j \neq i}$" entails that X_i is independent of Q, and X_i is independent of λ_j.

With these assumptions we can derive the conditional probability $P(\lambda_0 | X_0, \ldots X_n)$ for the failure rate at plant 0, given X_i failures observed at plant i, $i = 0$, $\ldots n$. This is sometimes called the posterior probability for λ_0.

2.1 Derivation of posterior probability for λ_0

We assume throughout that the plant of interest is plant 0. We seek an expression for

$$P(\lambda_0 | X_0, \ldots X_n) \quad (2)$$

We step through this derivation, giving the justification for each step. A more detailed exposition is found in [Cooke et al., 1995]. "$\propto$" denotes proportionality, "CI.1" means that conditional independence assumption i is invoked, $i = 1, 2$; "BT" denotes Bayes theorem: and "TP" denotes the law of total probability; and "FB" denotes Fubini theorem. The Fubini theorem (see [Cooke et al., 2002] for a more detailed discussion) authorizes switching the order of integration if the integrals are finite [Royden, 1968, p 269],

$$P(\lambda_0 | X_0 .. X_n) \propto_{BT} P(X_0 | \lambda_0, X_1..X_n) P(\lambda_0 | X_1..X_n)$$

$$\propto_{CI.2} P(X_0 | \lambda_0) P(\lambda_0 | X_1..X_n)$$

$$\propto_{TP,BT} P(X_0 | \lambda_0) \int_{\lambda_1..\lambda_n} \int_q P(\lambda_0 | \lambda_1..\lambda_n, q, X_1..X_n)$$

$$P(q, \lambda_1..\lambda_n | X_1..X_n) dq d\lambda_1..d\lambda_n$$

$$\propto_{CI.1} P(X_0 | \lambda_0) \int_{\lambda_1..\lambda_n} \int_q P(\lambda_0 | q) \quad (3)$$

$$P(q, \lambda_1..\lambda_n | X_1..X_n) dq d\lambda_1..d\lambda_n$$

$$\propto_{BT} P(X_0 | \lambda_0) \int_{\lambda_1..\lambda_n} \int_q P(\lambda_0 | q) P(X_1..X_n | q, \lambda_1..\lambda_n)$$

$$P(q, \lambda_1..\lambda_n) dq d\lambda_1..d\lambda_n$$

$$\propto_{CI.1,2;FT} P(X_0 | \lambda_0) \int_q P(\lambda_0 | q)$$

$$\int_{\lambda_1..\lambda_n} [\prod_{i=1..n} P(X_i | \lambda_i) P(\lambda_i | q) d\lambda_1..d\lambda_n] P(q) dq$$

$$\propto P(X_0 | \lambda_0) \int_q P(\lambda_0 | q) \quad (4)$$

$$\prod_{i=1..n} \int [P(X_i | \lambda_i) P(\lambda_i | q) d\lambda_i] P(q) dq$$

Expression (4) is normalized by integrating over all l0.

2.2 Summary of significant features

1. If $Q = q_0$ is known with certainty, then there is no influence from $X_1, \ldots X_n$ on λ_0. Indeed, in this case the posterior density in λ_0 is simply proportional to $P(X_0 | \lambda_0) P(\lambda_0 | q)$.
2. As the numbers X_i, T_i, $i = 1 \ldots n$ get large, $X_i / T_i \to \lambda_i$, then the Poisson likelihood $P(X_i | \lambda_i)$ converges to a Dirac measure concentrating mass at the point $X_i = T_i \lambda_i$. In the limit the "hyperposterior"

$$\prod_{i=1..n} [\int P(X_i | \lambda_i) P(\lambda_i | q) d\lambda_i] P(q) \quad (5)$$

becomes

$$[\prod_{i=1..n} P(\lambda_i = X_i / T_i | q)] P(q) \quad (6)$$

(6) corresponds to the situation where $P(q)$ is updated with observations $\lambda_1, \ldots \lambda_n$. Note that as the observation time increases, the number n does not change. If n is only modest (say in the order 10) then the effect of the hyperprior will never be dominated by the effect of observations (see [Cooke et al., 2002] for a more detailed discussion). We say that the hyperprior persists in the posterior distribution $P(\lambda_0 | X_0, \ldots X_n)$.
3. It is shown in [Cooke et al., 1995], [Hennings and Meyer, 1999] that improper hyperpriors $P(q)$ do not always become proper when multiplied by $[\prod_{i=1...n} P(\lambda_i = X_i / T_i | q)] P(q)$. In other words, the hyperposterior may well remain improper.

2.3 Selected literature review

Two-stage Bayesian models have been implemented by various authors. [Kaplan, 1983] used a log normal prior with a Poisson likelihood, which of course is not a natural conjugate. This method has been implemented by ZEDB. [Iman and Hora, 1989, 1990] and [Hora and Iman, 1987] proposed a natural conjugate gamma prior. [Vaurio, 1987] proposed a one-stage empirical Bayes approach, using other plants to determine the prior. The SKI data bank [1987] uses a two-stage model developed by Pörn [1990]. This model was reviewed in [Cooke et al., 1995], and further discussed in [Meyer and Hennings, 1999]. Recently [Hofer et al., 1997], [Hofer and Peschke, 1999] and [Hofer, 1999] have suggested that an incorrect chance mechanism underlies the two-stage models, and have proposed their own model. In this section we briefly review these developments.

In the two-stage Bayesian models considered here a Poisson likelihood is used. The prior is usually gamma or log normal. The second stage places a hyperprior distribution over the parameters of the prior gamma or log normal distribution. We briefly recall the definitions and elementary facts of the Poisson, Gamma, and log normal distributions in Table 1 below.

Using a gamma prior with parameters as above, the term $\Pi_{i=1\ldots n} \int P(X_i|\lambda_i)P(\lambda_i|q)d\lambda_i$ in (4) becomes after carrying out the integration:

$$\prod_{i=1}^{n} \frac{\Gamma(X_i + \alpha)}{\Gamma(X_i + 1)\Gamma(\alpha)}(\frac{\beta}{\beta + T_i})^\alpha (\frac{T_i}{\beta + T_i})^{X_i}$$

$$= \prod_{i=1}^{n} \frac{(X_i + \alpha - 1)!}{X_i!(\alpha - 1)!}(\frac{\beta}{\beta + T_i})^\alpha (\frac{T_i}{\beta + T_i})^{X_i} \quad (7)$$

Further calculation to solve equation (4) must be performed numerically. It is shown in [Cooke et al., 1995] that improper hyperpriors may remain improper after assimilating observations. The asymptotic behavior of the "hyperposterior"

$$P(\alpha, \beta|X_1\ldots X_n, T_1\ldots T_n) \propto P(X_1\ldots X_n, T_1\ldots T_n|\alpha, \beta)$$
$$P(\alpha, \beta)$$

will essentially be determined by the maximum of $P(X_1\ldots X_n, T_1\ldots T_n|\alpha, \beta)$. The significant fact is that $P(X_1\ldots X_n, T_1\ldots T_n|\alpha, \beta)$ has no maximum; it is asymptotically maximal along a ridge, see (Figure 2).

2.4 Vaurio

[Vaurio, 1987] proposed an analytic empirical Bayes approach to the problem of assimilating data from other plants. A simple one-stage Bayesian model for one plant would use a Poisson likelihood with intensity λ, and a $\Gamma(\lambda|a, b)$ prior. Updating the prior with X_i failures in time T_i yields a $\Gamma(\lambda|\alpha + X_i, \beta + T_i)$ posterior. Vaurio proposes to use data from the population of plants to choose the $\Gamma(\lambda|\alpha, \beta)$ prior by moment fitting. Any other two moment prior could be used as well. Data from other plants are not used in updating, hence, this is a one-stage model.

The model is consistent, in the sense that as $X_i, T_i \to \infty$, with $X_i/T_i \to \lambda_i$, his model does entail that $E(\lambda_i|X_i, T_i) \to \lambda_i$. Elegance and simplicity are its main advantages. Disadvantages are that it cannot be applied if all $X_i = 0$, or if the population consists of only 2 plants. Further, numerical results indicate that the model is non-conservative when the empirical failure rate at plant 0 is low and the empirical failure rates at other plants are high. A final criticism, which applies to most empirical Bayes models is that the data for the plant of interest is used twice, once to estimate the prior and once again in the Poisson likelihood.

2.5 Hofer

Hofer has published a number of articles [Hofer et al., 1997], [Hofer, 1999] and [Hofer and Peschke, 1999] in which the two-stage models are faulted for using a "wrong chance mechanism", and a new model is proposed. He does not explicitly formulate conditional independence assumptions, and does not derive the posterior by conditionalizing the joint as done above. Rather, the model is developed by shifting between the point of view of "observing λ_i" and "observing (X_i, T_i)". [Hofer, 1997] criticizes [Hora and Iman, 1990] for using the wrong order of integration of improper

Table 1.

	Poisson
Density	$P(X|T, \lambda) = \frac{(\lambda T)^X}{X!}e^{-\lambda T}, \lambda > 0, T > 0$
Mean	λT
Variance	λT
	Gamma
Density	$f(\lambda|\alpha, \beta) = \frac{\lambda^\alpha}{\Gamma(\alpha)}\beta^\alpha e^{-\beta\lambda}, \alpha > 0$
Mean	α/β
Variance	α/β^2
	Lognormal
Density	$f(\lambda|\mu, \sigma) = \frac{1}{\sqrt{2\pi}\sigma\lambda}e^{-\frac{(\ln\lambda-\mu)^2}{2\sigma^2}}, \sigma > 0$
Mean	$e^{\mu+\sigma^2/2}$
Variance	$e^{2\mu+\sigma^2}(e^{\sigma^2} - 1)$

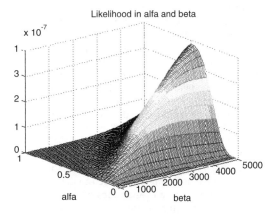

Figure 2.

integrals (see [Cooke et al., 2002] for a more detailed discussion). In later publications, a "deeper" reason is found to reside in the use of a "wrong chance mechanism". Hofer's model appears to result in a posterior of the following form:

$$P(\lambda_0|X_0..X_n) \propto P(X_0|\lambda_0) \int_{(q)} P(\lambda_0|q)$$

$$\int \int_{\forall \lambda_{1..n}} \frac{[\Pi_{i=1..n} P(\lambda_i|q)]P(q)}{\int_{(q)} [\Pi_{i=1..n} P(\lambda_i|q')]P(q')dq'}$$

$$[\prod_{i=1..n} P(\lambda_i|q)]P(\lambda_1...\lambda_n)d\lambda_1...d\lambda_n dq \quad (8)$$

Notice that this does not appear to have the form of (4).

Although Hofer does not explicitly formulate his conditional independence assumptions, he does use them. E.g. he uses CI.1 to derive the expression in the denominator (see equation (4) of [Becker and Hofer, 2001]). If CI.1 holds, then necessarily

$$\int_{(q)} [\prod_{i=1..n} P(\lambda_i|q')]P(q')dq'$$

$$= \int_{(q)} P(\lambda_1...\lambda_n|q)P(q)dq = P(\lambda_1...\lambda_n) \quad (9)$$

and (8) reduces to (4). If CI.1 does not hold, then the origin of the product $\Pi P(\lambda_i|q)$ is unclear. Hofer says that $P(\lambda_1...\lambda_n) = \Pi r(\lambda_i)$, where $r(\lambda_i)$ is a noninformative prior, which he takes to be constant. This entails that the λ_i are unconditionally independent. It is not difficult to show that if λ_i are unconditionally independent, and independent given q, that then λ_i is independent of q. Indeed, independence implies that for all λ_i, $P(\lambda_1, ... \lambda_n) = \Pi P(\lambda_i) = \Pi \int P(\lambda_i|q) P(q)dq$. By conditional independence; $P(\lambda_1,...\lambda_n) = \int P(\lambda_1, ... \lambda_n|q)P(q)dq = \int \Pi P(\lambda_i|q)P(q)dq$. The λ_i are identically distributed given q; take $\lambda_i = \lambda$; $i = 1, ...n$. These two statements imply $[\int P(\lambda|q)^n P(q)dq]^{1/n} = \int P(\lambda|q)P(q)dq$. Since the integrand is non-negative, this implies that $P(\lambda|q) = constant = P(\lambda)$ [Hardy, Littlewood and Polya, 1983, p 143]. This would make the entire two-stage model quite senseless. If $P(\lambda_1... \lambda_n) = \Pi r(\lambda_i) = constant$ in the numerator of (8), but not in the denominator, then (8) is not equivalent to (4), but rests on conflicting assumptions.

In any event, if (8) does not reduce to (4) then the assumptions CI.1, CI.2 do not both hold. Hofer does not say which assumptions are used to derive (8), in fact (8) is not derived mathematically, but is "woven together" from shifting points of view. The danger of such an approach is that conflicting assumptions may

be inadvertently introduced. This appears to be the case, as the λ_i are at one point assumed to be independent, and at another point are assumed to be conditionally independent given q.

2.6 Hyperpriors

Pörn (1990) introduces a two-stage model with a gamma prior for λ, similar to (Hora and Iman, 1990). He provides an argument for choosing the following non-informative (improper) densities for the parameters v, μ': $g(v) = 1/v$, $v > 0$, and $k(\mu') = 1/\sqrt{(\mu'(1 + \mu'))}$, $\mu' > 0$, where $v = 1/\alpha$ is the coefficient of variation and $\mu' = T\alpha/\beta$ is the expected number of failures at time T, given α and β. Assuming independence between these parameters and transforming back to the hyperparameters α and β, $\alpha = 1/v^2$ $\beta = T/v^2\mu$, a joint (improper) hyperprior density for α and β is obtained proportional to: $1/\beta\sqrt{(\alpha(\alpha + \beta/T))}$.

Another frequently used principle, called Jeffrey's rule, is to choose the non-informative prior $P(q)$ for a set of parameters q proportional to the square root of the determinant of the information matrix $\Phi_n(q)$:

$$P(q) \propto \sqrt{|\Phi_n(q)|}, \Phi_n(q) = E\{\frac{\partial^2 L}{\partial q_i \partial q_j}\} \quad (10)$$

and L is the log-likelihood function for the set of the parameters.

Hora and Iman (1990) apply this rule to the two-dimensional parameter vector (α, β) for the Gamma distribution of the failure rate λ. They get the (approximate) improper hyperprior: $1/\alpha^{1/2}\beta$, $\alpha, \beta > 0$.

3 ZEBD SOFTWARE VERIFICATION

Three data sets are used to check the concordance with the results from [Becker and Hofer, 2001]. Differences between our results and those of ZEDB reflect differences that may arise from an independent implementation based on public information. Although ZEDB recommends the lognormal model, both the lognormal and gamma models are supported, and both are benchmarked here. The data sets are shown in Tables 2–4.

3.1 Gamma model

The computation may be broken into three steps: Firstly, truncate the range of (α, β) to a finite rectangle. Then identify a range for λ_0 which contains all the "plausible" values. Finally, for every "plausible" value of λ_0, evaluate numerically the integrals over α and β, and interpolate to find the 5%, 50% and 95% quantiles.

Table 2. Data set 1 (4 in [Becker and Hofer, 2001]).

Nr. failures	Obs. time	Nr. failures	Obs. time
7	24000	0	24000
1	24000	0	24000
3	24000	0	24000
2	24000	2	24000
1	24000	0	24000
2	24000	0	24000

Table 3. Data set 2 (2 in [Becker and Hofer, 2001]).

Nr. failures	Obs. time	Nr. failures	Obs. time
1	20000	0	6000
0	2000	1	10000
0	4000	2	12000

Table 4. Data set 3 (3 in [Becker and Hofer, 2001]).

Nr. failures	0	0	1
Obs. Time	12000	2000	3000

Remark: The underlined field is the plant of interest.

The likelihood $P(X_1, \ldots X_n, T_1, \ldots T_n | \alpha, \beta)$ as a function of α and β (2.7) is presented in Figure 2. Values for $(X_1, \ldots X_n, T_1, \ldots T_n)$ are taken from data set 1. For uniform hyperpriors, this likelihood is proportional to the hyperposterior distribution $P(\alpha, \beta | X, T)$. Note that $P(\alpha, \beta | X, T)$ does not peak but "ridges". This means that a "natural" truncation for α and β cannot be defined; that is, we cannot define a finite rectangle for α and β which contains most of the hyperposterior mass. In our simulations, these ranges were chosen in a manner similar to [Cooke et al., 1995], using Pörn's heuristic. The inability to localize the hyperposterior mass for (α, β) means that we cannot localize the posterior mass

$$P(\lambda_0 | X_0, \ldots X_n) \propto P(X_0 | \lambda_0) \int \int_{\alpha, \beta} P(\lambda_0 | \alpha, \beta)$$

$$\prod_{i=1..n} [\int P(X_i | \lambda_i) P(\lambda_i | \alpha, \beta) d\lambda_i] P(\alpha, \beta) d\alpha d\beta$$

For each finite rectangle for α, β, the mass in λ_0 will be localized, but other choices for α, β could significantly shift the region in which λ_0 is localized. This means, of course, that the method of truncation in step 1 will influence the plausible values in step 2, and can have a significant effect on the results.

Figures 3 and 4 represent the hyperposterior distribution for Pörn's approach and Jeffrey's hyperpriors; also with $(X_1, \ldots X_n, T_1, \ldots T_n)$ from data set 1 below.

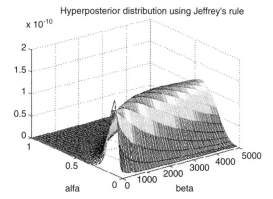

Hyperposterior distribution using Jeffrey's rule

Figure 3.

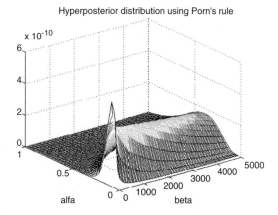

Hyperposterior distribution using Porn's rule

Figure 4.

Table 5. The 5%, 50% and 95% quantiles of the posterior distribution of λ_0 for data set 1.

	Uniform	Pörn	Jeffrey
5%	2.3971 E-5	2.8429 E-5	2.8665 E-5
50%	8.0511 E-5	8.4990 E-5	8.5678 E-5
95%	2.0012 E-4	2.0598 E-4	2.0670 E-4

Ranges	TUD	ZEDB
α:	0.033...1	0.002...351
β:	50...5000	582...232,219

Tables 5–7 compare our results for the uniform, Pörn and Jeffrey prior, and give the integration ranges for α and β for our computation and for the ZEDB results. Table 8 compares the TUD and ZEDB results. Note that the 5% quantile for dataset 3 is a more than a factor three lower in the TUD results. In dataset 1 the agreement is better, as there are more plants, more operational hours and more failures. These differences

325

Table 6. The 5%, 50% and 95% quantiles of the posterior distribution of λ_0 for data set 2.

	Uniform	Pörn	Jeffrey
5%	4.2371 E-5	3.9306 E-5	4.3845 E-5
50%	1.4603 E-4	1.4278 E-4	1.4908 E-4
95%	2.0012 E-4	2.0598 E-4	2.0670 E-4

	Ranges	TUD	ZEDB
α:		0.03…1.2	0.01…1926
β:		50…5000	789…350,098

Table 7. The 5%, 50% and 95% quantiles of the posterior distribution of λ_0 for data set 3.

	Uniform	Pörn	Jeffrey
5%	3.3976 E-5	3.6503 E-5	3.6394 E-5
50%	1.7514 E-4	2.1247 E-4	2.0711 E-4
95%	5.8914 E-4	6.6524 E-4	6.5640 E-4

	Ranges	TUD	ZEDB
α:		0.0154…0.3846	0.01…503
β:		50…5000	126…91385

Table 8. Comparison TUD (Pörn) and ZEDB for gamma model.

	Dataset 1	Dataset 2	Dataset 3
	Gamma	Model TUD	(Pörn)
5%	2.8429 E-5	3.9306 E-5	3.6503 E-5
50%	8.4990 E-5	1.4278 E-4	2.1247 E-4
95%	2.0598 E-4	3.3217 E-4	6.6524 E-4
	Gamma	Model	ZEDB
5%	3.2518 E-05	1.2220 E-04	1.2141 E-04
50%	6.9926 E-05	1.7247 E-04	2.5766 E-04
95%	1.3044 E-04	3.4473 E-04	7.4076 E-04

are consistent with the results reported in [Cooke et al., 1995], where "stress-testing" the gamma model by exploring the range of plausible choices for α, β resulted in differences up to a factor 5.

3.2 Lognormal model

ZEDB adopted the lognormal distribution as a prior, based on the maximum entropy principle invoked by [Jaynes, 1968]. The uncertainty over parameters μ and σ is expressed by hyperpriors. [Becker, 2001] takes into account four types of hyperprior distribution based on Jeffrey's rule. [Becker, 2001] proposes four different implementations of Jeffrey's rule. We caution against the multivariate implementation and version of the Jeffrey's rule when parameters of different kind e.g. location and scale parameters, are considered. In this case, as [Box and Tiao, 1974] suggested, it is wiser to choose parameters, which can be assumed independent

and then apply the one parameter version of the rule. This is done only in the first and the fourth case below (see [Cooke et al., 2002] for a more detailed discussion); case 1 is used by ZEDB.

– 1st case: Jeffrey's rule is applied to the parameters μ and σ: In this case the hyperprior has the well-known form $f(\mu, \sigma) \propto 1/\sigma^2$. The same result is obtained, if μ and σ are assumed to be independent, and Jeffrey's rule is applied twice.

– 2nd case: Jeffrey's rule is applied to the parameters $\alpha = E(X)$ and $CF = \sqrt{(VAR(X)/E(X))}$ (coefficient of variation). The resulting hyperprior in terms of μ and σ has the form

$$f(\mu, \sigma) \propto \sqrt{\frac{2e^{-2\mu - 3\sigma^2}(e^{\sigma^2} - 1)}{\sigma^6}}.$$

– 3rd case: Jeffrey's rule is applied to the parameters $\alpha = E(X)$ and $b = \sqrt{(VAR(X))}$. We recall that $VAR(X) = CF^2\alpha^2$. The resulting hyperprior in terms of μ and σ has the form

$$f(\mu, \sigma) \propto \sqrt{\frac{2e^{-4(\mu + \sigma^2)}(e^{\sigma^2} - 1)}{\sigma^6}}.$$

– 4th case: Jeffrey's rule is applied to the parameters $\alpha = E(X)$ and σ, assuming independence between them. In this case has we have:

$$f(\mu, \sigma) = \sqrt{\frac{e^{-2\mu - \sigma^2}(2 + 4\sigma^2)}{\sigma^4}}.$$

For each case the steps in the calculation are similar to those for the gamma model, except that in step 1, truncation is applied to the parameters of the lognormal density.

The likelihood $P(X_1, …X_n, T_1, …T_n|\mu, \sigma)$ as a function of μ and σ (7) is presented in Figure 5, with values for $(X_1, …X_n, T_1, …T_n)$ taken from case 1. For uniform hyperpriors, this likelihood is proportional to the hyperposterior distribution $P(\mu, \sigma|X, T)$. Note that, in contrast to Figure 2, $P(\mu, \sigma|X, T)$ does peak. This means that a "natural" truncation for μ and σ can be defined as any rectangle containing the peak. The choice which such rectangle will have negligible influence on the results.

Figure 6 shows the hyperposterior distribution. Again, the contrast with Figures 3 and 4 is striking. The mass is captured within the μ, σ rectangle containing the peak in Figure 5.

The results are presented in Tables 9–11 and for each hyperprior distribution discussed. For case 1 we include the effect of omitting the square root in the

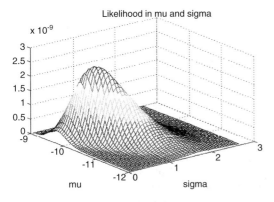

Figure 5.

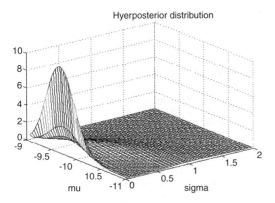

Figure 6.

Table 9. The 5%, 50% and 95% quantiles of the posterior distribution of λ_0 for data set 1.

Quantiles	5%	50%	95%
1st case	2.367 E-5	6.133 E-5	1.217 E-4
2nd case	2.803 E-5	5.805 E-5	1.067 E-4
3rd case	2.548 E-5	5.45 E-5	1.022 E-4
4th case	2.274 E-5	5.751 E-5	1.222 E-4
1st case ZEDB	2.01 E-5	5.91 E-5	1.44 E-4

Table 10. The 5%, 50% and 95% quantiles of the posterior distribution of λ_0 for data set 2.

Quantiles	5%	50%	95%
1st case	2.603 E-5	7.714 E-5	1.381 E-4
2nd case	7.838 E-6	4.817 E-5	1.246 E-4
3rd case	3.016 E-6	1.967 E-5	8.281 E-5
4th case	8.541 E-6	4.852 E-5	1.263 E-4
1st case ZEDB	2.539 E-5	7.710 E-5	2.058 E-4

Table 11. The 5%, 50% and 95% quantiles of the posterior distribution of λ_0 for data set 3.

Quantiles	5%	50%	95%
1st case	2.414 E-5	6.935 E-5	2.501 E-4
2nd case	2.179 E-5	3.794 E-5	1.357 E-4
3rd case	2.126 E-5	3.261 E-5	7.895 E-5
4th case	2.186 E-5	3.860 E-5	1.468 E-4
1st case ZEDB	1.12 E-5	6.228 E-5	3.398 E-4

Jeffrey prior. The corresponding ZEDB results are shown in each table. The differences are smaller than with the gamma model the differences noted above.

4 TRUNCATION

Using a gamma prior, the method of truncation seems to have a large influence on the posterior distribution of λ. It has been shown in section 2.3 that the likelihood in α and β has no maximum, but it is asymptotically maximal along a ridge. [Cooke et al., 1995] showed that different choices of truncation ranges can affect the median and the 95% quantile by a factor 5. In (4), the term $\Pi_{i=1...n} \int [P(X_i|\lambda_i)P(\lambda_i|q)d\lambda_i]$ cannot be calculated analytically when we have a lognormal distribution as prior for λ. Hence, we cannot study the asymptotic behavior of the "hyperposterior" $P(\mu, \sigma| X_1...X_n, T_1...T_n) \propto P(X_1...X_n, T_1...T_n|\mu, \sigma)P(\mu, \sigma)$ analytically. Performing numerical integration (Figure 9), one can see that a maximum occurs in the likelihood in μ and σ. Hence, if the parameters of the lognormal distribution μ and σ are truncated in a way that includes the bulk of mass around the maximum, then how they are truncated will not make a significant difference. To save time in the computation process, truncation is performed around the significant values of likelihood in μ and σ (Figure 5). Figure 7 shows this same likelihood, but a larger integration rectangle for μ and σ; integration over the larger rectangle produces effectively the same result. The intervals for integration over λ, were determined as in [Becker and Schubert, 1998]. One remark can be made: if the domains of integration are not large enough the posterior cumulative distribution of λ will not go to one. Using an iterative loop in software implementation, the natural interval of integration can be found.

The possibility of truncating the domains of integration so as to include the bulk of mass around the maximum of the prior is a significant argument in favor of the lognormal prior over the gamma prior. From Table 1 we see that if the variance of a gamma distribution is proportional to the mean, hence they go to zero at the same rate. For the lognormal the variance is proportional to the square of the mean, and hence the variance goes to zero faster than the mean. We can

327

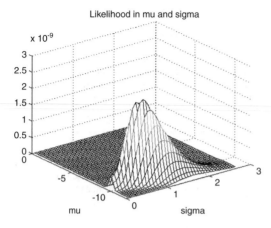

Figure 7. Likelihood μ, σ dataset 1; $\mu = -17.5...-3$, $\sigma = 0.1...4$, $\lambda = 2 * 10^{-6}...3 * 10^{-4}$.

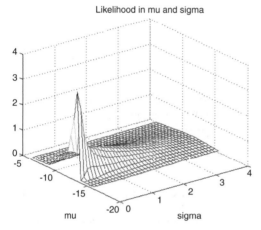

Figure 8. Likelihood μ, σ dataset 3: $\mu = -15.5...-6$, $\sigma = -0.1...4$, $\lambda = 2 * 10^{-5}...3 * 10^{-3}$.

anticipate concentration of mass near $\sigma = 0$ in such cases. In any event, with sufficient observation times failures will be observed and the lognormal prior will peak away from zero and hence admit a good truncation heuristic. For the gamma this is not the case. Figure 8 shows the likelihood of μ, σ for dataset 3. Here a peak is not visible at all, but mass seems to concentrate at $\sigma = 0$. These figures also give the integration ranges for μ, σ, and λ.

5 CONCLUSIONS

Two-stage models provide a valid method for assimilating data from other plants. The conditional independence assumptions are reasonable and yield a

tractable and mathematically valid form for the failure rate a plant of interest, given failures and operational times at other plants in the population. However the choice of hyperprior must be defensible since improper hyperpriors do not always become proper after observations. The lognormal model enjoys a significant advantage over the gamma model in that, as observation time increases, a natural truncation of the hyperparameters μ, σ is possible. In the context of a literature survey, Vaurio's one-stage empirical Bayes model is elegant and simple but will not work with zero observed failures or with a population of two plants. Moreover, Hofer's model appears to rest on shifting viewpoints involving conflicting assumptions. Consistent application of the standard conditional assumptions collapses his model into the form (4), which he criticizes as a "wrong chance model". Further discussion should wait until the conditional independence assumptions and mathematical derivation are clarified.

REFERENCES

Becker G. 2001. Personal communication.
Becker G. & Hofer E. 2001. *Schtzung von Zuverlssigheits-Kennwerten aus der Evidenz veschiedener Anlagen*, GRS-A-2918.
Becker G. & Schubert B. 1998. ZEDB – A practical application of a 2-stage Bayesian analysis code, ESREL '98 conference, June 16–19, Trndheim, Norway, *Safety and Reliability*, Lydersen, Hansen and Sandtorv (eds), Balkema, Rotterdam, vol. 1, pg. 487–494.
Box G. & Tiao G. 1974. *Bayesian Inference in Statistical Analysis*, Addison Wesley, New York.
Cooke R.M., Bunea C., Charitos T. & Mazzuchi T.A. 2002. *Mathematical Review of ZEDB Two-Stage Bayesian Models*, Report 02–45, Departement of Mathematics, Delft University of Technology.
Cooke R.M., Dorrepaal J. & Bedford T.J. 1995. *Review of SKI Data Processing Methodology* SKI Report.
Hofer E. 1999, On two-stage Bayesian modeling of initiating event frequencies and failure rates, *Reliability Engineering and System Safety*, vol. 66, pg. 97–99.
Hofer E., Hora S., Iman R. & Peschke J. 1997. On the solution approach for Bayesian modeling of initiating event frequencies and failure rates, *Risk Analysis*, vol. 17, no. 2, pg. 249–252.
Hofer E. & Peschke J. 1999. Bayesian modeling of failure rates and initiating event frequencies, ESREL 1999, Munich, Germany, *Safety and Reliability*, Schueller and Kafka (eds), Balkema, Rotterdam, vol. 2, pg. 881–886.
Hora S. & Iman R. 1987. Bayesian analysis of learning in risk analysis, *Technometrics*, vol. 29, no. 2, pg. 221–228.
Hardy G.H., Littlewood J.E. & Polya G. 1983. *Inequalities*, Cambridge University Press, Cambridge.
Iman R. & Hora S. 1989. Bayesian methods of modeling recovery times with application to the loss of off-site power at nuclear power plants, *Risk Analysis*, vol. 9, no. 1, pg. 25–36.

Iman R. & Hora S. 1990. Bayesian modeling of initiating event frequencies at nuclear power plants, *Risk Analysis*, vol. 10, no.1, pg. 103–109.

Jaynes E. 1968. Prior probabilities, *IEEE Transactions On System Science and Cybernetic*, vol. 4, no. 3, September, pg. 227–241.

Kaplan S. 1985. Two-stage Poission-type problem in probabilistic risk analysis, *Risk Analysis*, vol. 5, no. 3, pg. 227–230.

Meyer W. & Hennings W. 1999. Prior distributions in two-stage Bayesian estimation of failure rates, ESREL 1999, Munich, Germany, *Safety and Reliability*, Schueller and Kafka (eds), Balkema, Rotterdam, vol. 2, pg. 893–898.

Pörn K. 1990. *On Empirical Bayesian Inference Applied to Poisson Probability Models*, Linköping Studies in Science and Technology, Dissertation No. 234, Linköping.

Royden H.L. 1968. *Real Analysis*, Macmillan, Toronto.

T-Book Reliability Data of Components in Nordic Nuclear Power Plants, 1987, ATV Office, Vattenfall, AB, S-162, Vallingby Sweden.

Vaurio J.K. 1987. On analytic empirical Bayes Estimation of failure rates, *Risk Analysis*, vol. 17, no.3, pg. 329–338.

329

Safety and Reliability – Bedford & van Gelder (eds)
© 2003 Swets & Zeitlinger, Lisse, ISBN 90 5809 551 7

A non-parametric two-stage Bayesian model using Dirichlet distribution

C. Bunea, R.M. Cooke & T.A. Mazzuchi
Delft University of Technology, Delft, Netherlands

G. Becker
RISA, Berlin, Germany

ABSTRACT: As an alternative to standard two-stage Bayesian models, a non-parametric or Dirichelet two-stage model is presented. The analytic solution of the model and its clear interpretation are its main advantages over the classical models. A number of case study are simulated in order to check the robustness of the model. Three data sets from German project – ZEDB are also used to compare the results of the Dirichelet model with the results for standard one-stage and two-stage Bayesian models.

1 INTRODUCTION

Within the context of a recent review of a two-stage Bayesian model for processing data at a population of German nuclear plants, a nonparametric or Dirichelet two stage model was developed. This model has some advantages relative to the standard two stage models: it is analytically solvable, no numerical integration need to be performed, and it allows an intuitive interpretation of the (hyper)parameters – so called "equivalent observations". We check the robustness of this model with a simple numerical example. Preliminary calculations show some sensitivity of the model with regard to the number of cells that characterized the prior distribution and their end points. For the purposes of comparison with classical Bayesian model [Vaurio, Hofer, Becker], the results for three data sets are presented. Considering its apparent advantages, the authors may recommend that the Dirichlet model deserves further development, to qualify it for practical use in data base analysis.

2 BAYESIAN TWO STAGE HIERARCHICAL MODELS

A two-stage model is really nothing more than a joint distribution [Cooke et al 2002]. To be useful, however, we must derive conditional distributions. Typically we want to use data from "other plants" to make predictions about a given plant. This is very attractive in cases where the data from the given plant is sparse.

By specifying the model assumptions one can derive the posterior distribution $P(\lambda_0 | X_0, \dots X_n)$ for failure rate λ_0 at plant of interest 0, given X_i failures and T_i observation times at plant i, $i = 0, 1, \dots n$. First, we can identify the conditional independence assumptions in order to factor the joint distribution. The conditional independence assumptions met in the literature, with one possible exception [Hofer et al 1997][Hofer 1999], are stated below:

CI.1 Given Q, λ_i is independent of $\{X_j, \lambda_j\}_{j \neq i}$
CI.2 Given λ_i, X_i is independent of $\{Q, \lambda_j, X_j\}_{j \neq i}$,

where Q is the hyperparameter of the prior distribution from which the Poisson intensities $\lambda_1 \dots \lambda_n$ are drawn.

The expression "X_i is independent of $\{Q, \lambda_j, X_j\}_{j \neq i}$" entails that X_i is independent of Q, and X_i is independent of λ_j.

Giving the conditional independence assumptions, [Cooke et al 2002] derived the explicit form of the posterior distribution $P(\lambda_0 | X_0, \dots X_n)$ for failure rate λ_0:

$$P(\lambda_0 | X_0 .. X_n) \propto P(X_0 | \lambda_0)$$

$$\int_Q P(\lambda_0 | Q) \prod_{i=1}^{n} \int P(X_i | \lambda_i) P(\lambda_i | Q) d\lambda_i P(Q) dQ \qquad (1)$$

Assumptions must be made also regarding the fixed distribution types and the hyperprior Q. In the two stage Bayesian models considered here, the likelihood of the failure times from each plant i, $P(X_i, T_i | \lambda_i)$, given λ_i and given any information from other plants, are independent and follow a Poisson distribution with parameter λ_i.

Parametric two stage Bayesian models consider usually gamma or log normal as prior distribution $P(\lambda_i|Q)$. The second stage places a hyperprior distribution over the parameters of the prior gamma or log normal distribution.

Controversies arise over the choice of hyperprior. [Cooke et al 1995] showed that the noninformativeness is not a good criteria if it leads to improper distribution. Since improper hyperpriors do not always become proper after observation, they should be avoided if property cannot be demonstrated.

Another major criticism to parametric two stage Bayesian models is the non-analytically solution of the model. Numerical integration should be performed in order to obtain the posterior distribution for λ_0. [Cooke et al 2002] showed that the method of truncation seems to have a large influence on the posterior distribution of λ_0.

3   VAURIO MODEL

As an alternative to standard two-stage Bayesian models, [Vaurio, 1987] proposed an analytic empirical Bayes approach to the problem of assimilating data from other plants. A simple one-stage Bayesian model for one plant would use a Poisson likelihood with intensity λ and a *Gamma*$(\lambda|a, b)$ prior. Updating the prior with X_i failures in time T_i yields a *Gamma* $(\lambda|a + X_i, \beta + T_i)$ posterior. Vaurio proposes to use data from the population of plants to choose the *Gamma*$(\lambda|a, \beta)$ prior by moment fitting. Any other two moment prior could be used as well. Data from other plants are not used in updating, hence, this is a one-stage model.

We sketch Vaurio's model in the simple case that the observation times at all $n + 1$ plants are equal to T. The population mean and (unbiased) variance are estimated as:

$$m = (\sum_{i=0..n} X_i/T)/(n + 1)$$

$$v = (\sum i = 0..n(X_i/T - m)^2)/n$$

A shifted variance estimate, which is positive when at least one of the $X_i > 0$, $i = 0, \ldots n$; is defined as:

$$V = vm/nT$$

V and m are used to solve for the shape α and scale β of a gamma prior $G(\lambda_i|\alpha, \beta)$:

$$\alpha = m^2/V$$
$$\beta = m/V$$

Using the familiar gamma-Poisson one stage model, the posterior mean and variance for λ_i after

observing X_i failures in time T, are:

$$E(\lambda_i|X_i, T) = (\alpha + X_i)/(\beta + T)$$
$$Var(\lambda_i|X_i, T) = (\alpha + X_i)/(\beta + T)^2 \qquad (2)$$

The model is consistent, in the sense that as $X_i, T_i \to \infty$, with $X_i/T_i \to \lambda_i$, his model does entail that $E(\lambda_i|X_i, T_i) \to \lambda_i$. Elegance and simplicity are its main advantages. Disadvantages are that it cannot be applied if all $X_i = 0$, or if the population consists of only 2 plants. Further, numerical results in section 5 indicate that the model is non-conservative when the empirical failure rate at plant 0 is low and the empirical failure rates at other plants are high. A final criticism, which applies to most empirical Bayes models is that the data for the plant of interest is used twice, once to estimate the prior and once again in the Poisson likelihood. Thus, X_i occurs in (2) twice, once as X_i, and again in the estimate of α and β. This may contribute to the non-conservativism noted in section 5.

4   A NON-PARAMETRIC OR DIRICHLET MODEL

Given the problems with the approaches described above, we explore the possibility of a non-parametric Bayesian two stage model. Very roughly, in this model we select a number of points $L_0, L_1, \ldots L_k$. The parameters of our prior distribution are probabilities $q = (q_1, \ldots q_k)$, adding to unity, such that:

$$P(L_{i-1} < \lambda \le L_i|Q = q) = q_i; i = 1, \ldots k. \qquad (3)$$

The mechanism for doing this is the Dirichlet distribution. In this section we set up the model in simple terms, and examine its assumptions.

4.1   *Prior parameters*

The prior distribution $P(\lambda|q)$ is characterized by

- a fixed number of points: $L_0 < L_1 < \ldots L_k$
- a probability vector $q = (q_1, \ldots q_k); q_i \ge 0, \Sigma q_i = 1$
- a probability density $g(\lambda)$ defined on (L_0, L_k)

4.2   *Prior distribution*

The points $L_0, \ldots L_k$ and the density $g(\lambda)$ are chosen by the user in a manner discussed below, and are not uncertain. There is no hyperprior over these. Letting $C_i = (L_i - 1, L_i)$, the prior may be written as

$$P(\lambda|q) = \sum_{i=1..k} 1_{\{\lambda \in C_i\}}(\lambda)q_i g(\lambda)/ \int_{C_i} g(\lambda)d\lambda. \qquad (4)$$

Here, $1_A(x)$ denotes the indicator function taking the value 1 if $x \in A$, and zero otherwise.

We shall take $g(\lambda)$ to be the uniform density, then $g(\lambda)$ is constant and we may write this as:

$$P(\lambda|q) = \sum_{i=1..k} 1_{\{\lambda \in C_i\}}(\lambda) q_i/|C_i| \quad (5)$$

Note that a gamma distribution for $g(\lambda)$ will provide also an analytical solution of the model.

4.3 Hyperprior

The hyperprior is a Dirichlet distribution characterized by parameters $a_1 \ldots a_k$; $a_i > 0$:

$$P(q) = \prod q_i^{(a_i-1)}/D(a_1, \ldots a_k) \quad (6)$$

where $D(a_1, \ldots a_k)$ is the Dirichlet integral:

$$\int_{q_1..q_k} \prod q_i^{(a_i-1)} dq_1..dq_k = \frac{\prod(a_i-1)!}{(\sum a_i - 1)!} \quad (7)$$

where $q_1 + \cdots q_k = 1$.

For reasons explained below, the parameters are sometimes called "equivalent observations of $\lambda \in C_i$".

We will choose $a_i = 1$; $i = 1, \ldots k$. Then the $D(a_1, \ldots a_k) = 1/(k-1)!$, and may be absorbed into the normalization constant, so that the hyperprior becomes:

$$P(q) = 1 \quad (8)$$

4.4 Updating the model

4.4.1 Hyperposterior

Considering only plant i, after observing X_i failures at plant i, the hyperposterior becomes

$$\int P(X_i|\lambda_i) P(\lambda_i|q) d\lambda_i P(q)$$

$$= (1/X_i!) \int (\lambda_i T_i) X_i exp(-\lambda_i T_i) \sum_{h=1..k} 1_{C_h}(\lambda_i) q_h d\lambda_i$$

$$= (1/X_i!) \sum_{h=1..k} q_h \int_{C_h} (\lambda_i T_i)_i^X exp(-\lambda_i T_i) d\lambda_i \quad (9)$$

Since X_i is an integer, the integral in (9) can be evaluated explicitly. We find for $C_h = (L_{h-1}, L_h)$:

$$(1/X_i!) \int_{C_h} (\lambda_i T_i)_i^X exp(-\lambda_i T_i) d\lambda_i$$

$$= (1/T_i) exp(-L_{h-1}T_i) \sum_{j=0..X_i} (L_{h-1}T_i)^j/j!$$

$$-(1/T_i) exp(-L_h T_i) \sum_{j=0..X_i} (L_h T_i)^j/j! = A_{i,h} \quad (10)$$

4.4.2 Posterior

Substituting (10) into (9), and this into (1) we find:

$$P(\lambda_0|X_0, \ldots X_n) \propto$$

$$P(x_0|\lambda_0) \int \sum_{m=1..k} 1_{C_m}(\lambda_0) q_m \prod_{i=1..n} \sum_{h=1..k} A_{i,h} q_h dq$$

$$\quad (11)$$

The integral is taken over the set $\{q = q_1, \ldots q_k | q_i \geq 0, \sum q_i = 1\}$. We can write the integrand as a sum of products.

Let us consider one such term. We may write this term as

$$A_{1,h(1)} A_{2,h(2)} \ldots A_{n,h(n)} q_1^{r_1} q_2^{r_2} \ldots q_k^{r_k} \quad (12)$$

Where, since there are $n+1$ plants it total,

$$r_1 + r_2 + ..r_k = n+1$$

The terms $A_{1,h(1)} A_{2,h(2)} \ldots A_{n,h(n)}$ do not contain q, and we may evaluate this integral explicitly for each term, using the Dirichlet integral. We find

$$P(\lambda_0|X_0 \ldots X_n) \propto$$

$$P(\lambda_0|X_0) \sum A_{1,h(1)} A_{2,h(2)} \ldots A_{n,h(n)} \prod r_i! \quad (13)$$

Where the summation goes over all k^n terms, and $r = r_1, \ldots r_k$ is specific to each of the k^n terms. Note that (13) expresses $P(\lambda_0|X_0, \ldots X_n)$ completely in terms of $P(\lambda_0|X_0)$, L_j, $j = 0, \ldots k$; X_i, T_i, $i = 1, \ldots n$. In other words, this model is completely solvable analytically.

If n is, say 10, and k is, say 4, then we have $4^{10} = 10^6$ such terms, which is a feasible number.

4.5 Equivalent observations

The parameters $a_1, \ldots a_k$ of the hyperprior are sometimes called equivalent observations. Indeed, the Dirichlet distribution is the natural conjugate for the multinomial likelihood. Consider rolling a die with k faces M times, where the probability of seeing face i is q_i. The probability of seeing face "i" r_i times; $i = 1, \ldots k$ is

$$\prod q_i^{r_i} M!/(r_1!, \ldots r_k!) \quad (14)$$

The result of updating the Dirichlet prior

$$P(q) = \prod q_i^{(a_i-1)}/D(a_1, a_k)$$

with these observations is again a Dirichlet:

$$P(q|r_1..r_k) = \prod q_i^{(a_i+r_i-1)}/D(a+r_1..a_k+r_k) \quad (15)$$

This suggests that the parameters a_i of the original Dirichlet prior may be interpreted as if we started with the prior

P(q) 5 1/D(1, 1... 1)

and observed face "i" $a_i - 1$ times, $i = 1, k$. This yields a useful heuristic for interpreting the parameters of the hyperprior in the ordered Dirichlet two stage model. When we choose the Dirichlet hyperprior (7) with $a_i = 1$, we are adopting a (hyper)prior belief state in which we have not yet observed one λ falling in any cell $C_j, j = 1, \dots k$.

4.6 Choosing parameters

The Dirichlet model requires the user to choose parameter values which cannot be updated. These are

1. The number k of cells $C_1, \dots C_k$,
2. The points $L_0 \leq L_1 \leq L_2, \dots \leq L_k$
3. The values $a_1, \dots a_k$.
4. The density $g(\lambda)$.

We have already indicated that the parameters a_i are chosen equal to one, to reflect "no equivalent observations".

The number of cells should be chosen so as to guarantee that after a long period of observation at each of the $n + 1$ plants, the hyperprior is "forgotten". If we consider (15), we see that the hyperprior is forgotten when most of the terms r_i are greater than zero. This suggests that with, say 10 to 15 plants, the number of cells should not exceed, say, four.

Given that we will have k cells, we must choose the points L_j defining the cells $C_j = (L_j - 1, L_j)$. Since L is a measure of $\lambda_i = \lim_{T_i \to \infty} X_i/T_i$, the end points L_0 and L_k should be chosen such that the term $\exp(-LT_i)\Sigma_{j=0\dots X_i}(LT_i)^j/j!$ in (9), with $L_0 \leq L \leq L_k$, covers all the possible values. This corresponds to calculate the hyperposterior distribution in one cell scenario for all $i = 0, 1 \dots n$ plants. Hence, the lower and the upper bounds will be the minimum of L_0, respectively the maximum of L_k, found for every plant $i = 0, 1 \dots n$. Figure 1 shows the hyperposterior distribution as a function of L in one cell scenario, for a plant with one failure over 10000 hours of observation. Using the assimptotic properties of term $\exp(-LT_i)\Sigma_{j=0\dots X_i}(LT_i)^j/j!$, a natural band can be found as $L_0 = 10^{-6}$ and $L_k = 10^{-3}$.

The points L_j, $j \neq 0$ and $j \neq k$, should be chosen such that we expect, before performing observations,

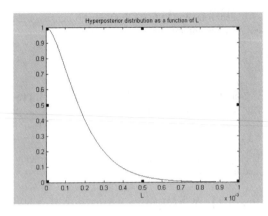

Figure 1. Hyperposterior distribution as a function of L in one cell scenario.

that the number of

$$\lambda_i = \lim_{T_i \to \infty} X_i/T_i$$

which, after a very long observation period at each plant, we will see, falling into each cell C_j is equal.

Finally we must choose the density $g(\lambda)$. We have chosen the uniform density as it is the least informative. Other choices could be made without sacrificing tractability.

4.7 Summary of significant features

We summarize the significant features of the Dirichlet model with the choices described above:

1. The model is solvable analytically, no numerical integration need be preformed.
2. The hyperprior has a clear intuitive interpretation.
3. The hyperprior is minimally informative in the sense of "no equivalent observations", and is proper.
4. The size and number of the cells C_j is chosen to insure that the hyperprior does not persist on observing $n + 1$ plants.

5 NUMERICAL EXAMPLE

We illustrate this model with a simple numerical example. These computations have all been performed on a spreadsheet, as no numerical integration is required. Nonetheless, for 7 plants the time required to compute the normalized posterior for λ_0 is about 10 minutes. We illustrate the model with 4 other plants, having between 1000 hours and 4000 hours operational time. Plant 0 is computed in three cases, namely with zero failures and 100, resp. 1000 resp. 10,000 operating hours. The

Table 1.

Plant	T	X	X/T	
1	1000	5	5.00E-3	
2	3000	20	6.67E-3	
3	3500	50	1.43E-2	
4	4000	100	2.50E-2	
0	100	0		

L0	L1	L2	L3	L4
1.00E-8	5.00E-4	5.00E-3	1.00E-2	1.00E-1

Ex w.o.	Ex w.	5%	50%	95%
4.08E-3	4.83E-3	7.57E-4	4.70E-3	9.46E-3

Prob	$(\lambda_i \in C_j)$			
Plant 0	0.04877	0.3447	0.23865	0.36783
Plant 1	1.42E-5	0.38402	0.54887	0.06709
Plant 2	−2.2E-16	0.08297	0.88174	0.03528
Plant 3	−2.2E-16	6.0E-11	0.00653	0.99347
Plant 4	3.3E-16	3.6E-15	−1.4E-15	1

Table 3.

Plant	T	X	X/T
1	1000	5	5.000E-3
2	3000	20	6.667E-3
3	3500	50	1.429E-2
4	4000	100	2.500E-2
0	10000	0	

L0	L1	L2	L3	L4
1.00E-8	5.00E-4	5.00E-3	1.00E-2	1.00E-1

Ex w.o.	Ex w.	5%	50%	95%
1.23E-4	1.23E-4	8.67E-5	7.22E-5	2.84E-4

Prob	$(\lambda_i \in C_j)$			
Plant 0	0.99316	0.00674	1.9E-22	3.7E-44
Plant 1	1.42E-5	0.38402	0.54887	0.06709
Plant 2	−2.2E-16	0.08297	0.88174	0.03528
Plant 3	−2.2E-16	6.0E-11	0.00653	0.99347
Plant 4	3.3E-16	3.6E-15	−1.4E-15	1

Table 2.

Plant	T	X	X/T	
1	1000	5	5.000E-3	
2	3000	20	6.667E-3	
3	3500	50	1.429E-2	
4	4000	100	2.500E-2	
0	1000	0		

L0	L1	L2	L3	L4
1.00E-8	5.00E-4	5.00E-3	1.00E-2	1.00E-1

Ex w.o.	Ex w.	5%	50%	95%
1.10E-3	1.16E-3	8.67E-5	8.88E-4	3.05E-3

Prob	$(\lambda_i \in C_j)$			
Plant 0	0.39346	0.59979	0.00669	4.5E-5
Plant 1	1.42E-5	0.38402	0.54887	0.06709
Plant 2	−2.2E-16	0.08297	0.88174	0.03528
Plant 3	−2.2E-16	6.0E-11	0.00653	0.99347
Plant 4	3.3E-16	3.6E-15	−1.4E-15	1

Table 4.

Plant	T	X	X/T	
1	1000	5	5.000E-3	
2	3000	20	6.667E-3	
3	3500	50	1.429E-2	
4	4000	100	2.500E-2	
5	2000	15	7.500E-3	
6	2500	15	6.000E-3	
0	100	0		

L0	L1	L2	L3	L4
1.00E-8	5.00E-4	5.00E-3	1.00E-2	1.00E-1

Ex w.o.	Ex w.	5%	50%	95%
4.08E-3	5.49E-3	8.93E-4	5.75E-3	9.54E-3

Prob	$(\lambda_i \in C_j)$			
Plant 0	0.04877	0.3447	0.23865	0.36783
Plant 1	1.42E-5	0.38402	0.54887	0.06709
Plant 2	−2.2E-16	0.08297	0.88174	0.03528
Plant 3	−2.2E-16	6.0E-11	0.00653	0.99347
Plant 4	3.3E-16	3.6E-15	−1.4E-15	1
Plant 5	1.8E-14	0.04874	0.79475	0.15651
Plant 6	5.2E-13	0.19397	0.78374	0.02229

results are shown in Tables 1, 2 and 3. Each table shows the operational data from the four other plants and the cells. This data is the same in each table. Also shown is the (unnormalized) probability that λ_i falls in cell C_j; this data differs for λ_0 in each table. Very small negative probabilities are caused by numerical errors in EXCEL. "Ex w." and "Ex w.o." denote updating with and without the data from plants 1 ... 4. Data from other plants has the effect of raising the posterior expectation.

For plant 0 with 1000 operating hours, Table 2 shows that the updating with other plants now has less effect on the posterior expectation.

For 10,000 operating hours and still zero failures at plant 0, the posterior expectations with and without other plants are practically the same. Now, after 10,000 hours, the expectation at plant 0 is determined only by the data at plant 0, and the other plants have almost no effect. Note that the probability for λ_0 is concentrated in cell C_1.

For purposes of comparison, Table 4 shows the results of updating with 6 other plants, when plant 0 has experience no failures in 100 operating hours. Plants 5 and 6 have empirical failure rates in the same order as the first 4 plants. We see that the results are a bit larger than those of Table 1.

Table 5 gives the results of the Dirichlet model for the Data set 2 of [Becker and Hofer 2001], and com-

Table 5.

Plant	T	X	X/T	
1	20000	1	5.00E-5	
2	2000	0	0	
3	4000	0	0	
4	6000	0	0	
5	10000	1	1.00E-4	
0	12000	2		

L0	L1	L2	L3	L4
1.00E-8	5.00E-6	5.00E-5	5.00E-4	1.00E-3

Ex w.o.	Ex w.	5%	50%	95%
2.32E-4	2.31E-4	7.45E-5	2.13E-4	4.32E-4
Gamma		1.22E-4	1.72E-4	3.45E-4
Lgnormal		2.54E-5	7.71E-5	2.06E-4
Vaurio		3.34E-5	1.14E-4	2.74E-4

Prob	$(\lambda_i \in C_j)$			
Plant 0	3.44E-5	0.02308	0.91492	0.06145
Plant 1	0.00468	0.25956	0.73526	0.0005
Plant 2	0.00993	0.08521	0.53696	0.23254
Plant 3	0.01976	0.16147	0.68339	0.11702
Plant 4	0.02949	0.22963	0.69103	0.04731
Plant 5	0.00121	0.08899	0.86937	0.03993

Table 6.

Plant	T	X	X/T	
1	10000	0	0	
2	10000	0	0	
3	10000	0	0	
4	10000	0	0	
5	10000	0	0	
0	10000	5	0.0005	

L0	L1	L2	L3	L4
1.00E-8	5.00E-6	5.00E-5	5.00E-4	1.00E-3

	Ex w.	5%	50%	95%
Dir	5.03E-4	2.03E-4	4.16E-4	9.37E-4
Vaurio	4.87E-4			

Plant	T	X	X/T
1	10000	5	0.0005
2	10000	5	0.0005
3	10000	5	0.0005
4	10000	5	0.0005
5	10000	5	0.0005
0	10000	0	0.0005

L0	L1	L2	L3	L4
1.00E-8	5.00E-6	5.00E-5	5.00E-4	1.00E-3

	Ex w.	5%	50%	95%
Dir	1.30E-4	1.81E-5	1.02E-4	3.17E-4
Vaurio	5.79E-5			

Plant	T	X	X/T
1	1000	0	0
2	1000	0	0
3	1000	0	0
4	1000	0	0
5	1000	0	0
0	1000	5	0.005

L0	L1	L2	L3	L4
1E-8	5.00E-6	5.00E-5	5.00E-4	1.00E-2

	Ex w.	5%	50%	95%
Dir	5.61E-3	2.54E-3	5.46E-3	9.05E-3
Vaurio	4.87E-3			

Plant	T	X	X/T
1	1000	5	0.005
2	1000	5	0.005
3	1000	5	0.005
4	1000	5	0.005
5	1000	5	0.005
0	1000	0	0.005

L0	L1	L2	L3	L4
1E-8	5.00E-6	5.00E-5	5.00E-4	1.00E-2

	Ex w.	5%	50%	95%
Dir	1.41E-3	4.39E-4	1.09E-3	3.39E-3
Vaurio	5.79E-4			

pares these with the results for standard Bayesian models [Cooke et al 2002]. We see that the results are of the same order, though a bit higher. The intervals (L_{i-1}, L_i) are chosen to bound the empirical rates (when failures are present).

Table 6 compares the Dirichlet results with Vaurio's estimator. To avoid numerical procedures, all observation times are equal. We see that there are significant differences. In general, Vaurio's estimate is closer to the empirical failure rate of plant 0. In those cases where plants 1..5 exhibit a high empirical failure rate, and the plant of interest, plant 0, has no failures, Vaurio's estimate is lower than the Dirichlet estimate by a factor 2. This feature is observed for short observation times (1000 hours) and long observation times (10,000 hours).

These results indicate that Vaurio's estimate displays a non-conservativism when the empirical failure rate at the plant of interest is much lower than the empirical failure rates of other plants in the population. The two-stage Dirichlet model (and presumably the other two stage models) are more sensitive to the empirical failure rates at 6 other plants.

Finally, we include the Dirichlet results for data sets 1 and 3 of [Becker and Hofer 2001]. The results are in the same order as those for classical Bayesian models [Cooke et al 2002], but tend to be a bit more conservative.

Figure 2 shows the density for λ_0 with and without updating from other plants. We see that the other plants have a (weak) tendency to lower the failure rate at plant 0.

6 CONCLUSIONS

The Dirichlet model enjoys two advantages relative to the other models discussed here. First, it is analytically

Table 7. Results for datasets 1 and 3.

Dataset 1	5%	50%	95%
Dirichelet	1.12E-05	7.79E-05	2.34E-04
Gamma model	3.25E-05	6.99E-05	1.30E-04
Lognormal model	2.01E-05	5.91E-05	1.44E-04
Vaurio's model	1.83E-05	6.84E-05	1.73E-04
Dataset 3	5%	50%	95%
Dirichelet	1.45E-04	4.75E-04	9.25E-04
Gamma model	1.21E-04	2.58E-04	7.41E-04
Lognormal model	1.12E-05	6.23E-05	3.40E-04
Vaurio's model	2.84E-05	2.19E-04	7.64E-0.4

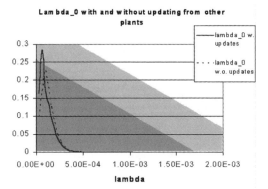

Figure 2. Dirichlet model, failure rates with and without updating from other plants, dataset 1.

solvable, no numerical integration need to be performed, and second, it allows an intuitive interpretation of the (hyper)parameters – so called "equivalent observation". It must be emphasized that this is a first implementation of this model. Additional testing should be performed before declaring the model fully operational. In particular, heuristics should be developed for choosing the number of cells C_j, and the values of their endpoints. Preliminary calculations show some sensitivity of the model in this regard.

REFERENCES

Becker G. & Hofer E. 2001. *Schtzung von Zuverlssigheits-Kennwerten aus der Evidenz veschiedener Anlagen*, GRS-A-2918.

Cooke R.M., Bunea C., Charitos T. & Mazzuchi T.A. 2002. *Mathematical Review of ZEDB Two-Stage Bayesian Models*, Report 02-45, Departement of Mathematics, Delft University of Technology.

Cooke R.M., Dorrepaal J. & Bedford T.J. 1995. *Review of SKI Data Processing Methodology* SKI Report.

Hofer E. 1999. On two-stage Bayesian modeling of initiating event frequencies and failure rates, *Reliability Engineering and System Safety*, vol. 66, pg. 97–99.

Hofer E., Hora S., Iman R. & Peschke J. 1997. On the solution approach for Bayesian modeling of initiating event frequencies and failure rates, *Risk Analysis*, vol. 17, no. 2, pg. 249–252.

Vaurio, J.K. 1987. On analytic empirical Bayes Estimation of failure rates, *Risk Analysis*, vol. 17, no.3, pg. 329–338.

Safety and Reliability – Bedford & van Gelder (eds)
© 2003 Swets & Zeitlinger, Lisse, ISBN 90 5809 551 7

Unavailability assessment of IFMIF target system

L. Burgazzi
ENEA, Italian Commission for New Technologies, Energy and Environment, Bologna, Italy

B. Giannone
APAT, Italian Agency for Environmental Protection and for Technical Services, Rome, Italy

F. Zambardi
APAT, Italian Agency for Environmental Protection and for Technical Services, Rome, Italy

ABSTRACT: In the frame of the safety assessment of IFMIF (International Fusion Materials Irradiation Facility), an analysis has been developed, aimed at evaluating an unavailability figure of the target system, and eventually at evaluating the expected performance of the plant, in terms of continuous and reliable operation. Starting points of the analysis have been the identification of functions relevant for plant availability and relationships between plant systems and functions and the definition of relationships among the systems themselves. Fault tree technique has been adopted to address the topic: despite the early stage of the design – at present many systems are not yet definitely established – systems boundaries and interfaces have been identified as well as the dependent failures between systems and components. Due to the novelty of the plant and its prototypical character, reliability data introduced in the numerical simulation underwent an accurate screening process among the current available databases, sometimes requiring the expert judgment assessment. Finally an uncertainty, importance and sensitivity analysis has been performed in order to add credit to the achieved results and to highlight reliability-critical systems and components: results are analysed to assess the compliance with the plant availability requirements and design criteria. Risk Spectrum code has been utilised for the system unavailability quantification.

1 INTRODUCTION

The IFMIF (International Fusion Materials Irradiation Facility) is an experimental facility consisting essentially of a particle accelerator and a target system to provide high energy neutrons for testing the radiation resistance of materials to be used in future nuclear fusion reactors. The design of the facility is jointly carried out by Japan, the United States, the European Union and the Russian Federation (JAERI, 2000).

Target facility unavailability during operation is evaluated as part of the safety analysis review foreseen in the development program of IFMIF (JAERI, 2000).

Because of its prevailing use in systems analysis, fault tree technique is adopted to evaluate all potential failure modes affecting the system during normal operation and to assess the system failure on the probabilistic standpoint. The construction of the fault tree is based on system success/failure criteria, with reference to the functions which is required to perform. System model

reflects as much as possible the current definition level of the design – at present many systems are not yet definitely established – it requires the identification of the relative boundaries and interfaces and the allocation of all the components required for system operation, support systems required for actuation and operation of the system components, and other components whose failure can degrade or fail the system. All the relevant and possible failure modes, previously identified by a well structured Failure Mode and Effect Analysis (FMEA) procedure by Burgazzi (2002), are evaluated: these include both component independent and dependent failures.

The system model construction and unavailability assessment are made with the support of Risk Spectrum® PSA Professional, a PC software package that performs risk and reliability analysis based on fault trees and event trees (RELCON AB, 2000) and the component failure rate data necessary for reliability assessment are selected from available data bases which are judged applicable.

Finally an uncertainty, importance and sensitivity analysis is performed in order to increment credibility to the fault tree model and to point out reliability-critical areas: results are analysed to assess the compliance with the plant design criteria and future operative requirements in terms of availability goals.

2 SYSTEM DESCRIPTION

A deuteron beam current delivered to a liquid lithium target by two 125-mA, 40 MeV accelerator modules operating in parallel, is at the base of the IFMIF concept. The IFMIF facility is subdivided in the following systems, as described by Martone (1996):

- Accelerator facility, which produces, accelerated deuteron beam;
- Lithium target facility, which produces a flowing stream of lithium to convert the impacting deuterons to neutrons;
- Test facility, which exposes specimens to neutrons;
- Auxiliary systems, such as vacuum system, lithium cooling, impurity removal, monitoring system, electrical power supply etc.

The lithium target may be divided into two basic subsystems. The first is the target assembly itself, which must present a stable lithium jet to the beam, where the kinetic energy of the deuteron beam is deposited and neutrons are produced. The second is the lithium loop, which circulates the lithium to and from the target assembly and removes the heat deposited by the deuteron beam. This heat is then transferred through heat exchangers to a secondary organic liquid cooling loop and a tertiary water cooling loop, and finally to a cooling tower acting as heat sink. The lithium loop also contains systems for maintaining the high purity of the lithium required for radiological safety and plant operational availability, in fact the reduction of impurities is essential to minimize the generation of activated substances and the corrosion of the loop structure induced by the hot flowing lithium.

A flow diagram of the IFMIF is shown in Figure 1. The vacuum system is interfaced to the accelerator and the lithium target: it is essential for the deuteron beam operation and for its interaction with the target itself. The major components in the main lithium loop are the target quench tank, the surge or overflow tank, the lithium dump tank, the main electromagnetic pump and one heat exchanger (to the secondary organic cooling loop), in addition to the valves. The piping and tanks are constructed of austenitic steel (either 304 or 316). There is, in addition, a trace heating system to maintain the temperature throughout the loop above the melting point of the lithium (180,6°C) at all times the metal is present in the loop, thermal insulation, valves etc.

The total estimated lithium inventory is estimated as $9\,m^3$, as reported by Burgazzi (2002).

3 METHODOLOGY FOR SYSTEM MODEL

Main steps of the methodology are the following:

3.1 Target facility functions and systems relationships

Starting point of the analysis is the identification of functions relevant for plant availability and relationships between plant systems and functions: the following Table 1 relates systems to the functions they perform.

3.2 Dependent failure analysis

The following types of dependency, to be modeled in the fault trees, are identified:

- Functional Dependencies, which consider the effects of the status of one system or safety function on the success or failure of another. These dependencies are explicitly accounted for as separate transfer events in fault trees, as regards the electrical power supply.
- Shared-equipment Dependencies, in the form of dependencies of multiple systems on the same components or subsystems. These dependencies have been found in the lithium-oil heat exchanger and in the oil-water exchanger and properly incorporated in the fault tree analysis, as separate events (damage of shell and pipes).
- Intercomponent Dependencies, also called common cause failures, due to shared root causes of failures in redundant components. Common cause failures are accounted for the lithium draining valves to dump tank and the related circuit breakers, as separate events.

3.3 Fault tree construction

The underlying success criteria in fault tree construction is the availability of the whole target facility which can be attained through the availability of each system whose failure in the requested mission contribues singularly to the loss of the target.

Therefore, fault tree models are constructed for each relevant target system as previously identified, as listed below:

- Loss of Primary Cooling System
- Loss of Secondary Cooling System
- Loss of Water Cooling System
- Loss of Cold Trap Cooling System
- Loss of Lithium Purification System

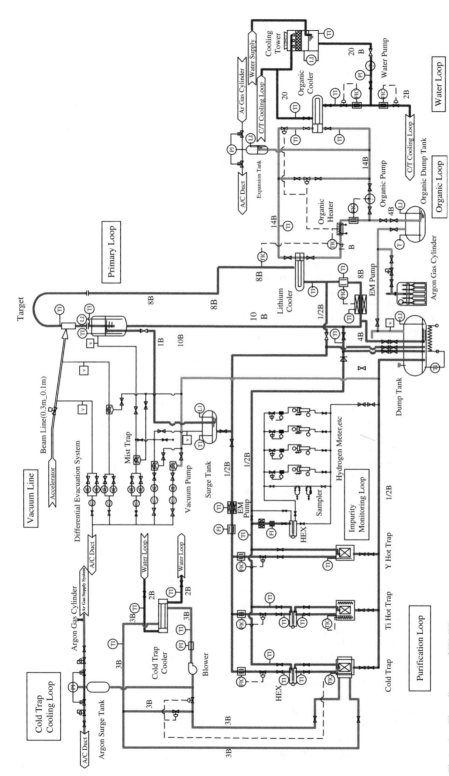

Figure 1. Flow diagram of IFMIF target system.

341

Table 1. Function/System relationship table.

Function	System
Target Cooling	Primary Cooling System
	Secondary Cooling System
	Water Cooling System
Target Radioactivity Removal	Cold Trap Cooling System
	Lithium Purification System
	Lithium Impurity Monitoring System
Target Assembly Vacuum	Lithium Target Vacuum System
Power Supply	Electrical Power Distribution System

- Loss of Lithium Target Vacuum System
- Loss of Electrical Power Distribution System

The monitoring system was not evaluated in the model because of lack of the failure data related to lithium contamination meters. In any case the system is assessed to have low weight in the probability prediction since reduction of contamination is made by purification loop.

Fault tree development involves the decomposition of a system into system segments or components and the development of system fault logic in terms of consequential faults due to failures in the segment/component; it has to be underlined that the boundaries (like switches, circuit breakers, mechanical interfaces and so on) and the physical interfaces of the various systems, required in fault tree development, are identified according to the current level of definition of the present design and the available system delineation related information.

In Figures 2 and 3 the whole target system fault tree and the primary cooling system fault tree are reported respectively.

4 UNAVAILABILITY MODELING

In Risk Spectrum model each system unavailability is calculated starting from the Boolean combination of failures of its components. All systems are analysed in condition of normal operation. Normal routine maintenance and periodic tests are not included in the model but they are expected to be performed between subsequent mission times (one week). Human errors, as well as planned and unplanned maintenance actions (i.e. repairs of hardware failures) and the impact of ageing, are not accounted.

Three reliability parameters are considered in the present probabilistic analysis:

- failure probabilities, in terms of failures per demand, for components that are activated and change state, such as valves;

- failure rates, in units of failures per hour, for normally operating components, such as pumps, electrical devices and pipes;
- mission time, that is the number of hours in one operating week (168 h).

In Risk Spectrum the following reliability models are associated to these parameters:

- constant unavailability: the unavailability "q" is the only parameter required in the model that is no time-dependent;
- fixed mission time unavailability: given a failure rate "λ" and a fixed mission time "TM", the probability of failure over a mission time is calculated with the formula:

$$Q = 1 - e^{-\lambda * TM}.$$

The beta model is selected for the common cause failure analysis with $\beta = 0.1$.

5 RELIABILITY DATA

Generally one of the major issue when performing a reliability analysis deals with the failure rates to be assigned to the components: this is particularly relevant in the present case, due to the prototypical character of the experimental plant. Consequently the reliability values to be introduced for numerical quantification are accurately selected, basing mainly on existing relevant fission data bases, including data developed for fusion plants and previous studies, when applicable, sometimes requiring the engineering judgment assessment, as related by Authors (2002).

The compilation of component failure rates were performed taking account of the following criteria, in order of importance:

1. IFMIF target system specific data
2. component specific data
3. most recent data
4. conservative values
5. recurrent values
6. component assimilation

The lognormal distribution is considered to describe the uncertainty of component reliability parameters; therefore a mean unavailability and an error factor are indicated for each component.

The list of evaluated components with the related failure modes, failure parameters and error factors is presented in Table 2.

6 RESULTS

All fault trees together with the relative results in terms of Minimal Cut Sets are detailed by Authors

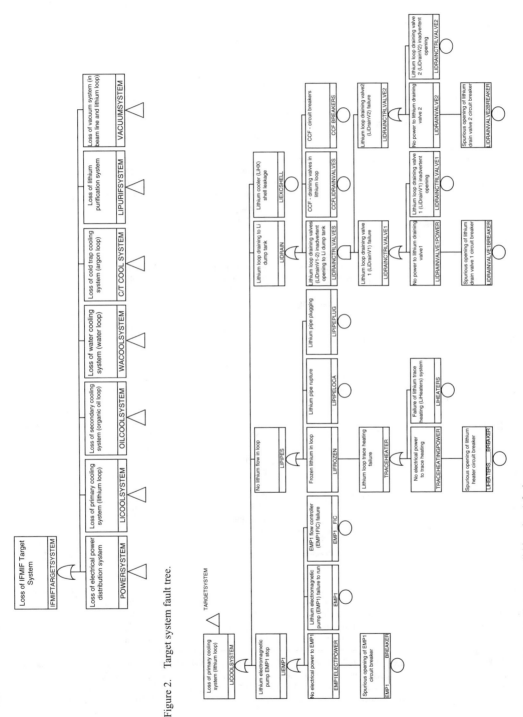

Figure 2.   Target system fault tree.

Figure 3.   Primary cooling system (lithium loop) fault tree.

Table 2. IFMIF component reliability data.

IFMIF Target System Component	Reference Component	Failure Mode	Failure Rate /h – Probability/d	EF	Source Ref. (*)
Electrical bus	Electrical bus	All modes	3.0E-8/h	3	II
Electrical transformer	Electrical transformer	All modes	1.0E-6/h	3	VI
Circuit breaker	Generic circuit breaker	Spurious change of position	1.2E-6/h	3	III
Flow controller	Flow controller	Failure to operate	1.0E-5/h	10	V
Pressure controller	Pressure controller	Failure to operate	3.0E-6/h	10	V
Temperature controller	Temperature controller	Failure to operate	5.0E-6/h	10	V
Lithium electromagnetic pump	Electromagnetic pump	Failure to run	1.0E-6/h	10	I
Lithium draining valve	Solenoid valve	Inadvertent opening	2.9E-6/h	10	CDA
Control valve in purification system	Pneumatic valve	Failure to operate	1.0E-3/d	10	I
Lithium piping	Liquid metal piping	Rupture/leakage	1.6E-8/h	30	I
		Plugging	1.6E-8/h	30	E.A.
Lithium trace heaters	Trace heater	All modes	1.5E-4/h	10	CDA
Lithium cooler	Liquid metal heat exchanger	Multiple pipe leakage	4.0E-6/h	10	I
		Multiple pipe plugging	5.7E-6/h	10	II
		Shell leakage	3.8E-7/h	30	I
Oil pump	Generic pump	Failure to run	2.9E-6/h	30	CDA
Oil control valve	Pneumatic valve	Failure to operate	3.0E-3/d	10	VI
Organic oil piping	Liquid piping	Rupture/leakage	1.6E-8/h	30	E.A.
		Plugging	1.6E-8/h	30	E.A.
Organic cooler	Liquid heat exchanger	Multiple pipe leakage	4.0E-6/h	10	E.A.
		Multiple pipe plugging	5.7E-6/h	10	II
		Shell leakage	3.0E-6/h	10	I
Organic oil heater	Generic heater	All modes	5.6E-7/h	10	II
A/C duct and gas cylinder control valve (oil loop)	Pressure relief valve	Failure to operate	1.0E-3/d	3	VI
Water pump	Water pump	Failure to run	3.0E-5/h	10	VI
Water control valve	Pneumatic valve	Failure to operate	3.0E-3/d	10	VI
Water piping	Water piping	Rupture/leakage	1.2E-8/h	3	I
	Liquid piping	Plugging	1.6E-8/h	30	E.A.
Cooling tower	Cooling tower	All modes	3.8E-6/h	30	CDA
Water supply	Water supply	Loss	1.2E-3/d	3	VII
Circuit breaker of water pump	Generic circuit breaker	Spurious change of position	3.0E-5/h	2	II
Argon blower	Blower/fan	Failure to run	1.0E-5/h	2	II
Argon control valve	Pneumatic valve	Failure to operate	3.0E-6/h	10	I
Argon piping	Gas piping	Rupture	3.0E-9/h	30	I
Cold trap cooler	Gas heat exchanger	Multiple pipe leakage	3.0E-6/h	10	I
		Multiple pipe plugging	5.7E-6/h	10	II
A/C duct and gas cylinder control valve (argon loop)	Pressure-relief valve	Failure to operate	1.0E-3/d	3	VI
A/C duct and gas cylinder self-operated valve (argon loop)	Pressure self-operated valve	Failure to change position	4.0E-3/d	6	II
Cold trap	Cold trap	Plugging	3.0E-5/h	30	E.A.
Hot trap (Ti and Y)	Getter	Getter saturation	3.0E-5/h	30	E.A.
Titanium trap heater	Generic heater	All modes	5.6E-7/h	10	II
Vacuum pump	Vacuum pump	Failure to run	1.0E-5/h	10	I
Mist trap	Strainer/filter	Plugging	3.0E-5/h	10	VI

(*) Reliability data sources; I. EGG-FSP-7922 (INEEL), INEL, 1988; II. IAEA-TECDOC-478, IAEA, 1988; III. Reliability Data Book for Components in Swedish Nuclear Power Plants, RKS-Ski, 1987; IV. NREP Reliability Data, NRC, 1983; V. LER data, E.W. Hagen, Nuclear Safety Vol.24, 1983; VI. NUREG/CR 2728 (IREP), NRC, 1983; VII. NUREG-75/014 (WASH 1400), NRC, 1975; CDA. Conceptual Design Activity report, Martone (1996); E.A. Engineering Assessment.

Table 3. Distribution of failure probabilities among the target systems.

System	Probability of failure Q_i	Importance % $(Q_i/Q_{tot}) * 100$
Vacuum system	2.79E-02	20.48
Primary cooling system (lithium loop)	2.72E-02	20.00
Lithium purification system	2.50E-02	18.35
Water cooling system	2.15E-02	15.78
Secondary cooling system (organic oil loop)	1.68E-02	12.32
Cold trap cooling system (argon loop)	1.09E-02	8.00
Power system	6.91E-03	5.07

(2002). The result of the plant fault tree analysis in terms of IFMIF target system probability of failure over a week mission time is:

$$Q_{TOT} = 1.29E\text{-}01$$

The uncertainty analysis, performed by the built-in Risk Spectrum model based on Monte Carlo simulation (number of simulations = 1000), gives the following results:

$$Q_{MEDIAN} = 1.03E\text{-}01$$
5th percentile = 5.71E-02
95th percentile = 2.75E-01

Notwithstanding the relatively high degree of uncertainty for some parameters, which implies a high error factor, the uncertainty analysis doesn't show a large spreading in the obtained results and the highest value is quite close to the expected one (i.e. mean value).

The importance analysis is developed at "system" and "component level". The importance results at "system level" are summarized in Table 3.

Concerning the importance analysis at "component level", it is possible to infer that:

- lithium heaters are the most critical components for probability of failure of the lithium cooling system (91.45%) and of the overall plant (19.34%);
- 3 kV breaker unavailability is predominant in failure of electrical power system (72.78%);
- cold trap, yttrium and titanium getters are predominant in failure of purification system (60.45%) and mist traps in vacuum system (54%).

The sensitivity calculations are carried out by Risk Spectrum in the following way:

- the value under consideration is set equal to the nominal value divided and multiplied by a sensitivity factor (SensFactor = 10);

- the new top event values are calculated and are indicated respectively with $Q_{TOP,L}$ and $Q_{TOP,U}$;
- the sensitivity value is calculates as:

$$S = Q_{TOP,U} / Q_{TOP,L}$$

The sensitivity analysis points out that the dependence of the results on lithium heater performance is three times the dependence evaluated for the other components.

7 CONCLUSIONS

In order to evaluate the unavailability of the whole target facility, the reliability assessment of most relevant systems belonging to IFMIF lithium target system are performed at "component level" applying the fault tree technique, by means of Risk Spectrum® PSA Professional code. Fault trees are built with the component reliability model, the component reliability data are assigned and the final unavailability is assessed through the quantification of the fault tree.

Results show a value, in terms of probability of failure, which lies around 1.0E-1: this value derives from the individual contributors relative to the target systems (i.e. cooling systems, lithium purification system, etc.) and pertains to the target probability to fail the required mission for a specified mission time (one week).

The analysis points out as reliability-critical areas, where modifications to the design could be required to reduce the probability of failure the following, as major contributors to the overall value: lithium target vacuum system, cooling systems and lithium purification system. Concerning the system architecture, it is considered that lack of sufficient redundancies involves relatively high probabilities of failure.

In addition, going into more detail, the study highlights as critical components the electrical heaters, placed around the lithium carrying components and designed to assure the lithium temperature above the melting point (180,6°C), the mist traps, required to trap the lithium mist in the vacuum system, and the getters of the purification loop.

At present there are not any requirements, in the fashion of unavailability figure, pertaining to the reliability of the system: however the allocable availability goal to be met calls for a high reliability degree (in terms of MTBF), to be attained by critical components redundancy, component overdesign, component derating and so on.

Implementation of the present reliability analysis is required as soon as a more detailed design is available because the level of definition of the current design is not sufficient to reach a final assessment.

REFERENCES

JAERI-Tech 2000–014, 2000. IFMIF International Fusion Materials Irradiation Facility Conceptual Design Activity Reduced Cost Report A supplement to the CDA by the IFMIF Team.

JAERI-Tech 2000–052, 2000. IFMIF International Fusion Materials Irradiation Facility Key Element Technology Phase Task Description.

Burgazzi, L. 2002. Hazard Assessment of Liquid Lithium Target. Proceedings of the Third Edinburgh Conference on "Risk: Analysis, Assessment and Management", Edinburgh (UK), 8–10 April 2002.

RELCON AB, 2000. Risk Spectrum® PSA Professional Version 1.20 – User's Manual.

Martone, M. 1996. ENEA Report RT-ERG-FUS-96-11 IFMIF Conceptual Design Activity Final Report.

Burgazzi, L., Giannone, B. & Zambardi, F. 2002. ENEA Report FIS-P127-002 IFMIF Lithium Target System Analysis.

Safety and Reliability – Bedford & van Gelder (eds)
© 2003 Swets & Zeitlinger, Lisse, ISBN 90 5809 551 7

Bayesian assessment of corrosion-related failures in steel pipelines

E. Cagno, F. Caron & M. Mancini
Department of Management, Economics & Industrial Engineering, Politecnico di Milano, Milan, Italy

A. Pievatolo & F. Ruggeri
IMATI-CNR, Milan, Italy

ABSTRACT: The probability of gas-escape from steel pipelines due to different types of corrosion is studied with real failure data from an urban gas distribution network. Both design and maintenance of the network are considered, identifying and estimating in a Bayesian framework the failure rate of the pipe depending on the characteristic of the gas distribution network and the type of corrosion. All over the proposed approach, particular attention is devoted to the elicitation of the experts' opinions obtained by pairwise comparisons deriving from the Analytic Hierarchy Process. We conclude that the corrosion process behaves quite differently according to the type of corrosion and that cathodically protected pipes should be installed in most cases.

1 INTRODUCTION

This paper deals with the assessment of the probability of gas-escape (from now on named «failure») from the steel pipelines of a gas distribution network, in order to give support to the estimation of its dependability and to the implementation of the maintenance and renewal policy. A case study of a urban low pressure network laid in a metropolis of 1.5 million citizens is presented. Steel pipelines have very strong mechanical properties (failures are very rare), but they are easy prey for corrosive agents unless they are correctly protected.

After a brief description of the technical problem in Section 2, a Bayesian analysis based only on the observed number of failures during the period 1978–1997 is presented in Section 3. There we focus on the probability that, given a failure, it occurs in a certain location and that is due to one among various types of corrosion processes.

This information can be used as a quick reference to decide the kind of protection needed by a pipe for design purposes.

In all cases, information provided by the experts is incorporated into the models according to the Bayesian paradigm.

During the period 1978–1997 the gas network has developed and the environment surrounding it may have changed. Thus it might not be appropriate to consider as a fixed quantity the probability that, given a failure, it has taken place in a certain lay location due to a corrosion of a given type. In fact this kind of data are all but repeatable. The Bayesian approach, having also had the possibility to collect experts' opinions, then comes as the natural one for this problem.

2 FACTORS RELATED TO THE PROBABILITY OF FAILURE

In the case under examination, the number of failures of the steel pipelines due to corrosion is very small (33 failures in the period 1978–1997 over a network of 275 kilometers as of 1997). The scarceness of historical data has led us to use a methodology that takes into account all possible kind of information. The principal source of information is the experts' knowledge (grown inside the company that manages the gas distribution network), and the Bayesian framework (Bernardo & Smith 1994) offers the tools to combine the historical data with the corporate memory. From the available historical data inside the company and the interviews with the experts, two principal elements that may be related to the failures of steel pipelines have been identified: the type of corrosion that led to the rupture of pipe, and the lay location of the pipe.

2.1 Type of corrosion

Three types of corrosion have been determined: natural, galvanic, and by interference. Each of them is characterized by specific causes and develops in fully different ways. This brings to their different relationship with the failure of the pipe.

2.1.1 Natural corrosion

It is principally caused by the aggressive properties of the ground. For example, very wet ground is a good conductor and as such facilitates the development of the electrolytic phenomenon on the steel.

2.1.2 Galvanic corrosion

A gas distribution network is very inhomogeneous because it is made up of very different materials (treated cast iron, spheroid graphite, traditional cast iron, polyethylene and so on). When a steel pipe is connected to a cast iron pipe, if the insulating material, which constitutes the principal protection of the pipe, is not perfectly efficient, a galvanic corrosion caused by the contact of the two materials can take place. In particular, there is a voltage difference between cast iron and steel, where the latter assumes the anodic behavior and corrodes, while the former assumes the cathodic one.

2.1.3 Corrosion by interference

It is also called corrosion by stray current, because it is due to the presence of stray currents in the ground coming from other electrical plants badly isolated (for example streetcar substations or train stations). These currents discharge themselves on the steel pipe increasing the corrosion rate by various orders of magnitude.

In other cities, also other corrosive agents hidden in the ground can result very important (especially near the sea), but not in the present case.

2.2 Lay location

The propensity to a particular type of corrosion is not the same all over the city, but changes according to the place where the pipe is laid. In fact in some places stray currents in the ground are significantly higher than in other places, which increases the propensity to a failure due to corrosion by interference. Therefore two different areas can be determined: zone A, characterized by the presence of streetcar substations or rail stations, and zone B, without them.

2.2.1 Closeness of streetcar substations

Near streetcar substations, the trunk of negative electric cables used to feed the streetcar line is hidden underground. The streetcar feeding current is generated in the substation, goes through the aerial line (positive cables) and is transformed into power by the streetcar; then it goes through the steel street-car tracks (negative cables) to close the circuit in the substation. The circuit is actually closed, in the last hundred meters, by underground cables connected to the steel streetcar tracks and to the generator of the substation. These cables generate stray currents because of the bad isolation.

2.2.2 Closeness of rail stations and lines

In the case of rail stations, the stray currents derive not only by the bad isolation of the tracks, but also by the strong electric field coming from the passage of the train. Therefore the train traffic is proportional to the mean intensity of the stray currents.

3 BAYESIAN ANALYSIS OF THE FAILURE DATA

In view of the above discussion, we determined two areas: zone A, where there is a trunk of negative electric cables used to feed the streetcar line or near rail stations or rails; zone B, all other areas. In order to compare the number of failures by zone and by type of corrosion, the number of observed failures should be divided by the lengths of the gas distribution networks in zone A and in zone B respectively. Since this datum is unavailable, we approximate it by the area (in squared kilometers) of the two zones. The observed number of failures by zone and by type of corrosion per squared kilometer (named here *failure rate*) is summarized in Table 1.

It is reasonable to expect that the failure rate by natural corrosion in zone A is comparable to that in zone B, but this is not the case for the observed failures, probably because of incorrect classifications made by the repairing teams. This will not affect the estimate of the marginal probability that, given a failure, it occurs in zone A, whereas it will bias the estimates of the marginal probabilities that a corrosion occurs of a given type, as well as the estimates of some conditional probabilities. Therefore some care must be taken in interpreting the result, even if the misclassification bias can be partially counterbalanced by the prior opinion of the experts.

3.1 Prior probability assessment: the analytic hierarchy process

From now on we will use the shorthand *zone probability* to stand for *the probability that, given a failure, it occurs in a given zone*. Experts prior estimates of

Table 1. Failure rate distinguished by zone and by type of corrosion; absolute frequencies in parentheses.

Failure rate	Natural	Galvanic	By interference
Zone A (12 km^2)	0.583 (7)	0.083 (1)	0.500 (6)
Zone B (88 km^2)	0.068 (6)	0.057 (5)	0.091 (8)

the zone probability have been achieved by means of the pairwise comparison technique of the Analytic Hierarchy Process (Saaty 1980, 1994, Cagno et al. 2000). The team of experts was formed by two technicians and two engineers: the technicians assess the conditions of the pipes after an excavation and decide what action to take; the engineers have an overall knowledge of the technical and information management problems that arise inside the company for each expert. Two types of assessments were requested: the probability of a failure in the two zones and the probability of a type of corrosion conditioned to the zone. All questions were phrased in terms of relative judgments about probabilities related to the two zones. For example, the first one was: «In your opinion, given a failure, is it more likely to happen in zone A or in zone B? How much more likely?».

Following AHP methodology (Saaty 1980), a linguistic judgment scale was used by experts to answer. The degrees of the linguistic judgment scale («equally likely», «moderately more likely», «significantly more likely», «strongly more likely» and «absolutely more likely») were referred to a numerical scale $\{1;3;5;7;9\}$ (eventually using intermediate values like 2, 4, 6, 8) and, using the eingenvalue method, the estimate of the $P(A)$ and $P(B)$ for each expert were obtained. The four expert estimates were combined calculating their mean (Saaty 1980) to obtain the subjective probabilities, $P(A)$ and $P(B)$ (also the standard deviation was calculated; see Table 2).

The experts answered considering that the areas near streetcar substations or rail stations are under close control. In spite of this, the subjective probability of failure in zone A was assessed about 4 times greater than that in zone B.

The second part of the questionnaire regarded the comparison among the three types of corrosion by zone. The typical question was: «In an area with (without) streetcar substations or rail stations is it more

likely to have natural or galvanic corrosion? How much more likely?».

From the questionnaire $P(N|A)$, $P(G|A)$, $P(I|A)$, $P(N|B)$, $P(G|B)$, and $P(I|B)$ were obtained for each expert (following the same above cited steps of the first part), while each expert's zone probabilities conditioned to the type of corrosion have been derived via Bayes formula:

$$P(c) = P(c|A)P(A) + P(c|B)P(B)$$

$$P(A|c) = \frac{P(c|A)P(A)}{P(c)}$$

where $c \in \{I, G, N\}$. To avoid confusion we remark that $P(A|c)$ is the probability that, given a failure which was caused by corrosion type c, it occurred in zone A, whereas $P(c|A)$ is the probability that, given a failure which occurred in zone A, it was caused by corrosion type c. The means and standard deviations of the experts' evaluations are shown in Table 2.

3.2 Posterior probability calculation

3.2.1 Posterior zone probability
Let us denote by p the probability that, given a failure, it occurred in zone A, so $P(A) = p$, and $P(B) = 1-p$. Denoting by n_A the number of failures in zone A, the observed data can be modelled by a binomial distribution with n trials and parameter p:

$$P(n_A \mid n, p) = \binom{n}{n_A} p^{n_A} (1-p)^{n-n_A}$$

where $n = 33$, which is the total observed number of failures. The observed value of n_A is 14. The Beta distribution is conjugate to the binomial, so we take $p \sim Beta(a,b)$ a priori, where a and b are chosen so that the mean and the variance of the Beta distribution match those of $P(A)$ in Table 2. We find $a = 2.582$ and $b = 0.671$. Then the posterior density of p is

Beta (a',b')
$a' = a + n_A = 16.582$
$b' = b + n - n_A = 19.671$

The maximum likelihood, prior, and posterior estimates of p are compared in Table 3.

The strong influence of the streetcar substations and rail stations on the probability of failure of the steel pipelines is unchanged a posteriori, which shows that little weight was attached to the prior, whose mean is considerably different from the MLE. According to the posterior mean, around 46% of the failures is concentrated on 12% of the city area.

Table 2. Subjective estimates of the probability that, given a failure, it occurred in zone A or B and of the probability that it was caused by corrosion type N, I or G.

	Mean	Standard deviation
$P(A)$	0.794	0.196
$P(B)$	0.206	0.196
$P(A|N)$	0.613	0.211
$P(A|G)$	0.622	0.217
$P(A|I)$	0.958	0.057
$P(B|N)$	0.387	0.211
$P(B|G)$	0.378	0.217
$P(B|I)$	0.042	0.057
$P(N)$	0.163	0.040
$P(G)$	0.277	0.130
$P(I)$	0.560	0.157

Table 3. Comparison of MLE, prior and posterior estimates of the zone probability. Standard error (for the MLE) and standard deviations in parentheses.

	Zone description	MLE	Prior mean	Posterior mean
p	Zone A (12 km²)	0.424 (0.086)	0.794 (0.196)	0.457 (0.082)
1 − p	Zone B (88 km²)	0.576 (0.086)	0.206 (0.196)	0.543 (0.082)

Table 4. MLE, prior and posterior estimates of the probability of types of corrosion. Standard error (for the MLE) and standard deviations in parentheses.

	MLE	Prior mean	Posterior mean
P(I)	0.424 (0.086)	0.560 (0.156)	0.493 (0.060)
P(G)	0.182 (0.067)	0.277 (0.130)	0.2302 (0.051)
P(N)	0.394 (0.085)	0.163 (0.040)	0.276 (0.054)

3.2.2 Marginal posterior probability of the type of corrosion

The progress of the electrolytic phenomenon depends strictly on the type of corrosion, thus it is necessary to study which type of corrosion is more frequent. Given a failure, a corrosion of a given type occurs according to the probabilities $p_I \equiv P(I)$, $p_G \equiv P(G)$, and $p_N \equiv P(N)$. The vector (n_I, n_G, n_N) of the number of failures by type of corrosion has a multi-nomial distribution with n trials and these probabilities:

$$P(n_I, n_G \mid n, p_I, p_G) = \frac{n!}{n_I! n_G! (n - n_I - n_G)!} \cdot p_I^{n_I} p_G^{n_G} (1 - p_I - p_G)^{n - n_I - n_G}.$$

The Bayesian procedure of the previous section is generalized by assigning a Dirichlet prior distribution to (p_I, p_G):

$$P(p_I, p_G; a_I, a_G, a_N) = \frac{\Gamma(a_{tot})}{\Gamma(a_I)\Gamma(a_G)\Gamma(a_N)} \cdot p_I^{a_I - 1} p_G^{a_G - 1} (1 - p_I - p_G)^{a_N - 1},$$

where $a_{tot} = a_I + a_G + a_N$. As c varies in $\{I, G, N\}$, these hyperparameters are again determined by matching the prior mean and variances of the three probabilities of corrosion type in Table 2 to those given by the Dirichlet prior:

$$E(p_c) = \frac{a_c}{a_{tot}}$$

$$Var(p_c) = \frac{a_c(a_{tot} - a_c)}{a_{tot}^2(a_{tot} + 1)}.$$

Solving for (a_{tot}, a_I, a_G, a_N), as c varies in $\{I, G, N\}$, we obtain three dependent equations for (a_I, a_G, a_N) and three equations for a_{tot},

$$a_c = a_{tot} E(p_c)$$

$$a_{tot} = \frac{E(p_c)(1 - E(p_c))}{Var(p_c)} - 1$$

so that three sets of solutions are possible, where all the prior means of the experts are matched, but only one variance is. As we choose which variance to match, the marginal prior variances given by the model share the same order of magnitude, increasing as a_{tot} decreases. We adopt the compromise solution of averaging the three values of a_{tot}, getting $a_{tot} = 34.398$, $a_I = 19.253$, $a_G = 9.518$, $a_N = 5.627$. Therefore we do not give too much or too little weight to the experts' opinions. The posterior distribution of (p_I, p_G) is again Dirichlet with parameters $a_c' = a_c + n_c$, $c \in \{I, G, N\}$.

The posterior mean of p_c is then $(a_c + n_c)/(a_{tot} + n)$, so that the a_c's play the same role of the actual number of observations for each corrosion type, and a_{tot} is contrasted to n. This is reminiscent of the "number of virtual observations" of van Noortwijk et al. (1992), although there this terminology is appropriate with reference to the "likelihood function of the experts". This is obtained after asking each expert to guess how many observations out of 100 will fall in each one of a predetermined number of classes.

The MLE, prior and posterior estimates are reported in Table 4.

Here the posterior means differ more markedly from the MLEs than in Table 3, especially in the balance between natural and galvanic corrosion.

3.2.3 Posterior conditional probabilities

When a new pipeline is to be installed in zone A or B it is useful to know the probability of each type of corrosion given the zone. We could proceed as in the previous section using only the failure in zone A or those in zone B. However, for consistency with the previous hyperparameters assignment, it is preferable to calculate the posterior distributions of $P(c|A)$ and $P(c|B)$, $c \in \{I, G, N\}$, using Bayes theorem:

$$P(c|A) = \frac{P(A|c)P(c)}{P(A)}.$$

We have found the posterior distribution of (P(I), P(G), P(N)) in section 3.2.2. The posterior distribution of P(A|c), as c varies, can be found in exactly the same way as section 3.2.1. Then, with a straight-forward

350

Table 5. Monte Carlo estimates of the probabilities of corrosion type conditioned to zone A.

	MLE	Prior mean	Posterior mean
P(I|A)	0.429	0.668	0.583
	(0.132)	(0.085)	(0.084)
P(G|A)	0.071	0.204	0.142
	(0.069)	(0.080)	(0.064)
P(N|A)	0.500	0.128	0.275
	(0.134)	(0.016)	(0.074)

Table 6. Monte Carlo estimates of the probabilities of corrosion type conditioned to zone B.

	MLE	Prior mean	Posterior mean
P(I|B)	0.421	0.094	0.377
	(0.113)	(0.024)	(0.095)
P(G|B)	0.263	0.546	0.342
	(0.101)	(0.110)	(0.093)
P(N|B)	0.316	0.360	0.281
	(0.107)	(0.118)	(0.085)

Monte Carlo simulation, we can calculate the posterior mean and standard deviation of $P(c|A)$. To insure that

$$P(I|A) + P(G|A) + P(N|A) = 1$$

during the simulation, we only simulate the numerator of the above Bayes formula and calculate the denominator from it *at each iteration*. We proceed similarly for zone B.

The results of our exercise for zone A are displayed in Table 5. Table 6 displays the results for zone B.

The posterior means in Table 5 suggest that cathodically protected pipes should be installed in zone A unless it is possible to ascertain that there are no stray currents. As regards zone B, the three types of corrosion are almost equally likely, so that more information would be needed about the lay location before a decision can be made.

4 CONCLUSIONS

A methodology for the assessment of the probability of gas-escape from steel pipelines has been presented to support the implementation of the maintenance and renewal policy of a urban gas distribution network. We identified the factors that influence reliability and focused on the type of corrosion and the lay location for design purposes. The developed methodology answered the requirement to combine, according to the Bayesian paradigm, scarce information derived from historical data with the experts' knowledge grown inside the company during the last running periods.

At a design stage, a simple Dirichlet-Multinomial model proved sufficient, showing that streetcar substations greatly increase the probability of a fast corrosion; therefore, considering that it is not easy to tell whether a corrosion by interference can take place in a given location, newly installed pipelines should be cathodically protected in most cases, unless some additional information is available.

The real case study presented emphasized the relevance of the practical solution, to be reused in other industrial cases.

REFERENCES

Bernardo, J.M. & Smith, A.F.M. 1994. *Bayesian Theory.* John Wiley & Sons, Chichester.

Cagno, E., Caron, F., Mancini, M., Ruggeri, F. 2000. Using AHP in Determining the Prior Distributions on Gas Pipeline Failure in a Robust Bayesian Approach, *Reliability Engineering & System Safety*, 67(3).

Saaty, T.L. 1980. *The Analytic Hierarchy Process*. McGraw-Hill, New York.

Saaty, T.L. 1994. *Fundamentals of Decision Making and Priority Theory with the Analytic Hierarchy Process*. RWS Publications, Pittsburgh, PA, USA.

Van Noortwijk, J.M., Dekker, R., Cooke, R.M. & Mazzuchi, T.A. 1992. Expert Judgment in Maintenance Optimization. *IEEE Transactions on Reliability, 41(3), 427–432.*

Safety and Reliability – Bedford & van Gelder (eds)
© 2003 Swets & Zeitlinger, Lisse, ISBN 90 5809 551 7

On the estimation of return values of significant wave height data from the reanalysis of the European centre for medium-range weather forecasts

Sofia Caires & Andreas Sterl
KNMI, De Bilt, Netherlands

ABSTRACT: The Peaks-Over-Threshold method is used to obtain return value estimates from buoy and reanalysis (ERA-40) significant wave height data. The large amount of data used in this study provides evidence for the significant wave height data having an exponential tail, which simplifies the estimation process. Comparisons between buoy and ERA-40 return value estimates show that the ERA-40 data can be reliably used to obtain return value estimates once a simple linear correction is applied. Further, we assess the effect of the significant wave height's space and time variability on the prediction of the extreme values. This is done by performing detailed global extreme value analyses using different subperiods of the 45-year long ERA-40 dataset.

1 INTRODUCTION

Currently, the European Centre for Medium-Range Weather Forecasts (ECMWF) is conducting ERA-40, a reanalysis of global atmospheric wind, temperature and humidity fields, stratospheric ozone, deep water sea states and soil conditions from 1957 to 2002. The reanalysis uses ECMWF's Integrated Forecasting System, a coupled atmosphere-wave model with variational data assimilation. This is the first reanalysis in which a wave model is coupled with the system used. Moreover, it will produce the most complete wave data set on a 1.5° by 1.5° latitude/longitude grid covering the whole globe. This data set can be used to study the climatology of ocean waves and to compute extreme wave statistics over the whole globe. Initial validations of the data reveal a generally good description of variability and trends (Caires et al. 2002), but some underestimation of high wave heights (Caires and Sterl, in press). The objective of this article is to use the ERA-40 data to compute global estimates of return values of significant wave height (H_s), paying particular attention to the effect the underestimation of the high H_s values has in the estimation.

Following Ferreira and Guedes Soares (2000) and Anderson et al. (2001), we compute the return value estimates using the peaks-over-threshold (POT) method rather than the widespread approach of fitting a distribution (such as the lognormal, Weibull, beta, etc.) to the whole dataset and extrapolating from it. Among the

arguments given by these authors against the latter method we mention the following:

- Due to dependence and non-stationarity, H_s series violate the independence and identity in distribution assumptions, which invalidates the application of the common statistical methods used (confidence intervals and tests) as well as the definition of return value;
- There is no scientific justification for using one particular distribution to fit to H_s data, and the usual goodness-of-fit diagnostics are not able (on the basis of realistic sample sizes and given the length of the required "prediction horizon") to distinguish data with type I (exponential) tail, say, from data with type II (heavier than exponential) tail. In contrast, if for example one concentrates on averages, maximum values, or excesses over a high threshold of very general variables, then statistical theory provides a scientific basis for the use of, respectively, the normal, extreme value and generalized Pareto distributions.

The POT approach consists of fitting the generalized Pareto distribution to the peaks of clustered excesses over a threshold, the excesses being the observations in a cluster minus the threshold, and calculating return values by taking into account the rate of occurrence of clusters. Under very general conditions this procedure ensures that the data can have only three possible, albeit approximate, distributions (the 3 forms of the generalized Pareto distribution), and moreover that observations belonging to different peak clusters are (approximately) independent.

Ferreira and Guedes Soares (1998) described an application of the POT method to predict return values of H_s The method was applied to 10 years of H_s data from one location off the Portuguese coast, and it was concluded that the exponential distribution (the generalized Pareto with shape parameter $\kappa = 0$) fitted the data well. Moreover, the closely exponential character of the data was explained theoretically: the H_s estimate of each sea state is a realization of a random variable whose approximate distribution is a mixture of Rayleigh distributions with an unspecified number of parameters, which has a type I or exponential tail. The exponential character of both ERA-40 and buoy H_s observations will be extensively assessed in this article using data from several locations. As we shall see, the establishment of the exponential distribution as an appropriate model for the peak excesses not only points to a general pattern but it also simplifies the problem of predicting extreme values, since one less parameter needs to be estimated.

The ERA-40 data available at the time of performing this study does not yet cover the whole 45-year period under study, but with the data available it is possible to define 3 decades: 1958–1967, 1972–1981 and 1986–1995. We will analyze the return values obtained from the data considering each decade separately and analyze the time and space variability of these estimates. We will not look at the effect that within-year variability and within-decade trends may have on the return value estimates. For a study of these effects the reader is referred to Anderson et al. (2001).

2 PEAKS-OVER-THRESHOLD METHOD

In the POT method, the peak excesses over a high threshold u of a timeseries are assumed to occur according to a Poisson process with rate λ_u and to be independently distributed with a Generalized Pareto Distribution (GPD), which is given by

$$
\begin{aligned}
F_u(x) &= 1 - (1 - \kappa x/\alpha)^{1/\kappa} & \kappa \neq 0 \\
&= 1 - \exp(-x/\alpha) & \kappa = 0
\end{aligned}
$$

where the range of x is $(0, \infty)$ if $\kappa \leq 0$ and $(0, \alpha/\kappa)$ for $\kappa > 0$. For $\kappa = 0$ the GPD is the exponential distribution with mean α for $\kappa < 0$ it is the Pareto distribution, and for $\kappa > 0$ it is a special case of the beta distribution. When $\kappa < 0$ the tail of the GPD, i.e., the function $x \to 1 - F_u(x)$ is heavier (i.e., decreases more slowly) than the tail of the exponential distribution, and when $\kappa > 0$ it is lighter (decreases more quickly and actually reaches 0) than that of the exponential. The GPD is said to have a type II tail for $\kappa < 0$, and a

type III tail for $\kappa > 0$. The tail of the exponential distribution is called a type I tail.

As said, the excesses over the threshold u of a timeseries $X_1, X_2, \ldots$ are the observations (called exceedances) $X_i - u$ such that $X_i > u$. A *peak excess* is defined as the largest excess in a cluster of exceedances, hence its definition depends on that of a cluster. In this paper we shall adopt the usual definition of a cluster as a group of consecutive exceedances, which implies that in the above mentioned Poisson process there is a one-to-one correspondence between clusters and peak excesses.

The model now outlined can be justified theoretically for a great variety of timeseries; see for example Leadbetter (1991) and the references therein.

One of the main applications of the POT method is the estimation of the m-year return value, which for a given threshold u is given by

$$
\begin{aligned}
x_m^{(u)} &= u + \alpha/\kappa(1 - (\lambda_u m)^{-\kappa}) & \kappa \neq 0 \\
&= \alpha \log(\lambda_u m) & \kappa = 0.
\end{aligned}
$$

If the number of clusters *per year* is a Poisson random variable with mean λ_u then the expected number of clusters/peak excesses in m years is $m\lambda_u$, and if the peak excesses over u are independently distributed with distribution function F_u then the expected number of observations exceeding is x is $m\lambda_u(1 - F_u(x))$. Setting this equal to 1 and solving for x gives $x_m^{(u)}$.

2.1 Estimation and testing

Once a threshold (u) has been selected and the peak excesses have been extracted from the timeseries, the scale (α) and shape (κ) parameters of the GPD can be estimated, for instance by the maximum likelihood method (see e.g. Hosking and Wallis (1987) or Davison and Smith (1990)).

The parameter λ_u, the yearly cluster rate, can be estimated by the average number of clusters/peak excesses per year. More generally, for yearly series with different numbers of observations, λ_u can be estimated by

$$
\hat{\lambda}_u = k^{-1} \sum_{i=1}^{k} N_i/p_i,
$$

where κ is the number of years considered, $p_i = n_i/n$, where n_i is the number of observations available in the i-th year, N_i is the corresponding number of peak excesses, and n is the maximum number of observations in a yearly series. Under the Poisson assumption, $E(\hat{\lambda}_u) = \lambda_u$ and

$$
\text{var}(\hat{\lambda}_u) = \lambda_u k^{-2} \sum_{i=1}^{k} p_i^{-1},
$$

354

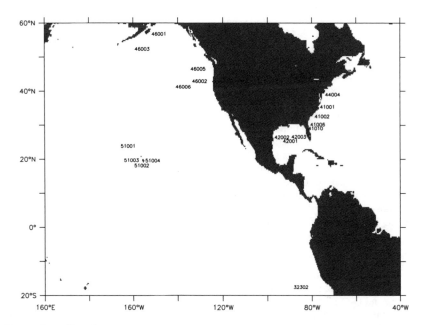

Figure 1. Buoy codes and locations.

and, since λ_u is relatively large, we have that $\hat{\lambda}_u$ is approximately normal with mean λ_u and variance $\text{var}(\hat{\lambda}_u)$.

Confidence intervals for the return value estimates can be estimated by using the delta method (see Ferguson (1996)); more precisely, the asymptotic variance of the estimated return values can be estimated as

$$\widehat{\text{var}(x_m^{(u)})} = d^T \Sigma d,$$

where d is the vector of derivatives of $x_m^{(u)}$ with respect to the estimated parameters (λ, κ and α if $\kappa \neq 0$, and λ and α if $\kappa = 0$) and Σ the asymptotic covariance matrix of the parameter estimates (in this case the inverse of the Fisher information matrix), both evaluated at the estimates of the parameters.

In order to test the exponentiality of the data we use two tests. The Gomes-van Monfort (Gomes and van Monfort 1986) test, which is intended for testing the exponential versus the GPD, and the Anderson-Darling statistic (see e.g., Stephens (1974)), which can be used for testing the exponential versus any other distribution. We use a 5% significance level. Under the null hypothesis of exponentiality, the Gomes-van Monfort statistic is asymptotically a standard normal variable, so the null hypothesis will be rejected whenever the modulus of the statistic exceeds 1.96. The 5% critical value of the Anderson-Darling statistics is 1.341.

3 VALIDATION

The buoy data to be used in this study come from the NOAA database (http://seaboard.ndbc.noaa.gov/). From all locations for which NOAA buoy data are available, we have selected a total of 18. The selection took into account the distance from the coast and the water depth. Only deep water locations should be taken into account since no shallow water effects are accounted for in the wave model, and the buoy should not be too close to the coast in order for the corresponding grid point to be located at sea. The locations of the buoys are shown in Figure 1. In order to compare the ERA-40 data with the buoy observations, time and space scales must be brought as close to each other as possible. The ERA-40 results are available at synoptic times (every 6 hours) and each value is an estimate of the average condition in a grid cell; on the other hand, the buoy measurements are local. They are available hourly from 20-minute long records. Time and space scales of the data are made compatible by defining a buoy observation as a 3-hour average. The reanalysis data at the 4 points surrounding a given buoy location, at the synoptic time around which the buoy measurements were averaged, is interpolated bilinearly to the buoy location.

In the following analysis, a "year" is defined as the period from October to September. This definition is preferable for the Northern Hemisphere data to that

355

of a calendar year; in this way the Winter period is not broken into two.[1]

We have started our analysis by trying to find the "good" threshold for the buoy and the ERA-40 time-series at different buoy locations. The criteria was the smallest value giving a reasonable fit of the peak exceedances by the GPD. The threshold is expected to depend on the location, the period considered and the data set (ERA-40 or buoy). The most important factor is supposed to be the location since locations at high latitudes will be exposed to more severe conditions than those at lower latitudes. From assessments of the ERA-40 data against buoy data (see Caires and Sterl, in press), the thresholds for the ERA-40 data are expected to be lower than those of the buoy data, since ERA-40 underestimates the high values of H_s. We have considered timeseries of 3 years and more (the longer one going from 1978 to 2000, the whole period for which the NOAA buoy data are available) and for different periods. We have tried to fix the threshold by looking at the fit of the GPD to the peak excesses using kernel density estimators and Q-Q plots at different thresholds, and at the stability of the estimates with relation to the threshold (see Ferreira and Guedes Soares (1998) and Anderson et al. (2001)). So far we have not been able to devise an automatic procedure to fix the threshold, but found that in most of the cases a threshold fixed at the 93% quantile of the whole data set (before computing the peaks excesses) gives a good adjustment. There were cases, however, where the threshold had to be set higher, up to the value of the 97% quantile. As noted in Ferreira and Guedes Soares (1998), the κ estimate of the fitted GPD often stabilized in the neighborhood of zero.

Fixing the threshold at the 93% quantile of the data we have applied the POT method to buoy data and to the corresponding ERA-40 data for the periods 1986–1990, 1994–1999 and 1986–1999. We have only considered buoy time series with small gaps so that the sample sizes of the corresponding buoy and ERA-40 timeseries are compatible.

Table 1 presents some results for the period 1986–1990. The 1st column gives the code of the data location (see Fig. 1), the 2nd column gives the number of points in each timeseries, the 3rd column the threshold used, the 4th column the λ_u estimates, the 5th and 6th columns the observed values of the Gomes-van Monfort and the Anderson-Darling statistics, respectively, the 7th column gives the α estimates of the exponential distribution, and the last 2 columns give,

respectively, the 20-year and 100-year return value estimates along with 95% confidence intervals calculated from the exponential distribution.

Looking at the test statistics, we see that the exponentiality of the data is rejected only in one case, for the ERA-40 data at the location 51001; the corresponding values of the statistics are underlined in Table 1. Since the tests are being done at a 5% significance level, one rejection in 30 tests is just about the expected proportion of rejections. However, there were also rejections with both buoy and ERA-40 data at other locations in the 2 other periods considered (results not shown), a total of 13 in 88 cases (15%), and these are relevant. We have analyzed each of the cases where the hypothesis was rejected and in all of these the hypothesis was not rejected once the threshold was increased. As to the rejection in Table 1, a threshold set at the 95% quantile no longer gives rejection. From the 13 rejected cases, all of them were rejected by the Anderson-Darling statistic and only 7 by the Gomes-van Monfort statistic, which indicates that when there is lack of fit the lack of fit also applies to the GPD. In view of this, and since the Anderson-Darling statistic is more powerful against deviations other than those in the direction of the GPD, we have opted for computing the values of the Anderson-Darling statistic for the POT analysis of the three 10-year periods of ERA-40 data. Figure 2 shows the values of Anderson-Darling statistic for the period 1958–1967. The regions where exponentiality is rejected at a 5% level are shaded. There are 15% of rejections; if the threshold is fixed at the 97% quantile of the data (at each point), the percentage of rejections drops to only 9%. Similar rejection percentages are obtained in the 2 other 10-year periods considered. Based on these results, we conclude that the exponential distribution is adequate for modeling the peak excess of both buoy and ERA-40 H_s data, and consequently will estimate return values using the fitted exponential distributions.

Comparing the buoy with the respective ERA-40 estimates we see that the ERA-40 threshold and the λ_u and α estimates are lower than those of the buoy data. Consequently, the ERA-40 return value estimates are lower. For the different periods considered in the buoy-ERA-40 comparisons, the λ_u and α estimates vary from location to location. At a given location, the confidence intervals of the λ_u and α for the different periods usually overlap and contain the estimates from the other periods. This says that the space variability of the extreme values is larger than the time variability; the values at a given point cannot be reliably estimated from data of a location further away. In other words, having local information is more important than the length and period over which the information was collected.

Figure 3 shows the ERA-40 versus the buoy 100-year return value estimates for the three periods

---

[1]The study of Hogg and Swail (2002), which was concentrated just in the Northern Hemisphere, defined a "wave-year" from July to June; this definition would have the same disadvantages in the Southern Hemisphere as the calendar year has in the Northern Hemisphere.

Table 1. Some results of the application of the POT method to buoy (codes ending with b) and ERA-40 (codes ending with e) data from 1986–1990. For an explanation see the text.

Buoy	n_t	u	$\hat{\lambda}_u$	GM	AD	$\hat{\alpha}$	$\hat{x}_{20}$	$\hat{x}_{100}$
32302e	7304	2.76	17	−0.91	1.04	0.25	4.22 (3.91, 4.53)	4.62 (4.22, 5.02)
32302b	6212	3.03	19	0.65	0.44	0.46	5.73 (5.13, 6.34)	6.47 (5.70, 7.23)
51001e	7304	3.20	16	−2.28	1.37	0.50	6.06 (5.41, 6.70)	6.86 (6.04, 7.68)
51001b	6700	3.87	22	−1.18	0.50	0.89	9.31 (8.26, 10.36)	10.74 (9.42, 12.06)
51002e	7304	2.78	15	−0.23	0.30	0.29	4.45 (4.06, 4.83)	4.92 (4.43, 5.40)
51002b	6084	3.47	24	−0.20	0.84	0.47	6.38 (5.75, 7.01)	7.14 (6.35, 7.93)
51003e	7304	2.86	15	0.00	0.33	0.39	5.09 (4.59, 5.60)	5.72 (5.08, 6.37)
51003b	6212	3.30	24	0.18	0.94	0.53	6.58 (5.93, 7.23)	7.44 (6.63, 8.26)
51004e	7304	2.83	14	−1.22	0.81	0.31	4.60 (4.19, 5.02)	5.11 (4.58, 5.64)
51004b	5852	3.40	22	−1.77	0.85	0.50	6.43 (5.75, 7.12)	7.23 (6.38, 8.09)
42001e	7304	1.66	24	−0.30	0.83	0.61	5.45 (4.76, 6.14)	6.44 (5.57, 7.31)
42001b	6944	2.20	29	0.13	0.31	0.72	6.76 (5.96, 7.56)	7.92 (6.91, 8.92)
42002e	7304	1.71	25	0.17	0.36	0.62	5.57 (4.88, 6.26)	6.57 (5.71, 7.44)
42002b	7180	2.33	32	−0.44	0.44	0.78	7.35 (6.55, 8.15)	8.60 (7.60, 9.60)
42003e	7304	1.59	20	0.64	0.25	0.65	5.48 (4.71, 6.25)	6.57 (5.55, 7.49)
42003b	6092	2.02	25	−1.35	0.93	0.67	6.20 (5.39, 7.02)	7.28 (6.26, 8.30)
41001e	7304	2.91	27	0.97	0.41	0.88	8.47 (7.53, 9.42)	9.89 (8.71, 11.08)
41001b	6564	3.93	33	0.81	0.66	1.04	10.69 (9.53, 11.86)	12.37 (10.92, 13.82)
41002e	7304	2.49	24	0.24	0.96	0.83	7.61 (6.69, 8.53)	8.94 (7.79, 10.10)
41002b	6460	3.46	28	0.12	0.82	1.07	10.26 (9.00, 11.51)	11.98 (10.42, 13.55)
41006e	7304	2.14	21	0.20	0.44	0.69	6.31 (5.51, 7.11)	7.41 (6.41, 8.42)
41006b	5348	2.90	25	−0.48	0.47	0.86	8.26 (7.06, 9.47)	9.65 (8.14, 11.16)
46003e	7304	4.85	26	0.14	0.43	1.04	11.38 (10.25, 12.50)	13.05 (11.64, 14.46)
46003b	6704	5.70	35	0.08	0.44	1.35	14.56 (13.14, 15.99)	16.73 (14.97, 18.50)
46002e	7304	3.97	20	−0.29	0.55	0.94	9.59 (8.46, 10.72)	11.11 (9.68, 12.53)
46002b	5480	4.74	24	−1.87	1.04	1.21	12.21 (10.55, 13.87)	14.16 (12.07, 16.24)
46005e	7304	4.38	24	−0.31	0.48	0.99	10.46 (9.36, 11.56)	12.04 (10.66, 13.43)
46005b	6452	5.07	27	−0.20	0.29	1.34	13.54 (11.94, 15.13)	15.70 (13.70, 17.69)
46006e	7304	4.25	21	0.95	0.30	0.95	10.00 (8.89, 11.11)	11.53 (10.13, 12.94)
46006b	5848	5.00	28	−0.02	0.62	1.22	12.74 (11.27, 14.21)	14.70 (12.87, 16.54)

357

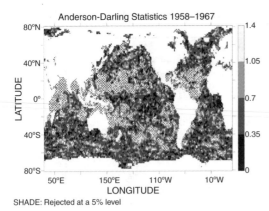

Figure 2. Spatial distribution of the values of the Anderson-Darling statistic; results of the application of the POT method to data from 1958 to 1967 using the 93% sample quantile as a threshold. In the shaded regions exponentiality is rejected at a 5% level.

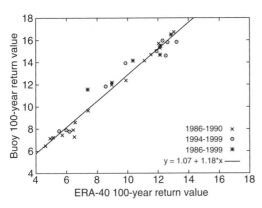

Figure 3. Illustration of the linear relationship between the 100-year return periods estimated from ERA-40 and buoy data.

considered. Clearly, the ERA-40 underestimation of the 100-year return value estimates can be reliably accounted for using a linear relationship. Using linear regression the following relation between buoy and ERA-40 data 100-year return values is found,

$$X_{100}^{buoy} = 1.07 + 1.18X_{100}^{ERA-40} \qquad (1)$$

Equation 1 is plotted in Figure 3.

We have performed similar comparisons between the 20- and 50-year return values. As in Figure 3, the scatter between the ERA-40 and the buoy estimates is small. The linear relationships are $X_{20}^{buoy} = 0.82 + 1.19\ X_{20}^{ERA-40}$ and $X_{50}^{buoy} = 1.00 + 1.18X_{50}^{ERA-40}$ for the 20- and 50-year return values, respectively.

4   ERA-40 ESTIMATES

Figure 4 presents global surface plots of the 100-year return values computed using the different 10-year periods and corrected by eq. (1). The storm tracks of the Southern and Northern Hemispheres can easily be identified; the highest return value estimates from all the decadal datasets occur in those regions. All datasets give low return value estimates for the tropics.

There are no significant differences between the return values estimated from the three different decades in the Tropics and Southern Hemisphere. On the other hand, significant differences do exist in the North Atlantic and Pacific storm tracks. These mirror the decadal variability in the Northern Hemisphere.

Several earlier studies have found high correlations between the North Atlantic wave pattern and the North Atlantic Oscillation (NAO) index[2]; see e.g., Lozano and Swail (2002). More precisely, the North Atlantic storm track varies according to the NAO index. During periods when the NAO index is positive the storms tend to move from North America in the direction of the Norwegian Sea. On the other hand, when the NAO index is negative the storms move in the direction of the Mediterranean Sea (see Rogers (1997, Fig. 3)), the wave conditions being milder (see Wang and Swail (2001)). From the beginning of the 1940's to the beginning of the 1970's the NAO index exhibited a downward trend, the index being negative from 1958 to 1967. From the beginning of the 1970's the trend was positive, the period between 1972 and 1981 being characterized by both positive and negative NAO index years. From 1986 to 1995 the index was always positive. The change in the pattern and intensity of the 100-year return values in the North Atlantic basin is completely in line with these decadal changes of the NAO index. The estimates from the period 1958–1967 are lower and more to the south of those of the later periods. The pattern of the estimates for the period 1972–1982 is characterized by 2 high lobes, one for the positive NAO index years and another for the negative NAO index years. The highest estimates, which are also those with the most northerly peak, are from the period 1986–1995, the period during which the NAO index was at its highest.

The plots of Figure 4 show a clear and strong increasing trend in the estimates of the 100-year return values from the 3 decadal periods in the North Pacific storm track. This is in line with the results of Graham and Diaz (2001). There is some discussion about the reasons for this increase. Graham and Diaz (2001) suggest the increasing sea surface temperatures in the western tropical Pacific as a plausible cause.

---

[2]A measure of the difference of the sea surface pressure between Reykjavik (Iceland) and Ponta Delgada (Portugal).

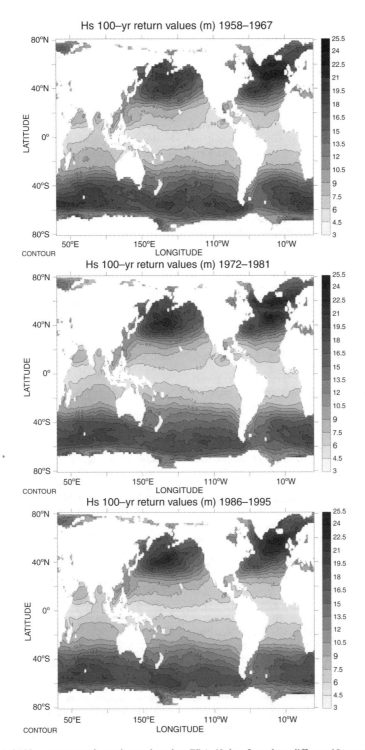

Figure 4. Corrected 100-year return value estimates based on ERA-40 data from three different 10-year periods. Top panel: 1958 to 1967. Middle panel: 1972 to 1981. Bottom panel: 1986 to 1995.

359

5 DISCUSSION

Based on the application of the POT method to buoy and ERA-40 data we have concluded that the tail of the significant wave height data is exponential. Using this knowledge, we have then obtained global 100-year return value estimates from the ERA-40 data. Since the ERA-40 data underestimates the high H_s peaks, it is necessary to apply a linear correction to the estimated return values. Corrected global estimates based on three different 10-year periods of ERA-40 data were given, and it was shown that estimates obtained from the different periods differ in the Northern Hemisphere storm tracks. These differences can be attributed to the decadal variability in the Northern Hemisphere, and we have linked them to changes in the global circulation patterns. In the Tropics and the Southern Hemisphere the estimates based on the different decadal datasets are compatible.

The establishment of the exponential distribution as the distribution of the H_s peak excesses can be very useful on a practical level. The reduction of one parameter in the GPD model is quite convenient, especially when considering small datasets, as is the case for altimeter data. On the other hand, the fact that most data sets lead to the exponential distribution can be used as a quality control check to data.

The fact that the distribution of peak excesses is well approximated by an exponential distribution suggests that the Gumbel distribution should also be a good model for the distribution of maxima over fixed periods. But to verify whether this holds with these data does not seem straightforward because the annual maxima method is comparatively wasteful of information. Hogg and Swail (2002) and Orelup et al. (2002) fitted the Gumbel distribution to H_s annual maxima and to samples obtained from a POT technique, respectively, in order to estimate H_s 100-year return values. We have compared the 100-year return values presented here with the ones plotted in Orelup et al. (2002) for the North Atlantic. Although there are some similarities in the patterns, the corrected ERA-40 estimates presented here are higher than those of Orelup et al. (2002). Since the estimates in Orelup et al. (2002) were obtained from model data, probably a calibration similar to the one applied to the ERA-40 return value estimates needs to be applied. Hogg and Swail (2002) noted that, at one buoy location, the 100-year return value estimate obtained from the dataset used by Orelup et al. (2002) underestimates the corresponding buoy estimate. The estimates are about 15 and 20 metres for the model and the buoy data, respectively. This suggests a correction of similar magnitude as the one obtained here.

There are some caveats on the return value estimates we present here:

- The choice of the threshold as the 93% percentile of the data has to be seen as preliminary. Also, we have not accounted for the influence that the choice of threshold may have on return value estimation (although some techniques are available for that). One of our next steps will be looking at ways of fixing the threshold automatically.
- The linear relation found between the ERA-40 and the buoy return value estimates is based on a small number of points, and no attention was paid to the influence that the location may have on the relationship. Further work also needs to be done in this front.
- Due to resolution, tropical cyclones are not resolved by the ERA-40 system. Therefore, the estimated return values for the regions of the tropical storms may be underestimated.

ACKNOWLEDGMENTS

We would like to thank the American National Oceanographic Data Center for the available buoy data, and the ECMWF ERA-40 team for the ERA-40 data and their prompt reaction to all our queries. We are indebted to José Ferreira for fruitful discussions and comments, and to Camiel Severijns for the Field library and technical support. This work was funded by the EU-funded ERA-40 Project (no. EVK2-CT-1999-00027).

REFERENCES

Anderson, C. W., D. J. T. Carter, and P. D. Cotton (2001). Wave climate variability and impact on offshore design extremes. Report for Shell International and the Organization of Oil & Gas Producers. 99 pp.

Caires, S. and A. Sterl. Validation of ocean wind and wave data using triple collocation. *J. Geophys. Res., in press.*

Caires, S., A. Sterl., J.R. Bidlot, N. Graham, and V. Swail (2002). Climatological assessment of reanalysis ocean data. In *7th Int. Workshop on Wave Hindcasting and Forecasting*, 21–26 October, Banff, Canada, pp. 1–12.

Davison, A. C. and R. L. Smith (1990). Models for exceedances over high thresholds. *J. R. Statist. Soc. B 52*(3), 393–442.

Ferguson, T. S. (1996). *A course in large sample theory.* Chapman & Hall.

Ferreira, J. A. and C. Guedes Soares (1998). An application of the peaks over threshold method to predict extremes of significant wave height. *J. Offshore Mechanics and Arctic Engineering 120*, 165–176.

Ferreira, J. A. and C. Guedes Soares (2000). Modelling distributions of significant wave height. *Coastal Engineering 40*, 361–374.

Gomes, M. Y. and M. A. J. van Monfort (1986). Exponentiality versus Generalized Pareto, quick tests. In *3rd Int. Conf. on Statistical Climatology*, 23–27 June, Vienna, Austria.

Graham, N. E. and H. F. Diaz (2001). Evidence of Intensification of North Pacific Winter Cyclones since 1948. *Bull. American Meteorological Society 82*, 1869–1893.

Hogg, W. D. and V. R. Swail (2002). Effects of distributions and fitting techniques on extreme value analysis of modelled wave heights. In *7th Int. Workshop on Wave Hindcasting and Forecasting*, 21–26 October, Banff, Canada, pp. 140–150.

Hosking, J. R. M. and J. R. Wallis (1987). Parameter and quantile estimation for the Generalized Pareto Distribution. *Technometrics 29*(3), 339–349.

Leadbetter, M. R. (1991). On a basis for "peaks over threshold" modeling. *Statistics & Probability Letters 12*, 357–362.

Lozano, I. and V. Swail (2002). The link between wave height variability in the North Atlantic and the storm track activity in the last four decades. *Atmosphere-Ocean 40*(4), 377–388.

Orelup, E. A., A. Niitsoo, and V. J. Cardone (2002). North Atlantic wave climate extremes and their variability. In *7th Int. Workshop on Wave Hindcasting and Forecasting*, 21–26 October, Banff, Canada, pp. 490–492.

Rogers, J. C. (1997). North Atlantic storm track variability and its association to the North Atlantic oscillation and climate variability of Northern Europe. *J. Clim. 10*, 1635–1647.

Stephens, M. A. (1974). EDF statistics for goodness of fit and some comparisons. *J. American Statistical Association 69*(347), 730–737.

Wang, X. L. and V. R. Swail (2001). Changes of Extreme Wave Heights in Northern Hemisphere Oceans and Related Atmospheric Circulation Regimes. *J. Clim. 14*, 2204–2221.

Safety and Reliability – Bedford & van Gelder (eds)
© 2003 Swets & Zeitlinger, Lisse, ISBN 90 5809 551 7

Graphical models for the evaluation of multisite temperature forecasts: comparison of vines and independence graphs

U. Callies
GKSS Research Centre, Geesthacht, Germany

D. Kurowicka & R.M. Cooke
Delft University of Technology, Delft, Netherlands

ABSTRACT: Vine and independence graphs are employed to extract conditional independence relations from multivariate meteorological data so as to construct a simple graphical model which adequately represents the interrelationships between observations and corresponding model results at different sites. The independence graph approach identifies partial correlations of maximal order. Statistically negligible partial correlations are set to zero. Iterative proportional fitting is used to find a maximum likelihood distribution satisfying the stipulated zero partial correlations. The deviance between the fitted distribution and the original distribution measures the goodness of fit. The vine approach constructs a regular vine in which negligible partial correlations are set to zero. No proportional fitting is required. Again, deviance is used to measure goodness of fit. The connection between vines and continuous belief nets, where an arc from node i to j is associated with a (conditional) rank correlation between i and j is presented.

1 INTRODUCTION

This paper investigates interaction structures between observed December mean temperatures at four European stations and corresponding 'forecasts' of a regression model based on sea-level pressure (SLP) fields as explanatory variables. With temperatures at different stations being correlated each local forecast can be expected to be informative to a certain extent about observations at other locations as well. The following three questions naturally arise: 1) Is the local forecast really more informative about local conditions than forecasts delivered for other sites are? 2) Knowing the local forecast, can additional information be obtained from consulting other forecasts? 3) If other forecasts are found to contain additional information, how much of this incremental information can be attributed to correlations between observations at different sites?

The statistical notion of conditional independence allows these questions to be formalized. Callies (2000) contrasted the concepts of conditional independence and sufficiency, which both refer to the complete statistical information embodied in the joint distribution of forecasts and subsequent observations. Brooks & Doswell (1996) contrasted the distribution-oriented approach to forecast verification with the more conventional approach of calculating global measures of correspondence between forecasts and observations. The purpose of this study is to explore the use of different techniques of graphical modelling for forecast evaluation. Throughout the paper a multinormal joint distribution of observations and corresponding forecasts is assumed.

2 DATA

Observations of local December mean temperatures at 14 European stations were taken from the World Monthly Station Climatology (WMSC) of the National Center for Atmospheric Research (NCAR). Corresponding diagnostic "forecasts" between 1900 and 1993 were produced by regressing local temperatures on monthly mean regional-scale atmospheric sea-level pressure distributions as represented by $5° \times 5°$ analyses (Trenberth & Paolino 1980) at 60 gridpoints covering the region 40°N to 64°N and 20°W to 25°E. Data are made available through NCAR. The regression scheme was calibrated for 1960–1980. This period was excluded from the analysis below.

Prior to regression, both regional and local data were filtered by standard principal component analysis to reduce the number of degrees of freedom and to avoid overfitting. In both data sets only four degrees of freedom were retained. Accordingly the complete amount of information contained in the forecasts is available after selecting any four of the predictions. The following (8×8) sample correlation matrix, S, embodies all information about the interactions between observations, θ, and the corresponding forecasts, F, at the four stations Geneva (G), Innsbruck (I), Budapest (B) and Copenhagen (K):

$$
S = \begin{pmatrix}
1 & .35 & .50 & .49 & .68 & .38 & .50 & .59 \\
 & 1 & .79 & .69 & .12 & .64 & .62 & .49 \\
 & & 1 & .72 & .18 & .61 & .58 & .43 \\
 & & & 1 & .05 & .46 & .47 & .43 \\
 & & & & 1 & .33 & .51 & .71 \\
 & & & & & 1 & .97 & .77 \\
 & & & & & & 1 & .90 \\
 & & & & & & & 1
\end{pmatrix}
\begin{matrix}
\theta^K \\ \theta^G \\ \theta^I \\ \theta^B \\ F^K \\ F^G \\ F^I \\ F^B
\end{matrix}
\quad (1)
$$

The values for the observation–forecast correlations vary from 0.43 for Budapest to 0.68 for Copenhagen.

3 FITTING INDEPENDENCE GRAPHS

The graphical modelling approach elaborated by Whittaker (1990) identifies genuine variable interactions that are not mediated through any third variable in the data set. For this purpose maximum order pairwise partial correlations given all remaining variables in the data set are analysed. The two independence graphs in Figure 1, for instance, are made up by only 11 edges. In a process of recursive model simplification, which

starts from the saturated graph, 17 edges have been discarded by setting corresponding partial correlations to zero. The two different graphs represent structural ambiguity caused by the very high correlation (0.97) of F^G and F^I. The maximum likelihood correlation matrix $V^{(a)}$ in Eq. (3) corresponds with graph (a) in Figure 1. It arises from a constrained fit for the non-zero partial correlations and minimises the following entropy measure (deviance) of the distance between S and V as function of V (N denotes sample size):

$$
\text{dev}(S, V) = N\left[\text{tr}(SV^{-1} - I) - \log \det(SV^{-1})\right] \quad (2)
$$

The graphical constraint manifests itself by the fact that all elements of $(V^{(a)})^{-1}$ corresponding with missing links in the graph must be zero. It turns out that only elements (given in bold type) of $V^{(a)}$ in the same position differ from the sample correlation matrix S (Whittaker 1990):

$$
V^{(a)} = \begin{pmatrix}
1 & .38 & .41 & .49 & .68 & .38 & .48 & .58 \\
 & 1 & .79 & .56 & .18 & .64 & .62 & .48 \\
 & & 1 & .72 & .12 & .50 & .48 & .37 \\
 & & & 1 & .05 & .35 & .32 & .23 \\
 & & & & 1 & .33 & .51 & .72 \\
 & & & & & 1 & .97 & .77 \\
 & & & & & & 1 & .90 \\
 & & & & & & & 1
\end{pmatrix}
\begin{matrix}
\theta^K \\ \theta^G \\ \theta^I \\ \theta^B \\ F^K \\ F^G \\ F^I \\ F^B
\end{matrix}
\quad (3)
$$

The total deviance of graph (a) in Figure 1 is 41.9 for sample size 63 (years with incomplete data have been discarded). All but three of these incremental deviances are larger than the total deviance of 41.9 due to 17 missing links. Therefore, the truncated interaction

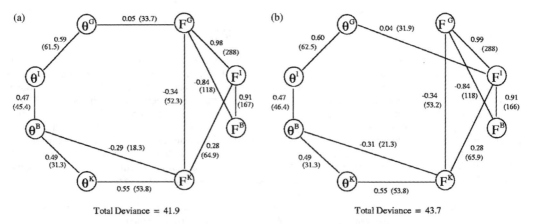

Figure 1. Two conditional independence graphs for observations, θ, and corresponding forecasts, F, at four locations. 17 edges have been discarded. Numbers indicate partial correlations and EEDs (in parentheses).

structure portrayed by the graph seems to provide a reasonable model for the data.

For each link the corresponding partial correlation and the increase of the graph's total deviance, dev (S, V) that would arise from the link's omission (Edge Exclusion Deviance, EED) are given. Note that the partial correlations θ^G–F^G in graph (a) and θ^G–F^I in graph (b), respectively, are very small due to the high correlation between F^G and F^I (cf. matrix S). Nevertheless large deviances (33.7 and 31.9, respectively) indicate that the corresponding links are relevant.

The matrices $V^{(a)}$ (cf. Eq. (3)) and $V^{(b)}$ have been specified numerically using iterative proportional fitting (Whittaker 1990). Generally, analytic solutions of the maximum likelihood problem are not available. This is related to the fact that partial correlations in the independence graph cannot be prescribed independently without violating the positive definiteness of the correlation matrix. Graphical vines are an alternative representation of correlation matrices, for which this is possible.

4 FITTING C- AND D-VINES

Vines have been introduced recently by Bedford & Cooke (2001, 2002). Basically, a *vine* on N variables is a nested set of trees, where the edges of tree j are the nodes of tree j + 1; j = 1, ..., N − 2, and each tree has the maximum number of edges (Cooke 1997). A *regular* vine on N variables is a vine in which two edges in tree j are joined by an edge in tree j + 1 only if these edges share a common node, j = 1, ..., N − 2. There are (N − 1) + (N − 2) + ⋯ + 1 = N(N − 1)/2 edges in a regular vine on N variables. Figure 2 shows a regular vine on 5 variables. The four nested trees are distinguished by the line style of the edges; tree 1 has solid lines, tree 2 has dashed lines, etc. The conditioned (before |) and conditioning (after |) sets associated with each edge are determined as follows: the variables reachable from a given edge are called the constraint set of that edge. When two edges are joined by an edge of the next tree, the intersection of the respective constraint sets are the conditioning variables, and the symmetric difference of the constraint sets are the conditioned variables. The regularity condition ensures that the symmetric difference of the constraint sets always contains two variables. Note that each pair of variables occurs once as conditioned variables.

We recall two generic vines, the *D-vine* D(1, 2, ..., n) and C-vine C(1, 2, ..., n), shown on Figures 2 and 3. Both C and D-vine are determined by the choice of the first tree.

In contrast with independence graphs vines are made up by partial correlations of varying order (the order of a partial correlation is the number of conditioning variables). Generally, the lower the order of an independence relation (i.e. partial correlation equals zero), the stronger this relation is. Whereas in an independence graph all partial correlations are all order N − 2, in a regular vine there is only one correlation of order N − 2, and N − 1 correlations of order zero. Hence, setting K partial correlations in a regular vine equal to zero imposes stronger independence than setting K partial correlations of maximal order equal to zero, as with independence graphs. Vines are derived from tree structures and therefore (again in contrast with independence graphs) assume a particular ordering of variables.

There is a one to one relationship between correlation matrices and partial correlation–vine specification so that the deviance defined in Eq. (2) can be used for the assessment of vines as well. Since the deletion of any partial correlation from a vine does not change other partial correlations no proportional fitting algorithm is needed to guaranty consistent correlation

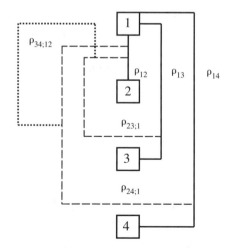

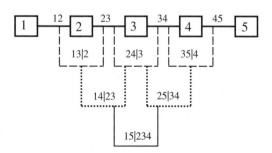

Figure 2. The D-Vine on 5 variables D(1, 2, 3, 4, 5) showing conditioned and conditioning sets.

Figure 3. The C-vine on 4 variables C(1, 2, 3, 4) with associated to the edges partial correlations.

matrices. However, additional efforts are needed to search for that permutation (ordering) of the set of variables, which allows for the most effective model simplification.

Generally to try all models with different degrees of simplification based on different orders of the variables is a tremendous task. Here it was assumed *a priori* that the number of edges being discarded from the vine should be 17 as it is in Figure 1. Then, for each of 40320 (=8!) permutations of variables EEDs of all individual links were calculated and the edge with the minimum EED was excluded from the graph (for the symmetric D-vines it was sufficient to test 8!/2 orderings of variables). This procedure was iterated until 17 links had been discarded. Even though other partial correlations are not affected by the removal of individual links, this is not true for the EEDs of the remaining links. Thus a strict optimisation would have needed the testing of all possible combinations of edges rather than using the sequential edge exclusion scheme. This, however, would have been too computationally demanding.

Figure 4 depicts the total deviances, which emerge from the 60 most successful orders of the eight variables when using C-vines and D-vines, respectively. In our example D-vines allow for a better fit than C-vines (note that the two panels are scaled differently), the deviance of the best D-vine is smaller than the deviances of the independence models in Figure 1. In addition, the optimum choice of the ordering of

variables is clearly better determined for D-vines. For C-vines the deviances tend to be more clustered and show a very flat minimum.

Figure 5 depicts the best fitting C-vine (one among four options with the same deviance) and D-vine, respectively, both being made up by 11 edges. It should be noted that the notation used in the graphs differs from the notations that was used in Figures 2 and 3. Instead of emphasizing the nature of a vine as a nested set of trees, links in Figure 5 indicate pairs of correlated variables. Numbers between variables specify marginal correlations. The C-vine allows also for marginal correlations between non-neighboring variables. These are indicated by dashed lines above the variable names. Partial correlations indicated below the variable names are to be interpreted differently for the two vines. In the C-vine partial correlations are conditioned on all variables to the left of the first variable connected by the respective solid line. In the D-vine the set of conditioning variables comprises all variables in between the two correlated variables.

Similar to the independence graphs both the C-vine and the D-vine employ three edges for establishing the link between observations and forecasts. Like the independence graph (b) in Figure 1 both vines include edges F^K–θ^K and F^I–θ^G.

In particular the D-vine tends to establish links conditioned on no or a small number of variables. However, (in contrast with the C-vine) no marginal correlation between observations and forecasts is maintained. The alignment of observations is the same as in the independence graphs.

Generally edges corresponding with strong partial correlations have been retained. There are, however, exceptions. In the C-vine the marginal correlation

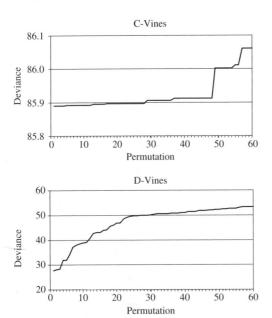

C-Vines

D-Vines

Figure 4. Deviances of graphical vine models with 17 missing edges as function of the 60 most successful permutations of variables.

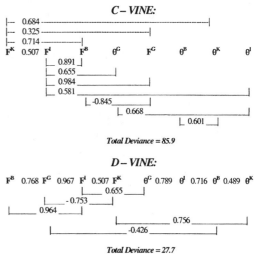

Figure 5. Optimized vines with 11 links (see text).

Table 1. Edge exclusion (inclusion) deviance for the C-vine in Figure 5. Existing links are written in bold. The first column ("Level") gives the number of variables that are held fixed for the respective partial correlation. The last column specifies the order in which the 17 links have been removed.

Level		EED	EID	Corr	
0	**F^K–F^I**	**3254.0**		**0.507**	
0	**F^K–F^B**	**651.3**		**0.714**	
0	F^K–θ^G		−2.36		(7)
0	**F^K–F^G**	**760.8**		**0.325**	
0	F^K–θ^B		0.24		(6)
0	**F^K–θ^K**	**55.8**		**0.684**	
0	F^K–θ^I		0.12		(8)
1	**F^I–F^B**	**272.0**		**0.891**	
1	**F^I–θ^G**	**60.3**		**0.655**	
1	**F^I–F^G**	**477.0**		**0.984**	
1	F^I–θ^B		14.31		(17)
1	F^I–θ^K		−9.29		(12)
1	**F^I–θ^I**	**43.8**		**0.581**	
2	F^B–θ^G		−0.72		(1)
2	**F^B–F^G**	**78.9**		**−0.845**	
2	F^B–θ^B		6.14		(15)
2	F^B–θ^K		1.59		(4)
2	F^B–θ^I		3.77		(11)
3	θ^G–F^G		0.30		(3)
3	θ^G–θ^B		0.91		(16)
3	θ^G–θ^K		−8.71		(13)
3	**θ^G–θ^I**	**37.5**		**0.668**	
4	F^G–θ^B		−0.76		(5)
4	F^G–θ^K		1.97		(10)
4	F^G–θ^I		0.09		(2)
5	**θ^B–θ^K**	**31.0**		**0.601**	
5	θ^B–θ^I		−2.02		(14)
6	θ^K–θ^I		4.70		(9)

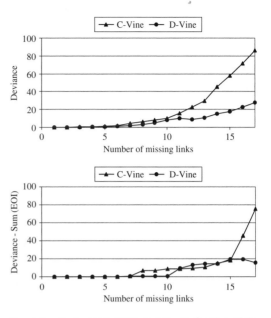

Figure 6. Upper panel: Total deviance as function of the number of edges discarded from the optimal C- and D-vine, respectively. Lower panel: Differences between total deviances and sums of individual edge inclusion deviances.

between forecasts F^K and F^G, 0.325, has been retained despite of its relatively small value. According to Table 1 the EED of this edge, 761, is much higher than, for instance, the EED of the link F^I–θ^G, 60, although the partial correlation for the latter pair of variables is 0.655. We may conclude once more that partial correlations do not necessarily provide information about the statistical significance of links in graphical models. A surprising result is that in contrast with independence graphs for vines the decrease of total deviance resulting from a link's inclusion (Edge inclusion deviance, EID) can be negative telling that the removal of a link would result in an improvement (!) of the model.

The upper panel of Figure 6 depicts for the two vines in Figure 5 the total deviances as functions of the number of discarded links. Obviously the total fit of the C-vine could be much improved by re-establishing the link F^I–θ^B, which has been removed last (according to Table 1 the total deviance would decrease by a value of 14.3). When doing so, however, the EID of the link θ^G–θ^B turns out to increase from its former value of 0.91

(cf. Table 1) to 13.4. The lower panel of Figure 6 reveals that for both kinds of vines the differences between the total deviances and the sums of individual EIDs increase with model truncation, i.e. the assessment of interaction structures becomes less local. Beyond 15 missing links, this non-locality increases dramatically for the C-vine while it still remains on the same level for the D-vine.

5 ASSOCIATING A D-VINE WITH A BELIEF NET

The idea of parameterizing the correlation structure with algebraically independent partial correlations, as in a regular vine, can also be extended to belief nets. In fact, the idea of capturing "influence" via conditional sampling leads to a natural homomorphism between regular vines and belief nets, which we now describe.

A belief net is a directed acyclic graph with nodes representing random variables and arcs representing "influence". In continuous belief nets we associate nodes in a belief net with continuous univariate random variables, and arcs with (conditional) rank correlations (Kurowicka & Cooke 2002). The following steps indicate how translate this into partial correlations on vines, and hence compute the entire correlation structure. We number the nodes in a belief net 1, ..., n.

Step 1 *Sampling order*

We construct a *sampling order* for the nodes, that is, an ordering such that all ancestors of node i appear before i in the ordering. A sampling order begins with a source node and ends with a sink node. Of course the sampling order is not in general unique.

Step 2 *Factorize joint*

We first factorize the joint in the standard way following the sampling order. If the sampling order is 1, 2, ..., n, write:

$$P(1, ..., n) = P(1)P(2|1)P(3|12)...P(n|1, 2, ..., n - 1).$$

Next, we underscore those nodes in each condition which are not necessary in sampling the conditioned variable. This uses (some of) the conditional independence relations in the belief net. For each term, we order the conditioning variables, i.e. the variables right of the "|", such that the underscored variables (if any) appear right-most and the non-underscored variables left-most.

Step 3 *Quantify D-vine for node n*

Suppose the last term looks like:

$$P(n|n - 1, n - 3, ..., \underline{n - 2, 3, 2, 1}).$$

Construct the D-vine with the nodes in the order in which they appear, starting with n (left) and ending with the last underscored node (if any).

If the D-vine $D(n - 1, n - 3, ..., 1)$ is given, the D-vine $D(n, n - 1, ..., 1)$ can be obtained by adding the edges:

$$(n, n - 1), (n, n - 3| n - 1), ..., (n, \underline{1} | n - 1, ..., \underline{2}).$$

For any underscored node $\underline{k}$, we have

$(n \perp \underline{k} |$ all non-underscored nodes $\cup$ any subset of underscored's not including k).

The conditional correlation between n and an underscored node will be zero.

For any non-underscored node j, the bivariate distribution

$(n, j |$ non-underscored nodes before j)

will have to be assessed. The conditioned variables (n, j) correspond to an arc in the belief net.

Write these conditional bivariates next to the corresponding arcs in the belief net. Note that we can write the (conditional) correlations associated with the incoming arcs for node n without actually drawing the D-vines. If the last factor is $P(5|1, 2, 3, \underline{4})$, we have incoming arcs (5, 1), (5, 2) and (5, 3) which we

associate with conditional correlations (5, 1), (5, 2|1) and (5, 3|12).

Step 4 *Quantify D-vine for node n − 1, for node n − 2 etc.*

Proceed as in step 3 for nodes 1, 2, ..., n − 1. Notice that the order of these nodes need not be the same as in the previous step. Continue until we reach the D-vine D(1 2) or until the order doesn't change in smaller subvines, i.e., if for node 4 the D-vine is D(4 3 2 1) and for node 3 it is D(3 2 1) then we can stop with node 4; or better, we can quantify the vine D(3 2 1) as a subvine of D(4 3 2 1).

Step 5 *Construct partial correlation D-vine (1, ..., n)*

As a result of steps 1–4 each arc in the belief net is associated with a (conditional) bivariate distribution. These conditional distributions do not necessarily agree with the edges in D(1, ..., n) since the orders in the different steps may be different. However, if the conditional bivariate distributions are given by partial correlations, then given D(1, ..., k) we can compute D($\pi(1)...\pi(k)$) where $\pi \in$ k! Is a permutation of 1, ..., k. Repeatedly using this fact, we compute the partial correlations for D(1, ..., n).

Since the values of conditional correlations on regular (sub)vines are algebraically independent and uniquely determine the correlation (sub)structure, the above algorithm leads to an assignment of correlations and conditional correlations to arcs in a belief net which are algebraically independent and which, together with the "zero edges" in the corresponding D-vines, uniquely determine the correlation matrix.

EXAMPLE 1

Sampling order: 1234
Factorization: $P(1)P(2|\underline{1})P(3|1\underline{2})P(4|23\underline{1})$
D-vine 4231:

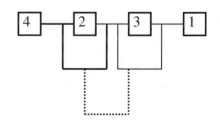

368

The dotted edge has partial correlation zero, the bold edges correspond to (4, 2) and (4, 3|2). These are written on the belief net and must be assessed.

We now consider the term $P(3|1\,\underline{2})$. The order is different than for the term $P(4|2\,3\,\underline{1})$. We construct $D(3\ 1\ 2)$:

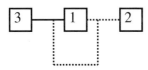

The dotted edges have partial correlation zero, the bold edge must be assessed, it is added to the belief net. The belief net is now quantified:

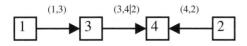

With the partial correlations in $D(3\ 1\ 2)$ we compute using the recursive relations $D(2\ 3\ 1)$. In fact, we find $\rho_{23} = \rho_{23,1} = 0$.

We now have $D(4\ 2\ 3\ 1)$ which corresponds to the belief net:

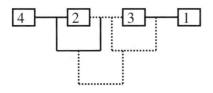

The distribution having specified univariate margins and satisfying rank correlation structure of the regular vine above can be obtained and sampled (Kurowicka & Cooke 2001).

6 ASSOCIATING A BELIEF NET WITH A D-VINE

Starting with a D-vine, we associate a belief net by identifying the sampling order of the belief net with the ordering of the vine. Let us assume that $D(n, n-1, ..., 1)$ is given, hence the sampling order is $1, 2, ..., n$. Start with the variable 1, draw arcs to all variables j such that partial correlation $\rho_{1j;2,...,j-1}$ on the D-vine is not equal to zero. Proceed the same way with the next variable in the ordering, that is 2 etc.

We illustrate this procedure using the D-vine in Figure 5. The ordering of this vine is: θ^K, θ^B, θ^I, θ^G, F^K, F^I, F^G, F^B. We start with θ^K. Both the marginal correlation θ^K and θ^B and the partial correlation θ^K and F^K given θ^B, θ^I, θ^G are non-zero. Therefore we

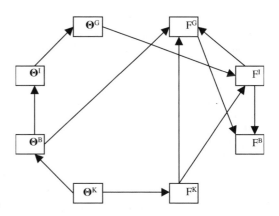

Figure 7. A belief net corresponding to the D-vine in Figure 5.

draw arcs from θ^K to θ^B and from θ^K to F^K. Similarly arcs must be drawn from θ^B to θ^I and from θ^B to F^G etc. Following this procedure the belief net in Figure 7 can be created that corresponds to the D-vine in Figure 5.

The graphs in Figure 1 and 7 are quite similar. Only the link θ^B–F^G is surprising since it was discarded in both cases in Figure 1. It must be emphasized, however, that in order to derive an undirected conditional independence graph from a directed belief net, generally new links must be added connecting all variables with a common child (moral graph; cf. Lauritzen & Spiegelhalter 1988, Whittaker 1990). This formal rule does not include a statement about whether or not the new links are statistically relevant, so that that the comparison of Figures 1 and 7 is not straightforward.

7 DISCUSSION

Numerical climate models are important tools for climate change studies. A general problem, however, is that global atmospheric models are not able to resolve the local scales at which most users require information so that statistical models must be employed to relate large-scale climate model output to regional scale observations (cf. Zorita & von Storch 1999). The temperature forecasts analyzed in this paper have been produced by such a "downscaling" method using observed large scale atmospheric pressure data as explanatory variables. A main objective of the study was to explore the application of two different techniques of graphical modeling for the assessment of the amount of genuine local information conveyed by the results of downscaling.

In the specific example it turned out that, for instance, the partial correlation between predictions for the stations Budapest, F^B, or Innsbruck, F^I, and observations at any of the four locations of interest are quasi zero given the values of forecasts for the two stations Geneva, F^G, and Copenhagen, F^K. This means that regression on forecasts F^G and F^K could be used as a surrogate for the forecasts delivered for Innsbruck or Budapest. In this sense the latter two forecasts, F^I and F^B, could be skipped without significant loss of information.

Another conclusion from the graphical independence model was that none of the four forecasts contained a significant amount of incremental information about temperature in Innsbruck as soon as the temperatures at Geneva and Budapest are known. This exemplifies how the analysis of interaction patterns can be used to identify sets of local stations, for which the downscaling scheme provides relevant independent information. Given the forecasts for these key stations observations at other stations might be simulated by an interpolating regression scheme.

A by-product of the graphical analysis is an estimate of the dimensionality of the interaction between the set of observations and the set of forecasts. Such information could also be achieved by considering, for instance, the number of pairs of correlated patterns that turn out to be relevant in conventional canonical correlation analysis. The main advantage of the graphical methods studied here is that they refer to the original set of variables and do not introduce derived artificial variables. Generally the exclusion of irrelevant derived variables (low correlation and/or explained variance) does not result in an reduction of the number of original variables the analysis is based on.

Graphical modeling intends to eliminate unimportant relations that may be present in empirical data in order to define parsimonious models with maximal explanatory power. Whittaker's method of independence graphs is currently the state of the art. The results of this study indicate that vines *might* offer improvements by imposing stronger independence relations (i.e. zeroing partial correlations of lower order) while deviating less from the original distribution. However, the computational burden is formidable.

Independence graphs discussed by Whittaker (1990) are defined by sets of pairwise conditional independence restrictions of maximum order. Non-zero partial correlations cannot be prescribed but must be estimated from data in a consistent way. In contrast, partial correlations in a vine can be prescribed independently so that the maximum likelihood estimate of non-zero partial correlations is trivial (simply choose the observed value). However, the occurrence of negative EIDs only for vines seems to be related to this fact that edges of vines are defined without adjustment of other links.

The main difficulty when fitting vines lies in the particular tree structure that each vine imposes onto the set of variables. It is an open question whether

methods can be designed that circumvent the need for testing all possible permutations of variables. Fitting independence graphs does not impose any ordering onto the set of variables, which makes fitting independence graphs computationally much less demanding.

The notation of D-vines is in closer correspondence with independence graphs than the notation of C-vines. Both use the very intuitive notation of context variables put in between those variables, the partial correlation of which is calculated. For instance, the orderings of observations in the graphical model and in the D-vine were found to be identical. In the present example D-vines performed clearly better than C-vines. This is an interesting result because due to its asymmetric topology for C-vines the number of different options for ordering variables is twice as large as for D-vines.

Fitting a C-vine, however, might be advantageous when a particular variable is known to be a key variable that governs variable interactions in the data set (maximum average correlation with other variables?). In such a situation one may decide to locate it at the root of the canonical vine.

All graphical models discussed in this study portrayed the mechanism of interactions between observations and forecasts in a consistent way. All methods used three edges for establishing interaction between observations and forecasts. Both the independence graph and the D-vine concluded that forecasts do not contain a significant amount of site specific information about θ^I, which cannot be replaced by using observations θ^B and θ^G as a surrogate. The C-vine with forecasts at its root has not the proper structure for portraying this fact.

ACKNOWLEDGEMENTS

For their open provision of data we are indebted to NCAR.

REFERENCES

Bedford, T. & Cooke, R. M. 2001. Probability density decomposition for conditionally dependent random variables modeled by vines. *Annals of Mathematics and Artificial Intelligence* 32: 245–268.

Bedford, T. & Cooke, R. M. 2002. Vines – a new graphical model for dependent random variables. *Annals of Statistics* 30(4): 1031–1068.

Brooks, H. E. & Doswell, C. A. 1996. A comparison of measure-oriented and distribution-oriented approaches to forecast verification. *Weather and Forecasting* 11: 288–303.

Cooke, R. M. 1997. Markov and entropy properties of tree and vines- dependent variables, Proceedings of the ASA Section of Bayesian Statistical Science.

Callies, U. 2000. Comparative forecast evaluation: Graphical Gaussian models and sufficiency relations. *Mon. Wea. Rev.* 128: 1912–1924.

Kurowicka, D. & Cooke, R. M. 2001. Conditional, Partial and Rank Correlation for Elliptical Copula, *Dependence Modeling in Uncertainty Analysis, Proc. of ESREL 2001.*

Kurowicka, D. & Cooke, R. M. 2002. The vine copula method for representing high dimensional dependent distributions: application to continuous belief nets. In E. Yücesan, C.-H. Chen, J.L. Snowdon & J.M. Charnes (eds), Proceedings of the 2002 Winter Simulation Conference, 263–269.

Lauritzen, S. L. & Spiegelhalter, D. J. 1988. Local computations with probabilities on graphical structures and their application to expert systems (with discussion). *J. Roy. Statist. Soc. B.* 50(2): 157–224.

Trenberth, K. E. & Paolino (jr.), D. A. 1980. The northern hemisphere SLP-dataset: trends, errors and discontinuities, *Mon. Wea. Rev.* 108: 855–872.

Whittaker, J. 1990. Graphical Models in applied multivariate statistics. John Wiley & Sons, Chichester.

Zorita, E. & von Storch, H. 1999. The analog method as a simple statistical downscaling technique: comparison with more complicated methods, *J. Climate* 12: 2374–2489.

Safety and Reliability – Bedford & van Gelder (eds)
© 2003 Swets & Zeitlinger, Lisse, ISBN 90 5809 551 7

Risk analysis methods to evaluate aluminium dust explosion hazard

D. Cavallero, M.L. Debernardi & L. Marmo
Politecnico di Torino, Dipartimento di Scienza dei Materiali e Ingegneria Chimica, Torino, Italy

ABSTRACT: A risk analysis method for factories at risk to explosion due to the presence of aluminium dust has been presented in this work. The companies that have been examined deal with the surface finishing of objects in aluminium through grinding and the dust that is produced is captured by suction plants and then subjected to dry or wet abatement. Each company has been divided into analysis environments and 4 study objectives have been identified for each of these environments.

1 INTRODUCTION

Following the introduction in Italy of Law 626/94, taken from Directives 89/391, 89/654, 89/655, 89/656, 90/269, 90/270, 90/394 and 90/679, an employer must carry out a risk analysis and identify the opportune prevention and protection measures in order to eliminate or reduce risks. No precise indications however exist on what the techniques are or on the modalities of carrying out the analysis, nor is there an explanation of the contents that the analysis should have. It therefore very often occurs that risk evaluations are incomplete and frequently carried out without the help of specific techniques or are even performed with the aid of rather general, and therefore not very precise, checklists. In the worst possible circumstances, an incomplete analysis could lead to the neglecting of very important risks or could even induce others to be created (secondary risks) as a consequence of the elimination or reduction of the existing ones.

During the first part of the nineties many Italian manufacturing industries were interested in the installation of vapour and dust capturing plants with the purpose of reducing the risks of the workers contracting illnesses caused by the inhalation of aerodispersed particles. In some cases, incomplete risk analysis did not highlight the secondary risks that these installations caused, in particular those of the explosion of combustible dusts, risks to which the sector involved in the finishing of aluminium objects is in particular exposed as many of the works that are carried out (grinding, polishing, etc.) produce explosive dusts.

Some industries in an industrial district in the north of Italy were involved, in the 1994–2000 period, in a series of very serious accidents that caused two deaths, several injuries and great material damage (Lembo et al. 2001). The first safety measure intervention is described in (Lembo et al. 2001). In this paper the continuation of the work, carried out to identify the methods and contents of the evaluation of the risks, is described, so that it could allow both primary and secondary risks to be identified.

Although we are dealing with small sized companies (2–15 workers), the risk analysis involves the study of very different situations (for example, a grinding machine, the dust abatement plant, the loading and unloading of materials) that require appropriate approaches. Furthermore, the quantification of risk itself differs according to the considered event: starting from the general concept of risk (Eq. 1)

$$R = F * M \tag{1}$$

where F is the frequency and M is the magnitude of the event. Two extreme situations can be considered on the basis of the magnitude M:

– M = death of one or more people,
– M = permanent/temporary invalidity of a worker.

Events characterised by such different values of M usually arise from different activities (e.g. explosion of a piece of apparatus or injury of a worker at a machine); these are more efficiently examined in different ways using specific techniques.

Table 1. The risk study fields.

Table 1. The risk study fields.

Field A: Machines and equipment with CE mark	Machines endowed with a CE mark and the relative documentation such as the use and maintenance manual conforming to Directive 89/392/CEE.
Field B: Machines and equipment without CE mark	Machines without a CE mark or with the mark but lacking either a part or all of the relative documentation as required by Directive 89/392/CEE.
Field C: Work sites	This usually includes all the areas where the personnel normally have access or can occasionally have access.
Field D: Complex plants	All those plants that perform articulated processes such as dust collection and abatement plants.

Table 2. The risk assessment topics.

1. Conformity of the machines, apparatus and procedures with the law provisions.
2. Suitability of the instruments and procedures.
3. Risks of "serious" accidents mostly related to malfunctioning or faults.
4. Risks the workers are exposed to.

own peculiarities that have to be considered when performing the analysis itself. On the basis of such peculiarities one or more Hazard Evaluation techniques were applied to each *Key Point*.

2 RISK ANALYSIS CRITERIA

The definition of residual risk contains explicit reference to that part of the risk, which is not nul, that persists in spite of the integral respect of the general and specific regulations, when they exist. The employer is always obliged to upgrade machines, equipment and plants according to the regulations in force; obviously the respect of this condition is the presupposition for the subsequent risk analysis. Plants and apparatus should conform with technical regulations, but these regulations unfortunately do not always present a sufficient degree of specificity (see for example EN 1127) and reference is often made to the "Best Available Technology" concept.

When the first part of this work began these obligations were not always generally respected by all the examined aluminium polishers and some situations involved technical levels that were much lower than acceptable standards while there was a decisive lack of knowledge as far as dust explosions was concerned.

A first intervention, extended to all the companies in the field, led to the production companies being upgraded to the minimum technical level that had been identified by a technical standard that was not harmonized (NFPA 651/98, Lembo et al. 2001, Lembo et al. 2001a). This work has started the study of the residual risk and has been carried out by dividing the companies that have been examined into 4 Fields of study, as shown in Table 1.

In addition, *4 risk assessment topics* were defined (see Table 2).

Each pair (*Field, Topic*) was defined as a *Key Point* in the evaluation of the risks (see Table 8); the analysis was extended to all the topics as each *Topic* has its

3 THE STUDIED COMPANY

The studied company performs polishing of boiler shells and coffee percolators for household use in aluminium alloy with semi-automatic machines. The investigated works are performed with different abrasive and brushing systems (corundum belt, disk grinders, rotating felt brush). The layout of the company is illustrated in Figure 1; it is located inside a 25×60 m industrial building which is divided into two spacious rooms. The machines are in room A (Fig. 1) and there is also an area used for the moving of goods while room B is used as a storeroom. The dust abatement plant is situated outside the industrial building and the three main branches of the suction plant, situated on the outside, converge in this abatement plant.

Production is estimated at $500 \div 700.000$ pieces per month, with an annual production of about 20 tons of aluminium dust.

The work procedures are carried out by partially automated grinding machines with grinding belts driven by pulleys with the operator only intervening at the initial stage, when he places the unprocessed piece on special supports, and at the final stage, when he removes the finished object. The machine is able to move the piece so that all the surfaces undergo the treatment.

The work procedures produce dusts of granulometry that varies according to the type of abrasive that is used, these being prevalently made up of shavings of sizes that vary from about 100 μm for the largest size to particles of the order of the μm and are mainly composed of metal under production with a slight addition of abrasive substances due to the wear and tear of the belts and traces of animal fats, which are sometimes used to treat the abrasive belts.

Each work procedure point is enclosed in a metallic casing (in aluminium to prevent sparks following knocks), equipped with an aspiration mouth (diameter equal to 148 mm) designed to intake 1300 m^3 of air per hour. The openings were made in such a way as

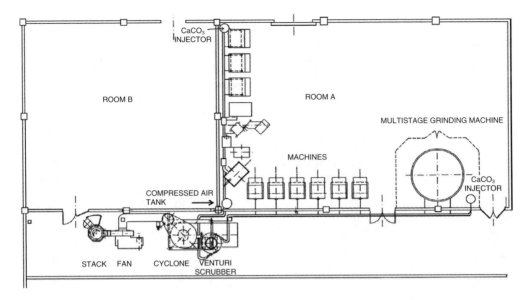

Figure 1: Layout of the rooms with the layout of the machines and the position of the capturing and abatement plant.

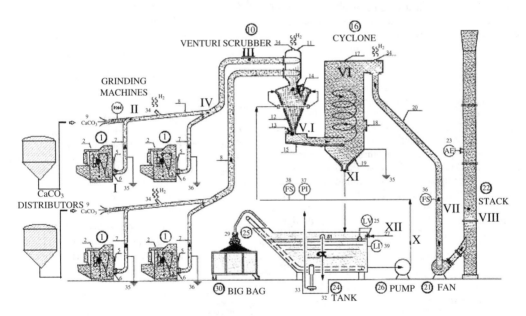

Figure 2: Scheme of the wet dust collection and abatement plant, with the relative controls and alarms. The nodes for the HazOp Analysis are also shown. [1 Grinding machines, 10 Venturi-Scrubber, 16 Cyclone separator, 21 Exhaust fan, 22 Exhaust chimney, 24 Washing water collection basin, 26 Washing water circulation pump, 30 Sludge collection container].

to prevent the stagnation of dust and are endowed with grills to avoid the accidental capturing of foreign bodies; the pulleys have been designed to prevent the stagnation of dust at the intrados.

The electric engines are class IP55 with an external temperature lower than 120°C, in normal working

conditions and lower than 165°C in overloaded conditions.

The dust collection plant (Fig. 2) consists of a pipe network that transports the dusts to the abatement plant. The air is aspired by a centrifugal fan that is placed downstream the abatement system.

The plant is endowed with three main intake manifold in galvanized steel plate, entirely lined in aluminium for anti-spark purposes (in the case of impact with foreign bodies on the internal surface) and these intake ducts enter into the venturi scrubber hood. The joints are of a head-to-head type that are inserted in the direction of the airflow and the curves are made in clipped sections with smooth internal surfaces. All the highest points of the intake manifolds have vent holes which are always open to disperse the hydrogen that could form when the plant is not working, in presence of humidity, into the atmosphere. Each manifold is supplied with a draft gage that sets off an alarm if the vacuum in the manifold falls below the threshold value; the power supply to the grinding machine is cut off.

Each manifold has a pneumatic inert dust (calcium carbonate) injector. Each of these has a tank in which there is a level sensor with an optical and acoustic alarm that intervenes when the carbonate descends below the minimum foreseen level; if the operator does not intervene within 15 minutes the energy to the grinder connected to the interlock manifold is immediately cut off. The distributor also has a second optical and acoustic alarm and a blockage system of the energy supply to the machines on the line whenever there is no supply of inert for 4 minutes.

Another optical and acoustic alarm causes the grinding machines to stop if the network pressure that activates the distributors and machines falls below 3 bar.

The dust abatement plant is a wet plant with a mixing hood, a Venturi scrubber and hydro-cyclones from which the air is sucked by the fan and introduced into the chimney; solids and liquids circulate in a water basin where the dusts settle. The water is recirculated to the hood and to the venturi throat by a centrifugal pump. The water level, the pressure and the recycling airflow are continuously measured and when necessary the optical and acoustic alarms are activated and the machines stop. The calcium carbonate and aluminum mud, which is extracted by a dredging machine, is stored in big bags and then sent to be disposed of.

4 DEVELOPMENT OF THE RISK ANALYSIS AND RESULTS

Each Key Point was examined through a series of analysis techniques. Some applications and the development method that was used are here given as an example.

4.1 Field A: Machines and equipment supplied with a CE mark

As far as the machines that fall into the Field A category is concerned, at least a verification of the

documentation (manuals, use, shortcomings) should be carried out in order to verify the suitability of the machine for the purposes it has to be used for (for example, in the case of grinding machines it should be clearly indicated that the machine can be used with combustible dusts and/or explosives). This approach could be sufficient as far as *Topics 1, 2* and *4* are concerned, according to which the machines do not require any further risk analysis. Further analysis can however be carried out as described for *Field B*. The effect of faults and malfunctioning of the machine on the operator and on the system was further analysed in order to illustrate how machine malfunctions can act as an event that provokes "serious" accidents (Topic 3). The FMEA, the What-If and the Checklist can be used for this purpose.

An analysis example carried out using the What-If and Checklist analysis techniques is shown in Table 3; for conciseness purposes the full list of questions is derived from standards in force (in the case this means NFPA 651).

4.2 Field B: Machines and equipment not supplied with the CE mark

It is first necessary to verify the correspondence to the provisions of the Laws in force (Topic 1) through specific standards type C regulations. In the case under examination a Checklist was developed starting from the project of type C standard prEN 13218. If there are no specific standards, it is possible to use regulations that pertain to type A and B (UNI EN 414, ISO 1127 etc.) even though these have less details.

The Checklist and the What-If were used for Topics 1, 2 and 3 while the FMEA resulted to be more suitable for Topic 3 (an example of this is shown in Table 4).

4.3 Field C: Work sites

Techniques such as the Checklist and the What-If developed according to the already presented criteria, allowing a complete analysis to be carried out.

4.4 Field D: Complex plants

In the considered company, this area is made up of the collection and abatement of the dust plant, as shown in Figure 2. The analysis of this environment varies considerably according to which Topic is studied.

4.4.1 Topics 1 and 2

The conformity to the provisions of Law was evaluated by developing Checklists (see Table 5). In order to guarantee that these Checklists were complete, they were derived directly from specific technical standards (NFPA 651/98). In the same manner, also the

Table 3. Extract from the What-If analysis – Key Point: Field A, Topics 1 and 2.

Reference NFPA 651/98	What-If	Consequences	Hazard	Preventive techniques	Integration and technical measures
4–2.1 Machines that produce fine particles of aluminium shall be provided with hoods, capture devices or enclosures that are connected to a dust collection system with a sufficient suction and capture velocity to collect and transport all the dust that is produced. The hoods and enclosures shall be designed and maintained so that the fine particles will either fall or be projected into the hoods and enclosures in the direction of the airflow	Insufficient pressure at the manifold mouth	Dust deposits; Accumulation of dust in the casing; Lowering of the suction velocity	Exceeding of the LEL in the casing	Velocity and pressure measuring devices in the pipes; Inactivation of the manifolds thanks to the injection of $CaCO_3$; Inactivation of the casing thanks to the manual introduction of $CaCO_3$	A minimum velocity of 20 m/s is guaranteed in the manifolds
	Obstruction at the grill	Accumulation of dust	Exceeding of the LEL	Velocity and pressure measuring devices in the pipes	

Table 4. Extract of the FMEA – Key Point: Field B, Topic 3.

Component	Function	Damage	Cause of damage	Consequences		Survey of damage	Adjustments	Observations
				Local effect	Global effect			
Pulley tightener	Tensioning of the grinding belt	Pulley not in position	Worn bearings	Oscillations of the belt Unravelling of the belt from the pulley	Objects badly ground Belt knocking the casing with the possibility of sparks	Oscillations of the belt	Substitute the bearings	The inside of the casing lined in aluminium as well as the mechanical parts
		Surface of the pulley uneven	Wear and tear	Unravelling of the belt from the pulley	Belt knocking the casing with the possibility of sparks	Oscillations of the belt	Regularly check the pulley surfaces	
		Pulley does not move	Bearing seizure	Belt does not move	Belt knocking the casing with the possibility of sparks	Belt not moving	Substitute the bearings Re-examine the maintenance procedures	

What-If analysis was applied (see Table 6) and again the list of points was derived from a specific standard.

4.4.2 Topic 3

The applied approach consists in carrying out recursive operability analysis (HazOp), the details of which are described in Piccinini et al. 1994, Piccinini & Ciarambino 1997, followed by the construction of Fault Trees. The HazOp method is proposed in the following Table 7. Since the HazOp method consists of the analysis of the deviation of the variables from their set point values. First of all the analyst should choose the variables that must be studied. In this work, the air velocity in the ducts and, as a consequence the

377

Table 5. Extract of the Checklist relative to point 4.3.5 of NFPA 651/98, Field D, Topics 1 and 2.

Reference NFPA 651/98	Requirements	Yes	No	Hazard	Notes
4.3.5.1 Ducts shall be constructed of conductive material and shall be carefully constructed and assembled with smooth interior surfaces and with internal lap joints facing the direction of the airflow	Are the aspiration ducts made of conductive material?			Accumulation of electrostatic loads and triggering of explosive mixtures	
	Are the internal surfaces of the ducts smooth and have the overlapping joints been mounted against the direction of the airflow?			Accumulation of combustible materials	

Table 6. Extract from the What-If method relative to point 4.2.1 of NFPA 651/98, Field D, Topics 1 and 2.

Reference NFPA 651	What-If	Consequences	Hazard	Preventive techniques	Integrations and technical measures
4–2.1 Machines that produce fine particles of aluminium shall be provided with hoods, capture devices or enclosures that are connected to a dust collection system with a sufficient suction and velocity to collect and transport all the dust that is produced	Damage to the dust capturing device	Accumulation of dust	Exceeding of the LEL	Draft gage in the manifold	
	Velocity of the suction diminishes	Insufficient suction at the collection mouth; Depositing of dust	Exceeding of the LEL	Flowmeter and draft gage in the manifold Inactivation of the dust in the manifolds	

connected variables such as dust concentration and vacuum were studied. Two top events were recognised thanks to the HazOp analysis: explosions in the casing and explosion of the capturing plant. It should be noted that some initiating events (e.g. the obstruction of the grill) derive from the analysis performed in Fields A and B.

The recursive structure of the HazOp analysis led to the immediate construction of the Fault Trees. This tool allowed the Minimal Cut Set (MCS) to be recognised. This analysis was performed with ASTRA software (Contini 1995), developed by the Joint Research Centre in ISPRA (Italy), and the minimal cut sets for the two Top Events were calculated:

I. (explosions in the casing), in which 2 MCSs of order 2 contribute (rupture of the I-II duct – trigger) and (obstruction of the grill and trigger), 10 MCSs of order 4 and others of even higher orders;

II. (explosion of the capturing plant), in which 16 MCSs of order 5, 120 MCSs of order 7 and others of even higher orders contribute.

As an alternative, the analysis can be developed according to the What-If method, as much better results can always be obtained if specific standards

are available from which the list of points can be obtained.

5 CONCLUSIONS

The analysis instruments that are the best to examine each Key Point are synthesised in Table 8. Obviously, a change in the Key Point could lead to a change in the most suitable instrument. This is part of the peculiarity of each instrument and confirms that it is useful to make use of different analysis methods.

5.1 Checklist

The Checklist and the What-If are techniques that can be widely applied, even though they are sometimes not efficient.

The Checklist is particularly efficient in the analysis of Topics 1, 2 and 4 when a specific standard is available from which the questionnaire can be derived. If there are no specific standards more general types of regulations can be used (EN 414). However the checklists are not suitable for the analysis of hazards that originate from the malfunctioning

Table 7. Extract from the HazOp method of the collection plant, Field D, Topic 3.

Deviations	Causes	Consequences	Warnings	Automatic protective means	Notes	TE
I. Low velocity	*I. Obstruction of the grill *Rupture of the I-II duct II. Insufficient pressure	Accumulation of dust in the I-II section I. Accumulation of dust				
I. Accumulation of dust	I. Low velocity	I. Explosive atmosphere				
I. Explosive atmosphere & *I. Triggering	I. Accumulation of dust	I. Explosion in the casing				
II. Insufficient suction	III. Low suction *Rupture of the II-III duct *The II-III duct obstructed	I. Low velocity	Optic and sound alarm (PXM)	Shut down of the line (PXM)		
III. Low suction	V. Low suction *Rupture of the III-V manifold *Rupture of another manifold	II. Insufficient suction & II-III. Depositing of dust				
V. Low suction	VII. Low velocity *VI. Obstruction	III. Low suction				
VII. Low velocity	Damage to the blower	V. Low suction & IV. Low separation efficiency	FS 36 LA 36	Shut down of the plant (FS 36)		

Table 8. Summary of the combinations of the objectives, environments and techniques that supplied the best results. The support techniques alone are given in italics.

Topics Fields	1 Conformity with the Law provisions	2 Suitability of the instruments and procedures	3 Risks of serious accidents	4 The risks the workers are exposed to
A: Machines and equipment supplied with a CE mark	Control of the documentation and of the procedures		What-If Checklist FMEA	What-If Checklist FMEA
B: Machines and equipment not supplied with a CE mark	What-If Checklist	What-If Checklist	What-If Checklist FMEA	What-If Checklist FMEA
C: Work sites	What-If Checklist	What-If Checklist	What-If Checklist	What-If Checklist
D: Complex plants	What-If Checklist	What-If Checklist	What-If HazOp FTA	HazOp What-If Checklist

of equipment or safety devices; it is generally advisable not to use Topic 3 except as an aid to HazOp.

5.2 What-If

As far as the What-If analysis is concerned, the same considerations can be made as for the checklists and it can be profitably adapted to the study of Topics 1, 2 and 4. It can be developed starting from specific standards (type C or not matched, such as NFPA 651) and can allow the estimation, though qualitative, of the consequences; it is therefore a more valid aid than the Checklist for FMEA and HazOp.

379

5.3 FMEA

This is extremely efficient for the study of incidents caused by faults to individual machines (Key Points of Fields A and B and Topics 3 and 4); in the case of production plants (Field D), the correlations between the events and protective systems not having been correctly evaluated, the FMEA analysis is not sufficiently complete. As it is a very demanding technique, its use is advised as an addition to significant problems on machines without the CE mark that have been identified with easier approaches such as Checklist. On the other hand, FMEA can supply indications on the initiating causes that have to be considered in the analysis (HazOp) which should be applied to the capturing plants.

5.4 HazOp

The recursive analysis has proved to be, beyond doubt, the most suitable technique for the reliable study of Field D, Topic 3 (it is often this Key Point that presents greater magnitude of danger) and can therefore be considered the best for the study of safety of dust capturing plants. This arises from the innate nature of the method, which was conceived to study the deviations of the variables of the system and is therefore particularly suitable for the study of production plants in which the deviations are protected by several systems that are designed to intervene in succession.

HazOp allows the Top Events that a system can reach following the malfunctioning of different components to be identified. It can be useful in the study of environment B, objectives 3 and 4 even though a technique like FMEA is preferable. It is also a powerful instrument for reliability analysis in that it allows the damage trees to be automatically constructed.

5.5 FTA

The damage tree is only useful for reliability purposes. In this sense FTA is the instrument it is necessary to resort to in order to determine the Minimal Cut Sets and for their successive probabilistic quantification, for which reliability data of the components are however necessary.

The damage tree is required where it is necessary to carry out the probabilistic quantification of an event, therefore, for example, in Field D, Topic 3. In the case under study, the reduction of the TEs to the MCSs identified through HazOP was carried out.

The construction of the FTA damage tree makes it easier to evaluate alternative safety measures in that it allows the consequences of a modification of the plant on the probability of a maximum level event occurring to be easily evaluated.

5.6 Topic 4

This topic cannot be studied independently, since the risks the workers are exposed to are a consequence of those individuated in Topics 1–3, hence no explicit reference to any technique was given in the paper. Column 4 of Table 8 summarises the tools by which Topic 4 is most properly studied but these are simply picked up from the previous studied topics. In fact the study of this topic is a sort of summary of the study of Topics 1–3.

ACKNOWLEDGEMENTS

This work has been carried out thanks to the co-ordination of Confartigianato of the Verbano Cusio Ossola Industrial Union with the financial contribution of the Verbano Cusio Ossola Chamber of Commerce and of the Verbano Cusio Ossola Province.

REFERENCES

UNI EN 414, Safety of machinery – Rules for the drafting and presentation of safety standards.

UNI EN 292/2, Machine safety – Fundamental concepts. General design principles – Techinical specification and principals.

NFPA 651, Machining and Finishing of Aluminum and the Production and Handling Aluminum Products, 1998.

Directive 98/37/EC of the European Parliament and of the Council of 22 June 1998 on the approximation of the laws of the Member States relating to machinery.

UNI EN 1127, Explosive atmospheres – Explosion prevention and protection – Basic concepts and methodology.

Contini S., 1995, *A new hybrid method for fault trees analysis*, Reliab. Engng System Safety 49: 13–21.

Demichela M., Marmo L., Piccinini N., 2002, Recursive operability analysis of systems with multiple protection devices, Reliability Engineering and System Safety 77 (3): 301–308.

Lembo F., DallaValle P., Marmo L., Patrucco M., Debernardi M.L., 2001, Aluminum airborne particles explosions: risk assessment and management at Northern Italian factories, in MG-Torino *eds, Proc. The European Conference ESREL 2001: Towards a Safer World*, p. 85–92, Torino, 16–20 September 2001.

Lembo F., Patrucco M., Debernardi M.L., Marmo L., Tommasini R., 2001a, Esplosione di polveri nei processi di finitura di manufatti in alluminio e leghe nella realtà produttiva ASL 14 VCO: analisi del rischio e misure di prevenzione, Centro Stampa della Giunta Regionale del Piemonte, 2001.

Piccinini N., Ciarambino I., 1997, *Operability analysis devoted to the development of logic trees*, Reliability engineering and system safety 55: 227, 241.

Piccinini N., Ciarambino I., Scarrone M., 1994, *Probabilistic analysis of transient events by an event tree directly extracted from operability analysis*, Journal of Loss Prevention in the Process Industry 7 (1): 23–32.

Safety and Reliability – Bedford & van Gelder (eds)
© 2003 Swets & Zeitlinger, Lisse, ISBN 90 5809 551 7

Optimization of testing and maintenance schedules

M. Cepin

Jozef Stefan Institute, Reactor Engineering Division, Ljubljana, Slovenia

ABSTRACT: Standby safety equipment in nuclear power plants is subjected to regular testing and maintenance, which are an important potential for risk and cost reduction. The method for optimization of schedule of testing and maintenance activities is described. The method integrates simulated annealing as an optimization method and probabilistic safety assessment models and results as the standpoint for determination of the objective function of optimization. The prerequisite for success of the method is modeling of standby safety equipment with probabilistic models considering parameters that have impact to testing and maintenance including time. Application of the method on a real example of probabilistic safety assessment model and its results, which correspond selected nuclear power plant, is discussed. The improvements of the method, which were discovered, when the method was tested by the real example, are identified. Results show that the optimization shows less improvement in optimal schedule versus initial schedule for more reliable components.

1 INTRODUCTION

Standby safety equipment in nuclear power plants is subjected to regular testing and maintenance, which are an important potential for risk and cost reduction.

Among the initiators in the field of improvement and optimization of testing and maintenance activities Jacobs, 1968, expressed empirically the unavailability formulas for standby safety equipment.

Wu & Lewins, 1993, embedded Monte Carlo simulation in the optimization of test intervals, so as to upgrade the previous calculations with the empirical formulas.

Then, many activities have been oriented to optimizing parameters of testing and maintenance (Vaurio, 1995, Cepin & Mavko, 1997, Munoz, Martorell et al., 1997, Vaurio, 1997, Cepin et al., 1999, Yang et al., 2000, Martorell et al., 2000, Giuggioli Busacca et al., 2001, Martorell et al., 2002b, Cepin & Martorell, 2002). The objective function in such optimizations was primarily a measure of risk, e.g. system unavailability, e.g. core damage frequency, e.g. large early release frequency, or in some cases a measure of costs.

Less emphasis was placed to the development of methods for scheduling of testing and maintenance activities (Harunuzzaman & Aldemir, 1996, Cepin, 2002a, Cepin, 2002b), but this issue is coming forward with increasing needs within the activities of development and application of risk monitors.

Methods for scheduling of testing and maintenance activities evolve in two directions:

– optimization of scheduling of testing and maintenance activities during specified plant operating modes (Harunuzzaman & Aldemir, 1996, Cepin, 2002a) and
– selection of the most appropriate plant operating mode for testing and maintenance, i.e. support of the on-line maintenance (Cepin, 2003).

The method for optimization of schedule of testing and maintenance activities, which was developed recently (Cepin 2002a, Cepin 2002b), is described.

2 METHODS

The simulated annealing methods (Metropolis et al., 1953, Press et al., 1992) and genetic algo-rithms (Marseguerra & Zio, 2000a, Martorell et al., 2000, Martorell et al., 2002a) have been found extremely useful for large and complex optimization problems.

Simulated annealing as the optimization method was selected as the standpoint for development of method for optimization of scheduling the testing and maintenance activities (Cepin 2002a, Cepin 2002b). The objective function for optimization was defined from results of probabilistic safety assessment

(PRA, 1982, Cepin & Mavko, 1997, Cepin & Mavko, 2002).

Probabilistic safety assessment is one of standardized ways of assessing safety in nuclear power plants (PRA, 1982). Its primary methods: fault tree and event tree analysis include a wide range of applications (Cepin & Mavko, 1997, Cepin et al., 1999).

Fault tree is a tool to identify and assess all combinations of undesired events in the context of system operation and its environment that can lead to the undesired state of a system (Roberts et al., 1981). Undesired state of the system is represented by a top event. Logical gates connect the basic events to the top event.

Basic events are the ultimate parts of the fault tree, which represent undesired events, such as component failures, missed actuation signals, human errors, contributions of testing and maintenance activities and common cause contributions.

Qualitative fault tree analysis identifies the minimal cut sets, which are the combinations of the smallest number of component faults that may cause the system fault. In other words: the minimal cut sets are combinations of the smallest number of basic events, which, if occur simultaneously, may lead to the top event.

Quantitative fault tree analysis includes calculation of the system unavailability (calculated as the fault tree top event probability) which is one of the main risk measures at the system and component level and which is based on probability of failure of safety system components.

Event tree is a tool to identify and assess possible scenarios (i.e. accident sequences) of safety systems functions responses (i.e. systems successes or system faults, which are further analyzed with the fault trees) to the initiating event (PRA, 1982).

The initiating event is an event, which may lead to the accident consequences. Safety system functions are the means to prevent the accident or to mitigate its consequences. Plant damage states are the end states of the scenarios.

Results of probabilistic safety assessment include several risk measures: e.g. system top event probability or e.g. safety function top event probability at the system or function level or e.g. core damage frequency or e.g. large early release frequency at the plant level (Vaurio, 1995, Cepin & Mavko 1997, Yang et al., 2000).

2.1 *Description of the optimization method*

Optimization method is described in more details in reference (Cepin, 2002a).

The function of core damage frequency, which is obtained from probabilistic safety assessment, and which is presented in equation 1, is minimized with the use of simulated annealing.

$$CDF_{mean} = \frac{1}{T} \sum_{t=1}^{t=T} CDF_n(\mathbf{Q_{Bj}(t, T_{pBj})})$$ (1)

where CDF_{mean} = mean core damage frequency over determined time interval; T = number of discrete time steps, which define time interval; $CDF_n(Q_{Bj}(t,T_{pBj}))$ = core damage frequency as a function of time dependent failure probabilities of basic events, which through minimal cut sets determine the core damage frequency; $Q_{Bj}(t,T_{pBj})$ = time dependent failure probability of equipment modeled in basic event B_j calculated by respective probabilistic model; t = time; T_{pBj} = test placement time (as the optimization parameter in the optimization method), which defines the time passed from time 0 to time of outage (e.g. due to testing or maintenance) of basic event B_j.

Probabilistic models for basic events are equations, which represent failure probability of component under investigation as a function of one or more parameters including: failure rate, test interval, test duration, repair duration, repair rate, test placement time and time.

2.2 *Application of the optimization method*

The real example of probabilistic safety assessment of selected nuclear power plant is investigated in this paper. Therefore, the core damage frequency was selected as the risk measure under investigation.

Prerequisite for obtaining the function of core damage frequency and its use in optimization method lays in development of the complete probabilistic safety assessment model and in its quantitative analysis.

2.3 *Limitations of the optimization method*

Limitations of the optimization method and its application are the following.

Probabilistic models for components are not yet developed in extent, which would cover all positive and negative aspect of testing and maintenance to the component failure probability.

Probabilistic safety assessment model is limited to single mode of operation, although activities in direction of integration of models for all modes of plant operation exist (Cepin, 2003).

Although the real example is considered in this paper, the number of considered minimal cut sets is small. The small number of selected minimal cut sets represents small partial contribution of the core damage frequency to the overall core damage frequency. It is possible to extend the number of considered minimal cut sets with rewriting of one of the subroutines of the optimization code.

3 MODEL

3.1 Characteristics of nuclear power plant under investigation

The real size probabilistic safety assessment model of a nuclear power plant is named NPP_S and represents the model of investigation.

The main characteristics of the nuclear power plant, which is modeled in the model NPP_S, are the following: 2 units (with 3 diesel generators: one for each of both units and third, which can be aligned to any of the units, one unit is under investigation), 3 loops, pressurized water reactor, sub-atmospheric containment and thirty years of successful plant operation.

3.2 Characteristics of probabilistic safety assessment model

The main characteristics of the probabilistic safety assessment model NPP_S for internal events as the standpoint for calculation are the following: 12 initiating events, 12 event trees, 41 functional events (event tree headings), 483 sequences, 105 fault trees, 461 basic events, 384 gates and 2 house events.

3.3 Characteristics of results of probabilistic safety assessment as an input to optimization method

Complete list of evaluated minimal cut sets for probabilistic safety assessment model NPP_S include 15549 minimal cut sets. Resulted core damage frequency is estimated to 3.48E-5/ry.

Equipment under investigation includes selected standby safety components modeled in probabilistic safety assessment model such as diesel generators, auxiliary feedwater pumps and safety injection pumps.

Selected number of minimal cut sets consisting of at least one basic event describing the fault of the equipment under investigation is 11175.

The first 30 minimal cut sets were used for evaluation among selected number of the minimal cut sets ordered by the core damage frequency from higher to lower value. Those 30 minimal cut sets represent 2.6% of the overall core damage frequency.

A larger number of minimal cut sets could be selected and evaluated. This would require manual preparation of larger input file for optimization, or alternatively, this would require integration of the optimization method and the probabilistic safety assessment code, which does not exist at the moment. Optimization code is separated from probabilistic safety assessment code at the moment and input preparation is not fully automated yet. Selected number of 30 minimal cut sets is large enough to evaluate the method and to direct possible improvements.

Selected 30 minimal cut sets include the following events, which can be grouped into three sets:

– 11 basic events, which represent standby components subjected to testing and maintenance; those components are modeled by probabilistic models, which are a function of several parameters including time; those components are the subject of the optimization,
– 14 basic events, which represent components that are modeled by constant failure probabilities,
– 4 initiating events, which are modeled by constant frequency.

11 basic events, which represent standby components subjected to testing and maintenance, are modeled in original probabilistic safety assessment by constant failure probabilities. For the optimization purposes those components have to be modeled with new probabilistic models, which may be a function of several parameters: test interval, test duration time, test placement time, failure rate, repair rate, repair time, time (Cepin & Mavko, 1997). Therefore, new probabilistic models were used for those 11 basic events and new data required for quantification of each of 11 basic events was determined.

Replacement of constant probabilistic models based on original probabilistic data with new probabilistic models with new data resulted in change of core damage frequency resulted from selected minimal cut sets.

The change is not important because the absolute value is not playing any role in optimization. It is important to keep in mind that initial core damage frequency obtained from selected minimal cut sets, which is presented on Figure 1 (2.4E-6/ry), differs for a factor of 2.5 from 9.21E-7/ry, which is 2.6% of original core damage frequency of 3.48E-5/ry. This difference is observed due to change in probabilistic models used and change in data required in respective probabilistic models.

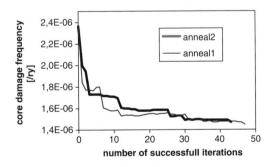

Figure 1. Convergence of the objective function (2 independent optimization runs: anneal1, anneal2).

4 PROCEDURE FOR OPTIMIZATION

Procedure for optimization starts with analyzing the model of probabilistic safety assessment qualitatively and quantitatively. Results of probabilistic safety assessment are obtained.

Minial cut sets are derived from obtained results (it is possible to consider all minimal cut sets or only the representative part of minimal cut sets, which includes minimal cut sets that include basic events of standby components that are under investigation).

Logical equation for minimal cut sets represents the standpoint for objective function of optimization. Subroutines are prepared, which quantifies the objective function. Finally, optimization algorithm is executed.

The practical steps of the procedure for optimization are the following:

1st step: carry out probabilistic safety assessment (PSA) for the plant under investigation,

2nd step: tag basic events under consideration (basic events connected with faults of standby equipment i: BESi),

3rd step: run analysis of selected PSA model,

4th step view results: minimal cut sets,

5th step: tag minimal cut sets, which include tagged basic events,

6th step: prepare input function (subroutine) for optimization: consideration of tagged minimal cut sets,

7th step: run optimization algorithm,

8th step: interpret the results.

5 RESULTS

5.1 Example results of selected optimization runs for real plant model

Several optimization runs were performed. Results of optimization were of secondary importance as of the primary importance was to investigate the usefulness of the method for the real plant examples and to direct the necessary improvements.

Figure 1 shows the convergence of the objective function for two independent optimization runs. Result of optimization is a set of test placement times for all 11 considered standby components that are subsequent to testing and maintenance, when the plant is at power operation ($T_{p1} = 121$ h, $T_{p2} = 239$ h, $T_{p3} = 189$ h, $T_{p4} = 186$ h, $T_{p5} = 261$ h, $T_{p6} = 213$ h, $T_{p7} = 152$ h, $T_{p8} = 231$ h, $T_{p9} = 218$ h, $T_{p10} = 138$ h, $T_{p11} = 114$ h) at core damage frequency contribution of considered minimal cut sets of 1.5E-6 [/ry].

Curve anneal1 on Figure 1 shows, that successful iterations sometimes (but rarely) result in increase of objective function. Such observations comply with the nature of simulated annealing. In simulated annealing the iteration (based on randomly selected values of the function parameters), which is better than temporary solution, replaces temporary solution with probability of 1. The iteration, which is worse than the temporary solution, replaces the temporary solution with a very small probability. And such small probability is decreasing exponentially with increasing number of iterations, but it is not 0. This is the feature of simulated annealing as the optimization method (Metropolis et al., 1953).

5.2 Specific features and identified improvements considering full size real nuclear power plant example

Experience shows that cases exist, where several basic events in a fault tree model their respective faults of the same component. Those basic events should be joned together into one in order that one component is modeled in optimization algorithm with one parameter of each kind (e.g. test interval, e.g. test placement time).

Some probabilistic models represent constant values in relation to optimization parameter. Those probabilistic models should be treated differently from probabilistic models, which represent functions of optimization parameter, in order to save memory and to decrease evaluation time.

The solution of this feature is enhanced notation, which is simplified in sense that threats differently basic events with constant failure probability and basic events with failure probability defined as a function of parameters.

The enhanced notation is achieved by joining selected minimal cut sets in sense to expose basic events modelling faults of standby equipment as it is schematically shown in equation 2.

$$TOP = \Sigma BES_i * (PMCS_m + ... + PMCS_M)$$
$$+ MCS_n + ... + MCS_N \qquad (2)$$
$$+ \Sigma MCS_z$$

where TOP = top event; BES_i = basic event connected with faults of standby safety equipment i (the respective probabilistic model represents time dependent function and function of optimization parameter); MCS_m = minimal cut set, which, among other basic events that are independent on time and on optimization parameters, include one basic event BES_i, which is exposed for the reason of enhanced optimization procedure; $PMCS_m$ = part of MCS_m without exposed basic event BES_i; MCS_n = minimal cut set, which include basic events independent of time and independent from basic events connected with faults of standby safety equipment under consideration; MCS_z = minimal cut set, which include more than one BES_j ($MCS_z = BES_p * BES_q * \cdots * BES_r$).

6 CONCLUSIONS

The application of the method for optimization of schedule of testing and maintenance activities of the standby safety equipment on the real example of probabilistic safety assessment is presented.

The improvements of the method, which were discovered by application of the real example, are identified and they include:

- update of probabilistic safety assessment models in sense that one component is modeled with one basic event, which considers all failure modes of the respective component, and
- enhanced notation of minimal cut sets as it is described in section 5.2.

The optimization shows less relative improvement of selected risk measure (i.e. core damage frequency) in optimal schedule versus initial schedule for more reliable components.

REFERENCES

Cepin M., 2002a, Optimization of Safety Equipment Outages Improves Safety, Reliability Engineering and System Safety, Vol. 77, pp.71–80.

Cepin M., 2002b, Risk Based Optimization of Schedule of Safety Equipment Outages, Transactions of ANS, Vol. 87, pp. 100–102.

Cepin, M., 2003, Development of a Method for Integration of All Modes of Probabilistic Safety Assessment, Proceedings of International Conference Nuclear Energy for New Europe 2003, Portoroz, 2003 (to appear).

Cepin M., A. Gomez Cobo, S. Martorell, P. Samanta, 1999, Methods For Testing And Maintenance of Safety Related Equipment: Examples from an IAEA Research Project, Proceedings of ESREL99: Safety and Reliability, Vol. 1, pp. 247–251.

Cepin M. & B. Mavko, 1997, Probabilistic Safety Assessment Improves Surveillance Requirements in Technical Specifications, Reliability Engineering and Systems Safety, Vol. 56, pp. 69–77.

Cepin M. & B. Mavko, 2002, A Dynamic Fault Tree, Reliability Engineering and System Safety, Vol. 75, No. 1, pp. 83–91.

Cepin M. & S. Martorell, 2002, Evaluation of Allowed Outage Time Considering a Set of Plant Configurations, Reliability Engineering and System Safety, Vol. 78, No. 3, pp. 259–266.

Giuggoli Busacca P., M. Marseguerra, E., Zio, 2001, Multiobjective Optimization by Genetic Algorithms: Application to Safety Systems, Reliability Engineering and System Safety, Vol. 72, pp. 59–74.

Harunuzzaman M. & T. Aldemir, 1996, Optimization of Standby Safety System Maintenance Schedules in Nuclear Power Plants, Nuclear Technology, Vol. 113, pp. 354–367.

Jacobs I. M., 1968, Reliability of Engineered Safety Features as a Function of Testing Frequency, Nuclear Safety, Vol. 9, No. 4, pp. 303–312.

Marseguerra M. & Zio, E., 2000a, Optimizing Maintenance and Repair Policies via a Combination of Genetic Algorithms and Monte Carlo Simulation, Reliability Engineering and System Safety, Vol. 68, pp. 69–83.

Martorell S., S. Carlos, A. Sanchez, V. Serradell, 2000, Constrained Optimization of Test Intervals Using a Steady-State Genetic Algorithm, Reliability Engineering and System Safety, Vol. 67, pp. 215–232.

Martorell S., A. Sanchez, S. Carlos, V. Seradell, 2002a, Simultaneous and Multi-Criteria Optimization of TS Requirements and Maintenance at NPPs, Annals of Nuclear Energy, Vol. 26 (2), pp. 147–168.

Martorell S., A. Sánchez, S. Carlos, V. Serradell, 2002b, Comparing Effectiveness and Efficiency in Technical Specifications and Maintenance Optimization, Reliability Engineering and System Safety, Vol. 77 (3), pp. 281–289.

Metropolis N., A. Rosenbluth, M. Rosenbluth, A. Teller, E. Teller, 1953, Equations of State Calculations by Fast Computing Machines, Journal of Chemical Physics, Vol. 21, pp. 1087–1092.

Munoz A., S. Martorell, V. Serradell, 1997, Genetic Algorithms in Optimising Surveillance and Maintenance of Components, Reliability Engineering and System Safety, Vol. 57, pp. 107–120.

PRA Procedures Guide, 1982, NUREG/CR-2300, US NRC, Washington DC.

Press W. H., A. S. Teukolsky, W. T. Vetterling, B. P. Flannery, 1992, Numerical Recipes in Fortran, Cambridge University Press.

Roberts N. H., W. E. Vesely, D. F. Haasl, F. F. Goldberg, 1981, Fault Tree Handbook, NUREG-0492, US NRC, Washington.

Vaurio J. K., 1995, Optimization of Test and Maintenance Intervals Based on Risk and Cost, Reliability Engineering and System Safety, Vol. 49, pp. 23–36.

Vaurio, J. K., 1997, On Time-Dependent Availability and Maintenance Optimization of Standby Units Under Various Maintenance Policies, Reliability Engineering and System Safety, Vol. 56, pp. 79–89.

Yang J. E., T. Y. Sung, Y. Yin, 2000, Optimization of the Surveillance Test Interval of the Safety Systems at the Plant Level, Nuclear Technology, Vol. 132, pp. 352–365.

Wu Y. F. & J. D. Levins, 1993, Mechanical System Surveillance Test Interval Optimization, Journal of Science and Technology, Vol. 30 (12), pp. 1225–1233.

Safety and Reliability – Bedford & van Gelder (eds)
© 2003 Swets & Zeitlinger, Lisse, ISBN 90 5809 551 7

An overview of maritime safety assessment trends in a stakeholder perspective

G. Chantelauve
Bureau Veritas, Paris, France
Ecole des Mines de Saint Etienne, Saint Etienne, France

ABSTRACT: Over the years, maritime safety has been under the spotlight of the public opinion. Nowadays, safety is one of the primary objectives of maritime transport and is regulated under the auspices of the International Maritime Organisation, which establishes and adopts international safety Conventions mainly of prescriptive nature. Recently, risk management philosophy has received an increasing recognition across the maritime community. Moreover, the importance of the stakeholders is now pointed out: stakeholders together with their safety needs/decisions and their safety initiatives are identified and involved in the safety improvement process. In order to create a sound understanding of the new emerging safety framework the interface between stakeholders, their needs and objectives on the one hand and the availability and accuracy of safety approaches on the other hand should be considered. This paper explores some trends in safety assessment approaches in light of their use by different and interrelated stakeholder networks. Both regulatory and industrial safety trends are investigated together with their implementation issues in a complex context. The paper concludes on the benefits and shortcomings of a suggested safety framework and outlines the key features of its evolution.

1 INTRODUCTION

Maritime transport is a sector that has long been governed by the freedom of the seas principle. However, Boisson (Boisson, 1999) underlines that: "nowadays, maritime safety is one of the essential objectives of the marine policy and justifies the principal attacks against the freedom of the seas principle".

Indeed, numerous efforts regarding maritime safety have been made during the second half of the 20th century. The International Maritime Organisation (IMO), a specialised agency of the United Nations created back in the 50's, acts as the global maritime regulator developing the international safety regulatory framework. Nevertheless, the recurrence of disasters, resulting in human life loss and environmental pollution, called into question the situation and the prescriptive regulatory framework. In 1992, following the Herald of Free Enterprise capsize, Lord Carver report (House of Lords, 1992) argued for a new vision and the need for a radical change in the safety regulation approach. Since then, safety initiatives have incorporated risk concepts and analysis. Moreover, the importance of the stakeholders, as the driving force for safety improvement, is acknowledged.

The present paper is a contribution to a better understanding of the interaction between risk analysis philosophy and decision-making. Theoretical work on the use of risk analysis in different stakeholder decision settings (Kørte, 2002) and specific research related to maritime stakeholders (THEMES, 2001), are combined. Having mapped maritime stakeholders into a generic typology of decision domains, a refinement of decision domains is proposed. Then, the implication on the use of risk analysis in different decision settings, with special emphasis on stakeholder needs, is considered. Finally, interactions and influences among stakeholders, within the waterborne transport sector, are theorised and discussed.

2 FOUNDATION

This work is mainly based on two existing research works. The first one (Kørte, 2002) describes constraints and potential for risk analysis and management in different decision settings. The second one (THEMES, 2001) focuses on stakeholders' stakes in waterborne transport.

2.1 Terms clarification

The word "decision" indicates, both, a decision-making process (systematic process comprising generally the problem formulation, collection and analysis of information, choice, execution and control of an alternative), and the result of that process. In the following, "decision-making" describes the process of making a choice between several alternatives according to specified objectives and available information. "Decision" describes the outcome of the process.

"Stakeholder" is defined as (THEMES, 2001) "any person, group, organisation or authority, which directly or indirectly either influences or is affected by the safety or cost effectiveness of ships and/or shipping."

In this paper, "safety approach" has a broad meaning, including both knowledge elements, such as data and models, and ethical elements in terms of values, objectives, etc.

2.2 "On the use of risk analysis in different decision settings" (Kørte, 2002)

Recognising risk as an important aspect of the decision-making process, Kørte presents a typology of decision domains (political, managerial, routine, analytical and crisis domains) using a two-dimension diagram (level of authority and proximity to hazard). Then, implications on risk analysis and management are described according to decision domains and constraints. Finally, normative ideas on interactions and influences among domains are presented. Kørte concludes, «This [theoretical] understanding needs to be tested and elaborated through empirical studies of the interplay between decision-making and risk».

2.3 THEMES "Thematic network for safety assessment in waterborne transport"

THEMES is a Thematic Network funded by the European Community, Maritime Directorate of DG TREN, under the 5th Framework Programme for Research, Technological Development and Demonstration. THEMES overall goal is to improve industrial safety and environmental protection in shipping, through support to and development of a pro-active safety culture. The objective is to establish a common knowledge base and a comprehensive framework of safety assessment and safety management for waterborne transport. It commonly recognises the importance of stakeholders. In particular, an exhaustive list of about forty stakeholders, together with their safety needs and decisions,

has been set up during the research activity (THEMES, 2001).

2.4 Transition

This paper applies the theoretical and generic work from Kørte to the specific waterborne transport domain. The present study proposes improvement on Kørte's ideas, and provides a better understanding of safety stakes within the maritime domain.

3 STAKEHOLDER MAPPING

First and foremost, the maritime stakeholders have been mapped into the two-dimension diagram proposed by Kørte (see Figure 1).

3.1 Two-dimension taxonomy refinement

The first dimension, proximity to hazard, is used to classify stakeholders from a sharp end setting to a blunt end setting. Stakeholders in a sharp end position are those physically close to hazard – either on board or in the surrounding of operation and accident consequences. Stakeholders in a blunt end setting are those not involved in an operational context or having limited direct contacts. Intermediate stakeholders are classified according to informational/communicational/operational distance. In order to have a good level of details, we have divided this dimension into six classes not presented here.

The second dimension is the level of authority. As described by Kørte, level of authority is conceived in formal term: "Stakeholder A has a higher level of authority than stakeholder B if Stakeholder A is entitled to give directives, orders, instructions to Stakeholder B, but not vice versa". Authority levels are political institutions (PI), regulatory institutions (RI), company (C), management (M) and staff (S).

In addition to these two dimensions, we have defined a last element as the External Constraint category. Stakeholders within this category have an impact and/ or an influence on the different levels of authority, but are not classified on the proximity to hazard axis.

3.2 Mapping and findings

3.2.1 Mapping

Figure 1 is a simplified mapping constrained by printing. Stakeholders selected are those commonly firstly identified. The whole mapping is composed of about forty stakeholders (see Annex for the list). In accordance with Kørte labelling, stakeholders close to the

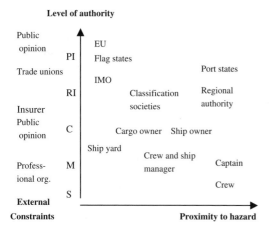

Figure 1. Simplified mapping of maritime stakeholders into the two-dimension model. EU: European Union. IMO: International Maritime Organisation. Political Institutions (PI), Regulatory Institutions (RI), Company (C), Management (M) and Staff (S).

hazard are positioned away from the origin of the "proximity to hazard" axis.

3.2.2 Complexity

The analysis is based on "generic" stakeholders, and gives a simplified description. However, trends indicating the increasing complexity of the maritime domain have been identified during the mapping.

A first trend is stakeholder networking. Traditional activity performed by a single stakeholder is now spread out among a network of new stakeholders. Ship-owner is an illustration of this networking with new functions such as ship manager, crew manager, etc. On the other hand, some stakeholders concentrate multiple roles and perspectives. For example, classification society performs classification mission, as well as statutory mission on behalf of Flag States, i.e. certification. The last trend is stakeholder shifting to the close end of hazard, this being due to external constraints and strategic stakes.

3.2.3 Decision domains

Kørte identifies five distinct decision domains: political (N.1), routine, crisis/emergency, managerial and analytical. The decision domains are displayed in the diagram, see Figure 2. As a result of the mapping, we have identified and defined a second "political domain", close to the hazard and at a high authority level (in grey in Figure 2).

The decision domains will be considered one by one in the next chapter. First, they are characterised by specific decision-making (process), decision (result), constraints, criteria and support. The main drawbacks

Figure 2. Classes of decision domains, adapted from (Kørte, 2002). Political Institutions (PI), Regulatory Institutions (RI), Company (C), Management (M) and Staff (S).

are also highlighted. Then, implications on the use of risk analysis accounting for specific maritime needs are discussed.

4 DECISION DOMAINS AND IMPLICATIONS ON RISK ANALYSIS

Kørte's findings on decision domains and implications on risk analysis are discussed, adapted and refined hereafter. This accounts for the waterborne transport context and for the safety needs identified within THEMES.

4.1 Political domain

4.1.1 Decision setting

In Kørte's approach, the political domain is defined by high level of authority and blunt end settings (political domain 1). Decision-making, resulting in law and regulation, is typically a decision-making by consensus supported by bureaucratic processing. Members of the group exert a role in a negotiable, discursive and deliberative manner. This is constrained by specific objectives and strategies and by potential conflict of interests. Within the maritime domain, this is reflected by Flag States deciding upon life safety and environmental issues through the IMO and its technical committee support. The second standardisation bodies are the classification societies issuing rules for the safety of ships.

From the stakeholder mapping, we have identified a second political domain (political domain 2) at the same level of authority but with the first lines of hazard. This results mainly from the double role and responsibility of states. On the one hand, Flag States develop and enforce international regulatory

389

instruments. On the other hand, Coastal and Port States are the location of the shipping traffic and potential pollution targets. The main concern is a kind of risk transfer not compensated by commercial issue, from countries having significant fleet and tonnage to countries having high waterborne traffic. Opposite from the decision-making by consensus – during so to speak "peace time" – decision-making by unilaterism takes place during "war time", i.e. accidents and disasters situation. Under growing public opinion pressure, politics should react and propose new measures in a shorter time horizon.

On the one hand, the difficulties in achieving a consensus on major change often dominate the decision-making by consensus. Changes rely on minor decisions based on limited analysis. Boisson pinpoints the main drawbacks of such a regime (Boisson, 1999): over-regulation and fragmentation of the maritime regulatory framework. This results from a composition effect, due to aggregation and combination of minor decisions. On the other hand, decision by unilaterism responds to a public demand, but could weaken international efforts. This kind of reactive action relies more on hasty judgement than on a rational analysis.

4.1.2 On risk analysis

Kørte claims «uncertainty and risk assessment should have an important place in informing public policy (decision) makers». This is stated in the context of decision by consensus aiming at developing regulatory instruments. We will consider hereafter the new political domain scene.

Within the first, removed from hazard, political domain, one of the objectives is regulatory instrument development. This domain is typically the one described by Kørte with decisions on safety regulations through IMO or by classification societies. For long time, decisions have relied on technical, i.e. analytical, support providing expertise in naval architecture, marine engineering and other scientific principles. However, the need for formal risk analysis supporting rule development is now recognised. Systematic and transparency properties of such analysis should give less room for discussion. Even if Kørte highlights that different groups could have variance in goal preferences, the prerequisite of such an analysis should therefore be an agreement on acceptance criteria and preferences. In case of high societal costs, cost benefit analysis (in the ALARP region) is recommended including a kind of stakeholders' involvement/analysis. Analytical support is required to perform this formal risk analysis. The analytical domain is also responsible for data collection and expert judgement techniques incorporation.

Safety needs validate this description. Beyond the primary need of policy setting, typical risk-related needs are described, such as information on/assessment of risk, database structuring and incident/accident information collection and archiving.

Achieving regulation improvement, balance between regulations and/or development of new ones using risk analysis depends merely on the structure of the regulatory framework. The current maritime safety regulatory framework is mainly prescriptive in nature, stating minimum equipment and competence standards. Within this situation, prior analysis by analytical domain should have an increased importance in supporting decisions. However, the evolution of the regulatory framework should also be considered, going from a prescriptive framework to a performance-based one. In the former, the regulator should be responsible for risk analysis. Therefore, no choice is left to the operator and compliance can be directly assessed. In the latter, managers/operators have to prove safety using risk analysis techniques according to a safety target specified by the regulatory body. As part of the new fire safety regulatory framework promoted by IMO, approval of an alternative design is possible. Equivalency to prescriptive-based document is assessed using fire safety engineering.

Finally, the enforcement of regulatory instruments, during both design and operational phases, is delegated to analytical and operational domains. Classification societies could provide this support, having high technical expertise and worldwide expert networks.

The second identified domain, close to the hazard, has in common the policy setting need. However, the proximity to the hazard – and the omnipresent pollution prevention and life protection needs – imposes an interweaving of political, routine and crisis domain properties.

This domain is the first line of defence in the enforcement and verification of safety standards. Knowledge of ship type specific risk and of individual ship should support a system targeting high-risk ship. Such a knowledge and information needs call for analytical processing and exchange of information at a maritime region level. The final acceptance/refusal of ship is based on spot check inspection. Assessment of ship type specific risk also supports checklist development, i.e. what to inspect in a short time horizon.

As part of a crisis domain, port and regional authorities should fulfil national/local requirements for contingency planning coupled with information on/assessment of environmental risk in certain areas. Moreover, being at the front line of hazard, accident and incident investigation is assumed. Finally, in a post-crisis situation, decision at a high level of authority, in a reactive and unilateral manner, occurs under public opinion pressure. As a modern decision-making process, such kind of decision should avoid hasty judgement and be supported by the analytical domain to perform rational risk-based decision-making.

4.2 Routine/operation domain

4.2.1 Decision setting

Routine domain is characterised by stakeholders at sharp end and low/medium level of authority. Rather that rational decision-making, Kørte lists three modes of action generation in Rasmussen and Reason's wake. Firstly, the skill based mode, where the behaviour is controlled by experienced behavioural patterns. The operator reacts to stimuli with little effort. Secondly, the rule based mode, where the behaviour uses readily available rules, procedures, etc. Finally, the knowledge based mode, where the behaviour is event specific and requires "higher level" cognitive process with a kind of analytical process (problem solving, goal setting and planning).

Moreover, on board routine decisions could not rely entirely on the first line operator. It therefore involves global communication and an increasing support from ashore stakeholders. It should be underlined that routine decisions are highly dependent on the managerial objectives.

4.2.2 On risk analysis

Kørte argues that risk is not an explicit issue as long as deterministic rules and trained skills exist. Skill and rule based modes could be programmed and formalised. Risk should be assessed and decision preprogrammed elsewhere with operator involvement and experience transfer. Decisions could be programmed insofar as they are repetitive and routine. Then, a standard or constant approach can been defined. On the other hand, if there is no matching between received information and skills/rules, un-programmed decision is made in unique, new and innovative situation. Judgement, creativity and experience are then required together with clear identification of roles and responsibilities.

Balance between programmed/un-programmed is more an art than a science. Several research efforts are performed on the problematic of formal vs. informal optimisation. As evidence, Perrow (Perrow, 1999) is astonished by the odd structure of sea collisions, so called "non collision-course collisions". Morel (Morel, 2002) highlights that many collisions result from an – incorrect – intuitive anticipation of the intention of the other. Collisions in initial non-collision course often result from a mixed use of the "left crossing" rule based decision and of a "right crossing" skill based decision accounting for distance between ships. A study (Global Maritime BV, 2002) on human HAZOP (Hazard and operability analysis) identifies little skill or rule based behaviour to be present on bulk-carriers; thus, the behaviour is predominantly knowledge based.

Kørte identifies also an alternative where information is provided on the safe operation boundaries. In other words, the objective is not to specify how to perform but to show boundaries between safe and unsafe ways to do the job.

The first line operators, together with their needs, are not well identified within THEMES. However, routine needs are encapsulated in managerial domain needs. Examples are training in routine operation – aiming at formalising skill – and International Safety Management Code compliance – aiming at formalising procedures. This is an evidence that routine domain is indeed highly dependant on managerial objectives and values. Above all, commercial and safety issues should be well balanced and prioritised.

4.3 Crisis and emergency

4.3.1 Decision setting

Kørte summarises crisis and emergency as situations relating "to an environment evolving dynamically with serious but uncertain consequences".

Narrow time limit, unknown/uncontrolled effects due to high information rate and problematic information trustworthiness, aggregation effects due to the high number of decision-makers at any level impose that decisions should above all limit negative consequences and avoid adverse outcomes. Kørte highlights that decision-maker should also limit anxiety and stress.

In this kind of situation, the number of stakeholders involved in decision(s) increases. Responsibilities and leadership change. One of the fundamental limitations towards effective decision is nevertheless different and contradictory objectives among the stakeholders.

4.3.2 On risk analysis

A characterising feature of disasters and crises is their unpredictability. In order to reduce the effects of disasters, people involved in emergency management have to be trained in those situations they can expect to encounter. However, for the emergency management organisation, it is impossible to think of every possible situations or events in advance. It is therefore important that the entire organisation can adapt flexibly and efficiently. In this context, the UK's Maritime and Coastguard Agency (Warsash Maritime Centre, 2001) makes a distinction between emergency and crisis management. Emergency management has been defined as a situation where decisions and actions are based on documented emergency procedures. Crisis management differs from emergency management in that decisions and actions do not have necessarily documented emergency procedures. In this case, there may not be a predefined response, or, if there are defined emergency responses, those responses may have conflicting requirements.

Crisis/emergency-related needs imply a high number of stakeholders in various domains as identified in THEMES. Managerial domain is responsible for training in emergency preparedness, and should comply with international emergency requirements. Regional and local authorities (political domain) and organisations (routine domain) should fulfil national/local requirements for contingency planning and provision of emergency services. Analytical domain should, firstly, provide solution for/education in emergency training and contingency planning, and, secondly, real time analysis and support.

According to the experience gained through enquiries, BEA-Mer – the French Marine Accident Investigation Bureau – (Courcoux, 2002) identifies five families of recurrent determining factors leading to accident. There are: "the decay (decrepitude) of the vessels and their operation outside of the by law operational limits"; "the shortage of communication between the vessels, and the shore and the vessels"; "the dilution of the responsibility between concerned operators"; "the difficulty to combine economics and technical constraints with consequences on the maintenance"; and, "the organisation of the work on board and the degradation of the look out".

This strengthens that co-ordination between on board, ashore and offshore efforts should be maximum. Ultimately, one of the most important factors leading to catastrophe is disagreement about objectives in crisis/emergency situation.

4.4 Managerial

4.4.1 Decision setting
Managerial domain is associated with high organisational authority level and no direct exposure to hazard. Their information process capability being finite – and their information rate high – and co-ordination not perfect – due to high number of problems to deal with –, Kørte states that managers apply a satisfying strategy. They decide upon option that is good enough according to the objectives and aspiration level.

Kørte describes two types of decision-making. The first one is a programmed decision-making, without formal reference to consequences, in accordance to rules and codes of conduct. The second one is un-programmed decision making in high or new risk situation. The decision-maker delegates to the analytical domain and approves the final decision.

4.4.2 On risk analysis
Kørte states that managers tend to apply accepted negotiated procedures according to some standards and accepted practices in order to limit data, information and analysis.

The current prescriptive regulatory regime leads to a compliance culture – strict rules application – within the maritime industry. However, new societal constraints impose that reliance on technical standards is no more a satisfactory strategy. Risk needs to

be stated/expressed and actively managed. This approach to safety could start from a responsible attitude – the manager targeting durability – or could also be "created" by a new regulatory framework and standards related to risk.

In the case of un-programmed decision, the support of analytical domain is encapsulated within the supervision process. Managers communicate values, preferences and ethical rules to the analyst in an interactive manner. The formal decision analysis aims at assessing consequences and risk with possible cost benefit analysis. Recommendations are in a form that limits the information amount. Measures of risk are summarised for final checking by the decision-maker.

The needs identified within THEMES reinforce this analysis. First of all, managerial stakeholders have a higher number of needs than stakeholders in other domains. Moreover, contacts and negotiation with regulatory, financial, customers groups, etc. are important. This means eventually high information rate and high number of problems to deal with.

Regulatory stake imposes compliance with standards. Strategic stake imposes evolution to a responsibility culture. This culture relies on an explicit management of safety, on global assessment, and also on development of code of good practices and other safety charters. Moreover, data collection and database development are now strategic and regulatory needs. For design manager, selection of option, optimisation, safety equivalency, incorporation of safety aspects as criterions and no more as constraints require risk analysis knowledge. Operational manager need to address crew training in routine and emergency, regulatory compliance, and accident investigation.

4.5 Analytical – bureaucratic

4.5.1 Decision setting
Stakeholders are far-removed from hazard with no direct level of authority.

On top of routine decision in their own account, an increased time limit for information collection and rational decision-making allows this domain to support stakeholders in other positions. Such a rational decision-making process is usually made of: objectives and scope definition, decision criteria development, identification of alternatives, assessment of the alternatives and choice of an alternative.

Even if the development of Western thought and science rises from a rationalist tradition, other ways of thinking exist and have emerged for a few decades such as systemic approach. Anticipating the future, the analytical process should not be established as the exclusive approach. It should be more accurate to speak of "modelling domain".

4.5.2 On risk analysis

Kørte identifies three decision categories. The first mode is decision made on their own account, applying rules or through satisfying against criteria. The second mode is risk-based decision support to decision-maker at higher levels. The form of this analysis, together with the form of the output, depends on the decision-maker request. Concerning the political domain expectations, we have identified, among others, supports for decision by consensus, for decision by unilaterism and for instrument enforcement. The last mode is risk analysis and pre-programming for sharp end function in terms of inspection, training, procedure, etc. both in routine and in crisis/emergency.

This domain is also in charge of gaining experience. This is expressed by database structuring accident, incident, inspection, near miss information, etc.

4.6 External stakeholders

External stakeholders are interest groups, associations, organisations, financial partners, end-users, etc. having an influence on all authority levels.

They form the environment of the other stakeholders and are therefore interested in overall assessment. This could be evaluation at a fleet level, development of safety campaign, development of code of good practices and advisory/persuasive activities. They are the driven forces for value and ethic evolution.

5 INTERACTION BETWEEN DOMAINS – ROLES AND RESPONSIBILITIES

From all this, it follows that it is possible to extent the discussion on decision domains to a description of the interactions between domains. Kørte proposes a framework of interaction between domains, roles and responsibilities. Firstly, he describes influences in terms of laws, value, strategy, experience data and process knowledge from all the domains towards the analytical domain. Secondly, he presents the analytical domain modelling. This domain processes information and provides recommendations, procedures, and programme requirements to the other domains.

5.1 How to relate the domains?

The cindynics (Kervern, 1999) offer the opportunity to formalise this framework of interactions. This discipline is indeed underlain by mnesic (data), epistemic (models), teleological (finalities), deontological (rules) and axiological (values) stakes. These five stakes are represented as the five dimensions of the so called "cindynic hyperspace", see Figure 3.

They are supposed to be the main aspects that frame the description of safety situations. According

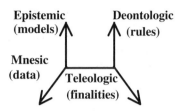

Figure 3. Cindynic hyperspace.

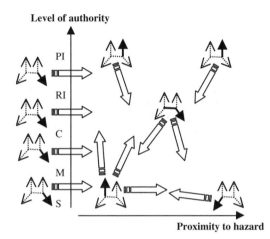

Figure 4. Influence between domains using the cindynic hyperspace representation. Main influence dimensions are in bold.

to Kervern, these five stakes could be identified for each stakeholder. This theory could be used for static description. In this case, cindynic stakes could be associated to each decision domain. Then, it is possible to work on the relations between domains. A framework of influences between domains could be developed, based on typical constraints, needs and implications on risk analysis.

5.2 Findings: influence framework

From the discussions above, some preferred cindynic dimensions emerge for each domain together with influence channels between domains (see Figure 4, the main dimensions, for each domain, are in bold).

Mnesic dimension is about the memory. More generally, it is the historic of an evolution. Data can be stored and organised in an experience feedback system. Information meets political, managerial and operational needs in term of accident, incident, near miss, inspection and audit data. These data are collected from the routine domain, stored and structured in database by the intermediary of the analytical domain.

393

Epistemic dimension is about models. Models are both risk methods and risk analysis results. This includes recommendations, procedures, checklists, training programs, etc. In the field of safety, the models are developed and used by the analytical domain to provide outputs to the other domains.

Teleological dimension is about objectives, finalities and goals. Accounting for this aspect is essential to consider the potentialities of evolution. Indeed, this dimension is the bridge between two approaches: a technical approach (the models and data) and an ontological approach (rules and values).

Deontological dimension is about rules, codes, standards, laws, etc. The main source is the political domain towards managerial domain. Managerial domain has also an important role to play in the development of a self-regulated safety culture.

Axiological dimension is about values. The analytical domain should account for values and preferences when performing analysis for higher decision-maker. Moreover, an important source of values is the external stakeholder category. This group has a strong influence on each authority level.

6 CONCLUSIONS

Starting from generic research on the use of risk analysis in different decision settings, and from a list of maritime stakeholders together with their safety needs, the combination of the two approaches has been performed. As a first result, the mapping of about forty stakeholders using a two-dimension taxonomy (proximity to hazard and level of authority) reveals some trends in term of maritime domain complexity. From a theoretical point of view, refinements of the taxonomy model – with a new category called "external constraints" – and of the five decision domains – political, managerial, routine, analytical and crisis, with the identification of a second political domain – have been performed. The descriptions of the decision domains, and of the implications on the use of risk analysis in these different settings, have been enriched by the use of identified maritime safety needs. Finally, the framework proposed by Kørte concerning influence between domains is supported by a conceptualisation using cindynic theory.

Such a work could be used as a pedagogic tool, and could also support thoughts about the development and implementation of a new pro-active maritime safety framework.

Future research is foreseen:

- Validation and improvement of the description using safety initiatives
- Identification of gaps and strengths of the current situation

- Identification of data, values, models, finalities and rules for each domain to improve the understanding of the influences.

This paper recognises the importance of interaction, ethical stakes, and knowledge development to go towards a modern safety culture and decision framework.

ACKNOWLEDGEMENT

The author acknowledges the work of Kørte and of its colleagues Aven and Rosness. The author wish to thank THEMES participants involved in stakeholder work.

DISCLAIMER

The opinions expressed in this paper are those of the author and under no circumstances should be interpreted as policy of BUREAU VERITAS or Ecole des Mines de Saint Etienne.

REFERENCES

Boisson, P. 1999. *Safety at sea – policies, regulations & international law*. Paris: Bureau Veritas.

Courcoux, L. 2002. *Maritime accident/incident investigation in France*. THEMES meeting in Windsor: BEA-Mer.

Global Maritime BV, 2002. *Formal safety assessment of bulk carriers- WP9A, High-level task inventory*. RINA.

House of Lords Select Committee on Science and Technology, 1992. *Safety Aspects of Ship Design and Technology*. London: HMSO.

Kervern, G.-Y. 1994. *Latest advances in cindynics*. Paris: Ed. Economica.

Kørte, J. et al. 2002. On the use of risk analysis in different decision settings. In ESREL2002, *European Conference on System Dependability and Safety proceedings, Lyon, 19–21 March 2002*. ESRA.

Morel, C. 2002. *Les décisions absurdes*. Pais: Gallimard.

Perrow, C. 1999. *Normal accidents*. Princeton: Princeton University Press.

THEMES, 2001. *D1.1 – Outline European Framework for the development of safety assessment in waterborne transport*. THEMES Thematic Network.

Warsash Maritime Centre, 2001. *Simulator training for handling escalating emergencies*. Maritime Coastguard Agency.

STAKEHOLDER LIST, FROM (THEMES, 2001)

Ship owners, Ship owner associations, Charterers, Ship managers, Ship yards, Classification societies, Financial organisations, P&I Clubs, Insurance companies, Ship sale and purchase brokers, Port authorities, Bridge/Lock Controllers (inland navigation), Flag states, Environmental authorities, National governments, Regional governments, Regional (waterway) authorities, Intergovernmental bodies, EU, IMO, Maritime research organisations, Seafarer training and other maritime educational establishments (including universities), Accident/incident database managers, Search and rescue (SAR) organisations, Consumer and environmental pressure groups, Political advisors (e.g. members of EU parliament), Pilots, Mooring masters, Trade Unions, Professional organisations, Crew managers, Tour operators, Stevedores/Shippers and professions related to cargo (forwarding agents, commissioners of transfer, export company, other carriers, etc.), Tug owners, Salvage companies, Immigration authorities, Customs authorities, Drug enforcement authorities, Fire department, Passengers.

Safety and Reliability – Bedford & van Gelder (eds)
© 2003 Swets & Zeitlinger, Lisse, ISBN 90 5809 551 7

New models for measuring the reliability performance of train service

Hsien-Kuo Chen (Frank)

Project Manager, Railroad Service Department, TÜV Rheinland Taiwan Ltd.

ABSTRACT: Reliability models as built can be used during the design stages with input data of field information from similar railway systems, to predict the lower margin of train service performances for the new railway system while the sublevel RAM targets could also be allocated. Two mathematical models for measuring the train service reliability and the train punctuality respectively were derived and verified based on the fundamental probability theory with comparing to several existing railway systems. Those failures that would interrupt revenue service or impair the daily timetable operation but not causing catastrophic event, e.g. derailment or collision, were taken into account. Uncertain of standard Reliability Block Diagram (RBD) to model the train service reliability and the punctuality has stimulated the development of these equations. Along with quoted references, the models have demonstrated the ability for handling the reliability problem of the complex redundancy configuration, especially the railway system.

1 INTRODUCTIONS

Generally, at the beginning of a project some overall performance requirements are set. In the case of a railway they might be expressed as the number of trains required to operate out of a total number of trains owned, over a specified time period with a specified level of reliability and punctuality. Such criteria can be specified in terms of reliability, availability and maintainability and, indeed, this has become the internationally recognized method. Owners of railways are also particularly concerned with the safety of people who use, operate, maintain or might otherwise be affected by the railway (e.g. power failures or data transmission causing trainsets to operate under the condition of no signaling).

This concept is embodied in the European Standard EN50126[1]. This Standard, which has been adopted worldwide as the procurement specifications, requires RAM target to be defined at an early stage in a railway construction project. To determine these RAM targets properly, some rationales of how to establish them need to be developed and hence this paper's main goal.

Actually, defining the Reliability, Availability and Maintainability (RAM) for the whole railway system helps to ensure that the pre-determined timetable will be achieved within affordable maintenance and logistics costs during revenue operation. Such reality reveals the importance of reasonable system wide RAM requirements. However, RAM targets generally deemed as part of systems inherent characteristics can only be achieved by the remedies of design and manufacturing processes at the design and build stage. The need of establishing practicable system wide RAM models is therefore undoubted.

When these system wide targets apportioned, RAM targets provide pre-determined performance requirements for contractors to satisfy, and goals for operation and maintenance for revenue services.

2 DEFINITIONS & ABBREVIATION

Reliability	The probability that an item can perform a required function under given conditions for a given time interval or distance.
Train Service	ReliabilityThe percentage of actual trainsets runs against the published timetable, including the backup trainsets if required.
Punctuality	The percentage of trainset arrivals at each stop within specific time against the published timetable.
SSRA	Shadow Strategic Rail Authority, UK

3   TRAIN SERVICE RELIABILITY

Train service reliability is generally determined by the combined reliability of the individual train and the associated subsystems, for example, signaling, communication, power supply, station's function, civil and depot etc. The model for establishing this reliability performance index therefore includes these respective elements according to their contribution to the service interruption.

Another commonly used parameter, which provides quite similar measurement on the capability of train service delivery is called the "Missed Trips". The train unreliability tends to the missed trip asymptotically when the cumulated number of days for revenue operation is increased and consideration of the train trips assigned to each trainset will be close as counted on a yearly basis. To simplify the modeling process for calculating the missed trips, it is assumed that the train unreliability is equivalent to the percentage of missed trips and would not induce significant biases.

A simple formula is defined below for counting the train reliability during the revenue operation (for example, start from 6 A.M. to 12 P.M. or 18 hours operating time):

Train Service Reliability

$$= \frac{\text{no. of actually operated trains}}{\text{no. of scheduled train per prior defined timetable}} \quad (1)$$

Knowing that such equation needs to be transformed into the equation of probability for predicting the possible achievement of train service reliability during the design stage, the corresponding mathematical model is therefore developed as follows.

The probability of failure for each subsystem, which could cause the service interruptions, shall be statistical independent. This can be explained by the definition of secondary failure[2] that were caused by the primary incident will not be counted, i.e. train derailed and collide adjacent items (equipment or facility) such failures of the damaged items will not be taken into account as its own failure (i.e. non-chargeable failure). In addition, such assumption will considerably reduce the complexity of the process to formulate the relevant equations unless there is evidence that such interaction cannot be neglected, e.g. EMI effect between signaling and power supply or communication. However, EMI has been broadly considered and tested[3] via proper design isolations. The concepts of isolation for the relevant subsystems, which comprise the system, to assure the minimum interferences between each subsystem are not limited to electrical/electronics applications. Much mechanical equipment has been designed to avoid escalation of failure consequence, e.g. protections, redundancy

and detection/warning etc. A typical application of isolation concept is the "Fail-Safe" requirement that has been broadly applied in the field of Reliability Engineering to deal with the safety concerned functions whenever single failure happened, e.g. alternative braking system.

To use such assumption properly in the model development, the failure categories should not contain those failures, which will lead to catastrophic accidents, e.g. derailment or collision. Hence, the definition of service interruption of train is listed below:

"A delay or cancellation of scheduled train arrivals, but not include catastrophic events"

In other words, the failed trainsets still are on the track but can't run under normal speed (equipment failure) or have to be removed from the mainline due to loss of mobilizing function (immobilizing failure).

According to the aforementioned, the generic model of train service reliability of the whole railway system can be properly expressed as follows:

$$R_T = f(R_{core}, R_{station}, R_{track}, R_{civil}, R_{depot}) \quad (2)$$

Where:

R_T = Train Reliability for whole railway system

R_{core} = Reliability of trainset system, including Rolling Stock, Signaling, Communication, Power Supply, and Wayside Equipment

$R_{station}$ = Reliability of Station's functions

R_{track} = Reliability of Track

R_{civil} = Reliability of Civil, including tunnels, bridges, viaduct, embankment, etc.

R_{depot} = Reliability of Depots, including main workshop, stabling yields and other depots

Equation (2) represented the whole system train reliability is a function of the reliability for its subsystems. When the assumption of statistical independent of failure between each subsystem is applied, this equation can be further simplified as follows:

$$R_T = R_{core} \cdot R_{station} \cdot R_{track} \cdot R_{civil} \cdot R_{depot} + a(w) \quad (3)$$

where the term $a(w)$ stand for uncertainties as may be brought by the interactions that existed between the subsystems. To simply the approach, this term is assumed to be vanished but will be exploded in next stage study.

In order to keep this report concise and readable, the final equation for calculation of train service reliability is given below directly:

$Rtr = Rrs \cdot Rnrs \cdot Rstation \cdot Rtrack \cdot Rcivil \cdot Rdepot + b(Rrs, Rnrs)$

$+ c(Rrs, Rstation) + d(Rrs, Rtrack) + e(Rtr, Rdepot) + f(Rtr, Rcivil)$

$+ Rt^n (1 - Rnrs \cdot Rstation \cdot Rtrack \cdot Rcivil \cdot Rdepot)(1 - rt / OT)PT$

$$(4)$$

where:

rt = time interval for the whole system out of service due to immobilizing failures (e.g. 3 hours);

PT = 95% probability of the whole system recovered within required maintenance intervals (e.g. 3 hours) when some part of the whole system failed (immobilizing failure) but excludes Rolling Stock;

R_{rs} – (see equation (5) for detail) is the daily reliability of Rolling Stock subsystem with support by backup train and includes the condition of train to be counted for backup service[4];

R_{nrs} – is the daily reliability of Non-Rolling Stock subsystems, that is, excludes the Rolling Stock from the R_{core} in the notes of equation (2);

$b(\cdot)$, $c(\cdot)$, $d(\cdot)$, $e(\cdot)$, $f(\cdot)$ – represent the interactions between rolling stock and other subsystems. This means that even some failure occurred due to the interference but will still have some successful train dispatches; and for sole rolling stock the following equation is used:

$$Rrs = \sum_i^n C_i^n (Rt)^i (1-Rt)^{n-i} (1-\frac{n-i}{OT}) PS \tag{5}$$

for $i = n-m,...,n-2,n-1,n$

where:

R_t is the reliability of a single train that has accomplished the required travel distances per day; n is the number of total required trains for daily operation considered for one particular year; m is the allowable number of breakdown trains (due to immobilizing failures) per day considered for one particular year; $PS = 0$ when $i = n$ or $PS = 95\%$ for $i \neq n$. The PS considers only the immobilizing failure (e.g. train axle locked or hot axle) caused a closure of the system temporary but can be recovered within a specific time duration (e.g. 3 hours). The total operation time (OT) normally is 17 hours per day for either MTR or long distance transportation railway.

4 PUNCTUALITY

As referred to in the previous section, the reliability of the train itself and the relevant subsystems (e.g. signaling, power, communication, station and track, etc.) again form the basis for development of the equation of train punctuality. From this point, the percentage of the scheduled train arrivals at the stations can be measured. Similar approaches and rules are adopted during this derivation with some additional notes and assumptions being made to simply the process of modeling.

As similar to the train reliability, a simple formula is defined in below for accounting the train punctuality

during the revenue operation:

$$\text{Punctuality} = (\text{no. of cumulated on time arrivals})/ (\text{no. of total scheduled arrivals}) \tag{6}$$

trains are punctual if they arrive at their destination within a specific time deviation (e.g. 10 minutes).

Due to equation (6) can't be used when no actual operation data is available for this new system, it must be expressed by the form of probability to predict the possible achievement of train punctuality during the design stages, the corresponding mathematical model is therefore shown below.

Generally, the average number of delayed train arrivals per day (Number of Delayed Arrival – DA) can be determined by the following equation:

DA 5 Total no. of scheduled train arrivals (ARt) – {Number of train arrivals on time for all subsystems operated under normal conditions (ARn) 1 Number of train arrivals on time when one or more service interruption failures occurred in rolling stock while others subsystems operated under normal conditions or Number of train arrivals on time when one or more service interruption failures occurred in other subsystem but not Rolling Stock(ARr)} (7)

The final equation for calculation of the train punctuality is described below for reference:
Punctuality = $100\% - (DA/ARt)$ or

$$\left[(\prod_{i=1}^{TP} R_i^{N_{i,j}} + \sum_{k=1}^{FT} g(k,j) \right] \bar{R}' + h(R_{rs}, R_{nrs}) + l(R_{rs}, R_{station}) +$$

$$p(R_{rs}, R_{track}) + s(R_{rs}, R_{civil}) + t(R_{rs}, R_{depot}) +$$

$$\left[(\prod_{i=1}^{TP} R_i^{N_{i,j}}) \right] (1 - \bar{R}')(1 - \frac{rt}{OT}) PQ \tag{8}$$

where:

R_i = Trainset reliability to complete a particular trip pattern i;

PQ = 95% represents the probability of recovering from a service interruption failure within specific time intervals (for immobilizing and equipment failures);

$\bar{R}'$ = $R_{nrs} \cdot R_{station} \cdot R_{track} \cdot R_{civil} \cdot R_{depot}$ (for service interruption failures);

TP = Total number of trip pattern per day;

$N_{i,j}$ = Number of train trip of day j for one particular trip pattern i;

FT = Number of train fault per day;

$h(\cdot)$, $l(\cdot)$, $p(\cdot)$, $s(\cdot)$, & $t(\cdot)$ – represent the number of successful train arrivals remained despite of some failures occurred due to interactions between rolling stock and other subsystems; and

$g(k,j)$ defining the number of possible delay of train arrival that could be encountered for different train trip patterns. These can be shown in equation (9) and (10) in the below:

$$g(k, j) = \sum_{m=1}^{TP} \left\{ C_1^{N_{m,j}} FS_{m,1,j} R_m^{N_{m,j}-1}(1-R_m) \prod_{\substack{q=1; \\ q \neq m}}^{TP} R_q^{N_{q,j}} \right\}, \text{ for } k = 1$$

$$(9)$$

$$g(k, j) = \sum_{m=1}^{TP} \left\{ C_2^{N_{m,j}} FS_{m,m,j} R_m^{N_{m,j}-2}(1-R_m)^2 \prod_{\substack{q=1; \\ q \neq m}}^{TP} R_q^{N_{q,j}} \right\} +$$

$$\sum_{\substack{m=1; \\ r \neq m}}^{TP} \sum_{r=1;}^{TP} \left\{ C_1^{N_{m,j}} FS_{m,r,j} R_m^{N_{m,j}-1}(1-R_m) C_1^{N_{r,j}} R_r^{N_{r,j}-1}(1-R_r) \prod_{\substack{q=1; \\ q \neq m \\ q \neq r}}^{TP} R_q^{N_{q,j}} \right\}$$

$$(10)$$

and when $k \geqslant 3$, $g(k,j)$ tends to zero due to

$$(1 - R_m)^k \to 0 \quad \text{when } R_m \to 1$$

The additional three functions: $FS_{x,y,z,}$ shown in equation (9) & (10) are defined as follows:

$FS_{m,1,j}$ – is the function for representing the possible remaining on-time train arrivals to the total daily scheduled train arrival when one of train trip (pattern m) fault occurred.

$FS_{m,m,j}$ – is the function for representing the possible of remaining on-time train arrivals to the total daily scheduled train arrival when two identical train trip (pattern m) faults occurred.

$FS_{m,r,j}$ – is the function for representing the possible of remaining on-time train arrivals to the total daily scheduled train arrival when two train trip (pattern m and pattern r) faults occurred.

The reasonable number for these functions: $FS_{x,y,z}$ can only be properly determined through conducting timetable simulation according to each train fault scenarios. Once the number is estimated, the train punctuality is obtained via these equations.

5 ANALYSIS AND DISCUSSION

Several input data can be obtained from the published data of similar railway systems[5] (website information), relevant RAM prediction data of each major

Table 1. Typical MTBF of major subsystem for model evaluations (for illustration only).

Subsystem	Train reliability (MTBF, hour)	Train punctuality (MTBF, hour)
Rolling Stock (for single train)	15,000	4,200
Signaling	1,300	250
Power Supply	4,400	2,700
Communication	–	50
Wayside E&M	500	250

subsystem and the operational plan for new railway system to make the equation (4) and equation (8) practicable for predicting the train service reliability and train punctuality respectively.

Substituting the necessary information gathered from the aforementioned assumptions for statistical independent and available data sources, the train service reliability for the system of the new railway system can be estimated, via the equation (4). For example, the Charter Standards as published by SSAR (i.e. R_{tr} = 0.99 or 99% *for Great Western and 99.2% for Gatwick Express*) and other available train reliability data in the public domain can be used as a benchmark to determine the achievable RAM performance targets for the new railway system. Failure information regarding the rolling stock, non-rolling stock as well as the others subsystems formed the basis for undertaking the calculations where these could be found in the references. To use this model is straightforward whenever the associated input data are available. Some typical data for major subsystems are listed in Table 1 as an example to describe how these models can predict the reliability performances before the system is to operate.

For the MTBF of Stations as used in equation (4), the probability of functional failure, either hardware or software within a station that could cause the closure of the track sections in that particular station or adjacent to it will be quite small. Such cases could only be a serious track flooding near the station platform or station fire. In this paper, the reliability for all stations along the can be obtained by the reliability of single station to the power of the number of stations. Track system normally contributes less than 10% to the total service interruption failures as refer to the field data from the reference railway system. The Civil subsystem is under steady state condition and the any failure of Civil & Structure should not affect the revenue operation as the normal maintenance regular deployed. Where depots are designed for only maintenance purposes and any failure of single depot equipment should not be allowed to cause the interruption of revenue service. Therefore these two subsystems have no numerical reliability requirements in terms of the train reliability for the models evaluation in this paper. However, if more detail information is available the effects of failure for Depot and Civil many bring the prediction more closely to the actual operational condition.

Based on the equation (8) and the same assumptions made for analyzing the train reliability, the calculated train punctuality for the system can be derived. This number is further round up to upper integer (e.g. 90% for long distance railway and 97% for MTR system) as a suggested target for the new railway system due to the consideration of worst case that has been adopted within the formulation. As compared with the statistics of train punctuality for the reference railway system, train punctuality of this new system's RAM performances seems reasonable. Attempting for escalating the number of train punctuality is, however, not recommended. It is due to considering the necessity of proper margin for future growth and the statistics collected from others similar railway systems, e.g. the train punctuality of Eurostar for only 89% within 15 minutes of delay at year 2000[6].

In developing the mathematical equations of train reliability and punctuality, one may think about the similar method, i.e. the reliability block diagram (RBD) as usually being applied among the reliability engineers. In this case, the RBD, however, cannot properly represent the system configuration, which combined the minimum number of required duty trainsets per the timetable with the standby trainsets. Other shortfall of RBD is that it cannot represent the conditions of possible train cancellations and train delays whenever some part of system failed. These subsequent effects can be properly reflected in the equation (8), (9) and (10) when the related parameters can be estimated properly.

Besides the aforementioned benefits, these two models can provide a basis (or reference) for conducting trade-off study for alternatives as it many be necessary to improve or rectify the poor quality part of the system whenever the significant incompliance RAM performances observed during any phases of the project.

6 CONCLUSION AND FUTURE STUDY

The less effectiveness of RBD to model the train service reliability and the punctuality has stimulated the development of the relevant equations. Along with quoted references, the models have demonstrated the ability for handling the reliability problem of the complex redundancy configuration, especially for the railway system.

Further development of the models will be conducted to incorporate the factors of interactions between each subsystem, if necessary, and the possible variations of quoted information in order to establish a standard procedure for evaluation of railway system reliability. This includes the parameters as shown in equation (4) & (8), i.e. function $b(\cdot)$, $c(\cdot)$, $d(\cdot)$, $e(\cdot)$, $f(\cdot)$, $h(\cdot)$, $l(\cdot)$, $p(\cdot)$, $s(\cdot)$ and $t(\cdot)$. Sensitivity of the models will also explore the uncertainties of the assessment during the applications. The raw data, being used in the models, will be further studied for their contributions to the system maintainability and availability. Additional scope of study will be to explore explicitly the interconnection between timetable simulation and the RAM performance for the trainset system, especially when serious failure occurs at normal operational conditions.

REFERENCES

[1] EN50126: 1999 "Railway applications – The specifications and demonstrations of Reliability, Availability, Maintainability and Safety (RAMS)".
[2] MIL-STD-785B "Reliability Program for Systems and Equipment Development and Production".
[3] MIL-STD-462 Revision D "Measurement of Electromagnetic Interference Characteristics".
[4] SSAR Explanatory Notes, Performance Standard and Statistics for 2000 year Annual Report, UK.
[5] JR-Central 2001 Annual Report (Website information).
[6] "Go by Train and Not by Plane?", Mr. Robinson, 31 October 2000 (www.btinternet/~mike.feeris/trains.htm).

Safety and Reliability – Bedford & van Gelder (eds)
© 2003 Swets & Zeitlinger, Lisse, ISBN 90 5809 551 7

Techniques for modelling PSAs for use in risk-informed PSA applications and risk monitors

A.J. Commandeur

Atkins, Woodcote Grove, Epsom

ABSTRACT: Probabilistic Safety Assessment (PSA) models often assume asymmetrical plant configurations in order to reduce modelling effort and to reduce model complexity. For example, in a four-loop reactor, a loss of coolant accident (LOCA) in the primary circuit is often assumed to occur in one specific loop. The analyst's judgement is used to determine which loop is considered to have failed; normally by considering the "worst case" scenario. Errors in the choices of "failed" or affected plant can be made as most facilities are never 100% symmetrical and the "worst case" is difficult to judge due to the complexity of some accident sequences and support system dependencies. Therefore where a choice for the most pessimistic option is thought to have taken place, in fact a more optimistic option may have been selected. Furthermore, the choice for the worst case is not always appropriate if the worst case is significantly worse than all the other cases, thereby skewing the risk for a particular fault.

For risk informed applications where investment or actions on specific items of equipment or plant are under consideration, the PSA model should be representative for the application. Furthermore, risk monitors are used to evaluate the risk of specific plant configurations where items of equipment are set to running or standby matching the plant configuration, and where several plant items have been taken out of service for maintenance. The resulting plant configuration does not normally match the plant configuration assumed in the PSA model.

General techniques are presented to reduce, as much as possible, the size and complexity of modelling a fully representative PSA. In addition our experience in developing the techniques for modelling symmetrical advanced gas cooled reactor (AGR) PSAs for risk monitor applications using RiskSpectrum® Professional software is discussed. In addition, due to special features available within RiskSpectrum® Professional, methods of simplifying the general techniques listed above are also presented.

1 INTRODUCTION

This paper presents techniques that have been applied to remove modelling asymmetries which are implemented in PSAs to simplify the modelling effort by assuming faults and maintenance occur in selected trains/circuits in the PSA model. The method presented in this paper to fully represent all faults has been developed on two PSAs for the British Energy AGR nuclear power plants at Heysham 2 (HYB) and Torness (TOR).

A risk (safety) monitor is used to determine the impact on risk of removing items of plant from service. In order to demonstrate the effect on risk of removing specific items of plant in a quadrant, or combinations of items of plant, from service, a risk model that could reflect such actual outages was required.

The application of the PSAs for living PSA and safety monitor applications therefore require that the original PSA assumptions be revisited and the systems modelling and top logic modelling be changed in order to allow a fault to occur in any of the trains/circuits it could occur in. This is to enable the complex configuration changes which occur during operation to be reflected directly in the PSA. This allows a more accurate assessment of the risk due to additional maintenance activities that may be desired, taking into consideration the actual plant availability and configuration.

British Energy are in the process of replacing the existing Risk Monitors at Heysham II (HYB) and Torness (TOR). British Energy did not wish to maintain two different models, the living PSA model for the safety case and a risk monitor PSA model. A PSA model was required which would form the basis of the safety case and could also be used directly in the risk monitor application. A prerequisite of the modelling techniques was therefore to ensure that the risk monitor and the living PSA would make use of the same PSA model.

An AGR reactor steam generation plant (boiler plant) typically consists of 4 quadrants. Faults affecting a single quadrant would typically be assumed to occur in a pre-selected quadrant based on judgement of which quadrant would be the most conservative. The PSA models revised now model the faults in all four quadrants and the revised models have been loosely termed 4-quadrant models:

- multiple initiating events for faults in all affected trains or loops
- use of initiating event fault trees to represent all affected trains
- use of initiating events instead of house events allowing all train faults to be represented in one event tree
- functional fault trees for faults that reduce system redundancy
- functional fault trees for actions required on fault trains/loops
- not logic in functional fault trees to remove disallowed combinations of initiating events
- modelling standby and running combinations for support systems
- modelling of maintenance house events for plant unavailability

The methods applied consist of the following steps to create a 4-quadrant model:

- Development of quadrant fault initiating events
- Development of functional fault trees to account for quadrant faults
- Development of event trees for quadrant faults
- Development of system fault trees
- Quantification of quadrant faults

2 DEVELOPMENT OF QUADRANT FAULT INITIATING EVENTS

Initiating event basic events which represent a fault in each of the trains/circuits in which the fault can occur have been developed as follows:

- use of multiple initiating events for individual faults in all affected trains or loops
- use of initiating event fault trees to represent all affected trains

The initiating event frequency for a quadrant fault determined for the initial PSA was sub-divided into individual quadrant initiating events for each applicable fault. In most cases this will mean dividing the fault frequency by a factor of four to get the individual quadrant fault frequency.

An initiating event fault tree is created to group the faults where the frequency is divided into four quadrants together. Four basic events representing the individual quadrant fault frequencies are placed under an OR gate. An example initiating event fault tree is presented in Figure 1.

3 DEVELOPMENT OF FUNCTIONAL FAULT TREES TO ACCOUNT FOR QUADRANT FAULTS

The most effort in the techniques described in his paper involved developing the functional fault trees to account for initiating events occurring in all of the four quadrants and is accomplished as follows:

- use of initiating events instead of house events allowing all train faults to be represented in one event tree
- functional fault trees for faults that reduce system redundancy
- functional fault trees for actions required on individual failed trains/loops
- not logic in functional fault trees to remove disallowed combinations of initiating events

The functional fault trees were developed to account for more than one initiating event being considered simultaneously in the tree. There are essentially two cases where the modelling is required to change from the situation where an event tree is developed for a single initiating event tree. The first case is where a quadrant fault renders a quadrant unavailable for mitigation

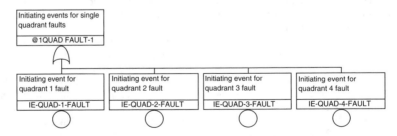

Figure 1.   Example fault tree with 4 quadrant initiating events.

post trip. The second case is where components specific to a particular quadrant need to be operated in order to mitigate an individual quadrant fault. An example event tree, which contains examples of FFTs for both cases mentioned above, is presented in Figure 2.

1. In the case where a quadrant fault renders a quadrant unavailable for mitigation post trip, the availability of the quadrant needs to be removed from the fault tree logic where four quadrants are normally modelled as available post trip to mitigate

the fault. This is usually accomplished by placing the individual quadrant mitigating systems under a top AND gate and using house events to switch out the contribution from a single quadrant for a given analysis. However as the event tree quantifies four faults simultaneously, the use of house events is not possible. A simple solution is to make use of the initiating event basic events to switch out the affected quadrant. An example of the functional fault tree logic required to achieve this is presented in Figure 3. For each quadrant, the initiating event

Single quadrant fault - frequency determined for all 4 quadrants	Feed to 1 out of 4 quadrants	Isolate the quadrant with the fault				
(1 QUAD FAULT)	(FEED 1/4 QUAD)	(ISOL FAULT QUAD)	No.	Freq.	Conseq.	Code
			1	1.00E+00	OK	
			2	1.00E-09	OK	(FEED 1/4 QUAD)
			3	1.00E-12	CD	(FEED 1/4 QUAD)-(ISOL FAULT QUAD)

Figure 2.   Example event tree.

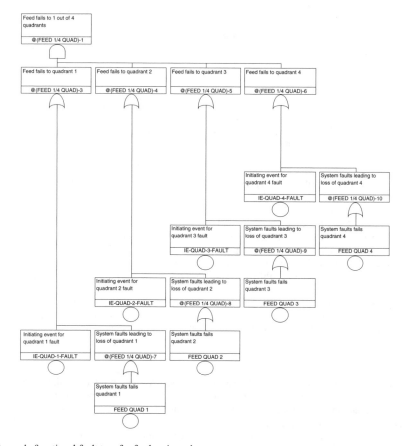

Figure 3.   Example functional fault tree for feed to 4 quadrants.

basic events and the respective quadrant system fault tree are placed under an OR gate. (In this example the system fault tree is represented by a single basic event.) This arrangement is shown on the left hand side of the FFT presented in Figure 3.

However, using this logic will result in combinations of the four initiating event basic events occurring in the same cut set. As the initiating events cannot occur simultaneously, these combinations need to be removed. This could be accomplished by placing the unacceptable combinations of initiating event basic events under a negated OR gate (NOT-OR or a NOR gate) under the top AND gate. This is illustrated by the branch on the right hand side of the functional fault tree in Figure 3. A NOR gate is where the gate is false ($=0$) for any combinations of the basic events under the gate occurring. The unacceptable combinations of basic events are determined by placing the four quadrant fault initiating event basic events under a 2/N gate. The 2/N gate returns combinations of any two initiating event basic events under the gate. Therefore any multiple initiating event basic event cut set combination will render the NOR gate false ($=0$). As the NOR gate is under the top AND gate the top AND gate will be false ($=0$) and therefore the combination will not be included in the minimal cut set results. The PSAs have been developed using RiskSpectrum Professional for Windows software. A useful feature of the software is that during determination of the sequence cutsets the software deletes any cutsets with more than one frequency event. As only the initiating events are the only frequency events defined in the models, the requirement to include the NOT logic above is removed.

2. In the case where a quadrant fault requires quadrant specific basic events/mitigating systems to mitigate the fault, the model requires that the functional fault tree needs to provide separate basic events/mitigating systems for each specific quadrant fault. This is normally accomplished by using house events to switch on or switch off logic given a specific initiating event or by modelling a simplified functional fault tree where only the basic events/mitigating systems required for that individual fault would be modelled. Four different functional fault trees would then be required for four separate individual quadrant faults. However as the event tree in the alternative method quantifies four single quadrant faults simultaneously, the use of house events is not possible. For the alternative method, an example functional fault tree is presented in Figure 4. The mitigating systems for each of the four quadrants are placed under a top OR gate. For each quadrant the mitigating system is placed under an AND gate together with a negated OR gate (NOT-OR or a NOR gate), the inputs of

which are the initiating events that are not mitigated by the system under consideration. Using this logic prevents combinations of the four initiating event basic events occurring in the same cut set. Therefore any non-relevant quadrant initiating event basic event cut set combination will render the NOR gate false ($=0$). As the NOR gate is under the quadrant AND gate the AND gate will be false ($=0$) and therefore the combination will not be included in the minimal cut set results.

4 DEVELOPMENT OF EVENT TREES FOR QUADRANT FAULTS

The existing event trees developed for the PSA were utilised for the 4-quadrant models. The event trees in general required linking of the initiating event fault trees developed and the redeveloped functional fault trees, which would have been developed for a specific fault group. In general the event tree branch points and sequences remained unchanged.

5 DEVELOPMENT OF SYSTEM FAULT TREES

The development of system fault are required to resolve asymmetries assumed in the PSA firstly for support system initiating event faults and secondly for support system asymmetries assumed in assigning specific duty or standby pumps and circuits.

5.1 Method used to resolve support system bounding fault tree asymmetries

The support system bounding fault tree asymmetries that have been identified are where a specific bounding fault assumes a specific pump (or train) within a support system is failed due to the fault. In original PSAs the bounding faults are assumed to occur in a specific train and the train is rendered unavailable for the bounding fault. In order to remove the support system bounding fault asymmetry the frequency is divided into the number of specific trains bounded by the bounding fault and assigned to initiating events basic events for each train.

An initiating event fault tree is created containing the initiating events basic events representing the faults in each train and the initiating event fault is assigned to the bounding fault event tree. In the original PSA a house event would be invoked that would render the support system train unavailable for mitigating the fault. However the method used in the 4-quadrant model uses a single event tree to quantify multiple bounding faults simultaneously and therefore house events cannot be used directly to switch in/out different fault tree branches for each of the bounding faults.

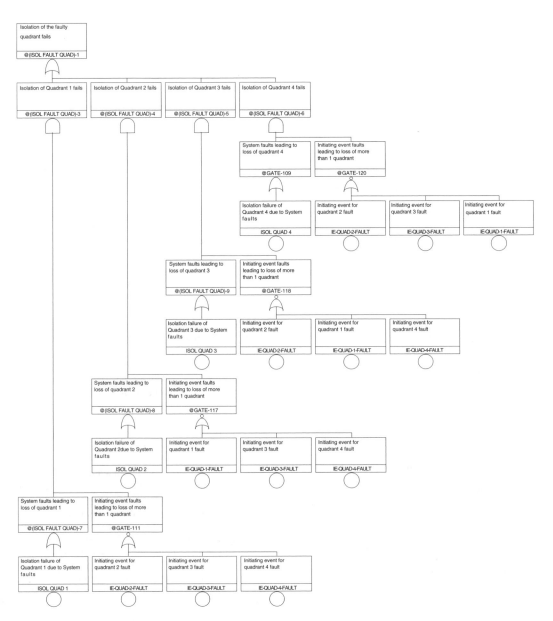

Figure 4.   Example functional fault tree for quadrant isolation.

However, in order to model the support system dependencies the fault house events are used to invoke the bounding fault initiating event that fail the relevant support system trains. This is replicated for the other support system trains failed due to each of the bounding faults.

Using this logic will result in combinations of the four initiating event basic events occurring in the same cut set. As the initiating events are assumed not to occur simultaneously, these combinations need to be removed. All the faults/initiating events are represented by basic events assigned a "Frequency" parameter type. Risk Spectrum Professional [4] does not permit cutsets where more than one event is assigned a "Frequency" parameter type and automatically removes such cutsets. Risk Spectrum Professional therefore solves the problem of more than one initiating event in a cutset automatically.

```
MINIMAL CUTSETS
     Top event frequency F = 1.000E-12
No.  Freq.   %   Q          Event          Description
1    2.50E-13 25  2.50E-01   IE-QUAD-4-FAULT Initiating event for quadrant 4 fault
                  1.00E-03   FEED QUAD 1     System faults fails quadrant 1
                  1.00E-03   FEED QUAD 2     System faults fails quadrant 2
                  1.00E-03   FEED QUAD 3     System faults fails quadrant 3
                  1.00E-03   ISOL QUAD 4     Isolation failure of Quadrant 4 due to System faults
2    2.50E-13 25  2.50E-01   IE-QUAD-3-FAULT Initiating event for quadrant 3 fault
                  1.00E-03   FEED QUAD 1     System faults fails quadrant 1
                  1.00E-03   FEED QUAD 2     System faults fails quadrant 2
                  1.00E-03   FEED QUAD 4     System faults fails quadrant 4
                  1.00E-03   ISOL QUAD 3     Isolation failure of Quadrant 3 due to System faults
3    2.50E-13 25  2.50E-01   IE-QUAD-2-FAULT Initiating event for quadrant 2 fault
                  1.00E-03   FEED QUAD 1     System faults fails quadrant 1
                  1.00E-03   FEED QUAD 3     System faults fails quadrant 3
                  1.00E-03   FEED QUAD 4     System faults fails quadrant 4
                  1.00E-03   ISOL QUAD 2     Isolation failure of Quadrant 2 due to System faults
4    2.50E-13 25  2.50E-01   IE-QUAD-1-FAULT System faults fails quadrant 2
                  1.00E-03   FEED QUAD 3     System faults fails quadrant 3
                  1.00E-03   FEED        Initiating event for quadrant 1 fault
                  1.00E-03   FEED QUAD 2     QUAD 4  System faults fails quadrant 4
                  1.00E-03   ISOL QUAD 1     Isolation failure of Quadrant 1 due to System faults
```

Figure 5.   Results for the simplified example event tree.

5.2   *Method used to resolve system fault tree asymmetries*

The original PSA models included a number of system fault tree asymmetries. Typically associated with the assumption that one of two (or more) pumps are nominated as a duty pumps and the other(s) as standby pumps. In order to model the contribution of failure to start for both pumps the branch modelling the failure to start contributors are placed under an AND gate with a basic event modelling the standby probability. This basic event in the case of a single running and single standby pump is assigned a probability of 0.5 and order to represent the 50/50 pump running or standby duty cycle. At the point where the two redundant pumps are placed under an AND gate, a Not-AND (NAND) gate is included which models the two standby probability basic events for each of the two pumps. This logic is used to remove any cutsets that include the two standby probability basic events.

6   RESULTS FROM APPLICATION OF THE TECHNIQUES ON PLANT PSAS

The results from the revised 4-quadrant models were benchmarked against the results obtained from the original PSAs. In many cases the results were broadly in line with the original PSAs, however for two groups of faults the result produced surprising results. In these two groups a "conservative" assumption was made in which the fault is assumed to occur and the initiating event multiplied by the number of quadrants to represent faults in the remaining quadrants. In the first case the quadrant chosen for faults was based on the most conservative quadrant. However the quadrant chosen, was a quadrant which was dependent on the unit transformer tap changer. The risk from this quadrant was discovered to be significantly higher than in any of the other quadrant. Therefore, due to the multiplier applied in the initiating event frequency the risk due to this fault was overestimated by a factor of 4. On a second group of faults the complex dependencies on cross-quadrant protection systems was not judged correctly and a quadrant with a significantly lower risk than other quadrants had been selected. The risk from this group of faults was therefore underestimated. On balance the combined risk from all faults was not significantly different, however the contributions to the risk from individual fault groups did change. The benchmarking and verification of the results demonstrated that the techniques applied in the paper were a cost effective in producing a PSA model suitable for the safety case as well as Risk Monitor applications.

7   CONCLUSIONS

The techniques applied did not require redevelopment of separate event trees for faults in each individual quadrant/circuit, and used the original event trees developed for the PSA.

The use of the techniques described above was estimated to have reduced the cost of modelling all faults by at least 30%, over traditional methods where each "Quadrant" Fault is modelled individually. The need for a large number of individual event trees and functional fault trees to represent the individual quadrant faults has therefore been omitted.

The method has also been particularly effective in modelling component specific transfer sequences and sequences where consequential faults need to be considered. E.g. Potential boiler tube leaks on boiler overpressurisation following spurious boiler stop valve closure.

REFERENCES

1. Relcon AB, Risk Spectrum Professional Version 1.20, Reference Manual, 1998.
2. Heysham 2 Power Station. Four Quadrant Probabilistic Safety Assessment Model Changes Report. WSA Report No. CJ0063-R1, Issue 4, October 2002.
3. Torness Power Station. Four Quadrant Probabilistic Safety Assessment Model Changes Report. WSA Report No. CJ0063-R2, Issue 4, October 2002.

Safety and Reliability – Bedford & van Gelder (eds)
© 2003 Swets & Zeitlinger, Lisse, ISBN 90 5809 551 7

Treatment of natural external hazards in the probabilistic safety assessment of a nuclear submarine refit programme

A.J. Commandeur
Atkins Defence Systems, Woodcote Grove, Epsom, Surrey, UK

R. Curry
Devonport Management Ltd., Devonport Royal Dockyard, Plymouth

ABSTRACT: The Royal Navy's fleet of Vanguard Class nuclear submarines are undergoing a major programme of refit and overhaul at Devonport Dockyard, Plymouth. Implementation of the programme is subject to the approval of a safety case by military and civil regulators. In support of the safety case, WS Atkins Consultants Ltd. (Atkins) has undertaken a Probabilistic Safety Assessment (PSA) of the parts of the refit programme relevant to the refuelling of the submarines' nuclear reactors.

This paper outlines the Probabilistic Safety Assessment (PSA) modelling work undertaken by Atkins and DML for the evaluation of natural external hazards. The Natural External Hazards Assessment (NEHA) is required in support of the refit programme and contributes to the 9 Dock LOP(R) pre operational safety case report (POSR). The natural hazards assessment (NEHA) has subjected the following potential natural hazards to detailed assessment: seismic, extreme tide, high wind, lightning strike and extreme temperature events. The NEHA has been fully incorporated into the pre operational safety case PSA model.

The ability of the Vanguard class submarine and associated 9 Dock facilities to withstand the above set of natural hazards has been assessed. For all hazards the PSA assessment considers hazard induced failures of the dock facilities and for the seismic hazard only, dock facilities and submarine components.

Traditionally, natural hazards are represented in a PSA at a design basis event (DBE), for example at a 10^{-4}/yr event and, in the case of seismic hazards, an additional seismic margin event (SME) (e.g. at 40% above DBE). This paper presents a novel "decade step" approach, in which the robustness of the dock facilities are examined for specific hazard events with return frequencies that decrease in decade steps e.g. seismic events with return frequencies of 1E-02, 1E-03, 1E-04, 1E-05, 1E-06 and 1E-07. As the submarine is being refitted in a Dockyard facility a wider range of hazards could potentially affect the submarine than a well sited civil nuclear facility. A decade step approach would therefore enable a comprehensive assessment of the risks and robustness of the facilities against the hazards at different hazard intensities and frequency levels.

1 INTRODUCTION

The Royal Navy's fleet of Vanguard Class nuclear submarines are undergoing a major programme of refit and overhaul at Devonport Dockyard, Plymouth. Implementation of the programme is subject to the approval of a safety case by military and civil regulators. In support of the safety case, WS Atkins Consultants Ltd. (Atkins) has undertaken a Probabilistic Safety Assessment (PSA) of the parts of the refit programme relevant to the refuelling of the submarines' nuclear reactors.

This paper outlines the Probabilistic Safety Assessment (PSA) modelling work undertaken by Atkins and DML for the evaluation of natural external hazards. The Natural External Hazards Assessment (NEHA) is required in support of the refit programme and contributes to the 9 Dock LOP(R) pre operational safety case (POSR) [1].

Following a Devonport Management Limited (DML) review of credible natural hazards [2], the following set of hazards was identified for detailed assessment:

1. Seismic (earthquake)
2. Extreme tide (flooding)
   - Extreme water levels
   - Tsunami/storm surge

3. Wind, Hurricane, Tornado
4. Other natural hazards
   - Air temperature
   - Lightning

The above natural hazards exist throughout all phases of the submarine refit process and could be considered to challenge the protection against criticality accidents, loss of primary coolant accidents or loss of heat sink accidents.

Traditionally, natural hazards are represented in a PSA by a design basis event (DBE), for example at a 10^{-4}/yr event and, in the case of seismic hazards, by an additional seismic margin event (SME) (e.g. at 40% above DBE). For comparison with the above criteria, the performance of the facility would only be considered for the design basis e.g. at 10^{-4}/yr events and at seismic margin events.

In the 9-Dock PSA the treatment of natural hazards was undertaken using a "decade-step" approach with the severity of the event based on natural hazard frequency data for the site, and the component fragility of SSCs (Systems, Structures and Components) to the event. The following paragraphs provide a description of the natural hazards assessment. However, specific details of the design, plant fragility and results of the assessment have not been included. An example has been provided to demonstrate the technique applied, and conclusions are drawn on the practical application of the technique.

1.1 Natural hazard frequencies

The treatment of natural hazards was undertaken using a "decade-step" approach in which the robustness of the dock facilities and the submarine were examined for specific natural hazard events occurring over a range of return frequencies, decreasing in orders of

magnitude (i.e. 10^{-2}, 10^{-3}, down to a frequency of 10^{-7} per year). For each order of magnitude the survivability of each of the safety provisions was assessed using withstand/fragility distributions and a failure probability was derived. Those judged to be capable of withstanding the hazard (i.e. having an extremely low additional failure probability) were retained in the model with their normal "operational" reliability.

Hazard Curves on the severity of the hazards affecting 9 Dock, for return frequencies of 10^{-1} down to 10^{-7} per year were assessed separately in Reference 2 and have not been detailed here. This data is summarised in Table 1 and is presented diagrammatically for the individual hazards in Figures 1 to 5.

1.2 Component withstands

The withstand capabilities, or "fragilities", of the various safety provisions with regards to natural hazards have been derived from the information yielded by other specialist studies conducted for DML.

Fragility Distribution Data – The fragility (i.e. probability of failure) of the 9 Dock structures/systems with regards to seismic, high wind, extreme tide, lightning and extreme temperature hazards have been assessed. In addition the fragility of the submarine components with regards to the seismic hazard have also been assessed. Submarine systems are considered to be shielded from the environmental effects for all but the seismic hazards.

To illustrate the "fragility distribution" approach, an example distribution based on seismic data is given in Figure 6.

1.3 Modelling of the Hazards in the PSA

By combining the Hazard Curve and Fragility Distributions, the probability of failure of a structure/

Table 1.

Natural hazard event	Confidence level	Exceedence Frequency						
		0.2E-01/yr (Normal design basis)	IE-02/yr	IE-03/yr	IE-04/yr (Extreme design basis)	IE-05/yr	IE-06/yr	IE-07/yr
Seismic (g horizontal)	Best estimate	0.017	0.026	0.094	0.230	0.420	0.700	1.060
Extreme tide – hamoaze – year 2000 (m ODN)	Best estimate	3.34	3.47	3.83	4.29	4.70	5.00	5.20
High wind – wind/hurricane – hourly average (m/s)	Best estimate	26.4	27.4	30.5	33.1	35.1	36.9	38.4
High wind – tornado (m/s)	63% Confidence	n/a	n/a	n/a	33	58	74	>84
Lightning strike – >1600 m² AO (kA)	Best estimate	0	0	0	68	168	200	200
Extreme minimum air temperature (deg C)	Best estimate	−8.9	−9.6	−11.6	−13.3	−14.6	−15.6	−16.4

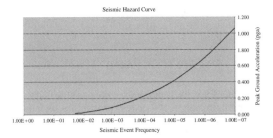

Figure 1. Seismic hazard curve.

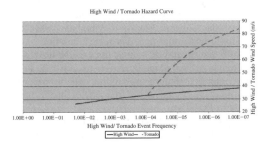

Figure 2. High wind hazard curve.

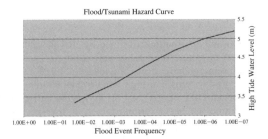

Figure 3. Flood hazard curve.

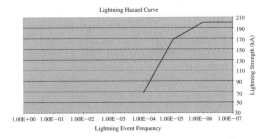

Figure 4. Lightning hazard curve.

system/equipment item for a given hazard can be expressed as a function of the return frequency of the hazard. The derived data, has been input to the PSA model alongside normal failure data via "exchange"

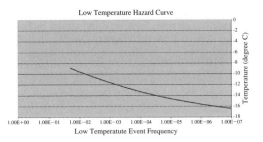

Figure 5. Low temperature hazard curve.

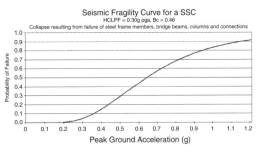

Figure 6. Sesimic fragility curve.

basic events. This has been done for the range of return frequencies from 10^{-2} down to 10^{-7} per year for the hazards.

The assessment of each of the natural hazards comprised the following steps:

1. Determination of component failure probabilities by:
   a. Identification of safety-critical components subject to the hazard, both at the 9 Dock facility and onboard the submarine (seismic only) considering both direct and indirect failure modes, e.g. direct = breakdown of equipment, indirect = building collapse onto safety-critical equipment.
   b. Determination of the likely loadings from the hazard (using Hazard Distributions and secondary response spectra).
   c. Consideration of the likely component failure modes under the hazard loading, where "component failure" is defined as the failure to perform or support the relevant safety function.
   d. Structural/component failure probabilities are estimated using Load vs. Withstand information and were then refined using Fragility Distributions.
2. Production of event trees to model event sequences following various hazard incidents occurring during each refit stage. The event trees model the responses of the dock, submarine, operators and assign consequential damage states.

3. The creation of "exchange" basic events in the PSA model's fault tree structure, using failure probabilities derived from the Fragility Distributions. The exchange events are used to represent the increased probability of components/structures failing, given the occurrence of a range of seismic events (return frequencies of the natural hazard events defined using the decade-step approach, i.e. from 10^{-2} to 10^{-7} per year).

The methodology used for the derivation of the component fragility distributions from the HCLPF and βc values, for each of the hazards, has followed the general EPRI-recommended strategy and is explained in more detail in Reference 3. The failure probabilities for each return frequency have been determined from the fragility distributions and entered into the PSA model as exchange events.

2 NATURAL EXTERNAL HAZARDS ASSESSMENT

The assessment for each of the 5 natural external hazards is described in this section.

2.1 Seismic hazard

The methodology used for the derivation of seismic fragility distributions from the High Confidence Low Probability of Failure (HCLPF) and composite log standard deviation values (βc) values has followed the general EPRI-recommended strategy. This is explained briefly below.

Lognormal distributions of component fragility under seismic loading have been determined separately for all 9 Dock structures and submarine equipment items. The fragility distributions have been derived based on information collected on walkdowns, design basis information, and experience collected from past events. The graphical representation of this log normally-distributed data is an S-shaped "fragility" distribution of failure probability vs. seismic loading (expressed in terms of horizontal peak ground acceleration). Component failure probabilities due to specific seismic events have been derived using the LOGNORMDIST functions in Microsoft Excel in conjunction with the following datasets:

- 9 Dock structures/systems – HCLPF values and βc.
- Submarine equipment – HCLPF margin values and βc.

In order to reproduce a cumulative lognormal distribution (i.e. an S-shaped curve), two pieces of information are required: the logarithmic standard deviation (β) and the median capacity (Cm). Starting with the

HCLPF value, the median capacity (i.e. the pga value at which there is a 50% probability of failure) is derived thus:

$$\ln(Cm) = \ln(HCLPF) + 1.645\,\beta \qquad (1)$$

The HCLPF value is defined as: the pga at which there is approximately a 5% or less probability of failure, said with 95% confidence. This value can be more easily estimated by engineers than the median capacity (x-axis value at which cumulative probability of failure is 50%).

The HCLPF margin values provided for submarine equipment are defined as the ratio of the HCLPF value against the Seismic Margin Earthquake (SME) of 0.35 ms-2 horizontal pga. Through tabulation of the HCLPF margin and βc values, the LOGNORMDIST function in MS Excel has been used to derive the structure/component failure probabilities for the relevant seismic events with return frequencies in the range 1E-02 to 1E-07/yr.

The component/structure failure probabilities for each event return frequency have been determined from the fragility distributions, and entered into the PSA model as "exchange" basic events.

2.2 High wind hazard

Lognormal distributions (defined by HCLPF and βc values) of failure probability versus hourly average wind velocity have been provided for all 9 Dock structures/systems. For high wind events with return frequencies of 2.0E-1/yr to 1.0E-4/yr, wind/hurricane conditions are the governing factor, with respect to wind velocities produced. However for return frequencies of 1.0E-05/yr and lower, wind velocities developed under tornado conditions become limiting and need to be treated separately.

Tornados need to be treated differently to high winds, as they are characterised by pressure differentials within the tornado as well as high windspeeds. In addition, there is the potential for tornado-borne missiles. It is thus probably too simplistic, and not necessarily conservative, to assume that the damage from a tornado at a certain return frequency is equivalent to that from a high wind at the same return frequency.

Mean high wind fragility distributions and corresponding event trees have been developed for 9 Dock structures for high wind event return frequencies in the range 1.0E-02 to 1.0E-07/yr. For return frequencies of greater than 1E-05 per year, the high wind speed was considered to be the limiting factor as the tornado windspeeds are significantly lower than the high wind speeds. For return frequencies of 1E-05 per year and lower, the tornado speed was considered to be the limiting factor.

2.3 Extreme tide/Flood hazard

A mixture of lognormal-, step function- and ramp function-distributions have been produced for component fragility to flooding, slow flooding of the dock and fast flooding of the dock, respectively.

Examination of the extreme tide hazard curve against the fragility data indicates that for tides of 10^{-4} per year, or more frequent, the water level will not rise above the cope edge. For the 10^{-5} per year case, there exists a probability of overflowing the cope edge in a slow dock flood scenario (step function) and for frequencies of 10^{-6} per year and lower there is a finite possibility of a fast dock flood (ramp function, 0.5 m increment). Thus, for extreme tide frequencies of 10^{-6} per year and less, it has been assumed that dock flood will arise as a fast dock flood.

Mean extreme tide/flood fragility distributions and corresponding event trees have been developed for 9 Dock structures for extreme tide/flood event return frequencies in the range 10^{-4} to 10^{-6}/yr. As it has been assumed that the hazard could result in fast dock flood at 10^{-6}/yr, the extreme tide at a return frequency of 10^{-7}/yr was considered to be bound by the 10^{-6}/yr event.

2.4 Lightning hazard

Lightning fragility data and lognormal distributions assumed for all 9 Dock structures/systems have been provided. An Effective Collection Area (Ao) of $>1600\,m^2$ is assumed for incident strikes at 9 Dock – Ao of the PCD/ACRC building alone is $>1600\,m^2$. Diffusion of strike currents throughout dock structures/systems has not been modelled to date, thus HCLPF strike current values are expressed in terms of a strike hitting any part of the 9 Dock facility, with general assumptions made regarding the percentage of the incident current that diffuses through to the structure/system in question, e.g. for electrical systems "it is probable that less than 10% of the incident strike current will be induced as a surge, by whatever mechanism." Thus, for electrical systems, the HCLPF value is given as ten times greater than the resultant surge current experienced by the system. The lowest fragilities have been attributed to C&I equipment throughout the dock. This may be overly conservative as credit has yet to be given for the action of their protection systems (if present).

The mean lightning fragility distributions and corresponding event trees model a single strike only – 28kA – with a return frequency of 3.91E-04 per year. This strike disables all Overside Services C&I equipment. The failure probabilities for C&I equipment at this particular return frequency (3.91E-04/yr) have been determined and entered into the PSA model as exchange events.

2.4 Extreme low temperature hazard

No credible extreme temperature initiated failures have been identified for the dock structure. Fragilities were assessed against different return frequencies (in the range 2E-02 per year to 1E-07 per year) of extreme minimum air temperature events. The following assumptions are made for mechanical pipework systems: The standby train and systems that act as back-up cooling water systems throughout the refit phases are assumed to contain static water, making them more susceptible to failure in freezing climatic conditions than running systems. The duty trains and running systems during the various refit nodes are modelled as running systems.

Lognormal distributions have been produced for all 9 Dock systems susceptible to low temperature events. The mean extreme low temperature fragility distributions and corresponding event trees have been developed for the affected 9 Dock systems for low temperature event return frequencies of 1.0E-02 to 1.0E-05/yr. As this fault results in the loss of all overside mechanical and electrical services at 1.0E-05/yr, frequencies below this were considered to be bound by the 1.0E-05/yr event.

3 EVENT TREE DEVELOPMENT

Integrated event trees have been developed for the NEHA PSA to represent all the natural hazard faults, using the Risk Spectrum software. The event trees model various factors that can influence the outcome of an initiating hazard fault by the use of event tree nodes. These include available protection and operator actions that can be claimed to militate against the fault. Some nodes have been represented by simple probabilities while others involve fault trees of varying complexity.

The use of exchange events to model explicitly the failure probabilities of SSCs allowed the assessment of the risk to core cooling to be evaluated using "standardised" event trees. Extensive modelling of damage states using house events were not required as the use of exchange events introduced the failure probabilities of the SSCs directly into the system fault trees. Additional events were required in some event trees to represent the failure of structures which under normal circumstances have a negligible failure probability, e.g. submarine cradle.

4 APPLICATION

The driver for using the decade step approach was that the initial results of the NEHA using the decade step approach could be fed back to the design teams.

Several vulnerabilities to hazards at frequencies of 1.0E-2 and 1.0E-3 per year were identified and design changes implemented to reduce the boat's vulnerabilities to these hazards. An example of where the decade step approach assisted in the design of the dock drainage systems and water services to the dock. The NEHA identified several water supply systems which were not seismically qualified for a 1.0E-04 per year event (Freshwater and Fire hydrant water mains). Simultaneous failure of these water supply systems could under specific conditions, result in the water entering the dock at a rate greater than which the drainage pumps could empty the dock. The NEHA was used to test various configurations and designs to find the optimum solution which satisfied multiple safety concerns. Seismically qualified isolation valves have been installed which will enable the operator to isolate any of the water supply systems given a rupture of the water services pipelines in the dock (i.e. seismically-induced or otherwise). This allows the water services to be fully available for mitigation of other more frequent faults e.g. Fire Fighting.

5 CONCLUSIONS

The PSA model results reflects the safety protection provided, against natural external hazards, for Vanguard class submarines undergoing a refit in 9 Dock. As a result, the NEHA provides a realistic representation of the capability of the safety provisions associated with the submarine and the facility to mitigate the effects of natural external hazards.

Using the method in presented in this paper, Hazard-Curves and Fragility Distributions have been combined to derive the revised failure probabilities of individual components/structures given the occurrence of a hazard at specific return frequencies separated by decade-step intervals, i.e. 10^{-2}, 10^{-3}, 10^{-4} per year etc. As the risk due to the natural hazard events for each return frequency has been assessed as an integral part of the PSA model, the authors claim that this approach allows a more comprehensive assessment of the risk presented by natural hazards.

The natural hazards PSA modelling could be further improved by applying the decade step approach to Lightning hazards, and by assessing in more detail the ability of the facility to withstand tornado events. As these hazards are bound by a higher frequency or greater severity event, the modelling of these faults is valid.

REFERENCES

1. SLG-PSC-290, 9 Dock Facility Pre-Operational Safety Report, Issue 4, November 2001.
2. SLG-HAZ-005, Generic Natural External Hazards PSA Data Derivation, Issue 1, May 2001.
3. EPRI, Methodology for Developing Seismic Fragilities, EPRI-TR-103959, Final Report, June 1994.

Safety and Reliability – Bedford & van Gelder (eds)
© 2003 Swets & Zeitlinger, Lisse, ISBN 90 5809 551 7

Project viability assessment for support of software testing via Bayesian graphical modelling

F.P.A. Coolen, M. Goldstein & D.A. Wooff
Department of Mathematical Sciences, University of Durham, Durham, UK

ABSTRACT: In a continuing collaboration with an industrial partner, we have developed a statistical method, the BGM approach, using Bayesian graphical models to support software testers and managers in dealing with the uncertainties in testing, including automatic design of test suites. There is a wide variety of benefits of using the BGM approach, which we briefly review in this paper, but there are also some costs involved in using it, e.g. in creation of the models and elicitation of the probabilities involved. In this paper, we discuss aspects that should be taken into account when assessing viability of this approach for particular projects, and we suggest general guidelines for deciding on the possible impact, benefits, and costs of implementing the BGM approach.

1 INTRODUCTION

In collaboration with an industrial partner, we have developed a statistical approach to support software testing, using Bayesian graphical models (Cowell et al. 1999, Jensen 2001). Details of this approach, which we refer to as the "BGM approach", can be found elsewhere (Rees et al. 2001, Wooff et al. 2002), some further aspects, in particular automated test design and the test–retest cycle, are described in Wooff et al. (2003).

In case studies, the BGM approach has proven to be a very promising method to support testers and managers in their complicated tasks of ensuring good testing for complex software. Nevertheless, adopting this approach also has some disadvantages, in particular with regard to up-front time for modelling and quantification of the testers' expert judgements. Hence, part of the collaboration addressed management aspects, in particular on project viability assessment.

In this paper, we describe general guidelines that can be used to decide on whether or not to adopt the BGM approach in individual projects. We restrict presentation to the main ideas, and use minimum numbers of classifications needed to illustrate the reasoning involved. In practical situations, more detailed classifications might be needed, depending e.g. on information available and relevance of the project to the company, but the general reasoning is along the same lines as presented here. First, however, we give a brief overview of the BGM approach.

2 BGM APPROACH FOR SUPPORT OF SOFTWARE TESTING

The BGM approach (Wooff et al. 2002) presents formal mechanisms for the logical structuring of software testing problems, the probabilistic and statistical treatment of the uncertainties to be addressed, the test design and analysis process, and the incorporation and implication of test results. Once constructed, the models are dynamic representations of the software testing problem. They may be used to answer whatif questions, to provide decision support to testers and managers, and to drive test design (Wooff et al. 2003). The models capture the knowledge of the testers and retain it for further use. We briefly summarize the main ingredients of the BGM approach, illustrating it via part of a substantial case study (Wooff et al. 2002).

Suppose that the function of a piece of software is to process an input number, e.g. a credit card number, in order to perform an action, and that this action might be carried out correctly or incorrectly. The various tests that we might run correspond to choosing various numbers and checking that the *software action* (SA) is performed correctly for each number. Usually, we will not be able to check all possible inputs, but instead we will check a subset from which we will, hopefully, be able to conclude that the software is performing reliably. Clearly, this involves a subjective judgement about the functionality of the software over the collection of inputs not tested, and the corresponding uncertainties. We must choose whether we explicitly quantify the uncertainties concerning further failures given our test results, or whether we are content to make a purely informal qualitative judgement for such uncertainties. In many areas of risk and decision analysis, uncertainties are routinely quantified as subjective probabilities representing the best assessments of uncertainty of the expert carrying out the analysis (Bedford & Cooke 2001). In the BGM approach, the testers' uncertainties for software failure are quantified and analysed using subjective probabilities, which involves investment of effort in thinking carefully about prior knowledge and modelling at a level of detail appropriate for the particular project. The rewards of such efforts include probabilistic statements on the reliability of the software taking test results into account, detailed guidance on optimal design of test suites, and the opportunity to support management decisions quantitatively at several stages of the project, including deciding on time and cost budgets at early planning stages.

The simplest case occurs when the tester judges all possible test results to be exchangeable, which implies that all such test results are judged to have the same probability of failing, and that observing each test result gives the same information about all other tests. Qualitatively, exchangeability is a simple judgement to make: either there are features of the set of possible inputs which cause the tester to treat some subsets of test outcomes differently from other subsets or, for him, the collection of outcomes is exchangeable. In our example, suppose that the software is intended to cope with credit cards with both short (S) and long (L) numbers, and that he judges test success to be exchangeable for all short numbers and exchangeable for all long numbers. For example, it might be the case that dealing with long numbers is newly added functionality. In addition, suppose that the tester judges test success to be exchangeable for numbers starting with zero (Z), and exchangeable for numbers starting with a non-zero digit (X).

Figure 1 is a BGM, in reduced form, reflecting such judgements. In such a BGM, each node represents a random quantity, and the arcs represent conditional independence relations between the random quantities (Cowell et al. 1999, Jensen 2001, Wooff et al. 2002).

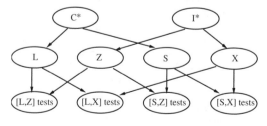

Figure 1. BGM (in reduced form) for a software problem with two possible failure modes.

We have four subgroups of tests, resulting from the combinations of number length and starting digit given the exchangeability judgements. For example, node "[L, Z] tests", with attached probability, represents the probability that inputs of the type [L, Z] lead to an output error when tested. Such error could be caused by three different problems, namely general problems for long numbers (node "L"), general problems for numbers starting zero (node "Z"), or problems specific for this combination. This last cause is not explicitly represented in this reduced BGM, as it would be a single parent node feeding into the node "[L, Z] tests", but for sake of simplicity we have deleted all nodes with a single child and no parents. The node C^* represents general problems related to length of the numbers, so problems in "L" can be either caused by problems in C^*, or in a second parent node of "L" (not shown), for problems specifically arising for long numbers. Similarly, the node I^* represents general problems related to starting digit. Finally, the actual test inputs are also not represented by nodes in this reduced BGM, they would be included in the form of a single node corresponding to each input tested, feeding off the corresponding node at the lowest level presented in Figure 1.

To build such a model, exchangeability assumptions are required over all relevant aspects, and many probabilities must be assigned, both for all nodes without parents, and for all further nodes in the form of conditional probability tables, for all combinations of the values of the parent nodes. For larger applications, which typically included several thousands of nodes, we have described methods to reduce the burden of specification to manageable levels (Wooff et al. 2002). For example, we chose to ask the tester to rank nodes according to "faultyness", and then we asked him to judge overall reliability levels to assign probabilities consistent with these judgements. For more details we refer to Wooff et al. (2002), where in particular further theoretical, modelling, and practical aspects are discussed.

The main case study presented in Wooff et al. (2002), where this method was applied in a project similar to management of credit card databases, involved a major update of existing software, providing new functionality and with some earlier faults fixed. We would call that case study "medium scale"; there were 16 separate major testable areas, and initial structuring by the tester led to identification of 168 different domain nodes (in Figure 1, there are 4 domain nodes, namely those at the lowest level presented), contained in 54 independent BGMs. A single test typically consisted of a variety of inputs spread over several of these BGMs. For example, one test could be "change of credit card", which would consist of a combination of delete an existing card from the database and add a new card to the database with appropriate credit limits.

418

The practical problem for the testers is optimal choice of the test suite, assessment of the information that may be gained from a given test suite is very difficult unless we use a formal method. In this case study, the tester had originally identified a test suite consisting of 233 tests. Application of the BGM approach revealed that the best 11 tests of these reduced the prior probability of at least one fault remaining by about 95%, and that 66 tests could not add any further information, according to the tester's judgements, assuming that the other tests had not revealed any failures. We also used the BGM approach to design extra tests to cover gaps in the coverage provided by the existing test suites. We managed to design one extra test which had the effect of reducing the residual probability of failure by about 57%, mostly because one area had been left completely untested in the originally proposed test suite. The value of this test was agreed by the senior tester.

In subsequent work (Wooff et al. 2003), we have developed a fully automated approach for test design. This leads to even better designs, and allows the inputs to be tested in sequences according to a variety of possible optimisation criteria, where for example aspects of test-fix-test can be taken into account, e.g. minimising total expected test time including time to fix faults shown during testing and the required retesting.

Although the BGM approach has vast potential, management decisions on whether or not it is viable for individual projects are non-trivial, due to the upfront investment required for the modelling. In the next section we discuss the general guidelines for such viability assessment.

3 PROJECT VIABILITY ASSESSMENT

We focus on an imaginary forthcoming project, and discuss what the assessment must cover. There are two aspects to consider:

(1) *General* – features which will be relevant to the assessment whichever testing approach is applied;
(2) *Specific* – features which depend on the approach that is used, one such approach being the BGM approach.

Each of these two aspects can be cross classified according to *cost* and *fault level*; the cost of overlooking faults for the project (*general*) and of carrying out testing (*specific*), and the level of fault in the software before testing (*general*) and after testing (*specific*).

3.1 *General assessments*

3.1.1 *Project cost*
There are many relevant considerations for the cost of overlooking faults in the software before release. This

is intimately linked to the general importance of the project for the company, and underpins all the effort worth taking to do a good job on the testing. Informally, we may classify the importance that the project performs appropriately as

(i) *High importance:* an important project at the heart of the company's profitability, for which it would be extremely costly (e.g. in time, effort, prestige) if the project developed major failures in the field, so that it is very important that the software quality is good.
(ii) *Medium importance:* a fairly typical project, for which poor performance will be "typically" costly.
(iii) *Low importance:* the kind of project where the software quality is not a major concern. For example, the project might still be in a trial phase, or concern areas with little consequence.

Of course such a classification is really a crude discretisation of (several) continuous quantities. If we were intending to carry out a full decision analysis, then such a quantification would doubtless be important. However, we are trying to illustrate the qualitative structure of the assessment. Further, as soon as we try to elicit precise quantitative effects, we will tend to face many practical, psychological, and political types of issues, whereas it is often quite quick and easy to get agreement on a broad categorisation.

Therefore, for each factor, here and below, we suggest three levels of importance: high, medium and low. By "*high importance*" we mean that improved testing is always a major issue. By "*low importance*" we mean that testing is not a major concern, according to the particular aspect under consideration. To implement the general recommendations presented here in a particular project, one may wish to use different classifications, but the general reasoning is as outlined here. If one wishes to refine such judgements, one might introduce a cost vector, e.g. representing the expected cost of each minor, major, and catastrophic fault found after release.

3.1.2 *Initial fault level of software*
The importance of testing will obviously depend on how many faults the software contains initially. Clearly, this is unknown, but an assessment can be made into the following categories.

(a) *High fault level:* the software is likely to contain many faults, so testing is very important.
(b) *Medium fault level:* the software is likely to contain several faults, so some testing is required.
(c) *Low fault level:* testing is mainly a formality.

The effect of this classification is to scale the probability of faults existing in the software before testing, and so likely to be remaining in the software after testing. We could further break down our information about the software according to considerations on, for example, whether the software is substantially new, or largely a routine rewrite of existing good quality software; difficulty of the application of the software; whether the software is long and elaborate, or short and simple; and whether the software is produced by a reliable team. We could build up a simple assessment procedure based on this type of consideration to give probabilities for each type of level of software deficiency. Again, one could introduce a probability vector for the different types of fault for each category.

3.2 Specific assessments

3.2.1 Test cost

We suppose that the company assigns a testing budget B, using their current methods of assessment. B has two components, money M and time T. Each of these is a prior expectation, which may be modified, but they are both set fairly early in the process, to some order of magnitude. For simplicity, we suppose that each element M and T may be classified as

(i) *High expense:* sufficiently expensive that reducing test cost is a major business concern.
(ii) *Medium expense:* there may be some scope for cost reduction, but this is not a driving consideration.
(iii) *Low expense:* cost reduction is not an issue.

Of course, these classifications are relative to the constraints of the problem. For example, a testing budget is expensive in time if the expected testing time gets very close to the promised release date of the software. Time enters the costings indirectly as a constraint. For example, if time cost is high and overall test quality is poor, then we will seek to use time more efficiently to improve testing.

3.2.2 Fault level of testing

With their current approach, suppose that the company now assesses how effective they believe their testing regime is likely to be. Presumably most testing is not planned to be poor, though time constraints and so forth will mean that full testing (whatever that might be) is not always possible. Again, to keep things simple for purpose of illustration, suppose that the company considers three categories of test deficiency. Note that we are concerned here with faults which could be caught by the testing process. Any further faults which can only be detected after release are beyond the scope of this analysis.

(a) *High deficiency:* there is large uncertainty as to whether testing will find all high priority faults. For example, some important areas of functionality might only be superficially tested, so improved testing is a priority.
(b) *Medium deficiency:* there is reasonable confidence that all highest priority faults will be found, but less

confidence that all major faults will be found, so there is scope for improved testing.
(c) *Low deficiency:* there is confidence that all highest priority and other major faults will be found, in sufficient time to fix before release. Improving test effectiveness is not considered important.

This classification is made by comparison with similar testing problems addressed by the company. In practice, this is essentially how budgets are assigned, in that expert testers make assessments of the resources required for careful testing, by analogy with previous testing experiences, and then by negotiation settle on a time frame and resource budget with which they are either happy (presumably as they feel this will allow low test deficiency) or which they accept as a compromise involving some degree of test deficiency. As with each of the other inputs, one could refine the categories by discussion with test managers for specific implementations of this approach.

Note that test fault level refers to lack of confidence in the *test procedures*, not in the quality of the software after testing. For example, if we did no testing at all, then we would have no confidence in the testing (obviously), but if we had high confidence that the software was of good quality then we would still be happy to release the software.

Effectively, we are dividing the belief specification problem into two parts. First, the chance of faults in the software (initial fault level), and secondly, the probability of finding such faults given that they are in the software (testing fault level). This decomposition is made to make it easier for the company to specify their judgements, and because software deficiency will be a common feature to the problem irrespective of the testing approach, but test deficiency strictly applies to what the company can achieve without the BGM approach. In the same way, in financial considerations, project importance is common to all approaches, but current test budget only relates to the company's choice without the BGM approach. Again, one may specify a probability vector for the ability to find each type of fault.

3.3 Use of assessments

We now consider the potential for the BGM approach to be worth considering for a new project. The financial considerations are summarised by the cost assessments, whilst the confidence in releasing good quality software is summarised in the initial and test fault levels. The advantage of starting with simple classifications is that for many cases the conclusion will be fairly obvious without detailed calculation, e.g. if all costs and fault likelihoods are assessed as high then we would expect the BGM approach to be very helpful, while if all are assessed as low then presumably we would expect much less benefit from more sophisticated approaches. In practice, we suspect that for many problems this cross-classification will quickly reveal whether the BGM approach should be explored, as the potential for improvement can be gauged from the number and nature of high values in the specification. If we do seek a quantification, the overall expected cost to the company is of the form

([Project cost] × [Initial fault level]
    × [Testing fault level]) + [Test cost],

suitably scaled, and possibly based on vectors of costs and probabilities. However, it is more fruitful to use the above assessments to direct the amount of effort we put into the BGM approach analysis, as follows.

3.4   Assessment of the BGM approach

Set against the various advantages of the BGM approach (some of these are not easily tangible in money terms, e.g. the availability to managers of detailed justifiable assessments of software reliability before and after testing), are the costs of implementing the approach. Essentially, the extra cost that is incurred is the additional effort required to construct the BGMs for the testing problem. Of course, how easy or difficult it is to construct the model depends strongly on the effectiveness of the tools that are put in place by the company to implement the approach. However, much of the work required to construct the BGM is required for any approach to construct the test suite, and, given good tools, the cost, in complicated testing problems, may actually be less than for the alternative approaches, as the BGM approach may be carried out either in full, or in cost-saving approximations, depending on the amount of time and thought that goes into the structuring and specification.

The BGM approach exploits the relationships between different test outcomes to transfer information from the outcome of one test to beliefs about potential failures elsewhere in the software. The more complex the diagram, the more the potential to exploit such transfers of information, but also the higher the cost in producing the diagram. Rather than making a somewhat arbitrary guess at the potential savings that a general BGM approach might yield, it seems more sensible to suggest how to match the amount of effort that should go into the BGM approach modelling with the requirements of the problem at hand, so that the BGM approach should always give good value.

In any case, we start with a description of the possible tests that can be run. This is essentially cost neutral, as we must make such an assessment whichever approach we take. We would imagine, however, that the general features of the problem (project cost, software quality) will influence the level of detail at which it is thought worth describing the observation space. The various mappings involved in mapping tests and inputs choices to observables are addressed elsewhere (Wooff et al. 2003).

Now we describe various choices for the level of detail for the BGMs that we shall construct. We identify three aspects to this, and for each aspect we describe three levels: full, partial and simple.

3.4.1   Outcome space description

The level of detail at which we *describe* the test outcome.

(1) *Full:* the finest level of detail that we consider relevant;
(2) *Partial:* an intermediate level of detail for the outcome space, aggregating outcomes which are almost equivalent;
(3) *Simple:* makes as few distinctions as possible.

As a general principle, the more important it is not to overlook faults, the more detailed we will want the outcome space description to be, as it will be important not to overlook fault types.

3.4.2   Model linkage

The level of detail of the *links* between different parts of the model.

(a) *Full:* makes all of the links between outcome nodes, through higher order structuring, that we can identify;
(b) *Partial:* keeps the main links, but drops link nodes which appear to have little value (as they are very unlikely, or are likely to have small influence);
(c) *Simple:* only adds unavoidable links to the diagram, minimises the number of parent nodes.

As a general principle, the more important it is to reduce the testing cost, the more detailed we will want the linkage to be, as this allows us to reduce probabilities in parts of the diagram based on tests in other parts of the diagram, and so reduce the number of tests.

3.4.3   Probability elicitation

The care with which we *elicit* probabilities for the model.

(i) *Full:* belief specification is made carefully for each node;
(ii) *Partial:* some use is made of exchangeability ("similarity") to simplify specifications which are almost exchangeable;
(iii) *Simple:* belief specification is based on a minimal number of exchangeable judgements and simple scaling arguments across parent nodes.

As a general principle, the more we suspect that we will not be able to carry out full testing, the more careful we will need to be in our prior specification, as we will not be able to drive all fault probabilities to near zero, so some of our assessments may be strongly influenced by prior values.

3.5 *Choosing an appropriate Bayesian approach*

We now describe how to put the various ingredients together. There are three essentially different ways that the BGM approach may be valuable.

1. For the same test budget as for an existing approach, it may provide a more effective test suite and thus reduce faults in software release.
2. It may achieve the same effectiveness as a test suite constructed according to current company practice, but using less test effort and so reducing test budget.
3. The company may judge that there is an intrinsic "good practice" value in following an approach which offers a more rigorous and structured approach to testing, which constructs models for the test process which can be maintained over a series of software releases, which provides probabilistic assessments of software reliability which can be fed directly into any risk assessment that must be made about software release, and which can be used to monitor and forecast time until completion of the testing process and so give advance warning of problems in meeting test deadlines.

Our main objective is to reduce faults when the project cost is high and to reduce budget when the budget cost is high. We suggest the following choices for setting the levels for the three aspects of the Bayesian specification:

A. Start with medium levels for outcome space description, model linkage and elicitation;
B. If the initial software fault level is high (low), then raise (lower) each of the three levels;
C. If the project cost is high (low), then raise (lower) the level of outcome space description;
D. If budget is high (low), then raise (lower) the level of model linkage;

E. If test fault level is high (low), then raise (lower) the level of detail of prior elicitation;
F. Consider raising the various levels to realise more fully the "good practice" value of the testing.

Obviously, we may suggest further guidelines, and add further criteria. However, this is intended to be illustrative of our intention to produce an approach which is appropriate to each problem, rather than a "use/don't use-rule" with some arbitrary cut-off. The approach outlined above is well suited for implementation, and can easily be adapted, by modifying or adding to the suggested rules, in the light of new experience or requirements.

ACKNOWLEDGEMENTS

This research has been supported by the UK Engineering and Physical Research Sciences Council, grant GR/M76775. We gratefully acknowledge our industrial collaborators, British Telecommunications plc, for the software testing case study material, with respect to which the probabilities and other results mentioned in this paper are illustrative and do not reflect the company's actual judgements.

REFERENCES

[1] Bedford, T. & Cooke, R. 2001. *Probabilistic Risk Analysis: Foundations and Methods*. Cambridge University Press.
[2] Cowell, R.G., Dawid, A.P., Lauritzen, S.L. & Spiegelhalter, D.J. 1999. *Probabilistic Networks and Expert Systems*. New York: Springer.
[3] Jensen, F.V. 2001. *Bayesian Networks in Decision Graphs*. New York: Springer.
[4] Rees, K., Coolen, F.P.A., Goldstein, M. & Wooff, D.A. 2001. Managing the uncertainties of software testing: a Bayesian approach. *Quality and Reliability Engineering International* 17: 191–203.
[5] Wooff, D.A., Goldstein, M. & Coolen, F.P.A. 2002. Bayesian graphical models for software testing. *IEEE Transactions on Software Engineering* 28: 510–525.
[6] Wooff, D.A., Goldstein, M. & Coolen, F.P.A. 2003. Bayesian graphical models for high reliability testing. *In submission*.

Safety and Reliability – Bedford & van Gelder (eds)
© 2003 Swets & Zeitlinger, Lisse, ISBN 90 5809 551 7

Challenges related to surveillance of safety functions

K. Corneliussen & S. Sklet

Dept. of Production and Quality Engineering, NTNU/SINTEF Industrial Management, Trondheim, Norway

ABSTRACT: One of the main principles for the safety work in high-risk industries such as the nuclear and process industry, is the principle of defence-in-depth that imply use of multiple safety barriers or safety functions in order to control the risk.

Traditionally, there has been a strong focus on the design of safety functions. However, recent standards and regulations focus on the entire life cycle of safety functions, and this paper focuses on the surveillance of safety functions during operations and maintenance. The paper presents main characteristics of safety functions, factors influencing the performance, a failure category classification scheme, and finally a discussion of challenges related to the surveillance of safety functions during operations and maintenance. The discussion is based on experiences from the Norwegian petroleum industry and results from a research project concerning the reliability and availability of computerized safety systems.

The main message is that there should be an integrated approach for surveillance of safety functions that incorporates hardware, software and human/organizational factors, and all failure categories should be systematically analyzed to (1) monitor the actual performance of the safety functions and (2) systematically analyze the failure causes in order to improve the functionality, reliability and robustness of the safety functions.

1   INTRODUCTION

One of the main principles for the safety work in high-risk industries such as the nuclear and process industry, is the principle of defence-in-depth or use of multiple layers of protection (IAEA 1999, Reason 1997, CCPS 2001).

The Norwegian Petroleum Directorate (NPD) emphasizes this principle in their new regulations concerning health, safety and environment in the Norwegian offshore industry (NPD, 2001a). An important issue in these new regulations is the focus on safety barriers, and in the first section of the management regulation, it is stated that "barriers shall be established which a) reduce the probability that any such failures and situations of hazard and accident will develop further, and b) limit possible harm and nuisance".

The IEC 61508 (IEC 1998) and IEC 61511 (IEC 2002) standards have a major impact on the safety work within the process industry, and describe a risk-based approach to ensure that the total risk is reduced to an acceptable level. The main principle is to identify necessary safety functions and allocate these safety functions to different safety-related systems or external risk reduction facilities. In IEC 61511 a safety function is defined as a "function to be implemented by a SIS (Safety Instrumented System), other technological safety-related system or external risk reduction facilities which is intended to achieve or maintain a safe state for the process in respect to a specific hazardous event". An important part of the standards is a risk-based approach for determination of the safety integrity level requirements for the different safety functions. IEC 61508 is a generic standard common to several industries, while the process industry currently develops a sector specific standard for application of SIS, i.e., IEC 61511 (IEC 2002). In Norway, the offshore industry has developed a guideline for the use of the standards IEC 61508 and IEC 61511 (OLF 2001), and the Norwegian Petroleum Directorate (NPD) refers to this guideline in their new regulations (NPD 2001a). Overall, it is expected that these standards will contribute to a more systematic safety work and increased safety in the industry.

Further, the NPD in section 7 in the management regulation (NPD, 2001a) requires that "the party responsible shall establish monitoring parameters within his areas of activity in order to monitor matters of significance to health, environment and safety", and that "the operator or the one responsible for the operation of a facility, shall establish indicators to monitor changes and trends in major accident risk". These requirements imply a need for surveillance of safety functions during operation. In accordance with these requirements, NORSOK (2001) suggests that "verification of that performance standards for safety and emergency preparedness systems are met in the operational phase may be achieved through monitoring trends for risk indicators. [...] Examples of such indicators may be availability of essential safety systems". Also IEC requires proof testing and inspection during operations and maintenance in order to ensure that the required functional safety of safety-related systems is fulfilled (IEC 2002).

In order to monitor the development in the risk level on national level, the NPD initiated a project called "Risk Level on the Norwegian Continental Shelf". The first phase of the project focused on collection of information about defined situations of hazard and accident (DSHA), while the second phase also focus on collection of information about the performance of safety barriers (NPD/RNNS 2002). According to this project, the performance of safety barriers has three main elements: (1) functionality/efficiency (the ability to function as specified in the design requirements), (2) reliability/availability (the ability to function on demand), and (3) robustness (ability to function as specified under given accident conditions).

The NPD uses the term safety barrier in their regulations. However, they have not defined the term, and in a letter to the oil companies as part of the project "Risk Level on the Norwegian Continental Shelf" (NPD/RNNS, 2002), they have referred to the definition proposed by ISO (2000): "Measure which reduces the probability of realizing a hazard's potential for harm and which reduces its consequence" with the note "barriers may be physical (materials, protective devices, shields, segregation, etc.) or non-physical (procedures, inspection, training, drills, etc.)". Accordingly, the NPD uses the term barrier in an extended meaning and is therefore similar to other terms used in the literature, such as defence (Reason 1997), protection layer (CCPS 2001), and safety function (as used by IEC). The term safety function is used in this paper.

Surveillance of safety functions during operations in order to meet the requirements stated by the NPD (NPD 2001a) and IEC (IEC 1998 and IEC 2002) is not a straightforward task, but is a challenge for the oil companies. Therefore, several oil companies have initiated internal projects to fulfill the requirements (see e.g. Sørum & Thomassen 2002). This paper focuses on the surveillance of safety functions during operations and maintenance. The paper presents main characteristics of safety functions, factors influencing the performance, a failure category classification scheme, and finally a discussion of challenges related to the surveillance of safety functions during operations and maintenance. The discussion is based on experiences from the Norwegian petroleum industry and results from a research project concerning the reliability and availability of computerized safety systems.

2 CHARACTERISTICS OF SAFETY FUNCTIONS

Safety functions may be characterized in different ways, and some of the characteristics influence how the surveillance of the safety function is performed. The following characteristics are further discussed in this section: type of safety function, local vs. global safety functions and active vs passive systems.

IEC 61511 (IEC 2002) defines a safety function as a "function to be implemented by a SIS, other technology safety-related system or external risk reduction facilities, which is intended to achieve or maintain a safe state for the process, in respect of a specific hazardous events". By SIS IEC means an instrumented system used to implement one or more safety instrumented functions. A SIS is composed of any combination of sensor(s), logic solver(s), and final element(s). Other technology safety-related systems are safety-related systems based on a technology other than electrical/electronic/programmable electronic, for example a relief valve. External risk reduction facilities are measures to reduce or mitigate the risk that are separate and distinct from the SIS. Examples are drain systems, firewalls and bunds.

A distinction between global and local safety functions is made by The Norwegian Oil Industry Association (OLF) (OLF, 2001). Global safety functions, or fire and explosion hazard safety functions, are functions that typically provide protection for one or several fire cells. Examples are emergency shutdown, isolation of ignition sources and emergency blowdown. Local safety functions, or process equipment safety functions, are functions confined to protection of a specific process equipment unit. A typical example is the protection against high level in a separator through the PSD (Process Shutdown) system.

CCPS distinguishes between passive and active independent protection layers (IPL) (CCPS 2001). A passive IPL is not required to take an action in order to achieve its function in reducing risk. Active IPLs are required to move from one state to another in response to a change in a measurable process property (e.g. temperature or pressure), or a signal from another source (such as a push-button or a switch). An active IPL generally comprises a sensor of some type (detection) that gives signal to a decision-making process that actuates an action (see Figure 1).

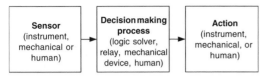

Figure 1. Basic elements of active protection layers (CCPS, 2001).

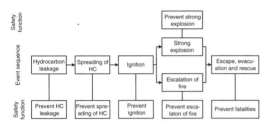

Figure 2. Event sequence for process accidents.

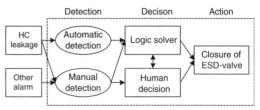

Figure 3. Safety function – prevent spreading of hydrocarbons.

be initiated automatically by the logic solver, or by a human operator pushing the ESD-button, or manually by a human operator closing the ESD-valve manually.

There should be an integrated approach for surveillance of safety functions that incorporates hardware, software and human/organizational factors.

3 SAFETY FUNCTIONS FOR PROCESS ACCIDENTS

The need for safety functions is dependent on specific hazardous events. Figure 2 gives a simplified illustration of the event sequence and necessary safety functions for "process accidents". The event sequence begin with the initiating event "leakage of hydrocarbons (HC)", and are followed by spreading of hydrocarbons, ignition, strong explosions or escalation of fire, escape, evacuation, and finally rescue of people. The main safety functions in order to prevent, control or mitigate the consequences of this accident are to prevent the hydrocarbon leakage, prevent spreading of hydrocarbons, prevent ignition, prevent strong explosion or escalation of fire, and to prevent fatalities. These safety functions may be realized by different kinds of safety-related systems. In this paper, we focus on the safety function "prevent spreading of hydrocarbons".

In principle, the safety function "prevent spreading of hydrocarbons" may be fulfilled in two different approaches (1) stop the supply of HC, and (2) remove HC. In this paper, we focus on the former approach in order to illustrate some of the challenges related to the surveillance of safety functions.

The main elements of the active safety function "prevent spreading of hydrocarbons by stopping the supply" are shown in Figure 3. Firstly, the leakage of HC must be detected, either automatically by gas detectors, or manually by human operators in the area. Secondly, a decision must be taken, either by a logic solver or a human decision. The decision should be followed by an action, in this case, closure of an ESDV (Emergency Shutdown Valve). The action may either

4 FAILURE CLASSIFICATION

For safety functions implemented through SIS technology (as in Figure 3), IEC 61508 and IEC 61511 define four safety integrity levels (SIL). The SIL for each safety function is established through a risk-based approach. To achieve a given SIL, there are three main types of requirements (OLF, 2001):

- A quantitative requirement, expressed as a probability of failure on demand (PFD) or alternatively as the probability of a dangerous failure per hour. This requirement relates to random hardware failures.
- A qualitative requirement, expressed in terms of architectural constraints on the subsystems constituting the safety function.
- Requirements concerning which techniques and measures should be used to avoid and control systematic faults.

The requirements above influence the performance of the SIS, and in this section we present a failure classification scheme that can be used to distinguish between different types of failure causes (hardware and systematic failures). The scheme is a modification of the failure classification suggested in IEC 61508.

The basis for the discussion can be traced back to the research project PDS (Reliability and availability for computerized safety systems) carried out for the Norwegian offshore industry (Bodsberg & Hokstad 1995, Bodsberg & Hokstad 1996, Aarø et al. 1989), and the still active PDS-forum that succeeded the project (Hansen & Aarø 1997, Hansen & Vatn 1998, Vatn 2000, Hokstad & Corneliussen 2000). The classification presented in this section is one of the results in the new edition of the PDS method (Hokstad & Corneliussen 2003).

According to IEC 61508 (Section 3.6.6 of part 4), failures of a safety-related system can be categorized either as random hardware failures or systematic failures. The standard also treats software failures, but we consider this as a subclass of the systematic failures (see Note 3 on p16 of IEC 61508-4). The standard makes a clear distinction between the two failure categories, and states that random hardware failures should be quantified, while systematic failures should not (IEC 61508-2, 7.4.2.2, note 1).

In IEC 61508-4 (Section 3.6.5), a random hardware failure is defined as a "failure, occurring at a random time, which results from one or more of the possible degradation mechanisms in the hardware". IEC 61508-4 (Section 3.6.6) defines a systematic failure as a "failure related in a deterministic way to a certain cause, which can only be eliminated by a modification of the design or the manufacturing process, operational procedures, documentation or other relevant factors".

The standard defines "hardware-related Common Cause Failures (CCFs)" (IEC 61508-6, Section D.2): "However, some failures, i.e., common cause failures, which result from a single cause, may affect more than one channel. These may result from a systematic failure (for example, a design or specification mistake) or an external stress leading to an early random hardware failure". As an example, the standard refers to excessive temperature of a common cooling fan, which accelerates the life of the component or takes it outside it's specified operating environment.

Hokstad & Corneliussen (2003) suggest a notation that makes a distinction between random hardware failures caused by natural ageing and those caused by excessive stresses (and therefore may lead to CCFs). The classification also defines systematic failures in more detail. The suggestion is an update of the failure classification introduced in the PDS project (Aarø et al. 1989), but adapted to the IEC 61508 notation, and hence should not be in conflict with that of IEC 61508. The concepts and failure categorization suggested by Hokstad and Corneliussen (2003) is shown in Figure 4.

Hokstad & Corneliussen (2003) define the failure categories as:

– Random hardware failures are physical failures, where the delivered service deviates from the specified service due to physical degradation of the module. Random hardware failures are split into ageing failures and stress failures, where ageing failures occur under conditions within the design envelope of a module, while stress failures occur when excessive stresses are placed on the module. The excessive stresses may be caused either by external causes or by human errors during operation.
– Systematic failures are non-physical failures, where the delivered service deviates from the specified

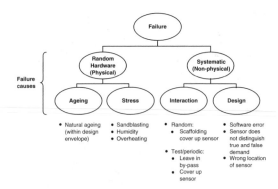

Figure 4. Failure categorization (Hokstad & Corneliussen 2003).

service without any physical degradation of the module. The failure can only be eliminated by a modification either of the design or the manufacturing process, the operating procedures, the documentation or other relevant factors. Thus, modifications rather than repairs are required in order to remove these failures. The systematic failures are further split into interaction failures and design failures, were interaction failures are initiated by human errors during operation or testing. Design failures are initiated during engineering and construction and may be latent from the first day of operation.

As a general rule, stress, interaction and design failures are dependent failures (giving rise to common cause failures), while the ageing failures are denoted independent failures.

To avoid a too complex classification, every failure may not fit perfectly into the above scheme. For instance, some interaction failures might be physical rather than non-physical.

The PDS method focuses on the entire safety function (Hokstad & Corneliussen 2003), and intends to account for all failures that could compromise the function (i.e. result in "loss of function"). Some of these failures are related to the interface (e.g. "scaffolding cover up sensor"), rather than the safety function itself. However, it is part of the "PDS philosophy" to include such events.

5 SURVEILLANCE OF SAFETY FUNCTIONS

This section discusses the surveillance of safety functions during operation related to the failure classification in the previous section.

The requirements for surveillance are related to the functional safety, and not only to the quantitative SIL requirements (see section 4). In IEC 61508-2, section 7.6.1 it is stated that one should "develop procedures to ensure that the required functional safety of the SIS is maintained during operation and maintenance", and more explicitly stated in IEC 61511-1, section 16.2.5, "the discrepancies between expected behavior and actual behavior of the SIS shall be analyzed and where necessary, modification made such that the required safety is maintained". In addition to the quantitative (PFD) requirement, systematic failures and changes in safety system/functions should be considered. Also changes not explicitly related to the safety function may influence the safety level (number of demands, operation of the process, procedures, manning, etc.), however such conditions will not be treated in this paper. The discussion is limited to the boundary outlined in Figure 3.

In operation or during maintenance the performance of the safety functions or part of the functions may typically be observed by means of a range of activities/observations, Table 1 illustrates the relation between the failure cause categories (as discussed in section 4) and the main types of activities/observations.

Not every failure encountered during the different surveillance activities may fit perfectly into the scheme, but it illustrates which failure categories that typically can be identified by use of different surveillance activities.

The actual demands of a function can potentially reveal both systematic and random hardware failures, provided that there is a systematic approach for registration of failures. The frequency of actual demands is, however, in most cases low, and it is therefore important that the organization focuses on the actions taken after an actual demand. As an example statistics from HSE (HSE 2002a) shows that gas detectors detected 59% of 1150 gas leakages reported in the period 1-10-92 to 31-3-01, while the remaining releases were mainly detected by other means, i.e., equipment not designed for the purpose (visual means, by sound, by smell, etc.).

In addition to the actual demands, the SIS functions must be tested, and there are two types of testing: 1) functional tests and 2) automatic self-tests. These tests are essentially designed to detect random hardware failures. However, no test is perfect due to different factors as the test do not reflect real operating conditions, the process variables cannot be safely or reasonably practicably be manipulated, or the tests do not address the necessary functional safety requirements (e.g. response time and internal valve leak) (HSE 2002b).

Components often have built-in automatic self-tests to detect random hardware failures. Further, upon discrepancy between redundant components in the safety system, the system may determine which of the modules have failed. This is considered part of the self-test. But it is never the case that all random hardware

Table 1. Different types of surveillance of safety functions.

Surveillance activity	Random hardware failures		Systematic failures	
	Ageing	Stress	Interaction	Design
Actual demand	x	x	x	x
Automatic self-test	x	x		
Functional test	x	x		
Inspection	x	x	(x)	
Random detection	x	x	(x)	

failures are detected automatically ("Diagnostic Coverage"). The actual effect on system performance from a failure that is detected by the automatic self-test may also depend on system configuration and operating philosophy.

Functional testing is performed manually at defined time intervals, typically 3, 6 or 12 months intervals for component tests. The functional test may not be able to detect all functional failures. According to Hokstad & Corneliussen (2003) this is the case for:

- Design errors (present from day 1 of operation), examples are: software errors, lack of discrimination (sensors), wrong location (of sensor), and other shortcomings in the functional testing (the test demand is not identical to a true demand and some part of the function is not tested).
- Interaction errors that occur during functional testing, e.g., maintenance crew forgetting to test specific sensor, tests performed erroneously (wrong calibration or component is damaged), maintenance personnel forgetting to reset by-pass of component.

Thus, most systematic failures are not detected even by functional testing. In almost all cases it is correct to say that functional testing will detect all random hardware failures but no systematic failures.

The functional tests may be tests of:

- The entire system/function typically performed when the process is down, e.g., due to revision stops.
- Components or sub-functions. Component tests are normally performed when the process is in operation.

Component tests are more frequent than the system tests due to less consequences on production. Experience do, however, show that full tests (from input via logic to output device) "always" encounter failures not captured during component tests.

In IEC 61511-1, inspection is described as "periodical visual inspection", and this restricts the inspections to an activity that reveals for example unauthorized modifications and observable deteriorations of the components. An operator may also detect failures in between tests (Random detection). For instance the panel operator may detect a transmitter that is "stuck" or a sensor left in by-pass (systematic failure).

6 DISCUSSION

The data from the various activities described above should be systematically analyzed to (1) monitor the actual performance of the safety functions and (2) systematically analyze the failure causes in order to improve the performance of the function. The organization should handle findings from all above surveillance activities, and should focus on both random hardware and systematic failures. The failure classification in PDS may assist in this work.

6.1 Performance of safety functions

As stated above, the performance of safety functions has three elements: 1) the functionality/efficiency, 2) the reliability, and 3) the robustness. The functionality is influenced by systematic failures. Since these failures seldom are revealed during testing, it is necessary to register systematic failures after actual demands or events that are observed by the personnel (inhibition of alarms, scaffolding, etc.).

Traditionally, the reliability is quantified as the probability of failure on demand (PFD) and is mainly influenced by the dangerous undetected random hardware failure rate (λ_{DU}), the test interval (τ) and the fraction of common cause failures (β).

The PDS-method (Hokstad & Corneliussen 2003), however, accounts for major factors affecting reliability during system operation, such as common cause failures, automatic self-tests, functional (manual) testing, systematic failures (not revealed by functional testing) and complete systems including redundancies and voting. The method gives an integrated approach to hardware, software and human/organizational factors. Thus, the model accounts for all failure causes as shown in Figure 4.

The main benefit of the PDS taxonomy compared to other taxonomies is the direct relationship between failure causes and the means used to improve the performance of safety functions.

The robustness of the function is defined in the design phase, and should be carefully considered when modifications on the process or the safety function are performed.

6.2 Analysis of random hardware failures from functional tests

Data from functional tests on offshore installations is summarized in a CMMS (computerized maintenance management system). The level of detail in reporting may vary between oil companies and between installations operated by the same company. Typically,

the data is presented as failure rates per component class/type independent of the different safety functions which the components are part of. This means that the data from component tests must be combined with the configuration of a given safety function in a reliability model (e.g. a reliability block diagram or PDS) to give meaning with respect to SIL for that safety function. Alternatively a "SIL budget" for detection (input), decision (logic) and action (output) might be developed. This can be advantageous since tests of the components are more frequent, and data from tests can be used to follow up component performance independent of safety functions.

It is important to have a historical overview of the number of failures and the total number of tests for all the functional tests in order to adjust the test interval, but it is equally important to analyze the failure causes to prevent future failures. This is particularly the case for dependent failures (i.e. stress failures). An example is sensors placed in an environment that results in movements and temperature conditions that further may lead to stress failures on several sensors. The functional tests will reveal random hardware failures but will not differentiate between independent (ageing) and dependent failures, and the fraction between independent and dependent failures must be analyzed.

Common cause failures may greatly reduce the reliability of a system, especially of systems with a high degree of redundancy. A significant research activity has therefore been devoted to this problem, and Høyland and Rausand (1994) describe various aspects of dependent failures.

For the β-factor model we need an estimate of the total failure rate λ, or the independent failure rate (λ_I), and an estimate of β. Failure rates may be found in a variety of data sources. Some of the data sources present the total failure rate, while other present the independent failure rate. However, field data collected from maintenance files normally do not distinguish between independent failures and common cause failures, and hence presents the total failure rate. In this case, the β, and λ_I will normally be based on sound engineering judgment. An approach is outlined in IEC 61508 for determining the plant specific $\beta(s)$.

The maintenance system (procedures and files) should be designed for assisting in such assessments, and it is especially important to focus on the failure causes discussed in this paper

The tests and calculated PFD numbers may be used as arguments for reducing the test interval or more critical, to increase the test interval. Such decisions should not be based on pure statistical evidence, but should involve an assessment of all assumptions the original SIL requirement was based on. OLF suggests an approach for assessment of the failure rate (OLF, 2001), but the oil companies have not implemented this approach fully yet.

428

6.3 Analysis of systematic failures

As described earlier, the systematic failures are almost never detected in the tests or by inspection, but it is important to analyze the systematic failures that occur in detail and have a system to control systematic failures.

Systematic failures are usually logged in other systems than the CMMS, but the information is normally not analyzed in the same detail as the data from functional tests. In particular, it is important to investigate the actions taken by the safety functions when an actual demand occurs. Systematic analysis of gas leaks is important for gas detection systems. Such analyses may indicate if the sensors have wrong location and do not detect gas leakages. In addition, other systems like incidents investigation, systems or procedures for inhibition of alarms, scaffolding work, and reset of sensors must be in place and investigated periodically. Another possibility that could be utilized more in the future, is to build in more detailed logging features in the SIS logic, to present the signal path when actual demands occur. This type of logging might give details about failed components and information about how the leak was detected.

6.4 Procedure/system for collection of failure data

Experiences from the failure cause analysis should be used to improve the procedures and systems for collection and analysis of failure data. A structured analysis of failures and events may reveal a potential for improvements in the actual maintenance or test procedures, or need for modifications of the safety-related systems to improve the functionality.

An important aspect regarding collection of failure data is the definitions of safety-critical failures. Ambiguous definitions of safety-critical failures may lead to incorrect registration of critical failures (e.g. failures that are repaired/rectified "on the spot" are not logged) or registration of non-critical failures as critical ones. The oil companies in Norway have initiated a joint project with the objective to establish common definitions of critical failures of safety functions.

6.5 SIS vs. other types of safety functions

Our case, "prevent spreading of HC by stopping the supply" is an active safety function, and we have not discussed challenges related to surveillance of passive safety functions. However, the functionality of passive safety functions is integrated in the design phase of the installation, and in practice, passive safety functions will be tested only during real accidents. Surveillance of passive safety functions may be carried out by continuous condition monitoring or periodic inspection.

The focus of this paper has been surveillance of SIS. However, surveillance of other safety functions as other technology safety-related systems and external risk reduction facilities is important to control the risk during operation. The failure classification and the surveillance activities presented above may also be used for other active, safety-related systems. Surveillance of some kinds of external risk reduction facilities in the form of operational risk reducing measures as operational procedures may require use of other kinds of surveillance activities.

7 CONCLUSIONS

Recent standards and regulations focus on the entire life cycle of safety functions, and in this paper we have focused on the surveillance of safety functions during operations and maintenance.

The main message is that there should be an integrated approach for surveillance of safety functions that incorporates hardware, software and human/organizational factors, and all failure categories should be systematically analyzed to (1) monitor the actual performance of the safety functions and (2) systematically analyze the failure causes in order to improve the functionality, reliability and robustness of safety functions.

Not all surveillance activities reveal all kind of failures, and a comprehensive set of activities should be used. Failures of safety functions should be registered during actual demands (e.g. gas leaks), testing (functional tests and self-tests), and inspection. The presented failure classification scheme can contribute to an understanding of which surveillance activities that reveal different types of failures.

REFERENCES

Aarø R, Bodsberg L, Hokstad P. *Reliability Prediction Handbook; Computer-Based Process Safety Systems.* SINTEF report STF75 A89023, 1989.

Bodsberg L, Hokstad P. A System Approach to Reliability and Life-Cycle Cost for Process Safety Systems. *IEC Trans. on Reliability*, Vol. 44, No. 2, 1995, 179–186.

Bodsberg L, Hokstad P. Transparent reliability model for faulttolerant safety systems. *Reliability Engineering & System Safety*, 55 (1996) 25–38.

CCPS, 2001. *Layer of Protection Analysis – Simplified Process Risk Assessment.* ISBN 0-8169-0811-7, Center for Chemical Process Safety of the American Institute of Chemical Engineers, New York, US.

Hansen GK, Aarø R. *Reliability Quantification of Computer-Based Safety Systems. An Introduction to PDS.* SINTEF report STF38 A97434, 1997.

Hansen GK, Vatn. *Reliability Data for Control and Safety Systems.* 1998 edition. SINTEF report STF38 A98445, 1999.

Hokstad P, Corneliussen K. *Improved common cause failure model for IEC 61508*. SINTEF report STF38 A00420, 2000.

Hokstad P, Corneliussen K. *PDS handbook*, 2002 Edition. SINTEF report STF38 A02420, 2003.

HSE 2002. *Offshore hydrocarbon releases statistics, 2001*. HID Statistics Report HSR 2001 002. Health & Safety Executive, UK.

HSE 2002b. *Principles for proof testing of safety instrumented systems in the chemical industry*, 2002. Contract research report 428/2002. Health & Safety Executive, UK.

Høyland A & Rausand M, 1994. *System Reliability Theory Models and Statistical methods*. ISBN 0-471-59397-4, Wiley-Interscience.

IAEA, 1999. *Basic Safety Principles for Nuclear Power Plants* INSAG-12, 75-INSAG-3 Rev. 1. IAEA, Vienna, 1999.

IEC, 1998. IEC 61508 1998. *Functional safety of electrical/ electronic/programmable electronic safety-related systems*. International Electrotechnical Commission.

IEC, 2002. IEC 61511-1 2002. *Functional safety: Safety Instrumented Systems for the process industry sector Part 1: Framework, definitions, system, hardware and software requirements*, Version for FDIS issue 8/1/02. International Electrotechnical Commission.

ISO, 1999. ISO 13702:1999. *Petroleum and natural gas industries – Control and mitigation of fires and explosions on offshore production installations – requirements and guidelines*. International Electrotechnical Commission.

ISO, 2000. ISO 17776:2000. *Petroleum and natural gas industries – Offshore production installations – Guidelines on tools and techniques for hazard identification and risk assessment*.

NORSOK, 2001. *Risk and emergency preparedness analysis*, NORSOK standard Z-013 Rev. 2, 2001-09-01, NTS, Oslo, Norway.

NPD, 2001a. *Regulations relating to management in the petroleum activities (The Management Regulations)*. 3 September 2001, The Norwegian Petroleum Directorate

NPD, 2001b. *Regulations relating to design and outfitting of facilities etc. in the petroleum activities, (The Facilities Regulations)*. 3 September 2001, The Norwegian Petroleum Directorate.

NPD/RNNS, 2002. *The Risk Level on the Norwegian Continental Shelf* (In Norwegain – Risikonivå på norsk sokkel), Oljedirektoratet, Stavanger, Norway.

OLF, 2001. *Recommended Guidelines for the Application of IEC 61508 and IEC 61511 in the Petroleum Activities on the Norwegian Continental Shelf*. The Norwegian Oil Industry Association.

Reason, J. 1997. *Managing the risk of organizational accidents*, ISBN 1 84014 105 0, Ashgate Publishing Limited, England.

Sørum M, Thomassen O. 2002. *Mapping and monitoring technical safety*. SPE paper 739230.

Vatn J. *Software reliability quantification in relation to the PDS method*. SINTEF report STF38 A0016, 2000.

Safety and Reliability – Bedford & van Gelder (eds)
© 2003 Swets & Zeitlinger, Lisse, ISBN 90 5809 551 7

Comparison between two organisational models for major hazard prevention

J-C. Le Coze, E. Plot & D. Hourtolou
INERIS, France

A.R. Hale
Safety Science Group, TU Delft, Netherlands

ABSTRACT: A clear difficulty today is to find practical organisational modelling dealing specifically with major hazard prevention and to be able to compare them in order to potentially enhance their relevance. This work is an attempt to do so. This paper provides a comparison of the differences, the common and complementary features of the I-RISK model (developed within the European I-RISK project) and the MIRIAM model (developed at INERIS, France).

1   INTRODUCTION

INERIS has been working for 2 years on a methodology named MIRIAM, which stands for "Maîtrise Intégrée des Risques d'Accidents Majeurs" (Integrated Control of Major Accident Risk). The MIRIAM method, based on an organisational model introducing a human factor approach, is aimed at implementing and assessing safety management systems. A next step in this work, through its development phase, is to compare it to other existing organisational models.

The model developed under the I-RISK research program (Integrated Risk) in the EU 4th Framework Program, had as objective the production of a probability of major hazard occurrence on a chemical site, weighted by human and organisational influences. The Safety Science Group of TU Delft, with Linda Bellamy of SAVE, developed the organisational modelling (Hale et al. 1999).

Both INERIS and Safety Science Group acknowledge the importance of the organisational aspects in major hazard prevention and agree on the necessity of comparing modelling to enhance the ability of modellers to introduce these important aspects into risk assessment.

In this paper, the result of the comparison of the two models is presented. These results will be particularly useful for the ARAMIS program. This program under the EU 5th Framework Program has the objective of producing a harmonised risk assessment methodology for Europe and includes the measurement of safety management effectiveness.

The paper is structured around 5 modelling areas under which the differences as well as the common and complementary aspects are discussed:

- a process approach to the organisational models,
- the interface between the technical and organisational models,
- the factors influencing the interface with the technical modelling,
- the conceptual approach to what is traditionally called the "human factor" in the risk field,
- the issue of management priorities,
- the modelling representation.

2   A PROCESS APPROACH

The MIRIAM model is based on a process approach that consists in decomposing the activities of an organisation. A process is seen as a system of activities that transforms inputs into outputs using resources and constraints. The final outputs in our case of safety management modelling are the safety related activities.

The I-RISK model used the SADT, (Structured Analysis and Design Technique) for the modelling of the safety management system, which is originally a technique that is useful for system planning, requirements analysis and system design. This type of decomposition has the same inputs, outputs, resources and constraints as the process approach used in MIRIAM (Figure 1).

The principle of this type of decomposition is to be functional. It is a "black box" approach that helps to focus not on "how it is done" but rather on "what it does". It is therefore relevant for anything that transforms inputs into outputs, meaning that any organisation can be described as a set of functions to be achieved, rather than in terms of the actual structure specified to achieve them. This process decomposition therefore allows a representation that suits various types of organisation and can be a support to comparing the effectiveness of their risk control – do they have the functions in place and working?

The underlying management science principle that is described through this decomposition is the PDCA, Plan-Do-Check-Act loop. It describes a feedback loop. This is a required feature of any goal-oriented systems. Safety management systems have a risk control objective, and must be moving towards this goal thanks to the application of this principle.

Both the MIRIAM and I-RISK models have this feedback and learning function that can be illustrated through the following cybernetic figure (Figure 2).

All these elementary components have to be defined and described through the process decomposition and all of them are equally important. For a system dedicated to major hazard prevention, these steps could be briefly described as follows:

2.1 The goal

The goal of a safety management system is the control or prevention of major hazard. For that, the hazards must be identified and the risks assessed, in order to know what is the type and level of risk generated by the activity. In both MIRIAM and I-RISK risk analysis is recognised as a core activity which provides the information about what the system has to control and monitor.

2.2 Decisions

From that risk analysis activity, decisions should be taken about the measures to be implemented for risk

control. These are technical measures as well as human/procedural and organisational ones. Technical measures include the implementation of safety devices, human measures include the use of competent people to carry out safety procedures or diagnose unsafe situations, whilst organisational measures include the maintenance planning of the devices, or the training of the operator on safety issues etc. At this stage the safety management system has set up its objectives to be reached through the implementation of the actions. In both MIRIAM and I-RISK this decisional aspect is represented using arrows that indicate information/decision/action flows.

2.3 Actions

From those decisions, actions are carried out in order to reach the objectives. These actions consist of implementing the training, implementing the maintenance planning, and all of the many other safety-related activities. In both MIRIAM and I-RISK these activities (operation, maintenance, inspection etc.) are explicitly mentioned.

2.4 Affected/observed variables

The actions taken must meet the objectives, and these objectives are the affected/observed variables. The choice of the affected/observed variables is crucial. Indeed these variables, in a risk control perspective, are the reference, and they are under the influence of the environmental disturbances. What should be the appropriate observed variables is not so well defined in the risk management literature. Many options are mentioned, including risk perception, safety culture or climate, procedures for safety related activities (i.e. management standard or guidance), power plays within the organisation, motivation etc. The very complexity

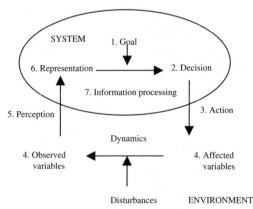

Figure 2. Basic components of a control system.

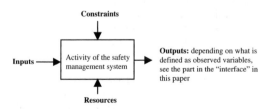

Figure 1. A process approach applied to major hazard prevention.

432

of human activity reveals numerous relevant factors from organisational life (from psychology, organisational sociology, ergonomics, politics, etc.). The difficulty is to define the relevant ones. In the next section the interfaces created with the technical model are discussed for MIRIAM and I-RISK and the affected/observed variables chosen each will be introduced.

The environment creates disturbances, or deviations from the desired goals. These can be internal perturbations like conflicts between people, the lack of commitment due to production pressure, but also the installation itself can generate a wide variety of unpredicted outcomes threatening safety. There are also of course external disturbances like economic constraints (less time and investment in safety related activities), public opinion, regulatory enforcement etc. These disturbances affect the variables that need to be assessed.

2.5 Perception

This relates to the ability of the organisation to observe the chosen affected variables and steer or improve their control of them. This could be based on auditing techniques, when it comes to evaluating the level of compliance with standards or procedures. Other ways of gathering information can be applied when the observed variables concern more "intangible" aspects like risk perception, safety culture, power plays etc. and would be ensured by methods like questionnaires, interviewing, observations. In both MIRIAM and I-RISK this perception aspect is operationalised with arrows that indicate these information flows.

2.6 Representation

The way this information is represented at all levels of the organisation is also fundamental. How is this represented information perceived by the system? Can it be quantitative, or is it better qualitative? What sort of indicators are possible when major hazard prevention imposes a no outcome (accident) objective and requirement (zero incidents/accidents does not mean the same thing as zero risk)? How to create a picture of the state of the variables? In both MIRIAM and I-RISK the exact content of these performance indicators is left open, because of the functional approach. This representational aspect is contained in the arrows. Management has to fill them in, but the arrows imply that connections between those functions of the management system are ensured. The quality of this connection depends on the quality of the information, and therefore part of this is its representation.

2.7 Information processing

Once this information is represented, how is it treated? How will good decision-making be ensured,

considering this information about the state of the observed variables? In both MIRIAM and I-RISK this information-processing activity is represented through the use of arrows that indicate information/decision/action flows. The quality of this activity relies on the ability of the information's receiver to interpret properly the information that is represented in a specific way, according to the affected variables that are chosen. (See 2.4 above)

3 THE INTERFACE BETWEEN THE TECHNICAL AND ORGANISATIONAL MODELS

Both models need to create an interface in order to somehow "plug" the organisational model into the technical one.

In MIRIAM, the interface is based on the prevention barriers. The technical model describes scenarios determined by risk analysis, that are prevented thanks to safety critical functions, represented by barriers chosen in the (organisational) design process. These barriers can be safety devices or safety activities, which fulfill the safety critical functions that need to be implemented at any time to ensure the level of risk control. Barriers are multiple, intervening at various stages in the development of the scenarios. This is the defence-in-depth approach. MIRIAM links to this approach by assessing the management system in terms of its ability to establish and maintain the quality of the barriers.

This interface is presented in Figure 3, part of the MIRIAM modelling.

I-RISK has an approach based on the implementation of a probabilistic method (QRA, Quantitative Risk Assessment). It focuses on initiating and base events as well as barriers, found in the calculation of the failure probability through the use of fault trees. This makes the interface more detailed than in MIRIAM. It is explicitly aimed at connecting with the QRA parameters, which are the following:

- Frequency of external events.
- Failure rate of unmonitored (standby) or monitored components.
- Time between testing.

Figure 3. MIRIAM modelling, interface based on barriers divided in activities and safety critical devices.

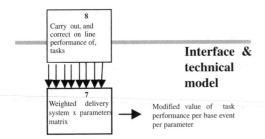

8
Carry out, and correct on line performance of, tasks

Interface & technical model

7
Weighted delivery system x parameters matrix

Modified value of task performance per base event per parameter

Figure 4. I-RISK modelling, Interface based on parameters used in QRA.

– Error in test and repair.
– Failure to detect and recover previous error in test and repair.
– Frequency of routine maintenance.
– Duration of unavailability due to routine maintenance.
– Duration of repair.
– Probability of error in operations or emergency.
– Probability of not detecting and recovering error.

The interface consists, therefore, in connecting these parameters with the process decomposition. This was found to be very complex, as it is really hard to decompose in practice all the activities that influence all of these parameters for all of the equipment and actions which can fail in a complex plant, as analysed by generic fault trees (Figure 4). I-RISK had to resort to grouping equipment, tasks and parts of the organisation in terms of the similarity (called "common mode") of their management, in order to reduce this complexity and make the task of auditing manageable. Rules were developed for this grouping, but it remained a difficult aspect of the modelling.

If there is no constraint imposed by the need to make a detailed quantitative, QRA link between technical and management models at the level of parameters, it may be appropriate to concentrate only on the barriers – the MIRIAM approach, in order to reduce the number of activities to be audited. The level of major hazard prevention would thus depend on the quality of the barriers, maintained by the quality of the management through the implementation of the appropriate activities. However, rules will still have to be developed in ARAMIS for grouping barriers into "common modes", based on the similarities of their management.

4 THE FACTORS INFLUENCING THE ACTIVITIES AND HUMAN BEHAVIOUR

The interface problem raises the question of the influence of the organisation on the elements, events, or parameters coming from the technical risk analysis.

How are the parameters coming from the technical analysis influenced?

In MIRIAM these parameters are the safety critical devices and the safety critical activities. To ensure the quality of these, one needs to consider how the safety critical devices are chosen, purchased, implemented and maintained, and therefore the related activities. Concerning the safety critical activities, one needs to analyse closely how they are carried out.

Considering this question leads the assessor to define how the management system influences these activities carried out by the people. Looking at the way activities are operated requires thinking about how people behave according to their social and physical context.

With a process approach, this becomes a question about the constraints and resources allocated to the people who perform a given activity that constitute the context in which activities, and therefore the work of individuals and group of individuals for any type of activities, are performed. These factors are defined as follows in MIRIAM:

4.1 Competence

Competence is the generic term that gathers all the abilities of an actor to take on his role and responsibilities and to use the technical support at his disposal. These abilities are covered by the three following dimensions:

- **Knowledge**: concerns all the concepts and rules known by the actors. Knowledge will help the actor understand a situation and a context but does not necessarily help make decisions and decide actions.
- **Know-how**: concerns the part of knowledge which is action-oriented. It is applicable to practical situations and very often already automated in the actor's mind. Know-how is intimately linked to *experience*.
- **Safety behaviour**: consists in working out careful solutions, in accepting the cost of these careful solutions and also in investing resources to develop new careful solutions. In our research, this dimension is very much linked to the *values*.

4.2 Co-ordination

Co-ordination is the generic term which allows work first to be divided into tasks and allocated to different human and material resources, and then for the various tasks to be co-ordinated into a coherent whole again. It provides the answer to the question *"Who does what, when, where and how?"*

Co-ordination in MIRIAM is represented by the three following dimensions:

- **Definition of roles and responsibilities**: this represents the two following aspects that need to be managed: *task planning* (What is to be done, where, when?) and *resource allocation* (Who does what?)

- **Modes of communication**: what channels and modes do people use for coordination?
- **Modes of decision-making**: how and where are decisions made?

4.3 *Technical support*

This generic dimension concerns the whole set of techniques (conceptual and material) that should enable an actor to achieve his objectives or effectiveness in his activity.

Technical support answers the question "*How* and *what* should a task be done *with*?" Technical support refers to:

- **Methods, operating procedures**: the answer to *how* question.
- **Tools, device, and man/machine interface**: answer to *what with* question?

These factors have been identified mainly through the body of knowledge found in organisational sociology.

In I-RISK, these influence factors are called delivery systems, to emphasise that they are delivered by management processes, linked often to specific staff functions of an organisation, which deliver resources and controls in the sense defined by SADT. These influencing factors (resources and constraints) have also been derived from the literature and previous modelling projects.

Competence: the knowledge, skills and abilities in the form of first-line and/or back-up personnel who have been selected and trained for the safe execution of the critical primary business functions and activities in the organisation (comparable to the *knowledge* influencing factor in MIRIAM).

Availability: allocating the necessary time (or numbers) of competent people to the safety-critical primary business tasks which have to be carried out.

Commitment: the incentives and motivation which personnel have to carry out their tasks and activities with suitable care and alertness, and according to the appropriate safety criteria and procedures specified for the activities by the organisation (comparable to the *safety behaviour* influencing factor in MIRIAM).

Interface and modifications: The ergonomics of all aspects of the plant which are used/operated by operations, inspection or maintenance (meets the *tools, device, man/machine interface* influencing factor in MIRIAM).

Spares: These are the equipment & spares which are installed during maintenance, which need to be the correct spares and in good condition for the replacement.

Internal communication and co-ordination: Internal communications are those communications which occur implicitly, or explicitly within any primary business activity, i.e. within one task or activity linking to a parameter of the technical model, in order to ensure that the tasks are co-ordinated and carried out according to the relevant criteria (comparable to the *modes of communication*, influencing factor in MIRIAM).

Conflict resolution: The mechanisms (such as management decision, supervision, monitoring, group discussion) by which potential and actual conflicts between safety and other criteria in the allocation and use of personnel, hardware and other resources are recognised, avoided or resolved if they occur (comparable to the *mode of decision making* influencing factor in MIRIAM).

Procedures, Output goals and Plans: Rules and procedures are specific performance criteria which specify in detail, often in written form, a "normative" behaviour or method for carrying out an activity (checklist, task list, action steps, plan, instruction manual, fault-finding heuristic, form to be completed, etc.). Output goals are performance measures for an activity which specify what the result of the activity should be, but not how the results should be achieved. Plans refer to explicit planning of activities in time, either how frequently tasks should be done, or when and by whom they will be done within a particular time period (month, shutdown period, etc.) (comparable to the *methods, operating procedures* influencing factor in MIRIAM).

From this description one can say that the two sets of influencing factors in the MIRIAM and I-RISK models are very close to each other. What differences there are represent some complementarity. Merging of the two lists provides a still more complete list of influences.

The next question is how these factors (resources and constraints) are combined in the model to produce a given unwanted outcome. We use as example the event "incorrect weld performed by an operator", found in a fault tree. A practical question is what factors lead to this error and how could/should they be avoided? Where does it appear in the models.

It falls under the maintenance activities, the welding that has to be performed. There are a number of possible reasons why the error could occur. These are represented in the influencing factors. It may be that the organisation should provide more training, because the welder is not competent. It may be that welders routinely violate procedures and the organisation should ensure more commitment. It could be that the welder did not receive clear instructions about what to do, because there is poor co-ordination between groups concerned, or that there is no procedure for checking work. Should the welding procedure be enhanced, better written? Perhaps the workplace in which the welding takes place is poorly laid out or physically awkward and should be redesigned. All these factors and more are covered under the influencing factors and their delivery systems. Both models provide the classifications and checklists of factors which allow the analyst to predict and discover all of these factors and to examine, in the specific company, how good the management systems are to prevent these shortcomings. How this evaluation is carried out will depend in detail on the conceptual framework they have in taking account of this human factor.

5 THE CONCEPTUAL APPROACH TO THE HUMAN FACTOR

Even if there is a similar vision on the influencing factors between the MIRIAM and I-RISK models, their conceptual frameworks for considering the human factor differ. For the question "how to explain observed or expected behaviours within the implementation of an activity", two types of practice, based on different human factor approaches, are described in the two models. MIRIAM goes into much more detail at this point to understand why the individuals behave in the way they do. I-RISK takes a more generic approach, emphasising the influences to manage the actions and errors. When represented in a fault tree, I-RISK limits itself to the questions shown in Figure 5.

In MIRIAM the influencing factors are to be understood through the conceptual framework of a strategy. From a strategic point of view, to act is to pursue an objective (to resolve problems) by making use of specific means (to find solutions).

This theory considers that all human behaviours can be analysed as individual strategies or as the result of aggregating collective strategies. The strategies are worked out by actors. Actors are individuals or groups. The collective strategies in an organisation can be analysed like the rules of a game. The game can be easily compared to a theatre play. Each actor plays his own role; his strategy is his interpretation of the role. Performing the whole play requires (explicit and tacit) collective bargaining in order for everyone to co-ordinate.

The notion of strategy also refers to the notion of reason. To analyse human acts as specific solutions to given problems is equivalent to acknowledging that actors always have good reasons to act as they do. These reasons are always at least subjectively good, even though they may not be objectively.

This is the theoretical point of view that MIRIAM provides to help the external observer to be able, through interviews, to uncover how activities and organisations function. The aim is to understand the influences on the strategies, in order to modify them. This is necessary when they do not comply with the essential safety requirements, or do not allow the actor(s) to take appropriate safety actions when facing, for example, unexpected situations or unplanned emergencies.

Looking at strategies means looking at the informal part of the organisation, a part that is not written or found in formal procedures, but which gives a complementary view of the organisational life. This informal part, corresponding to the aggregation of collective strategies, rules daily activities and forms what has come to be known as the safety culture of the organisation. This approach, consisting in revealing these unwritten rules, is a useful addition to the safety management assessment found in I-RISK.

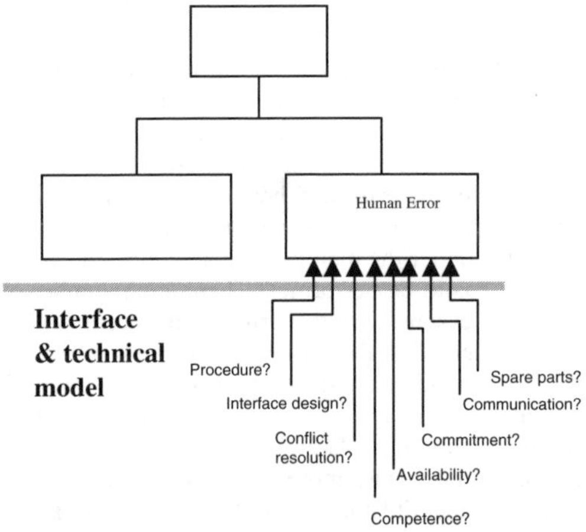

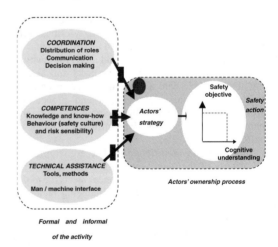

Figure 5.   How do the factors influence human error?

Figure 6.   Actors' strategy in MIRIAM.

436

In the MIRIAM modelling this has been represented as a separate part, where the strategy can be specifically addressed (Figure 6).

In I-RISK this aspect of culture is relatively under-represented. There are elements of "the actors" strategy in the delivery systems of "commitment" and "conflict resolution", but these are approached from an implicitly normative viewpoint.

6 EXPERT JUDGEMENT AND MANAGEMENT PRIORITIES

An issue which I-RISK has dealt with explicitly, but which remains implicit in MIRIAM is the question of how important each of the influences is on the ultimate achievement of the safety goals (the control of the parameters in I-RISK, or the barriers in MIRIAM). The approach relies on an expert judgement procedure to identify among all the influencing factors defined, which are the most important ones. Expert judgement consists in asking to experts in a specific field to answer questions about a subject, using systematic elicitation techniques (Cooke 1991). In a series of pilot studies (Hale et al. 2000) experts in maintenance management were asked to use a paired comparison technique to compare the effect of influencing factors, drawn from the 8 I-RISK delivery systems (see section 4 above), on maintenance parameters (see section 3 above). A mathematical treatment of the answers showed which areas generated a sufficient level of agreement for the results to be meaningful. This provided an empirical basis for priorities in auditing and proposals for change. The I-RISK approach is more empirical than MIRIAM in this respect, being driven by the requirements of quantification (QRA context).

7 THE MODELLING REPRESENTATION

The organisational modelling in both MIRIAM and I-RISK is a mental representation that helps an assessor to structure his/her approach to the organisation, for assessment purposes.

They are representations of a reality which is very complex and which can be observed from various angles and therefore be modelled in different manners. The modelling chosen depends on the constraints and objectives within which the modellers designed it.

The MIRIAM representation had as constraint that it must be transferable to industrialists. This constraint led the modellers to emphasise the visualisation of the model, through the use of colours and to simplify it, to have less elements as well as less visible connections between these elements.

Moreover the MIRIAM model separated the modelling into two parts, in order to emphasise the

conceptual human factor approach. The strategy is the core concept of that approach, and its application for safety purpose is highlighted by a second representation.

Because it had a more quantitative objective, I-RISK had to be as complete as possible about the factors which make a significant difference to the risk numbers. The expert judgement studies in section 6 above were aimed at filling the gaps in knowledge about which these are. In the absence of such data I-RISK had to make the assumption that all factors were significant. This drove the study to greater levels of detail than MIRIAM. This gives it more of an expert perspective. Communication to the user was an issue, which did lead to the abandonment of the full SADT analysis as representation, but was less of a constraint than for MIRIAM.

As a consequence, the difference between the two models is that MIRIAM looks simpler than I-RISK, which contains more visual information about the dynamic aspect of the organisation. The feedback loops are explicitly represented in I-RISK, in several recursive loops, but are only implicit in MIRIAM, though learning is acknowledged as a fundamental aspect of a safety management system there too.

It is not within the scope of this paper to create a new modelling out of the two previous ones. However the ARAMIS project will tackle that task and try to build on the strengths of both models to provide both clarity for outsiders and detail of the reality of the dynamic feedback loops of organisations, and integration of the strategic dimension of human behaviour.

8 CONCLUSION

The comparison of models is a fruitful exercise that allows modellers to exchange their views on complex issues. It is a long process where "paradigms" must be understood from both sides in order to integrate the best of each, depending on the purposes.

The MIRIAM and I-RISK models have been shown to be compatible with few conceptual differences. The models are both based on a functional process decomposition in order to represent the dynamic nature (PDCA, feed back loop) of organisations. They are both independent from the technical modellings (fault trees) for which a relevant interface has to be defined. The two different interfaces – base event parameters and safety barriers – are alternatives relevant for two different purposes. The parameters match existing QRA models better, but the barriers provide more appropriate insight into risk control options and their management.

The two interfaces describe influencing factors that are very similar. The main difference between the models lies in the conceptual approach to the human factor. In MIRIAM it is seen through the eyes of the body of knowledge of organisational sociology, and specifically through the strategy of the actors. In I-RISK the emphasis on this cultural aspect of organisations is less explicit and it is left to expert judgement to define the most relevant influencing factors. However I-RISK does, by this means tackle the issue of prioritising management influences.

Finally the question of the modelling representation is raised. The purposes of the modelling, qualitative or quantitative, for risk analyst, practitioner or manager, for assessment or improvement, must drive the representation used. I-RISK and MIRIAM are different in this respect, but their complementarity of content means than they can potentially be combined.

REFERENCES

Baechler, J. 2000. Nature et histoire. *Presses universitaires de France.*

Bellamy, L. & Van der Schaff, J. 1999. Major hazard management: Technical-management links and the AVRIM2 method, in Seveso 2000 – Risk Management in the European Union of 2000: The Challenge of Implementing Council Directive Seveso II. *Edited by EC-JRC-MAHB. Athens, Greece. 10–12 November 1999.*

Cooke, R.M. 1991. Experts in uncertainty – Opinion and subjective probability in science. Environmental Ethics and Science Policy. Oxford University Press. Oxford.

Duijm, N.J., Madsen, M., Andersen, H.B., Goossens, L. & Hale, A. 2003. ARAMIS project: Assessing the effect of safety management efficiency on industrial risk. *Proc. ESREL, Maastricht, Netherlands, 16–18 June 2003.*

Friedberg, E. 1993. Le pouvoir et la règle. *Edition du seuil.*

Hale, A.R., Goossens, L.H.J., Costa, M.F., Matos, L., Wielaard, P. & Smit, K. 2000. Expert judgement for the assessment of management influences on risk control. In: Cottam M.P., Harvey D.W., Pape R.P. & Tait J. (eds) Foresight & Precaution. 1077–1082. Balkema. Rotterdam

Hale, A.R., Guldenmund, F. & Bellamy, L. 1998. An audit method for the modification of technical risk assessment with management weighting factors. *Probability and Safety Assessment and Management. Springer London. 2093–2098.*

Hale, A.R., Guldenmund, F. & Bellamy, L. 1999. Management model. In Bellamy L.J., Papazoglou I.A., Hale A.R., Aneziris O.N., Ale B.J.M., Morris M.I. & Oh J.I.H. 1999. I-Risk: Development of an integrated technical and management risk control and monitoring methodology for managing and quantifying on-site and off-site risks. Contract ENVA-CT96-0243. Report to European Union. Ministry of Social Affairs and Employment. Den Haag. Annex 2.

Helighen, F. & Cliff, J. 2001 «Cybernetics and second order cybernetics» in R.A. Meyer (ed.) *Encyclopedia of physical science and technology, 3rd edition, academic press, New York.*

I-Risk: «Development of an integrated technical and management risk control and monitoring methodology for managing and quantifying on-site and off-site risk». Main report and annexes. Contract ENVA – CT96 – 0243.

Plot, E. & Lecoze, J.C. 2002. MIRIAM: an integrated approach to organize major risk control in hazardous chemical establishments. *ESREDA 23rd Seminar, Delft, Netherlands, 18–20 November 2002.*

Safety and Reliability – Bedford & van Gelder (eds)
© *2003 Swets & Zeitlinger, Lisse, ISBN 90 5809 551 7*

Industrial areas and transportation networks risk assessment

M.G. Cremonini, P. Lombardo, G.B. De Franchi, P. Paci
D'Appolonia, Genova, Italy

Terrinoni, C. Rapicetta, L. Candeloro
ENEA, Roma

ABSTRACT: There is a growing awareness and concern of governments, communities and industry about the risks to people from location and operation of hazardous activities. Production processes in some sectors of the chemical and petrochemical industry have, in fact, greater potential for fire, explosion, or catastrophic release of highly hazardous material (major accidents). Risk Analysis is well established for industrial plants and evaluates the frequency and severity of accidents, as well as their causes and consequences for people and the environment.

Road and rail dangerous substances transfer and transport are another important source of risk to people whose estimation and distribution on the territory is difficult to evaluate. Therefore, the identification, assessment and management of health and environmental risks due to transportation of hazardous materials are recognized as essential for orderly economic and social development and this requirement is strongly perceived.

A simplified risk analysis methodology, aimed to characterize the various transportation infrastructures, was developed specifically to evaluate the people safeguard conditions and represents an analytical tool to determine which industrial activities (for instance transportation of hazardous materials) can result in a "risk" not to be undervalued. This methodology, starting from the definition of possible accident scenarios and from the calculation of the probability of an occurrence of an accident, provided an estimate of the probability of occurrence of a significant accident.

Furthermore, the study provided useful indications for:

– data acquisition and processing strategies;
– adequate techniques for their subsequent entry in ad hoc databases;
– effective criteria for data analysis aimed at detailed mapping of risks (second level) related to industrial plants and transportation infrastructures.

The objective is the development of guidelines for the prevention, monitoring and control of risks linked to industry and transportation of hazardous goods and for managing possible emergency situations at local level.

1   INTRODUCTION

The paper is aimed at presenting a methodology developed to perform a risk evaluation of dangerous goods transportation at regional scale. Scope of the methodology is to characterize transportation network (i.e., roads, railways) located in an industrial region, and considered both as potential source of risk due to the volumes of dangerous materials/goods being transported, and as potential receptors because of the presence of industrial activities and storage of dangerous materials in their vicinity. The implementation of the methodology allows to identify and map areas of significant potential risk along the transportation corridors of an industrial region where the density of industrial

activities and storage of dangerous materials (the so called "critical areas" according to Italian law D.L. 334/99 and to European 96/82/CE) may create conditions of relatively high risk for the environment and the communities.

The main tasks of the study have been the following:

– design and generation of an ad hoc database compiling the existing data relevant to transportation of dangerous materials and characteristics of transportation network in Italy;
– development of a methodology to classify transportation infrastructures both as potential source of risk due to the volumes of dangerous materials/goods being transported, and as potential receptors because

of the presence of industrial activities and storage of dangerous materials in their vicinity;
- pilot application of the methodology to the industrial area of Ancona and Falconara Marittima in the Marche region, central Italy.

2 DATA BASE GENERATION

To quantify potential risks associated to industrial activities and transportation of dangerous goods, available information and data have been collected from the documents provided by the industries according to the Seveso directives, and from the national and regional public and private bodies in charge of the management of the primary transportation network (e.g., toll roads, national roads and highways, railways).

Road networks have been classified according to the following data:

- length of road segment;
- number of lanes;
- traffic data for both light (i.e., cars) and heavy commercial traffic (i.e., trucks);
- vehicle accidents occurred along the network, their location, involving for both light and heavy traffic.

Railway transportation and infrastructure data have been collected for the Italian network from the national railway company. The annual volumes of traffic of passengers and goods were calculated on the basis of the following parameters:

- traffic as number of trains/kilometer;
- volume of traffic expressed in tons/kilometer;
- incidence of railway accidents at national level expressed as overall rates for the whole network [1].

3 TRANSPORTATION NETWORK AS SOURCE OF RISK

The methodology follows the classification of dangerous substances into three broad categories (inflammable, explosive, toxic) in accordance with the standards set by ADR (European Agreement concerning the international carriage of Dangerous goods by Road), RID (Reglementations concernant le transport ferroviaire International des marchandises), and NST (Nomenclatura Statistica del Traffico), to which the statistics on traffic of goods refer.

3.1 Road network

The categorization of road network as a source of risk is carried out following two different approaches, based on specific data availability:

- approach 1: risk parameters obtained from specific data collected through monitoring;

- approach 2: where specific monitoring data are not available, the risk parameters are calculated using conservative indexes developed from the analysis of available data relevant to road segments with similar characteristics.

Each road is divided into segments of length Li (km) with homogeneous structural features and traffic volumes. For each segment "i" the expected incidence of accident Pi,j, is calculated as follows:

$$P_{i,j} = r_i^m \, U_{i,j}$$

where:

$P_{i,j}$ = incidence of accident per km for a dangerous substance belonging to the category "j" (inflammable, explosive, toxic) on the i-segment (event/year km);

r_i^m = rate of accident (event/vehicle km) occurring on the i-segment involving vehicles transporting goods;

$U_{i,j}$ = traffic of the dangerous substance j on the i-segment expressed as vehicles per year.

The segments where the following condition is verified is considered as "source of risk" and taken into account for the assessment of accident effects:

$$P_{i,j} \geq P_j^0 = r_j^0 \cdot U_j^0$$

where P_j^0 is a study-specific threshold which is defined for the incidence of accident involving the transportation of dangerous substances belonging to category "j", and below which the computed incidence of accident is assumed not to significantly impact the risk for the study area. Although it is noted that P_j^0 can be set by the analyst or the regulator at different values, P_j^0 was given, during the development phase and first application of the methodology, the following default values, according to the category of the substances transported on the given road segment:

Category of Dangerous Substances	j	P_j^0 (event/year km)
Inflammable	1	10^{-3}
Explosive	2	5×10^{-4}
Toxic	3	2.5×10^{-4}

Specific values of rim are not fully available for the Italian road network; rates of accident for the total traffic of goods are only available for the highways (toll roads). An approximate value of rim can be obtained from the annual average volume for the traffic of goods on a given road segment. Where specific data were not available, the overall rate of accident rit (event/vehicle km occurring on the i-segment involving all type of vehicles) was used by applying a multiplication factor equal to 2^1.

Where data on incidence of accidents are not available at all, and, therefore, values for rim and rit are unknown on a given road segment, an approximate value of rim was obtained with reference to a parametric approach developed on selected indicators: the traffic flow, a curve index ("indice di tortuosità", to characterize the presence of curves along the road) and a slope index ("indice plano altimetrico"). The values proposed[2] are presented in the following tables:

TRAFFIC FLOW (vehicles transporting goods/year)		CURVE INDEX
Low	<145000	>10
High	>145000	<10

(Estimated Rate of Accident) $\times 10^8$ (Traffic, Slope Index)	CURVE INDEX	
	Low	High
High		
Category 1	25.4	16.7
Category 2	32.4	23.7
Category 3	39.4	30.7
Low		
Category 1		
Category 2	25.3	17.9
Category 3		

The curve index T_i of the i-segment is calculated as follows:

$$T_i = 100 \cdot \left(\frac{L_{EFF}}{L_{LA}} - 1 \right)$$

where:

L_{EFF} = effective length (km) of the homogeneous i-segment;

L_{LA} = straight length (km) between the two ends of the homogeneous i-segment.

The slope index is defined by dividing the road segments into the following categories [2]:

– Category 1: segment running in a flat area;
– Category 2: segment running on a rolling terrain;

– Category 3: segment running in a rugged-mountainous area.

The traffic of the dangerous substance j on the i-segment $U_{i,j}$, can be evaluated from the total traffic of dangerous substances $U_{i,p}$ as follows:

$$U_{i,p} = f_{i,p} \cdot U_{i,m}$$

where:

$f_{i,p}$ = fraction of dangerous substances transported along the i-segment with respect to the overall goods traffic;

$U_{i,m}$ = effective traffic of goods on the i-segment.

The values of $U_{i,m}$ and of the fraction $f_{i,p}$ were obtained for the Italian network as the solution to an Operational Research problem using Linear Programming techniques on the basis of the following information:

– goods in/out from each region according to NST classification and type of transport (road, railway);
– origin-destination (O.D.) matrix for each type of transport at regional level;
– percentage of the NST categories per type of transport for each Italian administrative province;
– road networks.

$U_{i,j}$ for each category "j" of dangerous substances can be computed from $U_{i,p}$ as follows:

$$U_{i,j} = g_j \cdot U_{i,p}$$

where:

g_j = fraction of traffic of dangerous substances "j" transported along the i-segment with respect to the total traffic of dangerous substances.

The values of g_j, reported in the table below have been calculated following the approach proposed by [3] and have been used as average values for the Italian road transport.

ACCIDENT SCENARIO	A.D.R. CLASSIFICATION	j	g_j
Fire	3;4;5	1	0.85
Explosion	2;1	2	0.075
Toxic Release	6	3	0.075

3.2 Railway network

Similarly to the road network, the railway network has been divided into segments of length L_i (km) and each i-segment is associated to a traffic $T_{i,j}^c$ of trains, which transport dangerous substances belonging to

---

[1] This factor has been obtained from a statistical analysis of the accidents, occurred in the Italian highway network, which provided a r_i^m/r_i^t, 2 for the period 1997–98.
[2] From a statistical analysis of the accidents occurred in the Italian national road network in the period 1995–96.

441

category "j" (inflammable, explosive, toxic), expressed as number of trains per year.

The traffic $T_{i,j}^c$ of trains is obtained from the traffic of wagons computed as follows:

$$N_{i,j} = \frac{Q_{i,j}}{L_i C_j}$$

where:

$Q_{i,j}$ = quantity of all dangerous substances belonging to category "j" (ton km/year) transported on the railway segment "i";

C_j = loading capacity of a wagon for dangerous substances belonging to category "j" (ton/ wagon);

$N_{i,j}$ = number of wagons transporting the dangerous substances "j" on the railway segment "i" (wagons/year).

Once established that T_i represents the traffic of trains transporting goods per year on a given railway segment "i", the value of $T_{i,j}^c$ is calculated from the following criteria:

– if $N_{i,j} > T_i$, then $T_{i,j}^c = T_i$, assuming that the wagons transporting the dangerous substances "j" are uniformly distributed among the trains transporting goods;
– if $N_{i,j} \leq T_i$, then $T_{i,j}^c = N_{i,j}$, assuming that each train transporting goods has only one wagon transporting the dangerous substances "j".

As for road segments, the potential hazard (P_i) of the railway segment "i" can be expressed as follows:

$$P_i = \sum_{j=1}^{M} P_{i,j} = \sum_{j=1}^{M} r_{i,j} \cdot T_{i,j}^c$$

where:

$P_{i,j}$ = incidence of accidents for trains transporting a dangerous substance belonging to category "j" on each i-segment (events/year km);

$r_{i,j}$ = rate of accidents (events/train km) on the i-segment, which involved trains transporting a dangerous substance belonging to category "j";

$T_{i,j}^c$ = conventional train traffic expressed as the number of trains per year which transport a dangerous substance belonging to the category "j" on the i-segment.

The segments where the following condition is verified is considered as "source of risk" and taken into account for the assessment of accident effects:

$$P_{i,j} \geq P_j^0 = r_j^0 \cdot T_j^{c0}$$

where P_j^0, similarly to P_j^0 for the road network, was set at the following values:

Category of dangerous substances	j	P_j^0 (events/year km)
Inflammable substances	1	5×10^{-4}
Explosive substances	2	10^{-4}
Toxic substances	3	5×10^{-5}

Due to the lack of specific data to evaluate the parameter $r_{i,j}$, it has been conservatively assumed that the rates of accidents for the overall railway traffic (t) (both goods and passengers) are equal to the rates of accidents involving the transport of goods (m) and the transport of hazardous substances (j), that is:

$$r_{i,j} = r_{i,m} = r_{i,t}$$

where:

$r_{i,m} = r_{i,j} = 4.9 \times 10^{-7}$ (events/train km) for the rate of accidents involving trains on railway segments, including tunnels;

$r_{i,m} = r_{i,j} = 2.45 \times 10^{-6}$ (events/train km) for the rate of accidents involving trains in stations or travelling on railway segments with detour or intersection devices.

These values are average values of annual accidents per million of trains/kilometer considered "typical" by the UIC (Union Internationale des Chemins de Fer) and have been obtained from the data of the Italian railways company (FS) for the period 1992–1998 [1].

The parameter $Q_{i,j}$ which refers to the traffic of dangerous substances belonging to the category "j", is computed as follows:

$$Q_{i,p} = f_{i,p} \cdot Q_{i,m}$$

where:

$Q_{i,p}$ = annual traffic of potentially dangerous substances belonging to NST categories No. 2, 3 e 8, on the railway segment "i" (ton km);

$Q_{i,m}$ = annual average traffic of goods associated to all NST categories on the i-segment (ton km);

$f_{i,p}$ = fraction involving potentially dangerous substances with respect to the average annual traffic of goods.

The values of $Q_{i,m}$ and of the fraction $f_{i,p}$ were obtained as the solution of an Operative Research problem using Linear Programming techniques on the basis of available data concerning the annual traffic of goods and its distribution over the national territory.

$Q_{i,j}$ for each category "j" of dangerous substances can be computed from $Q_{i,p}$ as follows:

$$Q_{i,j} = g_j \cdot Q_{i,p}$$

442

where g_j represents the fraction of traffic of dangerous substances "j" compared to the traffic of all the potentially dangerous substances. The following table shows the average values of g_j derived from [3] and used as average value for the national railway network:

ACCIDENT SCENARIO	A.D.R. CATEGORIES	j	g_j
Fire	3;4;5	1	0.85
Explosion	2;1	2	0.075
Toxic Release	6	3	0.075

3.3 Areas of damage and areas of influence

Once selected road or railway segments have been identified as a source of potential risk, the "radius of damage" R_1 and the "radius of influence" R_2 surrounding these segments are computed through the association of a reference substance to each scenario. The computed radii generate corridors centered along the segments, which constitute the areas of damage and the areas of influence, where irreversible damage/casualties and reversible impacts are respectively expected in case of accident involving the reference substance. An example of the maps prepared for the pilot area is presented on Figure 1.

The following reference substance have been selected:

– gasoline for the scenario "Fire";
– propane for the scenario "Blast";
– chlorine and ammonia for the scenario "Toxic".

The areas impacted by each scenario have been calculated on the basis of typical loading capacities of vehicles (trucks and wagons) used for the transportation of dangerous substances and with reference to regulatory limits (DM Ambiente 15 Maggio 1996, Appendice III).

The following table shows the radii computed for the reference substances:

	DISTANCE R_2 (m)	DISTANCE R_1 (m)
FIRE		
Gasoline Tank Truck (30 t)	50	25
Gasoline Railway Wagon (55 t)	60	30
BLAST		
Propane Tank Truck (20 t)	450	200
Propane Railway Wagon (40 t)	580	230
TOXIC RELEASE		
Cl_2/NH_3 Tank Truck (24 t)	3140/838	–
Cl_2/NH_3 Railway Wagon (48 t)	3490/870	–

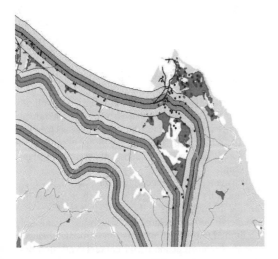

Figure 1. Example of map depicting areas of damage and influence along the transportation corridor for a blast scenario, superimposed to the population density and territorial vulnerable elements.

4 TRANSPORTATION NETWORK AS VULNERABLE RECEPTOR

The vulnerability of transportation infrastructures is calculated in terms of average linear density of population expressed as the number of persons per kilometer.

4.1 Road Network

The vulnerability assessment of a road is based on the calculation of the "average linear equivalent density" of population, which represents the average number of equivalent passengers per kilometer of a given homogeneous road segment. The parameters used to calculate the average linear equivalent density along a road segment are the total volume of traffic along the segment, the equivalent traffic (equivalent vehicles in a given period of time), the average speed of vehicles and road capacity.

The hourly equivalent traffic along a road segment is calculated as follows:

$$U_{eq} = \frac{k}{24} \cdot \left(Ug_p + Ug_m \cdot CE \right)$$

where:

U_{eq} = hourly equivalent traffic (equivalent vehicles/h);

Ug_p = average daily traffic of light passenger vehicles (cars/day);

Ug_m = average daily traffic of heavy vehicles transporting goods (trucks/day);

CE = coefficient of equivalence (assumed equal to 2.5 equivalent vehicles/heavy vehicle);

k = Peak Hour Factor.

The k factor defines the percentage of daily traffic on a given infrastructure attributable to the peak hours. Typical values of the k factor are between 2 and 2.5 according to the type of road [4].

The "average linear equivalent density" on a given road segment is calculated as follows:

$$D_{lm} = C_{RF} \cdot R \cdot \frac{U_{eff}}{v}$$

where:

D_{lm} = average linear equivalent density on a given segment (equivalent persons/km);

C_{RF} = coefficient for the reduction of peak hour vehicle traffic: non-dimensional factor which takes into account the different use of the road segment over the 24 hours;

R = average "filling coefficient" of the equivalent mean of transport (persons/mean of transport);

v = average speed of vehicles on a given segment (km/h);

U_{eff} = $(U_{gp} + U_{gm}) \cdot k/24$ = effective total peak hour traffic (vehicles/h).

4.2 Railway network

The assessment of the vulnerability of a given homogenous railway segment is based on the average number of passengers per kilometer of the segment (average linear density). The parameters necessary to calculate the average linear density along a railway segment are the peak hour train traffic (long distance trains and regional trains), the average commercial speed of trains and the number of passengers for each train.

The data, which are available at national level for each segment, are the annual total number of trains (divided into trains transporting goods, and regional and long distance trains transporting passengers) and the annual number of passengers (for both regional and long distance trains).

The hourly peak hour traffic T_{tot} is obtained as follows:

$$T_{tot} = \frac{k_1}{24} \cdot T_{lp} + \frac{k_2}{24} \cdot T_{loc}$$

where:

T_{lp} = daily traffic of long distance passenger trains (trains/day);

T_{loc} = daily traffic of regional passenger trains (trains/day);

k_1, k_2 = multiplying factors for the conversion from average daily traffic to peak hour traffic.

The "average linear density" D_{lm} (equivalent persons/km) for the two types of trains (T_{lp} and T_{loc}) on a given segment is determined as follows:

$$D_{lm} = \frac{C_{RF}}{24} \cdot \left[\left(R_{lp} \cdot k_1 \frac{T_{lp}}{v_{lp}} \right) + \left(R_{loc} \cdot k_2 \frac{T_{loc}}{v_{loc}} \right) \right]$$

where:

C_{RF} = coefficient for the reduction of peak hour vehicle traffic: non-dimensional factor, which takes into account the different use of the segment over the 24 hours;

R_{lp} = is the average filling coefficient for long distance passenger trains (persons/train);

R_{loc} = is the average filling coefficient for regional passenger trains (persons/train);

k_1, k_2 = multiplying factors for the conversion from average daily traffic to peak hour traffic for long distance and regional trains respectively;

v_{lp} = average commercial speed for long distance passengers trains (km/h) along the segment;

v_{loc} = average commercial speed for local passengers trains (km/h) along the segment.

The values of R_{lp} and R_{loc} are the ratios between the number of annual passengers and the annual number of trains along all the segments, averaged on the linear kilometers of all the railway lines.

5   CONCLUSIONS AND REMARKS

This current work represents a technique explicitly aimed to categorize on a relative basis the risk induced by the network arcs on the surrounding areas (or vice-versa) and to map this information merging it in a coherent way with the "classical" stationary plants risk data.

The main characteristic of this approach is its screening function on the whole network analyzed: using this method the arcs (roads, railways, water routes, etc) can be categorized as "relatively dangerous", using only the limited number of information immediately available in any normal industrial context, and then post processed with a more detailed and quantitative objective risk analysis, where considered necessary.

A comparison with available and well known small-scale standard risk analysis approaches, besides not being practicable at this level, would not add any value to the outcomes of the proposed methodology, from this point of view. In addition, it is important to remark that due to its nature, the risk here calculated is hardly comparable with those derived from other techniques in the absolute value, and only the "relative risk" among arcs belonging to the same transportation network can be used as representative value for any kind of comparison with other techniques.

REFERENCES

[1] Fermerci, 2000, Paper (In Italian) "Gli Standard della Sicurezza – Tra i migliori d'Europa", pagg. 23–24, Roma, Marzo/Aprile.

[2] Environmental System Research Institute (ESRI), Istituto Italiano di Statistica (ISTAT) e SEAT, 1997, User Manual "GeoStat", Roma.

[3] Ufficio Federale dell'Ambiente, delle Foreste e del Paesaggio (UFAFP), 1992, Manuel III de l'ordonnance sur les accidents majeurs OPAM, Dicembre.

[4] National Research Council (NRC), 1985, Highway Capacity Manual – Transportation Reseach Board, Washington, D. C.

Safety and Reliability – Bedford & van Gelder (eds.)
© 2003 Swets & Zeitlinger, Lisse, ISBN 90 5809 551 7

An Integrated Quantitative Risk Assessment of an Oil Carrier

R.B. Cross
American Bureau of Shipping Group Consulting, Houston, Texas, United States

J.E. Ballesio
American Bureau of Shipping, Houston, Texas, United States

ABSTRACT: Class Societies have traditionally taken an experience based approach to developing Class Rules that has provided a high level of safety for both personnel and the environment. With emerging technologies and advanced designs, however, the experience base is small. A risk-based approach to Classification is being developed by the American Bureau of Shipping (ABS) to ensure that the high level of safety that currently exists continues, and will be extended to new and novel designs. As part of this effort, an integrated Quantitative Risk Assessment (QRA) model for an oil carrier was developed. The basic design modeled is that of a double hulled, redundant propulsion/rudder vessel. This model is being used to evaluate the "value" of Class Rules, evaluate the relative benefit of redundancy, and the risk impact of maintenance while underway. This paper will present the basic design and methodology used to build the model, along with results from the applications performed.

1 INTRODUCTION

Classification Societies are independent organizations, most of them not-for-profit, that promote the security of life, property and the environment of ships and offshore structures. This is done through a procedure known as classification, by the establishment and administration of standards, known as rules, for the design, construction, and operational maintenance of marine vessels and structures. The American Bureau of Shipping (ABS) is one of the world's leading ship classification societies.

Classification societies have been investigating the use of risk based techniques applied to the classification of marine vessels and offshore structures. ABS undertook specific projects to support all three parts of the classification process: rule development, class assignment and maintenance of class (Ballesio & Diettrich 2002). The projects covered a wide spectrum of risk assessment techniques to support classification, from simple qualitative tools to full quantitative risk models. Specifically, two comprehensive Quantitative Risk Assessment (QRA) models were developed: one for a Floating Production, Storage and Offloading (FPSO) vessel and another for an Oil Tanker. The objectives for building these quantitative models are the following:

- to comprehensively represent major hazards associated with design, operation and maintenance
- to generate baseline risk estimates as independent benchmarks of performance
- to link the model to specific classification rules and activities

The models are expected to be used both internally within ABS for classification rule development and rule application, and externally by sharing it with clients to aid demonstration of acceptability for designs when they include alternatives to the prescriptive classification rules. These types of applications are described in more detail below:

1.1 Internal use by ABS

Two very important activities in the classification process are rule development and consistent application to the rules. Both of these activities can benefit from the use of risk assessment models.

For rule development, risk assessment can be used mainly for the following two objectives:

- to identify potential new requirements that can decrease risks not currently addressed by the rules (potential rule addition), and
- to identify current prescriptive requirements which can be shown not to have any impact on risk (potential for rule elimination or relaxation)

The quantitative models can be of use for the above objectives.

For the consistent application of classification rules, ABS has a procedure called "Rule Interpretations and Instructions" (I&I). The objective of the I&I Process is to facilitate consistency in the application of the rules. This procedure has been recently revised to include the use of risk assessment techniques to support the resolution of certain consistency issues as identified by ABS Engineers. The use of existing quantitative models is contemplated under this new procedure, when applicable. Using existing models throughout the company will help reduce subjectivity and variations in approach in assessing risks (e.g. among offices, countries, project to project, client to client)

1.2 External use

ABS has published information that outlines procedures the industry can follow to make use of risk assessment techniques as an alternative basis for classification approval. A range of risk assessment techniques can be used to support classification requests by demonstrating that the proposed designs offer an equivalent level of safety to that of traditional designs or related industry standards. Existing quantitative risk models are expected to be shared with the industry, to give a starting point in such equivalency proofs. The existing models in general will not directly provide a proof, but it can save considerable resources by using them as basis for further risk modeling in specific issues as needed.

Two cases are expected to be encountered:

- analyze designs presented by clients with specific alternatives to the more prescriptive classification rules
- support clients in the risk analysis of novel designs

This paper will describe the oil tanker model developed, including results and applications.

2 DEVELOPMENT OF THE MODEL

2.1 Selection of design and mission

The development of the model started with defining the tanker overall design parameters, and the location(s) where the tanker was assumed to operate. The selection of these parameters is important because of the potentially significant influence on the results and they also impact how widely applicable the "generic" model is for use in applications.

2.1.1 Overall design parameters

Based on industry trends and regulations, size and hull design were chosen as:

- Size – 100,000 – 150,000 DWT
- Hull – Double Hull

Many newbuild tankers fall within these parameters, and were therefore thought to be the most applicable looking forward to designs being considered by owners. A new trend in engine room design for some newbuild tankers is to build redundant engine rooms. This is beyond current regulatory requirements, but was noted as an emerging trend in some areas. In order to be able to provide the maximum analytical capability for the model, it was decided to model the engine room using a redundant design, with the capability to "turn off" one of the engine rooms and therefore simulate a single engine room. One caveat used in this decision was that both cases (single and redundant engine rooms) must meet Classification requirements for their respective designs.

2.1.2 Assumed mission

With the overall design selected, the next step was to select a mission for the tanker. The geographic area where a tanker operates influences risk through weather, duration and complexity of course in restricted waters, etc. Therefore, in selecting a mission, such factors need to be defined. A Trans-Alaskan Pipeline Service (TAPS) mission was chosen because operations in this area have a relatively high amount of data available. Specifically, the mission chosen for the model assumed is the following:

- Loading in Valdez
- Fully loaded transit to Washington (state) with a partial offload
- Transit to Long Beach to complete offload
- Transit to Valdez empty

In addition to the availability of data, this mission allows the analysis to examine all nominal loading conditions a tanker would experience.

2.2 Scope of analysis – hazards and consequences

2.2.1 Consequence selection

Hazards (through accident scenarios) and consequences are interrelated in that the potential consequence depends on the type of hazard. Therefore, a selection of consequences of interest to Class Societies and owners was performed in order to identify the appropriate hazards for the scope of the model. The consequences of interest were found to be:

- Personnel safety
- Environmental Damage
- Downtime
- Costs

Many hazards potentially lead to multiple consequences. For instance, an explosion could result in all of the above consequences. From the above potential consequences, personnel safety and environment were selected as the risk metrics to be analyzed. Tanker hazards vary significantly depending on the mode of operation and potential consequences. For instance, a contaminated bilge water discharge is a hazard since environmental damage could result, however, a drift grounding with subsequence severe weather would have much higher potential for environmental damage. For the scope of the model, the events considered of most interest were those with significant potential consequences.

2.2.2 Hazard selection

In order to assess the hazards associated with the consequences selected, a Hazard Identification (HazId) exercise was performed with tanker and risk assessment experts. Using a top down approach, hazard categories were identified that could result in fatalities or environmental damage. For tankers, four critical functions were identified that are necessary for tankers to be "fit for purpose." Failure of any one of these functions was considered as a hazard category as shown below.

- Buoyancy Failure – Any hazard that threatens the buoyancy or stability of the vessel. Examples include structural failure and internal flooding.
- Steering Failure – Any hazard or event that can lead to a steering failure such as loss of electric or hydraulic power.
- Propulsion Failure – Any hazard that threatens the propulsion capabilities of the vessel such as loss of electric power.
- Stationkeeping Failure – Any hazard such as mooring or anchor failure that impacts the ability to effectively keep stationed.

Categories of events that could lead to fatalities and/or possible environmental damage, without necessarily involving loss of one of the critical functions above were then identified. This resulted in the following five additional hazard categories shown below.

- External Hazards – Any hazard that is initiated outside the boundary of the vessel. Examples include severe weather, collision, etc.
- Personnel Hazards – Any hazard that impacts personnel safety, but is limited in potential threat to overall crew safety. Examples include maintenance incidents and dropped objects.
- Explosion – Any hazard or event that has the potential to lead to an explosive atmosphere, a release of stored kinetic energy, or disassembly of high energy equipment. Examples of these are failure of the inert gas system, overpressure of a hydraulic accumulator, or a diesel engine crankcase explosion.
- Fire – Any hazard or event that has the potential to lead to a fire such as fuel oil or other hydrocarbon release.
- Environmental Damage – Any hazard that leads to an environmental impact and does not threaten

personnel safety or the overall functionality of the ship. And example would be a contaminated bilge water discharge.

The HazId then proceeded and identified causes for each of these hazard categories. For instance, loss of electrical power, fuel system failure, lube oil failure, etc. were identified as causes for loss of propulsion. A risk matrix was developed to qualitatively rank the causes in order that they could be prioritized for modeling. The matrix used was a 3×5, with three consequence levels and five frequency levels. The matrix was consequence oriented in that an objective of the model was to focus on severe consequences rather than more frequent less severe consequences. The modeling priorities were assigned after each cause was ranked based on frequency and worst case consequence in the risk matrix.

2.3 Development of event sequences

2.3.1 Use of event sequence diagrams

Development of the model proceeded by detailing the possible event sequences for each hazard category considered (e.g. loss of propulsion). In order to do this, Event Sequence Diagrams (ESDs) were employed. These were used to describe critical events and potential consequences related to each hazard category. The ESDs identify critical systems and events to model with fault trees, as well as the logic required to be used in event tree development.

2.3.2 Loss of propulsion and steering

Events related to loss of propulsion and/or steering were the first to be addressed. Other events, such as fires, may lead to a loss of propulsion and/or steering so the applicability of this event sequence was developed in a global sense. The event sequences for propulsion and steering considered the following questions:

- Is the grounding likely to be powered or drift?
- Is the failure able to be repaired before grounding?
- Are tugs available to prevent grounding?
- Is anchorage an option to prevent grounding?
- What type of bottom is experienced when grounding?
- Is evacuation required?
- Are lifeboats available?
- Is helicopter rescue available?

A simplified version of the loss of propulsion and steering ESD is shown in Figure 1.

2.3.3 Fire

A "generic" ESD for fire event sequences was developed to be able to evaluate fires in various locations on the tanker. Questions considered in the development

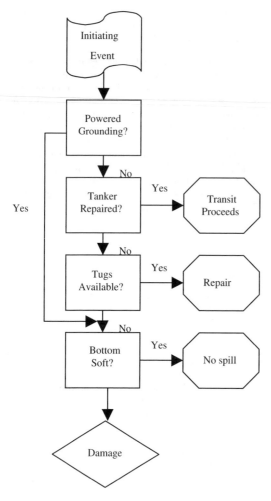

Figure 1. Simplified Loss of Propulsion and/or Steering Event Sequence Diagram.

of the fire event sequences are:

- What is the source of the fire?
- Does the fire have the potential to escalate?
- Is the fire detected?
- Are there personnel in the area of the fire?
- Can portable suppression be used?
- Is fixed suppression available?
- Does the fire escalate?

The outcomes from these questions along with the specific location of the fire were used to determine what equipment might be affected by the fire, if any fatalities may be expected, and if an evacuation was required. In some cases where loss of propulsion and/or steering was a result of a fire, the event sequences then consider the questions asked in the previous section.

2.3.4 Explosion

A "generic" ESD was also developed to define potential scenarios for cargo tank, ballast tank, and pump room explosions. The initiating event considered for these types of scenarios was assumed to be that an explosive atmosphere was present. The development of the ESD then proceeded by considering the following questions:

- Are there personnel in the area of the explosion?
- Is an ignition source present?
- Does structural damage occur?
- Is there a leak of hydrocarbons?
- Does the explosion cause flooding?

Like the fire event sequences, explosions were modeled to result in secondary effects such as fire, loss of propulsion, etc. In these instances, the scenarios considered the appropriate questions for the secondary events (e.g. fire).

2.3.5 Collision

Collisions were considered as initiating events and the ESD development considered the following questions:

- Is the ship docking or underway?
- What part of the tanker is affected?
- Does the collision penetrate the double hull?
- Is there a leak of hydrocarbons?
- Is there seawater flooding?

The distinction was made between docking and underway collisions, since docking collisions are generally lower in energy and are less likely to penetrate a double hull. The part of the tanker affected was assumed to be random and is based on geometric considerations. A high energy collision in the cargo block will cause a leakage of hydrocarbons and potential fire threat, while an impact in the engine room area may result in flooding and a loss of propulsion.

2.3.6 Structural failure

A relatively simple model was used to estimate structural failure since machinery events were the primary focus of the model. HECSALV™ (HEC 2000) calculations were performed at beginning of life and end of life of the tanker to estimate bending moments on the tanker. The end of life analyses considered the effects of maximum corrosion rates and very limited replenishment of steel. The conservative analysis showed that beginning of life of the tanker was very robust structurally. The calculated end of life bending moment was beyond yield strength of the deck plating. This was true, however, only for a severe wave environment. The structural failure probability was then estimated using historical structural failure data, based on the North Pacific area.

450

2.4 Consequence assignment

2.4.1 General consequence approach

As previously mentioned, the consequences of interest were selected as personnel safety and environmental damage. At the start of the model development, it was also noted that a second "level" of consequences existed that can occur before the ultimate consequences are realized. These "intermediate" consequences (e.g. loss of propulsion, loss of steering, etc.) are precursors to potentially significant events, and are generally based on design characteristics not environmental factors. Since, these consequences were considered when developing the ultimate consequence scenarios, the decision was made to include the ability to obtain results on both levels.

2.4.2 Intermediate consequences

Intermediate consequences for the model are based on the design of the ship and not on external influences such as helicopter rescue, human navigation errors, etc. It is recognized, however, particularly for structure, that environmental conditions can have a large impact on reliability. The sets of intermediate consequences modeled by this study are:

- Total loss of propulsion and steering – This is defined as the annual frequency of a loss of all propulsion and steering which includes a loss of normal shipboard power from the high voltage switchboards through the emergency switchboard.
- Total loss of the propulsion system– This is defined as the failure of the propeller to turn. This event only involves loss of propulsion and not propulsion and steering. For loss of propulsion cases with steering available, additional recovery time is assumed available relative to a total loss of propulsion and steering.
- Total loss of the steering system – This is defined as a loss of the ability to control the rudder(s).
- Fires – This is defined as the initiation of a fire in one of the areas identified in the HazId. Fire may have an immediate impact on personnel and could lead to loss of the critical functions above.
- Explosions – This is defined as the initiation of an explosion. Like fires, explosions may have an immediate impact on personnel and could lead directly to environmental damage or to loss of the critical functions.
- Structural Failure – Structural failure as an intermediate consequence is defined as one that can put the vessel in an unsafe condition to operate or a through wall crack that can lead to cargo leakage.

2.4.3 Environmental damage

The risk metric used to "bin" environmental damage was barrels of oil spilled. A combination of factors was used to estimate the amount of oil spilled for each scenario. The software HECSALV™ was used to calculate oil outflow from cargo tanks for various conditions. Key assumption used in the HECSALV™ calculations were that a grounding event would result in a loss of containment midway between the waterline and the tanker bottom, or on the bottom with an equal likelihood. For collisions, it was assumed that the loss of containment would manifest itself at the waterline. The program accounts for hull configuration, location of break, static head of water, etc. to determine the expected oil outflow. Weather conditions were then accounted for in terms of escalating the expected damage. For example, Table 1 shows how some of these factors combine to result in different consequence severity.

2.4.4 Fatalities

In terms of personnel safety, fatalities were chosen as the risk metric. Fatalities were estimated by considering two possibilities:

- Some events may result in immediate fatalities
- Evacuation (if required) may result in fatalities

In determining immediate fatalities, the type of event, location, and personnel manning was considered. To do this a Personnel On-Board (POB) distribution as shown in Table 2 was assumed.

Historical events were reviewed in order to determine the likelihood of experiencing immediate fatalities for events such as fire and explosion that are somewhat likely to lead to immediate fatalities. The historical data along with the assumed POB distribution were used to estimate the average number of immediate fatalities for specific events. Evacuation

Table 1. Estimated average environmental consequences for selected grounding scenarios.

Scenario	Weather	Bottom	Spill size (barrels of oil)
Drift Grounding, Fully Loaded	Calm	Mud	0
Drift Grounding, Fully Loaded	Calm	Rock	0
Drift Grounding, Fully Loaded	Moderate wave	Rock	30,000
Drift Grounding, Partially Loaded	Moderate wave	Rock	18,000

Table 2. Assumed Personnel On-Board distribution.

	Cargo Deck	Accommodation (including bridge)	Engine room areas
Days	4	10	6
Nights	0	20	0
Average	2	15	3

451

Table 3. Estimated average total fatalities from fires in the accommodations area.

Immediate fatalities	Evacuation conditions	Evacuation fatalities	Total fatalities
0.75	Evacuation not required	0	0.75
0.75	Evacuation required, rescue successful	1.7	2.45
0.75	Evacuation required, rescue fails	3.7	4.45

was assumed to be required for two conditions: fire escalation, and severe structural failure. For evacuation fatalities, a simple model based on Safety Case work done in the North Sea was used. This simple model considered lifeboat availability, liferafts availability, direct entry fatalities and helicopter rescue.

A simple event tree was made to estimate evacuation fatalities, which were then combined with immediate fatalities for each event. An example of this is shown in Table 3.

2.5 Data

2.5.1 Data requirements
In order to populate the model for quantification, necessary data was obtained from a variety of sources. In general, the types of data required fell into three categories: equipment failure rates, external event data, and initiating event data.

2.5.2 Equipment failure rates
Equipment failure rates were obtained from several commercially available sources. In many instances, data was available for a specific piece of equipment in multiple sources. In those cases, often there was good relative agreement. In cases with large differences, knowledge of the sources and analyst experience were used to determine the most appropriate value. Common cause factors were also used extensively in the analysis. The Multiple Greek Letter (MGL) method was used (USNRC 1998) with the MGL data parameters obtained from existing commercial nuclear power data.

2.5.3 External event data
External event data used in the model consists of items primarily related to estimating the consequences of events such as weather, tug availability, helicopter rescue, etc. A variety of means were used to develop data for these types of events. For weather (wind and wave), National Oceanic and Atmospheric Administration (NOAA) data was used from buoys in locations relevant to the analyses. At each location wind and wave were correlated to capture the dependence of wave

height on wind speed. Tug availability was estimated by reviewing factors such as tug locations versus time to grounding, weather, tug engine reliability, etc. These factors were then combined, and conditional probabilities developed as necessary for each specific location analyzed in the model. Some factors, such as bottom geology (sand or rock) where a tanker might ground was based on opinion after discussions with personnel familiar with the mission locations.

2.5.4 Initiating event data
Initiating event data was derived in two fashions. For machinery failure type events, fault trees were used with equipment failure rates as described in Section 2.6.2. For human error initiators leading to grounding and collision, structural failure, and fire events, data was obtained from United States Coast Guard (USCG 2000) databases. For the human error initiators, only data from the ports assumed in the mission were used because it was felt that the human error rate would be heavily dependent on location. Specific initiating event frequencies were developed for each port and sea passage when required. Only historical data for tankers was used to develop the model data.

3 RESULTS

Results from the model may be obtained at various levels including:

- Overall consequence frequencies (loss of propulsion, drift grounding, fire, etc.)
- Scenario rankings
- System level importance
- Component level importance

Additionally, all of these levels of results may be applied to a particular phase of the mission or consequence. Included in this paper, are only samples of each of the types of results in order that the reader may understand the detail and complexity involved in the model.

3.1 Intermediate consequence results

The high-level intermediate consequence results from the baseline model are shown in Table 4.

These results show by far the most likely intermediate consequence is a loss of propulsion. The result is expected since the majority of equipment normally in use during transit is there to support propulsion. Explosions, a major concern for tankers, were shown to be the least frequent. This is due, in large part, to the fact that during transit tank operations are only infrequently performed. Each of the items in Table 4 may be broken down into individual contributors. Because machinery events have similar consequences, the events experiencing loss of propulsion and/or steering can be evaluated as a group. Table 5 lists the top scenarios for these events.

As, expected, loss of propulsion is the consequences in all but one of the top scenarios. An examination of the results also shows that the major contribution to loss of propulsion comes from a variety of mechanical systems. If more scenarios were included, it would show that the loss of steering event is essentially all within the steering system itself. Losses of steering from supporting systems such as electrical power generally result in a concurrent loss of propulsion.

From these higher level results, contributions and sensitivities of contributors may be obtained on the system level and component level as discussed in Section 4.

3.2 Ultimate consequence results – environment

Ultimate consequence results related to the environment are shown in Table 6 and Table 7. These two tables are included here to show that the frequencies and consequences are not necessarily related. Collisions were found to be the most frequent event that could lead to environmental damage during transit, however, they are

only a small contributor to the expected risk. Sinking due to structural failure is a very low likelihood event, but because of the severity of the consequences, has a large impact to expected risk.

3.3 Ultimate consequence results – fatalities

Ultimate consequence results related to personnel safety are shown in Table 8 and Table 9. For fatalities, it is seen that the expected risk again does not mirror event frequency for those events that results in fatalities. Fire is estimated to be the leading contributor to expected risk for fatalities, while explosions during transit are estimated to be very low contributors for

Table 6. Estimated frequencies of events potentially resulting in environmental damage.

Event type	Frequency (events per year)
Collision	3.15E − 02
Powered Grounding	6.56E − 03
Drift Grounding	4.45E − 04
Structural Failure	1.76E − 04
Explosion	9.32E − 06
Total	3.87E − 02

Table 4. Single engine tanker estimated frequencies of intermediate consequences.

Consequence	Frequency (events per year)	One event per (years)
Loss of propulsion	5.04E − 1	1.98
Loss of steering	3.36E − 2	29.8
Fire initiation	1.25E − 2	80
Structural Failure	6.33E − 3	158
Loss of propulsion and steering	5.28E − 3	189
Explosion	9.37E − 6	107000

Table 7. Expected risk for environmental damage by event type.

Event type	Expected risk (barrels per year)
Structural Failure	6.96E + 01
Powered Grounding	5.97E + 01
Collision	5.87E + 00
Drift Grounding	2.08E + 00
Explosion	3.45E − 01
Total	1.38E + 02

Table 5. Top 10 intermediate consequence scenarios for machinery events.

Rank	Scenario	Frequency (per yr)	Consequence
1	Main Engine Fails	1.19E − 01	Loss of Propulsion
2	Freshwater Cooling Fails	2.54E − 02	Loss of Propulsion
3	Lube Oil System Fails	1.70E − 02	Loss of Propulsion
4	Fuel Oil System Fails	1.57E − 02	Loss of Propulsion
5	Seawater Cooling System Fails	1.43E − 02	Loss of Propulsion
6	Steering Gear System Fails	1.35E − 02	Loss of Steering
7	Power Take Off Generator and Ship Service Diesel Generator Fail	9.92E − 03	Loss of Propulsion
8	Main Engine and Ship Service Diesel Generator Fail, Ship Service Diesel Failure Has No Additional Effect	4.39E − 03	Loss of Propulsion
9	Main Engine and Emergency Diesel Generator Fail, Emergency Diesel Failure Has No Additional Effect	4.36E − 03	Loss of Propulsion
10	Failure of Both Sea Chest Intakes	1.65E − 03	Loss of Propulsion

Table 8. Estimated frequencies of events potentially resulting in fatalities.

Event type	Frequency (events per year)
Collision	3.13E − 02
Fire	1.22E − 02
Structural Failure	1.76E − 04
Explosion	9.33E − 06
Total	4.37E − 02

Table 9. Expected risk of fatalities by event type.

Event type	Expected risk (fatalities per year)
Fire	2.29E − 3
Structural Failure	8.78E − 4
Collision	1.46E − 4
Explosion	1.88E − 5
Total	3.33E − 3

the same reasons described in the previous section for environmental results.

4 APPLICATIONS

As mentioned in the Introduction, it is envisioned that the model will be applied both internally by ABS and externally by owners. One of the most significant internal uses being performed is an evaluation of Classification Rules. In this application, intermediate consequence results are used to study the tanker design and attempt to identify any weaknesses through use of system and component level results from the model. For example, the loss of propulsion and steering frequency as described in Section 3.1 may be broken down into its contributors. Table 10 shows the top contributions on a system level for a loss of propulsion and steering.

In Table 10, the Fussell-Vesely importance factor is used. This is a measure that simply identifies the fractional contribution of a system to a specific consequence. The sum of all Fussell-Vesely importance values over all systems can be greater than 1.0 since in some scenarios multiple systems are required to fail to realize a consequence. Some general conclusions can be drawn from the loss of propulsion and steering system level results:

– The prominence of the diesel generators, and small importance of the steering gear system itself (~0.03 Fussell-Vesely importance), results in the loss of electric power contributing to over 95 percent of the contribution to a loss of propulsion and steering.

Table 10. Fussell-Vesely system importance to loss of propulsion and steering.

System	Fussell-Vesely importance
Emergency Diesel Generator	0.955
Freshwater Cooling System	0.408
Ship Service Diesel Generator	0.287
Seawater Cooling System	0.23
Power Take Off Generator	0.171
Main Engine	0.082
Seawater Inlet	0.026

Table 11. Fussell-Vesely component importance to loss of propulsion and steering.

Component	Fussell-Vesely importance
Emergency Diesel Generator	0.91
Ship Service Diesel Generator	0.25
Freshwater Mixing Valve	0.16
Power Take Off Generator	0.11
Main Engine	0.11
Freshwater Piping	0.07
Emergency Generator Feeder Breaker	0.06
Freshwater Pump 1	0.04
Main Engine Power Take Off Clutch	0.03
Main Engine Power Take Off Gear Box	0.03

– Because of the high contribution of a loss of electric power to this consequence, the emergency diesel generator has a very high importance relative to other systems.
– Cooling systems such as freshwater and seawater have relative high importance because their failure may result in multiple failures of dependant systems such as the main engine and the ship service diesel generator. These dependencies make this type of event more vulnerable to these types of failures.

Results are also obtained on the component level. These are perhaps the most important for this application since many of the Classification Rules are on the component and subcomponent level. To continue with the loss of propulsion and steering example Table 11 shows the top ten contributors from a component perspective:

Based on the contributions of various components, Classification Rules may be studied to identify any weaknesses that may exist.

5 CONCLUSIONS

The oil tanker QRA model has been developed for use in the marine industry. This comprehensive QRA model is the first of its kind for an oil tanker, and will yield benefits for both the Class Society and tanker owners as a tool for evaluating risk tradeoffs, new designs, etc. The baseline results are based on a specific set of design and mission assumptions used to develop the model, however when compared against industry statistics they show good overall agreement. Analyses for specific ships may be tailored by revising those assumptions. Ongoing evaluations include:

– Evaluation of Class Society Rules
– Comparison of single and redundant engine designs
– Specific analyses such as evaluating the change in risk from removing the emergency diesel generator from the redundant engine design.

This tool and the associated analyses are an integral part of the overall plan to move toward integrating risk processes into the maritime and off-shore industries.

REFERENCES

Ballesio, J. & Diettrich, D. 2002. Risk and reliability applications to marine classification, *Proc. European Safety and Reliability Conference, ESREL 2002, Lyon, March 2002.*
HEC 2000. HECSALV 7.4.3, Herbert Engineering Corp., 2000.
USCG 2000. US Coast Guard *Marine Safety Management System (marine casualty and pollution database), 2000.*
USNRC 1998. US Nuclear Regulatory Commission, *NUREG/CR-5485: Guidelines for Modeling Common Cause Failures in Probabilistic Risk Assessments, 1998.*

Safety and Reliability – Bedford & van Gelder (eds)
© 2003 Swets & Zeitlinger, Lisse, ISBN 90 5809 551 7

Analysis of random inspection policies

A. Csenki
University of Bradford, UK

ABSTRACT: A single item subject to random failure is considered. The item is inspected at certain time instants to see whether it is still in working order. Each inspection carries a fixed cost, and there is also a cost per unit time associated with down times. For this scenario, there exist various optimum and near-optimum policies for *deterministic* inspection schedules. This paper is concerned with the case when the inter-inspection times are *random*, a situation that may arise in the practical implementation of deterministic inspection policies or in opportunistic inspection schedules. We use a technique which is known as the "Ross Scheme" to derive recurrence relations for the exact computation of the expected length of a renewal cycle and the moments of the total cost in a renewal cycle for random inspection policies. The item lifetime and the inter-inspection times are assumed distributed according to mixtures of Erlang distributions. A numerical example is used to compare the present policy with a (near-)optimal policy. The penalty for random (as opposed to deterministic) inspections is found fairly limited.

1 INTRODUCTION

Many papers have been published on inspection policies for single-item systems and this topic is still of ongoing interest; see, for example, the recent article of Leung [9] and the survey by Kaio, Dohi and Osaki [8]. We are concerned here with systems whose state can be established by an inspection only and which need to be in the up-state permanently. (Some weapons systems are instances thereof.) The situation considered here is as follows. The system is inspected at certain times. The duration of inspections is assumed negligible. Each inspection may find the system in one of the states *up* or *down*. If the system is found to be *down*, it is instantaneously replaced by a new one; otherwise it is inspected at the next inspection instant. Each inspection carries a fixed cost and system downtimes have a certain cost per unit time associated with them. Whereas most inspection policies examined in the literature are deterministic there is a need to be able to analyze policies whose inter-inspection times are random (the *random policies*). This is because in practice, for various reasons, the implementation also of deterministic policies may turn out to be random. (One reason for randomness may be, for example, by carrying out opportunistic inspections where the opportunities arise in conjunction with some random event.)

In the present paper we discuss a technique for the exact computation of the expected cost in, and the expected length of, a renewal cycle for a random policy. The technique is based on an extension of what is known as the "Ross Approximation", a method for establishing recurrence relations for the computation of the renewal function in Renewal Theory ([1], [10], [11]). As an application, we use this method to compare numerically, for a specific example, the cost of a random policy with that of a matching (near-)optimal policy.

The paper is organized as follows. In Section 2, we introduce some notation. In Section 3, pertinent formulae for the random inspection model are derived. In particular, it is shown that the quantities of interest are expressible in terms of the expectation $E(N(T))$ where T and $N(t)$ stand respectively for the system's (random) lifetime and the number of inspections in $[0, t]$. In Section 4, we show how $E(N(T))$ can be computed for various distributional assumptions on the inter-inspection times and that of the system's lifetime. Section 5 is devoted a numerical study. In Section 6, some directions for future work are indicated.

2 NOTATION

T	(random) system lifetime
$X_1, X_2, \ldots$	i. i. d. inter-inspection times
μ	mean of the inter-inspection time X
c_x^2	squared coefficient of variation of X
$N(t)$	number of inspections up to time t

$m(t)$	renewal function, $E(N(T))$
c_i	cost of a single inspection
c_d	cost per unit system downtime
C	(random) cost accruing in an inspection cycle for a random policy
C_{det}	(random) cost in an inspection cycle for a deterministic policy
L	(random) length of an inspection cycle for a random policy
L_{det}	(random) length of an inspection cycle for a deterministic policy
$\int f(t) P(X \in dx)$	integral of f with respect to the distribution of the r. v. X, $E(f(X))$
ERL (k, λ)	Erlang-k distribution with mean k/λ
$\Re$	Real line

3   A RANDOM INSPECTION MODEL

3.1   Model formulation

The system considered comprises a single item whose positive random lifetime is T. As long as the system is found to be in the up state, it is inspected regularly such that the inter-inspection intervals X_1, $X_2, \ldots$ are independent and identically distributed random variables, themselves independent of T. An inspection cycle finishes with the first inspection which finds the system in the down state, upon which the system is instantaneously replaced by a new one and a new inspection cycle commences. Because of the Renewal Reward Theorem (e.g. Tijms [14]) we may concentrate on the first renewal cycle. Then, the nth inspection will take place at time $S_n = X_1 + \cdots + X_n$, $n \geqslant 1$, and $N(t)$, the number of inspections during $[0, t]$, is defined by $N(t) = n \Leftrightarrow S_n \leqslant t < S_{n+1}$. The *random* cost C accruing during an inspection cycle can be written as

$$C = (N(T)+1)c_i + (S_{N(T)+1} - T)c_d, \qquad (3.1)$$

where c_i and c_d stand respectively for the cost of a single inspection and the cost per unit system downtime. The Renewal Reward Theorem tells us that the long term expected cost rate R (i.e. the expected cost per unit time) of this policy is $R = E(C)/E(L)$, where L stands for the (random) length of a renewal cycle; the latter can be written as $L = S_{N(T)+1}$.

3.2   Model analysis

The following proposition provides expressions for $E(C)$, $E(L)$, and thus also for $R = E(C)/E(L)$, in terms of $E(N(T))$.

Proposition 1. It is

$$E(C) = (c_i + c_d \mu)(E(N(T))+1) - c_d E(T), \qquad (3.2)$$

$$E(L) = \mu\{E(N(T))+1\}. \qquad (3.3)$$

Proof of Proposition 1. In the course of our discussion some elementary facts about conditional expectations and distributions will be used for which, say, the classic work by Breiman [3] may be consulted.

Taking expectations in (3.1) and because of the independence of T and N (which holds since the latter is a function of $X_1, X_2, \ldots$ only) the expected cost for a renewal cycle can be written as $E(C) = c_i\{E(N(T))+1\} + c_d \int E(S_{N(t)+1} - t)P(T \in dt)$. Let $\mu = E(X_1)$ denote the expected inter-inspection time. It is known that

$$E(S_{N(t)+1}) = \mu(m(t)+1) \qquad (3.4)$$

(e.g., Ross [11], Proposition 7.2), and thus the expected cost of an inspection cycle is

$$E(C) = c_i\{E(N(T))+1\}$$
$$+ c_d\{\mu(E(m(T))+1) - E(T)\}. \qquad (3.5)$$

But, again because of the independence of T and N, we have

$$E(N(T)) = \int E(N(T)|T = t)P(T \in dt) =$$
$$\int E(N(t))P(T \in dt) = \int m(t)P(T \in dt) =$$
$$\int m(t)P(T \in dt) = E(m(T)), \qquad (3.6)$$

from which (3.2) follows by (3.5). (3.3) is obtained by the independence of T and N, and by (3.4) and (3.6) since

$$E(L) = E(S_{N(T)+1}) = \int E(S_{N(T)+1} | T = t)P(T \in dt) =$$
$$\int E(S_{N(t)+1})P(T \in dt) = \mu\{E(m(T))+1\} =$$
$$\mu\{E(N(T))+1\}. \qquad \blacksquare$$

4   COMPUTING $E(N(T))$ BY THE ROSS RECURRENCE RELATIONS

Proposition 1 tells us that for model analysis we may concentrate on the computation of $E(T)$ and $E(N(T))$. This section is concerned with the latter problem.

The Ross Approximation ([1], [10], [11]) is a technique which has originally been conceived for the approximate computation of the renewal function

$m(t)$. An informal outline of the idea is as follows. Let $Y_1, \ldots, Y_r$ be independent exponentially distributed random variables, each with mean $1/\tau$. Then, by the *Law of Large Numbers*, for large r, most of the probability mass of the sum $Y_1 + \cdots + Y_r$ will cluster around $t = r/\tau$ and therefore we would expect $m(t) = E(N(t))$ to be closely approximated by $E(N(Y_1 + \cdots + Y_r))$. More importantly for us here, it turns out that the latter quantity is easily computed by means of recurrence relations, first established in Ross [10]. It is this latter observation which we want to pursue here for the computation of the quantity $E(N(T))$.

4.1 The original Ross scheme

As indicated earlier, the original Ross scheme assumes an Erlang distribution for T and an arbitrary inter-arrival distribution on the positive axis. T and the renewal process are assumed independent. (This latter assumption will be maintained throughout.)

Let us start with the simplest case where $T = Y_1$ is exponentially distributed with rate τ. Then,

$$\alpha_1 \underset{def}{=} E(N(Y_1)) =$$

$$\int E(N(Y_1)|X_1 = x) P(X_1 \in dx), \tag{4.1}$$

where, by the *Law of Total Probability*,

$$E(N(Y_1)|X_1 = x) =$$

$$E(N(Y_1)|X_1 = x, Y_1 < x) P(Y_1 < x \mid X_1 = x) +$$

$$E(N(Y_1)|X_1 = x, Y_1 \geq x) P(Y_1 \geq x \mid X_1 = x). \tag{4.2}$$

The first conditional expectation on the r. h. s. of (4.2) is clearly zero. And, for the second conditional expectation we have (due to the memoryless property of the exponential distribution)

$$E(N(Y_1)|X_1 = x, Y_1 \geq x) = 1 + E(N(Y_1)) \tag{4.3}$$

From (4.1)–(4.3) it follows that

$$E(N(Y_1)) = \int \{1 + E(N(Y_1))\} e^{-\tau x} P(X_1 \in dx)$$
$$= \{1 + E(N(Y_1))\} E(e^{-\tau x})$$

from which it is seen that

$$1 + \alpha_1 = 1/(1 - E(e^{-\tau x})) \tag{4.4}$$

Let us now examine the case where $T = Y_1 + \cdots + Y_r$ is a sum of $r \geq 2$ independent exponentially distributed random variables each with rate τ, i.e. T has an Erlang distribution ERL (r, τ). For the rest of Section 4.1, we shall be concerned with deriving an expression for

$$\alpha_r \underset{def}{=} E(N(Y_1 + \cdots + Y_r))$$

in terms of $\alpha_1, \ldots, \alpha_{r-1}$. First we observe that

$$\alpha_r = \int E(N(Y_1 + \ldots + Y_r)|X_1 = x) P(X_1 \in dx) \tag{4.5}$$

where, by the *Law of Total Probability*,

$$E(N(Y_1 + \ldots + Y_r)|X_1 = x) = \sum_{k=0}^{r} E_k P_k \tag{4.6}$$

with E_k and P_k respectively given by

$$E(N(Y_1 + \ldots + Y_r)|$$
$$X_1 = x, Y_1 + \ldots + Y_k < x \leq Y_1 + \ldots + Y_{k+1})$$

and

$$P(Y_1 + \ldots + Y_k < x \leq Y_1 + \ldots + Y_{k+1}|X_1 = x).$$

Now, we have for $k < r$ by the Renewal Argument (which applies because of the memoryless property of the distribution of Y_{k+1})

$$E_k = E\{1 + N(Y_1' + Y_2' + \ldots + Y_{r-k}')\} = 1 + \alpha_{r-k} \tag{4.7}$$

(Here, the sequence $Y_1' = Y_1 + \cdots + Y_{k+1} - x, Y_2' = Y_{k+2}, Y_3' = Y_{k+3}, \ldots$ has the same distribution as that of the original Ys.) Furthermore, it is

$$E_r = 0. \tag{4.8}$$

To evaluate P_k, we observe that the X and Y are independent and therefore the conditioning in the definition of P_k may be ignored. Thus,

$$P_k = P(Y_1 + \ldots + Y_k < x \leq Y_1 + \ldots + Y_{k+1})$$

$$= P(Y_1 + \ldots + Y_k < x) - P(Y_1 + \ldots + Y_{k+1} < x)$$

$$= \left(1 - \sum_{j=0}^{k-1} e^{-\tau x} \frac{(\tau x)^j}{j!}\right) - \left(1 - \sum_{j=0}^{k} e^{-\tau x} \frac{(\tau x)^j}{j!}\right)$$

$$= e^{-\tau x} \frac{(\tau x)^k}{k!}. \tag{4.9}$$

By (4.7)–(4.9) in conjunction with (4.6) we get

$$E\big(N(Y_1 + \ldots + Y_r)\big|X_1 = x\big)=$$

$$\sum_{k=0}^{r-1} E_k P_k = \sum_{k=0}^{r-1} e^{-\alpha} \frac{(\tau x)^k}{k!}(1 + \alpha_{r-k}). \qquad (4.10)$$

Substitute now (4.10) into (4.5) to see that

$$\alpha_r = \sum_{k=0}^{r-1} \frac{\tau^k}{k!}(1 + \alpha_{r-k}) E(X^k e^{-\tau X})$$

which is re-arranged to give

$$1 + \alpha_r = \frac{1 + \sum_{k=1}^{r-1} \frac{\tau^k}{k!}(1 + \alpha_{r-k}) E(X^k e^{-\tau X})}{1 - E(e^{-\tau X})}. \qquad (4.11)$$

In conjunction with (4.4), (4.11) is seen to hold *for all* $r \geq 1$. With a view to a computational implementation we note that (4.11) can be rephrased as an instance of the discrete convolution equation $g = a + g * b$, i.e.,

$$g_r = a_r + \sum_{k=1}^{r-1} g_k b_{r-k}, \ r \geq 1, \qquad (4.12)$$

with $g_r = 1 + \alpha_r$ and

$$a_r = \frac{1}{1 - E(e^{-\tau X})}, \ b_r = \frac{\tau^r}{r!} \frac{E(X^r e^{-\tau X})}{1 - E(e^{-\tau X})}. \qquad (4.13)$$

4.2 Supplementing the Ross scheme

There are two restrictions associated with the use of the Ross scheme for computing $E(N(T))$.

(a) For most inter-inspection distributions, the quantities a and b from (4.13) will not be available in a closed form. Since, however, finite mixtures of Erlang distributions (with even the same scale parameter) may be used to approximate any distribution on the positive axis arbitrarily closely (e.g. Tijms [14], Theorem 2.9.1), we may, in practical applications, work with inter-inspection times whose distribution is a mixture of Erlangians. For these then, a and b are easily computed. (See below.)

(b) The Ross Scheme works in the first instance with Erlangian T only. A corresponding scheme for $E(N(T))$ is easily derived for the case when the distribution of T is a finite mixture of Erlangians; the general case is dealt with by the familiar approximation argument again.

* * *

Proposition 2. If the distribution of the inter-inspection time X is ERL (k, λ), the sequences a and b in (4.12)–(4.13) are computed from

$$a_r = \frac{1}{1 - \left(\dfrac{\lambda}{\lambda + \tau}\right)^k}, \ r \geq 1, \qquad (4.14)$$

$$b_0 = \frac{\left(\dfrac{\lambda}{\lambda + \tau}\right)^k}{1 - \left(\dfrac{\lambda}{\lambda + \tau}\right)^k}, \qquad (4.15)$$

$$b_r = \frac{k + r - 1}{r} \frac{\tau}{\lambda + \tau} b_{r-1}, \ r \geq 1. \qquad (4.16)$$

Proof of Proposition 2. (4.14) and (4.15) hold since the Laplace transform of X is $\varphi(\tau) = \lambda^k (\lambda + \tau)^{-k}$. For $r \geq 1$ we have (4.16) by

$$b_r = \frac{\tau^r}{r!} \frac{E(X^r e^{-\tau X})}{1 - E(e^{-\tau X})} = \frac{\tau^r}{r!} \frac{(-1)^r \dfrac{d^r}{d\tau^r} \varphi(\tau)}{1 - \varphi(\tau)}$$

$$= \frac{\tau^r}{r!} \frac{\dfrac{(k + r - 1)!}{(k-1)!} \dfrac{\lambda^k}{(\lambda + \tau)^{k+r}}}{1 - \varphi(\tau)} = \frac{k + r - 1}{r} \frac{\tau}{\tau + \lambda} b_{r-1}.$$

∎

Note. To emphasize in the sequel the association of the b – sequence with the parameters of the specific distributional assumption of Proposition 2, an appropriate superscripting will be applied: $b = b^{(k,\lambda)}$.

Proposition 3. Let the distribution of the inter-inspection time X be a finite mixture of s Erlangians, i.e.,

$$P(X \leq t) = \sum_{i=1}^{s} c_i P(Z_i \leq t), \ c_1 + \ldots + c_s = 1,$$

with $Z_i \sim$ ERL(k_i, λ_i). Then the sequences a and b for X (from (4.13)) are given as follows. For all $r \geq 1$,

$$a_r = 1/(v_1 + \ldots + v_s) \text{ with } v_i = c_i\left(1 - \left(\frac{\lambda_i}{\lambda_i + \tau}\right)^{k_i}\right).$$

Furthermore, the sequence b is a convex combination of the b – sequences of the Z_i,

$$b_r = \sum_{i=1}^{s} w_i b_r^{(k_i, \lambda_i)} \text{ where } w_i = v_i/(v_1 + \ldots + v_s).$$

460

Proof of Proposition 3. It is

$$a_r = \frac{1}{1 - E(e^{-\tau X})} = \frac{1}{1 - \sum\limits_{i=1}^{s} c_i E(e^{-\tau Z_i})}$$

$$= \frac{1}{1 - \sum\limits_{i=1}^{s} c_i \left(\frac{\lambda_i}{\lambda_i + \tau}\right)^{k_i}} = \frac{1}{v_1 + \dots + v_s},$$

$$b_r = \frac{\tau^r}{r!} \frac{E(X^r e^{-\tau X})}{1 - E(e^{-\tau X})} = \frac{\sum\limits_{i=1}^{s} c_i E(Z_i^r e^{-\tau Z_i}) \dfrac{\tau^r}{r!}}{1 - \sum\limits_{i=1}^{s} c_i E(e^{-\tau Z_i})}$$

$$= \sum\limits_{i=1}^{s} c_i \frac{1 - E(e^{-\tau Z_i})}{1 - \sum\limits_{i=1}^{p} c_i E(e^{-\tau Z_i})} \cdot \frac{E(Z_i^r e^{-\tau Z_i}) \dfrac{\tau^r}{r!}}{1 - E(e^{-\tau Z_i})}$$

$$= \sum\limits_{i=1}^{s} w_i b_r^{(k_i, \lambda_i)}.$$

∎

Finally, the computation of $E(N(T))$ is fairly straightforward for when the distribution of the system lifetime T is a finite mixture of Erlangians. Assume that

$$P(T \leq t) = \sum\limits_{i=1}^{q} d_i P(T_i \leq t), \quad d_1 + \dots + d_q = 1,$$

with $T_i \sim \mathrm{ERL}(n_i, \tau_i)$. Then, by (3.6) it is seen that

$$E(N(T)) = E(m(T)) = \sum\limits_{i=1}^{q} d_i E(m(T_i))$$

$$= \sum\limits_{i=1}^{q} d_i E(N(T_i)).$$

4.3 Higher moments

The idea of the Ross Scheme can be employed also for the computation of higher moments of C and L. As the corresponding recurrence relations rapidly become rather cumbersome, the interested reader is referred to the Technical Report [4].

5 APPLICATION: COMPARISON OF RANDOM AND NEAR OPTIMUM INSPECTION POLICIES

As discussed in the Introduction, the *random* inspection policy introduced in Section 3 cannot outperform

an optimum policy which, as is known from Barlow and Proschan [2], may be assumed non-random. In this section, we compare our random policy with that of the near-optimum deterministic inspection policy of Kaio and Osaki, which together with a number of other inspection policies, is described in the recent review article by Kaio, Dohi and Osaki [8].

5.1 Deterministic near-optimal policy

The example to be considered and the pertinent figures for the said policy are taken from a numerical study by Kaio and Osaki [6]. (See also Kaio and Osaki [7].) Their assumed values are, $c_i = 20$, $c_d = 1$, $T \sim \mathrm{ERL}(r, \tau)$, $r = 2$ and $\tau = 0.01$. In Table 1 below, we show the policy's inspection, and inter-inspection times.

For a deterministic inspection policy with inspection times $t_1 < t_2 < t_3 < \dots$, the expected cost $E(c_{\mathrm{det}})$ and the expected duration $E(L_{\mathrm{det}})$ of an inspection cycle are respectively given by (see, e.g. [8])

$$E(C_{\mathrm{det}}) =$$

$$\sum\limits_{k=0}^{\infty} \left(c_i(k+1) + c_d t_{k+1}\right) P(t_k < T \leq t_{k+1}) - c_d E(T), \quad (5.1)$$

$$E(L_{\mathrm{det}}) = \sum\limits_{k=0}^{\infty} t_{k+1} P(t_k < T \leq t_{k+1}). \quad (5.2)$$

Table 2 below shows the pertinent quantities for the deterministic (near-optimum) policy from Table 1.

5.2 Random policy

It must be emphasized that in practice a random policy will hardly ever be chosen deliberately if an optimum

Table 1.

Inspection times t_i	Inter-inspection times $\Delta t_i = t_i - t_{i-1}$
114	–
195	81
271	76
344	73
415	71
485	70
554	69
623	69
691	68
758	67
825	67
892	67
959	67

Table 2.

$E(C_{\text{det}})$	$E(L_{\text{det}})$	$E(C_{\text{det}})/E(L_{\text{det}})$
95	241	0.394

Table 3.

c_X	k	$E(C)$	$E(L)$	$E(C)/E(L)$
0.4	7	109.7	240.8	0.456
0.3	12	106.4	238.3	0.446
0.1	100	102.7	235.4	0.436
0.01	10000	102.3	235.1	0.435
0	∞	102.0	235.0	0.434

or near optimum policy can be implemented. However, if circumstances so dictate, it may well be the case that we will have to content ourselves with a random policy as described in Section 3. The question for us arises here as to what random policy to consider (for comparison purposes) which reasonably matches the (near) optimum deterministic policy from above. In our quest for a distribution for the inter-inspection time X, we have been guided by the observation (e.g. Tijms [13], [14]) that mixtures of two Erlangian distributions of the form $\text{ERL}(k-1, \lambda)$ and $\text{ERL}(k, \lambda)$ are often used in practice to approximate distributions on the positive axis. These mixtures can be made to fit the first two moments of any given distribution provided that its squared coefficient of variation is less than unity. This situation fits Proposition 3 with $s = 2$, $k_1 = k-1$ and $k_2 = k$. Remain to be expressed the quantities $\lambda_1 = \lambda_2 = \lambda$, c_1 and c_2 in terms of μ and c_X^2, the latter being the squared coefficient of variation of X. The formulae sought are as follows (see, Tijms [13], [14]):

$$\lambda = (k - c_1)/\mu ,$$

$$c_1 = \frac{1}{1+c_X^2}\left[kc_X^2 - \left\{ k(1+c_X^2) - k^2 c_X^2 \right\}^{1/2} \right],$$

$$c_2 = 1 - c_1 .$$

It can be shown that c_X^2 lies between $1/k$ and $1/(k-1)$. Guided by Table 1, we put $\mu = 70$. Table 3 below displays the values of $E(C)$, $E(L)$ and $E(C)/E(L)$ for several parameter choices. (The last row in Table 3 is by application of (5.1) and (5.2).)

It is seen that the expected cost decreases as we approach the limiting, deterministic case. It approaches the value of a t-policy with $t = 70$. (A t-policy is a policy prescribing inspections at multiples of t time units). The "worst" random policy shown in Table 3 (top row) is more costly than the t-policy by a mere 7%, whereas the latter is more costly than the (near) optimum policy by around 8%. (The corresponding figures for the expected cost rate are 5% and 10%, respectively.) These data suggest that from a practical point of view not much is lost by employing either a t-policy, or, a random policy with a small coefficient of variation and a mean inter-inspection time which is close to the average inter-inspection time of the optimum deterministic policy.

The present numerical results have been obtained by an implementation in Haskell, a functional programming language suitable for concise coding and a problem-oriented programming style (Thompson [12]).

6 CONCLUSIONS AND FURTHER WORK

We have considered a random inspection model for a single-item system. The expected cost in an inspection cycle and also the expected length of an inspection cycle turned out to be expressible in terms of the quantity $E(N(T))$ which itself proved amenable to computation by what is known as the Ross Scheme, originally devised for the approximate calculation of the renewal function. We have extended the Ross Scheme to cater for a wide distributional assumption (mixtures of Erlangs). Finally, we have conducted a numerical study comparing the performance of the random policy with that of the optimum–, and t-policies. It appears that from a practical point of view there is not much gained by insisting on an optimum policy if for administrative or other reasons the latter two policies are easier to implement.

We are aware of work on models concerning the random inspection of more complex systems; a most recent one is [5], wherein the numerical inversion of Laplace transforms is employed. At a deeper (mathematical) level the Ross Scheme can be thought of as a kind of numerical Laplace transform inversion technique. (This has been pointed out by Ross himself in his original article [10].) An exploration of the interrelation of these two approaches for the analysis of inspection models and a practical application of the Ross Scheme to specific inspection models would be most welcome. In most cases, the "mixtures of Erlangians" – assumption is an approximation only (albeit satisfactory for practical purposes). It would be worthwhile (at least from the theoretician's point of view) to assess the error thereby committed. The work by Angus and Hong [1] may serve as a starting point for such studies.

REFERENCES

1. Angus, J.E. & Hong, X., On the rate of convergence of the Ross Approximation to the renewal function.

Probability in the Engineering and Informational Sciences, **10** (1996) 207–211.

2. Barlow, R.E. & Proschan, F., *Mathematical Theory of Reliability*. Wiley, New York, 1967.

3. Breiman, L., *Probability*. Addison-Wesley, Reading, Ma., 1968.

4. Csenki, A., Analysis of random inspection policies for single-unit systems. Technical Report, School of Computing and Mathematics, University of Bradford, December 2002.

5. Dieulle, L., Reliability of several component sets with inspections at random times. *European Journal of Operational Research*, **139** (2002) 96–114.

6. Kaio, N. & Osaki, S., Inspection policies: comparisons. In *Reliability Theory and Applications*, Proceedings of the China-Japan Reliability Symposium, September 13–25, 1987, Shanghai, Xian and Beijing, China (Osaki, S. and Cao, J. eds.). 140–147, World Scientific, Singapore.

7. Kaio, N. & Osaki, S., Comparison of inspection policies. *Journal of the Operational Research Society*, **40** (1989) 499–503.

8. Kaio, N., Dohi, T. & Osaki, S., Classical maintenance models. In *Stochastic Models in Reliability and Maintenance* (Osaki, S. ed.). 65–87, Springer-Verlag, Berlin, Heidelberg, New York, 2002.

9. Leung, F.K.-N., Inspection schedules when the lifetime distribution of a single-unit system is completely unknown. *European Journal of Operational Research*, **132** (2001) 106–115.

10. Ross, S.M., Approximations in renewal theory. *Probability in the Engineering and Informational Sciences*, **1** (1987) 163–173.

11. Ross, S.M., *Introduction to Stochastic Models*. Academic Press, London, 1993.

12. Thompson, S., *Haskell – The Craft of Functional Programming*. Addison Wesley, Harlow, 1999.

13. Tijms, H.C., *Stochastic Modelling and Analysis: A Computational Approach*. Wiley, Chichester, New York, 1986.

14. Tijms, H.C., *Stochastic Models – An Algorithmic Approach*. Wiley, Chichester, New York, 1994.

Safety and Reliability – Bedford & van Gelder (eds)
© 2003 Swets & Zeitlinger, Lisse, ISBN 90 5809 551 7

A modified moment method in structural reliability

M. Daghigh
Sharif University of Technology, Tehran, Iran

N. Shabakhty
Delft University of Technology, Department of Marine Technology, The Netherlands

ABSTRACT: Generally, transformation methods are utilized to find the failure probability by iterative procedures. Nonlinear performance functions would be solved with the aid of first and second expansion of Taylor
series. Alternatively, the failure probability may be obtained by direct utilizing the probability moment information of the performance function. In this study, two alternative methods have been adopted for this purpose.
In the first solution, the moments of performance function is calculated using numerical integration and application of FM-1 and FM-2 methods to correct result of the mean value. In the second approach, the distribution
of performance function is found by Monte Carlo simulation and then again using FM-1 and FM-2 methods to
correct these failure probabilities.

1 INTRODUCTION

This paper is concerned with the application of
moment methods in reliability analysis of structural
elements and systems. The main scope of the present
paper is to check the validity of the so-called moment
methods based on MVFOSM (Mean Value First
Order Second Moment) and comparing to AFOSM
(Advanced First Order Second Moment).

Among various methods of reliability analysis, the
first-order reliability method (FORM) is one of the
significant computational methods. Several research
studies have been conducted to improve the reliability estimates especially for the nonlinear limit states
and system reliability. These include the second
order reliability method (SORM), importance sampling techniques, response surface method and genetic algorithms.

Another route to improve the reliability estimates
with FORM method is the application of moment
methods. The moment methods overcome the difficulties in evaluation of failure probability when using
original FORM method.

When the performance function is not so complex,
moment methods are very efficient with respect to
either iteration or computation of derivatives require
to determine design point in reliability analysis.

The moment methods can be generally categorized
into transformation or simulation approaches.

Several researchers developed approximation of
moment method using Edgeworth and Cornish-Fisher
expansions technique, Zhao and Ono (2001). Furthermore, Winterstein and Bjerager (1987) proposed to
use hermit polynomial transformation technique
to calculate higher order moments of performance
function.

Grigoriu and Lind (1980) applied a first order with
the third moment reliability method (FOTM) and a
higher order moments standardization technique
(HOMST). These methods use the first order Taylor
series expansion of the performance function at the
design point with three and higher statistical moment
of performance function, respectively. Since calculation of the higher order moment is difficult and in
some situation is not possible, Grigoriu proposed to
apply Monte Carlo simulation technique.

For the approximation of moments, Hong (1996)
used a point estimate moment based reliability analysis method in which the failure probability is approximated by the Johnson family of distributions and the
concentrations in the point-estimates are obtained from
nonlinear equations. Zhao and Ono (2001) applied
point estimates in standard normal space without using
nonlinear equations. They illustrate application of this
method for several examples under conditions of independent variables.

In the present paper, moment methods are investigated for different problems. Two examples have been

excerpted from Zhao and Ono (2001) and the results have been compared with this study.

Furthermore, in this paper the application of moment methods for system reliability analysis of two types of series and parallel systems is illustrated in example 5.3.

It has been found that the moment method does not require iteration as usual for FORM routine and thus is convenient to be applied to structural reliability analysis.

2 HERMITE MOMENT AND STANDARDIZATION TECHNIQUE

The marginal distribution of virtually any non-Gaussian variable can be matched by applying an appropriate monotone function to a Gaussian process. A polynomial transformation including determinative coefficients is defined using third-moment standardization function for the standardized random variable z_u defined in the following form.

$$y = z_u + c z_u^2 \tag{1}$$

$$u = \frac{y - \mu_y}{\sigma_y} \tag{2}$$

when c is a deterministic coefficient, z_u is the normalized variable for a performance function $z = G(X)$.

$$z_u = \frac{z - \mu_G}{\sigma_G} \tag{3}$$

In order to make Equation 1 standard based on third moment, the skewness of y should be equal to that of the normal random variable. Then c can be determined by the following equation.

$$\alpha_{3y}\sigma_y^3 = (\alpha_{6z} - 3\alpha_{4z} + 2)c^3 + 3(\alpha_{5z}$$
$$-2\alpha_{3z})c^2 + 3(\alpha_{4z} - 1)c + \alpha_{3z} = 0 \tag{4}$$

Where α_{3y} and σ_y are the skewness and standard deviation of y, respectively. $\alpha_{3z}, \alpha_{4z}, \alpha_{5z}, \alpha_{6z}$ are the third, fourth, fifth and sixth dimensionless moments of z_u, they are equal to those of z respectively, according to the definition of probability moments.

According to the computational experience of Zhao and Ono (2001), c can be obtained by assuming $|c| \ll 1$ with the following equation.

$$c = \frac{\alpha_{3z}}{3(1 - \alpha_{4z})} \tag{5}$$

and

$$\mu_y = C \tag{6}$$

$$\sigma_y^2 = (\alpha_{4z} - 1)c^2 + 2\alpha_{3z}c + 1 \tag{7}$$

For other value of c, the quadratic equation of 4 can be used to determine the value of c, which involves fifth and sixth orders of performance function.

Now by substituting Equations 5–7 in Equation 2, the following expression for normalized variable u can be specified.

$$u = \frac{\alpha_{3G} + 3(\alpha_{4G} - 1)z_u - \alpha_{3G}z_u^2}{\sqrt{(5\alpha_{3G}^2 - 9\alpha_{4G} + 9)(1 - \alpha_{4G})}} \tag{8}$$

where α_{3G} and α_{4G} are third and fourth dimensionless central moment, i.e. skewness and kurtosis of $z = G(X)$.

The probability of failure can be determined by using performance function, z and the standardized variable z_u in the following format

$$Pr\,ob[z] = Pr\,ob[z_u \leq -\frac{\mu_G}{\sigma_G}]$$

$$= Pr\,ob[z_u \leq -\beta_{SM}] \tag{9}$$

where μ_G and σ_G are the mean value and standard deviation of $z = G(x)$, respectively.

The reliability index and failure probability based on the fourth-moment method can be obtained with

$$\beta_{FM} = \frac{3(\alpha_{4G} - 1)\beta_{SM} + \alpha_{3G}(\beta_{SM}^2 - 1)}{\sqrt{(9\alpha_{4G} - 5\alpha_{3G}^2 - 9)(\alpha_{4G} - 1)}} \tag{10}$$

$$P_{fFM} = \phi(-\beta_{FM}) \tag{11}$$

when α_{3G} approach to zero, Equation 11 shows that reliability index β_{SM} approach to β_{FM}. This formulation of failure probability is denoted as FM-1 reliability index where β_{FM} is the reliability index determined by using MVFOSM (Mean Value First Order Second Moment) procedure.

3 APPLICATION OF EDGEWORTH EXPANSION IN FOURTH MOMENT METHOD

If we use the Edgeworth expansion procedure up to the fourth statistical moment, the non-Gaussian distribution function can be determined with the following expansion.

$$F(z_u) = \phi(z_u) - \varphi(z_u).[\frac{1}{6}\alpha_{3G}.H_2(z_u) +$$
$$\frac{1}{24}(\alpha_{4G} - 3).H_3(z_u) + \frac{1}{72}\alpha_{3G}^2.H_5(z_u)] \tag{12}$$

where H_n is Hermite function from order n and is specified with:

$$H_2(x)=x^2-1 \qquad (13)$$

$$H_3(x)=x^3-3x \qquad (14)$$

$$H_5(x)=x^5-10x^3+15x \qquad (15)$$

Hence, the failure probability or reliability index can be obtained with the following expression.

$$P_{fFM} = \phi(-\beta_{SM}) - \varphi(\beta_{SM}).[\frac{1}{6}\alpha_{3G}.$$

$$H_2(-\beta_{SM})+\frac{1}{24}(\alpha_{4G}-3).H_3(-\beta_{SM})$$

$$+\frac{1}{72}\alpha_{3G}^2.H_5(-\beta_{SM})] \qquad (16)$$

$$\beta_{FM}=-\phi^{-1}(-P_{fFM}) \qquad (17)$$

The reliability index and failure probability determined with Equations 16 and 17 will hereafter be denoted as FM-2 reliability estimator, since it is determined based on first order of series expansion and fourth statistical moment of performance function.

In order to calculate the moments of the performance function, Zhao (2001) employed the point estimates. For this purpose, we define the fourth moments of the performance function using numerical integration or Monte Carlo Simulation (MCS). Having obtained the fourth moments of the performance function, the new reliability index is calculated without the need for any iteration.

4 SYSTEM RELIABILITY

The above procedure is also applied for the system reliability analysis of group components. The general idea behind this methodology is that the probability of failure of the whole system is obtained conditional to the failure probability of the correlated variables. In this method, the variables are categorized in three groups, x_1, x_2 and x_3. Let x_1 be the set of variables common to all individual failure events, $x_2^{(i)}$ be the set of variables common to the components of a subgroup i, and $x_3^{(ij)}$ be the set of variables belonging to component j of subgroup i and no others. Mathematically, the conditional probability of failure can be written for

both the series and the parallel systems. This conditional probability of failure is therefore assumed as a normalized random variable and the failure probability is then integrated with respect to the conditional random variables, Wen (1987).

Hovde (1995) has used such procedure to calculate system reliability of tension leg platforms and the application of this procedure in system reliability of jack-ups structure has been studied by Daghigh (1997). Mathematically, the conditional probability of failure can be written as

$$P_{f,sys}(x_1,x_2,X_3)=P_{f,sys}|(x_1,x_2)=$$

$$1- \prod_{i=1}^{i=m} \prod_{j=1}^{j=n} [1-P[g_{ij}(x_1,x_2^{(i)},X_3^{(ij)})\leq 0]] \qquad (18)$$

for series system, and

$$P_{f,sys}(x_1,x_2,X_3)=P_{f,sys}|(x_1,x_2)=$$

$$\prod_{i=1}^{i=m} \prod_{j=1}^{j=n} P[g_{ij}(x_1,x_2^{(i)},X_3^{(ij)})\leq 0] \qquad (19)$$

for parallel system. In this formulation i refers to the group number with maximum number of m and n refers to the number of elements in each group. Capital letters refer to stochastic variables, while small letters reflect the deterministic variables. The limit state function for whole system is calculated by the conditional failure probability as proposed by Wen (1987) and Bjerager (1988). The conditional system probability of failure can be determined from the following limit state function:

$$G_{sys}(X_1,X_2,x_{sys})=$$

$$x_{sys}-\phi^{-1}(P_{f,sys}(x_1,x_2,X_3) \qquad (20)$$

The performance functions of series and parallel systems can be obtained by substituting Equation 18 and 19 into Equation 20.

An alternative procedure to specify the performance function defined in Equation 20 is to perform a series of FORM analysis for component element of $i = 1, 2, ..., m$ and $j = 1, 2, ..., n$ with deterministic x_1 and x_2 and then to evaluate the final probability of failure from the numerical integration (Rieman integral).

$$P(G\leq 0)=P_{F,sys}=\int_{x1=-\infty}^{\infty} \int_{x2=-\infty}^{\infty}$$

$$[P_{f,sys}|(x_1,x_2)]f_{x1}(x_1).f_{x2}(x_2).dx_1.dx_2 \qquad (21)$$

5 EXAMPLES

In this section, three examples of single event and system reliability will be presented.

5.1 Simple R-S model (Excerpted from Zhao et al. (2001), Example 4)

This example is the Simple R-S reliability model with various probability distributions for random variables and the results have been compared with the study of Zhao (2001). Hence, the performance function is

$$G(X) = R - S \qquad (22)$$

when R is assigned to a resistance and S to load random variables. In the following investigations, the coefficient of variation of R and S are taken to be 0.2 and 0.4, respectively. In addition, R is considered to follow normal distribution and S lognormal distribution. The results of different reliability methods are summarized in Table 1.

As it is clear from Table 1, the reliability estimates are close for two different formulations of FM-1 and FM-2 and are less than those of numerical integration technique (last column β_{REAL}), FORM and SM.

5.2 Effect of formulation of the performance function (Excerpted from Zhao (2001), Example 6)

Another example is here taken in order to check the sensitivity of the moment method to the formulation of the performance function. Two simple performance functions are specified with

$$G(X) = x_1 - x_2 \qquad (23a)$$

$$G(X) = x_1^2 - 2x_2 \qquad (24a)$$

when in Equation 23a the random variables x_1 and x_2 are statistically independent and lognormally distributed with mean value 50 and 10, and standard deviation 10 and 4, respectively. Furthermore, in Equation 24a the variables x_1 and x_2 are statistically independent and normally distributed with the means 10 and 20, and standard deviations 2 and 5, respectively.

Table 1. Reliability indices of Simple R-S model.

Central factor of safety	β_{SM}	β_{FORM}	β_{FM-1}	β_{FM-2}	β_{REAL}
3.5	3.222	2.910	2.575	2.690	2.835
4.5	3.651	3.396	2.738	2.799	3.075
5.5	3.925	3.746	2.822	2.956	3.648
6.5	4.111	3.999	2.870	3.085	3.899

In order to investigate sensitivity of the moment method to the formulation of the performance function, Equations 23a and 24a have been rewritten with the following equivalent formulations.

$$G(X) = 1 - \frac{x_2}{x_1} \qquad (23b)$$

$$G(X) = 1 - \frac{2x_2}{x_1^2} \qquad (24b)$$

The reliability indices using the first fourth moments of the performance functions are listed in Table 2.

As it is clear from Table 2, the first four moments and the SM reliability index obtained from Equations 23a and 24a differ very much from those obtained from their corresponding equivalent formulations Equations 23b and 24b. However, FM-1 approach shows better result than SM.

Comparison with the results of Zhao (2001), it seems that using probability distribution functions for the Pearson system (FM-3), the reliability indices with FM-1 and FM-3 are almost insensitive to formulation of performance function.

5.3 System reliability for one group components with similar performance functions

For system reliability of group components, the hyperspherical limit state function is used. In order to check the results, the system reliability for series and parallel systems are found for different number of components and the results are compared with the previous study.

Assume the performance function of components are given by the hyperspherical performance function.

$$G(X) = \beta^2 - u_1^2 - u_2^{(j)^2} \qquad (25)$$

where β is the FORM reliability index, u_1 and $u_2^{(j)}$ are the standard normally distributed random variables. While the random variable u_1 is common to the

Table 2. Results of Example 5.2, number of simulations is 2×10^6.

	Eqn. (23a)	Eqn. (23b)	Eqn. (24a)	Eqn. (24b)
μ_G	40.18	0.779	64.06	0.538
σ_G	11.24	0.103	41.60	0.352
α_{3G}	0.383	-1.51	0.539	-153
α_{4G}	3.554	7.330	3.411	96080
β_{SM}	3.574	7.563	1.540	1.528
β_{FM-1}	3.873	4.875	1.610	1.643
β_{FM-2}	2.708	6.333	1.917	–
Zhao et al.	4.262	4.310	1.733	1.704

468

Table 3. System reliability indices β_{sys} for hyperspherical performance functions, number of simulations is 10000 and $\beta = 2$.

Number of Components	1	5	10
Series system, Case 1	1.565	0.6742	0.2015
Series system, Case 2	1.270	0.2687	−0.0533
Series system, Case 3	1.216	0.3903	−0.0798
Series system, FORM	1.691	0.8142	0.3254
Series system, exact	1.254	0.366	−0.1103
Parallel system, Case 3	1.259	2.640	2.972
Parallel system, FORM	1.691	1.993	2.000
Parallel system, exact	1.254	2.521	2.891

Table 4. Computational results of combined MCS and moment method, number of simulations is 10000.

Number of components	1	5	10
μ_G, series system	1.3829	0.3698	−0.2515
σ_G, series system	1.0954	1.2488	1.3671
α_{3G}, series system	−0.2083	−0.5786	−0.9186
α_{4G}, series system	3.2482	4.2045	5.1176
β_{SM}, series system	1.262	0.296	0.1839
β_{FM-1}, series system	1.257	0.3293	0.233
β_{FM-2}, series system	1.521	0.5246	0.4235
μ_G, parallel system	1.3829	4.4648	6.0082
σ_G, parallel system	1.0954	1.3293	1.7884
α_{3G}, parallel system	−0.2083	−0.6014	−1.500
α_{4G}, parallel system	3.2482	3.9448	6.700
β_{SM}, parallel system	1.262	3.359	3.359
β_{FM-1}, parallel system	1.257	3.064	3.247
β_{FM-2}, parallel system	1.521	2.365	2.405

performance functions of components, it is assumed that $u_2^{(j)}$ is different for any performance function (i.e. fully independent). The performance function for system of n components can be written such as Equation 26

$$G_{sys}(u_1, u_{sys}) =$$
$$u_{sys} - \phi^{-1}(1 - \prod_{j=1}^{j=n} \chi_1^2(\beta^2 - u_1^2)) \qquad (26)$$

for series system, and :

$$G_{sys}(u_1, u_{sys}) =$$
$$u_{sys} - \phi^{-1}(\prod_{j=1}^{j=n} (1 - \chi_1^2(\beta^2 - u_1^2))) \qquad (27)$$

for parallel systems.

In the performance functions, χ_1^2 refer to the chi-square distribution with one-degree of freedom. Results of solution techniques are presented in Table 3, which are system reliability by the application of FORM routine, the Importance Sampling Technique and the exact numerical integration. The Importance sampling density function is a rectangular function in which the lower and upper bounds are changed in three cases. Boundaries are given for Case 1; $[u_1] = [-1.5, 1.5]$, Case 2; $[u_1] = [-2.5, 2.5]$ and Case 3; $[u_1] = [-3.5, 3.5]$. Since the performance function of components are quadratic, the system reliability indices are different with the FORM reliability index for number of components $n = 1$.

The numerical integration is applied with lower and upper bounds of the integral in Case 3. As it is observed from Table 3, the efficiency of FORM method decreases when the hyperspherical performance function is solved for large numbers of n.

In addition, with respect to Importance sampling results, the choice of Importance sampling density

function and its boundaries are important and for this case a more wider band density function yields better accuracy in reliability index.

For the application of combined MCS and fourth moment method, the moments of performance function is determined by using random generators (MCS method) and by adopting two types of Fourth moment formulations (FM-1 and FM-2). The system reliability are calculated without iteration for these two approaches and is illustrated in Table 4.

It is also seen from Table 4 that even mean value results with better estimate of mean and standard-deviation will yield good results. However application of FM-1 and FM-2 fourth moment will yield to better prediction of failure probability compared to exact solutions. Due to nonlinearity of performance function, the results of combined MCS and moment method would be accurate than normal iterative FORM routine.

6 CONCLUSIONS

The fourth moment methods for structural reliability were investigated and compared with exact and iterative FORM method. Combined numerical integration and mean value FORM method or combined MCS, FM-1 and FM-2 formulations were presented and investigated through several numerical examples.

The variables in the examples were statistically independent but the application of this method for correlated variables is straightforward. For this purpose, using Hohenbichler approximation, the performance functions are transformed to other formulations with independent variables. For the new performance functions, the same methods for system or reliability analysis are used to predict the moments of any distribution.

The FM-1 reliability index which is obtained from HOMST (Higher Order Moment Standardization Technique) generally gives better results than other formulas of comparable simplicity. However, as also mentioned by Zhao (2001), when the skewness of the performance function is large the FM-1 reliability index produces significant errors.

The FM-2 reliability index, which is obtained from the Edgeworth Expansion, generally gives not suitable results, and is not recommended as well as FM-2 moment method.

Based on the examples studied in this paper, the lower bound and upper bound on the reliability index of series or parallel systems may be found by using FM-1 and FM-2 moment method. For systems with nonlinear performance function, the application of FM-1 and FM-2 moment method will yield to better estimation in safety results than the iterative *FORM* routine.

For simple performance functions with non-normal variables, the moment method will yield to a good approximation. For complex performance functions, a better estimate of reliability will achieve using Importance Sampling or normal MCS with less simulations and the application of moment methods described earlier.

Application of this method for correlated variables and system failures with ductile behavior is selected as an option for future studies.

ACKNOWLEGEMENTS

The authors thank Mr. Helal Makouie, Master of Science student of Sharif University of Technology to provide some modules of this study in Matlab environment.

REFERENCES

Bjerager P. et al. 1988. Reliability method for marine structures under multiple environmental load processes, Proceedings of the International Conference on Behavior of Offshore Structures, BOSS, Vol. 3, pp. 1239–1253.

Daghigh M. 1997. Structural system reliability analysis of jack-up platforms under extreme environmental conditions, PhD thesis, Delft University of Technology, The Netherlands.

Grigoriu M. & Lind N.C. 1980. Optimal estimate of convolution integral, J. Engrg. Mech., ASCE, Vol. 106, No.6, pp. 1349–1364.

Hong H.P. 1996. Point-estimate moment-based reliability analysis, Civil Engineering System, Vol. 13, pp. 281–294.

Hovde G.O. 1995. Fatigue and overload reliability of offshore structural systems considering the effect of inspection and repair, PhD thesis, Division of marine structures, Norway.

Wen Y.K. & Chen H.C. 1987. On fast integration for time variant structural reliability, Probabilistic Engineering Mechanics, 2:3, pp. 156–162.

Winterstein S. & Bjerager P. 1987. The use of higher order moments in reliability estimation, Proc. Int. Conf. Application of Statistics and Probability, 2, pp. 1027–1036.

Zhao Y.G. & Ono T. 2001. Moment methods for structural reliability, Journal of Structural Safety, 23, pp. 47–75.

Safety and Reliability – Bedford & van Gelder (eds)
© 2003 Swets & Zeitlinger, Lisse, ISBN 90 5809 551 7

Explosive Atmospheres – Implementation of ATEX 137 in the UK

P.A. Davies & J.H. Gould
ERM Risk, Manchester, England

ABSTRACT: Most, if not all EU members supported the need for the Explosive Atmospheres Directive 95 (ATEX 95). However, there was disagreement over the need for a supplementary workers protection directive (ATEX 137). Some members argued that the hazards to workers were adequately covered by existing domestic legislation. Furthermore, before committing to ATEX 137, there was an opportunity to wait and review the success of ATEX 95 and Seveso II. However, the directive was passed and becomes law in EU member states in January 2003 and July 2003 for new and modified operations, respectively. This paper summarises the implications of ATEX 137 in the UK, dismisses the widely held belief that it is little more than formalising hazardous area classification, and outlines the more onerous requirement to demonstrate that all reasonably practicable measures to reduce risk have been taken.

1 INTRODUCTION

In the UK it is generally agreed that most of the requirements of ATEX 137 (EU Directive 1999/92/EC) are covered by existing legislation and guidance. Certainly most of the risk assessment requirements are covered by the *Management of Health and Safety at Work Regulations 1999* (HSE 1999). More specific requirements currently exist under the *Highly Flammable Liquids and Liquefied Petroleum Gasses Regulations 1972* (HSE 1972). In addition, the Health and Safety Executive (HSE) has issued guidance to cover the safe use and handling of flammable liquids (HSE 1996), the storage of flammable liquids in containers (HSE 1998), and the storage of flammable liquids in tanks (HSE 1988). Furthermore, there is guidance on potentially reactive substances, chemical warehousing and the storage of packaged dangerous substances (HSE 1998b), and the design and operation of safe chemical reaction processes(HSE 2000).

However, the existing regulations and guidance lack the coherency of ATEX 137, enacted within the UK by the *Dangerous Substances and Explosive Atmosphere Regulations* (DSEAR) (HSE 2001). Although DSEAR introduces few new requirements, there is a need to demonstrate that all reasonably practicable measures have been taken to reduce the risk from potentially explosive atmospheres. As outlined below, simple "blanket" zoning to reduce the probability of ignition from electrical sources will not be sufficient.

DSEAR also implements the safety aspects of the *Chemical Agents Directive* (CAD) (EU Directive 1998). This is because there is much overlap between CAD and ATEX 137 and therefore, HSE has decided to combine the safety requirements into a single set of regulations. Only the safety aspects of ATEX are discussed in this paper.

2 IMPLEMENTATION OF ATEX 137 – DSEAR

At the time of writing (December 2002), DSEAR has not been enacted and one can only refer to the consultation document. However, whilst there may be minor amendments, it is expected that the requirements will remain largely unchanged.

In the past, HSE has been criticised for the issue of guidance long after regulations have become law. Experience has shown that guidance issued months after regulations are enforced can simply add further burdens and costs to those who have already made their own interpretations of the regulations. However, lessons have been learned and in addition to guidance within DSEAR a supporting Approved Code of Practice (ACoP) has been produced.

As with all guidance, organisations do not need to follow the ACoP nor guidance within the regulations. This is true provided the organisation can demonstrate that their interpretation of the regulations are equally as effective as the ACoP and guidance. This can be an extremely difficult task (especially in the event of an accident), and where an organisation chooses not to follow what HSE considers good practice, they had better have some good answers ready for when the Inspector calls!

The DSEAR ACoP states that *the risk assessment should include enough information to demonstrate that the workplace and the work are designed, operated and maintained with due regard for safety*. Hence, the risk assessment needs to contain this information to comply with the Regulations. Demonstration that the work and workplace is appropriately designed, operated and maintained is a function of the site's safety management system (SMS) and so reference to the SMS is needed. This of course has implications for the site's SMS. For example, it will most probably require updating to incorporate DSEAR.

The Regulations, its guidance and the ACoP amount to over 100 pages, and implementation will prove to be a significant task for any safety professional. This is true, even though there is very little in the way of new requirements. For example, all the risk assessments should have been completed under the *Management of Health and Safety at Work Regulations 1999* (HSE 199), and most of the measures to control risks and ignition sources are detailed in existing guidance and the *Highly Flammable Liquids and Liquified Petroleum Gasses Regulations 1972* (HSE 1972).

3 THE MAIN REQUIREMENTS OF ATEX 137 – DSEAR

There are some transitional arrangements on DSEAR's introduction. For example, organisations have a period of three years to classify zones, introduce signage to identify "at risk" areas, provide appropriate clothing, and co-ordinate with other occupiers. However, all the other requirements are mandatory immediately. This leaves employers with the task of:

1. marking all equipment containing substances that can give rise to an explosive atmosphere;
2. providing information, instruction and training to employees;
3. undertaking/updating risk assessments;
4. identifying and implementing control and mitigation measures; and hence
5. demonstrating that all reasonably practicable measures have been taken, and documenting this in an appropriate manner (e.g. in an Explosion Protection Document, as advised by ATEX 137).

As with Seveso II (EU Directive 96/82/EC), DSEAR focuses on the requirement for demonstration – by identifying the hazard, assessing the risks, and ensuring appropriate control and mitigation. A hierarchy of safety measures is specified: *elimination, control, and mitigation*.

3.1 Safety measures – elimination

Elimination can only start once hazardous substances which can give rise to an explosive atmosphere have been identified. The CHIP classification (SI 2002) is a useful guide and substances and preparations classed as dangerous under CHIP, or falling within the CHIP criteria will be dangerous under DSEAR. Other substances may be dangerous depending on how they are used. An obvious example is high flash point liquids held at high temperatures which if released could result in the formation of an explosive atmosphere.

Once the dangerous substances have been identified, the activities that do or could give rise to an explosive atmosphere can be listed. Identifying these activities and the possibility of ignition is fundamental to the demonstration that all reasonably practicable measures have been taken.

Elimination is regarded as the best solution, and involves replacing say, a highly flammable liquid, with a non-flammable liquid. In many cases this will not be practicable. For example, to replace a substance may require extensive and costly changes to a process, or it may require significant and lengthy research to identify an appropriate non-flammable substance.

Where elimination is not an option, it may be possible to reduce quantities or substitute with a substance of a "lesser" hazard (e.g. changing a low flash point solvent with a higher flash point one). This may be feasible for some processes, but safety gains in reducing batch sizes or using substitutes need to be carefully considered against rising costs and possible increases in the number and complexity of operations.

3.2 Safety measures – control

Control can be achieved through a series of measures. In order of importance, these include:

1. avoiding releases;
2. minimising releases;
3. preventing the formation of explosive atmospheres;
4. collecting and containing releases;
5. avoiding ignition sources;
6. avoiding adverse conditions;
7. segregating incompatible substances.

The guidance provides details on how these control measures may be implemented. It is significant that "avoiding ignition sources" is by no means the most "important" control measure, being fifth from seven. Prior to this guidance, there was a tendency to simply "zone" areas to control ignition sources (and ignore other controls) in the belief that this was sufficient. Under DSEAR, zoning will not be acceptable without investigating the possibilities for avoiding, minimising, preventing, and containing explosive atmospheres.

3.3 Safety measures – mitigation

Less guidance than expected is provided on appropriate mitigation measures. However, the most obvious and often forgotten measure of reducing the numbers exposed is included as the first of six measures listed in the Regulations.

Perhaps because the mitigation measures of explosion resistant plant, suppression, relief, and fire are specialised areas is the reason why HSE has not provided detailed guidance. Including detailed guidance on these areas would have repeated existing guidance and made the document even longer. However, HSE has missed an opportunity to cross reference HSE guidance and other standards that give good detailed advice on this subject.

Whilst most of the six mitigation measures listed in DSEAR are fairly obvious, the last one, *"providing suitable personal protective equipment"* is less so. This will most likely require the provision of fire resistant clothing to employees. This has the potential to save lives but also has a cost implication beyond purchasing fireproof overalls since most of these types of clothing require specialist cleaning.

3.4 Information, instruction and training

Provision of information, instruction and training adds little to what would be considered good practice and existing advice given by HSE. There is an absolute requirement to provide employees with details of the hazardous substances and the results of the risk assessment. Whilst this is very similar to Regulation 10 of the *Management of Health and Safety at Work Regulations 1999* (HSE 1999) it does go further in requiring the information, instruction and training to be adapted to take account of significant changes. Presumably, it will no longer be acceptable to have out of date operating procedures (an all too common feature found on many plants).

3.5 Emergency planning

The requirement for emergency plans is specified in Regulation 8 and detailed in Annex 8 of DSEAR. The guidance given, compliments existing guidance for Seveso sites where it is arguably set out in a more logical order. Experience in the offshore industry has shown that emergency arrangements have a major part to play in saving lives. Expertise developed for the offshore industry should be used to help employers develop and test emergency plans.

3.6 Shared workplaces

Shared workplaces have always been a "weak area" when trying to formulate mitigation measures. Whilst most employers are aware of their responsibility to protect nearby workers this can be met with an unsympathetic attitude by others who cannot see why they should be troubled with emergency arrangements because of activity outside their control. The new regulation and guidance adds little to the existing requirement under Regulation 11 of the *Management of Health and Safety at Work Regulations* 1999 (HSE 1999). Time will tell whether or not they will be more effective.

4 SAFETY MEASURES – A SIMPLE EXAMPLE

Consider a typical multi-purpose reactor found in industry.

Inside the reactor it would be classed as Zone 0 with flammable atmospheres expected to be present for long periods of time under normal operations. Assuming that the material could not be substituted for a less hazardous one, DSEAR requires employers to demonstrate that the flammable atmosphere could not be avoided by inerting, increased ventilation or operating above the upper flammable limit.

The opening around the top of the reactor would be classed as Zone 1 with flammable atmospheres expected for short periods of time in normal operations. DSEAR requires consideration of means to reduce the likelihood of a flammable atmosphere. For example, by means of ventilation, keeping the reactor under negative pressure, or fitting a "glove box" on top of the reactor so that the flammable atmosphere cannot disperse into the workplace.

It is common industry practice to zone the whole building as Zone 2 (or even Zone 1). Before this can be claimed as a suitable risk control measure, consideration will need to be given to eliminating practices that release flammable atmospheres. Additions of flammable liquids from drums may need to be performed in laminar flow booths, or eliminated altogether by using fixed pipework.

5 BLANKET ZONING AND RISK ASSESSMENT

Zoning of potentially flammable atmospheres is only a small part of the requirements of DSEAR, and as noted above the control of ignition sources should not be the first risk control measure. There is a requirement to consider eliminating or reducing the extent or duration of potentially flammable areas before considering ignition control.

The obvious place for demonstrating that all reasonably practicable measures have been taken is in the risk assessment. The risk assessment is referred to as an Explosion Protection Document in ATEX 137.

Advice from HSE indicates that this risk assessment can be combined with the *Management at Work* (HSE 1999) risk assessment. However, it is the authors opinion that this will only be the case for organisations with relatively "small" fire and explosion risks. Industries that traditionally use zoning as part of their risk control strategy will probably need to produce separate documents to show that they have considered risk control measures in the appropriate order detailed in the Regulations, ACoP and Guidance.

The risk assessment is likely to illustrate that blanket zoning is often unnecessary. Hence, over the coming years a reduction in the number and extent of zoned areas may well occur. However, where cost is not a factor and to simplify the management of electrical equipment, some organisations may well maintain a policy of blanket zoning.

6 CONCLUSION

Although ATEX 137 (DSEAR) introduces very few new requirements, the need to demonstrate that all reasonably practicable measures have been taken, places additional work and costs on industry. However, replacing the tendency for "blanket" zoning (to reduce the likelihood of ignition from electrical equipment) with a more robust risk-based approach should further reduce the risks from explosive atmospheres.

REFERENCES

EU Directive 1996. 96/82/EC of 9 December 1996 on the control of major-accident hazards involving dangerous substances, Official Journal L 010, 14/01/1997 p. 0013–0033.

EU Directive 1998. 98/24/EC of 7 April 1998 on the protection of the health and safety of workers from the risks related to chemical agents at work http://europa.eu.int/eurlex/pri/en/oj/dat/1998/l_131/l_13119980505en00110023.pdf

EU Directive 1999. 1999/92/EC of the European Parliament and of the Council of 16 December 1999 on minimum requirements for improving the safety and health protection of workers potentially at risk from explosive atmospheres. http://europa.eu.int/eurlex/pri/en/oj/dat/2000/l_023/l_02320000128en00570064.pdf

HSE 1972. Highly Flammable Liquids and Liquified Petroleum Gasses Regulations 1972.

HSE 1988. The Storage of Flammable Liquids in Tanks. HSG176.

HSE 1996. The Safe Use and Handling of Flammable Liquids. HSG140.

HSE 1998a. The Storage of Flammable Liquids in Containers. HSG51.

HSE 1998b. Chemical Warehousing, The Storage of Packaged Dangerous Substances. HSG71.

HSE 1999. Management of Health and Safety at Work Regulations 1999.

HSE 2000. Designing and Operating Safe Chemical Reaction Process. HSG143.

HSE 2001. Proposals for The Dangerous Substances and Explosive Atmospheres Regulations. Consultation Document CD180. http://www.hse.gov.uk/condocs/closed/cd180.pdf.

SI 2002. The Chemicals (Hazard Information and Packaging for Supply) Regulations 2002 (S.I. 2002/1689).

Safety and Reliability – Bedford & van Gelder (eds)
© 2003 Swets & Zeitlinger, Lisse, ISBN 90 5809 551 7

Adoption and enforcement: UK CIA guidance on protecting persons in buildings on chemical establishments

P.A. Davies & J.H. Gould

ERM Risk, Suite 8.01, 8 Exchange Quay, Manchester, England

ABSTRACT: Excluding the task of demonstrating appropriate protection, there are two principal issues associated with the UK Chemical Industries Association's (CIA's) guidance on occupied buildings: (1) adoption by industry; and (2) enforcement by the UK Health & Safety Executive (HSE). This paper discusses these two issues by reference to practical experience. Firstly, it covers industry's response to implementation of the guidance and subsequent implications. For example, the response of "do nothing until pushed" (i.e. receive HSE's request, park it and wait) compared with "a guide to help implement best practice and assure appropriate protection". Secondly, the paper outlines HSE's enforcement of the guidance. For example, in some instances, HSE has felt the need to inform companies that their delayed response to the guidance is excessive and as such, an Improvement Notice is imminent.

1 INTRODUCTION

In the UK, guidance to protect persons within buildings on chemical sites was published by the Chemical Industries Association (CIA 1998) in February 1998. This was largely in response to the fire at Hickson & Welch in September 1992 (HSE 1994 and Patterson 1995), which highlighted the need for guidance to cover all buildings on site and not just control and process buildings, for which guidance already existed (CIA 1979 and API 1990). Whether true or not, it is also widely rumoured that the CIA was keen to introduce guidance under its direction, rather than wait for the possibility of guidance or an Approved Code of Practice (ACOP) being issued by the Health & Safety Executive (HSE). Hence, in 1993 the CIA set up a Building Standards Task Force (BSTF) to develop this guidance (Davies & Patterson 2000).

Following publication, many chemical, oil, gas and pharmaceutical companies have used the guide to assess the "adequacy" of buildings to protect personnel. It is fair to say that in many cases use of the guide and the speed with which assessments have been completed has been driven by HSE. It is also fair to say that HSE's "enforcement" of the guide, whether true or not, has appeared inconsistent to many in industry. Furthermore, interpretation of the guide by individual companies has led to differing approaches and adoption strategies; from the "downright awful" to "over the top", dependent upon your point of view.

From this initial phase, lessons have been learnt by both industry and HSE, and the CIA is revising its guidance (especially that related to toxic hazards). With the above in mind, the following sections summarise the requirements of the CIA guidance, industry's adoption and HSE's enforcement, and the implications of satisfying the guidance. The summaries are largely based on a survey of the views of Risks Interest Group (RIG) members, discussions with industry and HSE, and work performed by ERM on behalf of chemical and pharmaceutical companies.

1.1 *The CIA guidance*

Essentially, the guide (CIA 1998) reinforces the importance of locating and designing site buildings to afford protection to occupants from accidental releases of flammable and toxic substances. The criteria against which to judge the appropriate level of protection is outlined. However, regardless of the level of risk, the guidance makes clear that the risk should be as low as reasonably practicable, and the design and operation of buildings and processes should follow the principles of inherent safety (IChemE 1995).

To assess the adequacy of protection from flammable releases both a hazard-based approach (HBA) and a risk-based approach (RBA) are outlined.

HBA simply relates the cumulative frequency of events that can attain a specified damage level at the building's location (e.g. 70 mbar, $6.3\,kW/m^2$) to a single numeric value. The guidance suggests a value of 10^{-4} per year (i.e. 1 in 10,000 years).

RBA relates occurrence frequency and the likelihood of exposure, to individual risk criteria. Based upon HSE's much published criteria for judging the "acceptability" of annual individual risk (IR) to persons, this translates to:

1 a practical IR target of less than 10^{-5} with a maximum of 10^{-4};
2 consideration of further risk mitigation measures between 10^{-4} and 10^{-6}; and
3 no requirement for further protection below 10^{-6}.

To assess the adequacy of protection from toxic only releases the guidance recommends a qualitative "checklist" approach. This covers practical approaches to limit building exposure, limit gas/vapour ingress, and aid emergency actions. However, it is the authors experience that for sites with many buildings (e.g. 50 or more), HBA or RBA is required to avoid unnecessary upgrades to buildings that cannot be impacted or where the risk to occupants is negligible. In a revision to the guidance, expected later this year (2003), CIA will clarify that adopting toxic refuges is an alternative to the potentially costly and impractical requirement to protect all occupied site buildings.

2 ADOPTION BY INDUSTRY

In many cases, awareness of HSE's expectations of industry's response to the CIA guidance has been via a formal letter. This letter has been issued in slightly different guises since late 1998 by local inspectors using a centrally prepared template. Essentially, the letter asks companies to provide information on those buildings that are in accordance with the guidance, and those that are not, and the corrective actions they plan to take (Davies & Patterson 2000). Initially, the information was required within one month, but following industry's objection that this was unreasonable, HSE has since stated three months. Others have been made aware of HSE's expectations via local inspectors visiting their site and verbally requesting information, or via attendance at industry forums and conferences where presentations by HSE have been made. At one of these presentations HSE made it clear that *"although this is industry guidance, HSE will treat this as if it were HSE guidance"* (Doolan 1998).

Responses to the letter or verbal request have varied considerably; from "take no action and hope that it goes away" to "issuing a report to HSE within a short time frame". One site on receiving the letter in March 1999 took no action and HSE did not follow-up. It was only upon receipt of HSE's review of the site's Seveso Safety Report in December 2001 that the company decided to respond; a delay of two years nine months. HSE concluded that the Safety Report was deficient, with one reason being that "occupied buildings" had not been addressed. A satisfactory response from the company was requested within three months otherwise an *Improvement Notice* would be issued specifically attributable to "occupied buildings"

By contrast, on receiving HSE's request most companies have responded, with many requesting and receiving further time to complete their reviews. This reflects not only the resources and expertise available to companies, but also, for some, the greater importance they attach to other safety concerns. Rightly or wrongly, if acknowledged by their local inspector, this latter point may well be the reason why some companies consider that "occupied buildings" has been treated appropriately by HSE compared to the views of some other companies.

The authors found two companies that had considered all onsite buildings well before issue of the CIA guidance. However, the companies acknowledged that the depth of analysis performed was less than that recommended in the guidance, and that further work was required.

The approaches to the review of "occupied buildings" have also differed markedly between companies. This reflects available resources, whether the hazards are flammable or toxic or both, perceived risk, site specifics, and expertise. For example, for fire and explosion hazards, some companies have undertaken consequence based screening analyses to identify buildings that can be "impacted", and then made a judgement as to the buildings requirements. Interestingly the criterion for impact (i.e. a specified level of thermal radiation and overpressure) has varied between these companies and that published by the CIA. Subsequently, and typically following work required for Seveso Safety Reports, event frequency has been appraised to further reduce the number of impacted buildings.

For toxic hazards, a number of companies have simply acknowledged or assumed that all buildings could be impacted, and the only solution is to ensure all buildings are in accordance with the guide. A number of companies have decided that this is impractical and have decided to provide toxic refuges. This approach reflects the expected revisions to the CIA guidance, that is, the number of buildings that need to be in accordance with the guide should reflect that necessary to provide appropriate protection of persons onsite. However, in a number of cases the refuges are not in accordance with the guidance, and arguments are put forward as to why they remain appropriate; in one case using cost benefit analysis to demonstrate that the cost of upgrades outweighs the benefits gained. In many cases, the potential frequency of toxic gas impact has been ignored. For sites with a large number of buildings and many people distributed over a large area, ignoring impact frequency is likely to overestimate the number of buildings impacted and the number of refuges required.

The reporting of the reviews has also varied, but has largely taken the form of: the production of standalone reports issued to, or available for inspection by HSE; formal letters to, or meetings with local inspectors summarising the approach, results and actions to be taken; and/or a summary in Seveso Safety Reports. Where reviews have been included in Safety Reports, a common response by HSE has been to state that this issue will be included in the Seveso inspection programme. Again, the differences between companies, reflects available resources, perceived risk, site specifics, and expertise.

Unsurprisingly, companies views on the guidance and HSE's enforcement of it are also diverse. With regards to the guidance, the views can be summarised by two extremes: "a framework to address a problem that has existed for some time – hence, a useful exercise"; and "of little assistance in improving site safety – provides a useful checklist". With regards to HSE's approach to enforcement there are those that feel (whether true or false) HSE has been "inconsistent and over zealous", and those that consider that their inspector has "treated the issue in the context of site safety as a whole".

2.1 Satisfying the CIA guidance – implications for industry

The time and effort expended in assessing the adequacy of occupied buildings can itself be a costly exercise, ranging from a few thousand to hundreds of thousands of euros. However, where buildings are found to afford unsatisfactory protection (i.e. they do not meet the guidance) the implications can be more serious both in terms of money and disruption.

For example, serious implications can include the re-location of personnel, protecting buildings from thermal radiation and explosion overpressure, through to changes to production processes. Less serious implications might well include the provision of hand-held respirators through to sealing cable and duct penetrations.

It is the authors experience that many of the implications in satisfying the regulations have been related to emergency preparedness. For example, changing alarms to distinguish between fire and toxic releases, re-locating fire and gas detectors, enlarging doorways so that they are suitable for egress wearing breathing apparatus, ensuring forced ventilation can be rapidly closed in the event of an accident, and providing alternative emergency assembly points and toxic refuges.

3 ENFORCEMENT BY THE HEALTH & SAFETY EXECUTIVE

Internal documents clearly illustrate HSE's commitment to, in its own words "tackle the risks to occupants of buildings". This commitment is based upon HSE's judgement that "a number of recent events … have clearly shown a lack of effective action by occupiers". In addition, the duty of occupiers to reduce this risk is "a long standing legal requirement under the Health & Safety at Work Act 1974" and "an issue which all prudent occupiers should be aware of, and have planned to deal with". From this, it is of no surprise that HSE has taken a firm line with enforcement.

The enforcement initiative was launched in October 1998, as part of a five year plan. This timescale has increased to six years, and may well extend beyond this. The reason for this, is a commitment to include not only top-tier Seveso sites (which have increased in number from 350 to 435 during transition from Seveso I to Seveso II) but also sites HSE considers present a similar hazard to occupants (e.g. sites manufacturing explosives, and many lower-tier sites). Furthermore, there is evidence to suggest that HSE delayed enforcement, on the assumption that "occupied buildings" assessments would be included in Seveso Safety Reports. Unfortunately, such inclusion has not been universal and many companies only refer to a commitment to undertake an assessment; far below HSE's expectations.

Training of HSE's staff in the approach to assessment and enforcement began in late 1998 and coincided with the distribution of internal guidance documents and the issue of the first letters to selected sites. This letter, which was drafted centrally as a "template", informed occupiers of HSE's expectations of them with regard to the CIA guidance. Although the letter has been modified a number of times, essentially its main theme has not changed. The letter asks whether an assessment has been performed, and what remedial measures have been taken or are planned. A response within one month is requested.

This formal request from HSE is referred to as a "pre-visit letter", and is sent to those sites selected for appraisal. Guidance is given on the selection of sites and targets set as to the number of sites to be appraised in each year. However, it is largely at the discretion of the local area to decide upon priorities and hence, the rate at which sites are contacted and follow-up action (if required) initiated. This may well be the reason for some in industry to perceive that HSE's enforcement has been inconsistent.

Following response to the pre-visit letter a meeting to appraise the assessment is made. Practice has shown that this meeting typically follows the request and receipt of an assessment report from the occupier, and that in some cases no site visit is considered necessary. Where a formal assessment is not completed (and/or no response received), enforcement action is taken to agree a reasonable period for completion. Guidance is given as to what constitutes a "reasonable period"; obviously it is site specific, but typically three to six months is considered adequate for most sites, although up to 12 months may be appropriate for "large" sites and multi-site occupiers.

As noted above, for top-tier sites, it was expected that "occupied buildings" assessments would be included in Seveso safety reports. In reality many assessments are absent or incomplete (e.g. often a plan for remedial measures is not given). In most, if not all of these cases, HSE has judged that the "all necessary measures" requirement has not been demonstrated, and hence, the Safety Report is deficient and enforcement action is required.

Contrary to the personal experience of the authors, it is not HSE's intention to appraise assessments in detail, but rather to concentrate on the suitability of the remedial measures identified to "secure compliance". In addition, there is no requirement to use the CIA guidance. Alternative assessment methods are acceptable, and internal guidance requests that these are judged against the applicable legal requirements (i.e. Seveso II and the Management of Health & Safety at Work Regulations 1999) and not the CIA guidance; essentially the onus is on the operator to demonstrate an equivalent standard of protection.

From discussions with Inspectors "in the field", there is a perception that the "occupied buildings" assessments have fallen below expectations, and industry implementation has been slow. Reasons cited for industry's "poor" performance include: lack of resources; lack of internal expertise in quantitative risk assessment techniques (i.e. with some justification this is unlikely to be considered a core skill required of onsite safety personnel); confusion over how to apply the guidance and its scope (e.g. assessment of toxics); and "waiting to see what happens". Interestingly, although Inspectors have "threatened" enforcement action, few have needed to resort to formal issue of *Improvement Notices*.

In summary, HSE considers that enforcement of the CIA guidance is an ambitious attempt to introduce a structured approach to the assessment of occupied buildings, and that it is succeeding in encouraging occupiers to address this issue. Furthermore, HSE acknowledges that implementation of the standards can be costly (especially at sites where insufficient consideration has been given to this issue in the past), but the indications are that companies are taking action in a proportional manner to improve safety and comply with the law.

All of the above is based upon "open government" internal HSE documents and discussions with Inspectors in the field (CHID 1998, 1999a, 199b, HID 2000, and Davies 2002).

REFERENCES

API 1990. Management of Hazards Associated with Process Buildings, API RP752.
CHID 1998. Circular 12-Nov-98, Location and Design of Occupied Buildings Inspection Project. SF/431.
CHID 1999a. Circular 10-Feb-99, Location and Design of Occupied Buildings Inspection Project – Overlaps with Land-use Planning Procedures, SF/431.
CHID 1999b. CHID 6, OPPG Unit, 5-Oct-99, "Below the Line Paper", The Location of Occupied Buildings on Major Hazard Sites, First Annual Review of Project, DMB/99/51.
CIA 1979. Process Plant Hazard and Control Building Design.
CIA 1998. Guidance for the Location and Design of Occupied Buildings on Chemical Manufacturing Sites.
Davies & Patterson 2000. Demonstrating the Adequacy of Protection Afforded by Occupied Buildings on Chemical Sites, Hazards XV, IChemE Symposium.
Davies 2002. Personal Communications, August 2002, Davies, P.A. (ERM) and HSE Inspectors.
Doolan 1998. Chemical and Hazardous Installations Division, presentation to the West Yorkshire CIA Responsible Care Cell.
HID 2000. Circular, 30-Nov-00, Location and Design of Occupied Buildings on Major Hazard Sites, National Inspection Project – Review and Continuation, SF/431.
HSE 1994. The Fire at Hickson & Welch Limited.
IChemE 1995. Training Package 027: Inherently Safer Process Design, Institution of Chemical Engineers.
Patterson 1995. The fire at Hickson & Welch, Castleford, UK, Major Hazards Onshore and Offshore, IChemE Conference.
RIG. Risks Interest Group, http://www.hazardview.com.

Safety and Reliability – Bedford & van Gelder (eds)
© 2003 Swets & Zeitlinger, Lisse, ISBN 90 5809 551 7

ARAMIS project: identification of reference accident scenarios in SEVESO establishments

C. Delvosalle, C. Fiévez & A. Pipart
Faculté Polytechnique de Mons, Major Risk Research Centre, Mons, Belgium

J. Casal Fabrega & E. Planas
Universitat Politecnica de Catalunya, CERTEC, Spain

M. Christou & F. Mushtaq
European Commission, Joint Research Centre, Italy

ABSTRACT: In the frame of the ESREL special session on ARAMIS project, this paper aims at presenting the work carried out in the first Work Package, devoted to the definition of accident scenarios. This topic is a key-point in risk assessment, and serves as basis for the whole risk quantification. A first part of the work aims at building a Methodology for the Identification of Major Accident Hazards (MIMAH), which is carried out with the development of generic fault and event trees based on a typology of equipment and substances. This work is coupled with an historical analysis of accidents. In a second part, influence of safety devices and policies will be considered, in order to build a Methodology for the Identification of Reference Accident Scenarios (MIRAS). This last one will take into account safety systems and lead to obtain more realistic scenarios.

1 INTRODUCTION

The identification of the possible accident scenarios is a key-point in risk assessment. However, especially in a deterministic approach, only worst cases scenarios are considered, often without taking into account safety devices used and safety policy implemented. This approach can lead to an over-estimation of the risk-level, and does not promote the implementation of safety systems.

In the ARAMIS European project (Hourtolou & Salvi, 2003), the idea developed in Work Package 1 is to identify major accidents (without considering safety systems), then to study deeply safety systems, causes of accidents and (qualitative) probabilities, in order to be able to identify Reference Accident Scenarios, these last ones taking into account safety systems.

In order to reach this objective, two main steps are defined.

The first objective is to define a Methodology for the Identification of Major Accident Hazards (MIMAH). On the basis of equipment considered and properties of chemicals handled, the methodology must be able to predict which major accidents are likely to occur.

The work is divided into several parts:

– the choice of a general approach (bow-tie method);
– the definition of a common vocabulary and a typology of equipment and hazardous properties of substances;
– the development of event trees and fault trees centered on critical events likely to occur on different kinds of equipment;
– an historical analysis of known accidents.

The second objective is to study the influence of safety devices and policies on scenarios identified by the MIMAH methodology, leading to the development of a second methodology (MIRAS: Methodology for the Identification of Reference Accident Scenarios).

2 OBJECTIVES, REQUIREMENTS AND STEPS OF MIMAH

2.1 *Objective*

The Methodology for the Identification of Major Accident Hazards (MIMAH) aims at defining the maximum hazardous potential of an installation. The term "Major Accident Hazards" must be understood as the worst accidents likely to occur on this installation, assuming that no safety systems (including safety management systems) are installed or that they are ineffective. The Major Accident Hazards will only depend on the equipment characteristics and the hazardous properties of the chemical handled in the equipment.

2.2 Requirements

Some requirements must be met according to the work to be done in the rest of the project:

– MIMAH must be generic in order to be applied to the different types of equipment likely to be present in a chemical plant;
– MIMAH must be built in an algorithmic form (with logical links), in order to allow the use of a methodical approach when safety systems will be considered;
– MIMAH will serve as a basic method to define MIRAS (Methodology for the Identification of Reference Accident Scenarios). MIMAH should then be build in such a way that the influence of the safety systems on the development of a scenario could be easily identified and further taken into account.

For all these reasons, the approach chosen in the ARAMIS project is the bow-tie approach. This one is a highly structured tool which offers the possibility to identify critical events (center of the bow-tie), to construct major accident scenario (event tree), to study the causes of accidents (fault tree), to estimate probabilities, to place safety barriers and to study how safety barriers influence the trees, for the definition of Reference Accident Scenarios.

Moreover, the bow-tie approach is considered as a very good tool to establish links with the work carried out by other partners in the ARAMIS project (e.g. Work Package 3 – safety management systems).

2.3 Steps in the work

As the methodology must be generic, the objective chosen is to define a method allowing to build an accident scenario only on the basis of three data: the equipment type, the hazardous properties of the handled substance and its physical state (solid, liquid, two-phase, gas/vapor).

For this purpose, the work is divided in several steps:

– Firstly, define an equipment typology and an hazardous substance typology, allowing to classify equipment and substances encountered on chemical plants.

– Secondly, study the center of the bow-tie (critical event), by drawing up a list of critical events likely to occur for each type of equipment.
– Thirdly, develop a systematic method to build an event tree. The approach chosen consists in defining logical links, on the one hand between critical events and secondary critical events, and on the other hand between secondary critical events and dangerous phenomena. This can be done by the way of matrices, and the result obtained is a tree giving the possible accident scenarios for each critical event studied. A selection of the scenarios obtained is then carried out according to the hazardous properties of the handled substance.
– Fourthly, develop a method to build generic fault trees.

3 THE BOW-TIE APPROACH

The bow-tie approach assimilates accident scenarios to a succession of events, as shown in Figure 1.

The bow-tie is centered on the critical event (CE), which is generally defined as a Loss Of Containment (LOC) for fluids or a Loss of Physical Integrity (LPI) for solids.

The left part of the bow-tie, named fault tree, identifies the possible causes of a critical event. Combinations of Undesirable events (UE) and Current Events (CuE) lead to Detailed Direct Causes (DDC) which, when combined, lead to Direct Causes (DC) which cause Necessary and Sufficient Conditions (NSC) provoking the Critical event.

The right part of the bow-tie, named event tree, identifies the possible consequences of a critical event. The Critical Event CE, such as a pipe failure, leads to Secondary Critical Events SCE (for example a pool formation, a jet, …), which leads to Tertiary Critical Events TCE (for example a cloud following a jet), which in turn leads to Dangerous Phenomena DP such as fire, explosion, dispersion of a toxic cloud, … Major Events (ME) are defined as the exposition of targets (human beings, structure, environment, …) to

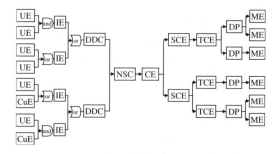

Figure 1. The bow-tie approach.

a significant effect due to the identified Dangerous Phenomena.

4 TYPOLOGY

As explained above, MIMAH aims at defining major accident scenarios associated with a given equipment containing a given substance. For this purpose, it is necessary to define an equipment typology and an hazardous substances typology.

4.1 Equipment typology

Equipment are classified in generic categories, according to their function and operating conditions. An essential rule has to be kept in mind all along the classification: equipment classified in the same category must generate the same generic bow-tie.

A first way to group equipment is the "unit", in which several types of equipment are described and defined. Four units and sixteen equipment types are defined. The list of selected equipment is presented here after:

– Storage equipment: mass solid storage (EQ1), storage of solid in small packages (EQ2), storage of fluid in small packages (EQ3), pressure storage (EQ4), padded storage (EQ5), atmospheric storage (EQ6), cryogenic storage (EQ7);
– Transport equipment: pressure transport equipment (EQ8), atmospheric transport equipment (EQ9);
– Pipes networks (EQ10);
– Process equipment: intermediate storage equipment integrated into the process (EQ11), equipment devoted to the physical or chemical separation of substances (EQ12), equipment involving chemical reactions (EQ13), equipment designed for energy production and supply (EQ14), packaging equipment (EQ15), other facilities (EQ16).

4.2 Substances typology

Hazardous properties of substances are classified in different ways. In Europe, the relevant regulation is the Directive 67/548/EC (Council Directive 67/548/EEC of 27 June 1967 on the approximation of laws, regulations and administrative provisions relating to the classification, packaging and labeling of dangerous substances). It should be noted that the hazard categories of the SEVESO II Directive are based on the risk phrases of the Directive 67/548/EC.

The ARAMIS project fits into the scheme of the SEVESO II Directive. This is why it was chosen to base the ARAMIS typology on the hazard categories of the SEVESO II Directive, coupled with a specific selection of risk phrases defined in the Directive 67/548/EC.

5 THE CENTRE OF THE BOW-TIE: THE CRITICAL EVENT

It should be reminded that the Critical Event is the center of the bow-tie. It can be defined as a loss of containment or a loss of physical integrity.

An important part of the work done in this research led us to define a concise list of 12 critical events likely to occur on equipment and to give a definition for each of them. The critical events retained are:

– Decomposition (CE1)
– Explosion (CE2)
– Materials set in motion (entrainment by air) (CE3)
– Materials set in motion (entrainment by a liquid) (CE4)
– Start of fire (LPI) (CE5)
– Breach on the shell in vapor phase (CE6)
– Breach on the shell in liquid phase (CE7)
– Leak from liquid pipe (CE8)
– Leak from gas pipe (CE9)
– Catastrophic rupture (CE10)
– Vessel collapse (CE11)
– Collapse of the roof (CE12)

A table linking equipment types and critical events was built. This table gives the critical events which could be observed on the different equipment types.

6 THE EVENT TREE

6.1 General overview of the MIMAH methodology

As explained above, the methodology must serve to build generic trees, in an algorithmic form (with logical links). It was chosen to work with matrices, which allow to give importance to the algorithmic and logical aspects. The methodology will be explained here according to its main principles.

Firstly, matrices are used in order to define which critical events must be associated with a given equipment containing a given substance. As explained previously (see section 5), there exists some possible associations between equipment and critical events, which will be expressed in a matrix crossing the equipment type (EQ) and the critical events (CE), as shown in Table 1. A sign "X" in a cell indicates that the association CE–EQ is possible, and, on the contrary, an empty cell indicates that the top-column critical event cannot be associated with the top-line equipment type.

Secondly, as said previously, the event tree must be built according to the equipment type and the substance handled in this equipment. Interesting substances properties in the frame of event trees are the physical state and the hazardous properties.

Hazardous properties will be studied with the issue of dangerous phenomena (DP). Concerning physical

Table 1. Example of matrix EQ–CE.

	...	CEi	CEj	CEk	CEl	CEm	...
EQa			X		X		
EQb		X	X	X			
EQc			X		X	X	
EQd		X		X		X	
EQe			X				
...							

Table 2. Example of matrix STAT–CE.

	...	CEi	CEj	CEk	CEl	CEm	...
STAT1			X	X	X		
STAT2		X		X		X	
STAT3		X	X			X	
STAT4			X	X	X		

Table 3. Association of CE with EQ and STAT.

	...	CEi	CEj	CEk	CEl	CEm	...
EQd		X		X		X	
STAT3		X	X			X	
EQd and STAT3		X				X	

Table 4. Example of matrix CE–STAT–SCE.

		...	SCEo	SCEp	SCEq	...
...						
CEi	STAT1		▨	▨	▨	
CEi	STAT2		X			
→ CEi	STAT3		X		X	←
CEi	STAT4		▨	▨	▨	
CEj	STAT1		X		X	
CEj	STAT2		▨	▨	▨	
CEj	STAT3			X	X	
CEj	STAT4				X	
CEk	STAT1		X			
CEk	STAT2				X	
CEk	STAT3		▨	▨	▨	
CEk	STAT4		X		X	
CEl	STAT1		X	X		
CEl	STAT2		▨	▨	▨	
CEl	STAT3		▨	▨	▨	
CEl	STAT4			X		
CEm	STAT1		▨	▨	▨	
CEm	STAT2		X		X	
→ CEm	STAT3			X		←
CEm	STAT4		▨	▨	▨	
...						

state, a given one cannot lead to all types of critical events. A matrix crossing the substance state (STAT) and the critical events (CE) is then defined, as shown in Table 2.

With these two tables (Table 1 and Table 2), it is possible to determine which critical events must be associated with a given equipment and a given physical state of the handled substance.

For example, assuming that an EQd equipment type is considered, handling a substance which physical state is STAT3, the following conclusions can be drawn:

- critical events CEi, CEk and CEm are compatible with EQd;
- critical events CEi, CEj and CEm are compatible with STAT3;
- thus, only critical events CEi and CEm are compatible with EQd and STAT3.

This can be better observed in Table 3, where the first line is for the equipment, the second one for the substance physical state and the third one for the combination of the equipment and the substance physical state. In this last line, a sign "X" is present for a given critical event only if it is present in the line of the equipment AND in the line of the substance physical state, for the same critical event.

Thirdly, it is useful to know which secondary critical event(s) occur(s) after a given critical event. This will depend on the physical state of the handled substance: a same critical event can give rise to different secondary critical events for different types of substances.

A matrix linking the critical events (CE), the substance state (STAT) and the secondary critical events (SCE) is thus built. An example is shown in Table 4. It can be observed that some cells are hatched: this means that the critical event and the physical state

concerned are incompatible, according to Table 2, and thus cannot lead to a secondary critical event.

In our example of definition of an event tree for the EQd equipment type containing a substance in physical state STAT3, it has be seen (Table 3) that the critical events CEi and CEm are possible. In Table 4, from the lines marked by arrows, it can be concluded that, for this physical state, the critical event CEi will lead to the secondary critical events SCEo and SCEq, and that the critical event CEm will lead to the secondary critical event SCEp.

So, the first part of the event trees can be drawn as in Figure 2.

Fourthly, it is necessary to define a matrix crossing the secondary critical events (SCE) with the tertiary critical events (TCE). The crossing is independent of the physical state of the substance. An example of such a matrix is presented in Table 5, and the event tree corresponding to our example is shown in Figure 3.

Fifthly, the same reasoning must be conducted about a matrix linking tertiary critical events (TCE)

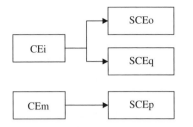

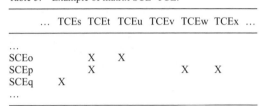

Figure 2.   Example of construction of an event tree (till SCE).

Table 5.   Example of matrix SCE–TCE.

	...	TCEs	TCEt	TCEu	TCEv	TCEw	TCEx	...
...								
SCEo			X	X				
SCEp			X			X	X	
SCEq		X						
...								

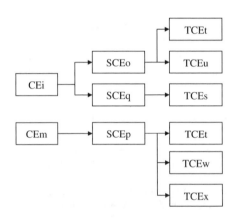

Figure 3.   Example of construction of an event tree (till TCE).

Table 6.   Example of matrix TCE–DP.

	DPa	DPb	DPc	DPd	DPe	DPf	DPg	DPh	...
...									
TCEs		X			X				
TCEt							X		
TCEu	X								
TCEv			X	X					
TCEw	X							X	
TCEx					X				
...									

and dangerous phenomena (DP). An example of such a matrix is presented in Table 6, and the event tree corresponding to our example is shown in Figure 4.

Lastly, the hazardous properties of the handled substance must be taken into account in order to

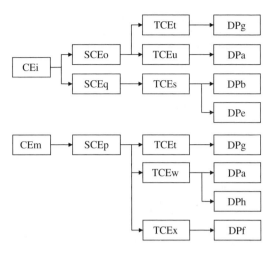

Figure 4.   Example of construction of an event tree (till TCE).

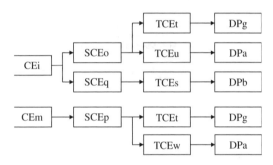

Figure 5.   Final event tree (including selection of DP according to hazardous properties).

select appropriate dangerous phenomena. This selection will lead to the deletion of some branches of the event tree. In our example, if the hazardous properties are only linked with dangerous phenomena DPa, DPb and DPg, the event trees will be less extended, as shown in Figure 5. In this final tree, all branches which do not end by the above-quoted dangerous phenomena (DPa, DPb or DPg) are deleted.

6.2   Summary of the methodology

With regard to the event tree construction, the MIMAH methodology can thus be summarized as shown in Figure 6.

6.3   Example

6.3.1   Data

The MIMAH methodology, explained above, is here applied on a storage vessel containing methyl alcohol.

483

The equipment type is "atmospheric storage" (EQ6). The substance physical state is "liquid" (STAT2). Risk phrases associated with methyl alcohol are:

- R11: Highly flammable
- R23/24/25: Toxic by inhalation, in contact with skin and if swallowed
- R39/23/24/25: Toxic: danger of very serious irreversible effects through inhalation, in contact with skin and if swallowed.

6.3.2 Choice of the critical events

Critical events likely to occur on atmospheric storage and those likely to occur with a substance in liquid state are given in Table 7.

The combination of these information gives as results that 6 critical events must be retained: start of fire (LPI); breach on the shell in liquid phase; leak from liquid pipe; catastrophic rupture; vessel collapse; collapse of the roof.

6.3.3 Construction of the event trees (without taking into account the risk phrases)

An event tree must be built for each critical event retained. For example, the event tree related to the critical event "leak from liquid pipe" will be explained.

First of all, the matrix CE-STAT-SCE (Table 8) must be used to choose the secondary critical events to be retained. It can be seen that one SCE must be selected: SCE3 "pool formation".

Secondly, the matrix SCE-TCE (Table 9) gives information about the TCE (tertiary critical events) retained. These are three: TCE4 pool ignited, TCE5 gas dispersion and TCE11 pool not ignited/pool dispersion.

Finally, the matrix TCE-DP (Table 10) gives the dangerous phenomena for each TCE:

- for TCE4: poolfire (DP1), toxic cloud (DP6) and environmental damage (DP11)
- for TCE5: VCE (DP4), flashfire (DP5), toxic cloud (DP6) and environmental damage (DP11)
- for TCE11: environmental damage (DP11)

The event tree presented in Figure 7 is then obtained. With the same analysis, event trees can be built for every critical event selected.

6.3.4 Construction of the event trees (taking into account the risk phrases)

As mentioned before, main hazardous properties of methanol are its high flammability and its toxicity by inhalation.

The MIMAH methodology gives rules in order to select appropriate dangerous phenomena. In this case,

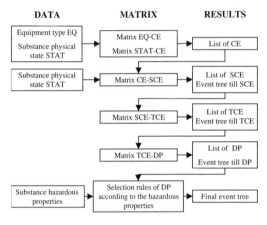

Figure 6. Summary of the steps followed by MIMAH (part "event tree").

Table 7. Choice of critical events.

	Atmospheric storage (EQ6)	Liquid (STAT2)	Results
Decomposition (CE1)			
Explosion (CE2)			
Materials set in motion (entrainment by air) (CE3)			
Materials set in motion (entrainment by a liquid) (CE4)			
Start of a fire (LPI) (CE5)	X	X	X
Breach on the shell in vapour phase (CE6)			
Breach on the shell in liquid phase (CE7)	X	X	X
Leak from liquid pipe (CE8)	X	X	X
Leak from gas pipe (CE9)			
Catastrophic rupture (CE10)	X	X	X
Vessel collapse (CE11)	X	X	X
Collapse of the roof (CE12)	X	X	X

Table 8. Matrix CE–STAT–CE (extract).

| Compatibility CE–STAT | Leak from liquid pipe (CE8) | | | |
| | X | | X | |
	STAT1 Solid	STAT2 Liquid	STAT3 Two-phase	STAT4 Gas/Vapour
Fire (SCE1)				
Catastrophic rupture (SCE2)				
Pool formation (SCE3)		X	X	
Pool inside the tank (SCE4)				
Gas jet (SCE5)				
Gas puff (SCE6)				
Two-phase jet (SCE7)			X	
Aerosol puff (SCE8)				
Explosion (SCE9)				
Materials entrained in air (SCE10)				
Materials entrained by a liquid (SCE11)				
Decomposition (SCE12)				

Table 9. Matrix SCE–TCE (extract).

	Pool formation (SCE3)
Fire (TCE1)	
Catastrophic rupture (TCE2)	
Pool ignited inside the tank (TCE3)	
Pool ignited (TCE4)	X
Gas dispersion (TCE5)	X
Toxic secondary products (TCE6)	
Gas jet ignited (TCE7)	
Gas puff ignited (TCE8)	
Two-phase jet ignited (TCE9)	
Aerosol puff ignited (TCE10)	
Pool not ignited / Pool dispersion (TCE11)	X
Explosion (TCE12)	
Dust cloud ignited (TCE13)	
Dust dispersion (TCE14)	

Table 10. Matrix TCE–DP (extract).

	Pool ignited TCE4	Gas dispersion TCE5	Pool not ignited/ Pool dispersion TCE11
Poolfire (DP1)	X		
Tankfire (DP2)			
Jetfire (DP3)			
VCE (DP4)		X	
Flashfire (DP5)		X	
Toxic cloud (DP6)	X	X	
Fire (DP7)			
Missiles ejection (DP8)			
Overpressure generation (DP9)			
Fireball (DP10)			
Environmental damage (DP11)	X	X	X
Dust explosion (DP12)			
Boilover and resulting poolfire (DP13)			

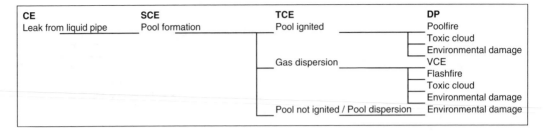

CE	SCE	TCE	DP
Leak from liquid pipe	Pool formation	Pool ignited	Poolfire
			Toxic cloud
			Environmental damage
		Gas dispersion	VCE
			Flashfire
			Toxic cloud
			Environmental damage
		Pool not ignited / Pool dispersion	Environmental damage

Figure 7. Event tree for the CE "leak from liquid pipe".

Table 11. Selection of DP (summary).

DP		Selected	Remark
DP1	Poolfire	Yes	
DP2	Tank fire	Yes	
DP3	Jetfire	Yes	
DP4	VCE	Yes	
DP5	Flashfire	Yes	
DP6	Toxic cloud	Yes IF …	Selected if the TCE is TCE5 (gas dispersion) or TCE14 (dust dispersion)
DP7	Fire (LPI)	No	
DP8	Missiles ejection	Yes IF …	Selected only after the SCE "catastrophic rupture"
DP9	Overpressure generation	Yes IF …	Selected only after the SCE "catastrophic rupture"
DP10	Fireball	Yes	
DP11	Dust explosion	No	
DP12	Boilover and resulting poolfire	No	

DP related to a fire phenomenon will be selected, so as DP related to the dispersion of a toxic cloud (but not the dispersion of toxic fumes). Moreover, the DP "missiles ejection" and "overpressure" resulting from the SCE "catastrophic rupture" must always be taken into account in the construction of the trees.

Rules given here are summarized in Table 11. With this table, it is possible to examine the event trees built without taking into account the risk phrases, and to delete some branches to give the new event trees with the influence of the risk phrases (shown in Figure 8).

7 THE FAULT TREE

The left part of the bow tie, or fault tree, identifies the possible causes of a critical event.

The objective sought in making these generic fault trees is to provide a basis for the next steps of the ARAMIS program, which is supposed to end up with a methodology for evaluating the probability and severity of accident scenarios and means of reducing both. Therefore, the fault trees must allow to make a link between the critical events and all the elements which could have an influence on their possible occurrence including management options as well as safety functions and barriers. Such trees should also provide basis for probability assessment. For these reasons, it was decided to keep a causal relationship between the events constitutive of the fault trees. This should also enable us to perform fault tree probability calculations in the next steps of the methodology.

Major difficulties encountered when building fault trees were to limit the number of levels, and to provide a sufficiently exhaustive analysis of each level. An equilibrium was to be found between the quantity of potential causes and the useful and exploitable information.

The structure of the fault trees is shown in Figure 1. For each level, a detailed analysis was carried out in order to obtain a reasonable but complete list of causes. The result consists in 12 fault trees (one for each critical event). It is unfortunately impossible to present the trees obtained in the frame of this paper.

8 HISTORICAL ANALYSIS

The Methodology for the Identification of Major Accident Hazards (MIMAH) is developed on a typology of equipment. A typology of accident scenarios,

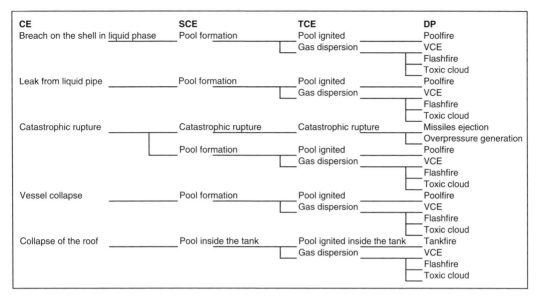

Figure 8.   Event trees, taking into account risk phrases (methyl alcohol).

causes, phenomena involved and their effects has also been defined. This typology is compared with an historical review of known major accidents. It is necessary to validate the theoretical approach by means of a comparison with historical accident data. For this purpose, various databases such as MARS (Major Accident reporting System from the EC), MHIDAS (Major Hazard Incidents Data Service) and HADES (HAzards Database and Effects Study) are consulted to determine the Most Often Observed Accidents (MOOA) as a function of the typology of accident scenarios and to identify, at least in qualitative terms, their probability of occurrence.

The objectives of the historical analysis are to verify the pertinence of the trees; to verify the lists of equipment types, critical events, dangerous phenomena, causes (have we forgotten an item ?) and to obtain statistics to define the MOOA (Most Often Observed Accident).

This part of the work is currently in progress.

9   CONCLUSIONS AND FUTURE DEVELOPMENTS

The development of the MIMAH methodology led us to obtain a tool able to identify major accidents likely to occur on an equipment item, starting from the basic causes and defining a succession of events leading to major effects. On the basis of the equipment type, the substance handled and its physical state, a matrix-based method allow to select appropriate critical events and to build event trees. On the other hand, for each critical event, generic fault trees have been built, reaching an equilibrium between the quantity of potential causes and the useful and exploitable information.

However, this methodology does not take into account the implemented safety systems. Future developments will allow to define safety functions and barriers in order to study the influence of safety systems on the fault and event trees. A quantification of probabilities will also be studied. This will give a second methodology called MIRAS (Methodology for the Identification of Reference Accident Scenarios).

REFERENCES

Hourtolou, D. & Salvi, O. 2003. ARAMIS Project: development of an integrated Accidental Risk Assessment Methodology for Industries in the framework of SEVESO II directive. *Proc. ESREL, Maastricht, Netherlands, 16–18 June 2003.*

Safety and Reliability – Bedford & van Gelder (eds)
© 2003 Swets & Zeitlinger, Lisse, ISBN 90 5809 551 7

Three years experience of the R&M case

Richard Denning
Ministry of Defence UK

ABSTRACT: For many years the UK MoD has sought to purchase Reliable & Maintainable equipment and systems to aid this process as with many other organizations, it has developed a series of standards. Historically these standards have been similar to many other R&M standards – requiring a programme of work to develop systems with good R&M characteristics. The standards also detailed the tasks required to achieve such a programme. In the mid 1990's it became apparent that this approach was not achieving the desired results and therefore it was decided to move from this prescriptive approach to a "case approach". This approach is similar to the safety case approach for assuring safety, although there was stronger emphasize on the progressive nature of the case. Since 1999 this is the approach that has been taken for new MOD contracts. The results have not always been what the authors originally intended.

1 BACKGROUND

The UK Ministry of Defence (MoD) spends approximately £11 (€18) billion per annum on purchasing and providing engineering support to equipment and systems. In order to achieve operationally effective systems, it includes R&M (Reliability & Maintainability) requirements in its contracts for the supply of new systems and equipment. In order to aid this process, as with many other organisations, it has developed a series of standards.

Historically these standards have been similar to many other R&M standards requiring a programme of work to develop systems with good R&M characteristics. The standards also detailed the tasks to be included in the programme.

In the mid 1990s it became obvious that this approach was not particularly suitable for many of the new systems that were being procured as these contained large amounts of software – to which the standard tools and techniques could not be easily applied. This resulted in the decision to take another approach for software intensive systems. It was while preparing this new guidance on the management of software reliability that the MoD came to the conclusion that the challenge presented by the need to demonstrate reliability was similar to the challenge to demonstrate safety. It may not be possible to prove absolutely that a system, equipment, sub-assembly or component is safe or reliable, but it is possible to build confidence by examining safety hazards and reliability risks and

then managing them. Thus the decision was made to adopt the approach of requiring that the supplier develop a "case" to give the MoD Assurance that the software reliability would meet the contracted requirements – this approach was incorporated in to the MoD standards on reliability in 1997 (MoD 1997).

In parallel with the development of this new approach, questions were being raised about the MoD's ability to tell suppliers how to do their job, the argument being that the supplier had the best understanding of their technology/processes and therefore should know the best way to develop solutions that meet the requirement.

Therefore it was decided to change the approach from a prescriptive – "do these activities and it will be accepted that the product is reliable", to a "put forward a reasoned argument that the product will be reliable and the argument will be assessed". After consultation a new standard was raised (MoD 1999a) with supporting guidance (MoD 1999b) these two documents formalized the new approach.

2 RELIABILITY CASE APPROACH

The new approach was based on the previous work to manage software reliability and was very similar to the safety case approach. The aim of the process is to build the confidence of all interested parties in the progress to, or the level of R&M achieved, making the best use possible of all available information. The process is called the Progressive Assurance of R&M where data, evidence, arguments and claims are brought together to form an R&M Case.

Under this process there are three objectives for all involved in the management of R&M (both customer and supplier):

1 Understanding the requirements;
2 Planning & Implementing a programme of activities to satisfy the requirements;
3 Generate assurance that the requirements are being/have been met.

Objectives 2 and 3 are important, but of limited value if the requirements are not defined or understood correctly.

At any given stage in a project, confidence that the R&M requirements will be/have been met can come from satisfaction that the three objectives are being met.

2.1 Setting R&M requirements

Confidence here comes from evidence that the customer's needs are fully and adequately documented and understood. Involvement of all stakeholders in the R&M of the system ensures that all aspects of the requirements are captured. This is a complicated task that must be completed as accurately as possible. The requirements must include all details concerning the intended use of the system. These include:

– How it will be used and pattern of usage.
– Where it will be used.
– Who will use it (skills, experience, expectations).
– Other systems it will have to interface or operate with.
– Who will maintain it.
– The maintenance policy.
– Etc.

To ensure traceability of the requirements, the evidence, assumptions and arguments employed in setting those R&M requirements should be brought together in a document (The Initial R&M Case). In this way anyone working on the project (especially any supplier trying to identify and produce a solution to meet the requirements) can fully understand the background to those requirements.

When these requirements are passed to a supplier, that supplier must ensure that they understand the requirements and their implications fully. This will involve a thorough analysis of them and lead to the identification and recording of R&M risks where the requirements are challenging or insufficient information is known or supplied by the customer.

There is always a likelihood that the requirements will change. If this happens careful configuration management, in addition to traceability, is essential.

2.2 Progressive assurance

Once the requirements have been set, the next step is to design a programme to select and develop a solution. The customer may include passing on the requirements to a supplier as part of that programme. The supplier has to produce a detailed programme in the form of a strategy and plan for the achievement of the R&M specified in the contract requirements.

A useful technique in developing the plan of the R&M activities is to base them on the plan for the management and mitigation of the R&M risks.

This programme must ensure that the strengths and weaknesses of all potential solutions are analysed and assessed against the requirements and that the planned development of a selected solution is appropriate to achieving those requirements. Progress in that direction is monitored, which must include a review and evaluation of the programme so that it can be amended if it appears that the requirements may not be achieved.

2.3 R&M Case Report

Because the volume of evidence used in the R&M case and the range of sources are likely to be large, it is unlikely that the Case could be brought together in a single document. In this instance a periodic summary document can be used that details progress to date in terms of the arguments and claims. It should refer out to data sources where necessary. This document and subsequent additions is called the R&M Case Report. The contract may define points in the programme at which copies of the R&M Case Report will be supplied to the customer.

Just as the requirement flows down the supply chain (Fig. 1), the R&M Case Reports should flow back up the chain, with each supplier incorporating the relevant parts of the lower level Cases in their own Case.

2.4 Review of the Case

On receipt of a Case Report the receiver should review it and consider if appropriate confidence is being generated, this may include drilling down to reports on

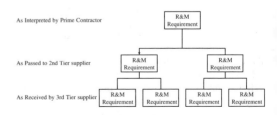

Figure 1. Flow of R&M requirements.

490

specific activities to see if the reviewer independently comes to the same conclusions as the originator of the document. In some cases it may be necessary to reproduce some of the work in order to make this judgment. Some larger projects have employed an independent body to undertake this review activity.

3   EXPERIENCE OF THE R&M CASE

One of the key requirements of the new approach is for the initiator of the requirement to capture the R&M requirement in some details, including the justification for the requirement and the consequences of not achieving that requirement. This is to enable informed tradeoffs to be made at a later date. Although previously much effort had gone into setting requirements, having to capture the justification was new and caused some problems. This was especially true for those systems which were similar to previous systems – custom and practice was to base the R&M requirement on the previous system's requirement without much thought about the relevance of the requirement or the practicality of meeting it. This problem has been exacerbated by industry not requesting this information. With the education of project teams on the new process, these problems are decreasing.

Although the new process was introduced in 1999 there are still many "legacy" systems being procured via the traditional approach. Due to the differing times for procurement, this has meant that we had (and still have) two processes for managing the R&M of systems. This has resulted in some confusion, both in-house and within industry, as people move from project to project. It should be noted that the time-scales of procurements range from a few months for a COTS (Commercial Off The Shelf) to several years for major platforms such as ships, therefore it is likely that these problems will continue for a few years yet. What is very positive is that for a number of the legacy projects, people have proposed changes to their programmes to take advantage of the case approach. This has resulted in the shortening of some procurement programmes and reduction in costs.

Related to this problem there are still requirements being produced by the "Cut'n'Paste" from previous requirements – this tends to be for the more minor projects and is being corrected on a case-by-case basis.

When developing the standard and the guidance document, a deliberate decision was taken not to provide a template R&M Case as it was felt (based on previous experience) this would become the standard solution. This would have been the opposite of our aim of allowing suppliers to develop a case that suits their product.

This lack of template has caused some problems within certain "areas" of industry which previously have had a "standard" document which they just modify for the next project – often this modification is a search and replace for the project name.

This has also meant that we have had to improve the training of our staff as it is no longer possible to assess a proposal or progress report by checking that all tasks are being done. There is now a need to consider if the justification for these tasks reasonable, or has the supplier missed something important.

The concept of accepting various types of evidence and merging these to come to a view on the R&M characteristics of a system, is very simple, we do it every day in our normal life. However, it has caused some difficulties in practice as it is much harder to justify decisions made when assessing the R&M Case rather than the older style pass/fail reliability demonstrations and tests. This is an area where more research is needed.

The underlying assumption to the approach was that "industry knows best". What has become clear is that this is not necessarily the case. There were a large number of skilled operatives who were very adept at applying the various standard techniques, but very few had had experience of deciding which techniques should be applied. They also lacked the experience of arguing the case for funding to undertake a task during a specific stage of a project. As a senior manager of a R&M team said, "When the contract specification required a R&M programme with specific tasks, I was asked how much will that cost? Now I have to argue the cost of not doing it – this is much more complex and we just do not have the information."

Related to this was the change from undertaking an activity, such as producing a fault tree to give a number which indicated that the system would be reliable, to one where the fault tree is produced and is wrapped in an argument of the form:

– The fault tree represents the structure of the system to an adequate resolution.
– The R&M data used is representative of the components that it is intended to use and of those components in the expected usage environment and under the expected usage profile.
– The fault tree will be valid for the life of the product (or if not when and why it will change).
– How the result of this fault tree impacts on the overall confidence that the system will achieve appropriate levels of R&M, and any changes required to the programme to maintain confidence that the characteristics will be achieved.

Although MOD had consulted widely with various relevant trade associations it has become obvious that this consultation process had not filtered down to all areas of the industry. This has resulted in the need to repeatedly explain the concept to the people involved with delivering R&M, in order that they understood the concept.

The major difference between Reliability and Safety Cases is the legal aspect. The law requires Safety Cases, whereas there is no legal requirement for Reliability Cases. This has meant that when introducing the new approach we have had to use persuasion rather than regulation.

4   LESSONS LEARNT

Having formally introduced this new approach 3 years ago it is possible to identify a number of lessons learnt which will improve the implementation of similar schemes in the future.

Having discussed the approach with the trade associations, it was considered that industry understood the concept and was in agreement with it. Therefore it would be possible to issue the new standard and it would be accepted and acted on, without any further input from the MoD. What became clear was that there was a need for a concerted programme of publicity to ensure all involved in the delivery of equipment and systems with good R&M characteristics were aware of and understood this new approach.

An issue that has arisen is that with down sizing of companies, many either subcontract their R&M activities to specialist consultancy companies or subcontract employees. This means that the traditional flow of information on new developments through companies is less efficient than it was.

After 3 years the R&M activities being undertaken are starting to improve, with much more thought being given to the effectiveness of the various elements of R&M programmes and how to evaluate the usefulness of tasks.

As discussed previously, there is a need for a more formal approach to combining different sources of information.

Moving from the traditional prescriptive approach required a step change in the R&M skill levels of all involved. This was not fully appreciated and prior to introducing the scheme an assessment should have been made of industry's ability to deliver the new approach & our ability to understand what was being delivered.

5   THE FUTURE

Having had 3 years experience of the R&M case, the guidance information is currently being modified to incorporate the lessons learnt and to clarify those areas which have been identified as being not as clear as originally believed. This work is due to complete spring 2003, with a period of public consultation prior to the issue of the new guidance. The new guidance will not change any of the underlying principles but it will change the emphasis slightly, putting more emphasis on:

1 The whole life nature of the R&M case.
2 The importance of setting the requirements.
3 The initial case.
4 The acceptance framework and hence the need for a clearer link from Risk to task to completion of task.

This approach is not limited to the UK defence sector, the concept being used in other areas in the UK and internationally. As the approach becomes the norm, many of the teething troubles should be overcome.

As an engineer this approach feels correct. It means that the skill levels of those undertaking R&M tasks have to increase. It also means that R&M can no longer be left to the specialist as the task is no longer to undertake an analysis and generate some numbers. The task is now to understand the R&M drivers of a product and therefore understand how a product will perform in-service.

6   CONCLUSIONS

This paper has briefly explained the MoD approach to R&M management and some of the experiences of the introduction of this new approach. The new approach is working and resulting in more consideration of programmes of R&M activities which give higher confidence that the systems being delivered will meet the required levels of Reliability and Maintainability. Similar approaches are being introduced in other countries and in other industrial sectors; the experience described in this paper is applicable to these areas, and appropriate consideration would ease the introduction in those areas.

Although this paper may appear negative, we are certainly in a much better position than we were five years ago, and things are getting even better – we now know how much we do not know.

The views expressed are those of the author and do not necessarily represent those of the Ministry of Defence.

© British Crown Copyright 2003/MoD – Published with the permission of the Controller of Her Britannic Majesty's Stationary Office.

REFERENCES

MoD 1997 Ministry of Defence, Defence Standard 00-42 (Part 2)/Issue 11 September 1997 – RELIABILITY AND

MAINTAINABILITY ASSURANCE GUIDES, PART 2: SOFTWARE

MoD 1999a Ministry of Defence, Defence Standard 00-40 (PART 1)/Issue 4 22 October 1999 – RELIABILITY AND MAINTAINABILITY (R&M) PART 1: MANAGEMENT RESPONSIBILITIES AND REQUIREMENTS FOR PROGRAMMES AND PLANS

MoD 1999b Ministry of Defence, Defence Standard 00-42 (PART 3)/Issue 1 22 October 1999 – RELIABILITY AND MAINTAINABILITY (R&M) ASSURANCE GUIDANCE, PART 3: R&M CASE

Safety and Reliability – Bedford & van Gelder (eds)
© *2003 Swets & Zeitlinger, Lisse, ISBN 90 5809 551 7*

Application of various techniques to determine exceedence probabilities of water levels of the IJssel Lake

F.L.M. Diermanse, G.F. Prinsen & H.F.P. van den Boogaard
WL\Delft Hydraulics, Delft, Netherlands

F. den Heijer
RIKZ, Den Haag, Netherlands

C.P.M. Geerse
RIZA, Lelystad, Netherlands

ABSTRACT: Three different methods are applied to derive exceedance probabilities of water levels of the IJssel Lake (the Netherlands). In the first method an extreme value distribution function is identified from observed water levels. In the other two methods a numerical model is used to simulate a series of synthetic events. Subsequently, computed maximum water levels and occurrence probabilities of the synthetic events are combined in a numerical integration procedure to derive exceedance probabilities of water levels in the IJssel Lake. The main difference between the second and third method lies in the schematisation of the physical processes involved. The second method uses random variables with a clear physical meaning, whereas the third method uses a principal component representation. Comparison of the three methods shows that resulting exceedance probabilities of water levels vary strongly.

1 INTRODUCTION

Without adequate flood protection, almost two-thirds of the Netherlands would be regularly flooded. Therefore, an extensive system of dikes and dunes has been constructed in order to prevent floods of the sea, rivers and lakes. The low lying part of the Netherlands is divided into 53 so-called dike-ring areas. Each dike ring area has its desired safety level, that is stated by law (Wet op de waterkeringen, 2002). This safety level, typically a small value in the order of 10^{-3} to 10^{-4} per year, is based on both the economic value of the protected area and the extent of the threat. The same law states that every five years an evaluation is executed to verify whether the flood defence structures still offer the desired safety level. As a consequence, exceedence probabilities of extreme water levels and wave heights in the major water systems in the Netherlands are regularly updated.

This paper deals with the derivation of exceedence probabilities of extreme water levels in the IJssel Lake, one of the primary water systems in the Netherlands. Currently, exceedence probabilities of water levels in the lake are derived from a fit of an extreme value

distribution function through a set of observed peak water levels. The main disadvantage of this type of method is that one looks for a regularity in measured water levels without taking into account the fact that these water levels are caused by highly non-linear and inhomogeneous processes. This observation has lead to the suggestion (e.g. Klemeš 1994, Kuchment et al. 1993) that progress in flood frequency analysis can only be expected from methods which explicitly take into account the involved physical processes.

In this study, two of such "physically based" methods are developed and applied to derive exceedance probabilities of extreme water levels of the IJssel Lake. The results of these two methods are compared with results of a purely statistical method.

2 SYSTEM AND MODEL DESCRIPTION

2.1 *The IJssel Lake*

With a total surface of $1,182\,\text{km}^2$, the IJssel Lake is the largest lake in the Netherlands. In the South East part of the lake the IJssel river, a distributary of the river Rhine, drains on average approximately 400 m³/s to the lake. Furthermore, a number of smaller rivers and canals (here referred to as regional contributaries) contribute an average accumulated discharge of about 165 m³/s to the IJssel Lake.

In the North, the lake is separated from the sea by a dike (the "Afsluitdijk"). Before the realisation of this dike in 1932, the area which now forms the IJssel Lake was still part of the sea. The Afsluitdijk is equipped with a number of sluices, enabling the drainage of surplus river water into the sea. The volume of water that discharges through the sluices depends on the sluice control, the water level in the lake and the sea level. The policy behind the sluice control is to keep the average water level at −0.40 m. +NAP in winter and at −0.20 m. +NAP in summer (where "m. +NAP" is the Dutch national reference level, which is approximately at mean sea water level).

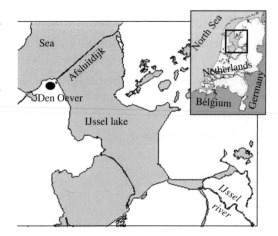

Figure 1. Location of the IJssel Lake and the IJssel river in the Netherlands.

2.2 High water events

High water levels in the IJssel Lake are the result of a period of 1 to 3 weeks during which the accumulated discharge of the IJssel river and the regional contributaries exceeds the outflow through the sluices of the Afsluitdijk into the sea. Peak water levels are generally observed a few days after a peak discharge on the IJssel river has occurred. High water events in the IJssel Lake occur almost exclusively in the winter half year since that is the season for high water events on both the IJssel river and the regional contributaries.

Besides the IJssel river and regional contributaries, wind (direction as well as velocity) and sea water level also significantly influence (maximum) lake water levels since they largely control the drainage of surplus water from the lake into the sea.

The maximum observed water level in the lake (since the start of measurements in 1932) is 0.50 m. +NAP and it occurred in October 1998. Extreme water levels like that in 1998 can only lead to floods if they occur in combination with a storm that causes wind set up in the Lake. However, the latter phenomenon will not be subject of this paper, i.e. only the dynamics of the area-average lake water level are considered. For an uncertainty analysis of water levels in the IJssel Lake which does include effects of wind set up, the reader is referred to Vrijling et al. (1999).

2.3 The WINBOS model

Two of the three methods presented in chapter 3 require a physically based numerical model to simulate water level dynamics in the lake during (hypothetical) events. For this purpose the model "WINBOS" is used. The model input consists of time series of processes that are relevant for the lake water dynamics, such as discharges of the IJssel river and the regional contributaries, wind direction, wind velocity and sea level. The model computes Lake-average water levels on the half-hourly time-scale. The model uses simple mass balance equations and user defined rules for sluice control. The WINBOS model has been validated and applied successfully in previous studies of the IJssel Lake dynamics (WL 1997, RIZA 1999).

3 METHODS

3.1 Introduction

This chapter describes the three methods applied to derive statistics of extreme water levels in the IJssel Lake. The first method consist of fitting an extreme value distribution function through the series of observed annual maxima and subsequently extrapolating this function to derive extreme value statistics.

In contrast to the first method, the other two methods explicitly take into account the physical processes that are responsible for the lake's water level dynamics. For a large number of synthetic inputs the WINBOS model is used to generate the corresponding water levels of the IJssel Lake. Combining the statistical description of the input processes and the associated water levels as computed with the model, the statistics of extreme water levels of the IJssel Lake are determined through numerical integration.

The main difference between the two numerical integration methods lies in the schematisation of the input processes. In the second method the schematisation is rather straightforward using a number of random variables with a "direct" and clear physical meaning (e.g. peak river discharge, sea water level, wind direction). In the third method the schematisation is much more complex and indirect. First, principal components of the multi-dimensional space of measured input variables are derived. Subsequently, linear combinations of the principal components are used to represent input variables of the synthetic events.

3.2 Method 1: Extreme value distribution functions

The first method is a purely statistically based fit-procedure of observed water levels. This type of procedure is still the most popular in extreme value analysis, mainly because it is relatively easy to apply.

The first step in the procedure consists of the selection of a subset of the available series of measurements. Since statistics of high water levels are to be derived, this subset should consist of relatively high water levels as well. Generally, either the series of annual maxima is selected or, alternatively, the series of peak water levels that exceed a user defined threshold (the so called peaks over threshold, or POT series).

Secondly, a prior estimate of the exceedance probability is derived for each observation of the selected subset. This estimate is based on the length of the period of observation and the number of times the concerning peak water level has been exceeded during this period. If the series of annual maxima has been selected, the prior estimate generally is derived as follows:

$$P = \frac{r - a}{N + b} \qquad (1)$$

where P is the prior estimate of the yearly exceedance probability, r is the rank number of the peak water level (1 = highest observed peak water level, 2 = second highest observed peak water level etc.), N is the total number of years of observation and a and b are user defined constants. If, for instance, Gringortens method is followed, a equals 0.44 and b equals 0.12 (Shaw 1988).

The third step consists of selecting a probability distribution function, preferably from the group of extreme value distribution functions (e.g. Pareto, Gumbel, Weibull).

The fourth and final step is to determine the parameter values of the selected distribution function that provide a "good fit" for the selected subset of observations. A number of procedures are available to derive the fit, of which the method of moments and the maximum likelihood method are most commonly applied.

The derived probability distribution function provides a relation between exceedance probabilities (or recurrence intervals) on one hand and related water levels on the other hand. This means the water level with a recurrence interval of, for instance, 1000 years can be derived from the distribution function, even if the available series of observations covers less than 100 years.

In the method described above the user has to select:

- a subset of observations;
- a method to obtain the prior estimate;
- a probability distribution function; and
- a fit procedure.

The prior estimate is used to visualise the goodness of fit of the selected distribution function. However, the prior estimates have no influence on the resulting parameters of the probability distribution function. In contrast, the other three choices mentioned above can have a strong effect on the final outcome, which is demonstrated in chapter 4.

3.3 Method 2: numerical integration

3.3.1. Introduction

In method 1 it is assumed that the derived probability distribution is valid for all water levels, even those that are much higher than the maximum observed water level. In doing so, the function is extrapolated outside the range for which it is proven to be valid. One of the consequences is that this method does not take into account the fact that the nature of the physical processes during extreme (unobserved) events may change in comparison with observed events. For instance the dominant role of floodplains during extreme events may not be reflected in the observed events.

Therefore, a second method is introduced that explicitly takes into account the physical processes responsible for water level dynamics in the IJssel Lake. In this method, hypothetical events are simulated with the WINBOS model (see section 2.3). The length of simulated periods is 30 days, which is sufficient to simulate the rise and fall of the water level in the lake. Such a hypothetical event is characterised by a quantitative description of the relevant physical processes. The physical features considered are:

- the discharge of the IJssel river;
- the accumulated discharge of the regional contributaries;
- the sea water level at Den Oever;
- the wind velocity; and
- the wind direction.

The entities described above are treated as stochastic variables. The following five steps are executed in order to derive exceedance probabilities of the water level in the IJssel Lake:

1. A relatively straightforward representation and quantification is defined for the stochastic variables.
2. For each stochastic variable the range of possible realisations is divided into a number of classes.

497

3. For each stochastic variable the available series of observations is used to derive the probability of occurrence of the classes defined.
4. For all possible combinations of realisations of the stochastic variables (which is limited as a result of the defined classes) the resulting water level in the lake is computed with the WINBOS model.
5. The probabilities as derived in step 3 and the water levels as computed in step 4 are combined to derive the statistics of extreme water levels of the IJssel Lake through numerical integration.

Some of the steps mentioned above are described in more detail below.

3.3.2   *Representation of stochastic variables*

The stochastic variables represent processes of a dynamic nature, i.e. with a strong temporal variability. (and in some cases some spatial variability as well). For practical purposes a relatively simple representation of these is required. It is vital, though, that this representation still contains the features that are relevant in the generation of high water levels in the IJssel Lake.

The discharge of the IJssel river is characterised by the peak discharge that occurs in the simulated event. The temporal variation of the discharge is represented by a standardised hydrograph, the shape of which is based on statistics of measured hydrographs. The assumed hydrograph is such, that the peak discharge occurs on day 15 of the simulated event (the simulated period is 30 days).

The sea water level during a single tide is modelled as a sine function with an amplitude of 0.75 m. The mean of the sine function is assumed to be constant for a period of 5 days. Since a period of 30 days is simulated this means a total of 6 stochastic variables is required to represent the temporal variation of the sea water level. The choice of a period of 5 days is a trade-off between an accurate representation of temporal variability on the one hand and the required number of model simulations on the other hand.

In contrast to the discharge of the IJssel river, the combined discharge of the regional contributaries is assumed to be constant over the simulated period of 30 days. The same assumption is done for the wind direction. Finally, the wind velocity is taken as a function of both wind direction and sea water level. This means the wind velocity is assumed to be constant over periods of 5 days, similar to the sea water level.

3.3.3   *Probability distributions of the stochastic variables*

The required statistical features of the IJssel river discharge, the sea level at Den Oever and wind were taken from previous studies (HKV 2001, RIKZ 2000). These statistics have been derived by fitting extreme value distribution functions through available data in

a similar manner as described in section 3.2. The available statistics of the sea level were converted from the tidal scale to the 5-daily scale, since the latter time-scale is used in the current analysis.

Statistics of the regional contributors were not readily available and consequently needed to be derived during the current analysis. Again, a data fit by an extreme value distribution was derived.

3.3.4   *Definition of classes*

In order to obtain a good coverage of the range of possible realisations, 6 classes were defined for the IJssel river discharge, 8 classes for the regional contributaries, 7 classes for the sea water level and 2 classes for the wind direction. This means the total number of realisations (and, consequently, required WINBOS runs) is equal to $8 \times 6 \times 7^6 \times 2 \approx 11,300,000$. In order to reduce the number of computations, the stochastic variables representing the sea water level of the first 5 days and the last 5 days of the simulated period of 30 days were replaced by deterministic values. This is allowed because the sea water level at the beginning and at the end of the simulated event has no influence on the maximum water level in the lake. Maximum water levels in the Lake generally occur around day 15 when the discharge of the IJssel river is at its peak. The number of required computations is now reduced to $8 \times 6 \times 7^4 \times 2 \approx 230,000$.

3.3.5   *Numerical integration procedure*

For a user-defined water level of the IJssel Lake the simulated events during which this water level is exceeded are identified. The probability of exceedance of the water level is equal to the accumulated probability of occurrence of these events. The available numerical integration tool assumes there is no correlation between the stochastic variables, the effect of which is discussed later on. The procedure is repeated for a range of water levels to obtain a probability distribution function.

A disadvantage of this procedure is the fact that the amount of model simulation runs required easily gets out of hand, as demonstrated in section 3.3.4. Depending on the computation time of a single model run one may be forced to reduce the amount of stochastic variables and output classes as much as possible. However, this may cause unacceptable loss of accuracy.

3.4   *Method 3: numerical integration based on principal component analysis (PCA)*

3.4.1   *Introduction*

The quantitative description of the physical processes responsible for the high lake levels plays a significant role in the method described in the previous section. Generally, a relatively simplified representation was used in order to limit the required model simulations.

The method presented in the current section offers an alternative that allows a much more dynamic representation of the processes involved. Furthermore it has the advantage that the (statistical) dependency between the processes is automatically taken into account.

The method is based on a mathematical technique called "principal component analysis" (PCA). PCA is a multivariate analysis or ordination technique that can be applied to detect common features in multiple sets of measurements. The available sets are represented by vectors of equal length, M, and as such form a multidimensional swarm of data points. The main idea of PCA is to find the directions in the M-dimensional space with the largest spread or scatter, i.e. the directions that explain the largest part of the total variance of the data set. For more detailed descriptions of PCA the reader is referred to Haan (1977) and Pielou (1984).

The PCA-analysis is used here to detect common features in observed patterns of the involved processes during high water events. This information is subsequently used to generate an ensemble of synthetic input series for the WINBOS model. As in method 2, the resulting water levels computed by the WINBOS model are used to derive exceedance probabilities of water levels through numerical integration. The method is further outlined below.

3.4.2 PCA analysis of pre-event histories
The first step in the analysis is to select a number (K) of high water events for which measurements of all relevant processes/stochastic variables (i.e. IJssel river discharge, sea water level, etc.) are available. For each event the measured values are "stored" in a single vector v:

$$v = \left(x_{-T}^{(1)},..,x_0^{(1)}; x_{-T}^{(2)},...,x_0^{(2)};....; x_{-T}^{(N)},...,x_0^{(N)} \right) \quad (2)$$

where $x_{-t}^{(j)}$ refers to the observed value of the jth stochastic variable at t time steps before the maximum water level in the lake occurs, N is the number of stochastic variables and T is the number of time steps in the observed period.

Since K events are selected we have K vectors in a $N \times (T + 1)$-dimensional space that offer a rather extensive representation of the pre-event histories. As these vectors are expected to be correlated to some extent, they are likely to contain redundant information. The main purpose of the PCA analysis is to reduce the dimension of these vectors in such a way that no vital information is lost. In doing so, the set of vectors is written as a linear combination of "base-vectors" φ and coefficients (or principal components) α:

$$v^{(k)} = \sum_{d=1}^{D} \alpha_d^{(k)} \varphi_d \ ; k = 1..K \quad (3)$$

where φ_d is the dth base vector and $\alpha_d^{(k)}$ is the dth coefficient of the kth vector $v^{(k)}$.

The base-vectors φ_d are selected as eigenvectors of the covariance-matrix of the K vectors $v^{(k)}$. If the full set of K eigenvectors is selected, each vector $v^{(k)}$ can be exactly reproduced by the linear combination in Equation (3) above by selecting the proper α-values. Furthermore, other combinations of α-values will result in "new" v-vectors which can be considered as mathematical representations of synthetic events. The α-values will therefore be treated as stochastic variables in order to generate a set of synthetic events. These synthetic events are subsequently applied in a numerical integration procedure, similar to the one presented in section 3.3, in order to derive exceedance probabilities of water levels in the IJssel Lake.

However, if K eigenvectors (equal to the number of selected observed events) are used, the number of stochastic variables is generally too large to handle in the numerical integration procedure. In order to reduce the number of stochastic variables, a subset of the eigenvectors is selected. First, the eigenvectors are ranked, based on the magnitude of the corresponding eigenvalues λ_k. The magnitude of λ_k is a measure of the fraction of the total variance of the set of vectors v which can be explained by eigenvector φ_k. Subsequently, the first D eigenvectors are selected, which means a fraction f of the total variance of the set of vectors v can be explained by equation (3), where f equals:

$$f = \sum_{d=1}^{D} \lambda_d / \sum_{k=1}^{K} \lambda_k \quad (4)$$

The value of D is chosen such, that f is satisfactorily close to 1.

3.4.3 Further reduction of the number of stochastic variables
Application of the PCA-analysis on the case study of the IJssel Lake showed that 20 eigenvectors are required to explain 90% of the total variance of the observed pre-event histories and 27 eigenvectors to explain 95%. Since the use of 20 or 27 stochastic variables is too much to handle in the numerical integration procedure, a modification of the described approach was developed.

Each pre-event history v is written as

$$v = \rho(v) \cdot v^* \quad (5)$$

where $\rho(v)$ represents the "norm" or "magnitude" of vector v and v^* is a scaled or normalised version of v, with $\rho(v^*) = 1$. Several alternatives have been considered for the definition of the norm $\rho(v)$, such as the L_1, the L_2 and the L_∞ norm, where L_n is defined as

$$\sqrt[n]{\sum_{d=1}^{D} |v_d|^n} \qquad (6)$$

For each of the considered alternatives the norms of observed pre-event histories were compared to the corresponding water level peaks of the lake. As a result, the L_1 norm was selected since it is a measure of the total volume of water which has drained into the IJssel Lake. This volume is a significant control on the (maximum) water level.

Subsequently, the L_1-norm of the input processes (IJssel river discharge, sea level, etc.) are treated as stochastic variables. In order to do so, a Weibull fit was derived for observed L_1 values of the input processes of 46 high water events. Then, a statistical description of the temporal variation of the input processes was derived from a PCA-analysis to the set of normalised pre-event histories v^*. The resulting principal components, or α-values, are also treated as stochastic variables, in this case with a Gaussian distribution. In this way the pre-event histories are described by a set of random variables of which one part governs the magnitude and the other part the temporal variation.

4   RESULTS

4.1   Introduction

The three methods described are applied on the series of observations of the period 1951–1996. The year 1951 was selected as start date because the available measurements of the relevant processes in previous years are insufficient. The resulting exceedance probabilities and corresponding levels in the IJssel Lake are presented and differences between the methods are explained.

4.2   Method 1: extreme value distribution functions

Two subsets of the available data of the period 1951–1996 were selected: the 46 annual maxima and the set of annual maxima (16 in total) which exceed a threshold of 0.12 m. +NAP. Gringortens method was used to obtain prior estimates of exceedance probabilities of observed water levels, which means $a = 0.44$ and $b = 0.12$ in Equation (1). The maximum likelihood method was used to obtain an optimal fit of the data for the Gumbel-distribution function.

Figure 2 shows the resulting fits for the two selected subsets. This figure clearly shows that the selection of the subset on which the data is fitted has a strong effect on the resulting exceedance probabilities.

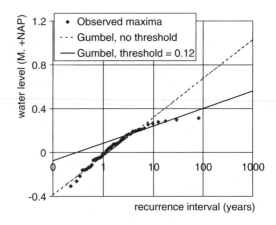

Figure 2.   Gumbel fit of (1) 46 annual maxima and (2) the set of annual maxima above a threshold of 0.12 m. +NAP.

For a recurrence interval of 4,000 years (the desired safety level for dikes around the IJssel lake) the difference in derived water levels is 0.6 m. The relatively large differences are explained by the change in the pattern of the data that occurs somewhere around the water level of 0.10–0.20 m. +NAP. If the full set of annual maxima is used the resulting Gumbel function largely follows the relatively steep pattern of the lower observations, whereas for the subset of annual maxima the relatively flat pattern of the highest observed values is reflected. Consequently, the introduction of the threshold leads to a poor fit of water levels with low recurrence intervals whereas without this threshold the fit for high recurrence intervals is poor. If other extreme value distribution functions are used, the same effect occurs.

An interesting question is whether the observed change in pattern has a physical cause and, subsequently, whether the pattern of the highest observations is characteristic for the extreme recurrence intervals of interest (i.e. >1,000 years). If this is the case, the Gumbel function resulting from the set of highest observed water levels is to be preferred. However, the change in pattern may also be a pure coincidence. In that case the use of the larger set of data is to be preferred, since it holds more information. The results of the more physically based methods (described below) may provide some answers.

4.3   Method 2: numerical integration

4.3.1   WINBOS results

As explained in section 3.3 a relatively simplified and straightforward representation of stochastic variables is used in order keep the number of model runs manageable. Realisations of the stochastic variables are used as input of the WINBOS model to compute

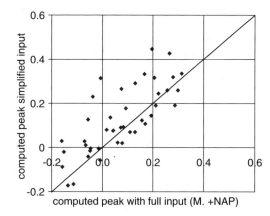

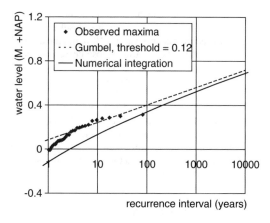

Figure 3. Derived model results, using [a] the full input information (x-axis) and [b] the simplified input representation (y-axis).

Figure 4. Results of the numerical integration (method 2) and the Gumbel fit of the set of annual maxima above a threshold of 0.12 m. +NAP (method 1).

(peak) water levels of synthetic events. Due to the simplified representations of stochastic variables an error is introduced in the computed (peak) water levels. In order to quantify the error, model simulations of measured high water events have been executed with [a] measured time series of the relevant input variables and [b] the simplified representation of these input variables.

Figure 3 shows computed peak water levels for 46 events which have been simulated with both input series. The simplification of input variables causes on average an overestimation of peak water levels of 5 cm. The error appears to be independent to the magnitude of the event. Therefore, in the numerical integration procedure a correction of 5 cm. is applied to the computed water levels, to assure the model is not biased.

4.3.2 Exceedance probabilities

Figure 4 shows the results of the numerical integration (method 2) in combination with the results of Gumbel-fit of the set of annual maxima (16 in total) above a threshold of 0.12 m. +NAP (method 1). For low recurrence intervals (<50 years) differences between the derived water levels of the two methods are substantial. This also means the reproduction of observed water levels by the numerical integration method is rather poor, as can be seen from Figure 4.

For larger recurrence intervals (>100 years) differences between the two methods diminish. For a recurrence interval of 4,000 years the difference in derived water levels is about 3 cm. However this close correspondence should not be considered as a sound validation for both methods. Differences with observed water levels for low recurrence intervals (even for method 1, which is a fit procedure) provide a warning in this respect.

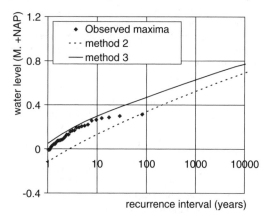

Figure 5. Resulting water levels and recurrence intervals of method 2 and method 3.

4.4 Method 3: numerical integration based on principal component analysis (PCA)

Figure 5 shows recurrence intervals and related water levels as derived with methods 2 and 3. The water levels derived for a given recurrence interval with method 3 are clearly higher than those derived with method 2. Differences are mainly due to the fact that method 3 accounts for the (statistical) dependency between the IJssel river discharge and the regional contributaries. As a consequence, the combined occurrence of extreme discharges on the IJssel river and the regional contributaries increases. Additional numerical integration runs show that for a recurrence interval of 100 years the total hydraulic load, i.e. the combined volume of the IJssel river and regional contributors increases with 230 million m^3.

This volume is the equivalent of an additional water level of approximately 20 cm. on the IJssel Lake.

For low recurrence intervals (<10 years) the derived water levels of method 3 are in close correspondence with observed water levels. For larger recurrence intervals this method appears to overestimate the water levels. However, in October 1998 a peak water level of 0.50 m. +NAP was observed (i.e. outside the period of observations as used in this studies). The recurrence interval of this event as derived with method 3 is 150 years which seems more likely than the recurrence intervals as derived with method 1 (500 years) and method 2 (750 years).

5 CONCLUSIONS AND DISCUSSION

Three different methods have been applied to derive exceedance probabilities of water levels of the IJssel Lake. Significant differences were found in the exceedance probabilities derived, using the three methods. Furthermore, the choices that have to be made within a single method (i.e. magnitudes of thresholds, number of stochastic variables) in some cases were also found to be of strong influence on the resulting exceedance probabilities.

The range of outcomes to some extent reflects the magnitude of the existing epistemic uncertainty (i.e. uncertainty resulting from a lack of knowledge of the system under extreme circumstances). Since both underestimation and overestimation of flood dangers can prove to be very costly, the range of possible outcomes preferably should be narrowed down as much as possible. In other words, some kind of validation procedure should be developed that enables the user to reject certain methods and/or ranges of outcomes.

A major problem in this respect is the length of the available series of observations, which is (much) too short to provide a sound validation of estimates of water levels with a recurrence interval of, for instance, 10^{-3} or 10^{-4} years. A verdict on the quality of predictions in any kind of (statistical) extreme value analysis is therefore partly based on good faith. A vital step in that respect is that the derived results should at least be physically realistic. Therefore, the second and third method presented here are believed to be the most promising, since they explicitly take into account the relevant physical features of the system (through application of the WINBOS model).

However, these two methods as currently applied still leave room for improvements, mainly due to the measures that were taken in order to reduce the required amount of model runs. For instance the number of classes in the second method and the number of principal components in the third method had to be reduced. Therefore, the first attempt to further improve the methods will consist of the application of alternative

methods for the numerical integration procedure. An approach based on the principle of Monte Carlo simulations is considered, since it is expected to reduce the required amount of model simulation runs significantly. The amount of computation time that may be saved in this way can then be invested in a more refined representation of the processes involved.

ACKNOWLEDGEMENTS

This research was funded by the Dutch Ministry of Transport, Public Works and Water management, for which the authors are grateful. Textual comments by Micha Werner are appreciated.

REFERENCES

Klemeš, V. 1994. Statistics and probability: Wrong remedies for a confused hydrologic modeler. In: V. Barnett and K.F. Turkman, *Statistics for the environment 2: water related issues*: 345–370. John Wiley & Sons Ltd.: Chisester, England.

Kuchment, L.S., Demidov, V.N., Motovilov, Yu.G., Mazarov, N.A. and Smakhtin, V.Yu. 1993. Estimation of disastrous floods risk via physically based models of river runoff generation, In: Z.W Kundzewics, D. Rosbjerg and S.P. Simonovic (eds), *Extreme hydrological events: Precipitation, floods and droughts* (Proceedings of the Yokohama symposium, July 1993): 177–182. IAHS Publ. 213.

Haan, C.T. 1997. *Statistical methods in hydrology*. The IOWA State University Press, Ames IOWA.

HKV, 2001. *Uitbreiding afvoerstatistiek Borgharen, Lith, Lobith en Olst*. HKV report, December 2001 (in Dutch).

Pielou, E.C. 1984. *The interpretation of Ecological Data. A primer on Classification and Ordination*. Wiley & Sons, New York.

RIZA, 1999. *Meerpeilstatistiek*. RIZA report March 1999 (in Dutch).

RIKZ, 2000. *Richtingsafhankelijke extreme waarden voor HW-standen, golfhoogten en golfperioden*, RIKZ report December 2000 (in Dutch).

Shaw, E.M. 1988. *Hydrology in practice, second edition*. Chapman and Hall.

Vrijling, J.K., van Gelder, P.H.A.J.M., van Asperen, L., van de Paverd, M., Westphal, R., Berger, R. and Voortman, H.G. 1999. Uncertainty analysis of water levels on Lake IJssel in the Netherlands. *Proceedings ICASP8*1999*. Sydney, Australia, 12–15 December 1999.

Wet op de waterkeringen, 2002. Wet, houdende algemene regels ter verzekering van de beveiliging door waterkeringen tegen overstromingen door het buitenwater en regeling van enkele daarmee verband houdende aangelegenheden. *Staatsblad 304* (in Dutch).

WL|Delft Hydraulics 1997. *Onafhankelijk onderzoek Markermeer*. Report, WL|Delft Hydraulics December 1997 (in Dutch).

Safety and Reliability – Bedford & van Gelder (eds)
© 2003 Swets & Zeitlinger, Lisse, ISBN 90 5809 551 7

Improving human factors analysis in maritime accident investigation: proposal for a new classification framework

A. Di Giulio, P. Trucco
Politecnico di Milano, Milan, Italy

M. Pedrali
D'Appolonia, Genoa, Italy

ABSTRACT: Despite the maritime transport is nowadays considered a strategic opportunity within the overall European transport system, a real throwing back of the sector requires a tragic improvement of safety standards and drastic reduction of accident at sea. In this context systematic accident analysis is a fundamental activity in order to build representative database which feeds quantitative models for risk analysis. Although the major role of human and organisational factors in maritime accident is widely recognised, a standardised taxonomy that allow their univocal and systematic classification does not exist yet. This paper summarizes a critical review of three taxonomies – IMO, MAIB and CASMET – of causal factors involved in accidents at sea. The review pointed out some requirements that have been used to develop a new classification framework with an event-based view of the accident. Finally, the new classification has been tested through its application in the analysis of an High Speed Craft (HSC) accident.

1 INTRODUCTION

The maritime transport of goods and people assumed a strategic role within the international logistic net. In Europe the maritime transport is responsible for more than 70% of the extra-community commerce and about 30% of the intra-community commerce. Starting from the '70 the maritime transport of goods in Europe is increased by 70% and the transport of passengers is increased by 110%. Nowadays, more than one billion tons of goods and about twenty millions of passengers travel every year through the European ports. It is estimated that the annual growth should continue to be about 2% (European Commission 2001).

However, a real throwing back of the sector requires a significant improvement of safety standards and drastic reduction of accident at sea. It is estimated that annually a hundred and forty people die for maritime accidents in European waters (1,4 deaths for 100 covered kilometres, more than twenty-five times the incidence of the deaths in the civil aviation, equal to 0,05 deaths for 100 covered kilometres; European Transport Safety Council 2001). The total cost associated to annual accidents (indemnities for the victims, damages to goods and ships, environmental damages) amounts to about 1.500 million Euro (European Transport Safety Council 2001, EUROSTAT 1998).

Despite the rapid technological development of ships and navigation support systems, the number of maritime accidents is not decreased according to the expectations.

Indeed, it is widely recognised that the human element plays the major role in most accidents involving modern ships. Thus, the Lord Carver report of the UK House of Lords summed it up succinctly when stating that it *"is the received wisdom that four out of five ship casualties … are due to human error …"*. Also national statistics (TSB 1998a) attribute the 74% of the accidents at sea to human errors and only the 1% to technical failures (Fig. 1). As shown in Figure 2, the 45% of the analysed cases admit as determinant cause the misjudgement (mistake) of the pilot or of the ship master; in the other 42% of cases human errors refer to: a) the incomprehension between the pilot and the master; b) to the inattention of the pilot and the officer of the watch (OOW); c) to the lack of communication between members of the crew.

Similar results are pointed out by a statistical analysis based on the data of the Lloyds Informative Maritime Service (Mathes et al. 1997) concerning

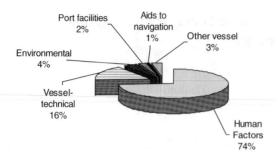

Figure 1. Incidence of the typologies of accident causes.

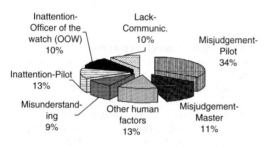

Figure 2. Typology of human factors causing an accident.

more than 15.000 accidents in a time span of ten years. Lloyds' statistics show that a wrong route and an excessive speed with respect to the traffic in the sea zone, are responsible of about 50% of all the maritime accidents and particularly of groundings. Moreover, 70–80% of the accidents are due to human mistakes or other events to be attributed to the human behaviour.

In this context systematic accident analysis is a fundamental activity in order to build representative database which feeds quantitative models for risk analysis. Those information and data can be collected through official accident investigation reports, or by the continuous recording of data during navigation. Although the role of human and organisational factors in maritime accident is widely recognized, a standardized taxonomy that allow their univocal and systematic classification does not exist yet.

The paper summarises a critical review of three taxonomies of causal factors involved in accidents at sea, focusing on causes that involve human and organisational factors: the International Maritime Organisation (IMO 2000), the Marine Accident Investigation Branch (MAIB 2001) and the Casualty Analyses Methodology for Maritime Operations (CASMET 1999). The review pointed out some requirements that have been used to develop and validate a new classification framework.

2 STATE OF THE ART OF HUMAN FACTORS ANALYSIS IN MARITIME ACCIDENTS

Taxonomies are amongst the most powerful tools for analysing the influence of human errors in accidents and investigating their root causes.

A lot of taxonomies have been developed across the transport industry with the main objective of classifying the causes of accident. As long as the maritime context is concerned, three classification systems can represent the state-of-the-art both at the international and European levels; these taxonomies are:

– the IMO taxonomy, which is the only system that approaches the status of being an international standard;
– the MAIB classification system, which has been developed and used extensively by a European maritime authority (namely the British);
– the CASMET classification system, which has been developed in a large-scale EC sponsored project but has not yet been put into operational use.

These taxonomies are supported both by an accident causation model and a method of analysis. As long as the analysis of human factors is concerned, Figure 3 and Figure 4 show the structure of these taxonomies.

IMO focuses on the accident as a whole and, after coding the descriptive information (e.g., time, location, type of vessel, etc.), it concentrates on the human element. In this respect, casual factors are described by three basic properties, namely: Human involvement (if a person was involved in the accident), Error mode (how this person made an erroneous action) and Underlying factors (why it happened); in addition, a large section of this taxonomy refers to casual factors related to fatigue. Error modes refer to Reason's distinction in Slips, Lapses, Mistakes and Violations (Reason 1990), whilst Underlying factors are subdivided according to the SHELL model (Edwards 1972).

The MAIB classification considers the casualty as a chronological list of pre-defined accident factors describing the sequence of failures leading to the accident; the focus is primarily the identification of "who" and "what" was involved in these failures, where and why these failures occurred. The "Underlying Accident Factors and Sub-factors" section addresses technical and human failures underlying the causal chain. As far as human factors are concerned (viz., why one person did an erroneous action), the following distinction between the categories (and sub-categories) is provided:

1 external bodies liaison;
2 company and organisation;
3 crew factors;
4 equipment;
5 working environment;
6 individual.

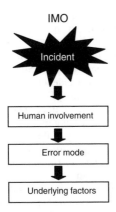

IMO

Figure 3. Structure of the IMO taxonomy.

CASMET looks at a casualty as an events chain. Two basic properties describe casualties, namely "Accidental Events" (what happened) and "Causal Factors" (why it happened). For causal factors, a distinction is made between those related to the operation of the vessel (Daily operations) and those taken by the land-based organisation (Management and resources).

The three classifications have already been compared quantitatively through an experimental study in order to identify possible shortcomings and propose a new one with satisfactory level of key attributes, such as reliability, accuracy, comprehensiveness and usability (Pedrali et al., 2002). Following the AHP methodology (Saaty 1980) the hierarchy of criteria has been set, as reported in Figure 5. Experts have been asked to express:

– judgements of the relative importance of criteria (level 1) in establishing the best taxonomy (level 0) by means of pairwise comparisons and using the basic scale of Saaty;
– judgements of the performance of each taxonomy (level 2) respect to criteria (level 1) using a six level evaluation scale.

Single expert judgements have been grouped using the geometric mean as grouping function.

As reported in Figure 6, the human factors section of MAIB turned out to have an average quality index higher than IMO and CASMET.

However, only a slight difference exists between the three taxonomies; variance between experts' judgement around the mean value are comparable, spanning from a maximum of 0.6 for IMO to a minimum of 0.5 for CASMET. Hence, the authors of the study concluded that the human factors section of these taxonomies are almost equivalent and can be indifferently selected for human factors analysis in maritime domain. These results have to be carefully handled because not

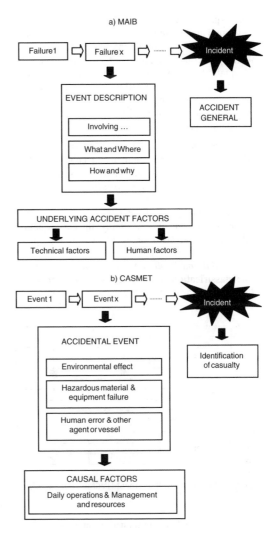

Figure 4. Structure of a) MAIB and b) CASMET taxonomies.

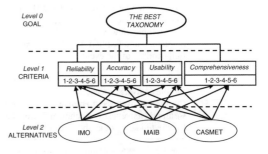

Figure 5. Hierarchical representation of the quantitative assessment (Pedrali et al., 2002).

505

all the experts participating at the study had the same background and, more important, not the whole structure of the three classifications has been tested, but only the part concerning human factors analysis.

A qualitative analysis of the three classifications has been carried out in parallel, according to a set of characteristics fundamental for exhaustive and efficient analysis of human involvement in casualty. These characteristics are reported hereafter.

– *Distinction between Technical and Human factors*: the classification offers human and technical factors sections for descriptive and analytical purposes.
– *Macroscopic view of the casualty*: the classification allows the analysis of an accident from the organisational point of view.
– *Microscopic view of the casualty*: the classification allows the analysis of an accident from the operator point of view.
– *Sequential and Explicit Accident causation model*: the classification is supported by a model that permits a reconstruction of the accident dynamics.

– *Fully coding of each event*: the classification provides a complete set of elements for coding the event.
– *Contextualisation of causal factors*: the classification allows to code the physical and organisational context where the accident occurred.
– *Identification of multiple causes per event*: the classification allows to trace back different factors and causal chain leading to the accident.
– *Simple analysis process*: the classification is supported by an straightforward analysis method.
– *Comprehensiveness of Human Factors categories*: degree to which the classification covers all the different kinds of human factors descriptions.
– *Comprehensiveness of Technical Factors categories*: degree to which the classification covers all the different kinds of technical factors descriptions.
– *Low redundancy of analysis results*: the classification does not use the same type of category in the coding phase.

3 PROPOSAL FOR A NEW CLASSIFICATION FRAMEWORK

A clear result of the study carried out by Pedrali et al. (2002) is the low importance that experts assign to comprehensiveness and usability of a classification system. Experts involved in the study stated that it is more important that the adopted taxonomy allows an accurate and repeatable (reliable) accident analysis instead of capturing a complete set of casualty factors.

Referring to the results of the qualitative assessment (Tab. 1), it is apparent that CASMET represent the most organised framework for accident analysis; on the other side, MAIB has the most complete and

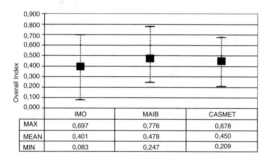

	IMO	MAIB	CASMET
MAX	0,697	0,776	0,678
MEAN	0,401	0,478	0,450
MIN	0,083	0,247	0,209

Figure 6. Quantitative assessment (Pedrali et al., 2002).

Table 1. Quantitative comparison.

	Classification systems		
Characteristics	IMO	MAIB	CASMET
Distinction between Technical and Human factors	*	***	***
Macroscopic view of the casualty	*	*	***
Microscopic view of the casualty	**	**	***
Sequential and Explicit Accident causation model	*	**	***
Fully coding of the event	*	**	***
Contextualisation of causal factors	/	/	/
Identification of multiple causes per event	/	/	*
Simple analysis process	*	**	***
Comprehensiveness of Human Factors categories	**	**	/
Comprehensiveness of Technical Factors categories	*	**	****
Low redundancy of analysis results	**	**	**

Legend: **** maximum correspondence.
* minimum correspondence.
/ lack of characteristics.

506

detailed classification of causes. Globally, both quantitative (Pedrali et al., 2002) and qualitative assessment point out that none of the existing taxonomies is clearly the best, but all of them have strengths and critical weaknesses.

In order to propose a new classification framework for maritime accident analysis, valuable outcomes of the quantitative and qualitative assessment have been turned into some recommendations termed as elements of synthesis and innovation.

As long as *synthesis* is concerned, a new maritime classification should have:

- apparent reference to an accident causation model inspired by CASMET's;
- the descriptive structure of each event should follow the CASMET's, but also integrated with categories derived from MAIB.

As long as *innovation* is concerned, the maritime classification under definition should:

- distinguish between human and technical factors at the event level rather than at the causal level;
- allow a microscopic approach for human event analysis;
- adopt also a macroscopic vision of the accident (following the Reason's approach);
- follow a cyclical and not hierarchical process during the analysis;
- use SHELL (Edwards 1972) model as a key to interpret/understand technical aspects of the accidents.

Figure 7 shows an overall picture of the new classification framework developed according to all the stated recommendations.

The classification framework is divided into two main sections. *Section one* is dedicated to the general description of the accident and of the context under which the accident occurred. The sequential event-based model of CASMET has been assumed to formally represent the accident. *Section two* deals with the analysis of the events composing the accident and the classification of their causes. Two different classification methods are set accordingly to the primary nature of a generic event, that could be technical (Fig. 7, section 2.1) or human (Fig. 7, section 2.2). Note that, unlike the approach of existing classification systems (IMO, MAIB and CASMET) the distinction between technical and human factors has been set at the event level and not at the cause level.

To characterise the human event a human factors list has been designed that assumes the CASMET's as a basis, but it has been completed adding some categories from the MAIB taxonomy. The basic structure of the list is reported in Table 2. The description of a technical event is carried out by a set of parameters as reported in Table 3.

Finally, *Tables of underlying factors* are provided which track the interactions between different causal factors. Those tables, directly derived from the CREAM classification (Hollnagel 1998), support the analyst in identifying root causes of a human error or a technical failure (indeed the causes of an event do not depend on his nature).

Table 2. Basic structure of the human factors list.

Attribute	Symbol	Number of Codes
Position	POS	27
Task affected	TSK	44
Performance mode	PERF	10
Error mode	ACWT (Action at the wrong time)	2
	ACWP (Action in wrong place)	1
	ACWY (Action of wrong type)	4
	ACWO (Action on wrong object)	1

Table 3. Categories for the description of technical events.

Category	Attribute
Hazardous material attributes (HAZ)	Material (MAT)
	Location (LOCQ)
	Hazard (HTYP)
	Failure type (TYPZ)
System/equipment failure parameters (FEQ)	System/equipment type
	Location (LOCQ)
	Failure type (TYPQ)
	Physical agent (PHY)

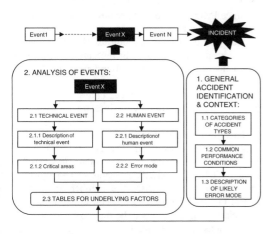

Figure 7. Framework of the proposed classification.

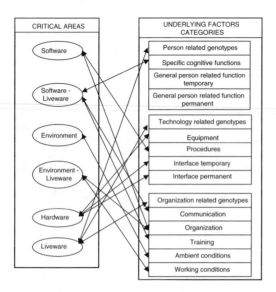

| CRITICAL AREAS | UNDERLYING FACTORS CATEGORIES |

Figure 8. Correlations between categories of critical areas and underlying factors for a technical event.

Since, it is likely that an event is produced by the concurrent contribution of a number of factors of different nature, the new classification lets the analyst to attribute several causes to a single event. Similarly, CASMET envisage two root causes of different nature for any event type and they are addressed by two different sections of the classification, namely "Daily operations" and "Management and resources". Nevertheless, unlike CASMET, the new classification framework supports the identification of several concurrent causes of different nature (human, organizational and technological) by means of an iterative classification process instead of a sequential one.

Two different types of correlations guide the analyst in identifying root causes: underlying factors vs. error mode (referring to human event) and underlying factors vs. critical areas (referring to technical event). The first type of correlations is established by the CREAM classification, whilst the second ones have been set establishing multiple links between the SHELL model, describing the critical areas of a technical event, and the categories of underlying factors (Fig. 8). The ADREP (ICAO 2000) taxonomy has been assumed as a reference to define the contents of the categories describing critical areas.

4   CASE STUDY: THE INCAT 045 AND LADY MEGAN II COLLISION

The collision between the fast ferry catamaran "INCAT 046" and the fishing boat "LADY MEGAN

II", taken place on September 4th, 1998 in the Yarmout bay in Nova Scotia (Canada), has been used as a case study for experimenting and validating the new proposed classification framework.

A ten years experienced investigator has been provided with the new classification and has been asked to classify human factors involved in ten major events previously selected referring to the official report of the collision (Tab. 4; TSB 1998b).

To validate the new classification, the results of the first analysis have been compared with the ones obtained by applying other existing taxonomies (i.e. IMO, MAIB and CASMET) to the same accident.

5   DISCUSSION

The first general consideration concerns the way the new classification method works: the preliminary evaluation of the accident context and the identification of a set of common performance conditions (CPC) allow to focalise a more detailed analysis. In such a way the new classification framework better supports the analyst in choosing the best fitting causal factors when considering single events.

The proposed classification have also proved to fully support the analyst in order to draw causal chains encompassing technical and human factors.

Indeed, unlike the other taxonomies, the new classification integrates the relationships between the root causes of technical failure (SHELL model of critical areas) and the categories of antecedents of human errors (genotypes of CREAM), allowing to highlight direct and indirect causes, i.e. active and latent errors, of a single event. In addition, providing both a microscopic and a macroscopic analysis of the accident, the new classification overcome some drawbacks of the IMO and MAIB classification schemes. Table 5 shows the results of the classification experiment.

Since IMO does not require any event decomposition of the accident, in order to make possible a comparison of the different taxonomies, the IMO scheme has been applied considering the single event as an independent occurrence (i.e. an independent accident cause).

Some significant differences emerged among the causal factors identified by the investigator using different taxonomies.

The proposed classification framework was able to classify all the considered 10 events (100%), where the IMO classified only 6 events (60%), CASMET 7 events (70%) and MAIB 8 events (80%). Furthermore, Table 6 summarises the number of different causal factors identified by a specific taxonomy. The new classification results to be the most accurate and comprehensive, as it classifies all the events, highlighting twelve different causes. On the other side, the

Table 4. Major events characterising the INCAT 046 and LADY MEGAN II collision.

Event	Text of the official accident report *(comments)*
1	… The visibility was restricted in thick fog…
2	… The navigation was primarily done by means of radar …
	(It is likely that the radar of the "LADY MEGAN II" was affected by the close quarters phenomenon of side lobes—making it difficult to determine the exact position, course and speed of the "INCAT 046")
3	… The "INCAT 046" was informed of the position of the fishing vessel and that it would hold off south of Bug Light until the "INCAT 046" had passed "LADY MEGAN II"…
	(The International Regulations for Preventing Collisions at Sea, 1972 prescribe the actions to be taken when two vessels meet in a narrow channel. Rule 9(b) of these regulations prescribes that a vessel less than 20 m in length shall not impede the passage of a vessel which can navigate only within a narrow channel or fairway. The draught of the "INCAT 046" would have permitted her to navigate outside the Main Channel. However, the size of the "INCAT 046" and the configuration of Yarmouth Harbour in the area of Bug Light, would have made such navigation unsafe. Because of this, the 17 m long "LADY MEGAN II" was required to not impede the passage of the "INCAT 046")
4	… The ebbing tide created a southerly current of about 2 knots …
	(The Canadian modifications of these regulations prescribe the actions to be taken when two power-driven vessels meet in a narrow channel or fairway where there is a current or tidal stream. Rule 9(k)(i) prescribes that the vessel proceeding with the current or tidal stream shall be the stand-on vessel and shall propose the place of passage and indicate the side on which she intends to pass. The next paragraph of the same rule prescribes that the vessel proceeding against the current or tidal stream shall keep out of the way of the vessel proceeding with the current or tidal stream and hold as necessary to permit safe passage)
5	…The skipper of the fishing vessel informed the crew of the ferry that the ferry was on his radar and that he had changed his intentions and he was now intending to go around Bug Light and wait on the east side of the Main Channel for the "INCAT 046" to pass by…
	(It's not allowed to change the passing agreement in a restricted channel once the two crossing vessels have come to an agreement)
6	…The skipper of the fishing vessel informed the crew of the ferry that the ferry was on his radar and that he had changed his intentions and he was now intending to go around Bug Light and wait on the east side of the Main Channel for the "INCAT 046" to pass by…
	(this change has reduced the distance between the two vessels at the passing agreement, from 87,5 to 45,5 metres)
7	… The engine setting of the fishing vessel was kept at the full speed ahead setting …
8	… A second deckhand entered the wheelhouse of the "LADY MEGAN II" at this time and was asked to look out for the "INCAT 046"…
9	… On the "LADY MEGAN II", the lookout suddenly sighted the forward white mast light of the "INCAT 046" dead ahead at close proximity …
	(The side lights of the "INCAT 046" are mounted at about mid-length of the vessel and 7.156 m above the waterline. From a position 100 m directly ahead of the vessel, the ship's sides do not allow an approaching vessel to see the "INCAT 046" port and starboard side lights)
10	… The collision occurred at about 2335:45, 111 m northwest of Bug Light on the western edge of the Main Channel at position …
	("LADY MEGAN II" took the wrong direction, moving West rather than East, as agreed)

existing taxonomies generate some redundancies in suggesting causal factors and particularly: "Violation of procedure" for MAIB, "Lack of knowledge" and "Selection/training of officers" for CASMET, "Less than adequate operative procedures and instruction" and "Design Failure" for IMO. A Coefficient of redundancy (COR) has been simply defined as the ratio between the total number of events describing the accident and the number of different root causes identified by means of a specific taxonomy (Tab. 6).

Since repeatable (reliable) classifications are assured by means of a multi step approach and fixed correlations among factors embedded in the taxonomy, the proposed framework results potentially more reliable than the other schemes; this feature came into sight also during the experiment.

6 CONCLUSION

The study started from the results of a quantitative assessment of the human factors sections of three classification schemes for maritime accident investigation (IMO, MAIB and CASMET), carried out by Pedrali et al. (2002), and a qualitative assessment of the same classification schemes based on a larger set of quality criteria. Valuable features as well as critical issues affecting those classifications have been

Table 5. Results of the classification experiment.

N.	IMO	MAIB	CASMET	NEW
			Causes	
1	/	*	/	Fog
2	Design failure	System defect	Inadequate tool or aid + inadequate testing	Inadequate reserves
3	Less than adequate operative procedures and instruction	Violation of procedures	Lack of knowledge + selection/training of officers	Violation
4	Less than adequate operative procedures and instruction	Violation of procedures	Lack of knowledge + selection/training of officers	Violation
5	Less than adequate operative procedures and instruction	Violation of procedures	Lack of knowledge + selection/training of officers	Insufficient knowledge
6	*	Perception of risk	Lack of knowledge + selection/training of officers	Insufficient knowledge + lack of knowledge + habit, experience
7	*	Perception of risk	*	Overlook preconditions + erroneous analogy + focus gambling
8	*	Management and supervision inadequate	Inadequate manning + selection/training of officers	Too short planning horizon
9	Design failure	Design inadequate	Inadequate tool or aid + design error	Inadequate quality control + design failure
10	Design failure	*	*	Temporary incapacitation + design failure: presentation failure

Legend: * The causal factor is not included in the proper category of the taxonomy.
/ The entire causal category is overlooked.

Table 6. Coefficient of redundancy (COR).

Taxonomies	IMO	MAIB	CASMET	NEW
Total number of events	10	10	10	10
Number of unclassified events	4	2	3	0
Number of different identified causes	2	5	6	12
COR	5	2	1,6	0,83

pointed out and used as reference for the development of a new classification framework.

Four main characteristics distinguish the new framework with respect to the state-of-the-art:

– the categories used to classify human erroneous actions and their root causes are basically taken from CREAM (Hollnagel, 1998), thus the proposed framework is based on a formal cognitive model;
– the distinction between technical and human factors is set at the event level instead of the cause

level, thus the human factors analysis has moved up within the investigation process;
– categories of root causes, grouped into tables of underlying factors, are shared both by technical event and human event, thus also the identification of latent causes is improved;
– the general description of the event and the context are turned into common performance conditions (CPC) and used to reduce the set of possible root causes, making the classification process more reliable.

A more consistent validation of the new classification framework may be carried out in the next future by means of an assessment study similar to the one performed by Pedrali et al. (2002); also a wider application of the framework in maritime occurrences investigation will offer a stronger field testing of its features.

Finally, the proposed classification can be seen as an analytical frame for retrospective and prospective applications, of which Formal Safety Assessment (FSA) is one possibility. The same classification approach is also extensible to minor events and operational mis-performance, in order to provide a common basis for data collection.

REFERENCES

Edwards, E. 1972. Man and Machine: system for safety. *Proceedings of British Airline Pilots Association Technical Symposium*. London: British Airline Pilots Association.

European Commission, 2001. *European Transport Policy for 2010*. http://europa.eu.int/comm/energy-transport/en/lben.html.

European Transport Safety Council (ETSC), 2001. *EU Transport accident, incident and casualty databases: current status and future needs*. http://www.etsc.be.

EUROSTAT 1998. *The relative safety of maritime transportation*.

Hollnagel, E., 1998. *Cognitive reliability and error analysis method: CREAM*. Oxford: Elsevier Science.

ICAO 2000, *ADREP 2000*.

International Maritime Organisation (IMO), 2000. *Reports on Marine Casualties and Incidents, Revised harmonised reporting procedures*. Reports required under SOLAS regulation I/21 and MARPOL 73/78 articles 8 and 12, MSC/Circ. 953 and MPEC/Circ. 372.

Kristiansen, S., 1999. *CASMET-Casualty Analysis Methodology for Maritime Operations*. Norwegian Marine Technologies Research Institute.

Marine Accident Investigation Branch (MAIB), 2001. *Marine Accident Investigation Branch – Database taxonomy*. Southampton. UK.

Mathes, S. Koch Nielsen, J. Engen J. & Haaland, E., 1997. *ATOMOS® II – Final Report*. Brussels: European Commission.

Pedrali, M., Andersen, H.B. & Trucco, P., 2002. Are Maritime Accident Causation Taxonomies Reliable? An Experimental Study of the Human Factors Classification of the IMO, MAIB and CASMET Systems. *Proceedings of 2nd RINA Conference on Human Factors In Ship Design and Operation*. October 2–3. London: RINA.

Reason, J., 1990. *Human error*. New York: Cambridge University Press.

Saaty, T.L. 1980. *The Analytic Hierarchy Process*. New York: McGraw-Hill.

Transportation Safety Board of Canada (TSB), 1998a. *Safety study of the operational relationship between ship master/watchkeeping officers and marine pilots*. http://www.bst.gc.ca.

Transportation Safety Board of Canada (TSB), 1998b. *Report Number M98M0061*. http://www.bst.gc.ca.

Safety and Reliability – Bedford & van Gelder (eds)
© 2003 Swets & Zeitlinger, Lisse, ISBN 90 5809 551 7

Towards a systematic organizational analysis for improving safety assessment of the maritime transport system

A. Di Giulio, P. Trucco & G. Randazzo
Politecnico di Milano, Milan, Italy

M. Pedrali
D'Appolonia, Genoa, Italy

ABSTRACT: Safety in maritime operations has become a serious issue in the last 10 years, following the occurrence of an increasing number of incidents and accidents at sea. In this paper we propose a systematic approach that supports safety analyses of the Maritime Transport System (MTS); particular emphasis is put on High Speed Craft (HSC). The approach is mainly based on the development of a functional model that provides a systemic vision of the MTS from the safety point of view; in the early design stage of its development, the model was inspired by a similar approach developed in the air transport system. The functional model supports the validation of Formal Risk Assessment, but can be also used as an aid for the optimisation of resources allocation within the MTS in order to reach a target safety level. An application of the functional model will be shown for the quantification in a more coherent way of the conditional probability of some Basic Event (BE) of an accident (collision) scenario represented as a Fault Tree (FT). Indeed, the potential contribution of human and organisational factors can be quantified by linking the BE of the FT to appropriate variables of the functional model.

1 INTRODUCTION

Despite the remarkable effort performed at different levels to achieve a safe Maritime Transport System (MTS), the occurrence of accidents and incidents at sea still increases. Statistics published by the European Transport Safety Council (2001) reveal that in Europe maritime accidents are responsible annually for 140 deaths and 1.500 MLN€ of goods loss and damages. Globally, the maritime transport system is responsible for 1,4 deaths every 100 km, 25 times more risky than the air transport system, that accounts for 0,05 deaths every 100 km. Grounding (32%), striking (24%) and collision (16%) are the most frequent occurrences and with the highest rate of casualties.

The official report concerning the Zeebrugge incident (capsizing of a passenger ship) (Rasmussen 1997) pointed out that it was not due to a coincidence of independent technical failures and human errors, but a systematic change in the organizational behaviour of operators under the influence of economic pressure in a strongly competitive environment. Thus, a systematic safety analysis of the MTS needs to be enlarged to include the interactions and effects of decisions taken by various actors of the MTS, the workplace and context conditions, including the economic pressure affecting the maritime sector.

Various actors (operators, shipyards, regulators and government) in their respective working contexts are very often involved in a sequence of events leading to an accident; this is the most critical issue in developing an effective risk or accident analysis. The error of the operator onboard a ship is only the final act of a long and complex chain of organisational and systemic errors (i.e., the so-called latent failures). Figure 1 (Rasmussen 1997) shows the conflicting interactions between actors in MTS, pointed out by the accidents analysis of oil tankers and ferryboats (Shell 1992; Estonia 1995; Stenstrom 1995). Similar critical relations were found out in the sixty within the civil aviation sector (Schiavo 1997).

The need for a systemic approach to analyse the MTS safety is therefore clear, not focused only on the mistakes and the violations of the operators, but aiming to find, if they exist, also the causes at the various levels of the socio-technical system which competes for determining the accidents.

The present work, following a similar approach developed for the civil aviation by the ASTER project (Pedrali, Roelen 2000), aims to the development of a

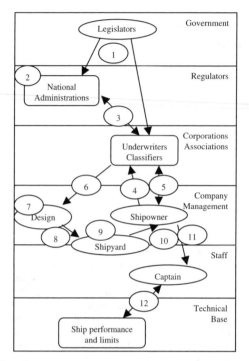

Legend:

1. & 2. The strategies for legislation appear to be inadequate during fast technological change.

3. Shipping industry's influence on legislators: Depressed shipping market of the 1980s leads to changes in its structure: Underwriters and National Administrations are neutralised by competition.

4. Ship owners influence classification societies.

5. Owners and classifiers co-operate and do not inform legislators adequately.

6. Communication from classifiers to designers inadequate.

7. Design based on established practice inadequate during period of fast pace of change.

9. Quality assurance of shipyard questionable.

10. Inadequate communication between design, manufacturing and operating communities.

11. & 12. Inadequate guidance to captain,learning by doing inadequate during fast pace of technological change.

Figure 1. Existing conflicts between the actors in the MTS (Rasmussen 1997).

functional model for the MTS. The principal purpose of the functional model is to provide a systemic vision of the MTS from the safety point of view.

According to a top-down modelling approach, the first step of the MTS analysis identified the principal actors; then, a more detailed analysis allowed to define the critical functions performed by each actor and the corresponding variables. Such a method was considered as the most suitable as it allows a selective description and modelling of the system; it was preferred to the bottom-up approach (i.e. starting from specific activities or actions that may lead to an error

or a technical failure) because the latter tends to underline detailed case-dependent descriptions of the cause-effect relations, where we need to identify systemic critical dynamics. Finally, a top-down approach allows to quickly identify the parts of the model which require different modelling techniques or a more detailed description.

Furthermore, we developed the functional model reducing at minimum the number of functions and variables, in order to make it more manageable by the safety analyst that will use it. Future developments of the research may lead to increase the level of detail – and thus the complexity – of the model, considering more detailed functions, an increased number of variables and dependencies or a more complex variable type.

2 FUNCTIONAL MODEL DEVELOPMENT

The model has been developed along three major steps: (1) Maritime Transport System analysis; (2) Qualitative Model Formulation; (3) Quantitative Model Formulation.

2.1 Maritime transport system analysis

In the first step, a wide-spectrum analysis of the MTS was led to identify the main actors and their critical functions, referring to an "ideal" situation in which "everyone follows the rules". This phase was carried out through interviews to maritime operators, shipyards and classification societies. Five main actors have been identified, namely: the Operator (that represents the maritime company), the Shipyard, the Port, the Regulator (that represents all the certification and regulatory bodies) and, finally, the Environment, considered also from a socio-technical point of view.

Secondly, functions performed by each actor were identified, under the hypothesis that the behaviours of the actors are perfectly pertaining to the provisions and the safety procedures. For instance, six critical functions have been identified for the actor "Operator", as listed in Table 1.

The description of the functions allowed to clarify and better understand the roles and the duties of each actor within the MTS, considered as a socio-technical system. Globally 18 critical functions have been identified (Tab. 2).

2.2 Qualitative model formulation

In the second step, a set of input and output variables were defined for each function. The current version of the functional model is built up with Boolean variables in order to limit the number of possible variable modes and thus the model complexity (binary variables are also coherent with the needs related to the

Table 1. Critical functions performed by the actor "OPERATOR".

N°	Critical function	Description
1	Personnel management	Modes of staff management in terms of distribution and assignment of the duties and management of shifts. Attribution of the work load according to the current disposals in matter, avoiding an excessive physical stress which could lower the attention level. Staff motivation and assignment of incentives to reward high performance. Promotion of a "safety culture" and near misses reporting among employees.
2	Fleet exploitation	Use and management of the fleet by the maritime company on the basis of the characteristics of the ships and their current state, with the purpose to use the most suitable means in terms of efficiency and reliability. Planning of ships' operations and maintenance sessions.
3	Ship maintenance	All the activities which have the purpose to take or keep the ship in conditions of navigability. Therefore, these activities include both the corrective maintenance and the preventive one.
4	Training	Technical training of the crews of the ships and the maintenance teams. Planning of the training sessions according to the current provisions so as to guarantee a certain frequency. Elaboration and updating of the operational and safety procedures.
5	Ship government	All the activities concerning the exercise of the navigation. Setting of the journey parameters (speed, route, etc.), use of the directional control for manoeuvring (rudder, command levers, etc.), use of the instrumentation (electronic nautical papers, radars, GPS, etc.) for the identifying of its position and that of other ships and the maintenance of the established route, monitoring and management of the systems (machinery, electric system, etc.) and communications.
6	Cargo and passengers handling	All the activities concerning the embarkation and the disembarkation of passengers, security controls, positioning of the equipment to allow the passengers embarkation. The control and the maintenance of all the embarkation equipment. Loading and unloading of goods and vehicles. Cargo disposition on the ship. Safety controls to verify the nature of the cargo.

Table 2. Critical functions included within the MTS functional model.

Actors	N°	Functions
Operator	1	Personnel management
	2	Fleet exploitation
	3	Ship maintenance
	4	Training
	5	Ship government
	6	Cargo and passengers handling
Shipyard	7	Ship design
	8	Ship development/Production/Testing
	9	Management of subcontracting
Regulator	10	IMO – Definition of international standards
	11	Classification Societies – Means of compliance
	12	Classification Societies – Provision of assistance
	13	Port State Authorities – Ensure of compliance with international standards
Port	14	Berthing, mooring and anchoring
	15	Traffic management
	16	Towage and pilotage
	17	Turnaround services
	18	Emergency and rescue services

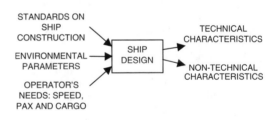

Figure 2. Input and output parameters of the function "'SHIP DESIGN".

presented application of the functional model, i.e. the estimation of some BE probabilities of a Fault Tree-shaped collision scenario). The input parameters (e.g. design requirements) affect how a specific function (e.g. ship design) is executed, whereas the output parameters represent the outcome of the same function (e.g. ship characteristics). The case of the function "Ship design" performed by the actor "Shipyard" can be considered to focalize the framework of the functional model (Fig. 2).

The design of the ship must be executed on the basis of the international standards (*Standards on ship construction*), on the basis of the environmental conditions in which the ship will have to work (*Environmental parameters*) and, obviously, considering the requests of the Operator included in the ship order (*Operator's needs: Speed, Passengers and Cargo*). Another independent variable of this function is obviously the design performance. On the basis of these input variables, the ship design function will return two output variables: the first related to the technical characteristics of the ship (e.g. propulsion system power, hold capacity) and the second related to the non-technical characteristics (e.g. workplace ergonomics, bridge layout).

The variables have been classified, according to the SHEL model of Edwards (1972), as *software variables* (e.g. sealing procedures), *hardware variables* (e.g. ship components and systems performances), *environmental variables* (e.g. weather parameters, time pressure and company safety culture) and *liveware variables* (e.g. human performance, crew decisions and crew communication).

The variables identification and definition represents the most critical issue in the development of the functional model, since these elements describe the relations inside the MTS and thus the causal links, throughout the entire system, that may lead to undesired events or accidents. In this phase the contribution of experts, coming from different areas of the MTS, is crucial.

2.3 Quantitative model formulation

Finally, in the third step the interactions between actors have been traced and quantified as functional relationships (deterministic) or organisational/systemic influences (probabilistic).

A functional relation exists when the output of a function is also the input of another function (e.g. the propulsion system, the directional control systems and the navigation and communication equipments, that are, the output parameters of the function "Ship construction", are also input parameters for the function "Ship government"; Fig. 3).

In a complex socio-technical system, not all the relationships are planned and explicit, but a number of constraints and relations may be identified that, in spite of being mainly latent, indirect or unplanned, are relevant in affecting the safety performances of the overall system. These relationships have been named *influences*. For example, the variable "Maintenance personnel performance" is an output parameter of the function "Ship maintenance" that influences the variables "Propulsion system", "Navigation system" and "Navigation and communication equipments" that are input parameters for the function "Ship government". Both these functions are performed by the "Operator". These influences show how the correct work of the ship components and systems depends also on the performance of maintenance personnel (Fig. 4).

Figure 3. Example of functional relations (deterministic).

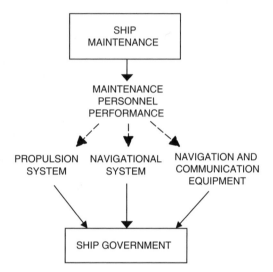

Figure 4. Example of influences between variables.

Notwithstanding with a Boolean definition of the variables (e.g. yes/no, good/bad, right/wrong), the quantification of a functional relation or an influence can be a very laborious process. It depends on the number of input variables of a function, or on the number of variables influencing another variable. Indeed, if N is the number of input variables of the function (or the number of influencing variables), under the hypothesis of independence of the input (or influencing) variables, the number of different combination that have to be estimated is 2^N. Therefore, to make the quantification of the functional model easier, the concept of macro-variable has been introduced, i.e. an aggregation of homogeneous variables whose state is dependent on the combination of modes of the basic variables. These macro-variables introduced some further nodes in the net, making the quantification easier and increasing the estimate precision.

The functional relations and the influences have been quantified by means of experts' judgements, concerning the modes assumed by each output variable for each combination of modes of the input variables. To this end a matrix format has been used, as reported in Tables 3–4. According to the deterministic

Table 3. Supporting matrix for the quantification of functional relations.

Input	Accidents data info			
	Exhaustive: Environmental parameters		Not exhaustive: Environmental parameters	
Output	Contemplated	Not cont.	Contemplated	Not cont.
Standards on collision avoidance				
Incorrect	1	0	1	0
Correct	0	1	0	1

Table 4. Supporting matrix for the quantification of the influences.

Influenced by	Manning			
	Good: Bridge layout		Bad: Bridge layout	
Influenced	Good	Bad	Good	Bad
Crew characteristics (temporary)				
Bad	0,90	0,70	0,30	0,10
Good	0,10	0,30	0,70	0,90

approach, a functional relation requires that for each combination of modes of the input variables, the output variable is able to assume just one state (e.g. correct or incorrect).

On the other side, the quantification of the influences between variables required the experts to express the value of the conditional probability for each cell of the matrix (Tab. 4). As a quantitative basis supporting the elicitation process, the experts have been provided with a study of AEA Technology Consulting (1996), concerning the statistical analysis of accidents happened to HSC in British ports between 1988 and 1992.

3 CASE STUDY: EVALUATION OF A COLLISION SCENARIO FOR HIGH SPEED CRAFT DESIGN

As a test application, the Functional Model (FM) has been used to estimate some BEs of a Fault Tree describing potential causes involved in a collision scenario. The FT has been developed in the context of the Safety at Speed (S@S) project. The purpose of the project, funded by the EU, is to develop a formal methodology for integrating safety performance in the design process of High Speed Craft (HSC). Thus, the project aims to promote the safety culture within

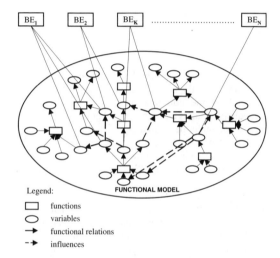

Legend:
☐ functions
○ variables
→ functional relations
⇢ influences

Figure 5. Relations between the BE of the Fault Tree and the variables of the Functional Model.

Figure 6. Direct causes of the Basic Event "DEFECTIVE RADAR".

the maritime transport sector, through the adoption of "Design for Safety" methods and techniques.

This application shows how the FM supports the quantification of some BE occurrence probability, making clear to the safety analyst its latent causes (Fig. 5).

Furthermore, the FM allows to identify those factors in the causal chain that have to be managed carefully in order to minimize the BE probability (Fig. 6). For

Table 5. State conditions for "Maintenance Personnel Performance" = HIGH.

	Maintenance planning and procedures	Maintenance equipment	Maintenance personnel training
1	Good	Adequate	Adequate
2	Good	Inadequate	Adequate
3	Bad	Adequate	Adequate

Table 6. State conditions for "Goods and services" = GOOD.

	Internal resources	External goods and services
1	High	Good
2	Low	Good

instance, the FM shows that the least occurrence probability for the BE "defective radar" (5,56 E-05) is assured by the combination Design parameters = "adequate", Goods and services = "good" and Maintenance personnel performance = "high".

"Maintenance personnel performance" is the dependent variable of the function "Ship maintenance", whose input variables are "Maintenance planning and procedures", "Maintenance equipment" and "Maintenance personnel training". According to the FM, as quantified by the experts (Tab. 5), the factor that assures a high performance of the maintenance personnel is the quality of the training programmes, while the equipment conditions and the maintenance procedures are less important. Thus the probability to have high personnel performance depends on the probability to have adequate training. Indeed, a well trained maintenance operator could overcome some lacks in the equipment set or in the maintenance procedures.

To make sure that the macrovariable "Goods and services" assumes the state "good" two combinations of the variables grouped in it are possible (Tab. 6).

In this case the experts evaluated that the necessary condition to assure the high quality of the goods and services is the good quality of subcontracted good and services, giving less importance to the internal resources of the naval yard.

"External goods and services" is the dependent variable of the function "Management of subcontracting", whose input variables are "Time constraints" and "Orders specification". According to the FM, as quantified by the experts (Tab. 7), there is only one combination of the states of the input variables that assures the good quality of the "External goods and services". Precisely, the time pressure have to be low and the specification of the orders have to be good.

The macro-variable "Design parameters" is constituted by the two variables "Technical characteristics"

Table 7. State conditions for "External goods and services" = GOOD.

	Time constraints	Orders specification
1	Low	Good

Table 8. State conditions for "Design parameters" = ADEQUATE.

	Technical characteristics	Non-technical characteristics
1	Adequate	Adequate

Table 9. State conditions for "Technical and Non-technical characteristics" = ADEQUATE.

	Operator's needs (speed, pax, cargo)	Designer performance	Standards on ship construction
1	High	High	Correct
2	Low	High	Correct

and "Non-technical characteristics", output variables of the "Ship design" function, which must necessarily assume the state "adequate" to make sure that the complex variable assumes the state "adequate" (Tab. 8). The input variables of the function "Ship design", are: "Operator's needs (speed, passengers and cargo)", "Designer performance", "Standards on ship construction". The combinations of modes that assures adequate technical and non-technical characteristics of the ship are Designer performance = "high" and Standards on ship construction = "correct" (Tab. 9).

In this case, the experts evaluated that the correctness of the standards on the ship building and the high performance of the designers are absolute conditions to have adequate technical and not technical parameters characterizing a safe HSC.

Finally, we have to take into account that the technical and non-technical characteristics have a strong influence on the variable *Orders specification* (Tab. 10).

Precisely, if the technical and non-technical characteristics of the ship were adequately planned, with a probability of 95%, also the specification of the external orders will be good, increasing the quality of external goods and services.

Besides, the mode of the variable "Time constraints" depends on the mode of the variables "Operator's needs (speed, passengers and cargo)" and "Designer performance" (Tab. 11).

In this case, the experts evaluated that the high performance of the designers is a condition to have low time pressure.

Table 10.   Influence of the modes of technical and non-technical characteristics on the mode of Orders specification.

Influenced by	Non-technical characteristics			
	Adequate: Technical characteristics		Inadequate: Technical characteristics	
Influenced	Adequate	Inadequate	Adequate	Inadequate
Orders specification				
Bad	0,95	0,5	0,5	0,05
Good	0,05	0,5	0,5	0,95

Table 11.   State conditions for "Time constraints" = LOW.

	Operator's needs (speed, pax and cargo)	Designer performance
1	Low	High
2	High	High

Summarising, the FM allowed to point out that to have a low occurrence probability of a "defective radar" the most critical organisational parameters are:

– correct standards on ship construction;
– high designer performance;
– an adequate maintenance personnel training.

Furthermore, the FM highlighted the direct and indirect influences of those variables, particularly with respect to the relevance of the technical and non-technical ship characteristics, the quality of orders specification and the time constraints of the HSC design and construction project.

4   DISCUSSION

Authors are well aware that the validation of the FM is heavily affected by the scarcity of empirical an historical data about accident and incidents involving HSC and therefore, that the only possibility for the quantification remains, for the moment, the use of expert judgements. Bayesian techniques will be used in the future to support the functional model with better quantified estimations of the causal relations, especially those related to human and organisational factors, for which data is much more sparse.

5   CONCLUSION

The FM is a useful support tool for the search of the human and organizational factors which cause the accidents and for the establishment of a hierarchy.

Future developments of the Functional Model follow two directions. The first concern the adoption of SADT (Ross 1977) as a technique for the formal representation of the model. Indeed, the introduction of this standardized "language" allows to distinguish two principal dimensions embedded in the network of relationships: the ship operation processes and activities (i.e. functions performed by the operator) and, transversally, the supporting processes, that provide the ship operation functions with technical equipments, human resources, control devices and criteria, knowledge, etc. The first dimension is directly affected by technical failures or human and organisational errors, the second dimension represents the way by which systemic vulnerabilities impact on the ship operations.

The second development concerns the quantitative formulation of the model: by means of the application of data models derived from the Bayesian statistics it will be possible to integrate historical data and experts' judgments to quantify the model relationships with higher degree of completeness and precision.

Another potential application of the FM concerns the support to the retrospective analysis of really occurred accidents.

REFERENCES

Edwards, E. 1972. Man and machine: systems for safety, in proceedings of british airline pilots association technical symposium. *British Airline Pilots Association*: 21–36. London.

Estonia 1995. Accident investigation report; part report covering technical issues on the capsizing on 28 September 1994 in the Baltic sea of the ro-ro passenger vessel MV ESTONIA. *The Joint Accident Investigasion Commission*. Stockholm: Board of accident investigation.

European Commission 2001. European Transport Policy for 2010. http://europa.eu.int/comm/energy-transport/en/lb-en.html.

Pedrali, M. & Roelen, A. 2000. Model of safety levels. *ASTER Technical Report WP2-TR2A*. Release 1.0. November.

519

Rasmussen, J. 1997. Risk management in a dynamic society: a modelling problem. *Safety Science* 27: 183–213.

Robinson, R.G.J. & Lelland, A.N. 1996. Marine incidents in ports and harbours in Great Britain, 1988–1992. *AEA Technology.*

Ross, D. 1977. Structured analysis (SA): a language for communicating ideas. *IEEE Transactions on Software Engineering:* 195–212. New Jersey.

Schiavo, M. 1997. Flying blind, flying safe. *Avon Books.* New York. *For revisions see TIME magazine, 31 March 1997: 38–48 e 16 June*: 56–58.

Shell 1992. A study of standards in the oil tanker industry. *Shell international marine limited, May.*

Stenstrom, B. 1995. What can we learn from the ESTONIA accident? Some observation on technical and human shortcomings. *In The Cologne Re Marine Safety: Seminar. 27–28 April.* Rotterdam.

APPENDIX A – THE FUNCTIONAL MODEL

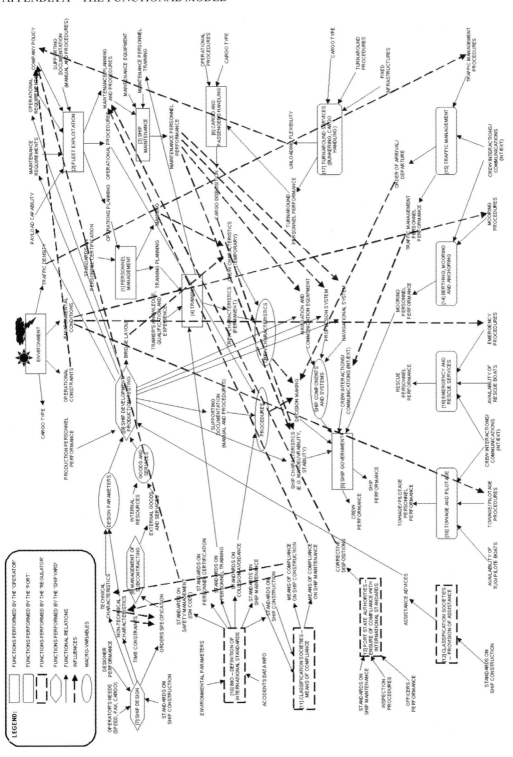

Safety and Reliability – Bedford & van Gelder (eds)
© 2003 Swets & Zeitlinger, Lisse, ISBN 90 5809 551 7

Design based safety engineering applied to railway systems, part II

E.M. Dijkerman & W. Kruidhof
Holland Railconsult B.V., Utrecht, Netherlands

J. de Boer & B. van der Hoeven
CMG Public Sector B.V., Den Haag, Netherlands

M. Uittenbogaard
ADSE B.V., Hoofddorp, Netherlands

ABSTRACT: What problems arise during a "real life" hazard analysis?

The BB21 Signalling System is being developed for the *Nederlandse Spoorwegen* (the Dutch railways) under the supervision of *Prorail-Railinfrabeheer*, the Dutch rail infrastructure provider. Instead of signals along the track, BB21 sends the signals by radio to a cab display in front of the train driver. Taking the BB21 Signalling System into use requires the approval of the Dutch railway safety authority.

The hazard analysis technique used for the BB21 Signalling System is based on Railtrack UK's "Yellow Book". The hazards to be analysed were identified using the process described in part I of this paper (Step 1 in the Yellow Book). Part II now describes the hazard analysis process itself and shows how the Yellow Book guidelines were adapted to the requirements of BB21. Where possible, the paper provides guidelines based on the authors' experience.

1 BB21 REQUIREMENTS

Introducing a new signalling system on a railway network is a gradual process of transition. It is important to avoid large differences between the behaviour of the new signalling system and that of the existing signalling system, as large differences make human error more likely.

A new signalling system such as BB21 therefore tends to use as many elements from the existing system as can be justified from a safety point of view.

The BB21 Signalling System has to satisfy the requirements of:

– The ERTMS standards (European Rail Traffic Management System). These are signalling standards issued by the European Union.
– The CENELEC standards, issued by the European Committee for Electrotechnical Standardization.
– The Dutch railway safety authority, IVW Rail (*Inspectie Verkeer en Waterstaat, divisie Rail*) (Ministry of Transport, Rail Division).

IVW Rail has defined specific areas in which the BB21 Signalling System has to provide a higher level of safety than the existing signalling system:

– Safety of track workers (implemented by assigning separate work zones to track workers).
– Safety on level crossings (implemented by a constant warning time, to avoid long waiting times that lead to drivers of road vehicles slipping across the level crossing if the actual arrival of the train takes longer than they expect).

For areas not explicitly mentioned, IVW Rail requires that the BB21 signalling system provide a safety level at least as high as the current safety level.

Because it is not possible to quantify the risks associated with the present signalling system adequately, IVW Rail has agreed to a qualitative risk assessment for the BB21 project.

2 SELECTION OF HAZARD ANALYSIS METHOD

The CENELEC standards allow different principles for demonstrating safety; including "ALARP" and "GAMAB". The above requirements mean that the principle applied will differ from one hazard to another:

- For some BB21 hazards, it will be necessary to demonstrate that everything has been done to provide a level of safety that is higher than the current level (ALARP: the risk must be As Low As Reasonably Practicable).
- For other hazards, it will be sufficient to demonstrate that the level of safety is at least as high as the current level (GAMAB, from the French *Globalement Au Moins Aussi Bon*, or "on the whole at least as good").

Consequently, the method used to assess the hazards and to specify the appropriate safety measures has to be able to cope with both principles – ALARP and GAMAB.

Railtrack UK's Yellow Book formed the starting point for analysing the hazards associated with the BB21 system.

The reasons for selecting the Yellow Book are that it is a well-documented hazard analysis guideline for railway systems and that it follows the CENELEC standards. To make using the method easier, all relevant BB21 project members attended a Yellow Book course.

The Yellow Book guidelines are designed to demonstrate ALARP. The Yellow Book allows both a qualitative and a quantitative approach. For the BB21 project, we adapted the Yellow Book guidelines to GAMAB and a qualitative approach.

3   YELLOW BOOK ADAPTATIONS

Table 1 shows which steps in the Yellow Book are applicable when demonstrating ALARP and which steps are applicable when demonstrating GAMAB. The "intermezzos" have been introduced by us. For

Table 1. Overview of steps.

Step	Step description	ALARP	GAMAB
1	Hazard identification	x	x
2	Causal analysis	x	x
3	Consequence analysis	x	x
4	Loss analysis	x	x
Intermezzo 1	Demonstration of GAMAB (1)		x
5	Options analysis	x	(x)
6	Impact analysis	x	(x)
Intermezzo 2	Demonstration of GAMAB (2)		(x)
7	Demonstration of ALARP	x	

the example shown in this paper, we followed the path indicated in grey (see explanation in Section 6).

Hazard analysis always starts with the first four steps of the Yellow Book, whether the criterion is ALARP or GAMAB. When it is sufficient to demonstrate that the risk in the BB21 signalling system does not exceed the risk in the current signalling system, and Steps 1 to 4 confirm that this is so, one has satisfied the GAMAB criterion.

There is then no need to execute Steps 5, 6 and 7 (defining options, assessing their impact and demonstrating ALARP).

However, in some cases the risk identified is higher than with the current signalling system. In that case, one can propose and analyse options for reducing this risk (Steps 5 and 6). There is therefore a second point at the end of Step 6 where one can determine whether the system meets the GAMAB criterion.

4   BB21 SIGNALLING SYSTEM DESIGN

We shall demonstrate application of the modified Yellow Book guidelines using an example from the BB21 hazard analysis – the revocation of a movement authority.

The BB21 system consists of the following subsystems:

Table 2.   BB21 subsystems.

Subsystem	Function
VPT	Man-Machine Interface with the signalman
Bev21	Safety system, consisting of: – Trackside: system core – Trainside: interface to driver
GSM-R	Radio connection between the trackside and trainside parts of the signalling system
25 kV	Overhead wire system

Abbreviations:
VPT: *Vervoer Per Trein* (transport by train).
Bev21: *Beveiliging 21* (safety system for the 21st century)
GSM-R: GSM for Railway applications.

In fact only a part of VPT is inside the scope of BB21. VPT also contains traffic management functions and planning functions that are outside the BB21 scope.

5   EXAMPLE: REVOCATION OF A MOVEMENT AUTHORITY

In the BB21 system, a train is given permission to move by a "movement authority", a message sent by radio that specifies how far the train can move and at what speed.

The subject of this hazard analysis is the revocation of a movement authority for a train because of a dangerous situation on or near the track ahead. In the Netherlands, revocations are done approximately 560 times a year.

Reasons for revoking a movement authority include the following:

- An unauthorized person on or near the track (referred to below as a "trespasser").
- Livestock on the track.
- A level crossing out of order.
- An obstacle on the track.

6   APPLICATION OF THE METHOD TO THE EXAMPLE

Because this was our first hazard analysis and the rules for using GAMAB were not yet established, we chose the ALARP criterion right at the start of the hazard analysis. Later, at Step 6, we realized that the GAMAB criterion was also permissible for this specific hazard and would be easier to apply. We therefore changed over to GAMAB.

6.1   Step 1: Hazard identification

6.1.1   Method
Please see part I of this paper for a discussion of the hazard identification method.

6.1.2   Example
The hazard is described as follows: "The signalman revokes the movement authority of a train, because of a dangerous situation. This revocation is not executed by the BB21 system."

6.2   Step 2: Causal analysis

6.2.1   Method
We drew up fault trees in accordance with the Yellow Book's recommendations on causal analysis. These fault trees show which subsystem is causing the hazard.

The causal analysis also involves determining the frequency with which each hazard will occur.

For BB21, the railway safety authority agreed that a qualitative assessment of the frequency of occurrence would be sufficient. The hazard frequency categories shown in Table 3 are taken from CENELEC 50126.

In practice, this proved quite difficult, because ideas differ as to what constitutes "frequent", "incredible", etc. We therefore agreed a system of quantification

with the railway safety authority, to ensure that categories were applied in a uniform manner throughout the project.

We used the concept of SIL (Safety Integrity Level) to estimate the hazard frequencies. CENELEC 50129 explains "Safety Integrity" as "the likelihood of a safety related function achieving its required safety features". Four discrete Safety Integrity Levels are specified, level 4 offering the highest safety, level 1 the lowest.

During the hazard analysis of the revocation example, we developed certain guidelines:

- The frequency with which a function is used can form the basis for estimating the frequency with which the hazard associated with this function will occur. For a little-used function, the frequency of a hazard may be so low that measures are not necessary.
- We can assume that the hazard frequency for the Bev21 trackside and trainside systems is "incredible", as the supplier's safety case will demonstrate that Bev21 is a SIL-4 system.
- We can assume that the hazard frequency for VPT is "remote" (100 times worse than "incredible"), as application of CENELEC 50128 will demonstrate that VPT is a SIL-2 system.
- If no information about failure rate is available, an estimate of the range is sufficient for the time being (e.g. "better than probable, but worse than incredible"). Whether more detailed information is required can be decided later.

Table 3.   Frequency categories.

Category	Description	Explanation
Frequent	Likely to occur frequently. The hazard will be experienced continually.	Almost daily.
Probable	Will occur several times. The hazard can be expected to occur often.	One or a few times every month.
Occasional	Likely to occur several times. The hazard can be expected to occur several times.	One or a few times every year.
Remote	Likely to occur sometime in the system life cycle. The hazard can reasonably be expected to occur.	Once every 10 to 30 years.
Improbable	Unlikely to occur but possible. It can be assumed that the hazard may exceptionally occur.	Once every 40 to 100 years.
Incredible	Extremely unlikely to occur. It can be assumed that the hazard may not occur.	"Never" occurs.

6.2.2 Example

The fault tree below shows the identified causes. See Tables 4 and 5 for further details of the causes.

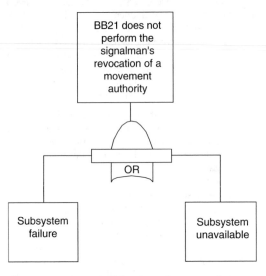

Figure 1. Fault tree for failure to perform revocation.

Table 4. Hazard occurrences caused by subsystem failures.

Cause description	Frequency of occurrence (category)	Reason
VPT fails	Remote	SIL-2
Bev21 (trackside) fails	Incredible	Bev21 safety case
GSM-R fails	Incredible	EURORADIO protocol ensures delivery of messages
Bev21 (trainside) fails	Incredible	Bev21 safety case

Table 5. Hazard occurrences caused by subsystem unavailability.

Cause description	Frequency of occurrence (category *)
VPT not available	Incredible
Bev21 (trackside) not available	Incredible
GSM-R not available	Remote
Bev21 (trainside) not available	Incredible

(*) The frequency of occurrence has been determined on the basis of the MTBF, taking into account the fact that only the first minutes of unavailability are relevant – after a few minutes, all movement authorities will have expired and all trains will have stopped.

6.3 Step 3: Consequence analysis

6.3.1 Method

We adopted the CENELEC 50126 method of classifying the severity of consequences (Tab. 6).

In line with the Yellow Book, we have used consequence trees to show the resulting accidents. However, we have not always extended the consequence trees to their limits; in general, we have only traced the paths to the most serious accidents. This is sufficient to identify the necessary measures. Furthermore, we have not included the frequency of accidents, as accident frequencies depend on many factors (number of trains, amount of road traffic passing a level crossing, ...). For each specific application of BB21 these factors are different.

We agreed these measures with the railway safety authority. Indeed, an important lesson learned during the consequence analysis was that showing early results to the railway safety authority helps to get their agreement on the required level of detail and therefore saves time.

6.3.2 Example

The diagram below shows part of the consequence tree for the example.

Table 6. Severity classes.

Class	Consequences for persons and environment
Catastrophic	Fatalities and/or multiple severe injuries and/or major damage to the environment
Critical	One person killed and/or one person seriously injured and/or considerable damage to the environment
Marginal	One person lightly injured and substantial threat to environment
Insignificant	Possibly one person slightly injured

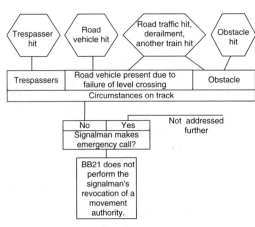

Figure 2. Part of the consequence tree.

6.4 Step 4: Loss analysis

6.4.1 Method
In line with the Yellow Book, we used a risk matrix to present the results of the loss analysis.

We have used the risk tolerability table in CEN-ELEC 50126:

Table 7.   Risk tolerability table.

Risk category	Actions to be applied against each category
Intol(erable)	Shall be eliminated.
Undes(irable)	Shall only be accepted when risk reduction is impracticable and with the agreement of the railway safety authority.
Tol(erable)	Acceptable with adequate control and with the agreement of the railway safety authority.
Negl(igible)	Acceptable with/without the agreement of the railway safety authority.

We based our risk evaluation on the example table for risk evaluation and acceptance from CENELEC 50126:

Table 8.   Risk levels.

Frequency of occurrence of a hazardous event	Severity levels of hazard consequences			
	Insignificant	Marginal	Critical	Catastrophic
Frequent	Undes	Intol	Intol	Intol
Probable	Tol	Undes	Intol	Intol
Occasional	Tol	Undes	Undes	Intol
Remote	Negl	Tol	Undes	Undes
Improbable	Negl	Negl	Tol	Tol
Incredible	Negl	Negl	Negl	Negl

6.4.2 Example
Having identified the consequences, we classified them as follows:

Table 9.   Classification of consequences.

No.	Consequence	Severity class
1	Collision with trespasser	Critical
2	Collision with road vehicle	Critical
3	Collision with road vehicle, followed by train derailment, followed by collision with another train	Catastrophic
4	Collision with obstacle	Critical

The resulting risk matrix is as follows:

Table 10.   Risk levels for revocation of a movement authority.

Frequency of occurrence of a hazardous event	Severity levels of hazard consequence			
	Insignificant	Marginal	Critical	Catastrophic
Frequent				
Probable				
Occasional				
Remote			1, 2, 4	3
Improbable				
Incredible				

(The numbers correspond to the consequence numbers in Table 9 above)

This combination of hazard frequency ("remote") and accident severity ("critical" or "catastrophic") classifies the risk as "undesirable" (see Table 8 in 6.4.1).

Note: the matrix only reflects hazard frequency and accident severity. It does not take accident frequency into account. This is in line with Step 3 above.

6.5 Intermezzo 1: Demonstrating GAMAB

6.5.1 Method
Demonstrating GAMAB does not form part of the Yellow Book guidelines.

However, if one wishes to demonstrate compliance with GAMAB, it is at this point in the analysis that one should determine whether the system satisfies the GAMAB criterion.

6.5.2 Example
In our example, the initial goal was to demonstrate ALARP, so we did not look into the possibility of demonstrating GAMAB at this point.

Even in retrospect, however, we can see that it would not have been possible to demonstrate GAMAB at this stage. This is caused by differences between the current ATB-system (Automatische Trein Beïnvloeding, automatic train protection) and the GSM-R-system. Unavailability of ATB at the moment of revocation leads to lower speed of the train (at most 40 km/h). Unavailibility of GSM-R leads to a train continuing at full speed. So a measure should be taken to solve this difference between BB21 and the current systems.

Therefore we continue with Steps 5 and 6.

6.6 Step 5: Options analysis

6.6.1 Method
We developed our own guidelines during this step:

– Try to eliminate the most likely causes of the hazard.
– Note all ideas circulating within the project.
– Do not estimate costs at this stage.

6.6.2 Example

The most likely causes of the accidents are failure of the VPT subsystem and unavailability of the GSM-R subsystem.

We identified the following options:

1. VPT: Improve the quality of the revocation function.
2. GSM-R: Improve availability.
3. GSM-R: Implement one of the possibilities that ERTMS offers in the event of loss of radio contact:
   - Train trip after contact has been lost. Train stops.
   - Apply service brake. Not applied in the Netherlands, so not an option.
   - No reaction. Does not solve the problem.

6.7 Step 6: Impact analysis

6.7.1 Method

The Yellow Book implies that when one has found a means of reducing a risk, one can move the accident to a position in the matrix associated with the new, lower risk.

In practice, this is not always easy to demonstrate. An alternative is then to try to show that the risk is comparable to the risk within the current signalling system, thereby demonstrating that the solution meets the GAMAB criterion. This is "intermezzo" 2 in Table 1.

6.7.2 Example

Now we consider the impact of the options for the example hazard.

Improving the quality of VPT would require considerable effort and would be very costly, because it would require upgrading to SIL-3 or SIL-4. In fact a major part of VPT has already been used for more than five years all over the Netherlands and has proved its quality in practice. For the demonstration of GAMAB it is sufficient to show that VPT has a quality comparable to the quality of the current VPT.

Improving the availability of GSM-R is also likely to be very expensive.

Implementing one of the ERTMS options for action in case of loss of radio contact does not result in a different risk matrix, since VPT still causes a remote number of hazards. However applying the first ERTMS option results in a higher safety level than that of the current system (see "intermezzo" 2).

6.8 Intermezzo 2: Demonstrating GAMAB

6.8.1 Method

While the GAMAB criterion is not part of the Yellow Book, we wished to verify compliance with GAMAB at this point in the analysis.

6.8.2 Example

Applying the first ERTMS option, i.e. forcing a train stop, demonstrates compliance with GAMAB. When the current system, ATB is unavailable, the train continues to run, albeit at a lower, "on sight" speed. The ERTMS option to force a train to stop is even safer.

It is however important to consider the consequences of this solution. A train trip occurs after a delay, the length of which depends on a timer. If this timer is set to zero, thousands of train trips will occur during a year, most of them unnecessarily. The timer will therefore be set to 20 seconds, which means that only serious unavailability of GSM-R is taken into account. That does not occur more than about 20 times a year at the most.

6.9 Step 7: Demonstrating ALARP and compliance

6.9.1 Method

In this last step it is shown:

- that the risk has been reduced such that it is in the tolerable region (see Table 8),
- that there exist no reasonable options which have not been implemented.

6.9.2 Example

We did not execute this step for the example.

7 CONCLUSIONS

7.1 General

In order to meet the requirements of the BB21 project, we successfully adapted the guidelines in Railtrack UK's Yellow Book to allow analysis according to the GAMAB principle.

We also showed that it is possible to take a qualitative approach to the occurrence of causes and consequences, and to the loss analysis. This is particularly important where the risk associated with the safety level of the existing safety system is considered acceptable, and when there is insufficient data to quantify this risk.

7.2 Recommendations

- Before starting detailed preparations and raising expectations that later might prove impossible to meet, decide whether the requirements for your project really allow you to follow the Yellow Book to its full extent or whether you will have to deviate from the Yellow Book.
- Show early results to the safety authority and make sure that your hazard analyses meet their expectations.
- Prepare a template for documenting the hazard analyses. This will help give the analysis a solid structure and will ensure a uniform presentation to the reader, i.e. the safety authority that has to approve the project safety case.

ACKNOWLEDGEMENTS

The authors wish to thank Henri van Houten and Henny Koppens (IVW Rail) for their very useful contributions during the application of the Yellow Book approach. They also wish to thank Gea Kolk, Ed Boekestijn and Erik Jeurissen for their valuable input.

Finally, the authors thank the BB21 programme for permission to use experience acquired during the project to write these papers.

REFERENCES

CENELEC 50126 (1999). *NEN-EN 50126, The specification and demonstration of RAMS.* Delft: Nederlands Normalisatie Instituut.

CENELEC 50128 (2001). *NEN-EN50128, Software for railway control and protection systems.* Delft: Nederlands Normalisatie Instituut.

CENELEC 50129 (2000). *NEN-EN50129, Safety related electronic systems for signalling.* Delft: Nederlands Normalisatie Instituut.ERTMS (2002). *System Requirements Specification, issue 2.2.2, Chapter 7, ERTMS/ETCS language, par 7.5.1.74.*

Railtrack (2000). *Engineering Safety Management – Yellow Book 3, Volume 1 and 2, Fundamentals and Guidance.* London: Railtrack PLC.

Safety and Reliability – Bedford & van Gelder (eds)
© 2003 Swets & Zeitlinger, Lisse, ISBN 90 5809 551 7

Effective risk management of design & construct contracts

J. Donders & P. Vermey
Grontmij Projects, De Bilt, Netherlands

ABSTRACT: In the last five years many infrastructure works in the Netherlands have been contracted out using a design & construct delivery system. The innovative aspect of the procedure and the increased value of the contract packages lead to a changed risk profile for both owners and contractors. Explicit risk management became an important instrument in assessing, managing and allocating project risks. Based on the authors experiences as consultants on both sides of the table, some "lessons learned" are defined and improved methodology is recommended in order to enhance the cost efficiency and benefits of risk management.

1 INTRODUCTION

In the mid 90's the Netherlands started some major investment programs in public infrastructure, such as such as cargo line Betuweroute, the High Speed Line South, the Rail 21 program and major road works. To manage the increase in workload through outsourcing, to improve project results, and to stimulate innovations and optimizations, many projects have used a design & construct delivery system. Compared with the traditional procurement system where design and construction are split, this lead to new risk profiles and risk allocation schemes. Explicit risk management was introduced, initiated by the public authorities. Methods and processes were developed, and risk management plans became an obligatory part of the bid documents. Most of these projects are now in the construction phase, and it is possible to evaluate the process. Is this form of explicit risk management an efficient and effective tool for project management?

The next paragraphs describe the risk management process (in short: RM) and perspective from the owners and contractors point of view, and the lessons learned and recommendations for the RM cycle, risk assessment, risk measures and risk allocation.

2 PROCESS

2.1 Perspective

A risk may be defined as an unplanned event leading to loosing an asset or failing to reach a target.

In a project environment, work packages are usually executed by different parties. Each party has a different risk domain, a different risk exposure and therefore a different perspective on risk.

In public works projects, main risk carriers in the chain of activities from initiative to commissioning are:

– Financier, or client
– Program manager, or owner
– Engineering firm
– (Sub-)contractors
– Insurance companies and banks (guarantees)

A party will only feel a risk, and therefore act on it, if it originates from his own domain and leads to exposure for himself. (A domain may be defined as all activities and circumstances for which he is responsible.) For example, a contractor's domain is limited to the scope and conditions of the contract, and his exposure is related to profit and continuity. The domain of the financier is larger: it consists of the contractor's domain plus all other activities in the project plus all other circumstances disrupting the planned process. Exposure of a public financier is related to project targets such as budget, planning and fit for purpose, as well as policy issues such as enhancing mobility or stimulating public transport. Exposure of a private financier is more related to return on investment during the operation phase.

A non-profit program manager (e.g. the rail authority) is the middle party. Exposure is related to professionalism and reliability: to what degree is the project delivered within time and budget constraints proposed by himself at an early stage, and to what degree does the delivered system represent best value for money. Further more, the owner often has a non-quantified responsibility towards society as a whole in the areas of

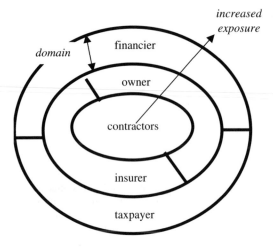

increased
exposure

domain

financier

owner

contractors

insurer

taxpayer

Figure 1.   Domain and exposure.

Table 1.   Risk attitudes in design & construct contracts.

Owner	Contractor
Invests in prevention	Prepares for correction
Pro-active: who controls	Reactive: who is liable
Contingency is buffer for not exceeding budget	Contingency is potential source of profit
Unlimited exposure	Limited Maximum Possible Loss (MPL)
Priority: fit for purpose after commissioning	Priority: costs before commissioning

environmental protection, acceptance, safety and aes-thetics. The domain is limited to the scope described in the project sanctioning document and the general mis-sion of the authority.

Differences in risk domain and exposure lead to a different attitude towards risk, which is sketched in Table 1.

These attitudes may change when the program manager is a one-time project director rather than an established (semi-) governmental authority. Based on statistical considerations, the one-timer may take more risks, and the contractor has only one opportunity to make a profit with this client.

2.2   *Process.*

A project follows three basic phases:

– Planning and development, resulting in project sanctioning
– Tendering, resulting in award of contract
– Execution, resulting in testing and commissioning

When using a d&c delivery system, risk management is a tool in each phase.

2.2.1   *Planning*
The owner uses RM to:

– Execute a continuous cycle of assessing and man-aging project risks;
– Establish the contingency in the cost estimate and slack in the planning;
– Define the acceptable RAMS levels in the specification;
– Define the contracting strategy and delivery systems based on the project risk profile and market risks.

2.2.2   *Tender*
The owner uses RM to:

– Execute the RM cycle;
– Define the pre-qualification requirements based on the risk profile of the contract packages;
– Define the risk allocation in the concept contracts;
– Define the risk management requirements in the tender documents;
– Evaluate the bids on compliance with the risk level in the RAMS specifications, the risk allocation and the risk management requirements;
– Negotiate the bid until there is a mutual perception and acceptance of the risk allocation in the defini-tive contract.

The contractor uses RM to:

– Execute the RM cycle during the bid-design process, with reference to the risk level in the RAMS specifi-cations and the risk allocation in contract;
– Set the contingency in the bid;
– Prepare the risk management plan as part of the bid documents;
– Negotiate bid the until there is a mutual perception and acceptance of the risk allocation in the defini-tive contract.

2.2.3   *Execution*
The owner uses RM to:

– Execute the RM cycle;
– Monitor risk management by the contractor, and to check claims and changes.

The contractor uses RM to:

– Execute the RM cycle;
– Check claims and changes.

3   RISK MANAGEMENT CYCLE

3.1   *Lessons learned*

The project risk management cycle is perceived as a paper tiger.

3.2 Recommendations

The RM cycle should be organized on the strategic level, not on operational level. The focus should be on those areas where explicit RM can add value to other management tools, and where the immediate effect on the operational results are visible.

In Figure 2 one can distinguish four management areas where explicit project risk management can be efficient and effective if used well.

- Area A – inefficiencies, events with a frequency >1 and low impact. No role for risk management.
- Area B – low impact risks, standard risks associated with craftsmanship. These can be managed through the quality management system. Risk management plays a role in identifying the critical objects or processes where standard risks may lead to a high impact due to the special nature of the project.
- Area C – major risks, obvious risks probably well known by the project manager. RM adds value by explicating the risk mechanism and potential measures, and by monitoring the residual risk.
- Area D – high impact low probability (HILP), the needle in the haystack, often capped in liability or insured. Assessing them is a major exercise and only relevant where new technology or processes are used, or where potential impacts can be extreme. RM can add value in identifying potential catastrophic losses or drivers without explicating the actual mechanisms.
- Area D – medium risks. This is the area where explicit RM can really add value.

The risk manager should report directly to the project manager with a balanced score card indicating the major items per area. Measures should then be focussed on checking the existing management systems of areas A, B and C, and on effective measures for area D including costs and a direct link to the contingency.

4 RISK ASSESSMENT

4.1 Lessons learned

Risk assessment is perceived as time consuming and costly. There is no standard in classifying or describing risk, which makes communication about risk difficult and perception of risks ambiguous. Assessment of probabilities and impacts are not understood well.

4.2 Recommendations

4.2.1 Generic model

On an abstract level, a project is a generic system of interrelated processes and objects. Risks are therefore also generic, and should only be specified where relevant using a structured approach.

A roadmap for such a structured approach contains the following elements

1. A standardized fault tree
2. A standardized process model of a project
3. Per process the standard risks based on input and output

This model can be applied to specify risks in relevant work packages or objects in the Work Breakdown structure.

The fault tree defines top down what type of consequences are assessed. It is one way of classifying risks. For example "quality risks" are risks which affect the quality in a negative way.

In Figure 4 some generic project processes are given. Standard risks can be assessed for each process individually. The more experience one has with a certain process, the more the risks are understood and managed. Experiences can than be fed back into the model, and used for the next project.

It is also a way of classifying risks. For example "design risks" are risks that originate from or affect the design process.

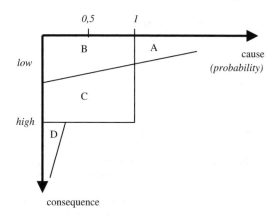

Figure 2.   Cause – consequence matrix.

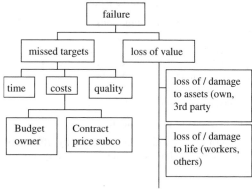

Figure 3.   Standard fault tree – example top level.

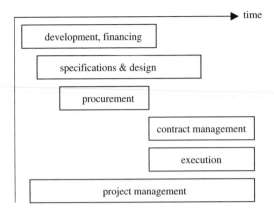

time

Figure 4.   Project processes (example).

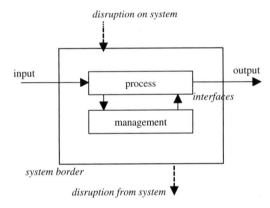

Figure 5.   Input/Output model.

Each process can be described as an input-output model, and standard risks can be defined. Examples:

The input failed when there is no input, or when the input is too late, of insufficient quality or too expensive. Input can be broken down into:

– Physical data and assumptions, e.g. type of soil
– Time depended data and assumptions, e.g. price of diesel in 5 years
– Conditions, e.g. date when permits are expected
– Output targets, e.g. scope, specifications, budget, milestones
– Baseline documents, e.g. final design

The blackbox or output failed:

– Internal organization, e.g. productivity, efficiency
– Resources and tools, e.g. quantity, quality or price

Interfaces failed:

– Impacts from other activities, e.g. planning constraint
– Impact on other activities, e.g. collateral damage

An extra advantage of using an I/O model is the possibility to allocate risks to the one who can control the input, the output or the interface.

An example of a structured approach: in the fault tree "time" is set as the impact to be considered. In the project the design process is investigated, more specifically the "final design of bridge X" (refer to WBS). One risk is now: "the project is too late due to the risk that the final design of the critical bridge X is too late due to the risk that the critical specifications Y (input) which were provided too late."

4.2.2   Probabilities and impacts

Risk can be measured using a ranking system of for example 1–5 for different type of impacts, and 1–5 for probabilities. The rankings are defined using standard definitions, for example: probability 3 means "medium" and represents a 10% probability.

This kind of ranking is easy and works well. However, it is not always clear to the estimators how to interpret it. For example, it should be clear that 10% means: "one in every 10 similar projects". Or an "impact of 100,000 euro" means that the residual risk after specified measures is 100,000 euro (refer to paragraph 5). More standardization and understanding will facilitate feedback of experiences.

5   RISK MEASURES

5.1   Lessons learned

Definitions of measures are fuzzy, lack commitment, and are not based on cost/benefit analysis.

5.2   Recommendations

5.2.1   Generic model

As risk assessments can be based on a generic model, so can risk measures. Generic measures can be defined for each input, output and interface risk. For example: standard measures to cover uncertainties in physical data (input) are:

– More tests
– Increase redundancy
– Set a contingency for corrective action, and define that action

5.3   Cost of risks

When measures are defined, the costs of risks can be estimated using standardized cost elements, e.g.:

– Investing in preventive measures such as redundancies in design, planning and resources; trade-off; data collecting studies; testing.
– Corrective measures to return a project back on track
– Corrective measures to repair or reimburse mistakes or damages

– Loss of assets
– Replacing defaulting contractor
– Costs of delay or inefficiencies, such as interest, overhead, non-productivity and liquidated damages
– Risk premium for transferring or insuring a risk
– Sureties, guarantees and bonds
– Financing costs of contingency for residual risks
– Risk management staff

Measures should be taken using a cost/benefit analysis, and the cost of residual risk should be linked directly to the contingency.

6 RISK ALLOCATION

6.1 *Lessons learned*

Risk allocation principles are too general. Definition of risk allocation in contracts is too formal and based on liabilities to stimulate efficient management.

6.2 *Recommendations*

6.2.1 *Risk allocation principle*
Often the principle used is that risks are allocated to the party who is best able to manage. This is too general and does not lead to an effective risk allocation. There is no separation in responsibility for causes and consequences, no difference between responsibility and liability, and it does not cover risk premiums or exposures.

Risks totally controlled by either contractor or owner can be allocated to that party. Risks which affect both, so called "gray" risks, should be managed by both. For these responsibilities, a risk premium should be paid rather than setting liabilities at a high price.

An enhanced set of principles can be: transfer a risk to the party:

– Best equipped to control cause and consequence of a risk, and
– Able to buy a risk for the lowest risk premium, and
– Solely exposed to the consequences, and
– Able to carry the possible impacts of a risk

If this is not possible, one should look into the details and find a solution of a common risk management alliance where risks and rewards are shared.

6.2.1 *Risk allocation mechanism*
During tendering the owner should specify which risks are clearly with him. The remaining risks are then up for sale. Contractors are requested to specify risks the "gray" risks, and price them, and to specify

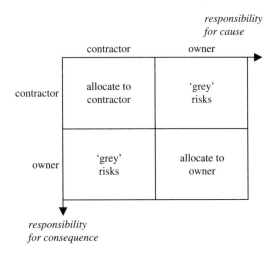

Figure 6.   Responsibility for cause or consequence.

which risks are solely with the contractor. These are then included in the total risk premium. Competition on the gray risks is a major element.

Award of contract is based on a design & construct contract including allocated risks and a risk management plan, and an alliance contract for the remaining "gray" risks.

If contracts are negotiated in a competitive market, legal wording of the risk allocation is usually based on two principles: clarity above fairness, and accountability above responsibility. It would be better to pay attention to the definition of risk, responsibility for causes and measures, liability to consequences, and finally and caps and guarantees.

7 FINAL NOTE

Some key points in effective risk management are:

– Focus on management, not on risk assessment or liability
– Optimizing the cost of risks, including measures and residual risks, should be the driving force for risk management
– Manage "gray" risks jointly using incentives

Over time, improvements in project risk management should lead to a reduction of the total costs, and an increase of profits and reliability. Otherwise risk management will become obsolete.

Safety and Reliability – Bedford & van Gelder (eds)
© 2003 Swets & Zeitlinger, Lisse, ISBN 90 5809 551 7

Risk-NL, The Netherlands quantitative risk analysis tool to quantify and analyse risks involved with storage of ammunition and explosives

Ph. van Dongen, G.H. Lodder & L.H.J. Absil
TNO Prins Maurits Laboratory, Rijswijk, The Netherlands

ABSTRACT: The risk of an accidental explosion is inherent in the handling and storage of ammunition and energetic materials. Depending on the nature of the explosion and the infrastructure, the consequences of an accidental explosion range from only minor structural damage to large-scale catastrophes in which many people can be killed. The EU SEVESO II Directive obliges companies handling dangerous goods to give authorities insight information into the hazards involved. Quantitative Risk Analysis (QRA) is the methodology to quantify and analyse the risks of storing and handling dangerous goods. With a QRA, process safety managers are able to take adequate risk-reducing organisational and constructive measures.

On behalf of the Netherlands Ministry of Defence, the TNO Prins Maurits Laboratory has developed the quantitative Risk Analysis software RISK-NL to quantify and analyse risks involved with storage of ammunition and explosives. The model was developed in the early eighties. In 1999, the latest definitions of Individual Risk and Societal Risk according to the Netherlands Ministry of Housing, Spatial Planning and Environment, which is responsible for external safety issues of dangerous goods in general, were implemented in the model. The risk is presented in iso-risk contours with regard to the individual risk and a so-called F/N-curve with regard to societal risk. The advantages and disadvantages of the imposed methodology for QRAs specifically on ammunition and explosives storage will be discussed.

1 SUMMARY OF RISK-NL PROCEDURE

The procedure to perform a Quantitative Risk Analysis with the model RISK-NL starts by defining an input data-file with relevant information regarding the storage site and its surroundings.

With this basic information, the maximum credible event per Potential Explosion Site (PES = each separate storage magazine on a site) is calculated including the effects of possible sympathetic detonation.

With the chosen initiation frequency of each PES and the calculated consequences for an unprotected (fictitious) individual, standing for 24 hours a day and all year long at an arbitrarily location, the Individual Risk is calculated. The risk for this fictitious person is the sum of risks of all PESs. This IR is presented in the form of iso-risk contours, which are lines with the same level of risk.

With the defined real exposed sites and their occupation, the Societal Risk is calculated. This SR is presented in the form of a so-called F/N-curve. Per possible explosion scenario (per PES), the expected number of persons being killed and the probability of that event are plotted in this curve.

The calculated IR and SR can then be compared with acceptance criteria set by the Netherlands Government. In case the risks are too high, an analysis of the risks follows and protective measures will be advised in order to decrease the levels of risk. In figure 1, a photo of a 1 ton detonation is presented.

Figure 1. Detonation trial of 1 ton explosives (view at 1300 m, with 300 mm telelens).

2 INPUT DATA

To obtain the Individual risk and Societal risk of a storage site, the expert-user of RISK-NL must define the input data-file. A set of choices have to be made regarding Potential Explosion Sites (PESs) and Exposed Sites (ESs).

2.1 Potential explosion sites

The size and orientation of a magazine is defined by its co-ordinates. The first two co-ordinates define the entry wall of the magazine (as being the most vulnerable side) while the third co-ordinate defines the side wall. Only three co-ordinates are necessary as the magazine is assumed to be rectangular. The types of magazines that are defined in RISK-NL are:

I7: Igloos designed for 7 bar;
I3: Igloos designed for 3 bar;
ID: Earth covered buildings with a headwall and door(s) resistant to high velocity projections if facing a Potential Explosive Site (PES);
IB: Earth covered buildings with a barricaded door facing a PES;
BR: Heavy-walled buildings (0.7 m concrete, brick or equivalent) with a protective roof (0.15 m concrete with suitable support) and barricaded door if facing a PES;
BD: Heavy-walled building and a barricaded door if facing a PES;
OB: On three sides barricaded open-air stack or light structure;
OT: On four sides barricaded open-air stack or light structure;
OP: Unbarricaded open-air stack or light structure.

For each PES on a storage site, the expert-user has to select a proper type of magazine. This choice has a main effect on the debris throw-out.

2.2 Exposed sites

Each exposed site is represented by one co-ordinate (x,y) except for public traffic routes. These are represented by two co-ordinates for two ends of the road. A road can be described as having several sections, therefore a road can be drawn using several co-ordinates to represent the different shapes of the sections.

With an (pre-determined) average number of people per unit, the total number of people involved with each exposed site can be tied up by a number of units coupled with each surrounding object. For public traffic routes, the 24-hour traffic intensities for cars and for cyclists/pedestrians must be known or estimated including their average speed.

The following are all Exposed Sites that can be used in RISK-NL:

HS: inhabited buildings;
HF: inhabited buildings with more than four floors;
RD: public traffic roads;
PL: POL facilities (Petroleum, Oils, and Lubricants).

For people in/near houses, the calculations are performed assuming the following: 90% of the people are in the house of which 5% are within a distance of 1.75 m of a window, and 10% will be outside the house. People in the open are assumed to be unprotected.

3 FREQUENCY OF EXPLOSION

When using RISK-NL, the frequency of an explosion must be determined. In the early eighties, TNO-PML had established a value of between 1.10^{-5} and 1.10^{-6} per magazine per year, which was based on available historical data of the Netherlands and NATO partners. For safety reasons, the worst case scenario was taken of 1.10^{-5} (per magazine per year).

In 1998 it was acknowledged by the NATO panel on explosion safety, AC/258 Storage Sub Group (STSG), that the methodology and techniques of quantitative risk analysis (QRA) should be a useful step forward in the area of explosion safety principles. This acknowledgement resulted in the establishment of the AC/258 STSG Risk Analysis Working Group (RAWG) which has the aim of examining the requirement for risk analysis input to the ammunition and explosive storage and transportation areas, the production of a NATO Risk Analysis Manual AASTP-4, and to make recommendations for technical risk analysis inputs to all other documentation produced by AC/258 and its Sub Groups. In the RAWG, the frequency of an accidental explosion in a storage magazine was a topic with high priority. It showed that NATO partners having an operational QRA model, used different approaches in obtaining data on initiation frequency. Instead of using one fixed value for initiation frequency, NATO partners made the value depending upon logical parameters, like the Compatibility Group of ammunitions, the amount of NEQ, the type of magazine, the type of activities and number of handlings per unit of time, etc.

At the moment, TNO-PML is investigating which foreign approach is best applicable in the Netherlands and has the intention to contribute to the NATO STSG Risk Analysis Working Group discussions on risk analysis.

4 CALCULATION OF EXPLOSION EFFECTS

4.1 Sympathetic detonations

In order to calculate the Maximum Credible Event of a certain explosion scenario, the model RISK-NL investigates firstly if sympathetic reactions inside adjacent magazines can occur (domino-effects). These calculations are based on the quantity-distance rules as given in the NATO publication AASTP-1. For each PES, the total mass involved in an ammunition reaction will be summed in order to calculate the total explosion effects and consequences for the surroundings.

4.2 External safety

4.2.1 Air blast

The blast parameters are deduced from hemispherical TNT charges detonated at ground level in the open air. These blast parameters are side-on and reflected blast, side-on and reflected impulse, positive phase duration, arrival time of shock wave and the shock front velocity. Blast parameters are valid for scaled distances ranging from 0.0674 to 40.0 m/kg$^{1/3}$. For each donor magazine with its total quantity of explosives, the side-on and reflected overpressure, the side-on and reflected impulse and positive phase duration are calculated from (NATO AASTP-1, 1997).

4.2.2 Debris and fragments

For each specific donor magazine with its total net explosive quantity, the fragment density and mass distribution as a function of distance is calculated with formulae derived from fitting experimental data of full-scale explosion trials. For the fragment density and mass distribution, fitting formulae of experimental data obtained from (Weals, 1973), (Weals, 1974) and (Feinstein, 1976) are used. These formulae were validated by comparing calculated results with more recent data from (Bowman, 1984), (Henderson, 1986) and (Henderson, 1988) and no significant contradictions were found.

First, the debris/fragment density is calculated. The following distinctions are made for HD1.1 ammunitions:

- density for non-igloo type magazines containing less than 3000 kg;
- density for non-igloo type magazines containing more than 3000 kg;
- density at the sides of an igloo magazine;
- density at the back of an igloo magazine;
- density at the front of an igloo magazine.

Secondly, a mass distribution is calculated for three mass classes:

- m1 < 0.1 kg;
- 0.1 kg < m2 < 4.5 kg;
- m3 > 4.5 kg.

As a function of the scaled distance, a mass distribution is defined, i.e. between scaled distances 4 and 5 (m/kg$^{1/3}$), the mass distribution between m1, m2 and m3 is 55%, 41% and 4% respectively.

The initial velocity of small fragments and debris are set at 1500 m/s when the object is situated at the door of an igloo type magazine, otherwise it is set at 1200 m/s. For fragments and debris of medium mass (m2), the initial velocity is set at 800 m/s and for heavy fragments (m3) at 300 m/s.

Impact velocities are calculated with formulas as given in the NATO manual AASTP-1 (NATO, 1997). Next, the ballistic limit velocities V50 for 22 cm brick (houses), skin or 1.5 mm mild steel (cars) of 0.1, 1.0 and 4.5 kg pieces of fragments and debris are calculated. With the impact velocities and the ballistic limit velocities, V50, the residual velocities after perforation is calculated.

4.2.3 Thermal radiation

For each donor magazine with its total quantity of explosives the radiation intensity and exposure time is calculated from (NATO AASTP-1, 1997).

5 CALCULATION OF CONSEQUENCES

With the quantified physical results found from the performed calculations, the consequences for surrounding exposed sites can be estimated. This is done using functions developed at TNO-PML from experimental data found in literature and from experiments performed by TNO-PML. These are also described in (Green Book, 1990) which contains vulnerability models for the effect of hazardous materials on humans. The general form of these "*probit functions*" are:

$$Pr = a + b.\ln S \qquad (1)$$

where a = a constant; b = a constant; S = a function with input parameters, e.g. air blast peak overpressure and impulse.

Depending on the value of S and the constants a and b, the magnitude of Pr determines the probability of lethality. The values of the probit values (PR) correspond to a certain probability of lethality given in percentages.

5.1 Airblast

The consequences of air blast on persons are given by the probit functions for the following phenomena:

- injury involving eardrum rupture (non-lethal);
- lethal injury involving lung damage;
- the probability of a house collapsing;
- the probability of structural damage to a house;
- the probability of light damage to a house;

- the probability of windowpane breakage;
- the probability of a lethality due to windowpane breakage for a person standing 1.75 m behind that window;
- serious/lethal injury by head impact;
- the probability of a lethality due to total body impact.

Assumptions made for blast criteria

- The probability of lethality due to a collapsing house is set at 35%, due to structural damage 5% and due to light damage it is set at 1%. These values have been taken from lethality data of collapsing buildings during an earthquake.
- It is assumed that 5% of the people in a house will be at a distance of <1.75 m from a window.
- The number of injuries due to the collapse of buildings is supposed to be twice the number of lethalities;
- The number of injuries due to fragment effects is supposed to be twice the number of lethalities.
- The number of injuries due to head and total body impact is supposed to be twice the number of lethalities These figures for the number of injuries are normally not used in the assessment of the individual and societal risk.
- For roads, an average value for the probabilities of lethality for cars and for cyclists/pedestrians is taken. The average velocity for a car is assumed to be 60 km/h while for a cyclist/pedestrian it is taken to be 15 km/h. The lethality of the object is the probability of lethality times the number of units times the average occupation of the object and where appropriate, the probability of lethality times the traffic intensity.
- From all this information, the effects of all donors from the storage site for all objects can be calculated.

The probit functions itself are given in (Green Book, 1990 and Absil, 1998). As an example, the probit function for lethal lung damage is given in the following:

$$Pr = 5.00 - 5.74.\ln S \qquad (2)$$

in which:

$$S = 4.2/Psc + 1.3/Isc \qquad (3)$$

and:

$$Psc = Ps/Po \qquad (4)$$

$$Isc = Is/(m^{1/3} \cdot Po^{1/2}) \qquad (5)$$

Ps = side-on peak over pressure (kPa); Po = ambient pressure (kPa); m = mass of person (average of 70 kg assumed); Is = side-on impulse (kPa.s).

When the probit, Pr, is calculated, the corresponding probability of lethality (given as a percentage) can be derived using the relationship for probabilities and probits (Green Book, 1990 and Absil, 1998).

5.2 Debris and fragments

To calculate the consequences of debris and fragment impact on persons, probit functions are derived by TNO-PML for masses m1, m2 and m3:

- for small masses (m1 is set to 0.1 kg), a skin perforation criterion is used (Absil, 1998);
- for medium masses (m2 is set to 1.0 kg), a kinetic energy criterion is used (Absil, 1998);
- for heavy masses (m3 is set to 4.5 kg), a skull fracture criterion is used (Absil, 1998).

5.3 Thermal radiation

The probability of lethality for unprotected human beings due to radiation heat effects is derived from carburetted hydrogen fires. NATO AC/258 criteria (NATO AASTP-1, 1997) for lethal and second degree burning are taken into account. The percentage of lethality depends on the heat radiation (kW/m^2) and the exposed time (s).

6 PRESENTATION OF RISKS

Knowing the possible explosion scenario's, their initiation frequencies, explosion effects and consequences, the risks can be calculated. To do so, the definitions of individual risk and societal risk are essential for the resulting figures.

6.1 Definition of individual risk

This is the probability per year that an unprotected individual standing at an arbitrarily location for 24 hours a day, all year long, will be killed by an unwanted event with dangerous goods.

The advantage of this definition is that the risk is only depending upon storage site specifications. The risk is not depending upon the surroundings, since we are dealing with a fictitious individual with constant specifications. As a result, using this definition of IR to assess the risk at all grid points around a site (resolution about 25 × 25 m^2 per grid cell), contours with the same level of risk can be drawn (iso-risk contours). The IR is graphically portrayed by iso-risk contours around the storage site with PESs, which are lines which joins all points of equal values of IRs. An example of iso-risk contours is given in figure 2. The black boxes in this figure are the PESs and the three iso-risk contours represent risk levels of 1.10^{-5}, 1.10^{-6} and 1.10^{-7}.

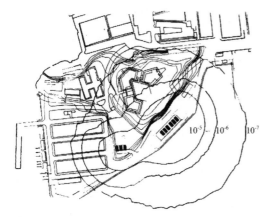

Figure 2.   Example of iso-risk contours projected on map (black boxes are PESs).

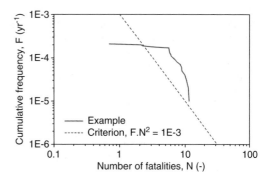

Figure 3.   Example of F/N-curve (dotted line is acceptance criterion).

6.2   IR acceptance criteria

The acceptance criteria set by the Netherlands Government are as follows:

- within the 10^{-5} contour only public traffic routes are allowed;
- between the 10^{-5} and 10^{-6} contour, buildings with short time occupation (e.g. offices, hotels, etc.) are allowed;
- outside the 10^{-6} contour, all other exposed sites, like dwellings, are allowed.

Also, a distinction is made for existing situations and new situations (e.g. new housing development). The criteria for existing situations are less stringent.

These acceptance criteria show the obvious disadvantage of the imposed definition of Individual Risk. The risk is assessed for a (fictitious) unprotected person in the free field standing all year long at a certain location, while the acceptance criteria include the time of presence (presence factor) and the conditions in which the persons are at the moment of the unwanted event (e.g. in a building or car). The definition of IR results in a worst-case figure, assuming that a person in the open is most vulnerable for the effects of dangerous goods. For HE-explosion, this is controversial. In the open at far field, debris and fragments throw out are less hazardous (probability of being hit is low) and air blast is far from lethal, but inside a building, window-pane breakage can form a serious threat to people.

However, the Societal Risk, which is defined to obtain figures on the total number of persons killed during an event, has to be derived for a realistic situation.

6.3   Definition of societal risk

The Societal Risk (SR) is the probability per year that one group of a certain size all die due to an accident

with dangerous goods. This includes persons in the vicinity of the PESs whereby the SR will be greater as the people near the PESs increase. For each explosion scenario, i.e. for each magazine or PES, the initiation frequency and the total number of victims expected are calculated. The SR is graphically depicted in an F/N-curve which is a graph showing the cumulative frequency of the accident scenario's and their subsequent number of victims. An example is given in figure 3.

6.4   SR acceptance criteria

The acceptance criterion for the Societal Risk is that an accident whereby ten deaths are expected should have a probability of occurring less than one in 100.000 per year. An aversion factor is included to account for the unwillingness of the society to accept events in which large groups of persons will be killed. The acceptance criterion is:

$$F.N^2 = 1.10^{-3} \qquad (6)$$

in which F = initiation frequency (per year); N = number of persons killed (-).

E.g. it is acceptable that an accident occurs with 100 persons being killed, if the probability that such an event takes place, is less than 1.10^{-7} per year.

7   CONCLUSIONS

In the early eighties, the TNO Prins Maurits Laboratory started with the development of the Quantitative Risk Analysis model RISK-NL. The model RISK-NL is based on the QRA methodology for dangerous goods in general in the Netherlands and the specific definitions of Individual Risk and Societal Risk.

The sub-models, describing the explosion effects and the consequences for persons in a variety of conditions are based on (inter)national research over the last decades and are frequently updated with state-of-the-art knowledge (e.g. through NATO AC/258 working groups) on explosion effects.

Advantage of the current model is that the sub-models are relatively detailed and well documented in NATO publication AASTP-1 (NATO, 1997) and national literature (i.e. Green Book, 1990). The probit functions which predict the probability of lethality for relevant explosion effects, are technically easy to adjust to the latest knowledge on consequence modelling.

Current disadvantage of the model RISK-NL is the lack of (Netherlands specific) data on initiation frequency or probability of an accidental explosion. NATO partners with comparable models put quite an effort in deriving more detailed (national specific) figures for this essential parameter. However, the NATO AC/258 STSG Risk Analysis Working Group is the proper panel to develop national QRA methodologies and a consistent NATO advise.

REFERENCES

Absil, Dr. L.H.J., Dongen, Ph. van, and Kodde, H.H. Inventory of damage and lethality criteria for HE explosions. TNO-report PML 1998-C21. NATO AC/258 informal working paper: NO(ST)UGS/AHWP IWP 2/98, dated March 1998.

Amelsfort, R.J.M. van. Risk calculations in storing explosives. TNO-PML 1993.

Bowman, F., Henderson, J., Walker, J., and Rees, N.J.M. Joint Australian/UK stack fragmentation trials, Phase I report. 21st Explosives safety seminar 1984.

Feinstein, D.I., and Nagaoka, H.H. Fragmentation hazard study, phase III: fragment hazards from detonation of multiple munitions in open stores. IIT Research Institute, Final report 76176, 1976.

Handreiking externe veiligheid (in Dutch). ISBN 90 322 7122 9.

Henderson, J., and Rees, N.J.M. Joint Australian/UK stack fragmentation trials, Phase I report. 23rd Explosives safety seminar 1988.

Henderson, J., Walker, J., Rees, N.J.M., and Bowe, R. Joint Australian/UK stack fragmentation trials, Phase I report. 22nd Explosives safety seminar 1986.

NATO Allied Ammunition Storage and Transport Publication AASTP-1. Manual on NATO safety principles for the storage of ammunition and explosives, 1992.

"Schadeboek" Commissie Preventie van Rampen door gevaarlijke stoffen (Green book). Directoraat-Generaal van de Arbeid van de Ministerie van Sociale Zaken en Werkgelegenheid. CPR, ISSN 0921-9633; 16, 1990.

Weals, F.H. ESKIMO I magazine separation test. DoD ESB, Naval Weapons Center, April 1973.

Weals, F.H. ESKIMO II magazine separation test. DoD ESB, Naval Weapons Center, September 1974.

Safety and Reliability – Bedford & van Gelder (eds)
© 2003 Swets & Zeitlinger, Lisse, ISBN 90 5809 551 7

Consequence assessment of large bomb explosions in urban areas

J.C.A.M. van Doormaal & L.H.J. Absil
TNO Prins Maurits Laboratory, Rijswijk, The Netherlands

ABSTRACT: At TNO Prins Maurits Laboratory a case study has been carried out to quantify the explosion effects and the consequences for a building at a certain distance from a large car bomb explosion in an urban area. Different urban configurations have been considered in order to obtain a range of relevant blast loadings. In addition, the vulnerability of different types of structures has been studied. Together, these data sets form a methodology to estimate the seriousness of a nearby explosion for a building in an urban environment and to determine the necessity for retrofit measures.

1 INTRODUCTION

1.1 Background

Terrorist attacks are usually committed in urban areas because of the location of the target. As a consequence not only the target may be exposed to the explosion load, but also the surrounding environment. In Oklahoma for instance, also the buildings surrounding the Federal Building suffered from the attack.

This shows that terrorism is not only a concern of those who may be a target, but also for organizations which are located close to such a possible target. Compared to the target building, the environment is, thanks to the standoff, better off in the way that the damage can often be categorized as light to medium and that protective measures are relatively easy to take.

A proper protection plan should be based on an analysis of the situation. Insight must be obtained in the possible explosion loading a building can be exposed to and the vulnerability of that building for this loading.

1.2 Case study

At TNO Prins Maurits Laboratory a case study has been carried out to quantify the explosion effects and the consequences for a building at a certain distance from a large car bomb explosion in an urban area.

The urban area is an essential parameter in this study. The buildings form obstacles for the expanding blast wave and thus influence the propagation of the blast wave. This is a very complex process of interaction between the moving blast wave and the obstacles. The blast loading on buildings and people in the neighborhood of an attacked building strongly depends on this interaction process. At locations where blast shielding occurs the overpressures may be relatively low. On the other hand high overpressures can arise in case of blast focusing by reflection of the wave in certain directions.

1.3 Objective

The objective of the case study was to develop a methodology that can support decisions on the necessity and usefulness of protective measures. The methodology should help to quickly estimate the seriousness of explosions for buildings in urban areas and the necessity and possibilities for protective measures.

1.4 Approach

The study has been approached in a stepwise procedure. The first step was to define a realistic explosion scenario, i.e. explosion source and distance. Next several street configurations have been defined to take into account the influence of the environment. For all configurations, calculations have been carried out to determine the explosion effects.

On the other hand, the vulnerability for blast loading has been determined for some typical structural elements of normal urban buildings. Combination of this vulnerability with the range of explosion loads enables the classification of a given structure in a given environment.

2 EXPLOSION SCENARIO

2.1 *Explosion source*

Theoretically, there is no limit to the size of a bomb. Practically however, the following categories can be defined based on the delivery devices (see for instance TSWG 1999):

1. small bombs on people or mail bombs, in the range of about 0.1 to 5 kg explosive material,
2. façade bombs (for forced entry purposes) and suitcase bombs in the range of about 5 to 25,
3. car bombs in the range of about 100 to 2000 kg charge weight, and
4. truck bombs in the range of 4000 to 20000 kg.

The bombs of category 1 and 2 will have only a small damage radius and are not relevant for the present study. Only large car bombs and truck bombs will be relevant for the environment of a target.

A 1000 kg ANFO-charge has been chosen as a realistic car bomb. In many international fora this size of car bomb is taken as a reference value for terrorist attacks. Examples like Oklahoma, Nairobi and Dar-es-Salaam, show that this amount is not too extreme.

ANFO is an explosive based on ammonium nitrate. Because the components can be easily obtained, it is often used by terrorists.

2.2 *Distance*

A first rough estimate of damage radii showed that the range between 100 and 200 m is the most interesting one. Closer to the explosion source the damage to be expected will be severe. Further away than 200 m the damage to expect decreases rapidly and the damage to the buildings will be very light.

From this rough estimate, it results that the most interesting distance is about 150 m.

2.3 *Environment*

The configurations for the environment have been chosen based on the following reasoning:

- the configurations should cover a broad range of situations, like densely or spaciously built-up areas, and low or high buildings;
- a worst case scenario should be studied;
- the configurations should be realistic (e.g. street widths and building heights);
- keep it as simple as possible.

Thus, the following configurations were defined:

1. A spaciously built-up environment, where any buildings present have an insignificant influence on the propagation of the blast wave, that means no focusing or shielding effects occur. The explosion can therefore be referred to as free field;
2. The explosion source and the building considered are in the same densely built-up street, which is 15 m wide. Two different heights for the buildings are considered, 12 m and 20 m. Any side streets are neglected, because the influence would be small. See Figure 4 for a schematic drawing.
3. The building under consideration is at the end of the street where the explosion occurs or just around the corner. Again a densely built-up area is taken with 15 m wide streets and 12 or 20 m high buildings. Any side streets are again neglected. See Figure 6.
4. The building under consideration is a high building (50 m) between lower buildings (20 m) in a spaciously built-up area, consisting of large blocks in wide streets (25 m). See Figure 8.

3 EXPLOSION EFFECTS

3.1 *Blast wave characteristics*

A blast wave arises as a consequence of the almost instantaneous release of the energy in an explosive. Inside the blast wave the pressure, density and particle velocities are very high.

A typical pressure history of a blast wave is shown in Figure 1. Since the blast wave travels faster than the sound speed, it does not give a warning of its arrival. One moment, the ambient is still undisturbed. The next moment, the pressure instantaneously increases to a maximum. Then the pressure decreases exponentially with time until atmospheric pressure is reached (positive phase). A negative phase follows with a pressure below atmospheric.

Particularly the positive phase is relevant for blast analysis and damage prediction. This phase is characterized by two parameters: the peak pressure and the impulse (i.e. the integrated pressure).

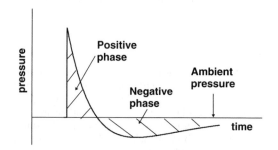

Figure 1. Typical blast pressure history.

3.2 Interaction with obstacles

Figure 2 gives an illustration of the influence of obstacles on the propagation of shock waves.

Reflection occurs when a blast wave finds an obstacle in its path. The movement of the particles in the shock front is abruptly stopped by the obstacle and reversed in direction. A new shock wave moves back through the medium. Locally the pressure is very high. This is called the reflected shock wave.

The reflected shock loading may be relieved by diffraction. Since the reflected overpressure is substantially higher than the overpressure in the immediate surroundings, there is a flow of air from the region of high pressure to the one at low pressure. This flow proceeds as a rarefaction wave over the front of the structure from the edges to the centre. At arrival of the rarefaction wave, the pressure drops back to the incident pressure. As a consequence the impulse of the reflected shock wave may be relieved. The reduction in impulse depends on the time the rarefaction wave needs to relief the whole front of the obstacle. The larger a building, the longer the relief process will take, and the less the reduction in overall impulse.

3.3 Calculations

The calculations have been carried out with two different computer programs. For the free-field case (case 1), the program ConWep (TM 5-855-1 1990) has been used to calculate the blast load. ConWep is a state-of-the-art semi-empirical code for Weapon Effects calculations (Bogosian et al. 2002). It can, however, not deal with complex situations, such as the influence of the urban area on the blast propagation.

For the other cases therefore, the software program BLAST (Van den Berg 1985) has been used. This program is an in-house developed program of TNO Prins Maurits Laboratory for simulation of blast wave-structure interaction.

3.4 The BLAST code

The TNO software BLAST is designed to simulate the gas dynamics of blast wave propagation in three dimensions by solution of the conservation equations for inviscid compressible flow (the Euler equations). The Flux-Corrected Transport scheme is used to capture and preserve shock phenomena. Flux-Corrected Transport employs the qualities of both, a first and a higher order finite difference scheme to obtain solutions with an optimum balance in accuracy and stability.

BLAST has proven its capability to model blast propagation. Comparison with experimental data showed good results (Absil et al. 1993, Van Doormaal & Absil 2001).

Figure 3 shows as an example the comparison of isobar patterns, obtained experimentally by means of

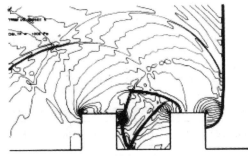

Figure 3. Comparison of interferogram (top) and calculated isobar pattern (bottom).

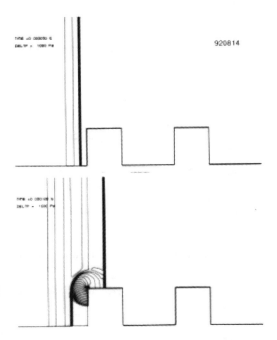

Figure 2. Propagation around obstacles.

interferometry and numerically with BLAST. Both patterns show the same features of reflection against the obstacles and diffraction around the obstacles. Also the pressure levels compare very well (see Absil, Van den Berg & Weerheijm, 1993).

3.5 Application of BLAST

When using BLAST a grid size must be chosen. This choice is a compromise between computation time and accuracy. Prevention of numerical blast decay requires a sufficient resolution. The 150 m distance however is quite large and with small cells the total mesh might become that large, that computer capacity problems may occur.

For these calculations the grid has been chosen as large as possible. The accuracy of the results has been checked by starting with a BLAST simulation of a free field 1000 kg ANFO ground burst. At different distances of the charge the blast characteristics have been determined and compared with the ConWep data base.

To further limit the computation time, use has been made of symmetry as much as possible. That means that the explosive charge has been assumed at the centre axis of the street. Although in real life, the charge would be expected to be off centre, the assumption of symmetry is allowed thanks to the large distance of 150 m for which the blast characteristics are to be obtained.

4   RESULTS BLAST LOADING

4.1   Case 1: free field

The blast loading for the free field case has been obtained with the ConWep program. It appeared that for this blast wave relief of the reflected shock wave can only be taken into account for small buildings (smaller than 4 m high and 8 m wide). Therefore, relief has not been considered here.

Table 1 summarizes the results.

4.2   Case 2: in narrow street

Figure 4 shows a schematic drawing of this case. The dashed lines are symmetry lines. The front façade of the building considered is the most interesting. Target points have been defined at this façade.

Table 1.   Blast loading characteristics for case 1: free field.

Distance = 150 m	Pressure (kPa)	Impulse (kPa.s)
Front face	18	0.39
Side and rear	8.7	0.21

The calculation with BLAST has been carried out twice: for 12 m and 20 m high infrastructure. Figure 5 shows the pressure–time curves for the 12 m high infrastructure. Table 2 summarizes the results.

In the narrow street, close to the explosion source reflections at both sides of the street will occur. At 150 m distance, these reflected shock waves have combined together to one propagating shock wave parallel with the street, as shown by the single peak in the curves in Figure 5. The propagating shock wave is relieved above the infrastructure. This is shown by the lower peak for the target points at 10 m height compared to the target points at 0 m height and by the reduction in peak pressure with increasing distance.

The channeling effect of the narrow street is clear from two observations:

1. Compared to the free field case the characteristics (pressure and impulse) of the skimming blast wave are high.

Figure 4.   Sketch of case 2.

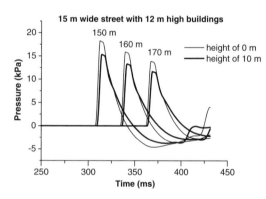

Figure 5.   Blast loading for case 2 with 12 m high infrastructure.

Table 2.   Blast loading characteristics for case 2: narrow street.

Distance = 150 m, H = 0 m	Pressure (kPa)	Impulse (kPa.s)
12 m high infrastructure	18	0.29
20 m high infrastructure	30	0.66

546

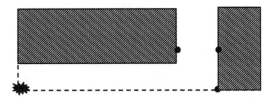

Figure 6. Schematic drawing of case 3: T-junction.

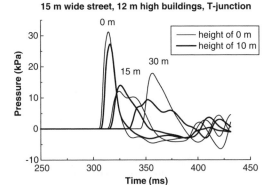

15 m wide street, 12 m high buildings, T-junction

Figure 7. Blast loading for case 3 with 12 m high infra-structure.

Table 3. Blast loading characteristics for case 3: T-junction.

Distance = 150 m	Pressure (kPa)	Impulse (kPa.s)
12 m high infrastructure		
Facing explosion	31	0.36
In side street	17	0.33
20 m high infrastructure		
Facing explosion	52	0.83
In side street	29	0.54

2. The blast characteristics are much more important for the 20 m high infrastructure than for the 12 m high buildings.

4.3 Case 3: T-junction

Figure 6 shows a schematic drawing of this case. The blast loading has been calculated at three positions: in view of the explosion source, and on two sides of the street around the corner.

Figure 7 shows the pressure pulses as calculated for the case with 12 m high infrastructure. The 0 m, 15 m and 30 m above the curves points out the distance from the symmetry axis.

Table 3 summarizes the main results.

Figure 8. Schematic drawing of case 4: high block between lower blocks.

Table 4. Blast loading characteristics for case 4: blocks.

	Pressure (kPa)	Impulse (kPa.s)
In street of explosion	13	0.28
In side street	17	0.43

Again the channeling effect can be observed. The reflected shock load on the façade facing the explosion is much higher than for the free field situation.

On the other hand, shielding effects can be observed in the results for the targets in the side street. The peak pressure for these targets is considerably reduced. The impulse however is still of the same order. This is due to additional reflecting shocks in the side street. As can be seen in Figure 7, some of the signals are clearly built up by a couple of shock waves.

Not only reflection in the street explains the number of shock waves, also diffraction over de buildings contributes to the total loading. This is shown by the curve for the 30 m off-axis target point at the height of 10 m. Here the diffraction wave over the buildings reaches the target point sooner than the blast wave from around the corner. That is why for this high target point the arrival time is shorter than for the target point at ground level.

4.4 Case 4: high block between lower blocks

For this case Figure 8 gives a sketch of the situation and Table 4 gives the main results.

Comparison with the other cases shows that focusing effects are less significant than for case 2 and case 3. This is not so strange, because of the larger width of the streets and the presence of broad side streets that provide relief.

5 POSSIBLE DAMAGE

5.1 General

The blast calculations provide a range of blast loadings which must be compared with the resistance of

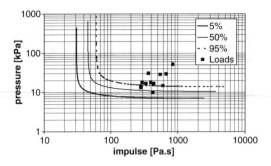

Figure 9. Probability of window failure of a 6 mm thick window with a size of 0.6 × 1.0 m; loads at 150 m distance for range of considered geometries.

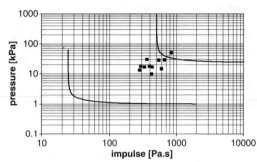

Figure 10. Blast resistance of masonry compared with the relevant blast loadings.

structures in order to get information on the possible damage. Similarly as for the environment, for structures there is a broad range of possibilities. It is impossible to cover all. Therefore a choice had to be made for some typical structural elements, which demonstrate the damage levels that may be expected.

5.2 Windows

One of the most vulnerable parts of buildings for blast loading is a window. It breaks easily and the glass shards will be thrown inwards. The glass fragments are usually sharp and have a high penetrating force which may be lethal for people present, depending on where it hits a person.

For the range of blast loadings relevant in this case study, window failure is certainly an issue. The diagram in Figure 9 compares the failure of a specific window (size 0.6 × 1.0 m, thickness 6 mm) with the range of blast loadings. The failure probability of this window ranges between 31% and 100%, depending on which loading case is considered. For most loading cases, the failure chance is 100%.

The failure probability of this window has been determined with use of the TNO-method for window failure (CPR 16E, 2000).

5.3 Walls

For walls three typical types were selected: lightweight cladding, masonry and concrete.

The cladding walls are known to be very vulnerable for blast loads, because of their low mass and low resistance. The sheets are often torn loose and spread around. Often these sheets are found outside the structure because they fail in the negative loading phase or in the rebound.

For masonry walls, it is not possible to give a clear level of blast resistance. This level strongly depends

on the quality and construction. Figure 10 shows the extremes which are possible for masonry. Weak masonry walls will certainly fail. Strong masonry walls may be able to resist the for this case relevant blast loadings.

Like for masonry, concrete walls can also be of very different quality and performance. Compared to masonry, the protection level of concrete is generally higher. The protection level of concrete walls would start from a little bit less than that of the stronger masonry walls. In this case study it would mean that only for the worst case blast loadings, a weak concrete wall could sustain too much damage.

Concrete thanks its higher capacity to its higher weight and its reinforcement. The reinforcement causes the concrete to fail in a more ductile way.

5.4 Other elements

Other parts of a structure that can be identified as vulnerable for blast loading are the roof and the ceilings. Lightly connected ceiling elements drop down very easily due to the vibrations in the structure. Roof elements, such as pane or corrugated sheets, often fail in the rebound phase and drop down in front of the building. Small pieces however may be thrown inwards.

6 RETROFIT

6.1 Danger for people

Fragments which are thrown inwards endanger people present the most. Not only the sharp pieces of glass with their high penetrating power can be lethal for persons present but also pieces of debris from the weak walls. These pieces also can fly with high velocities.

A second possible source of risk for people present is (partial) collapse of a structure. This may happen when a supporting wall fails. The loss of support causes the floors on top of the failing walls to collapse.

Although upon failure of parts of the structure blast pressure will leak into the building, this blast entering is not a serious risk for the people present. The rise in pressure is slowly and the maximum pressure inside is much lower than the peak pressure on the façades. (see for instance Mercx & Weerheijm 1991 or Van de Kasteele et al. 2002). The blast can only leak through small openings. During a relatively long time, the pieces of debris and glass are blocking the way for the blast to enter.

With the current set of relevant blast loadings, the pressure rise inside will cause minor damage to internal walls and objects, and no risk for people present. In fact, people can survive relatively high pressures.

6.2 *Retrofit for windows*

For windows, different types of retrofit measures have been developed. Three levels of protection can be distinguished.

A low level of protection is provided by the application of safety film to the glass. This is a very easy to add-on retrofit. The safety film does not or only slightly increase the strength of the window. The particular protection function of the safety film is that it holds the glass fragments together. The penetrating danger is then eliminated. Complete window panes however can still be thrown inwards.

Another category of retrofit is a catch system for the window. Possibilities are heavy curtains, mechanical fixation of the safety film to the surrounding structure, a second window or a cable catch system in combination with safety film. Such a cable catch system has been tested at TNO with use of the blast simulator shown in Figure 11.

The highest level of protection is provided by replacing existing windows by blast resistant windows.

Figure 11. Testing of cable catch system in TNO blast simulator.

This retrofit however needs a strong supporting structure, because high forces are transferred to the rest of the structure.

6.3 *Retrofit for walls*

For walls the principles of retrofit are similar as for windows. Catching the pieces of debris is sufficient when wall failure does not cause collapse.

For supporting walls however additional measures need to be taken. The walls need to be strengthened or it must be taken care of that the supporting function can be taken over by other elements.

Options for wall retrofits are listed below.

– Strengthen a wall with a layer of shotcrete on the inside (Whiting & Coltharp 1997).
– Span fabrics on the inside of the wall to catch the pieces of debris. Possible fabrics are geotextile or fiber reinforced plastics (Oswald & Chang 2001).
– Apply a membrane of spray-on, elastomeric polyurea on the inside of the wall, to both strengthen the wall and catch the debris (Knox et al. 2000).
– Cover the wall on the inside with steel sheet to catch the debris (Coltharp and Simmons 2001).
– Place columns behind the wall to strengthen the wall and to take over the supporting function (Coltharp and Simmons 2001).
– Add an extra wall.

7 CONCLUSIONS

A set of blast characteristics has been calculated for an explosion of 1000 kg ANFO at 150 m distance in different urban configurations. The scatter in the blast characteristics is considerable. This shows that the influence of the urban infrastructure on the propagation of the blast wave can not be neglected when predicting the blast load on a building at some distance from the explosion.

The set of blast loadings can be compared with vulnerability curves for typical structural elements, such as windows and walls. Thus the seriousness of a specific case can be estimated.

The combination of the data set on blast loading and the vulnerability curves provides the methodology to classify a structure. This will be a rough classification in categories like high, medium or low hazard. Based on this classification, grounded decisions can be taken concerning the next steps, whether and what retrofit measures are needed. The results may also give cause for a more detailed analysis.

Typical damage which may be expected for the studied case is window breakage and failure of weak walls. There are several options for retrofit in order to protect people in the building.

The obtained data set on vulnerability of urban buildings for a large bomb explosion may be a useful module in a risk-based program, which addresses risks of bomb explosions. This is however outside the scope of the current project.

REFERENCES

Absil, L.H.J., Kodde, H.H. & Mercx, W.P.M. 1993. The effectiveness of blast walls. *Proceedings of the 13th International Symposium on the Military Application of Blast Simulation, September 1993*, The Hague. Vol. 1, pp. 177–186.

Bogosian, D., Ferrito, J. & Shi, Y. 2002. Measuring uncertainty and conservatism in simplified blast models. *Proceedings of the 30th Explosives Safety Seminar, August 2002*, Atlanta, Georgia, Washington: Department of Defense Explosives Safety Board. On CD.

Coltharp, D.R. & Simmons, L. 2001. Blast response and retrofit of load-bearing masonry walls. *Proceedings of the 10th International Symposium on the Interaction of the Effects of Munitions with Structures. May 2001*, San Diego. On CD.

CPR 16E 2000. *Methods for the determination of possible damage to people and objects resulting from releases of hazardous materials*. The Hague: SDU uitgevers.

Knox, K.J. et al. 2000. Polymer materials for structural retrofit. *Proceedings of the 29th DoD Explosive Safety Seminar, July 2000*, New Orleans. On CD.

Mercx, W.P.M. & Weerheijm, J. 1991. Pressure development in a chamber due to an entering shock wave, *12th International Symposium on the Military Application of Blast Simulation, September 1991*, Perpignan.

Oswald, C.J. & Chang, K.K. 2001. Shock tube testing on masonry walls strengthened with Kevlar®. *Proceedings of the 10th International Symposium on the Interaction of the Effects of Munitions with Structures. May 2001*, San Diego. On CD.

TM 5-855-1 1985. *Technical Manual: Fundamentals of protective design for conventional weapons*. Washington: Department of the Army.

TSWG 1999. Technical Support Working Group. *Terrorist Bomb Threat Standoff*. Washington, DC.

Van den Berg, A.C. 1985. *BLAST – A code for numerical simulation of multi-dimensional blast effects*. Rijswijk: TNO Prins Maurits Laboratory.

Van de Kasteele, R.M., Absil, L.H.J. & Dirkse, M.W.L. 2002. *40 Tonne Donor/Acceptor Trial: Instrumentation and measurement results*. Rijswijk: TNO-report PML 2002-A15.

Van Doormaal, J.C.A.M. & Absil, L.H.J. 2001. Loading and response of brick houses by a 40 tonnes explosion. Moody F.J. (ed.). *Thermal hydraulics, liquid sloshing, extreme loads, and structural response – ASME 2001*. PVP-Vol. 421, pp. 31–37.

Whiting, W.D. & Coltharp, D.R. 1997. Retrofit measures for conventional concrete masonry unit buildings subject to terrorist threat. *Proceedings of the 8th International Symposium on the Interactions of the Effects of Munitions with Structures. May 1997*, Washington DC.

Safety and Reliability – Bedford & van Gelder (eds)
© 2003 Swets & Zeitlinger, Lisse, ISBN 90 5809 551 7

A distribution for modeling dependence caused by common risk factors

J. René van Dorp

The George Washington University, Washington D.C., USA

ABSTRACT: The cumulative distribution function (cdf) of a finite mixture of independent uniform random variables will be derived. The distribution is useful for uncertainty analyses in application domains such as, e.g., project risk analysis, decision analysis, finance, accident probability analysis and actuarial analysis, particularly when dependence between uncertain elements is present due to common risk factors. Use of the distribution reduces the number of dependence parameters that need to be assessed to specify dependence when compared to a correlation matrix approach. An example discussing the effect of dependence in the project risk analysis domain utilizing the mixture distribution is presented.

1 INTRODUCTION

The motivation for the construction of the cumulative distribution function (cdf) of a finite mixture of uniform random variables arose in the development of a risk analysis approach for project networks (see, Van Dorp & Duffey (1999)). The duration of the activities in such a network may be modeled as random variables. With the project network structure and an assumption of independence between these random variables the completion time of the project can be readily obtained using a combination of the Critical Path Method (see, e.g. Winston (1993)) and Monte Carlo methods (see, e.g. Vose (1996)). However, the independence assumption between these random variables may be specious (e.g. Duffey & Van Dorp (1998)) and one may resort to modeling statistical dependence between the random variables utilizing e.g. a correlation matrix approach via multivariate normal distributions. The long-standing issue of dependence between random variables has recently been discussed in application areas such as project risk analysis (see, e.g. Duffey and Van Dorp (1998)), accident probability analysis (see, e.g. Yi Bier (1998)), finance (see, e.g. Härdle et al. (2002)), decision analysis (see, e.g. Clemen and Reilly (1999)) and actuarial modeling (see, e.g. Frees and Valdez (1998)).

With n pre-specified marginal distributions a correlation matrix approach generally requires the specification of $\binom{n}{2}$ correlations. In project risk analysis project sizes of 100 activities are not uncommon and specification of $\binom{100}{2} = 4950$ correlations becomes a formidable task. When using engineering judgment to specify such a correlation matrix, inconsistencies occur as the correlation matrix needs to be positive definite, and often modifications to the engineering judgment are needed (e.g. Iman & Conover (1982)). Instead, one may develop an approach to model statistical dependence between these activity durations using common risk factors. The idea of *common risk factors* or *common causes* is not new and has already found wide appreciation in fault tree analysis for chemical and nuclear power plants (see, e.g. Haasl et al. (1981) or Zhang (1989)). Examples of possible common risk factors in a project risk analysis context are; weather, engineering change orders, productivity of workforce etc.

The dependence model in Van Dorp & Duffey (1999) uses common risk factors for modeling dependence, but is restricted to 1 common risk factor per disjoint subsets of activities. This, however, implies that only 1 risk factor influences the uncertainty in an activity duration which may be too restrictive for practical purposes. The cdf to be derived in this paper allows multiple common risk factors to influence the uncertainty in an activity duration and is thus more flexible than the dependence model in Van Dorp & Duffey (1999). Figure 1 displays the influence diagram representing the relaxed dependence model.

The authors Frees Valdez (1998), Duffey and Van Dorp (1998), Yi Bier (1998), Clemen and Reilly (1999) and Härdle et al. (2002) unanimously suggested the copula approach (see, e.g. Sklar (1959), Genest and McKay (1986) and Nelsen (1999)) for dependence modeling. An advantage of the copula approach

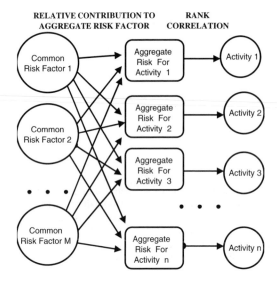

RELATIVE CONTRIBUTION TO   RANK
AGGREGATE RISK FACTOR   CORRELATION

Figure 1. A model for statistical dependence between activity durations due to common risk factors.

is that it utilizes the decomposition principle by separately describing the uncertainty aspect via the marginal distributions and dependence features between components via copula's. A one parameter copula is used to model the dependence between an activity duration and its associated aggregate risk factor indicated in Figure 1. The correlation between the uniform marginals of the copula is the rank correlation between the aggregate risk factor and its activity duration. The rank correlation has been proposed as an appropriate measure of positive statistical dependence (see, e.g. Joag-Dev (1984)).

An aggregate risk factor in Figure 1 is a combined measure of risk for a particular activity duration arising from the common risk factors between activities. A common risk factor may not have a natural attribute scale and different common risk factors are generally measured on different scales. Therefore, for convenience, each common risk factor i, $i = 1, \ldots, m$ is modeled as a uniform latent random variable U_i, where the lowest risk level for such a risk factor is transformed to 0 and the highest risk is transformed to 1. Latent random variable models have found wide application in the behavioral sciences (see, e.g. Bartholomew (1987)). Transforming the different risk factors to the same scale allows engineers to tradeoff risk factors using a tradeoff elicitation approach like e.g. *swing weights* as described in Clemen (1995). Using such an elicitation approach, relative contributions w_i of each risk factor pertaining to a particular activity may be obtained. The proposed aggregate risk factor Y for an activity is then calculated as

$$Y = \sum_{i=1}^{m} w_i U_i. \tag{1}$$

Note that Y is a mixture of uniform random variables U_i with mixture weights w_i, $i = 1, \ldots, m$. To be able to use the copula approach to model statistical dependence between an activities aggregated risk and the activity duration as in Figure 1, the random variable Y needs to be transformed into a uniform random variable. It is well known that the required transformation is $F(Y)$, where F is the cdf of the mixture of uniform random variables given by (1). The cdf of Y will be derived in the next section.

Note that for each activity m relative contributions of individual risk factors need to be specified to aggregate risk, in addition to a rank correlation between an activity's aggregate risk factor and the activity duration. Hence, the total number of parameters that needs to be specified, equals $m \cdot n + n$. With 5 common risk factors and 100 activities in a project network this amounts to 600 parameters as opposed to the 4950 correlations in a correlation matrix approach to build dependence between pre-specified marginal distributions. Also, no modifications to the dependence parameters are needed due to inconsistencies when engineering judgment is used to specify these parameters. The dependence model in Figure 1 has been successfully tested in Greenberg (1998). Multiple elicitation sessions with Naval Architects were used in Greenberg (1998) to specify; (1) the parameters for the uncertainty distribution of 254 activity durations in a project network and (2) the parameters for a dependence model as in Figure 1 with 5 common risk factors.

2 A MIXTURE OF UNIFORM VARIABLES

To use the copula approach to model bivariate dependence between a random variable X and its aggregated risk Y, both X and Y need to be transformed to the uniform marginals U and V of the copula. The required (integral) transformations of X and Y are $F(X)$ and $G(Y)$, where the functions $F(\cdot)$ and $G(\cdot)$ are the cdf's of X and Y, respectively. For a known marginal distribution for random variable X, $F(\cdot)$ is readily obtained either in closed form (e.g. in the case of a triangular distribution) or through numerical routines (e.g. for a beta distribution). The cdf of the linear combination Y (cf. (1)) is given by

$$Pr(Y \leq y) = G(y) = \sum_{v_1=0}^{1} \cdots \sum_{v_m=0}^{1} \tag{2}$$

552

$$(-1)^{\sum\limits_{i=1}^{m} v_i} \left\{ \frac{(y - \sum\limits_{i=1}^{m} w_i v_i)^m}{m! \prod\limits_{i=1}^{m} w_i} \right\} 1_{[0,\infty)} (y - \sum\limits_{i=1}^{n} w_i v_i)$$

(see, Mitra (1971) or Barrow & Smith (1979)). Unfortunately, their proofs – geared towards mathematically oriented readers – are very concise and somewhat difficult to follow. The proof discussed in the next section which seems to be new, is geometric in nature and is based on the time honored inclusion-exclusion principle

$$Pr\left\{ \bigcup_{i=1}^{m} A_i \right\} = \sum_{i=1}^{m} Pr(A_i) -$$

$$\sum\sum_{i<j} Pr\{A_i \cap A_j\} +$$

$$\sum\sum\sum_{i<j<k} Pr\{A_i \cap A_j \cap A_k\} - \dots$$

$$+(-1)^m Pr\left\{ \bigcap_{i=1}^{m} A_i \right\}, \qquad (3)$$

for arbritrary events $A_1, \dots, A_n$ (not necessarily disjoint). The geometric nature of the proof allows for an efficient algorithm to evaluate (2) and needed for the application of (2) in Monte Carlo based uncertainty analyses. The Appendix describes the algorithm in Psuedo Pascal .

2.1 Theoretical result

Let $C^m = \{\underline{u} \,|\, 0 \le u_i \le 1\}$ be the unit hyper cube in $\mathbb{R}^m$. Let $\underline{v} = (v_1, \dots, v_m)$, $v_i \in \{0, 1\}$ be a vertex of the unit hyper cube C^m and define the simplex $S_{\underline{v}}(y)$ at the vertex $\underline{v}$ as

$$S_{\underline{v}}(y) =$$
$$\{\underline{u} \,|\, \sum_{i=1}^{m} w_i u_i \le y, u_i \ge v_i, i = 1, \dots, m\}, \qquad (4)$$

where $w_i \ge 0$, $\sum_{i=1}^{m} w_i = 1$. Figure 2A displays C^3 and the simplex $S_{(0,0,0)}(y_1)$ (cf. (4)). Figure 2B displays C^3 and $S_{(0,0,0)}(y_2)$, $S_{(1,0,0)}(y_2)$, $S_{(0,1,0)}(y_2)$, $S_{(1,0,0)}(y_2)$, $S_{(1,1,0)}(y_2)$, $S_{(1,0,1)}(y_2)$ and $S_{(0,1,1)}(y_2)$. Our proof of (2) utilizes the following lemma.

Lemma 1: The hyper volume $V\{S_{\underline{v}}(y)\}$ of the simplex $S_{\underline{v}}(y)$ given by (4) equals

$$\frac{(y - \sum\limits_{i=1}^{m} w_i v_i)^m}{m! \prod\limits_{i=1}^{m} w_i} \cdot 1_{[0,\infty)} (y - \sum\limits_{i=1}^{n} w_i v_i). \qquad (5)$$

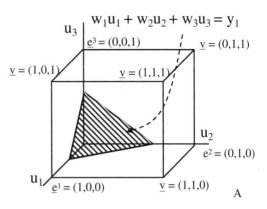

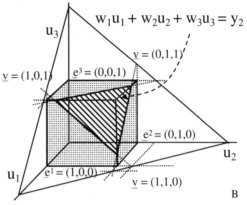

Figure 2. A: Evaluating $G(y_1)$ for $m = 3$; B: Evaluating $G(y_2)$ for $m = 3$.

Proof: From the definition of (4) it immediately follows that for $y \ge 0$

$$V\{S_{\underline{0}}(y)\} =$$
$$\int_{u_1=0}^{1} \int_{u_2=0}^{1-\sum\limits_{i=1}^{m-1} \frac{w_i}{y} u_i} \dots \int_{u_m=0}^{1-\sum\limits_{i=1}^{m-1} \frac{w_i}{y} u_i} du_m \dots du_1. \qquad (6)$$

Changing the variables of integration to $z_i = w_i u_i / y$, $i = 1, \dots, m$, integral in (6) becomes

$$V\{S_{\underline{0}}(y)\} =$$
$$\frac{y^m}{m \prod\limits_{i=1}^{m} w_i} \int_{z_1=0}^{1} \int_{z_2=0}^{1-\sum\limits_{i=1}^{1} z_i} \dots \int_{z_m=0}^{1-\sum\limits_{i=1}^{m-1} z_i} dz_m \dots dz_1. \qquad (7)$$

The integral in (7) is the hyper-volume of the unit simplex

553

$$S = \{\, \underline{u} \mid \sum_{i=1}^{m} u_i \leq 1, u_i \geq 0, i = 1, \ldots, m \,\}. \quad (8)$$

Realizing that the Dirichlet distribution (see, e.g., Kotz et al. (2000)) with density function

$$\frac{\Gamma(\eta)}{\prod\limits_{i=1}^{m+1} \Gamma(\eta \cdot \nu_i)} \cdot \left(\prod_{i=1}^{m} (u_i)^{\eta \cdot \nu_i - 1} \right) \left(1 - \sum_{i=1}^{m} u_i \right)^{\eta \cdot \nu_{m+1}} \quad (9)$$

where $\eta > 0$, $\nu_i > 0$, $\sum_{i=1}^{m+1} \nu_i = 1$ has support S (cf. (8)) and by setting its parameters in (9) equal to $\eta = m + 1$, $\nu_i = 1/m+1$, $i = 1, \ldots, m + 1$ it immediately follows from (8), (9) and the fact that $S_{\underline{0}}(y) = \varnothing$ for $y < 0$ that

$$V\{S_{\underline{0}}(y)\} = \frac{y^m}{m! \prod\limits_{i=1}^{m} w_i} \cdot 1_{[0,\infty)}(y). \quad (10)$$

Again changing variables $x_i = u_i - \nu_i$, $i = 1, \ldots, m$ we arrive utilizing (4) at

$$V\{S_{\underline{v}}(y)\} = V\{S_{\underline{0}}(y - \sum_{i=1}^{n} w_i \nu_i)\}. \quad (11)$$

The lemma now follows from (10) and (11). $\qquad \square$

Theorem 1: *The cdf of the weighted linear combination Y given by (1), where U_i are independent $[0, 1]$ uniform random variables is given by (2).*

Proof: The support of Y follows from (1) as $[0,1]$. Let $\underline{0} = (0,\ldots,0)$ be the origin vertex of the unit hyper cube C^m and let $\underline{e}^i = (e_1, \ldots, e_m)$, $i = 1, \ldots, m$, be the unit vertices of C^m (See, Figure 2), i.e. $e_i = 1$, $e_j = 0$, $j = 1, \ldots, m$, $j \neq i$. For illustration we shall consider the case $m = 3$ and evaluating $Pr(Y \leq y_1)$ for the value of y_1 indicated by Figure 2A and that of $Pr(Y \leq y_2)$ for the value of y_2 depicted in Figure 2B. Figure 2A displays C^3 and $S_{\underline{0}}(y_1)$ (cf. (4)). Figure 2B displays C^3, $S_{\underline{0}}(y_2)$, $S_{\underline{e}^1}(y_2)$, $S_{\underline{e}^2}(y_2)$ and $S_{\underline{e}^3}(y_2)$. From (1) and the independence of U_i, $i = 1, \ldots, m$ it follows that in Figure 2A $Pr(Y \leq y_1) = V\{S_{\underline{0}}(y_1)\}$. In Figure 2B the calculation of $Pr(Y \leq y_2)$ is somewhat more complicated. Figure 2B shows that

$$Pr(Y \leq y_2) = V\{S_{\underline{0}}(y_2)\} - V\left\{ \bigcup_{i=1}^{3} S_{\underline{e}^i}(y_2) \right\}. \quad (12)$$

Note that (12) also holds for the value in y_1 in Figure 2A as $S_{\underline{e}^i}(y_1) = \varnothing$, $i = 1, \ldots, 3$. Generalizing to $\mathbb{R}^m$ we obtain directly

$$Pr(Y \leq y) = V\{S_{\underline{0}}(y)\} - V\left\{ \bigcup_{i=1}^{m} S_{\underline{e}^i}(y) \right\} \quad (13)$$

The inclusion-exclusion principle (cf. (3)) yields

$$V\left\{ \bigcup_{i=1}^{m} S_{\underline{e}^i}(y) \right\} = \sum_{i=1}^{m} V\{S_{\underline{e}^i}(y)\} -$$

$$\sum\sum_{i<j} V\{S_{\underline{e}^i}(y) \cap S_{\underline{e}^j}(y)\} +$$

$$\sum\sum\sum_{i<j<k} V\{S_{\underline{e}^i}(y) \cap S_{\underline{e}^j}(y) \cap S_{\underline{e}^k}(y)\} -$$

$$\ldots + (-1)^m V\left\{ \bigcap_{i=1}^{m} S_{\underline{e}^i}(y) \right\}. \quad (14)$$

Utilizing (4) it follows that the intersections of the simplices $S_{\underline{e}}(y)$ in (14) are all of following form

$$\bigcap_{i \in I} S_{\underline{e}^i}(y) = S_{\underline{v}}(y) \quad (15)$$

where $I \subset \{1, \ldots, m\}$ and $\underline{v} = \sum_{i \in I} \underline{e}^i$. For example, $S_{(1,1,0)}(y_2)$ in Figure 2B is the intersection of $S_{(1,0,0)}(y_2)$ and $S_{(0,1,0)}(y_2)$. From (15), (14) and (12) we conclude that

$$Pr(Y \leq y) = \sum_{v_1=0}^{1} \cdots \sum_{v_m=0}^{1} (-1)^{\sum_{i=1}^{m} v_i} V\{S_{\underline{v}}(y)\}. \quad (16)$$

The proof of the theorem follows from Lemma 1. $\qquad \square$

From the proof it follows that an efficient method to evaluate the distribution in (2) for a particular value of y and a given set of weights $\underline{w} = (w_1, \ldots, w_m)$ is to develop a recursive algorithm enumerating all vertices $\underline{v}$ of the hypercube C^m and evaluate the hyper-volume of the simplex at each vertex $\underline{v}$ given by (5) when a vertex is visited by the procedure. An example discussing the effect of dependence in the Project Risk Analysis domain utilizing the cdf given by (2) is presented in the next section.

3 EXAMPLE – A CONTROVERSY IN PERT

Johnson (1997) proposed the triangular distribution to be used as an alternative to the beta distribution. Its parameters have a one-to-one correspondence to an optimistic estimate a, a most likely estimate m and a pessimistic estimate b of an activity duration T in a PERT network. Much earlier, Malcolm et al. (1959) fitted a four-parameter beta distribution by estimating a, m and b and used the method of moments to overcome difficulties involved with interpreting the beta parameters by setting

$$\begin{cases} E[T] = \dfrac{a + 4m + b}{6} \\ Var[T] = \frac{1}{36}(b-a)^2. \end{cases} \quad (17)$$

Solving for the beta parameters using (17) has been controversial (see e.g. Clark (1962), Grubbs (1962)) and its use is still subject to a discussion (see e.g. Kamburowski (1997)). Van Dorp & Kotz (2002) suggested the use of a Two-Sided Power (TSP) distribution, an extension of the triangular distribution, defined by the density

$$f_X(x|a, m, b, n) =$$
$$\begin{cases} \dfrac{n}{(b-a)} \left(\dfrac{x-a}{m-a} \right)^{n-1} & a < x \leq m \\ \dfrac{n}{(b-a)} \left(\dfrac{t-x}{b-m} \right)^{n-1} & m \leq x \leq b, \end{cases} \quad (18)$$

as a proxy to the beta, specifically in problems of assessment of risk and uncertainty (such as in PERT). For $n = 2$ in (18) the TSP density coincides with the density of a triangular distribution. The expressions for the mean and the variance for (18) result in

$$E[X] = \frac{a + (n-1)m + b}{n+1} \quad (19)$$

and

$$Var(X) =$$
$$(b-a)^2 \cdot \left\{ \frac{n - 2(n-1)\frac{(m-a)}{(b-a)}\frac{(b-m)}{(b-a)}}{(n+2)(n+1)^2} \right\}. \quad (20)$$

For a TSP distribution with $n = 5$, the mean values $E[T]$ in (17) and $E[X]$ in (19) coincide.

In the example to be discussed below, the effect of an assumption of independence between activity durations on the minimal completion time of a PERT network combined using the above setup (i.e. selecting either a beta, triangular or TSP distribution while utilizing the estimates a, m and b) will be compared to one associated with an assumption of dependence combined with triangular distributions.

3.1 Description

Figure 3 shows an 18-activity project network in the ship building domain from Taggart (1980). The uncertainty in each activity duration could be elicited through expert judgment via a lower bound a, most like estimate m and upper bound b as described in Table 1. Modernday ship production is a manufacturing domain in which innovative design and build strategies require special attention to risk factors that may impact cost and delivery time. Two major risk areas are the impact of Engineering Change Orders and crane unavailability.

Engineering changes may come from a variety of sources – such as owner-requested changes, inadequate design specifications, interface problems for vendorfurnished equipment, etc. Cranes are used to lift large prefabricated units and their unavailability due to outages may result in substantial project delays. The relative contributions of ECO and crane unavailability to aggregate risk and the rank correlation between the

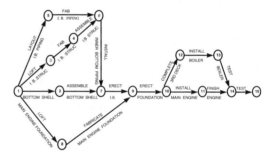

Figure 3.   Project network P for production process.

Table 1.   Parameters for modeling the uncertainty in activity durations for the project network in Figure 3.

ID	Activity name	a	m	b
1	Shell: Loft	22	25	30
2	Shell: Assemble	35	37	43
3	I.B.piping: Layout	19	22	29
4	I.B.piping: Fab.	4	5	10
5	I.B.structure: Layout	23	26	31
6	I.B.structure: Fab.	16	18	24
7	I.B.structure: Assemb.	11	14	20
8	I.B.structure: Install	6	7	12
9	Mach fdn. Loft	25	28	33
10	Mach fdn. Fabricate	33	35	40
11	Erect I.B.	27	30	37
12	Erect foundation	6	7	11
13	Complete #rd DK	4	5	9
14	Boiler: Install	6	7	10
15	Boiler: Test	9	10	15
16	Engine: Install	6	7	12
17	Engine: Finish	17	20	26
18	FINAL test	13	15	20

Table 2. Parameters for modeling the dependence between activity durations for the project network in Figure 3.

ID	Activity name	w_{ECO}	w_{Crane}	ρ
1	Shell: Loft	1.0	0.0	0.5
2	Shell: Assemble	1.0	0.0	0.5
3	I.B.piping: Layout	0.5	0.5	0.5
4	I.B.piping: Fab.	1.0	0.0	0.5
5	I.B.structure: Layout	1.0	0.0	0.5
6	I.B.structure: Fab.	1.0	0.0	0.5
7	I.B.structure: Assemb.	0.5	0.5	0.5
8	I.B.structure: Install	0.5	0.5	0.5
9	Mach fdn. Loft	0.5	0.5	0.5
10	Mach fdn. Fabricate	0.5	0.5	0.5
11	Erect I.B.	0.2	0.8	0.5
12	Erect foundation	0.2	0.8	0.5
13	Complete #rd DK	0.2	0.8	0.5
14	Boiler: Install	0.0	1.0	0.5
15	Boiler: Test	1.0	0.0	0.5
16	Engine: Install	0.0	1.0	0.5
17	Engine: Finish	1.0	0.0	0.5
18	FINAL test	1.0	0.0	0.5

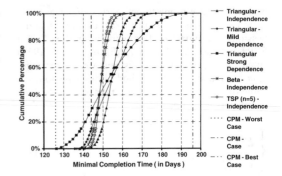

Figure 4. Comparison of distributions of minimal completion time for the project in Figure 3.

Table 3. Minimum, mean, maximum, standard deviation and range of the project completion time distribution using triangular, beta and TSP ($n = 5$) Distributions under an independence assumption and triangular distributions under a mild and strong dependence assumption. Results were generated utilizing Monte Carlo analysis involving 10,000 CPM calculations per case.

	Min	Mean	Max.	St. dev.	Range
Triang – Independence	138.11	155.15	172.49	5.04	34.38
Triang – Mild dependence	135.56	155.10	176.93	8.90	41.37
Triang – Strong dependence	126.29	154.94	192.29	14.64	66.00
Beta – Independence	136.27	150.01	164.14	4.06	27.87
TSP ($n = 5$) – Independence	140.63	149.85	160.78	2.96	20.15

activities and its associated aggregated risk are specified in Table 2. Note that due to similarity in exposure to ECO's and usage of the crane these parameters may not need to vary by activity, thereby further reducing the assessment of dependence parameters by pregrouping similar activities in terms of reliance on common risk factors.

3.2 Project completion time distribution analysis

To show the effect of mild dependence between the activity durations on the minimal completion time distribution of the project in Figure 3, the information in Tables 1 and 2 and the dependence model described above have been used. A rank correlation of 0.5 is assumed across the board and may be viewed as a mild form of dependence.

Amongst the TSP and beta distribution, the triangular distribution is the only distribution that is completely specified by a, m and b without additional assumptions. Hence, the minimal completion time distribution involving triangular distributions and an assumption of mild dependence is compared with the project completion time distribution assuming independence between the activity durations with a triangular form (cf. (18) with $n = 2$), a beta form (via (17) and employing the method of moments) and finally a TSP form (cf. (18) with $n = 5$). The Monte Carlo analysis results utilizing 10,000 CPM calculations per case are displayed in Figure 4. For robustness purposes of the dependence model herein, Figure 4 also contains the minimal completion time distribution of the project utilizing triangular distributions for activitaty durations involving complete dependence (the strong

dependence case in Figure 4). Complete dependence can be specified using the dependence model above by assigning rank correlations of 1 for all activities and mixture weights such that $w_{i*j} = 1$ and $w_{ij} = 0$, $i \neq i*$, $i = 1, ..., m$, $j = 1, ..., n$ for some $i* \in \{1, ..., m\}$. Finally, the minimal completion time of 126 days (CPM-Best Case), 144 days (CPM-Case) and 196 days (CPM-Worst Case) of a standard CPM analysis utilizing only the lower bounds, most likely estimates and upper bounds in Table 1, respectively, in are depicted by vertical lines.

The minimum, mean, maximum, the standard deviation and range of the project completion distribution for the five combinations are provided in Table 3. Comparing the first and fourth row in Table 3 it follows that with the independence assumption between beta activity durations, the use of (17) results in a significant reduction in the mean of the project completion time and a substantial reduction in its standard deviation when compared to utilizing trian-gular distributions whose parameters are directly specified by the three estimates a, m and b (See Table 1). Hence, the adoption of (17) may not be consistent with a conservative approach towards estimating project completion time and its uncertainty. Note that from the fifth row in Table 3 follows that when utilizing TSP distributions ((18) with $n = 5$), a similar mean shift occurs in the project completion time and even a larger shift in the standard deviation, providing an even more optimistic scenario.

The most notable results in Figure 4 and Table 3, however, follow from comparing the completion time distribution under an assumption of mild dependence and strong dependence (rank correlations equal to 1) with the distributions assuming independence. Although no mean shift occurs when comparing the first, second and third rows in Table 3, the standard deviation of the completion time of the project almost doubles (triples) when comparing the second (third) row to the first one. The same observation follows from Figure 4 where the distributions under the dependence assumptions possess much smaller slopes and appear to have a support that overlaps all of its counterparts. The latter observation is substantiated by comparing the minimum and maximum observed values for the minimal project completion time in the second and third row to those in the first, fourth and fifth row of Table 3. In addition, observe from Table 3 that the support of the distribution in the strong dependence case (third row) covers 94.29% of the possible range of 126 days (CPM – Best Case) to 196 days (CPM – Worst Case), whereas the accompanying independence case (first row) covers only 49.11%. The latter observation demonstrates the flexibility of the dependence model described in this paper, but also emphasizes the need for accurate assessment of the dependence parameter within the model.

Evidently, if the use of (17) and its resulting underestimation of project completion time and uncertainty were a reason for a long standing controversy (see e.g. Clark (1962), Grubbs (1962) and Kamburowski (1997)), the common assumption of independence between marginal distribution of activity durations should be subjected to the same level of scrutiny. Perhaps, such a level of scrutiny could lead to development of dependence models similar to the one described herein and the development of formal methods for assessing dependence parameters utilizing expert judgment elicitation.

Note also that it follows from Figure 4 that the probability of completing the project by 144 days calculated using the standard CPM method is less than 15% regardless of an assumption of mild dependence or independence. This result is due to the fact that the ingredient distributions of the activity durations are positively skewed. Positively skewed distributions were prevalent in the expert judgment used in Greenberg (1998). Such a prevalence may be explained by the existence of a motivational bias amongst experts resulting in optimism regarding the most likely value of activity completion. This fact, coupled with an independence assumption, could serve as an explanation for a low incidence of project success (on-time) when utilizing standard CPM analysis as a yard stick.

ACKNOWLEDGMENT

The author is indebted to Technical Program Board of the European Safety and Reliability 2003 Conference and to the referees. Their comments significantly improved the content and presentation of the first version.

REFERENCES

Barrow, D.L. and Smith, P.W. 1979. Spline Notation Applied to a Volume Problem. *American Mathematical Monthly* 86: 50–51.

Bartholomew, D.J. 1987. *Latent Variable Models and Factor Analysis*. New York: Oxford University Press.

Clark, C.E. 1962. The PERT Model for the Distribution of an Activity. *Operations Research* 10: 405–406.

Clemen, R.T. 1995. *Making Hard Decisions, An Introduction to Decision Analysis*. New York: Duxbury Press.

Clemen, R. and Reilly, T. 1999. Correlations and Copulas for Decision and Risk Analysis. *Management Science* 45: 208–224.

Duffey, M.R. and Van Dorp, J.R. 1998. Risk Analysis for Large Engineering Projects: Modeling Cost Uncertainty for Ship Production Activities. *Journal of Engineering Valuation and Cost Analysis* 2: 285–301.

Frees, E.W. and Valdez, E.A. 1998. Understanding Relationships using Copula's. *North America Actuarial Journal*, 2: 1–25.

Genest, C. and Mackay, J. 1986. The Joy of Copulas, Bivariate Distributions With Uniform Marginals, *The American Statistician*, 40(4): 280–283.

Greenberg, M. 1998. *An Activity Network Simulation Approach for Estimating the Production Schedule of a Convertible Container Ship*. Washington D.C.: The George Washington University (Masters Thesis).

Grubbs, F.E. 1962. Attempts to validate certain PERT statistics or a "Picking on PERT". *Operations Research* 10: 912–915.

Haasl, D.F., Roberts N.H., Vesely W.E. and Goldberg F.F. 1981. *Fault Tree Handbook*. Washington D.C.: Systems and Reliability Research, Office of Nuclear Regulatory Commission.

Härdle, W., Kleinow, T. and Stahl, G. 2002. Applied Quantitative Finance, Theory and Computational Tools, Springer, e-book.

Iman, R.L. and Conover, W.J. 1982. A Distribution Free Approach to Inducing Rank Order Correlation among Input Variables. *Communications in Statistics – Simulation and Computation* 11(3): 311–334.

Joag-Dev, K. 1984. *Measures of dependence*. Handbook of Statistics 4: 79–88. North Holland: Elsevier.

Johnson, D.G. 1997. The Triangular Distribution as a Proxy for the Beta Distribution in Risk Analysis. *The Statistician* 46(3): 387–398.

Kamburowski, J. 1997. New Validations of PERT Times. *Omega, International Journal of Management Science*, 25(3): 323–328.

Kotz, S., Balakrishnan, N. and Johnson, N.L. 2000. *Continuous Multivariate Distributions, Volume 1: Models and Applications, 2nd Edition*. New York: Wiley.

Malcolm, D.G., Roseboom, C.E., Clark, C.E. and Fazar, W. 1959. Application of a Technique for Research and Development Program Evaluation. *Operations Research* 7: 646–649.

Mitra, S.K. 1971. On the Probability Distribution of The Sum of Uniformly Distributed Random Variables. *SIAM Journal of Applied Mathematics* 20(2): 195–198.

Nelsen, R.B. 1999. *An Introduction to Copulas*, New York: Springer.

Sklar, A. 1959. Fonctions de répartition à *n* dimensions et leurs marges. *Publ. Inst. Statist. Univ. Paris*, 8: 229–231.

Taggart, R. 1980. Ship Design and Construction, The Society of Naval Architects and Marine Engineers. New York: SNAME.

Van Dorp, J.R. and Duffey, M.R. 1999. Modeling Statistical Dependence in Risk Analysis for Project Networks. *International Journal of Production Economics* 58: 17–29.

Van Dorp, J.R. and Kotz, S. 2002. A Novel Extension of the Triangular Distribution and its Parameter Estimation, *The Statistician*, 51(1): 63–79.

Vose, D. 1996. *Quantitative Risk Analysis, A Guide to Monte Carlo Simulation Modeling*, New York: Wiley.

Winston, W.L. 1993. *Operations Research, Applications and Algorithms*. New York: Duxbury Press.

Yi, W. and Bier, V. 1998. An application of Copulas to Accident Precursor Analysis, *Management Science*, 44: S257–S270.

Zhang, Q. 1989. A General Method Dealing with Correlations in Uncertainty Propagation in Fault Trees, *Reliability Engineering and System Safety*, 26: 231–247.

APPENDIX

The procedure $CalcCDF(\mathbf{G}, y, m, \underline{w})$ below evaluates the c.d.f. of Y given by (1). The algorithm uses functions

$ProdWeights(\underline{w}, m)$

to calculate $\Pi = \prod_{i=1}^{m} w_i$,

$SumElements(\underline{v}, m)$

to calculate $\Sigma = \sum_{i=1}^{m} v_i$ and

$SumProducts(\underline{v}, \underline{w}, m)$

to calculate $\psi = \sum_{i=1}^{m} w_i v_i$.

$VisitVertices(\mathbf{G}, y, i, \underline{v}, m, \underline{w}, \Pi)$;
Step 1: *if* $i < m$ *then* $v_i: = 0$;
$visitVertices(\mathbf{G}, y, i, \underline{v}, sum, m, \underline{w}, \Pi)$;
$v_i: = 1$;
$VisitVertices(\mathbf{G}, y, i, \underline{v}, sum, m, \underline{w}, \Pi)$;

Step 2:
$\Sigma: = SumElements(\underline{v}, m)$;
$\psi: = SumProducts(\underline{v}, \underline{w}, m)$;

Step 3: *If* $(y - \psi) > 0$ *then*
$G: = G + (-1)^{\Sigma} y - \psi/m! \cdot \Pi$

$CalcCDF(\mathbf{G}, y, m, \underline{w})$;
Step 1: *If* $y \leq 0$ *then* $G: = 0$; *Stop*;
Step 2: *If* $y \geq 1$ *then* $G: = 1$; *Stop*;
Step 3: $\Pi: = ProductWeights(\underline{w}, m)$;
Step 4: $VisitVertices(\mathbf{G}, y, 1, \underline{v}, m, \underline{w}, \Pi)$.

Safety and Reliability – Bedford & van Gelder (eds)
© 2003 Swets & Zeitlinger, Lisse, ISBN 90 5809 551 7

A general Bayes Weibull inference model for accelerated life testing

J. René Van Dorp & Thomas A. Mazzuchi
The George Washington University, Washington D.C., USA

ABSTRACT: This article presents the development of a general Bayes inference model for accelerated life testing. The failure times at a constant stress level are assumed to belong to a Weibull distribution, but the specification of strict adherence to a parametric time-transformation function is not required. Rather, prior information is used to indirectly define a multivariate prior distribution for the scale parameters at the various stress levels and the common shape parameter. Using the approach, Bayes point estimates as well as probability statements for use-stress (and accelerated) life parameters may be inferred from a host of testing scenarios. The inference procedure accommodates both the interval data sampling strategy and type I censored sampling strategy for the collection of ALT test data. The inference procedure uses the well-known Markov Chain Monte Carlo (MCMC) methods to derive posterior approximations. The approach is illustrated with an example.

1 INTRODUCTION

In the case of highly reliable items, e.g. Very Large Scale Integrated (VLSI) electronic devices, computer equipment, missiles, etc., mean times to failure (MTTF) exceeding a year is not uncommon. The use of these items, however, may still require reliability demonstration or verification testing, especially when these items are used for military or high risk public applications. With such MTTF's, it is often too time consuming and too costly to test these items in their use (or nominal) environment, as the length of time to generate a reasonable number of failures is often intolerable. If such is the case, it has become a standard procedure (see MIL-STD-781C) to test these items under more severe environments than experienced in actual use. Such tests are often referred to as Accelerated Life Tests (ALT's). Mann and Singpurwalla (1983) note that because of advancement in technology and increased reliability, ALT's are performed more frequently than ordinary life tests. There are two main problems associated with ALT's as: (1) optimal design of the ALT, and (2) statistical inference from ALT failure data. The focus of this paper is on the statistical inference problem, i.e. on how to make inference about the reliability in the use environment by obtaining information in the accelerated environments.

Typically inference methods have been developed assuming that: (1) the life time distribution in a constant stress environment belongs to a common family of distributions, and (2) the scale parameter of such a distribution is related to the stress environment via a parametric function known as a time transformation function (TTF) (see for example Mann et al. (1974)). In addition, most of the inference methods are based on the use of maximum likelihood estimation which may require large sample sizes for meaningful statistical ALT inference (see for example Nelson (1980)) In this paper, only the first assumption will be adhered to. Specifically, inference will be developed using the Weibull failure time model. The inference method is Bayesian in nature and will rely on the use of engineering judgment to specify prior distributions for the Weibull model parameters. While there is a host of literature in this area, the only Bayesian inference procedure developed for the Weibull model that we know of is presented in Mazzuchi et al. (1997) for constant stress ALTs in conjunction with the parametric TTF. The inference procedure herein will be developed for a wide range of ALT scenarios with no TTF assumption.

In Section 2 the general likelihood model is developed. In Section 3 the prior distribution for the shape parameter and scale parameters of the Weibull failure time model is outlined. The posterior inference is briefly discussed in Section 4. The approach is illustrated by an example in Section 5.

2 A GENERAL LIKELIHOOD MODEL

2.1 *Motivation*

A first step in any statistical inference procedure, whether classical or Bayesian involves developing the

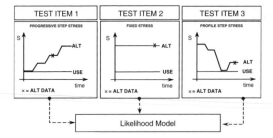

Figure 1.   A separate ALT design for each test item.

likelihood. The flexibility of the likelihood formulation drives the flexibility of the statistical inference procedure in terms of its applicability to different ALT scenarios. In this section, a likelihood model is developed that allows for a comprehensive representation of most ALT inference scenarios currently available to ALT practitioners, specifically, regular life testing, fixed-stress testing, and progressive step-stress testing. In addition, the likelihood model allows for profile step-stress testing and different ALT patterns for each test item as illustrated in Figure 1. Having such a flexible formulation of a likelihood model allows for the comparison of different ALT designs within a common modeling framework. In addition, allowing for such flexibility will increase the model's ability to represent ALT designs used by testing practitioners.

2.2   The failure rate over the course of an ALT

In developing a likelihood model, consider the following step-stress ALT setup. The ALT will consist of testing over a predetermined and fixed maximum number of test environments. An environment is defined by a combination of stress levels from the set of stress variables, e.g. temperature, vibration and voltage. Let K environments $E_1, ..., E_K$ be defined candidate test environments. Let E_ϵ denote the use environment, where $\epsilon \in \{1, ..., K\}$. The index e will be used to indicate a particular environment. Suppose that each of N test items will be subjected to a step-stress ALT with possibly a different step pattern per test item. The index j will be used to indicate a particular test item. The total length of each ALT may vary per test item, but each ALT will be subdivided into m steps. The index i will be used to indicate a particular step-interval within an ALT. Thus, for each test item j, m steps are defined by

$$0 \equiv t_{0,j} < t_{1,j} < ... < t_{m-1,j} < t_{m,j}$$

where the i-th step is defined as $[t_{i-1,j}, t_{i,j})$ and the ALT is terminated at time $t_{m,j}$ for test item j. A design matrix $A = \{a_{i,j}\}$ specifies the indices of the environments for each test item j in each step i. Thus during

the i-th step, test item j is subjected to environment $E_{a_{i,j}}$ where $a_{i,j} = 1, ..., K$. Note that this flexible formulation includes both regular life testing ($a_{i,j} \equiv \epsilon$ for all i) and fixed-stress ALT ($a_{i,j} \equiv z$ for all i and some particular environment E_z, $z \neq \epsilon$).

The approach to deriving the likelihood will be general and center around the failure rate, $\lambda_e(t)$, in a constant stress environment E_e. The failure rate for test item j over the course of the step-stress ALT, denoted by $h_j(t)$, is different from the failure rates in the constant stress environments, as the environments, and thus the failure behavior, vary over the course of the ALT stage. A generic expression will be derived for the failure rate $h_j(t)$ of test item j over the course of an ALT stage, conditioned on knowing the failure rate functions $\lambda_e(t)$ in the candidate test environments E_e, $e = 1, ..., K$.

The cumulative failure rate that test item j has accumulated up to $t_{i,j}$ in an ALT is given by

$$H_j(t_{i,j}) = \int_0^{t_{i,j}} h_j(u)du. \tag{1}$$

In a constant stress environment E_e, the cumulative failure rate would be given by

$$\Lambda_e(t_{i,j}) = \int_0^{t_{i,j}} \lambda_e(u)du. \tag{2}$$

Note that the operating environment in the ALT after $t_{i,j}$ equals $E_{a_{i+1,j}}$. It will be assumed that the change in environments at $t_{i,j}$ is assumed to be instantaneous. In addition, we assume that no additional failures are induced by the instantaneous change of environments $t_{i,j}$ through a shock effect.

Using (2), the cumulative hazard rate $H_j(t_{i,j})$ may be expressed using $\Lambda_{a_{i+1,j}}(t)$ for some value of t. Denoting this value t by $\tau_{i+1,j}$, and solving for $\tau_{i+1,j}$, yields

$$\tau_{i+1,j} = \Lambda_{a_{i+1,j}}^{-1}(H_j(t_{i,j})). \tag{3}$$

The time $\tau_{i+1,j}$ may be interpreted as the amount of time that would have elapsed to accumulate $H_j(t_{i,j})$ by testing in environment $E_{a_{i+1,j}}$ alone, starting at time 0 (see for example Figure 2).

Next, the failure rate function over the course of the ALT stage may be derived as

$$h_j(t) = \lambda_{a_{i+1,j}}(t - t_{i,j} + \tau_{i+1,j}), \tag{4}$$

for $t_{i,j} \leq t < t_{i+1,j}$, $i = 0, ..., m - 1$, where $\tau_{i+1,j}$ is given by (3) and $H_j(t_{0,j})$ is the initial cumulative failure rate of test item j prior to the ALT stage. In case

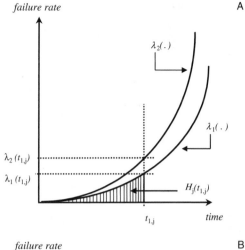

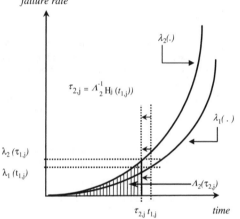

Figure 2. Failure rate construction using instantaneous Environment Changes. A: failure rate up to $t_{1,j}$; B: change Failure Rate to $\lambda(\tau_{2,j})$ at $t_{1,j}$ such that $\Lambda 2(\tau_{2,j}) = H_j(t_{2,j})$.

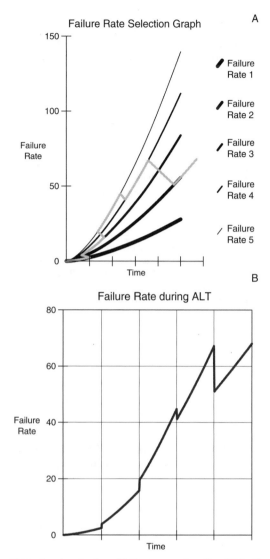

Figure 3. A: selection of failure rate sections from Weibull failure rate functions B: failure rate during a profile ALT by concatening the failure rate sections selected in A.

$H_j(t_{0,j})$ is a new test item, $H_j(t_{0,j}) \equiv 0$. However, the case where a test item has a history of operating hours in known environments may be easily accommodated. The jump in the failure rate at time $t_{i,j}$ follows as

$$\Delta h_j(t_{i,j}) = h_j(t_{i,j}^+) - h_j(t_{i,j}^-), \qquad (5)$$

where $h_j(t_{i,j}^-) = \lim_{t \uparrow t_{i,j}} h_j(t)$, $h_j(t_{i,j}^+) = \lim_{t \downarrow t_{i,j}} h_j(t)$.

It may be derived that the jump equals $\Delta h_j(t_{i,j})$ equals

$$\Delta h_j(t_{i,j}) =$$
$$\lambda_{a_{i+1,j}}(\tau_{i+1,j}) - \lambda_{a_i}(t_{i+1,j} - t_{i,j} + \tau_{i,j}) \qquad (6)$$

for $i = 1, \ldots, m - 1$. Figure 3 presents an example of the above construct for a profile ALT sequentially stepping through the enviroments E_1, E_3, E_5, E_4 and E_2 where a Weibull failure time distribution is assumed for each constant stress. Following the approach above, the current failure rate of a test item only depends on the current accumulated cumulative failure rate and the current stress. It can be shown that the *intrinsic time* failure rate construction approach above is equivalent to the assumption of the Linear Cumulated Damage (LCD) model (see for example Nelson (1980)).

The assumption that no additional failures are induced by the instantaneous change of environments between steps is an assumption which may be challenged, as an instantaneous change of environments may induce a shock effect, causing item failure. The above procedure, however, can be easily extended to the case of gradual environmental changes (see for example Van Dorp (1998)).

2.3 The likelihood given ALT test data

Using the formulation of the failure rate function over the course of an ALT-stage, the likelihood given ALT Test data may be derived for both interval and Type I censored data.

2.3.1 Interval data

Suppose failures can only be monitored at the end of a step interval $i + 1$ i.e. $[t_{i,j}, t_{i+1,j})$. The interval in which item j fails will be denoted by q_j. The probability of test item j surviving time $t_{i+1,j}$ given that it has survived up to time $t_{i,j}$ follows as

$$
Pr\{T_j \geq t_{i+1,j} | T_j \geq t_{i,j}\}
$$
$$
= exp\left\{ -\int_{t_{i,j}}^{t_{i+1,j}} h_j(u)du \right\}
$$
$$
= exp\left\{ -\int_{\tau_{i+1,j}}^{t_{i+1,j}-t_{i,j}+\tau_{i+1,j}} \lambda_{a_{i+1,j}}(v)\, dv \right\}, \tag{7}
$$

where $\tau_{i+1,j}$ is given by (3).

The probability of test item j failing before time $\tau_{i+1,j}$ given that it has survived up to time $t_{i,j}$ equals

$$
Pr\{T_j < t_{i+1,j} | T_j \geq t_{i,j}\}
$$
$$
= 1 - exp\left\{ -\int_{\tau_{i+1,j}}^{t_{i+1,j}-t_{i,j}+\tau_{i+1,j}} \lambda_{a_{i+1,j}}(v)\, dv \right\} \tag{8}
$$

The probability of test item j failing in interval q_j equals

$$
\prod_{i=1}^{q_j-1} \left\{ Pr\{T_j \geq t_{i,j} | T_j \geq t_{i-1,j}\} \right.
$$
$$
\left. \times Pr\{T_j < t_{q_j,j} | T_j \geq t_{q_j-1,j}\} \right\}, \tag{9}
$$

with the convention that $q_j = m + 1$ if the test item is censored at $t_{m,j}$ and $t_{m+1,j} \equiv \infty$. Substituting (7) and (8) in (9) yields

$$
p(q_j) \times
$$
$$
\prod_{i=1}^{q_j-1} exp\left\{ -\int_{\tau_{i+1,j}}^{t_{i+1,j}-t_{i,j}+\tau_{i+1,j}} \lambda_{a_{i+1,j}}(v)\, dv \right\} \tag{10}
$$

where $p(q_j)$ equals

$$
1 - exp\left\{ -\int_{\tau_{q_j,j}}^{t_{q_j,j}-t_{q_j-1,j}+\tau_{q_j,j}} \lambda_{a_{q_j,j}}(v)\, dv \right\} \tag{11}
$$

for $q_j < m + 1$ and is defined as 1 for $q_j = m + 1$

With (10) and assuming conditional independence between the failure times of the test items conditioned on knowing $\underline{\lambda}(\cdot) = (\lambda_1(\cdot), ..., \lambda_K(\cdot))$, it follows that the likelihood given interval data $(N, \underline{q}) = (q_1, ..., q_N)$, equals

$$
\mathcal{L}\{\underline{\lambda}(\cdot); (N, \underline{q})\} = \prod_{j=1}^{N} p(q_j) \times
$$
$$
\prod_{i=1}^{q_j-1} exp\left\{ -\int_{\tau_{i+1,j}}^{t_{i+1,j}-t_{i,j}+\tau_{i+1,j}} \lambda_{a_{i+1,j}}(v)\, dv \right\} \tag{12}
$$

where N is the number of test items in the ALT.

Though not specifically developed here, the previous equations may also be adjusted for the case where items begin the ALT with accumulated damage as in the case of retesting of items. This is considered in Van Dorp (1998).

To be able to perform inference with respect to the failure rates $\lambda_e(t)$ in each environment, it is convenient to reorder the product in (12) as a product over the environment index e instead of over the step interval index i. Given $\tau_{i,j}$ $i = 1, ..., m - 1$ via (3), such a reordering is possible. To accomplish such a formulation, let

$n_{e,j}$ = the number of times that test item j visits environment E_e during an ALT stage,

$vk_{e,j}$ = interval index for which item j visits E_e for the k-th time.

With the above notation (12), may be rewritten as

$$
\mathcal{L}\{\underline{\lambda}(\cdot); (N, \underline{q})\} =
$$
$$
\prod_{e=1}^{K} \prod_{j=1}^{N} \prod_{k=1}^{n_{e,j}} f(e, j, k \,|\, \underline{\lambda}(\cdot), \underline{q}) \tag{13}
$$

where

$$
f(e, j, k | \underline{\lambda}(\cdot), \underline{q}) =
$$
$$
\begin{cases} f_1(e, j, v_{e,j}^k | \underline{\lambda}(\cdot)) & v_{e,j}^k < q_j \\ f_2(e, j, v_{e,j}^k | \underline{\lambda}(\cdot)) & v_{e,j}^k = q_j \end{cases} \tag{14}
$$

and

$$
f_1(e, j, v_e | \underline{\lambda}(\cdot)) =
$$
$$
exp\left\{ -\int_{\tau_{v_e,j}}^{t_{v_e,j}-t_{v_e-1,j}+\tau_{v_e,j}} \lambda_e(v)\, dv \right\}, \tag{15}
$$

562

and $f_2(e, j, v_e | \Lambda(\cdot)) = 1 - f_1(e, j, v_e | \Lambda(\cdot))$ where $\tau_{i,j}$ is given by given by (3). Note that, f_1 (f_2) is the conditional probability of surviving the step (failing in the step) interval for which test item j visits environment E_e for the k-th time, conditioned on having survived up to the beginning of that step. When assuming a common family of life time distributions within a constant stress environment, i.e. specifying a functional form for $\Lambda(\cdot)$, the likelihood may be further derived using (13)–(15). Note, that in principle different failure models for different environments may be specified.

The interval data sampling strategy has the disadvantage that failure information is lost by only monitoring at the end of each step interval. In the type I censored sampling strategy, test items are continuously monitored over the course of the ALT.

2.3.2 Type I censored data

Suppose failures can be monitored continuously over the course of an ALT stage. In that case, the failure time r_j of test item j is known exactly if the test item fails in $[0, t_{m,j})$. It will be assumed that once an item has failed, it will be removed from testing in the same ALT Stage. Knowing the failure times r_j, the step intervals q_j in which the items failed may be easily derived. Using an analogous approach as in Section 2.3.1, the likelihood given the data $(N, \underline{r}, \underline{q})$, where $\underline{r} = (r_1, ..., r_N)$, and $\underline{q} = (q_1, ..., q_N)$, follows as specified in the expressions (17)–(20),

$$\mathcal{L}\{\Lambda(\cdot); (N, \underline{r}, \underline{q})\} =$$

$$\prod_{e=1}^{K} \prod_{j=1}^{N} \prod_{k=1}^{n_{e,j}} g(e, j, k | \Lambda(\cdot), \underline{r}, \underline{q}), \qquad (17)$$

$$g(e, j, k | \Lambda(\cdot), \underline{r}, \underline{s}, \underline{q}) =$$
$$\begin{cases} g_1(e, j, v_{e,j}^k | \Lambda(\cdot), \underline{r}) & v_{e,j}^k < q_j \\ g_2(e, j, v_{e,j}^k | \Lambda(\cdot), \underline{r}) & v_{e,j}^k = q_j, \end{cases} \qquad (18)$$

$$g_1(e, j, v_e | \Lambda(\cdot), \underline{r}) =$$
$$exp\left\{ -\int_{\tau_{ve,j}}^{t_{ve,j} - t_{ve-1,j} + \tau_{ve,j}} \lambda_e(v)\, dv \right\}, \qquad (19)$$

$$g_2(e, j, v_e | \Lambda(\cdot), \underline{r}) =$$
$$exp\left\{ -\int_{\tau_{v,j}}^{r_j - t_{ve-1,j} + \tau_{ve,j}} \lambda_e(v)\, dv \right\} h_j(r_j). \qquad (20)$$

Note that, g_1 is the conditional probability of surviving the step interval for which test item j visits environment E_e for the k-th time conditioned on having survived up to the beginning of that step. In addition,

note that g_2 is the conditional density at the time of failure in case the test item fails within the step interval for which test item j visits environment E_e for the k-th time conditioned on having survived up to the beginning of that step.

When assuming a common family of life time distributions within constant environments, i.e., specifying a functional form for $\Lambda(\cdot)$, the likelihood may be further derived using (17)–(20). Such expressions can for example be derived for the Weibull life distribution using $\lambda_e(t) = \lambda_e \beta t^{\beta-1}$.

3 PRIOR DISTRIBUTION

Given the ordering of the severity of the testing environments, it is natural to assume that

$$0 \equiv \lambda_0 < \lambda_1 < ... < \lambda_K < \lambda_{K+1} \equiv \infty, \qquad (21)$$

and, defining

$$u_e = e^{-c\lambda_e} \qquad (22)$$

for some constant c, it follows that

$$0 \equiv u_{K+1} < u_K < ... < u_1 < u_0 \equiv 1. \qquad (23)$$

The parameter c is chosen to insure numerical stability of the results (see Van Dorp and Mazzuchi (2003)). Rather than defining a prior distribution for λ exhibiting property (21), one may equivalently define a prior for $\underline{u} = (u_1, ..., u_K)$ exhibiting property (23). Concentrating on $\underline{u} = (u_1, ..., u_K)$, a prior distribution which is mathematically tractable, is defined over the region specified in (23), and imposes no other unnecessary restrictions on the u_e, is the multivariate Ordered Dirichlet distribution,

$$\Pi\{\underline{u} | \eta, \underline{\nu}\} = \frac{\displaystyle\prod_{e=1}^{K+1} (u_{e-1} - u_e)^{\eta \cdot \nu_e - 1}}{\mathbb{D}(\eta, \underline{\nu})}, \qquad (24)$$

where, $\eta > 0$, $\nu_e > 0$, $e = 1, ..., K + 1$, and

$$\mathbb{D}(\eta, \underline{\nu}) = \frac{\displaystyle\prod_{e=1}^{K+1} \Gamma(\eta \cdot \nu_e)}{\Gamma(\eta \cdot)}, \quad \sum_{e=1}^{K+1} \nu_e = 1. \qquad (25)$$

Analogous to the above, a beta prior distribution is specified for the transformed parameter $b = e^{-\beta}$.

$$\Pi\{b | \gamma, \kappa\} = \frac{(b)^{\kappa \cdot \gamma - 1}(1 - b)^{\kappa \cdot (1-\gamma) - 1}}{\mathbb{B}(\kappa \cdot \gamma, \kappa \cdot (1 - \gamma))}. \qquad (26)$$

The prior distribution of b is assumed independent of the prior distribution of $\underline{u}$.

Typically, to define the prior parameters, expert judgment concerning quantities of interest are elicited and equated to their theoretical expression for central tendency such as mean, median, or mode [see for example Cooke (1991)]. In addition, some quantification of the quality of the expert judgment is often given by specifying a variance or a probability interval for the prior quantity. Solving these equations generally leads to the desired parameter estimates. Specific quantities of interest for the problem at hand are the mission time reliabilities for each stress environment. An additional advantage of the Ordered Dirichlet distribution is that due to its mathematical properties, the incorporation of expert judgment is facilitated. From (24), for example, the prior marginal distribution for any u_e is obtained as a beta distribution given by

$$\Pi\{u_e\} = \frac{(u_e)^{\eta \cdot (1 - \nu_{e\cdot}) - 1}(1 - u_e)^{\eta \cdot \nu_{e\cdot} - 1}}{\mathbb{B}(\eta \cdot (1 - \nu_{e\cdot}), \eta \cdot \nu_{e\cdot})}, \quad (27)$$

where $\mathbb{B}(\cdot, \cdot)$ is the well known beta constant. This distribution can be used to make prior probability statements concerning mission time reliabilities at the different stress levels due to the one-to-one relationships of these quantities to u_e. Specifically,

$$\{R(t) | u_e, \beta\} = Pr\{X > t | u_e, \beta\} = (u_e)^{\frac{t^\beta}{c}}, \quad (28)$$

where $\{R(t) | u_e, \beta\}$ is the reliability of a test item exposed to environment E_e for a mission time t given u_e.

To obtain the prior parameter values, estimates of prior mission time reliabilities must be obtained. The focus is on mission time reliabilities rather than failure rates, as these may more easily be obtained through elicitation methods focussing on observable quantities. Specifically, for a specified mission length, an estimate of R_e^* the mission time reliability in environment e, a quantile estimate R_ϵ^L for the mission reliability at use stress, and an estimate of R_e^o mission reliability after G mission time durations at use stress is required. Given this information, the following problem is solved numerically to obtain the prior parameter estimates (see van Dorp (1998) for details)

$Solve\ \Theta = (c, \eta, \nu, \gamma, \kappa)\ from$

1. $Pr\{R_\epsilon(t) \leq R_\epsilon^* | \Theta\} = 0.50$

2. $Pr\{R_\epsilon(G \cdot t) \leq R_\epsilon^o | \Theta\} = 0.50$

3. $Pr\{R_\epsilon(t) \leq R_\epsilon^L | \Theta\} = 1 - q, q = 0.95$

4. $Pr\{R_e(t) \leq R_e^* | \Theta\} = 0.50,$

$e = 1, \ldots, K, e \neq \epsilon$

Thus with the exception of the quantile estimate, all prior reliability estimates are treated as median values.

4 POSTERIOR APPROXIMATION

The expression for the likelihood given ALT data was derived in Section 2 and resulted in expressions for interval and type I censored data. Rather than performing prior-posterior analysis using these expressions, one may perform prior-posterior analysis by expressing likelihood in terms of $\underline{u}$ and b instead of $\underline{\lambda}$ and β using (22) and using well known properties of the Ordered Dirichlet distribution.

The posterior distribution follows for interval data and type I censored data by applying Bayes Theorem to the prior and the appropriate likelihood expressions. The derivation of the posterior distribution of is intractable in most cases. It is therefore suggested to use the well known Markov Chain Monte Carlo (MCMC) method approach (see, e.g., Casella and George (1992)). Through the MCMC approach, a sample of the posterior distribution can be obtained. From the sample, approximations of moments and an approximation of the joint posterior distribution may be derived. The approximations of marginal posterior distributions and, using (28), that of the mission time reliability at any stress level may be derived by the estimation of their quantiles. These quantiles may be estimated up to a desired level of accuracy using order statistics arguments (see, e.g., Mood et al. (1974), pp. 513).

5 EXAMPLE

The following example is designed to show the flexibility of the Weibull ALT inference model. The use stress environment will be E_2 and different test items will be subjected to different step patterns. Assume that the following median mission time reliability estimates are available for a mission time of 1000 hours.

$R_1^* = 0.9; R_2^* = 0.8; R_3^* = 0.7; R_4^* = 0.6;$
$R_5^* = 0.5; R_2^L = 0.4; R_2^G = 0.4; G = 2,$

An approximate solution to the prior parameters may be solved from the above data (see Van Dorp (1998)) yielding

$c = 2917399.332; \eta = 747.06;$
$\nu_1 = 0.2110; \nu_2 = 0.1833; \nu_3 = 0.1568;$
$\nu_4 = 0.1312; \nu_5 = 0.1065; \nu_6 = 0.2110;$
$\kappa = 472.74, \gamma = 0.1308.$

In this example, 6 proof-systems are available for testing. The step data concerning environments in each step and step interval times are specified for each testing stage, f, by the following matrices A^f and T^f, $f=1, 2,$

$$A^1 = \begin{pmatrix} 1 & 2 & 3 & 4 & 5 \\ 1 & 2 & 3 & 4 & 5 \\ 5 & 4 & 3 & 2 & 1 \\ 5 & 4 & 3 & 2 & 1 \\ 1 & 3 & 5 & 4 & 2 \\ 1 & 3 & 5 & 4 & 2 \end{pmatrix},$$

$$T^1 = \begin{pmatrix} 100 & 100 & 100 & 100 & 100 \\ 100 & 100 & 100 & 100 & 100 \\ 100 & 100 & 100 & 100 & 100 \\ 100 & 100 & 100 & 100 & 100 \\ 100 & 100 & 100 & 100 & 100 \\ 100 & 100 & 100 & 100 & 100 \end{pmatrix},$$

$$A^2 = \begin{pmatrix} 1 & 3 & 5 & 4 & 2 \\ 1 & 3 & 5 & 4 & 2 \\ 1 & 2 & 3 & 4 & 5 \\ 1 & 2 & 3 & 4 & 5 \\ 5 & 4 & 3 & 2 & 1 \\ 5 & 4 & 3 & 2 & 1 \end{pmatrix},$$

$$T^2 = \begin{pmatrix} 100 & 100 & 100 & 100 & 100 \\ 100 & 100 & 100 & 100 & 100 \\ 100 & 100 & 100 & 100 & 100 \\ 100 & 100 & 100 & 100 & 100 \\ 100 & 100 & 100 & 100 & 100 \\ 100 & 100 & 100 & 100 & 100 \end{pmatrix}.$$

In the second testing stage failed items from the first stage are assumed minimally repaired. Items that survive the first stage are continued on test in the second stage. The mission-time of the system was set to 1000 hours. The test results over the ALT are summarized in terms of $\underline{r}^f$, $\underline{q}^f$ in Table 1. Note that $q_j^f = 6$ indicates that the test item has survived the ALT stage without failure.

A prior-posterior analysis for both interval data and type I censored data is presented. The Gibbs

Table 1. ALT test data in terms of $\underline{r}^f$, $\underline{q}^f$, $\underline{s}^f$.

Test item		1	2	3	4	5	6
ALT Stage 1	r_j^1	495	500	295	500	395	500
	q_j^1	5	6	3	6	4	6
ALT Stage 2	r_j^2	295	95	500	195	500	295
	q_j^2	3	1	6	2	6	3

Sampling Method was used to obtain posterior quantile estimates using test data obtained over 2 ALT stages for: (1) the scale parameters in each environment, and (2) the common shape parameter. The length of the Gibbs-Sequence generated was of length 100,000 and the Gibbs burn-in\Gibbs lag period was set to 25. MCMC diagnostics for this problem are discussed in Van Dorp and Mazzuchi (2003). Results are provided in Table 2.

Distributional results may also be obtained. For example, Figures 4 and 5 convey the prior and posterior

Table 2. Prior & posterior parameter estimates.

E_e	Prior λ_e^*	Posterior λ_e^*: Interval data	Posterior λ_e: Type I data
1	$8.11 \cdot 10^{-8}$	$8.22 \cdot 10^{-8}$	$8.18 \cdot 10^{-8}$
2	$1.72 \cdot 10^{-7}$	$1.73 \cdot 10^{-7}$	$1.73 \cdot 10^{-7}$
3	$2.75 \cdot 10^{-7}$	$2.77 \cdot 10^{-7}$	$2.76 \cdot 10^{-7}$
4	$3.93 \cdot 10^{-7}$	$3.95 \cdot 10^{-7}$	$3.95 \cdot 10^{-7}$
5	$5.34 \cdot 10^{-7}$	$5.36 \cdot 10^{-7}$	$5.36 \cdot 10^{-7}$
β^*	2.03	2.33	2.32

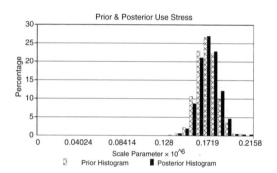

Figure 4. Prior & posterior scale parameter for environment 2 – Interval data.

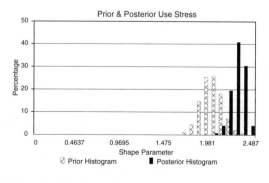

Figure 5. Prior & posterior shape parameter – interval data.

distribution for the shape and use stress scale parameter for the interval censoring case. Distributional results for the scale parameter or mission time reliability (for any specified mission time) at any stress level may also be generated.

It follows from Figures 4 and 5 that for this particular example, the greatest shift is observed in the distribution of the shape parameter rather than that of the scale parameter.

ACKNOWLEDGMENT

The authors are indebted to Technical Program Board of the European Safety and Reliability 2003 Conference and to the referees. Their comments significantly improved the content and presentation of the first version.

REFERENCES

Casella, G. and George, E. I. (1992). "Explaining the Gibbs Sampler". *The American Statistician*, Vol. 46, No. 3, 167–174.

Cooke, R. M. (1991). "Experts in Uncertainty". *Oxford University Press*.

Department of Defense Mil-Std-781C (1977). "Reliability Design Qualification and Production Acceptance Tests: Exponential Case", *United States Government Printing Office*.

Mann, N. R., Schafer, R. E. and Singpurwalla, N. D. (1974). *Methods for Statistical Analysis of Reliability and Life Data*, John Wiley and Sons.

Mann, N. R. and Singpurwalla, N. D. (1983). "Life Testing". *Encyclopedia of Statistical Sciences*, Vol. 4, John Wiley & Sons, 632–639.

Mazzuchi, T. A., Soyer, R. and Vopatek, A. (1997). "Linear Bayesian Inference for Accelerated Weibull Model". *Lifetime Data Analysis* Vol. 3, 63–75.

Mood, A. M., Graybill, F. A., Boes, D. C. (1974). "Introduction to the Theory of Statistics", *McGraw-Hill International Editions*, Third Edition.

Nelson, W. (1980). "Accelerated Life Testing – Step-stress Models and Data Analysis". *IEEE Trans. Reliability* Vol. R-29, 103–108.

Van Dorp, J. R. and Mazzuchi, T. A. (2003). "A General Bayes Exponential Inference Model for Accelerated Life Testing", To appear in *Journal of Statistical Planning and Inference.*

Van Dorp, J. R. (1998). "A General Bayesian Inference Model for Accelerated Life Testing", *Doctoral Dissertation,* The George Washington University.

Safety and Reliability – Bedford & van Gelder (eds)
© 2003 Swets & Zeitlinger, Lisse, ISBN 90 5809 551 7

Concepts of flood risk analysis methodology

A. Drab

Brno University of Technology, FCE, Department of Water Structures, Brno, Czech Republic

ABSTRACT: The aim of this paper is to indicate, in an intuitive form, the possible alternative system engineering methods which can be used in the process of flood risk management. The notion of system engineering is frequently encountered mainly in connection with the plans and requirements related to resolving problems of complex management of organizations, processes, extensive development projects, inter-personal, inter-organizational and international relationships. System engineering, its methods and techniques should simplify the process of resolving the emerging problems of modern society, economics and industry. The paper does not set out to present a complete theoretical explanation of the individual systems science procedures, rather it concentrates on the proposed concept of their use in protection against floods. In the introduction, the paper deals with defining the notions of object, system and model. It then goes on to briefly explain the procedures of system identification, system analysis, system design and system implementation.

1 INTRODUCTION

In the introduction to the paper we should start by outlining the notion of flood risk management as it will be considered in the following text. A clear idea of the notion is provided by Figure 1 illustrating all the stages of risk management. These include the stages of risk assessment, risk analysis and acceptable risk establishment and finally the risk control stage.

In order to continue our considerations it is necessary to *define the goal* of the risk management process. In this respect we shall consider *achieving an*

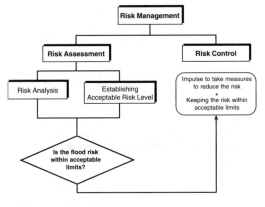

Figure 1. Risk management process diagram.

acceptable level of flood risk to be the main goal of the whole process. In addition, it is possible to formulate partial goals for each stage as well, such as:

- risk analysis goal – quantify flood risk;
- acceptable risk establishment goal – establish an acceptable flood risk;
- risk control goal – keep the risk within acceptable limits, or adopt measures to ensure the risk would reach acceptable limits.

Our task now is to achieve the above goals by applying system science methods. In order to do so, one needs to have an understanding of two basic notions – *the object and the system.*

2 REAL OBJECT CHARACTERISTICS

The current economic and decision-making practice at all levels of economic management requires that extensive data, creating a documentation image of the assessed object and its environment, is processed.

Prior to arriving at a decision, the decision-making subject needs to carry out a quantitative and qualitative analysis of the condition of the managed object. The image of its condition documents a large quantity of information which, although necessary for the decision-making, is not assessable by one individual. The need arises to create inter-linked information sets that are processed using suitable models and, having undergone transformation, facilitate decision-making concerning the current best possible development of the object.

In order to better grasp the complex processes of decision-making and control, we must be able to create appropriate models for the purpose of acquiring current information about the managed whole, we need to model the object using a suitable system.

Before introducing the notion of the system, we shall examine system properties of objects in order to advance nearer to the definition of the system considered in its application form.

In the real world, we are surrounded by large quantities of various types of objects. The method of investigating such quantities of objects requires an individual approach that abstracts their common properties, characteristics and changes in time.

Our consideration is based on one of the main premises of philosophy that in objective reality there exist various material objects with a wide variety of types of relationships of dependence and determinance between them. Material objects always have their own content and form.

Our principal goal from Section 1 above is achieving an acceptable flood risk in the *floodplain*. The floodplain is therefor the *object* of our interest. In the text below, we shall see that the floodplain as a whole meets all the preconditions of being an object from the system science viewpoint.

We shall call a specific concrete form of existence of an object the *expression of an object*. An expression of an object can be characterized by a set of characteristics identified on the object. The elements of the set of characteristics identified on an object will be called *components* of the expression of the object.

An interest in a given object is always motivated by a goal that makes it possible to create defined criteria to specify some of the significant components of the expression of the object determining a simplified variety as a specific reflection of the object scrutinized. From our viewpoint, the components of the expression of the object – floodplain – may include, for example, scope of deterioration or damage within the floodplain, space distribution of risk values, etc.

When scrutinizing the object we also want to identify the relationships between the components of the expression of the object as a whole. To this purpose it is convenient to break down the object into a collection of partial objects so that they do not *share common parts* and so that the expressions of the individual partial objects are more easily recognisable. The collection of the created partial objects comprises the original whole. The breaking down of an object into partial objects can be effected in a number of ways and the choice of a specific break-down is called the *recognition level* on the object.

The object – floodplain – that we have selected is part of objective reality and as such always exists in a specific environment (or surroundings) among other objects. There may be a specific relationship between the object and its environment. The expressions of an object towards its environment may be a consequence of the expression of the environment towards the object or vice versa. The components of the expression of an object on its environment will be called the *output components of the object* as they are directed towards the environment considered to be its organic part; at the same time the output components of the objects will be referred to as the input components to the environment of the said object. In the same way the environment affects the object. The individual components of the expressions of the environment on the object will be called the output components of the environment being at the same time the input components of the expressions of the environment to the object.

Identical impact of the environment on the object need not always lead to identical responses (expressions) of the object to the environment. An object consisting of partial objects which constantly interact happens to be in a specific *state* and the response of the object to a stimulus from the environment is a combined result of the immediate state of the object and the expression of the environment on the object. The state of the object therefore becomes a significant variable in the response of the object to its environment, to its surroundings. It varies depending on time, the previous state and the input from the environment. The state of the object in relation to the input components also affects the output components of the object, hence it is not possible to generally describe the behaviour of an object by its output components only, or its input influences.

We shall now try to analyze the object – floodplain – in terms of its internal structure. When scrutinizing it in greater detail we can distinguish independent parts – partial objects. As the scope of this paper does not allow us to attempt a detailed analysis of all the partial objects, we shall concentrate only on those that may significantly contribute to the external expressions of the object. In this we shall be helped by the so-called *flood hazard identification* which is part of risk analysis (see Figure 1). The cause of the hazard can be ascribed to flooding which in the wider sense of the word stands for water. The way by which water reaches the floodplain is a complicated process based on the hydrological cycle. From the moment the water hits the earth's surface in the form of precipitation its behaviour is influenced by many factors the investigation of which is the subject of hydrology and hydraulics.

Let us now introduce a partial object, calling it *water management*, and let us declare a set of variables known as the *flood progress characteristics (water depth, rate of flow, temperature, period of inundation, etc.*) as the components of its expression. The definition of the water management partial object is key as it effects, through the components of its expressions, all the other partial objects that we shall deal with. These include objects of a technical nature (structures, infrastructure, etc.), biological (animals, plants, etc.), social (population), formal (legislature, standards, etc.), and economic objects (financing, etc.).

On an object as a whole we are impossible to notice and investigate all the components of the expression of the object; in most cases we are only interested in some of them. The selection of the individual components of the expression of the object with a view to pursuing a goal is termed *object investigation viewpoint*. As the object investigation is subordinated to a particular goal we shall now proceed by restricting the scope of expressions of the object even further – choosing a specific subset of the expressions of the object. Further simplification is therefore achieved by determining whether or not the expression of the object is distinguishable.

The individual expressions may often occur under specific circumstances as expressions with a given probability of occurrence, expressions with specific values of components, expressions the values of which change depending on time, both in a discreet and continuous passage of time.

We have now created a simplified view of the object that is based on the premise that the above *simplifications are subordinated to the goal of investigating the object*.

3 THE CONCEPT OF A SYSTEM

Perhaps the most important argument in favour of studying the properties of systems is the knowledge that by implementing a system we define a class of properties of various types of objects. In the previous paragraph we studied the characteristics of an object. We shall now apply the knowledge directly to building up the concept of the system.

Investigating an object in terms of its internal behaviour and relationships with its environment leads to recognizing the system properties of the object.

Real objects under scrutiny often consist of a great number of parts with links between them and it is possible to establish many quantitative variables on them. The selected viewpoint then determines the level of detail of the investigation termed the *recognition level*. A process of investigating objects leads to implementing a system on the object.

The concept of a system is therefore a form of abstraction, it is a form of the object's reflection. The concept displays (reflects) all the system phenomena of the real world and applies to the abstract mathematical concepts as well. From the philosophical

viewpoint it represents a system of displaying the dialectical unity of the whole and its parts.

In everyday usage the term system often refers to a concrete object thus underlining its system qualities (such as a traffic system, enterprise system, procurement system, etc.). However, in scientific disciplines one has to consistently *distinguish between the system and the object* upon which the system is defined. In our imagination we cannot work with a real object, rather with its representation as a model. We work therefore with the model and its properties and then transpose them into a description of the behaviour of the object.

The notion real system will refer to a system implemented upon a concrete real object. The real system thus becomes a model created by the observer in the process of observing, investigating and describing the object under scrutiny. We have to bear in mind at all times that it is possible to define an infinite number of real systems upon the object under scrutiny.

Mathematical formulation of system definitions makes it possible to establish mathematical foundations for the modern systems theory. The principal parts of this theory (e.g. the theory of static, stochastic, determinist and dynamic systems) use mainly mathematical devices which facilitate a brief and adequately accurate expression of the general pieces of knowledge understandable to a wide range of users of the theory specializing in various fields of knowledge. The mathematical formulation, or the mathematical model, can then be implemented using computer technology.

Let us finally sum up the basic properties of systems that are generally recognized and are significant for our investigation.

The first component of the concept of the system is represented by the elements (parts) that make up the whole of the system (or more precisely the whole of the object). We concede that the whole may be created from parts of various types.

For example a system (i.e. the whole) defined upon an object – floodplain – is made up of structural elements (buildings, roads, water management structures, etc.), infrastructure network elements (water, gas, drainage distribution networks, etc.), economic elements (finances), social elements (age groups, nationality groups of the population, etc.), formal elements (legislature, methodologies, standards, etc.). The first basic property of the system derived from the first component of the concept of the system (i.e. from the elements) is the heterogeneity (diversity) of the system.

The second component of the concept of the system is represented by links, relationships and interactions between the parts of the system. In terms of the basic properties of the system it is quite obvious that the implementation of links represents the unification of the diverse elements of a type which is referred to as the homogenization of the heterogeneous parts.

The third component of the concept of the system is the system's purposeful behaviour. The whole viewed as a system exhibits changes directed towards a goal (note: the goal may be variable). The purposeful behaviour is, above all, an expression of the dynamic properties of the system. Here, we are facing, for the first time, the question: Where do the dynamics, i.e. the ability to change, arise in the system (determined by the elements and the relationships between them). We shall assign the ability to change, as a property, to elements, while the relationships will be the bearers of the results (activity outcomes), or of the backgrounds or initiators (activity inputs) of the implementation of the dynamic activities. At the same time we shall bear in mind that for some purposes it will be possible to depict the dynamic properties even formally, at a first glance, in an opposite way, i.e. we shall use the links to express an activity connecting specific states (activity outcomes and inputs).

In order to investigate the properties resulting from the dynamics of the system in greater detail we shall introduce the term *system state* as the purposeful behaviour is in fact represented by a sequence of states that the system development passes through. The simplest case from which we shall derive the system state is the description of the system by means of state variables onto which we can give values in time representing the value of the system state at that moment. Intuitively, we can introduce the notion of the system behaviour as a specific expression of dynamics in a specific environment, velocity and direction (assuming that the state values vary in specific directions and at a specific velocity).

4 FLOODPLAIN AS A SYSTEM

The system properties introduced above are, in general, easily recognizable in physical and biological objects. It is more difficult in economic and social objects or in those components of any type of objects. An example that we shall take to illustrate the system properties (in particular the homogeneity of heterogeneous parts) is the object – floodplain.

The system defined upon the object – floodplain – can be constituted, for example, from technical, biological, economic, social and formal elements. Links between the heterogeneous parts of the system (such as how the scope of damage to buildings affects the finances required for their repair) exist. However, their analysis or structure has some specific features. One of them is creating links between heterogeneous elements. The links need to be identified using the so-called "translational" (conversional) abilities built into the functions of the system elements. So, the functions of the economic element – *flood damage* – include the built-in residential building damage

which links this element to the technical element of residential build. If we fail to find or formulate the "translational" complementation of element functions, faults in the system function will occur.

5 SYSTEM IDENTIFICATION

In the sections above we have established formal means sufficient for expressing or depicting a system. We have studied the means based on the recognition of the properties of the objects that we want to control, manage or design better and more completely.

The intuitive approach to the study of identification processes indicates that it is, on the one hand, a question of the engineer, observer, his possession of knowledge and plans (observer state) and, on the other hand, a question of the object and its states into which the observer properties are projected using models and definitions. The relationships can be simply illustrated in greater detail in Figure 2.

If we sum up the identification process characteristics it is possible to state that the process of identification consists of the identification of the images, knowledge and plans of the engineer with the properties of the object (the original) expressed by means of models and definitions. The introductory consideration of identification can be concluded by saying that the objective of identification is to create a model of a specific (observed, designed) object under conditions specified by the observer, or engineer,

In order to start system identification it is essential that one has available means (tools) for acquiring information about the object understood in its complex state. The identification process itself has acquired, through engineering experience, the form of a procedure based on a specific sequence of steps. The steps need to be carried out in order to proceed to the record of the system model. In general, the identification process consists of the following steps:

– selecting recognition level;
– establishing elements;
– assigning functions to elements;

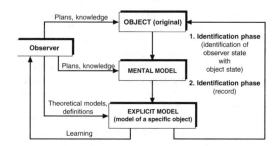

Figure 2.   System identification process diagram (Vlcek, 1984).

570

Table 1.   Language device characteristics (Vlcek, 1984).

Language	Characteristic				
	Description completeness and understandability	Semantic unambiguousness of language forms	Suitability for use in algorithms	Scope of applicability	Basic usage
Verbal description (natural or scientific language)	good	ambiguous	none	limited	Descriptions proper, explanatory comments
Tables	good	unambiguous	poor	wide	Descriptions of function inputs and outcomes
Drawings and block diagrams	good	unambiguous	limited	wide	Designing technical objects, logical structure of processes
Logical diagrams	satisfactory	unambiguous	limited	wide	Automated processes
Curves, graphs, nomograms, etc.	satisfactory	unambiguous	good	limited	Expression of dependence
Mathematical functions and models	poor	unambiguous	best	wide	Mathematically expressible processes, optimization

– identifying relationships between elements;
– identifying relationship parameters.

6   SYSTEM RECORD

In the previous section we dealt with the problem of system identification. In order to be able to continue working with the system we must express and record the results of the identification. The basic condition enabling us to record the identified properties of a specific object is the existence of a suitable language. Table 1 presents an overview of the characteristics of selected language devices.

7   SYSTEM ANALYSIS

The knowledge that the results of identification make it possible for us to record a system model of any object leads to the question of what we can do further with the system recorded in this way. Work with the system can be directed into two areas. We can either establish the existence of the system properties of objects or we can form the object so that it meets specific system properties. Both areas intermingle; the latter area uses the results of the former area. The first activity is referred to as systems analysis, the second activity is called systems engineering. The relationships can be illustrated using Figure 3.

Systems analysis is a collection of tasks and methods of their solution formulated upon an identified object. The objective of the methods is to establish or

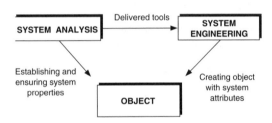

Figure 3.   Systems analysis and systems engineering.

ensure the system properties of the objects under scrutiny. The basic types of tasks in systems analysis include (Vlcek, 1984), for example, the task of the whole of the object, ensuring the system's existence, system's development progress, etc.

Systems engineering represents a substance of the constructive systems theory, as at its heart, if fulfils the purpose and objective of the constructive theory, i.e. represents an ordered set of pieces of information about creating objects with the properties of systems.

In terms of risk management the systems analysis covers the stage of risk assessment while risk control is more a problem of systems engineering (see Figure 1).

8   SYSTEM IMPLEMENTATION

System implementation fulfils the purpose and objective of the constructive theory of systems and, in the application field, the purpose and objective of systems engineering. The most important examples of implementing formal systems are management systems. Implementation will therefore refer to creating any object, either in terms of establishing its system properties or its creation with ensured system properties. However, system implementation is not just a mere conversion of the model into an object. It has problems of its own – as discussed partly in the previous sections. Their foundation is the objective existence of incompleteness, vagueness, difference between the model and its object.

In relation to system implementation one should raise a very important question of *systems reliability*. Solution of a reliability task is frequently based on generally recognized systems analysis tasks. We may mention the event tree method and the fault tree method (Blockley, 1992).

9 CASE STUDY

The theoretical methods mentioned in previous chapters will be applied within the design of a pilot locality situated near the town of Blatná in South Bohemia. This site has been chosen partly due to the floods which affected the Czech Republic in August 2002. The locality contains a large number of small ponds – reservoirs, which may be a potential source of flooding. A dam failure (rupture) of such a water body may create a flood wave with great destructive effects.

Assuming that individual reservoirs create a system (they are all connected), a so-called chain effect may occur; i.e. the flood wave will gradually travel over individual reservoirs while damaging them. This situation occurred in August 2002 when extreme rainfall caused overflow and ensuing rupture of the Melín pond dam (see Figure 4). The flood wave created

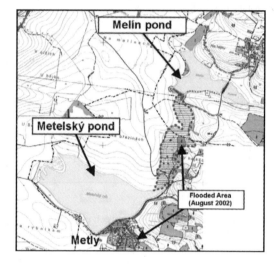

Figure 4.   Map of a part of the affected area.

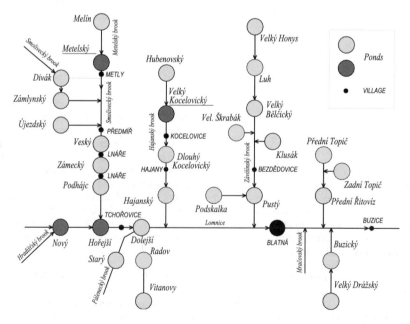

Figure 5.   Scheme of a part of the reservoir system.

travelled through the Metelský brook into the lower situated Metelský pond, the dam of which ruptured as well. As a consequence, the village of Metly and part of the village of Předmíř were swept away. The flood wave travelled further down the Metelský brook valley where it caused considerable damage to property. Figure 5 shows a scheme of chosen reservoirs situated in the locality in question. It is apparent that the mentioned water management system presents a relatively complex system. Although the reservoirs are relatively small, they should not be underestimated in terms of flood risk.

The first stage of the risk analysis – flood risk identification – is at present being carried out in the above mentioned locality. The data necessary for creating a Geographical Information System (GIS), which will form a basis for carrying out further stages of the risk analysis, are being collected. The flood risk identification also includes examination of the water management system (a system of ponds, water courses, water management structures, etc.). This will especially mean identifying individual elements of the system and analysing their mutual functional relations. Simulation of selected situations of the given system will then take place on the basis of these observations. Thus it will be possible to identify critical situations of the given system which are adverse in terms of flood risk, and to present eventual measures for their elimination.

10   CONCLUSION

The problem of protection against flooding, especially when related to the massive development of today's society, can no more be addressed by relying solely on the experience and intuition of experts. The necessity of applying systems disciplines to this area of human activity is obvious.

An essential feature of this is the co-operation of experts from many fields that touch on the problem of flooding. Systems engineering tools enable them to communicate with one another and effectively share information.

The system approach to resolving flood protection problems also opens the door to using other powerful tools, such as operation analysis.

The application of systems sciences to flood risk management offers many ideas for further research, in particular in the area of system identification. As was mentioned, in Section 5 above, system identification involves, in the first place, acquiring sufficient information about the object under scrutiny upon which we are going to define the system. Today it is possible to use the benefits of information systems from many areas to this purpose (Maidment, 2002). In connection to flooding one should not underestimate the role of

Geographic Information Systems (GIS). Other problem areas of system identification mainly concern developing methodology for selecting the recognition level, identifying elements, assigning functions to elements, identifying relationships between elements, identifying relationship parameters. The question raised in Section 4 relating to creating relationships between heterogeneous elements is also essential. This involves mainly data exchange between information systems related to objects in the floodplain.

The systems analysis methods mentioned in this article will be deployed for example in carrying out projects of the Grant Agency of the Czech Republic Nos. 103/02/0018 and 103/02/D100.

The main objective of project No. 103/02/0018 "Employing methods from probability theory, mathematical modelling, damage evaluation, and risk analysis in flood control measures" is to elaborate a methodology for evaluating the extent of risk in inundation areas, i.e. a process on the basis of which it will be possible to select, for a particular specification, a suitable method of risk analysis and corresponding instruments. The following activities are involved:

– Drafting event trees.
– Probability evaluation of individual events with methods from probability theory, mathematical modelling and stochastic modelling.
– Model evaluation of vulnerability and reliability of water management structures and buildings during flood events.
– Vulnerability analysis of inundation areas and proposal for their classification on the basis of an analysis of factors affecting the level of damage (water depth, current speed, time of inundation, temperature of water and air, etc.).
– Proposing methods for damage evaluation, constructing damage functions.
– Defining and quantifying risks.

Project No. 103/02/D100 "Employing mathematical modelling and GIS as instruments of inundation area risk analysis" deals with some problems connected to deploying 2D mathematical modelling and Geographic Information Systems (GIS) as instruments of risk analysis for inundation areas. Using 2D mathematical modelling of current of water in a risk analysis brings the necessity to consistently evaluate the level of uncertainty which affects the final risk estimation. Among the main sources of uncertainty in the case of 2D mathematical modelling are: selecting turbulence model, assessing roughness levels in an inundation area, and selecting a suitable numerical method of solution. Employing the second instrument – GIS, in connection with 2D mathematical modelling, requires the provision of the following: ensuring data exchange between GIS and the mathematical model, implementing verification methods, sensitivity analyses and calibration of 2D mathematical model and of risk analysis methods into the GIS environment.

The aim of the submitted project is to elaborate a methodology which would serve as a basis providing directions for carrying out risk analysis of inundation areas with a focus on application of 2D mathematical modelling of current of water and GIS. The content of the project is drawn up in such a way as to enable observations obtained to be applied in practice in drawing up a concept of flood control measures as well as in the detailed design of protection of individual areas at risk.

ACKNOWLEDGEMENT

The Research was supported by the Grant Agency of the Czech Republic, projects No.103/02/D100 and 103/02/0018.

REFERENCES

Blockley, D. 1992. *Engineering safety.* Cambridge: McGraw-Hill International.

Fukuoka, S. 1998. *Floodplain risk management*, Rotterdam: A.A.Balkema.

Maidment, D.R. 2002. ArcHydro: GIS for water resources. New York: ESRI.

Obona, J. 1990. *Systems and system analysis in practice (in Slovak)*. Bratislava: Alfa.

Robinson, B., & Prior, M. 1995. *System analysis Techniques*. Oxford: ITP.

Slijkhuis, K.A.H., & Van Gelder, P.H.A.J.M., & Vrijling, J.K., & Vrouwenvelder, A.C.W.M. 1999. On the lack of information in hydraulic engineering models. *Safety and Reliability, Vol. 1*: 713–718.

Van Gelder, P.H.A.J.M. 2000. *Statistical methods for the risk-based design of civil structures.* PhD-thesis: Delft University of Technology.

Vlcek, J. 1984. *System engineering methods (in Czech).* Praha: SNTL.

Vrijling, J.K., & Van Gelder, P.H.A.J.M. 1997. Societal risk and the concept of risk aversion, *Advances in Safety and Reliability, Vol. 1*: 45–52.

Vrijling, J.K., & Van Gelder, P.H.A.J.M. 2000. *An analysis of the valuation of a human life.* Esrel 2000 and SRA-Europe Annual Conference, Edinburgh, Scotland, UK.

Safety and Reliability – Bedford & van Gelder (eds)
© 2003 Swets & Zeitlinger, Lisse, ISBN 90 5809 551 7

Assessing the effect of safety management efficiency on industrial risk

N.J. Duijm, M.D. Madsen & H.B. Andersen
Risø National Laboratory, Systems Analysis Department, Roskilde, Denmark

A. Hale & L. Goossens
Delft University of Technology, Delft, Netherlands

H. Londiche & B. Debray
Ecole des Mines St. Etienne, France

ABSTRACT: One of the parts of the ARAMIS project concerns assessment of safety management efficiency and its effect on the calculation of the external risks of Seveso-establishments. The aim of safety management is to implement and maintain a series of barriers that prevent and mitigate accidents. The quality of safety management depends on the selection of the barriers and the organizational structures and processes that ensure that the barriers fulfill their purpose. The quality of these structures and processes can be reviewed by auditing them directly, but the safety culture and learning ability of the organization have an important influence on the effectiveness of the safety management and these and their mutual interaction are harder to assess directly. The ARAMIS project will pay equal attention to measuring all three aspects; safety management structures, processes and safety culture.

1 INTRODUCTION

Major hazard industrial sites in the European Union, covered by the so-called "Seveso-II" directive (European Community, 1996) are required to demonstrate implementation of a "Major Accident Prevention Policy" (MAPP) and a safety management system. The Directive and its national implementations provide prescriptions and requirements for the elements that need to be covered by the MAPP and the safety management system.

The Directive also requires a risk analysis to be performed and documented, demonstrating the possible development of major accidents and the likelihood of these accidents.

Until now, it has been difficult to discount the quality and effectiveness of safety management in the risk level of the industry, let alone to show satisfactorily how this should be accounted for in the risk analysis.

For example, the reference scenarios to be included in a risk analysis for land use planning assessment in France (DEPPR 1990) are prescribed regardless of the quality of the safety management at the industrial site in question.

Likewise, the "Purple book" with guidelines for the quantitative risk assessment to be performed in the Netherlands (CPR 1999), provides failure rate data not taking explicitly into account the quality of the management.

One of the goals of the "ARAMIS" project is to include the safety management efficiency in the risk assessment. The methodology should in principle be suitable to be used both in the context of a quantitative risk assessment (where risk is represented by contours of Individual (fatality) Risk and/or Societal Risk) and a qualitative risk assessment (where risk is represented by describing the consequences of a series of representative accidents). Depending on national implementations of the "Seveso-II" directive, both methods are applied in the European Union. However, it should be realized that safety management focuses on prevention and mitigation of accidents. Therefore, its efficiency can primarily be expressed by how much the *likelihood* of major consequences can be reduced, rather than by the *magnitude* of the (worst-case) consequences. In other words, in a qualitative risk assessment too, considerations regarding *probability* need to be included in some way or another.

In order to limit the complexity of the methodology, the goal is to develop a tool that allows an industrial site and its safety documentation (i.e. the safety report for the "high-tier" Seveso sites) to be analyzed within a couple of weeks.

The "ARAMIS" project runs from 2002 up to and including 2004. This paper describes the concept of the safety management assessment methodology as it has been developed during the first period of the project. The concept still needs to be made operational and it will be evaluated by means of 5 major case studies in large industrial establishments.

2 PREVIOUS WORK

Work on modeling safety management in relation to technical risk analysis began with the pioneering work on "Manager" (Technica 1988). This was followed in Europe by the studies that developed PRIMA (Hurst et al. 1996) and I-Risk (Bellamy et al. 1999), and in the USA by the development of the Work Process Analysis Method (WAPM) (Davoudian et al. 1994). All of these differed in the way in which the management model interfaced with the technical risk model. Manager calculated one management factor, which was used to weight the total risk figure from the technical risk analysis. PRIMA interfaced through eight generic management factors controlling significant failure scenarios, linking life cycle phases with characteristic problem areas (e.g. human factors in operations). I-Risk interfaced with the hundreds of initiating and base events in the generic fault trees and proposed eight systems delivering resources and controls to safety-critical tasks related to these parameters. WPAM also interfaced through work processes linked to the risk scenarios. The more points at which the management model links with the technical risk model, the more there is a risk of a combinatorial explosion in the quantitative analysis and lack of clarity in qualitative understanding of the functioning of the management system. I-Risk therefore introduced the notion of "common mode" management, whereby tasks were clustered for analysis and auditing purposes when they were operated by one management system, were susceptible to one type of human error, or were similar hardware components subject to common maintenance tasks etc.

3 APPROACH

The approach chosen for the ARAMIS project is based on the identification of functions to be fulfilled by safety management that together aim at developing, choosing, using and maintaining barriers or lines of defense against major accidents.

In order to be able to develop and choose these barriers, the industrial site's management needs to identify the possible accident scenarios. To support the identification of accident scenarios, the ARAMIS project is developing a number of generic fault trees that describe the routes from initiating events to a (potentially catastrophic) top event, often a Loss of Containment (LOC). Starting at the LOC, an event tree describes the development of alternative consequences, depending on external conditions and mitigation measures. The combination of fault trees and event trees are called the bow tie.

In the generic bow-tie descriptions, generic barriers can be identified. These barriers can consist of technical measures (safety devices) or procedural measures (e.g. operational instructions linked to behaviour and competence). These barriers must be put and kept in place and in a functional state, and their use must be ensured by organizational measures (requirements for e.g. maintenance, inspections, and qualifications). This approach relies on the "defence in depth" concept associated with the development of "safety barriers analysis".

3.1 The safety barrier principle

According to a typology, which is largely derived from a recent publication by Hollnagel (1999), the safety barriers can be classified regarding their nature, following five main categories:

- *Material or physical barriers* physically prevent an action from being carried out or an event from taking place. These barriers may also block or mitigate the effects of an unwanted event. One can distinguish between passive barriers such as buildings, walls, fences, railings, and containers, and barriers that need (automatic) activation such as fire curtains, etc.
- *Process instrumentation and control* can be considered as barriers which keep the process conditions within small intervals around set points and which are permanently active mechanisms. The process instrumentation and control ensure that the process proceeds in a "normal", desirable way within the design envelope.
- *Functional (active or dynamic) barriers* impede undesired actions from being carried out, for instance by establishing an interlock, either logical or temporal. These barriers set up preconditions that need to be met before the action can be carried out. These conditions are tested automatically without the need of a human interpretation. These safety systems have an exclusive safety function and therefore they should be independent of the control systems. A lock, for instance, can be a physical lock requiring the use of a key or a logical

lock requiring some password or identification device.

- *Symbolic barriers* need an interpretation by an "intelligent" agent in order to achieve their purpose. Signs and signals are symbolic barriers indicating a limitation on performance that may be disregarded or neglected. A typical example can be the reflective markers along a road.
- *Immaterial barriers* are not physically present in the situation, but their efficiency relies on the knowledge of the operator in order to reach their purpose. Typical immaterial barriers are: rules, guidelines, safety principles (safety culture), restrictions, and laws.

3.2   The safety function concept

A safety function can be defined as "a technical, organisational or combined function, which can reduce the probability and/or consequences of a set of hazards in a specific system". Human actions should also be considered as part of the organisational component.

In the framework of the ARAMIS project it has been proposed to classify the safety functions into five main categories described by five action verbs:

- *To avoid*: these functions aim at suppressing all the potential causes of an event by changing the design of the equipment or the type of product used. E.g. the use of a non-flammable product is a way to avoid fire.
- *To prevent*: these functions aim at reducing the probability of an event by suppressing part of its potential causes or by reducing their intensity. E.g. to prevent corrosion, a better steel grade can be used. This is probably not sufficient to suppress it but it may reduce its probability.
- *To detect*: the detection is not sufficient by itself to suppress an event or to limit its probability or its consequences but it is generally a necessary condition for action. The detection can be automatic or performed by humans.
- *To control*: the control functions aim at limiting the deviation from a normal situation so that it does not become an unacceptable one. A pressure relief system performs a control function. So does a computerized supervision system (even if its initial aim is not necessarily safety).
- *To protect*: once an event has occurred, it is necessary to protect the environment from its consequences. This holds for a critical event, which leads to undesirable effects such as heat or pressure wave. It holds also for normal situations e.g. the protection of pipes against corrosive products.

The safety functions are applied to the events that appear in the fault and event trees.

3.3   The generic safety barrier

A generic safety barrier is the combination of one of the barrier types proposed in 3.1 with one of the safety functions and the event to which it is applied.

For instance, a generic safety barrier could be "a material device to prevent corrosion". Each generic safety barrier can then be implemented depending on the context. The generic barrier "material device to prevent corrosion" could be implemented through a protective coating or through galvanic protection depending on the material and environment.

In many circumstances several different barrier systems are used together to achieve a common purpose.

The description of the action to be done by a safety function leads to a list of different safety barrier systems able to fulfil the action. All these systems are part of the same generic barrier. One of the major choices is whether, and if so how much, humans as opposed to hardware fulfill the function.

3.4   Managing the safety barriers

Choosing and positioning the appropriate barriers are two related steps in safety management; maintaining, ensuring the use of, and improving the barriers over time are additional, required steps.

The approach to be developed in the ARAMIS project consists of the identification of a generic set of management functions or actions that are required to both establish the appropriate barriers and to ensure and maintain their use.

This means that a reference list of generic fault and event trees, barriers, and safety management functions will be produced following the lines described in the sections above. For a given industrial site, the actual conditions can be mapped onto these generic data sets, i.e. a subset of the generic fault and event trees will apply, and, in order to speak about a reasonable or good level of safety management, a minimum number of the corresponding generic barriers and safety management functions, as identified, need to be implemented in practice with a certain level of management quality. So the ARAMIS safety management model will provide a definition of the components that need to be included in an ideal or reference safety management system and which elements link most strongly to which types of barrier. By comparing the actual implementation with this reference definition, both with respect to what components are omitted and what is added (and whether that addition is functional or unnecessarily complex), a judgment can be made about how effective the implementation is.

4 GENERIC SAFETY MANAGEMENT FUNCTIONS

Analysis of the safety management literature and comparison of the models proposed by the consortium partners – I-Risk from Delft TU (Bellamy et al. 1999), MIRIAM from INERIS (Plot & Le-Coze 2002), and the work of Risø – has led to a proposed set of nine generic management functions, which can be assessed independently to provide a qualitative profile of the management effectiveness and of where improvement needs to be made. The nine functions are designed to capture the necessary management structure and processes that have to take place to ensure the presence and use of the barriers and so control the risk scenarios. Linked to each structure/process element we propose to assess the "motor" in the organizational culture that can ensure a good or poor use of that process. This makes in total 18 elements to be assessed. These are:

Structure/Process	Culture
Risk assessment & evaluation & barrier selection process	Risk perception & priority
Allocation of tasks & responsibilities	Felt responsibility, causal attribution
Learning & change control system	Openness to learning, trust, just culture
Manpower planning and availability system	Response to stress, fatigue & pressure
Provision of competence & safety suitability	Perceived competence, fallibility, actor strategy
Commitment, incentive, compliance & conflict resolution systems	Job pride, ownership, felt responsibility, safety priority
Communication & coordination channels	Social control, trust, leadership style
Procedure, goal, rule management systems	Compliance, violations, flexibility, ownership of rules
Hard- and software, construction, maintenance, interface, ergonomics	Trust in hardware

Each of these elements will be assessed for the processes of design/choice, use and maintenance of the types of barrier. The notion of "common mode" developed in I-Risk will be used to group barriers into generic types according to the requirements to manage them. This will reduce the amount of assessments that have to be made to come to a complete management efficiency assessment.

The linking of structural elements of management with corresponding aspects of culture is an innovative aspect of the ARAMIS project. Further work in the project will determine whether it is a useful way of combining such disparate aspects that have been shown by earlier research to be relevant to good practice. The coupling does offer the prospect of using assessment of culture to modify the score given to elements of structure/process within a certain range.

Expert judgment studies will be carried out to assess the relative contributions of each of the elements to the successful management of different types of barrier in the bow tie diagram. This will provide guidance as to how the assessments should be combined quantitatively to modify the risk analysis, through their influence on the functioning of barriers.

5 MEASURING SAFETY MANAGEMENT EFFICIENCY

Following the structure of the ARAMIS model of safety management, three separate areas of focus can be distinguished when assessing safety management efficiency:

1. The quality of (continuous) identification and evaluation of (major) hazards and, as a consequence, the identification and implementation of adequate barriers against major accidents;
2. The appropriate identification and implementation of safety management functions (structures/ processes) to safeguard the use of the barriers;
3. The success of the safety management activities to integrate risk control in the operational practices of all levels of staff and to create a good culture relating to safety.

5.1 Measuring efficiency of the hazard evaluation

The efficiency of hazard evaluation and risk analysis can be measured by comparing and mapping the risk analysis, and specifically the identified accident scenarios, with the generic fault trees and event trees. Performing a risk analysis is an obligation for all industries covered by the "Seveso-II Directive". Judging the effectiveness of the choice of barriers requires expertise in the specific processes and the state of the art of prevention in that area.

5.2 Measuring efficiency of the safety management functions

The whole set of management priorities, goals, planned actions, and assigned resources that document the safety management functions can be considered to be the core of the safety management system. The commitment to planned goals and actions and the documentation of these goals and actions ensure that safety is maintained and improved over time and is not an ad-hoc activity.

Using audit techniques, it is possible to identify the functions actually being implemented on an industrial site. By comparing the actual implementation with the set of eight remaining generic management functions (the ninth is covered by 5.1 above), taking into account the actual barriers to be maintained, a ranking of the completeness and adequacy of the safety management can be obtained. As discussed before, the level of detail for auditing will not be the activities in respect of each single barrier, but it is assumed that a number of actions and functions have "common mode" properties, i.e. they are considered to be effective for several (types of) barriers.

Auditing must not only identify what functions are intended to be performed, but also reveal whether these functions and their actions and priorities are put to work in practice.

5.3   Measuring the status of the staff attitudes to risk control

Auditing the safety management's goals, plans and actions reveals the intentions of safety management, but it does not necessarily reveal to what extent safety is indeed made part of the everyday operational activities. In other words it does not tell how successful safety management has been so far or will be in the future.

Useful notions in this respect are the organization's safety culture and its ability to learn from both previous experiences, but also more abstractly, from the hazard evaluations.

Safety Culture Indicators

---

1. Management:
Attitudes and practice of company management with respect to safety management, safety priorities, safety commitment, responsibilities, communication style and error/blame perception

2. Employees:
Attitudes and practice of employees with respect to safety, motivation, trust and relations towards management, company communication, leadership style, responsibility and human performance limitations

3. Safety system:
Compliance with safety relevant operating procedures (SROPs); mechanisms/practices to update and revise SROPs; basic & recurrent training requirements. Housekeeping. Compliance/violations.

4. Learning system (safety culture maturity):
Ability of organization to adopt and exploit experience from incidents/accidents and successful achievements.

The organization can be represented by four interactive structural components as depicted in the above table. Safety culture is explicitly and tacitly reflected in the views and attitudes of management (component 1) and employees (component 2). The safety culture determines how management and employees handle and perceive the safety management system (component 3) and thus deal with violations and work-arounds.

Relevant indicators for safety culture in the framework of ARAMIS are the attributes presented along with the presentation of the 9 generic structure/ processes in section 4 above. The outcome of measurements of these attributes can modify the judgment of the effectiveness of the corresponding generic function. For example a high level of perceived stress of the employees means that the management function "Manpower planning and availability systems" is not implemented optimally.

Component 4 represents the integrated ability of the system to integrate operational experience, including that from incidents/accidents and translate this into successful learning.

5.3.1   Management
When assessing safety culture dimensions with respect to management this will also involve looking at middle management. Management and especially middle management play an important part in shaping safety practice. Equally, changes in safety culture cannot succeed unless management wholeheartedly supports initiatives all through the line. Indeed, a number of analysts argue that improvements in safety culture will not succeed unless management initiates them.

5.3.2   Employees
Employees' attitudes to and views on factors that impact on safety – sometimes referred to as "safety climate" – play a crucial role in defining the safety culture of an organization. Therefore, it is necessary to measure or assess employees' perceptions of, and attitudes to direct and indirect safety-related issues in order to identify points of strength and weakness of an organisation.

5.3.3   Learning capability indicators
Based on experience reported in the literature (Van der Schaaf et al. 1991) and on experience with surveys and interviews with pilots, air traffic controllers (Madsen 2002), hospital doctors and nurses (Andersen et al. 2002), and process industry safety engineers, we suggest that an organisation's capability of learning from negative as well as positive events is a useful safety culture indicator and, moreover, that it is an indicator of the maturity of the organisation in terms of safety culture. Organizations that have a high ability to learn will be at a higher level of safety cultural maturity than those that do not.

579

There are, of course, several sources of learning within an organisation. Within safety critical domains perhaps the most essential one is the ability to learn from surprises, errors, incidents and accidents. In order to capture the potential learning from these events, it is essential to have structures within the organization that can document critical incidents, analyse and disseminate the lessons learned, give feedback to the staff involved and follow up with possible improvements (Koornneef, 2000). To feed the learning process it is necessary that the employees are willing to tell about their own errors and mishaps. However, this type of candidness will often be absent, not least because employees will often be afraid of formal or informal sanctions if they report their own mistakes.

Finally, another and equally important source of learning is the ability to learn from "positive" events – from successes and achievements made within (or outside) the organisation – learning from good practice.

5.3.4 Measuring the structural components

Management and employee attitudes towards safety and the learning ability of the organisation can be determined using a variety of tools.

- Safety climate questionnaires – measuring the attitudes towards the cultural indicators corresponding with the nine generic management functions: risk perception, responsibility towards safety, causal attribution, openness to learning, including reporting of errors and incidents and reasons of not reporting, trust, justness and relationship with management, stress and pressure, perceived experience and competence, own fallibility, pride, communication;
- Interviews following the Critical Incident Techniques (CIT) – measuring actual practice with respect to accidents and incidents (Flanagan 1954);
- Audits reviewing not only records of compliance with procedures and transgressions of allowed levels but also mechanisms and practices of reviews of these
- Interviews with management – measuring attitudes and practice towards incident and error reporting and learning

6 DISCUSSION

The innovative approach of ARAMIS with respect to safety management is the integration of audits of the processes and structures of the (formal) safety management system with indicators of safety culture.

In the project, the model description is presently being defined and agreed by the partners.

The model elements and measurement techniques require testing in a series of case studies in industry.

A range of measuring techniques exists, but an appropriate combination has to be found, while additional instruments may be needed to fill some gaps.

Combination of data from measurements of single indicators has not been addressed so far. Weighting factors for the generic management functions, probably dependent on the type of barrier, need to be developed. Collection of expert judgment will play an important role in this activity.

Finally there is a need to collect benchmarking material, i.e. to develop a ranking for levels of good and bad safety management compared to an agreed reference level.

ACKNOWLEDGEMENT

The ARAMIS project is supported by the European Commission as part of the Fifth Framework Programme on Energy, Environment and Sustainable Development, Contract No. EVG1-CT-2001-00036.

The institutes co-operating on development of the ARAMIS safety-management efficiency index are Risø National Laboratory, Denmark, Delft University of Technology, Netherlands, Ecole des Mines St. Etienne, France, Ecole des Mines Paris, France, INERIS, France, and the Central Mining Institute, Poland.

REFERENCES

Andersen, H.B. et al. 2002. *Reporting adverse events in hospitals: a survey of the views of doctors and nurses on reporting practices and models of reporting.* In prep.

Bellamy, L.J., Papazoglou, I.A., Hale, A.R., Aneziris, O.N., Ale, B.J.M., Morris, M.I. & Oh, J.I.H. 1999. I-Risk: Development of an integrated technical and management risk control and monitoring methodology for managing and quantifying on-site and off-site risks. Contract ENVA-CT96-0243. Report to European Union. The Hague: Ministry of Social Affairs and Employment.

CPR. 1999. *Guidelines for quantitative risk assessment ("Purple Book").* Committee for the Prevention of Disasters. The Hague: Sdu Uitgevers.

Davoudian, K., Wu, J-S. & Apostolakis, G. 1994. Incorporating organizational factors into risk assessment through the analysis of work processes. *Reliability Engineering and System Safety* 45(1–2): 85–105

DEPPR. 1990. *Control of Urban Development Around high risk industrial sites.* Secretary of State to the Prime Minister for the Environment and the Prevention of major technological and natural risks. Industrial Environment Department France.

European Community. 1997. Council Directive 96/82/EC of 9 December 1996 on the control of major-accident hazards involving dangerous substances. *Official Journal of the European Communities.* No L 10/1.

Flanagan, J.C. 1954. The critical incident technique. *Psychological Bulletin* 51(4): 327–358.

Hollnagel, E. (1999). *Accident analysis and barrier functions*. Accidents and barriers Project TRAIN 99/09/30.

Hurst, N.W., Young, S., Donald, I., Gibson, H. & Muyselaar, A. 1996. Measures of safety management performance and attitudes to safety at major hazard sites. *J. of Loss Prevention in the Process Industry* 9(2): 161–172.

Koornneef, F. 2000. *Organised learning from small-scale incidents*. Delft University Press. Delft.

Madsen, M.D. 2002. A Study of incident reporting in air traffic control – moral dilemmas and the prospects of a reporting culture based on professional ethics. In Johnson, C. (ed.) 2002. *Proceedings of the Workshop on the Investigation and Reporting of Incidents and Accidents* (IRIA 2002), 17th–20th July 2002, University of Glasgow.

Plot, J. & Le-Coze, J.C. 2002. *An integrated approach to organise major risks control in hazardous chemical establishments*. Verneuil-en Halatte: INERIS.

Technica. 1988. *The Manager Technique. Management Safety Systems Assessment Guidelines in the Evaluation of Risk*. London: Technica.

Van der Schaaf, T.W., Lucas, D.A. & Hale, A.R. (eds). Near Miss Reporting as a Safety Tool. Oxford. Butterworth Heinemann. 1991.

Safety and Reliability – Bedford & van Gelder (eds)
© 2003 Swets & Zeitlinger, Lisse, ISBN 90 5809 551 7

Data analysis in a study on risk perception in product use

F.H. van Duijne, H. Kanis
TU Delft, School of Industrial Design Engineering, the Netherlands

ABSTRACT: Design supportive research that focuses on product use yields data in the form of anthropometric measurements and audio/video tapes of observed product use. Analysis of these data can be cumbersome. Results should outline observed user activities and explicate individual instances of user-product interaction. This paper describes the analysis of a study on risk perception in product use, and discusses scientific credentials to address the quality of the study. In addition, it discusses the use of software programs to facilitate data analysis of design supportive research that addresses product use.

1 INTRODUCTION

In order to get insight into perceptions of risk in user product interaction, studies of risk perception should focus on actual product use (Weegels & Kanis, 2000). In this respect, the collection of detailed information that provides accounts of the interaction process seems essential. Measurements and observations for studying user-product interaction include 1) observations of users' actions, 2) verbatim selfreports on perceptions and cognitions, 3) anthropometric measurements, and 4) background information on the context of product use (cf. Kanis, 1998). Data obtained by these research methods consist of information about user activities linked to featural and functional characteristics of the product that is being used. Analysis of these data should clarify the scope and variation in users' perceptions of risk and how these perceptions are related to product characteristics. This type of ergonomic/human factor study is a means to support design practice (Kanis, 2001, 2002).

These studies into user-product interaction yield raw data in the form of audio/video tapes and anthropometric measurements. The distinct use actions of the participants, their expressions that contain (subtle) perceptions of risks, and the specific contexts of product use can be distinguished on the tapes. Studies in the field of usability research address a number of cases, each of them providing such "rich qualitative data". This practice poses a problem for the researcher who wants to provide an overview of the observed actions and accounts, and to exemplify the specific character of each instance of user-product interaction simultaneously.

Analysing qualitative data from observational studies can be cumbersome. It is only recently that procedures for qualitative data analysis have gradually been clarified (Miles & Huberman, 1994). In sociological studies, guidelines for the coding and comparison of qualitative data are proposed (Miles & Huberman, 1994; Glaser, 1978; Glaser & Strauss, 1967). Given the manifold of approaches and strategies in social sciences, qualitative data analysis aims to understand structures and processes in social (interactive) settings. Analytic practices include (an iterative process of) affixing codes to unreduced data, making memos to reflect upon emerging understanding of the data, and sorting and organising codes in order to identify patterns grounded in data which leads to the formulation of hypotheses. Because qualitative data analysis is mainly developed to address social relationships, the existing procedures needed elaboration in order to be applicable to the study of risk in user-product interaction.

A way of coding, counting, and comparison needed to be formulated in order to analyse the nature of usability problems. The core of this approach to data analysis is to provide an overview of usability problems and to perform an in-depth analysis which will result in detailed conclusions and recommendations with a link to design practice (Rooden, 2001). The scientific criteria used to monitor the quality of this type of research are "amenability to repetition", made possible by a detailed description of the applied approaches, and "credibility", achieved by for instance the application of various methods and by presentation of outliers that question current propositions (cf. Kanis, 2001; Miles & Huberman, 1994).

Figure 1. Placing the cartridge in Camping Gaz 206N. This procedure of positioning can be dangerous. The cartridge is pierced before it is attached to the clamps. The lamp unit should be removed first.

This paper describes the analysis of data from a study into the risk of replacing pierceable cartridges of gas lamps used for camping purposes (see Fig. 1) (van Duijne & Kanis, 2002). Two research questions that were asked in the study and the answers provided by empirical work are presented here in order to illustrate the analytic approach. The research questions were: 1) how do featural and functional product characteristics influence users' actions and their perception of risk?; 2) are user activities related to users' capacities or individual conditions?

The paper follows an approach to data analysis that is intended to meet the requirements of design supportive research of user-product interaction. It outlines the use of software packages to support analysis of the diverse sorts of data pertaining to observational studies into user-product interaction.

2 DATA

The study was carried out at camping sites. Twenty-three users of gas products with disposable cartridges participated. Various sorts of data were obtained in this study. First, background information about participants was recorded that addressed the context of their camping unit (tent or caravan, number and generation of campers), type of gas product they owned, and their experience with this gas product. Second, participants were asked to demonstrate how they would replace the cartridge of three models of gas lamps that were selected for this study. This yielded video data on use activities. Interviews to clarify use activities and perceptions of risk produced verbatim recordings, provide a third source of data. Finally, some anthropometric measurements of the upper

extremities were recorded (grip force, hand length, and hand width) which created quantitative data.

A total of five field visit were made to collect the data. Between these visits, the data were transcribed and a preliminary analysis was done. This paper discusses the findings that refer to one of the models of gas lamps, 206N (Fig. 1).[1]

3 REPRESENTATION OF THE DATA

Design supportive research seeks to produce, on the one hand, overall descriptions of the observed phenomena, and on the other hand descriptions of the situated individuality of user-product interaction. The quantitative findings of this study provide frequencies of occurred phenomena and correlations of possible indicators of these phenomena. In-depth qualitative descriptions aim to identify how users give meaning to featural and functional product characteristics and what kind of actions they perform in relation to the product.

An important quality of the data analysis in this study is the necessity to inspect the unreduced data. This ensures preservation of case configurations and allows understanding of activities in the context of their production. In order to make the data accessible, all interviews were transcribed literally. To scroll the video tapes efficiently activities had to be coded by short descriptions and corresponding time slots.

Different representation formats facilitate qualitative and quantitative data analysis respectively in terms of making the data accessible for studying individual case configurations including actions, verbal accounts, and context characteristics. The data were loaded into three software programs as outlined in section 4. Some data were used in all of the software programs in order to examine relationships. The interview transcripts, and background information were used to make files in Atlas/ti, a software program for qualitative data analysis (www.atlasti.de). Anthropometrics, background information including experience, and a general description of the order of use actions for lamp 206 N formed the content of a data file in SPSS, a package for statistical data analysis. Codes of observed use activities were listed in Microsoft Excel.

Each of these software programs has different strengths. Although it may be possible to work with

---

[1] Gamping Gaz 206N consists of a lamp unit and a holder. At the base, the lamp unit has a pin, which is designed to pierce the cartridge, and a rubber seal. The holder has clamps to attach a gas cartridge. In order to replace an empty cartridge, the lamp unit should first be disassembled. It can be removed by screwing it off. Afterwards the old cartridge can be removed from the clamps and a new cartridge can be placed in the clamps. The cartridge has ridges that fix its position in the holder. Finally, the lamp unit can be screwed onto the blue casing which pierces the cartridge.

one or two software tools (or even to do it all by hand), we clearly benefited from the programs' different visual lay-outs of the data and from their analytic opportunities in terms of depth of analysis and manageability of the data.

4 ANALYTIC PROCESS

4.1 Atlas/ti

Qualitative data analysis using Atlas/ti was carried out as follows. A first step is to read the interviews closely and to make notes or "memos" as a preliminary analysis of the data. Memos helped to code episodes in the interviews, both thematically and by concepts emerging from the data. A thematical concept is determined by observations which describe participants' understanding of the order of actions for replacing the cartridge. They were coded by the name "order". In other parts of the transcripts participants mentioned the necessity to close the control knob when replacing the cartridge. Instances of data were coded by the name "control knob". Concepts emerging from the data were, for instance, expression of fear. The program attaches the code to the passage in the transcripts (or video or audio[2]) such that clicking on one of the codes in the code list (all code names developed in the analysis), causes the quotations to be highlighted. In this manner, the data specify and illustrate the meaning of the code names which are chosen by the researchers.

The individual files that represent each of the participants can be grouped into families. Subsets of participants can be formed that are differentiated by, for instance, experience with gas lamps, having children, or the order used to replace the cartridge.

A further analytic step is to cluster codes into families and to form networks of codes. Codes that describe participants' perceived need for caution such as "unpleasant", "insecure", and "panic" can be clustered into one code family. Networks can be used to outline a proposed relationship between codes. For instance, codes that reflect participants' understanding of product characteristics (clamps, ridges on the inside of a casing, the pin and the rubber seal, and a perceived lack of cues) "are part of" the code "usecues". Networks can also be used to describe and visualise a process: observed actions related to replacing the cartridges can be coded and linked in order to describe the process across time. In this manner, structures in the

data can be outlined. The contents and meaning of these structures can be scrutinised by using the query tool to find and read the quotations on which the structures are based.

To summarise our analysis in Atlas/ti, an extensive coding process of interviews and actions helps the researcher to get acquainted with the data and facilitates the search for evidence ever since. In addition, the program's tools for visualising properties and relations between objects support systematic inspection of the data. Codes can be linked by a network in order to visualise the narrative structure of verbatim and activities. By inspecting the quotes behind the codes, the networks are adjusted to improve their fit with the original data. Creating representations grounded in data facilitates the writing process of a results section that emphasises description of the locally produced actions and accounts.

4.2 SPSS

In addition to Atlas/ti, we built an SPSS data file that contains anthropometric data, information on the kind of product owned by participants and on the frequency of use, which can be seen as participants experience with gas products. The file also displays categorical variables that represent those actions of participants that are related to the replacement of the gas cartridges, and whether they used the instruction manual.

We used SPSS to produce frequency distributions and correlational analyses to outline the relationships between actions and user characteristics. These analyses provide insight into the numbers of participants who performed a certain action or expressed a certain understanding of risk. They describe the (absence of) relationships between users' actions and for instance experience and hand sizes. This can be visualised in tables and in plots. Their meanings can be outlined by providing a contextualised description of user activities, i.e. using the unreduced data made accessible by Atlas/ti and Excel.

4.3 Excel

This program was used to describe sequences in observed actions. When answering questions related to users' actions and users' perception and understanding of product characteristics, it appeared to be relevant to describe activities in ever more detail. We described segments of the activities and indicated whether this applied to participants. We added time slots that correspond with the original films. This yielded a matrix display. The horizontal axis represented the participants, the vertical axis the observed activities. Check marks and time codes indicated participants' actions. Similar to the application of the Atlas/ti, Excel was used to facilitate inspection of product use activities in order to analyse unreduced data.

---

[2] Atlas/ti also supports analysis of images and video fragments. However, the present state of technology poses limitations to the use of these options. Files that are too large should be avoided. This applies to the video recordings of our study.

5 RESULTS OF THE ANALYSIS

This section demonstrates the results of the data analysis for the research questions presented above. In order to clarify the contribution of these three software programs, their first letter indicates which findings stem from which part of the analysis; Atlas/ti (A), SPSS (S), and Excel (E).

5.1 *How do featural and functional product characteristics influence users' activities and their perception of risk?*

To answer this question we examined participants' actions in replacing the cartridge and their comments in which they discussed product characteristics.

First of all, participants were aware of the risk of highly flammable gas (A). They said that replacing the cartridge is not dangerous when doing it outdoors, at a distance from other people and away from fire which includes burning cigarettes.

Table 1 provides an overview of participants' actions when replacing the cartridge (S). Twelve participants removed the lamp unit. Eleven participants did not remove the lamp unit of which four changed their minds. Their experience with a gas product with similar characteristics is outlined in the table.

Taking a closer look at how people interpreted the gas lamp's product characteristics, the results show the following. The observations demonstrate that participants inspected the pin and the clamps and concluded that the cartridge should be placed between the clamps. The pin pierces the cartridge in this position (E). Some participants who unscrewed the lamp unit explained that pressing the cartridge directly into the holder can be problematic if it is placed askew, which causes the cartridge to leak. The participants who pressed the cartridge directly into the holder either did not know that the lamp unit had to be disassembled or they simply understood that the cartridge should be pressed into the clamps. One participant also did not unscrew the stove unit of her own product when replacing the cartridge. She did not know that it could be unscrewed. Another participant discovered this technique during the study. Before, he had pressed the cartridge directly into the pin and holder. Four participants changed their minds after attempting to press the cartridge into position and realised they had to disassemble the product. They then noticed that replacing by disassembling is easier and requires less force (A).

Fourteen participants had experience with the procedure of replacing the cartridge in 206N in terms of unscrewing the lamp/stove unit (Table 1) (S). However, these participants show similar variations in user activities to participants who owned a self-sealing product or a Coleman lamp and had no experience with replacing a pierceable cartridge in a gas product that had to be disassembled (E).

It appeared that many of the participants who owned a product of which the lamp/stove unit had to be removed used their product infrequently. As a consequence, they may have forgotten the procedure that they applied in the past. Eight out of fourteen participants did not remove the lamp unit when replacing the cartridge of the gas lamps under-study. Three of them changed their mind (S/E). Participants' memory of procedures is reflected in their comments: "Well, you have to pull this around it, press it or something. I don't remember, how was it? No, it must be turn this off, and then position it, and then turn this on." The observations and interviews suggest that participants' experience with products that have similar characteristics does not guarantee that they know the procedure for replacing the cartridge. Some participants strongly believed that they knew how to do it (A), relying on their experience (S), and still used an unintended procedure (E). One participant who did not use his stove this year claimed that "you have to press the cartridge in the holder resolutely. After that it is common knowledge" (A). These findings suggest that knowledge of procedures erodes if it is not applied frequently.

Participants considered disassembling "an additional action". Three of them indicated that they preferred not to do such extra tasks (A).

Further, the clamps of 206N gave the impression that the system is weak and unreliable. One participant said that it is "just a cartridge attached to a lamp". Fifteen participants blamed the clamps for their lack of stability and commented that the lamp may fall and cause accidents (A). Some participants who owned a Camping Gaz stove with similar clamps had improvised something to cope with this lack of stability. One

Table 1.   Technique used to replace the cartridge of 206N and the participants' experience with a gas product as to the removal of the lamp/stove unit.

| Observed action with 206N | Type of gas product owned | | |
	Removal of lamp/stove unit (similar to 206N)	Different system	Total
Lamp removed right at the start	6	6	12
Lamp not removed at all, replacement "succeeded"	5	2	7
Lamp removed after several attempts without removing lamp	3	1	4
Total	14	9	23

placed it on a bread board, one built a wooden case which fixes the stove (E).

5.2 Are use activities related to users' capacities or individual conditions?

To answer this question, we focused on the problems that occurred due to a mismatch between users' capabilities and forces required to replace the cartridge. The anthropometric data of the hand (length, width, and grip force) showed no correlation between users who do and users who do not disassemble the lamp unit (S). Individual case configurations demonstrate the absence of an association between physical characteristics and the order in which participants replaced the cartridge of 206N (Figure 2). These findings support previous research which demonstrated that human physical characteristics explain little of users' operational problems with consumer products (Kanis *et al.*, 1999).

We also relied on participants' self-reports to understand the problems related to users' capacities and individual conditions (A). Difficulties emerging from physical characteristics were salient in participants' self-reports expressed during use activities. The observations suggested that pulling the clamps around the cartridge requires considerable effort (E). When performing the replacement of the cartridge in the incorrect order, two participants complained that it was heavy to place the clamps around the cartridge. They mentioned that because this activity is heavy, it can be risky when one fails to place the clamps correctly. Three participants who changed their minds about the order of replacing the cartridge also said

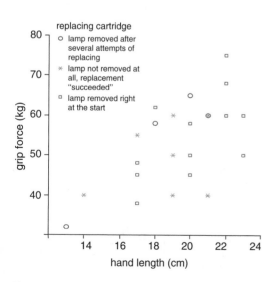

Figure 2. Hand length, grip force and the order for replacing the cartridge of 206N.

that pulling the clamps around the cartridge would require too much effort (A).

6 QUALITY OF THE RESEARCH FINDINGS

The reader may raise the question whether the findings of this study are credible and accurate and whether the process of collecting and analysing data in the study was consistent.

Scientific criteria that apply to this study can be classified as amenability to repetition and absence of indications for bias.

6.1 Amenability to repetition

The process of data collection and analysis needs to be described carefully in order to make the study amenable to repetition. Further, by describing the observed events as literal as possible, which includes citations of the various points of view, the researchers' conceptualisation of the problem under study is reflected in the report. Emphasis on description can be seen as a strategy to prevent early interpretation of the data (Wolcott, 1994; Glaser, 1978). Descriptions of observed events clarify the proposed interpretations. Analysis of unreduced data, activities and verbal reports, makes it possible to be sensitive to participants' expression of meaning. Reporting visual and verbal evidence ensures that the researcher makes inferences that take into account the data.

6.2 Bias?

The credibility of the data is demonstrated by presenting a range of different cases in order to inspect the natural variety in user activities. In this study, the authors studied 23 individual cases in five field visits, providing a rich source of variation and consistency in activities and verbal accounts. Further, the sample shows heterogeneity between participants (e.g. in their experience, anthropometrics and demographics) which does not explain differences in user activities. In this respect, the data do not point at idiosyncracies of the selected sample as being an unrepresentative sample.

Possible bias stemming from researcher effects can be recognised when analysing data from a first field visit. Improvements can be made in further field visits. Biasing tendencies of participants stemming, for instance, from discomfort due to the research setting and social environment, are anticipated by establishing a proper relationship by which participants are encouraged to tell about the subjects of interest. In sociological studies this is called "rapport". In addition, when we suspected participants' answers to be casual and unclear, we tried different questions to address the subject matter.

In order to prevent bias in the interpretation of data a fallabilistic approach involves inspection of presumed outlier cases (Seale, 1999). "Exceptions" in the data can test and strengthen findings. In this study, evidence from every participant was inspected when reporting on the findings. Actions and verbal accounts of participants were reported in order to emphasise individual cases of the multiple-voiced data. This strategy, which is essential in qualitative data analysis, implies that (early-) interpretations and conceptualisations are modified by reexamination of the data.

7 DISCUSSION

This field study on the replacement procedure of pierceable cartridges in gas lamps addressed various elements that constitute usage: featural and functional product characteristics, use actions, perceptions, cognitions, physical characteristics of users, and context elements. Using these observations and measurements on usage we aim to specify links between user activities and product characteristics, since these characteristics are a major component of design (Kanis, 2002). This implies that the research findings both summarise the observed activities, and report the "situatedness of product use", i.e. individuality of cases and their specific contextuals. Aggregation of data in terms of central tendencies is avoided, as this appears to provide little insight into the interactive processes in product use (Kanis, 2002).

The need to specify links between user activities and product characteristics conditions decisions on the data analysis. Various software packages were used to support qualitative and quantitative analysis of rich contextualised data. Each of them had different qualities and facilitated parts of the analysis. The need to report on particularities of user-product interaction implies that analysis needs to be carried out on unreduced data. For this purpose we used both Atlas/ti and Microsoft Excel. Verbal reports from the study were used as input for Atlas/ti. Through reading, coding, and grouping of data elements analytic rigour is achieved in the reported findings. Excel is used to create a coding scheme that describes the observed actions and facilitates inspection of the video tapes. The use of a coding scheme for actions is a tool for a systematic inspection and report of use actions.

In addition to the description of individual cases, research for design should report the scope and diversity of user activities in terms of actions, perceptions, cognitions, and experienced effort. The diversity of user activities can be described by counting the number of participants who perform certain actions. This can be done by using the code list in Atlas/ti. However, the statistical software package SPSS provides more extensive possibilities for studying frequency distributions and exploring relationships.

The variable structure in SPSS sets limits to the coding of observed actions. Actions, formulated as variables, need to be cut into a set of categories that represent all observed configurations. These "action variables" would have many categories because of the variety in observed activities. In addition, splitting up actions into variables reduces the possibilities to describe user-product interaction comprehensively. It prevents inspection of the data in the context of their production. Therefore, we used Excel to map the actions and add corresponding time codes in order to facilitate scrutiny of the original recordings.

The use of the three software packages enhanced the depth of analysis in this project. However, the use of one data base seems preferable when studying relationships between different sorts of data. Presumably, developments in software design will soon yield packages that support all aspects of our analysis sufficiently. Therefore, the present solution of dividing the data over three software packages can be seen as an interim solution that helped the analysis of problems in user-product interaction. In general, data analysis should not depend on software tools. Their role is to support the analytic process (c.f. Kelle, 1997).

Data analysis in design supportive research, such as the study of risk perception in product use, needs to have a firm qualitative basis. Counts and summarised data provide an informative but limited overview of findings. Designers need to know details of user activities in order to have focal points for design improvements (cf. Kanis, 1999). By means of the presentation of full contextualised data that emerge from observational studies this research has provided vivid explanations of problems in userproduct interaction that are grounded in identifiable local contexts. In this manner, the process of risk perception in product use can be further conceptualised and utilised by practitioners.[i]

---

[i] In a further step of this research project, conceptualisation of risk perception is approached by a qualitative analysis. Participants' self-reports and their activities are organised into networks that display the narrative structure of risk accounts and of activities. These networks address participants' accounts of the negative aspects of the products under study, which appear to describe perceived hazards. They also outline the (preventive) actions applied by participants. The networks try to relate discursive evidence of risk accounts in order gain insight into the concept risk perception and how it relates to activities. At this moment, an experiment is planned that will investigate how designers apply the findings of this field study on risk in product use in order to develop a redesigned product.

ACKNOWLEDGEMENT

The authors want to thank Prof. Dr. A.R. Hale and Prof. W.S. Green for constructive feedback.

REFERENCES

van Duijne, F.H. & Kanis, H. 2002. *Replacing pierceable cartridges in gas lamps.* Internal publication, Delft, University of Technology.

Glaser, B.G. 1978. *Theoretical sensitivity: advances in the methodology of grounded theory.* Mill Valley: CA: Sociology press.

Glaser, B.G. & Strauss, A.L. 1967. *The discovery of grounded theory: Strategies for qualitative research.* Chicago: Aldine.

Kanis, H. 1998. Usage centred research for everyday product design, *Applied Ergonomics* 29: 75–82.

Kanis, H., Weegels, M.F. & Steenbeckers, L.P.A. 1999. The uninformativeness of quantitative research for usability focused design of consumer products. In: *Proceedings of the Human Factors and Ergonomics Society 43rd Annual meeting,* 482–485.

Kanis, H. 2001. Scientific credentials for qualitative and quantitative research in a design context. In: M. Hanson (ed.) *Contemporary Ergonomics 2001,* 389–394, London: Taylor and Francis.

Kanis, H. 2002. Can design supportive research be scientific? *Ergonomics,* 45 (14): 1037–1041.

Kelle, U. 1997. Theory Building in Qualitative Research and Computer Programs for the Management of Textual Data. *Sociological Research Online,* 2 (2): <http://www.socresonline.org.uk/socresonline/2/2/1.html>.

Miles, M. & Huberman, A. 1994. *Qualitative data analysis: an explanatory source book.* Thousand Oaks, CA: Sage.

Rooden, M.J. 2001. *Design models for anticipating future usage.* Thesis, Delft University of Technology.

Seale, C. 1999. *The quality of qualitative research.* London: Sage.

Weegels, M.F. & Kanis, H. 2000. Risk perception in consumer product use, *Accident Analysis and Prevention,* 32 (3): 365–371.

Wolcott, H.F. 1994. *Transforming qualitative data.* Thousand Oaks, CA: Sage.

Safety and Reliability – Bedford & van Gelder (eds)
© 2003 Swets & Zeitlinger, Lisse, ISBN 90 5809 551 7

Organisational factors and safety performances

S. Duriez
*SNCF, Dir. Recherche et Technologie/Equipe de Recherche sur les Processus Innovatifs,
Paris, France*

E. Fadier
*Institut National de Recherche en Sécurité, Département Homme-Travail,
Laboratoire Ergonomie & Psychologie Appliquée à la Prévention,
Vandoeuvre, France*

O. Chery
*ENSGSI-INPL, Equipe de Recherche sur les Processus Innovatifs,
Nancy, France*

ABSTRACT: The professionals of the industrial world expect current researchers to provide them with lists of precursory criteria that would contribute to the prevention of counterproductive organisational failures. Conducted in the field of infrastructure maintenance in SNCF, this research project highlights the connections between the organisational aspects of a system and its level of safety. The approach is based on organisational sociology and uses semi-directive interviews to reconstruct a global picture of the organisation's daily activities. This study identified organisational factors and mechanisms that can generate hazardous organisational dysfunctions. It also brought crucial elements to the compilation of a list of the criteria one must monitor to integrate the idea of safety into the organisations' (re)design processes.

1   INTRODUCTION

Several studies have shown that organisational dysfunctions played a key role in such disasters as Chernobyl, Bhopal and the Challenger explosion. The term organisational accident as defined by Reason applies to those cases (Reason 1987). Various scientific schools have researched the links between "organisation and safety", either looking for the "secrets" underlying high risk organisations' safety, or trying to understand the causes of organisational unreliability. Essentially based on accident analysis, the investigations of Reason (1997), Rasmussen (2000), Vaughan (1996) and Turner (1978) have highlighted various relations between organisational aspects of the investigated systems and the occurrence of the accidents. Bourrier has produced an overview of related research and highlighted the following points (Bourrier 2001):

– *Researching the weaknesses of complex organisations when "everything works smoothly" attracts little interest from Senior Management; yet, social science have a definite contribution to make : instead*
of being restricted to retrospective analysis, (post-accidental enquiries), they could foster fledgling prospective endeavours.

– *"Organisational experts would like to set up a list of precursory signals as actual sources for organisational unreliability... Nevertheless, providing a comprehensive list of vulnerability sources and of their combinations would obviously be presumptuous."*

Produced within the context of a CIFRE PhD, this paper partially highlights the research work undertaken by the joint Research Teams of the Innovative Process Department of ENSGSI-INPL in Nancy and the Research and Technology Department of SNCF (French National Railway Company). Ongoing investigations within SNCF are based on the aforementioned prerequisites, i.e. their leading methodologies come from the social sciences and more specifically from Sociology of Organisations when identifying the causes of organisational vulnerability. This paper is based on an organisational analysis in the field of infrastructure maintenance within SNCF. The investigations bring to light diverging interpretations of the notion of safety within the organisation[1]. Results show how we can identify various elements of interest crucial to diagnosis and organisational changes for maintaining safety efficiency is concerned.

2 GENERAL PROBLEM

A central issue is the influence of an organisation's structural-functional parameters upon the safety level of a system. The industry's major demand is that current studies provide lists of precursory criteria that would contribute to the prevention of counterproductive organisational failures. The existing connections between organisational reliability and safety efficiency are often addressed. The concept of latent organisational failure, as defined by Reason (1995) in his causal pattern of organisational accidents, is a relevant example (Figure. 1).

On the one hand, current risk analysis methods that attempt to take the organisation into account appear to be hindered by their concern for quantification (Abramovici 1999); on the other hand, most studies committed to issuing recommendations on organisations usually are too general. The business world focuses on designing pragmatic solutions – no matter how global – that can associate safety components to each reengineering process required in organisational change. From there, it seems fundamental to provide a more thorough analysis of organisational failures and their impact on safety within the system by studying the organisation's day to day functioning. In order to provide a global modeling of those impacts, our researches are grounded on the investigations related to risk management from Reason (1997) and Rasmussen (2000) and try to conciliate those both approaches.

The method of analysis selected to that end is related to Sociology of Organisations. In practice, this method relies on semi directive interviews, and effectively highlights the connections between

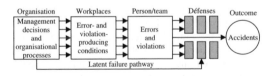

Figure 1. A model of organisational accident causation. *From Reason (1995).*

---

[1]In this article, the term "organisation" refers to formal organisation (structure, procedures, organisation of production, organisation of safety), as well as to informal organisation (actual activity, strategies used by actors, etc.).

structural-functional aspects of an organisation and its efficiency in terms of safety. This procedure will be detailed more extensively after a description of the research site.

3 BACKGROUND AND METHODOLOGICAL OPTIONS

3.1 Infrastructure maintenance in SNCF

Investigations detailed in this study relate to the field of infrastructure maintenance in SNCF (railways, signalling, etc.). Railway transportation is considered as a risky activity because the energies it develops and the presence of human beings can generate considerable danger (collisions, derailments, etc.). Furthermore, in the government-owned SNCF, organisation is a complex matter, for its numerous components (whether human, technical or organisational) are characterised by:

– their great diversity
– their interdependence
– the geographical distance between them (they are scattered all over France).

This complexity is illustrated in Figure 2, which provides an overview of the structural-functional organisation of SNCF and describes the flow of safety regulations from their design in central departments to their enforcement in production entities (see arrows).

These safety regulations are devised by the functional departments, disseminated among the relevant functional nodes in each of the 23 regions. The functional departments then send them down to the local

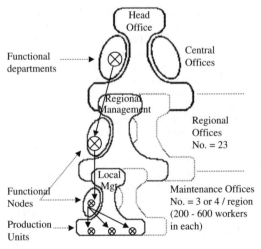

Figure 2. Safety regulations flow for infrastructure maintenance in SNCF. *Based on Mintzberg models (1989).*

functional nodes in the regions which in turn distribute them to the production units. At every step along the way, the safety regulations – applied to infrastructure maintenance – may be adapted to the local environment.

3.2   *The point of a sociological approach*

Even if accident analyses usually make up the core of the data used to foster research, they are too rare and often incomplete. However, crisis situations have the advantage of opening communication channels which would never have existed otherwise (Paries 1999). At best, they can help reduce the well-known phenomenon of "compartmentalisation". Stemming from labour division and geographical dispersion, this phenomenon exists in many large businesses such as SNCF, and may hide from view organisational dysfunctions attributable to a team, a section or even a department.

The systemic approach of sociological analysis has the advantage of allowing the reconstruction of a global picture often concealed by compartmentalisations. Thus, one may collect a greater amount of data by studying organisations in the course of their daily operations while simultaneously avoiding any compartmentalisation that would hinder a global understanding of the situation.

3.3   *Methodology*

Our point of departure is largely inspired by the work of Piotet & Sainsaulieu (1994) and Crozier & Friedberg (1977). It is composed of 3 successive phases:

1. Data collection through a series of semi-directive individual interviews,
2. Result analysis,
3. Reconstruction of a global picture bringing together all elements of interest.

3.3.1   *Data collection through semi-directive interviews*

As regards qualitative studies, the question of representativeness in the statistical sense of the term is irrelevant. The value of a sample depends on how well it corresponds to the objectives of the research… (Albarello et al. 1995); our objective being to illuminate organisational factors that may hinder results as far as the safety of the system is concerned. Thus, a sample of agents whose positions are directly linked to the safety of the system were interviewed initially, so as to identify the organisational constraints on their work.

The significant structural entities are:

– Production units (specialising in infrastructure maintenance) which work with the safety procedures and are directly exposed to the risks.
– The functional departments responsible for elaborating the safety systems and, more particularly, the regulations.

A sample of actors who depend upon those production units were interviewed, that is to say, those working in the functional nodes. Table 1 depicts the sample studied.

The interviews dealt with the actual position of the actors, their links with the other structural entities in the course of their activities, how they perceived the regulations, etc. Lasting an average 3 hours, the interviews were based upon 4 categories of open-ended questions:

– The first category sought to identify the tasks and position of the actor in his daily practice and his relationships with the other structural entities (example: "what exactly is your role?").
– The second and third categories were meant to show us where the actors stood in terms of risks and safety. (example, cat.2: "what do you think ensures a good level of safety?"; example, cat.3: "What do you think compromises safety or reduces the level of safety ?").
– The final category reveals the dysfunctions and organisational constraints that burden the tasks at hand. (example: "What problems do you encounter when working ? How do you resolve them ?").

These open-ended questions were deliberately vague so as to create entry points which could be enlarged upon later in the interview.

Table 1.   Sample studied.

Level	Structural entity	Positions	No. of interviews
Central management	Functional department	Regulation officer	5
Maintenance units	Functional node	Human resources manager	1
		Agent, safety node	1
		Agent, technical node (production)	5
	Production unit	Unit manager	4
		Assistant unit manager	2
Total			18

Table 2. Risk and safety perception.

	Risk and safety perception
Regulation authors	• Safety is the only preoccupation of the procedure design department. They define the general safety background (the safety system). • Observance of the regulations ensures a high level of safety. • Risks derive from non-observance of the regulations.
Regulation users	• Safety is seen as encompassed in a production/exploitation system. • Observance of the regulations ensures a high level of safety. • Risks derive from constraints burdening the activities and sometimes from difficulty in applying regulations.

3.3.2 Result analysis

The principal goal is to compare all the data collected on such specific themes as the perception of risks and safety, the constraints burdening the completion of tasks, the dysfunctions, etc. in order to:

– sort out all information regarding the different perceptions in terms of safety and risks contingent on the system
– see how the different actors in the system are interconnected
– isolate the resources and constraints contingent on the completion of the task and the fulfilment of the missions
– discover the solutions used to resolve the problems.

3.3.3 Reconstruction of a global picture

The diverse but complementary points of view of the several actors as regards the completion of their tasks are compared in order to reconstruct a global picture of the system. The next step is a charting of the essential processes linked to the activities directly related to results in terms of system safety. Not only does the sociological analysis provide a global vision of the organisation's day-to-day working, it also gives two types of information which are crucial to this study:

1. How the actors perceive the safety issues, and particularly the risks related to their activity. These data are vital because many informal risks are only temporary and thus are best known by those having an everyday, first-hand experience of the task (Laumont et Crevier 1986).
2. The nature of the organisational dysfunctions of every single structural entity and how they hinder results.

From these elements, one may draw links between the perceived risks and the organisational dysfunctions (that is to say, between the organisational dysfunctions and the system's results as regards safety). These links will illustrate which organisational factors are the most important as regards the system's safety efficiency.

4 SELECTED RESULTS

4.1 Perception of risks in terms of safety

First, it is to be remembered that, in SNCF, safety musters large organisational resources. Equipment, procedures and personal make up the "safety system", and SNCF's organisational culture, in observance of the regulations ensures a high level of safety.

The interviews show that risk perception and awareness depend on the actual position of the worker. There are two broad categories of perception: that of authors of regulations and that of users of regulations. Table 2 sums up the differences in safety awareness.

4.1.1 Regulation authors' perceptions

Usually, regulation authors see safety as being guaranteed by observance of the procedures. This dominant view derives from the fact that safety is legally defined: the law establishes general goals by compelling management to set up sufficient means to ensure safety (Seillan 1981). The main risk identified by regulation authors is the non observance of safety procedures, which, for them, amounts to errors and violations (this terminology recalls the notions defined by Reason (1990)). Nevertheless, regulation authors admit that in given situations, the procedures are not always easy to follow (they are aware of this fact probably because they, too, have been regulation users at a certain time in the course of their career). Regulation authors link organisational features to events entailing errors and violations (See Table 3).

Regulation authors see three main events at the source of errors and violations:

– Short deadlines, which, for them, are linked to time-related production constraints such as unpredicted events to be handled immediately (e.g: damaged railway tracks paralysing traffic) or a narrow time-limit to perform maintenance operations.
– Weak knowledge of the system and risks is related to human resources management, and more particularly, to training, "new on the job" and employees

Table 3. Main sources of risks and related organisational features (regulations authors' point of view).

Risks	Source events	Organisational features
Error/violation	– Urgent situation – Weak knowledge of systems and risks – Regulations unsuited to local environment	– Production constraints – Human resources management flaw – Unsuited conceptualisation – Lack of organisational resources – Burdensome formality

Table 4. Main sources of risks and related organisational features (regulations users' point of view).

Risks	Source events	Organisational features
Errors		– Production constraints
	– Urgent situation	– Lack of means
		– Planning problems
	– Weak knowledge of systems and risks	– Human resources management problems
	– Weak handling of the regulations	– Too formal or too complicated
Problems arising when performing maintenance	– Priority management as regards general maintenance	– Budgetary problems – Resource allocation problems (time)

turnover. They believe it is essential that knowledge and know-how be handled and transmitted properly.
- Organisational features related to unsuited procedures can take different aspects:

Unsuited conceptualisation
Lack of organisational resources: i.e. the connection between the means and the time allotted to perform a task. For instance, an insufficient number of agents to strictly follow a procedure, or too short a deadline for the agent to obey every step of the procedure.
Burdensome formality: the increasing complexity of the system and the tendency to formalise everything raises the question of the limit of assimilation of the regulations.

4.1.2 *Regulation users' perceptions*
Users do not question the general idea that observance of the regulations guarantees safety. Sometimes, they might complain about regulations when the procedures seem unsuited or entail further constraints that workers consider useless and burdensome in the completion of their task. Then again, the users wish their skills as regards risk management were better acknowledged, and that their know-how in terms of carefulness were considered a positive approach to safety (Brun 1997).

Their preoccupations as regards safety appear to focus more on how difficult it might be to reach their goals. The risks identified by the regulation users are the errors and problems arising while working on infrastructure maintenance. Table 4 sums up the users' point of view as regards the main events entailing the risks mentioned and the related organisational features.

Judging from these results, the users' perceptions seem to match significantly those of the regulation authors. Yet, there are several differences as to the perception of risks and the interrelation of source events with related organisational features. Indeed, the term "violation" is never used. When users describe their activities, they insist on the constraints related to the completion of the tasks (short deadlines, problems in getting the necessary means, unpredicted problems). *"when everything goes according to plan, there is no problem, but when something happens and you are running out of time, you have to find solutions…" (interview with a unit manager).*

The users know from first-hand experience what situations might get dangerous. These are expressed in terms of the essential criteria they need to complete their tasks and the constraints that get in the way of this completion (Rousseau & Monteau 1991). Regulation users agree with their authors: working on too short a deadline can be a source of error. However, users insist on the fact that emergency situations can be fostered by organisational problems. These elements will be further analysed in the next section through a study of maintenance operations and a closer look at all the organisational constraints identified by the users.

4.2 The specificity of infrastructure maintenance

In SNCF, infrastructure maintenance is carefully planned (De La Garza 1995). There are two general activities related to maintenance (Fadier & Mazeau 1996): "general maintenance" and "exceptional works". These usually consist in replacing large elements of infrastructure and can mobilise more than a hundred workers. Preparing these exceptional works is a crucial step as regards safety and involves attributing the necessary resources, defining on-site safety measures, and scheduling the various steps until the final completion.

- Resource planning aims at providing (on the required date) all the means necessary, that is to say:
  - human resources (in terms of number, skill, experience, knowledge of the system, knowledge of the regulations, etc)
  - technical means
  - information/organisational protocols
  - sufficient time to complete the works.
- Defining on-site safety measures: for the duration of the works, only technical trains are allowed on the railways being rebuilt. Relevant sections have to be closed, and safety equipment has to be modified to allow special technical trains to reach the restricted building site. At this stage, the general on-site safety rules are derived from the usual safety regulations. With particularly complex works, this can be quite a complicated matter.
- Scheduling means detailing the successive steps from the start to the completion of the work, and specifying the deadline for each operation. Time-related constraints are crucial, for every hour beyond the deadline will disrupt regular traffic.

The unit manager in charge of work preparation is supposed to coordinate all the operations, from resource planning to scheduling. The main problems they encounter are:

- they have very limited decision power as to the traffic-interruption periods needed to complete the work,
- they never know whether or not they will obtain enough resources at the right time.

When these difficulties become real, they put constraints on the actors' activity. The latter must then find solutions to resolve emerging problems and are sometimes obliged to develop informal behaviours, which may be at the limits for safe operation. This phenomenon is described by the notion of boundaries tolerated during use (Didelot 2001) and Neboit (2003). For example, when envisaged solutions require "last-minute" changes of work site safety regulations, the emergency nature of these changes presents a risk from the point of view of safety. Figure 3 illustrates this observation.

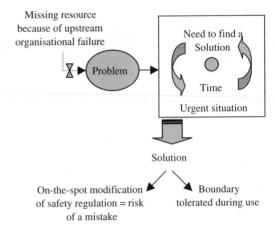

Figure 3. An example of the risk of inadequate behaviours caused by a missing resource.

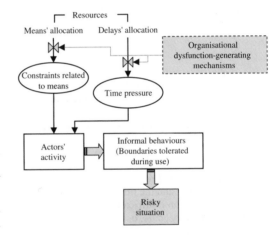

Figure 4. Synopsis of the impact of (organisational) dysfunction-generating mechanisms on the system's safety efficiency.

Solving the problems entails what Chabaud (1990) defined as « implicit requirements ». Indeed, the production constraints implicitly require the workers to solve the arising problems, since not reaching the goals would involve quite harsh consequences. For instance, not abiding by deadlines may cause traffic delays.

It seems obvious that the workers' inadequate behaviours may derive from organisational dysfunctions happening at the level of resource planning (See Figure 4).

Studying resource planning with sociological tools shows how we can identify various mechanisms that cause these organisational dysfunctions. One of the mechanisms, related to the planning of

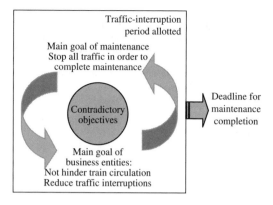

Figure 5. Mechanism of contradictory objectives in the planning of railways-shutdown periods.

This article focused on the connections between resources and the results in terms of safety.

The research project is going on: All data collected from the interviews should eventually help us to elaborate a global modelling of the links between organisational factors of a system and its safety performances. This representation could be used as a reflection framework by managers in order to better integrate the safety concepts in the processes for designing organisations.

The current investigation is related to the railway domain, this is a first step in research. In order to know if this approach could be used in other types of risky systems, the criteria and models which are developed in this context should be validated in other types of organisations. This would be a second step in order to provide conditions for generalising the models and criteria which are developed in our approach.

railways-shutdown periods (needed to work on these railways), arose in Figure 5.

This mechanism shows that the planning of railway-shutdown periods is negotiated between actors with contradictory objectives. On the one hand, the maintenance department needs to close railways long enough to complete the works, while on the other hand, commercial departments (under the pressure of economical constraints) want their trains to keep running.

Through this study, we showed that resources as an organisational feature may have an impact on the safety level of a system, and that resource planning may be hindered by various mechanisms that are liable to generate organisational dysfunctions (for lack of space, we deliberately omitted other factors and mechanisms we identified in the course of this study).

One should keep in mind that the on-site level of safety of a maintenance operation depends not only on the person in charge of the work, but also on the structural entities which planned and attributed the resources.

Therefore, when considering the safety of a system, it is crucial to study how resources are attributed, and more particularly how the attribution protocols could generate dysfunctions.

5 CONCLUSION

The sociological analysis we conducted highlighted the connections between organisational data and safety level of a system. Based on the interviews, we reconstructed a global picture of an organisation in the course of its daily activities. we highlighted the weight of several organisational features on the system's safety efficiency and identified various mechanisms that can generate organisational dysfunctions.

REFERENCES

Abramovici M. 1999. *La prise en compte de l'organisation dans l'analyse des risques industriels.* Thèse de Sciences de Gestion. Ecole Normale Supérieure de Cachan.

Albarello L., Digneffe F., Hiernaux J-P., Maroy C., Ruquoy D. & Saint-georges (de) P. 1995. *Pratiques et méthodes de recherche en sciences sociales.* Paris: Armand Colin.

Bourrier M. 2001. *Organiser la fiabilité.* Paris: L'Harmattan.

Brun J-P. 1997. Les savoir faire de prudence: une approche positive de la prévention. *Travail et Santé.* Vol. 13 No. 4. pp. 8–11. Décembre 1997.

Chabaud C. 1990. Tâche attendue et obligation implicite. in Dadoy M., Henry C., Millan B., de Tersac G., Troussier J-F. & Weill-Fassina A. *Les analyses du travail, enjeux et formes.* Paris: CEREQ. Collection des études, No. 54, pp.174–182.

Crozier M. & Friedberg E. 1977. *L'acteur et le système.* Paris: Seuil.

De la Garza C. 1995. *Gestions individuelles et collectives du danger et du risque dans la maintenance d'infrastructures ferroviaires.* Thèse de doctorat d'ergonomie. Ecole Pratique des Hautes Etudes.

Didelot A. *Contribution à l'identification et au contrôle des risques dans le processus de conception*, Thèse de doctorat, Institut National Polytechnique de Lorraine, 2001.

Fadier E. & Mazeau M. 1996. L'activité humaine de maintenance dans les systèmes automatisés: problématique générale. *RAIRO-APII-JESA.* Vol. 30 No. 10 pp. 1467–1486.

Laumont B. & Crevier H. 1986. Perception des risques dans une entreprise de chaudronnerie industrielle. *Archives des maladies professionnelles.* pp. 550–551.

Mintzberg H. 1989. *Mintzberg on Management, Inside Our Strange World of Organizations.* NY: Free Press.

Neboit M. 2003. A support to prevention integration since design phase: The concept of limit condition and limited activities tolerated by use. *Safety Sacience.* Special issue safety design Vol. 41 No. 2–3 E. Fadier guest editor.

Paries J. 1999. Intervention au séminaire "retours d'expérience, apprentissage et vigilances organisationnels: approches croisées". *Programme Risques Collectifs et*

Situations de Crise, Actes de la 4ème séance, 21 janvier. CNRS. Paris. pp. 109.

Piotet F. & Sainsaulieu R. 1994. *Méthodes pour une sociologie de l'entreprise.* Paris: Presses de la Fondation Nationale des Sciences Politiques & ANACT.

Rasmussen J. & Swendung I. 2000. *Proactive Risk Management in a Dynamic Society.* Risk & Environmental Department. Karlstad: Swedish Rescue Services Agency.

Rasmussen J. Risk management in a dynamic society: a modelling problem, *Safety Science* Vol. 27, No. 2/3, pp.183–213, 1997.

Reason J. 1997. *Managing the Risk of Organizational Accidents,* Aldershot: Ashgate.

Reason J. 1995. A system approach to organizational error, Ergonomics, Vol. 38, No. 8, 1708–1721.

Reason J. 1990. *Human error.* Cambridge: University Press.

Reason J. 1987. The age of organizational accident. *Nuclear Engineering International.* pp. 18–19.

Seillan H. 1981. *L'obligation de sécurité du chef d'entreprise.* Paris: Dalloz.

Turner, B. 1978. *Man-made disasters.* London: Wykeham.

Vaughan D. 1996. *The Challenger Launch Decision: Risky Technology, Culture, and Deviance at NASA.* University of Chicago Press.

Safety and Reliability – Bedford & van Gelder (eds)
© 2003 Swets & Zeitlinger, Lisse, ISBN 90 5809 551 7

Maintenance optimization scheme for a periodically tested automatic valve control system

S. Eisinger
Det Norske Veritas AS, Norway

M. Eid
CEA DEN DM2S SERMA, France

ABSTRACT: A maintenance optimization scheme is presented in this paper. It permits the description of the ageing behavior and the integration of the periodic testing. The scheme has been used to optimize the maintenance policy of an Automatic Valve Control System. The paper presents the first results of this work. Results are obtained and compared from both analytical-numeric methods and Discrete Event Monte Carlo Simulation methods. The work discusses some of the complications which arise when the often used exponential distribution approximation is replaced by more realistic models.

1 INTRODUCTION

Rational optimization of maintenance necessitates applying models that integrate the ageing behavior of the individual components and the impact of the periodic testing. Integrating components ageing in reliability-availability-maintenance (RAM) assessments is a major issue. On one hand, components ageing mechanisms are not yet well described in the mathematical sense. On the other hand, mathematical models that link between components time-behavior (local-time) and system time-behavior (global-time) are lacking. Very often, models are developed for given applications.

Components' ageing is very often described by a Weibull-like distribution. Weibull distribution allows modeling a large variety of time-dependent patterns. At specific values of the distribution parameters (α, β, η), Weibull distribution can even describe linear-time dependencies. Increasing field experience feedback as well as increasing knowledge about degradation mechanisms should allow the confirmation of the Weibull distribution choice and the validation of its specific parameters.

Weibull distribution will be used in this study to describe the ageing of some components included in the system.

Periodic-testing represents another aspect of the problem. What is the difference between a tested and a non-tested component? How may that difference be

described mathematically? Maintenance field-experience feedback proves that there is a difference between tested and non-tested components.

In this study, a time-dependent model has been developed and applied to the maintenance optimization of an Automatic Valve-Control System (AVCS). The AVCS is composed of three sensors that are actively redundant. They are independently connected to the same Central Processing Unit (CPU), which carries out a 2 out of 3 voting logic-testing. If the CPU receives a confirmed signal from the sensors (at least two out of three) it releases a signal to open two independent valves. The system fails if the 2 valves fail to open.

2 DESCRIPTION OF THE SYSTEM

The Automatic Valve-Control System (AVCS) is composed of three sensors that are actively redundant. They are independently connected to the same Central Processing Unit (CPU), which carries out a 2 out of 3 voting logic.

If the CPU receives a confirmed signal from the sensors (at least two out of three) it releases a signal to open two independent valves. The system fails if the 2 valves fail to open. The AVCS is schematically represented in Figure 1.

The sensors, the CPU and the valves are periodically tested. Each periodically-tested component becomes as good as new after each testing campaign. This is

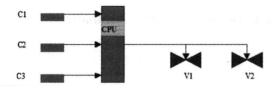

Figure 1. Schematic representation of the valve control system.

Table 1. Failure data.

	γ (failure to startup prob.)	$\lambda(t)$ (1/h)
Sensors	5e-4	(2.3/3.6e + 5)*(t/3.6e + 5)$^{1.3}$
CPU	1e-5	0.0
Valves	1e-5	1e-7

due to the fact that faulty components are immediately replaced/repaired during testing.

In spite of this policy of cyclic testing and immediate replace/repair of faulty components, the system unavailability increases with time in some situations. This is the case when some of the system components show ageing phenomena. To guarantee that the system overall unavailability will be kept below a given limit, the whole system will generally be replaced after a determined number of testing cycles. Maintenance policy optimization will be carried out using a maximum unavailability criterion. The criterion is that the system overall unavailability at any instant should be lower than 1e-4.

In our case, sensors show ageing and are supposed following a Weibull distribution, the valves have constant failure rate, while the CPU has a constant unavailability with time (hardware failures are either revealed immediately or will remain hidden, independent of tests). The failure data of the different components are given in Table 1.

3 COMPONENT TESTING MODEL

We recall that during testing faulty components are immediately replaced or repaired and become as good as new. The components replace/repair meantime is almost zero. Components' testing is periodic and has a periodicity of τ hours.

Under the above hypotheses, the component availability $p(t)$ is described by, (the model itself is described in a paper that will be issued soon):

$$p(t) = R(t) + \sum_{l=0}^{n-1} Q_l^- \bullet R(t - l\tau), \ (n-1)\tau < t < n\tau$$

(1)

where,

$$P_n^- = p(n\tau)$$

$$= R(n\tau) + \sum_{l=0}^{n-1} Q_l^- \bullet R((n-l)\tau), \qquad n > 1$$

$$Q_n^- = 1 - P_n^-, \text{ and } P_0^- = 1.$$

Where $R()$ is the component's reliability and n is the number of tests that the component passed through.

Component testing and replace/repair, if it was faulty, increases the component availability by:

$$\sum_{l=0}^{n-1} Q_l^- \bullet R(t - l\tau).$$

This increase in the component availability is a function of the test period τ and the total number of tests during the component-life $(n - 1)$.

If the component shows a Weibull-type ageing, its reliability $R(t)$ is given by:

$$R(t) = Exp\left(-\left(\frac{t}{\eta}\right)^{\beta}\right)$$

It has already been mentioned above that the sensors are supposed to show ageing, which follows a Weibull distribution, while the valves have a constant failure rate. Obviously, a component with a constant failure rate could be considered as a component having a Weibull distribution with a shape factor (β) equal to 1 and a scale factor η equal to $1/\lambda$. If a component has a constant failure rate (no ageing), its reliability $R(t)$ is described by:

$$R(t) = Exp(-t/\eta) = Exp(-\lambda t)$$

where, λ is the component failure rate.

If the component has a zero failure rate (CPU), its reliability $R(t)$ is assumed to be:

$$R(t) = 1.0$$

Equation (1) describes the impact of the periodic testing on the component unavailability time-profile. It allows distinguishing between ageing-components, time-independent failure and free-of-failure components. A detailed analysis of the mathematical model is beyond the scope of this paper. The paper is aimed at presenting a concrete application of maintenance optimisation that integrates ageing and periodic testing effects. However, a rapid apercu of some interesting aspects of the model is given in the following sections.

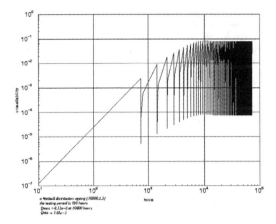

Figure 2. Unavailability profile of an ageing component ($\beta = 2.3$, $\eta = 1.e + 5$ hrs).

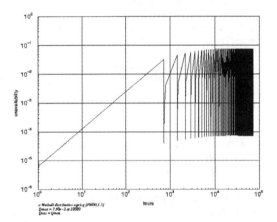

Figure 3. Unavailability profile of an ageing component ($\beta = 1.3$, $\eta = 1e + 5$ hrs).

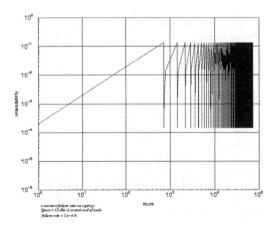

Figure 4. Unavailability profile of a non-ageing component ($\beta = 1$, $\eta = 1e + 5$ hrs).

In Figure 2, the unavailability profile of a periodically tested component ($\tau = 720$ hrs) is plotted against time. The component shows ageing following a Weibull distribution with $\beta = 2,3$ and $\eta = 1e + 5$ hrs. The unavailability increases steadily (after the 1st test) with the time (number of tests) in spite of testing and corrective actions. Around $2e + 4$ hours, the components' unavailability tends to its asymptotic value. The asymptotic value of the component unavailability is about 8e-2.

In Figure 3, the profile of the same tested-component is plotted assuming a Weibull distribution with a moderated shape-factor ($\beta = 1.3$). The unavailability profile tends slower to its asymptotic value (after the 1st test). The unavailability tends to its asymptotic value after lesser number of tests.

In Figure 4, the same component unavailability is plotted supposing a constant failure rate (no ageing). The testing policy and the corresponding corrective actions stop completely the unavailability increase (after the 1st test) with the time. Attending or not the asymptotic unavailability of a tested-component depends evidently on the report between the scale factor (η) and the test periodicity (τ).

4 RESULTS FROM ANALYTICAL CALCULATIONS

Using the model given above to describe the impact of periodic testing (with immediate replacement/repair of faulty component), the system overall unavailability has been calculated for 3 different testing policies.

These maintenance policies are: 1/testing all components once each 18 months, 2/testing all components once each 12 months, and 3/testing all components once each 6 month.

In all scenarios, if the system unavailability attends the value of 10^{-4}, a new system replaces the aged one. In fact, one replaces the valves and the sensors. The CPU unavailability is time-independent and need not be replaced. The value of 10^{-4} represents the maximum allowable overall unavailability.

In the scenario of 6 months testing policy, Figure 5, the system overall unavailability profile increases steadily. It becomes higher than the maximum allowable unavailability around 10^{+5} hours of operation. The system unavailability attends its asymptotic value of 6.7×10^{-4} after $3.4 \times 10^{+5}$ hours of operation. Subsequently, the whole system should be replaced before 10^{+5} hours of operation.

In the scenario of 12 months testing policy, Figure 6, the system overall unavailability profile attends the maximum allowable unavailability around $6.1 \times 10^{+4}$ hours of operation. The system unavailability attends its asymptotic value of 2.2×10^{-3} after $2.8 \times 10^{+5}$ hours

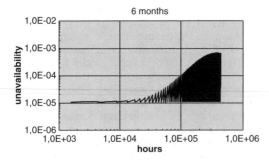

Figure 5.   AVCS unavailability time profile ($\tau = 6$ months).

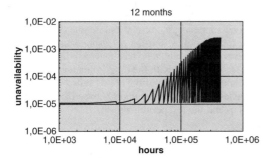

Figure 6.   AVCS unavailability time profile ($\tau = 12$ months).

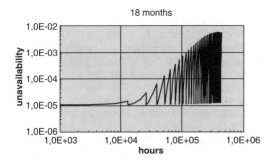

Figure 7.   AVCS unavailability time profile ($\tau = 18$ months).

of operation. Subsequently, the whole system should be replaced before $6.1 \times 10^{+4}$ hours of operation.

In the scenario of 18 months testing policy, Figure 7, the system overall unavailability profile attends the maximum allowable unavailability just before $5.2 \times 10^{+4}$ hours of operation. The system unavailability attends its asymptotic value of 5×10^{-3} after $5.5 \times 10^{+5}$ hours of operation. Subsequently, the whole system should be replaced before $5.2 \times 10^{+4}$ hours of operation.

Table 2.   Comparison between different maintenance policies.

Testing cycle (month)	Life cycle (no of test cycles)	Life cycle (months)	Unavailability (EoL)
18	4	72	1.28 e-4
	3	54	6.53 e-5
12	7	84	1.02 e-4
	6	72	7.31 e-5
6	23	138	1.00 e-4
	22	132	9.17 e-5

A synthetic comparison of these 3 scenarios is given in Table 2. That could be resumed as following:

A testing period of 18 months necessitates a standard replacement of the whole system each 3 testing cycles (54 months) in order to respect the maximum allowable unavailability criteria.

A testing period of 12 months necessitates a standard replacement of the whole system each 6 testing cycles regarding the maximum allowable unavailability criteria.

A testing period of 6 months necessitates a standard replacement of the whole system each 22 testing cycles, regarding the maximum allowable unavailability criteria.

5   COMPARISON WITH ANALYSIS USING DISCRETE EVENT MONTE CARLO SIMULATION

While analytical calculations often exhibit the advantage to be fast and relative insensitive to the choice of model parameters, they lack of a certain inflexibility concerning the classes of models which can be calculated. New model classes will often at least require a new tailored numerical tool, or turn out not to be tractable at all with analytical-numerical means. Simulation techniques are on the other hand more flexible with respect to the model classes which can be analyzed, such that existing tools can often be adjusted in a simple way when a new type of problem is at hand. Unfortunately, due to the stochastic nature of Monte Carlo Simulation, these techniques suffers often of the so-called rare event problem: If most events in a given model are uninteresting "intermediate" events and the interesting events are very rare, most of the simulation time is used up for uninteresting "noise", while the statistics for the interesting events remains poor at best. Safety systems have a tendency to fall into this class and can thus be difficult to treat using simulation techniques.

Also the system presented in Figure 1 exhibits the rare event problem: Maintenance events happen every about 6 month (dependent on the maintenance strategy),

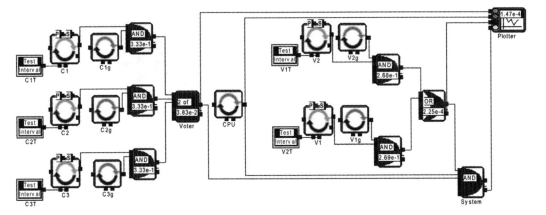

Figure 8. AVCS simulation model within EXTEND.

failures of a single sensor happen every about 450 months (mean value of the Weibull distribution) and a sensor system failure happens every about 4300 months.

The ratio between "noise" (maintenance actions and single sensor failures) and interesting events (sensor system failures) is thus nearly 1000. For the actuators, the picture is similar.

In a "brute force" simulation model all events of the system (maintenance and component failures) are simulated. For the problem at hand, it turns out that this way of modeling results in extensive simulation times, which renders the method in practice - impossible. Still, there are methods which solve, or at least reduce the rare-events problem. Unfortunately our model exhibits some specific properties, which make some of the methods hard to utilize:

- The components in the model are not statistically independent of each other. Through the common testing event, the components are coupled.
- The failure rate of the wear-out type failure modes are time-dependent, which means, that the unreliability as a function of time is not identical between testing intervals.

These properties render the modularization approach, presented in (Eisinger 1998) difficult to use. Still, a view back to the initial discussion in this chapter reveals a possible solution. The rare-event problem is, for this model, mainly caused by the short maintenance time. The "noise ratio" between single sensor failures and the sensor system failures is much less and should be tractable. Therefore, instead of the basic simulation approach (simulation of all events), a more sophisticated approach will be better suited for the problem at hand: Testing events are only generated when they are needed, i.e. when a sensor or an actuator has failed, the time for the next testing is calculated and processed as event. Implementing this

optimized model reduces the simulation time by about a factor 100 and makes simulation results with the desired degree of accuracy achievable.

An alternative approach would be the usage of other variance reduction techniques as they are presented in the literature (Rubinstein 1997). Such techniques have not been tried out for this paper, but would be worthwhile to test on this model.

Figure 8 shows the simulation model which is constructed with the off-the-shelf simulator EXTEND™ with an add-on "Reliability" library.

The test interval is modeled with the "Test Interval" blocks. The sensors have two failure modes each, the wear-out failure mode and the constant probability failure mode. All sensors are connected to the 2oo3-voter.

The CPU system consists of a single block modeling a constant failure probability. Similar to the sensors, the actuators are modeled with two failure modes and are connected with a logical OR. All three sub-systems (sensors, CPU and actuators) are connected with a logical AND.

Note, that a positive "reliability" logic is used when modeling the AVCS, meaning that the logical value of a working component is TRUE and a failed component receives a logical FALSE. The sub-system and the system unavailability are analyzed by the connected plotter block.

Figure 9 shows the results from the simulation run of the model presented in Figure 8. The testing time is chosen as 6 month. Clearly, the figure is very similar to Figure 5. One Million simulation runs are used to make stochastic variations as small as shown.

6 CONCLUSIONS

Different maintenance policies of an Automatic Valve-Control System (AVCS) have been assessed and

603

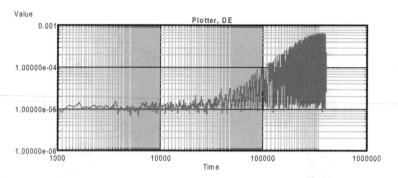

Figure 9. Results from the AVCS simulation with the testing time 6 months to be updated.

compared regarding a maximum allowable unavailability criterion.

The system is composed of 3 sets of components (sensors, CPU, valves). Each set of these components has its own failure pattern. The sensors show ageing following a Weibull distribution. The valves have a constant failure rate, while the CPU has a constant failure to respond probability. Maintenance policies based on testing period of 6 months, 12 months and 18 months were compared. The profile of the system unavailability should not exceed the maximum allowable unavailability level equal to 10^{-4}. A new system should replace the old one once the unavailability exceeds the value of 10^{-4}.

In this example, the CPU has a time-independent unavailability. Subsequently, one does not replace the CPU. Only, valves and sensors are replaced to obtain a new system.

The final selection between maintenance policies will be carried on after integrating the economic aspects related to the cyclic maintenance costs and the system replacement costs related to the system operational lifetime.

The results of the analytic calculation have been compared with the results from a Discrete Event Monte Carlo Simulation model. It has been shown,

that the simulation model can be readily constructed using an of-the-shelf simulation program with a graphical modeling interface. As in many safety systems, the rare-event problem renders the simulation nearly non-tractable. An analysis of the cause of the rare events helped in identifying an alternative way of modeling the system, which solved the rare-events problem, at least within the limits of the interesting model parameters.

In a later step, the economic considerations will be integrated to the model (maintenance cost, system standard replacement cost). It will also be interesting to use this rather simple model to study the usage of other variance-reduction methods.

REFERENCES

Eisinger, Siegfried 1998 *Safety and Reliability: Simulation of rare events using modularization techniques* Balkema, Conference proceedings ESREL'98.

Rubinstein, Reuven 1997 *Optimization of computer simulation models with rare events* European Journal of Operational Research 99 Elsevier.

Extend. *Professional Simulation Tools.* Imagine That Inc. www.imaginethatinc.com, Version 5, 2000.

Safety and Reliability – Bedford & van Gelder (eds)
© 2003 Swets & Zeitlinger, Lisse, ISBN 90 5809 551 7

Developments of explosion resistant doors and bulkheads

Andre van Erkel
TNO-PML, Rijswijk, the Netherlands

Leon F. Galle
RNLN, Den Hague, the Netherlands

ABSTRACT: The impact of an Anti Ship Missile is one of the most threatening scenarios for a naval platform. The accompanying warhead detonation will endanger crew, platform and its mission. A naval engineer has various options to reduce the vulnerability of his platform design, like smart arrangement and protection. One of the most important options is to increase the blast resistance of the longitudinal subdivision i.e. watertight (WT) bulkheads and doors. This paper will address the ongoing developments on blast resistant light or moderate weight steel structures at the TNO-PML. After an introduction on the threat and the followed approach, five structural products will be elaborated for two loading levels. Two blast resistant doors and three blast resistant bulkhead concepts have been developed and validated by full-scale experiments. Although the initiation of the development of these blast resistant structures is for Naval Defence purposes, there is a highly potential spin-off for in- and external explosions for bunkers, ammunition storage facilities and for applications in the offshore industry.

1 INTRODUCTION

1.1 *The above water threat*

The last decades, the threat of Anti Ship Missiles (ASMs) challenging warships has dramatically increased. ASMs have become more and more sophisticated in terms of velocity (high supersonic), agility, sensors, digital signal processing, joint smart attack and last but not least stealthy features.

1.2 *Blast loading phenomena*

Prior to the introduction of blast resistant structures, a short overview is given of the loading phenomena that are generated by an internal missile warhead or shell detonation. The explosion of such an HE warhead generates high velocity fragments and blast effects.

A wall in a compartment, where an HE charge detonates, will be exposed to a complex time-dependant pressure loading. The pressure-time history (at a given point) on the wall depends on the shape of the compartment, the location of the charge with respect to the wall and on the properties of the charge. A typical pressure recording is depicted in Figure 1.

Two phases in the pressure-time history can be distinguished, viz. a blast wave phase and a quasi-static overpressure (QSP) phase. For an internal explosion,

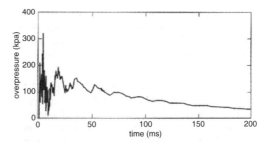

Figure 1. Typical pressure-time history from an internal explosion.

the blast wave having a very high pressure shock front, travels through the compartment and hits the wall. A reflection occurs at the wall of the compartment. The result of this reflection, the "reflected blast wave", returns towards the centre and reflects again. After several repetitions, this process damps out.

After the detonation, a volume (region) of hot particles results, which expands rapidly. As a result of the reflecting blast waves the hot volume with the chemical reaction products is forced to mix with the air in the confined structure. In general this causes a combustion or afterburning process. This combustion effect plays an important role in the creation of the before-mentioned second phase, the so-called Quasi-Static Pressure. The heat processes of detonation, blast wave dissipation and afterburning in combination with the formation of additional gases create a pressure, which is more or less equal in the whole compartment at a certain time, in contrast with the blast wave.

1.3 Improving the blast resistance of the RNLN LCF

In case of the newly designed RNLN LCF[1], the RNLN has opted for improvement of structures that make up the longitudinal watertight (WT) subdivision of the warship i.e. WT-Bulkheads & WT-Doors.

The installation of a longitudinal WT subdivision is already a classical design routine to secure residual buoyancy and stability after damage and consequent flooding. Focusing on the improvement of the blast resistance of this longitudinal WT subdivision has been initiated for three main reasons:

1. The Damage Control & Fire Fighting (DC/FF) Layout & Procedures of warships are mainly based on longitudinal separation;
2. In relation with present-day technology of blast resistant structures, it seemed impractical to go for improvement of the subdivisions in height i.e. the decks, because of the relatively small average detonation distances from the deck/deck-head and the large span width of decks.
3. The functional layout of systems is mainly based on longitudinal concentration of vital elements of "autonomous" systems within zones and longitudinal separation of redundant "autonomous" systems in zones.

Figure 2a shows the situation for the new RNLN Air Defence Command Frigate LCF, which has seven functional autonomous zones in terms of supply of chilled water, electric energy and data. In case a missile hits the ship with a High Explosive (HE) warhead, the detonation would at least result in failure of the four WT boundary structures i.e. forward bulkhead, aft bulkhead, deck and deckhead. Because of the failure of this structure, the damage will spread to the four adjacent compartments (physical damage). The damage in these compartments will kill the supply of chilled water, electric energy and data for these damaged zones. In the example case, zones 6 & 7 will be functionally killed.

In case blast resistant structures (WT-Bulkheads and WT-Doors) are installed at the zone boundaries, see Figure 2b, the physical damage can be limited to the hit zone, resulting in the functional damage being limited to only one zone (zone 6). The conclusion is

---

[1] Lucht Commando Fregat (Air Defence Command Frigate).

Detonation Compartment
Functional Damage

ZONE 7 / ZONE 6　ZONE 5 ZONE 4 ZONE 3 ZONE 2　ZONE 1

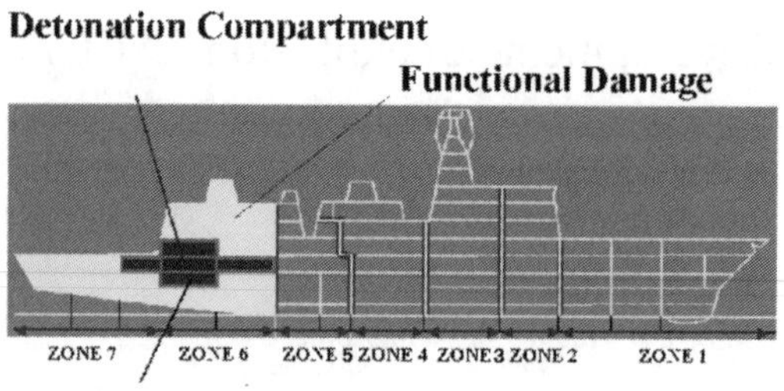

Figure 2a. Conventional design: functional damage exceeding the physical damage.

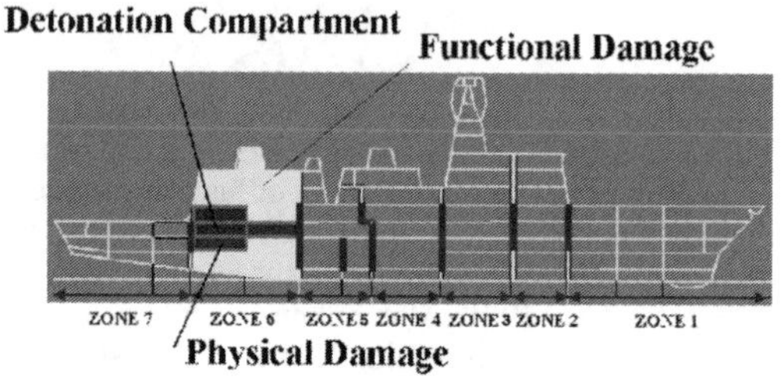

Detonation Compartment
Functional Damage

ZONE 7 / ZONE 6　ZONE 5 ZONE 4 ZONE 3 ZONE 2　ZONE 1
Physical Damage

Figure 2b. Design including blast resistant structures: functional damage limited to one zone.

that it is worthwhile to invest in the blast resistance of transverse WT-structures.

2 BLAST RESISTANT STRUCTURES

2.1 Principles

Here the principles for blast resistant structures are mentioned for both bulkheads and doors. In general, the challenge of improving the blast resistance can be solved in two ways: improving the bending resistance or exploiting the membrane mechanism.

In contrast with regular structures, blast resistant structures in a ship must be designed for the plastic strain realm. A structure can dissipate a lot of deformation energy in the plastic strain realm compared to the elastic strain realm only. Another general remark concerns large and not well designed stiffeners and brackets. For elastic designing, these elements are appropriate, however, for large blast loading they often cause cracks and as a consequence premature failure of the element.

When use is made of the bending resistance, the blast forces are taken by the internal bending moments, which is the usual mechanism for many structures. To increase the bending resistance of structures, particularly the section modulus must be increased, resulting in rather heavy structures that can withstand only

606

moderate pressures. To comply with the requirement of withstanding the blast effects in the explosion compartment, membrane structures are more suitable for an acceptable mass penalty.

The membrane mechanism is based on stretching a plate due to large deflections by preventing the in-plane movement of the edges of the plate. When the plate is loaded up to its yield strength the resistance of the bulkhead increases more or less linearly with its mid-deflection. The objective of the designer is to find measures that allow the highest possible deflection without failure of the panel. Designing in this way is unconventional as large reaction forces arise and a flexible connection to the regular ship structure is necessary.

In this paper the focus is on structures based on the membrane mechanism although also structures are presented based on the bending mechanism. Not all locations in the ship allow a membrane based structure to be applied. For instance, bulkheads with no adjacent bulkheads above show a lack of in-plane resistance and are less suitable for the membrane mechanism. In such cases it is inevitable to make use of both mechanisms (bending and membrane) for designing a blast resistant structure.

Blast improvement measures aim to avoid exceeding a critical level of local strain in the panel, particularly near the weld, see Figure 3.

So, locations of strain concentrations must be avoided as much as possible and inevitable strain concentrations must be kept as low as possible. Furthermore, the strain at these concentrations must be levelled to obtain a balanced concept.

The acceptable local strain level and the strain rate influence for the inevitable strain concentration is determined from experiments and theoretical research. TNO-PML and the Delft University of Technology have finished a PhD study on this subject that will be succeeded by another one. The parent material, the

weld material and the Heat Affected Zone material show different critical strains.

TNO-PML has performed several experiments on the influence of fragment holes on the critical strain level in the membrane plate of a bulkhead/door (not the strain concentration area). It appears that the decrease of the critical strain level largely depends on the density of the holes, particularly the spacing between the holes.

Due to the curvature of a deflecting panel a plastic hinge, thus a strain concentration, is formed at the edge-deck connection, see Figure 3. The tensile force from the membrane mechanism tends to concentrate this plastic hinge within a very small area and hence increases the local strain at the main horizontal welds at the top and bottom of the bulkhead. Three principles are available to cope with a plastic hinge at this location.

1. Avoid the rather brittle weld at the critical location to allow a large strain in the ductile parent material;
2. Give the part of the bulkhead connected to the deck a higher thickness to keep the ratio of the tensile force capacity of the middle plate over the capacity of this lower part below one.
3. Force the critical part of the bulkhead to follow a smooth curve in order to spread the strain.

All three principles can be used for blast resistant structures. Solutions based on a continuing bulkhead and a discontinuous deck are not considered for ships for reasons of fatigue. For solution 1 a careful welding procedure have to be applied to obtain a tough and strong weld, which is critical for the success of the bulkhead.

2.2 Levels of protection

Roughly speaking two levels of blast resistance can be distinguished for internal explosions:

1. Bulkheads and doors which must be able to resist the direct blast effects of a relatively small high explosive (HE) charge, or be able to withstand the venting pressure into a compartment adjacent to the explosion compartment. In some cases structures based on the bending mechanism can comply with this requirement.
2. Bulkheads and doors, which must be able to resist the blast effects in the explosion compartment for large HE charges or medium to large charge over volume ratio. For this requirement either the membrane mechanism has to be applied or a combination of bending and membrane.

The optimal (balanced) level of the blast resistance of a door will be to design it for the same level of blast resistance as the bulkhead in which the door is installed (matched design).

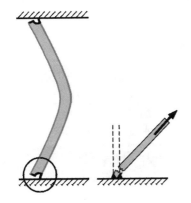

Figure 3.   Stress and strain concentration at the bulkhead & deck connection.

Two blast resistant doors (level I and II) and three blast resistant bulkhead concepts (level I and II) are dealt with below.

2.3 Blast & fragment resistant bulkheads

2.3.1 Short-fall conventional WT-bulkheads

Watertight bulkheads which are designed with conventional procedures are not able to resist the loading emanating from a missile warhead detonation, in case they are exposed to the direct blast effects. Conventional bulkheads adjacent to the explosion compartment might also fail depending on the charge mass and the location of the bulkhead. A lot of lessons on bulkhead response and failure were learned during three years of extensive RNLN/TNO-PML "Roofdier-Trials" on two decommissioned RNLN frigates. Figure 4 depicts a bulkhead that failed at the lower deck-connection and rotated around the deckhead into the adjacent compartment.

As explained earlier, to improve the blast resistance of the bulkhead, the connections between Bulkhead (BHD) and Deck/Deckhead have to be redesigned and if possible, the membrane mechanism must be exploited.

2.3.2 Blast resistant PriMa double bulkhead level II

A double bulkhead was designed, called the PriMa Double Bulkhead (PDB), based on the membrane mechanism. The double plating was put on a small horizontal part which was connected to a slender insert plate, see Figure 5. To cope with the plastic hinge at the deck connection the slender insert plate was carefully designed with FEM calculations based on LsDyna and Abaqus.

PriMa bulkheads present a high probability of survival against missile warhead detonation; thus preventing blast and fragments from propagating into and killing the next autonomous zone. This was shown with the in-house developed internal blast code DAMINEX, [Erkel et al. 1991]. A large number of explosion positions were simulated where the detailed outcome of the FEM calculations was used for the failure criterion.

The bulkhead, see Figure 6, was also thoroughly designed to be able to meet the operational requirements for flooding, buckling and fatigue. The PDB gives the designer the opportunity to install an additional ballistic material between the two steel plates to increase the fragment resistance.

The blast resistance of the PriMa bulkheads will be compromised by penetrations of piping and cables systems. This compromise has already been minimised for the new LCF, by decreasing the number of necessary penetrations, through concentration of system elements within one zone or even within one compartment. However where penetrations cannot be avoided, the insert plate element of the PriMa bulkhead is compensated in thickness.

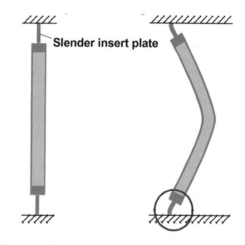

Figure 5. The principals of the PriMa double bulkhead.

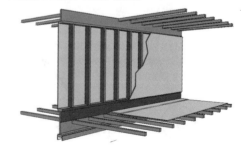

Figure 6. A schematic view of the PDB in the ship environment.

Figure 4. Failed conventional WT bulkhead during the roofdier trials.

Figure 7a. Experimental validation of the PDB under UK–NL cooperation program DERA.

Figure 7b. Application of PriMa BHDs o/b the RNLN LCF.

For those locations on the platform where there is a lack of membrane clamping (e.g. no adjacent bulkhead above) the PDB offers a very good bending behaviour as a result of the high section modulus and the prevention of stiffener tripping. The PDB has a large potential to withstand very high explosion pressures emanating from the largest threats. Recent FEM calculations show that by special shaping of the slender plate a reduction in the strain level of the plastic hinge by a factor two can be achieved.

2.3.3 Experimental validation and application

A PriMa bulkhead has been successfully tested in cooperation with UK DERA[2], Dunfermline, see Figure 7a.

---

[2] Defence Evaluation and Research Agency.

Figure 8. Recent validation tests on the PriMa single bulkhead.

PriMa BHDs are installed at the zone boundaries at the RNLN LCF (Figure 7b) and at positions were special functions have to be protected.

2.4 PriMa single bulkhead level I and II

After having developed the PriMa Double bulkhead TNO-PML was asked by the RNLN to develop a single plate blast resistant bulkhead. This was meant for level I and level II requirements, if possible. A single plate structure is a more regular ship structure, generally of lower mass and cheaper to fabricate.

A disadvantage of the single plate bulkhead is the lower efficiency of the single bulkhead for the locations in the ship with a lack of membrane clamping (e.g. no adjacent bulkhead above). Application of a single plate bulkhead in those locations will lead to larger deflections due to stiffener tripping and a lower section modulus compared to the PDB.

2.4.1 Principals PriMa single bulkhead

To cope with the plastic hinge, an insert plate was applied at the connection with the deck, in combination with a special shaped stiffener. The shape of the insert plate and the hole in the stiffener were designed carefully to force the deck connection of the stiffener to fail in a controlled way and to keep the concentration of strain low. This was done with the aid of numerous FEM calculations.

2.4.2 Experimental validation and application

Recently a test was carried out on two PriMa single bulkheads (PSB) in close cooperation with Germany. The test was very successful, both panels withstood the HE test without any crack, see Figure 8.

The deck connection performed as designed and tripping of the stiffeners was prevented due to the membrane strain. In several tests the influence of the girder and piping transits were determined.

The concept of the PSB can be applied for the level I and level II blast loading by keeping the principle and redesigning the dimensions.

The Spanish shipyard Bazán in close cooperation with TNO-PML has successfully presented the blast resistant PriMa single bulkhead as the best alternative for the new RNoN Frigate programme.

2.5 PriMa curvature limiters level I and II

It would be very worthwhile to have simple "add-on" techniques to improve the blast resistance of structures of already commissioned naval platforms, e.g. when they are upgraded. For these reasons TNO-PML investigated and developed, very simple, cost-effective solutions. A small add-on structure has been designed which is suitable for retrofitting and is called "curvature limiter". It will enlarge the membrane capacity of an existing bulkhead. By applying the curvature limiters at the bulkhead locations in warships between the zone boundaries (with specially designed blast resistant bulkheads) a further reduction in damage volume can be obtained.

The lightweight PriMa curvature limiters have been patented [Schilt] and have good opportunities for application in the offshore industry as well.

2.5.1 Principals PriMa curvature limiters

As mentioned earlier under the "principles of designing blast resistant structures", blast resistance improvement measures aim to avoid concentrated plastic hinges, particularly near the weld. The idea of the curvature limiter is based on the mentioned third principle of avoiding deformation concentration; i.e. spreading the curvature. By placing small wedges at the edges of the bulkhead the curvature is forced out of the welding area into the base material. By using wedges with a pre-defined radius, the curvature in the base material can be controlled and strain concentrations can be avoided, see Figure 9. This will lead to a larger allowable deflection for the panel and thus a higher resistance.

An accompanying advantage of this measure is that the high shear forces will act on a larger cross section.

2.5.2 Experimental validation

Some tests on the curvature limiters were performed in the TNO facility. A conventional bulkhead was provided with a curvature limiter at one side, the other side was regularly fitted. The heavily loaded panel ruptured completely along the regular side. The other side, with the curvature limiter underneath, remained intact, proving the resistance enhancing capability of the curvature limiter.

The curvature limiters have been also tested in the blast bunker at the Meppen proving ground in

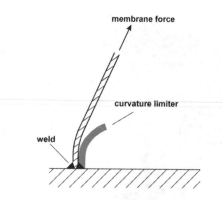

Figure 9.   The principals of the PriMa-curvature limiter.

Figure 10.   The large deflection of the conventional bulkhead without cracks, equipped with curvature limiters tested at Meppen (Germany).

Germany. Figure 10 shows the result of the test of a normal bulkhead retrofitted with the curvature limiter. It easily survived a loading twice the amount it normally would be able to endure. It showed a large deflection without cracks. So, it can be concluded that the curvature limiters perform very well.

2.6 Blast resistant door for conventional WT-bulkheads level I

2.6.1 Short-fall conventional doors

The most vulnerable link of "blast resistance" in the longitudinal watertight subdivision of conventional ship designs is the watertight (WT) doors. Conventional WT-doors are only tested and classified for static water pressure in flooded conditions. They are not classified for the highly dynamic loading as a result of blast effects. The result of this is that conventional WT-doors fail at very low internal explosion levels.

Figure 11. A Roofdier trial result; Four WT-doors in a row blown out by the detonation of a medium sized naval shell.

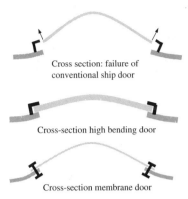

Figure 12. Different door concepts.

This fact has been proven during the RNLN/ TNO-PML "Roofdier-Trials" on decommissioned RNLN frigates of the Roofdier class. Even an internal detonation of a medium sized naval shell was able to blow out a large number of WT-doors over a long corridor distance, see Figure 11, exposing all involved compartments (personnel & materiel!), to overpressures and freeing the consequent spread of fire, smoke and water. The first step to improve the blast resistance of the longitudinal subdivision is improving the blast resistance of the WT-doors.

2.6.2 Principals for improvement of blast resistance WT-doors

In close cooperation with the RNLN, TNO-PML has carried out a number of tests on various existing ship door designs. A lot of lessons were learned, particularly on the failure mechanism. The relatively low blast resistance of conventional WT ship doors, is depicted in Figure 12, first picture. Already at low loading levels the doorplate will show a large mid deflection and particularly, distortions of the doorframe. Because of the axial shortening (in-plane movement) of the door, the door itself will, at low loading levels, "pop" through the frame.

Possible improvements of these bending based conventional designs can be divided into two categories. The first one deals with a loading direction against the frame and the second with a loading direction away from the frame. For a loading against the frame the frame length and the nearest stiffener must be designed properly as well as the edge of the frame. For a loading away from the door edge, the latches revealed to be the weakest part of the door and the designer must have a thorough look at (the shafts of) the latches.

2.6.3 Experimental validation and application of a level I design

The Dutch door manufacturer Van Dam, in close concert with TNO-PML and the RNLN, designed a bending door based on the above mentioned considerations. This door, of moderate weight, has experimentally been validated, with the defined dynamic loading, under a German–Netherlands Navy Cooperation Program. Doors of this type have been procured by the RNLN and will be installed on board the new RNLN LCF at all positions where use is made of conventional bulkheads. This level I door is commercially available. TNO-PML also designed a lightweight level I door based on an existing ship door design. This door presents a lower blast resistance than the bending door, but is still remarkably stronger than most of the conventional doors.

2.7 Membrane door for WT-bulkheads level II

The function of PriMa blast resistant bulkheads is to contain the loading of a missile detonation within the explosion compartment. This also holds for the installed WT-door. However a door of the already dealt with level I type will not suffice. Hence a new blast resistant and watertight door was developed by TNO-PML based on the membrane mechanism. This was performed in close cooperation with the manufacturer "van Dam" already mentioned for the level I door. The door has been patented recently, (see Schilt et al.) and is commercially available.

2.7.1 Principals blast resistant membrane door

To cope with the large blast loads in the explosion compartment without a large mass penalty, a special door must be designed. The RNLN defined an ultimate

611

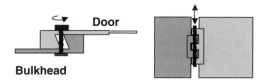  

Door

Bulkhead

Cross section door 1   **Front view door 2**

Figure 13.   Possible closure mechanisms for a membrane door.

Figure 14.   The PriMa membrane Mk3 door in the DERA blast test box, pre-trial and post-trial.

level for the mass of the door based on a sailor who must be able to operate the door during a certain sea state. Furthermore the manual operation of the door with one single lever is required.

Three different design concepts are drawn in Figure 12. Already at low loading levels, conventional ship doors show a large deflection and consequently 'pop' through the frame (Figure 12, first picture). Constructing a door with a high bending resistance can prevent this (Figure 12, second picture). However such a door is too heavy. A revolutionary approach is a door construction with its blast resistance based on the membrane stiffness, allowing a high deflection of the door (Figure 12, third picture). Unlike conventional doors, the doors turning part can be constructed relatively thin.

An additional advantage of a membrane door is the comparable strength for the two loading directions. However, there is a large difference with a bulkhead based on the membrane mechanism. The edges of the bulkhead are firmly connected to the decks that transfer the tremendous membrane force to the adjacent ship structure. This is contrary to the door which, of course, must be opened, but which also has to transfer the large membrane load in case a blast pressure. This faces the designer with a big challenge how to design the closure mechanism.

Several mechanisms have been designed and tested. Finally two favourable designs remain, of which the first one was selected for further testing and improvement, see Figure 13.

Due to the mass requirement, only a relatively small plate thickness of the door was allowed. After several simulations with DAMINEX it turns out that a membrane door based on a relatively small thickness was not able to withstand the blast load in the explosion compartment. Hence a special feature was developed based on the nature of the membrane mechanism which tells us that for the plastic strain realm the centre deflection of the plate is one of the governing parameters for blast resistance, irrespective of the strain level. So, folds were applied in the door which in the first stage of the response will be stretched like a spring and only after being "folded open" stretches the plate of the door, relieving that plate from high strain levels.

With DAMINEX a large number of simulations were carried out to find the optimum fold geometry in terms of number of folds, length, height and thickness of the fold. It appeared that an optimum for the fold geometry exists resulting in a minimum global membrane strain in the doorplate.

2.7.2   *Experimental validation and application of the PriMa membrane door*

The essential parts of a membrane door were validated during the Roofdier tests and later in the TNO test facility. In 1999, the PriMa door Mk2 has been tested in a large full scale ship section in close cooperation with DERA (UK). The door endured all tests satisfactorily. In December 2000, a final qualification blast test has been performed on the Mk3 design, see Figure 14, for the very high loading conditions of level II.

The Mk3 door incorporates a fully operational closure mechanism specially designed to protect the door from damage under large deflections. The mass of the rotating part of the door is about 120 kg.

The Mk3 design passed the qualification blast test without any trouble. The door deflected, introducing stretching of the folds and membrane strain of the door plate, and it survived the high explosion pressure without any crack or fracture, see Figure 14.

The RNLN and other navies are considering the procurement of the Mk3 blast resistant doors of the membrane type for application in the blast resistant bulkheads.

3   CONCLUSION

The importance of vulnerability reduction for warships has been stressed. Warships have various possibilities to avoid a hit. These susceptibility technologies have to be applied in a balanced cost-effective way to reduce the risk of a hit. However, they can never eliminate the chance of a hit with its successive loss of lives, casualties, loss of mission and materiel either by a high tech or a low tech threat.

Therefore vulnerability reduction technologies have to be incorporated in the design of fighting platforms. Improving the blast resistance of structures has proven to be one of the most effective solutions. The

RNLN, in concert with TNO-PML, has been focusing on the reduction of the longitudinal spread of damage after an internal detonation by improving the blast-resistance of longitudinal watertight subdivisions (i.e. bulkheads and doors).

TNO-PML has developed two blast resistant doors and three blast and fragment resistant bulkhead concepts for two levels of pressure loading. All five concepts have been validated very successfully by experiments. The door designs and the bulkhead concepts will be applied in the new Dutch LCF frigate and ships for other navies. The development and validation of these concepts was also founded on successful international co-operation with the German and British Naval Defence communities.

It must be noted that, although the initiation of the development of these blast resistant structures is for Naval Defence purposes, there is definitely a spin-off for in- and external explosions for bunkers, ammunition storage and for applications in the off-shore industry.

REFERENCES

Erkel, A.G. van & Haverdings, W. 1991 "Simulation of internal blast inside ships and its subsequent damage analysis", Presented at MABS-12, Perpignan.

Schilt, A. TNO-PML, *Curvature Limiters,* Patented at the Dutch patents office Nederlandsch Octrooibureau, under patent number 1013795, entitled "Explosiebestendige wandconstructie".

Schilt, A., Luyten, J.M. & Erkel, A.G. van Beheermaatschappij H.D. Groeneveld B.V. *Membrane door,* Patented at the Dutch patents office Nederlandsch Octrooibureau, under patent number 1010945 (International PCT/NL00/00003), entitled "Deurconstructie II.

Smit, C.S. "Ship combat survivability enhancement", Presented at IMDEX Asia 97, Singapore, may 1997.

Safety and Reliability – Bedford & van Gelder (eds)
© 2003 Swets & Zeitlinger, Lisse, ISBN 90 5809 551 7

Seismic risk of atmospheric storage tanks by probit analysis

G. Fabbrocino & I. Iervolino
Department of Structural Analysis and Design, University of Naples, Napoli, Italy

E. Salzano
Institute of Research on Combustion, CNR, Italy

ABSTRACT: The risk assessment of industrial facilities is based on availability of integrated procedures to quantify human, environmental and economical losses related to relevant accidents, thus including earthquake actions. Accordingly, results of seismic risk analysis has to be integrated into quantitative risk analysis either for single establishment or for entire industrial areas. Moreover, it's worth noting that the evaluation of the "domino effect" leads risk analysts to take into account the escalation of industrial accident even starting from a relatively minor natural event such as low-intensity earthquakes. In this paper, some considerations are given regarding the intensity and probability of occurrence of earthquakes and the vulnerability of atmospheric storage tanks subjected to seismic actions. Eventually, structural vulnerability and seismic hazard have been compared aiming at defining a simple and useful statistic tool as the "probit analysis". Some indications on the industrial seismic-related accidental scenarios are also given.

1 INTRODUCTION

The risk assessment of industrial facilities is based on availability of integrated procedures to quantify human, environmental and economical losses related to relevant accidents, thus including earthquake actions. Accordingly, results of seismic risk analysis has to be integrated into quantitative risk analysis (QRA) either for single establishment or for entire industrial areas [Lees, 1996]. Moreover, it's worth noting that in the mainframe of the Seveso II directive [Council Directive 96/82/EC], the evaluation of the "domino effect" has forced risk analysts to keep into account the escalation of industrial accident even starting from minor natural events such as low-intensity earthquakes. Eventually, interdisciplinary efforts between seismic, structural and chemical engineers are required to obtain reliable QRAs.

Risk analysis deals with the occurrence of individual failure events and their possible consequences on the analysed system [Kirchsteiger, 1999]. With reference to an industrial installation, aiming at providing a quantitative methodology for risk analysis (quantitative risk analysis, QRA, or Probabilistic Risk Assessment, PSA, or Probabilistic Safety Assessment, PSA), a deterministic or a probabilistic approach can be used.

When specifically the seismic risk is of concern, a deterministic approach uses the maximum "credible" earthquake event and "worst case" scenarios are considered for the evaluation of consequences.

If industrial installations are considered and people and/or equipment safety is of interest, this approach has to be coupled with another "deterministic" analysis, which keeps into account the evolution of the accidental scenario (the earthquake) starting from the material or the energy loss from the failed system of containment, i.e. the evaluation of consequences. Again, a worst case scenario should be considered. This approach leads to great over estimation of the total risk, often providing a risk grade which is both economically and politically not applicable, e.g. in the case of civil protection action. Moreover, the uncertainties on the initial conditions for either the seismic scenario or the evolution of the industrial accident scenario related to the earthquake itself, are often too large. This circumstance leads analysts to use a probabilistic approach, where uncertainties are explicitly taken into account and described through random variables, by their probability distribution.

Common measures for industrial quantitative risk include individual risk and societal risk [CCPS, 1989; Lees, 1996]. The individual risk for a point-location around a hazardous activity is defined as the probability that an average unprotected person permanently present at that point location, would get killed due to an accident at the hazardous activity.

The societal risk for a hazardous activity is defined as the probability that a group of more than N persons would get killed due to an accident at the hazardous activity [Bottelberghs, 2000]. The societal risk is often described as FN curve (frequency number curve), i.e. the exceedance curve of the annual probability of the event and the consequences of that event in terms of the number of deaths.

The practical evaluation of both individual and societal risk is a complex task which requires first the identification of all credible equipment failures, i.e. the "top event". The latest has to be coupled with related probabilities of occurrence, based on historical analysis or process related analysis (e.g. fault tree analysis).

At this point, the physical phenomena which are able to produce damages to people, e.g. fire or explosion or dispersion of toxic substance, and equipment (aiming at domino effect evaluation) and the related probability (e.g. through event tree analysis), for each of the possible top events, has to be modelled, in order to produce the temporal and spatial distribution of overpressure, heat radiation and concentration. Moreover, the relationship of this variables with human being has to be assessed.

Eventually, it's clear that QRA can only be considered as a rough evaluation of risk and that this instrument should only be used as a comparative tool, since arbitrary choices are often necessary, either on probability of occurrence of the top event or on the physical modelling of the entire accidental scenario or on the occurrence of damage. However, research efforts should be addressed to improve each of these aspects, in order to produce efficient "algorithms".

In this work, some considerations either regarding the intensity and probability of occurrence of earthquakes, or the vulnerability of equipment – specifically atmospheric storage tanks – to seismic action are given.

Structural vulnerability and seismic hazard quantitative results have been also compared aiming at defining the parameters of the simple statistic tool known as "probit analysis" [Finney, 1971; Vilchez, 2001].

2 THE INTENSITY AND PROBABILITY OF OCCURRENCE OF EARTHQUAKES

Measured earthquakes signals refer to seismic waves radiating from the seism epicentre to the gauge location and can be related to global characteristics of the earthquakes: magnitude, distance and soil type.

These quantities are mainly reflected in the frequency content of the motion.

Despite this simplification, earthquake signals carry several uncertainties and it is not even a trivial task to define a univocally determined "intensity" of earthquake, thus allowing comparison of records.

However, geophysicists and structural engineers use to classify earthquakes on the basis of two classes of parameters such as "ground parameters" and "structural dynamic affecting factors" [Chopra, 1995]. The choice of these intensity parameters is important since they summarize all the random features of earthquakes, including energy and frequency contents, which meaningfully affect the structural response of components [Eidinger, 2001].

Ground parameters refer to the peak values of variables experienced by the ground motion: the intensity of earthquakes is viewed as "ground shaking", characterized by a peak ground acceleration (PGA), or a peak ground velocity (PGV), or a response spectra (RS) at the site location of the component. Structural affecting factors usually refer to the dynamic amplification induced on a single degree of freedom system with the same period of the analysed structure (spectral acceleration). Although experimental investigations have demonstrated that different parameters are needed if the effects of earthquake on structures would be accurately reproduced by structural analysis.

For instance, in seismic analysis of piping system peak ground velocity is commonly used, while peak ground acceleration is more useful when steel storage tanks are under investigation. Hence, PGA is the global earthquake intensity measure used in the following.

This assumption is necessary in the light of simplification that is needed when complex risk assessment of industrial plants is performed. Regardless intensity definition, seismic pre-accident risk assessment needs the definition of a probability of occurrence of earthquake, given its intensity. To this aim, seismic hazard (H) is related to the time interval T – i.e. the service life of the structure – and to the seismological seismic intensity parameter (PGA):

$$H(T) = P(PGA > a|T) \qquad (3)$$

This relation gives the probability H that a given PGA exceeds the value of the constant a during T. Seismic hazard curves calculated for two equipment with a service life of respectively fifty and one year, evaluated at Benevento, located in south Italy, are reported in Figure 1.

Local authorities commonly produce the curves reporting the probability of occurrence of PGA both in Europe and US.

If different intensity parameters are used, all ground shaking parameters are related and can be found elsewhere [Claugh, 1982].

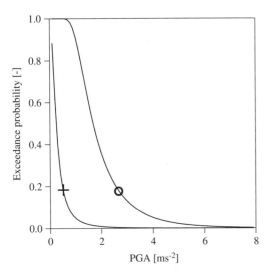

Figure 1. Hazard curves in terms of annual exceedance probability of Peak Ground Acceleration (PGA) for two equipment with service life of respectively: +: 50 year; O: 1 year.

Figure 2. Unanchored atmospheric tank subjected to "uplifting" and "elephant foot" buckling.

3 SEISMIC BEHAVIOUR OF ATMOSPHERIC STORAGE TANK

Starting from 1930 atmospheric storage steel tanks were fabricated as riveted, welded or bolted (especially for low values of height over radius ratio H/R); conversely in the last decades they were basically welded world-wide.

According to consolidated design and construction standards, these types of tanks exhibit strong structural similarities with water storage tanks. Nevertheless, the procedures [AWWA D100, 1996; AWWA D103, 1997; API Standard 620–650, 1998] provided by American Petroleum Institute and American Water Works Association do not prescribe any dynamic analysis, and the effects of earthquake actions are only evaluated in terms of overturning moment and total base shear.

Recently, Eurocode 8 (1998) has developed a more comprehensive and advanced guideline for the design of this type of facility from a structural standpoint.

The base plate of storage tanks is generally flat or conical shaped.

The tank shell consists of different steel courses approximately one meters and a half tall; their thickness decreases along the height and rarely exceeds two centimetres in the bottom course (large tanks reference value).

Shell thickness is calculated using empirical formulas (i.e. "one foot method") according to design guidelines and depends only on tank dimensions and content density. Shells include nozzles and openings and other piping connections. Roof can be shaped in many different ways as dome, conical or can be floating.

Roofs can be self-supported or columns supported in case of large diameters. International guidelines [API 620–650, 1998] provide minimum roof plate thickness and geometrical calculation (i.e. cone inclination, depending from diameter of tanks).

Tanks can be anchored or unanchored to the ground. Due to economical reasons; they are often simply ground or gravel bed rested; for large tanks and/or bad soil conditions concrete ring foundation can be effective.

Anchored tanks are more expensive and are generally recommended in seismic areas but their effectiveness is still under investigation.

A key issue in steel tank design is welding; indeed, welds are sensitive to corrosion and can lead to wide cracks during earthquake events, particularly in the shell/roof and shell/base plate joint zones.

Another critical aspect for the seismic behaviour of storage tanks is the foundation. The analysis of seismic damages pointed out the effects of foundation on collapse mechanisms and strength performances of the structure.

Assuming the same filling level and nominal dimensions, gravel rested tanks are subjected to uplifting and/or sliding motion, but the tearing of pipe connection can be activated in case of strong motions (Figure 2).

The dynamic behaviour of atmospheric storage tanks subjected to a earthquake is characterised by two predominant vibrating modes: the first is related to the mass that rigidly moves together with the tank structure (impulsive mass), the other corresponds to the liquid's sloshing (convective mass).

Liquid sloshing during earthquake action produces several damages by fluid-structure interaction phenomena and can result as the main cause of collapse for full or nearly full tanks.

Historical analysis and assessment of seismic damages of storage tanks has revealed that only full (or nearly full) tanks experienced catastrophic failures. Low H/R tanks only suffered cracks in conical roof connection, or damage by floating panel sinking.

The most common shell damage is the "elephant foot buckling" (EFB). For unanchored tanks and H/R < 0.8, EFB is not experienced but the base plate or the shell connection can fail causing spillage.

4 THE PROBIT ANALYSIS

The usefulness of probit analysis relies on the relatively simple integration of the probit function with QRA algorithms (e.g. ARIPAL [Spadoni, 2000]) which have been produced in the past aiming at the definition of industrial individual and social risks, as they are commonly defined in literature [CCPS, 1994; Lees, 1996]. Moreover, comparison of seismic hazard of tanks with different geometry and filling level can be easily obtained by means of probit coefficients. This tool has been widely used in hazard assessment since the first Canvey Report [HSE, 1978] and the Rijnmond report (1982), although only referred to person injury.

The probit variable (usually represented as Y) is a dose-response relationship and gives a measure of having certain damage as a function of the intensity of the variable V (the "dose") through a linear correlation with the logarithm of V:

$$Y = k_1 + k_2 \ln V \tag{1}$$

In this work, the dose has been considered as the seismic PGA whereas the effect is considered either as the structural damage of the tank or, more appropriately, the loss of containment of the tank subjected to a earthquake. The variable Y can be directly compared with the actual failure probability P by means of the integral [Vilchez, 2001]:

$$P(V) = \frac{1}{\sqrt{2\pi}} \int_{-\infty}^{Y-5} e^{-V^2/2} \, dV \tag{2}$$

Comparison of value of k_1 e k_2 gives direct and useful information on the gravity of the accidental events.

5 RESULTS AND DISCUSSION

A full stress analysis is certainly the more accurate way to design and to evaluate the risk of steel tanks under earthquake loads.

This approach leads to the direct computation of the interaction between shell deformations and content motion during earthquakes.

For base constrained and rigid tanks (anchored), a complete seismic analysis requires solution of Laplace's equation for the motion of the contained liquid. Solution of the latter equation has to be carried to obtain the total pressure history on the tank shell during earthquakes. When flexible tanks are considered, a structural deformation term must be also added to keep into account the "impulsive" and "convective" contributions.

Unanchored tanks are subjected to uplifting but also to sliding. Uplifting can crack base plate connection; besides it increases flexibility to the system isolating it. AWWA D-100 and API 650 focus their attention on base shear and overturning moment after Malhotra (2000) and provide methods to take into account of geometrical parameters of the tank and the earthquake zone classification factors.

Actually, when QRA of industrial installations have to be performed, the number of tanks and the complexity of the assessment of risk indexes does not allow the detailed analysis of interaction of all possible intensity of earthquakes with equipment.

Hence, in the light of simplification, statistical and empirical tools derived from post-earthquake damage analyses are needed, in order to define simple and general vulnerability functions.

Here, it's also worth noting that the similarity of seismic behaviour of water tanks and oil tanks, both operating at atmospheric condition, is certain, thus consistently enlarging the historical data set.

An extraction of data set used for the historical analysis of fragility of atmospheric storage tanks subjected to earthquakes is reported in Table 1. Here, and in the following, no separation between anchored and unanchored tanks has been taken into consideration.

Several studies [O'Rourke, 2000; Eidinger 2001] in the last decades have defined "damage states" (DS) in order to describe the seismic behaviour of steel tanks, starting from slight damage to the structures (DS2), to moderate damage (DS3), and finally to extensive damage (DS4) and total collapse of structure (DS5). The term DS1 refers to the absence of damage.

Table 1. An extraction of data set used for the assessment of vulnerability of atmospheric storage tank subjected to earthquakes.

PGA [g**] range	No. damaged tanks	Event*
0.17	49	Long Beach (1933)
0.19	24	Kern County (1952)
0.20 ÷ 0.30	39	Alaska (1964)
0.30 ÷ 1.20	20	San Fernando (1971)
0.24 ÷ 0.49	24	Imperial Valley (1979)
0.23 ÷ 0.62	41	Coalinga (1983)
0.25 ÷ 0.5	12	Morgan Hill (1984)
0.1 ÷ 0.54	141	Loma Prieta (1989)
0.35	38	Costa Rica (1992)
0.1 ÷ 0.56	33	Landers (1992)
0.3 ÷ 1	70	Northridge (1994)
0.17 ÷ 0.56	41	Others

* data from [Cooper, 1997; Wald, 1998; Haroun, 1983, Ballantyne and Crouse, 1997; Brown, 1995; Eidinger et al. 2001].
** g is gravity acceleration, ms^{-1}.

618

The DS values correspond to the classical limit states definition related to the economical loss to repair and restore the tank structure. Tables 2 and 3 report the damage analysis obtained using limit states, starting from the historical data set reported in Table 1, for the total number of tanks and for the tank whose filling level is greater than 50%, following the assumption that only highly filled tanks feel the effect of earthquake.

Indeed, structural analysis and empirical observation confirm that only filling level of 50% seems to be effective to vulnerability. Moreover, the choice of a filling level results useful when QRA on large storage area is performed and no detailed information on the average tank fill level can be obtained. Actually, in the mainframe of industrial risk assessment, the loss of hazardous substances from their system of containment is the main issue.

In fact, unless very catastrophic earthquakes are considered (often very rare and producing a complete destruction of the industrial installation), the loss of containment is a main consequence of earthquake-equipment interaction, thus providing the triggering for the escalation of the accident scenario.

Moreover, it should be considered that typical accidental scenario involve vapour cloud explosion (VCE), flash fire, pool fire or toxic dispersion; they all strongly depend on the total amount of substance [CCPS, 1994].

Hence, in the following, the data set has been re-organized in terms of three classes of damage or "risk state" for the atmospheric tank considered as a whole, including tube connections for loading, valves and general equipment.

The first class corresponds to an earthquake which slightly affects the structure of the tank thus resulting in the total absence of loss of containment, although post-accident analysis should be performed.

This class has been identified as RS1. Next, a structural damage of the shell or of an auxiliary equipment which gives rise to a "slight loss of content" is defined as RS2. Finally, a consistent and rapid loss of content has been identified as RS3.

The latest class identifies the damage related to an earthquake which affects the tank integrity, giving rise to a catastrophic accident and total loss of containment. Table 4 reports the damage analysis in terms of loss of containment for the data set of atmospheric storage tanks subjected to the earthquake as reported in Table 1. Also, the same table reports the damage analysis when isolating tanks whose fill level is higher than 50%. Figure 3 reports the fragility curve (the probability of getting the considered limit state) as derived from Table 4.

The results are practically coincident with those obtained by former studies [Eidinger, 2001]. Again, the fragility curve for tanks with 50% fill level has been reported for convenience.

Through the relationship given in Eq. 2 it's now possible to transform the fragility curve into a probit relationship with respect to PGA (Figure 4). Values of probit coefficient (see Eq. 1) as obtained from Figure 4 are reported in Table 5.

Table 2. Analysis of damage states for all the atmospheric tanks subjected to earthquake reported in the historical data set of Table 1.

PGA [g]	All	DS = 1	DS = 2	DS = 3	DS = 4	DS = 5
0.10	4	4	0	0	0	0
0.17	263	196	42	13	8	4
0.27	62	31	17	10	4	0
0.37	53	22	19	8	3	1
0.48	47	32	11	3	1	0
0.57	53	26	15	7	3	2
0.66	25	9	5	5	3	3
0.86	14	10	0	1	3	0
1.18	10	1	3	0	0	6
Total	532	331	112	40	25	16

Table 3. Analysis of damage states for the atmospheric tanks subjected to earthquake with filling level greater then 50%.

PGA [g]	All	DS =1	DS =2	DS = 3	DS = 4	DS = 5
0.10	1	1	0	0	0	0
0.17	77	22	32	12	8	3
0.27	43	16	12	10	4	0
0.37	22	3	11	4	3	1
0.48	25	12	9	3	1	0
0.57	48	22	14	7	3	2
0.66	15	4	2	3	3	3
0.86	10	7	0	0	3	0
1.18	10	1	3	0	0	5
Total	251	88	84	39	25	15

Table 4. Analysis of damage states in terms of loss of containment for the atmospheric tanks subjected to earthquake as reported in the historical data set of Table 1.

	Fill level [>50%]		Fill level [0–100%]	
PGA [g]	RS ⩾ 2	RS = 3	RS ⩾ 2	RS = 3
0.10	0	0	0	0
0.17	55	11	67	12
0.27	26	4	31	4
0.37	19	4	31	4
0.48	13	1	15	1
0.57	26	5	27	5
0.66	11	6	16	6
0.86	3	3	4	3
1.18	8	5	9	6

619

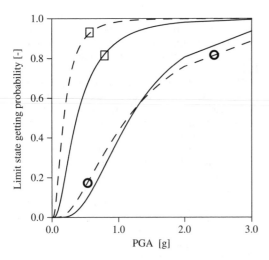

Figure 3. Experimental fragility curves for atmospheric steel tanks affected by earthquakes. Dotted line represents tank fill level >50%. □: RS2; ○: RS3.

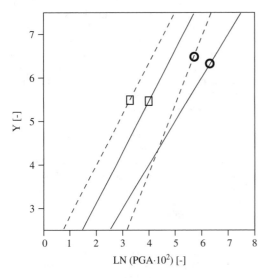

Figure 4. Probit analysis for steel tanks in seismic areas. Dotted line represents tank fill level >50%. □: RS2; ○: RS3.

Here, it's clear the similarity of behaviour of tanks with fill level greater than at 50% with the complete probit function.

The obtained results should be used in conjunction with the evaluation of the accidental scenarios which can derive from the loss of containment itself. Of course, some assumptions are necessary in order to provide a risk assessment.

First, here only atmospheric storage tank are considered, thus flammable or toxic liquid flow into catch

Table 5. Probit coefficient (Eq. 1) for atmospheric steel tank subjected to earthquake.

| | Fragility | | Probit | |
Risk state	μ	β	k_1	k_2
RS $\geqslant$ 2	0.38	0.80	0.43	1.26
RS = 3	1.18	0.61	−2.83	1.64
RS $\geqslant$ 2/fill level $\geqslant$ 50%	0.18	0.80	1.77	1.14
RS = 3/fill level $\geqslant$ 50%	1.14	0.80	−0.92	1.25

basins has to be analysed. Moreover, it is worth noting that ignition of fuel pool or flammable vapour cloud is very likely when seismic action is present.

In the case of low-intensity earthquakes, it is presumable that the response of operator and the safety procedures (e.g. sprinkler action) are able to prevent or at least to mitigate the risk of fire or explosion and to restore the plant normality within tens of minutes, at least for RS2 damages.

In this case, only toxic, flash fire (i.e. the fire of vapour cloud without the generation of destructive blast wave) and pool fire effects (in the close surrounding of tanks), should be considered, since vapour cloud explosions (a blast wave is produced in this case) need long term evaporation and fuel dispersion to give a potentially destructive homogenous flammable vapour cloud [CCPS, 1994].

Eventually, dispersion analysis has to be performed, either for the toxic dispersion or for the fuel air mixture within flammability limits (flash fire), and heat radiation effects on the structure and on the people has to be calculated, starting from the probability of occurrence of seism and from the probability of tank failure.

When RS3 damage occurs (and it's likely that several tanks are involved) or more generally structural damages induced by very catastrophic earthquakes are considered, the gravity of situation hardly allows the operator to take a full control even for the single equipment.

All the scenarios should be then considered: pool fire, flash fire, vapour cloud explosion and toxic dispersion. To this regard, the probability of having flash fire or vapour cloud explosion is strongly dependent on the fuel reactivity and on the geometry either of the accidental vapour cloud or of the industrial installation (specifically on the geometrical confinement and degree of congestion).

Moreover, the effect of pool fire should be always added to the first two type of non-localised fires. Description of the phenomena here reported are given elsewhere [Lees, 1996; Martin, 2000; Salzano, 2002].

Finally, the evaluation of domino effects should be always performed, particularly if pressurized tanks are in the nearby.

6 CONCLUSIONS

Risk assessment should always incorporate the probability of seismic occurrence, and its intensity, as a "top event", not only if industrial installation or areas lay on very seismic area but also when low intensity earthquakes are expected. Indeed, the loss of containment can trigger "domino effect" through several accidental scenarios which comprise explosion or fires.

In this work, a classification of damage state with respect to the peak ground acceleration (PGA), either in terms of structural effects, or in terms of simple loss of containment has been performed on the basis of a historical data set of atmospheric tanks.

Probit coefficient have been then statistically evaluated, in order to implement a seismic dose–effect relationship into algorithms, aiming at obtaining quantitative risk assessment.

Some indications on the risk assessment for different seismic-related scenario are given.

Further work should be addressed to the interaction of earthquake with other equipment, e.g. pressurized equipment, which often contain flammable or toxic substances and to the implementation of a fuzzy analysis starting from the results here reported.

ACKNOWLEDGMENTS

The authors would like to acknowledge funding and support received from the Geophysics and Volcanology Italian National Institute (INGV) within the research project "Reduction of Infrastructures and Environment Seismic Vulnerability" (VIA).

REFERENCES

API 620 – American Petroleum Institute, Design and Construction of Large, Welded, Low-Pressure Storage Tanks, Washington D.C., USA, 1998.

API 650 – American Petroleum Institute, Welded Steel Tanks for Oil Storage, , Washington D.C., USA, 1998.

AWWA D100-96 – American Water Works Association, Welded Steel Tanks for Water Storage, Denver, Colorado, USA, 1996.

AWWA D103-97 – American Water Works Association, Factory-Coated Bolted Steel Tanks for Water Storage, USA, 1997.

Ballantyne, D. and Crouse, C., Reliability and Restoration of Water Supply Systems for Fire Suppression and Drinking Following Earthquakes, prepared for National Institute of Standards and Technology, NIST GCR 97–730, 1997.

Bottelberghs P.H., Risk analysis and safety policy developments in the Netherlands, Journal of Hazardous Materials 71, 2000, 59–84.

Brown, K., Rugar, P., Davis, C., and Rulla, T., "Seismic Performance of Los Angeles Water Tanks, Lifeline Earthquake Engineering", in Proceedings, 4th US Conference, ASCE, TCLEE Monograph No. 6, San Francisco, 1995.

CCPS, Center for Chemical Process Safety of the American Institute of Chemical Engineers, "Guidelines for chemical process quantitative risk analysis", Center for Chemical Process Safety of the American Institute of Chemical Engineers, New York, 1989.

CCPS, Center for Chemical Process Safety of the American Institute of Chemical Engineers, "Guidelines for evaluating the characteristics of VCEs, Flash Fires and BLEVEs", Center for Chemical Process Safety of the American Institute of Chemical Engineers, New York, 1994.

Chopra, A.K., Dynamics of Structures, Prentice Hall, 1995.

Claugh, R.W. and Penzien J., Dynamics of Structures. McGraw-Hill, New York, 1982.

Cooper, T.W., A Study of the Performance of Petroleum Storage Tanks During Earthquakes, 1933–1995, prepared for the National Institute of Standards and Technology, Gaithersburg, MD, June 1997.

Council Directive 96/82/EC of 9 December 1996 on the control of major-accident hazards involving dangerous substances, Official Journal of the European Communities, L 10/13, Brussels, 1997.

Eidinger, J.M., Earthquakeic Fragility Formulation for Water Systems, ASCE-TCLEE, 2001.

Eurocode 8, Design of structures for earthquake resistance – Part 4: Silos, tanks and pipelines, UNI ENV 1998.

Finney, D.J., Probit analysis, Cambridge University Press, 1971.

Haroun, M.A., Implications of Recent Nonlinear Analyses on Earthquakeic Standards of Liquid Storage Tanks, in proc. of 5th US Conference on Lifeline Earthquake Engineering, TCLEE Monograph No. 16, ASCE Seattle, 1999.

HSE – Health Safety Executive, Canvey report: an investigation of potential hazards from operation in the Canvey Island/Turrock area, London, UK, Stationery Office, 1978.

Lees, F.P., Loss Prevention in the process industries (II ed.), Butterwoth-Heineman, Oxford (UK), 1996.

Malotrah, P.K., Wenk, T. and Wieland, M., Simple Procedure for Earthquakeic Analysis of Liquid-Storage Tanks, Structural Engineering International, 3, 2000.

Martin, R.J., Ali Reza, A. and Anderson, L.W., What is an explosion? A case history of an investigation for the insurance, Journal of Loss Prevention in the Process Industries, 13, 2000, 491–497.

O'Rourke, M.J., Eeri, M. and So, P., "Seismic Fragility Curves for On-Grade Steel Tanks", Earthquake Spectra, Vol.16, NY, USA, November 2000.

Rijnmond Public Authority, Risk analysis of six potentially hazardous industrial objects in the Rijnmond area, a pilot study, Reidel Dordrecht, The Netherlands, 1982.

Salzano, E., Marra, F.S., Russo, G. and Lee, J.H.S., Numerical simulation of turbulent gas flames in tubes, Journal of Hazardous Materials, 95:, 3, 2002, 233–247.

Spadoni, G., Egidi, D. and Contini, S., Through ARIPAR-GIS the quantified area risk analysis supports land-use planning activities, Journal of Hazardous Material, 71, 2000, 423–437.

Vilchez, J.A, Montiel, H., Casal, J. and Arnaldos, J., Analytical expression for the calculation of damage percentage using the probit methodology, Journal of Loss Prevention in the Process Industries, 14, 2001, 193–197.

Vrijling, J.K., van Hengel, and Houben, R.J., Acceptable risk as a basis for design, Reliability Engineering and System Safety, 59, 1988m 141–150.

Wald, J., Quitoriano, V., Heaton, T., Kanamori, H., and Scrivner, C., TRINET Shakemaps: Rapid Generation of Peak Ground Motion and Intensity Maps for Earthquakes in Southern California, Earthquake Spectra, Oct. 1998.

Safety and Reliability – Bedford & van Gelder (eds)
© 2003 Swets & Zeitlinger, Lisse, ISBN 90 5809 551 7

Considerations regarding the maintenance management for "k from n" structures

I. Felea
University of Oradea, Romania

N. Coroiu
Electricity Distribution and Supply, Electrica Transilvania Nord, Oradea Subsidiary, Romania

Fl. Popentiu-Vladicescu
University of Oradea, Romania

ABSTRACT: This paper belongs to the domain of reliability based maintenance. It presents a method for the coordination of the maintenance steps as well as for the operational succession which produces the maximum reliability gain regarding the "k from n" structures. For the particular cases (n = k + 1 and n = k + 2) and for the cases with practical interest (k = $\overline{1,10}$), it will deduce the indicators in terms of: time interval when the reliability gain reaches the maximum value, the safety time and reliability gain at that particular moment. The paper also presents the values of indicators which are established for structures with a large feasibility, part of electric power systems.

1   PRELIMINARIES

The structures of "k from n" type are often used in power systems. There are important research activities and applications that aim at structural (in design) and functional (in exploitation) optimization of systems of "k from n" type [1, 2, 3, 5, 6, 7].

The present paper connects with the category of reliability based on maintenance theory (RCM) having as object the preventive maintenance management of components within "k from n" systems. The maximal reliability gain is the basic criterion that determines the release moment of preventive maintenance actions, which justifies the including of this kind of approach in the category of RCM type. After the release of maintenance preventive actions, these will successively be performed for the "n" considered elements (Fig. 1).

The management of the preventive maintenance actions is approached in scientific papers [4, 5, 6, 8] according to the tackled principle with reference to the couple [structure of "1 from 2" type ≡ "1 + 1"].

When solving this problem, one can admit the hypothesis that all the elements from the analyzed structure have the same reliability level. Let "R" be the safety time (the reliability function) of an element.

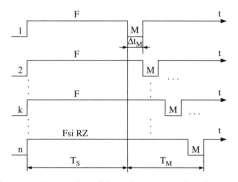

Figure 1.   Presentation of the maintenance actions for structures of "k from n" type. (F – operating; M – preventive maintenance; RZ – redundancy; T_s – the time interval within the reliability gain has an increasing evolution; Δt_M – the maintenance duration for one element; TM – total duration for preventive maintenance actions).

For this couple one can write:

- the time safety for systems with the two elements functioning (active):

$$R_2 = R^2 + 2R(1-R) = 2R - R^2 \tag{1}$$

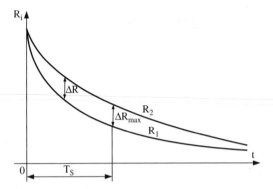

Figure 2.   Explanation of the safety time for a couple.

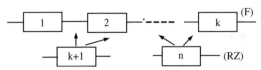

Figure 3.   ERD of "k from n" system.

The time safety for a system with "$n-j$" elements functioning (valid) can be expressed in a similar manner:

$$R_{n-j} = \sum_{i=k}^{n-j} C_{n-j}^i R^i (1-R)^{n-j-i}; j \leq n-k \qquad (6)$$

The reliability gain (time safety) between the two stages is expressed as:

$$\Delta R = R_n - R_{n-j} \qquad (7)$$

In order to determine the maximum value of the indicator "ΔR" and the value of T_s indicator, one can compute the derivative of "ΔR" with respect to "t". After some computations one obtains the following:

$$\frac{d[\Delta R(t)]}{dt} =$$

$$\frac{R'}{R} \left\{ n[R_n - R_{n-1}] - (n-j)[R_{n-j} - R_{n-j-1}] \right\} = 0 \qquad (8)$$

The analytical solving of the equation (8) is not possible even if the R indicator has the simplest expression possible, corresponding to the exponential distribution.

- the time safety for systems with only one element functioning (active)

$$R_1 = R \qquad (2)$$

- reliability gain

$$\Delta R = R_2 - R_1 = R - R^2 = R(1-R) \qquad (3)$$

The "ΔR" indicator has a time evolution of "normal characteristic" type.

In order to determine the value of the T_s interval beyond which ΔR reaches the maximum value (ΔR_{max}), one can apply the well-known mathematical procedure:

$$\frac{d(\Delta R)}{dt} = 1 - 2R \Rightarrow R(T_s) = \frac{1}{2} \qquad (4)$$

Based on this relation (4), the "T_s" indicator is analytically or numerically expressed, depending on the random variable distribution law of the random variable (t) and, therefore, on the expression of the R(t) indicator.

Starting from the above approach, one can generalize the procedure for structures of "k from n" type.

2   THE GENERAL CASE

The equivalent reliability diagram (ERD) for "k from n" system is presented in Fig 3.

The time safety for a system with "n" elements functioning can be expressed as follows:

$$R_n = \sum_{i=k}^{n} C_n^i R^i (1-R)^{n-i} \qquad (5)$$

3   PARTICULAR CASES

The "$n = k + 1$" and "$n = k + 2$" structures, which are often used in practical cases, will be analyzed in the following section of this paper. Similarly, structures with a known number of redundancy elements can be examined.

3.1   The "k from k + 1" structure

ERD is presented in Fig 4.

The realiability gain is expressed as follows:

$$\Delta R = R_n - R_k = R_{k+1} - R_k = k(1-R)R^k \qquad (9)$$

One obtains:

$$\frac{d(\Delta R)}{dt} = -kR^k + k(1-R)kR^{k-1} = 0 \qquad (10)$$

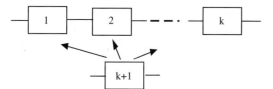

Figure 4. ERD of "k + 1" structure.

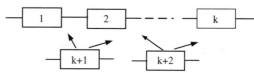

Figure 5. ERD of structure.

The time safety of an element at the moment when the reliability gain has its maximum value (beyond Ts) can be obtained using the equation (10):

$$R(T_s) = \frac{k}{k+1} \qquad (11)$$

In this case the analytical expression of T_s is:

- if the "t" variable has an exponential distribution:

$$R(T_s) = \exp(-\lambda T_s) = \frac{k}{k+1} \Rightarrow T_s = \frac{1}{\lambda}\ln\frac{k+1}{k} \qquad (12)$$

- if the "t" variable has a Weibull distribution:

$$R(T_s) = \exp\left[-\left(\frac{T_s-\gamma}{\eta}\right)^\beta\right] = \frac{k}{k+1} \Rightarrow$$

$$T_s = \gamma + \eta\left(\ln\frac{k+1}{k}\right)^{1/\beta} \qquad (13)$$

The meanings of $(\lambda, \gamma, \eta, \beta)$ parameters are those already established in [3, 5, 7].

3.2 The "k from k + 2" structure

ERD is presented in Fig 5.

Because this system has two redundant elements, the analysis can be extended (detailed) in terms of the state of one of the two redundant elements (e.g. k + 1) which can be active redundancy or passive redundancy. The other redundant element (e.g. k + 2) is supposed to be active.

3.2.1 The case when the two elements are active redundancy

In this case the reliability gain is expressed as follows:

$$\Delta R = R_n - R_{n-1} = R_{k+2} - R_{k+1} \qquad (14)$$

One obtains:

$$\Delta R = \frac{k(k+1)}{2}R^k(1-R)^2 \qquad (15)$$

Hence,

$$\frac{d(\Delta R)}{dt} = \frac{k(k+1)}{2}R^{k-1}(1-R)(k-kR-2R)=0 \qquad (16)$$

And this leads us to:

$$R(T_s) = \frac{k}{k+2} \qquad (17)$$

The analytical solutions for these two distributions are:

$$T_s = \begin{cases} \dfrac{1}{\lambda}\ln\dfrac{k+2}{k} & \rightarrow \text{ exponential distribution} \\[2ex] \gamma + \eta\left(\ln\dfrac{k+2}{k}\right)^{1/\beta} & \rightarrow \text{ Weibull distribution} \end{cases} \qquad (18)$$

3.2.2 The case when one element is passive redundancy

The reliability gain is expressed by:

$$\Delta R = R_n - R_{n-2} = R_{k+2} - R_k \qquad (19)$$

One obtains:

$$\Delta R = \frac{kR^k}{2}\left[(k+1)R^2 - 2(k+2)R + k+3\right] \qquad (20)$$

This implies that:

$$\frac{d(\Delta R)}{dt} = \left(k^2+3k+2\right)R^2 - 2\left(k^2+3k+2\right)R + k^2+3k=0 \qquad (21)$$

625

Table 1. The values and the expressions of the indicators for practical cases.

structure indicator		1	2	3	4	5	6	7	8	9	10
k from k+1	$R(T_s)$	1/2	2/3	3/4	4/5	5/6	6/7	7/8	8/9	9/10	10/11
	$\Delta R(T_s)$	1/4	8/27	$(3/4)^4$	$(4/5)^5$	$(5/6)^6$	$(6/7)^7$	$(7/8)^8$	$(8/9)^9$	$(9/10)^{10}$	$(10/11)^{11}$
	T_s E	$\frac{1}{\lambda}\ln 2$	$\frac{1}{\lambda}\ln\frac{3}{2}$	$\frac{1}{\lambda}\ln\frac{4}{3}$	$\frac{1}{\lambda}\ln\frac{5}{4}$	$\frac{1}{\lambda}\ln\frac{6}{5}$	$\frac{1}{\lambda}\ln\frac{7}{6}$	$\frac{1}{\lambda}\ln\frac{8}{7}$	$\frac{1}{\lambda}\ln\frac{9}{8}$	$\frac{1}{\lambda}\ln\frac{10}{9}$	$\frac{1}{\lambda}\ln\frac{11}{10}$
	T_s W $(\gamma=0)$	$\eta(\ln 2)^{1/\beta}$	$\eta\left(\ln\frac{3}{2}\right)^{1/\beta}$	$\eta\left(\ln\frac{4}{3}\right)^{1/\beta}$	$\eta\left(\ln\frac{5}{4}\right)^{1/\beta}$	$\eta\left(\ln\frac{6}{5}\right)^{1/\beta}$	$\eta\left(\ln\frac{7}{6}\right)^{1/\beta}$	$\eta\left(\ln\frac{8}{7}\right)^{1/\beta}$	$\eta\left(\ln\frac{9}{8}\right)^{1/\beta}$	$\eta\left(\ln\frac{10}{9}\right)^{1/\beta}$	$\eta\left(\ln\frac{11}{10}\right)^{1/\beta}$
k from k+2 and 2 RA (3.2.1)	$R(T_s)$	1/3	1/2	3/5	2/3	5/7	3/4	7/9	4/5	9/11	5/6
	$\Delta R(T_s)$	$4/3(1/3)^3$	$3/2(1/2)^3$	$8/5(3/5)^4$	$5/3(2/3)^5$	$12/7(5/7)^6$	$7/4(3/4)^7$	$16/9(7/9)^8$	$9/5(4/5)^9$	$20/11(9/11)^0$	$11/6(5/6)^{11}$
	T_s E	$\frac{1}{\lambda}\ln 3$	$\frac{1}{\lambda}\ln 2$	$\frac{1}{\lambda}\ln\frac{5}{3}$	$\frac{1}{\lambda}\ln\frac{3}{2}$	$\frac{1}{\lambda}\ln\frac{7}{5}$	$\frac{1}{\lambda}\ln\frac{4}{3}$	$\frac{1}{\lambda}\ln\frac{9}{7}$	$\frac{1}{\lambda}\ln\frac{5}{4}$	$\frac{1}{\lambda}\ln\frac{11}{9}$	$\frac{1}{\lambda}\ln\frac{6}{5}$
	T_s W $(\gamma=0)$	$\eta(\ln 3)^{1/\beta}$	$\eta(\ln 2)^{1/\beta}$	$\eta\left(\ln\frac{5}{3}\right)^{1/\beta}$	$\eta\left(\ln\frac{3}{2}\right)^{1/\beta}$	$\eta\left(\ln\frac{7}{5}\right)^{1/\beta}$	$\eta\left(\ln\frac{4}{3}\right)^{1/\beta}$	$\eta\left(\ln\frac{9}{7}\right)^{1/\beta}$	$\eta\left(\ln\frac{5}{4}\right)^{1/\beta}$	$\eta\left(\ln\frac{11}{9}\right)^{1/\beta}$	$\eta\left(\ln\frac{6}{5}\right)^{1/\beta}$

Table 2. The values of the T_s indicator for some equipment from Electricity Distribution and Supply, Electrica Transilvania Nord, Oradea Subsidiary, Romania structure.

structure equipment		1	2	3	4	5	6	7	8	9	10
k from k+1	TP$_1$ (W) [years]	3,5	1,5	0,84	0,56	0,39	0,28	0,23	0,19	0,16	0,13
	TP$_2$ (E) [hours]	9191	5377	3782	2959	2418	1968	1737	1562	1384	1264
	S$_{HV}$(E) [hours]	10289	6024	4237	3305	2707	2194	1945	1748	1549	1415
	S$_{MV}$(E) [hours]	6482	3792	2667	2087	1705	1388	1225	1101	976	891
k from k+2 and 2RA	TP$_1$ (W) [years]	7,55	3,53	2,12	1,4	1	0,83	0,66	0,54	0,46	0,39
	TP$_2$ (E) [hours]	14570	9192	6774	5377	4462	3815	3333	2959	2661	2418
	S$_{HV}$(E) [hours]	16311	10291	7584	6020	4995	4271	3731	3313	2979	2707
	S$_{MV}$(E) [hours]	10277	6484	4778	3792	3147	2691	2350	2087	1877	1705
k from k+2 and RA+RP	TP$_1$ (W) [years]	5,04	2,23	1,31	0,88	0,64	0,48	0,39	0,32	0,26	0,23
	TP$_2$ (E) [hours]	30843	6953	5919	3952	3261	2762	2407	2140	1908	1741
	S$_{HV}$(E) [hours]	34521	7780	6622	4425	3638	3088	2688	2405	2123	1945
	S$_{MV}$(E) [hours]	21749	4901	4172	2788	2292	1946	1693	1515	1338	1225

TP$_1$ – power transformer and autotransformer with S$_n$ ∈ [1;200] MVA; TP$_2$ – power transformer and autotransformer with S$_n$ ∈ [25;800] kVA; S$_{HV}$ – switchers with 110 kV; S$_{MV}$ – medium voltage switcher.

626

The solution of the equation (20) is:

$$R(T_s) = 1 - \sqrt{\frac{2}{k^2 + 3k + 2}} \qquad (22)$$

For the T_s indicator one obtains the expressions:

$$T_s = \begin{cases} \dfrac{1}{\lambda} \ln \dfrac{1}{R(T_s)} & \to \text{exponential distribution} \\[2ex] \left[\gamma + \eta \left[\ln \dfrac{1}{R(T_s)} \right] \right]^{1/\beta} & \to \text{Weibull distribution} \end{cases}$$

$$(23)$$

4 CASES OF PRACTICAL INTEREST

Based on the analytical expressions of the $[R(T_s)$, $\Delta R(T_s)$ and $T_s]$ indicators previously obtained, one can numerically express the indicators with regards to practical cases. One can see that, in practical cases, we often meet situations when $k \in \{1, 2, 3, 4, 5\}$, while situations in which $k \in \{6, 7, 8, 9, 10\}$ are quite rare.

Table 1 contains the values and expressions of the presented indicators for $k \in \overline{1,10}$.

For some electrical equipment from SEE Bihor structure, often used in "k from n" structures, the numerical values of the T_s indicator have been determined. For evaluation, one used the distribution functions (exponential, Weibull) which gave the best approximation of the empirical distribution, and the values of the parameters of the deduced and registered in [5] functions. The obtained values are written in Table 2.

5 CONCLUSIONS

An essential aspect of the maintenance management of "k from n" structures is that of identification of the releasing time of the maintenance actions, so as to the reliability gain is maximum. Such an approach belongs to the reliability based maintenance strategy.

The main indicators for the maintenance management of "k from n" systems, according to the presented criterion, are: the value of the time interval beyond which the reliability gain reaches its maximum value

(T_s), time safety $[R(T_s)]$ and the reliability gain $[\Delta R(T_s)]$ at time T_s.

The analytical expression of $[(T_s), R(T_s), \Delta R(T_s)]$ indicators is not possible in the general case (k from n), but only in particular cases often met in practical situations (e.g. k from k + 1, k from k + 2).

The analytical and numerical expression of $[(T_s), R(T_s), \Delta R(T_s)]$ indicators for cases of practical interest $(k \in \overline{1,10})$ and for large scale electrical equipment (table 1 and 2), shows the fact that:

- for all analyzed structures, the $R(T_s)$ and $\Delta R(T_s)$ indicators nonlinearly increase with respect to k (the indicators increase while the redundancy degree decreases).
- for all analyzed structures and equipment, the T_s indicator nonlinearly decreases with respect to k (it decreases while the redundancy degree decreases as well).
- Ş the values of $R(T_s)$ and $\Delta R(T_s)$ indicators depend only on the structure type and not on the reliability features of the equipment.
- the values of T_s indicator depend on the structure type and on the reliability features of the equipment as well.

BIBLIOGRAPHY

1. Billinton R., Allan R.N. Reliability Evaluation of Engineering Systems, PLENUM PRESS, New York and London, 1990.
2. Cătuneanu V., Popenţiu F. Optimization of Systems Reliability, Ed. Academiei, Bucharest, 1989.
3. Felea I. Reliability Engineering in Electric Power Engineering, EDP. Bucharest, 1996.
4. Felea I. a.s.o. Considerations Regarding the Implementation of Reliability Based Maintenance System of S.C. ELECTRICA S.A. Subunits with Applications at S.D. Oradea, Rev. Energetica nr.3., 2000.
5. Felea I., Coroiu N. Reliability and Maintenance of Electrical Maintenance, Technical Ed., Bucharest, 2001.
6. Monbray I. Reliability Centred Maintenance, Butterwarth, Heinemann, 1991.
7. Nitu V.I., Ionescu C. Reliability in Electric Power Engineering, EDP, Bucharest, 1980.
8. Stein M., ş.a. Contributions to RCM Application in Power Engineering Installations, Rev. Energetica, nr. 4, seria A, 1996.

Safety and Reliability – Bedford & van Gelder (eds)
© *2003 Swets & Zeitlinger, Lisse, ISBN 90 5809 551 7*

On some reliability approaches to human aging

M.S. Finkelstein
University of the Free State, Bloemfontein, South Africa

ABSTRACT: Two simple probabilistic models of biological aging are considered. The first one is based on the assumption that some random resource is acquired by an organism at birth. Death occurs when the accumulated wear exceeds the initial random resource. In the second model a nonhomogeneous Poisson and doubly stochastic Poisson processes of harmful events are considered. Each event results in death with a given probability or the damage caused by this event can be minimally repaired with a complementary probability. Similar to some demographic models, the mortality deceleration and subsequent leveling-off phenomenon is explained via the concept of frailty. Simple examples are considered.

1 INTRODUCTION

The literature on biological theories of aging is quite extensive. Various stochastic mortality models are reviewed, for instance, in Yashin *et al.* (2000). The nature of human aging is in some "biological wearing". There are different probabilistic ways to model wear. The simplest way of modeling is just to describe the corresponding lifetime random variable T by its distribution function (DF) $F(t)$. One of the standard demographic models is the Gompertz model for human mortality with exponentially increasing hazard rate $\lambda(t)$:

$$F(t) = 1 - \exp\left\{-\frac{\alpha}{\beta}[\exp\{\beta t\} - 1]\right\}$$

$$\lambda(t) = \alpha \exp\{\beta t\}, \quad \alpha > 0, \beta > 0. \tag{1}$$

The Weibull DF can be also used for this purpose:

$$F(t) = 1 - \exp\left\{-(\alpha t)^{\beta}\right\}$$

$$\lambda(t) = \alpha\beta(\alpha t)^{\beta - 1}, \quad \alpha > 0, \beta > 1. \tag{2}$$

It is well-known that along with mortality increase with age the subsequent mortality leveling-off takes place. Different explanations of this phenomenon are suggested in the literature. In the recent paper by Gavrilov and Gavrilova (2001) the following reliability model of human aging was discussed. It was assumed

that organisms consist of a series structure of redundant blocks. Each block in its turn consists of N initially "operable" components (in parallel) with a constant hazard rate λ. It was proved that when N is deterministic the shape of the corresponding hazard rate for the sufficiently small t can be approximately described by the power law (2). The authors considered Poisson and binomial distributions for describing the random redundancy. It was stated that for both cases the resulting hazard rate exponentially increases (as in (1)) for sufficiently small t and as $t \rightarrow \infty$, it asymptotically converges to λ from below. The latter fact is obvious, because as $t \rightarrow \infty$ (on condition that the block is operable) probability that only one component is operable tends to 1. It can be shown that the first statement is the consequence of the Poisson assumption (and of the Poisson approximation for the binomial distribution). It is easy to prove that the shape of $\lambda(t)$ can differ from exponential for other types of redundancy distributions. This problem can be viewed in a more general way like a problem of characterizing the hazard rate of objects with unknown (random) initial age. It is shown in Finkelstein (2002) that under certain assumptions the resulting hazard rate is initially decreasing even if the hazard rate of the baseline distribution is increasing.

The approach, based on some initial resource is discussed in Section 2. In Sections 3 and 4 the human mortality is modeled by the Poisson (doubly stochastic Poisson) process of "killing events". We also consider the model, when each event with a given probability can be "cured", which describes in some way the repair capacity of an organism. The models to be considered are probably oversimplified from the biological point of view, but they result in the general pattern of human mortality and describe the shape of the mortality rate function. In what follows we mostly use the conventional reliability terminology: a failure of an object corresponds to its death and the failure (hazard) rate, as usually, is the synonym for the mortality rate (the force of mortality).

2   MODELS BASED ON RESOURCE

Assume that in the process of production (engineering applications) or birth (biological applications) an object under consideration had acquired an initial unobserved random resource R with a DF $F_0(r)$: $F_0(r) = P(R \leq r)$. Suppose that for each realization of R the run out resource $W(t)$ (deterministic) to be called wear monotonically increases. Thus, the wear increment in $[t, t + dt)$ is defined as $w(t) + o(dt)$. Let additionally $W(0) = 0$ and $W(t) \to \infty$ as $t \to \infty$. Under these assumptions we arrive at the well-known in reliability theory accelerated life model:

$$P(T \leq t) \equiv F(t) = F_0(W(t)) \equiv P(R \leq W(t))$$

$$W(t) = \int_0^t w(u)du; \quad w(t) > 0; \quad t \in [0, \infty). \tag{3}$$

As follows from (3), the failure (death) occurs when the wear $W(t)$ reaches the random R. This equation describes a *deterministic* diffusion with a random threshold. It is natural to model a *random* wear, which is the case in reality, by a monotonically increasing stochastic process $W_t, t \geq 0$. Some appropriate models for $W_t, t \geq 0$ can be found in Lemoine and Wenocur, (1985) and Singpurwalla (1995).

Substituting the deterministic wear $W(t)$ in (3) by the increasing stochastic process $W_t, t \geq 0$ leads to the following relation (Finkelstein (2003)):

$$F(t) = P(T \leq t) = P(R \leq W_t) = E[F_0(W_t)], \tag{4}$$

where the expectation is taken with respect to $W_t, t \geq 0$. The DF $F_0(W_t)$ should be understood conditionally:

$$F(t|W_t) = P(T \leq t|W_t, 0 \leq u \leq t),$$

where $W_t, 0 \leq u \leq t$ is the history of this stochastic process.

Let, as previously, $\lambda(t)$ denotes the hazard rate (the *observed* hazard rate), which corresponds to the DF $F(t)$. Now consider firstly the case when the DF $F_0(t)$ is absolutely continuous. From relation (4), using

results of Yashin and Manton (1997) the following important relationship between the observed and the conditional hazard rates can be obtained:

$$\lambda(t) = E[w_t \lambda_0(W_t)|T > t], \tag{5}$$

where w_t denotes the stochastic rate of diffusion: $dW_t \equiv w_t dt$ and the hazard rate $\lambda_0(t)$ is defined by the DF $F_0(t)$. Equation (5) can be used for analyzing the shape of $\lambda(t)$, and this is very important in demographic studies, for instance. Specifically, it can show the deceleration in mortality for sufficiently large t (oldest-old mortality), as opposed to the exponential or power functions defined for $t \in [0, \infty]$ by (1) and (2) (see also examples 1 and 3 in Finkelstein and Esaulova (2001)). It is clear that a random initial age of an object: $R \equiv T_R$ is a specific case of the initial resource. As it was mentioned in Section 1, under certain assumptions the resulting hazard rate can initially decrease even if the hazard rate of the governing distribution (defined by $T_R = 0$) is increasing in accordance with the Weibull DF (2) (Finkelstein (2003)). The random variable T_R in this model can be also interpreted as some initial virtual age of an object.

Let $F_0(n) \equiv P(N \leq n)$ be now a discrete distribution A specific case was considered in Gavrilov and Gavrilova (2001), where $R = N$ is a random number of initially ($t = 0$) operable components (in parallel) with a constant hazard rate λ. The degradation process $W_t, t \geq 0$ for this setting is just a counting process for the corresponding process of pure death: when the number of events (failures of components) reaches $N = n$, the death occurs. Denote by $\lambda_n(t)$ the hazard rate for the time to death random variable for the fixed $N = n$, $n = 1, 2, \ldots$ ($n = 0$ is excluded, as there should be operable components at $t = 0$). Similar to (5) the mixture (observed) hazard rate is (see also Finkelstein and Esaulova (2001)):

$$\lambda(t) = E[\lambda_n(t)|T \geq t]. \tag{6}$$

Thus, due to this fact:

$$\lambda(t) \neq E[\lambda_n(t)] = \sum_{n=1}^{\infty} P_n \lambda_n(t), \tag{7}$$

where $P_n \equiv P(N = n)$, and the approximate equality takes place only for the sufficiently small t as in Gavrilov and Gavrilova (2001). Therefore, as follows from (7), $\lambda(t)$ is not necessarily approximately exponential (as it is, for instance, for the truncated Poisson distribution $F_0(n)$ when t is sufficiently small). To say more, similar to the continuous case, if the variance of N is sufficiently large, then $\lambda(t)$ can even initially decrease (Finkelstein 2002). Thus, the simple reliability "series-parallel system" model should be carefully specified in order to describe the human mortality.

3 MODELS BASED ON REPAIR CAPACITY

According to a number of authors (see Yashin *et al.* (2000) for references) the DNA repair capacity can be responsible for aging of humans. One can assume that in the absence of the proper repair spontaneous DNA mutation leads to the death of an organism (Yashin *et al.* (2000)). Thus, the repair mechanism can constitute the basis for the mortality modeling.

Let $P_t, t \geq 0$ denotes the nonhomogeneous Poisson process of harmful events with rate $\mu(t)$, which potentially can lead to death (diseases, disorders etc.). Another interpretation (Yashin *et al.* (2000)) is that $\mu(t)$ is the mortality rate in a given population without possibility of repair. If we assume, as in Vaupel and Yashin (1987), the possibility of n minimal repairs, then the new "better" mortality curve is obviously given by

$$\lambda_n(t) = \mu(t) \frac{M^n(t)}{n!\left(1 + M(t) + \frac{M^2(t)}{2} + ... + \frac{M^n(t)}{n!}\right)},$$

where $M(t) = \int_0^t \mu(x)\,dx$ and the minimal repair is understood in the usual reliability sense: it restores an object to a state it had immediately prior to the harmful event. If, on the other hand, in the line with the previous section we assume that the number of "new lives" is a random variable N, then relation (6) is also valid and the corresponding considerations on the shape of $\lambda(t)$ as well.

To make the model more realistic consider the following imperfect repair setting. Let the number of possible minimal repairs be unlimited but each event from the process $P_t, t \geq 0$ is minimally repaired with probability $1 - \theta(t)$ and is not repaired (thus resulting in death) with probability $\theta(t)$; $\theta(0) > 0$. Therefore, the function of age $\theta(t)$ describes the repair capacity of an object. It follows from Block *et al.* for instance, that the corresponding lifetime DF and the hazard rate are given in this case by

$$F(t) = 1 - \exp\left\{-\int_0^t \theta(u)\mu(u)\,du\right\}, \lambda(t) = \theta(t)\mu(t)$$
(8)

respectively. It is reasonable to assume that, as $\theta(t)$ describes the ability of the organism to perform a proper repair (or equivalently, to resist harmful events from $P_t, t \geq 0$), it should increase with t showing

deterioration, which reflects the corresponding aging. Due to the fact that the function $\theta(t)$ is bounded by 1 as t increases, it inputs in the mortality deceleration for sufficiently large t (as it will be discussed late, this factor is not the main which causes the deceleration).

Let $\theta(t) = \theta$ be constant in time for simplicity and assume that it is a random variable (independent of $P_t, t \geq 0$) with support in (0,1). In this way the population heterogeneity is introduced in the model. It follows from Yashin and Manton (1997) that, as relations (8) are valid conditionally on realizations of θ, the following formulas take place

$$F(t) = 1 - E\left[\exp\left\{-\theta \int_0^t \mu(u)\,du\right\}\right],$$
(9)

$$\lambda(t) = \mu(t)E[\theta \,|\, T \geq t].$$
(10)

Equations (9) and (10) define actually the mixture DF and the mixture hazard rate, respectively. The shape of $\lambda(t)$ in (9) (for the arbitrary continuous increasing function $\mu(t)$) can be already different from the shape of $\mu(t)$: it can even decrease for sufficiently large t (see later).

Another and maybe much more important source of heterogeneity in the model can be introduced by considering instead of the Poisson process of harmful events the corresponding doubly stochastic Poisson process $\hat{P}_t, t \geq 0$ (Cox and Isham (1980)) with a rate $\mu(t, \Psi)$, indexed by a random variable Ψ with support in $(0, \infty)$. Thus for the fixed $\Psi = \psi$ relations (8) are valid:

$$F(t \,|\, \psi) = 1 - \exp\left\{-\int_0^t \theta(u)\mu(u, \psi)\,du\right\},$$

$$\lambda(t) = \theta(t)\mu(t, \psi),$$
(11)

and similar to (9) and (10):

$$F(t) = 1 - E\left[\exp\left\{-\int_0^t \theta(u)\mu(u, \Psi)\,du\right\}\right],$$
(12)

$$\lambda(t) = \theta(t)E[\mu(t, \Psi) \,|\, T > t].$$
(13)

The existence of the exponential representation (11) in realizations (along with some minor assumptions) is a sufficient condition for (12) and (13). There can be different models for $\mu(t, \Psi)$, the multiplicative one being the simplest and the most appealing one: $\mu(t, \Psi) = \Psi\mu(t)$, where $\mu(t)$, as usually in the proportional hazards-type models, plays the role of a baseline (reference) hazard rate. Then, similar to (10), relation (13) turns to:

$$\lambda(t) = \theta(t)\mu(t)E[\Psi \mid T > t].\tag{14}$$

If, for instance $\mu(t)$ is an increasing failure rate Weibull DF and Ψ has a gamma distribution, then the function $\mu(t)E[\Psi \mid T > t]$ initially increases, reaches a single maximum and then decreases asymptotically to 0 as $t \to \infty$, thus exhibiting the dramatically different from $\mu(t)$ shape of the observed hazard rate $\lambda(t)$ (Finkelstein and Esaulova (2001)). Inclusion of the random θ in this model will contribute additionally to the heterogeneity:

$$\lambda(t) = \mu(t)E[\theta\Psi \mid T > t] \neq \mu(t)E[\theta \mid T > t]E[\Psi \mid T > t]$$

but the main effect on the shape of λ is caused in by Ψ.

It is worth noting that the doubly stochastic Poisson process effectively models the diversity among subpopulations (or on the individual level as well) in the rate of harmful events. Lifestyle, external factors, hereditary factors etc can define the corresponding covariates. The other noteworthy consideration is that inspite of the non-smooth shock-type influence of harmful events on lifetime variable T, the exponential-type formulas (9) and (12) take place. It can be shown, however, that, if for instance, a simplest dependence on history in the probability of failure (e.g., $\theta(t, z)$, where z is the time since the last survived event) is taken into account, then exponential formulas collapse and other methods should be used for analyzing $\lambda(t)$.

4 REPAIR CAPACITY DEPENDENT ON WEAR

Consider the stochastic process $(W_t, \hat{P}_t), t \geq 0$ where $W_t, t \geq 0$ is an increasing (tending to infinity) stochastic process of wear defined in Section 2 (right continuous with left-hand limits) and $\hat{P}_t, t \geq 0$ (independent of $W_t, \hat{P}_t), t \geq 0$) is a doubly stochastic Poisson process of the specific multiplicative form defined in Section 3. The process $(W_t, \hat{P}_t), t \geq 0$ is assumed to be unobserved in the sense that at time t only the information that our object is dead or alive is available. In the previous section the repair capacity $1 - \theta(t)$ was a time dependent decreasing function showing the deterioration of this ability with time. It is reasonable to assume that in reality it depends not directly on the chronological age of an object t but on the accumulated wear that can be actually considered as a characteristic of some "virtual age". Therefore, this model will be based on the wear dependent probability $\theta(W_t)$.

Though $W_t, t \geq 0$ is formally an internal process (Kalbfleisch and Prentice (1980)) the lifetime variable

T is not a stopping time for it and we can proceed with justification of the corresponding exponential-type formula for the survival function. The conditional hazard is defined in a following way

$$\lambda(u \mid w(u), 0 \leq u \leq t, \psi)$$
$$= \lim_{\Delta t \to 0} \frac{\Pr(T \in [t, t + \Delta t) \mid w(u), 0 \leq u \leq t, \psi; T \geq t)}{\Delta t}$$
$$= \theta(w(t)\psi\mu(t),\tag{15}$$

where $w(u), 0 \leq u \leq t$ is a realization of $W_u, 0 \leq u \leq t$ and ψ is a realization of Ψ for the multiplicative model $\mu(t, \Psi) = \Psi\mu(t)$. In this specific case the conditional hazard (15) constitutes the corresponding hazard rate process. It follows from (8) (and similar to (11)) that the exponential formula takes place for realizations of this hazard rate process. Therefore, as in the previous section, the conditional DF can be also defined as

$$F(t \mid w(t), \psi) = 1 - \exp\left\{-\psi \int_0^t \theta(w(u))\mu(u)du\right\},$$
$$\tag{16}$$

while the DF of T and the observed hazard rate $\lambda(t)$ are defined similar to (12) and (13), respectively:

$$F(t) = 1 - E\left[\exp\left\{-\Psi \int_0^t \theta(W_u)\mu(u)du\right\}\right],\tag{17}$$

$$\lambda(t) = E[\Psi\theta(W_t)\mu(t) \mid T > t],\tag{18}$$

where the operation of expectation is taken with respect to $W_t, t \geq 0$ and Ψ.

Usually one cannot perform an explicit integration in (17) and (18) and the corresponding numerical methods should be used, but the general analysis of the shape of the observed hazard rate can be still performed at least asymptotically. Assume firstly, that there is no heterogeneity caused by Ψ and harmful events occur in accordance with a nonhomogeneous Poisson process with rate $\mu(t)$. Then, as $\theta(t)$ is an increasing function and $\theta(t) \to 1$ as $t \to \infty$, it is clear that:

$$\lambda(t) - \mu(t) \to 0 \quad \text{as } t \to 0,$$

which means that the baseline and the observed hazard rates are asymptotically close and the effect of a random wear on the shape of $\lambda(t)$ can be seen only for smaller t. On the other hand, assuming deterministic $\theta(t)$, we arrive at (14), where the influence of the random Ψ plays the pivotal role in analyzing the shape of

$\lambda(t)$. Another illustration of this fact gives the following example:

Example. Let the baseline hazard rate be defined by the Gomperts model (1) (sharply increasing!) and let Ψ for simplicity be exponentially distributed with parameter ϑ and $\beta = 1$ in (1). Then integration in (18) gives (Finkelstein and Esaulova (2001)):

$$\lambda(t) = \theta(t)\left(1 + \frac{\alpha - \vartheta}{\alpha \exp\{t\} - \alpha + \vartheta}\right) \equiv \theta(t)\chi(t). \quad (19)$$

The initial value of the hazard rate is: $\lambda(0) = \theta(0)\alpha/\vartheta$. When $\alpha > \vartheta$, the function $\chi(t)$ is monotonically decreasing asymptotically converging to 1. Thus, depending on the shape of increasing $\theta(t)$, the hazard rate can be either also decreasing, or non-monotone, but converging to 1 in both cases. It is worth noting that condition $\alpha > \vartheta$ means that for the possible decreasing of $\lambda(t)$ it is necessary that the variance of the mixing distribution (DF of Ψ with the variance $1/\vartheta^2$) should be sufficiently large. This fact for a different setting was mentioned in Section 2.

When $k < \vartheta$ the function $\chi(t)$ is monotonically increasing asymptotically converging to 1 and so does the hazard rate $\lambda(t)$.

Thus, the random parameter Ψ under some reasonable assumptions on parameters of the model (17)-(18) and for the sufficiently large t actually defines the shape of the hazard rate $\lambda(t)$. As it was mentioned, this shape can dramatically differ from the shape of the baseline hazard rate $\mu(t)$. The decreasing of $\lambda(t)$ or its levelling-off can start at the non-attainable by humans age but the deceleration of mortality (compared with conventional models (1) and (2)) can be explained by our model for the realistic life span (see also Thatcher (1999).

5 DISCUSSION

The purpose of this paper is to suggest some reasonable probabilistic models, which could account for certain properties of human aging. Specifically, we are interested in the shape of the hazard rate (force of mortality). The conventional models (1) and (2) describe the sharp increase in mortality rates. However, as it was mentioned, the decline for oldest-old mortality was observed and the explanation of this fact is vital for modern demography and related sciences.

We explain this decline via several reliability-based models. The common feature of these models is the presence of unobserved heterogeneity, which seems to be realistic and reasonable assumption. The notion of heterogeneity in demographic models was considered before, but our approach differs from the conventional one, as it considers heterogeneity as a

part of the probabilistic model and not only as a tool for describing the final random variable of time to failure (death). The true nature of the decline in mortality is still hidden by nature, but as far as the model can explain certain effects, it has the right to exist. The validation of suggested models will be possible when the sufficient biological knowledge on mechanisms of aging and repair will be acquired. We hope that it will be soon, as the studies in this field are very intensive. In the rest of this section the considered above specific models will be discussed.

The model based on a Poisson (doubly stochastic Poisson) process of harmful events and the corresponding repair capacity is probably oversimplified from the biological point of view, but it gives a simple explanation of the deceleration in population aging by frailty in individual (or subpopulation) random factors. It is worth noting that the repair capacity can be also viewed as a characteristic of some biological anti-aging mechanism. Thus the deceleration in the oldest-old mortality in our model (like in some other models in the literature) is the consequence of the population heterogeneity.

The model based on the initial resource shows that even when the baseline hazard rate is increasing the population hazard rate can initially decrease. This interesting fact should be certainly taken into account in mortality studies. It is clear that both models can be easily combined into one approach.

The reasonable assumption of the minimal repair when, if repaired, the organism returns to the state (the state in our case is defined by the accumulated wear W_t, for instance) it had immediately prior to the harmful event, can be modified by taking into account some additional random wear V (independent of W_t, $t \geq 0$) as the result of the performed repair. It can be shown, however, that in this case simple exponential-type formulas similar to (12) and (17) are not valid, as the harmful events from the Poisson process do not have "a smooth influence" on the corresponding conditional survival functions in this specific case As the major impact on the shape of the hazard rate is from the heterogeneity factor Ψ, formulas (17) and (18) can be still used as approximate relations for this setting.

REFERENCES

Block, H.W., Borges,W., and Savits, T.H. Age-dependent minimal repair. *J. Appl. Prob.*, 22: 370–386.

Finkelstein, M.S., and Esaulova, V. 2001. Modeling a failure rate for a mixture of distribution functions, *Probability in the Engineering and Informational Sciences*, 15: 383–400.

Finkelstein, M.S. 2002. Modeling lifetimes with unknown initial age. *Reliability Engineering and System Safety*, 76: 75–80.

Finkelstein, M.S. 2003. A model of biological aging and the shape of the observed hazard rate. *Lifetime Data Analysis,* 1

Gavrilov, V. A., and Gavrilova, N.S. 2001. The reliability theory of aging and longevity. *Journal of Theoretical Biology,* 213: 527–545.

Kalbfleisch, J.D., and Prentice, R.L. 1980. *The Statistical Analyses of Failure time Data*: John Wiley.

Lemoine, A.J., and Wenocur, M.L. 1985. On failure modeling. *Naval Res. Log. Quart,* 32: 479–508.

Singpurwalla, N.D. 1995. Survival in dynamic environment. *Statistical Science,* 10: 86–108.

Thatcher, A.R. 1999. The long-term pattern of adult mortality and the highest attained age. *J. R. Statist. Society*. A, 162: 5–43.

Vaupel, J.W., and Yashin, A.I. 1987. Repeated resuscitation: how life saving alters life tables. *Demography*, 4: 123–135.

Yashin, A.I., and Manton, K.G. 1997. Effects of unobserved and partially observed covariate processes on system failure: a review of models and estimation strategies. *Statistical Science,* 12: 20–34.

Yashin, A.I., Iachin, I., and Begun, A.S. 2000. Mortality modeling: a review. *Mathematical Population Studies,* 8: 305–332.

Safety and Reliability – Bedford & van Gelder (eds)
© 2003 Swets & Zeitlinger, Lisse, ISBN 90 5809 551 7

Sensitivity studies for a probabilistic analysis of spacecraft random reentry disassembly

M.V. Frank
Safety Factor Associates, Inc., Encinitas, CA, USA

M.A. Weaver and R.L. Baker
The Aerospace Corporation, El Segundo, CA, USA

ABSTRACT: Sensitivity studies using a probabilistic paradigm for simulation of spacecraft random reentry disassembly have shown the importance of including epistemic uncertainties. The sensitivity of casualty area to spacecraft orientation and consequent free-flying ballistic element orientation is extreme. Casualty area standards for spacecraft that rely on deterministic estimates may not be providing the level of safety assurance that was intended.

1   INTRODUCTION

Reentering spacecraft break-up or disassemble by aero-thermal heating and aerodynamic stresses. Random reentry is the term used for orbiting objects that do not have the ability to de-orbit on a controlled trajectory that avoids populated areas. Controlled reentry is the term used when there is an attempt to direct the reentry such that the bulk of the spacecraft will land in water or known unpopulated land areas. This paper is concerned with random reentry. Much of a spacecraft's mass will burn-up or ablate in the atmosphere. The surviving pieces might strike people on the ground. A measure of surviving pieces used for randomly reentering spacecraft is called casualty area and spacecraft design standards that limit allowed casualty area have been promulgated in the United States (Gregory, 1995; OSTP, 2000). Casualty area may be thought of as the sum of the union of the areas presented by surviving pieces of reentering space vehicles with the target area that people present on the ground. The standards assume that casualty area of a randomly reentering spacecraft can be calculated relatively precisely. This is an assumption that we dispute and this paper provides evidence to support our claim that such calculations provide results that have significant uncertainty. Perhaps a contributing factor to the notion that deterministic standards are acceptable is that all prior calculation paradigms have been

deterministic (e.g. Refling, 1992; Montgomery, 1995; Fritsche, 1999; Lips, 2002; Rochelle, 2002).

Previously, the authors (Weaver, 2001; Frank, 2001) introduced a new probabilistic paradigm for the calculation of atmospheric breakup of randomly reentering spacecraft. We indicated that the detailed physical processes involved in spacecraft heating, disassembly and burn-up are not understood well enough for precise predictions to be made. Calculations, therefore, involve uncertainties in the models as well as in the physical properties and model parameters. These uncertainties are reflected in a large variation over casualty area. Our paradigm simulated thousands of trajectories, each with a different randomly selected set of variables and parameters, in order to calculate the uncertainty in casualty area and other intermediate outcomes such as disassembly altitude.

This work is concerned with attempting to characterize the effects of uncertainty on the calculated spacecraft disassembly process. This paper reports on sensitivity studies completed to understand the importance of the various uncertain parameters in the model for the purpose of focusing future efforts on improved modeling of the most sensitive parameters. As in our previous work, the disassembly of a Delta II second stage, focusing on the stainless steel propellant tank is used as the exemplar reentering spacecraft because propellant tanks have been recovered after reentry.

2 PROBABILISTIC APPROACH

We are interested in calculating the uncertainty in the breakup of reentering spacecraft, as measured by casualty area. The uncertainty is caused by inexact modeling capability, parameter variability, and parameter state-of-knowledge uncertainties. We recap the probabilistic paradigm introduced in Weaver, 2001 and Frank, 2001 in order to clearly explain the context of the uncertainties.

In essence, the approach attempts to replace assumptions with probability distributions that represent the range of knowledge and the inherent stochastic variability of the relevant phenomena. In this manner, the uncertainty in the range of possible casualty areas for a spacecraft consistent with current knowledge is revealed.

Simplified physical models are utilized to highlight the probabilistic aspects of the methodology. Reentry trajectories are described by the equations of particle motion over a spherical, non-rotating planet. A lumped-mass body temperature model provides for reentry heating.

The calculation begins with the initial velocity and angle of reentry of a space vehicle. Prediction of these quantities from randomly reentering orbital space vehicles is reasonably well understood. As with all previous reentry breakup algorithms, the deterministic application of the following equations assumes a spacecraft orientation (i.e. attitude). Later in this paper, we investigate the effects of accounting for orientation uncertainties. Given such initial conditions, the following describes the simplified set of heating, ablation and trajectory equations.

2.1 Heating

Temperature change for a lumped mass with convective heating and radiative cooling (for each lump) is governed by:

$$\frac{dT}{dt} = \frac{A}{mc_p}\left[B_2h(T_0 - T) - \varepsilon\sigma T^4\right] \tag{1}$$

where,

T = lump temperature
t = time
A = heating (wetted) area
m = mass
c_p = specific heat
B_2 = shape or shielding factor ($0 < B_2 \geqslant 1$)
h = heat transfer coefficient
T_0 = air total temperature

ε = emissivity
σ = Stefan-Boltzmann constant

The heat transfer coefficient for the convection term is obtained from the Detra-Kemp-Riddell correlation (Detra, 1957) which is valid only for hypersonic flow:

$$h = \frac{D}{\sqrt{R}(T_0 - 300K)}\left(\frac{\rho}{\rho_{ref}}\right)^{0.5}\left(\frac{V}{V_{ref}}\right)^{3.15} \tag{2}$$

where,

D = Detra-Kemp-Riddell coefficient
R = heating radius
ρ = free-stream mass density
ρ_{ref} = reference mass density
V = free-stream velocity
V_{ref} = reference velocity

2.2 Ablation

Once the material reaches melt temperature, T_m, mass change dM is determined by the material heat of fusion, h_{if}, according to:

$$dM = -\frac{\delta Q}{h_{if}} \leq 0 \tag{3}$$

where,

$$\frac{\delta Q}{dt} = A\left[B_2h(T_0 - T_m) - \varepsilon\sigma T_m^4\right] \tag{4}$$

2.3 Trajectory

Calculations with respect to orbital decay are able to provide initial conditions for density, flight-path angle, and velocity. The trajectory thereafter may be described as follows according to (Ashley, 1974):

$$\frac{d\gamma}{dt} = \frac{g\cos\gamma}{V}\left[\frac{V^2}{g(R_E + z)} - 1\right] \tag{5}$$

$$\frac{dz}{dt} = V\sin\gamma \tag{6}$$

$$\frac{dV}{dt} = -\left(\frac{V^2\rho_0 e^{-z/\alpha}}{2\beta} + g\sin\gamma\right) \tag{7}$$

where,

ρ_0 = exponential atmosphere reference density
α = exponential atmosphere scale height

z = altitude
γ = flight-path angle, positive for ascent
g = gravitational acceleration
R_E = Earth radius
V = free-stream velocity
β = ballistic coefficient

The purpose of the sensitivity studies described in this paper is to assess the relative significance of the models and parameters of Equations (1) through (7) as applied to both the reentering spacecraft and its pieces as it breaks up in the atmosphere. We do this to focus future research efforts on those uncertainties that most affect reentry risk estimates. Table 1 lists the uncertainties that we investigated and relates them to the above equations.

3 STRESS-STRENGTH INTERFERENCE

Our previously reported work used a simple thermal disassembly criterion. When a piece of a spacecraft, which we call an element, reached its melting point, it disassociated from the spacecraft and continued on an independent ballistic trajectory. We calculated the trajectory and heating of both the original spacecraft and the independently falling element. For the work reported herein, we introduced another disassembly criterion called a stress-strength interference model.

We use a numerical approach in which conditions such as spacecraft velocity and altitude, element temperature, and aerodynamic drag are calculated. We used the temperature dependent ultimate strength of 301 stainless steel from Mil-Hdbk-5H and concluded that two epistemic uncertainties are relevant. First, room temperature strength is uncertain owing to our lack of knowledge about grain direction, detailed specification, form, hardness, and measurement uncertainty. Second, the strength variation with temperature is uncertain as material characteristics are modified with temperature. We used very simplified stress calculations, which we called pseudo-stress, based on aerodynamic drag. The uncertainty in stress is assumed to vary with time because of the proportionality with velocity-squared and air density.

We wish to know the altitude at which an element (e.g. the propellant tank) disassociates from the spacecraft (e.g. the Delta II second stage). As the numerical solution marches forward in time, altitude, stress, strength, and element temperature are calculated at each time step. Us of an approximate stress calculation is acceptable for this work because we are interested only in the sensitivity associated with changes of its values. The disassociation altitude is taken in our calculations to be the higher of that altitude at which stress exceeds

strength or the element melt temperature is reached. This altitude is recorded as the disassociation altitude, along with trajectory angle, element temperature and velocity. These become the initial conditions for the element to continue on its independent ballistic trajectory.

These calculations are done within a Latin Hypercube simulation environment. Each trajectory from reentry to ground is considered one Latin Hypercube trial. All uncertain parameters are sampled at the beginning of a trial/trajectory and kept constant unless there is an intrinsic variation with a calculated variable such as temperature. For example, we have included strength and heat transfer parameters whose median values and uncertainties are functions of temperature that changes during the trajectory.

4 UNCERTAIN PARAMETERS

Table 1 shows the set of uncertain parameters used in this study along with the basis for estimating median values and uncertainties.

Some of the most significant unknowns are the trim orientations of the spacecraft and the independently falling elements. When we speak of orientation of a ballistic body, we mean the way it is moving on its axes and the way it is moving relative to the free stream velocity vector. For example, a cylinder-shaped tank may be spinning on its long axis (called broadside spinning), tumbling end over end on its short axis, spinning and tumbling and/or precessing, not spinning such that it presents a single face of the cylinder to the air-stream (either broadside stable or end-on stable), or angled end-on such that the stable trim orientation yields an acute angle between the plane perpendicular to the cylinder axis and the air stream. Deterministic methods can not account for the uncertainty in orientation and typically assume one (e.g. Portelli, 2002).

Sophisticated six degree of freedom calculations have been attempted but the results are highly sensitive to assumed initial conditions (Platus, 1994). We believe, therefore, that this uncertainty must always be a part of reentry disassembly calculations. We have tried to capture this uncertainty using the parameters for ballistic coefficient and heat transfer orientation shown in Table 1. The sampling procedure selects an orientation at the beginning of a trajectory/trial and then we simulate the trajectory heating and disassembly associated with this orientation. Because we use thousands of trials, we in effect simulate thousands of trajectories. This method provides a good estimate of the calculated effects of uncertain trim orientation and properly correlates ballistic coefficient with orientation factor.

Table 1. Uncertain parameters for disassembly calculations.

Parameter	Distribution	Median Basis	Standard Deviation or uncertainty basis
Specific Heat, c_p in (1)	normal	410 Stainless steel (460 J/kg/K)	10% of median
Melt temperature, T_m in (4)	normal	410 Stainless steel (1810 K)	10% of median
Heat of fusion, h_{if} in (3)	normal	410 Stainless steel (2.75E + 05 J/kg)	10% of median
Emissivity, ε in (1) and (4)	normal	Neither polished nor rough surfaced 410 stainless steel. 0.38 as interpreted from the data of (Touloukian, 1972)	Variations owing to rough or polished surfaces. 10% of median as interpreted from the data of (Touloukian, 1972)
Heat transfer coefficient, h in (1) and (4)	normal	Velocity dependent correlation from (Perini, 1975)	18.5% of median from (Perini, 1975)
Second stage ballistic coefficient, β in (7)	Triangular	Calculated for assumed spacecraft orientation	Variations owing to alternative orientations
Propellant tank ballistic coefficient, β in (7)	Triangular	Calculated for assumed tank orientation.	Variations owing to alternative orientations
Heat transfer orientation factor of element within spacecraft, B_2 in (1) and (4)	Triangular	Spacecraft acts as a spinning hemisphere, based on (Cropp, 1965)	Variations owing to possible alternative spacecraft orientations
Heat transfer orientation factor of independent element (cylindrical propellant tank), B_2 in (1) and (4)	Triangular	Tank spinning on long axis (broadside) based on (Cropp, 1965)	Lower bound is for random tumbling and spinning tank; upper bound is for broadside non-spinning stable tank. Based on (Cropp, 1965)
Room temperature material strength	normal	Based on 301 stainless steel in Mil-Hdbk-5H	Variations in Mil-Hdbk-5H
Temperature factor on material strength	normal	Based on 301 stainless steel in Mil-Hdbk-5H up to 1100 K then extrapolation	15% of median up to 1100 K, then based on variation due to alternative extrapolation models
Pseudo-stress	normal	Simple dynamic pressure model which varies over the trajectory	Three times median value

5 SENSITIVITY STUDIES

The purpose of these studies is to assess the relative significance of the uncertain parameters of Table 1 so that future efforts can be pointed toward reducing the epistemic uncertainties or improving the modeling in the most significant areas. We used two model results as the metrics of comparison. The first is the probability distribution of casualty area. This was an obvious choice because spacecraft must now meet deterministic casualty area standards. We found from our previous work that trends in casualty area are easier to explain if the altitude at which an element disassociates from the original spacecraft is known. Therefore, we also used as a metric the probability distribution of element disassembly altitude.

5.1 Effect of stress-strength interference disassembly criterion

Typically, a material under stress will fail owing to its reduced strength as temperature increases before it reaches its melting temperature. Therefore, we expected that including the stress-strength interference model described in Section 2 would have a significant effect. Figure 1 shows two cumulative distribution functions of the altitudes at which the propellant tank disassociated from the Delta II second stage, one resulting from using a melt temperature criterion and one resulting from the stress-strength interference criterion.

Each curve in Figure 1 is an ordered set of points. Each point on a curve is a result of a single simulated trajectory. Each trajectory is a result of a single Latin Hypercube trial. We see in the Thermal Criterion curve that approximately 60% of the simulated trajectories resulted in no disassociation of the element from the spacecraft. A zero altitude indicates that the element did not reach its melting point during the trajectory.

However, application of the stress-strength interference criterion indicates that 60% of the simulated trajectories resulted in disassociation of the propellant tank element at ~83 km or less. Indeed, in all trials the stress-strength criterion predicted disassociation of the propellant tank. We believe this to be, in some cases, a more realistic result than provided by the thermal criterion.

638

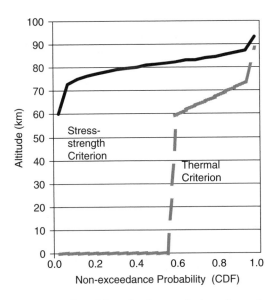

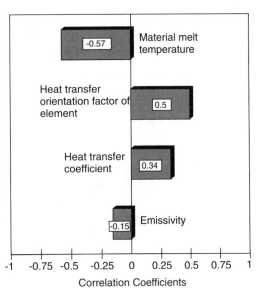

Figure 1. Effect of thermal and stress criteria on disassociation altitude.

Figure 2. Criteria on disassociation altitude.

5.2 Correlation of uncertain parameters

Our base case represents use of all the Table 1 probability distributions. We developed Spearman rank correlations (Lehmann, 1998) to determine those parameters that are most closely correlated with casualty area and disassociation altitude. Each trial served as a sample for the correlation calculation. Although correlation coefficients merely show whether or not there is a statistically significant correlation of casualty area and disassociation altitude with the parameters, in this case we argue there is also a physical cause and effect relationship.

Casualty area is inversely proportional to the ablation of a spacecraft element. We found it convenient to show the correlation coefficients in terms of ablation of an element. Figure 2 shows that the four parameters that most influence amount of an element that ablates are: element melt temperature, heat transfer coefficient, orientation factor, and radiative emissivity.

A falling element is exposed to aero-thermal heating and its temperature rises until it decelerates enough to allow radiative cooling to overcome aerothermal convective heating. During simulation trials in which a materials melt temperature is less than nominal, the melt temperature is reached when the element has more of the heating portion of the trajectory to transit, giving more opportunity for ablation. Therefore, ablation increases with reduced melt temperature and a negative correlation coefficient is reasonable. The heat transfer coefficient and the heat transfer orientation factor appear as multipliers in the

heating equations. Simulated trajectories in which these have larger values increase the amount of heat that reaches an element and ablation would, therefore, tend to increase as indicated by positive correlation coefficients. Conversely, increased emissivity increases radiative cooling of the element and we would expect ablation to decrease. This trend is indicated as a negative correlation.

Figure 3 shows the input parameters that most influence the altitude at which the propellant tank element disassociates from the Delta II second stage. An element disassociates from the spacecraft when the aero-thermal stresses exceed the temperature dependent strength of the material. Because they are multipliers that influence calculated spacecraft and element heating, higher values of heat transfer coefficient and heat transfer orientation factor increase element heating, reduce element strength and induce disassociation at a higher altitude, hence, a positive correlation in Figure 3. Similarly, reductions in material strength would induce disassociation at higher altitudes.

Parameters that induce cooling, such as emissivity, would have an opposite effect. That is increases in cooling would retard element disassociation and reduce the altitude at which it might occur. Simulated trajectories in which material specific heat is increased would have the effect of retarding the materials strength reduction because material temperature would rise more slowly. Therefore, increases in specific heat would cause disassociation to occur at lower altitudes.

Figures 2 and 3 show only the parameters to which casualty area and disassociation altitude are most

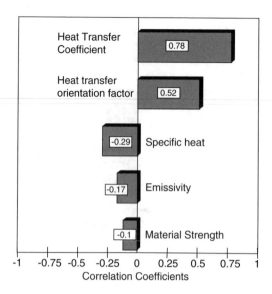

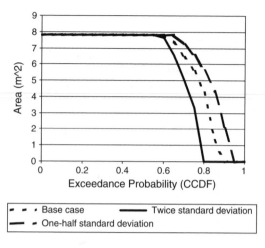

Figure 4. Sensitivity of casualty area to variations in the distribution of melt temperature.

Figure 3. Correlation of element disassociation altitude to input parameters.

sensitive. Variations in the parameters in Table 1 not shown in these figures are less important to the results. However, these figures do not show the magnitude of the sensitivities.

5.3 Magnitude of sensitivities

Our paradigm explicitly recognizes that important parameters are uncertain. The intention of this work is to determine how sensitive the results are to changes in the uncertainties. This is somewhat of a departure from traditional sensitivity studies which investigate the effect of single value changes on point estimate results. Our results are already uncertain and expressed as probability distributions. We are investigating the effect of probability distribution changes on the results.

We performed a sensitivity investigation for each of the parameters in Figures 2 and 3. The first result was that any parameter with a correlation coefficient less than an absolute value of 0.3 had a much smaller effect on results relative to the others. Thus, we show the sensitivity of results for melt temperature and heat transfer orientation factor relative to casualty area (i.e. amount of ablation); and the results for heat transfer coefficient and heat transfer orientation factor relative to disassociation altitude.

In Figure 4, the value of $7.8\,\text{m}^2$ corresponds to the casualty area of the Delta II second stage propellant tank element when there has been no ablation. An area of zero corresponds to complete ablation. This figure shows the probability on the x-axis that the area will have the corresponding y-axis area or

greater. That is, the x-axis shows the probability of exceedance of a casualty area. For example, the base case shows that approximately 60% of the simulated trajectories caused a casualty area corresponding to no ablation, approximately 80% of the simulated trajectories caused a casualty area $\sim 5\,\text{m}^2$ or more, and less than 10% of the trajectories caused the entire propellant tank to ablate.

The base case corresponds to the use of all Table 1 probability distributions which describe one set of beliefs about the uncertainties. The other two curves in Figure 4 show the effect of increasing and decreasing the uncertainty of melt temperature. The curve labeled "Twice standard deviation" uses 20% of the median value of melt temperature (362 K) as the standard deviation instead of the base case 10% (181 K) of the median value. The curve labeled "One-half standard deviation" uses 5% of the median value of melt temperature as the standard deviation. Notice that higher standard deviations produce lower casualty areas. If a trajectory occurs with a value of melt temperature near the central tendency ($\sim 1810\,\text{K}$) or higher, then ablation does not occur. As the uncertainty increases, more trajectories will occur with lower melt temperatures, which allow more ablation and lower casualty areas.

Figure 5 shows a more interesting variation. As noted in Section 3, the orientation of a spacecraft as well as the element when it is an independent ballistic body is poorly predicted. The base case assumed that the highest probability orientation of the propellant tank was broadside spinning corresponding to a heat transfer orientation factor of ~ 0.3 with an upper bound corresponding to a broadside stagnation orientation (~ 0.7) and a lower bound corresponding to

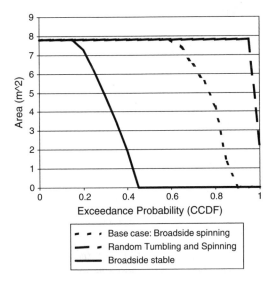

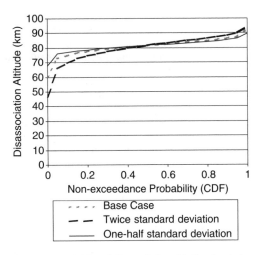

Figure 6.   Sensitivity of disassociation altitude to variations in heat transfer coefficient.

Figure 5.   Sensitivity of casualty area to variations in the element heat transfer orientation factor distribution.

random spinning and tumbling (∼0.2). We performed two sensitivity cases, one in which the central tendency corresponds to broadside stagnation with a standard deviation of 20% around that value; and one in which the central tendency corresponds to random spinning and tumbling with a standard deviation of 20% around that value. It is interesting to note that one of the recovered propellant tanks exhibited characteristics that indicate that it fell in near end-on stable orientation for at least part of its trajectory and may have changed orientation when a hole in the tank occurred (Baker, 1999).

A stable orientation allows aero-thermal heating to concentrate on a single side of the element. This is in contrast to either a tumbling or spinning element which distributes the heat and allows for more cooling. We expect, therefore, that the broadside stable configuration would exhibit more ablation and lower casualty areas than the other two. Similarly, an orientation of random tumbling and spinning provides more even distributed heating and cooling than the others. Note the dramatic effect of orientation. At exceedance probabilities of 50% and higher, the results span no ablation to complete ablation. This result is strong evidence that a deterministic approach in predicting casualty area may be grossly incorrect if it ignores uncertainties in orientation and the potential to change orientation during the trajectory.

Figures 6 and 7 illustrate the effect of changes on element disassociation altitude of the distributions for element heat transfer coefficient and the heat transfer orientation factor while within the spacecraft,

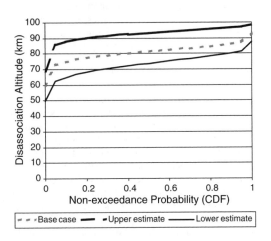

Figure 7.   Sensitivity of disassociation altitude to variations in heat transfer orientation factor of element within spacecraft.

respectively. The propellant tank element is attached to the Delta II second stage until the altitude at which disassociation occurs.

In Figure 6, the uncertainty is symmetric and increasing it has the effect of increasing the variation of disassociation altitudes about the median.

In Figure 7, the base case assumed that the highest probability orientation of the spacecraft was a randomly tumbling hemisphere corresponding to a heat transfer orientation factor of ∼0.3 with an upper bound corresponding to a combination of a stable broadside and spinning sphere (∼0.4) and a lower bound corresponding to a random spinning and tumbling cylinder

(~0.2). We performed two sensitivity cases, one (called upper estimate) in which the central tendency corresponds to stable broadside (~0.7) with a standard deviation of 20% around that value; and one (called lower estimate) in which the central tendency corresponds to random spinning and tumbling (~0.2) with a standard deviation of 20% around that value. We see the importance of the spacecraft orientation on the disassociation altitude. Orientations such as stable broadside that enhance heating over an area would tend to cause disassociation of an element from the spacecraft at significantly higher altitudes. At the median altitudes the difference between the upper and lower estimate is ~20 km.

6 RECAPITULATION

Before we introduced our probabilistic paradigm for randomly reentering spacecraft disassembly, all other such predictive calculation techniques were deterministic. Our method simulates thousands of trajectories, each of which uses samples from uncertain parameters in order to obtain the uncertainty in key results such as casualty area. Using relatively simple trajectory and heating models (Weaver, 2001), this work 1) investigated the variability of results owing to consideration of the epistemic uncertainties and 2) performed sensitivity studies on the probability distributions to estimate which most influence the casualty area (or element ablation) and disassociation altitude.

By use of the Spearman rank correlation process and actual sensitivity calculations, we found that 1) orientation of the element after disassociation and the melt temperature of the element most influence casualty area, and 2) orientation of the spacecraft before disassociation most influences the altitude at which elements disassociate or break-off from the spacecraft. Indeed, the sensitivity to free-flying element orientation of element casualty area is extreme, with median results varying between no ablation and complete ablation.

Based on these results we submit that any deterministic approach that does not account for uncertainties in our knowledge of spacecraft orientation, free-flying element orientation, and melt temperature may be grossly in error. Furthermore, because standards are based on such deterministic predictions, they might not yield the assurance for reducing casualty area that they are currently believed to have.

ACKNOWLEDGMENT

This work was funded by The Aerospace Corporation through the Center for Orbital and Reentry Debris Studies and by the SMC/AXF Acquisition Civil & Environmental Engineering, LAAFB, CA.

REFERENCES

Ashley, Holt, *Engineering Analysis of Flight Vehicles*, Dover Publications, 1974.

Baker, R.L., et al., "Orbital Spacecraft Reentry Breakup," *50th International Astronautical Congress*, Amsterdam, IAA-99-IAA.6.7.04, 1999.

Cropp, L.O., "Analytical Methods Used in Predicting Reentry Ablation of Spherical and Cylindrical Bodies," SC-RR-65-187, Sandia National Laboratory, Albuquerque, September 1965.

Detra, R.W., Kemp, N. H., and Riddell, F.R., "Addendum to 'Heat Transfer to Satellite Vehicles Re-entering the Atmosphere'," Jet Propulsion, Vol. 27, December 1957, pp. 1256–57.

Frank, M.V., Weaver, M.A., Baker, R.L., A Probabilistic Paradigm For Calculating Reentry Debris Casualty Area," *European Safety and Reliability Conference*, 2001, Torino, September 2001.

Fritsche, B., et al., "Spacecraft Disintegration During Uncontrolled Atmospheric Entry," *50th International Astronautical Congress*, Amsterdam, IAA-99-IAA.6.7.02, 1999.

Gregory, F.D., "Guidelines and Assessment Procedures for Limiting Orbital Debris," NASA Safety Standard 1740.14, NASA Headquarters, Washington, DC, August 1995.

Lehmann, E.L., and D'Abrera, H.J.M., *Nonparametrics: Statistical Methods Based on Ranks*, rev. ed. Englewood Cliffs, NJ: Prentice-Hall, pp. 292, 300, and 323, 1998.

Lips, T., et al., "Spacecraft destruction during re-entry - latest results and developments of the SCARAB software system," 34th Scientific Assembly of COSPAR, Houston, PEDAS1-B1.4-0030-02, October 2002.

Montgomery, R.M., and Ward, J.A., Jr., "Casualty Areas from Impacting Inert Debris for People in the Open," Research Triangle Institute, RTI/5180/60-31F, Cocoa Beach, Florida, 13 April 1995.

Office of Science and Technology Policy, "U.S. Strategy on Orbital Debris," Washington, DC, December 2000.

Perini, L.L., "Compilation and Correlation of Stagnation Convective Heating Rates on Spherical Bodies," *J. Spacecraft*, Vol 12, No. III, March 1975, pp 189–191.

Platus, D.H., "General Purpose Heat Source Module Hypersonic Reentry Attitude Stability," *AIAA Atmospheric Flight Mechanics Conference*, Scottsdale, AIAA 94-3462, 1994.

Portelli, C., et al., "BEPPOSAX Equatorial Uncontrolled Re-entry," 34th Scientific Assembly of COSPAR, Houston, PEDAS1 – B1.4-0027-02, October 2002, p. 4.

Refling, O., et al., "Review of Orbital Reentry Risk Predictions," The Aerospace Corporation, ATR-92 (2835)-1, 15 July 1992.

Rochelle, W, et al., "Modeling of Space Debris Reentry Survivability and Comparison of Analytical Methods," *50th International Astronautical Congress*, Amsterdam, IAA-99-IAA.6.7.03, 1999.

Touloukian, Y.S., and DeWitt, D.P., *Thermal Radiation Properties*, Vol. 7: Metallic Elements and Alloys, from *Thermophysical Properties of Matter*, Y.S. Touloukian and C. Y. Ho (editors), IFI/Plenum, New York, 1970–1972.

Weaver, M.A., et al., Probabilistic Estimation of Reentry Debris Area, *Third European Conference on Space Debris*, ESOC, Darmstadt, Germany , March 2001.

Safety and Reliability – Bedford & van Gelder (eds)
© 2003 Swets & Zeitlinger, Lisse, ISBN 90 5809 551 7

Experiences from application of model-based risk assessment

Rune Fredriksen & Bjørn Axel Gran
The Halden Reactor Project/Institute for Energy Technology, Halden, Norway

Ketil Stølen
Sintef Telecom and Informatics, Oslo, Norway

Ivan Djordjevic
Department of Electronic Engineering, Queen Mary University of London, UK

ABSTRACT: Risk assessment requires a firm, but nevertheless easily understandable, basis for communication between different groups of stakeholders. The CORAS Methodology for Model-based Risk Assessment for security critical systems developed by the IST-funded EU-project CORAS aims to facilitate this and summarises the experiences from testing major field trials on the use of this methodology in the telemedicine and the e-commerce areas.

1 INTRODUCTION

Risk assessment requires a firm, but nevertheless easily understandable, basis for communication between different groups of stakeholders. Graphical object-oriented modelling techniques are widely used for requirements capture and analysis. We assume that they might be equally well suited to support risk assessment. Class diagrams, use case diagrams, sequence diagrams, activity diagrams, dataflow diagrams and state diagrams represent mature paradigms used daily in the IT industry throughout the world. They are supported by a wide set of sophisticated case-tools, they are to a large extent complementary and, together, they support all stages in a system development.

The CORAS Methodology for Model-based Risk Assessment for security critical systems as defined by the IST-funded EU-project CORAS (CORAS 2000, Stamatiou et al., Raptis et al., Fredriksen et al., Winther et al.) is based on this paradigm.

The CORAS Methodology for Model-based Risk Assessment incorporates a documentation framework, a number of closely integrated risk assessment techniques and a risk management process based upon widely accepted standards. It gives detailed recommendations for the use of UML oriented modelling (OMG 2001) in conjunction with risk assessment in the form of guidelines and specified diagrams.

2 CONCEPTUAL MODEL

The CORAS framework is based on four main anchor-points,

- a risk documentation framework based on the Reference Model for Open Distributed Processing – RM-ODP (ISO/IEC 10746:1995),
- a risk management process based on the Australian standard for risk management (AS/NZS 4360:1999),
- an integrated risk management and system development process based on the Unified Process – UP (Jacobson et al. 1999),
- and a platform for tool-integration based on eXtensible Markup Language – XML (W3C 2000).

The risk documentation framework is used both to document the target of assessment as well as to record risk assessment results. The risk management process provides a sequencing of the risk management process into sub-processes for context establishing, risk identification, risk assessment, risk evaluation, and risk treatment.

The CORAS risk assessment methodology, as shown in Figure 1, is an integration of techniques and templates inspired by HazOp Analysis (Leveson 1995), Fault Tree Analysis – FTA (IEC 1025:1990), Failure Mode and Effect Criticality Analysis – FMECA (Bouti et al. 1994), Markov Analysis (Littlewood 1975, Storey

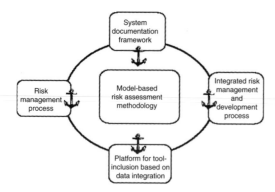

Figure 1. The CORAS Framework.

1996), and CCTA Risk Analysis and Management Method – CRAMM (Barber et al. 1992).

The CORAS risk management process provides a sequencing of the risk management process into the following five sub-processes:

1 Context Identification: Identify the context of the analysis. Describe the system and its environment, identify usage scenarios, the assets of the system and its security requirements.
2 Risk Identification: Identify the potential threats to assets, the vulnerabilities of these assets and document the unwanted incidents.
3 Risk Analysis: Evaluate the frequencies and consequences of the unwanted incidents.
4 Risk Evaluation: Identify the level of risk associated with the unwanted incidents and decide whether the level is acceptable. Prioritise the identified risks and categorise risks into risk themes.
5 Risk Treatment: Addressing of the treatment of the identified risks and how to prevent the unacceptable risks.

The Sub-processes 3 and 4 are also termed as the "Risk Assessment".

In addition, two concurrent sub-processes "Monitoring and review" and "Communication and consultation" are implicit, and running in parallel with the above five.

For each activity CORAS provides guidelines and recommendations on how to perform the activity in the sub processes. For a more detailed description of CORAS and the CORAS methodology for Model-based Risk Assessment see (Dimitrakos et al., Houmb et al.)

3 EXPERIENCES

The CORAS Methodology for Model-based Risk Assessment has been applied in a trial on the regional health network HYGEIAnet that links hospitals and public health centres in Crete (Stamatiou et al. 2003). The objective of the field trial was firstly to provide a security assessment of the Cretan healthcare structure that consists of a number of geographically separated healthcare centres in a hierarchical organization, and secondly to offer a process of identification and assessment of potential solutions.

The CORAS Methodology for Model-based Risk Assessment has also been applied in a field trial to the electronic retail market subsystem of an e-commerce platform, developed in another IST project (Raptis et al. 2002). The security assessment focused on the user authentication mechanism, the secure payment mechanism and on the use of software agents for accomplishing specialised purchasing tasks, offering a process for identifying and assessing potential solutions.

In the following we summarise the experiences from these field trials structured in accordance with the five sub-processes of the CORAS risk management process.

3.1 Sub-process 1: identify context

The aim of the first sub-process is to describe the system and its environment.

In the telemedicine trial an UML sequence diagram was used to model and assess a typical interaction scenario. The UML sequence diagram is an interaction diagram that emphasizes the time ordering of messages. Graphically, an UML sequence diagram as displayed in Figure 2, is a table that shows objects arranged along the X axis and messages, ordered in increasing time, along the Y axis. The UML use-case diagram, as displayed in Figure 3, proved useful in the process of describing the important scenarios. This process was performed in cooperation with both developers and users of the system (medical experts). The use of models proved useful in identifying the context of the system as they helped both risk assessment experts and other stakeholders to focus on the essential parts of the system. The identification of assets and stakeholders was performed in accordance with the recommendations in CRAMM.

In the e-commerce trial the same kind of modelling techniques were used as in the telemedicine trial. In addition a CORAS specific UML profile for risk assessment was used to describe more risk assessment specific documentation. For example, the results from a SWOT analysis (Strength, Weaknesses, Opportunities, and Threats) analysis was documented by a SWOT diagram as illustrated by Figure 4. UML component diagrams were employed to identify the platform environment, interfaces and architecture. The organisational context was described with a high level UML object diagram.

A more detailed description of the e-commerce platform was provided using UML class diagrams,

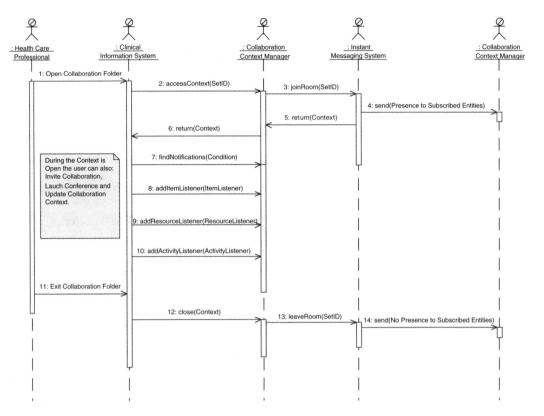

Figure 2. An UML sequence diagram from the telemedicine trial.

and UML sequence diagrams were employed to specify the dynamic behaviour.

3.2 Sub-process 2: identify risks

The aim of the risk identification is to identify the potential threats to assets, the vulnerabilities of these assets and document the unwanted incidents.

In both trials risk identification was performed using HazOp and FTA in order to identify possible threats for each of the identified assets. During the e-commerce trial it was decided to also employ FMEA for the bottom threats in the FTA.

The UML use case diagrams used in the context identification (sub-process 1) provided a suitable high-level of abstraction regarding input to the risk identification (sub-process 2). This was useful for supporting the identification of threats that could have effect on assets. For the more low-level and detailed risk identification, both UML activity diagrams and UML sequence diagrams provided the right level of abstraction.

Some new threats, compared to the preliminary assessment, were discovered during the process. HazOp

was useful to carry out to document the threats of the target system.

3.3 Sub-process 3: analyse risks

The aim of the risk analysis is to evaluate the frequencies and consequences of the unwanted incidents.

In the telemedicine trial, FTA was applied in order to analyze the least understandable of the threats identified by the HazOp. One of the advantages of FTA, observed and remarked by all participants in the trial, stems from the way it in a structured manner helps the risk assessment experts to communicate and explain the threats to non-experts (like doctors) and, likewise, helps non-experts to understand the causes of threats and the cause combinations that lead to the appearance of the threats. The FTA trees built during the risk analysis session also helped the network and system experts that participated to structure their view of the causes of the different threats. In order to conduct a more detailed analysis of some of the identified risks, FMEA (a variant of FMECA) was performed. FMEA was very time-consuming, but gave the opportunity for uncovering many interesting details

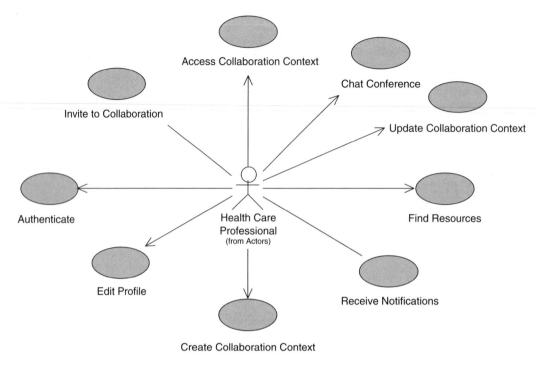

Figure 3.   An UML use-case from the telemedicine trial.

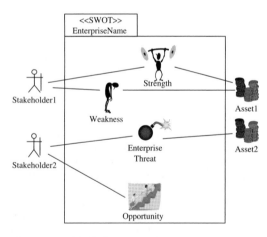

Figure 4.   A SWOT diagram template.

Table 1.   Risk classification from IEC61508.

Frequency	Consequence			
	Catastrophic	Critical	Marginal	Negligible
Frequent	I	I	I	II
Probable	I	I	II	III
Occasional	I	II	III	III
Remote	II	III	III	IV
Improbable	III	III	IV	IV
Incredible	IV	IV	IV	IV

about the analyzed threats. For these reasons FMEA is used only for the most security-critical parts of a system and it seems to require the participation of the experts on the specific aspect that is being analyzed.

3.4 Sub-process 4: risk evaluation

The aim of risk evaluation is to identify the level of risk associated with the unwanted incidents, decide whether the level is acceptable, prioritise the identified risks and categorise risks into risk themes.

In both trials a risk table was used to determine the level of risk. A risk table combines the values for consequence and likelihood into a description of the level of risk as suggested e.g. in (IEC61508:1998–2000) (see Table 1).

3.5 Sub-process 5: risk treatment

The aim of risk treatment is to address the treatment of the identified risks and how to prevent the unacceptable risks. The procedure of risk treatment is to match the risk levels against the risk evaluation criteria and to determine which risk treatment to prioritise.

In both trials the focus of risk treatment was to provide suggestions and communicate the possible solutions to improve the system according to the risk evaluation criteria. The UML models used previously in process for context identification and risk identification, was useful for communicating the necessary changes to the system to the stakeholders.

4 FURTHER WORK

Tool support for Model-based Risk Assessment will be provided through the CORAS Platform recently under development. The CORAS platform will be based on data integration implemented in terms of XML (eXtensible Markup Language) technology. The platform will be built around an internal data representation formalised in XML/XMI (characterised by XML schema). Standard XML tools are supposed to provide much of the basic functionality. Based on XSL (eXtensible Stylesheet Language), relevant aspects of the internal data representation may be mapped to the internal data representations of other tools (and the other way around). This allows the integration of sophisticated case-tools targeting system development as well as risk analysis tools and tools for vulnerability and treat management.

The CORAS Methodology for Model-based Risk Assessment for security critical systems incorporates a documentation framework and a number of closely integrated risk assessment techniques and a risk management process based upon widely accepted standards. It gives detailed recommendations for the use of UML-oriented modelling in conjunction with risk assessment in the form of guidelines and specified diagrams. However – the CORAS Methodology for Model-based Risk Assessment is based on a generic framework for risk assessment and should therefore be applicable on all types of risk assessments, e.g. for safety risk assessment.

5 CONCLUSIONS

The preliminary results from the CORAS trials indicate that it is possible to increase the communication and understandability of risk assessment by applying the CORAS Methodology for Model-based Risk Assessment.

The CORAS Methodology for Model-based Risk Assessment provides a firm, but nevertheless easily understandable, basis for communication between different groups of stakeholders. The use of UML models to assess and document the risk assessment seems to be a valuable approach.

ACKNOWLEDGEMENTS

CORAS is funded under the FP5 IST-program, IST-2000-25031. The CORAS consortium consists of eleven partners from four countries: CTI (Greece), FORTH (Greece), IFE (Norway), Intracom (Greece), NCT (Norway), NR (Norway), QMUL (UK), RAL (UK), Sintef (Norway), Solinet (Germany) and Telenor (Norway). Telenor and Sintef are responsible for the administrative and scientific coordination, respectively. The results reported in this paper have emerged through the joints efforts of the CORAS consortium.

REFERENCES

Australian Standard (1999): Risk Management. AS/NZS 4360:1999. Strathfield: Standards Australia.

Barber, B., Davey, J. (1992): Use of the CRAMM in Health Information Systems, MEDINFO 92, ed Lun K.C., Degoulet P., Piemme T.E. and Rienhoff O., North Holland Publishing Co, Amsterdam, pp 1589–1593.

Bouti, A, Kadi, A.D. (1994): "A state-of-the-art review of FMEA/FMECA", International Journal of Reliability, Quality and Safety Engineering, vol. 1, no. 4, pp 515–543.

CORAS: "A Platform for Risk Analysis of Security Critical systems", IST-2000-25031, (2000). (http://www.nr.no/coras/)

Dimitrakos, T., Raptis D., Ritchie, B., Stølen, K: Model based security risk analysis for web applications. To appear in Proc. Euroweb 2002.

Fredriksen, R., Kristiansen, M., Gran, B.A., Stølen, K., Opperud, T.A., Dimitrakos, T.: The CORAS framework for a model-based risk management process. In Proc. Computer Safety, Reliability and Security (Safecomp 2002), LNCS 2434, (2002), 94–105.

Houmb, S., Braber, F., Lund, M.S., Stølen, K: Towards a UML profile for model-based risk assessment. In Proc. UML'2002 Satellite Workshop on Critical Systems Development with UML (CSDUML'2002), pp 79–91, Munich University of Technology, 2002.

IEC 1025:1990 Fault tree analysis (FTA) (1990).

IEC 61508 (1998–2000): Functional Safety of Electrical/Electronic/Programmable Electronic Safety-Related (E/E/PE) Systems.

ISO/IEC 10746: (1995): Basic reference model for open distributed processing.

Jacobson, I., Rumbaugh, J., Booch, G: The unfied software development process. Addison-Wesley (1999).

Leveson, N.G.: SAFEWARE, System, Safety and Computers, Addison-Wesley (1995).

Littlewood, B.: A reliability model for systems with Markov structure. Appl. Stat. 24 (1975), 172–177.

OMG (2001): Unified Modeling Language Specification, version 1.4.

Raptis, D., Dimitrakos, T., Gran, B.A., Stølen, K.: The CORAS Approach for Model-based Risk Management applied to e-Commerce Domain, In Proc. Communication and Multimedia Security (CMS-2002), Kluwer, (2002) 169–181.

Stamatiou, Y., Skipenes, E., Henriksen, E., Stathiakis, N., Sikianakis, A., Charalambous, E., Antonakis, N., Stølen, K., Braber, F., Lund, M.S., Papadaki, K., Valvis, G. The CORAS approach for model-based risk management applied to a telemedicine service. To appear in Proc. Medical Informatics Europe (MIE'2003).

Stamatiou, Y.C., Henriksen, E., Lund, M.S., Mantzoranis, E., Psarros, M., Skipenes, E., Stathiakis, N., Stølen, K.: Experience from using model-based risk assessment to evaluate the security of a telemedicine application, To appear in Proc. Telemedicine in Care Delivery, (2002).

Storey, N. (1996): Safety-Critical Computer Systems. Addison-Wesley.

Winther, R., Gran, B.A., Johnsen, O-A. (2001): Security Assessments for Safety Critical Systems using HAZOPs, in SAFECOMP 2001 (LNCS 2187), U.Voges (Ed.), pp. (14–24), Springer-Verlag Berlin Heidelberg.

World Wide Web Consortium (2000): Extensible Markup Language (XML) v1.0, W3C Recommendation, Second Edition, 6 Oct. 2000.

Safety and Reliability – Bedford & van Gelder (eds)
© 2003 Swets & Zeitlinger, Lisse, ISBN 90 5809 551 7

Dynamic system acquisition and testing tradeoffs

D.P. Gaver & P.A. Jacobs
Department of Operations Research, Naval Postgraduate School, Monterey, California

E.A. Seglie
Office of Director, Operational Test and Evaluation, The Pentagon, Washington, D.C.

ABSTRACT: A multi-entity system of systems is to be designed, tested, re-designed if design faults appear during testing and eventually accepted for field usage after effectiveness/suitability/sustainability growth is confirmed by a success-run criterion. If operational flaws are uncovered in the field these are *either* tolerated/compensated – for operationally, by repair, *or* rectified by re-design, i.e., enhanced by upgrades, possibly gradually, i.e. upgrading by newly-designed spares, or under extreme circumstances, after a system recall. The latter tends to be quite costly if conducted after fielding, but is sometimes justified, and occurs for safety reasons. The challenge of technological obsolescence over time is explicitly factored into life-cycle determination, so late-in-life re-design or upgrading tends to be unattractive. The budgetary and operational availability and capability tradeoffs and control issues encountered in life-cycle budgeting of development, the pre-fielding operational test-fix-test (TFT) (stopping) scenario, plus the test-fix-test and field-fix-field operational periods and associated penalties are described, balanced, and appraised in a dynamic atmosphere of uncertain operational requirements. Implications for finite test asset (range and range time) requirements are addressed.

1 TESTING PROTOCOLS

Consider a system that has a characteristic mission time, m. A reconnaissance vehicle (e.g. a helicopter or UAV) or an intermediate armored vehicle are examples. Subject it (precisely, a physical version or realization of the current design) to *trials* of length $rm = \tau$ (e.g., $r = 3$, or 6, where r is a decision variable). One option is to perform a test as a sequence of trials; stop the test when, for the first time, a *trial* (of r missions) results in *no unit failure*, i.e., the first successful (sequential) trial. Note that since $\tau = rm$ this qualitatively resembles the r-Success Run criterion adopted in previous work (Gaver, *et al.* (2002)): the test terminates only after occurrence of r successful missions, run end-to-end. This is more stringent than our previous run criterion. Another option is to perform test missions in parallel, using different units that embody the system design as replicates. For instance, a *trial* may now consist of exercising r parallel missions, or missions in blocks r (3 or more); the system is accepted when the first successful trial (all missions succeed) terminates.

1.1 Test model: Sequential r-mission-success stopping rule

Assume initially that the number of design defects in the system is Poisson with mean μ. Let the probability that any defect activates (a failure occurs) during a trial be, $p(r) = p$, independently from trial to trial and across defects (for the present); $\bar{p} = 1 - p$, the probability of defect trial survival. All failures are assumed rectifiable/removable (alternatively, p is the net probability of failure occurrence and successful removal; we do not here allow for the real possibility of "faulty re-design", i.e., the addition of new defects).

From well-known properties of Poisson processes and by conditioning we can write expressions for the generating function and/or moments of these random variables:

N : number of trial on which the first r-mission success occurs;

T : test length $=$total missions (times m time units);

D_F : number of design defects removed during the test;

D_R: number of design defects remaining at test termination (after which the system will be fielded).

The joint generating function of these is

$$g(z_F, z_R, z_N) \equiv E\left[z_F^{D_F} z_R^{D_R} z_N^N\right]$$

$$= e^{-\mu p} e^{-\mu \bar{p}[1-z_R]} z_N$$

$$+ \sum_{n=2}^{\infty}\left[\prod_{i=0}^{n-2}\left[e^{-\mu \bar{p}^i p[1-z_F]} - e^{-\mu \bar{p}^i p}\right]\right.$$

$$\left. \times e^{-\mu \bar{p}^{n-1} p} e^{-\mu \bar{p}^n[1-z_R]} z_N^n\right]. \qquad (1)$$

Expressions for means and other moments result from differentiation at $z = 1$; an expression for the probability that, $D_R \equiv D_R(N) = 0$, no defects remain at test termination results by setting $z_N = z_F = 1$ and $z_R = 0$. Here below are some results for means ("population averages") of important random variables in this model:

$$P\{N = 1\} = e^{-\mu p}$$

$$P\{N = n\} = \prod_{i=0}^{n-2}\left[1 - e^{-\mu(\bar{p})^i p}\right] e^{-\mu \bar{p}^{n-1} p} \text{ for } n \geq 2$$

$$E[N] = \sum_{n=1}^{\infty} n P\{N = n\}. \qquad (2)$$

$$E[D_F] = \mu p$$

$$+ \sum_{n=2}^{\infty} \mu \bar{p}^{n-1} p \prod_{i=1}^{n-1}\left(1 - \exp\left\{-\mu \bar{p}^{i-1} p\right\}\right) \qquad (3)$$

$$E[D_R] = E[E[D_R | N]] = \mu E\left[\bar{p}^N\right] \qquad (4)$$

1.2   Cost parameters

c_T: Cost per trial per unit time

c_F: Cost to fix design defect (DD) exposed during trial

c_R: Cost to repair DD in field

c_D: Cost to redesign the system to remove one DD in field, ("fix in field")

c_U: Cost per item, per design defect redesign

c_M: Cost to manufacture one basic system, prior to later upgrade costs

B : Total budget

L : Time until obsolescence

λ_f: Rate of occurrence of DD in field

λ_t: Rate of occurrence of DD in testing

ϕ : Rate of occurrence of missions in the field per platform per unit time

ρ : Probability a redesign effort after the system is fielded removes a DD

m : Length of one mission

The probability a DD survives a mission of length m during testing is $\exp\{-\lambda_t m\}$; the probability of survival in the field is $\exp\{-\lambda_f m\}$. Other assumptions are possible.

1.3   Cost of testing

The cost incurred during test is sum of the cost of test time plus the cost of re-design:

$$C_T = c_T Nrm + c_F D_F. \qquad (5)$$

2   THE FIELD PHASE OF SYSTEM MANAGEMENT

In advance of fielding, following test termination, suppose that D_R design defects (DD) remain in the system, the ith of which has activation/failure rate $\lambda_f(i)$.

Suppose M units are fielded. For initial simplicity, suppose each unit has equal exposure to failure (this is heroic, but can be modified and generalized, certainly using Monte Carlo simulation). Since each unit has all of the D_R DDs, any particular DD has an occurrence/failure rate $\lambda_f M$. Thus, any DD may expose itself by failing in least one of M design replicas.

Assume that the time and identity of a DD failure (DDF) is known. In field usage it would be plausible to repair/replace (RR) the component causing DD. Alternatively, the component can be re-designed (RD) and all fielded copies of the original design be withdrawn from service (the system is recalled or "grounded") until the re-designed component is replaced; or undergo "modernization/upgrading through spares"; after the (usually costly) decision to RD is made, then as each new DD of a unit occurs, it is replaced by the RD, presumably with a lower failure rate (for simplicity here the re-designed failure rate is taken to be zero). Many systems will have a planned finite lifetime, L, so RD may well be unwise if too near L. In this paper we consider the first and last options, omitting recalls.

2.1   Examples: Some fault response options and their cost

Consider a representative DD that remains after testing. Let its activation/failure rate be λ_f. Consider monetary cost (in $) of these (limited) alternatives:

(A) Each activation/failure of the DD is simply repaired, e.g., using a standard spare, at cost c_R, where typically $c_R \ll c_D$. We neglect the time and facilities required, as well as any logistical delays; these important issues are left for attention elsewhere.

(B) As soon as the *DDF* occurs the *DD* cause is redesigned and the defect removed with probability ρ; the cost of this is c_D and is relatively high (we neglect removal time for the present); if the re-design is ineffective, future failures from that source are repaired. In reality, both such actions may have variable costs; so c_D and c_R are merely illustrative values.

In view of the finite lifetime of the systems, L, it may well be unwise to adopt (B), *RD*, if the failure occurs quite late, i.e., after time L_0, $L_0 < L$. This option is not studied here.

2.2 Expectations of performance and cost properties of fielded items bought under (A) and (B)

Assume missions for each fielded vehicle occur according to a Poisson process with rate ϕ.

(A) Repair only

Given the number of trials until have the first success, estimate the number of *DD*s remaining by $E[D_R|N]$. The expected cost of repairs in the field for one system during its lifetime L is $E[D_R|N]\,\phi\,c_R$ $[1 - \exp\{-\lambda_f m\}]L$. The conditional expected number of systems bought given N is (the numerator = budget – (test costs); the denominator is estimated lifetime unit field costs)

$$E[M_R\mid N]$$
$$= \frac{B - c_T Nmr - c_F E[D_F\mid N]}{c_M + E[D_R\mid N]\phi c_R\left[1 - e^{-\lambda_f m}\right]L} \qquad (6)$$

The conditional expected cost incurred due unit purchase and repairs given N is

$$E[C_R\mid N] = E[M_R\mid N]$$
$$\times \left[c_M + c_R E[D_R\mid N]\phi L\left[1 - e^{-\lambda_f m}\right]\right] \qquad (7)$$

(B) Redesign and replace by spares

Given the number of the trial of the first success, N, estimate the number of *DD*s remaining after test completes by $E[D_R|N]$. An estimate of the number of systems bought given N satisfies

$$M_D(N) = \left[c_M + c_U E[D_R\mid N] e^{-\phi L\left[1 - e^{-\lambda_f m}\right]} \right.$$
$$+ c_R(1-\rho)E[D_R\mid N]\phi L\left[1 - e^{-\lambda_f m}\right]\right]^{-1}$$
$$\times \left[B - c_T Nmr - c_F E[D_F\mid N] - c_D E[D_R\mid N]\right]$$
$$\left.\times \left[1 - \exp\left\{-\phi M_D(N)L\left[1 - e^{-\lambda_f m}\right]\right\}\right]\right] \qquad (8)$$

M_D (N) can be solved for by successive substitution starting with

$$M_D(N;0)$$
$$= \left[B - c_T Nmr - c_F E[D_F\mid N] - c_D E[D_R\mid N]\right]$$
$$\times \left[c_M + c_U E[D_R\mid N] e^{-\phi L\left[1 - e^{-\lambda_f m}\right]}\right.$$
$$+ c_R(1-\rho)E[D_R\mid N]\phi L\left[1 - e^{-\lambda_f m}\right]\right]^{-1} \qquad (9)$$

The conditional expected cost due to purchase, redesign, and repair by sparing incurred given N is

$$E[C_D\mid N]$$
$$= \left\{ c_D E[D_R\mid N] \atop \times \left(1 - \exp\left\{-M_D(N)\phi L\left[1 - e^{-\lambda_f m}\right]\right\}\right) \right\}$$
$$+ M_D(N)c_M$$
$$+ M_D(N)$$
$$\times \left[c_U \rho E[D_R\mid N] \atop \times \left[1 - \exp\left\{-\phi L\left[1 - e^{-\lambda_f m}\right]\right\}\right]\right]$$
$$+ M_D(N)$$
$$\times \left[c_R(1-\rho)E[D_R\mid N]\phi L\left[1 - e^{-\lambda_f m}\right]\right] \qquad (10)$$

Finally, the condition on N is removed by weighting by $P\{N = n\}$ and summing over n.

Example: The table below displays results from the analytical models. The estimated number of systems bought after testing is the integer part of (6) and (8). The costs are in $1000 units, so the budget for the

Trial size, r	E[Cost of test]	E[Number fielded (Repair)]	E[Mission success probability] if only repair	E[Field cost due to purchase and repair of failures]
3	126.8	14.5	0.8	816.2
4	148.2	26.0	0.9	824.5
5	164.3	41.0	0.9	812.0
6	177.6	58.9	1.0	814.1
7	189.1	78.3	1.0	797.7
8	199.5	98.0	1.0	792.1
9	209.2	116.3	1.0	786.3
10	218.3	132.8	1.0	777.9
11	227.1	149.1	1.0	770.9
12	235.6	165.4	1.0	761.8

entire project, including (estimated) costs when fielded is $1M.

$B = 1000$, $c_M = 3$, $c_T = 5$, $c_F = 20$, $c_R = 1$, $\mu = 5$, $\lambda_f = -\ln(0.8) = 22$, $T = Nmr$, $c_U = 2$, $c_D = 160$, $\phi = 50$, $L = 10$, $\rho = 0.75$.

Trial size, r	E[Number fielded (Re-Design)]	E[Mission success probability] after 1 time period	E[Field cost due to redesign]	E[Field cost due to purchase repair by sparing, and repair of failures]
3	40.0	1.0	141.9	853.7
4	66.8	1.0	82.6	844.5
5	94.2	1.0	51.0	831.0
6	120.4	1.0	32.9	816.8
7	144.0	1.0	21.9	809.6
8	163.6	1.0	14.8	798.5
9	180.0	1.0	10.2	788.3
10	194.4	1.0	7.0	779.0
11	207.3	1.0	4.8	771.9
12	217.6	1.0	3.3	763.6

2.3 Discussion

The present model is greatly simplified in that it studies only one class of homogeneous design faults, with no stagewise blocking as in Gaver et al. (2001). However, it analyzes a plausible sequential test stopping rule and projects its mean consequences through assumed system lifetime. It thus provides a test planner with quick perspective on the value of operational testing when the end-to-end consequences are accounted for. Model refinements and data acquisition effort should be stimulated by such an initial effort, as well as those referenced. Simulations reveal agreement, but exhibit variability of fiscal and operational outcomes (risk) likely to be encountered in practice.

REFERENCES

Block, H.N. & Savits, T.H. 1997. Burn in. *Statistical Science* 12(1): 1–13.

Gaver, D.P. & Jacobs, P.A. 1997. Testing or fault-finding for reliability growth: A missile destructive-test example. *Naval Research Logistics* 44: 623–637.

Gaver, D.P., Jacobs, P.A., Glazebrook, K.D. &.Seglie, E.A. 2001. Probability models for sequential-stage system reliability growth via failure mode removal. Naval Postgraduate School Operations Research Technical Report.

Gaver, D.P., Jacobs, P.A. & Seglie, E. 2002. Balancing "test-fix-test" and released system reliability under a fixed budget. In H. Langseth & B. Lindqvist (eds). *Communications 3rd intern. Conf. on Mathematical Methods in Reliability, Trondheim, Norway, 17–20 June, 2002*: 251–254. Norwegian University of Science and Technology.

Quigley, J. & Walls, L. Cost-benefit modeling for reliability growth. Department of Management Science, University of Strathclyde, Working paper, 2002–06.

Seglie, E.A. 1998. Reliability growth programs from the operational test and evaluation perspective. In K.J. Farquar & A. Mosleh (eds). *Reliability Growth Modeling: Objectives, Expectations, and Approaches*. The Center for Reliability Engineering, Univ. Maryland, College Park, MD.

Thomke, S. & Bell, D.E. 2001. Sequential testing in product development. *Management Science* 47: 308–323.

Safety and Reliability – Bedford & van Gelder (eds)
© 2003 Swets & Zeitlinger, Lisse, ISBN 90 5809 551 7

Hydrogen fuel cell bus: operating experience in Italy

R. Gerboni & E. Ponte

Energy Department, Politecnico di Torino, Italy

ABSTRACT: This paper describes and comments the safety related aspects that have been experienced during the design, test and delivery of a prototype of Hydrogen fuel cell Bus. Risk analysis has been identified as the best tool to be adopted for the evaluation, not only qualitative but also quantitative, of the hazards connected to the hydrogen application on a fuel cell bus for an urban public transport line and its refuelling station in Turin (Italy). The automatic approval to the circulation of the bus was not possible due to the lack of specific codes and standards in Italy about this innovative technology. Thus, the risk analysis seemed to be a necessary support for the final authorisation to the operation in the urban context. The paper reports an overview of the project and the main achievements of the analysis: the definition of the main sources of risk (resulted acceptable) and the possible accidental events; the draft of new guidelines about hydrogen use and storage in the transport sector.

1 INTRODUCTION

Global transport keeps on spreading throughout the last decades, involving an over-increasing quantity of polluting substances in the atmosphere. Following up a sustainable mobility, it is necessary to adopt specific intervention strategies that enable the reduction of the global CO_2 emissions, to limit the presence of benzene and fine dusts and to reduce the levels of acoustic pollution, in particular in the urban areas.

In this context, remarkable sources have been spent in the development of "green" technologies for the transport sector and in particular the hydrogen fuel cell vehicles represent a promising alternative for the future.

Following the footsteps of other transport companies in Chicago, Vancouver and Munich, also in Turin (Italy) a project has been developed, regarding the planning, realisation and management of an hydrogen fuel cell bus for an urban public transport line and its refuelling station. This project, promoted by the local public transport company, Gruppo Torinese Trasporti S.p.A. (GTT) and financed by the Ministry for the Environment, partnered by Irisbus Italia S.p.A, Sapio S.r.l, Compagnia Valdostana delle Acque S.p.A. (CVA), Ansaldoricerche S.r.l, Ente per le Nuove tecnologie, l'Energia e l'Ambiente (ENEA) and has the following partners in the realisation: TÜV Bau und Betrieb GmbH, International Fuel Cell (IFC) and Centro Ricerche Fiat (CRF).

The project, first example in Italy, involved the study of several safety problems concerning hydrogen storage and use in the transport sector.

Because of the innovative technologies applied to the bus, it was not possible to proceed to an automatic approval for the vehicle circulation. Besides, the approval of a vehicle with nine cylinder of compressed hydrogen (200 bar) onboard represented a complex problem due to the lack of specific codes and standards in Italy. Thus, the risk analysis has been identified as the best tool to be adopted for the evaluation, not only qualitative but also quantitative, of the danger connected to the hydrogen application on the bus and it seemed to be a necessary support for the final authorisation to the operation in the urban context.

The safety design studies related to the equipment and installations are being developed by the German TÜV. The Energy Department of the Politecnico di Torino gave has been involved for a Preliminary Risk Assessment concerning the potential release of Hydrogen in the urban context and to assess the status of standards and codes referring to Hydrogen application technologies.

2 THE PROJECT AND ITS AUTHORIZATION

The authorization of a transportation mean passes through the verification that operating conditions do not affect the safety of the passengers, officers and maintenance personnel.

Up to now, Italian laws and standards have discussed and regulated in detail the use and application of compressed natural gas on vehicles, while the adoption of hydrogen has been neglected.

As a general concept, to allow the free circulation of any transportation mean in a city, these three steps should be taken:

– verification the conformity to technical procedures;
– normative requirement fulfillment;
– approval procedure;

At an international level, the activity in order to achieve the standardization in the use of hydrogen as an automotive propeller is very lively.

The most important standardization bodies, ISO for the general aspects and IEC for the electrical ones; they have constituted several working groups about the different aspects on safety for hydrogen applications.

2.1 ISO working groups

Two working groups were set up by ISO: the Technical Committee n. 197 (*Hydrogen technologies*) which aims to the standardization of the systems and devices for the production, storage transport and use of hydrogen, and Technical Committee n. 22 (*Road vehicles*) which is concerned with all questions of standardization concerning compatibility, interchangeability and safety, with particular reference to terminology and test procedures for evaluating the performance of many types of road vehicles and their equipment. TC 22 has a subcommittee (n. 21) which is especially concerned with electric vehicles (www.iso.ch).

Table 1 lists the ISO TC 197 working groups.

2.2 IEC working groups

The International Electric Committee was primarily involved in the standardization process of the hydrogen related technologies and in particular of fuel cells performances. A technical committee (IEC TC 105 – *Fuel cell technologies*), with its specific working groups, is analyzing the several safety aspects of the installation and operation of stationary and portable fuel cells (www.iec.ch).

Table 2 lists the IEC working groups.

2.3 Other working groups

The United Nations Economic Commission for Europe (UNECE) originated a Working Party related to regulate the transport area: the WP.29 (*World Forum for Harmonization of Vehicle Regulations*). At its inside, a subgroup (*GRPE: Working Party on Pollution and Energy*) is concerned with the use of compressed natural gas (methane) on vehicles and, as a consequence, it is going to deal with hydrogen components. An ad hoc group inside GRPE was created to

Table 1. ISO working groups.

Working group	Activity
WG 1	Liquid hydrogen – Land vehicles fuel tanks
WG 2	Tank containers for multimodal transportation of liquid hydrogen
WG 4	Airport hydrogen fuelling facility
WG 5	Gaseous hydrogen blends and hydrogen fuels – Service stations and filling connectors
WG 6	Gaseous hydrogen and hydrogen blends – Land vehicle fuel tanks
WG 7	Basic considerations for the safety of hydrogen systems
WG 8	Hydrogen generators using water electrolysis process
WG 9	Hydrogen generators using fuel processing technologies
WG 10	Transportable gas storage devices – Hydrogen absorbed in reversible metal hydride

Table 2. IEC working groups.

Working group	Activity
WG1	Terminology
WG2	Fuel cell modules
WG3	Stationary fuel cell power plants – Safety
WG4	Performance of fuel cell power plants
WG5	Stationary fuel cell power plants – Installation
WG6	Fuel cell system for propulsion and auxiliary power units (APU)
WG7	Portable fuel cell appliances – Safety and performance requirements

write down a proposal for regulation (www.unece.org/trans/). As ISO is involved on the same topics, an integration between the two bodies is foreseen and, for some aspects, already active.

The GRPE Working Group takes advantage of the work done in the framework of the EIHP project (*European Integrated Hydrogen Project*) whose main partners are the major European carmakers brands. The EIHP report is constituted by two parts: one referred to specific components of motor vehicles using liquid (or gaseous) hydrogen and the second one referred to the vehicles with regards to the installation of specific components for the use of liquid (or gaseous) hydrogen. These draft provisions are based of the existing regulations for natural gas with some variations due to the different nature of hydrogen.

Also SAE (Society of Automotive Engineers) started to work about regulations for hydrogen fuelled vehicles. Many of the SAE members (Toyota, Ford, General Motors, Nissan, Peugeot, Honda, Daimler Chrysler, Renault, etc.) are also developing fuel cells cars and this results in a strong interest in the development of a proper standard not only for the fuel cell test procedures but also for the electric interface with the rest of the automotive system.

Six working groups formed and several guidelines and provisions have already been published (www. sae.org).

3 FINAL CONSIDERATIONS ON THE NORMATIVE ASPECTS

The work of so many groups and standardization bodies would be very welcome if it were co-ordinate and not overlapping. Italian representatives are involved in some of the most important bodies, but no national regulation is in form of draft as yet. This results in a delay in publicly operating hydrogen technology applications, such as the bus, even though technical problems are overcome and the vehicle has passed the trial session.

The innovative character of the technology applied to the bus, proved to be interesting from the scientific point of view, puts the Authorities in the difficult position of not having the legal bases to proceed to an automatic approval. The risk analysis has been identified as the ideal tool for the assessment of safety in order to support the final authorization for the circulation on public roads.

4 BUS DESCRIPTION

The bus called CityClass-Fuel Cell is powered by an electric engine charged by the fuel cell and by a storage battery. The fuel production is obtained electrolytically from water, using hydroelectrical energy, therefore the "zero emission" is ensured for the whole cycle of production, both for the vehicle and the fuel.

The fuel storage is provided by 9 gas cylinders placed on the top of the vehicle, in which hydrogen is kept at a pressure of 200 bar; each bottle has a shut-off valve, a flux-relief valve and a fuse that intervenes in case of external fire, enabling the release of high pressure gas. Each gas cylinder is connected to the fuel manifold, which makes possible:

– to fill the gas cylinder during the fuel charging loading phase by means of flexible pipes in which hydrogen goes from the gas station to the gas cylinders; the process is kept safe by positioning tools that verify that hoses are joined to the vehicle. Three pipe are foreseen: the first one carries the hydrogen from the gas station to the gas cylinders; the second one enables the controlled release of hydrogen present into the bus before refueling; the last one charges the necessary nitrogen for

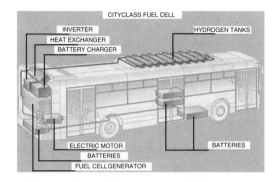

Figure 1. Bus main systems (www.comune.torino.it/atm/idrogeno.htm).

pneumatic valves that work during the refueling phase and to assure pipe cleaning at the end of the loading.

– to enter fuel into the fuel cell: hydrogen is high pressure stored (200 bar), so it is necessary to make a triple reduction of pressure using valves, in order to maintain a low pressure in the cell.

Fuel arrives to the fuel cell system that is constituted by:

– A compressor and a filter to feed the process air
– A ventilation system of the cell cabin
– A fresh water cooling system of the cell, made up of a feeding tank and a pump with filter
– A gas blow out system and a water separator that separate hydrogen from the condensate.

There are many tools that monitor the installation activity in most part of the vehicle (valve area on the top of the bus, hydrogen refueling area, fuel cell discharge area).

5 SAFETY RELATED PROBLEMS IN PROJECT DEVELOPMENT

Development and realization of the Turin Hydrogen Bus project led to several safety problems due to hydrogen dangerousness and related to the specific use and storage conditions.

Hydrogen, which is colorless and odorless, is not autoigniting, not oxidizing, not toxic, not corrosive, not radioactive, not carcinogenic; it presents a wide range of inflammability and a little ignition energy at open air. Hydrogen burns with an invisible flame with a very little heat radiated from the flame. It has significantly narrower detonation limits in air than explosion limits: when ignited early, it burns before detonation limits are reached.

Hydrogen is lighter than air and vanishes rapidly upwards: it has an high diffusion coefficient (four times more than methane) and dilutes rapidly in air.

If comparing hydrogen flammability characteristics with methane and propane ones, it's clear that hydrogen is more dangerous and has worst energy properties.

Besides, the use of hydrogen as fuel for vehicles implies the storage onboard as compressed gas at high pressure or liquefied (other techniques of storage, like nanotubes or metal hydrides are still under study).

As the above considerations suggest, the main safety aspects that emerged during the project development have been:

- operation of the vehicle. The most important safety aspects are related to: the risk of collisions with other vehicles and the loose roads pavement, that cause an increase of the frequency of a possible hydrogen release; the storage system onboard and the technical characteristics of the cylinders; the presence of a high number of people onboard and in the bus operation area, the urban center, that is characterized by a remarkable vulnerability.
- bus in depot. The safety aspects concerning the bus depot, maintenance and management phases are mainly related to the human agency and to consequence associated to a hydrogen release in a confined space.
- refueling of the bus. Also in this case, important factors for safety are the operator interventions and the consequence of an accidental event, depending on the refueling station location.
- refueling station. The most important safety problems about the station concern the quantity of gas stored at high pressure (200 bar) and the operator

Table 3. Comparison among hydrogen, methane and propane main safety characteristics.

Flammability characteristics	H_2	Methane	Propane
Flammability limits (% vol)	4.0–74.5	5–15	2.2–9.5
Ignition minimum energy (mJ)	0.017	0.28	0.25
Autoignition temperature (1 atm, °C in open air)	570	580	480
Flame temperature in air (°C)	1430	1957	1980
Lower heating value (kWh/Nm³)	3.00	9.97	25.89
Flame Velocity in air (m/s)	2.65	0.40	0.51

agency in the management of the plant. Besides, it is also important to consider that the station location, in the GTT depot, is near to the area reserved to the diesel and methane propelled buses and the methane refueling stations.

6 TOOLS AND METHODS

Because of the lacks of the national and international standards and codes, above mentioned, the safety aspects detailed in paragraph 5 were studied under two different points of view: the first one is the risk evaluation and the second one is the risk management. In particular, frequency and consequence magnitude of accidental events during vehicle in operation, in depot, refueling and during hydrogen production can be examined by risk analysis techniques (Hazard Identification, Fault Tree Analysis, Event Tree Analysis and Consequence Assessment); besides, the risks determined by these analyses can be manage with the elaboration of technical rules and procedures, specifically elaborated for the management and maintenance of the bus in the depot and for the fuelling, phases for which human intervention is particularly significant.

The following paragraph will deal with the phases and the results of the preliminary risk analysis performed on the operating vehicle: particular attention has been dedicated to the bus circulation because of the particular context in which the bus must operate.

7 BUS CIRCULATION PRELIMINARY RISK ANALYSIS

The preliminary risk analysis aims mainly to the achievement of a provisional risk evaluation of the vehicle and the definition of possible improvements of the system. The last goal of the study was to verify the adequacy of this kind of analysis to the problem.

The preliminary risk analysis is organized into the following steps:

- Hazard Identification (Historical Analysis, Failure Mode Effect and Criticality Analysis, Hazard and Operability Analysis);
- Event Tree Analysis;
- Consequence Assessment.

7.1 Hazard identification

This first phase of the study aims to determine the possible system sources of damage, the initiator events.

For each applied method, the obtained results are briefly described.

656

7.1.1 Historical analysis

Because of the innovative technology applied to the vehicle really recent, existing accidental databases do not relate specific records about accident involving hydrogen vehicles. Thus, the historical study led to examine accidents that involved different applications.

With reference to the following databases: MHI-DAS, MARS, CIRC (United States Chemical Safety), LPPI (Loss Prevention in the Process Industries, F.P. Lees, 1996), 111 accidental events, involving hydrogen, have been analyzed and catalogued depending on causes (planning mistake, human error, damage or break of components, domino effects, etc.) and consequences (flash fire, fires, BLEVE, releases, UVCE, VCE, explosions, fireballs, etc.) of the accident.

The following results are pointed out:

– Accidents during the transportation of hydrogen tanks are more than 10% of the total.
– More than 25% of total accidents concern compressed hydrogen and its storage in cylinders.
– The most common causes of accident are the leakage of compressed hydrogen and the human error. Component leaks are mainly related to valves and pipes; the erroneous intervention of the operators are more frequent during the maintenance than in management of the plant, in normal conditions.
– Explosions represent the most probable evolution of an accident.

7.1.2 FMECA e HAZOP

The operative phase taken into account is Bus operation, i.e. the hydrogen supply from the cylinders storage on the top of the bus to the fuel cell.

The systems considered during the analysis are essentially composed by the three following elements:

– Nine cylinders of compressed hydrogen, located on the top of the bus;

– Refilling pipe for the connection of the cylinders with the fuel cell;
– Fuel cell.

All these macro-components are equipped with their control and safety systems.

Analysis concerning the fuel cell component couldn't be exhaustive due to the confidentiality associated to this new technology.

The most critical accidental event is the break of the pipes that transport hydrogen at the pressure of 200 bar: because of external events, like collision, corrosion, thermal and mechanical stress, the break of the pipe can happen and produces the release of all the gas stored on the bus, present in the different pipes and cylinders. This is due to the lack of shut-off systems able to isolate each single cylinder.

The design of a shut-off system and its localization to prevent the release of the total mass of hydrogen is the first result obtained by the risk analysis.

7.2 Event tree analysis

This probabilistic technique has been applied to the Initiating Event (IE) described above, that is the release of hydrogen from one of the main pipes, considering both the possibilities of the presence and of the lack of a shut-off system that allows to limit the quantity of hydrogen released.

With reference to the OREDA 92 and EGIG 1997 Databases, it has been possible to estimate the frequency of the two accidental events, that involve the release of the total mass of hydrogen content in the nine cylinders (Initiating Event A, without shut-off system) and the release of the contents of two cylinders (Initiating Event B, supposing the presence of a shut-off system). The occurrence frequencies of the possible evolution of the accidents are reported in Table 4.

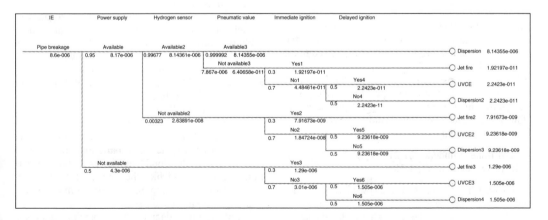

Figure 2.   Event tree for the iniating event B.

Table 4. Obtained occurrence frequency for A and B initiating events.

	A IE (ev/year)	B IE (ev/year)
Jet-fire	1.3×10^{-6}	1.3×10^{-7}
UVCE	3.0×10^{-6}	1.5×10^{-7}
Atmospherically dispersion	3.0×10^{-6}	1.5×10^{-7}

Table 5. Results from the release model.

	Max flow rate (kg/s)	Outflow duration (s)
A IE	1.22	94
B IE	1.22	21

Table 6. Results from the Jet-fire model.

Jet-fire length (m)	11
Max jet-fire width (m)	4
Distance (Heat radiation = $12.5\,kW/m^2$) (m)	13
Distance (Heat radiation = $5\,kW/m^2$) (m)	15

The main lack of this probabilistic study is related to the unavailability of reliable data about the break of the particular pipes used for hydrogen transport; besides, the contribution to this frequency due to the particular location and use conditions of the pipes could be estimated only by expert judgment.

7.3 Consequence analysis

The study concerning the modeling of physical phenomena has a great importance within the risk analysis because the bus activities interest an urban center, area characterized by a high vulnerability.

The phenomena simulation, aimed to the evaluation of the heat irradiation and overpressure values following a fire or an explosion; it was performed by using simplified physical-mathematical models, described in the Yellow Book of TNO and implemented in Effect 4.0 Software.

The release flow rate has been estimated for both the hypothetical IE, described above, considering the release of a quantity of hydrogen equal to the contents of all the nine cylinders (A) and to the contents of two cylinders (B). The consequence of both outflows have been evaluated, considering the happening of a jet-fire (if hydrogen is ignited early) and of a UVCE (if hydrogen is ignited later).

7.3.1 Release

The critical outflow values (A and B IE) have been obtained by the application of the TNO Model for gas release from vessel through hole in pipe linked to the vessel itself. The hypotheses made for these simulations are very conservative: the release has been considered to happen from the pipe (diameter equal to 12.7 mm) which works at the highest pressure; outflow of the greatest possible quantity of hydrogen has been studied, equal to:

– IE A (nine cylinders): 21.2 kg
– IE B (two cylinders) : 4.7 kg

Analysis has been performed considering two different size of the hole in the pipe, equal to 50% and 100% of the diameter; the results here related concern the worst case, 100%.

The results from the release model are reported in Table 5.

7.3.2 Jet-fire

The study of this phenomenon is based on the Chamberlain Model, described in the Yellow Book of TNO.

The obtained results have been the same for the two Initiating Events (A and B), since the maximum flow rate value is the same in both releases: the flow rate, in fact, depend only on the storage pressure and on the size of the hole and not on the quantity of substance released.

Outputs of jet-fire model are affected by a particular uncertainty due to the great influence on the results of the wind velocity, introduced as input data. In fact, the bus speed influences this parameter: the value to insert in the model has to consider both factors: wind velocity (and its direction as to bus one) and vehicle speed.

Table 6 shows the results obtained analyzing an horizontal jet, a wind speed of 2 m/s and a rupture of the 100% of the pipe diameter.

7.3.3 Dispersion

Dispersion modeling was performed using a gaussian model for neutral gas dispersion in atmosphere (Yellow Book of TNO); since hydrogen is a gas lighter than air and vanishes rapidly upwards, model outputs are not so realistic but very conservative. An alternative could be the application of a turbolent model for turbolent free jet.

Dispersion model introduces a large number of uncertainties in the analysis because the used simplified models don't allow to accurately describe the geometry of the systems, in particular the level of confinement of the cloud that is an essential information in this case, in which the context of the accidental event presents complex geometry (urban context with high buildings and narrow roads).

Using as input data in the dispersion models a wind speed equal to 1 m/s and the Pasquill-Gifford

Table 7. Results from the dispersion model .

	IE A	IE B
Total explosive mass (kg)	7.5	2.5
Maximum distance of source to LEL (m)	91.8	60.3
Maximum distance of source to UEL (m)	0.0	18.2
Maximum height to LEL (m)	3.4	3.4
Maximum height to UEL (m)	0.8	0.0

Table 8. Results from the UVCE model.

	IE A	IE B
Distance to the source-$\Delta p = 0.3$ bar	26 m	17 m
Distance to the source-$\Delta p = 0.07$ bar	74 m	48 m

stability class F, the following parameters have been calculated at time t, when the most important consequences occur.

7.3.4 UVCE

The not confined explosion of the hydrogen cloud has been modeled by the TNO Multienergy Model. Since this analysis is a preliminary study, an Unconfined Vapour Cloud Explosion (UVCE) has been analyzed, although the hydrogen quantity is small.

7.4 Vulnerability and risk evaluation

The vulnerability analysis has been performed considering the bus route on the urban center of Turin, route characterized by a high density of people (due to the presence of buildings, public schools, commercial activities), narrow streets, refueling stations or industrial plants and by the presence of possible trigger sources, like electrified tramlines.

Risk evaluation of the operating bus has been performed in three different situations: vehicle in motion (case 1), vehicle standing at the bus stop (case 2), vehicle passing near to a local market (case 3).

Thus, the damage areas due to the jet-fire and UVCE phenomena are applied to zones characterized by different vulnerability degrees.

Besides, some general hypotheses for the damage calculation have been done: in the event of an explosion, it was supposed that the 5% of the people hit by an overpressure wave higher than 0.3 bar it dead. This hypothesis is conservative: in fact Lees suggests a death probability minor than 1% about overpressure inferior to 1–2 bar. In the event of a jet-fire, vulnerability has been considered equal to 100% for people directly reached by the flame, while it has been supposed to be 5% for the people interested by a heat irradiation equal or superior than 12.5 kW/m² (also this hypothesis is conservative, if it is considered that Lees

Table 9. Obtained social risk values.

	End of	Risk value (dead/year)		
IE	sequence	Case 1	Case 2	Case 3
A	Jet-fire	1.4×10^{-7}	3.0×10^{-6}	2.9×10^{-7}
	UVCE	1.8×10^{-7}	1.3×10^{-6}	9.0×10^{-7}
B	Jet-fire	1.3×10^{-8}	2.8×10^{-7}	2.7×10^{-8}
	UVCE	8.0×10^{-9}	6.0×10^{-8}	4.2×10^{-8}

proposes a lethality equal to 1% for a 10.2 kW/m² heat radiation continuous for 45.2 s or more.

Case 1, vehicle in motion: the value of average density of population has been considered equal to the density of inhabitants of the urban center of Turin.

The occurrence frequencies of the accidental events, calculated by Event Tree Analysis and reported in paragraph 7.2, have been modified depending on:

– time of bus circulation, excluded time at bus stops;
– daily service hours (stop in the bus depot time is about 10.5 hours per day);

Case 2, vehicle standing at the bus stop: considering a presence at the bus stop or near it equal to 30 people, a density of 0.4 person/m² has been obtained.

Also in this case, the occurrence frequencies resulted by the event tree analysis were corrected and by:

– probabilities that the vehicle is operating and stopping at the bus stop at the accidental time;
– probability of the jet in the stop direction.

Case 3, vehicle passing near to a local market: the accidental event has been considered to happen on holiday (day of greatest crowd for the market) and the presence has been estimated in 700 people in a 2400 m² area.

The factors that modify the occurrence frequency in this case are:

– probability of the bus presence near the market square;
– probability that the vehicle is operating at the accidental time;
– probability of the jet in the market direction.

The risk values obtained as product of frequency and damage of accidental events are below related.

The risk value, so determined, is the social risk. The individual risk, that interests the single inhabitant of the city, has been determined too. Individual risk can be calculated dividing the social risk by the number of people that can be really interested by the phenomenon when it happens.

Table 10. Comparison between obtained risk values and threshold values.

	Calculated value	Threshold value
Individual risk (dead/year · person)	1.35×10^{-9}	8.8×10^{-5}
Social risk (dead/year)	2.26×10^{-6}	5.0×10^{-6}

Comparing the obtained values with the respective threshold values for acceptable risk, it is possible to evaluate the risk acceptability.

In Table 10, the worst case is reported.

Threshold value for the individual risk has been calculated with reference to risk associated to bus road accidents in the United Kingdom, that is not very different from the road accident risk of every kind of vehicle in Italy.

Threshold value for the social risk has been determined with reference to the frequency-damage diagram elaborated by the Italy region Friuli Venezia Giulia to estimate the industrial risk, not existing in Italy any codes or standard to define an acceptable risk.

8 CONCLUSION

The results obtained by the preliminary risk analysis allowed to conclude that the vehicle would circulate safely, also in the area characterized by an high vulnerability, as the local market area.

The preliminary risk analysis allowed also to find out the possible improvements to the system, as the arrangement of pneumatic valves to intercept each single cylinder just upstream the manifold, the arrangement of a safety valve in the highest pressure system part and the replacement of the two components for the pressure reduction with only one valve.

The preliminary analysis, although it allowed to perform a first evaluation of the bus risk, presents several lacks and approximations; some considerations about this work will be able to direct the future studies:

– the definition of the occurrence frequency of pipe rupture for the particular working condition and location of this component requires a further investigation, in order to reduce the associated uncertainties;

– since accurate information about the fuel cell aren't available, it was considered as a "black box" in the study. It could be useful to examine this component more carefully, although the fuel cell works with hydrogen at low pressure (2–3 bar) and it is expected that the worst possible bus accidents concern hydrogen at high pressure;

– the physical-mathematical models used for the consequence study are simplified: in consideration of the complexity of the context in which the bus works, it would be better to use more accurate models, as CFD and 3D codes;

– the Domino Effects possibility (although improbable) that can happen, e.g. because of the presence of several gasoline refueling stations along the bus course, should be investigated.

ACKNOWLEDGEMENTS

The authors thanks the Gruppo Torinese Mobilità S.p.A (GTT), particularly Dr. A. Santel, Eng. M. Morza and G. Eandi; SAPIO S.r.l, particularly Eng. S. De Sanctis; Irisbus Italia S.p.A, to Compagnia Valdostana delle Acque S.p.A. (CVA), that all believe in and develop this innovative project in Italy.

REFERENCES

3rd EGIG-Report 1970–1997. 1998. *Gas pipeline incidents*, Doc. number: EGIG98.R.0120

Gatto, Domenico 2002. *Uso dell'idrogeno come propellente di un autobus urbano: valutazione del rischio, Tesi di Laurea*. Politecnico di Torino.

Lees, F.P. 1983. *Loss prevention in the process industries*. Londra. Butterworths.

OREDA Participants 1992. *Offshore reliability data handbook*. Hovik (Norvegia). Veritec.

TNO 1998. *Methods for the calculation of physical effects (yellow book TNO)*. Voorburg (Olanda).

Safety and Reliability – Bedford & van Gelder (eds)
© 2003 Swets & Zeitlinger, Lisse, ISBN 90 5809 551 7

Risk management in the pre-design phase of civil structures

P.J. Van Gestel & E.C.J. Bouwman
Delta Pi, Duiven, Netherlands

ABSTRACT: The city council of Amsterdam has planned to build a road traffic tunnel called "De Bongerd tunnel". Holland Railconsult has the assignment to perform the pre-design activities resulting in a set of requirements and a preliminary design. To ensure that the set of requirements is complete, a risk management cycle was established with the first step being a preliminary risk analysis. From this risk analysis it was also possible to derive measures in order to prevent unexpected risks to occur in the next design stages. To support the risk management cycle a risk register was created for use during the design, build and operation phases of the project. The conclusion was unanimously: risk management proves to be a successful methodology to guarantee a complete set of requirements, to foresee risks and to build a risk register. It is also expected to be useful to manage the risks in the future phases of the project. In the paper the methodology and the process followed will be explained. The added value of the authors approach consist of a more structured risk analysis on projects using the FMECA methodology known from Systems Reliability. Because of confidentiality reasons the results are presented in general terms.

1 GENERAL RISK MANAGEMENT APPROACH IN PROJECTS

1.1 The Risk management steps

Risk management keeps a structural overview of subjects, like financial issues or throughput times that are important for the satisfactory progress of projects. Risk management in projects is a process aimed at the structuring of project activities and the managing of important project risks. Five steps are identified:

1. Risk analysis
2. Definition of control measures
3. Implementation of control measures
4. Evaluation of control measures
5. Update of the risk analysis

The risk management process is characterized by a cycle and a layered structure. In the beginning of a project the management will be confronted with great uncertainties and a lack of clarity. At that moment, risk management has a structuring role in the creation of a project plan and the identification of the most important bottlenecks. A rough risk analysis is performed. As the project develops, the lack of clarity will diminish and the existing uncertainties and risks can be analyzed in more detail.

A risk analysis captures a moment in time. This is repeated during the whole project. The depth of

Figure 1. Overview risk management process.

analysis depends on the extent of the risks, the necessary counter measures and the phase of the project.

1.2 Risk analysis

In order to perform a risk analysis of a project, a Plan of Action is required. This plan should at least describe all the analyzing activities that have to be performed, the planning of those activities and the budget available for the complete project. This analysis can be split up into several phases.

1.2.2 List of risks

The first phase of an analysis is to compile a comprehensive list of risks. This can be done in discussions and brainstorming sessions with those involved. Unwanted occurrences that do not directly belong to the common uncertainties of project management are registered in a database: a risk register.

It is recommended to work from coarse to fine. In that way risk management gradually provides better insight and overview. Based on this, priorities are set, which can be adjusted by a more refined analysis.

1.2.2 Quantification

In order to distinguish important risks from less important ones, it is necessary to perform some kind of quantification. If exact data are not available, a method can be used in which the probability of occurrence and the severity of the effect can be classified into 4 to 5 categories, ranging from small to large. The risk is quantified by multiplying the probability of occurrence with the severity of the effect.

Special attention is required for those unwanted occurrences with a small probability and large effect.

1.2.3 Modeling and simulation

To make the right decisions a good foundation is required. Modeling and simulation can add to the depth of the analysis. Using these techniques, it is possible to accurately predict the risk of exceeding available time or budget. In addition, these and/or possible other dominant risks will become clearly visible.

1.2.4 Risk management cycle

After performing the initial risk analysis the effect of the risks on the objectives of the project will become visible. There are two reasons why risks should be managed in time.

On the one hand risks change as the project develops. Risks are added up, some disappear; uncertainties become less as time progresses. The actual status of potential risks should be kept up to date regularly. During the phase transitions, frequently new risks are introduced.

On the other hand the project organization takes steps to reduce the probability of occurrence of unwanted events and their effects. The formulation and implementation of these steps are triggered via control measures. The effects of these actions have to be evaluated. They may change the risk profile of the project to such an extent that the risk manager has to initiate a periodic update of the risk analysis.

Despite the control measures, some risk frequently remains: the residual risk. An accurate judgment of the size of the residual risk heavily depends on the reliability of the control measures: their actual implementation and their effectiveness. The risk manager is responsible for judging this risk. The project manager determines if the residual risk is acceptable and whether it is in line with the objectives of the company and the aims of the project.

1.3 Definition of control measures

Risk management can be defined as a pro-active control of risks. This implies that in advance, considerable effort is put into possible measures to be taken. Such measures can be preventive or corrective. Preventive measures are aimed at decreasing the probability of occurrence or at limiting the expected negative effects. When an unwanted event manifests itself, corrective measures are applied. They are always aimed at reducing the effects or repair activity.

Prior to the decision on which measures should or should not be applied, criteria must be set to indicate which risks are acceptable. Is the measure conclusive and will it limit the risks to an acceptable level or is it not possible to find conclusive measures and will the company have to find other solutions?

It is not discussed here how relevant criteria can be defined. These criteria have to be defined at the beginning of the project. They should be recorded in a risk matrix in close consultation with the top management.

As a rule, measures have their price. Evaluations will have to be made to compare the costs of a measure against the possible advantages. During such evaluations the HILP events (High Impact Low Probabilities) will play an important role. Often, it is reasonable to cover these risks by insurance. For each measure a person should be nominated, who takes care of the execution and is responsible for the results.

1.4 Execution of control measures

The actual execution of the actions resulting from the risk analysis is as important as the definition of those actions. The awareness of the risks will be a disclosure to many, but the project manager is ultimately responsible for the execution of the actions and the required risk reduction. The risk manager has only a facilitating and advisory task. In addition he should guard the progress of the different control measures.

It is important to appoint a responsible person for each deviation in order to ensure that the measures are really carried out. When risk measures are formulated, this person is the first one responsible to pursue the progress of the measures being carried out. The risk manager will support and question him on a regular basis on the status of the risks and the accompanying measures.

Due to measures formulated or carried out the risk manager must periodically keep the amendments to the risks in the risk register up to date.

1.5 Evaluation of control measures

The aim of the evaluation of the control measures is to determine if the control measures had the desired effect and if they were efficient (costs versus assets).

In the case of risks with a low probability of appearance it is difficult to prove the effectiveness of the proposed measures, let alone their efficiency. A retrospective analysis is easier. For such an analysis, only the information available at the time the decision was made has to be taken into account. It is therefore of the utmost importance that this information is well documented.

During the project, the evaluation of control measures should be performed on a periodical basis. In the long term, an evaluation after project closure is advisable (lessons learned).

1.6 *Update of the risk analysis*

During a project more, new, risks can surface. These risks are integrated into the risk management cycle. The effect of measures may also lead to risk reduction. In this case, an update of the risk analysis is appropriate. Sometimes it is sensible to completely re-do a risk analysis. This will occur when many changes have been made to the original concept. It is the task of the risk manager to recognize this need. Also, a phase transition may require a completely new analysis to be carried out.

2 THE "BONGERD" APPLICATION

2.1 *Description of the technique applied*

After defining the surroundings of the project, the actors, the phases and the tasks of all actors in the different phases were established. The different phases are:

- Design
- Build & Construct
- Exploitation

With this information it is possible to carry out the first risk analysis. A kind of FMECA (Failure Modes Effects and Criticality Analyses) technique was used. This method is originally designed for the technical domain and is meant to identify in a systematic manner as many risks as possible. By using "guide words" all possible deviations (failure modes), the accompanying causes and effects will be investigated for every defined task. Examples of such "guide words" are: not, incomplete, incorrect, inverse, too early or too late. Further, for every identified deviation the probability of occurrence has been determined on the scale "Credibility" of 1 to 5, where 1 means "unlikely" and 5 means "high".

The effects of the deviations are categorized in:

1. Technique
2. Judicially and Planning Procedures
3. Licenses
4. Safety
5. Co-operation (internally and externally)
6. Political decision making
7. Finance
8. Maintenance and Exploitation
9. Underground Infrastructure
10. Environment
11. Soil conditions
12. House-building
13. Surrounding
14. Other

To determine the degree of seriousness of the deviation in every category, the deviations of the original objectives should be investigated. In the "Bongerd" Case the scale "Misery" of 1 to 5 has been used. The criticality is determined by multiplying the probability of occurrence with the degree of severity of the effect. This criticality factor is a measure of the risk for a deviation. All unwanted events can be plotted in a criticality graph for each category and for all categories.

As it is impractical to determine the measures that can be taken to decrease the criticality for all the deviations, only the most dominant deviations are considered. To be able to determine these, a limiting value of probability, criticality or severity of effects should be defined. All deviations higher than or equal to these limits will have to be investigated in more detail:

- Credibility > 2
- Misery > 2
- Credibility * Misery > 9

After determination of the risks and criticalities measures need to be defined to decrease the risks. If risk management is started up in the early stages of a project, it should be relatively easy to define measures that will decrease the probability of occurrence. Nevertheless, there are always risks that are difficult to be influenced by project management or cannot be influenced at all. This is especially true for deviations caused by third parties such as the government or other external parties.

The criticality graph can be recalculated after defining the measures.

2.2 *The "Bongerd" results*

Risk management started in a very early stage of the pre-design phase. In a brainstorm session all the stakeholders were invited to sum up all kinds of risks and possible measures. This overview became the basis for further analysis. In the pre-design phase focus was on the set of requirements. Important issues were legislation and regulations. The needs and the nice to have subjects were investigated and distinguished.

In the second step of the risk approach an overview of activities was made. Not only the activities in the pre-design phase have been taken into account, but also the activities in the other phases (Build & Construct and Exploitation) were listed. It was important to understand that the risks gathered from the brainstorm session would have different impact in the different phases, so different measures were needed. The project activities were analyzed with the FMEA technique and the results were coupled to the results of the brainstorm session. The initial set of risks and measures was updated. More possible risks and measures were found. Above all the risks were more structured and placed in a time path.

In the third step the Criticality Analysis was prepared. First of all a good description of the different severities was needed. The so-called Misery Index ranges from 1 to 5, but what is the meaning of the numbers? The result of the discussions was:

1. Extra time or money, but within project margins
2. Delay and small exceeding of the budgets
3. Social complications, political anxiety, severe delay or economical losses
4. Accidents or Environmental damage with social consequences, budget problem with political consequences
5. Fatalities, national political consequences, no start or immediate stop of building the tunnel

No discussion has taken place concerning the probability of occurrence the so-called Credibility Index from 1–5: Nil/very small, unlikely, likely, real and big.

In step four all stakeholders were interviewed. For the second time the risk analyses was updated not only qualitatively, but also quantitatively.

In step five the results of the risk analyses were used to analyze the initial set of requirements.

In step six the results were reported. 138 unwanted events spread over 14 categories were identified. 86 events had a risk higher than 8 (see table 1). To every unacceptable event one or more measures were coupled in the three different phases. At the time of the report 25 measures were already taken. 39 measures were initiated, but not ready and 68 measures had to be taken in future phases.

For all categories a risk matrix can be drawn. In this paper the combined risk matrix is given in figure 2. The actual risk matrix given the measures taken is depicted in figure 3.

It is expected that the other proposed measures will lower the risk under the criticality of 9 except for:

- Traffic accidents in and near the tunnel (risk 20 → 9)
- Fire in tunnel (risk 20 → 15)
- Fire near tunnel (risk 12 → 9)
- Explosion in tunnel (risk 15 → 10)

It is possible to lower the risk concerning fire in tunnel from 15 to 9 by building a separate fire lane. This option is still open.

Table 1. Number of unacceptable events spread over categories.

Category/Risk number	9	10	12	15	16	20	Total
Technique	2		4				6
Judicially and Planning Procedures	6		5				11
Licences	4		5		2		11
Safety	1	5	2	2	1	3	14
Co-operation (internally and externally)	5	1	2		3		11
Political decision making	1						1
Finance	2				2		4
Maintenance and Exploitation			2	1			3
Underground Infrastructure	1						1
Environment	1		6				7
Soil conditions	1		2				3
House-building	2		2		1	1	6
Surrounding	2	1					3
Other	2				2	1	5
Total	30	7	30	3	11	5	86

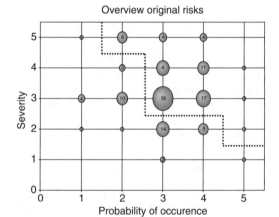

Figure 2. Overview original risks and acceptance criteria.

2.3 Actualization of the risk analysis using the risk register

From the point of view of risk management it is important to follow the risk assessment in time, specially concerning:

- The effect of the measures taken
- The milestones concerning the end of measures taken
- The decision concerning the measures not taken yet (yes or no, when?)

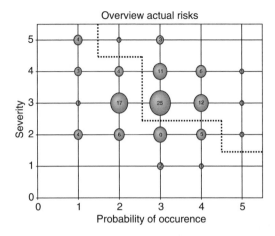

Overview actual risks

Figure 3. Overview of actual risks and acceptance criteria.

It is important to actualize the risk analysis after a considerable period of time, when a new phase is entered or when the project has been considerably redefined. A database, the so-called risk register, can be used in order to have an overview of all risks well organized and quickly available. Moreover, it enables the project manager to follow the history of the risks management process. All information regarding each risk, its causes, effects, probabilities and seriousness are filed here. In addition, for each risk the status is stored according to the measures defined, even if they have not yet been executed at that time. If a risk disappears because of one or more measures taken, this fact is also stored in the database.

3 CONCLUSIONS

Risk Management is a good approach to keep control over the risk in this project and projects in general, especially when started in a very early phase.

The effort spent contributes to the quality of the product. Above that it permits an easy control over the progress in the project (are targets met?) and over the permitted budget (with the resources allocated?) .

The added value of the use of the FMECA methodology to derive risks is threefold:

a) It provides a more structured framework for future expansion and adaptation of the Risk Analysis (and therefore can be reused to a large extent in other similar projects)

b) The use of guidewords triggers people to recognize risks that otherwise may not have been noticed (and therefore a more complete list of risks will be generated)

c) The use of the guidewords facilitates the standardisation of the description of Risks (and therefore minimize the number of double entries of the same risks)

REFERENCES

Methods for determining and processing probabilities ("Red Book"), CPR 12E, Committee for the Prevention of Disasters, Second edition 1997, ISBN 9012085438, Chapter 7.6.

Safety and Reliability – Bedford & van Gelder (eds)
© *2003 Swets & Zeitlinger, Lisse, ISBN 90 5809 551 7*

Processing method of microtremors for local site effect estimation

G. Ghodrati Amiri & R. Motamed
Department of Civil Engineering, Iran University of Science & Technology, Tehran, Iran

A. Ghalandarzadeh
Department of Civil Engineering, Faculty of Engineering, University of Tehran, Tehran, Iran

ABSTRACT: The aim of this study is to show how to process microtremors effectively. Commonly two different methods of spectral ratios H_s/H_r and H/V are used for analyzing microtremor data. Regardless of the type of employed method (H_s/H_r or H/V), special emphasis is given to the processing procedure of microtremors as signals. Evaluation of different methods of windowing; Hamming, Hanning, etc. and different windowing lengths are investigated. Five different window lengths of 10, 15, 20, 24 and 30 seconds and four overlap values of 10%, 30%, 50% and 70% of window length are used. The results are compared in terms of fundamental frequency and amplification factor. Statistical study is performed to show the consistency of the calculated values regarding parameters such as frequency resolution and standard deviation. This study shows that the data processing method has a significant effect on the final results and by applying suggested methods, obtained results will be more reliable.

1 INTRODUCTION

Damages occurred in the recent earthquakes showed that local site conditions have a significant effect on ground motions. Therefore, site response studies have an important roll in seismic microzonation studies. The application of microtremors to determine dynamic characteristics (predominant period and amplification factor) of soil layers was pioneered by Kanai and Tanaka (1954, 1961). Nowadays microtremor measurements are generalized in site characterization due to its simplicity, low cost and minimal disturbance to other activities. Investigation of microtremors for engineering needs leads to a complex task both for the problems related to the nature and the characteristics of microtremors, and the insufficiently standardized methods used. This is the main reason why there are differences in opinions, biases in results and the insufficient reliability of results. The existing methods are not adequate regarding the characteristics of microtremors (Stojkovic, 1998). Therefore, in this study main emphasis is given to the optimum processing methods from the viewpoint of Digital Signal Processing (DSP) (Oppenheim et al., 1995). First, the time window is evaluated and after introducing the suitable one, the most accurate overlapping value is determined. Finally, effect of different windowing methods is considered and the most suitable one is introduced. Results are concluded based on the dynamic soil

characteristics (fundamental frequency and amplification factor) and statistical methods. In this study, microtremor measurements carried out in Urmia city in Iran, are used. A computer program was developed in MATLAB, using Signal Processing Toolbox for analyzing microtremors, and H/V spectral ratio method (Nakamura, 1989) was applied. Calculated H/V spectral ratio curve is assumed to be the transfer function of the soil layer and the first peak is treated as the fundamental frequency and its amplitude is an estimation of the amplification factor at that site.

2 NATURE AND MAIN CHARACTERISTICS OF MICROTREMORS

Based on many experimental investigations, it was concluded that microtremors represent a permanent process of soil ambient vibration. These vibrations are induced by the complex different factors acting in the surrounding of the investigated place which are not known, but their effect is always variable. Considering the complex nature of these factors, they can't be predicted reliably. Therefore, based on consideration of similar events induced by unknown factors, it was concluded that microtremors represent a random stationary process of stochastic nature. The random characteristics of microtremors are due to their, always changing, excitation, wherefore each microtremor observation has different and non-repeatable record, while their characteristics as a stationary process are due to the effect of the characteristics of the local soil medium which makes microtremors of specific characteristics at each place (Stojkovic, 1998).

However, the effect of soil medium is changeable in the course of time depending on the permanent modification of the amplitude and frequency content of microtremor excitation, wherefore it may be expected, with a certain probability, that the stationary part of microtremors will be able to reflect the predominant period and the vibration characteristics of soil. This makes microtremors of a stochastic character. (Stojkovic, 1998)

3 EFFECTIVE PARAMETERS ON MICROTREMOR PROCESSING

Some parameters affect the results obtained from microtremors. These parameters have significant role in site characteristics like predominant period and amplification factor. In this study, the effect of these parameters is investigated.

3.1 Conditions of microtremor observation

All the microtremor records should be carried out at a certain place under corresponding and identical ambient conditions of excitation. In this case, the effect of the soil medium upon the stationary part of microtremors shall be reflected more thoroughly and uniformly (Stojkovic, 1998). For example, all the observations should be done at night which the noise like traffic and human activities are minimum. In addition, record length must be long enough to detect the site dynamic characteristics. Whatever the record length is long, the site fundamental frequency is represented in the record and after transferring into frequency domain, it will appear more manifestation. Stojkovic (1998) related the record length to the predominant period of site that can be estimated according to the well-known expression $T = 4H/V_s$, H is the soil layer thickness and V_s is the shear wave velocity of soil, based on general geological properties of the investigated area. The record length should be 4 to 6 times greater than the expected predominant period, in order to be more certain that the expected predominant period which will be reflected in the spectrum. Considering these points, 8 to 10 records of 120 seconds length were recorded in Urmia city.

In order to detect the main frequency content of signal, the sampling frequency rate should be at least two times greater than f_{max} which is the maximum frequency in signals. The minimum sampling frequency, $f_s = 2f_{max}$, is called the Nyquist frequency (Solnes, 1997).

Regarding the fact that interested frequencies in engineering practice are 0.1 to 10 Hz, in this study, sampling frequency was 100 Hz that satisfies the mentioned criterion.

3.2 Window length

As window length becomes longer, the frequency resolution of calculated spectrum increases. This matter leads to a non-smoothed curve that fundamental frequency cannot be recognized easily. In addition, spectral leakage can diminish by limiting the window length. This is a trade-off between requested frequency resolution and window length (MATLAB User's Guide, 1999).

In order to find the best window length, ten microtremor measurement sites in Urmia city were chosen. Selected sites were situated in different geological conditions. Five different window lengths of 10, 15, 20, 24 and 30 seconds were considered in this study. In each site, calculations were done and H/V spectral ratios were plotted. As shown in Figure 1, longer window length results in irregular curves and dynamic site characteristics can be recognized hardly. Figure 2 shows the variation of amplitude with window length. As mentioned before, longer window length results in higher amplitude which increases the uncertainty inherent microtremors for evaluating the soil amplification factors.

It is shown that the window length of 20 seconds is the best choice. In this case, spectral ratio curves have a well bell-shape, sufficient resolution. Also, fundamental frequency and amplification factor can be easily distinguished.

3.3 Overlap Value

It is possible to reduce the variance of the spectrum by breaking the signal into non-overlapping sections and averaging the spectra of these sections. The more

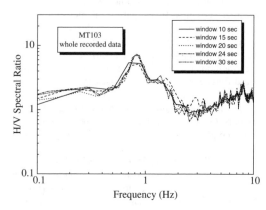

Figure 1. Transfer function calculated by five different window lengths.

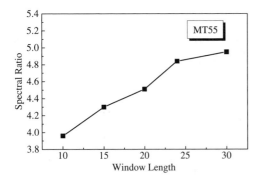

Figure 2.   Relation between window length and spectral ratio.

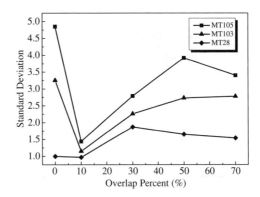

Figure 4.   Standard deviation values at resonance peak.

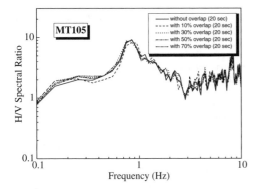

Figure 3.   Transfer function calculated by five overlap values.

Table 1.   Different windowing methods in DSP (MATLAB User's Guide).

Windowing methods	
Rectangular	Hamming
Triangular	Kaiser
Hanning	Chebyshev

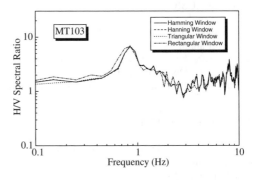

Figure 5.   Transfer function calculated by different windowing methods.

sections are averaged, the lower variance will be achieved. However, the signal length limits the number of sections. In order to obtain more sections, the signal is broken into overlapping sections. There is a trade-off between the number of sections and the overlap rate (MATLAB User's Guide).

There is no general rule for selecting overlap value. It depends on data nature (Satoh et al., 2001). In this study, five overlap values of 0, 10, 30, 50 and 70 percent were considered. Resulted transfer functions are shown in Figure 3 and can be seen that different overlap values result in almost same curves.

Therefore, calculated fundamental frequency and amplification factor are similar. If standard deviation at the resonance peak is considered, it can show that standard deviation in the case of 10% overlapping value is minimum (Figure) 4.

3.4   Windowing method

Another way to improve the spectrum is to apply a nonrectangular window to the sections before computing the spectra. This reduces the effect of section dependence due to overlap, because the window is tapered to zero on the edges. Also, a nonrectangular window diminishes the side-lobe interference or "spectral leakage" while increasing the width of spectral peaks. With a suitable window (such as Hamming, Hanning, or Kaiser), and optimum overlap rates, the calculated transfer function has lower variance. Also, nonrectangular windows can reduce the Gibbs effects. Different windowing methods in Digital Signal Processing (DSP) concern are shown in Table 1. Application of different windowing methods depends on the nature of signal and window properties. In order to find an optimum window, four common window types were selected; Rectangular, Triangular, Hanning and Hamming. Analysis was done in each case and resulting transfer functions were plotted (Figure 5). Results

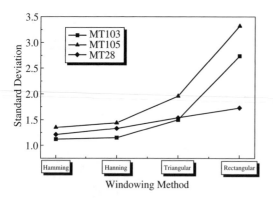

Figure 6. Standard deviation at resonance peak through different windowing types.

show that calculated transfer functions are very similar in different cases. But standard deviation in the resonance peak of curves shown in Figure 6 indicate that rectangular window gives higher standard deviation values and Hamming type gives the lower value.

4 CONCLUSIONS

It is concluded that time window length of 20 seconds, with overlapping value of 10% and Hamming window type are the best choices in processing of microtremor data. Obtained fundamental frequency and amplification factor in this case are more reliable.

ACKNOWLEDGMENTS

The assistance of Mr. S. Soleimani of University of Arak is highly appreciated. Thanks are also extended to Mr. H. Nouri for his great help.

REFERENCES

Kanai, K. and Tanaka, T. 1954. Measurement of the microtremor. *Bulletin of Earthquake Research Institute*, Tokyo University, 32: 199–209.
Kanai, K. and Tanaka, T. 1961. On microtremors. VIII. *Bulletin of Earthquake Research Institute*, Tokyo University, 39: 97–114.
Oppenheim, A.V., Schafer, R.W. 1995. *Discrete-Time Signal Processing*. Prentice-Hall International Inc.
MATLAB User's Guide. 1999. Mathworks Inc.
Nakamura, Y. 1989. A method for dynamic characteristics estimation of subsurface using microtremor on the ground surface. QR of RTRI, 30(1): 25–33.
Satoh, T., Kawase, H. and Matsushima, S. 2001. Differences between site characteristics obtained from microtremors, S-waves, P-waves, and Codas. *Bulletin of the Seismological Society of America*, 91: 313–334.
Solnes, J. 1997. *Stochastic Process and Random Vibrations*. John Wiley and Sons, England.
Stojkovic, M.B. 1998. Methodological aspects of investigation of stationary characteristics of microtremor ground motion. *11th European Conf. on Earthquake Engineering*, France.

NOTATION

H_s/H_r: Spectral ratio of horizontal component of recorded microtremors on sediment to rock outcropping.

H/V: Spectral ratio of horizontal to vertical component of recorded microtremors on sediment.

Safety and Reliability – Bedford & van Gelder (eds)
© *2003 Swets & Zeitlinger, Lisse, ISBN 90 5809 551 7*

Seismic risk analysis: a method for the vulnerability assessment of built-up areas

S. Giovinazzi
DICEA, University of Genoa, Italy

S. Lagomarsino
DISEG, University of Genoa, Italy

ABSTRACT: The principal reason for loss life and injury due to earthquakes is the collapse and damage of vulnerable buildings located in highly seismic regions; it is fundamental to identify the most potential danger-ous situations for mitigating seismic losses through an optimised allocation of funds to be used for prevention strategies and emergency management. The proposed method represents a useful tool to reach this aim as it allows the vulnerability analysis of ordinary buildings at a territorial scale employing data with different origins and precision. The method makes reference to EMS 98 typology classification; an index of average vulnerabil-ity is associated to each typology, which may be refined on the basis of behaviour modifiers. This index allows the identification of an analytical relationship between seismic input and damage through which all possible scenarios can be assessed such as collapsed or uninhabitable buildings, casualties, homeless and economic loss.

1 INTRODUCTION

1.1 Risk and scenario analyses

Carrying out a vulnerability analysis means assessing the consistency of a built-up zone in a given area, both in quantitative and qualitative terms, and in particular estimating its propensity to being damaged by an earthquake. A methodology for vulnerability analysis must therefore define how to carry out the survey of the built-up area and of its characteristics and define suitable models that correlate the severity of the seis-mic motion with the effects in terms of physical dam-age and losses, both economic and social.

Thus, having carried out the vulnerability analysis and noted the seismic risk of the area in question, or the characteristics of the seismic motion expected in the region, possibly differentiated to take into con-sideration the effects of local amplification (micro-zoning), it is possible to estimate the distribution of damage to the built-up area. Whenever the study of risk is conducted in probabilistic terms, the structural and economic consequences will also be expressed probabilistically: this approach represents that which is called a risk analysis. Instead, in the case in which the seismicity is studied on a deterministic basis, or extracting one or more significant earthquakes from a

catalogue of historical seismicity or simulating with theoretical-numeric models the mechanisms of source and the propagation of waves in the earth's crust, one carries out a scenario analysis, or rather one assesses the effects on the territory following a specific seismic event.

The choice between risk analysis and scenario analysis depends on the aims of the study. In the case of a study of the territory for preventive purposes, a risk analysis is preferable in that it brings together the effects of all the potential seismic sources of the area and supplies a comparable evaluation between all the different communities interested by the study. Instead, to analyse the aspects of management of the emer-gency linked with the Civil Defence, a scenario analy-sis is the most significant, in that it reproduces a realistic distribution of the effects on the territory, a fact that allows elaboration of strategies for the post-earthquake; nevertheless, the risk in some areas of the study is underestimated, in that it is referred to a single precise event. In both cases, however, the same mod-els can be used for the vulnerability analysis; for this reason, in this work we will not come back to this dis-tinction. The vulnerability analysis on an ordinary built-up area on a territorial scale requires evaluations of a large number of samples; the use of structural calculation models cannot be proposed both due to the difficulty of identifying simple but reliable models, and for the quantity of data that would be necessary to collect in the field. The methodologies available must therefore base themselves on a few empirical parameters and their validation cannot disregard the observation of the damage produced by real earthquakes.

1.2 *Vulnerability methods in use*

According to different authors (Corsanego & Petrini 1994, Dolce et al., 1995, Sandi 1982, UNDP/UNIDO 1985) vulnerability methods can be classified as follows: typological (categorization methods, statistical methods), inspection and rating methods indirect, expert judgment) and mechanical methods (analytical).

Typological methods group buildings by their material, structural characteristics and constructive techniques into vulnerability classes; for each vulnerability class a curve or a damage probability matrix establishes a relationship between the earthquake severity and the damage. Inspection and rating methods assign a vulnerability index to each building measuring its ability to hold the earthquake; a relation between the earthquake and the damage is established for each vulnerability index. Typological methods and rating methods can be also referred as observed vulnerability methodologies as their formulation is based on the damage assessment after earthquakes.

Mechanical methods (HAZUS 1999) are not yet widespread in Europe because they need an experimental validation, at least on the traditional masonry construction in Europe. A new trend in the field of vulnerability methodologies has been to consider hybrid techniques, in order to reduce the great uncertainties connected with a statistical model; certain recent experiences have already considered the possibility of a mixed typological-rating approach both in Italy (Faccioli & Pessina, 1999) and in Europe (Dolce et al., 1995).

The proposed method can be considered as one of these attempts in fact, nevertheless being a typological methodology, the vulnerability evaluation may be carried out taking into account other characteristics of the buildings considered as behaviour modifiers.

2 PROPOSAL OF A NEW VULNERABILITY MODEL

2.1 *The European Macroseismic Scale EMS 98*

The natural way to define a typological methodology is to start from the observed vulnerability, by analysing the damage assessment after different intensity earthquakes. However this is not possible in all European countries and should lead to a heterogeneous definition

Table 1. Attribution of vulnerability classes to different building typologies.

Typologies	A	B	C	D	E	F
MASONRY						
Rubble stone	■					
Adobe (earth brick)	■—|					
Simple stone	|••	■				
Massive stone		|—	■	••|		
Unreinforced masonry		|••	■	••|		
U. Masonry with R.C. floors		|—	■	••|		
Reinforced or confined masonry			|—	■	••|	
REINFORCED CONCRETE						
Frame in R.C without E.R.D.		|••••	—■	••|		
Frame in R.C moderate E.R.D			|•••• —	■—|		
Frame in R.C high E.R.D.			|••••	—■—|		
Shear walls without E.R.D.		|••	■—|			
Shear walls moderate E.R.D			|••	■—|		
Shear walls high E.R.D			|••	■—|		
STEEL STRUCTURES				|••••	—■—|	
TIMBER STRUCTURES			|••••	—■—|		

Situations: ■ Most probable class; | — Possible class; | •• Unlikely class (exceptional cases)

of the vulnerability building sets (vulnerability classes). The authors have derived a method, applicable to the European territory, making reference to the EMS-98 (Grunthal 1998), which implicitly contains a model of vulnerability as its aim is to measure the severity of an earthquake from damage observation.

In EMS-98 some typologies are considered for masonry, reinforced concrete, steel and timber buildings; in spite of the detailed distinction of each type of building, it is recognized that the seismic behaviour of buildings, in terms of apparent damage, may be subdivided by, at least, six vulnerability classes (Table 1). Thus, different types may behave in a similar way (see, for example, massive stone and unreinforced masonry with r.c. floors); on the other hand, it emerges that even if each type of structure is characterized by a prevailing vulnerability class, it is possible to find buildings with a better or worse seismic behaviour, depending on their constructional characteristics.

The vulnerability table is used to assign the macroseismic intensity to an earthquake in a particular location, in the post-earthquake investigation. To this end, the matrices of seismic effects on buildings belonging to different vulnerability classes may be used (Table 2); the quantity definition of building suffering each damage grade is given in a vague way (few, many, most).

2.2 *Completing EMS 98 vulnerability model*

The vulnerability model provided by EMS-98 is incomplete and vague.

Table 2. EMS 98 damage description for Class A.

Damage grades	1	2	3	4	5
Intensity V	Few				
VI	Many	Few			
VII			Many	Few	
VIII				Many	Few
IX					Many
X					Most
XI					
XII					

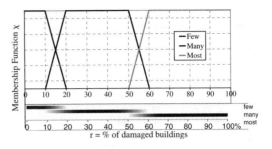

Figure 1. Percentage ranges and membership functions χ of the quantitative terms Few Many Most.

A proper probability distribution of damage grades (a discrete distribution) may be used in order to complete the DPM. The binomial distribution has been successfully used for the statistical analysis of data collected after 1980 Irpinia earthquake (Italy) (Braga et al., 1982); but the simplicity of this distribution, which depends on only one parameter, does not allow defining the scatter of the damage grades around the mean value.

Sandi (Sandi, 1995) observes that the dispersion of the binomial distribution is too high, when you consider a detailed building classification; this may lead to overestimate the number of buildings suffering serious damages, in the case of rather low values of the mean damage grade. The distribution that better suits the specific requirements is the beta distribution (also used in ATC-13, 1985):

$$PDF: p_\beta(x) = \frac{\Gamma(t)}{\Gamma(r)\,\Gamma(t+r)} \frac{(x-a)^{r-1}(b-x)^{t-r-1}}{(b-a)^{t-1}} \qquad a \leq x < b \qquad (1)$$

$$\mu_x = a + \frac{r}{t}(b- \qquad (2)$$

where a, b, t and r are the parameters of the distribution; μ_x is the mean value of the continuous variable x, which ranges between a and b.

In order to use the beta distribution, it is necessary to make reference to the damage grade D, which is a discrete variable, characterized by 5 damage grades plus the absence of damage; for this purpose, it is advisable to assign value 0 to the parameter a and value 6 to the parameter b. Starting from this assumption, it is possible to calculate the probability associated with damage grade k (k = 0,1,2,3,4,5) as follows:

$$p_k = P_\beta(k+1) - P_\beta(k) \qquad (3)$$

Following this definition, the mean damage grade, mean value of the discrete distribution, and the mean value of the beta distribution (2) can be correlated through a third degree polynomial:

$$\mu_x = 0,042 \cdot \mu_D^3 - 0,315 \cdot \mu_D^2 + 1,725 \cdot \mu_D \qquad (4)$$

Thus, by (2) and (4), it is possible to correlate the two parameters of the beta distribution with the mean damage grade:

$$r = t(0.007\mu_D^3 - 0.0525\mu_D^2 + 0.2875\mu_D) \qquad (5)$$

The parameter t affects the scatter of the distribution; if t =8 is used, the beta distribution looks very similar to the binomial distribution.

2.3 Fuzzy interpretation of EMS 98 linguistic damage definition

Having solved the matter of incompleteness by the discrete beta distribution, in order to derive DPM for the EMS vulnerability classes (Table 1), it is necessary to tackle the problem of vagueness of the quantity definitions (few, many, most). As there is little point in translating the linguistic terms into a precise probability value, they can be better modelled as bounded probability ranges. The fuzzy sets theory (often proposed for seismic risk assessment methods: Dong 1988, Sanchez-Silva 2001) can offer an interesting solution to the problem, leading to the estimation of upper and lower bounds of the expected damage (Bernardini, 1999).

Figure 1 shows the range of percentage corresponding to the quantitative terms (few, many, most) according to EMS-98: it emerges that while there are some definite ranges (few, less then 10%; many, 20% to 50%; most, more than 60%), there are situations of different terms overlapping (between 10% and 20% can be defined both few and many; 50% and 60%, both many or most).

These qualitative definitions are interpreted through membership functions χ, which are used in the fuzzy set theory (Dubois & Parade, 1980). If you consider a parameter, the membership function defines the belonging of single values of the parameter to a specific set; the value of χ is 1 when the degree of belonging is plausible (that is to say almost sure), while a membership between 0 and 1 indicates that

673

Table 3. Damage distributions and mean damage values related to the upper and lower bounds of plausibility and possibility ranges for class A.

CLASS A	1	2	3	4	5	μ_D
Intensity VI	Many	Few				
A$^+$ Upper Plausible	32.0	**10**	1.9	0.2	0.0	0.68
A$^-$ Lower Plausible	**20**	4.3	0.6	0.0	0.0	0.43
A^{++} Upper Possible	40.6	**20**	5.5	0.7	0.0	1.81
A^{--} Lower Possible	**10**	1.6	0.2	0.0	0.0	0.25

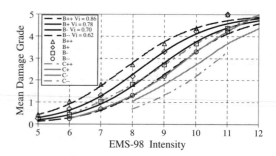

Figure 2. Class B and C plausibility and possibility draft curves and their interpolation.

the value of the parameter is rare but possible; if χ is 0, the parameter does not belong to the set.

Using the fuzzy sets theory and starting from EMS-98 definitions (Table 2), DPM are built, through the discrete beta distribution (3).

Let us consider the vulnerability class A and the macroseismic intensity VI (Table 3): many buildings should suffer a damage of grade 1 (slight) and a few a moderate damage (grade 2).

The plausible values of the parameter μ_D are the ones for which all EMS-98 quantity definitions must be respected in a plausible way; in other words the plausible range of the membership functions χ cannot be exceeded (damage grade 1 = Many = 20% ÷ 50%; damage grade 2 = Few = 0% ÷ 10%). The range of plausible values of μ_D is defined by two plausibility bounds, obtained when $p_2 = 10\%$ (upper bound) and when $p_1 = 20\%$ (lower bound).

The possible values of the parameter μ_D are the ones for which all EMS-98 quantity definitions are plausible or possible, with at least one which is only possible. The ranges of possible values are adjacent to the plausible range, being defined by two possibility bounds, obtained when $p_2 = 20\%$ (upper bound) and when $p_1 = 10\%$ (lower bound).

In Table 3 the upper and lower bounds of the mean damage grade, related to plausibility and possibility, are shown for the vulnerability class A. The corresponding distributions of the damage grades are shown: the value that determines the bound is in bold character.

Repeating this procedure for each vulnerability class and for the different intensity degrees it is possible to obtain, point by point, the plausible and possible bounds of the mean damage. By connecting these points, draft curves may be obtained, which define the plausibility and possibility areas for each vulnerability class, as a function of the macroseismic intensity (Fig. 2).

2.4 Vulnerability Index and Vulnerability Curves

Observing the diagram of Figure 2, it stands out that there is a plausible area for each vulnerability class

and intermediate possible areas for contiguous classes. In other words, the area between B+ and B− is distinctive of class B, while there is a contiguous area in which the best buildings of class B and the worse of class C coexist (the B− curve coincides with the C++ one; the B−− curve coincides with the C+ one). Another important outcome of the analysis above presented is that curves in Figure 2 are, more or less, parallel; this is because the damage produced by an intensity to buildings of a certain vulnerability class is the same caused by the following intensity degree to buildings of the subsequent vulnerability class. On the base of these considerations a conventional Vulnerability Index V_I, defined inside the Fuzzy Set Theory, represents the belonging of a building to a vulnerability class. Vulnerability Index numerical values are arbitrary as they are only a scores to quantify in a conventional way the building behaviour (represent a measure of the weakness of a building to the earthquake); for the sake of simplicity a 0–1 range has been chosen, allowing to cover all the area of possible behaviour, being values close to 1 those of the most vulnerable buildings and values close to 0 the ones of the high-code designed structures.

Thus, the membership of a building to a specific vulnerability class can be defined by this vulnerability index (Fig. 3); in compliance with the fuzzy set theory they have a plausible range ($\chi = 1$) and linear possible ranges, representing the transition between two adjacent classes.

The fuzzy definition of the Vulnerability index (Fig. 3) is perfectly coherent with the considerations previously done relatively to the vulnerability curves; the membership functions of the six vulnerability classes have the same shape and are translated of the same quantity, according to the parallelism and the constant spacing between the curves (Fig. 2). For the operational implementation of the methodology it is particularly useful to define and analytic expression (6), capable of interpolating the curves in Figure 2; in the proposed formula (6) the mean damage grade μ_D is given as a

674

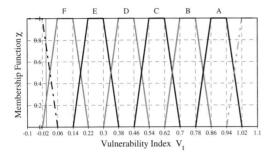

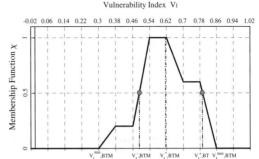

Figure 3. Vulnerability Index membership functions for EMS 98 vulnerability classes.

Figure 4. Membership Function of V_I for Massive Stone typology and its representative values.

function of the macroseismic intensity I, only depending from the parameter V_I (vulnerability index):

$$\mu_D = 2.5 \left[1 + tanh\left(\frac{I + 6.25 \cdot V_I - 13.1}{2.3} \right) \right] \quad (6)$$

2.5 Vulnerability curves for building typologies

In order to make clearer the proposed vulnerability model application, it is useful to refer directly to the typological classification instead of to the vulnerability classes. This is allowed as the behaviour of different typologies is represented in term of an only one parameter, the vulnerability index.

EMS-98 vulnerability table (Table 1) expresses the different belonging of a typology to the vulnerability classes through linguistic terms well interpretable inside the Fuzzy Set Theory. It is reasonable to attribute to "most likely", "probable" and "exceptional case" terms the percentage values respectively near to 100%, 60% and 20%.

It is possible to define the membership function of each building type, as a function of the vulnerability index V_I, through a linear combination of the vulnerability class membership functions (Fig. 2). As an example, the membership function of the massive stone masonry (M4) is shown in Figure 5 and so defined:

$$\chi_{M4}(V_I) = \chi_C(V_I) + 0.6 \cdot \chi_B(V_I) + 0.2 \cdot \chi_D(V_I) \quad (7)$$

For the membership function of each typology, five representative values of V_I have been defined (Fig. 4) through a defuzzification process (Ross, 1995):

- V_I^* most probable value V_I for the specific building type;
- V_I^-; V_I^+ bounds of the plausible range V_I for the specific building type;
- V_I^{min}; V_I^{max} upper and lower bounds of the possible values of V_I for the specific building type.

These values are represented in Figure 4 for Massive Stone masonry typology and reported in Table 4 for all

the EMS-98 buildings typologies; replacing the values of V_I^*, V_I^-, V_I^+ inside the given analytical function (6) vulnerability curves are found, representative, respectively, of the "average" behaviour believable and of a meaningful range of behaviour for the same typology.

In order to validate the method, the vulnerability curves have been compared with Italian observed

Table 4. Meaningful values for the building typologies.

Typology	$V_{I\,min}$	V_I^-	V_I^*	V_I^+	$V_{I\,max}$
MASONRY					
Rubble stone	0.62	0.81	0.873	0.98	1.02
Adobe (Earth brick)	0.62	0.687	0.84	0.98	1.02
Simple stone	0.46	0.65	0.74	0.83	1.02
Massive stone	0.3	0.49	0.616	0.793	0.86
Unreinforced masonry	0.46	0.65	0.74	0.83	1.02
Un. M. with RC floors	0.3	0.49	0.616	0.79	0.86
Reinforced/ confined M.	0.14	0.33	0.451	0.633	0.7
REINFORCED CONCRETE					
R.C Frame without E.R.D.	0.3	0.49	0.644	0.8	1.02
R.C Frame moderate E.R.D	0.14	0.207	0.447	0.48	0.86
R.C Frame high E.R.D.	−0.02	0.046	0.287	0.48	0.7
Shear walls without E.R.D.	0.3	0.367	0.544	0.67	0.86
Shear walls moderate E.R.D.	0.14	0.487	0.384	0.207	0.7
Shear walls high E.R.D	−0.02	0.047	0.284	0.35	0.54
STEEL STRUCTURES	−0.02	0.047	0.286	0.48	0.7
TIMBER STRUCTURES	0.14	0.207	0.447	0.64	0.86

675

vulnerability methods (Giovinazzi & Lagomarsino, 2001), with the observed and deduced vulnerability data reported by different authors for many building typologies (Balbi & Giovinazzi, in prep.) and with simplified mechanical methods (Giovinazzi et al., 2003).

3 FACTORS ABLE TO MODIFY THE SEISMIC BEHAVIOUR

The approach followed up to here has been typological based, however there are a number of different factors that affect the overall vulnerability of a structure, besides the building type, relative both to the variety of the constructive methods and to the different vulnerability in a region, which is related to the structural details, materials, etc. The proposed method allows taking into account all this parameters.

3.1 Regional vulnerability factor

The range bounded by V_I^-, V_I^+ values is quite large in order to be representative of the huge variety of the constructive techniques used all around the different European regions.

A Regional Vulnerability Factor is introduced to take into account the typifying of some building typologies at a regional level: a major or minor vulnerability could be indeed recognized due to some traditional constructive techniques of a region.

According to this Regional Vulnerability Factor it is allowed to modify the V_I^* typological vulnerability index on the base of an expert judgment or on the base of the historical data available. The first case is achieved when precise technological, structural, constructive information exist attesting an effective better or worse average behaviour in regard to the one here proposed. The second one occurs when data about observed damages exist; the average curve ($V_I = V_I^*$ in Formula 6) can be shifted in order to obtain a better approximation of the same data (Fig.5).

3.2 Behaviour modifiers

The seismic behaviour of a building does not only depends on the behaviour of its structural system but it involves other factors such as the quality of the construction, the height, the irregularity and its maintenance. The main vulnerability factors characteristic of masonry and reinforced concrete buildings can be considered inside this approach; modifying scores, V_m, able to decrease or increase the typological vulnerability index V_I^* are proposed for them (Tables 5–6).

The behaviours modifiers identification and the score attribution has been made empirically, on the basis of the observation of typical damage pattern, taking into account also what suggested by several

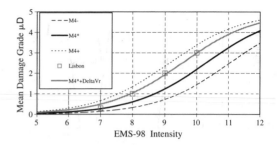

Figure 5. Reinforced Masonry shows a better behaviours than the average in Lisbon: a $\Delta V_R = 0.12$ is applied.

Table 5. Proposal of scores for the vulnerability factors for RC buildings.

REINFORCED CONCRETE

Vulnerability Factors		Code level		
		Low	Medium	High
State of preservation:	Bad	+0.04	+0.02	0
Number of floors:	Low	−0.04	−0.04	−0.04
	Medium	0	0	0
	High	+0.08	+0.06	+0.04
Plan Irregularity:	Shape	+0.04	+0.02	0
	Torsion	+0.02	+0.01	0
Vertical irregularity:		+0.04	+0.02	0
	Short column	+0.02	+0.01	0
	Bow windows	+0.04	+0.02	0
Foundations:	Connected beams	−0.04	0	0
	Beams	0	0	0
	Isolated footing	+0.04	0	0
Aggregate buildings: Insufficient aseismic joints		+0.04	0	0

Inspection Forms (ATC 21 1988, Benedetti & Petrini 1984, UNDP/UNIDO 1985).

The overall score that modifies the typological vulnerability index V_I^* is evaluated summing the contribution of the recognised vulnerability factors:

$$\Delta V_m = \sum_k V_{m,k} \qquad (8)$$

4 OPERATIVE ASPECTS FOR THE METHOD APPLICATION

According to the proposed method the vulnerability index is evaluated summing the contributions of three parameters:

$$\overline{V_I} = V_I^* + \Delta V_R + \Delta V_m \qquad (9)$$

Table 6. Proposal of scores for the vulnerability factors for masonry buildings.

MASONRY Vulnerability Factors	Parameters	Scores
State of preservation:	Good	−0.04
	Bad	+0.04
Number of floors:	Low	−0.02
	Medium	+0.02
	High	+0.06
Structural system:	Wall thickness	
	Distance between walls	−0.04
	Connection between walls	0.04
	Connection walls- horizontal structures	
Soft story:	Demolition – Transparency	+0.04
Plan Irregularity:	Shape and Torsion	+0.04
Vertical Irregularity:		+0.02
Superimposed Floors:		+0.04
Roof:	Roof weight and thrust	−0.08
	Roof connections	+0.08
Foundations:	Different levels	+0.04
Retrofitting intervention:		
Aseismic devices:	Buttresses, Barbican	−0.04
Aggregate buildings:		
Position:	Middle	−0.04
	Corner	+0.04
	Header	+0.06
Vertical irregularity:	Staggered floors	+0.04
	Different heights	+0.02

where V_I^* is the Typological Vulnerability Index, ΔV_R is the Regional Vulnerability Factor, ΔV_m represents the contribution given by the behaviour modifiers. The final vulnerability index has to comply with the possibility bounds recognized for each typology (Table 4):

$$\bar{V}_I = Min\left(\bar{V}_I ; V_{I\,max}\right) \tag{10}$$

$$\bar{V}_I = Max\left(\bar{V}_I ; V_{I\,min}\right) \tag{11}$$

4.1 Processing of the available data for the vulnerability index evaluation

In order to evaluate these parameters any available database related to buildings must be taken into consideration, classifying the information contained both from a geographic and a qualitative point of view. From the geographic point of view, the minimum unit, on the territory, for which data are available leads up to the minimum unit to make reference to for the analysis performing; this can be, indifferently, a single building, a building aggregate or a macro area containing several buildings (such as a district or a census section). From the qualitative point of view, it is necessary to recognize, inside the available data, the ones useful for the building typology identification and the

Table 7. Example of a category definition.

CATEGORY Age	Urban Context	Typologies	Percentage
1919	Rural Area	Rubble stone	50%
1945		Simple stone	30%
		Unreinfoerced masonry	20%

ones helpful for a deeper knowledge of the building behaviour. If no data are available at all, a field survey must be performed carried out through a rapid visual inspection; the first aim of the field survey must be the typological identification; secondly it is very important to collect as many vulnerability factors as possible (Tables 5–6), according to the availability of money and time.

4.1.1 V_I^* Typological vulnerability index
The building typology is univocally recognized once the vertical and horizontal structure type is known; if the available data are not enough for a direct typological identification, it is necessary to make reference to more general building categories, identified combining less detailed data such as the age, the height and the urban context (Table 7).

The vulnerability index value V_I^* for a building typology is univocally attributed according to the proposed table (4); otherwise the vulnerability index of a category is evaluated as the weighted average of the assessed typological composition:

$$V_{I\,cat\,i}^* = \sum_t p_t \cdot V_{I\,typ}^* \tag{12}$$

where p_t is the ratio of buildings inside the category C_i supposing to belong to a certain typology.

If the analysis unit is a small geographic area, the evaluation is referred to a set of building rather then to a single building; in this case the V_I^* evaluation is so performed:

$$V_{I\,typ}^*(set) = \sum_t q_t \cdot V_{I\,typ}^*(sb) \tag{13}$$

$$V_{I\,cat}^*(cat) = \sum_c q_c \cdot V_{I\,cat}^*(sb) \tag{14}$$

where q_t = ratio of buildings inside the set supposing to belong to a certain building typology; q_c = ratio of buildings inside the set supposing to belong to a certain building category.

If no other data are available $V_{I\,typ}^*$ and $V_{I\,Cat}^*$ represent the final vulnerability evaluation.

4.1.2 ΔV_R Regional vulnerability factor

Any available knowledge about observed vulnerability or traditional constructive techniques can be considered inside the ΔV_R parameter, as previously explained (see section 3.1) either for a typology or for a category. If we are referring to a set of building rather than to a single building:

$$\Delta V_R = \sum_t r_t \Delta V_{R\,t} \qquad (15)$$

where r_t = ratio of buildings recognized as belonging to a specific typology t affected by the recognized $\Delta V_{R\,t}$.

4.1.3 ΔV_m Behaviour modifiers factor

If further data allowing a deeper knowledge of the building behaviour are available, the final vulnerability index is refined through the ΔV_m. Its evaluation is done according to the (8) for a single building while, for a set of building:

$$\Delta V_m = \sum_k r_k V_{m,k} \qquad (16)$$

where r_k = ratio of buildings characterized by the modifying factor k, with score $V_{m,k}$.

4.2 The physical damage evaluation

Evaluated the final vulnerability index V_I the proposed analytical functions (6) allows to estimate a mean damage grade for the forecasted scenario intensity. Once the μ_D is evaluated, the corresponding damage distribution may be obtained through the beta distribution.

It is useful to remember that to each value of parameter μ_D (having definitely assumed a = 0, b = 6, and for a fixed value of t), through (5) and (3) a damage grade distribution corresponds.

The value t = 8 is advised, anyway, if data about damage distributions are available for some typology or category, a better approximation can be find inside the meaningful range t = 2 ÷ 16. Figure 6 shows a good approximation of damage, observed in stone buildings after Aegion earthquake (1995), obtained through the beta distribution with t = 2 (Penelis et al., 2002).

Another method to represent the vulnerability of a building is through the fragility curves. They express the probability that the expected damage of a structure will exceed a fixed damage grade during the ground shaking (in this case given in terms of macroseismic intensity).

$$P(D \geq D_k) = \sum_{j=k}^{5} p_j \qquad (17)$$

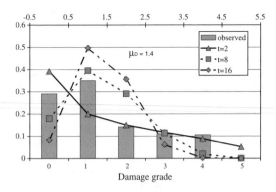

Figure 6. Observed Data interpolation through beta distribution.

Table 8. Correlation between the damage grade distribution and the effects on building and population.

Unusable buildings:	40% of buildings with damage grade 3 +100% of buildings with damage grade 4 and 5
Collapsed buildings:	100% of buildings with damage grade 5
Homeless	100% of the population living in unusable buildings – casualties and severely injured
Casualties and severely injured	30% of the population living in collapsed buildings

where p_j =probability associated with damage grade j (j = 0,1,2,3,4,5).

The fragility curves for the damage grades may be obtained in analytical form from equation (6) and are given directly by the cumulative probability beta distribution:

$$P(D \geq D_k) = 1 - P_\beta(k) \qquad (18)$$

4.3 Losses and consequences evaluation

The consequences to people (casualties, severely injured and homeless) and the social impact (unfit for use buildings, economic losses) may be evaluated from the damage grades probability distribution; statistical correlation between physical damage and consequences may be established on the basis of the data collected during the damage assessment in the emergency. Table 8 summarizes what observed after the last earthquakes in Italy (Bramerini et al., 1995).

The damage can also be measured in economic terms through a parameter, the damage index (D_I), defined as the ratio between repair cost and

Table 9. Correlation between the different values of D and D_I.

Damage Grade	0	1	2	3	4	5
D_I	0	0.01	0.1	0.35	0.75	1

reconstruction cost (the latter corresponds to the value of the building):

$$D_I = \frac{repair\ cost}{building\ value} \qquad 0 < D_I < 1 \qquad (19)$$

The correlation between damage and D_I have been obtained by processing the data of funding necessary for the repair and rebuilding of damaged structures (Table 9).

5 CONCLUSIONS

The method is not constricted by any specific requirements in terms of data and analysis units. It can be easily applied using existing data at little cost. As every step can be calculated automatically in GIS, the use of the results for risk reduction becomes an effective tool: the possibility of a constant updating of the data and the rather fast computational operation allows the decision-makers to simply construct different scenarios and to act on risk reduction strategies.

ACKNOWLEDGEMENTS

A significant part of this research has been done and funded within the 5th Framework European Commission Project RISK-UE: An advanced approach to earthquake risk scenarios with applications to different European towns – Contract: EVKT.

REFERENCES

ATC, 1985. *Earthquake Damage Evaluation Data for California*. Redwood (CA): Applied Technology Council report ATC-13.

ATC, 1988. *Rapid Visual Screening of Buildings for Potential Seismic Hazards: A Handbook* (CA): Applied Technology Council report ATC-21.

Benedetti, D. & Petrini, V. 1984. On seismic vulnerability of masonry buildings: proposal of an evaluation procedure. *The industry of constructions*. Vol. 18, pp. 66–78.

Bernardini, A. 1999. *Seismic Damage to masonry Buildings. Proc. of the workshop of seismic Damage to masonry Buildings*. Rotterdam: Balkema.

Braga, F., Dolce, M. & Liberatore, D. 1982. A statistical study on damaged buildings and an ensuing review of the M.S.K.-76 scale. *Proceeding of the 7th* European Conference on Earthquake Engineering.

Bramerini, F., Di Pasquale, G., Orsini, G., Pugliese, A., Romeo, R. & Sabetta, F. 1995. *Rischio sismico del territorio italiano. Proposta per una metodologia e risultati preliminari* Rome: SSN/RT/95/01. (In Italian).

Corsanego, A. & Petrini, V. 1994. Evaluation criteria of seismic vulnerability of the existing building patrimony on the national territory. *Ingegneria Sismica* Vol. 1, pp. 16–2.

Dolce, M., Zuccaro, G., Kappos, A. & Coburn A. 1994. Report of the EAEE Working Group 3: Vulnerability and risk analysis. *Proceeding of 10th* European Conference on Earthquake Engineering. Vol. 4, pp. 3049–3077.

Dong, W., Shah, H. & Wong, F. 1988. *Expert System in Construction and Structural Engineering*. London and New York: Chapman and Hall.

Dubois, D. & Parade, H. 1980. *Fuzzy Sets and Systems*. New York: Academic Press.

Faccioli, E., Pessina, V., Calvi, G.M. & Borzi, B. 1999. A study on damage scenarios for residential buildings in Catania city. *The Catania Project*. Rome: CNR-GNDT.

Giovinazzi, S. & Lagomarsino, S. 2001. Una metodologia per l'analisi di vulnerabilità sismica del costruito *Proceeding of 10th* Italian Conference on Earthquake Engineering.

Giovinazzi, S., Logomarsino S. & Penna, A. 2003. *Seismic Risk:scenarios: typological and mechanical approachesin vulnerability assessment*. Nice: EGS-AGU-EUG Joint Assembly.

Grunthal, G. 1998. *European Macroseismic Scale 1998*. Luxembourg: Cahiers du Centre Europèèn de Géodynamique et de Séismologie.

HAZUS 1999. *Earthquake Loss Estimation Methodology*. Washington: Federal Emergency Management Agency.

Penelis, G., Kappos, A., Stylianidis, K. & Lagomarsino, S. 2002. *Statistical Assessment of the vulnerability of Unreinforced Masonry Buildings*. Bucharest: International Conference on Earthquake Loss Estimation and Risk Reduction.

Ross, T.J. 1995. *Fuzzy Logic with Engineering Applications*. New York: McGraw Hill.

Sanchez-Silva, M. & Garcia, L. 2001. Earthquake Damage Assessment Based on Fuzzy Logic and Neural Network. *Earthquake Spectra*. Vol. 17, N. 1, pp. 89–112.

Sandi, H. & Floricel, I. 1995. Analysis of seismic risk affecting the existing building stock. *Proc. of the 10th* European Conference on earthquake Engineering. Vol.3, pp.1105–1110.

UNDP/UNIDO Project RER/79/015, 1985. Post Earthquake damage Evaluation and Strength Assessment of Buildings under Seismic Condition (Vol.4) UNDP, Vienna.

Safety and Reliability – Bedford & van Gelder (eds)
© 2003 Swets & Zeitlinger, Lisse, ISBN 90 5809 551 7

Reliability prediction using degradation data – a preliminary study using neural network-based approach

T. Girish, S.W. Lam & J.S.R. Jayaram
Design Technology Institute, National University of Singapore

ABSTRACT: One of the most critical aspects of degradation-based estimation of reliability lies in the modeling of degradation data. A variety of techniques ranging from mixed-effects non linear regression to techniques like Neural Networks have been applied for capturing the behaviour of the degradation paths. In this work, we propose to study the behaviour of the distributional parameters of degradation signals across time using a Neural Networks approach. We used data generated from an underlying process of a Gaussian distribution. A neural network is trained to recognize the variation pattern of the distributional characteristics of the process. The neural network is then used to predict these characteristics' values at future time intervals. These predictions were found to be comparable to the theoretical values and were subsequently used to arrive at reliability estimates.

1 INTRODUCTION

There are a variety of reasons why reliability prediction through degradation data modeling presents an attractive alternative in the assessment and improvement of reliability of components, sub-systems and systems from design stage in the product development process to the usage environment during a product's lifecycle.

The complex and highly-reliable products needed to satisfy increasing customers' demands and retain customers' loyalty in today's competitive marketplace require components with extremely high reliability. However, with shortening product development cycle, the component reliabilities for such systems may not be adequately estimated using traditional reliability life tests. This is so because it has become increasingly difficult to test these components to failure within the available time frame as new products become highly reliable. Degradation tests with their accompanying modeling and data analysis presents an alternative to predict failure of the product without testing till failure. This can be done with an appropriate definition of component failure in terms of a specified level of degradation and the estimation of the time-to-failure distribution from the relevant degradation measurements.

In addition, degradation measurements and modeling presents useful information regarding component reliability (Chinnam, 1999) that provides pre-emptive

forecasts on component failures. Hence, appropriate preventive maintenance policies can be generated to improve overall system's reliability and availability during the product's lifecycle.

Most failures can be traced to at least one underlying physical process that results in degradation. Such degradation processes can sometimes be observed and monitored through some quality characteristics. These are sometimes termed degradation characteristics. Examples of such degradation characteristics for monitoring degradation processes include the wear of tires and luminosity of light bulbs. Although this paper deals with the modeling and analysis of only a single degradation characteristic, the technique proposed is applicable to the study of multiple degradation characteristics as well.

Figure 1 shows three general shapes of degradation characteristics in terms of the amount of degradation over time. Such models of degradation characteristics over time are known as degradation paths. In the prediction of component failures, there is a need to define a particular critical level of degradation beyond which the component is deemed to have failed. This is represented by the horizontal line known as the Critical Failure Limit, *D*, in Figure 1.

The degradation paths shown in Figure 1 represent three general types of degradation commonly found in physical systems. They are linear, convex and concave degradation processes. Linear degradation arises in some simple wear processes such as automobile

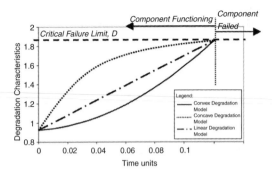

Figure 1. Possible shapes for degradation process of a single degradation characteristic.

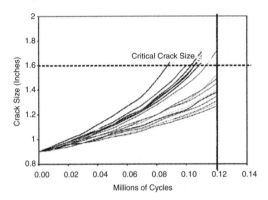

Figure 2. Fatigue Crack Growth Data (Bogdanoff and Kozin, 1985) showing variability in degradation paths for different samples.

tire wear (Meeker and Escobar, 1998). Convex degradation are used to model physical degradation processes where the degradation rate increases with the level of degradation (eg. growth of fatigue cracks) whereas concave degradation are used to model degradation processes that depends on the presence of certain chemical compounds which decrease over time, thereby reducing the rate of degradation. An example is described by Meeker and LuValle (1995).

In the real manufacturing and operating context, variability amongst degradation path inevitably exists (Meeker and Escobar, 1998). The possible degradation paths of the multiple samples taken from a population are shown in Figure 2. The sources of variability can be due to unit-to-unit variability contributed by initial accumulated degradation of the samples, varying material properties, geometrical dimensions or variability within the sample. Sources of variability can also arise from the operating and environmental conditions under which the particular sample is subjected to. Hence, the challenge is to propose a modeling scheme that would capture the statistical essence of such degradation

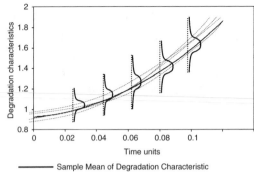

Sample Mean of Degradation Characteristic
Sample Degradation Paths

Figure 3. Sample degradation path and graph of sample means against time for an underlying process that is gaussian distributed.

paths and to also use that model to predict these statistical essence so that an inference on reliability can be made. For example, in the case of normally distributed data, this statistical essence may be captured by the estimates of the distribution parameters, namely, the sample mean and the sample variance.

In order to capture the variation of degradation characteristics with time, Lu and Meeker (1993) proposed linear and non-linear degradation models consisting of fixed and random effects and proposed methods for the analysis of degradation data to determine the failure distributions. In order to broaden the range of application to a more general model, Lu and Meeker (1993) assumed a vector of random effects, Θ, or some appropriate re-parameterization, $\theta = H(\Theta)$, that follows a multivariate normal distribution with a mean vector μ_θ and variance-covariance matrix, Σ_θ. The assumption of normality allows them to summarize the information in the sample paths with only the mean vector and variance-covariance matrix without loss of substantial information (Lu and Meeker, 1993). In our paper, we present an artificial neural network (ANN) based approach to model the degradation paths in terms of the sample mean and sample variance for degradation data that follows a univariate normal distribution at every time instant.

Consider the sample degradation profile shown in Figure 3. At any time, t, the degradation data form an associated distribution due to both inherent and environmental sources of variability. In addition, through the degradation process over time, the degradation characteristic changes as shown in Figure 3. This change of degradation characteristic translates to either a change in the form and/or the parameters of the distribution. In this paper, we assume that the form of the distribution remains the same and is either Gaussian in nature or one that is transformable to a Gaussian distribution. This implies that the mean and variance

Table 1. Example of degradation data from fatigue crack growth data of Bogdanoff and Kozin (1985).

| Path | Million of cycles | | | | | | | |
	0.00	0.01	0.02	0.03	0.04	0.05	0.06	0.09
1	0.9	0.95	1	1.05	1.12	1.19	••	1.64
2	0.9	0.94	0.98	1.03	1.08	1.14	••	1.47
3	0.9	0.94	0.98	1.03	1.08	1.13	••	1.46
4	0.9	0.94	0.98	1.03	1.07	1.12	••	1.43
5	0.9	0.94	0.98	1.03	1.07	1.12	••	1.43
6	0.9	0.94	0.98	1.03	1.07	1.12	••	1.41
7	0.9	0.94	0.98	1.02	1.07	1.11	••	1.41
8	0.9	0.93	0.97	1	1.06	1.11	••	1.39
:	:	:	:	:	:	:	:	:
:	:	:	:	:	:	:	:	:
21	0.9	0.92	0.94	0.97	0.99	1.02	1.04	1.14
Sample mean	0.90000	0.92952	0.96524	1.00143	1.04048	1.08238	1.13143	1.32381
Sample variance	0.00000	0.00009	0.00028	0.00065	0.00128	0.00208	0.00381	0.01635

are jointly sufficient to completely describe the distribution of the data. Having established this, we propose an ANN based approach to model the data in terms of the sample mean and variance.

Table 1 shows a typical set of degradation data taken from Bogdanoff and Kozin (1985). The sample mean and sample variance for the degradation data at each time interval is obtained from the 21 samples. This sample mean and sample variance are jointly sufficient statistics that specify entirely the distribution of the data at any given time instant for a univariate normal distribution. The same holds for any other data that can be transformed to the normal space using a single transformation for the entire data set.

Likewise it may be argued that for any other valid distributional assumption such as Weibull, if the form of the distribution remains consistent across time, variation of the distributional parameters with time may be modeled to predict the characteristics of the distribution at any future instant.

In this paper, the distributional parameters of the degradation paths over time have been modeled by ANNs. In the following section, a short review is given on the ANN methodology used. This is followed by a discussion on the results obtained for this preliminary study. The idea of modeling the distributional parameters of the sample degradation paths using ANN can similarly be extended to a multivariate case where multiple degradation characteristics are observed. This would be a topic of future research.

2 NEURAL NETWORK METHODOLOGY

2.1 Outline

As explained earlier, in this study, we have attempted to map the trend of the estimates of distribution

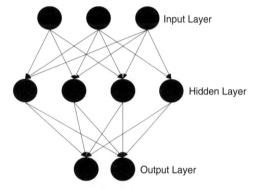

Figure 4. Multilayer feedforward network.

parameters – namely the mean and variance under an assumption of data normality (or data transformed to normality) – using an approach based on Artificial Neural Networks. Neural Networks are now common place in engineering applications and have been found very useful due to their ability to learn from data. Neural Networks are said to be non-parametric in nature because they can usually be used without any assumptions regarding the functional form of the input output model. Although it is often aptly argued that neural network techniques are not to be seen entirely divorced from the conventional statistical methods, they do come in handy in many difficult situations and offer a good set of tools that have been studied extensively.

One of the most commonly used neural networks, is the Multilayer Feed Forward Network (FFN) which is also known as the Multilayer Perceptron (Figure 4). A single hidden layer of neurons is said to be sufficient for a multilayer perceptron to compute an

683

approximate realization to a continuous function with arbitrary precision. (Haykin, 1999). The Radial Basis Function (RBF) Network, usually with a single hidden layer, is another commonly used network architecture. In an RBF Network, the neurons in the first layer use the Euclidean or Mahalanobis distance as opposed to the inner products used by typical Multilayer feedforward network algorithms. It may be stated that multilayer feedforward networks, along with the popular error backpropagation learning algorithm form a sizeable bulk of neural network-based solutions available today. Useful reference books by Haykin (1999) and Bishop (1995) give general conceptual and architectural descriptions of artificial neural networks.

In a typical setting for function approximation in the supervised learning paradigm, the available input data along with known results (targets) are used for training the network (Reed and Marks, 1999). "Training" refers to the process by which the connecting weights of the network are determined (in other words, the data is "learnt") based on performance criteria and objective functions such as minimizing the mean squared error.

In training a neural network several factors such as the network structure, objective function, learning algorithm and learning rate come into play and have to be optimally chosen. The issue lies in choosing the network weights so as to minimize the sum of squares of the error in predicted values compared with the true values of the output. The trained network is later used for predictions after sufficient testing and validation.

In our work, a multilayer feedforward network was used for mapping the estimates of the distribution parameters of the degradation process. Degradation signals typically tend to be autocorrelated in nature. Typically degradation data tend to exhibit non-stationarity (Chinnam, 1999). However, under the assumptions that the form of the distribution does not change, the values of the estimates of the distribution parameters at different instants of time may also be analyzed as time series data as explained in the previous section. Therefore, in this work, the estimates of the mean and variance are separately studied under the learning paradigm of neural networks. Appropriately chosen networks would be used for predicting the values of the estimates of the distribution parameters at a future instant of time.

2.2 Input data

In order to test and validate the proposed approach, a two-stage data generation process was used. Initially, an autoregressive AR(1) process was used to generate a stationary series. This process may be represented as:

$$Y_t = aY_{t-1} + \varepsilon_t \tag{1}$$

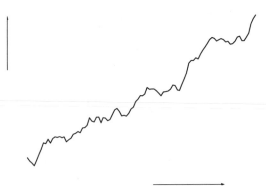

Figure 5.   One realization of the input data model.

where t is the time variable, Y_t refers to the signal measurement at any instant and ε_t represents the error that is normally distributed with mean μ_ε and variance σ_ε^2.

Then this stationary data was made non-stationary by the addition of a deterministic non linear component to every data point, so as to arrive at a process similar to the degradation process. The non linear component that was used is of the form:

$$Y = e^{kt} \tag{2}$$

where k is any constant value. A realization of the data is shown in Figure 5. The expectation and variance for such a process can be obtained analytically as:

$$E(Y_t) = \mu_\varepsilon \left(\frac{1-d}{1-a} \right) + e^{tk} \qquad |a| < 1$$
$$= \mu_\varepsilon t + e^{tk} \qquad |a| = 1 \tag{3}$$

$$Var(Y_t) = \sigma_\varepsilon^2 \left(\frac{1-a^{2t}}{1-a} \right) \qquad |a| < 1$$
$$= \sigma_\varepsilon^2 t \qquad |a| = 1 \tag{4}$$

If there are many realizations of such a process, statistics of those realizations such as the mean and variance, at any instant, would be estimates of the theoretical values given in Eqns. (3) and (4).

2.3 Network design and training issues

A feedforward network structure along with error backpropagation algorithm was designed and tested for feasibility in modeling. Typically, the choice of the network architecture depends upon the problem under consideration. Figure 4 shows the schematic layout of the feedforward error backpropagation network architecture being used here.

Table 2. Details of network architecture.

Item	Network for mean	Network for variance
Input size	4	
Hidden units	10	
Activation function	Linear	
Backpropagation training algorithm	One step secant method	
Performance (MSE)	0.00214	0.0604

In the present case, we have used two different neural networks, one for the estimate of the mean and the other for the estimate of the variance. There are 4 input variables consisting of time (t) and the lag terms Y_{t-1}, Y_{t-2}, Y_{t-3} . The latter three refer to the three most recent observations prior to the next observation, Y_t. This input structure applies both for the estimate of the mean as well as that of the variance. Here 3 lag terms were chosen to begin with. It may be noted that in many situations, it may be necessary to try different number of lag terms to obtain a good fit. Time is also considered as another input variable. The objective is to be able to map the series patterns which depend on the autocorrelation as well as the dependence upon time so that prediction at any future instant is made possible. An illustration of this methodology can be found in the work reported by Stern (1996).

The FFN used here consists of one hidden layer, the number of hidden units for mean and variance are shown in Table 2 along with other network information. These network choices were arrived at after some initial attempts to get good training results.

The One Step Secant method is one of the training algorithms available for training backpropagation feedforward neural networks (Battiti, 1992). This method was found to be efficient for the data generated and has been used for both the networks corresponding to the mean and the variance estimates.

Both the time series data consists of 297 observations. The training samples are derived from these observations, each set of which consists of 3 consecutive observations $(Y_{t-1}, Y_{t-2}, Y_{t-3})$ and time t as inputs. The known target value or response for the sample Y_t is also supplied in training the network. Training results for the mean and the variance after a few initial trials are shown in Figures 6 and 7 and are satisfactory.

In Figures 6 and 7, the original data as well as the network-fitted data are shown together for comparison. The fit for the mean-estimate data was found to be better than that of the variance-estimate, as is also indicated by the performance values in Table 2. It may be due to the fact that variance is more sensitive to sampling variations than the mean.

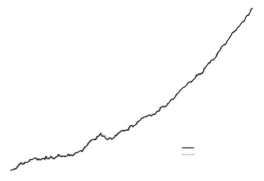

Figure 6.   Training fit for the mean estimate.

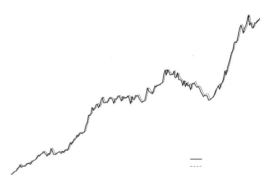

Figure 7.   Training fit for the variance estimate.

3   RESULTS AND DISCUSSIONS

3.1   NN testing

A testing data set is normally used to verify how well the neural network has learnt from the training data and helps to ensure that the NN has not over fitted the data. In this case, another new set of 300 observations were derived from the simulations of the same process for the purpose of testing the network. As before, 297 sample testing sets were derived from these observations. Each of these testing sets consisted of 3 most recent observations and time. At every stage, the neural network predicted output is compared to the sample output. These results of testing are illustrated in Figures 8 and 9 for the mean and variance respectively and it appears that the test set predictions are satisfactory.

3.2   Prediction

Prediction of mean and variance at any future time instant becomes possible once the network architecture has been arrived at. In this case, since the neural

685

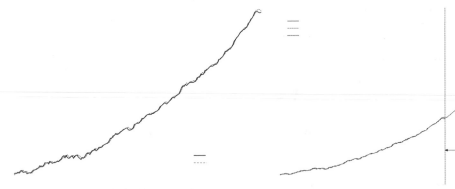

Figure 8. Testing results for the mean estimate.

Figure 9. Testing results for the variance estimate.

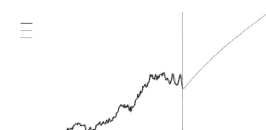

Figure 10. Iterative predictions for the mean estimate from $t = 200$ to $t = 300$.

Figure 11. Iterative predictions for the variance estimate from $t = 200$ to $t = 300$.

networks require the previous few lag terms as input, the idea would be to iteratively generate the inputs for subsequent predictions. In other words, if the degradation signal data is available until a time instant t, i.e., the lag terms for time t, $t - 1$, $t - 2$, etc. are known, then the first prediction is made for the $t + 1$ instant using the network. Having obtained the prediction for $t + 1$ instant, it is now possible to predict in a similar manner for $t + 2$ using the lag terms corresponding to $t - 1$, t, $t + 1$ time instances. This process may be continued until the distribution parameters for a desired time instant can be obtained. It may be noted that a similar method is suggested by Chang et al. (1999) for step stress data.

The predictions using such a process is shown in Figures 10 and 11, where the values from the real testing input data for time instances $t = 298, 299, 300$ are taken as inputs and the predictions are made from $t = 301$ up to $t = 400$. This is indicated by dotted lines in the prediction phase in the figures. It was found that the values thus obtained using the network were comparable to the theoretical values given by Equations (3) and (4).

3.3 Error bounds

It would be of significant interest to study the error bounds of the degradation distribution parameters predicted in the manner described above. In Figures 12 and 13, the network predicted estimates for mean and variance can be compared to the equivalent estimates using simulated data (using the same sample size as in the training of the network). This comparison can be shown using the 5th and 95th percentiles of a large number of simulated estimates of mean and variance using the data generating process described in section 2.2.

3.4 Reliability calculation

Under settings of normality (or transformed to normality), reliability calculation becomes simple. In Figure 14, the distribution of the degradation signal values at any given time is shown. Since failure is defined as reaching a critical degradation level, the

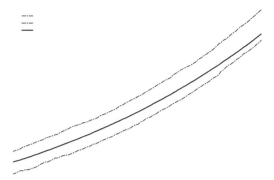

Figure 12. Error bounds for mean estimate predictions.

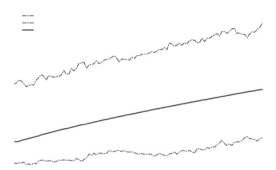

Figure 13. Error bounds for variance estimate predictions.

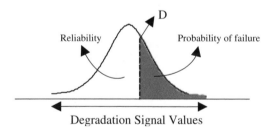

Figure 14. Probability of failure from distribution of degradation signals.

reliability is simply the cumulative probability of the degradation signal being less than a critical value, D. This is shown by the unshaded portion in Figure 14 for a "smaller the better" problem.

The estimates of the mean and variance provided by the neural network prediction can be used to estimate this normal distribution and this information can therefore be used in the prediction of reliability.

In the case of normality or in situations where the distribution has a closed form expression, this task is simple. In other situations such as those when there are multiple degradation signals with interdependence, Monte Carlo simulations would come in handy for a similar calculation of reliability.

4 CONCLUSIONS

Degradation signals have been analyzed in the past using statistical nonlinear regression techniques. Although they are proven to be highly beneficial, it often has the disadvantage of having to assume an underlying form of the non linear model in situations where the real underlying physics is not known. However, a neural networks approach coupled with the idea of modeling the distribution parameters of degradation signals is free from such assumptions and offer an alternative perspective of looking at the issue of degradation signal modeling for components. This proposed approach under assumptions of normality has been demonstrated in this paper. Although normality assumption may be a constraining one, it is felt that the general methodology would be useful for as long as the form of the distribution itself does not change with time. For other situations, there is ample scope for further theoretical development.

This is essentially a preliminary study that shows workable results for reliability predictions using neural networks. There are other issues that are currently being considered in making the reliability predictions. For example, in cases where the data cannot be approximated using the simple architectures discussed here, other improved network models such as dynamic recurrent networks may prove to be more applicable.

REFERENCES

Battiti, R. 1992. First and second order methods for learning: Between steepest descent and Newton's method. *Neural Computation* 4(2): 141–166.

Bishop, C.M. 1995. *Neural Networks for Pattern Recognition.* Oxford: Oxford University Press

Bogdanoff, J.L., and Kozin, F. 1985, *Probabilistic Models of Cumulative Damage.* New York: John Wiley.

Chinnam, R.B. 1999. On-line Estimation of Individual Components Using Degradation Signals. *IEEE Transactions on Reliability* 48(4): 403–412.

Chang D.S., 1999. Modeling and Reliability Prediction for the Step-Stress Degradation Measurements using Neural Networks Methodology. *International Journal of Reliability, Quality and Safety Engineering* 6(3): 277–288.

Haykin, S. 1999. *Neural Networks: A comprehensive foundation.* Upper Saddle River, New Jersey: Prentice Hall, Inc.

Lu, J.C. and Meeker, W.Q. 1993. Using Degradation Measures to Estimate a Time-to-Failure Distribution. *Technometrics* 35(2): 161–174.

Meeker, W.Q. and Escobar, L.A. 1998. *Statistical Methods for Reliability Data*, New York: John Wiley.

Meeker, W.Q. and LuValle, M. J. 1995. An Accelerated Life Test Model Based on Reliability Kinetics, *Technometrics*, 37:133–146.

Reed, R.D. and Marks, R. J. II. *Neural Smithing: Supervised Learning in Feedforward Artificial Neural Networks*. Cambridge, MA: MIT Press

Stern, H.S. 1996. Neural Networks in Applied Statistics. *Technometrics* 38(3): 205–215.

Safety and Reliability – Bedford & van Gelder (eds)
© 2003 Swets & Zeitlinger, Lisse, ISBN 90 5809 551 7

Comparative quantitative risk assessment of railway safety devices

Louis H.J. Goossens
Delft University of Technology, Delft, the Netherlands

Chris M. Pietersen & Margy den Heijer-Aerts
TNO Safety Solutions Consultants, Apeldoorn, the Netherlands

ABSTRACT: Railways are complex systems and currently much more effort is put into quantification of risks of changes in the system, and of new safety devices. In particular, when the risk assessment largely depends on human handling and communication between track workers and train traffic control, a more detailed analysis is required. The risk comparison approach presented in here will be illustrated for out-of-service situations of railroad tracks. The approach taken applies the following techniques: (1) fault trees are used (or developed) to provide an in-depth analysis of the technical and human failure modes. Communication errors are part of the fault tree basic events; (2) for both new and existing safety devices, the mutual differences in the branches of the fault tree are identified: some branches remain the same (or change minimally), other branches are largely different; (3) Hierarchical Task Analyses are applied to specifically identify the potential communication error modes relevant for the *design phase*, and to identify whether communication errors may lead to critical situations. The fault trees are validated with the results so far; (4) in the *operational phase* human handling is further analysed with the quantitative human error technique HEART, in order to identify erosion of work (and communication) procedures. Observations during operations are used to identify differences in types of human error with time when the application of the safety device becomes more mature; and (5) a comparative risk assessments results from the above steps in the analysis.

1 INTRODUCTION

Railway systems are complex systems requiring complex risk assessments as well. Nowadays, quantitative risk criteria have been defined for existing railway use and particularly also for new developments like the new high-speed trains in the Netherlands (Frijters et al., 1998) or the Betuwe-line for (hazardous) materials transport from Rotterdam harbour to Germany. The most used quantitative risk criterion is the Individual Risk (IR) which specifies risks in terms of fatalities per number of persons in a category of users per year. The categories used are railway employees, track workers, passengers, residents living nearby railway tracks, and so on.

Due to its complexity, risk assessments have to be structured to a certain level of detail. If needed detailed risk assessments are performed for specific in-depth analyses. The broad approach to risk assessment in the railway industry follows what has been laid down in Cenelec 50126 (Cenelec, 1996). This safety case approach identifies hazards, evaluates associated risks

and weighs acceptance of the risks case by case. Hazard identification is performed through HAZID-sessions bringing required expertise together. Risk evaluation is done with the Risk Matrix or Risk Graph techniques. The Risk Graph technique – as described in IEC 61508 (IEC, 1997) – is normally applied to identify the required additional risk reduction in terms of Safety Integrity Levels.

Fault trees and event trees are used to the required level of detail in order to quantify failure events and consequences to people. This paper describes the in-depth comparative risk assessment of two safety devices for the protection of track workers, in which the human factor plays a major role. The risk assessment methodology will be emphasised, the safety devices are considered to be examples.

Tracks are traditionally put out-of-service with the help of self-signalling short circuiting lances (in this paper called *lance*). A novel device currently taken into use, is the working zone switch (in this paper called *switch*). To install a lance track workers have to go in between the rails without other protection than a safety man while the tracks are still operational. The switch, however, is controlled from a key-box next to the tracks and prevents track workers from going into operational tracks presupposing a safer working environment for track workers.

2 LANCE VS SWITCH

2.1 *The self-signalling short circuiting lance*

The Work safety leader has a written Work safety instruction in which all activation and deactivation activities of the working zone(s) are mentioned including names and phone numbers of all personnel to execute the activities. One of these activities is placing lances. In the preparatory phase a designated track worker goes to the place where the lance has to be put in place, taking with him his communication devices and other necessary means. He is accompanied by a safetyman. The Work safety leader specifically assigns the placing of the lances. The lance short-circuits both rails and the resulting safe situation is similar to that of having a train on the tracks.

The Train service leader observes this phenomenon, but cannot act as it is beyond his responsibility. It is the responsibility of the Work safety leader. According to procedures there is no interaction between both leaders, but in practice they often communicate.

2.2 *The working zone switch*

The working zone switch is an instrument to activate working zones from a distance. Functionally this means that after turning the switch all signals leading to the working zone tracks are in position "stop", and they cannot be taken out of this position. Now incoming trains cannot enter the working zone. Track points are not put in diverting positions. This entails that the Train service leader can steer various elements of the system while the working zone is activated. This is different from the situation with lances (see section 2.1). If a Train service leader wants to control whether the right working zones are put out-of-service, he can give way for a train through that working zone. The signal should then stay in position "stop".

At shunting yards the working zone switches are present in lockable cupboards. The cupboards can be opened with a unique key. Each switch is locked with a unique lock. The person who operates the switch (the switch operator) gets the key of this unique lock from the Work safety leader. After opening the lock, the switch can be turned. Once the switch has been turned, it will be locked again with the unique key to prevent unintentional de-switching. The cupboard will also be locked. After turning the switch, all relevant controlled signals will be "red-blinking". If no signals are available, a working zone lamp will give a signal. The "red-blinking" position of the controlled signals is only active if the conditions are fulfilled (all signals towards the working zone are in position "stop" and cannot be taken out of that position). Also a control lamp put on the switch gives feedback that the safety functions are controlled and that all conditions prior to activation of the working zone are satisfied.

3 COMPARATIVE RISK ASSESSMENT APPROACH

The railway infrastructure company was faced with the question whether the risks for track workers would change and how much in cases they replace lances with switches. This required a quantitative approach whereby the risk analysis primarily aimed at comparing the risks of the two safety devices with each other. To some extent, this approach simplifies the analysis, as it merely should focus on where the differences in risks appear.

The comparative risk analysis took the following steps:

1. Fault trees (without quantitative data of the basic events) were already available for both safety devices. The fault trees showed the technical and human failure modes. It was obvious from the start that the human failure modes were dominating the basic events of both safety devices. In particular, communication errors are largely part of the basic events.
2. For both lance and switch, the differences in the branches of the fault trees were identified: some branches are similar or change minimally, some branches differ largely and one branch was not present in the fault tree of the switch. The major question was which branches dominate the top events of both safety devices.
3. The Hierarchical Task Analysis technique (HTA) was applied to specifically identify the potential communication error modes found relevant already in the *design phase*, and whether communication errors could lead to critical situations. If so, those communication errors should end up as basic events in the fault trees.
4. In order to be able to properly quantify all human handling failure modes in the fault trees, observations of work processes of the track workers were performed. Communication errors and other human handling errors were identified. Moreover, communication errors were assessed on their ability to lead to a potential critical situation. The HEART technique was applied to identify erosion of communication and work procedures in the *operational phase*.
5. A comparative risk assessment resulted from the above mentioned four steps.

4 FAULT TREES AND DIFFERENCES

This section describes the fault trees and there differences: steps 1 and 2. Table 1 summarises the results of the fault trees in quantitative terms. Table 2 describes the events and the (sub)-branches.

The top events "collision of train with track crew" of both safety devices are connected through an OR-gate with two relatively comparable branches (branches 1 and 2) and one branch (branch 3) only relevant for the lance fault tree ("collision with track worker while putting the lance in place on the tracks"). This last activity is particularly dangerous as the track cannot be put out-of-service at that time. A safetyman accompanies the track worker and checks and warns for incoming trains. In discussing the comparative risk analyses of the results, the contribution of the placing of lances (branch 3) will be taken into account.

Branch 1 of the top event "train in activated working zone" is built up of 3 (in the switch fault tree) or of 4 (in the lance fault tree) sub-branches connected through an AND-gate. The three sub-branches present in both fault trees are fully identical. The additional sub-branch 4 of the lance fault tree might decrease the probability of a train being in the working zone with a factor of 10.

The differences are found in branch 2 of the top event "track crew in insufficiently activated working zone". This branch resembles branch 1 as it is also built up of 3 (switch) or 4 (lance) sub-branches. Sub-branch 1 "working zone not activated for crew" differs slightly, but the other two or three sub-branches are exactly similar for both this branch 2 and the previous branch 1 "train in activated working zone". The additional sub-branch 4 of the lance fault tree might decrease the probability of a working zone being not activated with a factor of 10 also.

Sub-branch 1 with the differences between both safety devices consists of an OR-gate with three sub-branches under it:

1. Sub-branch a describes the track crew entering the activated working zone too early (not being activated yet), or entering the wrong working zone, or leaving a de-activated working zone too late: "human error at entering/leaving activated working zone". For both safety devices this sub-branch is practically similar, with a minor difference of about ten percent decrease of the sub-branch probability in favour of the lance.

2. Sub-branch b describes the failure to properly activate the working zone. This is a key branch of the fault tree as it shows how the safety device malfunctions. Designers of both safety devices had anticipated the importance of failures occurring during activation of a working zone. For that reason, they backed up the activation procedure with a check with signals. This sub-branch is therefore built up from two sub-branches ("activation of working zone failed" and "check of signals failed") connected through an AND-gate. Due to this AND-gate, the contribution of sub-branch b as a whole appears to be orders of magnitude (2 to 3) lower than the other two sub-branches. Even the small differences in sub-branch b of both safety devices do not change this picture significantly: eventually the contribution of the lance sub-branch is about three times larger than for the switch sub-branch.

3. Larger differences appear in sub-branch c, which describes the intentional and unintentional de-activation of the safety device. For the switch these are (a) technical failures and short-circuiting,

Table 2. Explanation of events and (sub)-branches from Table 1.

Event description
Top event: "collision of train with track crew"
Top event has OR-gate with 　Branch 1: "train in activated working zone" 　Branch 2: "track crew in insufficiently activated working zone" 　Branch 3: "collision with track worker while putting the lance in place on the tracks"
Branch 2 has AND-gate with 　Sub-branch 1: "working zone not activated for crew" 　Sub-branch 2: "train present" 　Sub-branch 3: "work safety leader fails to act" 　Sub-branch 4: "train service leader fails to act"
Sub-branch 1 has OR-gate with 　Sub-branch a: "human error at entering/leaving activated working zone" 　Sub-branch b: "failure to activate working zone" 　Sub-branch c: "(un)-intentional de-activation of safety device"

Table 1. Summary of fault tree quantification for the switch and the lance respectively.

Event description	Failure probability of switch	Failure probability of lance
Top event	1.9 * E-4	1.2 * E-4
Top event has OR-gate with		
Branch 1	7.8 * E-5	7.8 * E-6
Branch 2	1.2 * E-4	1.3 * E-5
Branch 3	N/A	1.0 * E-4
Branch 2 above has AND-gate with		
Sub-branch 1	1.2 * E-2	1.3 * E-2
Sub-branch 2	1.0 * E-1	1.0 * E-1
Sub-branch 3	1.0 * E-1	1.0 * E-1
Sub-branch 4	N/A	1.0 * E-1
Sub-branch 1 above has OR-gate with		
Sub-branch a	1.1 * E-2	1.0 * E-2
Sub-branch b	6.2 * E-6	2.0 * E-5
Sub-branch c	2.0 * E-4	2.6 * E-3

(b) switch unintentionally put in a de-activation mode, and (c) switch put back in a de-activation mode by an unauthorised person. For the lance these are (a) technical failures, (b) lance unintentionally removed by an authorised person, (c) "frozen track" opened for trains, and (d) lance intentionally removed by an unauthorised person.

5 HIERARCHICAL TASK ANALYSIS (HTA) AND QUANTIFICATION OF HUMAN ERRORS

Quantifying basic events, which are human error modes, has largely supported the quantification of the fault trees. Parts of the human error modes are communication errors. It must be stressed that not every communication error leads to a critical situation. Therefore the failure probabilities of communication errors had to be adjusted to the sole contribution of critical communication errors.

With the HTA technique (Kirwan and Ainsworth, 1992) the sequence of tasks is identified. Also loops in the task sequence are shown. The HTA technique is particularly applied to analyse whether the task sequence of activation and de-activation activities makes sense as they are originally *designed*. It is clear that the designer of these tasks recognised the possibility of communication errors as a large contributor to failures of the execution of the task. Redundancy has been built in the task sequence whereby the activation procedure for both safety devices has been backed up by checking of control signals. This redundancy leads to a relatively low contribution of that particular sub-branch b of sub-branch 1 (see Table 1).

Observations were made of incidents in the task sequence of all activations of the switch at shunting yards nearby railway stations (during a period of one year after the switches had been newly installed). 57 incidents were recorded during 133 activations of the switch, of which 51 fall under sub-branch b (Table 2), for which the switch designer had already anticipated (communication) errors. Out of the 51 incidents, 33 incidents represent activation errors and 18 incidents represent control errors (checking by signals).

The observations also reveal that about one in 30 human error incidents could have led to a critical situation. In the other 29 cases the communication error was made, but the situation was not critical. For example, if the Work safety leader forgets to name the switch number and the Train service leader does not ask for it, it is not critical in cases where the switch was the right one anyway. For the quantification of critical communication errors (represented as basic events in the fault trees) the corresponding overall human error rate has been adjusted with a factor of 1 in 30: 0.033.

Table 3. Generic human error rates (Kirwan, 1994), abbreviated description, HEP = Human Error Probability.

#..	Description	HEP
#1	General rate for very high stress	0.3
#2	Complex non-routine task (stress)	
#3	Supervisor does not recognize error	0.1
#4	Non-routine operation, with other duties	
#5	Operator fails to act within 30 minutes	
#6	Errors in simple arithmetic, self-check	0.03
#7	General error for oral communication	
#8	Failure to reset manual valve	0.01
#9	Operator fails to act within few hours	
#10	General error of omission	
#11	Error in routine operation, with care	
#12	Error of omission in a procedure	0.003
#13	General error rate for incorrect act	
#14	Error in simple routine operation	0.001
#15	Selection of wrong switch	
#16	Selection of key-switch over non-key	0.0001
#17	Human performance limit: one operator	
#18	Human performance limit: operator team	0.00001

Table 4. Overview of relevant human error modes for switch and lance.

Description of human error modes (HEM)	HEM #.
Communication error	HEM #1
Mistake of Work safety leader	HEM #2
Incorrect permission to set switch	HEM #3
Misinterpretation of Train service leader	HEM #4
Incorrect notification of crew position in tracks	HEM #5
Incorrect admission of crew to enter tracks	HEM #6

For the quantification of human errors a generic list of human error (Kirwan, 1994) rates has been used (Table 3). The numerical values are judgement-driven and are assumed to be reasonable estimates for operator errors in the nuclear and chemical industries. The numerical values should not be treated as an absolute "true" number, but they indicate the order of magnitude of human errors and provide a means to compare human error probabilities of different human actions.

All basic events of the fault trees associated with human errors were mapped on the generic human error rates of Table 3. Moreover, it was identified whether the thus found human error rate had to be adjusted for critical situations only. Table 4 shows the relevant human error modes. Table 5 presents the human error rates used for the quantification of the human error probabilities of the basic events. Notice that some HEP values are corrected for critical situations only applying the factor of 0.033. These human error modes are only relevant when the system is in a critical situation.

Table 5. Human error probabilities of human error modes of Table 4.

HEM # from Table 4	HEP # from Table 3	HEP from Table 3	HEP corrected for critical situations
HEM #1	HEP #7	0.03	0.001
HEM #2	HEP #6	0.03	0.001
HEM #3	HEP #11	0.01	0.00033
HEM #4	HEP #15	0.001	0.001
HEM #5	HEP #12	0.003	0.003
HEM #6	HEP #14	0.001	0.001

Misinterpretations and incorrect admission of crews are critical under all circumstances.

6 OBSERVATIONS DURING OPERATIONS

The analysis so far used estimates related to the system as *designed*. It is well known that during *operations* procedures are not always well followed. This may in particular be the case for communication errors.

Among the various human error techniques the HEART technique has been chosen for its readiness to apply in these situations. HEART stands for Human Error Assessment and Reduction Technique – and has been developed by Williams (cited in Kirwan, 1994). Basically the technique recognises 7 generic human error probabilities, which must be adjusted for error producing conditions (EPC) to comply with the real situation during operations (as, for instance, observed). The EPCs increase the generic human error probabilities with a specific multiplication factor different for each EPC. The human error probabilities during operations may increase sometimes with orders of magnitude.

From the observations of activations of switches at shunting yards near railway stations it was clear that communication tasks were sensitive to inadequate performance. The communication errors most identified during the field observations can be associated with the HEART task "fairly simple task performed rapidly or given scant attention" having a generic human error probability of HEPgen = 0.09. In 133 observations one then expects 12 observed incidents with communication errors. In fact, 51 incidents have been reported being over four times higher.

Out of the 26 EPCs available in the HEART technique the most appropriate EPCs for communication errors of switch operations appear to be

- "the need to transfer specific knowledge from task to task without loss" with a multiplication factor of 5.5
- "a shortage of time available for error detection and correction" with a multiplication factor of 11, and

- "a means of suppressing or overriding information of features which is easily accessible" with a multiplication factor of 9.

The human error probability during operations is calculated with

$$HEP = HEPgen * [(EPC - 1) * EFF + 1]$$

The factor EFF represents the effectiveness in the railway company, i.e. if a certain EPC is strongly influencing the resulting HEP in that particular company, one tends to use EFF = 1, if the influence is moderate one tends to take EFF = 0.4. For the railway company an average effectiveness has been assumed, so EFF = 0.7.

Given 51 incidents instead of the expected 12 incidents over 133 observations the operational human error probability equals HEP = 0.38. Taking HEPgen = 0.09 results in a multiplication factor of 5.7, which is reasonably in agreement with the first mentioned EPC above (with a multiplication factor of 5.5), which appeared to be dominant in the observed incidents (about half of them).

In the last quarter of the observation period a strong increase of the second EPC mentioned above was noticed. This means that the time necessary for detection and control was often diminished too much.

7 DISCUSSION

In this paper a study is presented of a comparative risk analysis of two safety devices in the railway industry: the self-signalling short circuiting lance and the working zone switch (the latter being a recent development).

The final step 5 of the analysis shows that there are differences in risks (Table 1). The overall difference appears to be marginal mostly depending on the risks caused by putting the lance in place prior to its effective operation. If that estimate can be reduced by a factor of 10 the lance appears to be safer than the switch.

The risks of the switch are almost equally caused by both intermediate events "train in activated zone" and "track crew in insufficiently activated zone" (branches 1 and 2 in Table 2).

The basic events of both fault trees are dominated by human error modes. Quantification of these error modes has been achieved using generic human error probabilities corrected for critical situations, if required. The factor to adjust for critical situations (0.033) has been obtained from field observations at railway shunting yards near railway stations. The Hierarchical Task Analysis technique has been applied to check the fault trees against the tasks to be performed as originally designed.

The HEART technique has been used to identify whether in reality the procedures are followed or not. In particular, communication errors and later on shortage of time for detection and control of errors were dominantly contributing to not following up procedures.

In conclusion, the risk levels of both safety devices are comparable. If more than one lance has to be placed at an out-of-service situation, the risk of the use of lances might increase. On the other hand, it has been observed that loosening of the application of the procedures may increase the risk level of the switch. Communication appears to be a weak element in using working zone switches that would require management attention.

ACKNOWLEDGEMENT

The authors want to acknowledge the Railinfrabeheer (RIB) company responsible for track work on the Dutch railways for their financial support of the study. The authors also want to thank Mrs. Y. van der Ven and Mrs. N. Delsing for their comments and support, and Mr. J. van Lokven for performing the observations.

REFERENCES

Cenelec. 1996. Standard prEN 50126, Railway applications: The specification and demonstration of dependability-reliability, availability, maintainability and safety (RAMS), Brussels, Belgium.

Frijters, M.P.C., van Hengel, W., Houben, R.J. An integral safety plan for the high speed train link in the Netherlands. In: Lydersen, Hansen, Sandtorv (eds). Safety and Reliability (ESREL 1998). Balkema, Rotterdam P. 73–77.

IEC, 1997. IEC 61508-5, Functional safety of electrical/ electronical/programmable electronical safety-related systems. Part 5. Examples of methods for the determination of safety integrity levels. Geneva.

Kirwan, B. 1994. *Human error reliability*, Taylor & Francis Publishers.

Kirwan, B., L.K. Ainsworth. 1992. *Guide to task analysis*. Taylor & Francis Publishers.

Safety and Reliability – Bedford & van Gelder (eds)
© 2003 Swets & Zeitlinger, Lisse, ISBN 90 5809 551 7

Integrating risk assessments and safety management systems

J.H. Gould, S.A. Caruana & P.A. Davies
ERM Risk, Manchester

ABSTRACT: Ever since the introduction of the goal setting regulations, risk assessments have become commonplace throughout industry. Risk assessments provide a means of identifying a site's hazards and the operators control measures that ensure the risks are properly controlled. Across high hazard industries, risk assessments have proved to be valuable tool in targeting resources to where they will be most effective in reducing the risks. Safety management systems are also widely used throughout industry. They are the systems by which the safety of the site is maintained. Not only are safety management systems a key requirement within hazardous installations, they are the principle means by which an employer implements the aspirations of the safety policy to produce a safe system and place of work. The relationship between risk assessments and safety management systems is not clearly defined. Not only is there a place for both risk assessments and safety management systems but the two systems should work together to ensure that the controls identified in the risk assessments are properly implemented by the safety management systems. This paper describes the relationship between risks assessments and safety management systems and how they should support each other to ensure the safety of the undertaking.

1 INTRODUCTION

The Seveso II Directive has given new importance to linking control measures to risk assessments. The UK Health and Safety Regulators, Health and Safety Executive (HSE) assess Seveso II safety reports against published criteria, HSE (2003). One criterion is the report's ability to describe how risk assessment is used to make decisions about the measures necessary to prevent accidents and mitigate their consequences. Many of these measures refer to engineering measures where the link with risk assessment is well established. However, Seveso II emphasises the importance of management systems and many of the measures will be the softer management systems rather than engineering ones.

Whilst demonstrating the link between risk assessment and Safety Management Systems (SMS) is important, these two important aspects of safety should be working together to improve safety. In essence, risk assessment is an integral part of an SMS that helps to direct it in managing the hazards.

2 SAFETY MANAGEMENT SYSTEMS

Safety management system is a broad term that includes any system to manage safety. The UK regulator HSE has produced guidance (HSE 1997) on safety management that has proved to be an important influence on how much of UK industry approach safety management.

HSE use the POPMAR framework for managing health and safety. POPMAR refers to the key elements of a safety management system: Policy, Organising, Planning and implementation, Measuring performance, Audit and Review. Figure 1 shows how these elements are arranged.

Whilst it is beyond the scope of this paper to explain HSE's guidance on safety management it necessary to describe the six aspects in order to illustrate its relationship with risk assessment.

2.1 Policy

The policy is the foundation stone for the whole SMS. It sets the organisation's philosophy and aspirations and provides the basis for shaping the organisational culture and the way in which safety is managed throughout the organisation.

2.2 Organisation

The aspirations in the policy will remain aspirations unless some organisation is put into place to make things happen. Organisation is about getting people to

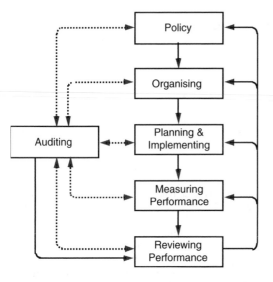

Figure 1. Key elements of successful health and safety management.

do things. Organisation is broken into the four "C"s of safety culture: Control, Communication Competence and Co-operation.

Without control within a company, nothing is ever started. In order to exercise control it is important to communicate a positive safety message top down from senior management and also for mangers to listen to employees concerns. Yet, there is no point in correctly controlling and communicating an organisation's aspirations if the person or persons are not competent to carry out the actions expected of them. Competence of the workforce ensures that the positive management intentions are carried out competently at the frontline. Finally, without co-operation at all levels all the good intentions described in the policy will remain intentions, rather than translated into an effective proactive safety culture.

2.3 Planning and implementation

With the policy and the organisation in place, a company is ready to take the actions necessary to ensure that its activities are completed safely. It is necessary to decide what to do and then to actually do it.

The HSE guidance describes three levels of a health and safety management system:

– management arrangements;
– risk control systems;
– workplace precautions.

The management arrangements is the POPMAR structure described here, the risk control systems are the discrete systems in place to ensure that the risk

control systems function, e.g. emergency plans and incident reporting systems, and the workplace precautions are the specific actions to control risks, e.g. machine guards and PPE.

2.4 Measuring performance

Measuring performance is an extensively researched topic. It enables a company to know when it is achieving the desired outcome. The most common performance measure is the number of accidents or accident rate. This is often entrenched in the policy with the objective of having "zero" accidents. These are lagging indicators of performance, in that they measure the failures that have occurred. However, it is important to emphasise that the reporting of incidents should be encouraged, not avoided in order to keep accident numbers down, as it is from the reporting that lessons can be learnt. If a more proactive safety culture is to be developed, it is important to use leading indicators also, such as the reporting of good safety performance. Yet, given that the whole point of managing safety is to stop harm, it is difficult to escape the use of lagging indicators of the performance of safety management systems.

2.5 Audit

This element of a safety management system is often confused with measurement. Checklists of safety equipment or a shop floor walk through are often described as audits, yet these types of activities are part of the measurement to see if the arrangements are functioning.

Audit is the structured process of collecting independent information on the efficiency and reliability of the total health and safety management system and drawing up plans for corrective action. The output from the audit is key to the review process.

2.6 Review

Arguably, the most important part of the whole system, the review, is where the measurements are examined. Two important questions are asked:

– Are the systems fulfilling their function?
– Is the function contributing to the overall safety policy?

If the answer to these questions are "no", then the review needs to take action to improve the system. Even if the answers are "yes", then the review must look for areas for improvement and set the performance targets for the following year. This aspect is usually set in the policy that expects continual improvement. The review is the part of the management system that closes the circle, taking the output and feeding it back into the management systems to ensure continued improved performance and prevention of the whole safety management system slipping into decay.

3 RISK ASSESSMENTS

From the above description, risk assessments only play a small part in the overall SMS. They are used to identify the workplace precautions. Yet, risk assessment is the keystone behind regulating health and safety in the UK. Experience in developing safety management systems for Seveso sites in the UK has shown the importance of using the results of risk assessments to guide the risk control systems.

4 EXAMPLE RISK ASSESSMENT

The following example is a simplified risk assessment for a flammable material (ethanol) being transferred from 205 litre drums to a mixing vessel. Drums of ethanol are brought from the drum store by a forklift truck (FLT) to a room that houses the mixer. The mixer is brought under a vacuum and connected to a hose via an exchange. The drums are opened and standpipes are inserted. The hoses are fitted to the top of the standpipes and the valves opened to the mixing vessel. The ethanol is drawn into the mixer. When the drum is empty, the valves are closed and a FLT takes away the drum.

The example shows how risk assessment can be used to drive the safety management system and is not intended to show a comprehensive risk assessment of such an operation.

4.1 Identifying the hazards

The hazards are identified using a checklist. For this example only fire explosion hazards are considered, those being related to the ignition of flammable atmosphere:

- in the drum causing a fire or explosion;
- in the drum area causing a fire or explosion;
- and ignition of a pool causing a poolfire.

4.2 Identifying the consequences

The consequences, if a flammable atmosphere should form and be ignited, depending on the location. There are three possible outcomes:

1. Ignition of vapours inside the drum would lead to an explosion in the drum affecting the operator and up to five people nearby by the production of missiles.
2. If an explosive atmosphere is formed around the filling area and ignited there is a possibility of a flash fire or an explosion in the mixing area affecting both operators and up to five people nearby.
3. If liquid is spilt and ignited, any flash fire or explosion would be followed by a poolfire and the potential for escalation and the building to be destroyed.

4.3 Identifying the cause and measures

Normally the hazards, consequences, causes, control and mitigation measures are listed on a single A3 sheet. For this paper, it has been split up into a list and two tables.

The causes and measures in place to reduce their likelihood have been identified and are listed in Table 1.

Table 1. List of causes.

Cause	Control measure
Flammable atmosphere inside the drum	
Partially full drum under vacuum when opened drawing in air.	Drums stored at same temperature as user area.
	Partly used drums inerted before storage.
Air blown into drum from the standpipe.	Air is not used on the mixer vessel.
	Stand pipe not inserted until mixer is under vacuum.
Ignition of flammable atmosphere inside the drum	
Above plus ignition from static.	Drums earthed before opening.
Above plus ignition from rf radiation.	Mobile phones are not allowed in process areas.
Above plus ignition from third party.	Ignition sources restricted by standing orders on igniters and permit-to-work system.
Vapour released from drum forming flammable cloud	
Partially full drum under pressure when opened.	Drums stored at same temperature as user areas.
	Partly used drums inerted before storage.
Drum pressurised via standpipe.	Stand pipe not inserted until mixer is under vacuum.
Liquid released from drum forming a pool and a flammable cloud	
Drum leaks due to corrosion.	Drum condition is inspected on receipt.
	Drum storage area covered. Stock rotation system. Policy on stockage. Standard instruction to deal with leaks and not bring them into the process areas.
Drum leak due to internal conditions.	Procurement quality control.

(Continued)

697

Table 1. (*Continued*)

Cause	Control measure
Drum leak due to impact.	Competent FLT drivers.
	Site traffic movement controlled.
Spill due to drum overturning.	Area flat and clear of obstructions.
Spill due to siphon.	Flexible hose arranged so it always above the level of the drum.
Ignition of flammable vapour	
Spark due to electrical equipment.	All electrical equipment in Zone 1 within 1 meter of transfer area and Zone 2 for other areas.
Ignition due to hot surfaces.	All equipment certified for Zone 1 within 1 meter of transfer area and Zone 2 for other areas.
Sparks due to friction.	All equipment certified for Zone 1 within 1 meter of transfer area and Zone 2 for other areas. All spark generating operations controlled by permit-to-work system.
Ignition form static.	All plant is earth bonded. Drums earthed before opening. All personnel wear conductive clothing and footwear. Static is considered in permit-to-work system.
Ignition from external sources	All work on site controlled by permit-to-work system. Policy to stop all work and make safe if fire occurs in nearby factories. Policy to stop all work and make safe if fire occurs in nearby factories. Building protected lightening conductors.
Ignition from nearby work or people.	All work controlled by permit-to-work system. Site policy not to carry igniters. Maintenance of zoned areas. FLTs are certified for Zone 2.

Table 2. Consequence and control measures.

Consequence	Mitigation measure
Explosion in drum affecting operator and up to 5 people nearby.	Access to process area restricted.
	Part empty drums emptied as single drums rather than on pallets with other materials. Emergency procedures to deal with casualties.
Flash fire or small explosion affecting both operators and up to 5 people nearby.	Process room ventilated at a rate of 5 air changes an hour to dilute flammable clouds.
	Flammable gas detectors that 1. sound alarm to evacuate room; 2. shut off power supply; 3. increase ventilation rate.
Pool fire with potential to escalate and destroy process area.	All liquids drained to central drain that is kept wet to dilute any spilt ethanol. Amount of flammable and combustible in the building is kept to minimum. Heat detectors linked to a fire alarm. Emergency procedures to evacuate the buildings. Fire fighting team.

4.4 *Identifying the mitigation measures*

A similar exercise is performed to identify the mitigation measures. Table 2 shows the consequences and the measures in place to mitigate the effects.

4.5 *Rating the risks*

To complete the risk assessment the likelihood has to be estimated and a judgement made that no further control or mitigation measures are reasonably practicable. This is often done by judgement supported by a risk matrix.

4.6 *Relating measures and control systems*

The risk assessment has identified the workplace precautions (measures). However, these measures are part of key risk control systems. Each workplace precaution is managed by one (or occasionally more) risk control system. A table can easily be made aligning the measures and risk control systems that shows immediately the most important risk control systems for this activity. Table 3 shows the measures.

Once the measures are linked with the risk control systems they can be used to help define their scope and objectives. From the above example it is clear that the emergency plan needs to deal with casualties, organising the fire fighting teams, and evacuation in an emergency. As all the risk assessments are combined and the measures linked to the key risk control systems a full picture of their requirements can be produced.

5 ADVANTAGES OF INTEGRATION

Aligning the measures and risk control systems immediately allows identification of the most important risk control measures for this activity. Perhaps more importantly this alignment produces a list of the

Table 3. Workplace precautions (measures and their risk control systems).

Risk control system	Measure
Operational procedures	Drums stored at same temperature as user area.
	Partly used drums inerted before storage.
	Stand pipe not inserted until mixer is under vacuum.
	Drums earthed before opening.
	Mobile phones are not allowed in process areas.
	Ignition sources restricted by standing orders on igniters and permit-to-work system.
	Drum condition is inspected on receipt.
	Stock rotation system.
	Policy on stock age.
	Standard instruction to deal with leaks and not bring them into the process areas.
	Area flat and clear of obstructions.
	Policy to stop all work and make safe if fire occurs in nearby factories.
	Site policy not to carry igniters.
	Access to process area restricted.
	Part empty drums emptied as single drums rather than on pallets with other materials.
	Amount of flammable and combustible in the building is kept to a minimum.
Maintenance	Drums stored at same temperature as user area.
	All plant is earth bonded.
Maintenance continued	All equipment certified for Zone 1 within 1 meter of transfer area and Zone 2 for other areas.
	Building protected lightening conductors.
	Maintenance of Zoned areas.
	FLTs are certified for Zone 2.
	Process room ventilated at a rate of 5 air changes an hour to dilute flammable clouds.
	Flammable gas detectors that
	1. sound alarm to evacuate room;
	2. shut off power supply;
	3. increase ventilation rate.
	Heat detectors linked to a fire alarm.
Management of change	Drums stored at same temperature as user area.
	Air is not used on the mixer vessel.
	Drum storage area covered.
	Area flat and obstruction free.
	Flexible hose arranged so it always above the level of the drum.
	All equipment certified for Zone 1 within 1 metre of transfer area and Zone 2 for other areas.
	All plant is earth bonded.
	Building protected lightening conductors.
	FLTs are certified for Zone 2.
	Process room ventilated at a rate of 5 air changes an hour to dilute flammable clouds.
	Flammable gas detectors that
	1. sound alarm to evacuate room;
	2. shut off power supply;
	3. increase ventilation rate.
	All liquids drained to central drain that is kept wet to dilute any spilt ethanol.
	Heat detectors linked to a fire alarm.
Procurement procedures	Procurement quality control.
Training	Competent FLT drivers.
	Site traffic movement controlled.
	All personnel wear conductive clothing and footwear.
Permit to work system	All spark generating operations controlled by permit-to-work system.
	Static is considered in permit-to-work system.
	All work on site controlled by permit-to-work system.
	All work controlled by permit-to-work system.
Emergency plan	Policy to stop all work and make safe if fire occurs in nearby factories.
	Emergency procedures to deal with casualties.
	Emergency procedures to evacuate the buildings.
	On site fire fighting team.
	Civil fire fighting team.

aspects that need to be managed that are key to managing safety.

This process has been applied using Windows based databases and spreadsheets. Whilst this simplifies the data handing, the simplicity of the process allows the big picture to be retained.

Perhaps the biggest advantage is a focusing of the key risk control systems that allows realistic performance targets and measurement of the hazards that the system should be controlling. This is in contrast to many targets for risk control systems that reflect the efficiency of the system rather than its effectiveness. Coupled with good communication and co-operation, this system also allows the reasoning behind many of the rules that on the surface and sometimes may appear to be pointless, to be highlighted.

6 CONCLUSIONS

Integration of risk assessment and safety management systems should not be an add-on to the way industry manages safety. They are both essential to making the workplace safer and they both work better when considered together rather than as two separate activities.

The integration, whilst applied to Seveso II sites in this example, is applicable to all workplaces. The integration described in this paper is not the result of sophisticated technical advances, rather it the application of common sense to produce a simple and elegant methodology to ensure the risk control systems address the key aspects of safety.

REFERENCES

HSE 1997. Successful Health and Safety Management, Health and Safety Executive, (HSG65) ISBN 0 7176 1276 7.
HSE, 2003. COMAH safety report assessment manual, Health and Safety Executive's web site http://www.hse.gov.uk/hid/land/comah2/index.htm.
Seveso II, Council Directive 96/82/EC on the control of major-accident hazards involving dangerous substances, *OJ No L10/13, 1997.*

Safety and Reliability – Bedford & van Gelder (eds)
© 2003 Swets & Zeitlinger, Lisse, ISBN 90 5809 551 7

The Halden open Dependability Demonstrator

B.A. Gran & R. Fredriksen
The Halden Reactor Project/Institute for Energy Technology, Halden, Norway

R. Winther & S. Mathisen
Østfold University Collage, Halden, Norway

ABSTRACT: The Halden open Dependability Demonstrator (HoDD) is established in co-operation between Østfold University College and the Halden Reactor Project. The objective of HoDD is to provide an open-source test bed for testing, teaching and learning about risk analysis methods, risk analysis tools, and fault tolerance techniques. HoDD is operated by student projects. The Inverted Pendulum Control System (IPCON) was the first system to be established as a part of HoDD. The main task of IPCON is to keep a pendulum balanced and controlled. The risk assessment of the IPCON made use of the CORAS methodology for model-based risk assessment. From the assessment of IPCON, we believe that the framework is also applicable in a safety context. This paper presents HoDD and summarizes some of the experiences from the risk assessment of the first test system.

1   INTRODUCTION

Increasing dependence on programmable equipment (or Information and Communication Technology, ICT) is a well-known fact. Systems used in e.g. nuclear plants; transportation systems and process control systems involve exposure to the risk of physical injury and environmental damage. These are referred to as safety related risks, and there are several examples of software failures that have caused serious accidents:

– Total loss of Ariane 501 due to an operand error in a non-critical routine.
– Deaths of several patients treated with the Therac 25 computer controlled radiation therapy machine.

The increased use of ICT-systems, in particular combined with the tendency to put "everything" on "the net", gives rise to serious concerns also regarding security, not just in relation to confidentiality, integrity and availability (CIA), but also as a possible cause of safety problems. With the increasing dependence on ICT-systems saboteurs are likely to use logical bombs, viruses and remote manipulation of systems to cause harm (Winter et al. 2001). Some simple examples illustrate the seriousness:

– The result of a HIV-test is erroneously changed from positive to negative due to a fault in the medical laboratory's database system.
– The next update of autopilot software is manipulated at the manufacturers site.

– The corrections transmitted to passenger airplanes using differential GPS (DGPS) as part of their navigation system is manipulated in such a way that the airplane is sent off course.

It should be noted that security might be a safety issue whether the system is real-time or not, and that it is not only the operational systems that need protection, but also the support, development and back-up systems. An attacker could also target systems under development and software back-ups.

It is also quite clear that

– there is a need for implementing security and safety cost-effectively,
– and that trust in the solutions is essential.

There is therefore a need for experimental investigation of methods both for the development, as well as the assessment, of software-based systems. A well-known problem when developing methods is the lack of a sufficient test-system. Relevant systems are rarely open-source, making it difficult to use such systems for research or educational purposes.

An objective of the research activities within risk assessment of total digital systems at Østfold University College and the Halden Reactor Project is to increase the objectivity of methods for dependability assessment. HoDD is organised by Østfold University College (HiØ), and supported by the OECD Halden Reactor Project (HRP). Other companies are also welcome to become partners and sponsors in HoDD.

2 DESCRIPTION OF HoDD

2.1 Objective of HoDD

The HoDD project has two main objectives:

- to provide test beds for risk analysis methods, risk analysis tools, and fault tolerance techniques; and
- to make all relevant information available using open-source principles.

The test bed shall make it possible to study safety and security issues, real-time control-systems and web-applications as well as systems involving all these characteristics. Management and continuous development of HoDD will be an integral part of the master degree program in safety-critical programmable systems at Østfold University College.

By applying open-source principles the aim is to attract researchers from other institutions to make use of the HoDD test bed systems and publish the results on the HoDD web site or in appropriate open publications.

2.2 HoDD sub-systems

The HoDD project is planned to consist of a number of sub-systems, and to be continuously extended with new sub-systems to provide a range of test-bed systems. One of the sub-systems is somewhat special since its purpose is twofold: It will be subject to research and teaching activities, as well as representing the front-end of HoDD providing free access to HoDD-generated information. This system is called the HoDD Information System (HoDD-IS).

Each sub-system is planned to go through 5 main stages, which might be performed in an iterative process:

1. Establishment of requirements/specification and design documentation for the system.
2. Dependability analysis and assessment of the system.
3. Implementation of the system.
4. Operation of the system.
5. Maintenance of the system.

The Inverted Pendulum Control System (IPCON) was the first system to be established as a part of HoDD. Several other systems, such as robots, flying platforms and web accessible databases, are currently under development or planning.

3 IPCON

3.1 Description of IPCON

The main operational task of IPCON is to keep the pendulum balanced and controllable within a given balance interval, which may be changed during a run.

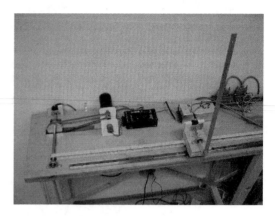

Figure 1. A picture of the IPCON.

The system's hardware components are rail, cart, pendulum, motor, servo-amplifier, sensors that observe the cart's and the pendulum's positions, and a PC, as shown in Figure 1.

Through an interface the user may give commands to start or stop the run, move the cart during a run or execute an emergency stop to immediately stop the cart. The user may also manipulate the signals (input and output) to simulate failures.

The first IPCON design was based on the work of two groups of students from HIØ suggesting requirements and models for IPCON as part of a student project the spring 2002. These designs was united and refined by the use of UML (UML 2000) by another student. These descriptions provided the basis for further work on IPCON, including the risk assessment by applying the CORAS methodology for model-based risk assessment (CORAS 2000, Fredriksen et al. 2002, den Braber et al. 2003).

3.2 Relevance of IPCON

It might be interesting to note that also others (Sha et al. 1998) have used the inverted pendulum to test and demonstrate methods for safety-critical software.

The IPCON system can be characterized by the various modes of operation that are relevant:

- Pre start-up routines: Setting of configuration parameters (e.g. length of rail), confirmation of configuration parameters, checking of external conditions to ensure that conditions are within limits (e.g. no obstacles on rail, hardware responds as predicted, etc.).
- Start-up: Controlled activation of physical system (raising of pendulum) while continuously ensuring the system is within operational limits (e.g. inside allowable part of rail) and that it reaches a normal operational state within a specified time interval.

- Continuous control with strict timing requirements: Pendulum will fall unless the system responds fast and correctly to the gravitational pull on the pendulum. By e.g. changing the physical characteristics of the system the timing requirements can be manipulated.
- Automatic handling of critical deviations: Various incidents might occur threatening to put the pendulum out of control or cause potentially dangerous situations and the system must be able to handle these adequately through e.g. automatic or operator initiated emergency shut-down procedures.

Comparing this to e.g. process control systems it is easy to see the similarities. Although the physical characteristics of a process system comprise such things as temperature, pressure and flow, while an inverted pendulum is characterized by the movement of its cart and pendulum, this essentially only means that the physical laws underlying the control algorithm are different. As the safety and reliability of the control system is not directly related to the actual physical laws, but rather the structure into which these are put, we see that even a simple system like IPCON involve many of the same concerns as "real" systems.

4 THE CORAS METHODOLOGY FOR MODEL-BASED RISK ASSESSMENT

The concept of model based risk assessment (MBRA) has been a research topic since early 80-ies, and builds on the concept of applying systems modelling when specifying and describing the systems to be assessed as an integrated part of the risk assessment. Some important works are by Garrett et al. (1995) and Jalashgar (1997, 1998 & 1999).

The CORAS methodology for model-based risk assessment is the main result from the IST-funded EU-project CORAS (CORAS 2000). CORAS aims to adapt, refine, extend, and combine methods for risk analysis, semi-formal description methods – in particular, methods for object-oriented modelling, and computerized tools to build a specialized RM-ODP (ISO10746:1995) inspired framework targeting risk analysis of security critical systems.

Model-based Risk Assessment incorporates a documentation framework and a number of closely integrated risk assessment techniques and a risk management process based upon widely accepted standards. It gives detailed recommendations for the use of UML oriented modelling in conjunction with risk assessment in the form of guidelines and specified diagrams.

4.1 The CORAS framework

The CORAS framework is based on four main anchorpoints, (1) a risk documentation framework based on the Reference Model for Open Distributed Processing, RM-ODP, (2) a risk management process based on the Australian standard for risk management (AZ/NZS1999:4360), (3) an integrated risk management and system development process based on the Rational Unified Process, RUP (Booch et al. 1999), (4) and a platform for tool-integration based on XML (W3C 2000).

The risk documentation framework is used both to document the target of assessment as well as to record risk assessment results. The risk management process provides a sequencing of the risk management process into sub-processes for context establishing, risk identification, risk assessment, risk evaluation, and risk treatment.

The CORAS risk assessment methodology is a careful integration of techniques and templates inspired by HazOp Analysis (Redmill et al. 1999), Event Tree Analysis (ETA) and Fault Tree Analysis (FTA) (Andrews & Moss 1993), Failure Mode and Effect Criticality Analysis (FMECA) (Bouti & Kadi 1994), Markov Analysis (Littlewood 1975) as well as CCTA Risk Analysis and Management Method (CRAMM) (Barber & Davey 1992).

4.2 The CORAS risk management process

The CORAS risk management process provides a sequencing of the risk management process into the following five sub-processes:

1. Context Identification: Identify the context of the analysis. Describe the system and its environment, identify usage scenarios, the assets of the system and its security requirements.
2. Risk Identification: Identify the potential threats to assets, the vulnerabilities of these assets and document the unwanted incidents.
3. Risk Analysis: Evaluate the frequencies and consequences of the unwanted incidents.
4. Risk Evaluation: Identify the level of risk associated with the unwanted incidents and decide whether the level is acceptable. Prioritize the identified risks and categorize risks into risk themes.
5. Risk Treatment: Address the treatment of the identified risks and how to prevent the unacceptable risks.

In addition, two concurrent sub-processes "monitoring and review" and "communication and consultation" are implicit, and running in parallel with the above five.

4.3 Experiences on applying CORAS

The CORAS methodology for model-based risk assessment has been applied in a trial case on the regional health network HYGEIAnet that links hospitals and public health centres in Crete. It has also been applied in a trial case to the electronic retail market subsystem of an e-commerce platform. (Fredriksen et al. 2003).

5 RISK ASSESSMENT OF IPCON

5.1 Limitations of the assessment

The following limitations were assumed for the assessment:

– The system is assumed not to be online. All content is local. We will not evaluate the risk of being hacked or network problems.
– Mechanical failure in HW, as loose parts, loose cables etc will not be considered.
– Misuse will not be considered.
– SWOT (Strength, Weaknesses, Opportunities, Threats) – analysis, as defined in the CORAS methodology for model-based risk assessment was not performed in its full form.

As can be seen from these limitations we were primarily focusing on the software of the system, the Inverted Pendulum Control System (IPCON).

5.2 Identification of assets and stakeholders

Based on textual descriptions of the IPCON system, its environment and context, a set of stakeholders and assets were identified. In order to provide easy communication of the results an asset-stakeholder diagram was designed as shown in Figure 2. The diagram is based on the UML-profile suggested by CORAS (den Braber et al, 2003).

5.3 HazOp analysis based on use case diagram

Based on a UML use-case diagram, as outlined in Figure 3, it was triad to perform a HazOp analysis. This did not succeed, due to the fact that the use-case was on a too high abstraction level, for the HazOp team members to identify possible deviations in the system. It was therefore decided to split the HazOp team in two.

5.4 HazOp analysis based on an extended use case diagram

One of the groups refined the use-case diagram in Figure 3 to contain the use-cases: {set parameters, check parameters, raise, control, stop}. This resulted in the identification of several deviations with respect to how parameters are set, checked and used in the system. The latter was not within the initial description of the system, and points to the situation that a user both can select to set parameters and apply stored parameters in the storage. However, it was also concluded that the review use-case could not have been made, without an assessment of the use-case in Figure 3.

5.5 HazOp analysis based on state diagram

The other team was assigned a UML state diagram, as outlined in Figure 4. By applying a set of traditional guidewords the HazOp analysis was performed. After a few HazOp meetings it was concluded that a small set of the guidewords: {no, as well as, early, late, other than, part of} were descriptive enough to identify both expected and unexpected deviations from normal behavior. Out of a number of 225 possible deviations, 24 were related to unwanted incidents in the context of danger.

5.6 Event Trees based on diagram

As a result of the HazOp meetings there was also identified two Event Trees based on the initiating events: "Run OK" and "Run not OK". The event trees contained events such as: parameters are set correct,

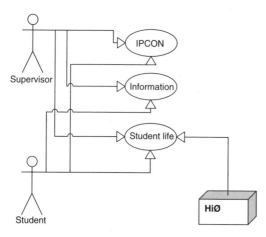

Figure 2.   Asset-stakeholder diagram for IPCON.

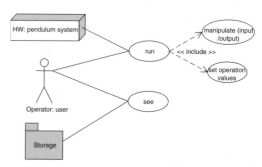

Figure 3.   The first use-case diagram for IPCON.

704

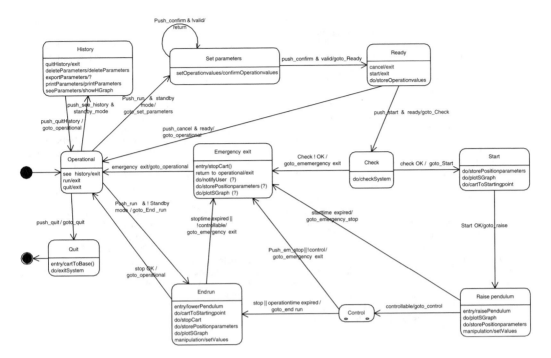

Figure 4.   UMS state diagram for IPCON.

parameters are checked correct, environment is checked correct, raise is initiated, the pendulum is controllable, the control algorithm concludes that the pendulum is controllable, and the emergency stop is working correct.

By inserting worst case consequences at the outcomes it was discovered, without assessing any frequencies at the events, that a safe operation of the inverted pendulum is depending upon safety barrier and controls related to all the events listed. This was not as assumed by the system provider, which had assumed that the main focus had to be on the raise and control algorithms. The set parameters and check parameters algorithms may be equal important.

6   CONCLUSIONS

This paper has presented the Halden open Dependability Demonstrator (HoDD) which is established in co-operation between Østfold University College and the Halden Reactor Project. The objective is that HoDD will provide an open-source test bed for testing, teaching and learning about risk analysis methods, risk analysis tools, and fault tolerance techniques. Therefore other companies, organizations and universities are invited to become partners and sponsors in HoDD.

The Inverted Pendulum Control System (IPCON) was the first system to be established as a part of HoDD. The risk assessment of the IPCON made use of the CORAS methodology for model-based risk assessment, and found it promising. In particular we gained good experiences with performing HazOp analysis with the help of state diagrams. This should be exploited further. On the other hand it was concluded that the use-cases needs some content to be effective. The assessment also resulted in some unexpected results.

REFERENCES

Andrews, J.D. & Moss, T.R. 1993. *Reliability and Risk Assessment*, 1st Ed. Longman Group UK.
Australian Standard AS/NZS 4360:1999 Risk Management. Strathfield: Standards Australia.
Barber, B. & Davey, J. 1992. Use of the CRAMM in Health Information Systems. In ed Lun K.C., Degoulet P., Piemme, T.E. & Rienhoff, O. (eds), *MEDINFO 92*, 1589–1593, Amsterdam: North Holland Publishing Co.
Booch, G., Jacobson, I. & Rumbaugh, J. 1999. *Unified Software Development Process*, Addison-Wesley.
Bouti, A., & Kadi, A.D. 1994. A state-of-the-art review of FMEA/FMECA, *International Journal of Reliability, Quality and Safety Engineering*, Vol. 1, No. 4, 515–543.

den Braber, F., Dimitrakos, T., Gran, B.A., Stølen, K., & Aagedal, J.Ø. 2003. Model-based risk management using UML and UP. To appear as chapter in book titled *UML and the Unified Process*. IRM Press.

CORAS 2000. A platform for risk analysis of security critical systems. IST-2000-25031 (http://www.nr.no/coras/)

Fredriksen, R., Dimitrakos, T., Gran, B.A., Kristiansen M., Opperud, T.A., & Stølen, K. 2002. The CORAS Framework for a model-based risk management process. In Andersson, S., Bologna, S., and Felici, M. (eds), *Computer Safety, Reliability and Security, Safecomp 2002 (LNCS 2434)*, 94–105, Berlin Heidelberg: Springer-Verlag.

Fredriksen, R., Gran, B.A., Stølen, K., & Djordjevic, I. 2003. Experiences from application of Model-based Risk Assessment. Paper accepted for ESREL'2003.

Garrett, C.J., Guarro, S.B., Apostolakis, G.E. 1995. The dynamic flowgraph methodology for assessing the dependability of embedded software systems, *IEEE Trans. on Systems, Man, and Cybernetics*, Vol. 25, No. 5, 824–840.

ISO/IEC 10746:1995. Basic reference model for open distributed processing.

Jalashgar, A. 1997. *Identification of Hidden Failures in Process Control Systems Through Function-Oriented System Analysis*. PhD Thesis Risø-R-936, Denmark.

Jalashgar, A. 1998. Identification of Hidden Failures in Process Control Systems Based on the HMG Method, *International Journal of Intelligent Systems*, Vol. 13, 159–179.

Jalashgar, A. 1999. Goal-Oriented Systems Modelling: Justification of the Approach and Overview of the Methods, *Reliability Engineering and System Safety Journal*, Vol. 64, No. 2, 271–278.

Littlewood, B. 1975. A Reliability Model for Systems with Markov Structure, *Applied Statistics*, 24(2), 172–177.

Redmill, F., Chudleigh, M., & Catmur, J. 1999. *Hazop and Software Hazop*. Wiley & sons.

Sha, L., Goodenough J.B., & Pollak B. 1998. Simplex Architecture: Meeting the Challenges of Using COTS in High-Reliability Systems. *CROSSTALK. The Journal of Defense Software Engineering*, April 1998.

UML proposal to the Object Management Group, Version 1.4, 2000.

Winther, R., Gran, B.A. & Johnsen, O.A. 2001. Security Assessments for Safety Critical Systems using HAZOPs. In Voges, U. (ed.), *Computer Safety, Reliability and Security, SAFECOMP 2001 (LNCS 2187)*, 14–24, Berlin Heidelberg: Springer-Verlag.

World Wide Web Consortium, 2000. *Extensible Markup Language (XML) v1.0*, W3C Recommendation, Second Edition.

Safety and Reliability – Bedford & van Gelder (eds)
© 2003 Swets & Zeitlinger, Lisse, ISBN 90 5809 551 7

Tripod: managing organisational components of business upsets

J. Groeneweg
Leiden University, Leiden, The Netherlands

G.E. Lancioni
Bari University, Bari, Italy

N. Metaal
Maastricht University, Maastricht, The Netherlands

ABSTRACT: This paper will discuss the role of human error in the accident causation process and indicate what the most effective way is of human error prevention. Human error is an important contributing cause in up to at least 90% of all accidents. Consequently the elimination of human error should be a most promising target for accident prevention. Tripod concentrates upon systemic factors and the way in which management decisions can propagate into unsafe conditions at the work place. It attempts to help the organisation to control the accident causation process and not to focus mainly on the individual worker or the negative outcomes like accidents and incidents. Optimal control of the controllable environmental conditions that cause human error makes an organisation maximally intrinsically safe. Managing these conditions is the next step in getting as close to zero accidents as economically and logistically feasible.

1 INTRODUCTION

Since the publication of Human Error (Reason, 1990) a consistent trend in the interest in the contribution of human error to industrial accidents can be noticed. The common factor in this trend is the theory that prevention of human error is most effectively gained by controlling the working environment instead of focusing at the individual who "failed" (Cullen, 1990, Wagenaar, 1992). Safety does not, as many experts believe, depend on the number of sprinklers and hydrants installed, but a high proportion of accidents and catastrophes are the obvious result of management error (Brauner, 1991). According to Rasmussen (1998) accidents are the result of lack of control: "A closer look at major accidents indicates that the observed coincidence of multiple errors cannot be explained by a stochastic coincidence of independent events. Accidents are more likely caused by a systematic migration toward accidents by an organisation operating in an aggressive, competitive environment. [...] Safety is a control problem." World wide a variety of different methods are utilised to reduce human error and Tripod is one of them. To prevent human error a range of techniques are available, some more effective than others. So, why is a new approach to managing safety like Tripod still necessary, with a whole new package of actions to be taken, if other initiatives as Accident Analysis, Unsafe Act Auditing, Qualitative Risk Assessment and Technical Safety Auditing are in place? In many companies these techniques are applied to increase safety. These techniques may be necessary but are not yet sufficient to further decrease the number of accidents. Essential in trying to improve the safety state of individuals is to acquire insight into the situations that lead to accidents and how those specific situations can be avoided. These factors are not only present at the work floor but also at other supervisory and managerial levels. The most successful ones focus on the managerial responsibility in identification and elimination of adverse conditions at the workplace. Due to the complexity of dynamic organisations, management cannot develop fail-safe long term solutions. They should therefore not focus on the complete elimination of human error and the corresponding dynamics of human behaviour by enforcing strict compliance with procedures but on the soundness of their organisation. They have to control the processes they initiate to remedy deficiencies in the structure of the organisation.

Due to the complex nature of organisations and consequently the inability to formulate a fail-safe long term strategy that will prevent all accidents in the future. Concepts can be used from complexity theory. This theory provides a way of thinking about the successes and failures of organisations. This part will conclude that, ultimately, it is not the outcome of the process that should be subject to managerial control but the process itself. Central to complexity theory are some core-ideas: non-linearity, self-organisation and emergence. Findings from complexity theory suggest that it is impossible to predict the future behaviour of complex dynamic systems, like large. This is not what most managers believe: their common assumption is that part of their job is to decide where the organisation is going and to take decisions designed to get there. According to complexity theory this is a dangerous delusion. Management, afflicted by increasing information overload and complexity, can react by becoming quite intolerant of ambiguity. Factors, targets, organisational structures all need to be nailed down. Uncertainty is ignored or denied. The management task is seen to be the enunciation of mission, the determination of strategy and the elimination of deviation.

Traditionally, in the "ideal" organisation there is a Chief Executive Officer (CEO) presiding over a cohesive management team with a vision or strategic intent supported by a common culture. The organisation should stick to its core business and competencies, build on its strengths and keep its eyes focused on the bottom line. This top-down strategic initiatives approach is a recipe for organisational disaster. Even the US army is evolving from just following orders. Since the Gulf War a practice called "directional intent" has been used in which commanders set up units with broad objectives, and the units make decisions semi-autonomously and learn as much from each other as from central command. Leaders and managers should aim at developing conditions which allow self-organising behaviour to flourish and blossom. They need to create adaptive organisations with flexible structures, skills, processes and information flows, rather than hierarchically imposing change.

Although they had many ancestors, it was primarily in the 17th century Newtonian science and Cartesian philosophy that fathered those kind of "outdated" management theories, giving rise to the machine metaphor. It declared that the universe and everything in it, whether physical, biological or social, could only be understood as clock-like mechanisms composed of separable parts acting upon one another with precise, linear laws of cause and effect. That metaphor has dominated thinking in the whole of western society, and increasingly the rest of the world, to an extent few fully realise (Giere, 1999).

Since then managers have structured society in accordance with that belief, propagating this with ever more increasing scientific knowledge: ever more specialisation, ever more technology, ever more linear education, ever more law and regulation, ever more hierarchical command and control, ever more efficiency. We could learn to engineer organisations within which to pull levers at one place and get precise results at another and know with certainty which lever to pull for which result, regardless of the fact that human beings must be turned into resources and made to behave as cogs and wheels in the process. (Hock, 1999). The CEO is seen as the watchmaker of the organisation. The developments in complexity theory make clear that long-term planning is impossible, that visions become illusions, consensus and strong cultures become dangerous and statistical relationships become dubious. Cause and effect of actions are not close in time and distance as well as symptoms and causes of problems are sometimes very distant and in a non-linear world their relation is far from evident.

It is essential to manage the strategic fundamentals related to the soundness of an organisation itself. Even when all the relevant factors determining the soundness of an organisation are identified, taking the best decision is a difficult, even bewildering problem. This paper focuses directly on this issue: finding the right strategies to take the most effective decisions based on the parameters that determine the soundness of the corporate immune system.

2 TRIPOD: THE CONCEPT

The Tripod theory and methodology has been developed at Leiden and Manchester University. It started as a research project investigating ways of preventing human error initiated by the Royal Dutch/Shell Group in 1986. The project resulted into an instrument that is now applied in settings ranging from the Nuclear Energy Authority in the U.K., chemical companies like DSM, Unilever and Shell Chemicals, oil and gas production on the North Sea, the Traffic Control and Safety Units of the Dutch National Railway Company (Groeneweg, 2002).

According to the Tripod philosophy accidents are always caused by one (or more) substandard act(s). Not all substandard acts result in accidents. What an organisation needs to prevent are the "operational disturbances" that precede accidents and incidents (Groeneweg, 1993). If disturbances of the desired way of operating still take place, the organisation has to put barriers in place to prevent these disturbances from turning into an accident or incident. When they are breached or not present at all, an accident or incident occurs. An accident is seen as an operational disturbance followed by its consequences. In between the operational disturbance and the accident barriers are (possibly) located, but they failed, were breached, or circumvented.

708

Substandard acts are by no means random events. They have their immediate origins in psychological states of mind, or patterns of reasoning, which are called psychological precursors. In turn these psychological precursors are elicited by the physical and organisational working environment of people. This can be the way the work is organised, the way the equipment and tools are designed, but also the ergonomics of the work place (Groeneweg, 1998). Environmental conditions that cause the psychological precursors of substandard acts are called latent failures. Latent, because they are present long time before a specific substandard act or accident has occurred and remain hidden without a specific local trigger. These latent failures are usually the result of fallible decisions made by the upper level in systems, such as decision-makers, legislators, designers, managers, inspectors. The Tripod theory recognizes 11 parameters that are critical to the level of control in an organisation. These determinants are called BasicRiscFactors (BRFs). The level of control of these BRFs is indicative for the quality of the management of all business/production processes in different types of organisations (Table 1).

The Tripod theory has served as basic concept for a method to measure the performance, the level of control on processes, of an organisation. This method is called Tripod Delta Survey and is designed to detect weak areas in the environment in which people are operating. The survey uses questionnaires to collect data relating to factual verifiable operational experiences.

In order to adapt the survey to the practical requirements of the end users, the relative abstract BRFs are each subdivided into 4 aspects: Drivers, Resources, Methods and Output. Each individual aspect is addressing a specific organisational level responsible for the quality of the aspect concerned (Figure 1).

Drivers
The "Drivers" represent the "intentions" of the control process: the standards and organisational policies. These represent the most stable control aspect. Standards and policies usually last for a long period of time, making or changing these is a time and effort consuming process. Initiating a process of creating and maintaining standards and policies is usually a task performed on the management level.

Resources
Determination of the "Resources" required is depending on the contents of the defined standards and procedures. Time, money, people, tools and equipment should be made available according to the requirements defined by the drivers. The organisational level which is usually associated with the handling of the resources is of a supervisory nature. If the required "Resources" are not available, not suitable or not fit for purpose, this level is expected to remedy this substandard condition. The time frame in which changes in "Resources" are established is usually between weeks and months.

Methods
The "Methods" determine what type of operating practices is utilised. This aspect comprises the way the work is planned, organised, executed and supervised. Again, usually the people on the supervisory level will deal with this part of the control process. The time frame in which changes in "Methods" are established is usually between weeks and months.

Output
"Output" is what the operational staff experience as their work environment. They have to face the consequences of failures in the previous stages of the control process. Usually the operational staff are not sufficiently authorised to change the situation themselves; they can only detect a substandard condition and raise alarm. "Output" related conditions can be changed relatively quickly: it is merely a matter of minutes and hours rather than weeks or longer.

The BRFs are the categorisation each of these four aspects into the eleven business process fundamentals. Deficiencies in these fundamentals can propagate

Table 1. The eleven BRFs.

Specific for a branch	Generic
Design (DE)	Procedures (PR)
Hardware (HW)	Training (TR)
Maintenance (MM)	Communication (CO)
Housekeeping (HK)	Incompatible Goals (IG)
Error Enforcing Conditions (EC)	Organisation (OR)
Defences (DF)	

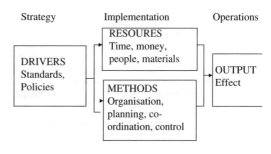

Figure 1. Aspects of the BRFs.

709

through the business process into output problems and business upsets like accidents and incidents (Groeneweg, 2002).

These mutually independent BRFs determine the safety state of an organisation with respect to level of control over human error. Optimal control of the BRFs makes an organisation maximally intrinsically safe and is, therefore, a prime indicator of management performance. By eliminating the latent failures within the various BRF categories, the future causes of accidents are eliminated (Pearce, 1996 and Wagenaar & Van der Schrier, 1997).

Survey results are presented in a quantitative, a graphical profile as well as in a qualitative way: a textual explanation about what aspect is weak/strong.

3 FIELD EXPERIENCE

During an extensive (4 years) field test period a question library (DeltaBase) has been compiled and calibrated to ensure the required level of reliability and validity of the survey results. The method has shown to be of added value to organisations that are interested in their level of organisational vulnerability. Tripod Delta Surveys have been successfully utilised in different settings (different types of industries in various geographical areas). The application of Tripod Delta Surveys in organisations has shown that survey results presented on aspect level give clear and tangible information to strategic as well as operational management. This information has proven to be a sufficient basis for the people who are responsible to develop an improvement plan for a particular area of operation, which has shown to be relatively weak. The results can also be used to avoid overspending of money in relative "strong areas".

4 CONTROLLABLE TARGETS

By the pro-active nature of the method it enables organisations to determine controllable performance targets, without having to rely on the results of accident analyses. This approach enhances "preventative thinking" instead of the more traditional mode of "corrective thinking". Achieved results of interventions can be measured over time to verify improvement as well as for benchmarking. So people are held accountable the quality of the results of planned interventions (improvement plans). By this means people are able to control their performance by the way they conduct their planned actions. This enhances workforce involvement for all staff positions.

The method addresses the latent "underlying" factors of operational processes. As such Tripod Delta

Surveys are considered as an organisation focused add-on to existing inspection, audit and other existing evaluation techniques.

The know-how of the developing partners of this survey is vested in the Stichting Tripod Foundation. Exploitation is delegated to Tripod International BV and its Providers, consultancy firms accredited by Tripod International to conduct Tripod Delta Surveys.

A strict QA/QC protocol, supervised by Leiden University serves as safeguard over the quality of survey results. All survey reports include a quality statement related to the reliability and validity of the survey results.

5 THREE CASE STUDIES

In order to reduce the time needed to complete questionnaires these are split into different sub lists containing a limited amount of questions. Sub lists cover a reduced number of BRFs. By a careful design all sub lists together cover all 11 BRFs, one BRF is represented on all sub lists and serves as "anchor BRF" to verify whether there are no significant differences in response patterns over the population. The sub lists are randomly distributed over the response group, e.g. operational staff of a factory. In case the design is right and the groups are truly representative for the organisation surveyed, the profile scores of any randomly selected sub group (batches) from this population should be similar within acceptable tolerances. This should also be the case with the cumulative distribution of the profile scores of the batches in question, which means that the different batches have about the same amount of respondents returning low profile scores and the same amount of respondents returning high profile scores. These two conditions, similar profile scores and similar distribution, are both required in order to state that all sub lists distributed over these random selected batches are truly indicative for the state of affairs in the organisation surveyed.

In the case of company "A" the questionnaire was split into 4 sub lists which were randomly distributed over the group "operational staff". Every sub list contained the 25 questions of BRF Procedures, the "anchor BRF". The survey covered a group of 240 people working on 15 different work sites controlled by one organisation.

The profile scores did not differ significantly (75 ± 2, n.s.). Also the distribution of the scores of the different batches does not differ significantly. Reduction of response time by splitting up Tripod Delta questionnaires, respecting certain statistical requirements, will not reduce the quality of the survey results. Test-retest reliability of a survey is established by comparing profile scores of the same population obtained by different checklists.

The effects of Tripod interventions (validity) can be measured in two ways:

- No difference in profiles, in case there have been none or unsuccessful interventions
- Improved scores in those areas where interventions have been successful.

Company "B" has been surveyed the first time in 1996. Based on the survey results an ambitious plan was designed to improve the score on the weakest BRF: Maintenance Management, MM.

After 3 years the management requested a second survey to verify whether planned interventions had positive effects on the company's performance. The re-survey showed no significant improvement in the area of maintenance (46 to 48, n.s.). In depth investigation, triggered by the outcome of the survey, showed that the potentially effective improvement plan had been kept securely in a drawer after it had been presented to the management. Tripod Delta Surveys also indicate when improvement plans are not implemented.

Company "C" was surveyed in 1997 and invested a substantial amount of money in Maintenance Management as the survey showed this BRF as being the weakest area. To verify the effect of this investment in Tripod terms the company decided to conduct a re-survey on this BRF only, using a questionnaire with twice as much (50) questions; a set of 25 questions identical to the first survey, and another set of 25 questions which were never asked in this company before. All questions were drawn from the calibrated Tripod Delta question pool, the Delta Base, and mixed into a single questionnaire. As such this survey design complies with the requirements to test the test–retest reliability of the survey method. A comparison of the 1997 and 1999 surveys show the effect of the investment: a significant increase of 15 points (52 to 67, $p < .05$) in scores measured by both question sets. The high level of test–retest reliability of the Tripod Delta Survey was established by comparing the "MM Question set 1" and "MM Question set 2". Both lists give equal profile scores, showing that survey results are independent of the specific set of questions used and indeed indicative of the level of control an organisation has over this BRF.

6 CONCLUSIONS

The scientific approach of the Tripod Delta survey method eliminates the influence of the subjective views of individual auditors or inspectors. Survey results are determined by straight forward statistical processing of data without otherwise inevitable human bias. This makes Tripod Delta a reliable method to conduct benchmark studies between different organisations.

Benchmarks have been established already in the railways as well as in the drilling industry. At the time this paper is written an extensive survey is conducted in the shipping industry. The method has repeatedly shown to be valid and reliable. The results presented in this paper confirm an earlier study by Shell Expro in Aberdeen. In this study it was shown that Tripod Delta Surveys provide useful recommendations at a fraction of the costs of "traditional" audits (Pearce, 1996 and 1997). The results of a Tripod Delta Survey form a sound basis for interventioins aimed at improving the level of management control in a company.

Tripod can evaluate actual performance of an organisation with regard to the control of human error inducing situations, regardless of the Safety Management System an organisation has put in place. It is fully in line with the ISO 9000 and 10000 requirements. This tool cannot replace all other existing techniques: it is an add-on to techniques already in use in many organisations. It is focused on a specific part of the accident causation process and "only" aimed at the prevention of human error (Pearce, 1997).

Tripod Delta can also measure actual performance of an organisation regardless of size of the company, structure, complexity level or activity. Management and leadership performance are not defined in terms of directly assessable qualities of managers and leaders. Accident counts are not valid parameters to measure individual management performance, but are indicators of organisational weaknesses. It is management's responsibility do something about these weaknesses in the process without trying to eliminate the self-organising properties of the organisation. Managing safety is done most efficiently and effectively by managing the start of the accident causation process. This is equivalent to managing an organisation properly: lack of safety is the result of inadequate management of deviations in the intended organisational processes. Controlling the organisation is a management responsibility.

Faced with an increasingly complex organisation, managers can be tempted to reduce the accompanying induced uncertainty by imposing strict rules and procedures for every action at every level in the company. They can require strict protocols for communication, feeling the need for being completely informed. They will try to "freeze" the company into a stable and uniform entity. Because safety is an emergent property of a self-organising system, taking away the dynamics by introducing uniformity and strict compliance has devastating effects. The self-organisation will disappear and organisations are at the mercy of managers who believe that they can run their organisations like some kind of clockwork mechanism. Because of the fundamental impossibility to predict future behaviour of their own organisation as well as demands of the "outside world" they will be effectively steering on noise. The organisation will not be able to self-adapt to changing environments rendering management's task even more difficult by introducing the need for a perfect diagnosis of the present situation to make the necessary changes. This should be avoided at all cost. Management should be held accountable for creating a sound self-organising company where their role is limited to guarding the business process. Therefore, they should be audited by determining the soundness of that organisation and on the process that they initiate to remedy deficiencies in that organisation.

There is no leader or management team that can force organisations into a rigid structure where all deviations from "desired" behaviour are eliminated. It is neither possible or desirable to rule out every form of variability in behaviour by producing more and more procedures in a fruitless and counter-productive attempt to eliminate "human error".

Managing safety is done most efficiently and effectively by managing the start of the accident causation process. This is equivalent to managing an organisation properly: lack of safety is the result of inadequate management of deviations in the intended organisational processes. Controlling the organisation is a management responsibility. Companies need dynamics to self-organise in a stable and sound and competent organisation. In audits, management should monitor their own control over the business process as the way to make the seemingly intangible qualities of management and leadership visible. The Tripod Delta-tool was developed to audit the business process. The relevant parameter that can be audited is process related: the number of latent failures introduced into the organisation and the (dis-) ability to manage them is indicative of quality of management and leadership.

Acknowledging what has already been achieved, a further understanding of accidents and focusing in on their history is the next step in getting as close to zero accidents as economically and logistically feasible and thus to prevent as many people as possible from being injured.

REFERENCES

Bagulay, P. 1994. *Improving organisational performance. Handbook for managers.* London: McGraw-Hill Book Company.

Brauner, C. 1991. *Estimating risks in the petrochemical industry.* Zurich: Swiss Reinsurance Company.

Giere, R.N. 1999. *Science without laws.* London: University of Chicago Press.

Groeneweg, J. 1993. Risk prevention strategies: an alternative approach. In: Bernard Moncelon (Ed.) *Matriser le risque au poste de travail.* Nancy : Press Universitaires de Nancy.

Groeneweg, J. 2002. *Controlling the controllable, the management of safety, 5th revised edition.* Leiden: Global Safety Group.

Groeneweg, J. & Roggeveen, V. 1998. Tripod: Controlling the human factor in accidents. In: Lydersen, S., Hansen, G.K. & Sandtorv, H.A. (Eds) *Safety and Reliability. Proceedings of the 1998 ESREL Safety and Reliability Conference, Trondheim.* Rotterdam: Balkema.

Hock, D.W. 1999. *The birth of chaordic organisations. Human resources or resourceful humans?* Paper presented at the Chaordic Alliance conference, Boca Raton, Florida.

Pearce, A.J. 1996. Tripod in Shell Expro or learning from what hasn't gone bang yet. In: *Proceedings of the 1996 Production and Maintenance conference,* London: IIR.

Pearce, A.J. 1997. Tripod in Shell Expro. Creating an open safety culture with full workforce involvement. In: *Proceedings of the IIR Conference on Behavioural Measurement and Management Conference.* London: IIR.

Rasmussen, J. 1998. Major accident prevention: What is the basic research issue? In: Lydersen, S., Hansen, G.K. & Sandtorv, H.A. (Eds) *Safety and Reliability. Proceedings of the 1998 ESREL Safety and Reliability Conference, Trondheim.* Rotterdam: Balkema.

Reason, J.T. 1990. *Human error.* Cambridge: Cambridge University Press.

Wagenaar, W.A. & Groeneweg, J. 1987. Accidents at sea: Multiple causes and impossible consequences. *International Journal of Man-Machine studies* (27): 587–598.

Wagenaar, W.A. & Schrier, J.H. van der, 1997. Accident analysis. The goal and how to get there. *Safety Science,* Vol. 26, 1/2, 25–33.

Safety and Reliability – Bedford & van Gelder (eds)
© *2003 Swets & Zeitlinger, Lisse, ISBN 90 5809 551 7*

Models of ship's traffic flow for the safety of marine engineering structures evaluation

L. Gucma
Maritime University of Szczecin, Poland

ABSTRACT: The paper presents an introduction to modeling of vessels traffic used for determination of accident probability of ships with fixed offshore structures. The models are based on Monte Carlo simulation principles. They can be used for evaluation of safety of any fixed offshore structures as underwater pipelines, offshore drilling rigs, offshore wind farms. The case study of offshore wind farm localization safety evaluation is also presented for demonstration purposes.

1 INTRODUCTION

The evaluation of complex marine systems safety demands creation of various models that include parameters like: ship's traffic, meteorological conditions and other relevant navigational features. Most of these parameters have random nature and analytical models are not fully suitable to describe of such complex and non-deterministic systems, especially if there is necessity to include also human error possibility. Evaluation of such systems safety can be performed by simulation methods using particularly Monte Carlo simulation method. The paper presents the methodology of creating simulation models for safety determination of fixed engineering structures located on the sea with consideration of its impact on shipping safety. The presented methodology is exampled by the model created for determination of offshore wind farm localization safety. The basic advantage of applied models is fact that with use of them it is possible to obtain more quickly a large amount of simulated data from a wide space of time and changes in the input parameters are very simple (Merrick et al. 2001).

It should also be noticed that presented model takes into account the human factor related with any errors and decisions to be made (for example: dropping the anchor decision) which is not easy to include in analytical methods. The presented method take also account of physics and hydrodynamics of modelling processes like ships behaviour in different meteorological conditions and under anchor.

The presented models can be applied after making several changes for determination of safety of such structures like offshore wind farms, offshore drilling rigs and other objects that can be considered as dangerous for surface navigation.

2 GENERAL CONCEPT OF SIMULATION MODELS

It is should be consider that even the good controlled ships traffic vessels are tends to go off-route and collide with objects or go aground. Mainly blame for such accidents are caused by navigation mistakes due to human or technical failure (or mostly combination of them).

There are also several factors that can influence collision of ships with fixed offshore structures (Fancourt 1991, Gucma & Materac 2002, Hansen & Simonsen 2000, Karlsson et al. 1998, Wennink 1988). The main of them can be presented as follows:

1. Technical failure of ship:
   - technical failure of main ship's equipment such as main engine, steering gear or generators,
   - combined with unfavorable meteorological conditions,
   - no prevent action available: external such as tugs, internal such as ship's anchors.
2. Navigation failure of ship (mainly due to human error):
   - chart data error (inadequate or uncorrected charts),
   - course keeping error (failure of detection action on ship),
   - no course change on waypoint arriving,
   - position error (failure of detection action on ship),

- often combined with unfavorable meteorological conditions.
3. Prevention system failure (if such exist, such as guarding ships, lights, etc.).

Presented models are to consider also physical process of ships behavior in special conditions such as drifting, leeway, anchor dragging, etc. The accidents can be also classified by the part of ship which causes the accident and can be necessary to determine the scale of consequences:

- ship's hull (drilling rigs, wind turbine),
- ship's anchor (pipelines, cables),
- indirectly due to propeller stream or waves generated by passing ship.

The analysis and observation of complex marine systems in safety aspect can lead us to distinguish several kinds of problems which can be solved with use of simulation models in aspect of accident. Typical area of models application is to assess the safety of:

- offshore drilling rigs,
- offshore wind farms,
- elements of fixed navigational marking,
- underwater pipelines and cables,
- other fixed offshore engineering objects.

3 BASIC MODEL PARAMETERS

There are several important parameters which affect the models. The most significant of them are modeling the ship's traffic, meteorological conditions, technical failures and human reliability.

3.1 Ship's traffic

The traffic of ships along a definite route is considered as a process affected by numerous factors changing with time, as well as the route length. These factors make the traffic a random process, and probabilistic methods are used for its description. The flow of the ship's stream at the water area can be presented on a time axis, where the moments of ships passing through a given point are random occurrences. The random stream of ships can be analyzed by examining the distribution of:

- the number of ships passing through a given point in time Δt,
- time intervals between ships,
- local ship's speed; and the scatter of ship positions from the assumed (mean) trajectory.

The number of ships passing through a given route point in the case of free movement, that is such movement where the ships are free to choose the speed and maneuvers, can be considered as a random process

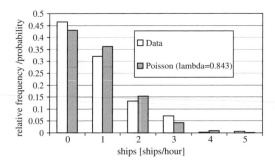

Figure 1. Sample of ship's entering and leaving Swinoujscie Port fitted to Poisson distribution.

described by Poisson distribution, where the probability of the appearance of $X = n$ units in the time space Δt is equal to (Figure 1):

$$P(X = n) = \frac{(\lambda \Delta t)^2}{n!} e^{-(\lambda \Delta t)} \qquad (1)$$

where λ – traffic intensity [ships/h].

The distances between vessels thus have exponential distribution. In case when there is a minimal safe distance between two ships two-parameter shifted exponential distribution can be used in research.

The ship's speeds in free traffic can be described by normal or lognormal distribution (Gucma 2002). When normal distribution is considered the following density function can be used for generation of simulated ships speed:

$$f(v_s, \delta_v) = \frac{1}{\delta_v \sqrt{2\pi}} e^{-\frac{(v-v_s)^2}{2\delta_v^2}} \qquad (2)$$

where v_s – mean speed of a given class of ships; δ_v – standard deviation of ship's speed.

It is assumed that the speed determined in this way is limited on the upper end by the speed possible to obtain with the propulsion applied on the ship. Ships parameters such as length, breadth, drought, and type can be easily calculated with use of distribution function usually obtained from statistical data (Figure 2).

3.2 Course and position of ships

The offset of ship positions from the chosen trajectory (route course) can be described by normal distribution, and in the case of a navigational obstruction by asymmetric distributions (e.g. Rayleigh's). Usually for modeling the ships traffic in open waters the combination of normal and uniform distributions are applied. It is often assumed that 1% of traffic is uniformly distributed and normal distribution is therefore modifying accordingly (Karlsson et al. 1998).

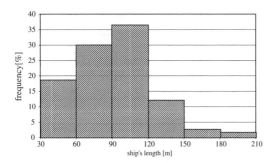

Figure 2.   Distribution of ship's length of vessels passing to and from Szczecin-Świnoujście ports complex to Eastern Baltic Sea.

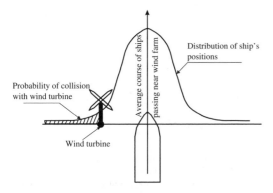

Figure 3.   Calculation of ship-wind turbine accident probability based on ship's position distribution function.

Method of calculation of ships accident probability due to course offset is presented in Figure 3.

3.3   Weather conditions

Effect of wind, current and waves are crucial in presented models. The wind and current affecting behaviour of disabled ships (drift speed and direction) and can be very useful for modelling the consequences of an accident. The wind vector is usually modelled with use of available long term statistical data. It should be noted however that wind speed is correlated with its direction and thus distribution of conditional probability of wind from given direction should be applied. Statistical data bases are usually build for meteorological stations located on land and special corrected factors for wind speed at open sea should be therefore applied.

3.4   Technical failures

Technical reliability (influence of possible breakdown of some navigation devices) can be taken into account for the devices such as main engines, steering gears, auxiliary engines, generators, radars etc. It can be estimated by the technical reliability functions. In order to calculate the reliable operation of the above appliances, there is used an intensity function of damage at time, which is a density function of damage occurrence on the condition that no damage has taken place so far. After multiplication of probabilities with assumption of its independence the overall failure technical probability can be estimated. For commercial vessels it's approximately on level 10^{-4} failure per hour of operation. The calculation of technical reliability can be performed with assumption that an examined machine has been operating failure-free for some time. Then we can consider the machine safety reliability at its stable operating stage. During the stable stage of operation the risk function $\lambda_e B(t)$ does not depend on time and is can be considered as constant. On this basis with assumption that the times between successive technical failures of ships can be presented by exponential distribution with failure intensity λ_e. The probability of reliable operation can be expressed by this equation:

$$P(t) = e^{-\lambda_e t} \qquad (3)$$

In presented models the technical failures are calculated with use of exponential distribution taking into account the time of sailing near examined structure. In more sophisticated models the repair time can be estimated depending of the kind of damaged equipment and other conditions (for example weather). The data for such models can be taken from statistical data or be estimated by general ship engineering judgment.

4   EXAMPLE RESULTS (CASE STUDY)

The example application is presented for one kind of model designed for accident probability determination of offshore wind turbine farm in consideration of disabled ships. Presented example model is based on Monte Carlo simulation methodology with an implemented human decision algorithm of anchor dropping and a physical model of anchor working with the effect of wind and current acting on the vessel. In the study due to simplification only disabled ships accident has been modelled. The model does not consider accidents due to errors in course keeping and accidents due to steering gear failures in close vicinity of wind farm. Probability of these accidents are however much more smaller than considered accident caused by disabled ship (Gucma & Materac 2002).

The prevent action on ship (dropping the anchor) and time of rescue action from shore was taken into account. The most common type of accident is accident, where the ship, due to extremely unfavourable

715

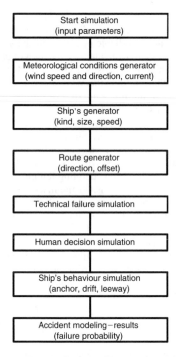

Figure 4.   Basic diagram of presented model.

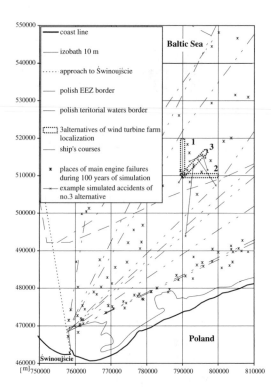

Figure 5.   Three alternatives of wind turbine farm localisations. Examples of simulated places of technical failures during 100 years of simulation. Example accidents scenarios of no. 3 wind turbine alternative.

conditions, will not be able to hold on to the dropped anchor and will drift towards wind farm. The possibility of breaking off the anchor chain is also considered in the model which is especially important for large ships. Human error such as neglecting to drop the anchor is also considered. Simplified diagram of model is presented in Figure 4.

Three possible alternatives of offshore wind farm localisation in close vicinity of Port of Świnoujście have been evaluated with consideration of ship-wind turbine accident probability (Figure 5). The Figure 5 presents simulated examples of places where technical failures occur (e.g. breakdown of the main engine) during 100 years of the simulations. An analysis of the results presented in this figure confirms the self-evident thesis that technical failures occur mostly in places of the highest traffic intensity. Figure 5 also presents several chosen ship-wind turbine accidents and the dropping anchor way immediately preceding it. Most frequent breakdowns are caused by dragging anchor in bad meteorological conditions. The distance of anchor dragging is different for various accidents, but it does exceed 10 km as a rule, and the direction is corresponding with the wind direction.

The most important result obtained with use of presented model are concerned with investigated accident probability. In case of systems with relatively long time between accidents the time of running simulation should be sufficient to get stable results (about one million years in presented case). After main simulation, sensitivity analysis should be performed with consequently changes in the crucial model parameters.

Other important safety factor are time distributions between successive simulated ship-wind turbine accidents. The obtained in simulations times between accidents were fitted to the probability distributions of some random variables. It was found out in accordance with previous supposition that the best fit gives exponential distribution with estimated parameter for considered alternatives: $\lambda_1 = 906$ years, $\lambda_2 = 667$ years and $\lambda_3 = 630$ years (Figure 6). By means of this distribution it can be estimated that an accident will occur in shorter time than the expected lifetime of the wind turbines (approximately 30 years). This probability amounts consequently: 3.2%, 4.4% and 4.6% for the investigated three alternatives of wind farm.

It should be remembered that this analysis does not consider other relevant factors that are important for shipping safety like influence on existing navigation marking and proper operation of electronic equipment installed on the ships.

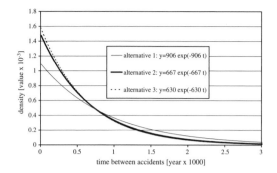

Figure 6. Estimated time distribution between simulated successive ship-wind turbine accidents of investigated wind farm alternatives.

5 CONCLUSIONS

Presented methodology of safety modeling of can be applied for determination of ship collision with offshore structures accident probability. Furthermore the marine risk assessment could be performed with application of consequence modeling. The further steps on this field should be concerned with verification of basic model parameters and creating more adequate models of human decision in different conditions and models of physical ship's performance before and after accident.

REFERENCES

Fancourt, R. 1991. Fixed and Floating Structures – Maritime Risk Assessment and Desiderata for Safe Navigation, *The Journal of Navigation* Vol. 44.

Gucma, L. & Materac M. 2002. Risk of collision of ships with maritime offshore wind farms in aspect of its localization (in polish). *Influence of Offshore Wind Power Plants Location on Navigation Safety, Wind Power – Planning and Realization; Proc. Intern. Conf.*, Sopot 2002.

Gucma, L. 2002. Distributions of ship's on Szczecin-Świnoujście waterway. *The Role of Navigation in Support of Human Activity on the Sea; Proc. Intern. Conf.*, Gdynia 2002.

Hansen, P.F. & Simonsen, B.C. 2000. GRACAT: Software for Grounding and Collision Risk Analysis. *Collision and Grounding of Ships; Proc. Intern. Conf.*, 2000, Copenhagen.

Karlsson, M. & Rasmussen, F. & Frisk, L. 1998. Verification of Ship Collision Frequency Model, In H. Gluver & D.Olsen (eds), *Ship Collision Analysis*. Rotterdam: Balkema.

Merrick, J.R.W. et al. 2001. Modelling Risk in Dynamic Environment of Maritime Transportation. *2001 Winter Simulation Conference; Proc. Intern. Conf.*, Washington 2001.

Wennink, J. 1988. Offshore Platform Collision Exposure to Passing Ships, *The Journal of Navigation* Vol. 41.

Safety and Reliability – Bedford & van Gelder (eds)
© 2003 Swets & Zeitlinger, Lisse, ISBN 90 5809 551 7

Applications of safety and reliability approaches in various industrial sectors

C. Guedes Soares
Instituto Superior Técnico, Portugal

N.K. Shetty
Atkins, United Kingdom

O. Hagen
Det Norske Veritas, Norway

A.P. Teixeira
Instituto Superior Técnico, Portugal

L. Pardi
Autrostrade, Italy

A. Vrouwenvelder
Technical University of Delft, The Netherlands

E. Kragh
COWI, Denmark

K. Lauridsen
RISOE, Denmark

ABSTRACT: This paper compares the state of application of safety and reliability approaches in various industries in different countries. In particular it deals with the Power, Process, Oil and Gas, Maritime, Highways and Building sectors. It draws the similarities and key differences between the sectors and aims to identify where the best practice can be found on various technological areas. It deals with the regulatory framework and the status of the available Code standards and guidance documents. The present position with regard to a number of key technical aspects involved in the use of safety and reliability are discussed. Based on this a technology maturity matrix is presented. Barriers to the wider use of risk and reliability methods in design and operation of installations are identified and possible ways of overcoming these barriers are suggested.

1 INTRODUCTION

An important initiative at European level, involving about 70 institutions, addresses the questions related with the assessment and management of safety and reliability of industrial products, systems and structures. It covers in principle all types of industrial plant, equipment, structural systems, buildings and other civil engineering facilities. The focus will be on safety-critical systems in industrial sectors of oil & gas production facilities, process industries, power (nuclear, conventional, renewable energies), shipbuilding and maritime transport, surface transport (road and rail), and buildings.

It aims at reviewing the state-of-the-art to identify emerging ideas and solutions which hold promise for practical implementation and also the current practice in different countries and different industrial sectors with a view to identify the "best" practices.

It intends to identify gaps in practice in different industrial sectors and countries considering the industrial problems for which there are no satisfactory technical solutions, the gaps between practice and research, the gaps in the level of technology used between countries, and the gaps in the level of technology used between industrial sectors.

The follow-up is to promote ways and means for addressing the identified gaps by developing best practice guidance documents for application to industrial problems, stimulating transfer of technology between industrial sectors and countries, identifying priority areas for research to respond to industrial needs, and promoting the initiation/coordination of research projects in these areas, providing recommendations to code committees and regulatory bodies, developing systems and technical information for use in training and education, and dissemination the results through industrial workshops, discussions with industry forums, publications, web site, newsletters and other similar initiatives.

One of the initial activities in this project was to compare the status of maturity of different industrial sectors in the various counties as concerns the application of safety. Each of the individual industrial sectors was reviewed, including the characteristics of major hazards. This was followed by an outline of the Regulatory framework among some member countries in Europe and the status of available codes, standards and guidance documents. The present position with regard to a number of key technical aspects involved in the use of safety and reliability approaches were discussed. Based on this review a Technology Maturity Matrix was synthesized and barriers to the wider use of risk and reliability methods in the design and operation of installations are identified and possible ways for overcoming these barriers are suggested.

This paper focuses on drawing similarities and key differences between the sectors and aims to identify where best practice can be found on various technological areas. Based on this priorities for research are identified. Opportunities for transfer of technology between industrial sectors are identified.

2 NATURE AND COMPOSITION .OF INDUSTRIAL SECTORS

The nature and composition of each industrial sector including the characteristics of major hazards are described in detail in the respective Position Papers. This covers aspects such as:

– Types of installation/facilities/structures,
– Characteristics of the sector,
– Major hazards and their characteristics,
– Operational and environmental loads,
– Deterioration mechanisms,
– Accidental events,
– Gross errors due to human and organisational influences.

The types of installation range vary very widely from buildings, bridges, tunnels and roads to ships,

offshore oil & gas installations to industrial plants used by chemical/process industries and the more complex nuclear power plants. Within each sector a range of systems are used, for convenience these can be broadly grouped into three types:

– Structural systems whose primary function is to support imposed loads – both operational and environmental. Examples include buildings, bridges, tunnels, pavements, offshore structures, ships, pressure vessels, pipelines, moorings, etc.
– Plant & Equipment whose primary function is to facilitate certain mechanical or chemical operations. Examples include pumps, compressors, separators, drilling/cutting/shaping equipment, heating/cooling systems, cranes, pressure vessels, process piping, etc.
– Safety & Control Systems whose primary function is to ensure safe operation of the installation. These may include a range of detection and monitoring systems and instruments, emergency isolation, shut-down and pressure/temperature relief systems, mitigation systems (e.g. water sprinklers), physical barriers, communication systems, and escape, evacuation and rescue systems.

A diverse range of actions and influences can pose a threat to the safety and integrity of the above systems, for convenience these can be broadly grouped into the following hazard sources:

1. *Operational loads* – gravity loads, live loads, pressure, temperature, etc.
2. *Environmental actions* – wind, waves, snow, floods, earthquake, ice loads, landslide, etc.
3. *Deterioration* – corrosion, fatigue, settlement, erosion, scour, chemical attack, etc.
4. *Accidental events* – fire, explosion, collision, capsizing, toxic release, radiation release, etc.
5. *Gross errors due* to human and organisational influences, which can occur during design, construction, operation, maintenance or de-commissioning.

3 REGULATORY FRAMEWORK, CODES AND STANDARDS

3.1 *Regulatory framework*

The regulatory framework for health, safety and environmental management is well developed and comprehensive in most European countries. Safety is regulated through a combination of European Directives such as the Seveso II, the ATEX, etc. combined with national laws and regulations.

Major hazards industries such as nuclear power, offshore oil & gas and the railways have the more stringent set of safety regulations.

The regulations have evolved generally from a "prescriptive" approach to "goal setting" and "risk-based" approaches. They require that all hazards to an installation are identified, their risks quantified and measures taken to reduce the risks to as low as reasonably practicable. There is a recognition that absolute safety cannot be guaranteed and that the effort spent in risk reduction should be commensurate with the benefits that can be obtained.

The risk-based regulations provide the necessary legal framework for the use of risk-based methods and for assessing the acceptability of risks once they are quantified.

The regulations predominantly focus on ensuring safety and do not generally deal with inspection and maintenance aspects, which in turn can affect safety.

3.2 Codes and standards

Codes and standards are well established for the design of structures and the production of various types of products, materials and their testing for quality. Codes are produced by national standards bodies and increasingly these are being replaced by international bodies such as the European Committee for Standardization (CEN) and the International Organization for Standardization (ISO).

In most cases, codes have the status of "guidance" only but in some cases they are explicitly referred to by regulations in which case they achieve a legal status. Nevertheless, codes represent what is considered by the profession as "good practice" and as such are deemed to satisfy relevant regulations.

Many structural codes are based on the principles of structural reliability theory embodying the concepts of uncertainty, variability, limit states and acceptable levels of safety. However, in their application they are deterministic in nature employing single characteristic values of variables combined with one or more partial safety factors. The recent Eurocodes for bridges and buildings, the draft ISO code for offshore structures, the NORSOK code represent the most advanced set of codes.

A code of practice for structural design is not necessarily appropriate to assess an existing structure. The code provisions and partial factors for design generally need to be reviewed for use in assessment in the light of knowledge of the existing structure. Codes and standards for assessment of existing structures are generally few in number and are not particularly well developed.

4 PRESENT POSITION

4.1 Risk analysis

Techniques for Quantitative Risk Analysis (QRA) are well established in the nuclear power, process and oil & gas sectors. The supporting data and guidance are also well developed in these sectors. These methods are beginning to be used in other sectors, particularly in the area of fire risk analysis of buildings and tunnels.

Risk analysis methods have been used mainly for accidental events such as fire, explosion, collision, toxic release etc. arising from operational activities (e.g. welding), equipment malfunction, fluid leaks from pipes or valves or loss of containment. The influence of human and organisational factors to risk is taken into account generally in a qualitative way.

The conventional application of QRA relies on historical failure data for system components, which are sometimes combined with fault-tree and event-tree techniques to produce the probability of outcome events. This approach is adequate for systems containing mass-produced mechanical or electrical components such as pumps, valves, switchgears, etc. for which good data on failure rates is available. For structural components and systems, which are quite unique in their geometry, strength and service load levels, such failure data are not readily available. For this reason, the use of risk analysis methods for structural failures is not well developed, however structural reliability methods can be used for this purpose.

4.2 Risk acceptance criteria

The results of QRA are assessed in terms of various quantitative measures such as Individual Risk Per Annum (IRPA), Fatal Accident Rates (FAR), F-N curves, risk contours etc. In some countries the regulatory authorities specify acceptable values for these while in other countries the acceptance criteria are left to be specified by the operator of the facility following some guidelines.

There are considerable differences in the way safety requirements are specified and the acceptable risk levels used in the different countries. In most European countries, the ALARP (As Low As Reasonably Practicable) approach is used whereby the risks within the ALARP zone are considered to be tolerable only if the effort involved in further risk reduction is considered to be grossly disproportionate to the benefits obtained.

For structural failure events, the above risk measures cannot be readily applied where the failure probability is evaluated using structural reliability analysis methods. For structural systems target reliability levels have been recommended by a number of codes and standards and guidance documents, usually related to consequences of failure.

4.3 Treatment of human and organisational factors

Although the role of human and organisational factors (H&OF) in accident causation is well recognised most industries do not at present have formal methods for the quantification of their effects on risk. Human and organisational errors (HOE) are mainly controlled through traditional Quality Assurance and Quality Control measures. This approach may be adequate for simpler systems and operations but are clearly not sufficient for managing major hazards in complex operations such as a process plant or a nuclear power station.

Industries such as nuclear, maritime and oil & gas are now researching this field, particularly with the aim of providing quantitative methods for evaluating HOE probabilities and their effects. Experience has shown that the majority of errors manifest themselves during the operating phase; however, many of these may have origins in design and/or construction errors. Therefore greater attention should be paid to minimising errors at the design stage by the use of appropriate quality management systems.

A proper understanding of HOE and their causation helps in developing an effective Safety Management System (SMS) to minimise these errors and thereby improve safety. The SMS should cover aspects such as safety policy; role definition and competency management; performance measurement and management; work control and change control processes; and quality assurance and quality control systems.

4.4   Reliability analysis for design and assessment

Reliability analysis for mechanical/electrical/electronic systems has hitherto followed a different approach to that used for structural systems. Fault-tree technique based on component failure rates is the common technique used for the former. For structural systems structural reliability analysis methods are used which aim to model the underlying uncertainties in the loading, resistance and other variables using probability distributions and evaluate the probability of failure of individual components or the system as a whole.

For industrial plant and equipments, reliability analysis is mainly used for control, monitoring and emergency systems, and for criticality based maintenance planning. Process plant reliability and availability is sometimes calculated using proprietary software, in order to confirm that the plant has adequate spares to achieve the target availability or planned product deliverability.

For structural systems, although the structural reliability analysis methods are now well developed and a number of commercial software packages for reliability calculation are available, their use during design is limited to very few specialist and major structures such as long span bridges and novel systems for offshore oil & gas production. Most common use of reliability analysis has so far been in the derivation of partial safety factors in design codes. The recent Eurocodes have recognised the direct use of reliability analysis methods for structural design and this should provide further impetus for the use of these methods.

4.5   Reassessment and life extension

The assessment of existing structures is usually less formalised than the design of new structures. It is generally recognised that for assessment there is a need to use more advanced methods of structural analysis combined where feasible with reliability analysis. Reliability based assessment of existing structures is gaining considerable interest and there have been a number of practical applications in the areas of offshore structures, bridges and ships. The main motivation for this stems from the need to make proper use of additional information (general impression, results of visual inspection, results of measurements) and to adopt an appropriate level of target reliability considering the importance of the structure, the consequences of failure and the cost of repair.

A number of guidance documents are now available for the reliability-based assessment of existing structures, for example from ISO, JCSS and DNV.

Although the structural reliability analysis methods are equally applicable, their use in the assessment of industrial plant components such as pressure vessels, process piping etc. has rather been limited. The nuclear sector is beginning to take greater interest in the use of these methods.

4.6   Reliability based inspection and maintenance planning

The use of Reliability Centred Maintenance (RCM) has gained considerable interest in recent years in many industrial sectors while scepticism remains about its merits. For this method to be successful it should be based on sound historical data on plant performance and component failure rates and a sound understanding of deterioration mechanisms and the uncertainties involved. Judicious use should be made of established techniques such as Failure Mode and Effect Analysis (FMEA), Availability, Reliability and Maintainability (ARM) and structural reliability analysis depending on the criticality of the component/system and the nature of degradation/failure.

For structural systems, there has been a growing interest in recent years for the use of structural reliability analysis methods for the rational planning of in-service inspection and maintenance for offshore structures, pipelines, ships and bridges. However, these applications have primarily been carried out by specialists and the wider structural engineering profession has not yet adopted these techniques. There

Table 1. Technology Maturity Matrix.

Topic area	Oil & gas	Maritime	Highway	Building	Process	Power
Regulatory Framework for Safety	5	4	3	4	5	5
Reliability basis of structural codes	3–4	3	4	4	4	3
Standards for reliability analysis	3	2	4	4	1	2
Standards for risk analysis	4	4	3	2	4	2–3
QRA Methodology	5	3	3	2	5	3–5
Data for QRA	3–4	2	3	2	3	4
HRA Methodology & Data	2–3	2	2	1	2	2–3
Integration of QRA, HRA & SRA	1–2	1	1	1	1	2
Reliability based design	3–4	2	3	2	1	2
Assessment of existing structures	3–4	3	3	3	3	3
Risk-based inspection/ maintenance	4	3	3	2	3	2–4
Training & Education	4	2	3	2	3	3–4

is a general lack of guidance in this area. Structural reliability analysis methods also hold promise in the risk-based inspection and maintenance planning of pressure parts in process and power plants but their application so far has been limited.

5 MATURITY AND SCOPE FOR TRANSFER OF TECHNOLOGY

Based on the review of current position, the maturity of safety and reliability technology in the different industries is evaluated allowing the identification of the scope for transfer of technology from one sector another sector.

Table 1 shows the Technology Maturity Matrix in which the maturity in each topic area is assessed on a 5-point scale defined as:

1. basic research or early stages of development,
2. applied research,
3. pilot applications,
4. growing trend but used only by specialists,
5. well established in practice.

The regulatory framework for safety is well developed in the power, process and maritime sectors and there may be some opportunities for harmonisation between these sectors. The framework in the building and highway transport sectors, although less stringent, may be appropriate given the nature and severity of hazards. However, greatest benefits may be possible if the regulatory framework for safety and environmental risk management can be harmonised across different countries within Europe. However, this is a complex and challenging area, as it needs to address political, cultural and social dimensions besides the technical aspects.

The methods for Quantitative Risk Analysis (QRA) are well established in the nuclear power, offshore and process industries and could be readily transferred to other sectors. On the other hand, the methods for Structural Reliability Analysis (SRA) have been well developed for civil structural systems such as buildings, bridges and offshore structures and their application to pressure parts in process and power plants should be considered as a matter of priority as they are likely to provide the greatest benefits in these cases.

Techniques for the assessment of human and organisational factors are in early stages of development across all sectors and some coordinated effort in this would be very useful.

There is a considerable scope for developing a consistent and unified framework for the risk assessment and maintenance management of structural systems and plant and equipment across industrial sectors.

Availability of good quality data on the performance of various types of industrial components and incident data on various hazards is vital to the successful application of risk and reliability methods. Cross-industry and pan-European initiatives in this area are likely to provide the greatest benefits.

Within the Power sector, there is scope for transfer of technology from the more developed nuclear and conventional power sectors to other sectors such as renewable energy.

6 BARRIERS TO PROGRESS

There are a number of barriers, which have hitherto limited the wider use of formal risk and reliability methods in industrial practice, and these must be overcome to promote their use. These are identified and discussed below.

6.1 General/organisational

In sectors such as Buildings and Process which are not well organised and have multitude of owners and stakeholders and where there is a lack of influential industry forums it is generally difficult to develop standardised methods and tools for risk and reliability.

Where industries and owners have a regional or national focus, this becomes a barrier for harmonisation and transfer of technology.

Poor safety culture within plant owners, lack of understanding among key stakeholders of risk and reliability issues can also pose a major barrier.

6.2 Regulatory framework

Regulatory framework has a determining influence on the use of risk and reliability methods. Where safety requirements are of a prescriptive nature (e.g. building sector) this provides little incentive for the owners to invest in expensive risk/reliability studies.

Safety regulations at present are largely within the domain of national authorities with only a few directives from the European Commission. This poses a major barrier for harmonisation of regulations. In some cases, even within one country, safety in different industrial sectors is regulated by different authorities, leading to differences in approaches and potential overlaps between regulations.

6.3 Availability of codes, standards and guidance documents

Availability of codes, standards and good guidance documents is essential for the wider application of risk and reliability methods by practicing engineers. Otherwise their use will be limited to specialists alone. At present there is paucity of standards in structural reliability analysis and techniques used for maintenance management.

6.4 Availability of software tools

Availability of software tools makes it easier for more people to use formal risk and reliability techniques. The status is reasonably good with regard to many of the traditional techniques such as fault-tree, simulation and ARM. However, for structural reliability analysis the available tools need to be more closely integrated with structural analysis and design software to make them readily accessible to structural engineers.

6.5 Availability of data

This is still a major problem in many industrial sectors and for many applications. Lack of good quality and commonly recognised risk/reliability data makes it very difficult to use advanced risk and reliability techniques and also raises scepticism about their efficacy.

In addition, there seems to be a general lack of structured databases, which hold the basic information on asset inventory, condition, performance, maintenance intervals, replacement cycles etc. in many industrial sectors. Industry specific efforts are needed to define data structures and to develop comprehensive database systems for this purpose.

6.6 Training and education

There is a general shortage of suitably qualified and experienced personnel for the application of risk and reliability techniques. Education at universities in these subjects is only offered at post-graduate levels and there is a general lack of facilities and programmes for training practicing engineers in these areas. As a result risk and reliability assessments are carried out only by few specialists. This limits the opportunities for the wider use of these techniques in practice.

6.7 Economic/logistical constraints

High costs and longer timescales needed for applying rigorous risk and reliability analysis techniques often preclude their use for routine applications. This is combined with the difficulty in quantifying the benefits that can be obtained for the costs incurred. There is also considerable scepticism about the merits of these methods where the data and models on which these techniques are based are of a poor quality.

7 KEY ISSUES AND INDUSTRIAL NEEDS

Based on the review of current position in the different industrial sectors and an understanding of the barriers, key issues and industrial problems that should be addressed are discussed below.

7.1 Regulatory framework, codes and standards

This effort should focus on the specification of safety requirements in regulations, codes and standards. As mentioned earlier, since safety is regulated at national level and codes and standards tend to be industry specific, there is a multiplicity of approaches used at present for specifying safety requirements. The different approaches should be reviewed to identify best practice(s) and recommendations made to regulatory bodies and code and standards committees for rationalising and harmonising safety specifications.

7.2 Risk analysis and risk acceptance criteria

There is an acute need for developing a unified methodology for risk analysis by integrating Quantitative Risk Analysis (QRA), Structural Reliability Analysis (SRA) and Human & Organisation Factor Analysis (H&OFA) techniques. At present these techniques are applied to different hazard types and lack a consistent theoretical basis. In practice however there is a good deal of inter-action between the different hazards covered by these techniques, which should be taken into account for a rational assessment of the risks.

In view of the above, there is also a need to formu-late risk acceptance criteria, which is consistent with the integrated risk analysis approach. At present risk acceptance criteria have been established based largely on historical experience and in response to cat-astrophic events, which have caused adverse public opinion. The acceptance criteria should be formulated based on a rational treatment of costs and benefits of risk reduction measures taking also into account the social dimension and public perception of risk. Recent approaches such as the Life Quality Index should be explored further to assess their feasibility for applica-tion to a wide range of safety problems.

7.3 Risk management

There seems to be a lack of established practice and guidance for risk management in many of the indus-trial sectors. Besides risk analysis and assessment, risk and safety management needs to address other softer issues such as safety culture, business processes, roles and responsibilities, training and competency, quality and performance management, etc. which ultimately have a major impact on the safety of operations.

In certain sectors such as the nuclear power, oil & gas and railways where there is a requirement for safety cases as in the UK, good progress has been made in the development of integrated systems for health, safety and environmental management. These developments should be reviewed to identify best practice and har-monise approaches where possible. There is also a need for these systems to take account of business risks of an organisation as commercial pressures always tend to have a negative impact on safety.

7.4 Treatment of human and organisational factors

In order for the human and organisational factor assessments to be credible they need to consider all the life cycle phases of an installation starting from concept design – through operations and maintenance – to decommissioning. The data and methodology used for the evaluation of likelihood of errors and severity of their consequences should be able to take into account the influence of the following factors:

- Regulatory regime
- Safety culture within the industry and within the organisation
- Economic climate and commercial pressures within which the organisation operates
- Safety/Risk Management System (formal or informal)

- Organisational maturity
- Systems for knowledge management, communica-tions and change management
- Performance management and incentive systems
- Training and competence of staff within the Operator of the facility and the entire supply chain which contributes to the operations
- Procurement procedures (risk-reward mecha-nisms) and supply chain management
- Quality assurance and control (QA/QC) processes
- Technological maturity or obsolescence of the hardware and software systems

7.5 Structural reliability analysis and integration of techniques

Methods for structural reliability analysis are now well developed and a number of commercial software packages for reliability calculation are available. Some further developments in the areas of system reliability analysis, time-variant methods and deterio-rating systems would be useful. However, the meth-ods are still seen as complex, time-consuming and requiring specialist knowledge. In order to overcome barriers to the wider use of these methods efforts should be focussed on the following:

- Standardisation of probability distributions for basic variables
- Integration of Structural Reliability software with advanced structural analysis software
- Use of real time structural monitoring data
- Integration of Quantitative Risk Analysis (QRA), Structural Reliability Analysis (SRA) and Human & Organisation Factor Analysis (H&OFA) techniques to evaluate overall risks
- Guidance documents which offer practical help

7.6 Reassessment and life extension

For structural systems, the key issues for reassess-ment and life extension are:

- Incorporating service data on structural perform-ance and knowledge of extreme loading events
- Methods for determining the number of tests needed to collect structure specific information
- Methods/tools to predict future demand on the structure (e.g. traffic load on a bridge)
- Modelling the accumulated damage when good records do not exist about the previous service (e.g. ships, bridges, etc.)
- Databases for durability and service lives of mate-rials and components
- Risk acceptance criteria for existing structures vis-a-vis new structures
- Guidance documents for reassessment and life extension

For industrial plants in process, power and oil & gas industries, reassessment may include the need to upgrade the use of the entire plant e.g. to higher pressure or temperature, or to other products, and not just decisions on life extension of structural systems (vessels, piping etc.) alone. For this reason, the WP 6 should address reassessment and life extension of the entire facility.

7.7 Reliability based inspection and maintenance planning

The main drivers for maintenance management of industrial and infrastructure systems are safety, availability, reliability and sustainability. The objective is to ensure that the asset base meets the required performance requirements with the minimum whole life costs.

For systems such as buildings, bridges, highways and railways much of the maintenance spending is on civil/structural systems and for these risk-based inspection and maintenance techniques using structural reliability methods should be more widely applied. The key issue here is to ensure practicability of these methods. While these methods can be applied directly for complex one-off systems such as offshore floating production systems, this approach is not feasible for thousands of common structures such as bridges. In this case, reliability based optimal maintenance strategies need to be developed for groups of similar structures.

For transportation systems such as highways and railways, which form an integrated network, it is important for the maintenance management to address the performance of the entire network and not the individual assets in isolation.

For industrial plant in the process, oil & gas, maritime and power sectors, much of the maintenance spending is devoted to plant and equipment and not structural systems. For these sectors, therefore, the maintenance management should be targeted at these systems. This would require an integrated approach combining Failure Mode and Effect Analysis (FMEA), Availability, Reliability and Maintainability (ARM) studies with Reliability-Centered Maintenance (RCM) and Risk-based inspection Maintenance (RBIM) techniques.

Inspection and maintenance are also critical to ensuring the safety of operations. The effort in this case should be focussed on safety critical systems within a plant.

8 CONCLUSIONS

This paper synthesises the current position on the applications of safety and reliability approaches in Oil & Gas, Process, Power, Highways, Maritime and Building sectors. This report focuses on drawing similarities and key differences between the sectors and identifies where best practice can be found on various technological areas. The following key conclusions can be drawn from this study:

– There is a need to define a generic risk management framework which addresses all aspects which have an influence on safety. This framework can then be customised to individual sectors at a level which is appropriate.
– Integration of Quantitative Risk Analysis, Human and Organisational Factor Analysis and Structural Reliability Analysis is critical for the evaluation of total risk to an installation and for addressing all hazards on a consistent basis.
– The approach for setting risk acceptance criteria need to be consistent with the above methodology for integrated risk analysis.
– Recommendations need to be developed for regulatory bodies and code and standards committees to harmonise specification of safety requirements and for deriving risk acceptance criteria.
– The methodologies for risk and reliability based maintenance management need to recognise that there will be typically many hundreds of components within a plant and many thousands of assets within a transport network.
– Besides safety, the main drivers for maintenance management are cost, availability, reliability and sustainability. For industrial plants in process, power and oil & gas sectors the maintenance spending is governed by active components and therefore every effort should be made to develop a consistent framework for the maintenance management of both active and passive (structural) systems.
– Decisions on reassessment and life extension need to take account of all systems in an installation and hence a consistent framework is needed to address both active and passive systems.

Further work is necessary for identifying research priorities and putting in place appropriate measures to facilitate transfer of technology between countries and between industrial sectors.

ACKNOWLEDGMENTS

The work presented was performed within the EU – Project "*Safety and Reliability of Industrial Products, Systems and Structures*" (SAFERELNET), which is described in http://mar.ist.utl.pt/saferelnet, and has been funded by the European Commission under the contract number G1RT-CT2001-05051.

Safety and Reliability – Bedford & van Gelder (eds)
© *2003 Swets & Zeitlinger, Lisse, ISBN 90 5809 551 7*

Bayes inference for a bounded failure intensity process

M. Guida
Dept. of Information Engineering and Electrical Engineering, University of Salerno, Fisciano, Italy.

G. Pulcini
Dept. of Statistics and Reliability, Istituto Motori – CNR, Napoli, Italy.

ABSTRACT: The failure pattern of repairable mechanical equipment subject to deterioration phenomena sometimes shows a finite bound for the increasing failure intensity. A non-homogeneous Poisson process with bounded increasing failure intensity is then illustrated and its characteristics are discussed. A Bayes procedure, based on prior information on model-free quantities, is developed in order to allow technical information on the failure process to be incorporated into the inferential procedure. Posterior inference on the model-free quantities, on the model parameters and on the expected number of failures is provided, as well as prediction on the future failure times and on the number of failures in a future time interval. Finally, a numerical example is given to illustrate the proposed inferential procedure.

1   INTRODUCTION

The failure pattern of repairable mechanical equipment, subject to deterioration phenomena and undergoing minimal repairs, sometimes shows a finite bound for the increasing failure intensity. This can occur when, as a consequence of the repeated application of repair actions and/or substitutions of the failed parts, the equipment becomes composed of parts with a randomized mix of ages and its failure intensity becomes a constant value. Typical examples are given by automobiles, that can show a constant failure intensity after about 100,000–150,000 km (Ascher 1986), off-road vehicles and trucks. Thus, if in such a situation a model with unbounded increasing intensity, such as the Power Law process (Crow 1974) is used to describe the observed failure pattern, pessimistic estimates of the equipment reliability will arise.

A non-homogeneous Poisson process (NHPP) with bounded failure intensity was first proposed by Engelhardt & Bain (1986), but its mathematical simplicity makes it inadequate to describe actual failure data in many cases. More recently, Pulcini (2001) proposed a two-parameter NHPP, called the Bounded Intensity Process (BIP), for which the failure intensity $\lambda(t)$ is an increasing bounded function with the operating time:

$$\lambda(t) = \alpha\,[1 - \exp(-t/\beta)] \qquad \alpha, \beta > 0 \qquad (1)$$

Pulcini (2001) provided point maximum likelihood (ML) estimators for the BIP parameter, as well as approximate interval estimates based on the asymptotic results. A test procedure for testing the absence of a time trend against the BIP alternative was also illustrated and "exact" percentiles of the test statistic were evaluated through a large simulation study for selected sample sizes and significance levels. Unfortunately, since the exact distribution of the ML estimators of the BIP parameters is not available, both exact interval estimates and prediction on observable quantities, such as the future failure time or the number of failures in a future time interval, cannot be derived.

In this paper a Bayes procedure is proposed to allow technical information on the failure process to be incorporated into the inferential procedure. In particular, the proposed procedure is based on prior information on two quantities: *a)* the asymptotic value λ_∞ of the intensity function, and *b)* the ratio r of the failure intensity at a given time τ over λ_∞, say $r = \lambda(\tau)/\lambda_\infty$. Such quantities possess a clear physical meaning and are defined irrespectively of the functional form of the failure model (*model-free* quantities). In addition, the prior information on λ_∞ and r, which appears to be generally available to the analyst, can be easily converted into the prior density on model parameters, thus producing tractable mathematics.

Posterior inference on such model-free quantities, on the model parameter β and on the expected of failures is developed. Prediction on the future failure times and on the number of failures in a future time interval is also proposed.

Finally, a numerical example, referring to actual failure data, is given to illustrate the proposed model and the inference procedure.

2 THE BOUNDED INTENSITY PROCESS

The BIP model is an NHPP whose failure intensity is of the form (1), which is an increasing bounded function with the operating time t, equal to 0 at $t = 0$, and approaching an asymptote of α as t tends to infinity. The slope of the intensity (1):

$$\frac{d\lambda(t)}{dt} = \frac{\alpha}{\beta} \exp(-t/\beta) \tag{2}$$

is positive for any $t > 0$ and any α and β value in the parameter space.

The parameter α then represents the asymptotic value of the intensity function: $\alpha \equiv \lambda_\infty$, and the parameter β is a measure of the initial increasing rate of $\lambda(t)$: the smaller β is, the faster the failure intensity increases initially until it approaches α. In addition, the parameter β is the time for which the intensity function should reach the asymptotic value for the cases in which $\lambda(t)$ increases linearly with slope equal to the initial value α/β. Hence, β also represents a measure of the strongly time-dependent phase of the failure process.

The expected number of failures up to a generic time t is equal to:

$$M(t) = \alpha\{t - \beta\,[1 - \exp(-t/\beta)]\} \tag{3}$$

and it approaches asymptotically the straight line $y = \alpha\,(t - \beta)$. In Figure 1 the behavior of $\lambda(t)$ is depicted for $\alpha = 2$ and selected β values.

For t values much smaller than β, $\lambda(t)$ and $M(t)$ are approximately equal to $\alpha t/\beta$ and $\alpha t^2/(2\beta)$, respectively, and hence the BIP model evolves initially as the PLP with shape parameter equal to 2. Then, it converges asymptotically to the homogeneous Poisson process (HPP) with constant intensity equal to α.

In addition, since the quantity $[1 - \exp(-t/\beta)]$ approaches 1 as β tends to zero, it follows that:

$$\lim_{\beta \to 0} \lambda(t) = \alpha \qquad \lim_{\beta \to 0} M(t) = \alpha\,t \tag{4}$$

Thus, the HPP is a limiting form of the BIP.

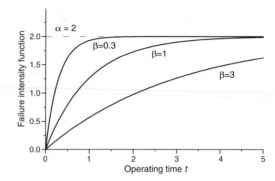

Figure 1. Behavior of the intensity function for the BIP model.

3 THE BAYESIAN PROCEDURE

Let $t_1 < t_2 < \cdots < t_n$ denote the first n times to failure observed until $T \geq t_n$. The likelihood function relative to the above data $\boldsymbol{t}$ is:

$$L(\alpha,\beta\,|\,t) = \alpha^n \prod_{i=1}^{n} U_i(\beta)\ \exp[-\alpha\,Z(\beta)] \tag{5}$$

where:

$$U_i(\beta) = 1 - \exp(-t_i/\beta) \qquad (i = 1,\ldots,n)$$
$$Z(\beta) = T - \beta\,[1 - \exp(-T/\beta)]$$

For failure truncated sampling: $T \equiv t_n$.

The proposed Bayesian procedure is based on prior information on quantities which possess a clear physical meaning and are defined irrespectively of the functional form of the chosen model (*model-free* quantities), except that its failure intensity is a bounded function.

The main advantage is given by the fact that, in general, the prior information the analyst possesses is actually independent of the probabilistic model which he will use to analyze the observed data. In particular, we assume that the analyst is able to anticipate technical information on:

a) the asymptotic value $\lambda_\infty = \alpha$ of the intensity function, and
b) the ratio r of the failure intensity at a given time τ over λ_∞, say $r = \lambda(\tau)/\lambda_\infty$.

Prior information on the asymptotic failure intensity, or equivalently on the parameter α, is formalized through a Gamma density:

$$g(\alpha) = \frac{b^a}{\Gamma(a)}\alpha^{a-1}\exp(-b\alpha) \qquad a,b > 0 \tag{6}$$

728

and, if the analyst formulates his prior knowledge on α in terms of a prior mean μ_α and a standard deviation σ_α, then the Gamma parameters a and b follow:

$$a = (\mu_\alpha / \sigma_\alpha)^2 \qquad b = \mu_\alpha / \sigma_\alpha^2 \qquad (7)$$

Prior information on the ratio r, which is positive and restricted to be no greater than 1, is chosen to be represented by a Beta density:

$$g(r) = \frac{1}{B(p,q)} r^{p-1}(1-r)^{q-1} \qquad p,q > 0 \qquad (8)$$

whose parameters are easily related to the prior mean μ_r and standard deviation σ_r by:

$$p = (\mu_r / \sigma_r)^2 (1 - \mu_r) - \mu_r \qquad (9)$$

$$q = p(1 - \mu_r) / \mu_r \qquad (10)$$

Since:

$$r = \frac{\lambda(\tau)}{\lambda_\infty} = 1 - \exp(-\tau / \beta) \qquad (11)$$

the prior density (8) can be easily converted into the prior density of the BIP parameter β:

$$g(\beta) = \frac{\tau [1 - \exp(-\tau / \beta)]^{p-1} \exp(-q\tau / \beta)}{B(p,q) \beta^2} \qquad (12)$$

Under the assumption of prior independence, the joint prior density of the BIP parameters is the product of (6) and (12), so that the joint posterior density of α and β results in:

$$\pi(\alpha, \beta \,|\, t) = \frac{\alpha^{n+a-1}}{D} I(\beta) \exp\{-\alpha[b + Z(\beta)]\} \qquad (13)$$

where:

$$I(\beta) = \prod_{i=1}^{n} U_i(\beta)\, \beta^{-2} \exp(-q\tau / \beta)\, [1 - \exp(-\tau / \beta)]^{p-1}$$

$$D = \Gamma(n+a) \int_0^\infty I(\beta) [b + Z(\beta)]^{-n-a}\, d\beta$$

3.1 Posterior inference on the BIP parameters and on the ratio r

By integrating the joint density (13) over β, the marginal posterior density of α, or equivalently of the asymptotic failure intensity, results in:

$$\pi(\alpha \,|\, t) = \frac{\alpha^{n+a-1}}{D} \int_0^\infty I(\beta) \exp\{-\alpha[b + Z(\beta)]\}\, d\beta \qquad (14)$$

From (14), the kth marginal moments and the cumulative posterior function easily follow:

$$E\{\alpha^k \,|\, t\} = \frac{\Gamma(n+a+k)}{D} \int_0^\infty I(\beta) [b + Z(\beta)]^{-n-a-k}\, d\beta \qquad (15)$$

$$\Pi(\alpha \,|\, t) = \frac{\Gamma(n+a)}{D} \int_0^\infty I(\beta) [b + Z(\beta)]^{-n-a}$$
$$\times IG\{\alpha[b + Z(\beta)]; n+a\}\, d\beta \qquad (16)$$

where:

$$IG(x; a) = \int_0^x z^{a-1} \exp(-z)\, dz / \Gamma(a)$$

is the incomplete Gamma function. By integrating analytically the joint density (13) over α, the marginal posterior density of β is:

$$\pi(\beta \,|\, t) = \Gamma(n+a)\, I(\beta)\, [b + Z(\beta)]^{-n-a} / D \qquad (17)$$

By changing variables $\beta = -\tau / \ln(1-r)$ in (17), the marginal posterior density on the ratio r results in:

$$\pi(r \,|\, t) = \frac{\Gamma(n+a)}{D\tau} \prod_{i=1}^{n} U_i(r)\, r^{p-1}(1-r)^{q-1}[b + Z(r)]^{-n-a} \qquad (18)$$

where:

$$U_i(r) = 1 - (1-r)^{t_i / \tau} \qquad (i = 1, \ldots, n)$$

$$Z(r) = T - \tau\, [(1-r)^{T/\tau} - 1] / \ln(1-r)$$

From the marginal densities (17) and (18), the marginal posterior moments and cumulative functions of β and r, respectively, can be obtained via numerical integration.

3.2 Posterior inference on the expected number of failures

By making the change of variables:

$$\alpha = M_t / C_t(\beta) \qquad (19)$$

where:

$$C_t(\beta) = t - \beta \, [1 - \exp(-t/\beta)]$$

in the joint posterior density (13) and integrating over β, the marginal posterior density of the expected number of failures in the time interval $(0, t)$, say $M_t \equiv M(t)$, follows:

$$\pi(M_t \mid t) = \frac{M_t^{n+a-1}}{D} \int_0^\infty I(\beta) \, C_t(\beta)^{-n-a}$$

$$\times \exp\{-M_t[b + Z(\beta)]/C_t(\beta)\} \, d\beta \qquad (20)$$

From (20), the kth marginal posterior moments and the cumulative posterior function of M_t easily follow:

$$E\{M_t^k \mid t\} = \frac{\Gamma(n+a+k)}{D}$$

$$\times \int_0^\infty I(\beta) \, [C_t(\beta)]^k \, [b+Z(\beta)]^{-n-a-k} \, d\beta \qquad (21)$$

$$\Pi(M_t \mid t) = \frac{\Gamma(n+a)}{D} \int_0^\infty I(\beta) \, [b+Z(\beta)]^{-n-a}$$

$$\times IG\{M_t[b+Z(\beta)]/C_t(\beta); n+a\} \, d\beta \qquad (22)$$

3.3 Posterior prediction on the future failure time

Given the failure times observed until T, we are interested in predicting the time of the next failure, say $t_{N(T)+1}$. If n failures have been observed until T, then $t_{N(T)+1} \equiv t_{n+1}$.

In general, large attention should be addressed to prediction procedures rather than to statistical inference referred to model parameters and expected number of failures. In fact, in the latter case, we are not able to known how these unobservable quantities have been estimated, whereas the accuracy of prediction, both of failure times and number of future failures, can be easily verified. This allows the analyst to assess the accuracy of selected model and the correctness of prior information.

The conditional density function of t_{n+1}, given that n failures have been observed until T, is:

$$f(t_{n+1} \mid T; \alpha, \beta) = \alpha \, [1 - \exp(-t_{n+1}/\beta)]$$

$$\times \exp[-\alpha \, Y_{t_{n+1}}(\beta)] \qquad t_{n+1} \geq T \qquad (23)$$

where:

$$Y_{t_{n+1}}(\beta) = (t_{n+1} - T) - \beta \, [\exp(-T/\beta) - \exp(-t_{n+1}/\beta)]$$

By making the product of the posterior density (13) and (23) and integrating it analytically over α, the predictive posterior density of t_{n+1} results in:

$$\pi(t_{n+1} \mid t) = \frac{\Gamma(n+a+1)}{D} \int_0^\infty I(\beta) \, [1 - \exp(-t_{n+1}/\beta)]$$

$$\times [b + Z(\beta) + Y_{t_{n+1}}(\beta)]^{-n-a-1} \, d\beta \qquad (24)$$

This predictive density represents the current prediction on the value of t_{n+1} taking into account both the uncertainty about the value of model parameters, and the residual uncertainty about t_{n+1} when α and β are known. The cumulative function is:

$$\Pi(t_{n+1} \mid t) = 1 - \frac{\Gamma(n+a)}{D}$$

$$\times \int_0^\infty I(\beta) \, [b + Z(\beta) + Y_{t_{n+1}}(\beta)]^{-n-a} \, d\beta \qquad (25)$$

from which both the median and a prediction interval of a given probability content γ can be obtained through an iterative procedure.

3.4 Prediction on the number of future failures

The number of failures in the future time interval $(T, T + \Delta)$, say $N_\Delta = N(T, T + \Delta)$, is a Poisson random variable with mean:

$$E\{N_\Delta\} = \alpha \, X(\beta, \Delta) \qquad (26)$$

where:

$$X_\Delta(\beta) = \Delta - \beta \, \{\exp(-T/\beta) - \exp[-(T+\Delta)/\beta]\}$$

so that the probability that m failures will occur in the future interval $(T, T + \Delta)$ is given by:

$$Pr\{N_\Delta = m\} = \frac{[\alpha \, X_\Delta(\beta)]^m}{m!} \exp[-\alpha \, X_\Delta(\beta)] \qquad (27)$$

By using the joint posterior density (13) and integrating analytically over α the product of (13) and (27), the posterior predictive probability of N_Δ results in:

$$Pr\{N_\Delta = m \mid t\} = \frac{\Gamma(n+a+m)}{D \, m!} \int_0^\infty I(\beta)$$

$$\times [X_\Delta(\beta)]^m \, [b + Z(\beta) + X_\Delta(\beta)]^{-n-a-m} \, d\beta \qquad (28)$$

A conservative γ upper credibility limit for N_Δ is obtained by finding the smallest value of m_u which satisfies:

$$\Pr\{N_\Delta \leqslant m_u \,|\, t\} \geqslant \gamma$$

and the posterior expectation of N_Δ is:

$$E\{N_\Delta \,|\, t\} = \frac{\Gamma(n+a+1)}{D}$$

$$\times \int_0^\infty I(\beta)\, X_\Delta(\beta)\, [b+Z(\beta)]^{-n-a-1}\, d\beta \tag{29}$$

From (28), the system reliability, i.e. the probability that no failure will occur in the future time interval $(T, T + \Delta)$, is given by:

$$R(T, T+\Delta \,|\, t) = \frac{\Gamma(n+a)}{D}$$

$$\times \int_0^\infty I(\beta)\, [b+Z(\beta)+X_\Delta(\beta)]^{-n-a}\, d\beta \tag{30}$$

4 NUMERICAL APPLICATION

Consider the failure data given by Ahn et al. (1998) relative to six automobiles (1973 AMC Ambassador) once owned by the Ohio state government. In particular, we analyze the subset #4, which consists of 18 times (in days) to failure for all causes of failure. Such data are given in Table 1. In absence of other information, we treat them as a failure truncated sample.

Pulcini (2001) analyzed these data in the framework of the maximum likelihood method by using both the PLP and the BIP model. He concluded that: (a) there is statistical evidence of trend with the operating time, and (b) the BIP model fits the data set somewhat better than the PLP model.

Suppose that the analyst is able to anticipate a prior mean $\mu_\alpha = 0.02$ failures per days and a standard deviation $\sigma_\alpha = 0.01$ for the asymptotic intensity $\lambda_\infty \equiv \alpha$, so that he uses the Gamma density (6) with parameters $a = 4$ and $b = 200$. In addition, from previous experiences, the analyst possesses a vague belief that the failure intensity at the time $\tau = 500$ days is nearly half its asymptotic value. Then, he formalizes his

Table 1. Times to failures (in days) for the automotive application.

202	265	363	508	571	755	770
818	868	999	1054	1068	1108	1230
1268	1330	1376	1447			

prior knowledge on the ratio $r = \lambda(\tau)/\lambda_\infty$ through the Beta density (8) having mean $\mu_r = 0.5$ and standard deviation $\sigma_r = 0.15$, so that: $p = q = 5.056$.

Combining the above prior information with the observed failure data, the marginal posterior densities of the model parameters α and β and of the ratio r are obtained. In Figures 2–4 these posterior densities are

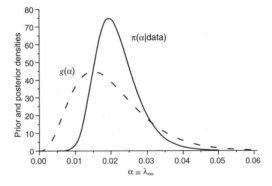

Figure 2. Prior and posterior densities of the parameter α in the automotive application.

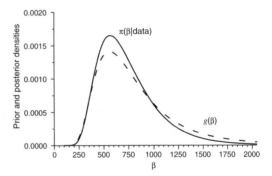

Figure 3. Prior and posterior densities of the parameter β in the automotive application.

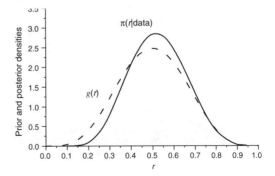

Figure 4. Prior and posterior densities of the ratio $r = \lambda(\tau)/\lambda_\infty$ in the automotive application.

731

compared with the corresponding prior ones, in order to show how the observed failures have updated the knowledge on α, β and r.

In Table 2 the prior and posterior mean and standard deviation of α and r are given. We note that the observed data have reduced the uncertainty about the true value of the model parameters, especially that of the asymptotic intensity.

In Table 3 the posterior mean and the 0.95 credibility interval of α and β are compared to the corresponding ML estimates given in Pulcini (2001). We note that the Bayes means are close to the ML point estimates and that the use of technical knowledge on the failure process has significantly reduced the uncertainty on the value of the model parameters with respect to the ML results.

Figure 5 compares the observed number of failures to the posterior mean of its expected value $M(t)$.

Credibility limits with a probability content of $\gamma = 0.90$ are also given. We observe that the posterior mean fits very well the observed points (t_i, i) and that the credibility bound contains the larger part of data points.

In order to assess if the BIP model describes adequately the observed failure pattern and if the prior knowledge is correct, we use the first $n = 14, \ldots, 17$ failure times to make prediction on the time t_{n+1} of the next failure. Of course, the posterior estimates of model parameters now slightly differ from the estimates based on the whole data set. Using the predictive posterior density (24), we compute both the median and the 0.80 prediction interval. By comparing the predicted failure times, given in Table 4, to those actually occurred, we observe that the actual data are very close to the median values and are included into the prediction intervals.

Another model check consists in comparing the predicted number of failures which will occur in a future time interval with the actual value. At this end, we use the first 14 failure times to make prediction on the number N_Δ of failures that will occur in the time interval (t_{14}, t_{18}).

From (28), the posterior prediction of N_Δ, where $\Delta = t_{14} - t_{18} = 217$ days, is depicted in Figure 6. The conservative $\gamma = 0.90$ upper credibility limit for N_Δ

Table 2. Prior and posterior mean and standard deviation of model parameters.

	Mean value		Standard deviation	
	α	r	α	r
Prior	.0200	.500	.0100	.150
Posterior	.0217	.524	.0060	.131

Table 3. Bayes and maximum likelihood estimates of α and β.

	Point estimate		Interval estimate	
	α	β	α	β
Bayes	.0217	743.	(.0123 –.0358)	(328.–1534.)
ML	.0199	596.	(.0052 –.0765)	(38.–9307.)

Table 4. Comparison of occurred failure times t_{n+1} ($n = 14, \ldots, 17$) with posterior prediction.

n	Actual t_n	Actual t_{n+1}	Posterior median	0.10 Lower limit	0.90 Upper limit
14	1230	1268	1271	1236	1370
15	1268	1330	1308	1274	1404
16	1330	1376	1370	1336	1465
17	1376	1447	1415	1382	1508

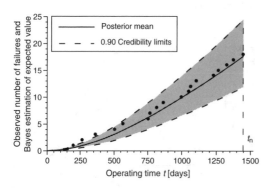

Figure 5. Observed number of failures and posterior estimate of the expected value $M(t)$ in the automotive application.

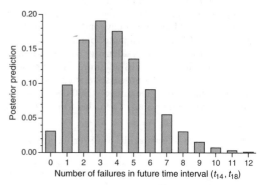

Figure 6. Prediction of the number of failures in the future time interval (t_{14}, t_{18}) in the automotive application.

is equal to 7 failures and the posterior mean is equal to 3.84 failures, very close to the number of failures, namely 4, actually occurred in (t_{14}, t_{18}).

Finally, from (30), the system reliability in the time interval (t_{14}, t_{18}) is equal to 0.031.

5 FURTHER APPLICATIONS

The BIP model, as well as other processes showing a bounded intensity function, can find application in the analysis of repairable systems that operate for a very long time period and are repaired at failure, rather than substituted with a new unit, even when they have experienced a large number of failures. In this section, two other applications of the BIP model and of the proposed inferential procedure are briefly illustrated.

Let consider the failure times of a superheater of a lignite fired unit given by Tibor (1992) and showed in Table 5.

The above failure data have been combined with:

(a) a Gamma prior density (6) on the asymptotic intensity $\lambda_\infty \equiv \alpha$ with prior mean $\mu_\alpha = 0.06$ and standard deviation $\sigma_\alpha = 0.02$, and

(b) the Beta prior density (8) for the ratio $r = \lambda(\tau)/\lambda_\infty$ (with $\tau = 300$) having mean $\mu_r = 0.25$ and standard deviation $\sigma_r = 0.05$.

The point posterior estimates of the BIP parameters α and β are equal to 0.0577 and 1099, respectively, and the 0.95 credibility intervals are given by (711, 1667) and (0.0354, 0.0869), respectively.

In Figure 7 the observed number of failures is depicted together with the posterior estimate of the expected number of failures $M(t)$. We note that the point estimate of $M(t)$ fits adequately the observed data and that the 0.90 credibility limits include almost all the data points.

The last application consists of the failure data (collected for two years) of the hydraulic system of a load-haul-dump machine (denoted as LHD17) deployed at Kiruna mine and given by Kumar & Klefsjö (1992). Such data, treated as a failure truncated sample, are reported in Table 6.

The prior information on the asymptotic intensity $\lambda_\infty \equiv \alpha$ and on the ratio $r = \lambda(\tau)/\lambda_\infty$ (with $\tau = 1500$) has been formalized through a Gamma density (6)

on $\lambda_\infty \equiv \alpha$ with mean $\mu_\alpha = 0.015$ and standard deviation $\sigma_\alpha = 0.005$, and a Beta prior density (8) on r having mean $\mu_r = 0.6$ and standard deviation $\sigma_r = 0.2$. Combining such prior information with the failure data of Table 6, the expected number of failures $M(t)$ has been estimated and compared in Figure 8 with the observed number of failures.

The posterior estimate of $M(t)$ fits well the observed data and the credibility bound contains the larger part of the data points.

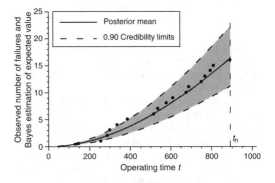

Figure 7. Observed number of failures and posterior estimate of the expected value $M(t)$ in the superheater application.

Table 6. Times to failures (in days) of the hydraulic system of the LHD17 machine.

401	437	455	614	955	1126	1150
1500	1572	1875	1909	1954	2278	2280
2350	2407	2510	2521	2526	2529	2673
2753	2806	2890	3108	3230		

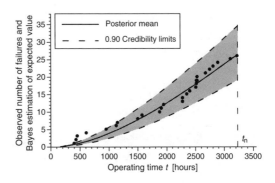

Figure 8. Observed number of failures and posterior estimate of the expected value $M(t)$ in the LHD17 application.

Table 5. Times to failures of the superheater of a lignite fired unit.

256	290	301	337	389	517	536
577	610	678	688	751	768	796
810	891					

6 CONCLUSIONS

The bounded intensity process is a suitable alternative to the well-known power law process when technical considerations on the failure mechanism and on the applied repair policy suggest that the failure intensity approaches a constant value.

On the other hand, when technical knowledge on the failure mechanism is available, Bayes inference constitutes a useful tool to obtain more accurate estimates and predictions with respect to the maximum likelihood method. In this paper a Bayes procedure based on prior information on *model-free* quantities, which appears to be easily available and produces tractable mathematics, has been discussed. The posterior distribution of a number of unobservable and observable quantities, such as the expected number of failures and the failure times in a future time interval, has been obtained.

The estimation procedure has been applied to several case studies in the mechanical field. The BIP model showed to be in a good agreement with the observed failure patterns. Moreover, it was shown that the uncertainty on quantities of interest can be significantly reduced even when a not too strong technical knowledge on the failure process is available.

REFERENCES

Ahn, C.W., Chae, K.C. & Clark, G.M. 1998. Estimating parameters of the power law process with two measures of failure time. *Journal of Quality Technology* 30(2): 127–132.

Ascher, H. 1986. Reliability models for repairable systems. In J. Møltoft & F. Jensen (eds), *Reliability Technology – Theory & Applications*: 177–185. Amsterdam: Elsevier Science.

Crow, L.H. 1974. Reliability analysis for complex repairable systems. In F. Proschan & R. Serfling (eds), *Reliability and Biometry*: 379–410. Philadelphia: SIAM.

Engelhardt, M. & Bain, L.J. 1986. On the mean time between failures for repairable systems. *IEEE Transactions on Reliability* 35(4): 419–422.

Kumar, U. & Klefsjö, B. 1992. Reliability analysis of hydraulic systems of LHD machines using the power law process model. *Reliability Engineering and System Safety* 35(3): 217–224.

Pulcini, G. 2001. A bounded intensity process for the reliability of repairable equipment. *Journal of Quality Technology* 33(4): 480–492.

Tibor, C. 1993. Some parameter-free tests for trend and their application to reliability analysis. *Reliability Engineering and System Safety* 41(3): 225–230.

Safety and Reliability – Bedford & van Gelder (eds)
© 2003 Swets & Zeitlinger, Lisse, ISBN 90 5809 551 7

Simulation of industrial systems

F. Gustavsson & S. Eisinger
DNV Consulting, Oslo, Norway

A.G. Kraggerud
Hydro AS, Oslo, Norway

ABSTRACT: Simulation techniques are widely used in the reliability area, this paper will illustrate methods for modeling large systems and running sensitivities. Uncertainties have been given special attention, presentation of results and practical aspects are discussed. Some simple examples are included, showing the importance of a flexible tool where the reliability specialist is not limited to predefined logic.

1 INTRODUCTION

This paper describes modern simulation tools used in offshore field development projects and how the simulation output is factored into the concept selection process. The paper is focusing on the Ormen Lange field development and the use of simulation tools to assess the production availability. The Ormen Lange field, which is the first deepwater discovery to be developed in offshore Norway, is located 130 km from the West Coast of Norway at a depth of about 1000 meters. Developing Ormen Lange represents a major challenge with a combination of large water depth, extremely rough seabed conditions, long tie-back distance and demanding weather conditions. Concepts that were evaluated cover a full range of innovative solutions – ranging from a relatively novel "Subsea to Land" to a more traditional deepwater concept with an offshore processing facility with wet and dry wells. In order to give proactive decision support the response time from a design change to valuable availability results has to be minimized, effective work procedures, simulation tools and an easy communication interface is of utmost importance.

Traditionally, reliability block diagrams or fault trees are used to assess the regularity of systems. Unfortunately, these modeling paradigms are rather inflexible and require approximations to be implemented, which are often uncontrollable and may reduce confidence in the analysis results. One simple example may be given by dependencies between two sub-systems, for example a steam turbine, which is dependent on the functioning of gas turbines delivering the

steam power. Even if exact models can be designed in some cases, these become complicated and difficult to maintain.

Therefore, early in the project, Discrete Event Monte Carlo Simulation was chosen as the tool to perform the various regularity calculations. Simulation technology offers a degree of flexibility unknown to analytical alternatives. This flexibility was used throughout the project, for example

- to specify complex flow rules for gas, condensate and water
- to model storage of condensate and tanker offloading
- to implement dependencies between component failures
- to implement complex production rules
- to deliver very detailed results, normally not feasible with analytical calculations

Since the aim of the project was decision support for the concept choice, it was in addition important to design flexible, modularized models of the various sub-systems and to maintain a vast amount of simulation cases and their results. The project demonstrates that modern simulation tools are fully capable of performing well under all these requirements and that simulation times are no problem with standard computers, even for huge industrial systems and the simulation of many cases.

The paper presents an overview over the model building process, including examples for a few sub-systems. Emphasis is put on non-traditional modeling challenges and the chosen solutions. Some main results are presented, with the aim to give an overview over the variety of relevant regularity measures becoming accessible by simulation technology.

2 SIMULATION TECHNIQUES

The simulation technique used is a next event Monte Carlo Simulation, with full array of building blocks, animation, graphical interface, unlimited hierarchical decomposition, full connectivity and interactivity with other programs, library based, build in compiled C-like programming language, to mention some of the most important features. The flexible method enables modeling of multi purpose tasks like system availability, on demand failure probability, multiple product flow, loading/storage capacities and spare part optimization. In the Ormen Lange project all these features were used during the concept selection process. The interface between the simulation tool Extend (www.imaginethatinc.com) and Excel for input data, made it easy to structure data, extract data, running sensitivities and customize results presentation.

The simulation tool is built of hierarchical blocks where blocks can be constructed by the user, both regarding interface (pictures etc.) and programming the logic of the block. This feature is used to build up the model as a flow network with similar interface as a PFD (process flow diagram). The flow network is intended to simplify the comprehension of the availability model for design engineers and management. Although some customers used to reliability block diagrams (RBD), prefer this interface for transparency, showing component redundancy instead of product flow.

2.1 Excel input sheets

All input in terms of MTTF, MTTR, capacities and redundancy are listed in Excel sheets with connection to Extend. Also results are reported back to Excel via the link shown Figure 1.

The Excel sheet is divided into columns where each column gives specific input, the columns directs data into cells in the Extend model via a hierarchical path structure. Any input changes in the Excel sheets are updated in the model by re-initialization of the case block.

The Excel interface makes data treatment easy especially regarding running sensitivities. In the Excel sheet a base case is defined and virtually any number of named sensitivity cases. Only changes from the

base case have to be repeated in the sensitivity. Failure data and repair data are stored in another excel sheet and linked in to the simulation cases through look-up functions. In this way the data become traceable and easy to maintain.

3 STRUCTURE OF LARGE SYSTEMS

To cover a system consisting of subsea components, risers, topside facilities, export lines and an onshore terminal, requires a model broken down into subsystems and sub-sub-systems. A hierarchal structure gives an easy overview, of the system. By clicking on the hierarchal icon all under laying components are viewed. There are no limitations in the possible number of levels in the hierarchal structure.

3.1 Recorder

The availability model is structured into subsystems consisting of maintainable components. The size of a subsystem is decided by the need for duplications. For example a cluster consisting of 4 wells was used to build up the entire subsea layout consisting of total 24 wells. A recorder is used to transfer data from a subsystem (well) to the main model (subsea system). The recorder logs data from a subsystem in a text file. In the case of the wells, the data consists of flow levels and time intervals.

The recorder in the main model calls for the specific text file via a file name. For all sensitivities a text file was created and stored, sensitivities could be different no. of wells and/or capacities.

Using more than one recorder in the same model requires independency between recorders calling for the same text file. The independency is ensured by a

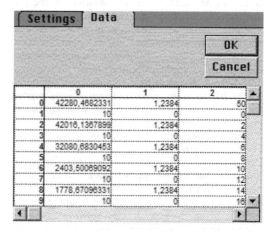

Figure 2. Result window in a recorder.

Figure 1. Excel extend interface.

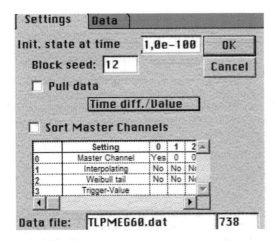

Figure 3.   Input to a recorder calling-up a text file.

different random seeds value for each recorder. The tool automatically generates seed numbers to each block in the model, in order to keep components independent. Using recorders requires a large population of data points to ensure a good statistical basis when the recorded data is re-sampled in another model. The amount of data points required depends on the failure rate and repair times. The population must be especially large when the model involves extremely rare events with extreme consequences. If the new model is to be used in another model even more data points are required.

3.2   Simulations of large systems

Large systems are normally split into sub-models to get an easier model to analyze. For the Ormen Lange concepts a subsea-, riser-, topside-, export- and onshore-model were built to get overall availability figures. With a sophisticated simulation tool it is possible to combine all these sub-models in a large main model either by direct integration or through the usage of stored data (recorder), which are used in the main model. If the availability figures for each subsystem are multiplied according to equation (1), the overall figure will be conservative. This is shown in a simple example below.

$$A(t) = a(t)_1 \times a(t)_2 \times a(t)_3 \qquad (1)$$

Two cases below shows the difference between a simulated result and a result obtained by multiplying sub models.

Sub-System A   Sub-System B
$a_1 = 1$ *fullflow* $a_4 = 1$
$a_2 = 0.5$ *halvflow* $a_5 = 0.5$
$a_3 = 0$ *noflow* $a_6 = 0$

Example 1 (multiplying sub-models)
$$A_{total} = SystemA \times systemB =$$
$$= (a_1 \times 0.5a_2) \times (a_4 \times 0.5a_5) = \qquad (2)$$
$$= a_1a_4 + 0.5a_1a_5 + 0.5a_2a_4 + 0.25a_2a_5$$

Example 2 (simulating sub models in a main model)
Flow level 1 = a_1a_4
Flow level 0.5 = $a_2a_5 + a_1a_5 + a_4a_2$
Flow level 0 = 1 − (Flow level 1) − (Flow level 0.5)

$$A_{total} = a_1a_4 + 0.5a_2a_5 + 0.5a_1a_5 + 0.5a_4a_2 \qquad (3)$$

This simple calculation shows that if both systems are running on reduced capacity, only a simulated result gives the correct output. The simulated result Example 2 combines the flow levels for all systems at any time, taking the lowest level supplied from any system. When multiplying each subsystem as in Example 1 all simultaneous capacity reductions will lead to a too high availability reduction, with a conservative overall availability result.

Using the features with recorder to gather data and connect data series to a main model, makes it possible to simulate very large systems. On a normal notebook computer, the simulation times ranged from 1 minute to 5 minutes for a whole field model.

4   MODELLING DEPENDENCIES AND RELATIONS

The simulation technique has been widely used for multi purpose tasks, trying to imitate the real world. In reliability engineering a number of tools have been developed to imitate the real world as good as possible. To do this, logical rules, shut off functions and simple if rules help the reliability engineer to set up a realistic models.

The tool used for the Ormen Lange availability studies had the flexibility of programming logical relations directly in the model. A C dialect, called ModL is used to define complicated dependencies. One example is the gas export trains topside, which were designed with 3 gas turbines and one steam turbine driven on heat recovered steam from the gas turbines.

Figure 4 shows an if-code controlling the relationship between gas turbines called x0, x1 and x2, the steam turbine called x3 and the heat recovery system called x5. Modeling this relationship has been done in close connection with experts in the mechanical department, providing energy and capacity curves. A range of different combinations of gas and steam turbines have been tested to investigate potential availability improvements.

Other smart blocks are flow logics used in the topside and onshore model. These models separate

737

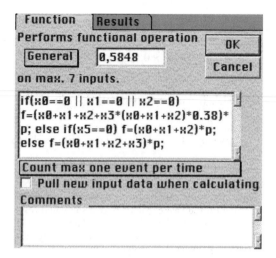

```
Function   Results
Performs functional operation          OK
General    0,5848
                                     Cancel
on max. 7 inputs.

if(x0==0 || x1==0 || x2==0)
f=(x0+x1+x2+x3*(x0+x1+x2)*0.38)*
p; else if(x5==0) f=(x0+x1+x2)*p;
else f=(x0+x1+x2+x3)*p;

Count max one event per time
☐ Pull new input data when calculating
Comments
```

Figure 4. Logic between gas turbines and a steam turbine, ModL code.

multiphase flow into condensate, gas and water. Routing flow is done with percentage, first in first out, priority, minimum or batch flow rules, which mix different ingredients and route the flow in the desired routes. One example is the recompression trains. If the flow is blocked downstream, flow could automatically be routed to flare. Using the priority block flow gets automatically re-routed from the primary to the secondary route, which in this case is short term flaring. By connecting a counter to the flare route, the number of flaring events can be measured to identify environmental impact.

5 GENERATING RESULTS

Results can be presented in a number of ways, for example unavailability bars, pie-charts, contributors, fraction of time per flow rate and availability distributions. In this chapter results will be discussed, both how they were produced and presented.

5.1 Producing results

As discussed above, results can be transferred between different models to minimize the size of the overall model. Both results from the main model, capturing all sub models, and results from each sub model have been of interest in forms of:

- Availability figures
- Fraction of time per flow rate
- Contributors to flow loss
- No. of shut downs
- Maintenance costs

To generate these results a Plotter block and a generalized cut set block are used. These blocks will be further discussed in this chapter.

5.1.1 Plotters

The plotter is used to view different results such as availability distributions and fractions of time per flow rate. The plotter records and stores all changes (next event Monte Carlo Simulation) during the simulation, plotting either

- Mean trace
- P(X <= x): the cumulative distribution of the input values
- Value histogram on time intervals

The entire life cycle could be viewed, with the possibility to zoom in specific time periods of interest. This is helpful for quality checks and model validation.

The plotter can either run continuously during the simulation run or "pop up" at the end of each simulation. Up to four input (flow) values can be connected to the plotter, differentiated with colors. The graphs can for example illustrate the cumulative distributions for exceeding a certain flow rate for up to four subsystems. The fraction of time per flow rate model included recorders from all subsystems. The plotter was used to analyze the contribution to unavailability from each subsystem at different flow levels.

5.1.2 Sets

The Sets block is used to find contributors to production unavailability, i.e. set of components for which the system performance is reduced. The Sets block thus finds generalized (minimal) cut and path sets and calculates a number of statistical values of these sets.

The component with highest contribution to unavailability has caused the highest gas volume loss. The volume loss is calculated with the formula:

$$\text{Loss} = (\text{Capacity reduction}) \times (\text{Acc. time in state}) \quad (4)$$

where the *accumulated time in state* is the sum of the down times for the set of components. A 3 x 33% system loses 33% upon a failure, while a 3 x 50% system exhibits no capacity loss if one component fails.

The Sets block also defines the remaining capacity at a failure event called *performance*. Other statistical measures of the Sets block are

- no. of events when a set of components have failed during the simulation time
- accumulated performance
- % of total time in the state
- accumulated time in the state

Having access to this information a wide range of interesting results can be generated just by filtering the data. For example the number of shut downs with duration longer than 10 hours is important and can suit as a simplified approach to estimate the effect of pipeline packing. The accumulated time delivering below market demand over one year, is another interesting result easily filtered from data in the Sets block. For the Ormen Lange case loss contributors were ranked after highest percentage to total loss. The number of events below demand and the number of shut downs per year were also calculated.

All events are logged in the Sets-block. Even component failures which do not lead to any system performance drop are logged. Often sets of components give high contribution with a high capacity loss but lower frequency than single components. A single component with high frequency and high capacity loss gives of course a high contribution.

5.2 Uncertainties

Treating uncertainty in availability analysis is an opportunity for state of the art simulation methodologies. The uncertainties can for example be grouped into the categories:

- regarding input data uncertainties
- time variations

5.2.1 Uncertainties in input data

The input determines the quality of the output. This is a well known phrase and still very important. To highlight the variations regarding input data gets very important when dealing with new technology, with no or limited amount of operational experience. In the Ormen Lange project one major task was to identify systems and components with unexpected high reliability uncertainties.

This was mainly done by studying a wide range of data sources, recently logged failures not yet recorded in any data base, expert judgments and test results. The availability analysis covered the entire system, from subsea to land. It was managed to rank systems regarding uncertainty in reliability. This is valuable information for estimating the size of the qualification process needed, which also influences the decision process regarding time schedules and cost.

After having identified systems and components with uncertainties the task is to quantify these uncertainties. Both the MTTF and MTTR figures could be subjected to uncertainty. Which of the two is dominant? The MTTR for subsea components located at a depth of 1000 meters could vary depending on local conditions and climate. Three categories of repair times[i] are introduced: minimum, most probable and maximum repair time. These sensitivities are applied to all subsea components with the purpose to get a picture of the

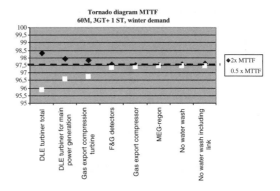

Figure 5. Tornado diagram.

availability variations. The results show considerable variation when no overcapacity is included, including over capacity the minimum, most probable and maximum down time show marginal effects to overall results.

It was discussed if the model should consider simultaneous repairs or if each failure event should be treated individually. Trying to keep the model as simple as possible it was decided to neglect effects of simultaneous repairs with the argument that this would not be any case differentiator. In practise interventions/repairs may not be initiated when the failure occurs, but postponed until e.g. the system is not capable to deliver market demands or an intervention/repair campaign is carried out during the summer season. Not considering simultaneous repairs gives a conservative estimate of the availability, although postponing repair activities causing reduced redundancy leads to a lower availability. Postponed repairs and intervention/campaign is planned to be included in the model during spring 2003.

From the most probable estimate, both regarding MTTF and MTTR, a top ten list of contributing components for each subsystem are listed.

Each component was varied individually to assess the variations on the overall subsystem availability.

The MTTF parameters were multiplied and divided by a factor of two, based on the availability archived a tornado diagram could be drawn.

The MTTR parameter was varied with minimum and maximum repair time and a tornado diagram was drawn based on the overall variations. Establishing the tornado diagrams required 4 simulations per component, 10 components leads to 40 simulations. Sensitivity analysis is of importance in the process of quantifying uncertainty.

5.2.2 Availability distributions

This chapter focuses on variations between each year. In year 1 we have a specific availability but in year 2

---

[i] Repair time includes mobilization time, wait on weather, active repair time and start up time

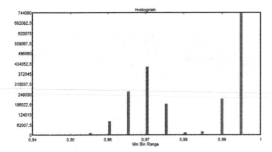

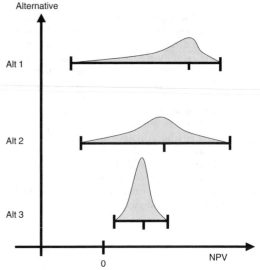

Figure 6. Winter and summer availability.

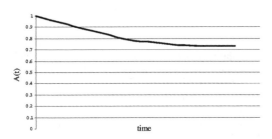

Figure 7. Availability of a components with exponential life- and repair time.

Figure 8. Decision under uncertainty.

it could be different. To some extent we need to quantify how much the availability can fluctuate from year to year. The variation in availability is a natural process, caused by failures occurring randomly from year to year. By simulating over 1000 years and documenting each run in a bar diagram, it is possible to get a distribution curve over the availability for each year. In this case we have done simulations over a summer demand and winter demand leading to two availability curves. The summer curve shows a higher mean availability and a narrower distribution. This is caused by a lower market demand rate where many failures do not effect the availability. The winter and summer availability curves are combined in a simulation model using recorders for sub-models and an equation block combining the summer and winter curves.

The width of the distribution is very much dependant on the simulation period. Decision making based on mean values and standard deviation should consider contractual aspects. If the operator signs an agreement, promising 96% deliverability each day, the distribution curve needs to reflect simulation runs lasting for 24 hrs. The spread will be larger for a short simulation period than a long period.

If σ is the standard deviation of a statistical variable, the variance of the mean over n samples is:

This means that the standard deviation is expected to decrease with the square root of n. The standard deviation of a simulation run over 30 years is expected to be $\sqrt{30}$ smaller compared with a simulation over a one year period.

MTTF and MTTR based on exponential distributions require a time period to converge towards a stable value. It is therefore important to have a long simulation run. To over-come convergence problems it is preferred to take readings of the availability results during a time increment Δt equal to the contractual assignments. The contractual assignments or control intervals are often a one day interval. If the simulation is restarted after each Δt, it is important to check that the availability has converged towards a stable value.

Using distributions gives additional information from RAM analysis. But there still are some learning required among customers for them to feels comfortable with this kind of presentation.

REFERENCES

1. System reliability theory, Arnljot Høyland and Marvin Rausand.
2. Ormen Lange Case selection, DNV report 2002-0136.
3. Extend™ user guide.
4. Probability & statistics for engineers, Richard A. Johanson.

$$\frac{\sigma^2}{n} \qquad (5)$$

Safety and Reliability – Bedford & van Gelder (eds)
© 2003 Swets & Zeitlinger, Lisse, ISBN 90 5809 551 7

A methodology to investigate heavy gas dispersion by water-curtains

Karin Hald

von Karman Institute, Department of Environmental and Applied Fluid-dynamics, Rhode-St-Genèse, Belgium Ecole des Mines d'Alès, Laboratoire Génie de l'Environnement Industriel, Alès, France

Aurelia Dandrieux & Gilles Dusserre

Ecole des Mines d'Alès, Laboratoire Génie de l'Environnement Industriel, Alès, France

Jean-Marie Buchlin

von Karman Institute, Department of Environmental and Applied Fluid-dynamics, Rhode-St-Genèse, Belgium

ABSTRACT: In the chemical industry, risk assessment methods are needed in front of accidental gas releases. Nowadays, the water-curtain is recognized as a useful technique to mitigate a heavy gas cloud. Models already exist to study their efficiency and they are often based on experimental results. The concept of validating these models by different methods is presented, and each part is explained. They consist of medium-scale field tests, Wind-Gallery tests and numerical simulations. Both experimental methods in different scales and numerical methods are discussed. Preliminary results of field-tests and spray characteristics are given.

1 INTRODUCTION

In chemical industry the safety related around gas releases always remains an important task. In the case of a toxic or flammable gas or pollutant release, the consequences of an accident depend on the concentration. Therefore, methods to mitigate the hazard of a toxic gas release must be studied. Among these methods, water-spray curtains are recognized as a useful technique of mitigation. A water-curtain consists of one or several pipelines equipped with a uniform distribution of nozzles that interfere with each other after some distance to constitute a vertical curtain. It can affect a heavy gas cloud by three different means. First, the mechanical effect of the gas entrainment in the water spray leads to a forced dispersion of the cloud. Next, depending of the gas solubility in the curtain liquid, a mass-transfer in the liquid (physico-chemical absorption) leads to a dilution of the gas cloud. To increase this feature for low water solubilities, it is possible to add substances in the water of the watercurtain as (Schatz 1990) and (Molag et al. 2000). Finally, a cloud warming by two-phase heat transfer leads to buoyancy effects. The system consists of a gaseous phase interacting with a water-curtain under certain meteorological conditions (wind, relative humidity, temperature...). It is a complex multiphase flow where the three controlling parameters are the mass-flow of the gas-source, the features of the water-curtain and the wind velocity. Large field-tests as the Buxton test series (Moodie 1981), the Gold-fish project (Blewitt et al. 1981), and the Hawk series (Schatz 1990) involve carbon dioxide and hydrogen fluoride, respectively. They investigate the effect of the controlling parameters. However, these tests are expensive and not easy to operate but are a base for the first models. Models to design water spray curtains and estimate their efficiency have been proposed (Buchlin 1994), (McQuaid, J. & Fitzpatrick, R. D. 1981), (Fthenakis, V. M. & Blewitt, D. N. 1995), (St-Georges et al. 1992), (Griolet et al. 1995). However they are not systematically used in the process industry because of unavailability, unfamiliarity and not full validation (Molag et al. 2000).

As an outcome of the European RIDODO and ASTRRE projects the engineering code CASIMIRE has been developed (Malet et al. 1995), (Uznanski, D. T. & Buchlin, J.-M. 1998). It is made to design waterspray curtains for the mitigation of the consequences of accidental situations such as gas releases or radiative sources (fire) in chemical industry. Considering gas releases, it evaluates the dilution factor in the case of a gas dispersion interacting with a watercurtain with respect to the curtain configuration (number of nozzles per meter, type of nozzles, upward or downward mode etc.). The code CASIMIRE contains databases of nozzle-types and chemical products. It has been validated only by laboratory tests conducted in the VKI Wind-Gallery (St-Georges et al. 1992), (Griolet et al. 1995), (Malet et al. 1995). The aim of the present project is to provide new set of data based both on medium-field test and CFD simulation with the objective to extend the applicability range of the CASIMIRE code. The validation by field-test measurements is needed to include three-dimensional effects in the code. Only the part consisting of the forced dispersion in the mitigation will be evaluated. The complexity of the system leads to a separate investigation of the different part mentioned above. Concerning field-test, the forced dispersion is the more user-friendly part to investigate. This paper presents results from the first field-tests and a description of the other parts of this project, which consists in further Wind-Gallery tests and numerical simulations.

2 METHODOLOGY

The methodology of the study is presented through the description of the different parts of the project. First, field-tests are performed to investigate a heavy gas dispersion in a more 'real case scenario'. Also, the use of industrial nozzles needs to be validated in the code CASIMIRE. Next, Wind-Gallery experiments will consist in repeating the field-tests at a smaller scale. Of course, this approach is two-dimensional but the control of all parameters is valuable. Then, CFD simulations will be made to compare with the experimental results. CASIMIRE will be tested in all experimentally tested configurations. At last, comparisons between experiments of different scales and different numerical simulation will lead to a complete validation and qualification of the engineering code. The different parts are described below. For the moment, only the first step is partly achieved.

3 FIELD-TESTS

The experimental set-up is described in details, presenting the different parts involved in such tests. The configuration on the field consists of the gas source, the water-curtain and its nozzles, the distribution of the measurement points, and the meteorological measurements.

3.1 The field

The field-tests are conducted on a large flat terrain in a military base camp. The present results are measurements taken during autumn and winter 2002–2003.

The area including the source, the water-curtain and the measurement points is of the order of $500 \, \text{m}^2$. For safety reasons, the experiments are controlled by safety services and a security area is defined during the trial.

3.2 The source

The gas chosen for the investigation of the forced dispersion is chlorine. It is a heavy gas (vapor density of 2.48) with low solubility in water. It disperses slowly at the level of the ground.

Liquefied chlorine bottle is used. It consists of 24 kg of pressurized gas at 10 bar in order to generate a relatively constant release. It is a typical small storage container of chlorine. The emission lasts 4 minutes, which allows a gas flow-rate of 5 kg/min for one bottle. The release is moderate for safety reasons, but also such that the thermal aspects are negligible compared to the forced dispersion of the spray. A manometer is used to control the release while the total discharge is measured by weighting the bottle before and after the release and timing the duration. The position of the source is usually placed 50 cm above the ground.

3.3 Water-curtain & nozzles

The water-curtain consists of a 5 m long and 2 m high pipeline equipped with a uniform distribution of pressure nozzles every 40 cm (with possibility to change to a factor of 0.2 m). The pressure difference at the nozzle exit breaks up the liquid jet and produces a discrete phase of droplets characterized by a droplet distribution. While separated at the top, the sprays interfere with each other after 20 cm to constitute a vertical curtain. The maximum water-flow-rate is 1000 l/min but it is typically in the tests 200 l/min. It is possible to rotate the rack in order to make an upward water-curtain. The nozzles are then 40 cm above the ground.

Different nozzles can be mounted on the water-curtain in order to investigate their efficiency with respect to their characteristics. Until now, only the full-cone nozzles presented in section 4.2 are used for the field tests.

The water-curtain is placed at 4 m from the gas-source in the angle of gas release. At that distance, the cloud has still a heavy gas behavior. The nozzles are spraying in 90° angle such that the surface impacting on the ground is located between 3 and 5 m from the source. With the effect of the wind, the latter is shown to increase.

3.4 Measurement points & technique

The measurement points are distributed within in a circular mesh downwind of the source in different lengths from the source at 5, 10, 15, 20 and 25 m as sketched in Figure 1. To measure the gas concentration

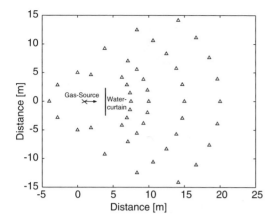

Figure 1. Sketch of the field with the representation of the measurement points positions with respect to the source and the water-curtain.

Figure 2. Water-curtain used in field-test.

the mixture of air and chlorine is absorbed and bubbled in a solution of soda (0.1 M) in order to trap the chlorine. The bubbling starts 30 sec after the beginning of the gas release and lasts till the end. Previous tests in laboratory demonstrated that more than 95% of the chlorine was trapped by the bubbling. Under these conditions, concentrations are average concentrations over 3.5 min. The chlorine concentration is then deduced directly by UV spectrophotometer of hypochlorite ions formed during the reaction between chlorine and soda. The lowest detectable value is 11 ppm.

3.5 Meteorological conditions

The meteorological conditions are measured during the experiments by two different means, a vane propeller anemometer and an ultrasonic anemometer. The wind velocity and direction are measured at 2 and 10 m above the ground. In addition the temperature, relative humidity of the ambient air are measured.

3.6 Experimental procedure

Just before starting the experiment, the gas-source is aligned and water-curtain is oriented perpendicular to the wind. The water-curtain is switched on as the gas-source is opened and interacts directly with the water. A picture of a field-test is given in Figure 2.

To evaluate the dilution factor of the water-curtain (which is one quantity predicted by CASIMIRE) tests with similar gas mass-flows are conducted with and without water-curtain. In this manner the dilution factor DF is defined by the following relation

$$DF = \frac{Concentration_{Free\ dispersion}}{Concentration_{Forced\ dispersion}}. \qquad (1)$$

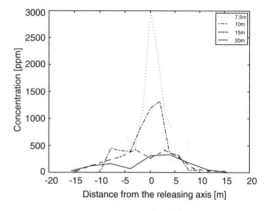

Figure 3. Concentration in function of the axe of gas-release.

3.7 Preliminary results

Concentration results are given from the absorbance of chlorine in the soda solution with respect to the volume of the solution and the volume of air bobbled through the solution. While the estimated error in the measurements is 10%, the uncertainty of the concentration is close to 15%. The results are presented in ppm value.

As Figure 1 shows, measurement points are placed mainly downwind of the water-curtain. However, some points are located on a circle at 4 m behind the source. These points give very constant results with mean values around 40 ppm for gas releases between 2 and 5 kg/min. On the sides of the source, at 5 m, the concentration depends strongly on the wind conditions. For low wind velocities, mean values are there 100 ppm.

An example of concentration measurements are presented in Figure 3 for a test with downward curtain. The wind velocity is 2.9 m/s, the mass-flows of the source and the water-curtain are 1.8 and 300 kg/min, respectively. It is a mass-flow ratio of 170. The figure points out the distribution of the concentration.

At the present time an estimation of a dilution factor from these preliminary tests is not available due to the lack of comparable tests. Further tests will be performed in order to determine the dilution factor DF of tests conducted at similar conditions. The influence of the flow ratio water-curtain/source will also be investigated as well as the meteorological stability class.

4 WIND-GALLERY & SPRAY CHARACTERISTICS

In order to check an eventual scaling factor between large-scale tests and Wind-Gallery experiments, similar experiments will be reproduced in the VKI-Wind-Gallery facility (Figure 4). The code CASIMIRE has been validated with tests of this kind. However, the question mark to be solved is the scaling effect. Therefore, industrial nozzles used in the field will be appropriately scaled down for Wind-Gallery test.

4.1 Wind-Gallery

The Wind-Gallery test section is 1 m high, 1.3 m wide and 7 m long. The wind speed can vary from 0.25 to 1.5 m/s, which is equivalent to 5 to 27 km/h at full scale. It can simulate, at scale, a pollutant leak that is absorbed and dispersed by a liquid curtain. Upward and downward pointing curtains can be tested. The air flow is produced by a battery of four ejectors mounted at the back end, thus producing a low pressure that keeps gas leaks towards the inside of the test section. The gallery has demonstrated very uniform velocity profiles and turbulence levels. Wind speed, pollutant source and water-curtain can be monitored. Concentrations and temperature can be measured, both in the gas and in the liquid.

To investigate the effect of forced dispersion, SF_6 have been previously used (Uznanski & Buchlin, 1998). Concentration profiles have been measured using a hot-wire anemometer especially developed for this purpose. The principle relies on the heat transfer response of the hotwire at constant velocity but to a change of the gas properties.

The geometrical scaling between the field-tests and the Wind-Gallery experiments is 4:1. That is easy to fulfill since the field-test were designed just for this purpose. In addition, a momentum scaling has to be defined. For this, a similarity of the momentum ratio of the curtain-to-wind R_M has to be respected. The ratio curtain-to-wind momentum R_M is defined by

$$R_M = \frac{m_u \cdot U_{d0}}{\rho \cdot V^2 \cdot H} \qquad (2)$$

Figure 4. VKI Wind-Gallery.

where m_u [kg/s/m] = mass-flow per unit length of the water-curtain; U_{d0} [m/s] = velocity at nozzle orifice; ρ [kg/m^3] = gas cloud density; V [m/s] = wind velocity; and H [m] = water-curtain heigth. The ratio R_M can not be easily fixed in field-tests because of the freedom of the wind velocity.

The Wind-Gallery tests are actually under preparations and will be conducted during spring and summer 2003.

4.2 Spray characteristics

The main hydrodynamic characteristics to evaluate the efficiency of a spray for industrial hazards are the entrained gaseous flow-rate and the interfacial area of the droplets. The gas entrainment quantifies the ability of the spray to dilute a toxic cloud by mechanical action and by thermal heating. The spray characteristics such as the interfacial area of the droplets and their velocities are the key properties for these actions. To determine these quantities, experiments are conducted in the VKI-Water-Spray facility. The set-up is composed of a hydraulic circuit supplying a single nozzle with a maximum flow rate of 1 l/s at 800 kPa. The pulverized water is collected in a 12 m^3 pool. The droplet size distribution in the spray and the entrained gas velocities are measured using Phase Doppler Anemometry (PDA). The influence of pressure and position in the spray are studied. Different types of nozzles, (full-cone and flat fan) can be tested. In this paper, only results for the full-cone nozzles are presented, as the field-tests are performed with this type. The full-cone is a swirl nozzle, the diameter of the orifice is 5.1 mm, its flow-number is $4.76 \cdot 10^{-4}$ kg/s/$\sqrt{Pa}$ Since the field-tests consist in evaluating the mechanical action of the spray, the air entrainment properties of the spray are fundamental. These can be estimated by measuring the smallest droplets of the spray, of the order of 20 μm assuming

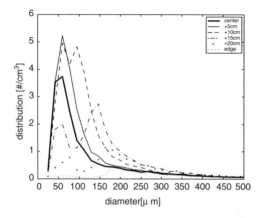

Figure 5. Droplet size distribution through a spray at 30 cm from the nozzle orifice for $\Delta P = 300\,kPa$.

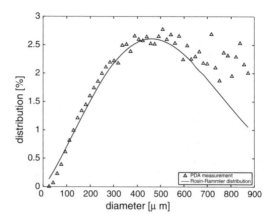

Figure 6. Comparison of the experimentally measured droplet distribution with Rosin-Rammler distribution.

they behave as tracers. As the PDA makes local measurements, profiles of the spray characteristics are acquired by measuring every 5 cm through a horizontal plane. Figure 5 presents the droplet size distribution through a horizontal section of the spray at 30 cm from the orifice for an operating pressure of $\Delta P = 300\,kPa$. The figure clearly states the effect of air entrainment in a spray. It shows that the edge is charged with few, but big droplets, while the center is charged with numerous small droplets. In between, for 5, 10, 15 and 20 cm we can observe how the droplets increases in diameter and decrease in quantity. However, in the center point, a small decrease of the droplet concentration is observed compared to the position at 5 and 10 cm. It is worth noting that experimental data fits very satisfactorily the Rosin-Rammler distribution as presented in Figure 6. The spray characteristics are significant in this project as we are trying to couple field-tests measurements to an engineering model CASIMIRE, Wind-Gallery tests and numerical simulations. The scaling of the nozzles in terms of flow-rate is easy to achieve, however, differences in droplet distributions is important to evaluate.

5 NUMERICAL SIMULATION

Multidimensional CFD simulation would be performed with the general-purpose code Fluent/UNS v. 6.1. The averaged Navier-Stokes equations coupled to the RNG k-ϵ model are solved. The experimental boundary conditions are satisfied.

The droplet flow is described by Langrangian approach: the particle velocity is calculated by solving the motion equation, taking into account the drag and gravity forces. The actions of the particles on the fluid flow and vice versa is ensured by a two-way

coupling via a source term added to the momentum equation of the gas phase. Upward and downward configurations of the spray-curtain may be reproduced by injecting the droplets at the floor or at a given height, respectively. The droplets trajectories are defined by a Rosin-Rammler distribution as the spray characteristics demonstrated a good similarity in Figure 6.

The objective is twofold: first, the numerical predictions will allow the preparation and the interpretation of the laboratory and field tests measurements (adjusting boundary conditions and scaling factor). Indeed, such a CFD simulation provide detailed information of the flow and concentration fields which are difficult and expensive to get in experimental approach (few measurement points in the field-tests, especially in the vertical direction). Once validated, the CFD simulation will be also regarded as numerical experiments to develop new models to be implemented in CASIMIRE. For instance, recirculation zones observed experimentally and numerically upstream or downstream of the water-curtain according to the operating mode should be modeled within CASIMIRE code. Indeed, they affect significantly the effect on the dispersion efficacy of the spray curtain.

6 COMPARISONS & VALIDATIONS

The comparisons between the field-tests and the Wind-Gallery experiments will be defined by geometrical and mass-flow scalings. Like (Hall, D. J. & Walker, S. 1997) defined scaling rules for his comparisons between field-tests and wind-tunnel experiments for the study of gas dispersion, scaling rules will be adapted for this project. Respect to the positioning of the measurement points and the source to water-curtain will be fulfilled.

The experimental and numerical test cases, will be compared with respect to similarity conditions and geometrical scalings. Then, the effect of the scaling factor will be assessed.

As mentioned earlier, CASIMIRE evaluates the dilution factor of a water-curtain with respect to the curtain-to-wind momentum ratio R_M. Therefore, these results are the one to be compared for all the different cases for the final validation.

CASIMIRE will be run in all test cases, and a new database (for industrial nozzles) will be added. The effect of scaling will finally be introduced and improve the engineering code to yield for real dimensions on industrial plants.

7 CONCLUSIONS

Different methods to study the mitigation effect produced by water-spray curtains on a heavy gas cloud are presented. The methodology of the project consists in comparing medium-scale field-tests with laboratory experiments and multidimensional CFD simulation. The objective is to validate an existing engineering code to design in an optimal way industrial water spray curtains. Emphasis is given to the forced dispersion effectiveness of the curtain. Aspects related to scaling and similarity effects in the experimental parts are introduced. Preliminary hydrodynamics diagnostic of industrial nozzles and examples of concentration results from field-tests are presented.

REFERENCES

Blewitt et al. (1981). Effectiveness of water sprays on mitigating anhydrous hydrofluoric acid releases. pp. 155–180. International Conference on Vapour Cloud Modelling.

Buchlin, J.-M. (1994). Mitigation of problem clouds. *Journal of Loss Prevention in Process Industries* (7), 167–174.

Fthenakis, V. M. & Blewitt, D. N. (1995). Recent development in modelling mitigation of accidental releases of hazardous gases. *Journal of Loss Prevention in Process Industries* (8), 71–77.

Griolet et al. (1995). Mitigation of accidental releases of toxic clouds by reactive fluid curtain. *Loss Prev. Saf. Prom. Proc. Ind. 1.*

Hall, D. J. & Walker, S. (1997). Scaling rules for reduced-scale field releases of hydrogen fluoride. *Journal of Hazardous Materials* (54), 89–111.

Malet et al. (1995, July). An experimental and numerical study of the mechanical and thermal actions of water spray curtain for toxic cloud mitigation. Volume 1, Toulouse, France. 15th Annual Conference on Liquid Atomization and Spray Systems.

McQuaid, J. & Fitzpatrick, R. D. (1981). The uses and limitations of water spray barriers. *IChemE N.W. Branch Papers* (5).

Molag et al. (2000). The use of fluid curtains to mitigate gas dispersions. EPSC. Rugby.

Moodie, K. (1981). Experimental assessment of full-scale water-spray barriers for dispersing dense gases. *North Western Branch Papers, Institution of Chemical Engineers* (5), 5.1–5.13.

Schatz, K. W. (1990). Water spray mitigation on hydrofluoric acid releases. *Journal of Loss Prevention in Process Industries* (3), 222–233.

St-Georges et al. (1992). Fundamental multidisciplinary study of liquid spray for absorption of pollutant or toxic clouds. *Loss Prev. Saf. Prom. Proc. Ind. 2*(65).

Uznanski, D. T. & Buchlin, J.-M. (1998, May). Mitigation of industrial hazards by water spray curtain. Washington, D.C., USA. Fifth Annual Conference of the International Emergency Management Society.

Safety and Reliability – Bedford & van Gelder (eds)
© 2003 Swets & Zeitlinger, Lisse, ISBN 90 5809 551 7

The collaboration between Arbo-specialists in the Netherlands

Andrew R. Hale

Safety Science Group, Delft University of Technology, Netherlands

Henk J.L. Voets

ABSTRACT: Under the Dutch Working Conditions Act the Dutch government stipulates that the employer shall have or shall have a contract with a certified arbo-service, as complying with a number of requirements about organisation and expertise. Given the different professional backgrounds of the experts and the autonomy and power of professionals their co-operation cannot be taken for granted. The paper presents the results of a research project which was carried out over a number of years, to see how the co-operation takes place in practice.

1   INTRODUCTION

Under the requirements of the Dutch Working Conditions Act (NL 1993), which translate the European Framework Directive (EC 1989), all employers are required to have a working conditions (in Dutch "arbo") service to advise them and conduct such compulsory tasks as approving a risk evaluation, holding periodic medical examinations of designated workers and organizing a regular medical surgery hour. This service can be internal, or the employer may have a contract to an external service to supply some or all of the expertise required. The service must be certified by a third party certifier. One of the requirements for this is that the service employs at least four core experts: occupational physician (OP), occupational hygienist (OH), safety professional (SP) and work and organization specialist (WO). Under the Working Conditions law it is stipulated that these four specialists must ensure that they collaborate to give the employer and employees (works council) an integrated advice. In addition to the compulsory tasks required by law, the employer may give his internal service, or contract from his external service additional tasks such as advising on prevention, on plans to reduce risk, or on specific arbo-problems.

This paper looks at the principles and practice of collaboration between the specialists through 12 case studies of the working of a range of different types of arbo-service.

2   WORKING CONDITIONS SERVICES & COLLABORATION

The organisations which offered their services to fill the requirement for compulsory working conditions services came from a number of different backgrounds.

1. *Internal services.* Large companies (over 500 employees) in manufacturing and harbour work had long been required to employ a safety practitioner and an occupational physician. These had often been in different services, the safety service often, particularly in chemical companies, combined with an environmental service, the occupational physician often linked to personnel services. Some internal services of this nature were combined and expanded to include the other required professionals, all employed within the company. Other internal services expanded the safety and environment service with an occupational hygienist and outsourced the occupational physician and work and organisation specialist functions. A few large companies outsourced their whole service, so that it could serve not only their own needs, but attract other clients to help spread the overhead.

2. *Existing external occupational health services.* There had been a number of regional, or industry-based external services offering the legally required professional services, particularly of occupational physicians. They were usually dominated by

occupational physicians and employed a handful of other professionals to do contract advisory work flowing from those links. Many of these expanded their professionals to cover all the four required. In addition there was a major flood of mergers leading to a few nationally based services.

3. *Insurance companies.* As a result of the major reorganisation of the social insurance market, which paralleled the setting up of working conditions services, a number of companies working in that sector set up services to exploit the new market.

4. *Independent consultancies.* Some consultant organisations moved into the new market, which they saw as a large, new guaranteed source of income.

The last two categories of service tended to be much more commercially oriented and to conduct aggressive marketing, whilst the first two types tended to retain the form of a professional bureaucracy for many years.

Some 200 services were certified in the first period after the introduction of the new law. A cutthroat period of competition resulted in this number being reduced in the intervening period to under 100.

2.1 History of the professionals

Two of the designated specialists, occupational physician and safety practitioner, have a long history in the Netherlands, having been required to be employed in at least large industrial companies for more than 50 years. Their professional associations date from the 1940s. The other two specialists have a more recent history, both as profession and as required specialist. The occupational hygienist professional body in the Netherlands was established in 1983 whilst the work and organisation specialist only became an integrated profession with its own professional body in 1997, after the law designated them as required specialists.

Whilst there is no clear and agreed formulation of the areas of expertise of the four groups (Hale 1990, 2000, Hale et al. 1997), we can designate the focus and background of each group. The occupational physician centres on the individual's capacity to work and the chronic harmful effect of work on him or her; the methods and models are medical in origin and hence individual in focus; the practitioners are exclusively graduates. The safety practitioner focuses on acute harm, accidents, and technical and, increasingly, organizational analysis and solutions and must have a technical education, which can be at graduate, but is often at higher, or middle technical level; the approach is now based on systems thinking and pragmatic engineering and problem solving. The occupational hygienist concentrates on assessing exposure to and harm from chronic workplace hazards and also has a scientific education at either graduate or higher technical level; the approaches are analytical and based on chemistry and the health sciences. The work

and organisation specialist has a more diverse focus, on problems of stress and on organizational malfunctions leading to poor well-being, but also on organizational change and leadership; most work and organisation specialists are graduates with a social science, group oriented, management approach. This short sketch seems to indicate both the need for and the problems of collaboration.

2.2 The need for collaboration

The requirement for the four experts to collaborate seems obvious. Most employers are faced with a range of hazards covering safety (SP), health (OP, OH) and well-being (WO). The hazards do not come isolated from each other, but together in processes, which need to be modified in an integral way technically (SP, OH) and organizationally (WO) so that they accommodate to individual's capacities (OP) and inclinations (WO). However, we should point out that this analysis assumes that the knowledge and skill to analyze, understand and control these hazards requires a deep expertise in these areas, which can only come from specialization (Hale 1999); and that generalists cannot be trained to understand and advise on the majority of hazards and solutions, with recourse to specialists with a deep knowledge only in the second line. In other words there are more possible models for deploying arbo-expertise than the one currently required in the Dutch law. We can illustrate a number of these alternatives in the figure 1.

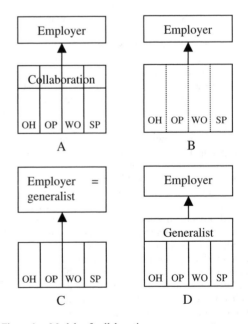

Figure 1. Models of collaboration.

A is the currently proposed format with the requirement for collaboration being laid on all of the experts. B sees a fading of the disciplinary barriers with all specialists becoming more generalist and able to incorporate enough of others' expertise into their work to ensure integral coverage. C sees the employer as sufficient of a generalist to do the integration. D sees a generalist employed by the arbo-service doing the integration with the four specialists as second line experts. Such a generalist could be the head/elite of the service growing from one or all of the experts (e.g. from model B), or could be specially trained as a broad employee (c.f. the general practitioner in medicine).

2.3 Problems for collaboration

The distinctly different educational backgrounds and theoretical and professional approaches of the four experts have been sketched in 2 above. These differences provide barriers to mutual comprehension between the professions, which still do not have a completely shared vocabulary and set of concepts to talk about their work. The problem of equal collaboration between professionals with a different level of education, particularly between graduates and middle technical personnel is also an issue. We need to add to this the intrinsic difficulties for professional groups in the same area to collaborate (v.d. Krogt 1981), when their professional associations may be in disagreement about the boundaries which they claim. Professions are by their nature protective of their knowledge areas and impose entry requirements in the form of training and experience on their members. The stronger a profession is, the more it expects others to be subordinate to it in its claimed area of expertise. Given the dominance of the medical profession in society generally, and the numerical dominance of the occupational physician in the Dutch arbo-services (see v. Amstel & v. Putten 2002 for recent figures), we can expect to meet these problems in practice, despite the existence of some evidence (Hale et al. 1997) that the majority of the different arbo-professionals hold each other in reasonably high regard and are keen on developing collaboration.

3 THE RESEARCH PROJECT

The authors, in collaboration with Prof. J.T. Allegro and graduates of the work and organisation psychology course in Leiden, have carried out a research project over a number of years, to see how the co-operation between the experts in the arbo-services takes place in practice. The study also looked to see whether there appeared to be differences in this respect between the traditional arbo-services which grew out of the older occupational health services and newer arbo-service organisations, which have been set up to make use of the newly created market for certified arbo-services. It also looked at the differences between internal company services and external contract arbo-services.

3.1 Methodology

We selected four graduate students. Two carried out their study in internal company services and two in external arbo-services. They all worked in the organisation for their fieldwork/graduation project for six months. During that period they concentrated on the functioning of the organisation itself. For their doctoral thesis cases (projects or problems potentially suitable for collaboration between specialists in the services) were selected and studied which were put forward by the organisation where they worked during their fieldwork period. In this way the actually occurring co-operation between the professionals within these organisations could be studied more in depth.

During the whole period the students made observations and studied written material with regard to the functioning of the professionals, the organisation and how the cases were carried out. Their study was completed by interviews with the professionals, their superiors, as well as their clients. So the students made participatory observations in the organisation as a whole, as well as in the projects that were carried out.

The sample of services studied is small and, hence, we cannot draw general conclusions from it. However, we believe that the picture we sketch is not untypical (see v. Amstel & v. Putten 2002)

3.2 Results

3.2.1 Internal arbo-services
3.2.1.1 Organisation: Both internal arbo-services studied are part of a military organisation. One is the central arbo-service for the Air force in the Netherlands, the other for the Land forces in the Netherlands. Both arbo-services are certified, which means that they have all the professionals required by law, and in both cases even more than that. From our contacts with other internal arbo-services it became clear that this situation is becoming more and more exceptional. Several organisations which used to have a fully equipped internal arbo-service centre decided during our research period to outsource the service of one or more professionals (usually the occupational physician) and hire them in from an external arbo-service. The argument for it was always the same: the costs of having a fully equipped internal service were considered to be too high as compared with the costs of hiring these services from an external arbo-service. This organisational separation of the professionals adds one more difficulty to the collaboration.

The members of the arbo-service are a part of the military organisation but they are a staff function. The military organisation is spread over several locations which have their own internal arbo-service. But these services are usually the (former) medical centres, now supplied with the other professionals from the central arbo-service, who work there on a part-time or sometimes even permanent basis (depending on the number of people working on the location). The commander of the local organisation can also call upon the central organisation for help when there is a problem which cannot be solved by his own (local) arbo-service, as happened in the cases which were studied by our students. The internal arbo-services studied are not representative in so far as most internal arbo-services are not a part of a military organisation. But they are alike in having a staff postion and having to cope with the pressure of the (local site) directors.

3.2.1.2 Collaboration between the professionals: All the professionals required by law were active within the internal arbo-services. In both centres the director of the service was a professional with a work and organisation specialist background. Because of the high position and rank of the directors within the military organisation the other professionals had to take orders from their director. Working at a specific location they also had to take orders from the commander of the location. This situation is comparable in some ways to the line responsibility of the arbo-professionals to the site manager in internal services in other industries, but perhaps more binding, as it is military.

The collaboration between the specialists can best be typified by model D in section 2.2. Although both directors were the work and organisation specialist, they acted more as a generalist, based on experience and high status. The co-ordination of the work was mostly done in regular meetings of all the members of the arbo-centres.

The work in the central arbo-service of the Air Force was split up into projects. For all projects it was made clear which specialist should participate in the project and who was in charge of the project. And also at what moment(s) the director should be informed about the progress made. Files were kept of all the projects.

The decision to involve one or more professionals was made by the director, but usually these matters were discussed beforehand in a regular meeting of all the professionals.

In the arbo-centre of the Land forces the director had a more informal way of working, relying more on direct face-to-face contact. Also the project teams were formed informally.

The professionals did not have much contact with each other apart from in the regular meetings. Most of the work was done by one professional, though sometimes by two. Whenever the director decided that the advice of more specialists was necessary, he would ask for it himself. Sometimes the professionals did not even know that another colleague had been asked to give his/her advice. But the professionals were not unhappy with that situation. They agreed that it was the task of the director/generalist to make such decisions and to ask for their professional opinion. Collaboration was thus via the director and seldom directly with other professionals.

3.2.1.3 Cases: For both arbo-services we selected a problem that in our view needed the involvement and collaboration of more professionals.

Case A. (Air force). The guarding of the bases. The problem for the guards is that the work itself is not very interesting and there is also not much variety. So the workers get bored and de-motivated. In the past the problem was "solved" by making use of soldiers who would do the job for some months and then leave the organisation. But now the work has to be done by professional soldiers the problem has got more acute. There were car accidents reported due to lack of attention and it was difficult to find enough good personnel for the job. Nevertheless the case had been lying around for some time and had not been picked up by the team. It was agreed that this project would be studied as an example of collaboration. However, what happened was that our student was asked to interview a representative group of guards and to make a report of the situation. The results were reported in a staff meeting but the arbo-service centre took no further action. The case had been delegated to our student as "generalist", rather than forming the basis for collaboration.

Case B. (Land forces). Work in the stores for surplus military goods (from small items to complete tanks). The surplus items are taken apart in the stores, if necessary, and sold or shipped for disposal. This could involve a great physical effort. The occupational physician reported a high percentage of sickness absenteeism. This was chosen as topic to study the collaboration, but again the "solution" chosen by the service was to ask our student to interview a representative group of workers and make a report of the situation. The local commander then studied the results and, without feeling the need to consult with the arbo-professionals, carried through a reorganisation. That was the end of the story, since the problem was defined as "solved".

Both cases are good examples of problems which could, and in our view should, have been dealt with by a group of professionals from the arbo-centres. Now they were dealt with by our students, who interviewed the different specialists. They were used by the services as "barefoot" generalists, a version of model D in section 2.2 above.

The excuses for not handling the problem among the professionals themselves were in case (A) lack of time and in case (B) the action taken by the local commander, because he informed the director that there was no further need for action from his centre. It is clear that as a result of this the problems that were diagnosed and that had existed already for many years, were not necessarily solved.

As mentioned before the internal arbo-services studied are not fully representative. But a lot of the problems mentioned are also found in the literature with respect to staff organisations in general (Joseph L. Massie 1970, see also Wolf Heydebrand 1990).

3.2.2 External arbo-services

3.2.2.1 Organisation: For the fieldwork/graduation project of the students we selected two external arbo-organisations: one was a service which had grown out of the insurance system as a new player in the market (service A), the other was part of the arbo-service formed from the merger of regional services growing out of the old occupational health services (service B). For their doctoral thesis we selected projects from different organisations which were clients of these two services.

3.2.2.2 The collaboration between professionals: Our students studied 6 cases which were all "Risk Inventory and Evaluation" projects (RIE's). The RIE is a legally required document for all companies in the Netherlands. It summarises the risks which are present in the organisation and forms the basis for a "plan of campaign" to control them. The RIEs may be carried out by the company itself, or by the arbo-service. In the former case, the arbo-service, as independent body, must legally approve the RIE. This means that most companies, to save time, let the arbo-service also carry it out.

3.2.2.3 Cases: The RIE must cover all types of risk, to safety, health and well-being. All the specialists required by law (the occupational physician, occupational hygienist, safety practitioner and work and organisation specialist) have to approve each RIE. It is therefore an obvious candidate for collaboration between the different professionals. Hence we selected it as topic for both arbo-services.

Case A. (Service A). The RIE in this service is made according to a standard procedure. First there is a meeting with the managers of the organisation where the RIE will take place to inform them about the procedure. Secondly questionnaires are sent to a group of workers, randomly chosen from the organisation. These include questions about all the types the risk mentioned before. In the meantime the occupational

hygienist and safety practitioner specialists make an inspection tour in the organisation observing and talking to the work-floor staff. These tours may be, but often are not, conducted together. When at least 65% of the interviews have been received the work and organisation specialist makes a summary. This summary is discussed with the works council and the director(s) of the organisation. On the basis of the summary and the discussion the work and organisation specialist makes his/her report and puts it together with the reports of the occupational hygienist en safety practitioner specialists. This is the concept RIE report. The concept report is again discussed with the works council and the director(s). On the basis of these discussions the final RIE report is made. The last step is that the work and organisation specialist makes a plan for further action on the basis of the final report and discusses that with the organisation.

Case B. (Service B). This organisation also has a standard procedure for a RIE, but it is more flexible. Also here it starts with an introduction of the RIE in the organisation. The service also has a standard questionnaire which is used in all organisations, but, in addition to the written questionnaire, interviews are also part of the procedure. The interviews are flexible: they may focus on different types of risk and different groups of workers, according to the situation. Only one of the specialists makes an inspection tour in the organisation. The report of the work and organisation specialist on the basis of the interviews and the report of the other specialist who did the inspection tour are put together in a concept report which is discussed with the works council and the director(s). The final RIE report is made on the basis of the outcome of the discussions. Also here the last step is a plan for further action made by the work and organisation specialist and discussed with the organisation.

In both of the cases the contact between the different professionals is relatively limited. They work on the same project, but their interaction is largely via the paper of the report and its recommendations. To save time and money it is not usual for them to visit the company together for either the RIE or the discussion. Even in the preparation of the report there is not much direct contact. The work of the specialists is more as disciplines adding to each other than really collaborating. Here, as in the internal services the work and organisation specialist has something of an integrating role.

3.2.3 *The comparison between internal and external arbo-services*

From our case studies it was clear that the internal services resembled model D and that the model for the external services did not clearly fit any of our

751

models. It resembled a cross between A without real collaboration and C, in which the integration is left to the employer, with some minimal indication of D, with the work and organisation specialist acting as coordinating generalist.

The idea that internal services may profit from the fact that they have a better knowledge of the organisation itself in designing their intervention strategies was not clearly confirmed by our research, but the topics chosen for the cases did not, by their nature, make it easy to check that hypothesis. A RIE is a diagnostic, rather than a change instrument. Only when it is turned into an operational plan does it start to produce change. That step was not taken with the RIE's we studied in the external services.

4 DISCUSSION AND CONCLUSION

It is clear from our case studies that collaboration between the professionals in arbo-services seldom occurs in practice. Model A from section 2.2 seems not to be present in the cases and arbo-services we studied. There may be meetings where all the professionals are present and where some projects are discussed. However, we found little evidence of different professionals working closely together on a particular project. Sometimes the collaboration was limited to a signature on the bottom of a report, which each specialist had contributed to and read independently. In effect this amounts to model C in section 2.2. The integration is left to the employer. This is the model which operates in, for example, the United Kingdom, where the employer is simply required to employ competent persons to advise him, without the law specifying that the competence has to reside in named types of professionals, or that the professionals have to collaborate.

What we did find evidence of was the existence of Model D, particularly in the internal services, but also to an extent in the external services. The director, or the work and organisation specialist acted as generalist on the basis of his experience and status. It may not be chance that it should be a work and organisation specialist who did this. The work and organisation specialist qualifying course is the one which has the most emphasis on collaboration and provides the most information about what the other professionals do (Hale 1999).

What we also saw in the cases was the (internal) service using our students as "generalists", but then more in the terms of the "barefoot doctors", people with limited training and experience used to solve the simple multidisciplinary problems. It may be a solution for companies with relatively simple problems to use generalists with a smattering of knowledge across the broad field, but that will hardly resolve the complex multidisciplinary problems which also exist in the field in the more complex organisations.

Our study technique was not really suitable to see whether Model C from section 2.2 operated. We would have needed to go much more deeply into the way in which each professional worked to see if it was the case that the experienced ones had gradually acquired sufficient knowledge of the adjacent disciplines to be able to resolve more multidisciplinary problems on their own. If we look at the qualifying courses for the professionals, there is some evidence that each has some input about the work of the fellow professionals. However this is only significant between safety practitioner and occupational hygienist, and between occupational physician and work and organisation specialist (Hale et al. 1997, Hale 1999). As stated above the work and organisation specialist training gives the most time to this basis for collaboration.

In conclusion we have demonstrated that the requirement in Dutch law that requires collaboration between professionals is not working in practice in our sample. Although our sample may not be fully representative of all arbo-services, we have no indication that the situation in other organisations is significantly better (v. Amstel & Putten 2002). We think it is time to go back to first principles and to re-examine the reasons for that collaboration. These are that the problems in practice do not come in the neat disciplinary packages which form the basis of the professional disciplines. There are many multidisciplinary problems which require, for their real solution, a combination of knowledge about different types of risk arising from the same processes (safety, health and well-being), which require a fine nuance in their solution so as not to exacerbate one problem whilst solving another.

Hale has long advocated (e.g. Atherley & Hale 1975, Hale et al. 1986, Hale 1990) that we should go back and look more fundamentally at the basic training courses of the professionals and their disciplinary boundaries, that date from a time when the discipline was determined by the result of the hazard and not by the causes of it (injuries for the safety practitioner, environmental exposure for the occupational hygienist, ill health for the occupational physician, and stress for the work and organisation specialist). We should seek to merge the common features necessary for a broad solution of working conditions problems. These could form a common basis for training, which could then go on to give specialisations in the specific areas now seen as separate disciplines. A further merging of the safety practitioner and occupational hygienist training would be the most simple to carry out, taking many elements of ergonomics also into account. An open question is whether the occupational physician training could be linked to this, given the strength of the medical profession in controlling its qualification processes.

REFERENCES

v. Amstel R. & v. Putten D. 2002. Ervaringen met de dienstverlening: arbodienstenpanel; rapportage van de vierde peiling (Experiences with service provision: arbo-service panel; report of the fourth assessment). TNO Arbeid. Hoofddorp report 1010007.

Atherley G.R.C. & Hale A.R. 1975. Prerequisites for a profession in occupational safety & hygiene. Annals of Occupational Hygiene v18 pp. 321–34.

EC. 1989. Directive concerning the execution of measures to promote the improvement of the safety and health of workers at their work and other subjects (Framework Directive). Official Journal of European Community. 12 June 1989.

Frick K. 1994. From "sidecar" to integrated health and safety work – Work environment control as a management problem in Swedish manufacturing industry. Swedish Centre for Working Life. Stockholm.

Hale, A.R. 1990. The training of professionals in prevention. Proceedings of the Conference of the International Social Security Association, Paris, 31 May–2 June 1989 on Education and Training in Prevention. ISSA. Paris.

Hale A.R. 1999. Why occupational health and safety experts? Keynote address. Proceedings of the International Symposium of the International Social Security Association: Training of occupational health and safety experts; issues at stake and future prospects. Mainz 30 June–2 July. ISSA Section on Education & Training. INRS. Paris.

Hale A.R. 2000. Why occupational health and safety experts? Proceedings of the International Symposium of the International Social Security Association: Training of occupational health and safety experts; issues at stake and future prospects. Mainz 30 June–2 July. ISSA Section on Education & Training. INRS. Paris.

Hale A.R., Piney M.P. & Alesbury R.A. 1986. The Development of Occupational Hygiene & the Training of Health & Safety Professionals. Annals of Occupational Hygiene v30 (1) pp. 1–18.

Hale A.R., Heming B., Musson Y. & v.d. Broek B. 1997. De veiligheidskundige professie: kansen en bedreigingen (The safety profession: chances and threats). Utrecht. NVVK.

Heydebrand, Wolf. 1990. The Technocratic Organization of Acacdemic Work, in: Craig Calhoun et al. (eds), Structures of power and constraint, Cambridge University Press.

v.d. Krogt, T.P.W.M. 1981. Professionalisering en collectieve macht; een conceptueel kader.

Massie, Joseph L. 1970. Management Theory, in: James G. March (ed.), Handbook of Organisations, Rand McNally & Company, Chicago (3rd Printing).

NL 1993. Aangepaste Arbeidsomstandighedenwet (Modified Working Environment Law) 1993. Staatsuitgeverij. Den Haag.

Safety and Reliability – Bedford & van Gelder (eds)
© 2003 Swets & Zeitlinger, Lisse, ISBN 90 5809 551 7

Improved vessel safety in offshore supply services

L. Hansson
Norwegian Marine Technology Research Institute

ABSTRACT: The supply services in the North Sea are a working place with a high level of personnel risk mainly in occupational accidents. Supply vessels, anchor handling vessels and stand-by vessels are operating in a harsh environment in an integrated logistic chain including offshore installations, mobile installations and shore bases. An ongoing research project for a Norwegian oil company addresses the risk for the vessels in the supply services and will develop a risk model that aims at giving decision support in the process of selecting the correct risk reducing measures. This paper describes available methodology for risk analysis and their application in the maritime and offshore industry. Further it discusses the similarities and differences between the vessels in the supply services and the rest of the maritime and offshore industry with regard to traditions, working conditions, operations and regulations. Finally, it gives some guidelines for choosing a risk methodology for the vessels in the supply services. The methodology will be based on task analysis for the most exposed operations on the supply and anchor handling vessels.

1 INTRODUCTION

The number of injuries in offshore supply services has increased year by year since the reporting started in 1996. The increasing number of injuries is the background for the work reported in this paper and the work is based on the injuries reported in the oil company Statoil. Statoil has made considerable efforts to stop this negative development and the choice of risk-reducing measures has been based on common knowledge, experience and accident investigations. One measure is employee or user involvement. The captains operating the vessels participate in a seminar every year were safety aspects are the main topic. These seminars have given positive effects, and the number of injuries in year 2002 has decreased drastically compared to the previous year. A systematic approach to assist the choice of the most effective risk-reducing measures could be helpful.

The events and injuries on the vessels in the supply services are first of all so-called occupational accidents. Occupational accidents happen during work operations, there is a direct connection between the cause and the consequence and the accidents often involve only one or a few people. The consequence can be serious; from first aid injuries to death but the material damage is usually limited. If a person falls down a ladder and breaks his leg during a work operation, this is an occupational accident. The other type

of accidents is the major accidents were the chain from the cause to the accidental event can be long, many people can be involved and the material damage can be substantial. One relevant major accident in the supply services is collision between vessel and installation. In 2000, 12 collisions (or contacts) were reported in Statoil.

The risk model described in this paper is developed in order to give decision support for the choice of risk-reducing measures. The second objective for this risk model is to give enhanced knowledge and understanding about risky situations. The next sections give an overview over risk analysis, the different methods available and a detailed description of the model developed for use in the offshore supply services. The last section of this paper discusses the model that has been developed in relation to the objectives. It also outlines ideas for further work.

2 THEORY AND METHODOLOGIES

2.1 *Risk analysis and available methods*

Risk analyses have been used in all types of industries over the past decades. One major objective of risk analysis is to measure risk for the purpose of risk control. In the development phase of a project, risk analysis is used as a design tool. Risk analysis plays an important role in the task of developing safety related rules and regulations in maritime and offshore industries. Formal safety assessment applied to the maritime industry has this as its primary aim. Risk analysis can be used for educational purposes as well. According to the British sociologist Robert Moore "*hazards and risks are often identified and controlled most effectively by those involved in the work tasks by a process of constant monitoring or risk evaluation from below. In order for operational-level safety intelligences to be effective, workers need to be involved in the process of risk assessment*" (Flin, 1996).

Several methods exist for use in risk analysis. While the first generation of risk analysis covered only technical aspects it has been realized that the whole story cannot be told without including and understanding the influence of human and organizational factors. Some of the available methods are Fault trees, Event trees, FMEA (Failure Mode Effect Analysis), Risk Assessment Matrix, Influence Diagram, HRA (Human reliability assessment) and HAZOP (Hazard Operability Analysis).

The area of risk analysis is mainly developed and used for major accidents. Major accidents have complicated cause-consequence chains and a potential of ending up in a catastrophe. The other category of accidents is the occupational accidents, events happening during work operations and with a direct connection between cause and consequences. The consequences are limited to one or a few people, but the damage or injury to the exposed person may be serious and in worse case fatal. Some of the methods used for major accidents can be used for occupational accidents as well but there are also some additional methods for occupational accidents (Harms-Ringdahl, 1993); Deviation Analysis, Action Error Analysis, Energy Analysis, Job Safety Analysis, MORT and Change Analysis.

2.2 Human error and occupational accidents

Kirwan (Kirwan, 1990) points out that one of the future challenges within risk assessment and human reliability analysis is to cover individual accidents (occupational accidents). "*The HRA (Human Reliability Analysis) field has mainly concerned itself with the high risk, high technology industry sector, including nuclear power plants and chemical plants. There exist however, a large number of other lower technology sectors, e.g. mining, which often incur a high risk via a large number of 'small' accidents (one or two fatalities), rather than (high technology) industries where the high risk is caused by a very small probability of an accident with many and serious consequences. This is clearly an area where applied human reliability should be able to help reduce risk.*"

2.3 Influence diagram and Bayesian network

The influence diagram is an important part of the developed model and this method is described more in detail. Mathematically, fault trees are probabilistic influence diagrams. (Barlow, 1998) The fault tree approach was based on engineering considerations and was invented by mechanical engineers. Influence diagrams provide an excellent graphical tool for understanding probabilistic conditional independence. The influence diagram is a graphical representation of the relationships between random quantities that are judged relevant to a real problem. The fault tree is the more useful representation for analysing system failure events. For decision problems the influence diagram is the more useful representation.

Influence diagrams have a sound mathematical basis in Bayesian probability theory. According to Rettedal (1997) the Bayesian approach is considered attractive since it does not break down in the absence of experience data and allows a systematic integration of expert opinions.

According to Øien (2001) there are specific advantages with the influence diagram (or Bayesian network) technique. It provides an intuitive representation of the causal relationships linking the organizational factors to the quantitative risk model. This intuitive representation is essential when communicating with experts. The relations as well as the states may be represented probabilistically, and there is no limitation on the number of states (thus we are not restricted to binary representation). Interaction between factors can be explicitly taken into account, e.g. the effect of poor training and poor procedures at the same time may be worse than the added individual effects of these factors being in a bad state.

The number of weights that have to be assigned is rather large even for moderately complex models. The propagation of the rates and the weights are an inherent part of the influence diagram techniques, and are no longer a problem even for large models. It has recently been solved by the development of "clever" algorithms. What really constitutes the practical challenge is the assignment of weights, that is the conditional probabilities given all possible combinations of states. Usually some kind of expert judgment procedure is proposed in order to establish these weights, but also data-driven approaches have been suggested.

3 CASE AND RISK PICTURE

3.1 The case

The offshore supply services cover the transport of goods to the oil and gas installations in the North Sea by supply vessels. Further it covers anchor-handling activities performed by special equipped vessels and the emergency preparedness which is taken care of by the stand-by vessels. The supply vessels transport food, chemicals, piping, equipment and spare parts from the shore base to the installations offshore and empty cargo back to the shore base. The stand-by vessels are located in a position close to the installation they are serving. Their main aim is to assist the installation in an emergency situation. To be well prepared they repeatedly rehearse, which may be a risky situation in itself. The anchor handlers are involved when the mobile drilling rigs move from one location to another. Their task is to fasten and unfasten the drilling rig anchors. The case to be discussed in this paper focuses on the supply services in Statoil.

Statoil operates the supply services but the vessels are contracted from different shipowners. All crew in the supply services have a shift with 4 weeks on duty and 4 weeks off duty. Traditionally the crew were recruited among seamen or fishermen, and for these groups the supply services were an attractive work place. Today there is a recruitment problem, one reason is that the conditions for the personnel onboard the vessels are worse the conditions for the crew onboard the installations. For the offshore workers on the installation the wages are higher, the working hours are shorter (they are on duty for 2 weeks and off duty for 3 or 4 weeks) and in addition or partly due to this, the offshore worker has a higher social status than the vessel crew.

The supply services operate in accordance to rules and regulations from the Norwegian Maritime Directorate and partly from the Norwegian Petroleum directorate. Vessels are operating partly according to Statoil procedures and partly according to shipowners' procedures. The captain is in charge onboard the vessel but is reporting to the Statoil Traffic control.

3.2 Risky operations

The main risky operations identified for the vessels in the supply services are:

- loading and unloading from supply vessels
- anchor handling activities
- maintenance
- navigation along installations

3.3 Risk picture of the vessels in the supply services compared to merchant vessels

"Normal accidents" by Perrow (1984) dedicates a section to marine accidents. The accidents described by Perrow are related mainly to merchant vessels, and according to Perrow the marine system is an "error-inducing system". In an error-inducing system, the components promote error inducement and it does not help to change one component. The tendency to attribute blame to operator error is prominent in an error-inducing system. Perrow predicts that the main reason for the marine system being error-inducing, is that the victims of an accident have a low status, third party victims of pollution and toxic spills are anonymous and the effect of pollution is delayed and the federal presence is minor.

It can be argued that the vessels in the supply services constitute a less error-inducing system than the merchant vessels. The vessels in the supply services operate closer to shore and in close contact with the other actors in the offshore industry. The safety level in the supply services is compared to the safety level at the oil and gas installations and this makes the accidents more visible. As a collision between vessel and installation is a threat to the installation itself, the system is less error inducing because the victims have a high status, the victims are not anonymous, the effect is not delayed and the federal presence is high. Some conditions affecting safety according to "Normal accidents" (Perrow 1984), are discussed below. The statements marine accident is given in italic.

Captains and crew can be on duty for 40–50 hours without sleep. This has been and is still a problem in the offshore business, on the installations as well as on the vessels in the supply services. For the vessels, especially anchor-handling activities can result in long hours. Relocation of a mobile drilling rig is a cost intensive operation and it is important to avoid delay.

The captains are avoiding radio contact. This is not described as a problem for the captains in the supply services. They are in frequent contact with the other actors in the services. Communication problems may however be addressed as a problem and can partly be blamed on the seamen tradition.

The captain is in supreme command and problems may stem from incompetent captains. If one person has unquestioned, absolute authority over a system, a human error by that person will not be checked by others. The competence of the captains on the vessels has not been questioned, but the average age of the captains is high, and the industry is concerned about the recruitment situation in the future.

4 DESCRIPTION OF THE DEVELOPED MODEL

4.1 Model objective and choice of methods

The aim of the developed risk model is first of all to give decision support to the process of prioritizing between risk reducing measures. A second aim is to enhance competence and knowledge about risky situations. Employment involvement is one important way to reduce the risk and the developed model will also fulfill this aim. The development process itself may serve as a process that increase the level of skill. Based on the fact that the model will be semi-quantitative, will be used for occupational accidents and decision support and for educational purposes it was decided to use a combination of the influence diagram and operational task analysis.

4.2 Limitations

The model has been developed for the loading and unloading operations from a supply vessel. The loading and unloading operations can be divided into three categories dependent on the cargo handled; deck cargo, bulk and casing. SYNERGI is the reporting system (database) used by Statoil and other companies for accidents and near-misses. The database contains reports on injuries on the vessels from 1996 until today. There have been 31 injuries during unloading/loading. The number of injuries reported on supply vessels in this period is approximately twice as high. The second half of these events happened during maintenance, in the machine room or in the kitchen.

The reporting of the injuries in SYNERGI includes a short verbal description and different coding describing the event, the causal chain and the personnel involved. Coding is defined for direct as well as the organizational factors. The organizational factors are however most often not described.

4.3 Influence diagram

An influence diagram has been drawn for the reported injuries. The first level represents the different event types, such as "hit by" and "hit against". The second level represents the direct causes and the third level the organizational factors. The influence diagram presented in Figure 1, includes the number of events and direct causes.

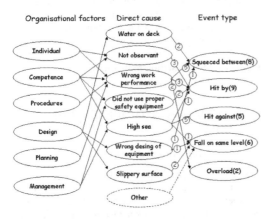

Figure 1. Influence diagram for events related to loading/unloading from supply vessels.

4.4 Organizational factors in the influence diagram

The organisational factors are approximately the same factors as described by Øien (2001) that are representative for the accidental event "leak of hydrocarbons" on offshore installations. The factor "individual" was described but not used by Øien (2001). For the supply services with a risk picture characterised by occupational accidents, the "individual" factor seems rather important.

The organizational factors are:

- **Individual factor** refers to slips and lapses. Loss of motivation and fatigue are examples on individual factors.
- **Competence** refers to the training and competence that is necessary for the operating personnel to carry out their jobs.
- **Procedures** refer to all written and oral information describing how to perform the operational and maintenance tasks in a correct and safe manner.
- **Planning** refers to the preparation being necessary before execution of tasks.
- **Design** refers to the physical construction of equipment. The design must be such that operation can be performed in a safe manner.
- **Management** refers to management on the vessel, in the shipowner's office and in the Statoil organization.

The connections between the organizational factors and the direct causes are made based on common knowledge. These are hardly indicated in accident reporting. In the resulting model an expert team of people with more experience in the field, should draw these connections. The relative importance between the organizational factors and probabilities should be included for the model to be used in decision support.

4.5 Direct causes in the influence diagram

The direct causes are identical to those used in the reporting system SYNERGI. The list of direct causes is long and these are reported for most of the events. As we can see the list of direct causes consist of a mixture of causes and of factors influencing on the consequences. "High sea" is a direct cause but needs to be combined with some sort of risky operation or lack of observation from the personnel. The cause "high sea" occurs frequently and this can indicate that loading/unloading operations have been done under too rough conditions.

"Did not use proper safety equipment" is rarely a direct cause. The event does not happen because the personnel does not use safety equipment such as helmets and glasses, but the consequence of the event is dependent on the use of the safety equipment. It can be argued that, if safety equipment was used in the reported injuries, these events would not have been reported at all because the personnel would not been injured.

The causes "water on deck" and "slippery surface" are quite similar. This is a general problem for the supply vessels and the vessel deck becomes slippery when the sea flushes over the deck. This problem can be prevented with a good vessel design.

4.6 *Event types in the influence diagram*

The event types used in the influence diagram are also taken from the reporting system SYNERGI. In the event type "hit by", a moving object is hitting a person. While in the event type "hit against" the person itself is moving and is hit by something. "Squeezed between" is used both when describing a person being squeezed between two containers and when a finger is squeezed in the bulk hose coupling. "Overload" has been used a couple of times describing the situation were a foot is twisted. "Hooked" has been used only once for an incident in which the wire was splintered and hooked in a person's hand.

4.7 *Task analysis*

One of the objectives with the risk model is to increase the understanding and applying competence in knowledge to the risky situations and for this purpose; a set of task analysis was drawn. The three categories of operations (loading of deck cargo, loading of casing and loading of bulk) were described in three separate task analysis. It was obvious from the reported events that the injuries could be connected to a limited number of the subtasks and based on this it was decided to keep the task analysis on a course level. Some of the injuries could not be connected directly to the task analyses and were defined as "general deck work". Figure 2 presents the task analyses for the loading/unloading operations of deck cargo. The shaded subtask, fasten/loosen the hook on the cargo, indicates where most of the injuries are happening.

The task analyses describe the main tasks carried out by the captain, the crane operator and the deck hand; the main actors in the loading/unloading process. The captain decides if the sea state and the weather conditions are good enough to start the operation and gives the start signal. The deck hand checks the load manifest and makes the first cargo ready for loading. The operations continue as a close cooperation between these three actors until the cargo is placed on the installation deck or the return cargo on the vessel deck.

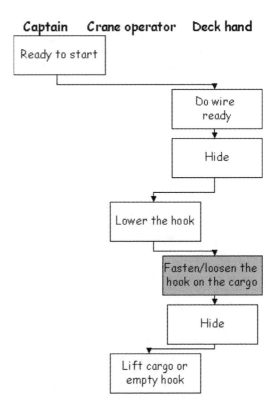

Figure 2.   Task analyses for unloading of deck cargo.

5   APPLICATION OF THE DEVELOPED MODEL

5.1   *Decision support for risk reducing measures*

A number of different risk reducing measures are identified but the problem is to prioritizing between these measures. The influence diagram will support these decisions. More time for rest, reduced noise on the vessel and improved quality of safety equipment are some examples on risk reducing measures. The first step in the analysis is to discuss the effect of these measures on the technical and organisational factors on the left side in the influence diagram. This discussion should be performed in an expert panel and the documentation of the discussion is essential. The second step is to quantify the influence and thereafter the resulting effect of the number of unwanted events can be calculated.

One result from the model is that the number of injuries can be reduced by for example 5% by the introduction of a specific risk reducing measure. Held together with the cost of the measure, this give a cost benefit estimate. The other result, which is as important, is the discussion about why and how this measure affects the safety level.

The influence diagram in Figure 1 indicates that "hit by" and "hit against" are two important types of events ending in injuries. The most prevailing causes of these events are that loading and unloading is performed in sea states that are probably too harsh ("high sea"). If we look at "best practice" and procedures we find some limitations given by conditions when the operation should be stopped. In real life it is difficult to give an exact measure of the sea state and the wind speed. The organizational factors applicable for these events are "competence", "management" and "procedures". The captain is the one to decide if operations are to continue or stop, and to be able to do so the captain needs the right competence and experience. He is under pressure by the crew on the installation because they need the cargo and he is under pressure from the rig owner company because they will have a good reputation when they are to re-negotiate the contract with Statoil. In addition the captain has to consider the safety of his crew. If the conditional limits for safe operations were more conservative, these limitations would have been a support for the captain under these circumstances.

5.2 *Educational purposes*

In order to use the influence diagram directly for educational purposes, examples can be presented in the influence diagram by marking out the events and causes and supporting this by a verbal description. One example is given in Figure 3. The event described in the influence diagram in Figure 3 happened during unloading of a container from the installation deck to the vessel. The crane had been used for lifting of casing on the installation deck previous to this operation and was therefore equipped with a double "forerunner". The high weight of this forerunner made it difficult for the deckhand to loosen the hook and he squeezed his finger between the hook and the container. The result was a broken finger. The direct cause in this event is "wrong work performance"; the forerunner should have been removed before starting the container lifting. The vessel crew should not have accepted to start operation with this double forerunner. The organizational factors are partly "competence" and partly "procedures". The reason for not removing the forerunner could be that the crew did not know the procedures or they could have known the procedure but left the forerunner on to save some time. The influence diagram should be verified and quantified by support from expert panels and by expert judgment.

The task analyses presented in Figure 2 cannot be used for educational purposes on their own. If we combine the task analysis and the influence diagram this will give us some useful information, see Figure 4. This example describes the operation loading/unloading of deck cargo. About half of the events could be connected to one specific sub-task in the task analyses

Figure 3.   Influence diagram illustrating one given event.

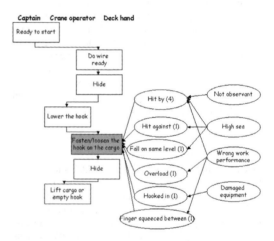

Figure 4.   Combined task analysis and influence diagram for loading/unloading of deck cargo.

while the rest could be classified as "general deck work". As the model in Figure 4 indicates, the division in different task analyses result in a low number of events connected to each of the operations, and as these are described by many different causes the result is more a description of some specific events than a general model.

The stories from concrete events should be used in order to give an increased understanding and knowledge. If the reported injuries could be told as a story, using animation, text or video this could be attached to the task analysis.

If the same task analyses also give a link to "best practice", this will be a good basis for a training package. This idea is presented in Figure 5. The model in the influence diagram will be used to choose some of the most common events. The influence diagram

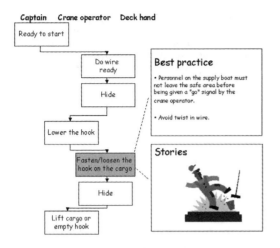

Captain   Crane operator   Deck hand

Ready to start

Do wire ready

Hide

Lower the hook

Fasten/loosen the hook on the cargo

Hide

Lift cargo or empty hook

Best practice

• Personnel on the supply boat must not leave the safe area before being given a "go" signal by the crane operator.

• Avoid twist in wire.

Stories

Figure 5.   Task analysis used for educational purposes.

and the description of events should be updated in a continuous process as more events are reported. In addition, the influence diagram can be used in training about how to report accidents. This will illustrate what "happens" with the reported material.

5.3   Best practice

Crane operations are regarded as one of the most high-risk offshore activities. Statoil operates around 15 cranes on 15 installations and an effort to improve operation and safety has been made through the "best practice" work, (Hepsø, 2001). The development of this common practice was made to create a collective reflection process among the 400 crane operators and banks men in Statoil. A task force spent considerable time in discussing the values of the work practice via search conference seminars and these seminars discussed what is required to further improve safety in crane and lift operations with given safety targets? What are the elements of a safety culture? How do we communicate with those involved in crane and lifting operations? And what are the skills and demands expected from those working in this domain? The resulting written and explicit practice does not describe how crane and lifting should be conducted in detail. It includes tips on important issues, how to maintain the crane, prepare and execute crane operations, how to handle critical situations, how to load cargo with what straps, how to communicate during crane and lifting operations, provide guidelines for special lifts, the transport of personnel and internal transport on the installation. Examples of formulations related to crane and lift operations:

"All lifting operations are high risk. A good practice for each person is to think through the whole lifting operation and evaluate if all necessary efforts for safe operations are taken".

"Everybody involved in the loading/unloading operation must be equipped with UHF communication equipment that have a headset and an integrated microphone".

"A safety zone must be defined before the operations start".

"Personnel on the supply boat must not leave the safe area before being given a "go" signal by the crane operator".

The idea with the "best practice" is that the conversation shall be kept going, meaning reflection and action as a continuous activity. Best practice has been made for the anchor handling activities as well.

These are supported with animated stories describing some of the serious accidents.

6   CONCLUSION

The influence diagram is suitable for the purpose of decision support. The developed influence diagram includes the most important events happening during loading/unloading operations from supply vessels. These represent about half of the events reported on the supply vessels. The rest of the events happened during maintenance, in the machinery room and in the galley. It should be considered to include all these events and events occurring during anchor-handling activities and on the stand-by vessels in the same influence diagram.

For educational purposes it should be considered using the stories from concrete events. The overall influence diagram gives a good overview of the most relevant events. A combination of task analysis, influence diagram, best practice and stories should be developed for training purposes.

The implementation of the risk model in the safety management system should be illustrated showing the process from accident reporting, through risk reducing initiatives to risk statistics. The developed model should be built on the same categories and definitions as the reporting system SYNERGI, to secure the possibility of this implementation. The safety management system should also include the training aspects and feedback of experience from the accident reporting to the crew, management and to procedure development.

REFERENCES

Barlow, R.E., (1998), *Engineering Reliability*, University of California, Berkeley, California.

Flin, R., Slaven, G., (1996), *Managing the Offshore Installation Workforce*, Penn Well Publishing Company ISN 0-87814-396-3.

Harms-Ringdahl, L., (1993), *SAFETY ANALYSIS principles and practice in occupational safety*, Elsevier Science Publishers LTD., London.

Hepsø, V., Botnevik, R., (2001), *Improved Crane Operations and Competence Development in a Community of Practice*, Paper, Statoil, Norway.

Kirwan, B., 1990, Evaluation of Human Work, Chapter 28 in *Human Reliability Assessment, Taylor* & Francis Ltd, ISBN 0-85066-480-2.

Øien, K. (2001), *Risk Control of Offshore Installations*, NTNU Report 200104, Trondheim, Norway.

Perrow, C. (1984), *Normal Accidents, Living with High-Risk Technologies*, Basic Books, New York.

Rettedal, W.K., (1997), *Quantitative Risk Analysis and structural reliability analysis in construction and marine operations of offshore structures*, Stavanger University College, Stavanger, Norway.

Safety and Reliability – Bedford & van Gelder (eds)
© 2003 Swets & Zeitlinger, Lisse, ISBN 90 5809 551 7

Investigation of barriers and safety functions related to accidents

Lars Harms-Ringdahl

Industrial Economics and Management, Royal Institute of Technology, Stockholm, Sweden, and
Institute for Risk Management and Safety Analysis, Stockholm, Sweden

ABSTRACT: A new approach to investigating accidents has been developed and tested. In any such investigation, barriers and safety functions related to an accident or near-accident are identified and classified. The method is described, and a case study of an incident at an electricity distribution company is used to illustrate the type of results obtained. One finding was that a large number of safety functions failed in relation to the incident. In particular, the analysis indicated low efficiency of functions related to higher organisational levels. Another experience was that people and work groups quickly obtained an intuitive understanding of the safety function concept, and its practical application.

1 INTRODUCTION

When an accident has taken place, the course of events leading to its occurrence is usually the main target of investigation. A question is also how it could happen. Especially in systems with large hazards, there are several safety features in place to prevent accidents from occurring. Accordingly, an essential complementary aim of any investigation is analysis of why the safety system failed.

Concepts and terminology related to safety features vary considerably (e.g. Harms-Ringdahl, 1999; Hollnagel, 1999). One common term is "barrier", which often has a concrete technical meaning related to energies. The term can also be used more generally to encompass organisational aspects (e.g. Johnson, 1980). Other commonly used terms are "barrier function" (Svenson, 1991), "defence" (Reason, 1997), and "protection layer" (CCPS, 1993).

There are a number of accident-investigation approaches that focus on barriers and the like. Management Oversight and Risk Tree (MORT) – the classical method (Johnson, 1980) – can be used for accident investigations, while the Accident Evolution and Barrier Function Method (Svenson, 1991) is specifically designed for analysis of accidents and incidents. Also, Safety Barrier Diagrams (Taylor, 1994) offer a way of presenting and analysing barriers to accidents.

One of the starting points for this paper lay in an interest in exploring whether the generic concept of "safety function" could be used to advantage in accident investigations. The author had favourable experiences of applying the concept in safety analysis (Harms-Ringdahl, 2000; 2001).

The general purpose of the study presented here was to test the extent to which "safety function" can be a valuable concept in practical accident investigations. A more specific aim was to obtain a summary of safety features in the workplace under study.

2 APPROACH

2.1 The concept of safety function

Safety function (SF) is a rather common term, but there are no clear definitions available in the specialised literature. Even the "Standard on Functional Safety" (IEC, 1998), where the term is used several times, does not contain a definition. Accordingly, the term may be used in divergent senses in different applications.

A general definition of safety function has been proposed by Harms-Ringdahl (2001). This is worded as follows:

> *A safety function is a technical, organisational or combined function that can reduce the probability and/or consequences of accidents and other unwanted events in a system.*

Human actions are here regarded as part of the organisational component. Safety function is a broad concept and, in specific applications, requires more concrete characterisation. This can be achieved using a set of "parameters". For example, the same reference proposes the following:

(a) Level of abstraction.
(b) Systems level.
(c) Type of safety function.
(d) Type of object.

2.2 Study approach

This case study was performed in four broad steps:

– Selection of accident to investigate.
– Deviation Investigation of the event.
– Identification of safety functions (SFs).
– Analysis of SFs.

Thus, the study involved two specific methods, which are briefly described below.

2.3 Deviation investigation

The Deviation Investigation method (e.g. Harms-Ringdahl, 2001) contains three major steps:

– Identification of deviations prior to the accident.
– Evaluation of the importance and seriousness of the deviations.
– Proposals for potential improvements.

The identification of deviations is supported by means of a checklist of technical, human and organisational functions. The identification results in a list, which is sometimes quite extensive. There is a need to select the most important deviations, for which proposals are then developed to improve the system. This procedure is supported by simple systematics.

2.4 Safety function analysis

Based on the concept of safety function, a methodology called Safety Function Analysis (SFA) has been developed (Harms-Ringdahl, 2000; 2001). The goals of a Safety Function Analysis (SFA) are to achieve:

– A structured description of a system's safety functions.
– An evaluation of their adequacy and weaknesses.
– Proposals for improvements (if required).

In principle, SFA has two general applications. The first is to review a system (e.g. a workplace) and its hazards and safety features. The second is to conduct an accident investigation, designed to draw conclusions about SFs and their weaknesses on the basis of an accident or near-accident event.

The major steps in an SFA are:

– Identification of SFs.
– Classification and structuring of SFs, based on the parameters (a)–(d) in Section 2.1 above.

– Evaluation of the SFs.
– Generation of proposals for improvements (optional).

3 THE CASE STUDY

3.1 The investigated incident and workplace

The case study was based on an incident at an electrical power distribution station in direct connection with a hydropower station. Part of the electricity net had been disconnected when servicing was performed. When the service was finished, one of the technicians (according to task schedule) reconnected the station to the electric power line.

By mistake, he went to a wrong coupling booth, adjacent to the correct one. When he made the connection, a high voltage line was connected to earth. Nobody was injured, but the error caused a disruption to electricity supply across a large region.

A number of companies were involved in service operations in the workplace. Each of them had a specific role; one was concerned with the hydropower station, one with the power lines, and two with service tasks. This organisational structure was the product of having split up one original company into several smaller ones, and also an outsourcing process.

This particular incident was chosen from among a few others because it appeared, at least at first sight, to provide a simple test of the method. A prior investigation had been made by the company in question. It concluded that a time delay had occurred that had placed stress on technicians. They deviated from normal work procedure, which gave rise to the mistake. The proposed counter-measure was that technical staff should be told always to follow formal rules, even when time is scarce.

3.2 Deviation investigation of the incident

Deviations were identified in interviews with three persons, and through study of relevant documentation. The result was a list of around 40 deviations.

For the case study, an ad-hoc group was created, representing both the companies and the trade-union involved. The group evaluated and discussed the results. First, the list was scrutinised, and somewhat modified. Second, the deviations were evaluated on a four-point scale in order to select the ones that needed better prevention or management. There was complete agreement on all deviation estimates except one.

The final step was to propose ideas for essential improvements. Table 1 summarises the number of deviations and proposed measures. The investigation found 42 deviations prior to the incident, with the majority related to management issues. For example, one of the ideas for improvement was formulated as a "need to analyse the whole organisation". A number of specific deviations had to be considered in this analysis. These are counted as separate proposals in Table 1.

3.3 Identified safety functions

In this example, the identification of safety functions was based entirely on the Deviation Investigation protocol. The text of the protocol was studied, and any issue interpreted as involving a safety function noted down. Both deviations and ideas for improvements were studied.

The SFs were classified into two groups: technical and organisational. The latter included three human actions, which in this case meant that a person had been supposed to perform a certain action. The SFs were then structured and grouped as in the list below. The first two were technical, and the rest organisational. The structuring was also based on different systems level, from workplace (3) to societal functions (7):

1. General technical SFs.
2. Local technical SFs.
3. Local organisational SFs.
4. Company management.
5. Co-ordination between companies.
6. Corporate safety management.
7. Societal SFs.

In this case, the SFs were evaluated according to whether or not they had performed their intended

Table 1. Summary of results from the Deviation Investigation.

Type of results	Number
Identified deviations	42
Deviations evaluated as essential	27
Proposals for safety improvements	29

function during the incident. The SFs related to a suggested improvement are shown in a separate column. Table 2 provides a summary of the SFs and their evaluation.

In all, 40 SFs were recorded and structured under 7 main headings. As an illustration Group 6, "Corporate safety management", comprised 5 safety functions, none of which were in satisfactory operation. The SFs were:

- Corporate policy for safety (out of date, inadequate).
- Information and distribution of policy (hard to find).
- General agreement between companies expected to regulate their co-operation on safety (out of date, unsatisfactory).
- In commercial agreements, clarify how competing goals shall be handled (proposal).
- In commercial agreements, clarify how policy shall be applied (proposal).

Another way of illustrating the results is shown in Figure 1. It is intended to provide a clearer hierarchical view, and it provides other examples of functions at a more detailed level.

4 DISCUSSION

4.1 The incident and the company's safety work

The incident was first thought to be "simple", and the event sequence was envisaged as uncomplicated. However, the analysis found many deviations prior to the incident. A natural question is whether all the deviations were relevant or whether the identifying step exaggerated problems and the situation.

This question was answered during the evaluation round, when most of the deviations were regarded as essential (Table 1) by the ad-hoc group. The evaluations were made virtually in consensus. The group also made a substantial number of suggestions for improvements.

Table 2. Summary of safety functions at incident at electric power station.

Safety function	Σ	Performed intended function				OK (%)
		Yes	Partly	No	SI*	
1 General technical SFs	1	0	0	0	1	0
2 Local technical SFs	6	2	1	1	2	33
3 Local organisational SF	11	4	1	5	1	37
4 Company management	6	0	0	5	1	0
5 Co-ordination between companies	10	1	1	1	7	10
6 Corporate safety management	5	0	0	2	3	0
7 Societal SFs	1	1	0	0	0	100
Total	40	8	3	14	15	20
Share (%)	100	20	7	35	38	–

*SI = A safety function noted as suggested improvement.

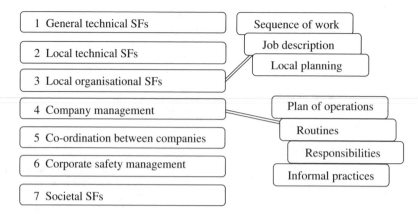

1 General technical SFs	Sequence of work
2 Local technical SFs	Job description
3 Local organisational SFs	Local planning
4 Company management	Plan of operations
5 Co-ordination between companies	Routines
6 Corporate safety management	Responsibilities
7 Societal SFs	Informal practices

Figure 1.   Summary of safety functions related to the incident, with examples of functions at a more detailed level.

This can be interpreted as a large discrepancy between how the people involved would like planning and safety procedures to operate, and what was observed in the investigation. The final list of proposed improvements covered 29 issues.

A further analysis showed that as many as 40 SFs were recorded as directly related to the incident. A larger number of SFs would have been obtained if a complete summary had been the aim, but this was not the case here.

Of the recorded SFs, a majority did not actually operate satisfactory, and only 20% worked as expected (the OK column in Table 2). Thus, the analysis indicated low efficiency of functions. It should be remembered, however, that the aim was to identify problems, not all SFs.

Nevertheless, it is essential also to look at absolute numbers. As many as 32 SFs did not work or needed improvement. Most were at higher organisational levels (4, 5 and 6 in Table 2), where there were 20 unsatisfactory SFs out of 21.

According to the findings, deficiencies were greatest at company and corporate management levels, which is somewhat surprising in light of the long tradition of safety work in the electricity distribution field. At the time of the incident, some of the companies had quality systems in place, and there were preparations to implement ISO 14000, which here was intended to also cover safety issues. Obviously, these systems were not adequately working in the safety arena.

One of the explanations for the deficiencies lay in the series of organisational changes that had taken place. Electric power distribution has well established routines, which continued to work rather well during the period of organisational change. Safety work appeared to be preserved by co-operation between people who used to belong to one and the same company.

Such informal co-operation (a kind of SF) emerged as essential during discussions with the work group. There were also some suggestions to strengthen co-operation of this kind, e.g. by considering it carefully in the education of technicians.

4.2   About the methodology

The investigation in this case study had two distinct parts. The first – Deviation Investigation – is concerned with uncovering facts about an incident and proposing improvements to a system. The second part – Safety Function Analysis – structures the results obtained from Deviation Investigation for evaluation and presentation.

The result of a Deviation Investigation is a list, usually a fairly long one, that is not necessarily arranged in cause-consequence sequence. The main reason for this is that there is not always strict coupling between a specific deviation and the final event. It is most common for any one deviation to affect just the *probability* of other events and deviations occurring. An advantage of not having to show all direct connections is that it makes the analysis rather quick. In this case it took slightly more than two days to perform the Deviation Investigation.

Here, the intention was to use SFA in a limited way so as to obtain a structured and evaluated summary of existing SFs. One experience was that the people involved quickly gained an intuitive understanding of the safety function concept and its practical application.

From another perspective, SFA can be regarded as a way of presenting the results of an investigation performed using another method. The results appear in the format of a fairly simple table and illustration (see Table 2 and Figure 1), which are easily communicable. In particular, the role of management and its related problems become more obvious through this type of analysis.

In this study, SFA was used in a limited way by just re-arranging existing material, but more steps could be included in an SFA analysis. One additional part might be to obtain a fuller picture of existing SFs. In addition to those identified, there were several other obvious SFs that had not been clearly entered into the protocol. One further step might be to suggest improvement that would give better coverage or efficiency to a system of SFs.

4.3 General comments

What is special (and new) about SFA compared with other barrier-oriented methods is that it is not based on a predetermined set of SFs, such as in MORT (Johnson, 1980). Instead, the functions are based on what is observed in the system.

Another characteristic of SFA is that there is no demand for the establishment of causal relationships between the different SFs and the chain of events. By contrast, this is the case, for example, in the Accident Evolution and Barrier Function Method (Svenson, 1991) and in Safety Barrier Diagrams (Taylor, 1994). Identification and structuring is simple in SFA.

An additional feature of SFA is that it provides an opportunity to work at various levels of abstraction and different system levels. It is also of value to be able to include problems and potential improvements on the same "dimension".

These characteristics give considerable flexibility, but the analysis is supported by a simple principle, which enables it to be performed in a consistent manner.

The evaluation of the SFs, based on whether or not they were satisfactorily operating during the incident or not, was simple but sufficient in this case.

Looking at the findings of the investigation, it was surprising to see that the role of management showed so many deficiencies. This indicates that the methodology might be relevant to specialised investigations where management systems have high priority.

5 CONCLUSION

Investigation of a simple incident identified a large number of deficiencies, especially concerning higher management.

The approach involved the adoption of a new method, called Safety Function Analysis (SFA). The safety function concept proved to work, in the sense that it was easily understood by the people involved.

The concept also provided a foundation on which to present the results of investigation in a consistent manner.

The experience of this case was generally positive. It appears that safety function might be an interesting concept to work with further in accident investigations. The analysis methodology could be developed to include further steps so as to enable more thorough identification and evaluation of safety functions.

ACKNOWLEDGEMENT

The project was supported by the National Centre for Learning from Accidents, which is a part of the Swedish Rescue Services Agency, the Swedish Council for Working Life and Social Research, and Ångpanneföreningen's Foundation for Research and Development. The interest and co-operation of the people in the companies involved in the case study are greatly appreciated.

REFERENCES

CCPS (Centre for Chemical Process Safety), 1993. *Guidelines for Safe Automation of Chemical Industries.* New York: American Institute of Chemical Engineers.

Harms-Ringdahl, L. 1999. On the modelling and characterisation of safety functions. In Schueller, G.I. and Kafka, P. (eds), *Safety and Reliability ESREL'99.* Rotterdam: Balkema.

Harms-Ringdahl, L. 2000. Assessment of safety functions at an industrial workplace – a case study. In Cottam, M.P., Harvey, D.W., Pape, R.P. and Tait, J. (eds): *Foresight and Precaution, ESREL2000.* Edinburgh: Balkema.

Harms-Ringdahl, L. 2001. *Safety analysis – Principles and practice in occupational safety* (2nd edition). London: Taylor & Francis.

Hollnagel, E. 1999. Accident Analysis and Barrier Functions. Kjeller (Norway): Institute for Energy Technology.

IEC (International Electrotechnical Commission), 1998. Functional safety: safety related systems (IEC 1508) Geneva: IEC.

Johnson, W.G. 1980. *MORT Safety Assurance Systems.* Chicago: National Safety Council.

Reason, J. 1997. *Managing the risks of organizational accidents.* Aldershot: Ashgate Publishing.

Svenson, O. 1991. The accident evolution and barrier function (AEB) model applied to incident analysis in the processing industries. *Risk Analysis* 11(3): 499–507.

Taylor, J.R. 1994. *Risk analysis for process plan, pipelines and transport.* London: E & FN Spon.

Safety and Reliability – Bedford & van Gelder (eds)
© 2003 Swets & Zeitlinger, Lisse, ISBN 90 5809 551 7

Psychological rewards and bad risks: the roles of prosocial, antisocial, moral and immoral attitudes, values and beliefs as determinants of individual risk-taking behaviour: a framework for further research

Joan Harvey & George Erdos
Newcastle University, United Kingdom

ABSTRACT: This paper examines some of the psychological factors which may lead an individual to take decisions which are neither optimal nor beneficial to them in terms of risk. The particular psychological factors under consideration are: altruism vs. selfishness, heroism, saving/giving/losing face, anger and retaliation. It uses findings from the authors' own work and research in other domains to present a framework which attempts to account for behaviours which may be known to be relatively risky and with potentially poor outcomes for the individual but which are nevertheless selected over more rational behaviours. The notion of prosocial behaviours and morality have been incorporated into models explaining many areas of behaviour, including that at the workplace, from absenteeism to job satisfaction, but less so into models of risk. This paper evaluates the evidence for these factors influencing risk and decision making and proposes different types of drivers of behaviour as part of a framework for incorporating them into future risk research.

1 INTRODUCTION

This paper examines some of the psychological factors which may lead an individual to take decisions which are neither optimal nor beneficial to them in terms of risk but which are nevertheless selected over more rational behaviours. Risk in this context means personal risk to an individual, i.e. the likelihood of suffering loss, injury or death. The particular psychological factors under consideration here are: altruism and selfishness, heroism, saving/giving/losing face, annoyance/anger and retaliation.

There are many anecdotal stories of people making apparently "bad" decisions which involve them in additional risk for no apparent personal gain and even in a clear personal loss. These stories cover a diverse range of situations and issues. For example, there is evidence in the consumer field to show the extent to which people will pursue a costly retaliatory or revengeful strategy rather than accept the current situation or lose face. There is also research attempting to explain road rage in terms of frustration, expression of anger, traffic density, regulation of behaviour according to contingencies and pressures. A highly moral behaviour, heroism, has been explained as a form of sensation-seeking, altruism, citizenship and

bravery and as a desirable adaptive response in Darwinian terms.

Risk taking as an activity is inherently dynamic rather than static. Not only can it change from situation to situation but certainly changes through one's life span (insurance premia are a good indication of this); it is also linked to personality factors.

People operate in a "phenomenological universe"– the world according to I – and sometimes they are unaware of the risk; heroism may be viewed this way and is often defined retrospectively.

The paper now explores some of the psychological factors which can clearly influence risk-taking behaviour, followed by consideration of a framework, measurement issues and finally a model is proposed.

2 HEROISM

Heroism is apparently non-adaptive in Darwinian terms, so why does it exist? Historically, risk-taking and heroism have been considered to be predominantly male tendencies. Research shows that women prefer risk-prone brave males to risk-averse non-brave males, and that men are aware of this preference (Kelly & Dunbar, 2000). The Kelly and Dunbar study also found that altruism has value only over the longer term, and is not perceived as important in the evolution of heroism as bravery and a demonstrated willingness to take risks.

Heroism may not only have a direct value to the individual in moral and self-esteem terms: admiration for heroes can form the basis for a civic identity, fostering active citizenship (Kelly, 1997). In other words, heroism endows the society as well as the individuals involved with positive values and culture. One form of heroism is "courageous resistance", defined as a form of voluntary selfless behaviour in which there is significantly high risk or cost to the individual and possibly their family and associates. Not all altruistic people will exhibit this, and those who do are not always courageous (Shepela, 1999).

Are heroes different to the rest of us, or is it just the circumstances at the time? Sensation-seeking behaviour is associated with heroism: it played a significant role in both performance during war and in relation to long term adjustment subsequently – those high on sensation-seeking behaviour were able to adjust in later years (Neria et al., 2000). It is also likely that heroism is also associated with higher moral values and ethics.

The most obvious examples of heroism in recent years have been the actions of emergency services in various disasters such as "9/11" in New York.

3   ALTRUISM/SELFISHNESS

Altruism has been described as cooperation and helping behaviour and may indeed be associated on occasions with heroism, as described earlier. Margolis (1982) suggests that within each of us, there are two "selves"– one selfish [S] and the other group-orientated [G], and the individual follows a Darwinian rule for allocating the resources between the selves. He describes how the balance between G and S changes during what he terms "social actions" such as for example a disaster. The fact that someone may put their own life seriously into jeopardy in order to save a stranger is explained by an increase in G such that other group members might perceive his behaviour and the example it sets as being beneficial to the development of the group and society. Thus, S is almost overpowered by G at the instant that the life-saver makes the decision to act so altruistically.

Low levels of altruism have been associated [along with high levels of sensation-seeking, irresponsibility, anxiety and aggression, and more likely to be men than women] with being identified as high-risk groups in traffic, and such high risk groups were least likely to appreciate traffic safety campaigns (Ulleberg, 2001).

Examples of altruism can be found in the animal world, where [to use the Margolis terminology] some behaviours which are G- orientated, such as signalling threats to the whole group, have no cost to the individual and considerable benefit to the group. Other G-behaviours enable animals to hunt in packs. However in humans, some altruistic behaviours are clearly not in the best interests of the individual and are more difficult to explain in a rational sense, unless one places a value on the altruistic behaviour itself as, for example, good citizenship [see for example, the concept of communitarianism by (Etzioni, 1996)].

Although not strictly altruism or selfishness, self-sacrificial decisions for the sake of fairness have been investigated and clearly demonstrate that the fairness motive and moral concerns can influence decisions that have economic impact; these decisions may involve both punishment and reward of someone who has demonstrated a prior intent to be either fair or unfair to another person (Turillo et al., 2002).

Thus, altruism can be described as being potentially able to induce very risky behaviour such as heroism in extreme circumstances but is normally associated with low levels of risky behaviour and high values placed on citizenship and morality. However where it links to perceptions of justice and fairness it can alter decisions so that the chosen course of action may be more or less risky than the most rational one.

4   RETALIATION AND ANGER

Retaliatory behaviour has been studied in recent years from a consumer perspective; it can be defined as aggressive behaviour, but the type which is done with the intention to get even, thus making it an equity issue (Huefner & Hunt, 2000). It has been observed that retaliation is a natural aspect of behaviour, performed when people lack better means of restoring equity (Demore et al., 1988). It is possible to distinguish aggressive behaviour from assertive behaviour, the former probably being unidimensional. Richins & Verhage (1987) found the two to be statistically independent but did find a correlation between pleasure in seeking redress and aggression.

The desire to take retaliative action can be inherently risky– in the consumer sphere, retaliation can take the form of vandalism and trashing, stealing, loss-creating or spoiling behaviour, negative word of mouth, and so on; in the traffic sphere, we hear of road rage, air rage and other behaviours which may be described as relating to anger management and aggression. Huefner & Hunt (2000) expanded the Exit-Voice-Loyalty model (Hirschman, 1970 & 1986) to include retaliation, which for dissatisfied consumers effectively reduces to Exit-Voice-Retaliation since loyalty is an unlikely outcome in such cases. Whilst Exit and Voice accounted for many of the responses, the retaliations which were implemented when Voice proved non-viable or unsatisfactory invariably involved illegal behaviours. As with the high-risk traffic groups, the group most likely to use retaliation as a response were young men.

Huefner & Hunt were able to relate a number of emotional and personality states to the different retaliatory responses and go on to state: "The creativity and anger that went into many of the retaliations were amazing". They suggest that retaliatory behaviours brought emotional release to the retaliators but have no potential to improve their well-being. This latter suggestion is questionable: retaliation can be described as what Freud would have termed "cathartic" behaviour, which has little utility but makes you feel good!

In order to test the likelihood that people would respond in a retaliatory manner in terms of consumer behaviour, Harvey & Erdos (2001) asked members of the public how likely they would be to respond in a variety of ways after receiving poor goods or service; they found that whilst 87% would be prepared to retaliate by damaging the organizations' reputation by telling other people, only 7% said they were prepared to take "drastic action" such as wilful damage. This might be contrasted with the studies showing the incidence of risky behaviour in a traffic situation, where 20% of drivers reported red light running (Porter & Berry, 2001) and 30–40% (according to age) of self-selected respondents admitted to getting into confrontations with other drivers, with 30–45% admitting retaliatory driving behaviour (roadragers.com, 2001).

The association of retaliatory behaviours and related risky behaviours with personality and emotions is clear in the Huefner & Hunt's research into consumer behaviour and the studies of driver behaviour. The difficulty here is that personality is relatively stable and enduring, and probably hard-wired, whilst emotions are transient and variable.

5   LOSING, GIVING AND SAVING FACE

The notions of losing face, saving face and giving face have been an important part of some cultures for hundreds of years, and indeed is part of the rituals of interpersonal behaviours and a necessary part of filial piety in far eastern cultures (Child, 1994). In Europe and the USA, face has been investigated in several areas such as family therapy, as well as how groups and teams operate and the role of the leader. In negotiations and interpersonal or intergroup conflict, saving face can become the primary motivator of behaviour, to the point where it is a major determinant of the conflict resolution (Worchel & Lundgren, 1991) and may also be an important part in explaining the "groupthink" in policy fiascos (Janis, 1972). What is proposed in this paper is that saving face can be a sufficiently powerful motivator that it can drive

individuals and groups to behaviours which serve an ego-protective function but which could even be considered bizarre by others. To the best of our knowledge, the extent to which the threat of losing face motivates different or more extreme risk-taking behaviour has not yet been measured.

Giving face, often apparent in Chinese and Japanese cultures, serves the purposes there of filial piety, harmony and respect, and is a low-risk behaviour as it would seem to involve no negative outcomes; moreover, when necessary it has the positive value of allowing others to save face and thus avoid humiliation (Chuang, 2000).

6   MEASUREMENT ISSUES

There already exist a number of psychometric measures covering the variables considered above or others related to them which may be of value in predicting risk-taking behaviours. These include: sensation-seeking scale (Zuckerman, 1979), uncertainty avoidance (Hofstede, 2001), emotional intelligence (e.g. Goleman, 1998; Higgs & Dulewicz, 1999), measures of risk attitudes and safety climate or culture (e.g. Cheyne et al, 2002; Harvey et al, 2002).

Although there is evidence that anger, aggression and anxiety are associated with high-risk behaviours, there is little evidence to show a clear-cut risk-orientation through the "big five" personality factors. Measuring aggressiveness directly is arguably not socially acceptable as part of risk research studies, and requires more specialised measures, such as the Psychopathic Personality Inventory (Lilienfield & Andrews, 1996).

7   ASSEMBLING THE VARIABLES AND ESTIMATING THE RISKS

The different variables considered above have little in common with each other save the fact that they are all potential psychological influences upon the tendency to take risks and the nature of the risky decisions made, and they may all act upon risk taking behaviour in a way that is neither rational nor beneficial to the individual, though the "group" or "society" (whoever they may be) may benefit.

Much of the research into the above areas, whilst considering the behaviours themselves, has not addressed the extent to which they may influence perception of risk and risk-taking behaviour. The first step in doing this is to attempt to establish the nature of the outcomes or consequences, both positive and negative, of the various behaviours; the second step would be to establish what prompts or triggers the behaviour. This approach is summarised in Table 1,

Table 1. Personal propensity to act and trigger points in relation to risk.

Behaviour	Trigger	Nature of behaviour	Positive outcomes	Risk or negative outcomes
Heroism	Catastrophe or accident	Positive/group/self	Self-esteem, feeling of worth	Life threat
Altruism	Perceived opportunity to help	Positive/group	Advantage to group self esteem	Disadvantage to self lost progress
Desire to protect	Threat to group	Positive/group	Feeling of worth	Life threat
Authoritarianism	Desire for power	Negative/self	Power and influence	Hatred/attack
"Cut corners"	Save time	Negative/self	Stimulant saved time	Loss of job personal injury
Selfishness	Greed, inequity	Negative/self	Achievement	Loss of social support
Selfishness of others	Perceived inequity	Negative/other	Perceived justice	Loss of position
Retaliation	Unfairness, anger, threat to personal space, alcohol	Negative/other	Temporary emotional release	Lowered well-being, punishment,
Injustice to others	Perceived inequity	Negative/other	Restorative justice, personal esteem	Isolation from group
Saving face	Conflict, threat	Positive/self	Self esteem, status	Humiliation, stress
Giving face	Culture, desire for harmony	Positive/group	Closer relations	None

A three-dimensional representation of risk-taking

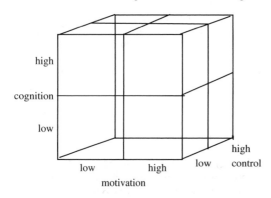

where the variables considered in this paper plus some other risk-related organizational behaviours [cutting corners and authoritarianism] are displayed. In the table, we are proposing that various behaviours are "triggered" by either an event and/or a psychological state (most often affective rather than cognitive). Once the behaviour is triggered, individuals are extremely unlikely to think through the consequences or the probabilities associated with them, otherwise they would not, on a pure utility basis, behave in such a way.

This "trigger" approach, paralleled in motivation theory by unsatisfied needs as triggers, proposes that a combination of an affective state and a trigger point

prompts action which is sub-optimal and not beneficial to the individual, regardless of the consequences or probabilities involved. However, underlying all these variables would seem to be three main elements which form the context in which the affect and trigger combine to generate action. Firstly, there is cognition– the extent to which people are aware or unaware of the risk. Secondly there is control – the extent to which the risk-taking behaviour is perceived to be under the individual's control or impulsive and environment-contingent. Thirdly, there is the motivation to act and its value to the individual – the extent to which the individual is prepared to put emotional and personal commitment into the act. Simply classifying each of these three elements into high or low would provide eight different scenarios, which may be organized as a cube.

The psychological variables considered earlier fit into this model easily. For example, heroism which is life threatening would be low on cognition, high on control and high on motivation, but retrospective justification would involve acquiring the knowledge of the risks taken and thus moving the cognitive element from low to high as part of the post hoc attribution process. Another example might be that behaviour in a decision-making meeting might be low on cognition, that one was relatively unaware of the risks, high on motivation because of loss of face, low on control as it was a relatively impulsive response to the actions of others. A final example in terms of consumer purchasing behaviour might be low on cognition, high on motivation and low on control – such would have been the case with consumers purchasing beef before the BSE crisis loomed.

We propose that this three-way classificatory system helps in understanding why people behave as they do in some circumstances when they might have been predicted to behave otherwise and why they retrospectively might describe the situation and justify their behaviour very differently.

8 CONCLUSIONS AND IMPLICATIONS

One major conclusion to be gained from this is that several of these behaviours occur without much warning, are relatively unpredictable, and contain some emotional or affective responses to a particular situation; this can be seen in heroism, some altruism, retaliation and rage and some selfish behaviours. They may also be considered to be opportunistic in some ways. However, these features inherently make them difficult to predict, manage or influence. There is a growing literature on emotionality in the workplace and elsewhere; emotional intelligence and similar concepts that attempt to quantify some of the elements that may be relevant to risk models. Similarly there are some psychometric measures, such as uncertainty avoidance and sensation-seeking which can clearly shed light directly on risk-taking behaviour, in particular risk-seeking and risk-avoiding behaviours. Slovic (2002) demonstrated just how heavily loaded risk perception is with affect and emotion.

The three-dimensional model proposed here incorporates emotion inherent in much risk-taking into the broader picture which includes the cognitive and control elements.

REFERENCES

Cheyne, A., Oliver, A., Tomas, J.M. & Cox, S. 2002. The architecture of employee attitudes to safety in the manufacturing sector. *Personnel Review.* Vol 31 Nos 5–6: 649–670.

Child, J. 1994. *Management in China during the Age of Reform.* Cambridge: Cambridge University Press.

Chuang, T. 2000. Personal communication.

Demore, S.W., Fisher, J.D. & Baron, R.M. 1988. The equity-control model as a predictor of vandalism among college students. *Journal of Applied Social Psychology.* Vol 18: 80–91.

Etzioni, A. 1996. Positive aspects of community and the dangers of fragmentation. *Development and Change.* Vol 27 No 2: 301–314.

Goleman, D. 1998. *Working with Emotional Intelligence.* London: Bloomsbury.

Harvey, J. & Erdos, G. 2001. *Consumers' responses to poor goods or service.* Unpublished paper.

Harvey, J., Erdos, G., Bolam, H., Cox, M.A.A., Kennedy, J.P. & Gregory, D. 2002. An analysis of safety culture in a highly regulated environment. *Work and Stress.* Vol 16 No 1: 18–36.

Higgs, M. & Dulewicz, V. 1999. *Making Sense of Emotional Intelligence.* Windsor: NFER-NELSON.

Hirschman, A.O. 1970. *Exit, Voice and Loyalty: Responses to Decline in Firms, Organizations and States.* Cambridge, MA: Harvard University Press.

Hirschman, A.O. 1986. *Rival Views of Market Society and Other Recent Essays.* New York: Viking.

Hofstede, G. 2001. *Culture's Consequences.* California: Sage

Huefner, J.C. & Hunt, H.K. 2000. Consumer retaliation as a response to dissatisfaction. *Journal of Consumer Satisfaction, Dissatisfaction and Complaining Behaviour.* Vol 13: 7–29.

Janis, I.L. 1972. Victims of Groupthink. Boston: Houghton Mifflin: 197–198.

Kelly, C. 1997. Rousseau's case for and against heroes. *Polity.* Vol 30 (2): 347–366.

Kelly, S. & Dunbar, R.I.M. 2000. Who dares, wins – Heroism versus altruism in women's mate choice. *Human Nature-An Interdisciplinary Biosocial Perspective.* 12 (2): 89–105.

Lilienfeld, S.O. & Andrews, B.P. 1996 Development and preliminary validation of a self-report measure of psychopathic personality traits in noncriminal populations. Journal of Personality Assessment. Vol 66 No 3: 488–524.

Margolis, H. 1892. *Selfishness, Altruism & Rationality: a theory of social choice.* Cambridge: Cambridge University Press.,

Neria, Y., Solomon, Z., Ginzburg, K. & Dekel, R. 2000. Sensation seeking, wartime performance, and long-term adjustment among Israeli war veterans. *Personality & Individual Differences.* 29 (5): 921–932.

Porter, B.E. & Berry, T.D. 2001. A nationwide survey of self-reported red light running: measuring prevalence, predictors and perceived consequences. *Accident Analysis and Prevention.* Vol 33 No 6: 735–741.

Richins, M.L. & Verhage, B.J. 1983. Assertiveness and aggression in marketplace exchanges. *Journal of Cross-Cultural Psychology.* Vol 18: 93–105.

Roadragers.com. 2001. *Analyze Your Driving Style: Results* www.roadragers.com/test/stats.htm

Shepela, S.T. 1999. Courageous resistance – A special case of altruism. *Theory & Psychology.* Vol 9 (6): 787–805.

Slovic, P. (2002) *Risk as analysis and risk as feelings: some thoughts about affect, reason, risk and rationality.* Paper presented to the SRA Annual Conference, New Orleans.

Turillo, C.J., Folger, R., Lavelle, J.J., Umphress, E.E. & Gee, J.O. 2002. Is virtue it's own reward? Self-sacrificial decisions for the sake of fairness. *Organizational Behavior and Human Decision Processes.* Vol 89, No 1: 839–865.

Ulleberg, P. 2001. Personality subtypes of young drivers: the relationship to risk-taking preferences, accident involvement and response to a traffic safety campaign. *Traffic Psychology and Behaviour.* Vol 4 No 4: 279–297.

Worchel, S. & Lundgren, S. 1991. The nature of conflict and conflict resolution. In K.G. Duffy & J.W. Grosch (eds) *Community mediation: A handbook for practitioner and researchers.* New York: Guilford Press.

Zuckerman, A. 1979. *Sensation-seeking: beyond the optimum level of arousal.* New York: Wiley.

Safety and Reliability – Bedford & van Gelder (eds.)
© 2003 Swets & Zeitlinger, Lisse, ISBN 90 5809 551 7

The relationship between attitudes and behaviour in terms of smoking and responses to food scares

Joan Harvey, George Erdos & Amy Strong
University of Newcastle, United Kingdom

ABSTRACT: Attitudes to risk associated with health scares have been investigated over many years and there remains some debate concerning the extent to which attitudes in one domain generalise to others, or whether specific attitudes are only predictive of specific behaviours. This study investigates attitudes to and perceptions of smoking and smoking behaviour and whether or not respondents have changed their purchasing and eating habits as a result of the BSE and GM issues. The method used was a questionnaire of Likert-style items plus consumption and demographic questions administered to shoppers. The findings confirm that smokers are less likely to be concerned with other health risks than non-smokers and that attitudes to smoking are also predictive of people changing their habits in relation to other health risks. The findings are discussed in light of the current theories.

1 INTRODUCTION

Smoking and health-protective attitudes and behaviours have been topical areas for investigation for many years, especially following a number of well-documented health scares (Uzogara, 2000; Harvey et al., 2001). It is likely that attitudes and behaviours in one domain generalise to others, although this is difficult to explain in terms of the attitude-behaviour relationship, which tends to be strongest when the attitudes and behaviours are more specific.

This research investigates the relationship between smoking attitudes and behaviour and health-related attitudes and in particular those in relation to BSE and GM foods. This is an interesting area because the risks associated with smoking have been well-defined over many years, and smoking as a behaviour can be described as being largely under voluntary control with risk increasing with the amount smoked; BSE on the other hand is a more recent problem, is one where damage may already have been done by behaviour not thought to be risky at the time; GM foods present, from the health point of view, a completely unknown area of risk.

1.1 Models of health perceptions and attitudes

Coping with a perceived health threat such as smoking or an unsafe food supply involves behaviours, cognitions and perceptions to minimise the impact of the threat. The Health Belief Model (HBM) offers a single-dimension explanation of why, even when symptoms are absent, people engage in behaviour to protect their health: for example, people experiencing a threat are likely to use a problem-solving approach, such as seeking more information or changing habits, whereas those who perceive the threat to be less are more likely to endorse faith, wishful thinking, fatalism, avoidance and trust ((Becker & Maiman 1975; Rosenstock 1990; Schafer et al., 1993). The single dimension explaining the behaviour is the perceived threat of contracting a health condition, but the HBM also incorporates perceived benefits; however, in terms of the BSE scare, these may well be more general health gains associated with eating less red meat, and for smoking there may be perceived social benefits, both consequences which the HBM handles less well than the theory of reasoned action (Ajzen, 1991).

The HBM has inadequacies such as not incorporating past behaviour as an integral element and not considering how attitudes from other health domains might influence particular attitudes or behaviours. Although the theory of reasoned action perhaps offers the best explanatory power so far, it still accounts for a relatively small amount of the variance within attitudes and behaviour, even taking into account the important predictive role of past behaviour.

1.2 Attitudes

Attitudes and behaviour have been causally linked (e.g. Pratkanis & Santos, 1993; Ajzen, 1991; Connor, 1994; Slovic 2000; Flay et al., 1994). It may be that attitudes also have a indirect effect, for example affecting evaluation of food which in turn influences purchase behaviour (Connor, 1994). There is increasing evidence, particularly based on the theory of reasoned action, showing the relationship between intention, attitudes and habit with consumption of foodstuffs (Worsley & Skrzypiec, 1998; Saba & Di Natale, 1999), and the importance of safety-related meat attitudes and their impact on consumption in relation to the BSE crisis has been emphasised recently (Verbeke & Viaene, 1999). Although Foxall & Bhate (1993) argue that a person's concern about what others think of them renders attitudes of only limited value in the prediction of consumer behaviour, it is now generally agreed that attitudes measured more specifically rather than generally correlate well with specific behaviours and that attitudes to risk in one domain may generalise to another (Davidson & Jaccard, 1979).

Behaviour in response to a perceived threat to health from information provided may reflect cognitive dissonance, insofar as the denial of the threat, or it's displacement into wishful thinking are the result of the imbalance between knowledge and conflicting behaviour (Festinger, 1957). The effects of cognitive dissonance are very marked in studies of personal risk evaluation: for example, in relation to smoking, McMasters and Lee (1991) found that smokers and non-smokers had similar amounts of factual knowledge about health risks due to smoking, but smokers, whilst estimating that smokers had higher risks of ill health, considered their personal risk as lower than for the 'average' smoker.

Attribution theory might also have some explanatory power; for example as time passes with no direct experience of anyone with Creuzfeldt-Jacob disease, the perceived threat of BSE will reduce and denial of the threat by attributing the cause to non-consumed inferior products or the Government's scare mongering will increase (Hewstone, 1989); in the case of smoking, a similar pattern of denial of the threat may be heightened by citing examples of life-time heavy smokers who have no lung or chest illnesses at all.

In relation to attitudes and beliefs specifically about health, food risks and smoking, there are clear demographic differences. Women show more positive health-related behaviours and attitudes than do men (Charlton, 1984; Kann et al., 1993; Bord & O'Connor 1997; Miller et al., 1996; Figueroa et al., 1999; Harvey et al., 2001). Socio-economic status and age have also been found to be correlated with smoking and other risk behaviours (Graham, 1994; Williams and Rucker 1996).

Smoking and risks due to BSE and GM foods are likely to occur in different domains of the two-factor space of known and dread risks (Fischoff et al., 2000), smoking being perceived as more voluntary and

known than BSE and GM, the latter being completely unknown at present. However in terms of dread risk, BSE would probably score as certain to be fatal, whereas smoking would occur in the middle and GM foods as not certain to be fatal.

The GM debate in Europe is interesting because any associated health threats are unknown at present, whereas for BSE there are many clear case examples in cattle and humans. Yet in the United Kingdom there was a public change in attitudes and consumption of various food products associated with this debate, probably because of the previous BSE crisis (Harvey et al., 2001). Existing models may well be inadequate to deal with something where the risk is entirely perceived rather than real. The intuitive rationalising of perceptions, attitudes and behaviour in relation to GM foods may be explained by attribution and cognitive dissonance theories, although the conventional explanation that perceived threats reduce as the time without health consequences increases may not be correct as the GM threat seems to be more complex than just disease-related (Festinger 1957; Hewstone, 1989).

1.3 Hypotheses

Hypothesis 1: that those who exhibit stronger risk-taking attitudes and behaviours in relation to smoking will also do so in relation to BSE and GM foods and that these are more likely to be exhibited by men. (Derived from Charlton, 1984; Kann et al., 1993; Bord & O'Connor 1997; Miller et al., 1996; Figueroa et al., 1999; Harvey et al., 2001)

Hypothesis 2: that there are sex and smoking vs. non-smoking differences in attitudes to food risk, health and smoking. (Derived from Pratkanis & Santos, 1993; Ajzen, 1991; Connor, 1994; Slovic 2000; Flay et al., 1994)

2 METHODOLOGY

2.1 Method

The method used to investigate the hypotheses was a questionnaire constructed using primarily Likert-style items derived from existing questionnaires previously constructed by the authors. The questionnaire was designed to be completed in 4 to 5 minutes in order to maximise the response rate. The 6-point scales, ranging from strongly agree to strongly disagree, covered attitudes to the health dangers of active and passive smoking, smoking compared to other risks, social aspects of smoking, advertisement of smoking, cost of smoking and causes of increased or decreased smoking activity; eight of these items related to smoking behaviour of the respondents themselves and thus would only be completed by those who currently smoked. Ten items covered non-smoking issues, covering attitudes to risk generally, food safety, health and fitness. In all, the questionnaire comprised 39 attitude items.

The remaining questionnaire items addressed frequency of smoking, demographic variables [age, sex, SES] and whether or not [in yes/no format] people had changed their purchasing and eating behaviours as a result of the BSE crisis and the GM food controversy. Copies of the questionnaire are available upon request from the authors.

2.2   Sample

The sample was 150 shoppers selected at random by approaching every 3rd person leaving a large city-centre department store. In order to obtain a good cross-section, people were approached at various times of day, on varying days of the week. The success rate of this way of obtaining respondents was high with 65% of those approached agreeing to participate. Significantly more men smoked than women ($\chi^2 = 10.67$, p = .0011, 1 d.f.), which is consistent with national data: Table 1 shows the frequencies of smokers by sex. Included in the table are χ^2 tests [showing degrees of freedom and probabilities based

on contingencies] which demonstrate the frequencies of smokers and non-smokers who changed their eating and shopping habits.

2.3   Procedure

The questionnaires were prepared in large type and presented on six laminated sheets, with the demographic questions first, followed by the consumption questions and finally the attitude items, with the questioner recording the verbal answers. People leaving the department store were asked if they would agree to participate in an anonymous survey, conducted by the University, which would take 4 or 5 minutes of their time.

3   RESULTS

In a sample of 150, one would expect approximately 7 or 8 to identify themselves as not eating beef or other meats (MINTEL, 1997). This study found 14, but these were significantly higher in proportion for non-smokers. Both smokers and non-smokers claimed to have changed their eating or shopping habits as a result of the BSE crisis, although the result for GM was not significant. Results are given in Table 2.

3.1   Factor analysis of items

Since smokers were asked 8 more questions than non-smokers, two orthogonal factor analyses with varimax rotations were conducted, one for each group; the screen plot suggested 5 factors for each group. Because the question sets were different, it was not possible to conduct a confirmatory analysis, so the two factor structures were compared manually in

Table 1.   Frequencies of male and female smokers and non-smokers.

	Smoker	Non-smoker	Totals
Male	33	31	64
Female	22	64	86
Totals	55	95	150

Table 2:   Frequencies of food activities and smoking.

	Smokers	Non-smokers	Total
Eat beef	$\chi^2 = 4.179$, 1 d.f., p < .0409		
Yes	51	84	135
No	3	11	14
Total	54	95	149
Changed where bought beef products	$\chi^2 = 3.353$, 1 d.f., p < .0671		
Yes	3	14	17
No	48	70	118
Total	51	84	135
Changed beef products eaten	$\chi^2 = 3.145$, 1 d.f., p < .0762		
Yes	16	42	58
No	35	48	83
Total	51	90	141
Changed shopping or eating since GM	$\chi^2 = 1.091$, 1. d.f., n.s.		
Yes	8	20	28
No	47	73	120
Total	55	93	148

Table 3. Factor loadings for first three factors.

Questionnaire item	Factor loadings				
	Smokers	Non-smokers	Smokers	Smokers	Non-smokers
Factor 1: *tolerance of smoking*					
15 Restaurants should be non-smoking	.756	.630			
17 People should only smoke where it doesn't offend	.556	.709			
18 If someone smokes, other people can open windows	−.689	−.331			
32 Inhaling other people's smoke is bad for health	.482	.618			
34 I try to avoid places where people are smoking	.701	.610			
35 People smoking near me doesn't bother me at all	−.787	−.557			
36 People should be allowed to smoke where they like	−.762	−.594			
39 People should not smoke in front of children	.574	.710			
Factor 2: *smoker; tries to reduce risks*					
1 Stick to the same brand of cigarettes or tobacco			−.582		
3 The amount I smoke doesn't vary from year to year			−.665		
6 My smoking increases when I am stressed			.627		
7 The amount I smoke has reduced over the years			.601		
9 If another food scare, would stop eating the food			.678		
14 I reduced beef consumption following the BSE scare			.608		
Factor 3: *attitudes to smoking adverts*					
19 The government should ban all smoking advertising				−.710	.589
24 Bothered when cost of smoking rises above inflation				−.733	.326
29 I notice health ads about smoking				−.501	.591
30 There is too much advertising of cigarettes				−.694	.799

order to establish whether it was possible to assemble robust factors by only using items which loaded highly on to factors in both cases. The first factor accounted for 21.5% of variance for smokers and 21.6% of variance for non-smokers and was clearly common to both groups. This factor, which may be termed '*tolerance of smoking*', is shown in Table 3.

A second factor, relating to reducing smoking and reducing other risks was clearly identified for smokers only and accounted for 9.6% of the variance; a third factor was identified with some common elements for both smokers and non-smokers which can be termed 'attitudes to advertising' and it accounted for 7.5% of the variance; these factors are shown in Table 3.

Other factors were clearly identifiable for each group included one which related to the awareness of smoking dangers and another which related to complacency; however, it was difficult to find enough common items to identify them clearly.

Thus only the three factors identified above were used to investigate further the link between attitudes and behaviour: the factor scores were calculated by aggregating the scores for items in each factor.

The aggregated means were compared using T-tests for the yes/no questions concerning whether people had changed their eating/shopping habits and for smokers/non-smokers. The findings are summarised in Table 4; factor 2 analysed for 'changed where bought beef' as N for smokers who had changed their behaviour was too low.

2-way ANOVAs were performed on all three smoking factors to test for sex and 'changed beef products eaten' differences. For change beef products, the ANOVA for factor 1 yielded F = 12.07, p = .0007 and for factor 2, F = 12.27, p = .0010; the F value for factor 3 was not significant. For sex differences, the ANOVA for factor 1 yielded F = 10.24, p = .0017 and for factor 2, F = 3.13, p = .0833; the F value for factor 3 was again not significant. All of these were in the predicted directions, of men being less likely to change their buying and eating habits and the amount smoked being inversely related to likelihood of changing habits following food scares, and thus can be treated as one-tailed hypotheses.

3.2 Summary of findings

Results confirm that smokers would seem to take more food risks in relation to BSE than non smokers, with sex differences in predicted directions. Further analyses of attitudes as predictors of smoking behaviour suggest that smokers employ a variety of dissonance reduction strategies such as taking more 'fatalistic' views about smoking and justify it as a response to stress. Thus, both hypotheses are supported.

A factor analysis of the attitudes items yielded 3 main factors which could be used for further analysis: 'tolerance of smoking', 'risk-reducing' smokers and attitudes to advertising. The first two of these factors were associated with changed behaviour related to

Table 4.   Comparison of smokers and non-smokers and changed shopping and eating habits on the three smoking-related factors.

Factor	$\bar{x}$ yes	$\bar{x}$ no	T	Probability
Smokers?				
Factor 1: tolerance of smoking	28.49	36.64	−7.83	<.0001
Factor 3: attitudes to advertising	13.26	13.77	−1.06	n.s.
Changed where bought beef*				
Factor 1: tolerance of smoking	37.82	31.61	2.76	.0066
Factor 3: attitudes to advertising	12.76	13.47	−.95	n.s.
Changed beef products eaten				
Factor 1: tolerance of smoking	36.52	31.61	4.07	<.0001
Factor 2: smokers reducing risks	21.87	17.06	3.54	.0009
Factor 3: attitudes to advertising	13.26	13.58	−.70	n.s.
Changed since GM				
Factor 1: tolerance of smoking	37.46	32.69	3.21	.0016
Factor 2: smokers reducing risks	20.37	18.13	1.72	n.s.
Factor 3: attitudes to advertising	13.57	13.62	−.09	n.s.

Notes: For FACTOR 1 tolerance of smoking, higher scores indicate less tolerance; for FACTOR 2 reducing risks, higher scores mean greater risk-reducing orientation, and for FACTOR 3 attitudes to advertising, higher scores mean a wish to reduce advertising.
* Factor 2 had too few smokers who changed their purchasing behaviour to analyse.

food scares, and the first factor showed very clear differences between smokers and non-smokers.

4   DISCUSSION

These results suggest that even relatively specific smoking-related attitudes are not only linked to smoking behaviour but also to other health-related behaviours, particularly BSE and GM foods. The finding for changing behaviour in response to the GM debate is complicated because of the so far lack of evidence showing any personal health risks associated with genetic modification of foods (Uzogara, 2000). The HBM (Becker & Maiman, 1975; Rosenstock, 1990) would explain these findings in terms of the general health motivation factor, although this is still problematic since it tends to frame severity and benefits in terms only of health-related considerations for particular diseases. Additionally, the inclusion of psychosocial factors in the HBM might help to explain social smoking behaviour. However, the theory of reasoned action (Ajzen, 1991) perhaps offers a better explanation because it specifically allow for non-health benefits which in the case of smoking would be relevant; examples of non-health benefits might include those associated with social smoking, status and smoking to alleviate nerves or to aid dieting.

At a time when the public are bombarded with many health-related issues and larger threats such as terrorism, the motivation to avoid health dangers from specific threats may well be swamped by these larger

elements, allowing a variety of rationalising belief systems to be constructed. This would help to explain the 16 respondents who changed their eating habits in relation to beef- such changes could be made without social and other personal costs and without having to 'give something up', even though the health-related logic of such a behaviour is contradictory to smoking behaviour; attribution theory might help to explain this situation (Hewstone, 1989) and cognitive dissonance might explain the smokers' attitudes as being consistent with their personal perceptions that they are at less risk than other smokers (McMasters & Lee, 1991; Festinger, 1957).

The sex differences found are consistent with other research into health risks (Charlton, 1984; Kann et al., 1993; Bord & O'Connor 1997; Miller et al., 1996; Figueroa et al., 1999). Women who had changed their beef product eating habits showed, of all the groups, the least tolerance of smoking and were most risk-reducing in orientation.

The finding that none of the factors related to perception of advertisements is interesting since the current political climate in the United kingdom is one which is debating the complete withdrawal of tobacco advertising. However, it does suggest that smokers are no more or less sensitised to advertising than non-smokers, although it is likely that these are perhaps for different reasons, since the non-smokers' tolerance of smoking is so much lower than that of smokers.

The three health issues investigated here – smoking, BSE and GM foods – present very different cognitive issues to the public. They differ in terms of the amount

779

of control each individual has over the situation (Fischoff et al., 2000); for many years people consumed beef without any knowledge at all of the risk, which cannot be said of smoking in the last 30 years. They differ in terms of the perceived consequences and the perceived probability that those consequences will actually happen. Although we have no evidence of people's assessment of risk, they differ in the perceived and real personal vulnerability.

5 CONCLUSIONS

This study has identified strong attitudes and behaviour linkages associated with smoking, BSE and GM foods. We believe that these data lend support to the notion that attitudes and behaviour are causally related, and that attitudes in one domain can and do influence attitudes and behaviours in other domains.

Attempting to explain shifts in behaviour and attitudes in terms of various models of health behaviours has been only moderately successful and we propose that the theory of reasoned action offers the most potential in this area because it incorporates non-health benefits which are clearly relevant to smoking behaviour. However, even this theory is pressed to explain some of the inconsistencies in behaviour and attitudes, such as smokers seeking to reduce other health risks.

In psychological terms, cognitive dissonance abounds through all three of the health issues examined here. For example, the justification for avoiding GM foods in order to avoid cognitive dissonance over the longer time period could involve a variety of processes including denial, selective processing of information, risk perception etc. These same mechanisms could be brought to bear during the current debate and if that were so, may exacerbate the propensity to reduce or cease GM food consumption. But this is hardly comparable with the much larger amount of cognitive dissonance that smokers must justify in the circumstances where they can hardly avoid knowing the risks. How are these dissonances brought together into one set of consistent risk-related beliefs? The effects may well require more potent psychological explanations concerning cognitive processes in risk-taking and decision-making to justify behaviour. There is clearly scope for more research in this area.

REFERENCES

Ajzen, I. 1991. The theory of planned behavior. *Organizational Behavior and Decision Processes,* Vol. 50: 179–211.

Becker, M.H. & Maiman, L.A. 1975. Sociobehavioral determinants of compliance with health and medical care recommendations. *Medical Care.* Vol. 13: 10–24.

Bord, R.J. & O'Connor, R.E. 1997. The gender gap in environmental attitudes: The case of perceived vulnerability to risk. *Social Science Quarterly,* Vol. 78, No. 4: 830–840.

Charlton, A. 1984. Smoking and weight control in teenagers. *Public Health.* Vol. 98: 277–81.

Connor, M.T. 1994. Accounting for gender, age and socioeconomic differences in food choice. *Appetite.* Vol. 23: 195.

Davidson, A.R. & Jaccard, J. 1979. Variables that moderate the attitude-behaviour relation: results of a longitudinal survey. *Journal of Personality and Social Psychology,* Vol. 37: 1364–1367.

Festinger, L. 1957. *A theory of cognitive dissonance* Evanston Ill: Row, Peterson.

Foxall, G.R. & Bhate, S. 1993. Cognitive style and personal involvement as explicators of innovative purchasing of "healthy" food brands. *Journal of Economic Psychology,* Vol. 14, No. 1: 33–56.

Fischoff, B., Slovic, P., Lichenstein, S., Corrigan, B. & Combs, B. 2000. How Safe is Safe Enough? A Psychometric Study of Attitudes Toward Technological Risks and Benefits In P Slovic 2000. *The Perception of Risk:* 80–103. London: Earthscan.

Figueroa, J.P., Fox, K. & Monir, K. 1999. A behaviour risk factor survey in Jamaica. *West Indian Medical Journal.* Vol. 48, No. 1: 9–15.

Flay, B.R., Hu. F.B., Siddiqui, O., Day, L.E., Hedeker, D., Petratis, J., Richardson, J. & Sussman, S. 1994. Differential influence of parental smoking and friends' smoking on adolescent initiation and escalation of smoking. *Journal of Health and Social Behaviour.* Vol. 35: 248–65.

Graham, H. 1994. Gender and class as dimensions of smoking behaviour in Britain: insights from a survey of mothers. *Social Science and Medicine.* Vol. 38, No. 5: 691–698.

Greatorex, M. & Mitchell, V.W. 1994. Modelling consumer risk reduction preferences from perceived loss. *Journal of Economic Psychology.* Vol. 15, No. 4: 669–685.

Harvey, J., Erdos, G. & Challinor, S.C. 2001. The relationship between attitudes, demographic factors and perceived consumption of meats and other proteins in relation to the BSE crisis: a study in the United Kingdom. *Health, Risk and Society.* Vol. 3, No. 2: 181–197.

Hewstone, M. 1989. *Causal Attribution: from cognitive processes to collective beliefs.* Oxford: Blackwell.

Kann, L., Warren, W., Collins, J.L., Ross, J., Collins, B. & Kolbe, L.J. 1993. Results from the national school-based 1991 Youth Risk Behavior Survey and progress towards achieving health objectives for the nation. *Public Health Reports.* Vol. 106, No. 1: 47–55.

McMasters, C. & Lee, C. 1991. Cognitive dissonance in tobacco smokers. *Addictive Behaviors.* Vol. 16: 349–53.

Miller, B.E., Quatromoni, P.A., Gognon, D.R. & Cupples, L.A. 1996. Dietary patterns of men and women suggest targets for health promotion: the Faringham nutrition studies *American Journal of Health Promotion,* Vol. 11, No. 1: 42–52.

MINTEL (1997) *Red Meats.* Mintel Food and Drink July 1997

Pratkanis, A.R. & Santos, M.D. 1993. The psychology of attitude change and social influence. *Contemporary Psychology,* Vol. 30, Pt 1: 16–18.

Rosenstock, I.M., Strecher, V.J. & Becker, M.H. 1988. Social learning theory and the health belief model. *Health Education Quarterly,* Vol. 15, No. 2: 175–183.

Saba, A. & Di Natale, R. 1999. A study on the mediating role of intention in the impact of habit and attitude on meat consumption, *Food Quality and Preference*, Vol. 10, pt 1: 69–77.

Schafer, R.B., Shafer, E., Bultena, G. & Holberg, E. 1993. Coping with a health threat: a study of food safety. *Journal of Applied Social Psychology*, March Vol. 23, No. 5: 386–394.

Slovic, P. 2000. Do Adolescent Smokers Know the Risks? In P Slovic 2000. *The Perception of Risk*: 364–371. London: Earthscan.

Uzogara, S.G. 2000. The impact of genetic modification of human foods in the 21st century: A review. *Biotechnology Advances*, Vol. 18, No. 3: 179–206.

Verbeke, W. & Viaene, J. 1999. Beliefs, attitude and behaviour toward fresh meat consumption in Belgium: empirical evidence from a consumer survey. *Food Quality and Preference*, Vol. 10, pt 6: 437–445.

Viscusi, W.K. 1990. Do smokers underestimate risks? *The Journal of Political Economy*. Vol. 98, No. 6: 1253–1269.

Williams, D.R. & Rucker, T. 1996. in Kato PM et al., *Handbook of diversity issues in health psychology* New York: Plenum Press.

Worsley, A. & Skrzypiec, G. 1998. Do attitudes predict red meat consumption among young people? *Ecology of Food and Nutrition*. Vol. 37, pt 2: 163–195.

Safety and Reliability – Bedford & van Gelder (eds)
© 2003 Swets & Zeitlinger, Lisse, ISBN 90 5809 551 7

Systematic derivation of safety rules for railway operations

Heijer T., Hale A.R. & Koornneef F.

Safety Science Group, Faculty of Technology, Policy and Management, Delft University of Technology

ABSTRACT: Rail operations involve a large number of different actors (driver, conductor, train controllers, maintenance crews, etc.), who are often in poor real-time communication with each other. The majority of communication takes place via the infrastructure and its signals, etc. This means that the rail system operates to a considerable extent in open-loop mode. This places very high demands on the strict adherence of all persons to rules, since people in the system can only predict the behaviour of others by assuming that they will follow the rules perfectly. Devising and managing an effective system of safety rules under such constraints is no easy task. This paper describes the work of a Dutch project to develop better ways of managing safety rules against this background. It describes the rail system and assesses its susceptibility to regulation, evaluates the literature on safety rule management and provides a draft protocol of good practice for developing and managing safety rules. The paper is illustrated from two case studies of sets of rules for Dutch rail operations (train drivers and track maintenance).The paper also gives an introduction to a larger European project (SAMRAIL) which will expand the work on this topic to other countries in the context of developing guidance for safety management systems for all European railways.

1 INTRODUCTION

Every technology and activity has safety rules, which are usually formulated explicitly, taught to those operating in the system and imposed on them. These rules are based on the nature of the activities and their systematic context. The project therefore contained three steps:

- A first step to draw together what is known of good and bad practice in this area, particularly in deciding what rules should be explicitly formulated and imposed
- The second one to establish the system dynamics and structure of the communication paths of the railway system
- The third step to apply the findings of the first step to those of the second step to reach conclusions about the current state of affairs and make recommendations for improvement. This step was still under way at the moment this paper was written and therefore falls outside the scope of this paper.

2 GOOD AND BAD PRACTICE REGARDING SAFETY RULES

Safety rules are seen as central to all attempts to prevent accidents and achieve safety. These rules and

attitudes towards their use and violation are generally considered as significant dimensions of safety culture (see also Hale et al. 2002, Swuste & Guldenmund 2002).Yet it is surprising how little literature there is about how to generate and manage safety rules effectively. The standard safety management texts only rarely (e.g. de Reamer 1980, Geller 2001) have a chapter on safety rules. The main emphasis is on discipline and enforcement, not on the positive contribution that rules make and how that can be maximised. Those operating in the system are considered actors who should simply obey.

Yet this view is being increasingly challenged. The revolution in regulation, starting in the 1960s and 1970s, questioned the imposition of strict rules at government level and advocated the development of framework rules formulated as goals, to be filled in lower down in the system hierarchy with detailed action rules. The 1980s saw an extensive study of rule violations conducted within the framework of cognitive theory by Reason followers (e.g. Reason & Free 1993, Free 1994, Reason et al. 1995, 1998). These showed comparable disutilities in detailed rules imposed at organisational level, which lead to enforced violation to get the work done in real life. It made clear what was bad in current practice, but got no further than generating general guidelines for the formulation of safety rules. At the same time research of high reliability organisations (HROs) (e.g. Roberts 1990, Rochlin 1989) showed that these were characterised by a lack of detailed written safety rules, but the presence of highly trained experts and by intense consultation, mutual checking and communication about safe actions.

Current knowledge about safety rules and their management still suffers from this fragmented approach. Detailed rules have a bad name; thick rule books are seen as symptoms of system pathology (Hale 1990), yet rules are still seen as essential ingredients for controlling system safety. The cry is for better rule use and management, but there are still not very clear ideas of how and of how to avoid the main pitfalls in producing workable safety rules.

2.1 A simple framework

This paper deals only with rules that are imposed from outside the individual and which, by definition, are limitations on individual freedom of choice.

We consider an ideal rule management system to be an adaptive process that consists of a number of logical steps (see figure 1).

Initially the processes to be controlled must be modeled as a set of system objectives and functions in a coherent process or set of processes, allowing for all relevant steps in the life cycle. This step indicates the crucial system transitions, the functions to be carried out and the system actors who conduct them (and hence the communication necessary). The accident scenarios, which are known or considered credible, can then be derived from this model, together with the barriers and controls which prevent them from developing (see Johnson 1980). It is these barriers and controls that form the basis for defining the desired behaviour of the system and its actors.

The next step is the most controversial and one where there is the least guidance, the decision as which of these "rules" needs to be explicitly defined

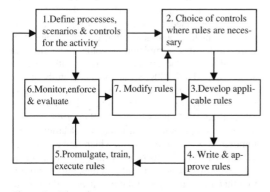

Figure 1. The rule management process.

and imposed to one degree or another on the system actors. Hopwood (1974, see also Reason et al. 1998) defines four ways of imposing such behaviour:

- Hard-wired rules, which automatically achieve defined system behaviour by means of hardware. These are effective unless the hardware fails.
- Technical forcing functions, through design, layout, etc., which prevent or elicit specific behaviour. Ergonomics has studied these extensively in the form of stereotypes, triggers and warnings.
- Administrative standardisation in the form of defined and imposed rules. Here we will use the categorization of Hale and Swuste (1998) that distinguishes output or goal rules, procedural rules which tell you how to arrive at a decision on what to do, and action/state rules which tell you exactly how to do it (see also also Dien 1998).
- Self control through expertise and competence based on practice and acquired knowledge (or internalised administrative rules), which may not be explicit or accessible to the person possessing it. Full self-control includes self-monitoring against the rules.
- Social group regulation, in which the rules reside at the group level, more or less articulated, but in any case imposed by group pressure. In the case of new situations the group debates and decides the course of action.

Once it has been decided what rules are appropriate, the rules must be developed, written (or formulated in another way which can be communicated) and approved for application. Here must be decided who is involved in these processes and how, and what expertise and experience is brought to bear in what way in testing the appropriateness and applicability of the rules. The form of the rules must be decided, including how rigid they are and how and by whom deviation from the rules can be decided, worked out and sanctioned.

The next steps are to inform and train those who have to carry out the rules and to ensure that everyone is aware of them and of any changes made to them. Here decisions have to be made about what to include in training, what in rule books and other means of access to the total set of rules and what media to use for promulgation.

Once the rules are introduced there is a phase of operation in which the rules must be monitored, deviations dealt with, new insights into the working of the rules gathered and decisions made about what may need changing. Here the decisions are about supervision, dealing with situations not envisaged by the rules and requiring adaptations and action after deliberate or inadvertent deviation from the rules in cases where the outcome is either positive or negative for the system or the individual.

The final step to complete the circle is the modification process in which the activity, processes or technology may change, in which the lessons from experience may suggest that rules for previously unregulated situations are necessary, or that rules can or should be removed or relaxed, or simply changed for better rules. Here the issues are again of who is involved, what processes are needed to approve a change and how to avoid the processes of rule accretion and calcification described in the next section.

2.2 Rules, violations and enforcement

A number of studies, particularly by Reason and his co-workers (Reason et al. 1998, Free 1994, Lawton & Parker 1999), have looked at the circumstances under which people violate rules. They have identified three main types of violation:

- Routine violations; usually short cuts that have developed to provide individual advantages
- Situational violations , that occur when the rule cannot be carried out in particular circumstances
- Exceptional violations in exceptional circumstances in knowledge-based operations

The first two are both concerned with mismatches between rules and reality.

The factors which encourage violations have been summarised by Hudson et al (undated) in a booklet for Shell managers:

1. People who expect that they will not be able to perform the activity if they follow all the rules
2. People who consider they have the skills and knowledge to work out for themselves how to operate outside the rules
3. Opportunities to take short cuts with impunity

4. Inadequate planning and resources, so that people are brought into situations in which they have improvise.

Several authors have extensively researched rule formulation, enforcement and violations. In this study we applied Hale's propositions (Hale & Swuste 1998) that indicate that the hierarchy of rule types (from goal, through procedural to action/state rules) should parallel the organizational levels in a system: goals should be set high in the organisation translated in a level below in procedures for arriving at decisions and, where necessary, as close as possible to the work floor.

2.3 Some guidance for an ideal system

The pathology of bad safety rule systems can be turned on its head to formulate guidance about ideal rule systems (see table 1).

These general rules for guidance fill in the various steps in figure 1 with more detailed criteria for assessing how good a safety rule management system is.

We turn in the next section to the case study of safety rule management in railways. This section concentrates on the first three steps in figure 1 and discusses further what are the constraints within the technology and organisation of the railway system which determine what the current use of safety rules is, and how these constraints make it difficult to implement a number of the ideal characteristics of a good safety rule system.

3 RAILWAYS: A CASE STUDY IN SAFETY RULE USE

This case study is based on interviews in diverse organisations operating in the Dutch railway system

Table 1. Ingredients for an ideal rule system.

Regular & well managed involvement of rule users in deciding on and evaluating rules	Hierarchy of types of rule (goal, procedural, action/ state) matched with level of formulation in the organisation
Rules limited to essentials – predictable, clear cut situations at rule-based level	Indication of status of each rule in a hierarchy from compliance to advisory
Meta-rules (procedural) for situations for which existing rules cannot apply	Assessment of violation potential & cost-benefit of correct & incorrect compliance
Agreed system for approving exceptions or deviations from agreed rules	Clear, unambiguous language. Linked to task/function steps
Structured way of learning from rule violations and coping with changes to the system	Clear specification of objectives, area of application and ownership of each rule. No complex cross-referencing
Regular review of rule system to weed out unnecessary or contradictory rules and review total compatibility of rule set	Clear structure & indexing of rule book/ computer retrieval system, so that rules are easy to find when on-line use essential
Clear system for monitoring rule compliance and applying sanctions	Resources & planning available to formulate good rules and to work according to rules

in passenger and goods transport, in infrastructure provision and maintenance and in regulatory roles. It is part of an ongoing European study regarding increasing interoperability and privatisation. This has led to the introduction of many new companies into railway operations and maintenance and to the international operation of companies used only to one national set of rules.

The interviews showed that the safety rule system in use in Dutch railways has many of the hallmarks of a system which has grown by accretion over a long history of trial and error, some of it through major accidents. Despite some attempts to rationalise the system (e.g. Keijzer 1999), there is still no explicit use of the steps set out in the framework in section 2 of this paper. In particular the processes, critical life cycle transitions, accident scenarios and risk control barriers have not been specified in a comprehensive and systematic way. The information about them is largely present in the system, but scattered, sometimes only present in the heads of experienced staff, and, above all, is not used as an explicit basis for deriving safety rules. This means that rules as now formulated are not comprehensive or tested for their compatibility or violation potential.

The research team has conducted an analysis along the lines of section 2 of this paper for part of the system, as part of the study. This shows that railway operations have a number of particular characteristics distinguishing them from many other high hazard, high technology systems:

1. Whilst the objectives of the system can be simply stated as the safe and efficient transport of goods and passengers over the infrastructure provided (or to be built), using the power supply and under control of a traffic control system, the achievement of the objectives takes place in a system with many boundaries and life cycle transition steps which are concentrations of vulnerability.
2. Many of the accidents happen at boundaries with other systems: suicides, trespassers and animals on the track invade the system for their own individual purposes; other transport systems cross the railway track at bridges and more particularly level crossings; vandals can use these crossing points to throw objects onto the track; passengers entering and alighting from trains at stations are more at risk. This means that the railway system has to negotiate risk controls with other systems, or impose them on them.
3. The railway system has been designed and built with only a limited concern for the maintenance and modification phases of the life cycle. Maintenance is not incorporated into the initial design of the train timetable and has to be dealt with as disturbance to it. The sharing of parts of the

system between operations and maintenance and the transfer from one system state to another (also non-nominal operations and emergency situations) lead to many potential accident scenarios which have to be controlled by rules.

4. Above all the communications between the various parties to the system is quite limited and largely intermittent. Figure 2 shows this. The train controller, the train drivers in his section of the system and the contractors working on the track are the main protagonists in on-line risk control. There is no direct communication between any of them and the external parties threatened by the system (motorists at level crossings, trespassers, suicides, etc). It is only through the physical infrastructure that their behaviour is controlled (crossing controls, fences, barriers). This limitation is perhaps to be expected. What is more striking is that the main protagonists also "communicate" largely through the physical infrastructure (signals, lookout men, ATB) or through off-line means (the information sheets issued to drivers daily about work in progress, restrictions).

   For example, the controllers do not know exactly where any train is, only in what block it is present; there is no direct contact between train driver and maintenance crew.
5. By its nature the system is distributed, with several regional train controllers dividing the country, many trains operating with little communication between driver, train crew and controller. The staff in the trains is constantly traveling and there is only limited group communication and cohesion because of the lack of opportunities to meet with colleagues or for supervisors to monitor behaviour or communicate with their subordinates.
6. The main regulator of the system is the timetable, filled out with the maintenance planning. If all goes according to this, within certain margins, the system functions well, but only provided everybody complies very strictly with the expected behaviour and the signals.
7. There has been a strong resistance to increasing the local communication in the system (telerail is a limited and not very reliable system, mobile phone use has been discouraged, communication between train and track workers is seen as a dangerous loss of central control).
8. Because of all this the system works to a significant extent on open loop (feed forward control), with a considerable delay in feedback. If anything happens which causes significant deviation from the timetable and expected planning the system rapidly reaches overload, because the communication demands suddenly increase and the communication means are very limited. In order to

guarantee the margins for predicted operations and to cope with deviations, there are wide margins of space and time built in between trains. The basis of the rules for much of the operation is also that actions or movements are only permitted if there is positive confirmation through the infrastructure that the system state is safe – lack of communication means that this cannot be checked otherwise with other parties. This means that a major disturbance can often only be coped with by stopping that part of the system and resetting and restarting it. With increasing pressure on the system capacity, through government policy to transfer more road traffic to public transport, and higher demands on system performance, these margins are coming under increasing pressure.

All of this means that rules and strict adherence to them form a central prerequisite to the safe functioning of the system. Only if the rules are followed exactly is it predictable what will happen and can the open-loop system work with sufficient certainty. The system has been designed on this basis since its inception. It has always had paternalistic and even militaristic characteristics, whereby rules are devised and imposed from top down and unquestioning obedience is expected and has in the past been enforced. Obedience is therefore assumed, leading to great surprise and alarm (and strong disciplinary action) when it is shown that violations have occurred. So much is obedience expected that there are some surprising gaps in technical support. For example Dutch trains have indicators in the cab showing that they have gone through a yellow light; however there is no indicator that a red has been passed. The reasoning seems to be that this is forbidden, hence it will not happen and one does not need to indicate it. The automatic train system "sees" this violation and responds, provided that the train is travelling at more than 40 kph. Below that it does not work. The result is that a moment's inattention or attention directed at the wrong signal below this speed can result in an error which cannot be recovered, since there is no record of it for the driver to see a few seconds later.

The distributed nature of the system and the constant traveling of the majority of staff means that there is no such thing as a cohesive work group. The limitations imposed on communication between the main protagonists, added to the recent development that they are now all working for different companies rather than the old monopolist, makes it impossible for any cohesive relationship to develop between them. Indications from the interviews suggest that the privatisation has even led to more conflict in these relationships, since drivers are judged on performance criteria which emphasis punctuality, whilst their

progress is governed by train controllers working to other targets of safety and capacity. The coherent communication necessary for the consensus on defining rules and the social control in carrying them out is therefore not easy to find. The task of communicating any changes in the rules, or organising refresher training is also a major logistic exercise.

Our conclusion from this stage of the analysis of the railway system and its use of safety rules is that there are a number of characteristics of the technology and the way it has been designed and managed up to now, which make it difficult to apply a number of the ideal characteristics of a rule system as set out in section 4.

It would be possible to gather together the existing fragmented information about the safety critical processes, scenarios and state transitions and to make them the explicit basis for identifying the barriers and control important for safety. These can then be used as an explicit basis for deriving the required behaviour of the various actors in the system. Use of the criteria defined for deciding which of these potential rules should be explicitly formulated as a safety rule can then provide a far more coherent basis for operational, task-oriented rules. Identification of the actor(s) responsible for each critical task and decision, particularly in the state transitions or during non-nominal or emergency states, can form the basis for the rules about responsibility and authority. This will improve the coherence and transparency of the rule system. What is clear is that the efficacy of the rules and their ability to control risk situations will only be radically improved if the means and possibilities for communication between the on-line actors in the system are improved. The developments in the ERTMS and in traffic control which are taking place are an opportunity to make this improvement, but its implementation will require a major change in the operating philosophy to give a far more active role to the "front-line" risk controllers, the track workers and train drivers.

When it comes to deciding which rules should be made an explicit part of the safety rule system and to formulating and writing the rules, the major problems to be solved are the involvement of staff. There are currently increasing attempts to feed in this expertise (Keijzer 1999), but more needs to be done to create the feeling of ownership of the rules and a common attitude towards them by all the on-line risk control protagonists. This will involve a shift of emphasis from rules written in the office by specialists to a more distributed process of consultation and approval. To be effective in the dynamic environment of a rapidly changing and expanding railway system, this needs to include processes for regular review and adaptation of the whole set of rules appropriate to given activities.

REFERENCES

Åberg L. 1998. *Traffic rules and traffic safety.* Safety Science 29(3), 205–216.

Baram M. 1993. *The use of rules to achieve safety: introductory remarks. Paper to the 11th NeTWork Workshop: The use of rules to achieve safety.* Bad Homburg. 6–8 May.

Battman W. & Klumb P. 1993. *Behavioural economics and compliance with safety regulations.* Safety Science 16(1), 35–46.

de Brito G., Pinet J. & Boy G. 1998. *Etude SFACT sur l'utilisation de la documentation opérationelle dans les cockpits de nouvelle generation: Etude de l'interface Homme – Machine (Study for SFACT on the use of operational documentation in new generation cockpits: study of the man – machine interface).* EURISCO report T-98-o54. European Institute of Cognitive Sciences and Engineering. Toulouse.

Bourrier M. 1998. *Elements for designing a self-correcting organisation: examples from nuclear plants.* In Hale A.R. & Baram M. Safety management: the challenge of change. Pergamon. Oxford.

Dien Y. 1998. *Safety and application of procedures, or how do "they" have to see operating procedures in nuclear power plants?* Safety Science 29(3), 179–188.

Elling M.G.M. 1991. *Veiligheidsvoorschriften in de industrie (Safety rules in industry). PhD Thesis. University of Twente.* Faculty of Philosophy and Social Sciences Publication WMW No.8. Netherlands.

Fleury D. 1998. *Reinforcing the rules or integrating behavioural responses into road planning.* Safety Science 29(3), 217–228.

Free R. 1994. *The role of procedural violations in railway accidents.* PhD thesis. University of Manchester.

Geller S. 2001. *The Psychology of Safety Handbook.* Lewis Publishers. Boca Raton.

Grote G. 2002. *Safety and autonomy – a necessary contradiction?*

Guldenmund F. 2000. *The nature of safety culture: a review of theory and research.* Safety Science 34(1–3), 215–257.

Hale A.R. 1990. *Safety rules OK? Possibilities and limitations in behavioural safety strategies.* J. Occupational Accidents. 12, 3–20.

Hale A.R. & Swuste S. 1998. *Safety rules: procedural freedom or action constraint?.* Safety Science 29(3), 163–178.

Hale A.R. & Waterbeemd H., Potter B., Heming B.H.J., Swuste P.H.J.J., Guldenmund F.W. 2002. *Safety culture assessment in a steelworks: using diverse data sources to develop an effective diagnosis for safety improvements.* In: P.R. Mondelo, W. Karwowski & M. Mattila. Proceedings of the 2nd International Conference on Occupational Risk Prevention. Gran Canaria February 20–22.

Hopwood A.G. 1974. *Accounting systems and managerial behaviour.* Saxon House. Hampshire.

Hudson P.T.W., Verschuur W.L.G., Lawton R., Parker D. & Reason J.T. Undated. *Bending the Rules II: why do people break rules or fail to follow procedures and what can you do about it.* University of Leiden.

Johnson W.G. 1980. *MORT Safety Assurance Systems.* Marcel Dekker, Inc., New York.

Keijzer R. 1999. *Information and communication: drafting user-friendly rules and procedures.* Paper to the UIC World Conference on Occupational Health and Safety. Sept 22–24 Paris. UIC.

Lawton R. 1998. *Not working to rule: understanding procedural violations at work.* Safety Science 28(2), 77–96.

Lawton R. & Parker D. 1999. *Procedures and the professional: the case of the British NHS.* Social Science & Medicine 48, 353–361.

Leplat J. 1998. *About implementation of safety rules.* Safety Science 29(3), 189–204.

Mager Stellman J. (ed.). 1998. *Encyclopaedia of Occupational Health and Safety.* ILO/WHO. Geneva.

Norros L. 1993. *Procedural factors in individual and organisational performance.* Paper to the 11th NeTWork Workshop: The use of rules to achieve safety. Bad Homburg. 6–8 May.

Perin C. 1993. *The dynamics of safety: the intersections of technical, cultural and social regulative systems in the operations of high hazard technologies.* Paper to the 11th NeTWork Workshop: The use of rules to achieve safety. Bad Homburg. 6–8 May.

Rasmussen J. 1993. *Rules: how to do things safely with words? Paper to the 11th NeTWork Workshop: The use of rules to achieve safety.* Bad Homburg. 6–8 May.

Rasmussen J. 1997. *Risk management in a dynamic society: a modelling problem.* Safety Science 27(2/3) 183–213.

de Reamer R. 1980. *Modern Safety & Health Technology.* John Wiley. New York.

Reason J.T. 1990. *Human error.* Cambridge University Press. Cambridge.

Reason J.T. & Free R. 1993. *Bending the rules: the psychology of violations. Paper to the 11th NeTWork Workshop: The use of rules to achieve safety.* Bad Homburg. 6–8 May.

Reason J.T., Manstead A.S.R. & Stradling S.G. 1995. *Driving errors, driving violations and accident involvement.* Ergonomics 38, 1036–1048.

Reason J.T., Parker D. & Lawton R. 1998. *Organisational controls and safety: the varieties of rule-related behaviour.* Journal of Occupational and Organisational Psychology. 71, 289–304.

Roberts K. 1990. *Some characteristics of one type of high reliability in organisation.* Organisation Science. 1(2), 160–176.

Rochlin G.I. 1989. *Informal organisational networking as a crisis-avoidance strategy: US naval flight operations as a case study.* Industrial Crisis Quarterly. 3(2), 159–176.

Simard M. & Marchand A. 1997. *Workgroups' propensity to comply with safety rules: the influence of micro-macro organisational factors.* Ergonomics 40(2), 172–188.

Swuste P.H.J.J., Guldenmund F.W. & Hale A.R. 2002. *Steel industry, safety culture and the effectiveness of safety interventions.* In Proceedings of the 2nd International Conference on Occupational Risk Prevention. Mondelo P.M., Karwowski W & Mattila M.. (Eds).

Williams J.C. 1997. *Assessing the likelihood of violation behaviour – a preliminary investigation.* Paper to the Institution of Nuclear Engineers Conference COPSA 9.

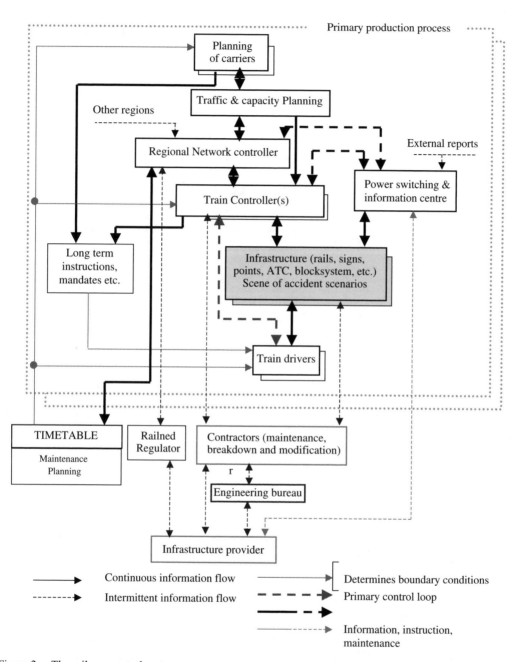

Figure 2.   The railway control system.

Safety and Reliability – Bedford & Van Gelder (eds)
© 2003 Swets & Zeitlinger, Lisse, ISBN 90 5809 551 7

Availability optimization by sensitivity analysis of fiber optical network systems

M. Held & Ph.M. Nellen
EMPA Swiss Federal Laboratories for Materials Testing and Research, Duebendorf, Switzerland

L. Wosinska
KTH Royal Institute of Technology, Kista, Sweden

ABSTRACT: A method for availability optimization of complex optical network systems is presented. This method can be especially useful when the underlying reliability data of components or sub-systems is uncertain. Sensitivity analysis is applied by systematically varying the input data, mainly failure rates of critical components. Monte Carlo simulation is used for availability analysis of an optical cross connect, which can be represented by a reliability block diagram. The possible states of a generalized telecommunication network are shown in a diagram of state transition probabilities and availability is calculated using a Markov process. As a result the system parts dominating the overall system availability can be identified and their influence quantified to some extent.

1 INTRODUCTION

1.1 *Availability requirements*

Telecommunication systems and networks must fulfill high quality of service (QoS) requirements. An important criterion is the availability of the physical transmission equipment. Depending on the demanded services availability values of 95% up to 99.999% are specified by telecommunication service providers.

Optical networks basically consist of optical or electro-optical nodes where traffic is added, dropped or routed, and spans – implemented with optical fiber – connecting the nodes. Availability requirements can be matched by using highly reliable components (comprising hardware and software), redundancy for failed components, and failure handling capabilities including repair resources. On the hardware side redundancy in network nodes is implemented by multiplying critical equipment with the possibility to switch from failed to redundant equipment in short time. Spans in network are usually protected against failure by the allocation of spare transport capacity and the according rerouting scenario. Availability is defined as the ability of an item to be in a state to perform a required function at a given instant of time. In telecommunication practice availability can be defined on all levels of a network, e.g. for a piece of equipment, a transmission path comprising several nodes and spans, or the network as a whole.

1.2 *Network resilience*

Resilience of a network is a general term used to group all mechanisms to recover the network in case of failure. Network survivability, i.e. the ability of a network to maintain or restore an acceptable level of performance during network failures is used as a measure of resilience. A general distinction of resilience mechanisms is made between protection and restoration [Maier et al., 2002].

Protection involves the static reservation of backup resources and the use of protocols to bring these into use to recover from a failure. Protection techniques use a fixed assignment of protection resources to working lines, and will handle failures only up to some designed level. However, when failures occur, protection results in deterministic behavior, with very fast traffic recovery times (50 ms in synchronous digital hierarchy SDH networks).

In contrast, restoration involves the dynamic allocation of network resources to replace either an end-to-end circuit, or some portion of a circuit, when it experiences a failure. Recovery can therefore be done without any pre-assignment of backup resources, using the overall pool of network resources as backup. The result is a more efficient use of resources and a higher degree of service survivability, i.e. the ability to continue providing service in the presence of failures. However, since restoration takes place at the service layers, traffic recovery times are generally longer (seconds to several minutes) than those for protection techniques. Since span failures are generally assumed to be more probable than node failures, both mechanisms – protection and restoration – are usually designed to provide a 100% restorable network in the case of a single span failure. Few investigations have treated multiple failures, e.g. [Clouqueur & Grover, 2002, Choi et al., 2002].

Due to high cost for redundant equipment and repair actions these resilience mechanisms aim at the optimization of performance, availability, and cost.

1.3 Availability, failure and repair rates

Any availability analysis is based on failure rate λ and repair rate μ of the involved equipment, respectively mean time to failure MTTF $= 1/\lambda$ and mean time to repair MTTR $= 1/\mu$ when failure free operating times and repair times are exponentially distributed. The availability function A(t) is defined as the probability of an item being in an up state at the instant of time t. Asymptotic (steady state) availability is generally expressed as the ratio of mean up time MUT to the mean time between failures MTBF that is the sum of MUT and mean down time MDT (i.e. MTBF = MUT + MDT), thus:

$$A = \lim_{t \to \infty} A(t) = \frac{MUT}{MUT + MDT} \qquad (1)$$

When failure free operating times and repair times are exponentially distributed and assuming continuous operation and no preventive maintenance actions the asymptotic availability of an item can be directly calculated using the resulting time independent (constant) failure and repair rates:

$$A = \frac{\mu}{\lambda + \mu} = \frac{MTTF}{MTTF + MTTR} \qquad (2)$$

Asymptotic unavailability U is then U $= 1 - A$ and if $\mu >> \lambda$ holds, approximated by U $= \lambda$ MTTR. Failure rates for components of telecommunication systems are provided by prediction models [Telcordia SR-332, UTE RDF 2000], field data analysis of operators, and results of accelerated stress tests of vendors and research institutes.

2 ANALYSIS METHODS

For many years networks were analyzed using formal approaches of graph theory. However, calculating reliability and availability of telecommunication networks was found to be difficult, with nominally all problems of interest in this field classified as NP-hard problems [Spragins, 1986], even with the simplifying assumptions and omissions that were commonly made, e.g. non-repairable systems, perfect nodes etc. For engineering purposes reliability and availability can be analyzed by describing the systems with reliability block diagrams or diagrams describing system state transitions. But also with this approach analytical calculations for characteristic reliability and availability parameters such as mean time to failure or average availability for complex, repairable systems containing redundancies rapidly become costly and intractable. Monte Carlo simulation and Markov process calculations can therefore be very useful for availability analysis of complex systems. Results of simulations and calculations show sufficient accuracy and high flexibility for sensitivity analysis [Held et al., 2002].

2.1 Reliability block diagram

A reliability block diagram RBD is an event diagram and answers the question: which elements of the considered item are necessary for the fulfillment of a clearly defined required function and which can fail without affecting it? The elements necessary for the required function are then connected in series, while elements that can fail with no effect on the required function are connected in parallel and represent a redundancy. From the operating point of view, one can distinguish between active, warm, and standby redundancy. For active (also called hot) redundancies with identical elements, both elements are subject to the same load and have identical failure rates. The redundant element in standby (cold) redundancies is not loaded and its failure rate is often assumed to be zero. In warm redundancies the redundant element is partly loaded and its failure rate is generally lower than the one of the operating element.

2.2 Monte Carlo simulation

When performing a Monte Carlo simulation, a random series of simulations are performed on the RBD. These simulations are test runs through the system from a start node through an end node to determine if the system completes its task or fails. For illustration purpose the steps of a Monte Carlo availability simulation of a two component redundant repairable system are given. This requires the simulation of operating-repair cycles for each component:

1. Generate a random number
2. Convert this number into a value of operating time using a conversion method on the appropriate times-to-failure distribution, e.g. exponential with given failure rate λ
3. Generate a new random number
4. Convert this number into a value of repair time using a conversion method on the appropriate

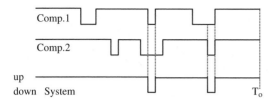

Figure 1. Sequence of component and system up/down states.

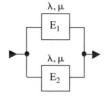

Figure 2a. Reliability block diagram.

times-to-repair distribution, e.g. exponential with given repair rate μ

5. Repeat steps 1–4 for a desired operating life T_o
6. Repeat steps 1–5 for each component
7. Compare the sequences for each component and deduce system up and down times, see Figure 1
8. Repeat steps 1–7 for the desired number of simulations N

The results of the Monte Carlo simulation are then the statistics over the whole series of N simulations. As a general rule, the more simulations performed, the more accurate the results become.

2.3 Markov process

Many physical phenomena observed in everyday life are based on changes that occur randomly in time. Examples are the arrivals of calls in a telephone exchange, radioactive decay, or the occurrences of failures in technical equipment. Markov processes can mathematically describe these phenomena. They are characterized by the property that for any time point t their future depends on t and the state occupied at t, but not on the history up to time t, hence a behavior without memory. Time homogeneous Markov processes, THMP, describe processes where the dependence on time t also disappears, so that the future of the process, i.e. the next state depends only on the current state and its state transition probabilities. THMP are often used to describe the behavior of repairable systems consisting of components with time independent (constant) failure and repair rates.

A given system is considered, at any instant in time, to exist in one of several possible states. A state transition diagram defines the operational and failed system states and the transitions between these states. After certain properties are assigned to states and the transitions between states, these diagrams contain sufficient information for developing equations describing the system behavior. Figure 2a) depicts the RBD of a 1-out-of-2 repairable, hot redundancy with two identical elements E_1 and E_2. In the state transition diagram depicted in Figure 2b), Z_0 and Z_1 are up states (system operational) and Z_2 is the down state (system failed).

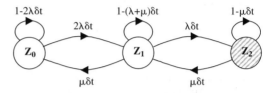

Figure 2b. Diagram of state transition probabilities in $(t, t + \delta t]$ for availability analysis for a 1-out-of-2 hot redundancy with identical elements $E_1 = E_2$, constant failure rates λ and repair rates μ, one repair crew, arbitrary t, $\delta t \downarrow 0$, Markov processes.

Provided that failure rate λ and repair rate μ are time independent the system can be described by a system of differential equations for the state probabilities $P_0(t)$, $P_1(t)$ and $P_2(t)$, with $P_i(t)$ = probability {process in Z_i at time t}, i = 0, 1, 2. From the analytical solution of the differential equations the parameters of interest can be obtained, e.g. steady state availability. In this work, computational support based on numerical calculation of differential equations, i.e. Runge-Kutta approximations are used to analyze more complex systems consisting of a larger number of elements.

3 SENSITIVITY ANALYSIS

The results of availability calculations obviously depend on the failure and repair rates as input data. The difficulty of such calculation is the lack of failure rate data mainly for new components and the uncertainty of the available data, i.e. a variety of sources provide diverse failure rates about components without specifying tolerances, confidence levels or statistical background. This observation coincides with the plainly different results of failure rate prediction models for electronic components [MIL-HDBK-217, Telcordia SR-332, UTE RDF2000, IEC 61709]. In contrast repair rates are less critical for the evaluation of availability because reasonable values have emerged from practical experience, hence MTTR ≈ 6 h for node equipment and MTTR ≈ 24 h for spans are widely accepted. Table 1 shows a selection of failure rates (given in FIT) for components and systems based on prediction

Table 1. Failure rate values.

Component	Range [FIT]	Typical value [FIT]
Fiber optical cable	5–500 FIT/km	100 FIT/km
Optical cross connect	–	10 kFIT
Optical receiver	100–400 FIT	250 FIT
Optical connector	5–100 FIT	50 FIT
Optical line amplifier	–	2000 FIT
Line equipment	–	2000 FIT
Network node	1 k–100 k FIT	10k FIT

1 FIT =1 failure in 10^9 operating hours.

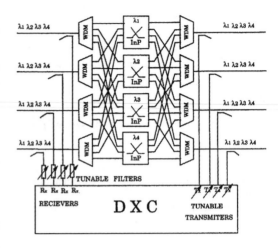

Figure 3. OXC architecture, λ_1–λ_4 designate wavelengths.

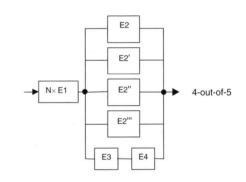

Figure 4. OXC reliability block diagram for *one link* (N = 5) and *four links* (N =8).

models and field data analysis (Telcordia SR-332, UTE RDF, 2000, Mikac, 2002, De Maesschalck, 2002).

Nevertheless, availability analysis based on typical failure rate values or a failure rate range as indicated in Table 1 could give valuable results in the sense of identification of weak system parts of a network and their possible influence on the overall reliability and availability. Thus, the sensitivity of the achieved results on the input data represents the added value of this type of reliability investigation.

Sensitivity analysis means that a set of calculations is done by systematically varying the input data, especially failure rates of critical components. As a result the system parts dominating the overall system availability can be identified and also be quantified to some extent. This gives valuable input for possible system optimization comprising technical and economical aspects.

4 RESULTS

4.1 *Optical cross connect*

Figure 3 shows the architecture of an optical cross connect OXC capable of switching 16 wavelength channels described in [Wosinska, 1993]. It is based on 4 pieces of 4×4 InGaAsP/InP laser amplifier gate space-division switches, further designated as GSDS. The digital cross connect DXC contains four receivers and four tunable transmitters and can be used for regeneration of the optical signals or as the opto-electrical interface in the case of originating or terminating traffic. Furthermore it represents a redundancy for failures of one of the four 4×4 GSDS.

A reliability block diagram RBD, depicted in Figure 4, is derived for the OXC considering two required functions a) or b):

a) The one link approach considers *one of the four input fibers, carrying four input channels to four output channels.* The incoming optical signal has to be demultiplexed by the according input wavelength division multiplexer WDM, cross-connected by the four GSDS,

and multiplexed by the four output WDMs. A failure of one of the five necessary WDMs leads to at least one channel not present at the output. This is considered as a system failure. Therefore the first part of the RBD is a series structure of N = 5 WDMs (5 × E1). E3 and E4 represent the tunable filter TF and the DXC, respectively, in a series structure. They build a redundant path in case of a failure of one of the GSDS (E_2, E_2', E_2'', E_2'''; therefore together they form a 4-out-of-5 hot redundancy in the RBD. In this approach the GSDS have a failure rate of 10'250 FIT.

b) The four links approach considers all *four input fibers, i.e. 16 input channels are switched to 16 output channels.* The incoming optical signal at each input fiber have to be demultiplexed by in total four input WDMs, cross-connected by the four GSDS, and multiplexed by the four output WDMs. A failure of one of the 8 necessary WDMs leads to at least three channels not present at the output (one wavelength channel on a failed input WDM could be rerouted

794

Table 2. Designations and failure rates of OXC system elements.

Element	E1 (N×)	E2–E2‴		E3	E4
Function	WDM	GSDS		TF	DXC
Failure rate	$\lambda 1$	$\lambda 2$		$\lambda 3$	$\lambda 4$
Value [FIT]	100	10'250*	24'000**	400	3'500

* one link requirement ** four links requirement.

through the DXC). This is considered as a system failure. The first part of the RBD in this case is a series structure of N = 8 WDMs (8 × E1). Again the TF and DXC represented by E3 and E4 in a series structure – serve as a redundant path in case of a failure of one of the GSDS. Therefore the 4-out-of-5 structure remains as second part in the RBD. In this approach the GSDS have a failure rate of 24'000 FIT.

Failure rates λ_1–λ_4 and designation of the respective elements are given in Table 2. Elements having very small failure rates such as couplers, splitters, and connectors are neglected [Wosinska 1993]. The according repair rates μ_1–μ_4 are $1/6\,h^{-1}$ for all elements, thus mean time to repair (MTTR = $1/\mu$) is 6 h.

4.1.1 General assumptions

In order to investigate reliability and availability of complex repairable structures the following assumptions are defined:

a) *Independence*: failure free operating times and repair times are statistically independent.
b) *Continuous operation*: each element of the system is operating when not under repair.
c) *No further failures after system down*: at system down the system is brought into an up state by repair. Failures during system down time are not considered.
d) *Online repair*: redundant elements can be repaired without interruption of system operation.
e) *As good as new*: a repaired element is considered to be as-good-as-new.
f) *Repair resources*: One repair crew. Last-in/first-out strategy applied, i.e. priority of repair on the last failed unit. Spare parts availability is unlimited. Preventive maintenance is neglected.

4.1.2 Availability results

Since a reliability block diagram is available for the OXC configurations a Monte Carlo simulation is chosen for availability analysis. Since there is no general rule a wide range of test runs between 10'000 and 10'000'000 is simulated and the convergence of the obtained results observed. Simulation is done without specified random number seed for the pseudo random number generator and with a predefined confidence level $\gamma = 95\%$ and the following settings: online

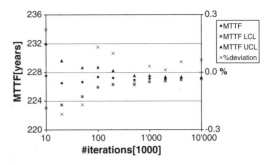

Figure 5. MTTF as a function of iterations.

Table 3. Comparison of analytical and simulated results.

	Analytical one link	Monte Carlo simulation one link*	Monte Carlo simulation four links*
MTTF (LCL/UCL$^{\nabla}$) [years]**	227	226.6/227.5	136.6/137.1
Steady state unavailability [-]	$3.015 \cdot 10^{-6}$	$3.012 \cdot 10^{-6}$	$4.93 \cdot 10^{-6}$
Mean down time [min/year]	1.56	1.56	2.56

$^{\nabla}$ lower/upper confidence limit, * 1'000"000 iterations, $\gamma = 95\%$, ** 1 year = 360 days = 8640 h as in [Wosinska 1993].

repairs are allowed, i.e. system continues to operate while a redundant element is under repair; and components do not operate after system failure. MTTF with a lower confidence limit LCL and an upper confidence limit UCL, steady state unavailability and mean down time are calculated as parameters of interest. Figure 5 shows the convergence of the simulated MTTF with upper and lower confidence limits UCL, LCL as a function of the number of iterations and the deviation in % of the respective mean MTTF from the analytical value of 227 years. In Table 3 the results are compared to the analytical solution given in [Wosinska, 1993].

4.1.3 Result of OXC sensitivity analysis

The result of a sensitivity analysis on the OXC system for the *one link* approach is shown in Figure 6. Reference point A shows the result with parameters λ_1–λ_4 and repair rate μ as given in Table 2. The impact of variation of the single parameters over a range of a factor 100 (0.1–10) on the total down time of the system is calculated using Monte Carlo simulation (100'000 runs, 95% confidence level) and is given as down time variation factor. It is clearly shown that the WDMs with

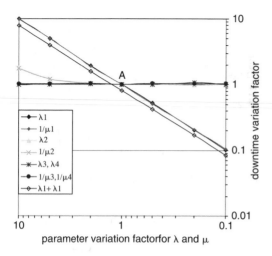

Figure 6. Sensitivity analysis for the optical cross connect, *one link* requirement.

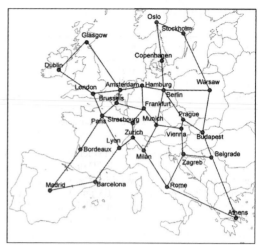

Figure 7. Topology of a proposed pan-European network.

λ_1 and μ_1 dominate system availability with an approximately linear dependency. In contrast, down time is almost completely insensitive to variation of λ_2, λ_3, λ_4 and their appropriate repair rates.

A decrease in system mean down time can therefore be achieved by a lower failure rate λ_1 or an increased repair rate μ_1 (shorter MTTR). The difference between these two possibilities is that a lower λ_1 will lead to higher MTTF, while MTTF remains constant when varying μ_1. This must be considered for spare parts provisioning.

Due to this result and the fact that WDMs are single points of failure in the reliability block diagram it is worth to consider redundancy for this element. However, implementing a 1-out-of-2 hot redundancy for the input WDM in the *one link* consideration results in a down time decrease of only a factor of approximately 0.8, depicted as $\lambda_1 + \lambda_1$ in Figure 6.

4.2 Pan-European network

4.2.1 System description
A pan-European long haul network is analyzed in order to investigate the sensitivity of the network availability on failure rates of spans and nodes. Different mesh topologies for such networks were studied within the European COST action 266 [De Maesschalck et al., 2002]. The topology shown in Figure 7 is one of them and consists of 28 nodes and 41 spans.

In many investigations availability of paths with defined start and end nodes within networks is analyzed in order to check their suitability for specific services requiring a defined availability and quality of service along this path [Wosinska & Svensson, 2000].

This work however aims at availability analysis of the complete network, i.e. the probability that a defined number of nodes and spans are available. In combination with a sensitivity analysis the influence of failure rates can be weighted as well as possibilities to improve network derived.

The network is therefore generalized containing 28 identical nodes and 41 identical spans with the following characteristics:

Nodes The key systems in nodes of future optical networks are optical cross connects OXCs and optical add drop multiplexers OADMs. In any case nodes consist of a large number of components with different failure and repair rates. In this work they are considered as a black box characterized by a constant failure rate λ_N with a typical value of $10'000$ FIT and a mean time to repair $\text{MTTR}_N = 1/\mu_N$ of 5 hours.

Spans A span of length L connecting two electro-optical nodes consists of line equipment (booster amplifier and receiver, λ_{LE}) within the two nodes, fiber optic cable (λ_F), and a number N_{OA} of optical amplifiers (λ_{OA}) along the span. A single span is therefore represented by a series reliability block diagram.

Thus, a span has a failure rate of $\lambda_S = L \cdot \lambda_F + 2 \cdot \lambda_{LE} + N_{OA} \cdot \lambda_{OA}$. The mean time to repair MTTR for a span can be approximately expressed as:

$\text{MTTR}_S \approx 1/\lambda_S(L \cdot \lambda_F \cdot \text{MTTR}_F + 2\lambda_{LE} \cdot \text{MTTR}_{LE} N_{OA} \cdot \lambda_{OA} \cdot \text{MTTR}_{OA})$. For simplification the $\text{MTTR}_S = 1/\mu_S$ for a span is assumed to be 20 h, even though it is a little shorter due to the line equipment MTTR_{LE} of 5 h.

The average span in a generalized network derived of the topology in Figure 7 has a length L of 625 km and is equipped with $N_{OA} = 8$ optical amplifiers.

The availability of the proposed network can then be analyzed using a THMP as described in section 2.3. The detailed model however would have 2^{N+S} states, which is not solvable in practice. This state space explosion is contained by truncation of the analyzed states to represent no more than K concurrent failures. When component failure rates are orders of magnitude smaller than component repair rates, the truncation has negligible effect on the accuracy of the network availability results [ITU-T G911].

Figure 8 shows the diagram of state transition probabilities truncated at 4 concurrent failures.

4.2.2 General assumptions
In order to investigate the availability of the network as a whole assumptions a), b), d), e) as in

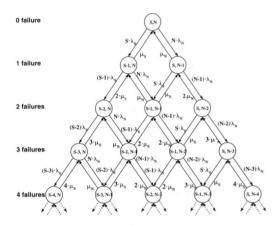

Figure 8. Diagram of state transition probabilities in $(t, t + \delta t]$ for an generalized network consisting of N nodes with identical, constant failure rate λ_N and identical, constant repair rate μ_N, and S spans with identical, constant failure rate λ_S and identical, constant repair rate μ_S. All transition rates must be multiplied by δt. Designation of states: number of operating Spans and Nodes. Unlimited repair crew, arbitrary t, $\delta t \downarrow 0$, Markov process. Probabilities to remain in a state are not shown for clarity of Figure.

section 4.1.1. are applied. Assumptions c) and f) however are:

c) *Further failures after system down*: if the system is in a down state, further failures up to a predefined number of failures can occur.

f) *Repair resources*: Unlimited repair crew. Spare parts availability is unlimited. Preventive maintenance is neglected.

4.2.3 Network availability results
Network availability is calculated using the typical failure rates and repair rates in Table 4, thus $\lambda_S = 82'500$ FIT, $\lambda_N = 10'000$ FIT, $\mu_S = 0.05$, $\mu_N = 0.2$, $S = 41, N = 28$. Using the Markov state transition diagram in Figure 8 enables a flexible definition of network failure by assigning up and down status to each state. The following scenarios are considered with designation of states as in Figure 8 and up states of the network given in brackets showing the number of operating spans and number of operating nodes N; for all other states the network is considered down, results are given in Table 5.

a) Network unprotected. Up: $\{S, N\}$
b) One span failure protected. Up: $\{S, N\}, \{S\text{-}1, N\}$
c) Two span failures protected. Up: $\{S, N\}, \{S\text{-}1, N\}, \{S\text{-}2, N\}$
d) Three span failures protected. Up: $\{S, N\}, \{S\text{-}1, N\}, \{S\text{-}2, N\}, \{S\text{-}3, N\}$
e) One span and one node failures protected. Up: $\{S, N\}, \{S\text{-}1, N\}, \{S, N\text{-}1\}, \{S\text{-}1, N\text{-}1\}$
f) Two span and one node failures protected. Up: $\{S, N\}, \{S\text{-}1, N\}, \{S\text{-}2, N\}, \{S, N\text{-}1\}, \{S\text{-}1, N\text{-}1\}, \{S\text{-}2, N\text{-}1\}$
g) Two span and two node failures protected. Up: $\{S, N\}, \{S\text{-}1, N\}, \{S\text{-}2, N\}, \{S, N\text{-}2\}, \{S\text{-}1, N\text{-}1\}, \{S\text{-}1, N\text{-}2\}, \{S\text{-}2, N\text{-}1\}, \{S\text{-}2, N\text{-}2\}$

4.2.4 Results of network sensitivity analysis
The sensitivity of network availability on the two parameters assumed to have the highest uncertainty – failure rate of fiber λ_F and failure rate of node λ_N – is investigated for protection scenarios b) *one span failure protected*, and e) *one span failure and one node failures protected*. Results of the network down time

Table 4. Failure and repair rates of network elements.

Function	Node	Fiber optical cable	Line equipment	Optical amplifier
Failure rate	λ_N	λ_F	λ_{LE}	λ_{OA}
Failure rate range	0.1 k–100 k FIT	1–1000 FIT/km	–	–
Typical failure rate	10 k FIT	100 FIT/km	2000 FIT	2000 FIT
Repair rate	μ_N	μ_F	μ_{LE}	μ_{OA}
MTTR $=1/\mu$	5 h	20 h	5 h	20 h

Table 5. Network availability and down time for $\lambda_N = 10'000\,\text{FIT}$ and $\lambda_F = 100\,\text{FIT/km}$.

Protection scenario	Network availability	Down time [hours/year]
a) unprotected	0.93333	584
b) one span	0.99647	30.9
c) two spans	0.99856	12.6
d) three spans	0.99860	12.3
e) one span and one node	0.99787	18.7
f) two spans and one node	0.99995	0.4
g) two spans and two nodes	0.99996	0.39

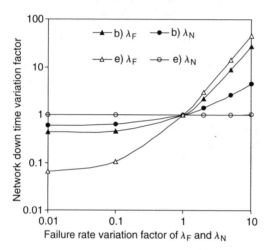

Figure 9. Network down time variation factor as function of failure rate variation factor of λ_N and λ_F for protection scenarios b) and e).

variation factor in Figure 9 are referenced to the respective down times calculated for typical values in Table 5.

Network down time and therefore loss of traffic in both protection scenarios is more sensitive to fiber failure rate λ_F, respectively the resulting failure rate of spans $\lambda_S = L \cdot \lambda_F + 2 \cdot \lambda_{LE} + N_{OA} \cdot \lambda_{OA}$. For the protection scenario b) which is usually considered in telecommunication, the variation of fiber failure rate λ_F over a range of 1000 (0.01–10) will vary network down time by a factor of approximately 69, whereas the same variation on node failure rate λ_N yields a factor of 7.5, thus roughly 9 times less.

This influence is even stronger for protection scenario e) where the variation of λ_N has practically no influence on network down time, in contrast to λ_F which will vary network down time by a factor of more than 700.

It is therefore clear that efforts to obtain more accurate failure rates as well as possible improvements in network availability must be targeted on spans and their components. However, this does not imply that node failures can be neglected because the cost of node restoration presumably can be very high.

5 CONCLUSION

We have shown that sensitivity analysis gives a valuable input for overall system availability optimization even if and especially when the underlying failure rates are uncertain. The main benefit of such analysis is not the accuracy of the obtained results but the identification of weaknesses, i.e. components or system parts whose failure or repair characteristics have the highest impact on availability at system level. Results of sensitivity analysis can therefore give a contribution to sustainable systems in the sense of the optimization of performance, availability, and cost.

ACKNOWLEDGMENT

The authors thank the *Swiss Federal Office for Education and Science BBW* for financing the project *RAMON Reliability, Availability and Maintainability of Optical Networks (BBW project C01.0087)*.

REFERENCES

Maier G., Pattavina A., De Patre S., Martinelli M., Optical network survivability: protection techniques in the WDM layer, *Photonic Network Communications* 4:2/4, pp 251–269, 2002.

Clouqueur M., Grover W.D., 2002. Availability Analysis of Span-Restorable Mesh Networks. IEEE Journal on Selected Areas in Communications. Vol. 20, No. 4, pp 810–821, May 2002.

Clouqueur M., Grover W.D., Mesh-restorable networks with complete dual failure restorability and with selectively enhanced dual-failure restorability properties, *OptiComm* 2002, Boston, MA, USA, July 29–Aug. 2, 2002.

Choi H., Subramaniam S., Choi H.-A., On double-link failure recovery in WDM optical networks, *IEEE INFOCOM*, pp. 808–815, 2002.

Telcordia SR-332, Reliability Prediction Procedure for Electronic Equipment, Telcordia Technologies, 2001.

UTE RDF 2000, Reliability Data Handbook: A universal model for reliability prediction of electronics components, PCBs and equipment. Union Technique de lÖElectricite. 2000.

Spragins J.D. et al., 1986. Current telecommunication network reliability models: a critical assessment, *IEEE Journal on Selected Areas in Communications,* (SAC-4)/7, 1986.

Held M., Nellen P.M., Wosinska L., Availability Calculation and Simulation of Optical Network Systems. *Proceedings*

SPIE Photonics Fabrication Europe 2002, Brugge, Belgium, in press, 2003.

MIL-HDBK-217, Notice 2, Reliability Prediction of Electronic Equipment, DoD, 1993.

IEC 61709, Electronic components reliability: Reference conditions for failure rates and stress models for conversion, 1996.

Wosinska L., 1993. Reliability Study of Fault-Tolerant Multiwavelength Nonblocking Optical Cross Connect Based on InGaP/InP Laser-Amplifier Gate-Switch Arrays, *IEEE Photonics Technology Letters,* Vol.5, No.10, 1993.

Wosinska L., Svensson T., 2000. Analysis of Connection Availability in All-Optical Networks. *Proceedings NFOEC,* 2000.

Mikac B., 2002. Modelling and Availability Evaluation of Optical WDM Networks. *COST 270 Meeting*, Graz, April 8–9, 2002.

De Maesschalck S., Colle D., Lievens I., Pickavet M., Demeester P., Mauz C., Jaeger M., Inkret R., Mikac B., Derkacz J., Pan-European Optical Transport Networks: an Availability-based Comparison. Submitted to *Photonic Network Communications,* 2002.

ITU-T G911. Parameters and calculation methodologies for reliability and availability of fibre optic systems. International Telecommunication Union, 1997.

Safety and Reliability – Bedford & van Gelder (eds)
© 2003 Swets & Zeitlinger, Lisse, ISBN 90 5809 551 7

A dynamic risk model for evaluation of space shuttle abort scenarios

E.M. Henderson
National Aerospace and Space Administration (NASA), Lyndon B. Johnson Space Center, Houston TX, United States

G. Maggio, H.A. Elrada & S.J. Yazdpour
Science Applications International Corporation, New York, NY, United States

ABSTRACT: As the Space Shuttle ascends to orbit it transverses various intact abort regions evaluated and planned before the flight to ensure that the Space Shuttle Orbiter, along with its crew, may be returned intact either to the original launch site, a transoceanic landing site, or from orbit. An intact abort may be initiated due to a number of system failures but the highest likelihood and most challenging abort scenarios are initiated by a premature shutdown of a Space Shuttle Main Engine (SSME). The potential consequences of such a shutdown vary as a function of a number of mission parameters but all of them may be related to mission time for a specific mission profile.

   This paper focuses on the Dynamic Abort Risk Evaluation (DARE) model process, applications, and its capability to evaluate the risk of Loss Of Vehicle (LOV) due to the complex systems interactions that occur during Space Shuttle intact abort scenarios. In addition, the paper will examine which of the Space Shuttle subsystems are critical to ensuring a successful return of the Space Shuttle Orbiter and crew from such a situation.

1   INTRODUCTION

The Dynamic Abort Risk Evaluation (DARE) model is a dynamic risk assessment model that evaluates the risk of intact Space Shuttle abort scenarios, namely, Return To Launch Site (RTLS), Transoceanic Abort Landing (TAL) and Abort To Orbit (ATO).

   The DARE model was developed by Science Applications International Corporation (SAIC) under the sponsorship of the NASA Johnson Space Center (JSC). DARE is being used to:

- Assess the risks of each of the abort scenarios and identify their major risk contributors
- Identify the abort scenarios with the least risk in the event that one of three Space Shuttle Main Engines (SSME) benignly shuts down during ascent
- Perform risk-informed design and operational trade studies

2   SPACE SHUTTLE INTACT ABORT OPTIONS

As the Space Shuttle ascends to orbit, it transverses various intact abort regions planned before the flight to ensure that the Space Shuttle Orbiter, along with its

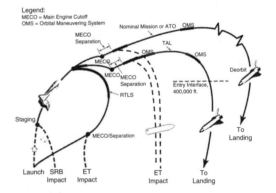

Figure 1.   Space shuttle intact abort options.

crew, may be returned to either one of the following: the original launch site (RTLS), a transoceanic abort landing site (TAL), or from orbit (ATO) in the event that an abort is initiated. Each of these options is shown in Figure 1. If a failure should occur late in the trajectory, mission control may opt to simply continue on to the planned orbit (Press to MECO – PTM). If a significant failure during an intact abort occurs, then a contingency abort is executed.

As the Space Shuttle climbs to orbit, the consequences of a single engine shutdown become less severe. More specifically, as the vehicle gains momentum and altitude it is better situated to maneuver and conduct aborts. The most challenging of the aborts available to a Shuttle pilot are the return to the launch site (RTLS) or a transoceanic abort landing (TAL). Both of these abort modes require the Shuttle to have a minimum altitude and velocity providing sufficient energy (kinetic and potential) to enable the spectrum of vehicle state vectors. This spectrum is necessary to allow for the successful completion of either of these difficult landing approaches. These minimum energy profiles require that the remaining two engines continue firing for the length of time necessary to ensure that the proper profile is attained to attempt an abort.

3 DARE MODELING CONCEPT

The DARE modeling concept is shown in Figure 2. It begins with an initiating event, and terminates with an end state. This methodology was designed to handle time-dependent conditional failure probabilities for the pivotal events, and conditional failure probabilities based on pivotal events interdependencies (i.e. failure of the Orbital Maneuvering System/Reaction Control System (OMS/RCS) to dump propellant would affect the weight of the Orbiter at landing and thereby increase the Loss Of Vehicle (LOV) risk). For the

Figure 2. DARE modeling concept.

DARE model, pivotal events represent failures of systems or conditions that might result in a failure to successfully abort a mission and land the Orbiter intact.

4 DYNAMIC MODEL DEVELOPMENT

Figure 3, shown below, represents the outline for the DARE model development process. The DARE modeling approach involves a disciplined time dependent analysis of scenarios.

The process begins with the Dynamic Event Diagram (DED) and ends with producing the report of results. A brief description of the steps that were taken to develop DARE is outlined in the following subsections.

4.1 Dynamic event diagrams (DED)

In a conventional probabilistic risk assessment (PRA) practice, an event sequence diagram (ESD) is developed to represent the successes/failures of the systems required for a successful abort. Reaching beyond the limits of the ESD, the DED was developed to show the *time* at which these systems are required to initiate and function. DEDs, developed for each abort scenario (RTLS, TAL and ATO), represent the systems that must function to accomplish a successful abort. These systems can be considered pivotal events as defined above. Simply, the DED is used as a map to further investigate the initiating events, the abort region, and the systems to be modeled and analyzed. Figure 4, shown below, represents a non abort-specific DED.

4.2 Identification of abort initiators

An intact abort may be initiated due to a number of system failures, but the most likely abort scenarios are

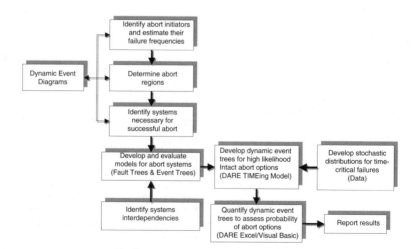

Figure 3. DARE methodology process.

initiated by a premature benign shutdown of one of three SSME (Maggio G. et al. 2000). Therefore, this was the abort-initiating event that was chosen for further in-depth dynamic probabilistic risk assessment.

4.3 Determine abort region

As shown in Figure 5, different intact abort options are available and chosen depending on the time a benign engine shutdown occurs. The dynamic probabilistic risk model for intact aborts uses abort region estimates from flight dynamicists that include the time of a benign engine shutdown and its associated available abort options. The actual abort boundaries are calculated prior to each flight based on a number of complex mission time-dependent parameters such as vehicle weight, orbital inclination, and available landing sites, etc.

There are cases where these abort regions might overlap. If this occurs (where performance loss is the only factor), the next available abort option would be chosen (i.e. if both are available, a TAL is always chosen over an RTLS) when the initiating event is a benign engine shutdown.

Exercising any one of the abort options will depend upon the time at which the first engine benignly shuts down. For example, an RTLS due to an engine out at lift-off is selected at the earliest time, approximately two minutes, twenty seconds into the mission (after solid rocket booster separation). The probability of a benign shutdown of the second engine or the probability of a catastrophic failure of either one of the

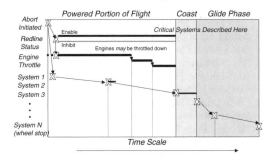

Figure 4.   Dynamic event diagram (DED) example.

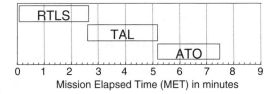

Figure 5.   Typical abort boundaries used in DARE.

remaining two engines will vary depending on when the first engine shutdown occurred.

The STS101 and STS111 are the missions that are currently being evaluated by the DARE model. The baseline DARE model uses 104.5% as its throttle level of the Block II SSMEs. However, the model has the capability to evaluate the risk of elevated power levels as well as evaluating the LOV risk for various other planned missions.

4.4 Identification of systems necessary for a successful abort

The following is a list of the modules that make up the DARE model. These modules represent the systems that must function during an abort in order to accomplish a safe landing.

SSME: As the Space Shuttle ascends to orbit, the severity of the consequences of a single engine shutdown changes with time. The DARE model evaluates the LOV risk due to a second SSME shutdown and the LOV risk due to catastrophic failure of one of the remaining two SSMEs.

External Tank (ET) Separation: When the SSMEs are shutdown, the ET is jettisoned and it breaks up as it enters the Earth's atmosphere. Successful ET separation depends on the success of the Reaction Control System (RCS) to perform its intended function.

ET Debris Hit: During a TAL abort, subsequent to Orbiter/ET separation, the ET may be in relatively close range of the Orbiter when it ruptures during entry. This increases the risk of the Orbiter being struck by ET debris.

Powered Pitch Around (PPA): The PPA maneuver is performed only during an RTLS abort. The PPA maneuver changes the Orbiter's attitude from heads-down going away from the launch site to heads-up pointing toward the launch site by an approximately 60 to 70 degree maneuver Failure to perform this maneuver within a specific time period would lead to landing short of the runway or missing it completely which was assumed to result in LOV.

Powered Pitch Down (PPD): This is an RTLS specific critical maneuver. During the powered portion of an RTLS abort, in order for altitude and flight path angle constraints to be met at Main Engine Cut Off (MECO), a positive angle of attack ($\sim$30 degrees) is required. However, an angle of attack of $-$ 2 degrees is necessary to ensure a safe ET separation. This transition is what is referred to as powered pitch down and must be completed rapidly by properly gimbaling the engine thrust to avoid large sink rates, which may cause overheating and overstressing. Failure to perform the PPD maneuver was quantified by modeling the inability to gimbal one of the remaining two SSME.

Control Surfaces: Control surface failures were assumed the driving risk factor in maintaining control of the vehicle during the glide phase of RTLS and TAL intact aborts. A streamlined model based on aircraft parts reliability was developed to account for this risk.

Thermal Protection System: The failure modes of the TPS during a TAL abort are assumed similar to those that might occur during a nominal mission. These failure modes include: debris damage with subsequent tile debonding (external tank, right solid rocket booster nosecone or orbital debris) and debonding from other sources (heating loads, high temperatures, aero-acoustic loads, cycle degradation of bonding materials and maintenance errors).

Touchdown Associated Failures: This includes the risk due to: excessive sink rate, tire failure(s) or runway under/overshoot, misalignment.

Orbital Maneuvering System/Reaction Control System (OMS/RCS) Dumps: Similar to the MPS, the OMS/RCS is modeled in DARE for its ability to dump propellant to reduce the likelihood of LOV. The amount of OMS propellant onboard the Shuttle during launch is mission-specific. During an abort, the OMS propellant is dumped by burning it through both OMEs and possibly through the twenty-four aft RCS thrusters. This improves the performance during the abort and enables the Orbiter to achieve an acceptable landing weight and center of gravity (C.G.) location.

Main Propulsion System Dump: The MPS is modeled in DARE for its ability to dump propellant to reduce the likelihood of LOV. If the MPS fails to dump enough propellant, it would result in violating the Orbiter C.G. envelope, which in turn, could lead to loss of control of the vehicle.

4.5 System interdependencies

A number of interdependencies exist among the modules listed above. For example, failure of the OMS/RCS to dump propellant during an abort would increase the risk at touchdown because an excessively heavy Orbiter will have too high a sink rate, possibly causing the Orbiter to slap down onto the runway. Furthermore, a high sink rate could increase the likelihood of landing gear collapse and/or tire blowout.

4.6 Develop and evaluate models for abort systems

The DARE modeling process included the development of risk models for each of the systems discussed above. Some of these risk models used conventional risk methodology that include fault tree and event tree type models. For example, the PPD was modeled and quantified using fault tree analysis and was later incorporated into the DARE model. The OMS/RCS

uses an event tree to both display and quantify the different failure scenarios of the OME and RCS jets. Some of these sequences resulted in end states that violate the Orbiter's C.G. limits affecting flight stability and were assumed to result in LOV. Other systems, such as the SSME time dependent risks are modeled using dynamic event trees and quantified using MS Excel and Visual Basic programming.

4.7 Data analyses

Statistical distributions representing the pivotal event failures were constructed and used as input into the DARE model to quantify the abort risks. For example, lognormal distributions for both the likelihood of the orbiter getting hit by debris due to the breakup of the external tank in the Earth's atmosphere and the mean time-between-failure for the benign shutdown and catastrophic failure probabilities for the SSMEs were developed and used to evaluate the LOV risk. The Weibull distribution was used to evaluate the time dependent SSME risks.

4.8 Quantification

The DARE model has a graphical user interface (GUI) which employs complex MS Excel and Visual Basic macros to perform a Monte Carlo simulation to evaluate the abort risks.

The DARE model uses a set of inputs provided by NASA flight dynamicists. These parameters are mission specific and include the following items:

- Abort Landing Site
- Inclination (deg)
- First Engine Shutdown Time (sec)
- Solid Rocket Booster Staging (sec)
- OMS Assist On and Off (sec)
- Abort Initiation Time (sec)
- RCS Ignition Time (sec)
- OMS/RCS Dump Stop (sec)
- OMS Dump Duration Time (sec)
- RTLS Turnaround Time (sec)
- Powered Pitch Down Time (sec)
- Main Engine Cutoff Time (sec)
- External Tank Separation Time (sec)
- Orbiter center of gravity at lift off (x and y (in))
- Second stage lift off weight (lbs)
- Orbiter lift off weight (lbs)

The results for each specific trajectory are generated and output within the MS Excel environment.

The DARE model evaluates the LOV risks due to failure of any of the systems described above. During the quantification process, DARE selects conditional failure probabilities for the systems components based on: (1) the time of the first engine shutdown; and (2) other system failures during a specific abort scenario.

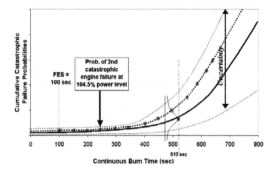

Figure 6. Estimation of Catastrophic Failure (conceptual)

An example of how the DARE model evaluates the dynamic risk is illustrated below in Figure 6. This figure represents a conceptual diagram for the evaluation of the LOV risk due to a catastrophic failure of either one of the two remaining SSMEs.

Given a first engine shuts down at 100 seconds, the probability of a second engine catastrophically failing is given by the dotted line (with "stars"). The dotted line is the Weibull function randomly selected from the family of possible Weibull distributions within the range of uncertainty shown. One Weibull distribution is chosen using a lognormal distribution around a baseline Weibull (dark center line). The selected Weibull distribution provides the cumulative probability with operation time of one of the remaining engines. At each time interval (shown by the stars on the dotted line), the probability of one of the remaining engines failing catastrophically is calculated by taking the difference between the cumulative probabilities at that time minus the cumulative at the previous time segment. Furthermore, this family of Weibull distributions represents the SSME failure probability given that the power level is at 104% following the first engine shutdown. The family of Weibull distributions may change if the power level varies from this baseline case.

5 APPLICATIONS OF THE DARE MODEL

The primary application of the DARE model is to evaluate the Loss of Vehicle (LOV) risk due to an abort given a benign engine shutdown. The DARE model has been used to perform sensitivity analyses and trade studies that include:

- Evaluating the abort risk of the SSME Block II vs. Block IIA
- Assessing the risk of being in a black zone given a second engine shutdown
- Assessing the probability of crew bailout given a second engine shutdown

- Assessing the probability of having to perform an East Coast Abort Landing (ECAL) given a second engine shutdown
- Evaluating the abort risk given higher SSME throttle levels

The DARE model is being used, and will continue to be used in the decision-making process to assess and improve the Space Shuttle operation, maintenance, and emergency procedures for the nominal and the abort scenarios. For instance, the DARE model may be used in conjunction with a PRA to consider the addition of another segment to the Shuttle's Solid Rocket Booster (SRB) thereby adding additional failure modes to the system for the benefit of eliminating a relatively high risk RTLS abort mode.

The DARE model may also be used to conduct risk trade studies to consider abort options for other types of failures besides those for which abort options exist today. For instance, one may postulate abort options for the type of TPS damage believed to have caused the Columbia accident and determine whether based on limited information a decision to abort the mission by conducting an RTLS, which has a relatively benign heating profile, is prudent as compared to the risk of continuing the mission. These types of trades are not possible without a probabilistic abort risk modeling capability that allows one to compare the risks between abort and mission continuance.

6 CONCLUSIONS

This paper discussed the development and application of a Space Shuttle Dynamic Abort Risk Evaluation (DARE) model that may not only be utilized to assess the relative risk of existing abort options but also to study the benefits and effectiveness of proposed abort scenarios. In addition the model, in conjunction with a PRA of nominal flight scenarios, may be implemented to balance the risk of continuing a mission with degraded systems versus conducting an abort to return the vehicle to a landing site as rapidly as possible. Lastly the DARE model also offers NASA the capability to determine the overall impact on both nominal and abort scenarios of significant changes to the Space Shuttle design. Changes that while increasing nominal risk may decrease the likelihood or risk of abort scenarios. Conversely, some design changes may increase the likelihood of aborts while diminishing overall flight risk by transforming potential catastrophic events into benign abort situations.

In summary, the DARE model is a potential tool in NASA's tool box to thoroughly evaluate design and operations decisions for the comprehensive set of mission scenarios that include the benefit or implications of mission abort.

ACKNOWLEDGEMENTS

The development of DARE has been a true multidisciplinary, multi-organizational undertaking. Although the core team consisted of a few NASA and SAIC managers and engineers there have been contributions made by many others throughout NASA and their contractors.

The authors would like to extend a special thanks to Jan Railsback and Richard Heydorn; without their initial impetus and vision this effort would not have been possible. Barbara Conte who despite her commitment to Space Shuttle mission planning also took the time to provide the team with specialized abort trajectories for our analyses. Finally, Andy Foster, who's insight into Space Shuttle abort protocols and engineering knowledge allowed the team to digest and comprehend the complexities of abort situations.

REFERENCES

Fragola J.R., Maggio G., et al. 1995. Probabilistic Risk Assessment of the Space Shuttle, A Study of the Potential of Losing the Vehicle during Nominal Operation, SAIC/NY 95-02-25, New York.

Heydorn, R., Railsback J. and Nguyen C. 1998. *Shuttle Abort Probabilities for Redline Limits Management*, JSC White Paper.

Maggio G. and Fragola J.R. 1995. Combining Computational Simulations with Probabilistic Risk Assessment Techniques to Analyze Launch Vehicles, Reliability and Maintainability Symposium Proceedings.

Maggio G., et al. 1997. *A Dynamic Probabilistic Assessment of the Premature Shutdown Risk for the Space Shuttle Main Engine*, SAIC/NY 97-12-01, New York.

Maggio G., Heydorn R.P. and Railsback J.W. 1998. *A Risk-Based Assessment of the Space Shuttle Main Engine Redline Management Philosophy*, AIAA 98-3207.

Maggio G., Railsback J.W., Heydorn R.P. and Safie F., 2000. *A Dynamic Risk Assessment of Space Shuttle Intact Abort Scenarios*, Probabilistic Safety Assessment and Management (PSAM5) Proceedings.

Safety and Reliability – Bedford & van Gelder (eds)
© *2003 Swets & Zeitlinger, Lisse, ISBN 90 5809 551 7*

Reliability analyses and observables

A. Hjorteland & T. Aven
Stavanger University College, Stavanger, Norway

ABSTRACT: Although reliability theory and analysis is a well-established discipline in engineering applications, many analysts often find it challenging to conduct reliability analysis in practice. Seldom the theoretical models and methods are directly applicable. There are usually a number of ways of approaching a reliability problem. The reliability theory offers a number of possible advanced models and techniques, but what is the proper approach in a specific case is not easily extracted from the theory. The general impression is that a probability model needs to be specified, and parameters estimated. The focus of the analysis is then the parameters, quantities that are often difficult to interpret.

In this paper we point at the need for a rethinking of this tradition. An approach where focus is observable quantities would make the analysis easier to understand, and the message from the analysis is more direct. In the paper we present and discuss such an approach, within a predictive Bayesian framework.

1 INTRODUCTION

As a part of a modification task on an offshore installation several design options for a gas compressor system are to be considered. The alternative system configurations are to be analysed and evaluated to support decision-making. The basis for the evaluation is some sort of performance in the operational phase. The first issue to be discussed is how to measure this performance. There are several indices to choose from, and they represent different starting points of the reliability analysis.

To conduct the reliability analysis, a team is established, comprising a reliability analyst and an engineer who is familiar with the various systems being analysed. Suppose a brainstorming within the team gives the following list of possible indices of system performance for further evaluation:

- Y, representing the number of failures of the system in the time interval $[0, T]$,
- A prediction of Y,
- λ, defined as the expected (mean) number of failures, $\lambda = EY$,
- An uncertainty distribution $P(\lambda') = P(\lambda < \lambda')$,
- A distribution of Y, $P(Y < y)$.

Now, what indices should we use, and what effect would the various indices have on the analysis? The question is what type of performance measures we should look for. Basically we see two categories:

1. Observable quantities such as number of failures, production volumes, number of times production is below a certain number, etc.
2. Probabilities and expected values representing some average performance of similar units to the one being studied.

Depending on the selected category, different approaches for the reliability analysis is necessary, and uncertainty is reflected to variable degree. The purpose of this paper is to address these issues and give some guidance on how to think.

Our recommended thinking is in line with the Bayesian paradigm. The essential feature of the Bayesian thinking is that probability is a measure of expressing uncertainty about the world, seen through the eyes of the assessor and based on some background information (knowledge). It is however not obvious how such a thinking should be implemented in practice, in a decision-making context, and different frameworks can be established.

We believe that the focus of the analysis should be observable quantities; that is quantities that express states of the "world" or the nature, that are unknown at the time of the analysis but will (or could) become known in the future. The purpose of the analysis is to predict these quantities and assess uncertainties. Probability is used as a measure of uncertainty related to the true values of these quantities.

Our recommended approach is referred to as the predictive Bayesian framework, cf. Aven (2003). See also Aven (2001). This framework put emphasis on observables quantities and predictions of these, in contrast to a more traditional Bayesian approach, which put attention to unobservable parameters. The predictive Bayesian approach has to large extent been implemented in the Norwegian oil and gas industry for analysis of production systems (NORSOK (1998)).

The framework is in line with the modern, predictive Bayesian theory as described in e.g. Bernardo & Smith (1994), Barlow (1998) and Barlow & Clarotti (1993), cf. also Lindley (2000) and Bedford & Cooke (2001).

2   OBSERVABLE QUANTITIES

In the introduction section we divided the performance measures into two categories, probabilities representing relative frequencies and observable quantities.

In this section we will look closer into this categorization and discuss in more detail what we mean by saying that a quantity is observable. In particular we discuss the conditions required to define a relative frequency as an observable quantity.

When using observable quantities focus is placed on quantities expressing states of the "world"', i.e., quantities of the physical reality or the nature, that are unknown at the time of the analysis but will, if the system being analysed is actually implemented, take some value in the future, and possibly become known. This is not always the case when using relative frequencies.

Consider the system reliability analysis example presented in the introduction section. The basis for the evaluation is some sort of observable performance measure in the operational phase. To make the discussion more clear, we will in the following study only one of the system concepts.

The first step is to identify suitable performance measures, expressing the goodness of the system. Three specific performance measures will be discussed to explain the meaning of an observable quantity. The system analyzed is in the design phase and has not been observed in operation. Suppose the systems operational lifetime is 10 years, and Y_i represents the performance measure in year i, defined by the number of system failures occurring in this period, $i = 1, 2, ..., 10$, see Figure 1.

We need to define precisely what a system failure means. And according to such a definition, we would have only one correct value of Y_i. The fact that there could be measuring problems in this case, some failures are not reported, does not change this. The point is that a true number exists according to the definition

and if sufficient resources were made available, that number can be found. This example illustrates that observable quantities include cases where we better could describe the quantities as potential observable quantities.

Based on the Y_is, we can define an accumulated, observable quantity by the sum, representing the total number of failures for the ten-year period. And from this quantity we may compute the mean value $\bar{Y}_{10}$, which is of course also an observable quantity.

Next, suppose that we add observations from similar systems, to the one studied. We may for example have observations from four other gas compressor systems, for five previous years. The observations are from the same type of system as studied, from other installations. See Table 1.

Let $\bar{Y}_{20}$ denote the mean of these 20 years of observations and $\bar{Y}_{30}$ the mean of the 30 observations, adding the 10 observations $Y_1, Y_2,..., Y_{10}$. Furthermore, let P_{20} and P_{30} denote the corresponding empirical distribution functions obtained from these observations. Clearly, these quantities are all observable quantities.

The reliability analysis may now proceed along two different paths, either

1. Use the 20 observations to predict Y_i and $\bar{Y}_{10}$, and assess uncertainties in these values.
2. Extend the population by a thought construction of the 30 observations to an infinite population of similar units, and introduce the limiting quantities P and E expressing the relative frequency interpreted probabilities and expectations. The aim of the reliability analysis is then to estimate these limiting quantities.

We will discuss these two paths in the following section. Here we will just reflect on to what extent P and E are observable quantities.

In general we conclude that if the infinite population of "similar systems" is just a thought experiment,

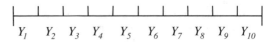

Figure 1. Y_i represents the number of system failures in year i, $i = 1, 2, ..., 10$.

Table 1.   Observed number of failures.

System \ Year	1	2	3	4	5
System					
"Similar" sys. 1	4	5	3	2	6
"Similar" sys. 2	7	4	6	5	2
"Similar" sys. 3	6	1	4	3	6
"Similar" sys. 4	2	4	1	2	3

it is fictional, then the relative frequencies are not observable. We will not be able to observe the relative frequency in the future; it is not a state of the world.

Now in a practical context one may argue that there is not much difference between $\bar{Y}_{30}$ and P_{30} on the one hand, and E and P on the other. If we can ignore the error by going from 30 to the limits, P and E are approximately observable. What is a sufficient large population needs to be determined in each application. Ensuring large populations of say 30 observation means that one often has to extend the meaning of similar to a broad category, with for example quite different operating conditions. Consequently, a non-relevant population is established.

The important point is however not whether the quantities should be called observables or not, but what should be the focus of the analysis. Should the analysis express knowledge related to Y_i and $\bar{Y}_{10}$, or P and E. In the latter case, the data becomes more or less relevant for the system being studied. This is a key issue, which we will discuss in the following section.

3   THE RELIABILITY ANALYSIS

Again, consider the system reliability analysis example in the introduction section. For the sake of simplicity, the system being studied is a $1 \times 100\%$ system configuration as illustrated in Figure 2.

In the following we will discuss the two different reliability analysis paths, according to Section 2. Depending on these, the analysis approach will be different. This relates to:

- How the introduced models are used and understood,
- The treatment of uncertainty,
- How to assess the unknown quantities of the model.

3.1   *Analysis approach with focus on observable quantities*

As a basis for the decision to be taken, it is of interest to obtain information about relevant performance measures, in this case Y_i and $\bar{Y}_{10}$. Now in the planning phase Y_i and $\bar{Y}_{10}$ are unknown, thus we are led to prediction of these quantities. Later we can accurately measure the performance; they are observables. The prediction can be done in different ways. We may compare with similar systems if available, or we could develop a more detailed model of the system reflecting the various equipment and subsystems of

the concept of interest; a reliability model. Using the observations for the four additional systems of Table 1, a prediction of Y_i and $\bar{Y}_{10}$ equal to 3.8 is obtained. Furthermore, uncertainty distributions are derived for these observable quantities. This assessment presumes that the analyst judges the data to be relevant.

Regardless of the approach taken, we will use all relevant information to predict Y_i and $\bar{Y}_{10}$, and assess uncertainties in these values.

A model is a simplified representation of a real world system. The general objective of developing and applying models in this context is to arrive at performance measures based on information about related quantities and a simplified representation of the real world phenomena. Since models are used to reflect the real world, they only include descriptions of relationships between observable quantities.

When adopting the predictive Bayesian approach to reliability analysis, modelling is a tool for identifying and expressing uncertainty and thus also means for potentially reducing uncertainty. The uncertainty can be identified by including more system-specific information in the analyses, in terms of an expanded information basis for uncertainty statements and in terms of the model structure itself. Furthermore, modelling gives flexibility to the reliability analysis since it allows us to express uncertainty in the format found most appropriate to obtain the objectives of the analysis.

To model the gas compressor system, let X_t represent the state of the system; $X_t = 1$ if the system is functioning at time t and $X_t = 0$ if the system is not functioning at time t. We assume $X_0 = 1$. Let T_m, $m = 1, 2, \ldots$, represent the positive length of the mth operation period of the system, and let R_m, $m = 1, 2, \ldots$, represent the positive length of the mth repair time of the system. The system performance is illustrated in Figure 3. Here Y represents the number of system failures in $[0, t]$. Clearly, Y is a function of the T_ms and R_ms, and through this model a relationship between Y and the T_ms and R_ms is established.

A topic closely related to the use of models, which is widely discussed in the literature, is model uncertainty. Several approaches to interpretation and quantification of model uncertainty are proposed in the literature, cf. Nilsen and Aven (2003) and the references therein. In our setting, the model is a function of

Figure 2.   System of $1 \times 100\%$ system configuration.

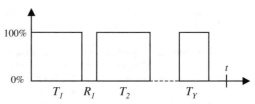

Figure 3.   Illustration of uptimes and downtimes.

809

the T_ms and R_ms. It provides a framework for mapping uncertainty about the observable quantities of interest, from expressions of epistemic uncertainty related to the observable quantities on a lower system level, and does not in itself introduce additional uncertainty. What is interesting to address is the goodness or appropriateness of a specific model to be used in a specific reliability analysis and decision context. Clearly, a model can be more or less good in describing the world. No model reflect all aspects of the world, per definition it is a model. In this setting the model is merely a tool judge useful for expressing partial knowledge of the system.

Returning to the reliability analysis, the quantities T_m and R_m are unknown and we express our uncertainty related to what will be the true values in the future. Now, how do we express uncertainty about the observable quantities? We need a measure to quantify our uncertainty, and probability is our answer. When focusing on observables we introduce probabilities, as tools for expressing our uncertainties about observable quantities.

Probabilistic expressions reflect epistemic uncertainty or lack of knowledge related to the values of such quantities. The reference is a certain standard such as drawing a ball from an urn. If we assign a probability of 0.10 for an event A, we compare our uncertainty of A to occur with drawing one specific ball from an urn having 10 balls. The assignments are based on all available information and knowledge.

When expressing uncertainty, all probabilities are conditioned on the background information (knowledge) that we have available at the time we quantify our uncertainty. This information covers historical system performance data (as in Table 1), system performance characteristics (such as policies, goals and strategies of a company, type of equipment to be used, etc.), knowledge about the phenomena in question (gas leakage, corrosion, human behaviour, etc.), decisions made, as well as models used to describe the world. Assumptions are an important part of this information and knowledge. Based on the available knowledge, the probability distributions are established.

To proceed the reliability analysis we express our uncertainty related to the unknown quantities in the model, T_m and R_m. To express our uncertainty related to T_m, we use a probability distribution F for all uptimes, where $F(t)$ is an exponential distribution $(1 - e^{-\lambda t})$, with 1/20 as the failure rate (λ). The repair times, R_m, are judged to be a constant equal to 1. These judgements are of course a rather strong simplification; we judge all uptimes and repair times to be independent and ignore learning when observing future lifetimes and repair times. But in some cases the background information is such that we could justify the use of independence; when an observation would have marginal influence on the assessed quantity. For further

consideration on justifying independence, see Aven (2003) and Aven & Kvaløy (2002). These references also show how to implement a full Bayesian analysis when independence cannot be justified.

Now, based on the assigned distributions of T_m and R_m, we deduce the distribution of Y. Since the repair times are small compared to the uptimes, we see that the distribution of Y, when using independence, will be an approximate Poisson distribution with mean λt, i.e.

$$Y \approx Poisson(\lambda t) \tag{1}$$

We refer to Aven (2003) for a detailed discussion on the relevance of the Poisson distribution in this case.

In this simplified case study, the uncertainty distribution is easily established, but in general it may be difficult to compute the uncertainty distributions for Y_t and approximation formulas need to be used, cf. Aven & Jensen (1999). It is also common to use Monte Carlo simulation, cf. Aven (2003).

3.2 Analysis approach with focus on probabilities and expected values

In this framework reliability is supposed to be an objective characteristic or property of the system being analyzed, expressed by probabilities and statistical expected values of random variables such as the number of system failures for a given period time. To be more specific we now draw attention to P and E, and the estimation of these true values from the system reliability analysis example in Section 2. The reliability analysis provides estimates of the true reliability for the system, i.e., probabilities and expected values.

To carry out the analysis according to this approach, the previously observed data from the four "similar systems" in Table 1 are used directly to estimate E and P. The mean number of system failures per year of operation considering the observed data is found to be 3.8. When it comes to the probability of, say maximum four failures occurring in a one year period, an estimate of 0.65 is established.

Uncertainty is in this framework interpreted as the gap between the estimated value and the true value. In this example where no system modeling is done, uncertainty is mainly related to the input data. This contribution to uncertainty is related to what extent the data is representative for the system of interest, lack of knowledge about the system being analyzed and statistical variation.

By using quantities like standard deviation, or confidence interval it is possible to express statistical variation based on the observed data. These methods are often seen in practical applications of reliability analysis. It is, however, difficult to quantify other sources of uncertainty.

In the previous section we discussed the meaning of a model when focusing on observable quantities. Now, when focusing on P and E, the interpretation of models is somewhat different. Also here the model represents a simplification of the world, but it also is a tool to generate the true values E and P. The model in this case covers the probabilistic quantities, the exponential and Poisson distributions. In this setting we have two levels of uncertainty; the estimates of lower system quantities and choice of the model. See Aven (2003) for further details.

4   DISCUSSION AND CONCLUSION

To conduct a reliability analysis is not straightforward. A clear understanding of the concepts to be assessed is required. The key point is to determine whether the analysis should have a focus on observable quantities or theoretical underlying quantities expressing relative frequencies in a population of similar units to the one studied. We find that the former approach is most attractive in a practical reliability context as the observable quantities are the key quantities of interest and they are simple to understand, whereas the statistical concepts are hard to explain the meaning of.

Of course, if an infinite (or large) population of similar situations can be defined, then the statistical concepts (relative frequencies, parameters of a distribution) represent a state of nature – they are observable quantities – and we can speak about our uncertainty of these quantities. The question is whether such a population of similar situations should be introduced. In our view, as a general rule, the analyst should avoid introducing fictitious populations. In the example considered in this paper, it may not be obvious how to define an infinite population of similar situations, and without a precise understanding of the population, the uncertainty assessments become difficult to perform and it introduces an element of arbitrariness. What is a fictitious population and what is a real population is a matter for the analyst to decide, but the essential point we are making here is that the analyst should think first before introducing such a population.

REFERENCES

Aven, T. 2003. *Foundations of Risk Analysis*. New York: John Wiley & Sons, to appear.

Aven, T. & Hjorteland, A. 2003. A predictive Bayesian approach to multistate reliability analysis. *International Journal of Reliability, Quality and Safety Engineering*, to appear.

Aven, T. & Jensen, U. 1999. *Stochastic Models in Reliability*. New York: Springer.

Aven, T. & Kvaløy, J.T. 2002. Implementing the Bayesian paradigm in risk analysis. *Reliability Engineering and System Safety*, 78, 195–201.

Barlow, R.E. 1998. *Engineering Reliability*. Philadelphia: SIAM.

Barlow, R.E. & Clarotti, C.E. & Spizzichino, F. 1993. *Reliability and decision making*. London: Chapman & Hall.

Bedford, T. & Cooke, R.M. 2001. *Probabilistic Risk Analysis: Foundations and Methods*. Cambridge: Cambridge University Press.

Bernardo, J.M. & Smith, A. 1994. *Bayesian Theory*. Chichester: Wiley & Sons.

Lindley, D.V. 2000. The philosophy of statistics. *The Statistican*, 49, 293–337.

Nilsen, T. & Aven, T. 2003. Models and model uncertainty in the context of risk analysis. *Reliability Engineering and System Safety*, 79, 309–317.

Safety and Reliability – Bedford & van Gelder (eds)
© 2003 Swets & Zeitlinger, Lisse, ISBN 90 5809 551 7

Industrial evaluation of an adaptive elicitation process to capture engineering knowledge about product design reliability

R. Hodge, J. Quigley & L. Walls
University of Strathclyde, Glasgow, Scotland

M. Balderstone
Smiths Industries, Cheltenham, England

D. Lumbard
Agusta Westland, Yeovil, England

J. Marshall
Goodrich, Birmingham, England

ABSTRACT: During the product development process it is important to use engineering knowledge to identify potential reliability problems within the proposed design. A process, developed to elicit engineering concerns about new designs and to estimate the chance concerns may occur as failures, has been applied to several industrial cases. The translation of the elicitation theory to practice is examined. Mapping techniques are used to better capture engineering reasoning about design concerns. The data gathered as part of the evaluation of the approach allows recommendations for adapting the elicitation process for different engineering cultures.

1 INTRODUCTION

Reliability assessment is an important activity during product design since decisions made at this stage will impact the potential performance and cost of the product downstream (Denson, 1998). It is usual to base reliability assessment on designs, not yet available as hardware, upon the considerable knowledge that exists in the experience base of engineers responsible for the family of designs (Kerscher et al., 1998). To harness this knowledge requires formal procedures.

While reliability tools, such as FMEA, do aim to capture engineering understanding of the ways in which designs will fail, such methods have been criticised because they are used too late in the product development cycle. Therefore the information gathered is not used to impact design decisions (Marshall and Newman, 1998). Yet there is considerable material describing formal processes for eliciting expert knowledge. In risk assessment the seminal text is Cooke (1991) and recent papers include Hokstad et al. (1998), Cagno et al. (2000), Rosqvist (2000). The theory and practice reported in risk assessment has informed the elicitation process designed to capture engineering knowledge in design and development. For example,

see Walls and Quigley (2001), Hodge et al. (2001) who describe a methodology for eliciting probabilities about design concerns and report some preliminary test results.

These findings highlight the dangers of designing processes that are scientifically controlled but may be difficult to implement in practice. Therefore we adopt the approach of Eden and Huxham (2002) who stated the need for more research in the practical arena, since it represents typical situations for problems in probability assessment to support decision making. We also believe that it is important to allow users of the process to assess its usefulness and applicability to the products for which it was designed. Design focused elicitation will never gain acceptance in industry until applications to real life problems can be demonstrated and there is evidence that the process provides useful data to the design team.

This paper reports the application of an elicitation process to several cases from the aerospace industry and an evaluation of the effectiveness of the approach within the different design environments. One goal of the research project supporting this investigation is the provision of elicitation guidance that will allow it to be implemented successfully within different aerospace company cultures and design environments. Hence, as part of our evaluation we have captured data about the engineering culture and we make some tentative analysis of how this affects the implementation of the elicitation process.

Section 2 overviews the elicitation process and outlines the role of mapping techniques to capture and structure engineering knowledge. Section 3 describes the industrial cases, while Section 4 presents guidance for adapting the process to different design environments.

2 ELICITATION PROCESS

2.1 Overview

The elicitation process developed for REMM is an adaptation of the procedures proposed in Walls and Quigley (2001) for eliciting a prior distribution to support Bayesian reliability growth modeling. Both approaches are informed by the logical framework of the SRI model (Merkhofer, 1987). Figure 1 presents an overview in the stages of the REMM elicitation process and shows how they relate to the aforementioned approaches.

2.2 Mapping engineering knowledge

Details about each stage in the elicitation process are given in Walls and Quigley (2001) and procedures are given in REMM Elicitation Guidance. One major change to the original process that arose through initial test application was the need for mapping techniques (Fortune and Peters, 1995). These were found to provide an excellent way of engaging engineers, capturing understanding and providing records of beliefs and probabilities. These replace the original spreadsheet tables of concern descriptions and probabilities.

A form of spray diagram is used during interviews. A paper template is prepared where information is captured by design change and the concerns that arise. The reasoning behind concerns are also noted as sub-concerns and the linkages between elements are drawn.

3 PLAN FOR CASE RESEARCH

3.1 Research questions

The goal of the research is to increase the effectiveness of the elicitation process across engineering cultures within the UK aerospace sector.

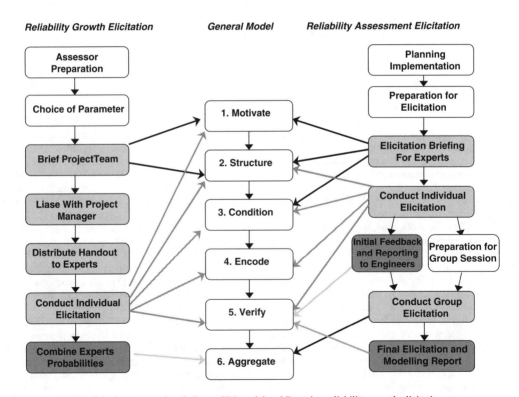

Figure 1. REMM elicitation process in relation to SRI model and Bayesian reliability growth elicitation process.

This translates to the following aims and objectives:

A. To evaluate the application of the REMM elicitation process to real-life projects
B. To understand the effect of engineering culture on the effectiveness of the elicitation
C. To provide guidance on adapting the process given specified implementation environments.

3.2 Research methodology

A comparative multiple case study approach has been selected as the primary means of meeting the research objectives. The comparison across implementation environments will facilitate the evaluation of the generic applicability of the process whilst the use of case study research allows an understanding of the nature and complexity of the process in practice (Voss et al., 2002). This research tactic was regarded as the most effective method of meeting the need for explaining the complex nature of elicitation whilst maintaining a holistic perspective.

The cases have been selected to ensure a representative set of engineering products and processes typical of the UK aerospace industry. Table 1 provides a summary.

However the case study approach does present certain challenges. Conducting case studies, particularly multiple cases is time consuming and requires care to be taken when drawing generalizations from limited data. To ensure relevant data on process effectiveness is captured, from the perspective of all stakeholders, shown in Figure 2, various data collection methods are used. For example, interview, both semi-structured and unstructured, as well as personal observation and informal conversations aim to allow triangulation of findings (Easterby-Smith et al. 1991). This should provide confidence in the findings of each case assuming there is no conflict across data sources. Conflict would identify indifference, that is no compelling evidence either way, therefore no inference would be possible.

The evaluation framework is embedded within the elicitation process interacting at key points as illustrated in Figure 3. A balance is sought between allocation of resources for elicitation and evaluation based on the trade-off with expected benefits. In order not to detract from the primary elicitation, much of the evaluation data collected from direct industrial stakeholders has been conducted retrospectively. This has the advantage of allowing company personnel to provide opinions in a holistic manner having had time to reflect on the process.

Throughout implementation the REMM facilitator and other researchers recorded personal observations. These provide useful data such as cross-company comparisons, hunches about relationships, anecdotes and informal observations on engineers' acceptance of the process. Although it does raise the potential problem of researcher bias since the researchers starts from a perspective of belief in the elicitation process. Such personal bias can shape what is heard, observed and recorded. However the collation of observations from multiple stakeholders, employing multiple data collection techniques, should minimise the risk of bias, although we remain conscious of this issue.

3.3 Measuring engineering culture

Engineering culture is considered as the collection of the attributes and characteristics of the environment in which the elicitation will be applied. The engineering culture encompasses all variables in the implementation process. These variables can be grouped together under headings such as organisational culture, the design team profile, the product under investigation, the champion (or principal informant), the experts being interviewed and the facilitator(s) involved in

	Direct Involvement	Indirect Involvement
Research Areas	Elicitation facilitator	REMM Modellers
Project Team	Project facilitator	Reliability and design experts

Figure 2.   Elicitation stakeholders.

Table 1.   Summary of industrial cases.

Case	Industrial partner	Product	Changes from earlier design
1	Goodrich	Electronic engine controller for medium engine	Materials, environment, platform
2	Goodrich	Electric actuation of primary and secondary control surfaces	Design, environment, material, manufacturing process
3	Agusta Westland	Top cover assembly of main rotor gearbox	Material, manufacturing process
4	Smiths	Electrical power management	Platform, environment

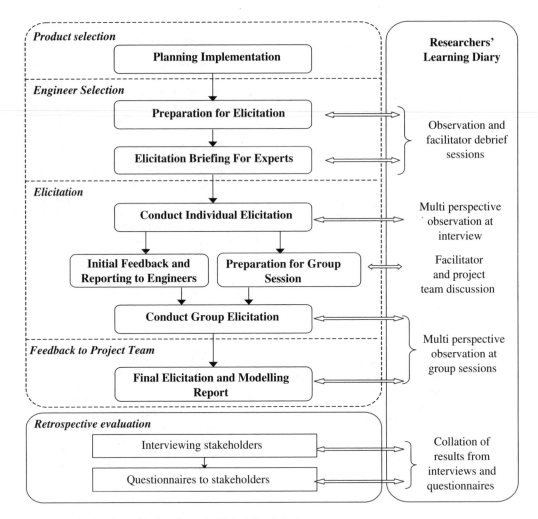

Figure 3.   Evaluation data collection plan embedded within elicitation process.

interviewing. Table 2 shows the attributes that will be measured and used to characterise engineering culture. The measures noted are to be supplemented with descriptions.

4   CASE HISTORIES

4.1   *Case 1*

The design is an EEC for a business jet, which will spend a high proportion of life in storage. The design is a replacement for a hydra-mechanical controller and the motivation for change is the use of new technology. The base circuit design being modelled on that for the controller of a large commercial jet. Therefore the new design is a variant of two existing products – the system to be replaced that provides

information about the operational profile and the technology to be incorporated from a related system.

The organisation responsible for the design has a close working relationship with their direct customer and reliability tasks are negotiated. Responsibility for reliability lies with one the design engineers within the project team and is consistent with the project based structure of the organization. The design team is between 10–15 people depending on development stage and includes manufacturing engineers in an effort to better integrate design and manufacturing activities. The design team are colocated. The elicitation is championed by the lead electronic engineer and is facilitated by a member of the REMM project team.

The elicitation has been implemented on two occasions for this project – in late concept and early detailed design stage. However only part of the

Table 2. Selected measures of engineering culture.

Group	Criteria
Organisation	Position of company in the supply chain
	Closeness of working relationship with supplier
	Closeness of working relationship with top level customer
	Autonomy on reliability tasks
	Barriers between engineering groups
	Role of reliability within team
	Company structure
	Number of times elicitation has been conducted in company
Product under investigation	Product type/technology
	Novelty of design
	Aircraft type
	Location on aircraft
	Stage of design
Champion	Who was the champion (i.e. the person with whom the elicitation facilitator worked with during implementation)
	Design project membership of champion
	REMM membership of champion (knowledge of REMM process)
	Knowledge domain of champion
	Knowledge of in-service characteristics of product
	Knowledge of REMM elicitation process at time of implementation
	Understanding of reliability tools and techniques
Experts	Knowledge of experts
Facilitator	Experience in elicitation
	Multiple researchers

procedures outlined in section 2 have been implemented. Namely, the initial elicitation conducted the process to individual interview stage only, although a group meeting was held to feedback data and review initial modelling. The second elicitation followed up the initial concerns to assess which still existed and what other issues had arisen as the design had matured.

Initially considerable time was spent communicating the purpose and procedures of elicitation to the champion and in return for the facilitator to learn more about the system design. This was necessary to ensure the elicitation was grounded in the project requirements. The first elicitation involved seven engineers whose domains of knowledge were mutually exclusive. The second elicitation involved 13 engineering, only 5 of whom were involved in the first elicitation. The reason for the changes were necessity, due to personnel changes on project, as well as a desire to include multiple experts per domain, especially to augment the experience base of the project team. Interviews

followed the standard elicitation procedures and lasted about one hour per engineer. Those engineers interviewed a second time were given a shortened version that began with a review of their earlier concerns since last review, however the reduction in the process was felt to diminish the quality of the data collected. Another change in the second elicitation was the decision to use two facilitators, the lead being taken by the original REMM facilitator and assisted by a company facilitator. This allowed the company facilitator the gain first hand experience of the process and to contribute more detailed questioning about design concerns based on his better knowledge of the products.

Overall feedback on the elicitation process suggests that it is regarded as an effective way to gain insight into design concerns and the design team were willing to contribute their views and experience to make the process constructive. The decision to widen the scope of engineering experts and to include a company facilitator is considered to have lead to better quality data, although in part this may also be due to the REMM facilitator moving up the learning curve.

The most challenging aspect has been managing expectations of the project management, who have been keen to use the elicitation data as a means of reliability demonstration. This is positive insofar as it indicates acceptance of the approach within the reliability programme and the links to existing company risk assessment processes have been made explicit. A change in design team culture has been observed with an increase in awareness of reliability and its impact throughout the product lifecycle. Although the elicitation process was novel to the engineers, its best outcome has been to encourage them to review the impact of their design decisions on product reliability.

4.1 Case 2

The design is a secondary flight control system, which contains a significant amount of novel technology, its environment is expected to be more harsh than usual for control systems and the system should be in stand-by mode for most of its life. The elicitation has been conducted in early detailed design. The design team, which comprises 20–25 engineers, are grouped by sub-system function and so there is common general engineering knowledge of design as well as specialist expertise. The nature of the system means that the team is quite internalised and does not have access to extensive historical data apart from certain sub-systems. The relationship with the customer is less close than for case 1 and the team have more autonomy in deciding an appropriate reliability programme. Again one the design engineers has responsibility for reliability within the project. The lead project engineer was the main champion within the team, while a technology consultant helped to facilitate the elicitation.

Twelve engineers were selected for interview as they were considered to cover the appropriate knowledge domains. However these engineers did not possess a deep understanding of the issues arising with the family of control systems from which aspects of the new design will be inherited. Therefore it was decided to match design engineers from the new and existing designs to provide a richer source of data and to encourage learning between projects. All but one of the engineers had been interviewed previously therefore again briefing was an important first stage.

The REMM facilitator managed the interviews, lead discussions and recorded the data, while the company facilitator followed through questioning to explore issues in more technical details and to seek clarification and completeness of concerns. Two company observers were also present during most interviews as they wanted to learn about the process first-hand. Despite initial reservations about the potential to introduce bias into the data collection, it appears that their presence did not influence discussion in any negative way. As before, the scope of issues considered was extended to cover programme risks as well as reliability concerns. Only the latter is of direct interest to subsequent reliability analysis. As it was not feasible to schedule group sessions, the concerns noted were reviewed by the company facilitator who identified duplicates and therefore produced a final list of unique concerns for the design.

Again the co-operation of the engineers to contribute to the process has been beneficial. The multiple facilitators worked well as it permitted the REMM facilitator time to manage and record, while the company facilitator de-mystified in-house jargon and procedures. The company facilitator also played a vital role in customizing the generic questions for the design and so enhanced the interview output, although it did sometimes lengthen interviews leading to delays in schedules and increasing facilitator and expert fatigue. For this case, the mapping approach used to capture and record data was felt to have worked well.

Although useful data was output about design concerns, there were several hiccups in implementing the process. At the beginning the REMM facilitator did not possess a real understanding of the design to be examined and this had a knock-on effect to the plan for customizing the elicitation. Engineering experts were not fully briefed because the introductory group session was held too far in advance of the interviews and was only partially attended due to competing project demands. This in turn led to concern over what data was being collected in interviews and how it would be used. Also the need to brief within interviews also lengthened them and reduced the time available for constructive elicitation. Time-keeping, or lack of, became a major problem.

The interviews were scheduled in order of convenience, however on retrospect it would have made more sense to have scheduled systems engineer first as they possess overarching knowledge of the design, then electronic designers who have specialist knowledge of functions and specialists who provide technical support across all sub-systems. Also, in future manufacturing experts should also be included in the latter group. The decision to double up in some interviews by combining project engineers with experienced designers of the related family of control systems did not work as well as expected. The reason was the impact of the seniority of the control engineers that made project engineers more reticent. Although a group feedback session was conducted to gain some closure on the process, in future the results of commonality assessment of concerns should be available in time for the group session.

4.2 Case 3

The design is a mechanical transmission assembly for a multi-role helicopter undergoing an upgrade programme. The modification of the gearbox topcover aims to mitigate known operational reliability problems as well as improving the weight gain capacity. The product is near the end of detailed design and the preliminary design review had already been conducted.

The company designing the modification is a systems integrator and the design team is a fully integrated project team member of the customer. The champion for this elicitation is a reliability engineer who advises the project team. The team is quite small, only 10 engineers, and has developed a long-term working relationship. Considerable overlap in experience is anticipated although each engineer has a specialized role within the team. The novelty of the design is low and there is considerable field experience of the base system.

The REMM facilitator led the elicitation process, while another REMM team member was present as an observer, and the company champion provided vital support in integrating the elicitation with the design team's objectives and schedule. The implementation of the elicitation was almost directly in line with the planned procedures. Only 6 engineers were interviewed and these were on time and constructive. De-brief sessions were held between interviews to review how well the process was working. Reports were issued within days of the sessions being completed and a group workshop held within a week to discuss the results. This allowed debate about the concerns and their commonalities as well as about the probabilities allocated. Agreement was reached to assign nominal probabilities in some cases for initial modeling and where views differed and could not be reconciled, the variation in probabilities was noted and their impact investigated through what if reliability modeling. While most of the concerns raised were known to the design team as a result of earlier reviews, there were insights generated about their chance of occurrence.

Overall this case worked well, primarily because many lessons had been learnt from the previous two cases and great effort was made not to repeat mistakes. For example, the facilitators were provided with background information about the design well in advance and so had time to develop an understanding of the available data and so prepare the elicitation plan better. Interviews were scheduled for longer time periods to provide some slack within interviews for briefing and to allow time between interview to reflect on how well interview went and to prepare for next. The order of interviews starting with systems engineers and then moving to mechanical engineers and other specialists worked well. The system engineer helped set the scene by considering the breadth of the system function and the potential weaknesses while the specialists were able to explore issues in more detail.

The design team was highly motivated and experienced in conducting risk assessment. However they felt the elicitation process provided a systematic way of encouraging independent thinking about issues and their chance of occurrence as failures. The completeness of this process also permitted further development of the data reporting system from elicitation to provide details of the uncertainty in issues and probabilities.

This elicitation did raise a number of limitations. For example, the selection of this project was risk averse and so had less impact on design decisions than might be expected of an elicitation conducted earlier in design. There remain concerns that the previous exposure of the design team to H/M/L scoring systems and brainstorming of the concerns has led to anchoring on previous information. Also, the group elicitation phase is still immature as the engineers were observed to anchor on 50% as a default probability.

5 EVALUATION SUMMARY

The evidence collected from the cases suggests that the design of the elicitation process is satisfactory, in principle. However there is no doubt that the guidance documentation must be expanded to provide better examples of how the process should be adapted for a given context. This presents one of the natural outputs of the evaluation activity.

From the case studies we have managed to gather some evidence about the impact of engineering cultural factors on the effectiveness of implementation. While not feasible to measure scientifically, we present our subjective inferences.

Those factors that directly affect the design team, such as team size and structure, role of reliability specialist, degree of design novelty, product reliability requirements, do appear to have a direct impact on implementation. This is not surprising since these factors influence the motivation and practical organisation of elicitation implementation. For example, use of data, reporting systems, interview duration and schedule. We have seen how proper recognition of these can lead to successful implementation, while lack of attention can make the process less efficient, even though effort has been expended to ensure effective data are collected.

On the other hand, factors related to the wider engineering culture, such as position in supply chain, organisational structure, type of technology, have an indirect impact, if any, on the elicitation. This is positive as we would wish the process to be robust to such variance in environments.

The biggest impact factor has no doubt been the degree of experience of the facilitation. As the facilitators' learning has increased the elicitation has progressed more effectively and efficiently. The challenge now is to capture that experience in the guidance documentation so that company facilitators can assume full responsibility for leading elicitation and so benefit from the experience gained through these evaluation cases.

Most discussion has focused upon the operationalisation of the elicitation process and little has been said about the probability assessment. In part this is due to the practical nature of the early cases and in part due to the users primary interest in feedback in the form of engineering descriptions. However, the third case did serve to illustrate the usefulness of probability as a means of encouraging debate about the importance of concerns. Reviewing the details of the elicitation process post-cases and comparing it with the original specification, as well as the underlying principles described in Walls and Quigley (2000), we find that the changes introduced have corrupted some of the probability assessment.

For example, mapping concerns has led to debate about whether the probabilities required were P(failure/concern exists) or P(failure/fault). In actual fact the probabilities of interest are P(fault exists in design/concern exists). Another problem has been the introduction of a H/M/L scale in initial probability assessment that anchors experts on intervals rather than point estimates. The intent behind this system was simply to provide a means of checking consistency of values since the previous approach used in growth modeling where the expert was presented with the prior distributions of number of concerns based on fractile probability estimates and direct rating was not accepted as meaningful by the design engineers.

6 CONCLUSIONS

The goal of this research has been to improve the effectiveness of the elicitation process. The case studies have allowed us to identify many shortcomings of the process and to suggest, and demonstrate, means of improvement for some. The lessons learnt have been used to produce a revised version of the elicitation guidance that is based on experience as well as theoretical principles.

ACKNOWLEDGEMENTS

This work has been conducted under the auspices of the REMM project and sponsorship of the UK DTI CARAD programme.

REFERENCES

Cagno, E., Caron, F., Mancini, M. and Ruggeri, F. 2000. Using AHP in determining the prior distribution on gas pipeline failures in a robust Bayesian approach, *Reliability Engineering and System Safety*, 67, 275–284.

Cooke, R.M. 1991. *Experts in uncertainty*, John Wiley.

Denson, W. 1988. The history of reliability prediction, *IEEE Transactions in Reliability*, 47, 321–328.

Easterby-Smith, M., Thorpe, R. and Lowe, A. 1991. *Management research: An introduction*, Sage.

Eden, C. and Huxham, C. 2002. Action research in *essential skills for management research*, D Partington (Ed), Sage.

Fortune, J. and Peters, G. 1995. *Learning from Failure: The Systems Approach*, Wiley.

Hodge, R., Evans, M., Marshall, J., Quigley, J. and Walls, L. 2001. Eliciting engineering knowledge about reliability during design – lessons learnt from implementation, *Quality and Reliability Engineering International*, 17, 169–179.

Hokstad, P., Oien, K. and Reinertsen, R. 1998. Recommendations on the use of expert judgement in safety and reliability engineering studies, *Reliability Engineering and System Safety*, 61, 65–76.

Kerscher, W., Booker, J., Bennet, T. and Meyer, M. 1998. Characterising reliability in a product/process design assurance program, *Proc Reliability and Maintainability Symposium*, Los Angeles.

Marshall, J. Newman, R. 1998. Reliability enhancement methodology and modeling for electronic equipment – the REMM Project, *Proc. ERA Avionics*, 4.2.1–4.2.13.

Merkhofer, M. 1987. Quantifying judgemental uncertainty: methodology, experiences, and insights, *IEEE Trans. On Systems, Man. and Cybernetics*, 17, 741–752.

Rosqvist, T. 2000. Bayesian aggregation of experts' judgements on failure intensity, *Reliability Engineering and System Safety*, 283–289.

Voss, C., Tsikriktsis, N. and Fohlich, M. 2002. Case research in operations management, *International Journal of Operations and Production Management*, 22, 195.

Walls, L. and Quigley, J. 2001. Building prior distributions to support bayesian reliability growth modelling using expert judgement, *Reliability Engineering and System Safety*, 74, 111–128.

Safety and Reliability – Bedford & van Gelder (eds)
© 2003 Swets & Zeitlinger, Lisse, ISBN 90 5809 551 7

Use of risk acceptance criteria in Norwegian offshore industry: dilemmas and challenges

P. Hokstad & J. Vatn
SINTEF Industrial Management, Trondheim, Norway

T. Aven
Stavanger University College, Stavanger, Norway

M. Sørum
Statoil, Stavanger, Norway

ABSTRACT: Risk analyses have been used in the Norwegian offshore petroleum industry for more than two decades. The analyses have been closely linked to the use of risk acceptance criteria (RAC). This paper gives a short review of the motivation and use of RAC in the management of risk at offshore installations. Focus is on the present situation and challenges in the Norwegian petroleum industry. The paper will discuss some recommended principles for arriving at risk acceptance. The relation between the acceptability of risk and the cost/benefit of risk reducing measures is also considered, mainly in the context of the ALARP (As Low As Reasonably Practicable) principle.

1 INTRODUCTION

Today the use of risk acceptance criteria (RAC) is an integrated part of risk management in the Norwegian offshore industry. It is a requirement that specific considerations shall be undertaken to decide on which risks are tolerable ("acceptable") in order to initiate an activity. Companies shall formulate RAC stating the overall acceptability of the total risk related to the activity of an installation. But more generally, RAC can be used in various ways to control and manage risk; both in the design phase, to assist decisions of operational/management personnel, and by modifications.

However, there is an ongoing discussion on the role of RAC in the risk management of offshore installations. This discussion is partly related to the level of detail and accuracy of the risk analyses, which are carried out to verify that the risk is acceptable. Are these analyses of a format so that the resulting assessment of the risk level is really credible, that serve the primary goal of both verifying that risk will be tolerable/acceptable, and be a tool in the continuous work to reduce risk? Or has this analysis now more become an exercise (that to some extent can be "manipulated" to give the "correct" result), and could

a closer follow up of risk during operation be a better way to achieve the overall goals?

The present situation for offshore installations on the Norwegian continental shelf is that the specified RAC are often met with good margins. It may be considered risky to start the design process with small margins, in particular since changes are expected, and these are simpler to include if the margins are good. So it could be risky from an economic point of view to specify too ambitious RAC. Then risk analyses become a verification tool, and the RAC does not give a drive for improvements. Of course, other means exist for obtain such drive, but in our view there is a need for rethinking of the RAC regime presently being used for the Norwegian sector. We are not using the tool of risk assessment and cost/benefit analyses in the best possible way.

The paper will discuss these issues, based on the experience in the Norwegian offshore industry. First a short overview of the use of RAC in Norwegian offshore industry is given, in particular based on Aven et al. (1998) and the new regulations of the Norwegian Petroleum Directorate, NPD (2001, 2002). Next, the paper will discuss various aspects of the use of RAC and normative (ethical) issues, e.g. see Fishhof et al. (1981), Vatn (1998). A major question concerns the relation between cost/benefit and the acceptability of risk, e.g. use of the ALARP principle, see Melchers (2001), Schoefield (1998). Finally some recommendations are formulated.

2 USE OF RISK ASSESSMENT AND RAC IN NORWEGIAN OFFSHORE INDUSTRY

The Norwegian safety regime reflects the basic principle of the licensees' full responsibility for ensuring that the petroleum activity is carried out in compliance with the conditions laid down in the legislation. The safety regime has since 1985 been founded on internal control, meaning that the authorities' supervisory activities are aimed at ensuring that the management systems of the licensees are catering adequately for the safety and working environment aspects in their activities.

The initial petroleum legislation from the 1970s was technically oriented with detailed and prescriptive requirements to both safety and technical solutions. The authorities with the Norwegian Petroleum Directorate (NPD) in a key role have gradually changed the legislation to a functional or goal orientation.

The NPD regulatory guidelines for concept safety evaluations (CSE) were introduced in 1980. The guidelines introduced a quantified cut-off criterion related to the impairment frequency for nine types of accidents that could be disregarded when defining dimensioning accidental loads, the so-called 10^{-4} criterion, i.e. a maximum probability of 10^{-4} per year for loads exceeding the dimensioning level for each accident type. These guidelines contributed in a positive manner to using formalised techniques for analysis of risk in the industry, and encouraged the industry and authorities to communicate regarding risk and acceptable risk. However it also had some unfortunate effects, as it could seem that "number crunching" exercises could divert attention from concentrating on the real issues. Too much emphasis was placed on the methodology and the "magic" 10^{-4} target.

New NPD regulations regarding implementation and use of risk analyses came into force in 1990, and new regulation on emergency preparedness appeared in 1992.

The 1990 regulation had a focus on the risk analysis process. The purpose of the risk analyses is to provide a basis for making decisions with respect to choice of solutions and risk reducing measures. According to the regulations the operator shall define safety objectives and risk acceptance criteria. The objectives express an ideal safety level. Thereby they ensure that the planning, maintaining and the further enhancement of safety in the activities become a dynamic and forward-looking process. Accidental events must be avoided (the realisation of any accidental event is unacceptable). This means that risk is kept as low as reasonably

practicable (ALARP), and attempts are made to achieve reduction of risk over time, e.g. in view of technological development and experience. The need for risk reducing measures is assessed with reference to the acceptance criteria. The acceptance criteria and the basis for deciding them are to be documented and auditable.

New NPD Regulations relating to management in the petroleum activities came into force in 2001. It is a requirement that QRAs shall be carried out giving a nuanced and as far as possible complete picture of the risk. Necessary sensitivity calculations and evaluations of uncertainty shall be carried out. The effect of risk reducing measures shall be calculated. Specific analyses on emergency preparedness shall also be carried out. Risk analyses of major accident is also required, and may be quantitative or qualitative. (Major accident means an accident involving several serious personal injuries or deaths or an accident that jeopardises the integrity of the facility.)

This NPD management regulation states some of the "old" 10^{-4}/year requirements, but also that operator shall formulate acceptance criteria (upper limit of acceptable risk) relating to major accidents and to the environment. The acceptance criteria shall be used by evaluation of results from the various QRAs and shall be given for

a) personnel on the installation as a whole, and for personnel groups that are particularly exposed to risk
b) loss of main safety functions
c) pollution from installation

In order to fulfill the requirements and acceptance criteria for major accidents the NORSOK Z-013 standard should be used.

3 RAC AND THE DECISION PROCESS

The discussion on RAC cannot be separated from the decision process leading to an acceptable solution. Figure 1 illustrates elements of such a decision process.

1) Stakeholders. The stakeholders are here defined as people, groups, owners, authorities etc that have interest related to the decisions to be taken. Internal stakeholders could be the owner of the installation, other shareholders, the safety manager, the union, the maintenance manager etc, whereas external stakeholders could be the safety authority (NPD), environment groups (Bellona, Green Peace etc.), the public society, research institutions etc. Only the internal stakeholders will take part in the formal discussions even though external stakeholders will play a role in e.g. the public domain (press etc.). The stakeholders will express their values, goals, criteria and preferences in some

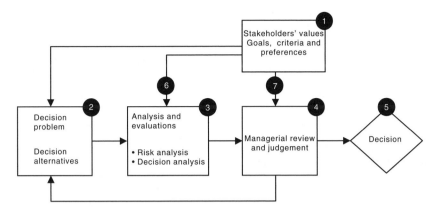

Figure 1. Basic structure of the decision-making process (Aven 2003).

kind of normative document, e.g. "Agreed RAC and value tradeoffs for company X".

2) Decision problem and decision alternatives. The starting point for the decision process is typically a demand from one or more stakeholder to improve aspects of an activity, start a new activity etc.

3) Analysis and evaluation. The different alternatives and options are subject to risk analysis and decision analysis. Risk analyses of the different alternatives are carried out, and the risk and costs are compared to the RAC and other value tradeoffs. This leads to a set of recommendations and risk statements.

4) Managerial review and judgement. Usually a management decision is required to make the final decision. The main input to this final decision process is the recommendation and risk statements from the analysis group. In Figure 1 we have also indicated that the stakeholders also may influence the final decision process 7 in addition to their stated RAC and value tradeoffs 6.

The main normative issues in the decision process could be categorized in three major aspects:

- Which *dimensions* of risk to include, e.g. safety, safety for whom, major accidents, environment, etc.
- Setting RAC for the risk dimensions considered.
- Preferences and tradeoffs between dimensions (e.g. the challenging question of quantifying the value of a statistical life).

It is obvious that a discussion is needed to define those dimensions of risk to include. For example are we only dealing with "average values", or do we focus on high risk groups, potential for major accidents etc.

We are assuming a *risk informed* approach, and there are three options:

1) Only using RAC and no value tradeoffs.
2) Only using preferences and value tradeoffs. To be pragmatic we will denote this approach a

"cost/benefit approach" and not differentiate between the important distinction made e.g. by Fischhoff et al. (1981) between a cost-benefit approach and decision analysis approach.

3) Combined use of RAC and the cost/benefit approach.

Of course there is a possibility of deciding to have a fully "discursive management strategy", see Klinke and Renn (2001). They suggest that such strategies could apply when there are major uncertainties related to frequency or consequences of potential events, making a risk based/informed approach unfeasible. Then the decisions are fully taken based upon discussions/evaluations, without reference to fixed acceptance targets. In the present paper we do not evaluate this as a general strategy.

4 USE OF RAC AND THE NORMATIVE DIMENSION OF THE DECISION PROCESS

4.1 *Use of RAC*

The decision situations could be related e.g. to *i)* building a new installation on an old field, *ii)* development of a new field where there are new environmental conditions (weather, depth, etc), *iii)* introduction of new technology, *iv)* new operational philosophy (say unmanned installation), *v)* modification, life extension, and end of life problems, *vi)* maintenance, etc. Of course both the decision process and the type of persons involved can be different for the various decision situations.

The quantitative RAC are typically given in terms of upper limits for the probabilities of unwanted events, yearly probability of loss of life etc. These RAC are related to performing a full QRA or analyses related to specific equipment or processes. The RAC shall be defined by the operator, and thus are not identical throughout the industry. A typical requirement could be that FAR (Fatal Accident Rate) should be less than 10 for all personnel on the installation, and say FAR ⩽ 50 for high risk workers.

In addition there may be various RAC at a "lower" level, related e.g. to specific annual gas leak frequencies, annual overpressure frequencies, etc.

Safety requirements for the acceptance of the various safety functions (e.g. related to the PSD and ESD systems) are based on the OLF Guidelines (2001). This is based on the IEC standard (2000), but contrary to the standard, provides deterministic requirements for standard safety functions, (without performing a prior risk analysis).

The criteria on individual risk (e.g. on FAR) are founded in the company's responsibility for its workers. If risk reduction was the only goal, other measures of risk will be used as the expected number of fatalities directly. There will always be discussions on which risk should be included and which risk to be focused; the criteria often set the agenda.

The rationale behind the use of RAC in the process could be

1) Better risk control: Contribute to undesired consequences of the planned activity being properly evaluated and controlled to a level that is acceptable to all affected parties.
2) Improved efficiency of decision process (reduce the workload of the decision making): Contribute to the decision processes being carried out in a more efficient way. It may simply be an efficient way to structure the tasks of the decision process. Further, *to some extent* decisions are automated; it is not necessary to repeat all arguments every time; even if the RAC also may be tailored to the specific situation.

In the first point we indicate that use of RAC may contribute to more focus and involvement regarding safety issues for affected parties; but unfortunately this is not necessarily the case. The use of RAC could also lead to rather "automatic" decisions.

The second point should be a very relevant argument in many decision situations. For instance, at a "low level" of the system (e.g. Safe Job Analysis), risk matrices may used to take decisions regarding the need for risk reducing measures.

One argument against using RAC at a "lower level" (e.g. for decisions on maintenance) is that they may lead to sub optimal solutions. i.e. resources may not be used in an optimal way to reduce overall risk.

Further, it is noted that a successful implementation of RAC (without additional incentives) requires that these criteria should be difficult to meet, such that they represent a direction for improvement. Otherwise, the criteria become a way of justifying doing nothing, and risk analysis becomes a verification tool, with no

important role to play in the system development. However, setting more ambitious RAC could also be difficult, as that could be too costly.

This demonstrates a problem by using RAC. Setting a target does not give drive for improvements beyond this level. Even if the RAC are difficult to meet, the creative process of finding even better arrangements and measures is in practice limited to meeting the criteria. Consequently, the RAC do not play an active role in the risk management process.

However, the RAC could in the design phases function as a guide to how much resources should be given to technical safety measures, (as opposed to resources spent in later phases). There are indications there are more to be gained on follow up and operational measures, as stated in a Norwegian white paper, (Stortingsmelding 7, 2001). But in the design it is more a question on getting more safety out of the resources being spent. In this context the RAC seem to function well in the design phase.

A preliminary conclusion is that Option 1) of Section 3, i.e. "only using RAC and no value trade-offs" seems a rather unattractive solution.

4.2 Principles for establishing RAC and examples of use

Assume there is given a situation/application where RAC is found appropriate (in combination with additional incentives). Further preferences and values are identified; how should now the "internal stakeholders" (see Section 3) proceed to determine the "acceptable/tolerable limit". There exist some suggested principles/approaches to "determine" the actual limit for acceptable risk. Some are listed below, also indicating some applications in the offshore industry.

1) "Comparison criteria" of NORSOK Z-13 standard, (NORSOK 2001) which is essentially the same as the French GAMAB ("Globalement Au Moins Aussi Bon") principle. This is primarily used when "non-standard" solutions, e.g. new technology, are to be implemented. Then the acceptance will require that the solution shall give at least as low risk as the present accepted practice/solution. This has been applied (apparently with some success) to a quite large extend, e.g. related to the introduction of HIPPS (High Integrity Pressure Protection Systems) in the Norwegian offshore industry. The HIPPS represents an unconventional solution with quick closing valves for pressure protection in the gas/oil process, and the authorities will require comparisons with the performance of conventional systems to give acceptance. The acceptance criteria is typically related to the value of PFD (Probability of Failure on Demand), see IEC 61508, 2000.

2) "*Additional risk*" *criteria*, which can be seen as a version of the (German) MEM (Minimum Endogenous Mortality) principle. Roughly speaking, this principle starts from an existing "basic" risk. Then a new activity shall not significantly increase this. By "knowing"/specifying such an underlying basic risk, we are assisted in specifying a RAC (e.g. require that the increase in risk shall be less than a certain percentage). This has been used on the Norwegian Continental Shelf e.g. by process extensions/modifications, relating the acceptance criteria to increase in the frequency of overpressure events (e.g. pressure above test pressure) for the module/plant.

3) *Safety Integrity Level.* The IEC 61508 standard specifies an approach for identifying the SIL (Safety Integrity Level) requirements for safety systems. This standard is suggesting a risk graph approach for assessing the risk, and thereby imposing requirements on the safety system of a safety function. Thus, the suggested approach of the standard indirectly provides the RAC (i.e. a SIL requirement) on the safety system, based on the risk assessment to be carried out on the EUC (Equipment Under Control). Now this fully risk based approach has in the Norwegian offshore industry been experienced as rather time consuming and difficult to use, see Hauge et al. (2001). The approach does not utilise the existing experience and current practice. Further, the risk graph is experienced to be very "flexible" (low reproducibility) in producing results. Thus, the Norwegian offshore industry has provided a recommended practice in the OLF Guideline (2001), which is also referred to by the NPD. In this Guideline fixed SIL requirements are given for "standard functions", whereas a risk based approach is required to specify the SIL requirements for "nonstandard" functions. So, the OLF Guideline determines the actual RAC for the "standard functions". These RAC were the result of broad discussions by a team of experts/representatives from the industry and some research institutions.

Example 3 above (use of SIL) is an example of a more general setting, where acceptance is based on performing a comparison of the risk (assessed say by a risk graph or a risk matrix) and the available control measures. In these cases there seem often to be a lack of true normative discussions, as the actual risk acceptance limit may be rather vaguely defined.

It is felt that the examples indicated above, also demonstrates that there are situations where the formulation of RAC could be a helpful tool in the decision process.

It is quite possible to investigate two or more of the above approaches, before deciding on the RAC limit. In general the *Comparison Criteria* seems most

applicable, for instance when new installations shall be built, introduction of new technology, new operational philosophy and modifications. But as indicated above, there are cases where the other two seem more relevant. Whatever principle is chosen this should be used in combination with other principles such as (to be decided by "internal stakeholders", see Section 3):

- Principle of continuous risk reduction.
- Principle of justice for all affected parties.

It is noted that the following are in general useful input, when an actual RAC limit shall be specified:

- "Historical risk data" and acceptability of risk in similar activities; (i.e. utilise accumulated knowledge).
- Assessment of perceived risk and willingness to accept the risk by involved parties (and voluntariness, control).

It is also noted that RAC have to be calibrated. If we use risk analysis and RAC for the first time for a new type of system or activity we need to see what is a "normal" risk level for this system, before we can specify the RAC.

Further, the explicit RAC limit should always be specified relative to a given context of analysis methodology and input data. The risk analyst cannot be free to choose this independent of the RAC being specified.

We also make the following note: RAC should be revised periodically (e.g. to reflect a vision about no accidents). For example every five year such a revision process could be run. Thinking in the lines of continuous improvement this will lead to more strict RAC as a basic approach. However, we may have situations where introduction of new technologies, new operation philosophy, new environmental conditions etc is likely to increase the risk. Could we then relax on the RAC? If we do not want to reduce the advancement in the industry to much we believe that we in some situations need to relax on the RAC. Some principles should however be followed:

- The risk for high exposed groups should not exceed fixed limits.
- The RAC for average risk should not be increased beyond similar levels 10 years ago. This means that we would not relax on the RAC in the long run (10 years period or more).

Example of a situation where relaxation of RAC could be considered: We want unmanned platforms. The PLL (Probability of Loss of Life) will decrease as there are less people, but the remaining personnel may experience a higher risk, e.g. FAR = 20, which could be acceptable compared to other high risk groups?

4.3 Requirements regarding the use of RAC in the decision process

Used in a proper way, we suggest that use of RAC in combination with other incentives will for several decision situations prove useful for the decision process. Key words are involvement and risk reduction. A main message is that if a "RAC-approach" shall be beneficial there must exist an ethical foundation, securing that safety is not compromised. Use of RAC must not be a pretext for doing nothing; if applied it must be seen as a true means to reduce risk. However, we make the some requirements in order that an "RAC-approach" should be recommended.

- The group specifying the RAC ("internal stake-holders") should be broad (e.g. including both safety management and representatives of labour union), so that all affected parties are represented in the discussion. It should be a clear objective that these parties arrive at a consensus on the RAC. The ideal should be a "discursive process" (i.e. promoting participation and involvement of all, all with the same possibilities to promote their view, and with a common objective to arrive at consensus). Of coarse there must be a balance; decision on acceptance e.g. regarding modifications that are considered minor may be carried out in a smaller group.
- The group should decide whether the risk is such that a *risk informed* approach is appropriate (implying that a RAC shall be specified), or whether the situation requires a *precautionary* or *discursive* management strategy, cf. Klinke and Renn (2001). They should also evaluate other value issues, e.g. whether there are special (highly exposed) groups or operations requiring specific attention.
- There should be a clearly formulated ambition of risk reduction. The argument is that there may be several factors working towards a negative trend (minor unnoticed changes, increased complexity, "dullness", …), and accepting that status regarding safety is OK may in practice lead to degradation.
- The acceptance of a (risky) activity should be followed up (during operation) by formulating and following up goals that could be defined regarding trends of various risk performance measures ("conditional acceptance").

With this in mind, use of RAC should in several cases assist to make a more structured decision process.

5   COST/BENEFIT ANALYSES, ALARP AND THE ACCEPTABILITY OF RISK

How should we structure the decision-making process such that "good" decisions can be taken; reflecting in a balanced way the accident risk and other relevant factors? In the offshore industry this balance is carried out by means of separate analyses of relevant attributes, such as accident risk, investment costs, operational costs, cost due to loss of reputation, and risk perception. Then cost/benefit or cost/effectiveness analyses integrate the values of the separate attributes.

The idea of the cost/benefit (cost/effectiveness) analyses is to assign monetary values to relevant burdens and benefits, and summarise the "goodness" of an alternative by the expected Net Present Value (NPV) of costs and equivalent monetary values of benefits and burdens.

The standard procedure when analysing the goodness of a risk reducing measure, is to compute the accident risk and the expected NPV gain by implementing this measure, or alternatively by a simple cost/effectiveness index; the cost per expected saved life. If the expected NPV is positive or the cost/effectiveness index is relatively small, the measure should be implemented. This is in line with the ALARP principle that is adopted for the offshore activities, cf. e.g. Pape (1995), UKOOA (1999), which in practice means that some type of cost/benefit or cost/effectiveness analyses need to be carried out to compare relevant options. Using such analyses, there is seemly no need for risk acceptance criteria. Acceptance is not linked to one particular attribute – what is acceptable is the option – one accept all the attributes of the option.

In practice however there is often a need for simplification in the value judgements related to safety – the above approach has too many "free variables" – and constraints, such as risk acceptance criteria, are introduced. The standard approach when using the ALARP principle is to consider three regions; 1) The risk is so low that it is considered negligible 2) The risk is so large that it is intolerable (unacceptable) 3) An intermediate level where the ALARP principle applies.

In most cases in practice risk is found to be in region 3 and the ALARP principles is adopted and cost/benefit or cost/effectiveness analyses are used. In Norway a risk acceptance criteria typically lower than the intolerability limit is normally used, and this is the way of application discussed in the previous sections of this paper.

In UK the ALARP principle applies in such a way that the higher or more unacceptable a risk is, the more, proportionately, employers are expected to spend to reduce it. At the point just below the limit of tolerability a considerable effort is spent even to achieve a marginal reduction in risk. In our opinion this is a reasonable approach. Generally, higher risks call for greater spending. More money should be spent to save a statistical life if the risk is just below the intolerability level than if the risk is "far away" from this level.

As an example, for individual risk, it is common to use a probability of accidental death of less than 10^{-6} as negligible and higher than 10^{-3} as intolerable. The basis for these numbers are comparisons with other activities. In practice most systems would produce risks that are in the ALARP region so we will not use much time on discussing these boundary values. But why use RAC at all? Why not restrict attention to the use of some type of cost/benefit or cost/effectiveness analyses, consistent with the region 3 of the ALARP principle?

Now, applying the cost/benefit analyses is not straightforward either. Cost/benefit analyses has been subject to strong criticism, due to both philosophical objections and formal treatment of values and weights, see e.g. Adams (1995), Bedford and Cooke (2001), Fischhoff et al. (1981) and Melchers (2001). The main problem is related to the transformation of non-economic consequences such as (expected) loss of lives and damage to the environment, to monetary values. What is the value of a (statistical) life?

This is a difficult issue and has been devoted much attention in the literature. There are no simple answers. The cost/benefit analyses often just focus on certain consequences and ignore others. Nevertheless, we find that this type of analyses provides useful insight and decision support in many applications. We are, however, sceptical to a mechanical transformation of consequences to monetary values; in many cases it is more informative to put attention on each consequence separately and leave the weighting to management and the decision maker, through a more informal review and judgement process, see Figure 1.

Cost/benefit analysis is a decision-making tool. It provides input to the decision-maker, not the decision. By presenting the results of the analysis as a function of the value of a statistical life, we can demonstrate the sensitivity of the conclusions of the analysis. We should acknowledge that decisions need to be based on managerial review and judgement. The decision support analyses need to be evaluated in the light of the premises, assumptions and limitations of these analyses. The analyses are based on a background information that must be reviewed together with the results of the analyses. Considerations need to be given to factors such as

- The decision alternatives being analysed.
- The performance measures analysed.
- The fact that the results of the analyses represent to a large extent conducted by experts.
- The difficulty of assessing values for costs and benefit.
- The facts that the analysis results apply to a model, i.e., a simplification of the world, and not the world itself.

The weight the decision-maker will put on the results of the analyses depends on the confidence he has to the analyses and the analysts. Important issues are; what competence do the analysts have, what methods and models do they use, what is their information basis, what quality assurance procedures have they adopted in the planning and execution of the analyses, are the analysts influenced by some motivational aspects?

In our setting the analysis provide decision support, not hard recommendations. Thus we may for example consider different values of a statistical life, to get insight into the decision. Searching for a correct objective number is meaningless. The value of a statistical life used in the analysis is a value that represents an attitude to risk and uncertainty, and that attitude may vary and depend on the context. When using a one-dimensional scale, uncertainty of observable quantities is mixed with value statements about how to weigh the different assessed uncertainties. Then, of course, we cannot expect to obtain consensus about the recommendations provided by the cost/benefit analysis, as there are always different opinions about how to look at risk in a society. Adopting a traditional cost/benefit analysis, an alternative with a low expected cost is preferred to an alternative with a rather high such value, even if the latter alternative would mean that we can ignore a probability of a serious hazard, whereas this cannot be done in the former case. In traditional cost/benefit analysis it is also common to discount the values of statistical lives, the results being often that negligible weight is put on consequences affecting future generations. Clearly, it is of paramount importance that the cost/benefit analyses are being reviewed and evaluated, as we cannot replace difficult ethical and political deliberations with a mathematical one-dimensional formula, integrating complex value judgements.

6 CONCLUSIONS

As seen from the previous discussion there are several (conflicting) interests to consider in order to reach a good decision process regarding risk acceptance.

Some preliminary recommendations, presented below, are based on some evaluation criteria; the decision process should:

- promote risk reduction
- promote cost-effectiveness in the choice of risk reducing measures
- secure a sufficient degree of involvement; i.e. the relevant parties are heard and should be satisfied with the result reached. So there should be an arena to ease/support free discussions to arrive at consensus on what is acceptable
- promote an efficient process to arrive at acceptance.

The overall recommendation is that the process should be guided by a *combined* use of RAC and some sort of cost/benefit analysis/considerations for choosing amongst solutions that are found acceptable. The RAC can provide a useful support in the decision process, but will not by itself give a sufficient drive to risk improvement.

1) The group of "internal stakeholders" shall perform normative discussions. This group needs to take a stand to the normative issues, and RAC are formulated based on a broad discussion including e.g. risk perception, historical risk, need for risk reduction and justice to all (incl. highly exposed groups).
2) The evaluation of various options satisfying RAC (cf. the ALARP step) should be based on comparing costs/benefits, without necessarily performing a "strict cost/benefit analysis. For instance it should be "allowed" to use greater spending to save a statistical life, given the risk is high. Further, the options could be evaluated also considering for instance (qualitatively) simplicity, robustness and flexibility.

So there should be a drive to reduce risk below RAC, and thus also to explore concepts giving lower risk. However, it is not seen as imperative to require a full ALARP approach, e.g. also specifying value of life (VOL). Cost/benefit values should be provided to assist decisions without having to be compared to a fixed "acceptable" value.

Finally, management commitment is of course required. If management do not want to prioritise high safety, the approach allows it do so, by specifying non-ambitious risk acceptance criteria and using the further risk reduction (ALARP principle) with no enthusiasm and commitment.

ACKNOWLEDGEMENTS

The authors are indebted to NFR for the funding of the work and their parent organisations for the permissions to publish this paper.

REFERENCES

Adams J. 1995. Risk. London: UCL Press.
Aven T. and Pitblado R. 1998. On risk assessment in the petroleum activities on the Norwegian and UK continental shelves. Reliab Engn Syst Safety 61: 21–29.

Aven T. 2003. How to approach risk and uncertainty to support decision making (to appear). New York: John Wiley & Sons Ltd.
Bedford T. and Cooke R. 2001. Probabilistic risk analysis. foundations and methods. Cambridge: Cambridge University Publishing Ltd.
Fischhoff B., Lichtenstein S., Slovic P., Derby S. and Keeney R. 1981. Acceptable risk. New York: Cambridge University Press.
Hauge S., Hokstad P. and Onshus T. 2001. The introduction of IEC61511 in Norwegian offshore industry. Proceedings of ESREL 2001: 483–490.
HSE 1992. *The tolerability of risk from nuclear power stations*, Health and Safety Executive, London.
IEC 61508 2000. "Functional safety of electrical/electronic/programmable electronic (E/E/PE) safety related systems", parts 1–7.
Klinke A. and Renn O. 2001. Precautionary principle and discursive strategies: classifying and managing risk. Journal of Risk Research 4 (2): 159–173.
Melchers R.E. 2001. On the ALARP approach to risk management. Reliab Engn Syst Safety 71: 201–208.
NORSOK 2001. NORSOK Standard Z-013. Risk and Emergency Preparedness Analysis.
NPD 2001. Norwegian Petroleum Directorate. Regulations relating to Management in the Petroleum Activities (the Management Regulations); see http://www.npd.no/regelverk/r2002/frame_e.htm.
NPD 2002. Norwegian Petroleum Directorate. Guidelines to Management Regulations.
OLF 2001. OLF guideline 070 on the application of IEC 61508 and IEC 61511 in the petroleum activities on the Norwegian Continental Shelf, OLF, Rev 01; see http://www.itk.ntnu.no/sil.
Pape R.P. 1997. Developments in the tolerability of risk and the application of ALARP. Nuclear Energy 36(6): 457–463.
Schofield S. 1998. Offshore QRA and the ALARP principle. Reliab Engn Syst Safety 61: 31–37.
Stortingsmelding 7. 2001. Helse, miljø og sikkerhet i petroleumsvirksomheten. Arbeids og administrasjonsdepartementet, Oslo. (in Norwegian).
UKOOA 1999. A framework for risk related decision support – Industry guidelines. UK offshore Operators Association.
Vatn J. 1998. A discussion on the acceptable risk problem. Reliab Engn Syst Safety 61: 11–19.

Safety and Reliability – Bedford & van Gelder (eds)
© 2003 Swets & Zeitlinger, Lisse, ISBN 90 5809 551 7

ARAMIS Project: development of an integrated accidental risk assessment methodology for industries in the framework of SEVESO II directive

D. Hourtolou & O. Salvi
INERIS, Verneuil-en-Halatte, France

ABSTRACT: The ESREL conference welcomes a special session on ARAMIS European project. This session represents halfway workshop of the project, which started in January 2002. The aim is to disseminate first results and collect comments from the public. This article is one of the five papers constituting the session and it presents the overall frame of ARAMIS project. ARAMIS objective is to build up a new integrated risk assessment method that will be used as a supportive tool to speed up the harmonized implementation of SEVESO II Directive. The proposed method results in an integrated risk index composed itself of three independent indexes. Index 1 assesses the consequence severity of first defined reference scenarios. Index 2 evaluates Safety Management effectiveness and accounts thus for the scenario probability. Index 3 estimates the environment vulnerability. Efforts have been made to disseminate work progress and results from the start. A dedicated web-site has been created and a review committee gathering industry risk experts and decision-makers from EU competent authorities periodically monitors the project.

1 INTRODUCTION

ARAMIS project was accepted for funding in February 2001 by the European Commission, in the 5th Framework Programme for Research and Technological Development, in the field of "Energy, Environment and Sustainable Development", chapter entitled "Fight against major natural and technological hazards". This three-year project started in January 2002. The project has been built in particular on the conclusions and results of ASSURANCE and I-RISK, two other European projects funded in the 4th Framework Programme.

ASSURANCE stands for ASSessment of the Uncertainties in Risk Analysis of Chemical Establishments. This project was a benchmark exercise, which aimed at improving the understanding of the sources and types of the uncertainties connected with risk analyses. As a rough conclusion, the project stated that the benchmark exercise revealed noteworthy variation in the final results. Discrepancies were present both in the assessment of frequencies and in the assessment of consequences. The different results would have obviously affected the relevant risk-informed decisions, mainly land use planning, emergency planning and acceptability of risk.

The initial statement of I-RISK project-I for integrated was the idea that Quantitative Risk Assessments (QRA) and safety management audits were so far two separate tools and that would be valuable to integrate both to address major hazard management. In this respect, the main objective was first to develop a management model for risk control and monitoring, then to implement this model into a dynamic QRA tool. Conclusions of the project point out that the integrated technical and management model was very robust and helped audit organizations in a new way. However it also turned out that a full-scale site integrated QRA was too time and detail demanding, so not currently practical or relevant.

From both projects, but also from everyone's experience in his own country, it emerges the need for a methodology giving consistent rules to identify accident scenarios and taking into account both prevention and mitigation measures peculiar to each plant operator. Those safety measures are obviously controlled in a safety management system.

There is also an underlying need for a risk assessment method that could reach a consensus amongst risk experts from both Industry and Competent Authorities and then be used with reduced uncertainty to make risk-informed decisions. The ARAMIS project has been set up to propose solutions to both latter requirements.

The paper first recalls the context of major accident hazards prevention in the EU. ARAMIS overall objectives are then expressed and the work contents are described in details. As a conclusion, the expected impacts of such a methodology are addressed.

2 CONTEXT OF THE SEVESO II DIRECTIVE

The annual report from the European Environment Agency (1999) indicates, among others, that the trend in notified accidents has been constant over the last twenty years. This statement alone shows that many of the often seemingly simple "lessons learned" from accidents have not been yet enough implemented in industry's standards. There is unfortunately no doubt that disasters will continue to occur throughout the EU. Some will be due to technology, some to the hazards of nature. The problem of low-probability, high-consequence accidents is likely to remain a key issue in terms of risk management. Nevertheless, hazards have to be managed and risks can be reduced.

The most significant EU Directive to help protect people and the environment from major accident hazards is the SEVESO II Directive. This applies to industries that use "significant amounts of hazardous substances". Their operators must demonstrate in particular they apply a major accident prevention policy and they implemented appropriate prevention and mitigation measures controlled and monitored in a safety management system.

The SEVESO II Directive sets very clear objectives relating to major hazard management, but the remaining question is: how to reach them and to control they are reached? For instance, there is no harmonized definition of the scenarios to be considered for risk assessment. The ASSURANCE project showed in this way that the chosen scenarios (BLEVE, full bore rupture or small leakage, amount of substance caught in an explosion, etc.) are different according to the experts judgement and experience, and according to the deterministic or risk-based approach of the Member State applying the Directive. Moreover, constraints in land-use planning (Cassidy & Amendola, 1999) sometimes urge the operators to consider reduction of the safety zones by choosing "realistic" scenarios and accounting for the effectiveness of dedicated safety devices. Actually the lack of rules for identifying scenarios and carrying out risk assessment makes often the expert's job tricky and largely too subjective to base transparent risk-informed decisions on it.

In addition to uncertainties in risk analyses, differences of culture among the Member States result in a multiplicity of methods and approaches (Kirchsteiger, 1999). At a recent JRC International Workshop (Kirchsteiger & Cojazzi, 2000), most participants agreed that comparative risk assessment along harmonized procedures would significantly help the decision understanding. A harmonized risk assessment methodology would thus ensure that risk-based decision making provides the necessary transparency and the right balance between scientific understanding and principle of precaution.

Proposing a harmonized methodology for risk assessment is a tough and tricky task. However, deterministic and probabilistic approaches should not be opposed since they are often complementary (Libmann, 1996). From a historical point of view, deterministic methods first allow to check the safe design of an installation. Probabilistic methods help then evaluate the residual risk of the installation. Both approaches have their strengths and weaknesses (Hourtolou, 2002). A first basic idea of ARAMIS is to take advantage of each approach's strengths and to develop upon it an alternative semi-quantitative method based on the evaluation of the safety barriers – lines of defense-peculiar to each site.

3 OBJECTIVES

The main objective of the ARAMIS project is to create a new integrated risk assessment methodology by combining the strengths of different methods currently used in European Countries. Accordingly, the method should be flexible enough to account for different national cultures like deterministic or risk-based approaches, in order to become a recommended tool used by risk experts and endorsed by risk decision makers in the whole EU.

The proposed method in ARAMIS should allow to characterize an integrated risk index composed itself of three distinct and independent indexes. Index 1 is to assess the consequence severity of first defined reference scenarios. Index 2 is to evaluate prevention management effectiveness, which allows thus to account for the reference scenarios probability in a semi-quantitative manner. Index 3 is to estimate the environment vulnerability by evaluating the sensitivity of potential targets located in the vicinity of a SEVESO plant.

The project has been set up (Figure 1) to reflect the logical construction of the risk index and has been divided accordingly into work packages:

1. First goal is then to develop a method to identify "reference" accident scenarios. These scenarios are consensual "realistic" scenarios to be used in SEVESO II safety report and taking account of some prevention and mitigation measures of the site according to their effectiveness.
2. Second task is to build up the integrated risk index made up of the three distinct indexes, i.e.:
   – consequence severity evaluation,
   – prevention management effectiveness,
   – environment vulnerability estimation.

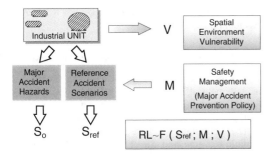

Figure 1. ARAMIS methodology representation.

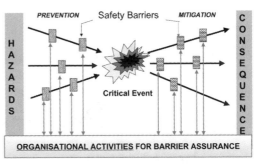

Figure 2. Bow-tie approach for scenario identification.

4 DESCRIPTION OF WORK

The work plan of the project has been built according to the logical construction of the final resulting risk index and it has been presented the same way in this article.

Halfway through the life of the project, the newly built methodology is to be tested on three SEVESO industrial sites in Europe. At this stage, two new partners from eastern-European countries will join the consortium and will also test the full method as totally unbiased end-users, each in one additional test site of their own country.

Moreover and since ARAMIS is intended to be a supportive tool to promote harmonized risk assessments throughout Europe, the project leaves from the start an extensive part to large exchanges with potential end-users of the method. Identified end-users are both industrial companies and Competent Authorities in charge of enforcing SEVESO II Directive. For that purpose, a dedicated Work Package deals with valorization and dissemination of project progress and results. Industrial partners have also been directly included in the consortium and a Parallel Review Team gathering potential end-users has been constituted.

4.1 WP1: Scenario identification

Identification of the possible accident scenarios is a key-point in risk assessment (Amendola et al., 2002). However, especially in a deterministic approach, worst case scenarios are considered, often without taking into account existing safety devices and implemented safety policy. This approach can lead to an overestimation of the risk level and does not promote the implementation of safety systems.

The aim of this Work Package (Delvosalle et al., 2003) is first to identify major accidents without considering safety systems. A second step is then to study in depth safety device effectiveness and safety management efficiency, which will allow to assess-qualitative-probabilities, in order to identify finally Reference Accident Scenarios taking into account some of the implemented safety systems.

The first objective is to define a Methodology for the Identification of Major Accident Hazards (MIMAH). On the basis of considered equipment and properties of handled chemicals, the methodology must be able to predict which major accidents are likely to occur. Properties of substances are found out thanks to Directive 67/548/EEC (substance classification and labeling) and their own conditions of use (pressure, temperature, flow, etc.).

The work has been divided in several parts. Firstly, it was necessary to select a general approach. The bow-tie method was chosen (Bellamy & Van der Schaff, 1999) because it is a highly structured tool and it is considered as a very good way to establish links with other parts of the project and especially Management Efficiency (Figure 2). Secondly, a special effort was made to develop a common typology of equipment and hazardous substances. Thirdly, event trees and fault trees centered on critical events have been built, and above all a methodology able to build generic trees was created. Critical events are defined as "Loss of Containment" or "Loss of Physical Integrity" event.

At this stage, the MIMAH methodology is able to predict which major accidents are likely to occur on a given equipment. With the help of a deep study of safety systems, causes of accidents and a historical analysis of known accidents, the objective of the work to be done is now to place lines of defense-safety functions, safety barriers on the different branches of the trees. This will lead to a second Methodology of Identification of Reference Accident Scenarios (MIRAS) which has to take into account some of the safety systems according to their effectiveness. Therefore, the Reference Accident Scenarios will use results from the work performed on the prevention management effectiveness (Figure 3). These scenarios are used afterwards as an input to evaluate the severity index, i.e. the hazard potential.

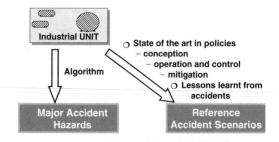

Figure 3. Scenarios identification process.

4.2 WP2: Severity of the consequences

The objective of this task is to define a severity index S characterizing the possible consequences of accident scenarios (Casal et al., 2003). In this respect, only the physical characteristics of the phenomena involved in accidents are studied in order to evaluate the severity of both major scenarios and reference scenarios identified in WP1.

First task of WP2 was the selection of the most suitable models for the calculation of the effects of the various dangerous accidental phenomena. Thus, a survey of the existing models for the calculation of the effects due to explosions (overpressure and missiles), fires (radiation), toxic clouds (concentration), BLEVE-Fireballs (overpressure, missiles and radiation), pollutant plumes into the water (concentration), soil pollution and domino effects is now available at this stage of the project.

The Severity Index must be independent of the other two indexes. It is thus constructed in such a way that every dangerous phenomenon has a corresponding specific sub-index. The contribution of each dangerous phenomenon to the global index S is strongly related to the probability of occurrence of the phenomenon associated to each critical event (e.g. probability of ignition) and identified in the WP1 event trees (Figure 2).

Each specific sub-index associated to the various physical phenomena takes into account in its construction the following parameters:

- the effect area concerned with the phenomenon, e.g. a disc in case of an explosion, a plume surface for gas dispersion;
- the kinetic of the phenomenon: rapid for an explosion, much slower for a fire;
- the potential of generating domino effect: fragment emission, delayed phenomena triggered off.

The severity index S is therefore a function of parameters only associated with physical phenomena. All the identified scenarios should then be evaluated and ranked in this way according to the calculation of

S_o for Major Accident Hazards and S_{ref} for Reference Accident Scenarios.

4.3 WP3: Prevention management effectiveness

This work package deals with the assessment of safety management efficiency and its effect on the calculation of external risks for SEVESO plants (Duijm et al., 2003).

The methodology is based on the identification of initiating events and direct causes of the accident scenarios (bow-tie approach). Safety barriers are then related to generic fault and event trees representing all possible accident scenarios leading to critical events (Figure 2). The safety organization includes both the adequacy and completeness of technical and managerial barriers (lines of defense) that are implemented to prevent these accidents and the management system that ensures that these barriers are maintained and adjusted properly.

The methodology recognizes a number of dimensions of safety management (delivery systems), derived from previous work on safety management modeling, notably the I-Risk (Hale, 1998) and MIRIAM (Plot, 2002) models. These are made explicit in specific functions that need to be executed to maintain a safety barrier. Examples of these delivery systems are: ensuring good competence and commitment of employees, manpower availability, communication, procedures, plans, hardware and human-machine interfaces.

Currently, the focus is on developing instruments to measure the set of dimensions, using a combination of audit, questionnaire, interview and observation techniques. The combination of measurements ensures that not only the implementation of functions, but also its conditions and outcome (e.g. good safety commitment of the employees) are taken into account.

The measurement techniques address in particular the specific safety functions (M_{LOD1} and M_{LOD2}, Figure 4) found in a given establishment. However measurement will also be carried out in a generic way onsite (M_{SMS}, Figure 4), assuming then the quality of the dimensions represents a common mode for the quality of safety barriers maintenance. The efficiency of the barriers can then be adjusted according to the measurement scores to select the final set of Reference Accident Scenarios.

The assessment of technical barriers effectiveness follows the principles described in the norms IEC61508 and IEC61511 (Functional safety: safety instrumented systems for the process sector). Among these principles, effectiveness is analyzed through the definition of "Safety Integrity Levels" linked to device characteristics (design, reliability, maintainability, testability...) and also through criteria upon the activities in charge to maintain them.

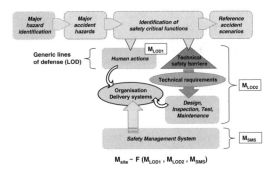

$$M_{site} \sim F (M_{LOD1} , M_{LOD2} , M_{SMS})$$

Figure 4.   Structure of index M.

Challenges in the development of the methodology include the need to calibrate the scores in the measurement of dimensions, as well as the need to determine the efficiency of technical barriers as a function of these scores.

4.4   *WP4: Environment vulnerability*

An installation handling dangerous substances is hazardous only according to the potential of vulnerable targets liable to be affected. In assessing the overall risk level of a plant, it is therefore quite relevant to characterize the spatial vulnerability of the environment surrounding the plant. Such is the aim of this work package (Tixier, 2003).

Vulnerability of the surroundings depends on the features of the environment that are potential targets (human, environmental and material) and on the type of impact due to hazards (fire, explosion, and toxic release). It also strongly depends on the considered target area.

To address this issue, the area of interest in the vicinity of a plant has been first divided into meshes and the potential targets have been identified and localized with the support of Geographic Information Systems (GIS). The major difficulty is then to rank and prioritize the various sensitivities of the targets according to the various expected impacts.

A suitable solution has been found out by applying a multi-criteria ranking approach, such as SAATY methodology which allows to define priorities from complex situations. At this stage of the project, SAATY method has been extensively applied to the concern of vulnerability estimation.

First step was to describe and classify potential targets, hazards and impacts in adequate typologies. Following step involved expert judgement. Through experts' answers to specific questionnaires, SAATY method helped build up mathematically the generic coefficients of target vulnerability.

Final step will be to test and validate the calculated coefficients and resulting index through full-scale case studies.

Thanks to the combined use of SAATY method and GIS, index V should be represented as vulnerability maps of a plant surroundings, and should become in this way a powerful tool for risk-informed decision making.

4.5   *WP5: Risk level integration and validation*

4.5.1   *Characterization of final risk level RL*
The risk level RL of an installation in its environment is a function depending on the severity index S, the vulnerability index V and management effectiveness index M:

$$RL = \lambda \times \frac{S^{\alpha} \times V^{\beta}}{M^{\alpha}} \qquad (1)$$

The objective of this phase is to study the relation between S, M and V to characterize final risk level.

It will be decided at this stage whether the risk level should remain characterized by 3 separate indexes or whether the 3 indexes could be aggregated into one single index.

4.5.2   *Case studies*
Halfway through the life of the project, five case studies will be carried out with the collaboration of five different SEVESO establishments throughout Europe in order to test and validate or improve the new methodology.

To select the test sites, it has been assured that both countries with consequence-based and probabilistic approaches would be represented. Moreover two case studies out of five will take place in Slovenia and Czech Republic. Two institutes from these countries will indeed join the consortium at this stage and test the full method with the test sites. Both of them will thus act as totally unbiased end-users since they were not involved in the method development.

After these full-scale exercises, the ARAMIS methodology will be improved again and give rise to its last version in the project.

5   WP6-7: VALORIZATION, DISSEMINATION

Since ARAMIS is intended to be a supportive tool to promote harmonized risk assessments throughout Europe, the project leaves from the start an extensive part to large exchanges with potential end-users. Identified end-users are both industrial companies and Competent Authorities in charge of enforcing SEVESO II Directive.

In the valorization process, industrial end-users are directly represented in the consortium through an association of European industrial companies. This helps the consortium to relay information about the project progress and results, and to convince plants for the case studies.

A Parallel Review Team gathering risk experts from industry and EU competent authorities has also been constituted. This review team gathers every six months to monitor the project and thus to ensure needs from end-users are indeed fulfilled and the final approach will be widely accepted. In this respect their main comments concern the applicability and usefulness of achieved results.

In the dissemination process, a dedicated web site has been set up first: please visit http://aramis.jrc.it. The web-site aims at promoting the project towards the public and also works as a quick communication tool among the partners. An electronic newsletter is also released every six month on the web-site, in order to get the public informed of work progress.

Two workshops were also planned during the project. This article is part of the halfway workshop held at the ESREL conference. A final workshop will also take place at the end of the project to disseminate main achievements to all relevant stakeholders. Proceedings of the workshop will be made available on the web-site.

6 CONSORTIUM DESCRIPTION AND INVOLVEMENT

The consortium consists of twelve organizations expert in the field of risk analysis (Table 1). Nine partners represent mostly research centers throughout Europe. The last three institutes represent Newly Associates States from Eastern Europe.

INERIS is the coordinator of the project. It has a European expertise in the field of major accident prevention. It works as technical support for the French Ministry of Ecology in charge of SEVESO II Directive application. In particular, it manages with a steering committee the risk index aggregation and validation phase. INERIS is also deeply involved in the valorization and dissemination process.

EC-JRC-IPSC and especially MAHB has a recognized international expertise in the field of major accident prevention. It animated EU Working Groups dealing with the application of SEVESO Directives and is also experienced with accident databases and GIS tools at a European level. MAHB is WP leader of the dissemination activities and also coordinates the Parallel Review Team.

FPMs-MRRC has a great experience in the application of SEVESO Directives, and already developed methodologies and tools in the field of domino effects and source term/consequence modeling. MRRC is

Table 1. Description of partner organization.

Organization full name	Short name	Country
Institut National de l'Environ-nement Industriel et des Risques Accidental Risk Division	INERIS	France
European Commission – Joint Research Centre – Institute for the Protection and Security of the Citizen – Major Accident Hazard Bureau	EC-JRC-IPSC-MAHB	Italy
Faculté Polytechnique de Mons – Major Risk Research Center	FPMs-MRRC	Belgium
Universitat Politecnica de Catalunya – Centre for Studies on Technological Risk (CERTEC)	UPC	Spain
Association pour la Recherche et le Développement des Méthodes et Processus Industriels	ARMINES	France
Risø National Laboratory System Analysis Department	RISØ	Denmark
Universita di Roma Dipartimento Ingegneria Chimica	UROM	Italy
Central Mining Institute Safety Management and Technical Hazards	CMI	Poland
Delft University of Technology – Safety Science Group	TUD	Netherlands
Institution of Chemical Engineers European Process Safety Centre	IChemE-EPSC	U.K.
Jozef Stefan Institute – Department of Inorganic Chemistry and Technology	IJS	Slovenia
Technical University of Ostrava – Energy Research Centre	VSB-TUO	Czech Republic

WP1 leader concerning scenario identification and also brings its experience to WP2.

UPC-CERTEC has a recognized expertise in the evaluation of accident consequences: dispersion, explosion, fire modeling. UPC is WP2 leader.

ARMINES is a consortium of research centers from French school of mines. Three different research teams take part in the ARAMIS project.

"Pôle Cindyniques" from Mines de Paris has built up a debriefing and interview methodology to learn better from both technical and organizational incidents. By this means, Mines de Paris contributes to develop ARAMIS organization model and management effectiveness index.

"SITE" Center from Mines de Saint-Etienne has a long experience in risk analysis and environmental management system. In ARAMIS, this team mainly focuses on developing generic "bow-ties" and an assessment method for technical barrier effectiveness.

"LGEI" Laboratory from Mines d'Alès has developed a methodology based on GIS and multi-criteria SAATY approach to study risks in transportation of hazardous substances. This partner is WP4 leader concerning environment vulnerability estimation, which intends to use the same competence.

RISØ is experienced with both SEVESO Directives. It also coordinated the ASSURANCE project. Furthermore, it is experienced in applying function-oriented modeling to analyze the effectiveness of an organization and its safety culture. RISØ is WP3 leader dealing with Prevention management.

UROM is experienced in risk analyses and area risk studies, linked in particular with the use of GIS. Its activities in ARAMIS mainly concern the interfaces to build up between the model developed for V index and the use of GIS to represent this index.

CMI has a long experience both in fire and explosion modeling and in the use of safety management standards. In this respect, CMI works in WP2 to list and select appropriate models for consequence modeling, and in WP3 to analyze how the common management standards fit in the ARAMIS model.

TUD was a major partner of I-RISK project. It brings its expertise to the project in safety management modeling and auditing, and also in scenario identification with the use of bow-tie approach.

IChemE-EPSC only participates in the dissemination process to other member companies or associates from EPSC. It also cares about Industry participation into the Parallel Review Team.

IJS is the largest Slovenian research organization. VSB-TUO has been involved for long in Industrial Environment Research, and has enlarged since 1995 its focus towards technological and natural risks. Both institutes have been chosen to test the full ARAMIS method in companies of their own country and also to compare it to commonly used approaches in their respective country.

7 CONCLUSIONS

ARAMIS project supports the European Research Area concerning knowledge improvement, encouragement of the Science-Industry dialogue and harmonization in decision-making process related to hazardous establishments.

The resulting method should indeed be proposed as a recommended and harmonized tool used by risk experts and recognized by risk-informed decision-makers of EU competent authorities. Harmonizing industrial risk assessment in Europe would significantly contribute to the European Commission's overall efforts to establish harmonized policies following the SEVESO II Directive.

For both Competent Authorities and Industry, such a harmonized risk assessment procedure would constitute first a useful comparison tool for industrial sites, which integrates the strengths of both probabilistic and deterministic approaches. The risk assessment procedure would at last be linked to the setup of progress plans within the framework of a safety management system. It would also allow to reach a consensus in the selection of accident scenarios that takes into account plant-specific safety devices and safety management effectiveness, i.e. suitable frequencies for the scenarios as required in a safety report for risk control demonstration.

The participation of potential end-users in the project from the start, in particular through the Parallel Review Team ensures that the ARAMIS project will contribute on a very practical level to EC research objectives and to consistent implementation of European policies in major hazard prevention.

ACKNOWLEDGEMENTS

The ideas presented in this publication are developed in the frame of EU project ARAMIS "Accidental Risk Assessment Methodology for IndustrieS", contract n°EVG1-CT-2001-00036 funded by the Energy, Environment and Sustainable Development Program in the 5th Framework Programme for Science Research and Technological Development of the European Commission.

The project is coordinated by INERIS but also includes a consortium of fifteen efficient institutions representing ten countries which all contributed to the results presented during this workshop.

The Parallel Review Team represents also fifteen organizations either from Industry or from EU competent authorities, which are also kindly acknowledged for their advice and contribution during the periodic reviews.

REFERENCES

Bellamy, L. & Van der Schaff, J. 1999. Major Hazard Management: Technical-Management Links and the AVRIM2 Method, in Seveso 2000 – Risk Management in the European Union of 2000: The Challenge of Implementing Council Directive Seveso II. *Edited by EC-JRC-MAHB. Athens, Greece. 10–12 November 1999.*

Casal Fàbrega, J., Planas, E., Delvosalle, C., Fiévez, C., Pipart, A. 2003. ARAMIS project: The severity index. *Proc. ESREL, Maastricht, Netherlands, 16–18 June 2003.*

Cassidy, K. & Amendola, A. 1999. Special issue: The SEVESO II Directive (96/82/EC) on the control of major accident hazards involving dangerous substances, *Journal of Hazardous Materials*: Vol. 65, N°1–2. Elsevier Science.

Delvosalle, C., Fiévez, C., Pipart, A., Casal Fàbrega, J., Planas, E., Christou, M., Mushtaq, F. 2003. ARAMIS project: Identification of reference accident scenarios in SEVESO establishments. *Proc. ESREL, Maastricht, Netherlands, 16–18 June 2003.*

Duijm, N.J., Madsen, M., Andersen, H.B., Goossens, L., Hale, A. 2003. ARAMIS project: Assessing the effect of safety management efficiency on industrial risk. *Proc. ESREL, Maastricht, Netherlands, 16–18 June 2003.*

European Environment Agency. 1999. Environment in the European Union at the turn of the century, Chapter 3.8. Copenhagen, Denmark.

Hale, A.R., Guldenmund, F., Bellamy, L. 1998. An audit method for the modification of technical risk assessment with management weighting factors. *Probability and Safety Assessment and Management. Springer. London. 2093–2098.*

Hourtolou, D. 2002. ASSURANCE – Assessment of the uncertainties in risk analysis of chemical establishments. U.E. Project U.E. ENV4-CT97-0627. Rapport final opération a DRA-07. *Ref. INERIS-DHo-2002-26824*

Kirchsteiger, C. 1999. Special issue on international trends in major accidents and activities by the European commission towards accident prevention, *Journal of Loss Prevention in the Process Industries*, Vol. 12, N°1. Elsevier Science.

Kirchsteiger, C. & Cojazzi, G. (ed.) 2000. Promotion of technical harmonization on risk-based decision-making, *Proc. intern. workshop, Stresa, Italy, 22–24 May 2000.*

Lauridsen, K., Kozine, I., Markert, F., Amendola, A., Christou, M., Fiori M. 2002. The ASSURANCE project. Final summary report.

Libmann, J. 1996. Eléments de sûreté nucléaire. Institut de Protection et de Sûreté Nucléaire.

Plot, E. & Lecoze, J.C. 2002. MIRIAM: an integrated approach to organize major risk control in hazardous chemical establishments. *ESREDA 23rd Seminar, Delft, Netherlands, 18–20 November 2002.*

Tixier, J., Dusserre, G., Dandrieux, A., Bubbico, R., Luccone, L.G., Mazzarotta, B., Silvetti, B., Hubert, E. 2003. ARAMIS project: Assessment of the environment vulnerability in the surroundings of an industrial site. *Proc. ESREL, Maastricht, Netherlands, 16–18 June 2003.*

Safety and Reliability – Bedford & van Gelder (eds)
© 2003 Swets & Zeitlinger, Lisse, ISBN 90 5809 551 7

Cognitive prerequisites for safety-informed organizational culture

K. Hukki & U. Pulkkinen
VTT Industrial Systems, Espoo, Finland

ABSTRACT: The paper introduces a methodological approach which has been developed for identification of the cognitive prerequisites of safety-informed organizational culture. A framework of analysis has been created to identify the requirements and the sufficiency of the organizational support for developing an understanding of the safety-critical significance of one's work. In taking the prerequisites, instead of the culture itself, as the starting point, the approach represents a different orientation when compared with investigations on safety culture. The approach has been developed in the context of nuclear waste management. The focus of the analysis is on multidisciplinary expert work, typical of this field. The paper introduces the methodological approach and the framework and the results of the analysis concerning the company responsible for the research and development and later also for the implementation of final disposal of spent nuclear fuel in Finland.

1 INTRODUCTION

The concept of safety culture has been adopted in nuclear power generation and other hazardous technologies to describe safe working practices and management. The commonly accepted attributes of safety culture include good organizational communications, good organizational learning, and senior management commitment to safety (Sorensen 2002).

The concept and measurable attributes of safety culture are not, however, clear. According to INSAG (IAEA 1991) safety culture is defined as: "…that assembly of characteristics and attitudes in organizations and individuals which establishes that, as an overriding priority, nuclear plant safety issues receive the attention warranted by their significance." This, like other traditional definitions of safety culture, are abstract and general, and thus they seem insufficient to take into account the essence of practical work and the characteristics of the target organization's activities (cf. e.g. Reiman & Oedewald 2002). In addition, the internalization of values, which is often considered when speaking about safety culture, is difficult to indicate, due to the implicit nature of the underlying values and tacit assumptions that determine the working practices (Schein 1985). The connection between the safety values and the work tasks is not always easy to discover, at least in complex expert work including uncertainties. Respectively, practical criteria for indicating the level of safety culture are not easily provided by adopting the internalization of values as the starting point in the definition of safety culture.

Instead of the value-based analysis we have taken an approach with the objective to develop an understanding of the *prerequisites* of safety culture provided by the organizations. The approach has been developed in the context of nuclear waste management. Safety is a central requirement encountered at multidisciplinary expert work which is typical of this field. In spite of this, the investigation of safety culture in nuclear waste management has been started only recently. The concept of safety culture itself is relatively unknown in this area, being associated rather to nuclear power plants.

There is no need to consider organizational culture and safety culture as separate from each other because the safety of the organization's activities depends on the ways of action evolved in the organization (see e.g. Reiman & Oedewald 2002). Therefore, the term safety-informed organizational culture is used in the present paper, instead of safety culture.

2 METHODOLOGICAL APPROACH

The roots of the approach lie in the psychological and decision analytic approach developed in VTT Industrial Systems. The psychological analysis of the demands of expert work (e.g. Hukki & Norros 1998, Norros & Nuutinen 2002) is based on ecological study of work and is near to Vicente's cognitive work analysis (Vicente 1999). The integration of the psychological and the decision analytic view on expert work (Holmberg et al., 1999) has helped in developing integrated models of decision making from the contextual and systemic points of view.

Also studies on organizational culture have been carried out on the basis of the VTT approach (e.g. Reiman & Norros 2002, Reiman & Oedewald 2002, Oedewald & Reiman 2002). In these studies the problems concerning the abstract and general character of traditional studies on safety culture have been tackled by connecting the aspects of organizational culture with the critical content of work. The approach to be presented in this paper has a common interest with these studies in that orientation and a common emphasis on the case-specific character of organizational culture. The present approach differs in having its focus on the analysis and evaluation of the prerequisites of safety-informed organization culture, instead of the culture itself.

The worker attitudes toward safety are considered relevant widely but the mechanism by which they affect the safety of operations is not clear (Sorensen 2002). The internalization of safety values can be expected to manifest itself as commitment to work. Commitment is a central concept e.g. in the INSAG's definition of safety culture (IAEA 1991). Especially commitment to the values and goals of the organization is essential (Senge 1990). In the present approach, instead of paying attention to values and attitudes, another direction has been taken by asking *whether the organization offers sufficient prerequisites for commitment*. In addition, the consideration has been focused on the *cognitive* basis of commitment to work. In the evolvement of motivation and commitment it is essential to be able to comprehend the purpose, object and nature of one's work. The possibility to understand issues which are relevant and necessary for performing the tasks is the fundamental prerequisite of work. According to Antonovsky (1988) the sense of manageability is contingent on the sense of comprehensibility. Comprehensibility and manageability are important contributors to the sense of meaningfulness, which, for its part, contributes to commitment.

It can, therefore, be expected that the understanding of the safety-critical significance of one's working practices is an important aspect of commitment in safety-critical organizations.

Taking the cognitive prerequisites for developing an awareness of the safety issues inherited in one's work as the main focus of the approach helps in making concrete the way the work is related to safety. At the same time a general view to orgnizational culture is provided, because the need for this awareness is valid for every safety-critical organization. What is not general is the comprehension to be gained because it depends on the domain of the expert work and also on the characteristic features of the individual organization.

According to the suggestion by IAEA (1998) the characteristics and attitudes mentioned in the INSAG's definition of safety culture (see Chapter 1 in this paper) should be commonly held which means that there is a core of key attitudes and values that are acknowledged by the majority. The present approach emphasizes the importance of every worker to be aware of the safety-critical significance of his/her work and of a sufficiently consistent understanding of the work-related safety issues inside the organization. The organization should support the workers by providing them with relevant knowledge in this sense. Here the ways of interaction and communication are important because they can promote or prevent the understanding. This is essential particularly in multidisciplinary expert work.

3 FRAMEWORK OF THE ANALYSIS

The approach has been developed by utilizing interviews made at Posiva Oy. Posiva is responsible for the research and development and later also for the implementation of final disposal of spent nuclear fuel in Finland. It has an own staff of about 30 experts of nuclear waste management, but the major part of the work is carried out by external contractors from universities, research institutes and conculting companies representing different kinds of expertise in science and technology. The work is characterized by efficient utilization of expertise from several disciplines, such as geology and other earth sciences, physics, chemistry, mathematical modelling and computing. In addition to this, the work of Posiva includes technical design and construction of the waste repository and the waste encapsulation plant.

The aim of the study was to discover, with the help of expert interviews, the central cognitive requirements of safety-informed organizational culture at Posiva and to see how well they can be met by the company.

The interviews were carried out at Posiva in 2001 and 2002. The interviewees, 23 altogether, either worked for Posiva or one of its subcontractors or were representatives of the regulatory authority.

A framework, based on the adopted theoretical assumptions, was developed for the design of the interviews and for the analysis of the material obtained through the interviews. The study, which was intended to support the R & D at Posiva, included aspects related to the cognitive, motivational and emotional aspects of safety-informed organizational culture. Due to the focus of this paper only the cognitive aspects are considered here.

The analysis can be described as contextual, systemic, subject-centered, inter-subjective and problem-based. This means, in short, that expert work was considered by taking the domain-specific and multidisciplinary characteristics of the work and the mutual dependencies between the tasks into account, and, at the same time, by focusing on the experts' possibilities to form an idea of the safety-critical requirements of their work in the organization.

The steps of the analysis were:

1. the identification of the challenges and difficulties of the expert work at the company
2. the identification of the optimal cognitive prerequisites of safety-informed organizational culture in nuclear waste management
3. the identification of the development needs and the practical development procedures concerning the organizational support to expert work at the company

In the first phase of the analysis the organization-specific working and communicating practices were identified by considering the expert work from the substance and interaction points of view. The former means finding out difficulties which can be encountered while performing one's work task and which are caused by the domain-specific features of work. The latter means finding out difficulties in communication between the experts.

In the second phase the optimal cognitive requirements of safety-informed organizational culture were identified on the basis of the comprehension gained in the first phase. The expert work was considered from the comprehensibility point of view, by identifying the types and significance of knowledge which are necessary in order to be able to comprehend the safety-critical significance of one's work and to achieve a shared understanding of the safety issues in the organization.

In the third phase the development needs of the organizational support to expert work were identified by comparing the current cognitive prerequisites at the company with the optimal ones. The expert work was considered from the comprehensibility point of view, by identifying the difficulties in discovering one's contribution as part of a whole and in relation to safety assessment. On the basis of the considerations the practical ways of eliciting and utilizing necessary knowledge recognized in the second phase of the analysis were identified.

4 RESULTS OF THE ANALYSIS

4.1 Challenges and difficulties of multidisciplinary expert work at Posiva

The expert work at Posiva has, until the year 2001, been concentrated on the selection of the site for waste disposal and on the analysis of the long-time safety of waste disposal solutions to pave way for the policy-decision by the Finnish parliament ("Decision-in-Principle"). In this connection, safety is more or less an abstract concept, and the experts must consider the safety of the designed disposal facilities over very long time scales. Safety is connected to the decisions made in the daily work only indirectly which makes the recognition of the safety-informed working and communicating practices difficult. The safety-critical questions of the work have been such as the selection of scenarios and phenomena for analysis, the transparency of arguments behind these decisions, and the sufficiency and completeness of the analyses. Thus, safety manifests in the work in two ways, firstly as the safety of the technical or scientific solutions produced in the work, and, secondly, as the inherent quality of the expert work procedures, i.e. in the ways of action in performing the work tasks and in communicating with the other experts.

Now, after the Finnish parliament has given the principal permission for waste disposal, the nature of the work has changed. The work is becoming more concrete, including the design of the disposal facilities and the start of subsurface investigations with the construction of the underground research facilities. The work orientation changes from theoretical considerations towards more practical studies. At the same time, the interaction with new types of expertise, such as construction engineers, brings the organization into a new situation. The research work must take into account the requirements and the constraints of the practical construction project, and vice versa. The expert work associated with underground investigations of the geological site is sequential, and the decisions on how to continue investigations are made on the basis of the findings of the research. The decision making in this kind of "observational working method" is difficult because it is not easy to obtain all the necessary information, or to take into account the needs and constraints of different parties, i.e. of underground investigations, technical design and construction.

Due to the complex and abstract nature of the phenomena to be investigated and the multidisciplinary character of the work it may be difficult to comprehend one's contribution as part of a whole. Besides, in the communication with experts from other disciplines, one should be able to understand the basic reasoning principles characteristic to these disciplines. The integration of knowledge produced by experts from different domains is, therefore, difficult, due to the differences in the working practices and in the ways of thinking.

The organisation of Posiva is small, and a significant part of the work is commissioned from subcontractors. Posiva is responsible for planning the work and collecting and interpreting the results. The company has built a network of subcontractors, and trusted in long-time co-operation. In their expert role the consultants are acting fairly independently of the contracting agency. The work of single experts is very important and their expertise has a lot of power in the safety related decisions. The subcontractors are mainly research institutes or engineering consultants, and although they follow the usual quality assurance practices it is evident that their working practices are different. This creates requirements for the Posivas own personnel, which have the final responsibility for the quality of the subcontractors' contributions.

The safety assessment of nuclear waste disposal, of which Vieno et al. 1999 is an example, requires knowledge from several disciplines, which look at the problem from different perspectives, using different methods and models. It has to bring these, in some cases even contradicting approaches, to one well-structured and well-argumented analysis. Safety assessment is based on defence-in-depth thinking, and describes how the different barriers prevent propagation of the harmful phenomena. The work of each expert can, in principle, be identified as support for the analysis of certain safety barrier. Safety assessment collects knowledge of the phenomena related to each safety barrier to a holistic view on safety of the whole waste disposal solution. The role of safety assessment is thus central for the company's work but due to its holistic and reduced character it may be difficult to understand for the experts from other domains.

4.2 Optimal cognitive prerequisites of safety-informed organizational culture in nuclear waste management

It is important that the experts in the field of nuclear waste management are able to understand what the concept of safety means in relation to one's work. In addition, it is essential that safety is interpreted in a sufficiently consistent way in the organization. In order to be able to form an idea of the safety-critical significance of one's work and develop a shared understanding of the safety-related aspects of work in the organization the experts should be aware of the most central issues concerning the companys's work and of the way one's task is related to these issues and to the other experts' tasks. This means that scientific and technical knowledge to be adopted at work should be comprehended in a holistic and integrated way in this sense.

The fulfillment of these demands requires, firstly, knowledge of the principles and significance of safety assessment, due to its central role in the company's work. It is important to provide this knowledge by training.

Secondly, there is need for knowledge which is related to substance issues but is not scientific-technical by nature. This knowledge, here called *supplementary knowledge*, is needed for developing an awareness of the context of one's work. It is essential, from the safety point of view, to understand the connections between one's own and the other experts' tasks. It is also important to recognize the knowledge-related dependencies between the tasks and the way the results, obtained from other experts in order to be utilized in performing one's own task, have been produced.

Supplementary knowledge can be divided to two subtypes, contextual and background knowledge. The first concerns one's own contribution and the latter the other experts' contributions.

Contextual knowledge concerns e.g. (1) the connection of one's task with the main goals of the company, (2) the connection of one's task with safety assessment, (3) the connections and knowledge-related dependencies between one's and the other domain experts' tasks. The last subtype concerns (a) the significance of the knowledge, resulting from one's own contribution, in other experts' tasks, (b) the way this knowledge is utilized in other domain experts' tasks (which may be influenced by discipline-specific practices and ways of thinking, and constrained by discipline-specific boundary conditions), and (c) the way this knowledge is used in safety assessment.

Background knowledge concerns e.g. (1) aims and motives behind the other experts' contributions, (2) the grounds of the selection of their working methods, (3) the grounds (choices, interpretations, inferences etc.) underlying their scientific and technical judgments, (4) the uncertainties related to their judgments and (5) other relevant factors related to their judgments (such as circumstances during the underground measurements).

Contextual knowledge makes it easier to comprehend the safety-critical significance of the results of one's contribution and of the communicating practices in the interaction with other experts. Background knowledge makes it possible for the domain experts to take the significance of the background factors concerning the other experts' results better into account in their work. Also the experts on safety assessment benefit from this type of knowledge in making their analyses. Both types of supplementary knowledge enhance mutual understanding between the experts and, therefore, facilitate the integration of the results produced in different domains and the reconciliation of the requirements set by the different parties.

It is important to recognize the role of supplementary knowledge in the development of safety-informed ways of action and to mediate this knowledge in the organization. Supplementary knowledge enhances the transparency of communication. It is possible, with the help of it, to present scientific and technical knowledge in a holistic and integrated way. This contributes to better awareness of the systemic nature of the multidisciplinary expert work which, for its part, facilitates taking account of the requirements set by safety of nuclear waste management.

4.3 Development needs of the organizational support to expert work at Posiva

Although people working for Posiva recognize the importance of safety in their work and are committed to the work, comprehension of one's contribution as part of a whole, in connection with the other experts' tasks and in relation to safety assessment is not always self-evident. The understanding of the safety-critical significance of the work may not be sufficiently shared among the experts working for the company. The communicating practices need development in order to mediate the necessary information in a safety-informed way. These findings have led to the start-up of practical development procedures in the organisation.

The company can support multidisciplinary expert work by taking the significance of the ways of action in task performance and in communication as the object of discussion in the organization. This kind of collective conceptualization of each work process should be inherent in the work of the organization.

Collective recognition and definition of central issues concerning the company's work and of the significance of the connections between the experts' tasks makes it possible to enhance one's awareness of the context of his/her work and develop a shared understanding on safety-informed working and communicating practices in the organization.

The importance of shared models and collectively accepted concepts in multidisciplinary expert work was emphasized by Holmberg et al. (1999). The models are important, not only as explicit descriptions of certain issues, but as tools for the identification of the sense, implications and uncertainties related to the results obtained during the work. In order to perform in an appropriate way, the organization must have a common acceptance and control over the decision criteria, and common tools to deal with uncertainty. Thus, adequate models and practices for decision making, possibly even rather formal decision analyses, are needed. The models can be presented as charts, road maps, tables etc.

The purpose of modelling is to make the knowledge-related dependencies between the tasks explicit and to recognize the instances where supplementary knowledge is needed. This means recognition and definition of the type and content of supplementary knowledge necessary for safety-informed working and communicating practices. Discussion on these issues helps in planning accurate development procedures.

At least the following types of models would be of relevance here: (1) models describing the physical situation, like the bedrock model, (2) models describing the process of investigation, (3) models describing the actual dependencies between underground investigations and (4) models supporting practical decisions made during the construction of the underground investigation facilities.

It is especially important to develop the models in co-operation. Every expert whose tas k belongs to the work process should participate in modelling. Collective selection of the object to be modelled,

identification of central issues and the related problems, selection of the way of modelling, analysis of the knowledge-related dependencies between the tasks in relation to central issues and specification of the way of updating the created models improves mutual understanding. Moreover, viewing a work process from different angles enhances learning and motivation. Especially the less experienced experts benefit from common discussions.

At its best, safety assessment could support development of shared models by providing them with a common perspective. As a holistic view over all analyses and as a collection of safety arguments concerning different disciplines it can be an important tool for understanding the significance of the safety-critical aspects of the work. It is necessary to make sure that it does not remain an abstract and inpenetrable model, the purpose and principles of which the domain experts are not able to comprehend. It is, therefore, important that the experts are provided with basic knowledge concerning safety assessment.

Supplementary knowledge identified as the result of collective modelling can be utilized in the communication between the experts and in knowledge management. It facilitates the experts' interaction by enhancing the informativeness of the requests for knowledge and of feedback. In knowledge management the knowledge to be mediated can be made more informative by e.g. the following procedures: (1) in commissions by making the connections with the other commissions explicit, (2) in reporting by including knowledge which makes the connections with other tasks visible, by considering the significance of the produced results as part of a whole, and by investing in reports in which the results produced in different disciplines are considered together in an integrated way, and (3) in intranet by offering shared models and databases for the scientific-technical work. These procedures facilitate documentation and communication. Moreover, the enhanced transparency of the internal processes of the organization leads to enhanced transparency of communication with the external stakeholder groups

The role of the management is very important because the creation of cognitive prerequisites for expert work is on their responsibility. The level of safety-informed organizational culture depends on the attitudes of all the members of the organization but safety-informed attitudes develop and strengthen on the basis of the adequacy of organizational support.

It is also important to consider the requirements concerning the organizational support to expert work as part of requirements management, to include the development procedures in the management system and to ensure by auditing that sufficient cognitive prerequisites of safety-informed organizational culture are maintained.

5 CONCLUSIONS

The basic idea of the introduced methodological approach is that safety of activities in nuclear waste management depends highly on the understanding of the safety-critical significance of the ways of action in performing tasks and in communicating.

In order to be able to develop this understanding it is necessary to be aware of the knowledge-related dependencies between the tasks and of the effect the way of producing the results has on the integration of the results in multidisciplinary expert work. There is, therefore, need for supplementary knowledge which is necessary for this awareness. This knowledge is related to substance issues but is not scientific-technical by nature.

In practice the experts' awareness of the context of their work and of the significance of the connections between their tasks can be enhanced by collective conceptualization of actual work processes. Collective modelling, together with other experts participating in the same work process, makes the knowledge-related dependencies between the tasks explicit. This makes it possible to recognize case-specific need for the type and content of supplementary knowledge necessary for safety-informed working and communicating practices.

Safety assessment, due to its central role in nuclear waste management, and as a holistic view over all analyses seems to provide a common perspective to the modelling of the work processes. Collective conceptualization and basic knowledge concerning the purpose and principles of safety assessment, provided by training, make it easier for the experts from other domains to understand its significance and nature.

Consideration of work processes improves learning, motivation and mutual understanding. Discussion on the significance of the ways of action in performing one's task and in communicating with other experts provides a basis for collective definition of safety-informed working and communicating practices. Shared models and the presentation of scientific and technical knowledge in a holistic and integrated way, with the help of supplementary knowledge, improves the transparency of communication. These procedures help in enhancing the experts' awareness of the safety-critical aspects of their work and the consistency of safety concepts in the organization.

By taking the cognitive prerequisites for the experts' work as the central requirement for the safety-informed organizational culture the introduced approach represents a different orientation when compared with investigations on safety culture. Investing time and energy in the development of adequate cognitive support to expert work provides a way of developing safety-informed working and communicating

practices which is easier to measure and indicate to the regulatory authorities than safety values and attitudes.

The conceptual framework has been developed in the context of nuclear waste management but it can be utilized also in other safety-critical domains.

ACKNOWLEDGMENT

The authors are indepted to Posiva Oy, Finland, for their financial support of the work.

REFERENCES

Antonovsky, A. 1988. *Unraveling the Mystery of Health. How People Manage Stress and Stay Well.* San Francisco: Jossey-Bass Publishers.

Holmberg, J., Hukki, K., Norros, L., Pulkkinen, U. & Pyy, P. 1999. An integrated approach to human reliability analysis – decision analytic dynamic reliability model. *Reliability Engineering and System Safety.* 65, 239–250.

Hukki, K. & Norros, L. 1998. Subject-centred and systemic conceptualisation as a tool of simulator training. *Le Travail Humain*, 4, pp. 313–331.

IAEA, Safety Series No. 75-INSAG-4. 1991. Safety Culture. Vienna: International Atomic Energy Agency.

IAEA, Safety Reports Series No. 11. 1998. Developing Safety Culture. Practical Suggestions to Assist Progress. Vienna: International Atomic Energy Agency.

Norros, L. & Nuutinen, M. 2002. The Core-task concept as a tool to analyse working practices. In: Borham, N., Samurcay, R., Fischer, M. (eds.). *Work Process Knowledge.* Routledge.

Oedewald, P. & Reiman, T. 2002. Maintenance Core Task and Maintenance Culture. Paper presented at IEEE 7th Conference on Human Factors and Power Plants. Scottsdale, Arizona, USA, September 2002.

Reiman, T. & Norros, L. 2002. Regulatory Culture: Balancing the Different Demands of Regulatory Practice in Nuclear Industry. In: Kirwan, B., Hale, A.R. & Hopkins, A. (eds.), *Changing Regulation – Controlling Risks in Society.* Oxford: Pergamon.

Reiman, T. & Oedewald, P. 2002. The Assessment of Organizational Culture – a Methodological Study. VTT Research Notes 2140. Otamedia, Espoo.

Schein, E.H. 1985. *Organizational Culture and Leadership. A Dynamic View.* San Francisco: Jossey-Bass.

Senge, P.M. 1990. *The Fifth Discipline.* New York: Doubleday Currency.

Sorensen, J.N. 2002. Safety culture: a survey of the state-of-the-art. *Reliability Engineering and System Safety* 76 (2002) 189–204, Amsterdam: Elsevier.

Vicente, K. 1999. *Cognitive Work Analysis. Toward Safe, Productive, and Healthy Computer-Based Work.* London: Lawrence Erlbaum.

Vieno, T. & Nordman, H. 1999. Safety assessment of spent fuel disposal in Hästholmen, Kivetty, Olkiluoto and Romuvaara – TILA-99. Report POSIVA 99-07.

Safety and Reliability – Bedford & van Gelder (eds)
© 2003 Swets & Zeitlinger, Lisse, ISBN 90 5809 551 7

Sensitivity of assessed bridge safety to the method of analysis and site data

D. Imhof & C.R. Middleton

Engineering Department, Cambridge University, United Kingdom

ABSTRACT: When the safety of existing structures is assessed, it is often not adequate to use the design codes which may be rather conservative in nature. There is less uncertainty compared to the design stage, as the structure is already in place. More precise information can be gathered concerning the resistance of the structural elements, but also of the applied loads. This paper examines the sensitivity of the safety of two example concrete slab bridges to the inclusion of updated strength information, to site-specific loading and to the method of structural analysis.

1 INTRODUCTION

Extensive bridge assessment programmes have been undertaken or are in progress in many nations around the world. These have highlighted the need to distinguish between the approach used with the *design of new structures* and the *assessment of adequate safety in existing structures*. In particular, since design codes must be appropriate for an entire population of generic bridges there will inevitably be a level of conservativeness incorporated in them since for simplicity's sake they make little or no allowance for differences in statical systems, resistance parameters and loading. As a result, codes of practice specifically targeted at assessment are evolving to exploit the inherent reserves of strength that are known to exist in bridges. Over the last 15 years the four national overseeing organisations in the UK have developed a very extensive suite of assessment codes which aim to reduce the level of conservativeness and incorporate the additional information known about existing structures into the assessment procedure to more realistically model the safety of the bridge stock.

Two specific ways in which assessment codes achieve this are the modification of partial safety factors on both load and resistance variables and also the change in the actual parameter values themselves to reflect the better estimate of magnitude and reduced uncertainty involved when dealing with existing structures where in-situ data can be collected on material properties, geometry and the applied loading. In the UK, the use of *Worst Credible Strength* values in assessment as compared with characteristic strength values in design is widely adopted. *Bridge specific loading models* have also been permitted for use in assessment. Similarly reduced partial safety factors are also employed when the in-situ evidence warrants it.

Although widely adopted in practice there is little in the literature which quantifies the effects of these modifications. In this paper a sensitivity analysis of two typical concrete bridge decks is undertaken to quantify the effects on safety of different modifications to partial factors and key load and resistance parameters in accordance with current practice in the assessment codes.

A third and possibly even greater influence on the derived estimate of safe load capacity is the choice of method of analysis and failure criterion adopted in the assessment procedure. In particular, the convention in current practice is to adopt elastic analysis to determine stress resultants and failure of the structure is defined in terms of failure of an individual element. It is shown that a substantial increase in estimated load capacity is derived by adopting system failure or global collapse as the appropriate criterion for assessment in circumstances where there is sufficient ductility and redundancy to warrant such an approach. This would allow engineers to represent more realistically the actual load carrying capacity of a bridge rather than using some (hypothetical) elastic limit, which may give very little indication of the margin to the true collapse load.

With the evolution of an advanced computer program for plastic yield-line analysis such collapse analyses are now possible. This paper also examines the sensitivity of the assessed capacity/safety to the method of analysis adopted.

2 ASSESSMENT EXAMPLES

2.1 Assessment procedure

The assessment procedure consists of different steps with progressively increasing levels of precision. The assessment effort is increased with each step in order to reduce the likely need for costly intervention (e.g. Fig. 1).

Generally bridges are first assessed with the current design codes and no further analysis is required if the structure is shown to be adequate. Design rules are rather conservative for checking the safety of existing structures. They are based on generic rather than structure-specific material and structural data. If this first assessment suggests the need for strengthening, then the procedure should continue, and later stages should seek to remove excessive conservatism in the calculations.

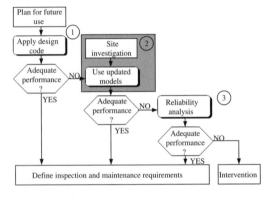

Figure 1. The evaluation procedure in Switzerland (adapted from Imhof (2001)).

In a second step updated resistance and load models are used together with site-specific data. If this assessment should still be inadequate, some organisations and researchers suggest a full reliability analysis to be undertaken, although this approach raises many questions as well. This third step is not considered in this paper.

In this paper the improvement due to the different options of step 2 compared with the design code situation is quantified and recommendations are made. Two short-span concrete bridges from the UK road network are used as examples. The UK design (BS5400 1990) and assessment codes (BD21/01) are applied to the examples.

2.2 Example bridges

Sandhole bridge is a rectangular, simply-supported, 2-lane reinforced concrete slab bridge. It was apparently constructed in two stages as indicated by a full-length longitudinal construction joint (Fig. 2). The site investigation report indicates variations in the concrete strength and the amount of steel on either side of this joint (Middleton & Ibell 1995). Only bottom reinforcement is provided. The longitudinal steel lies perpendicular to the abutments and the transverse lies parallel to it.

Allt Chonoghlais bridge is a 3-span, skewed, continuous, 2–lane reinforced concrete slab bridge. Each end is shown on the plans to be cast integrally with the abutments however, without any continuity of the deck reinforcement, it has been assumed that the deck will act as if simply supported at these locations. Alternate bottom longitudinal reinforcing bars are turned up near the piers, to provide hogging steel in this region. There is a single layer of transverse bottom reinforcement running perpendicular to the sides of the bridge deck, rather than parallel to the abutments. The spacing of the longitudinal bars is reduced in the region of the kerbs, producing stronger edge regions.

3 ANALYSIS METHODS

To estimate the load effects in a bridge and compare them to the resistance, a bridge can be examined with different analysis methods. These methods differ in

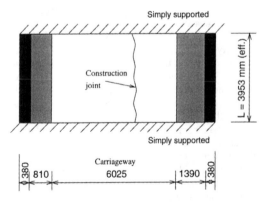

Figure 2. Geometry of Sandhole bridge (deck thickness: 330 mm).

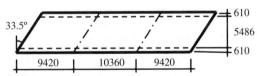

Figure 3. Geometry of Allt Chonoghlais bridge (slab thickness: 533 mm, thickness of edge beams: 762 mm).

simplicity and precision (completeness). In this paper three commonly used methods have been compared: grillage, finite-element and yield-line analysis. Grillage and finite-element analysis are elastic methods. Assuming a load effect diagram in equilibrium with the applied loads and not exceeding the plastic yield limit can be found, they provide a lower bound of the ultimate load capacity. The estimated capacity is thus a safe one. Yield-line analysis is an upper bound plastic method, in which compatible failure mechanisms are postulated to find an upper bound of the ultimate load capacity.

3.1 Grillage analysis

Grillage analysis is probably the most popular analysis method for bridge decks, because of its simplicity and wide availability of software packages. It has been shown to be accurate (Hambly 1991). In this technique the physical deck is idealised by a grid structure of rigidly connected longitudinal and transverse beams, each with a bending and torsional stiffness representative of the region of slab the members model. At each junction of the grillage beams, deflection and slope compatibility equations can be set up.

The load effects are calculated for each grillage beam element and the highest local values compared to the resistance of this beam.

3.2 Finite-element analysis

Traditionally grillage analysis was favoured over finite-element analysis (FEA) which often was only used for the most complex problems. With today's availability of inexpensive, high-speed computers and user-friendly programs the finite-element method has begun to replace the grillage method, even for more straightforward bridge decks. Finite-element analysis

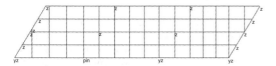

Figure 4. Grillage model of Allt Chonoghlais bridge and support conditions (grillage members along direction of reinforcement).

Figure 5. Finite element model of Allt Chonoghlais bridge.

is relatively easy to use and gives in most cases accurate results. But there is the risk that inexperienced users may analyse complicated structures without understanding the true behaviour of the structure. In the FEA the slab is modelled as a finite number of discrete segments of slab. These elements are connected together at the "nodes".

The load effects are calculated at defined points of a finite element. Using interpolation the effects per unit width can be obtained for every point of the structure. The highest load effects are then compared to the local resistance.

3.3 Yield-line analysis

This plastic method is usually less conservative than elastic methods and therefore significant savings can be made when assessing concrete structures. Today the use of yield-line analysis by practising engineers is not widespread, possibly because it is not well known. Also because of being an upper bound technique it may give unconservative results.

The yield-line method requires the engineer to postulate an appropriate collapse mechanism for the slab being considered. In the program used here (COBRAS), the user selects appropriate failure mechanisms from a pre-defined library of typical slab failure mechanisms. The work done in the assumed yield-lines is then equated to the work done by the applied loads to find the critical load. A stepwise optimisation of the mechanism geometry is performed to find the critical failure mode geometry. In comparison with elastic analysis a more realistic ultimate load capacity can be obtained. But as an upper bound method, there is always a possibility that another, more critical, failure mechanism may govern the actual collapse behaviour of the structure. Also, it must be emphasised that in conventional yield-line analysis only flexural failure is considered.

For the Allt Chonoghlais bridge both the central span and one end span were analysed to determine which part would govern the load capacity of the bridge. These two spans are assessed independently and the lowest capacity used to determine the overall bridge safety. At the two interior pier supports the deck is considered continuous providing fully fixed supports for the centre span.

4 IN-SITU INFORMATION

4.1 Characteristic strength

The initial assessments are made using the characteristic strength values tabulated in the design code. In the UK those values correspond to the 5%-fractile of the strength distribution.

4.2 Worst credible strength

The concept of "Worst credible strength (WCS)" has been introduced with the British assessment standard BD 44/95. It should take into account the actuarial material strengths of the structure. It is defined as the worst value of strength which the engineer believes could be obtained in the structure under consideration. This value may be greater or less than the characteristic value. The WCS-value may be obtained by extracting samples from the structure. The number of samples should be as high as possible, but it is not possible to take too many samples as the structure might be damaged critically. The advice note BA 44/96 calculates the WCS from n samples ($n \geqslant 3$) with strengths f_i:

$$WCS = \frac{1}{n}\sum_{i=1}^{n} f_i(1 - \frac{0.2}{\sqrt{n}})$$ (1)

In statistical terms this formula corresponds to the lower bound of the estimate of the population mean, μ_{low}, in the case where the population standard deviation, σ, is known:

$$\mu_{low} = m_x - z_\alpha \frac{\sigma}{\sqrt{n}} = m_x(1 - \frac{z_\alpha \sigma}{m_x \sqrt{n}})$$ (2)

where m_x = mean of the strength test results and z_α = coefficient of the normal distribution for a confidence level $(1 - \alpha)$. Hence, for $(1 - \alpha) = 95\%$ and thus $z_\alpha = 1.64$ equation (1) is only acceptable if σ/m_x is smaller than 0.12. Based on typical σ-values (Middleton & Hogg 1998) and applied to real bridge examples, it can be shown that this ratio may be at least twice as big.

Since the WCS-value eliminates some of the uncertainties associated with the use of characteristic strengths, the code permits partial safety factors for material, γ_m, to be reduced (Table 1).

4.3 Eurocode characteristic value

In the case of a normally distributed random variable X with an unknown coefficient of variation, the final draft of Eurocode prEN 1990:2001 finds the characteristic value, X_k, by using the lower bound of the prediction interval (Tamhane & Dunlop 2000):

$$X_k = m_x(1 - t_{n-1,\alpha}\sqrt{1 + \frac{1}{n}} \cdot V_x)$$ (3)

where $t_{n-1,\alpha}$ = coefficient of the Student-distribution with α = probability of being exceeded and n = number of tests and V_x = coefficient of variation of the sample. In the case where the coefficient of variation of the population is known, the characteristic value is based on the normal distribution instead of the Student distribution. For this case the Eurocode is statistically not totally correct, as it replaces V_x by the known coefficient of variation, which is $V = \sigma/\mu$ and not σ/m_x. As usual in statistics, the standard deviation should be used rather than a coefficient of variation.

4.4 Bayesian characteristic value

The derivation of a characteristic value from tests may take into account not only the statistical uncertainty associated with the number of the tests and the scatter of the test data, but also prior information based on the assessing engineer's knowledge and experience. The ISO 2394-Code contains a method to calculate the characteristic strength including prior information and the level of confidence in them.

Another approach is considered in this paper. In the WCS approach above a lower limit for the mean is found. But the engineer is less interested in estimating parameters of the distribution and more concerned about where the individual observations may fall. He would like to determine a "characteristic value for assessment" rather than a confidence interval for the mean. An exception is yield-line analysis where the failure criterion is a global one, and thus using a mean value of the strength makes sense. But for elastic analysis we suggest to use the same principle as for design, namely to use the 5%-fractile as the characteristic value, and additionally include the results of the tests. Using Bayesian updating a posterior distribution can be found and from this distribution we get the 5%-fractile for assessment: the Bayesian characteristic value (BCV).

Based on typical distribution parameters for the strength variables (e.g. Middleton & Hogg 1998) the prior distribution parameters, μ_0 and σ_0, corresponding to the characteristic value of design can be found. For the concrete strength, σ_0 for design has been taken as 5.96 MPa. The standard deviation for design is clearly bigger than for assessment because for design allowance for variance between concrete plants, between batches and within the batch has to be made. For the upstream section of Sandhole bridge,

Table 1. Partial safety factors for material.

	Char. value (BS 5400)	WCS (BD 44/95)
Concrete	1.5	1.2
Shear in concrete	1.25	1.15
Reinforcement	1.15	1.10 (1.05*)

* If depth of reinforcement measured.

the concrete volume is so small that it is likely that only one batch of concrete has been put in place. So a lower σ_1 of 2.38, consistent with FHWA 1999, was assumed.

In the following a strength variable X (e.g. concrete strength or steel yield strength) is considered normally distributed and the conditional density $f_x(x|\mu_1)$ has only one uncertain parameter M, which is the mean value, and a constant variance σ_1^2. The density of M, $f_M(\mu_1)$, is also normally distributed with constant mean value $\mu_2 = \mu_0$ and constant variance μ_2^2. We then have the prior density of X normally distributed from the law of total probability:

$$f_X(x) = \int_{-\infty}^{\infty} f_X(x|\mu_1) f_M(\mu_1) d\mu_1 \propto N(\mu_0; \sigma_0) \quad (4)$$

where $\sigma_0 = \sqrt{(\sigma_1^2 + \sigma_2^2)}$. Hence, from the sitespecific σ_1 we can derive σ_2. Given a sample of strength tests $x_n = (x_1, x_2, \ldots, x_n)$, the posterior conditional density $f_M(\mu_1|x_n)$ is normally distributed (Tamhane & Dunlop 2000):

$$f_M(\mu_1|x_n) \propto N(\frac{nm_x\sigma_2^2 + \mu_2\sigma_1^2}{n\sigma_2^2 + \sigma_1^2}; \frac{\sigma_1\sigma_2}{\sqrt{n\sigma_2^2 + \sigma_1^2}}) \quad (5)$$

where m_x = sample mean. The posterior density of the strength variable X becomes:

$$f_X(x|x_n) = \int_{-\infty}^{\infty} f_X(x|\mu_1) f_M(\mu_1|x_n) d\mu_1 \propto N(\mu_2''; \sigma_0'') \quad (6)$$

where μ_2'' and σ_2'' are the parameters of $f_M(\mu_1|x_n)$ and $\sigma_0'' = \sqrt{(\sigma_1^2 + \sigma_2''^2)}$. The BCV-value corresponds to the 5%-fractile of the posterior distribution (Fig. 6).

This method allows to calculate a *characteristic value* and not only a lower bound for the *mean*. The updating process however, is also executed on the mean. The prior density function describing the design strength distribution is represented as the integral of the product of a density conditional on the mean and the density of the mean itself. Assuming a site-specific standard deviation for the conditional probability, the prior distribution of the mean can be found. Using Bayesian statistics the density of this mean can be updated when additional information (test results) becomes available.

4.5 Comparison of strength values

Table 2 shows the comparison of the different strength values obtained with the methods described above and for the concrete strength tests undertaken on Sandhole bridge. The strength distributions were assumed either normally or lognormally distributed. Three cores were tested resulting in the following values: 27.5, 22.5 and 34.5 MPa. The design characteristic value was 20 MPa.

For the remainder of this paper the assessment strength has been calculated with the WCS-value as this was the highest value for the strength and thus the biggest effect on updating could be achieved. Although the WCS-concept is included in the official UK assessment code, the authors believe that at least for elastic analysis using a BCV would be more appropriate. The same safety philosophy as in design would then be applied for assessment.

5 IMPROVED LOAD MODEL

5.1 Introduction

The traffic load model contained in the design or assessment codes has to be applicable for all kinds of

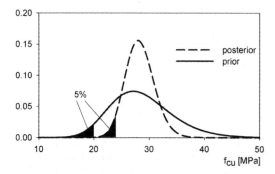

Figure 6. Prior and posterior distribution of the Sandhole concrete strength using a lognormal distribution, 5%-fractiles defining the design and bayesian characteristic value.

Table 2. Concrete strength used for assessment [MPa].

	Without prior	With prior information ($\sigma = 5.64$ MPa)
Minimum strength	22.5	22.5
WCS	24.9	24.9
Eurocode normal	7.9	17.5
Eurocode lognormal	2.7	19.0
ISO Bayesian normal*	7.8	20.2
BCV normal	–	23.8
BCV lognormal	–	23.9

* almost deterministic prior information ($n' = 50$, $m' = 29.8$ MPa, $s' = 5.64$ MPa, $v' = 50$ and probability of being exceeded: 0.05).

847

bridges, independently of their spans, cross-section and geometry. Thus, they are rather conservative for some types of bridges. A more accurate prediction of the load effects that really occur in the structure considered can often help to prove that a bridge is safe enough. Here two methods have been used to find the most severe realistic load effects: a deterministic loading based on extreme axle loads and a simulation-based model. The maximum of the two has then been used for the safety evaluation.

5.2 Type HA loading

The UK standard loading per lane is composed of a uniformly distributed load (UDL) that acts over the "loaded length", and a knife-edge load of 120 kN (KEL) uniformly distributed across the lane width. The "loaded length" is defined as "the base length of that area under the live load which produces the most adverse effect at the section being considered". The magnitude of the UDL is given by the following equation:

$$W = 336(\frac{1}{L})^{0.67} \qquad (7)$$

where W is the UDL in kN per metre length of lane of width 3.65 m and L is the loaded length in meters. For assessment, a reduction factor K is applied to the UDL and the KEL. The reduction factor K, defined as the ratio between assessment live loading and Type HA Loading, is a function of the traffic flow and the road surface condition. The load effects obtained are multiplied by the partial safety factor for live load, $\gamma_{fL} = 1.5$ and by γ_{f3}, a factor taking account of inaccurate assessment of the effects of loading. For elastic analysis methods γ_{f3} is 1.1 and paradoxically for plastic methods $\gamma_{f3} = 1.15$ (BD 44/95), even though the later aim to more realistically model the actual failure behaviour.

5.3 Deterministic critical loading

The load effects of short span slab bridges, like the two analysed in this paper, are often governed by individual vehicle axles rather than a traffic jam load situation. Individual vehicles have therefore been placed on the structure such as to induce the most severe effect at the location considered. No detailed vehicle statistics are available for the UK. As the legal axle weight limits in the UK, Europe and Switzerland are nearly the same, the axle loads have been based on the values measured in France and Switzerland (Imhof 2001) assuming the same overloading of the vehicles. The axle loads used correspond to the 99%-Fractile of the beta distribution approximating the measured histogramme (e.g. Fig. 7).

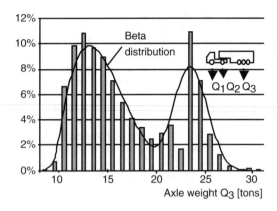

Figure 7.   Probabilistic description of rear axle weight.

5.4 Simulation-based load effects

In the computer programme QSIM (Bailey 1996) random traffic actions were generated for defined traffic types and the load effects calculated. The static load effect of the vehicles in a traffic lane is the sum of the effects of all axle groups, which are considered as an axle stream. These streams consist of random vehicles which have random weights and geometries. Probabilistic distributions for axle weights, axle spacings and the distance between vehicles have been based on traffic surveys made in France and Switzerland (Imhof 2001). As an example Figure 7 shows the probabilistic approximation of the rear axle of an articulated vehicle by a bi-modal beta distribution.

The traffic has been separated in different vehicle types depending on their axle configuration. The percentages of each vehicle type have been taken from the UK Transport Statistics 2001. To include all possible distances between vehicles three different traffic flow conditions have been looked at: free-moving (500 vehicles/hour/lane), congested (30 km/h) and at-rest (2 km/h). In each case the distance is described with a different probabilistic distribution.

The influence lines of each traffic lane have been determined using the analysis methods above. As it is not possible to calculate an influence line with yield-line analysis, traffic effects could not be evaluated using simulation with this type of analysis.

The traffic stream is stepped over a bridge by incrementing the axle positions. At each step the calculated effects are compared to the current maximum effect which is updated if necessary. The final maximum effect is achieved once the maximum number of vehicles, N, has been passed over the bridge. N is derived from the annual average daily traffic on the road considered and the remaining service lifetime, t_{rsl} (in this paper $t_{rsl} = 40$ years).

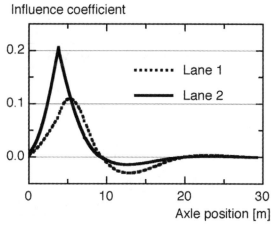

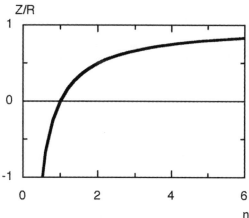

Figure 8. Influence lines for the endspan sagging moment in traffic lane 2 (upper lane) of Allt Chonoghlais bridge, calculated by finite element analysis.

Figure 9. Relationship between the factor of safety and the safety margin.

A dynamic impact factor is then applied to the static load effects to include the additional effects exerted by a moving vehicle. It follows the recommendations of Cantieni, 1984. Finally, the safety factor γ_{f3}, which takes into account inaccurate assessment of the effects of loading, is applied ($\gamma_{f3} = 1.1$).

6 RESULTS

6.1 Measure of safety

In the literature the factor of safety, n, is often used when the safety of different structures or structural elements are compared. For elastic analysis methods n is the ratio between the resistance, R, and the load effects, S:

$$n = \frac{R}{S} \quad (8)$$

For the yield-line method n is the ratio between the energy dissipated in the yield-lines, E_{YL}, and the work done by the loads, W_Q:

$$n = \frac{E_{YL}}{W_Q} \quad (9)$$

Often the safety margin $Z = R - S$ is preferred as it is a measure of the capacity reserve. The relation between those two safety measures is shown in Figure 9. For higher safety values, doubling n is not the same as doubling Z as Z remains almost constant.

In the examples studied in this paper the safety factor was always found between 0.5 and 1.4. In this region of Figure 9 n and Z have similar significance.

Here it was decided to use n as the parameter to compare safety. The resistance R has been calculated according to the British codes. It is worthwhile to add that neither of these two safety measures includes the uncertainty of R and/or S.

6.2 Sandhole bridge

The load effects have been calculated in the upstream section (left side of joint in Fig. 2) where there is less reinforcement and the concrete strength is lower. The sagging moment at midspan and the shear force at the abutment near the construction joint have been considered. The WCS-values for the concrete strength are based on three core tests in the upstream section. No tests were undertaken for the yield stress of the reinforcement, nevertheless a WCS value of 250 MPa has been taken to show the effect of updating. The initial design characteristic value was 230 MPa. Updated values of the reinforcement depth have also been used, but no improvement was achieved as the updated depth was almost equal to the design value. As expected for such a short span, the simulation-based loading was always giving lower load effects in the slab than the deterministic critical loading. In the yield-line analysis, the critical failure mechanism was found to be a full-width transverse sagging yield-line located at midspan.

Table 3 shows the factor of safety n varying with the different assessment options. In this table the safety factor corresponds to the lower value of the safety factors obtained for bending and for shear. The assessment code n-values in the table correspond to the bending safety, whereas for the improved loading shear is critical (in yield-line analysis shear is not considered however). The low values of n for the

849

Table 3. Factor of safety, n, for the Sandhole bridge corresponding to different assessment strategies.

Analysis method	Design code: Char. value	Assessment code (bending governs)				Improved loading			
		Char. val.	WCS f_{cu}	WCS f_y	WCS $f_{cu} + f_y$	Char. val.	WCS f_{cu}	WCS f_y	WCS $f_{cu} + f_y$
Grillage	0.61	0.66	0.68	0.77	0.80	0.63	0.76	0.63	0.76
Finite-element	0.59	0.63	0.65	0.74	0.76	0.71	0.84	0.71	0.84
Yield-line	0.77	0.82	0.85	0.97	1.00	1.12	1.13	1.31	1.33

Table 4. Factor of safety, n, for Allth Chonoghlais bridge corresponding to different assessment strategies.

Analysis method	Design code: Char. value	Assessment code				Improved loading			
		Char. val.	WCS f_{cu}	WCS f_y	WCS $f_{cu} + f_y$	Char. val.	WCS f_{cu}	WCS f_y	WCS $f_{cu} + f_y$
Grillage	0.56	0.59	0.62	0.69	0.74	0.73	0.77	0.85	0.91
Finite-element	0.70	0.74	0.84	0.74	0.95	0.88	0.93	0.91	1.10
Yield-line	0.82	0.85	0.87	0.99	1.03	0.98	1.02	1.16	1.20

improved loading can be explained by very high (local) shear stresses when a deterministic critical loading (extreme axle load) is considered. Updating the concrete strength, f_{cu}, has a big impact when shear is considered, but much less for bending. Updating the reinforcement yield strength, f_y, has a big influence on the bending resistance. The finite element results give slightly smaller factors of safety than the grillage analysis. This is principally because of the higher, but very localised, stresses in the structure.

The example of Sandhole bridge shows that a bridge that fails using the assessment code and including in-situ information can be shown to be safe if plastic analysis together with an improved load model is used.

6.3 Allt Chonoghlais bridge

The endspan has been found to be the critical part of the bridge. The sagging moment has been calculated in the central region of the upper traffic lane, the hogging moment and shear force at the centreline of the bridge near to the left support pier. The WCS-value for the concrete compressive strength (33.5 MPa) is based on core test results and is considerably higher than the characteristic value (20 MPa). The steel characteristic value (230 MPa) can be increased to 250 MPa when the three tested steel bars are used with the WCS-equation. Only when the shear force was considered, the deterministic critical loading induced bigger load effects than the simulation-based loading.

Table 4 contains the minimum factor of safety. This factor corresponds in most cases to the sagging moment and sometimes to the shear force evaluation. The differences between grillage and FE analysis come from the modelling of the pier supports. In the grillage analysis

they were modelled with only three point supports (Fig. 4). Interestingly yield-line analysis does little improve the assessed safety in this case. In this analysis, the critical yield-line mechanism is a full-width mechanism with a sagging yield-line near midspan. Note that subsequent analysis has shown that the point of curtailment near the abutment would be a more critical section than this one at midspan. But to be comparable with the elastic results, the critical yield-line pattern has been constrained to be in the central span region. In reality, all the safety factors would be lower than those in Table 4 since this other location governing failure.

7 CONCLUSIONS

The sensitivity of the assessed safety to different assessment strategies has been shown. Grillage and finite-element analysis give approximately the same results. Yield-line analysis can predict the ultimate load capacity more accurately.

If bending is critical, updating the steel yield-strength is worthwhile; if shear is critical it is better to collect some core samples to get concrete strength data. A new method has been introduced to determine a characteristic value for assessment to be used with elastic analysis. For plastic analysis however, a lower bound on the mean is more appropriate. More typical data (e.g. limits on strength variance) should be made available to determine the prior strength distributions.

The factor of safety can also considerably be increased by using a bridge-specific load model. A simulation-based model relying on a bridge's influence surfaces has been suggested. Unfortunately this model is not applicable for yield-line analysis.

ACKNOWLEDGEMENTS

The financial support of the Swiss Academy of Engineering (SATW), Mott MacDonald Ltd. and King's College Cambridge is greatly appreciated. The authors would also like to thank Roland Morgan of the Scottish Executive Development Department.

REFERENCES

BA 44/96 1996, The assessment of concrete highway bridges and structures, *Design manual for roads and bridges*, Volume 3, Section 4, Part 15, London: HMSO Publications Centre.

Bailey, S.F. 1996, *Basic principles and load models for the structural safety evaluation of existing road bridges*, Thesis N° 1467, Lausanne: Swiss Federal Institute of Technology.

BD 21/01 2001, The assessment of highway bridges and structures, *Design manual for roads and bridges*, Volume 3, Section 4, Part 3, London: HMSO Publications Centre.

BD 44/95 1995, The assessment of highway bridges and structures, *Design manual for roads and bridges*, Volume 3, Section 4, Part 14, London: HMSO Publications Centre.

BS 5400 1990, Code of practice for design of concrete bridges, *Steel, concrete and composite bridges*, Part 4, London: British Standards Institution.

Cantieni, R. 1984, Dynamic load testing of highway bridges, *IABSE Periodica* 3: 57–72.

FHWA 1999, *Performance related specifications for PCC pavements*, Publication FHWA-RD-98-171, McLean: US Department of Transportation, Federal Highway Administration.

Hambly, E.C. 1991, *Bridge deck behaviour*, 2nd edition, London: E & FN Spon.

Imhof, D. 2001, *Modèle de charge (trafic 40t) pour l'évaluation des ponts-routes à deux voies avec trafic bidirectionnel*, Research report 533, Zurich: Swiss association of road and traffic experts (VSS).

ISO 2394 1998, *General principles on reliability for structures*, Geneva: International Organization for Standardization.

Middleton, C.R., Hogg, V. 1998, *Review of basic variables for use in the reliability analysis of concrete highway bridges*, Report CUED/D-STRUCT/TR.172, Cambridge: Engineering Department, Cambridge University.

Middleton, C.R., Ibell, T.J. 1995, *Collapse Analyses of A83–150 Sandhole Bridge and A82–620 Allt Chonoghlais Bridge*, Report for the (then) Scottish Office, Cambridge: Campbell R. Middleton, Consulting Engineer.

prEN 1990 2001, *Eurocode – Basis of structural design*, Final draft, Brussels: European Committee for Standardization.

Tamhane, A.C., Dunlop, D.D. 2000, *Statistics and data analysis*, Upper Saddle River: Prentice-Hall, Inc.

Transport Statistics Great Britain 2001, London: Department for Transport.

Safety and Reliability – Bedford & van Gelder (eds)
© 2003 Swets & Zeitlinger, Lisse, ISBN 90 5809 551 7

Using HAZOP for assessing road safety measures and new technology

H.M. Jagtman, T. Heijer & A.R. Hale
Safety Science Group, Faculty of Technology, Policy and Management, Delft University of Technology

ABSTRACT: The introduction of new technologies in traffic produces a range of unknown deviations in the desired traffic process. These developments require a renewal of the ex-ante assessment procedures for measures which will be implemented in the traffic system. In this paper the HAZOP methodology is applied to road traffic measures to provide scenarios based upon predicted deviations and problems with new, mainly in-vehicle technologies. To make HAZOP applicable for road safety purposes analysis of the expectations of road users is added to the traditional approach. In this paper some results are shown for speed reduction measures. Besides the results, their dependency on the membership of the HAZOP team is discussed.

1 INTRODUCTION

The HAZOP technique has long been used in the chemical industry for assessing designs. In recent years its area of application has increasingly been extended to other industries and technologies. In the Safety Science Group in Delft these applications have included road maintenance work, tunnel building and more recently driving (e.g., Swuste & Heijer, 1999, Swuste & Wiersma, 2000). This paper de-scribes the approach taken in the last area and the results which have been achieved in assessing the potential safety effect and effectiveness of both conventional road features, such as speed humps, and new technologies, such as intelligent speed adaptation (ISA).

This study forms part of a larger research into the proactive assessment of intended and unintended effects on safety of proposed new technology (both in-car and roadside). The HAZOP approach has been tested using road users and local policymakers as experts. In particular this paper will show the adaptations to the methodology that were necessary in order to apply it to road traffic operations. These included decisions about the representation to be used, the experts to be involved and the parameters and guidewords to be chosen.

Special attention was given to the dependency of the HAZOP results on the composition of the HAZOP team. The results of two different groups are compared to show to what extent the identification of problems and the interpretation of these depend on the experiences of the HAZOP team.

2 NEED FOR PROACTIVE SAFETY ASSESSMENT IN TRAFFIC

The use of ICT (Information & Communication Technology) based technology in vehicles has been increasing since the 1980s (e.g., ETSC, 1999). Whereas the first applications were mainly based on providing various sources of information to drivers (e.g., RDC-TMC for regional traffic information and navigation systems), nowadays systems are on the market that are able to influence driving tasks directly. An example is adaptive cruise control (ACC), a system designed to keep a minimum headway to a vehicle in front. Although car manufacturers sell these systems as so-called "comfort extensions" functional safety problems might occur when using these systems. Furthermore the systems might influence one's driving behaviour and through this the safety of the traffic system as a whole. Jagtman et al. (2001) discuss the current knowledge on safety effects of ACC-like systems. They showed a gap in the types of effects that were incorporated in safety studies of these systems. The Ex-ante studies performed involved safety problems relating to the desired process that was defined for a system (e.g., keeping a safe headway) and often did not incorporate safety problems resulting from deviations from the desired process, such as malfunctioning of the system or driver adaptation.

Recently, Carsten and Nilsson (2001) have argued that a generic safety assessment for driver warning and vehicle control systems is lacking. They concluded that a standardised safety performance test will not be feasible and that a process-oriented approach is necessary. Part of this approach is the definition of possible test scenarios. In order to assess the safety of driving support systems these scenarios should at least contain the normal and desired process and plausible deviations from this process. The scope of possible deviations should be known before defining the scenarios containing deviations. The complexity resulting from the implementation of all kinds of new technology increases the need for a method to identify the test scenarios.

Elliott & Ownen (1968) described similar problems during the design of new chemical plants. They tried to provide a systematic approach to think about not only the process and its predictable deviations but also to try to take into account unknown deviations (Swann & Preston, 1995). The need to adopt such methods in the field of road traffic increases with the introduction of (complex) technologies that may decrease the possibilities of the road users to adapt their behaviour to the situation they find themselves in.

3   HAZOP FOR TRAFFIC PURPOSES

The fields in which HAZOP studies have been used (e.g., chemical process engineering, food process engineering, nuclear power and programmable electronic systems, [Kletz 1997, 1999]) have in common the investigation of a process that is delimited both in space and in the number of different operations. The road traffic system has much in common, but differs in some ways:

– It deals with a larger number of people who moreover do not all have similar experiences or expectations in participating in the traffic process
– The number of participants and the variety in these participants (car drivers, pedestrians, etc.) within the traffic process is variable in time and in space/location
– Within the traffic process individual participants cannot be assumed to have the same goals as each other or as the authorities.

As a result of these differences the tasks of different participants are harder to describe in a generic process description. However, as the HAZOP approach seems to show great promise for the identification of test scenarios for ICT-based new technologies in traffic, the approach has been adapted to traffic processes.

3.1   Traffic process

The desired traffic process discussed in a HAZOP is defined as the regulators' goals specified for the location under study. As these goals comprise both achieving a sufficient capacity, depending on the function of the network under study (for instance through-flow for a motorway or access in residential areas) and minimising the number and/or severity of traffic accidents, a HAZOP in traffic can study both the operability of the system and the hazards in it, as is the case with the original field of application (Kletz, 1997). Environmental impacts resulting from the traffic process could be studied in a similar way, but are not further discussed here.

A general description of the desired traffic process is defined as (Jagtman, 2002a): *Offering a particular capacity in a defined area under conditions of minimum loss.* The process is worked out in detail based upon the specific goals such as:

– Priority for specific road users (e.g., vulnerable road users, heavy traffic)
– Speed reduction during specific time slots (e.g., day time, lunch break)
– Set time windows for goods distribution (heavy traffic is not allowed at the location outside these time windows).

3.2   Representation

The traffic process under study is characterised by means of the physical layout, including information on the surroundings. The area information includes possible special destinations in the neighbourhood of the study object, such as: hospitals, schools, playing fields, shops, housing, offices, and industries. This information might be of importance for the events (including incident or accident events) that could occur.

The road (safety) measures included in the HAZOP are represented using standard descriptions from handbooks, where available (e.g., CROW 1996, 2002), or using the information available from pilot studies for the measures which use new technologies.

3.3   Basic assumption in the traffic HAZOP

The basic assumption for the Traffic HAZOP is that a safe traffic process (for a particular location) is established when all traffic participants involved have both sufficiently similar expectations about how to resolve the situations in that traffic process, and the means to effectuate that behaviour. "Sufficiently similar" means that expectations do not have to be exactly identical but that they should at least not conflict with each other. The complexity of expectations has been shown for different junction settings by Heijer and Wiersma (2001). All measures implemented in the traffic system, either in-vehicle, in the infrastructure or in regulation, could influence these expectations. For instance, a driver with a detection system for other traffic (of which ACC is a preliminary system) will possibly put less effort into observing the other traffic and overlook a pedestrian near a pedestrian crossing. Moreover, this modified behaviour of the drivers or vehicles with such ICT-based systems could also interact with the expectations of other road users without these systems. An example could be the pedestrian on a pedestrian crossing expecting the driver to have an automatic system to notice him/her and hence to give priority.

In order to cope with the differences as described above, discussion of the traffic participants' expectations is added to the standard HAZOP approach. The following steps are therefore added to the Traffic HAZOP:

1 Discuss whether or not the road users' expectations are likely to be sufficiently similar
2 Discuss if the use of a given measure will affect the expectations of road users at other locations than the one discussed during the HAZOP
3 Conclude whether the users' expectations fit the desired process at the location as defined by the regulator.

3.4   Traffic parameters and guidewords

For the application of the HAZOP methodology parameters and relevant guidewords have to be chosen for the specific field under study (Redmill et al. 1997). Safety problems in traffic result from deviations caused by a single road user as well as from deviations caused by situations in which several road users are involved. Consequently, we included traffic parameters both for single users (marked with "a" in Table 1) and parameters for traffic situation with multiple users involved (marked with "b"). A detailed explanation of these parameters and guide words can be found in Jagtman (2002a).

4   EXAMPLE: SPEED REDUCTION IN BUILT-UP AREAS

The HAZOP methodology has been applied to study the use of speed humps in a built-up area with a speed limit of 30 km/h. The 30 km-zone is located in the neighbourhood of a school. The location under study was rebuilt in 2001 to improve priority given to pedestrians on the pedestrian crossing. This priority was emphasised as part of the desired traffic process. In general the process is defined as "adherence to the formal rules" by all participants. At this location these rules include: "give priority to pedestrians on the pedestrian crossing", "give priority to traffic on the main road" and "do not block the junctions". Notice that these rules may lead to conflicts and problems. For instances for drivers from the "A" direction in Figure 1 the driver first crosses the junction having priority, then has to stop for pedestrians, if present, during which he blocks the junction, and finally to cross the speed hump.

Besides the current layout, the use of Intelligent Speed Adaptation (ISA) at the same location instead of the speed hump was analysed. ISA, as it is now in development, consists of vehicle-based technology

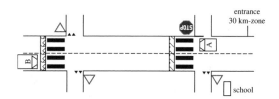

Figure 1.   Visualisation of location and the infrastructure measure (speed humps).

Table 1.   Parameters and guidewords for traffic-HAZOP.

	None/ no of	(Too) high	(Too) low	Wrong	Fail	Part of	Unknown	Unexpected
a. speed								
a. direction								
a. location								
a. focus								
a. attention								
a. travel time								
b. speed difference								
b. distance								
b. road users								
b. number of road users								
b. violations								
b. flow rate								
a. expectation				this cell is discussed at the end of the HAZOP				

that assists motorised vehicles in adhering to the local speed limit. The assistance can be provided in various manners, such as information on a display, auditive warning signals and physical feedback. We studied an ISA application called the "active gas pedal" in which the driver experiences a counter force through the gas pedal at the moment the vehicle exceeds the local speed limit. GPS positioning and a digital map are used to determine the local speed limit. The driver is able to overrule the assistance by kicking down on the gas pedal. (Further details on the system: Biding & Lind 2002).

4.1 Results

The HAZOP was performed twice for the two versions of this example: one HAZOP team consisted of road users and the other team consisted of local and regional policy makers. A selection of the meaningful cells for the HAZOP matrix was made by eliminating the unfeasible cells (the blank cells in Table 1). Both sessions resulted in over 140 useful deviations of which a selection will be presented in this section. An extended review of the results of the HAZOP performed by the road users can be found in Jagtman & Heijer (2002).

We will show the results in a framework with four levels of safety problems in traffic. The basis for this framework was set by Carsten et al. (1993) and Morello (1995) to evaluate the use of new in-vehicle technologies. They both distinguished a functional level, a driver level and a traffic level at which effects on safety in traffic systems might occur. The traffic safety level is divided into the interaction between a group of road users, who meet at the same location at the same time and the traffic flow as a whole. The four levels are defined as follows (Carsten et al., 1993, Morello, 1995, ETSC, 1999, Jagtman et al., 2001):

- Functional Safety Level: covers safety problems that result from the hardware and software design of the measure, in particular the technical reliability, probability of system failures and the potential of it getting into a dangerous or unexpected mode.
- Driver Safety Level: focuses on the interaction between the user and the measure under study. This level covers the appropriateness of the design, possible distraction of the user and adequate support in performing a safe trip.
- Safety of Interactions: focuses on the interaction of drivers and of the measure under study with their close environment, including other road users, vehicles and the infrastructure.
- Traffic Safety Level: covers the effects of the measure under study on safe operation of the traffic system. This system incorporates macro safety effects on the whole network.

4.1.1 Speed humps
A selection of deviations, problems and conflicts for the implementation with speed humps resulting from the HAZOP sessions is shown in Table 2.

4.1.2 Intelligent speed adaptation ISA
In Table 3 a selection of the results for the ISA system is shown. The percentage of cars equipped with the ISA system was not fixed for the discussion. If specific deviations result from mixed traffic, in which ISA and none-ISA cars meet each other, this is made

Table 2. Deviations, conflicts & problems for speed hump implementation.

FUNCTIONAL SAFETY
- Vehicle breakdown caused by engine, lack of fuel, flat tyre;
- Driving with too high speed caused by wrong estimation of speed required to smoothly pass the speed hump.

DRIVER BEHAVIOUR & SAFETY
- Misjudgement of situation or road signs;
- Slowing down for pedestrian, but paying no attention to other road users;
- Distraction by events in the vehicle (children, mobile phones, radio);
- Short cuts: crossing main road diagonally, short cut while turning left. This increases when road users are in a hurry (for instances if they are almost late for school);
- Speed up between speed humps;
- Pedestrians not crossing at pedestrian crossing;
- Impatient with waiting and as a result taking priority;
- Driving slowly and/or carefully because the location is unknown to drivers or they are inexperienced;
- Complexity of the situation leads to performance of sequential, partial solutions;
- Mainly focus on the speed hump ahead instead of on possible other road users.

SAFETY OF INTERACTIONS
- Groups of pedestrians/cyclists take priority;
- Blocking part of the road and/or the junction (removal lorries, delivery vans, cars who stop to pick up someone);
- Speed up to get to pedestrian crossing before pedestrian does;
- Rear-end shunts caused by short headways and sudden braking by vehicle in front;
- Pedestrians expect to get priority when they cross at the pedestrian crossing;
- Vehicles stop across junction to give pedestrians the possibility to cross the road at the pedestrian crossing;
- Motor vehicles on the main road will not give priority to pedestrians in order not to block the side roads.

TRAFFIC SAFETY
- Speed hump increases travel time for emergency vehicles;
- Change of public transport routes because of discomfort caused by the speed hump for driver and (standing) passengers;
- Traffic from side road cannot cross or enter main road because of blocking of the junction in congested circumstances.

Table 3. Deviations, conflicts & problems for ISA.

FUNCTIONAL SAFETY
- Feedback of the ISA system;
- Wrong speed limit used by the ISA system caused by a positioning error;
- Wrong speed limit used by the ISA system caused by old maps, temporary limits at road construction sites;
- Difference in location of speed limit boundary between the road signs and the feedback of the ISA system;
- Two ISA vehicles both driving on the main road may suffer from calibration problems resulting in different boundary for the vehicles and possibly slightly different speed limits.

DRIVER BEHAVIOUR & SAFETY
- Relying on ISA for keeping to the speed limit although a lower speed might be required for the actual situation (extreme: use of ISA as cruise control);
- Immediate speeding up when getting into a zone with higher speed limit/unexpected fast accelerating by ISA vehicle caused by pressing down the gas pedal totally;
- Not noticing the 30 km/h sign and suddenly experiencing the counter force by the ISA system;
- Drivers experiencing ISA for the first time mainly pay attention to the new system;
- ISA users forget to change gear after the speed limit is passed;
- Check the ISA system by (more than normally) observing the speed on the speedometer;
- Trust in the ISA system resulting in less checking of the speedometer;
- Irritation at the ISA system if the speed limit is inappropriate or incomprehensible (e.g. night time);
- Kick-down of the ISA system provides a way of release for the irritation at the system, without immediate speeding by the vehicle;
- Non-ISA drivers will probably drive faster because there are no speed humps at this location.

SAFETY OF INTERACTIONS
- Expecting vehicles to slow down when entering the lower speed zone may lead to problems if a vehicle without ISA is approaching (high number of ISA vehicles in the traffic flow);
- Trusting the ISA system to slow the vehicle down when entering a lower speed limit zone could result in problems if a leading vehicle starts braking before the zone boundary;
- Non-ISA users, pedestrians and cyclists will anticipate on (smaller) gaps to cross or enter the main road;
- Mopeds, which are not equipped with ISA, will overtake the ISA vehicles in the 30 km zone;
- Irritation of non-ISA users because of the ISA users drive slowly in the 30 km zone.

TRAFFIC SAFETY
- Too high speeds of ISA vehicles at the pedestrian crossing caused by wrong/awkward boundary set in the ISA system;
- Compensate for unexpected delays by driving at or beyond the speed limit in areas where ISA is not activated/supported;
- ISA may result in less acceleration and deceleration and as a result show a smoother traffic stream (perhaps even provide a greater capacity).

clear in the table. Similarly, it is expressed in the description of the deviation if this occurs in a fully equipped traffic flow. Note that the deviations in Table 2 not directly concerning the speed humps may also apply when using ISA.

4.2 Expectations of road users

In addition to the traditional HAZOP, the expectations of the different road users were discussed. As all road users do not act according to one pattern, the teams discussed likely behaviours and the expectations that would lead to this behaviour. The results are summarised in this section.

4.2.1 Speed hump

The motorised traffic approaching from direction A in Figure 1 will be focussed on the speed hump and might not observe other traffic, including pedestrians, as a result. They could block the side roads while waiting for a pedestrian to cross or continue without giving priority to the pedestrian, in order not to block the intersection. Pedestrians will not assert their priority more compared to a situation without the speed hump, because it is located wrongly for traffic in this direction and offers them no protection from speeding traffic.

Vehicles from the opposite direction (B) have to slow down for the speed humps before crossing the pedestrian crossing and the junction. Since they will already have reduced their speed for the humps they will be more likely to give priority to the traffic from the side road than will vehicles from direction A, although the traffic on the main road has priority. This is reinforced by the side traffic that will be inclined to accept smaller gaps before entering the main road. In this case the pedestrians' resolve to assert their priority will be strengthened by the presence of the speed humps.

Cyclists will not change their behaviour as a result of the speed humps. In the case of dense traffic on the main road pedestrians will use the special crossing more intensely than in low traffic intensity. On the other hand groups of road users, such as a group of pupils from the school, may act as a single powerful entity and take (and get) priority at the crossing location and elsewhere around the intersection.

When driving at another location with speed humps drivers will have learned from earlier experiences to cross the hump slowly because higher speeds might be uncomfortable. After experiencing the hump once the road user knows how to pass it the next time. Further use will hardly affect their focus on vulnerable road users.

4.2.2 ISA (all motorised traffic equipped)

When all vehicles are equipped with the ISA system the behaviour of all drivers of motorised vehicles will be more uniform and there will be no distinction between directions A and B on the main road (speed humps no longer required when ISA is used). The traffic on the main road will move steadily with a speed of about the maximum allowed speed (30 km/h). As result of this low speed the pedestrians and the side traffic will probably neglect the priority set at the intersection and cross or join the traffic stream when they think is appropriate. Moreover the pedestrian crossing will be seen as pointless and the pedestrians may cross anywhere.

Mopeds, depending on the feasibility to install ISA in such motorised bicycles, may cause problems. If these are not equipped with Intelligent Speed Adaptors they will most likely drive faster than the motorised traffic stream that is equipped.

4.2.3 ISA (mixed traffic allowed)

Depending on the traffic intensity, clusters of vehicles may form behind one or multiple ISA-equipped vehicles. If the intensity is low and drivers without ISA are irritated by the system, they may start over-taking the slower vehicles.

In the case of clusters of vehicles, the ease of crossing the street or entering the main stream will increase. If, however, the number of overtaking manoeuvres increases, this will increase the risk to pedestrians, cyclists and (motorised) traffic from the side roads. Road users have difficulty estimating speed outside a range of expected speeds. Especially when the use of ISA becomes more common, exceptional speeds (of vehicles without ISA) will become less and less expected.

The benefits of ISA and speed limits needs to be understood by the road users in order for them not to be irritated by these kind of systems, for example in quiet hours such as at night. This problem will increase in mixed traffic situations in which not every motorised vehicle is equipped.

Finally the use of ISA requires both users and other traffic to learn. The system will not only work locally but elsewhere in a network. Road users may experience the assistance of the system and/or meet vehicles that are equipped with the system at different points in the network and will need to learn how to recognise which regime is applicable.

4.2.4 In summary

Building expectations upon the behaviour of the other road users in the case of speed humps is problematic, because the options are variable. Will the vehicles on the main road slow down or stop to give priority to pedestrians and, for direction A, where will they stop? What is the behaviour of the side traffic, will they take priority where the traffic on the main road stops? We measured the behaviour of the main road traffic in our study by video observation and showed that it is

independent of the situation. Hence, it is hard for pedestrians and side traffic to estimate if and where these vehicles will slow down (Jagtman, 2002b). These findings confirm the current badly chosen design of the intersection.

Estimation of the speed of ISA equipped vehicles is easier. Where ISA is used, the possibility to estimate other road users' behaviour will depend on whether or not all motorised traffic is equipped, whether or not ISA vehicles can be distinguished from non-equipped vehicles. The speed behaviour in general in the Swedish pilots (Bidding & Lind, 2002) showed a lower average speed and a reduction in the spread of speeds. On 30-km roads in local networks the Swedish study found speed reductions in mixed traffic of 0.3–0.9 km/h.

5 THE EFFECTS OF THE COMPOSITION OF THE HAZOP TEAM

Lawley (1974) stated that the team performing a HAZOP should be carefully chosen to provide sufficient knowledge and experience to lead to adequate results. Though the team should be small enough to be efficient, it should contain sufficient spread of skills and disciplines to cover all aspects of the process under study.

In the field of road traffic everybody has some experience, at least as a participant. The HAZOP methodology was applied with two teams with different experience. One team consisted of road users and the other team of local and regional policy makers. The similarities and differences in the kinds of deviations that were identified in the workshops are described in this section.

The participants did not have knowledge of the HAZOP methodology beforehand, except for the leader of the workshops, but this did not present any problems. Lawley (1974) and Kletz (1999) indicate that it is not necessary for all participants to be familiar with the method.

5.1 Differences in types of deviations noted

The two workshops generated a variety of deviations. Tables 4 and 5 show the number of deviations mentioned for each of the levels explained in section 4.1. for each measure. The participants were asked in the session over ISA only to indicate additional problems to those which would occur with the humps. They indicated that over 95% of the deviations in respect of the speed humps are also applicable to the ISA implementation.

In both workshops and for both speed-reducing measures, the largest category of deviations was that relating to the level of the individual driver. The relative importance of the other levels differed between

Table 4. Types of deviations, conflicts & problems (road users).

	Functional	Driver	Interaction	Traffic	Total
Hump	14	38	24	21	97
ISA	12	18	9	9	48
Total	26	56	33	30	145

Table 5. Types of deviations, conflicts & problems (policy makers).

	Functional	Driver	Interaction	Traffic	Total
Hump	3	52	36	15	106
ISA	11	28	9	10	58
Total	14	80	45	25	164

the speed humps and the use of ISA. The share of the functional level for an ISA measure was about 19–25%. For the speed humps this was less than 15%. The level of interaction between road users was mentioned relatively more often for speed humps than for the use of ISA (25–34% versus 16–19%). The percentage at the traffic safety level was higher for speed humps than ISA for the team of road users. For the team of policy makers this was opposite.

The distribution of the deviations over the four levels is more even for the team of the road users than for the team of policy makers. The latter team has about 75% at the driver safety and safety in interaction levels.

5.2 Subjects of deviations noted

This section discusses the subjects of the deviations noted in more detail.

5.2.1 Infrastructure and ISA

It is the task of the policy makers to decide about the design and maintenance of the infrastructure and measures used to regulate traffic using that infrastructure. The team with policy makers came up with a number of problems for the ISA case:

– Changes of speed limits made by the regulator have to be changed in the digital maps as well
– Awkward definition of boundaries results in inappropriate speeds of the ISA-equipped vehicles
– Temporary speed limits at road construction sites are often not accompanied by temporary regulations; the digital map used by ISA does not know about those temporarily changed limits
– Incorrect data in the digital maps.

For the team of road users, this aspect is not part of their task. They did not go into the problems of failing

to set the temporary regulation but they did mention problems at roadworks, where the speed limit is temporarily changed. They also discussed errors in the database of the digital map caused by outdated or incorrect data. Not understanding the change of a speed limit was also discussed by this team. Moreover, they also got into the problems that arise if an ISA user experiences feedback from the ISA system that does not correspond to the road signs. Although the road users have another perspective than the policy makers they got into most of the same issues, even if they did not express them in a policy maker's specific details.

5.2.2 Road users' behaviour

All HAZOP participants, including the team of policy makers, are road users in addition to their possible professional involvement in road traffic. Up front we expected these groups to come up with more or less the same deviations in relation to road users' behaviour. The commonly occurring "problems" like short cuts, crossing at other locations than the pedestrian crossing, impatience and irritation at slow drivers were mentioned during both workshops.

For ISA similar problems in behaviour of users were mentioned, for example:

– Testing the possibilities of a new system in your vehicle, which may result in less attention for the situation in traffic, or even show-off behaviour
– Use ISA as a cruise control (without getting speed tickets) by pressing the gas pedal until the counter force of the system is felt
– Instead of slowing down when noticing a road sign with a lower speed limit, drivers could wait to be slowed down by the counter force
– Driver in the wrong gear (particularly in too high gear) because the system is slowing you down.

Although the two groups generally mentioned similar problems, one or two problems were noted in only one of the workshops.

5.2.3 Violations in traffic

Besides "classical" violations such as speeding and not giving priority, some other violations or deadlock problems were mentioned in both workshops. In relation to the location of the pedestrian crossing it was mentioned that a vehicle could stop on the junction itself to let pedestrians cross, while blocking the road for the side traffic. It was noted that irritation with the system might lead the user to try to tune the ISA. One violation (mentioned by only one group) involved a situation in which all vehicles are equipped with the ISA system except for mopeds. In that case moped drivers are likely, as they commonly do in Dutch traffic, to drive beyond the limit of 30 km/h in urban areas, while the other motorised traffic keeps to this limit.

859

5.3 Importance of composition of the team

In the two workshops the variety of deviations found and the way in which the teams discussed subjects did not vary so much. Based on these results the composition of the team does not seem to be the deciding factor for detecting deviation and defining conflicts and problems when applying a HAZOP in road traffic. For the interpretation of the results from the HAZOP however, special knowledge will be necessary. Ranking of the problems will most likely require in-depth experience in traffic design and operation. Both HAZOP teams were rather inexperienced in the field of new technologies used in road traffic. To find out if knowledge about the new technologies will influence the results the findings of these two groups will be compared to a team with researchers, who are involved in the ex-ante evaluation of these systems.

6 CONCLUSIONS

This paper describes the adaptations made to the HAZOP methodology to apply it to traffic processes, in particular those relating to the traffic process and the traffic parameters. To cope with the divergent goals of road users, that moreover may conflict with the desired process set by the regulator, the discussion of expectations of all road users was added to the standard HAZOP procedure.

Application of the traffic HAZOP was shown for two speed-reducing measures in the neighbourhood of a school. This resulted in a large number of possible deviations at this location. After discussion of the expectations of the road users at the study location it was concluded that the current speed hump implementation does not assist in building compatible expectations. A video observation confirmed this conclusion. For the use of a new technology to assist the drivers of motorised vehicle in keeping to the local speed limit the expectations of road users will depend on whether all vehicles are equipped with the system, if ISA vehicles can be distinguished from the other traffic and the acceptation of the system itself.

The deviations discussed during the two workshops were divided over four levels of which safety and behaviour of one road user and safety of an interactions between a number of road users covered about 75% of the problems.

Although the road users participating in the workshop do not have the experience of policy makers in the design and operation of local and regional road networks they came up with almost the same deviations. They differed in the depth of explanation of the origin of the problems concerning design and maintenance of the network. We conclude, at least for the moment, that the composition of the HAZOP team

will probably not be crucial for identifying problems, though specific knowledge will be required to interpret the results of a HAZOP in traffic. In a third workshop this issue will be further examined in respect of differences in the knowledge of new technologies.

The cases studied so far show that the HAZOP methodology can be used very effectively to study traffic processes. The participants found the structured approach of HAZOP particularly useful when thinking of infrastructure design for safety. To prove the value of performing HAZOP studies in the design of new technologies to be used in road traffic purposes, our next step will be to perform a HAZOP to set an evaluation framework for a large-scale pilot with an ISA system.

ACKNOWLEDGEMENTS

The authors would like to thank the local policy-makers and students who participated in the two workshops described in this paper.

The study described is part of a PhD project sponsored by the Dutch Institute for Road Safety Research (SWOV).

REFERENCES

Biding, T. & Lind, G. 2002. *Intelligent Speed Adaptation (ISA), Results of large-scale trials in Borlänge, Lindköping, Lund and Umeå during the period 1999–2002*, Börlange: Vägverket 2002:89E

Carsten, O.M.J. et al. 1993. *Framework for prospective traffic safety analysis, HOPES*, deliverable 6, Leeds

Carsten, O.M.J. & Nilsson, L. 2001. Safety assessment of driver assistance systems. In H. Priemus (eds.), *European Journal of Transport and Infrastructure Research*, 1 (3): 225–243 Delft University Press

CROW 1996. *ASVV 1996 – aanbevelingen voor verkeersvoorzieningen binnen de bebouwde kom*, nr. 110, Ede: CROW (in Dutch)

CROW 2002. *Handboek Wegontwerp: wegen buiten de bebouwde kom*, nr. 164x, Ede: CROW (in Dutch)

Elliott, D.M. & Owen, J.M. 1968. Critical examination in process design. In J.M. Pirie (eds.) *The chemical engineer* no. 233: CE377-CE383, Institute of Chemical Engineers

European Transport Safety Council (ETSC) 1999. *Intelligent Transportation Systems and Road Safety*, Brussels: ETSC

Heijer, T. & Wiersma, E. 2001. A model for resolving traffic situations based upon a scenario approach. In *ITS Transforming the Future: Proceedings of the 8th world conference on Intelligent Transport Systems*, Sydney 30 September–4 October. ITS Australia

Jagtman, H.M., Marchau, V.A.W.J. & Heijer, T. 2001. Current knowledge on safety impacts of Collision Avoidance Systems (CAS). in P.M. Herder & W.A.H. Thissen (eds.), *Critical Infrastructures – 5th International Conference on Technology, Policy and Innovation*, paper 1152, Utrecht: Lemma

Jagtman, H.M. 2002a. Dealing with deviations from intended operations of intelligent transport systems. In *ITS: enriching our lives, Proceedings of the 9th world congress on intelligent transport systems*, Chicago 17–17 October 2000, paper 2215, ITS America

Jagtman, H.M. 2002b. The "Traffic HAZOP" – an approach for identifying deviations from the desired operation of driving support systems. In E.F. ten Heuvelhof (eds.) *Proceedings of 1st International Doctoral Consortium on Technology Policy and Management*, 17–18 June 2002, Delft, ISBN: 90-5638-094-x, pp. 89–101

Jagtman, H.M. & Heijer, T. 2002. Applications of HAZard and OPerability studies (HAZOP) to ISA and speed humps in a built-up area. In *e-Safety Congress 2002*, Lyon 16–18 September, paper nr. 2175, Ertico

Kletz, T.A. 1997. Hazop – past and future. In P. Leclerq & C. Guedes Soares (eds.) *Reliability Engineering and System Safety* 55 (3): 263–266, Elsevier Science Ltd.

Kletz, T.A. 1999. HAZOP and HAZAN – *Identifying and Assessing Process Industry Hazards*, fourth edition, Rugby: Institution of Chemical Engineers

Lawley, H.G. 1974. Operability Studies and Hazard Analysis. In J. Howe (eds.) *Chemical Engineering Progress*, 70 (4): 45–56, American Institute of Chemical Engineers

Morello, E. 1995. Evaluation framework for driver assistance applications. In *Proceedings of the First World Congress on Applications of Transport Telematics and Intelligent Vehicle Highway Systems*, Artech House, Boston, pp. 1639–1646

Redmill, F., Chudleigh, M.F. & Catmur, J.R. 1997. Principles underlying a guideline for applying HAZOP to programmable electronic systems. In P. Leclerq & C. Guedes Soares (eds.) *Reliability Engineering and System Safety*, 55 (3): 283–293, Elsevier Science Ltd

Swann, C.D. & Preston, M.L. 1995. Twenty-five years of HAZOPs. In P.F. Nolan (eds). *Journal of Loss Prevention in the Process Industries* 8 (6): 349–353, Elsevier Science Ltd

Swuste, P. & Heijer, T. 1999. *Project: Onderzoek (on)veiligheid wegwerkers – rapportage van het onderzoek.* Amsterdam: Arbouw (in Dutch)

Swuste, P. & Wiersma, E. 2000. *De arbeidsomstandigheden bij het boren van tunnels in Nederland.* Project N800 Amsterdam: Arbouw (in Dutch)

Safety and Reliability – Bedford & van Gelder (eds)
© 2003 Swets & Zeitlinger, Lisse, ISBN 90 5809 551 7

Evaluation of tunnel safety and cost effectiveness of measures

S.N. Jonkman
Delft University of Technology, Faculty of Civil Engineering, Section of Hydraulic Engineering, Delft, Netherlands
Centre for Tunnel Safety, Civil Engineering Division, Rijkswaterstaat, Ministry of Transport, Public Works and
Water Managment, Utrecht, Netherlands

J.K. Vrijling, P.H.A.J.M van Gelder B. Arends
Delft University of Technology, Faculty of Civil Engineering, Section of Hydraulic Engineering, Delft, Netherlands

ABSTRACT: Aim of this paper is to propose a framework for the evaluation of tunnel safety. Two methods for the analysis of tunnel safety, the probabilistic and the deterministic approach, and their characteristics have been described. Probabilistic criteria are proposed for the judgment of personal, societal and economic risks of tunnels. Furthermore some methods to analyse cost effectiveness of (additional) safety measures are discussed. The application of the aspects and methods discussed is illustrated with experiences and results from some practical tunneling projects.

1   INTRODUCTION

Some large accidents in tunnels in recent years, such as the fires in the Mont Blanc, Gotthard and Tauern tunnels, have lead to an increasing attention for the subject of tunnel safety. Many countries have announced additional investments in existing tunnels and the initiation of extensive studies to improve the knowledge on tunnel safety. However, absolute safety does not exist, and the possibility of a serious tunnel accident can never be completely excluded. Safety criteria have been suggested for individual tunnel projects, see for example (Geyer, 1996). And although some general target safety levels are proposed by Diamantidis et al. (2000), no commonly applicable framework is available to support safety discussions. This problem is reflected in the complicated decision making processes in many large tunnelling projects. Key points in these safety discussions are the determination of acceptable risk levels and assessment of the required investments in risk reduction measures to ensure an "optimal" safety level.

Aim of this paper is to propose some guidelines for the evaluation of tunnel safety to support these often difficult design processes. The suggested methodology can be considered as an advice to the decision makers (the politicians) from a technical point of view. The suggested criteria will deal with the problem of acceptable risk, and the optimisation of investments in safety measures in relation with economic and societal demands. The application of the framework will be illustrated with some experiences from large tunneling projects in the Netherlands.

Firstly, two approaches for the analysis of tunnel safety, the probabilistic and deterministic approach, are analysed in section 2. In section 3 a framework for the evaluation of tunnel safety is proposed. Section 4 focuses on the analysis of cost effectiveness of life saving measures. Some of the aspects that are brought forward are illustrated with results from some practical case studies in section 5. Section 6 contains the conclusions of this study.

2   PROBABILISTIC VS DETERMINISTIC APPROACH OF TUNNEL SAFETY

In the analysis of tunnel safety both probabilistic and deterministic analyses are carried out.

The **probabilistic analysis**, or the quantitative risk analysis (QRA), is based on an inventory of all possible accident scenarios. The potential accidents can be shown in a so-called event tree, which outlines possible events after occurrence of the accident. Probabilities and consequences (e.g. fatalities, economic damage) of all these scenarios have to be estimated and combined in a risk number, for an example an FN curve or an expected number of fatalities. If the costs of measures to reduce the probability or consequences of an

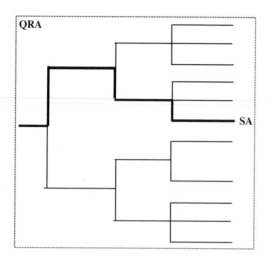

QRA

SA

Figure 1. Schematic difference between probabilistic (QRA) and deterministic analysis (SA).

accident are known, an implicit or explicit optimisation can lead to a decision on the level of protection and consequently to accepted level of risk.

The **deterministic** or scenario analysis (SA) focuses on the analysis of processes during one accident scenario and the outcomes of this scenario are used as input for design and decision making. A detailed analysis of accident scenarios (as is carried out in a deterministic analysis) will provide more insight in the processes during the accident, the possibilities for the rail/road user to escape in the event of a disaster (self-rescue), as well as the provision of aid by the public emergency services.

The difference between the probabilistic and deterministic analysis is shown in Figure 1. The whole imaginary event tree in the figure forms the basis for a probabilistic analysis. Each branch of the event tree represents one accident scenario. The bold line shows one (random) accident scenario which will be analysed in a deterministic analysis. In the deterministic analysis the probability of occurrence of the accidents is generally not taken into account.

A possible problem of the use of solely the deterministic analysis in deciding on desired levels includes the selection of an overly risky (or safe) scenario. This will provide overly negative (or optimistic) information to the decision maker and will force him to choose extreme (or too little) safety measures. The result of this decision making can be that society discovers during the implementation or use of the structure that indeed extreme measures have been chosen and that the costs are out of proportion. The final optimisation will then be the result of a trial and error process, in which policy making will be triggered by accidents, and not of thinking in advance.

Therefore the probabilistic analysis is most suitable for the evaluation of tunnel safety. It takes into account all relevant scenarios and thus ensures a balanced decision making. The benefit of the risk based approach is that different kinds of measures can be judged for their safety merits. The reduction of accident probabilitis caused by preventive measures can be compared with the reduction of consequences resulting from mitigating measures. The main problem in the application of the probabilistic analysis is the determination of acceptable risk levels. However, also the deterministic analysis of accident scenarios can prove a useful additional tool for a more detailed consideration of certain aspects such as the possibilities of self rescue and the demands for emergency services.

3 SUGGESTED FRAMEWORK FOR THE EVALUATION OF TUNNEL SAFETY

In the previous section it has been stated that the evaluation of tunnel safety should be mostly based on probabilistic evaluation of safety levels. A set of probabilistic criteria is suggested in section 3.1. Also the value of a deterministic approach has been recognized and section 3.2 discusses some aspects of the deterministic approach, such as the choice of a representative design scenario and the need for additional safety measures.

3.1 *Probabilistic risk assessment framework*

Based on previous research (Vrijling, 1998) a set of rules is presented for the judgment of personal, societal and economic risk for tunnels. The three approaches should all be investigated and presented. The most stringent of the three criteria should be adopted as basis for the "technical" advice to the political decision makers.

3.1.1 *Personally acceptable level of risk*
The first criterion is concerned with the personal level of risk. Although many, slightly different definitions are in use for the personal or individual risk, they are all concerned with probability for the individual of losing one's life. In the case of tunnels two types of parties at risk can be distinguished. Internal parties are persons who are at risk in the tunnel, the users for road tunnels, the passengers and employees for railway tunnels. External parties are the persons living in the area of tunnel. Since all these parties will have different relations with and various attitudes towards the hazards resulting from the presence of the tunnel, different risk levels can be considered acceptable for them.

A criterion for the acceptable individual risk (IR) is proposed by Vrijling et al. (1998), which takes into account the degree to which the activity is voluntary, and the benefit perceived.

$$IR < \beta \cdot 10^{-4} \left(yr^{-1} \right) \tag{1}$$

Table 1. Proposed β values for different parties involved in tunnel safety.

	Party	β
Internal	Employees (rail)	1
	Passenger or user	0.1
External	Persons living near the tunnel	0.01

where:

β policy factor, varies according to the degree to which participation in the activity is voluntary and with the perceived benefit.

In Table 1 β values have been suggested for the parties involved in tunnel safety based on Vrijling (1998).

3.1.2 Socially acceptable risk

While the former section deals with risks from point of view of the individual, it should also be considered from a social point of view. Societal risk is concerned with the probability that a whole group of people will be killed due to an accident with a certain probability of occurrence.

Societal risk is often represented graphically in a FN-curve. This curve displays the probability of exceedance as a function of the number of fatalities, on a double logarithmic scale. Also the expected values of the number of fatalities (which equals the surface under the FN curve) is often used. An important aspect in the societal judgment of hazardous activities are the "small probabilities – large consequences" accidents. The expected value is generally very low for this type of accidents (as are tunnel accidents) and therefore does not seem a good risk measure for tunnels. The standard deviation of the number of fatalities is relatively high for these types of accidents. Therefore, the so-called characteristic value is proposed as a suitable measure for societal risk. It consists of the expected value of the number of fatalities and the standard deviation, which is multiplied by a risk aversion factor k, for which a value of $k = 3$ is proposed based on the analysis of several activities in Vrijling (1995):

$$E(N) + k \cdot \sigma(N) \tag{2}$$

The total risk takes a risk aversion index k and the standard deviation into account and is therefore called risk averse. The following limit, which again takes into account the policy factor β, is proposed to limit risks on a national level:

$$E(N) + k \cdot \sigma(N) < \beta \cdot 100 \tag{3}$$

It has been shown (Vrijling, 1998) that this national criterion for acceptable risk can be translated into a standard for a single (tunnel) location. This criterion

has the typical form of a FN limit (with a quadratic steepness):

$$1 - F_N(x) < \frac{C}{x^2} \tag{4}$$

where:

$1 - F_N(x)$ probability of more than x fatalities per year

C constant that determines the position of the FN limit line

Suppose that the expected value of the number of fatalities is much smaller than its standard deviation (which in general is true for accidents with low probabilities and large consequences) and assume a Bernoulli distribution of the number of fatalities. The factor C can now be written as a function of the number of installations on a national level (N_A), the risk aversion factor (k), and the policy factor (β):

$$C = \left[\frac{\beta \cdot 100}{k \cdot \sqrt{N_A}} \right]^2 \tag{5}$$

For applications for tunnel safety again the distinction can be made between internal users (or employees) and external parties. Considering the differences between these parties, different standards should be applied for these parties, and different β's are applicable for them, see Table 1 for suggested values. The finally chosen height of the limit will also depend on the number of installations. In the derivation of the local limit, first the acceptable risk should be set on the national level. Consequently, the acceptable risk should be distributed over the tunnel locations. Since no risk limit for tunnels has been established on a national level yet, often safety criteria applied in other tunnelling project are used as reference values (see also section 5 for an overview).

3.1.3 Economic optimisation

The derivation of the (economically) acceptable level of risk can also be formulated as an economic decision problem, as has been shown by van Danzig (1956) for flood defences. According to the method of economic optimisation, the total costs in a system (C_{tot}) are determined by the sum of the expenditure for a safer system (I) and the expected value of the economic damage ($E(D)$). In the optimal economic situation the total costs in the system are minimised:

$$\min(C_{tot}) = \min(I + E(D)) \tag{6}$$

With this criterion the optimal probability of failure of a system can be determined, provided investments (I) and the expected economic damage ($E(D)$) are a function of the probability of failure. For tunnels investments can be done in safety measures to prevent or mitigate damage D. The economic damage can consist of the direct losses of the tunnel construction and it's installations (i.e. the investments to rebuild them) and indirect damage due to the loss of the transport connection of which the tunnel is part. When the probabilities of the different accident scenarios are known the expected economic damage can be assessed, and an economically optimal level of protection can be derived. In addition to the determination of the optimal level of protection it should be investigated whether the project generates economic benefits for society, i.e. whether the total costs in the optimal situation are smaller than the total costs in the economic optimum.

In tunnel safety an important consequence is the potential loss of life for certain accident scenarios. It is possible to take the value of human life into the economic optimisation. Assume that a certain scenario will result in N fatalities and that every person has an economic value d. For the valuation of human life for example the present value of the nett national product per inhabitant is proposed. The economic optimum can again be found by minimising the total costs:

$$\min(C_{tot}) = \min(I + E(D + N \cdot d)) \tag{7}$$

The valuation of human life may raise numerous ethical and moral questions, because some people consider life invaluable. It can thus be easily understood that not taking into account the economic value of human life in the economic optimisation will lead to lower expected damages and thus to lower optimal safety levels. Experience shows that the influence of loss of life is relatively limited in an economic analysis. Therefore separate criteria for limitation of the risk of loss of life have been proposed above.

3.2 Deterministic approach: representative design scenarios and additional measures

Based on the probabilistic approach a tunnel design can now be chosen applying the rules as described above. The tunnel design will now comply with the standards for individual and societal risk. Furthermore the design will be optimised from an economic point of view.

In addition to the probabilistic approach also a scenario analysis should be performed. This deterministic analysis provides more insight in the accident processes and focuses on the optimisation of self rescue and emergency assistance. Also standards can be formulated for deterministic scenarios: for example the allowable time for all passengers to have left the tunnel in case of a train fire.

A problem in the deterministic analysis is the choice of a representative scenario. In section 2 it has been described how the selection of a representative design scenario without consideration of the probability of

occurrence can lead to inconsistent decision making. Theoretically, a design scenario can be derived with the method of economic optimisation, as will be shown in a (simplified) example. Assume a tunnel in which one type of accident can happen with probability p_1, causing damage D_1. In calculation of the economic risk the discount rate r is applied. Assume a function in which investments increase as a function of the negative logarithm of the probability of that accident – $ln(p_1)$ with a steepness of I'_1. The total costs (C_{tot}) can now be written as:

$$C_{tot} = I'_1 \cdot -\ln(p_1) + p_1 \cdot D_1 / r \tag{8}$$

The optimal system failure probability (p_{opt}) can be derived as follows.

$$dC_{tot} / dp_1 = 0 \quad p_{opt} = I'_1 \cdot r / D_1 \tag{9}$$

The representative design scenario is the scenario which occurs with a probability of p_{opt}. When two types of accidents (type 1 and 2) can occur independently, the total costs are (equation 10):

$$C_{tot} = I'_1 \cdot -\ln(p_1) + I'_2 \cdot -\ln(p_2) + p_1 \cdot D_1 / r + p_2 \cdot D_2 / r \tag{10}$$

The optimum can be found as follows.

$$dC_{tot} / dp_1 = 0 \qquad p_{opt1} = I'_1 \cdot r / D_1$$
$$dC_{tot} / dp_2 = 0 \qquad p_{opt2} = I'_2 \cdot r / D_2 \tag{11}$$
$$p_{opt} = p_{opt1} + p_{opt2}$$

When the two scenarios are fully dependent the following optimum will be found:

$$p_{opt} = \max(p_{opt1}, p_{opt2}) \tag{12}$$

In these cases it is recommended to choose scenarios which dominate this optimum as representative design scenarios (i.e. the scenarios which have the highest contribution to the optimal probability). Also it is possible to select more representative design scenarios, as is often done in practice. However, in practice the optimisation functions and the determination of the representative design scenario will be more complex. It is easily understood that the assumption of a linear investment function is quite unrealistic. Costs will for example increase dramatically when the decision is made to construct a second tunnel tube to prevent frontal collisions. Also the assumption of independence of scenarios can be questioned. The large disastrous accidents initially start with a small disruption of the regular situation, and thus the various scenarios will be linked. Also the effect of preventive measures will not be limited to one type of accidents, but result in risk reduction for more types of accidents.

Consider a tunnel which complies with the probabilistic safety criteria as mentioned in section 3.1. A possible outcome of the deterministic analysis could be that extra measures are advisable to limit the impacts of certain accidents. These are defined here as additional measures. Experiences from tunnelling projects in the past have shown that based on deterministic considerations measures have been chosen in the design which would not be necessary from a strictly probabilistic point of view. To facilitate decision making concerning additional measures some principles are proposed here. First of all the costs of additional measures should be reasonable in comparison with the total project costs. This criterion can be fulfilled by requiring that the investments on additional measures should amount no more than a certain percentage of the total project costs. Further study on the expenses on additional measures for different tunnelling projects could reveal the order of magnitude of this fraction.

Furthermore the effectiveness of the additional measure should be explicitly taken into account in the decision making processes. This requires that is not merely said that a tunnel is to be designed with optimal efficiency. Costs of investments and their risk reducing effects should also be explicitly quantified. The subject of cost effectiveness of measures in relation with the reduction of risk of human life loss is discussed in the next section.

4 COST EFFECTIVENESS OF SAFETY MEASURES

A large part of the investments in tunnels is concerned with the reduction of potential loss of life, either by reducing the probabilities of accidents or by reducing the consequences of accidents. The subject of cost effectiveness of investments in relation with reduction of fatalities is thoroughly studied for many sectors in literature, however relatively little is known of this subject for tunnels. One important approach relates the value of human life to the investment made and to the number of prevented fatalities. The cost of saving an extra life (CSX) expresses the investment made for saving one extra (statistical) life. The investment is generally related to the (reduction of) the expected number of fatalities:

$$CSX = I / \Delta E(N) \qquad (13)$$

It has been shown by Vrijling and van Gelder (2000) how the cost of saving a human life per year (related to the expected value) can be determined from the economic optimisation. The costs of saving an extra life year (CSXY) can be calculated by involving life expectancy in this method. An extensive study of CSXY values in various sectors, carried out by Tengs et al. (1995), showed that CSXY values vary widely across different sectors.

However, in the case of rare accidents with large consequences the expected value of the loss of life will be small. Involving this low expected value in the calculation of cost effectiveness does not reflect the social aversion against these types of accidents (e.g. airplane crashes, large tunnel accidents). A characteristic of this type of accidents is that standard deviation will be relatively high. Therefore it can be considered to use the characteristic value (which includes the standard deviation) in determination of cost effectiveness (instead of the expected value):

$$CS_{E(N)+k\cdot\sigma(N)} = I / \Delta(E(N) + k \cdot \sigma(N)) \qquad (14)$$

In the two cases mentioned above statistical or probabilistic information is used. However, a deterministically oriented decision maker will be more interested in the reduction of loss of life for a certain accident scenario, regardless of the probability of occurrence. From a deterministic point of view, cost effectiveness can now be related to the number of fatalities prevented for a certain accident scenario.

$$CS_N = I / \Delta N \qquad (15)$$

Consider now the following example: It is proposed to install a sprinkler system in the tunnel. The costs of the sprinkler are Euro 20 million and it will prevent the occurrence of a serious accident (for example an explosion) which has a probability of occurrence of 10^{-5} per year and which will cause an estimated 100 fatalities. The probabilistic decision maker (for example the risk analyst) relates the investment to both expected and characteristic value.

$$E(N) = p \cdot N = 10^{-5} \cdot 100 = 10^{-3} fat / yr \quad \sigma(N) = \sqrt{p} \cdot N = 0,31 fat / yr$$
$$TR = E(N) + k \cdot \sigma(N) = 0,95 fat / yr$$
$$CSX = I / \Delta E(N) = 20 \cdot 10^9 Euro / fat / yr$$
$$CS_{E(N)+k\cdot\sigma(N)} = I / \Delta(E(N) + k \cdot \sigma(N)) = 21 \cdot 10^6 Euro / fat / yr$$

The risk analyst concludes that when the expected value is considered, the sprinkler should not be installed. When risk aversion is considered, as is done with the characteristic value, the investment might be an option, but still one at high cost. The final decision will depend on the available budget, and the preferences of the final decision maker.

The deterministic decision maker will relate the investment, regardless of the probability of occurrence of the accident to the number of fatalities prevented. From this (deterministic) point of view it can be concluded that this investment might be a good option.

$$CS_N = I / \Delta N = 200.000 Euro$$

Although, this example concerns a theoretical case, it is a good illustration of the discussions which have occurred between different stakeholders in some large tunnelling projects in the Netherlands.

4.1 Application of cost effectiveness of measures in decision making

Before performing an analysis of cost effectiveness, it should first be analysed whether the tunnel complies with the decision criteria presented in section 3.1. A cost effectiveness analysis should be carried out to decide on the additional level of measures which can be achieved at reasonable cost. The analysis can thus used in applying the well known ALARA principle. Since the additional measures are linked to risk levels, determination of the cost effectiveness of measures in relation to life safety should be based on a probabilistic analysis. Costs of additional measures can be weighed against the risk reduction. The typical relation between investments and risk levels is plotted in Figure 2. The figure shows that the incremental costs of reducing risk increase as the risk becomes smaller (Bohnenblust, 1998).

Depending on the preferences of the responsible decision maker expected value ($E(N)$) or characteristic value (or another probabilistic risk measure) can be chosen as the unit for risk. It should be noted that the function is plotted as a continuous line, but in practice it will have a more stepwise form. Consider for example decisions such as the construction of second tube or the installation of a sprinkler installation. The optimal risk reduction curve is formed by the measures resulting in the largest risk reduction with the smallest investments. This method offers the possibility to compare various alternative measures for their cost effectiveness (for example: heat resistant lining vs. sprinkler). Also a comparison between different projects and sectors can be made. Economic efficiency requires that the marginal benefits per Euro spent should be equal for various investments across different sectors (Tengs et al., 1998).

5 CASE STUDIES: APPLICATION OF THE FRAMEWORK

This section will illustrate the application of the framework presented in section 3 and the analysis of cost effectiveness as discussed in section 4. Some information from recent tunnelling projects in the Netherlands has been investigated, and where no figures were available some fictive examples have been given.

5.1 Individual and social risk

Criteria for individually and socially acceptable risk are commonly applied in tunnelling projects in the Netherlands. An overview of the standards applied for the judgment of internal risks for a few tunnels in the Netherlands is given in Table 2.

The personally acceptable risk can be presented as the probability of death per kilometre travelled per year or the probability of losing life for an average user or employee. It is calculated as the probability of losing one's life for the "average" user or employee of the tunnel. For the judgment of societal risks both FN limits and limits for the characteristic value have been proposed.

For the external risks near tunnel locations an acceptable personal risk of 10^{-6} per year and an acceptable societal risk of $10^{-2}/N^2$ are applicable in the Netherlands (Tweede Kamer, 1996).

5.2 Economic optimisation

Although "optimal" safety is often stated to be an important aim in design processes, no direct practical application of the method of economic optimisation can be presented. Therefore a case study is presented below.

Figure 2. Relation between investments in safety measures and risk levels.

Table 2. Overview of safety standards applied in some projects (Molag, 2002).

		Tunnel		
		HSL	Western Scheldt	Betuwe
PR	Users/ passengers	$1.5 \cdot 10^{-10}$ /km/yr	$1 \cdot 10^{-10}$ /km/yr	–
	Employees	$5 \cdot 10^{-5}$/yr	–	$5 \cdot 10^{-5}$/yr
GR	F_N	$4 \cdot 10^{-2}$ /N²/yr/km	10^{-2} /N²/yr/km	10^{-2} /N²/yr/km
	Char. Value	2.3 fat/yr	–	–

Consider the following simplified example of the design of a tunnel system. A decision maker has to choose between two design alternatives: a single and a double tube tunnel. Two types of accidents can occur: a frontal collision (only in the single tube) and an accident in which a truck catches fire within the tunnel system. The following decision tree can now be made as shown in Figure 3.

The economic optimisation is applied to decide on the necessity of certain measures. Assume that the probability of a truck fire is known for both alternatives: $p_2 = 0, 1$ and that $r = 0.05$. The total costs can now be determined as follows for both tunnels:

$$C_{tot1} = 1000 + p_1 \cdot 100/r + p_2 \cdot 200/r$$
$$= 1400 + p_1 \cdot 2000$$

$$C_{tot2} = 2000 + p_1 \cdot 0 + p_2 \cdot 100/r = 2200$$

Thus only if the probability of a frontal collision $p_1 \geq 0.4$ per year, a double tube tunnel is economically optimal. However, also human life is involved in these accidents. This value can be included in the economic optimisation as is shown in section 3.1. Assume a value of the human life of 1 million Euro. Again a similar analysis can be carried out.

$$C_{tot1} = 1000 + p_1 \cdot (100 + 10)/r + p_2 \cdot (200 + 20)/r$$
$$= 1440 + p_1 \cdot 2200$$

$$C_{tot2} = 2000 + p_1 \cdot 0 + p_2 \cdot (100 + 10)/r = 2220$$

If the loss of human life is included then a probability of frontal collision of $p_1 \geq 0.35$ per year, will already point to a double tube as economically optimal. Experience shows that the influence of loss of life is relatively limited in an economic analysis.

Western societies place however implicitly a greater value on human life than explicitly. This leads to separate reasoning for the acceptable levels of individual

and group risk and a relatively strong preference for double tube tunnels.

5.3 Cost effectiveness

An estimation of cost effectiveness has been made for two practical cases in the Netherlands. The first is a tunnel which is constructed near the city of Roermond as part of the A73 highway. It is constructed as a double tube tunnel (2×2 lanes) with a length of 3260 m. The second example considers a feasibility study which is undertaken on the construction of a seven kilometre long double tube tunnel (2×3 lanes) to connect the two Junctions of the A6 and A9 highways near Amsterdam. Both are so-called category 0 tunnels, which means that no limitations for the transport of dangerous goods are applicable.

Now the cost effectiveness of the installation of a sprinkler system is investigated for these two tunnels. Many international studies have been published on the effectiveness of sprinkler systems. In the simplified approach in this study the risk reducing effects of the sprinkler system are estimated as follows. It can prevent a so-called "hot BLEVE", a BLEVE (gas explosion) which occurs after some time when the contents of a fuel tank have been heated sufficiently. The sprinkler system can not prevent an instantaneous explosion or BLEVE. Since little is known about the mitigating effects of sprinklers for tunnel fires, this effect is not taken into account (in fact: some authors argue that the use of sprinklers in case of an accident can increase the risks since large amounts of steam will be developed). Based on experience data the costs of the sprinkler system are roughly estimated at 10 million Euro/km tunnel.

The risk reducing effects of the installation of the sprinkler system have been analysed with the Tunprim model (de Weger, 2001). With this quantitative risk analysis model the internal risks for the users have been assessed. The FN curves for both tunnels, with and without sprinkler system, are shown in Figure 4. Moreover expected value ($E(N) =$ area under the FN curve) and characteristic value have been determined. The figures only show the FN curves for the small probability – large consequences events, such as BLEVEs and large fires. Experience shows that expected value is to a large degree determined by collision accidents without fire of explosion. In Table 3 cost effectiveness is related to the expected value and the characteristic value.

The presented risk calculations are just indicative results for these examples, and can not be seen as officially determined risk levels. However, from the figure and the table it can be seen that the sprinkler does have a minor influence on FN curves and expected value of the number of fatalities. It can also be concluded that the investments in a sprinkler installation for these cases

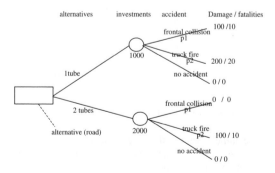

Figure 3. Decision tree for the tunnel example (costs in millions of Euros).

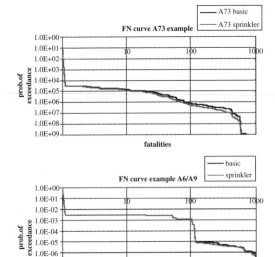

Figure 4. FN curves with and without sprinkler for the A73 example and the A6/A9 example for the small probability large consequences accidents.

Table 3. Cost effectiveness figures for the installation of a sprinkler system.

			Expected value (fat/yr)	
Tunnel	Length	Estimated cost of sprinkler	Basic	Sprinkler
A6/A9	7000	$7 \cdot 10^7$	1.9189	1.9179
A73	3260	$3 \cdot 10^7$	0.1103	0.1101

	Characteristic value (fat/yr)		Cost effectiveness (Euro/(fat/yr))	
Tunnel	Basic	Sprinkler	CSX	$CS_{E(N) + k.\sigma(N)}$
A6/A9	16.1	15.7	$7 \cdot 10^{10}$	$1.75 \cdot 10^8$
A73	1.44	1.32	$1.5 \cdot 10^{11}$	$2.5 \cdot 10^8$

are not preferable when cost effectiveness is considered from a probabilistic point of view. From the two examples it can be seen that the cost effectiveness of the sprinkler system are in the same order of magnitude: about 10^{11} Euro/fatality per year when related to the expected value and about $2 \cdot 10^8$ Euro per fatality per year when related to the characteristic value. This can partly be explained by the fact that the same QRA model with the same assumptions is used in the calculations for the two tunnels. An investigation for other tunnels and measures should reveal whether cost effectiveness figures are still in the same order of magnitude for other cases.

6 CONCLUSIONS

Aim of this paper is to propose some guidelines for the evaluation of tunnel safety. Two methods for the evaluation of tunnel safety have been analysed, the probabilistic and the deterministic approach. It can be concluded that the application of solely a deterministic analysis can lead to inconsistent decisions and that the probabilistic risk analysis should be the basis for the design of the tunnel. The deterministic or scenario analysis can be applied as an additional tool to provide more insight in the accident processes, the possibilities for self rescue and emergency assistance.

A (probabilistic) framework has been proposed for the judgment of personal, societal and economic risks of tunnels. The three criteria should all be investigated and presented. The most stringent of the three should be adopted as basis for the "technical" advice to the political decision makers. Cost effectiveness of measures should be analysed based on probabilistic information. Some methods to analyse cost effectiveness from different points of view have been discussed.

An analysis of the application of elements of the framework proposed in practical situations has shown that standards for limitation of personally and socially acceptable risk are commonly applied in tunneling projects in the Netherlands. A simplified example of economic optimisation of a tunnel shows that a method can be applied to determine the optimal level of safety in the tunnel, from an economical point of view. Two case studies for an existing and a planned tunnel have shown that cost effectiveness of the installation of a sprinkler system is very low for the studied examples. The derived cost effectiveness numbers for the two examples are of the same order of magnitude. Further study should indicate whether the cost effectiveness figures for other tunnels also show a similar pattern.

Although this paper does not provide full solutions for the complicated safety discussions in tunnelling projects, it is the hope of the authors that these ideas might contribute to a more rational and efficient decision making on the investments in tunnel safety.

ACKNOWLEDGEMENTS AND DISCLAIMER

The authors want to thank Jelle Hoeksma for his contributions to this paper. Any opinions expressed in this paper are those of the authors and do not necessarily reflect the position of the Dutch Ministry of Transport, Public Works and Water Management.

REFERENCES

Bohnenblust, H. 1998. Risk-based decision making in the transportation sector; In: Jorissen, R.E., Stallen, P.J.M. (eds), *Quantified societal risk and policy making*; Kluwer academic publishers.

Danzig D. van, 1956. *Economic decision problems for flood prevention*, Econometrica 24, p. 276–287.

Diamantidis, D., Zuccarelli F., Westhauser A. 2000. Safety of long railway tunnels, *Reliability engineering & System Safety* 67, p. 135–145.

Geyer, T.A.W. & Morri, M.I. 1996. Channel tunnel safety case: Quantitative risk assessment methodology, In: *Proceedings of the 1996 annual meeting of the society for risk analysis Europe.*

Molag, M. 2002. *Overzicht normen spoor* (an overview of risk limits for railway tunnels), TNO MEP documents.

Tweede Kamer. 1996, *Risiconormering gevaarlijke stoffen*, vergaderjaar 1995–1996, 24 611 nr. 2.

Tengs, T.O., Adams, M.E., Pliskin, J.S., Gelb Safran D., Siegel, J.E., Weinstein, M.C., Graham, J.D. 1995. Five-hundred live saving interventions and their cost effectiveness, In: *Risk Analysis*, vol. 15, no.3, p. 369.

Vrijling, J.K. et al. 1995. Framework for risk evaluation, *Journal of Hazardous Materials* 43, p. 245–261.

Vrijling, J.K., Hengel, W. van, Houben, R.J. 1998. Acceptable risk as a basis for design, *Reliability Engineering and System Safety* 59, p. 141–150.

Vrijling, J.K. Gelder, P.H.A.J.M. van. 2000. An analysis of the valuation of a human life, In: M.P. Cottam, D.W. Harvey, R.P. Pape, & J.Tait (eds) *ESREL 2000 – "Foresight and precaution"*, Volume 1, p. 197–200, Edinburgh, Scotland, UK.

Weger, D. de, Kruiskamp M.M., Hoeksma J. 2001. Road tunnel risk assessment in the Netherlands, TUNprim: A Spreadsheet Model for the Calculation of the Risks in Road Tunnels, In: E. Zio, M. Demichela, N. Piccinni (eds) *ESREL 2001*, International conference, Torino, Italy September 16–20.

Safety and Reliability – Bedford & van Gelder (eds)
© 2003 Swets & Zeitlinger, Lisse, ISBN 90 5809 551 7

Inspection and maintenance decisions based on imperfect inspections

M.J. Kallen
Delft University of Technology, Delft, Netherlands

J.M. van Noortwijk
HKV Consultants, Lelystad and Delft University of Technology, Delft, Netherlands

ABSTRACT: The process industry is increasingly making use of Risk Based Inspection (RBI) techniques to develop cost and/or safety optimal inspection plans. This paper makes use of the gamma stochastic deterioration process to model the corrosion damage mechanism. This model is successfully extended to update prior knowledge over the corrosion rate with imperfect wall thickness measurements. This is very important in the process industry as current non-destructive inspection techniques are not capable of measuring the exact material thickness, nor can these inspections cover the total surface area of the component. The model is illustrated by examples using actual plant data.

1 INTRODUCTION

In order to illustrate the requirements, which are necessary for the development of a suitable model, we start with a general introduction on current practices in the process industry concerning the use of decision models for inspection planning.

1.1 Risk based inspection

Since the late 1980s, numerous companies and organizations have developed several qualitative and quantitative models to aid plant engineers with the prioritization of component inspections. The average chemical process plant or refining installation will have thousands of components like pipelines, columns, heat exchangers, steam vessels etc. which operate under significant pressure. This pressure, combined with the corrosive nature of the chemicals inside the systems and the exposure to the weather on the outside, will degrade the quality of the construction material of the components. Among the most common degradation mechanisms are internal and external thinning (e.g. corrosion), cracking, brittle fracture and fatigue.

The highest uncertainties in the decision model are associated with the rate at which these mechanisms reduce the resistance of the construction material. Inspections are used to reduce this uncertainty, but since current wall thickness or crack length measurement techniques are not perfect, the measurements contain (small) errors. Some inspection techniques, e.g. ultrasonic wall thickness measurements, are highly accurate, but there is always the spatial variability in the measurements. This means that the quality of the material is not uniform for the whole component. Inspections have to be carried out such that the most critical spots are covered and such that the average measurement is representative for the complete item. Decision models should take these measurement errors into account.

1.2 Current methodologies

A common approach in tackling the inspection problem, is to start with a prioritization. This means that usually a more qualitative model is used to determine the components which constitute the highest risk. The assumption is that 80% of the risk is generated by only 20% of the components. This prioritization is then followed by a detailed quantitative analysis of these high-risk items, in which the remaining lifetimes are estimated and the consequences of failures are modelled. The American Petroleum Institute (API) has published a large document (American Petroleum Institute 2000), which describes the methodology developed in cooperation with an industry sponsor group. This document has become a methodology in itself and is used by many companies as a basis for further development.

Typically, structural reliability methods are used to estimate the failure probabilities of the components. The condition of the construction material is described by a state function g, which is represented by a resistance (R) minus stress (L) model:

$$g = R - L. \tag{1}$$

The uncertain variables in this model have normal densities with suitable parameters assigned to them. The failure probability is then approximated by using a reliability index method (e.g. FORM: First Order Reliability Method), which makes use of the limit state $g = 0$ or $R = L$. Many of the methods currently applied in the process industry, also make use of a simple discrete version of Bayes' theorem to update prior knowledge, or an estimate for the rate of degradation, with inspection results. The likelihood in Bayes' formula represents the likelihood of an inspection correctly identifying the state of the component.

In the following sections we will take a different approach to this problem. We will use the gamma stochastic process to model the degradation and we will update the parameters of this process using inspection data. This is inspired by the successful application of this stochastic process to the problem of optimally inspection large civil structures like dikes and storm-surge barriers, e.g. see (van Noortwijk 1996). It is not claimed that this is the best approach to this problem. The purpose of this paper is to illustrate the versatility of the gamma process and to show that there is great potential for its use in the process industry. Characteristically, in the development of most current models, many assumptions and simplifications are applied in order to keep the models easy to use. We will use the same approach in order to avoid large amounts of input and to keep the necessary input as simple as possible.

2 MODELLING THE DETERIORATION

In this paper we will only consider degradation due to corrosion, but the application of this model is not restricted to this damage mechanism. For this purpose we will first define the corrosion state function, after which the gamma process with suitable parameters is introduced.

2.1 Corrosion state function

The state function for corrosion is taken from (American Petroleum Institute 2000) and is defined as:

$$g(t) = \underbrace{S\left(1 - C\frac{t}{x_0}\right)}_{Resistance} - \underbrace{P\frac{d}{2x_0}}_{Load}, \tag{2}$$

where S [MPa $= 10$ bar], C [mm/yr] and P [bar] are random variables representing the material strength, corrosion rate and operating pressure respectively. The component diameter d [mm] and the initial material thickness x_0 [mm] are assumed to be known. Next to the variables are the dimensions which are used throughout this paper. Note that the amount of wall loss at time t is given by $w(t) = C \times t$ [mm].

The component is assumed to have failed at time t when $g(t) < 0$. If t' is the time at which failure occurs, i.e. $g(t') = 0$, then $w(t') = C \times t'$ is the maximum amount of wall thickness which can be lost until the component fails. If we call this the safety margin m, then we can calculate this amount from the state function (2):

$$m = w(t') = x_0 - P\frac{d}{2S}. \tag{3}$$

Each component will usually have a so-called corrosion allowance c_{max} associated with it. This is a value given by the manufacturer of the component and represents the maximum amount of wall loss up to which the component is assumed to be able to function safely. It should hold that $0 \leqslant c_{max} \leqslant m$, which means that we can write the corrosion allowance as a percentage of the safety margin:

$$c_{max} = \rho m, \quad 0 \leq \rho \leq 1. \tag{4}$$

We will use this value as the replacement level, i.e. we will preventively replace the component if the wall loss is more than the corrosion allowance.

2.2 Gamma deterioration process

Instead of using a reliability index method, we will model the cumulative wall loss with a gamma process. We will use the following definition for the gamma density with shape parameter $\alpha > 0$ and scale parameter $\beta > 0$:

$$Ga(x|\alpha, \beta) = \frac{\beta^\alpha}{\Gamma(\alpha)} x^{\alpha-1} \exp\{-\beta x\} \text{ for } x \geq 0 \tag{5}$$

The gamma process with shape function $at > 0$; $t \geqslant 0$ and scale parameter $b > 0$ is a continuous–time process $\{X(t) : t \geqslant 0\}$ with the following properties:

1. $X(0) = 0$ with probability one,
2. $X(\tau) - X(t) \sim Ga(a(\tau - t), b)$ for $\tau > t \geqslant 0$,
3. $X(t)$ has independent increments.

Let $X(t)$ denote the amount of deterioration at time t, then the probability density function of $X(t)$ is given by

$$f_{X(T)}(x) = Ga(x|at, b) \tag{6}$$

In essence, we replace the amount of wall loss $w(t)$ with the process $X(t)$. There are a number of advantages to using the gamma process. Most interestingly, the increments are always positive, therefore increases in wall thickness are not allowed for. The fact that increments are independent and therefore exchangeable fits well with the physics of corrosion and deterioration in general.

Using moment generating functions, it can be proven that the expectation and the variance of the process $X(t)$ are given by:

$$\mathbb{E}(X(t)) = \frac{a}{b}t \quad \text{and} \quad \text{Var}(X(t)) = \frac{a}{b^2}t. \quad (7)$$

Assuming that the expectation and variance are linear in time, i.e.

$$\mathbb{E}(X(t)) = \mu t \quad \text{and} \quad \text{Var}(X(t)) = \sigma^2 t,$$

we find that the parameters of the process $X(t)$ are defined as:

$$a = \mu^2/\sigma^2 \quad \text{and} \quad b = \mu/\sigma^2, \quad (8)$$

where μ is the expectation for the average corrosion rate and σ^2 is the variance of the process. These two parameters are uncertain and assessing both variables for each individual component is too much work. In order to keep the method practical and easy to use for the plant engineer, we will fix the standard deviation σ relative to the mean μ through the use of a coefficient of variation ν (this coefficient is often referred to as the COV):

$$\sigma = \nu \times \mu \Longrightarrow \nu = \sigma/\mu. \quad (9)$$

This approach is used in many of today's models: the variances of the uncertain variables are predetermined by expert judgment and subsequently fixed in the model. Using (9), the density for $X(t)$ in (6) reduces to

$$f_{X(t)}(x) = \text{Ga}\left(x\left|\frac{t}{\nu^2}, \frac{1}{\mu\nu^2}\right.\right). \quad (10)$$

Now we are only left with the uncertain variable μ for the average corrosion rate. In the following section we will define a suitable prior density for this variable and we will discuss the consequence of fixing σ relative to μ.

3 INSPECTION UPDATING

Due to the fact that most components are usually at most inspected every 2 years, with the average inspection interval being about 6 to 8 years, there is not enough data available for a statistical method like regression analysis. Using Bayesian updating we can efficiently incorporate the measurements and the engineer's prior knowledge or estimate of μ.

3.1 Choosing the prior

Due to the large number of components and the usually limited experience of the plant engineer with probability theory, it is not feasible to ask this engineer to assess a suitable density for μ. In line with current practices in the process industry, we will only ask for an average corrosion rate over which a default density will be placed. The API (American Petroleum Institute 2000) uses the simple discrete prior density, which is shown in Figure 1. The idea behind this density is that the model has 50% confidence in a corrosion rate which is less than or equal to the rate assessed by the plant engineer. The other 50% is divided into 30% between 1 and 2 times the assessed rate and 20% between 2 and 4 times the engineer's estimate.

For our gamma process model we will not use this discrete prior, but a continuous inverted gamma density. The definition of this density is given by

$$\text{Ig}(x|\alpha, \beta) = \frac{\beta^\alpha}{\Gamma(\alpha)}\left(\frac{1}{x}\right)^{\alpha+1}\exp\left\{-\frac{\beta}{x}\right\} \quad (11)$$

for $x \geqslant 0$. Notice that a random variable X is inverted gamma distributed if $Y = X^{-1} \sim \text{Ga}(\alpha, \beta)$ with shape parameter $\alpha > 0$ and scale parameter $\beta > 0$. With suitable choices for the parameters, we can again define a default prior density for μ as is shown in Figure 1. Most confidence is placed between the engineer's

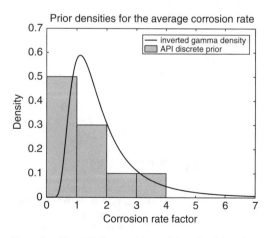

Figure 1: The API discrete prior and the related (continuous) inverted gamma density.

assessment and 2 to 3 times this value. The prior density in Figure 1 is arbitrarily chosen for the purpose of demonstration. In practice, the parameters of this density should be determined using expert judgment. One could also define a number of different default priors from which the practitioner can select the one which best represents his own confidence/uncertainty in the degradation rate.

In the following sections we will use Bayes' theorem to update the prior density with the likelihood of the inspection measurement to obtain a posterior density. The continuous version of Bayes' theorem is given by

$$\pi(\mu|x) = \frac{l(x|\mu)\pi(\mu)}{\int_{\mu=0}^{\infty} l(x|\mu)\pi(\mu)d\mu}, \tag{12}$$

where $\pi(\mu)$ is the prior density for μ and $l(x|\mu)$ is the likelihood of measurement x given μ.

3.2 Case 1: perfect inspections

The choice for the inverted gamma prior becomes clear when we only consider perfect inspections. In other words, we first assume that we can measure the exact wall thickness. We can then precisely determine the amount of wall loss x at time t and it can be proven that the posterior $\pi(\mu|x)$ is given by

$$\pi(\mu|x) = \text{Ig}\left(\mu \,\middle|\, \alpha + \frac{t}{\nu^2}, \beta + \frac{x}{\nu^2}\right), \tag{13}$$

where α and β are the parameters from the prior density. The fact that the product of the inverted gamma density and gamma distributed likelihood is again proportional to a inverted gamma density is also used in (van Noortwijk et al. 1995).

The equivalent of (13) for $n \geq 1$ inspections is

$$\pi(\mu|x_1, \ldots, x_n)$$

$$= \text{Ig}\left(\mu \,\middle|\, \alpha + \sum_{i=1}^{n} \frac{t_i - t_{i-1}}{\nu^2}, \beta + \sum_{i=1}^{n} \frac{x_i - x_{i-1}}{\nu^2}\right).$$

which reduces to

$$\pi(\mu|x_n) = \text{Ig}\left(\mu \,\middle|\, \alpha + \frac{t_n}{\nu^2}, \beta + \frac{x_n}{\nu^2}\right), \tag{14}$$

if we assume that $x_0 = 0$ at $t_0 = 0$. In other words, because we fixed the standard deviation σ relative to the mean μ, only the last inspection is needed to calculate the posterior when using perfect measurements. Besides the fact that perfect measurements do

not exist, this result will be hard to sell to any plant engineer or regulator.

3.3 Case 2: imperfect inspections

This is where we present the main feature of this paper, namely the extension of the gamma process updating model with uncertain measurement data. Similar to (Newby and Dagg 2002), we consider a new process $Y(t)$, which includes the original process $X(t)$ together with a normally distributed error ε:

$$Y(t) = X(t) + \epsilon, \quad \epsilon \sim \mathcal{N}(0, \sigma_\epsilon). \tag{15}$$

Here we assume the error has a mean 0 and a standard deviation σ_ε. Taking a mean different from zero would mean that the inspection tends to over- or underestimate the actual wall loss. The likelihood of the measurement y given the corrosion rate μ is now determined by the convolution:

$$l(y|\mu) = f_{Y(t)}(y) = \int_{-\infty}^{\infty} f_{X(t)}(y - \epsilon)f_\epsilon(\epsilon)d\epsilon, \tag{16}$$

where $f_{X(t)}(y - \varepsilon) = \text{Ga}(y - \varepsilon|at, b)$ is the likelihood of the gamma increment $X(t)$ as given by (10). We immediately go over to the likelihood for more than one inspection:

$$l(\mathbf{y}|\mu) = \prod_k l_{Y(t_k)-Y(t_{k-1})}(y_k - y_{k-1}|\mu), k > 1, \tag{17}$$

where $\mathbf{y} = \{y_1, \ldots, y_k\}$ are the wall loss measurements. In (17), we have used the fact that the increments are independent. Now we introduce some notation: the increment of $X(t)$ between two inspections is defined as $D_k = X(t_k) - X(t_{k-1})$, the difference between two measurements is $d_k = y_k - y_{k-1}$. Using this notation and the integral convolution as in (16), the likelihood (17) can be rewritten as

$$l(\mathbf{y}|\mu) = \int_{-\infty}^{\infty} \cdots \int_{-\infty}^{\infty} \cdots$$

$$\cdots \prod_k f_{D_k}(d_k - \delta_k)f(\delta_1, \ldots, \delta_k)d\delta_1 \cdots d\delta_k, \tag{18}$$

where $\delta_k = \varepsilon_k - \varepsilon_{k-1}$. Clearly the δ's are not independent since every δ_k depends on δ_{k-1}. We are left with two options: we calculate the covariances between the δ's and analytically solve the likelihood using the joint distribution of the δ's or we simulate the ε_k's and approximate the likelihood. Since the first option will complicate matters considerably, we will use the simulation approach. The likelihood (18) can be formulated as an expectation, which in turn can be approximated by the average of the product:

$$l(\mathbf{y}|\mu) = \mathbb{E}\left\{\prod_k f_{D_k}(d_k - \delta_k)\right\} \approx$$

$$\approx \frac{1}{N}\sum_{j=1}^{N}\prod_k f_{D_k}\left(d_k - \delta_k^{(j)}\right) \text{ as } N \longrightarrow \infty. \quad (19)$$

Here we use the law of large numbers to perform a so-called Monte Carlo integration. For each inspection k we

1. sample $\varepsilon_k^{(j)}$ for $j = 1, 2, ..., N$ and
2. calculate $\delta_k^{(j)} = \varepsilon_k^{(j)} - \varepsilon_{k-1}^{(j)}$, then finally
3. calculate (19).

Since the gamma distributed $f_{D_k}(x)$ is not defined for $x < 0$, we need to make sure that the argument is non-negative. We have worked our way around this problem by using the minimum function as follows:

$$l(\mathbf{y}|\mu) \approx \frac{1}{N}\sum_{j=1}^{N}\prod_k f_{D_k}\left(d_k - \min\left\{\delta_k^{(j)}, d_k\right\}\right) \quad (20)$$

as $N \longrightarrow \infty$, where

$$f_{D_k}\left(d_k - \min\left\{\delta_k^{(j)}, d_k\right\}\right) =$$

$$= \mathrm{Ga}\left(d_k - \min\left\{\delta_k^{(j)}, d_k\right\} \middle| \frac{t_k - t_{k-1}}{\nu^2}, \frac{1}{\mu\nu^2}\right).$$

Equation (20) can now be substituted in Bayes' formula (12), which can then be solved by discretization (i.e. simple numerical integration) to obtain the posterior $\pi(\mu|y_1, ..., y_n)$.

The choice for use of simulation to determine (18) greatly reduces the efficiency of the model, but we also see that the choice for the prior distribution is then no longer restricted to the inverted gamma density. Also, we can easily introduce correlated measurement errors (ε_k) using a multivariate normal distribution in order to see the effect of inspection dependence on the end result. Due to the absence of the required data, we have not included this here.

4 DECISION CRITERION

We now have a posterior density over the average corrosion rate, which incorporates the plant engineer's prior knowledge and all the available inspection data. Using this density and the densities for the uncertain variables S (material strength) and P (operating pressure), we can calculate the probability of failure or a preventive replacement of the component as a function of time.

Inspections in a process plant are very expensive due to the fact that the process often has to be stopped during the inspection. Also, because many components contain highly corrosive and/or toxic chemicals, they have to be flushed and cleaned before an internal inspection can be done. The waste resulting from this rinsing needs to be treated, which brings added costs to the whole procedure. We therefore would like to make a cost optimal decision on when to inspect the components. This means that we want to maximize the time interval between two subsequent inspections. On the other hand, we have to ensure that the component operates safely and therefore we also need to make sure that we do not inspect at too large intervals. For this purpose we use a cost based criterion, suggested by (Wagner 1975), called the expected average cost per time unit. This cost criterion is derived using renewal theory and uses the concept of the component life cycle. The length of this cycle is the time from the service start until a renewal, which is either a preventive replacement or a corrective replacement due to failure. The expected average costs per time unit is given by the ratio of the expected costs per cycle over the expected cycle length, see e.g. (van Noortwijk et al. 1997):

$$C(\theta, \Delta k) = \frac{\sum_{i=1}^{\infty} c_i(\theta, \Delta k)p_i(\theta, \Delta k)}{\sum_{i=1}^{\infty} ip_i(\theta, \Delta k)}, \quad (21)$$

where c_i and p_i are respectively the costs incurred during time unit i and the probability of renewal during this time unit. The ratio is a function of the replacement and failure levels, which is represented here by $\theta = \{m, \rho\}$, and we calculate this ratio for each inspection interval Δk. The failure probability also depends on the three uncertain variables S, P and C. For the sake of legibility, they are not included in the above equation. We refer to (Kallen 2002) for the details of the implementation of this particular model, which can be obtained by request from the authors.

In order to include the uncertainty over the material strength, operating pressure and the corrosion rate, we apply the same technique as in equation (19). We need to determine the expectation of (21), therefore we sample a large number of sets for the three uncertain variables: $\{s^{(j)}, p^{(j)}, c^{(j)}\}, j = 1, 2, ..., N$. For a sufficiently large N, the average of the calculated (21) will approximate the expectation. The corrosion rate can be sampled from the posterior $\pi(\mu|y_1, ..., y_n)$, which we have determined in section 3.3.

5 CASE STUDY: INSPECTING A HYDROGEN DRYER

We will illustrate the gamma process decision model with a case study on a hydrogen dryer. Table 1 summarizes the operational and measurement data, which is taken from an undisclosed plant in the Netherlands. The cost numbers are fictive. The material strength is determined using the tensile (TS) and yield (YS) strengths with the equation:

$$S = \min\{1.1(TS + YS)/2, TS\}.$$

The values for these strengths can be looked up in (ASME 2001). The uncertainty in S comes from the fact that these values represent the minimum requirements for this particular material type and therefore the material could be stronger. On the other hand, the material loses strength when it is processed by the manufacturer of the component, therefore it could be weaker than indicated.

Note that we will disregard the measurement taken in 1982, because it is too inaccurate to be considered for this analysis. This is not uncommon practice in the process industry, as older measurements (older than 10 to 15 years) are not considered to be reliable. This is due to the fact that the quality of the inspection techniques used in those days is not comparable to the accuracy of current techniques. Our gamma process model can incorporate this measurement only if the uncertainty distribution of the error term is chosen such that the probability of material growth is very small.

In the Netherlands, the inspection of pressurized vessels is regulated by the Dutch rules for pressure vessels (Stoomwezen 1997). All types of components are categorized and for each category there exists a fixed mandatory inspection interval. For the hydrogen dryer in our example the fixed interval is 4 years, which can be clearly observed in the measurement dates. If a plant engineer wants to extend this interval based on the results of a RBI analysis, then he will have to submit an application to the proper authorities for acceptance. The rules require all components to have had at least one regular inspection, according to the fixed intervals, before this application can be made. For items with a fixed interval of 2 years, this minimum is two inspections. There will thus always be at least one set of measurement data available for the model to be applied.

Using the Bayesian updating mode, which we have discussed in section 3, we can calculate the posterior density for the corrosion rate given the measurement data in Table 1. The results are shown in Figure 2. For comparison purposes, we have also included the posterior density, which we would get if we assumed that the measurements were exact. In this case only the last measurement is of importance, as we have determined

Table 1: Operational, material and inspection data for a hydrogen dryer.

Component type:	vertical drum
Material type:	carbon steel
Service start:	1977
Tensile strength (TS):	413.69 MPa
Yield strength (YS):	206.84 MPa
Operating pressure:	32 bar(g)
Drum diameter:	1180 mm
Init. material thickness:	16.8 mm
Corrosion rate (est.):	0.1 mm/yr
Corrosion allowance:	4.5 mm

Ultrasonic wall thickness measurements:

1982:	15.0 mm
1986:	15.6 mm
1990:	14.6 mm
1994:	14.2 mm
1998:	13.8 mm

Costs for different actions:

Inspection:	10,000 $
Preventive replacement:	50,000 $
Failure + replacement:	1,000,000 $

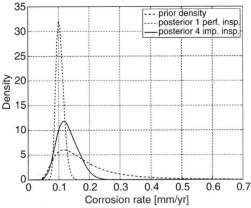

Figure 2: Posterior densities for 1 perfect and 4 imperfect inspections compared to the prior density.

in equation (14). The difference between the perfect and imperfect inspections is quite large. The assumption of perfect inspections clearly increases the confidence in the estimated corrosion rate considerably compared to the assumption of imperfect inspections.

The next step is to use the posterior density for the corrosion rate, the densities for S and P, and the cost criterion from section 4 to determine the optimal inspection period until the next inspection.

In Figure 3, we can see that the expected average costs per year will be high when we inspect too often

878

Figure 3: Expected average costs per year for the hydrogen dryer.

and when we inspect too little. The result for the optimal inspection interval length is $\Delta k = 37$yrs, but since the column was taken into service in 1977, we have to deduct its age of 25 years from this result. Therefore, the optimal period until the next inspection is 12 years. This is a very acceptable result, because it is less than the absolute maximum of 50 years and no more than 4 times the regular prescribed inspection interval of 4 years for this type of component. If this next inspection increases the confidence in the corrosion rate, then this model can be used again to determine the optimal inspection interval starting from the service start. A risk-averse decision maker can decide for a shorter inspection interval based on the result in Figure 3. A choice for Δk between 29 and 37 years will not significantly increase the expected costs per year. With current models, the decision maker will determine the failure probability as a function of time using the state function (2). Subsequently he will either choose to inspect before the fifth quantile of this distribution or he will define a risk criterion to determine the time before which an inspection should take place.

Future development of this model will include the calculation of the standard deviation of the expected costs per year. Interested readers can take a look at (van Noortwijk 2003), which discusses in great detail the required formulas for this purpose.

To finish, we note that the model will also be applicable when no inspection data is available. In other words: even when the component is new, we will be able to determine a responsible inspection interval with this decision model. The high uncertainty in the prior density for the degradation rate will initially make the optimal inspection interval quite short. As the number of measurements increase, so will the time between inspections increase.

6 CONCLUSIONS

There are a number of conclusions to be drawn from this research. First, we have determined that the gamma process is a suitable stochastic process to model the uncertain reduction of wall thickness due to corrosion. In line with what is currently done in the process industry, we have considerably simplified the parameters of this process by fixing the variance to the mean of the process. Together with the assumption of a fixed prior density for the average corrosion rate, this results in a model with minimum input requirements. This is a very desirable feature, as we are typically dealing with hundreds, maybe even thousands, of components in the average plant.

Next, we have created a simple extension to the Bayesian updating model, such that the model can incorporate the results from inaccurate measurements. In this step we have lost a lot of efficiency, because we have taken the path of simulation to solve the resulting equations. If we also consider other variables to be uncertain, e.g. the material strength or operating pressure, then the computational effort also becomes larger. The efficiency of the model is where the greatest improvements can be made in the future.

In order to make both cost optimal and safe inspection decisions, we have used the cost criterion of the expected average costs per year. Not only does this criterion fit well with our requirements, it also results in a graph which is easy to interpret by the plant engineers. This will ensure that the model will have some transparency and it will be less of a black box to the practitioners. Experience shows that the presentation of a single optimal value often leaves practitioners and regulators with more questions than they started out with.

The case study on a hydrogen dryer showed encouraging results of the whole model. We conclude that the use of the gamma Bayesian stochastic process is a viable alternative to the structural reliability methods which are currently used in the process industry.

ACKNOWLEDGMENTS

The research in this paper was performed by the author during an internship at Det Norske Veritas B.V. in Rotterdam, the Netherlands. We would like to thank Chris van den Berg and his colleagues at DNV for their support and for the plant data which is used in the example.

REFERENCES

American Petroleum Institute (2000). *Publication API581* (2nd ed.). American Petroleum Institute.

ASME (2001). Boiler and pressure vessel code. Technical report, American Society of Mechanical Engineers, New York, USA.

Kallen, M. J. (2002). *Risk based inspection in the process and refining industry*. Master's thesis, Technical University of Delft, Faculty of Information Technology and Systems (ITS), Delft, the Netherlands.

Newby, M. and R. Dagg (2002). Optimal inspection policies in the presence of covariates. In *Proceedings of ESREL 2002*, Lyon, France.

Stoomwezen (1997). *Rules for pressure vessels*, Volume 1–3. The Hague, the Netherlands: Sdu Publishers.

van Noortwijk, J. M. (1996). *Optimal maintenance decisions for hydraulic structures under isotropic deterioration*. Ph.d. thesis, Technical University of Delft.

van Noortwijk, J. M. (2003). Explicit formulas for the variance of discounted life-cycle cost. *Reliability Engineering and System Safety*. Article in press.

van Noortwijk, J. M., Cooke, R. M. and M. Kok (1995). A bayesian failure model based on isotropic deterioration. *European Journal of Operational Research 82*(2), 270–282.

van Noortwijk, J. M., Kok, M. and R. M. Cooke (1997). Optimal maintenance decisions for the sea-bed protection of the eastern-scheldt barrier. In R. M. Cooke, M. Mendel, and H. Vrijling (Eds), *Engineering Probabilistic Design and Maintenance for Flood Protection*, pp. 25–56. Kluwer Academic Publishers.

Wagner, H. (1975). *Principles of operations research* (2nd ed.). Prentice–Hall.

Safety and Reliability – Bedford & van Gelder (eds)
© 2003 Swets & Zeitlinger, Lisse, ISBN 90 5809 551 7

A proposal for managing and engineering knowledge of stochastics in the quality movement

Ron S. Kenett
KPA Ltd

M.F. Ramalhoto
Instituto Superior Tecnico, Lisbon

John Shade
Advanced Integrated Technologies Group

ABSTRACT: Many authors in books and papers have discussed the role of statisticians in the quality movement. We go further, and in the context of an extended quality movement: (1) extend the quality movement to other relevant areas including reliability, maintenance, innovation etc'…, (2) use profiles and pattern models, (3) present a database in the context of a modern e-library organized to provide roadmaps, diagnostic tools and solutions for problems in industry, business and governmental institutions. We define this extended quality movement, SQM, Stochastics for the Quality Movement. Having also in mind that the successful business executive is forecasting; purchasing, producing, marketing, pricing and organizing, all activities requiring SQM assistance and therapies. This proposal is a very demanding in terms of human resources and funding. Some progress towards SQM has been made in the context of ENBIS, the European Network for Business and Industrial Statistics (www.enbis.org).

1   INTRODUCTION

We think there is a need for the consulting community to mimic an approach adopted by the medical community, as indicated in Ramalhoto (2001).

Physicians first perform a diagnosis of an illness and only then prescribe the necessary treatment. Management consultants take a similar approach. Assessments are followed by discussions of various organizational remedies which are then translated into detailed implementation plans. Consultants are not always involved in the implementation phase. On the other hand, physicians are always supposed to assume some responsibility for the follow up of the prescribed treatment. They can even assume official responsibility for a patient's death if it is proven to be due to malpractice. Assuring the ethical dimension of doing medicine is crucial for their business reputation.

Physicians are organized regionally and worldwide in order to be able to meet these responsibilities towards their customers.

For example, a model of a disease (failure or problem of some sort) is created based on medical literature and advice from experts in the diagnosis and treatment of the disorder. This model is usually introduced as a Clinical Practice Guideline (CPG) under the supervision of the Institute of Medicine (IM).

In this paper it is proposed that for the community of statistical consultants (in business or academia) to assume responsibilities similar to that of physicians, there is a need to identify and create diagnostic tools and treatments of the CPG type, as well as to build a large network of specialists that can work cooperatively.

The business performance gains possible through enhanced products and process quality control have become well established, and are available for those companies willing to follow the lead of the pioneers. Substantial further gains are being pursued by them through lean manufacturing and more effective maintenance planning. These gains in industrial competitiveness can come from more effective "asset utilization" that result from scientifically based product, process and maintenance planning. The recent interest in combining Six Sigma, Design For Six Sigma (DFSS) with Lean Manufacturing are a result of this evolution. Keeping in mind that Reliability is Quality over time and Maintenance is a Reliability Function, SQM, Stochastics for the Quality Movement, an "extended quality movement" integrating the reliability and the maintenance cultures makes sense, where stochastics plays a much more relevant role.

In this paper we introduce and discuss the "SQM Library" concept, in order to identify and create diagnostic tools and best practice methods in managing and engineering stochastics for the extended quality movement.

Methodology leaders in business, such as master black belts in six sigma programmes, should benefit from a clear and comprehensive assessment of what their statistical practitioners might tackle, and how.

SQM embraces all the non-deterministic quantitative methods relevant to industry, business and governmental institution practices. It includes Statistical Process Control (SPC), maintenance, reliability and quality improvement, experimental design, risk analysis, as well as decision support systems, simulations, management statistics, graphics, key performance indicators, measurement systems, and the various interactions relevant to improvement and innovation. The SQM concept has been first introduced by Ramalhoto (1999). Its emphasis is on interdisciplinary and synergistic problem-solving.

The SQM library has been first introduced and developed by Ramalhoto (1999 and 2002), Kenett and Shade (2001), Shade and Kenett (2001) and Kenett (2002) within ENBIS, the European Network for Business and Industrial Statistics, and Pro-ENBIS, a Fifth Framework Thematic network funded under the Growth program.

The SQM Library will be organized in an intelligent and updated manner in order to be easy to use, very informative and complete. It will also include SQM research, education and training sites, computation, applications, statistical consultancy, quality improvement and innovation, Six-Sigma, free software, standards, data banks, relevant European Commission Directives etc'…

The first part of the SQM library consists of: a) Profiles of Industry/Business/Governmental institutions, b) Statistical consulting pattern models, c) Case-Studies of lessons learned and success stories. d) SQM Software links and resources, and e) an SQM survey of companies, consultants and academics. The next sections provide an outline, with examples, of the SQM Library.

2 INDUSTRY PROFILES

Industry profiles in an SQM library are illustrated in this section by the semiconductors industry. Similar profiles can be made available for other industries.

2.1 An example: The semiconductors industry

Manufacturing of semiconductors is a process that goes from raw silicon at one end to encapsulated microchips at the other. There are dozens of stages and hundreds of tasks involved. See "http://www.sematech.org/public/news/mfgproc/mfgproc.htm" for some details of the process.

The process generally takes several weeks, and for well-established products, process yields of over 90% can be achieved. Given that a chip can contain millions of elements and that failure in any one of these can lead to rejection of the entire chip, the need for extremely high quality levels at each stage becomes apparent.

Many chips are manufactured simultaneously on circular wafers of silicon. The general trend has been to increase the diameters of these wafers and increase the density of components on chips so that more chips can be produced as the wafers move through the various stages.

Since a dust particle can wreck a chip, much of the processing takes place in clean-rooms, where the air is filtered to achieve very low levels of dust. The improvement of these clean-rooms is an important ongoing task.

Improvements in all the stages are of course of interest, e.g. further reducing dust content in clean-rooms. Major themes are increasing uniformity of surfaces in terms of smoothness and of thickness of deposited layers, increasing precision of etching (when selected areas on the wafer are removed) and of doping (when materials are forced or diffused into the wafer to achieve required conductivities).

Some recent announcements suggest that 300 mm diameter wafers with sub-90 nanometer features are moving closer to production. Getting these to high levels of yield is a challenge for managers and engineers.

The design of the chips themselves, including the materials they use, is an active area of research. For example, this year the Semiconductor Research Corporation is running a competition (along with some companies such as IBM) for new designs using Germanium with Silicon (the "SiGe Design Challenge"). In 2002 IBM, Sony, and Toshiba have committed to a joint venture to develop new process technologies for chips with features as small as 50 nm on 300 mm wafers. A large part of IBM's new 300 mm wafer facility in New York State (East Fishkill) will be used for this. Other collaborations announced in 2002 include:

Taiwan Semiconductor Manufacturing Company is teaming with the Canadian ATMOS Corporation to develop high density memory chips using 90 nm process technology.

Canon Inc is collaborating with Applied Materials Inc to develop a 300 mm wafer manufacturing process in Sunnyvale, CA.

DuPont, Infineon, and AMD Inc have major development plans for sites in Dresden to supply photomasks for next-generation semiconductor developments in the 65 to 90 nm range and below.

In May 2002, semiconductor equipment maker SEZ Group opened its first 300 mm applications lab (in Phoenix, AZ).

Intel is resuming construction of a plant in Ireland that is scheduled to go into production in 2004 to produce 300 mm wafers using 90 nm design rules. This may be the largest single construction project in Ireland (est. £2 billion cost).

2.2 References

See "http://www.semi.org/report/index.html" for a report on the semiconductor industry (also downloadable as a pdf file). Statistical techniques have been widely applied in semiconductor manufacturing, not least due to the efforts of SEMATECH: http://www.sematech.org. For more details and perspective on projected developments over the next 5 to 7 years see: http://www.sematech.org, http://public.itrs.net./Files/2001ITRS/ExecSum.pdf

For news and information on conferences see: http://www.semiconductoronline.com http://www.sst.pennet.com, http://www.semi.org http://public.itrs.net, http://www.semiconductor2k.com

3 STATISTICAL CONSULTING PATTERNS

3.1 Background to statistical consulting patterns

Statistical consultants use some kind of procedures for planning the details of a consulting assignment, for reviewing data sets and detecting missing or suspicious values, for continuous improvement using the Deming-Shewhart PDSA cycle, for structuring an intervention in some process, and so on. Most consultants have a set of such techniques that they bring to bear almost automatically in the appropriate situation. Six Sigma programs popularized the acronym DMAIC (Define-Measure-Analyze-Improve-Control) as a structure for a wide range of projects aimed at process or product improvement. There are many such prescriptions for problem-solving and management, see Kenett and Shade (2001) and Shade and Kenett (2001).

Imagine a resource, possibly a database of some kind, which would allow statistical consultants to obtain some suggestions as to what ideas/practical tips/solutions etc might help in a given situation. Suppose further that this resource was created by statistical consultants to reflect what they saw as successful solutions that could be applied time and again to various recurring situations. We will use the word

"patterns" to refer to such ideas or "solutions" (Alexander, 1979).

3.2 Overview of patterns

The software development community is a useful source of examples of pattern use, in particular by those advocating object-oriented approaches and re-use.

What these efforts have in common is in their attempt to exploit knowledge about *best practice* in some domain. Best practice knowledge is constructed in "patterns" that are subsequently used as the starting point in the programming, design or analysis endeavors. Patterns therefore, are not invented but rather they are discovered within a particular domain with the purpose of being useful in many similar situations. A pattern is useful if it can be used in different contexts.

Alexander defines a pattern as describing "a problem which occurs over and over again in our environment and then describes the core of the solution to that problem, in such a way that you can use this solution a million times over, without ever doing it the same way twice" (Alexander, Isihikawa et al., 1977). Here, the emphasis is put on the fact that a pattern describes a recurrent problem and it is defined with its associate core solution.

According to Alexander, what repeats itself is a fabric of relationships. For example, when a statistical consultant is first approached by a customer and gets a first look at some data, or a detailed description of it, the statistical consultant is initiating a basic investigation to understand the context of the problem. This example represents a structural pattern which is repeatable in many different settings; for example in a troubleshooting assignment in manufacturing, or a market research study. Aligned to this structural pattern, there is a pattern of events which is also repeatable, in our example the basic investigation preceding the statistical analysis takes place time and time again within a company.

It is important to note that a pattern relates a problem to a solution.

3.3 Pattern templates

A pattern is more than just a description of some thing in the world. A pattern should also be a "rule" about when and how to create that thing. Therefore, a set of desirable properties for a pattern may be the following:

- A pattern should be made explicit and precise so that it can be used time and time again.
- pattern should be visualizable and should be identifiable, so that it can be interpreted equally well by all who might share the pattern. In this sense "visualization" may take the form of "statements in natural language", "drawings" conceptual models' and so on.

A pattern is explicit and precise if:

- It defines the problem (e.g. "we want to improve yield in a manufacturing process") together with the forces that influence the problem and that must be resolved (e.g. "managers have no sense for data variability", "collaboration of production personnel must be achieved" etc.). Forces refer to any goals and constraints (synergistic or conflicting) that characterize the problem.
- It defines a concrete solution (e.g. "how should basic problem investigations be done"). The solution represents a resolution of all the forces characterizing the problem.
- It defines its context (e.g. "the pattern makes sense in a situation that involves the initial interaction between the statistical consultant and his customer"). A context refers to a recurring set of situations in which the pattern applies.

In the literature there are many different proposals for the description of those desirable properties e.g. Alexander, Isihikawa et al., 1977, Coad 1992, Gamma, Helm et al., 1995. An example from Alexander, Isihikawa et al., 1977 is the following:

Name: it should be short and as descriptive as possible,

Examples: one or several diagrams/drawings illustrating the use of the pattern,

Context: it focuses on the situation where the pattern is applicable,

Problem: it is a description of the major forces/benefits of the pattern as well as its applicability constraints,

Solution: it details the way to solve the problem, it is composed of static relationships as well as dynamic ones describing how to construct an artefact according to the pattern. Variants are often proposed along with specific guidelines for adjusting the solution with regards to special circumstances. Sometimes, the solution requires the use of other patterns.

The structure of the pattern includes information describing the pattern, examples of use along with information describing the relationships that the pattern has with other patterns and so on.

Here is a simple example:

Name: Robust Design.

Examples: Choosing the dimensions and material composition of an engine component.

Context: Making products that minimize losses associated with deviations from ideal or target values of performance characteristics.

Problem: Multiple sources of variation can knock a manufactured product off target, and

some compromise over design may be desirable to produce products robust to uncontrolled sources of variation such as environmental conditions.

Solution: Identify control and noise factors, and conduct multi-factor experiments to identify impacts on location and spread of response variables, and in due course identify optimal product specifications and/or process settings.

3.4 Examples of statistical consulting patterns

Listed below is a sample of seven more elaborate statistical consulting patterns that are part of the SQM Library.

*Name: **CS_Data Analysis** (Cox and Snell)*
Context: A client presents a data set for analysis, perhaps with a request for a specific type of analysis.

Problem: The data set may be not ready for analysis, e.g. in a binary file from automated measuring equipment. The data set may be flawed, with missing values, suspect values, inadequate labeling or context information e.g. on randomization. The requested analysis may not be feasible or necessary.

Solution: Phases of analysis according to Cox and Snell (1981, see page 6): "(i) Initial data manipulation (the assembling of the data in a form suitable for detailed analysis and the carrying out of checks on quality). (ii) Preliminary analysis (the intention is to clarify the general form of the data and to suggest the direction which a more elaborate analysis may take), often best done by simple graphs and tables. (iii) Definitive analysis (intended to provide the basis for the conclusions). (iv) Presentation of conclusions in an accurate, concise and lucid form, that leads usually to a subject-matter interpretation of the conclusions"

As the authors remark, "While this division is useful, it should not be taken too rigidly. Thus an analysis originally intended as preliminary may give such clear results that it can be regarded as definitive. Equally, an analysis intended as definitive may reveal unexpected discrepancies that demand a reconsideration of the whole basis of the analysis. In fields in which there is substantial experience of previous similar investigations one may hope largely to bypass preliminary analysis."

*Name: **H_Data Analysis** (Hare)*
Context: You have helped design an investigation (e.g. experiment or survey) and data collection is soon to begin.

Problem: Data may not be collected in the right manner, nor in the right quantities. Further analysis of the data may be mishandled (not done, postponed, conducted despite problems with the data).

Solution: Phases according to Hare (2000): (i) Get real and get involved. Be present while samples are generated and participate, if possible, in their generation. You'll gain credibility and understanding, which will protect you from doing irrelevant things during the data analysis stage. Take ownership of the success of the project. (ii) Defend your client. Carry out the analysis and interpretation of the data, keeping project objectives in mind. When the time comes to present results to management, use graphs and charts to demonstrate the value to be had in following your recommendations. (iii) Defend yourself. "There's no limit to what can be accomplished as long as it doesn't matter who gets the credit." This phrase is written on a sign that used to hang in my office. Sadly, I took the sign down because during these days of corporate downsizing it does matter who gets the credit. To assure that you are protected, and to assure that your organization is not deprived of a statistical resource by someone who is misinformed about your worth to the organization, take don't grab credit for your work. Follow up in six months to assess the effort's value, and publicize your contributions to your own management team. Keep a cumulative record of your achievements in anticipation of the next passing of the downsizing grim reaper.

Name: *D_Surveys and Experiments (Deming)*

Context: You have been invited to play a major part in the design and analysis of a survey or experiment.

Problem: Client may not have a clear idea of what you will do, what you will deliver, and what the limits of your contribution may be.

Solution: Responsibilities of the statistician according to Deming (1960) – "The statistician's responsibility in a survey or experiment is for its statistical aspects, specifically: (i) To assist the client to formulate his problem in terms that will indicate whether and how statistical information might be useful on the problem. To formulate a choice of possible statistical surveys or experiments, if it appears that new information might be worth more than its cost. To explain to him the statistical procedure, cost, and utility of these plans. (ii) To explain to him the use and importance of the frame; to explore with him the various frames that might be suitable; and to explain to him that the results of any survey or experiment may be limited if a proposed frame and experimental conditions fail to include all the materials, areas, methods, levels, types, and conditions concerning which he desires information. (iii) To design a sample of specified material, covered by the frame that the client certifies as necessary and sufficient for study, to reach precision agreeable to the client; or to design an experiment to each significance in the tests of materials, methods, levels, types, and

conditions that the client certifies are necessary and sufficient. (iv) The statistician's instruction will include detailed procedures for use in the office and in the field for the delineation, definition, classification, and serialization of the sampling units. (v) The statistician will furnish statistical plans in writing when they are finally fixed, and the client thereafter will make no changes in procedure without further instruction from the statistician so long as the statistician's responsibility remains in force. (vi) The statistician will explain to the client the effect of departures from instructions. (vii) The statistician will furnish at the client's request statistical tests to use as an aid in supervision of the work, to attain more uniform performance in testing or in the interviewing and in coverage than would be possible otherwise. (viii) The client will arrange for the statistician to have direct access at any time to the people that carry out the preparation of the sample, the testing or the interviewing, the supervision, the coding, and the computations. (ix) The statistician will explain to the client, when the tabulations are finished, the meaning of the results of the survey in terms of their statistical significance. However, the statistician will not recommend that the client adopt any specific administrative action or policy. The uses of the data obtained by a survey or experiment are in the end entirely up to the client."

The above paragraphs give a bare outline of the statistician's responsibilities. They do not say how he should do his work.

Name: *H_Design (Hare)*

Context: You have been invited to a team meeting which is addressing a problem for the solution of which designed experiments are likely to be very useful.

Problem: You think DOE is a good idea, others may not. Many solutions may be presented. In fighting your corner, you may oversell the benefits and likely contribution of DOE.

Solution: Phases according to Hare (2000): (i) Understand the business process and value to the company. (What value will the current effort add if successful? What total resource will be required to gain that value? What structure exists, organizationally, to support a recommended change based on experimental outcome? Who on the management team will champion the change?) (ii) Listen actively. (Sit patiently for at least the first half of the meeting and let the extroverts vent. Only when the ideas have emerged and the real objectives have been made clear should the statistician speak. Summarize the experimental objectives and continuously make sure that the team agrees.) (iii) Ask the naive questions. (This is a statistician's luxury; Take advantage of it and enjoy it! Clients will never see statisticians as subject matter experts. Well-placed questions often cause a fresh look at objectives and key variables.) (iv) Avoid promising more than can be delivered. (The strong desire to be everyone's friend can lead to promising more than the design (or the statistician) can deliver. Marketing pressures, when left unchecked, can reduce development time to a matter of days. Be sure to allow adequate time for data analysis and interpretation.) (v) Defend the client. (Statistics don't dictate the experimental design; objectives do. Statisticians lead without authority; they are the clients' advocates. Their ability to guide is derived from their knowledge of statistical thinking and methods.)

The design must involve the right amount of the right kind of data to support experimental objectives. Of course, replicate and randomize. A good check on the appropriateness and adequacy of the design – an idea of statistician Stu Hunter – is to plot the data before the experiment is run. This is not as impossible as it sounds; clients can often anticipate responses that might correspond to treatment combinations. An analysis of anticipated responses often lends insight to design suitability.

*Name: **KZ_Change management** (Kenett & Zacks)*
Context: You have been brought in as a statistical resource, or are to lead an initiative calling for substantial changes in the way an organization operates, for example a Six Sigma program. The organization made several attempts to implement SPC and DoE without experiencing concrete bottom line results. These efforts were considered "academic exercises".

Problem: The organization needs to match statistical methods with management approaches. A fire fighting approach will not provide actual benefits to SPC or DoE implementations.

Solution: An increasing number of organizations have shown that the apparent conflict between high productivity and high quality can be resolved through improvements in work processes and quality of designs. The different approaches to the management of industrial organizations can be summarized and classified using a four steps **Quality Ladder** proposed by Kenett and Zacks (1998). The four management approaches are (1) Fire Fighting, (2) Inspection, (3) Process Control and (4) Quality by Design. To each management approach corresponds a particular set of statistical methods. Managers involved in reactive fire fighting need to be exposed to basic statistical thinking. Managers attempting to contain quality and inefficiency problems through inspection and 100% control can have their tasks alleviated by implementing sampling techniques. More proactive managers investing in process control and process improvement are well aware of the advantages of control chart and process control procedures. At the top of the quality ladder is the quality by design approach where up front investments are secured in order to run experiments designed to impact product and process specifications.

Efficient implementation of statistical methods requires a proper match between management approach and statistical tools. For example if a fire fighting manager tries to implement control charts he might have the workforce maintain charts covering the walls but the early warning signals provided by the charts will go unnoticed, grossly reducing the effectiveness of control chart or any other process control procedures. In order to take full advantage of the benefits of proactive process control and improvement, fire-fighting management habits have to change.

*Name: **K_Change Management** (Kotter)*
Context: You have been brought in as a statistical resource, or are to lead an initiative calling for substantial changes in the ways people handle their work, for example a TQM or a Six Sigma program.

Problem: Statistical thinking alone will not be enough (put all experimental design issues to rest) Part of the organizational challenge is enculturating statistical thinking, and inspiring others to pursue your approach.

Solution: Phases according to Kotter (1996). Kotter's eight steps for leading change – (1) Establishing a sense of urgency. (2) Creating a guiding coalition. (3) Developing a vision and strategy. (4) Communicating the change vision. (5) Empowering broad based action. (6) Generating short-term wins. (7) Consolidating gains and producing more change. (8) Anchoring new approaches in the culture.

The first four, Kotter says, appear to be missing from many change efforts, causing many such efforts to fail. In fostering more extensive use of the right kind of experimental design and statistical methods in general, we must ask ourselves what we have done to create a sense of urgency, cultivate a (formal or informal) guiding coalition, or develop and communicate vision and strategy.

Example: Ron Harris, the leader of research at Nabisco Inc., does wonders to foster and reward the creative use of experimental design. He helps create a sense of urgency by tracking the time to market, and he emphasizes the role of experimental design at every research gathering. To build the guiding coalition, Nabisco formed a cross functional council of non-statisticians who understand the value of statistical thinking and statistical methods and who form strategies for their furtherance. As a result, statistics is represented on several corporate councils including a process council, a quality council and a council devoted to applying technology to the solution of operational problems. Nabisco also assures that statisticians are directly involved in major projects and are available to provide consultative support to other projects (Hare, 2000).

*Name: **B_Management Consultancy** (Block)*

Context: You are at the initial stage of a consultancy opportunity, and want to pursue it in an orderly and structured way.

Problem: Every opportunity is different. How can you add some structure and clarity of purpose to your efforts at each stage?

Solution: Phases according to Block (1981): (i) Phase 1. Entry and contracting. (It includes setting up the first meeting as well as exploring what the problem is, whether the consultant is the right person to work on this issue, what the client's expectations are, what the consultant's expectations are, and how to get started.) (ii) Phase 2. Data collection and Diagnosis. (Consultants need to come up with their own sense of the problem. This may be the most important thing they do. The questions here for the consultant are: who is going to be involved in defining the problem, what methods will be used, what kind of data should be collected, and how long will it take?) (iii) Phase 3. Feedback and the Decision to Act. (The consultant is always in the position of reducing a large amount of data to a manageable number of issues. There are also choices for the consultant on how to involve the client in the process of analyzing the information. In giving feedback to an organization, there is always some resistance to the data (if it deals with important issues). The consultant must handle this resistance before an appropriate decision can be made about how to proceed.) (iv) Phase 4. Implementation. (This involves carrying out the planning of the previous step. In many cases the implementation may fall entirely on the line organization. For larger change efforts, the consultant may be deeply involved.) (v) Phase 5. Extension, Recycle, or Termination. (Phase 5 begins with an evaluation of the main event. Following this is the decision whether to extend the process to a larger segment of the organization. Sometimes it is not until after some implementation occurs that a clear picture of the real problem emerges. If the implementation was either a huge success or a moderate-to-high failure, termination of further involvement on this project may be in the offing. There are many options for ending the relationship and termination should be considered a legitimate and important part of the consultation. If done well, it can provide an important learning experience for the client and the consultant, and also keep the door open for future work with the organization.)

4 CASE STUDIES AND SQM SOFTWARE LINKS AND RESOURCES

4.1 *Case studies*

ENBIS is currently collecting case studies in Designed Experiments for the benefit of its members and the associations and organizations linked to ENBIS. Submission of case studies follow specific guideline rules: (1) Introduction and background. (2) Main objectives of the experiment. (3) List of factors participating in the experiments. (4) List of responses. (5) Issues and constraints. (6) Initial model. (7) Initial design. (8) Final design. (9) Modeling and data analysis. (10) Conclusions and points for discussion.

For each case study provided, it is necessary to make available the full data or, when not possible, a masked version of it. The case study reader should be able to replicate the analysis and/or present alternative approaches.

4.2 *Software*

In the SQM Library, links to web-sites with downloadable documents, presentations and software resources will be made available. Examples include web sites referred to in the European Network for Business and Industrial Statistics web site (Kenett (2002), www.enbis.org). Similar resources in the context of reliability and safety are available on the ESRA Newsletter and web site.

5 SQM SURVEY

What is the state of the art in industrially implemented SQM, what exists, and what needs to be developed? In order to understand and correctly answer these questions Ramalhoto (2002, Rimini) introduced the SQM survey concept. Pro-ENBIS is currently running an SQM survey with companies, consultants and academia. The first version of the survey questionnaire is now in a pilot phase and consists of: (a) Company profile. (b) Lists of reliability models and techniques in use, ready to use, needed. (c) Lists of maintenance management models and techniques in use, ready to use, needed. (d) Lists of control charts models and techniques in use, ready to use, needed. (e) Lists of acceptance sampling models and techniques in use, ready to use, needed. (f) Lists of economical design models and techniques in use, ready to use, needed. (g) Lists of cost-benefit models and techniques in use, ready to use, needed. (h) Lists of inventory/supply-chain models and techniques in use, ready to use, needed. (i) Lists of data quality techniques in use, ready to use, needed. (j) Lists of quality improvement and innovation models and techniques in use, ready to use, needed. (l) List of scientific and technical periodicals that exist in the company and are of practical or theoretical interest. (m) List of the Conferences, workshops attended. (n) List of SQM web-sites or other technical sites visited. (o) Type of long-life training/ learning attended (including distance learning, for instance, internet and open university courses).

There is also an electronic Forum linked to the SQM questionnaires to discuss: (1) The most relevant problems reported by industry so far. (2) The already identified problems to close the gap between industry and academia.

6   CONCLUSIONS AND FINAL REMARKS

Many authors in books and papers have discussed the role of statisticians in the quality movement (Bisgaard, 1991, Hahn and Boardman, 1985, Hogg, 1985, Joiner, 1985). We go further and propose the development of an SQM library in the context of an extended quality movement whose aim is to establish a profession using medicine as a role model.

This proposal is very demanding in terms of human resources and funding. An initial attempt at its implementation is achieved through ENBIS and Pro-ENBIS. Physicians have an enormously wealthy industry behind them, operating now for several decades. The statistical community could significantly gain from adopting such a model.

ACKNOWLEDGEMENT

This paper is supported by funding from the "Growth" program of the European Community and was prepared in collaboration by member organizations of the Thematic Network – Pro-ENBIS- EC contract number G6RT-CT-2001-05059.

REFERENCES

Alexander, C. (1979), *The Timeless Way of Building*, Oxford University Press, New York, 1979.

Alexander, C., Ishikawa, S., Silverstein, M., Jacobson, M., Fiksdahl-King, I. and Angel, S. (1977), *A Pattern Language*, Oxford University Press, New York, 1977.

Bisgaard, S. (1991), "Teaching Statistics to Engineers", *The American Statistician*, 45, pp. 274–283.

Block, P. (1981), *Flawless Consulting*, Pfeiffer and Company.

Coad, P. (1992), Object-Oriented Patterns, Communications of the ACM, Vol. 35, No. 9, 1992, pp. 152–159.

Cox, D.R., and Snell, E.J. (1981), *Applied Statistics*, Chapman and Hall. ISBN 0 412 16570 8.

Deming, W.E. (1960), *Sample Design in Business Research*, John Wiley & Sons – Classics Library Edition published in 1990. ISBN 0 471 52370 4.

Gamma, E., Helm, R., Johnson, R. and Vlissides, J. (1995), *Design Patterns: Elements of Reusable Object-Oriented Software*, Addison-Wesley, Reading, MA, 1995.

Hahn, G.J. and Boardman, T. (1985), "The Statistician's Role in Quality Improvement", *AmStat News*, March (113), pp. 5–8.

Hare, Lynne, B. "You Don't Have To Be Awful To Be a Statistician, but It Helps", ASQ, 2000.

Hogg, R.V. (1985), "Statistical Education for Engineers: An initial task force report", *The American Statistician*, 39, pp. 168–175.

Joiner, B. (1985), "The Key Role of Statisticians in the Transformation of North American Industry", *The American Statistician*, 39, pp. 224–227.

Kenett, R.S. and Zacks, S. (1998), *Modern Industrial Statistics: Design and Control of Quality and Reliability*, Duxbury Press.

Kenett, R.S. and Shade, J. (2001), The Statistical Consulting Patterns Models, ENBIS Report, Statistical Consulting Working Group.

Kenett, R.S. (2002), Web Resources for Statistical Consulting, ENBIS Report, Statistical Consulting Working Group (www.enbis.org).

Kotter, J. (1996), *Leading Change*, Cambridge, MA: Harvard University Press.

Ramalhoto, M.F. (1999), The Research and Education Global Network in Stochastics for the Quality Movement, European Journal for Engineering Education, Vol. 24, No. 4, pp. 405–413.

Ramalhoto, M.F. (2001), Some R&D Issues in Safety, Risk, Reliability and Maintenance: A Trans-disciplinary Approach. In "Safety & Reliability", Editors: E. Zio, M. Demichela and N. Piccini, Publisher: "Poiltecnico di Torino" Press, Vol. 2, pp. 1077–1082.

Ramalhoto, M.F. (2002), Notes #1, Pro-ENBIS Kick-off Meeting, Brussels, February 11–12, and Notes #2, Pro-ENBIS Meeting, Rimini, September 22–23, WP4-Reliability, Safety and Quality Improvement Work Package.

Shade, J. and Kenett, R.S. (2001), An Initial Repository of Statistical Consulting Patterns Model, ENBIS Report, Statistical Consulting Working Group.

Comparison of data representation tools for the analysis of complex time-dependant systems

C. Kermisch & P.E. Labeau
Université Libre de Bruxelles, Belgium

ABSTRACT: Petri nets and event sequence diagrams are two graphical data representation tools used in the modelling of complex time-dependent systems. Their application in the dynamic reliability framework has also been proposed. The capabilities of these methods are compared on simple examples: a time-dependent problem with operational constraints and no dynamics, and a dynamic transient. Criteria are suggested with the objective of assessing the potential applicability of these two methods to real industrial situations.

1   INTRODUCTION

In this paper we present the results of a comparative study – based on a set of criteria – of two graphical data representation tools used in the modelling of complex time-dependent systems. Some of these problems resort to dynamic reliability techniques, which study in an integrated way the discrete evolution of a plant between components states and the continuous evolution of the process variables in the different states (Labeau, Smidts & Swaminathan 2000). The two methods we propose to investigate and compare are Petri Nets (PN) and Event Sequence Diagrams (ESD), respectively.

As it is well known, PNs (section 2) have been used for several years as support tools for the Monte Carlo assessment of dependability characteristics of time-dependent systems. ESDs (section 3) have served as a technique helping the analyst in the construction of event trees. Recent works have attempted to adapt these techniques to dynamic reliability problems (Swaminathan & Smidts 1999, Châtelet, Chabot & Dutuit 2000, Chabot, Dutuit & Rauzy 2001). Their capabilities and performances are therefore assessed on simple examples: a time-dependent problem with operational constraints and no dynamics, and a dynamic transient. This comparison (section 4) is realized according to criteria we have defined. Finally, further developments (section 5) based on the results of this work are suggested.

2   PETRI NETS

2.1   *Basic principles*

For information about basic stochastic PNs applied to dependability problems, the reader can refer e.g. to Dutuit, Châtelet, Signoret & Thomas (1997).

2.2   *Example 1: time-dependent system*

Figure 1 presents a PN describing a two-component system with active redundancy and only one repair crew. The first component failing is the first one repaired. The PN corresponding to this system is made of two identical subnets, each one associated to one of the two components.

Let us consider on the left hand side of Figure 1, the sub-Petri net corresponding to component n°1. It is made of three places that correspond to the three possible states of the component: working (place Pl_1), failed (place Pl_3), being repaired (place Pl_5); of three transitions (Tr_1, Tr_2, Tr_5) between these places; and of a message (RD) corresponds to the availability of the repair crew, and that forbids the repair of the second failed component as long as the first one is not repaired and vice-versa. The right hand side sub-Petri net, corresponding to component n°2, is similar to the one which has been described here above.

As it can be seen, the components of this system are modelled separately. This PN rests therefore on a

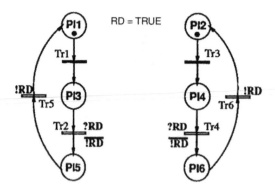

Figure 1. Example 1 modelled by a PN (Leroy & Signoret 1992).

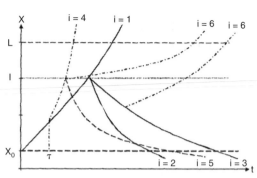

Figure 2. Transient evolutions of the process variable $x(t)$ (Chabot, Dutuit & Rauzy 2001).

component-based construction of the system model. The interactions between the components are introduced afterwards, by the message « RD ». This message governs therefore the order of the component repairs: if the first component failed before the second one, RD forbids the repair of the second component as long as the first one is not repaired.

In a more general way, it can be said that the PNs allow a component-based representation. Indeed, the analyst can first model each component (or subsystem) without having to account for the system in its totality. At this stage, each component is represented by a subnet. After that, the dependences between these subnets can be introduced thanks to messages, additional arcs, inhibitor arcs, etc …

The coupling between the evolutions of the subnets marking and the dependences between these subnets give birth to scenarios at the level of the whole system. The analyst does not have to carry about the burden of integrating the behaviour of the different parts of the system in order to determine which scenarios can occur at the level of the system as a whole. Indeed, from the different components of the sub-systems and their dependences, the evolution of the PN marking leads to provide potentially "all" the possible scenarios. Therefore, it can be said that PNs allow an automatic generation of the scenarios.

2.3 Example 2: extension of PNs to dynamic reliability

Let us consider the example proposed in Figure 2 (Labeau 2000). A vessel is described by one process variable x (the pressure), with its steady state value x_0. At time t_0, a transient is initiated in state 1, and the evolution of x follows an exponential law, possibly until $x > L$, when the system fails (vessel rupture). When level $x = l$ is reached, however a protection device is solicited (component C_1). The latter can either fail

(with a probability p_0, state 1), work correctly (with a probability of $1 - p_0 - p_1$, state 2), or be imperfectly triggered (with a probability of p_1, state 3).

In the last two cases, x starts to decrease, and a safe shutdown is reached as soon as $x < x_0$.

We add to this description the possibility of an additional component failure (component C_2; failure rate λ) that accelerates the transient, i.e. a transition from state i to state $i + 3$, $i = 1, 2, 3$. This more severe transient can still be mitigated in case of a perfect working of the protection device C_1, but it is only slowed down in case of partial triggering. In this last case, the scenario ends with delayed rupture.

This quoted above problem has been modelled by PNs in (Chabot, Dutuit & Rauzy 2001). The continuous evolution of the process variable $x(t)$ and the discrete behaviour of components C_1 and C_2 can be modelled by means of the Petri Net shown in Figure 3.

Two different kinds of transitions can be distinguished in this PN model:

– Deterministic transitions. On one hand, there are failures on demand transitions ("sol x y z", x, y and z representing the probabilities of the different outcomes of the transition) with zero delay that appear when level l is reached. On the other hand, there are Dirac transitions ("Drc" with a numerical value corresponding to the data given in (Chabot, Dutuit & Rauzy 2001) with constant delay. The transition between places 4 and 1 can be considered for example. Indeed, this transition corresponds to the time necessary to the vessel rupture in case both C_1 and C_2 fail. Note that the values of these constant delays are analytically calculated from the pressure trajectories.

– Hybrid transitions ("HYB$_x$") correspond to special laws that provide the times of occurrence of the physical process events by taking into account the effect on this continuous evolution of the change of

890

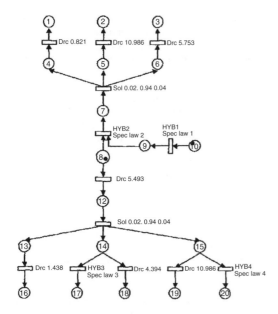

Figure 3. Example 2 modelled by a PN (Chabot, Dutuit & Rauzy 2001).

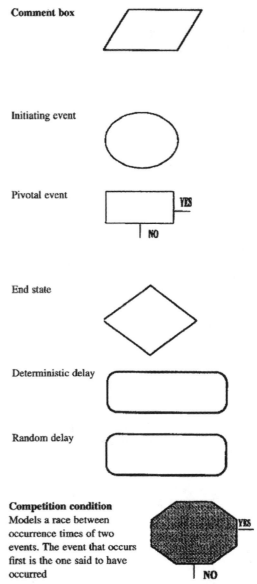

Comment box

Initiating event

Pivotal event

End state

Deterministic delay

Random delay

Competition condition
Models a race between occurrence times of two events. The event that occurs first is the one said to have occurred

Figure 4a. Events and conditions in the ESD framework (Labeau, Smidts & Swaminathan 2000).

discrete state of the system. They allow to analytically determine the delays depending on the pressure trajectories, which are deterministic per sections. So, they possibly take into account other delays that have been sampled before. An example of such a transition is the one between place 14 (C_1 working) and place 17 (failure of C_2 after the solicitation of C_1 that delays the pressure decrease), which shows the interaction between a discrete transition (C_2 fails) and the continuous process (the decrease of the pressure is delayed).

The interested reader can refer to (Chabot, Dutuit & Rauzy 2001) for the detailed description of this PN.

It is important to mention that it has been possible to build this PN only thanks to the a priori knowledge of the qualitative solution, i.e. of all the scenarios. Indeed, such a PN does not help the analyst in the scenario identification process. On the contrary, its building requires that the analyst identifies the next event at each step of the PN construction. The exhaustivity and the correctness of the scenarios in such a PN model rely thus only on the analyst. In order to complete a dynamic reliability study by means of such PNs it is necessary to have the qualitative solution to build the PN on which the calculation of the quantitative solution is based. Moreover, the special laws are equivalent to inverting analytically the dynamic laws in the different states. As a consequence, this procedure becomes quickly prohibitive.

3 EVENT SEQUENCE DIAGRAMS

3.1 *Basic principles*

ESDs are visual representations in the form of flowcharts of dynamic scenarios (Labeau, Smidts & Swaminathan 2000). In order to explicitly represent these dynamic situations, all the components listed in Figures 4a, 4b can be used.

3.2 Example 1: time-dependent system

The first example considered in section 2.2 is studied here in the ESD framework. Figure 5 represents the whole system evolution, unlike PNs, which represent the system component by component.

At $t = 0$, the components A and B start working. After a random delay, one of these components fails, which constitutes a competition: did A or B fail? Assume that component A has failed (upper part of the scheme). Note that the process would exactly be symmetric if B had failed (lower part of the scheme). The current situation is thus "A failed and B working". After another random delay, the system state changes again: either A is repaired, or B fails, which induces another competition.

This example allows us to draw the following conclusions.

- The EDSs initially envisioned in accident sequences ending in absorbing states are also able to deal with scenarios presenting cycles. Thus, this formalism can also be used in availability studies even if other representations seem to be more appropriate.
- They represent the evolution of the whole system state, it is thus impossible to model the system component by component as it was the case with PNs. The ESD is built while the analyst identifies step-by-step the possible scenarios. Therefore, the ESD is not able to provide the analyst with scenarios that he/she would not have identified during the construction phase.
- As a consequence, the size of this type of ESDs considerably increases with the number of components. However, more generally, it is the number of events that defines the size of the problem.

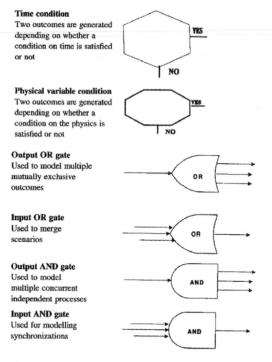

Time condition
Two outcomes are generated depending on whether a condition on time is satisfied or not

Physical variable condition
Two outcomes are generated depending on whether a condition on the physics is satisfied or not

Output OR gate
Used to model multiple mutually exclusive outcomes

Input OR gate
Used to merge scenarios

Output AND gate
Used to model multiple concurrent independent processes

Input AND gate
Used for modelling synchronizations

Figure 4b. Conditions and gates in the ESD framework (Labeau, Smidts & Swaminathan 2000).

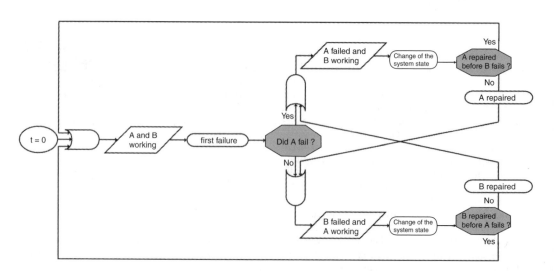

Figure 5. Example 1 modelled by a ESD.

3.3 Example 2: extension of ESDs to dynamic reliability

Consider in the ESD framework the second example treated in section 2.3. The diagram corresponding to this problem is shown in Figure 6.

The initiating event is followed by a competition: does C_2 fail before pressure reaches level l or not? Let us analyse both cases:

- C_2 fails before the pressure reaches l. After some delay, C_1 is solicited. As it is mentioned in section 2.3, three situations are then possible.
- C_2 does not fail before pressure reaches l. The interpretation of the end of this ESD is easy to read and intuitive. Thus, it will not be explicited here.

This example shows that this kind of graphical representation is well fitted for situations where the system evolves towards an absorbing state.

4 COMPARISON BETWEEN PNS AND ESDS

Based on the observations drawn from the two examples, we propose here four criteria in order to compare these two data representation tools.

4.1 Approaches of the system treatment

PNs allow to analyse the system component by component, or, more generally, event by event. This property is of paramount importance. Indeed, the representation of a complex system can be done in a modular way if the interactions and dependences between the subnets are correctly taken into account.

Unlike the PNs, ESDs need a vision of the whole system because they directly represent the possible scenarios at the system scale. The correctness of the representation rests on the capability of the analyst to mentally perform the integration of all aspects of the problem.

4.2 Readability

The graphical formalism of PNs is reduced: the same symbols are used to represent very different kinds of situations. Their readability can thus be highly restricted, especially when there are a lot of interactions between the subnets.

ESDs offer a wider graphical diversity, while keeping a reasonable number of symbols. In this point of view, ESDs should be preferred to PNs

4.3 Representation of the time evolution of the system

In the PN framework, the time evolution of the system is represented by the time evolution of the marking.

This marking is essential: it allows the analyst to know the situation of the whole system. Without this marking, it would be impossible to know the subnets states and the possible evolutions of the whole system via the subnets evolutions.

This marking does not have any equivalence in the ESD framework. However, it is not a major concern because the system is analysed as a whole: the branches of an ESD sequentially provide the different situations that the whole system can lie in.

4.4 Scenario generation

Let us consider a system taking into account only discrete states in the PN framework. In this case, the analyst does not need to know all the possible evolution modes of the system considered as a whole. Indeed, PNs are able to generate new scenarios by the evolution of their marking. In this case, PNs present significant advantages on other forms of representation. However, if the dynamics of a system has to be integrated in a PN (see section 2.3), the analyst determines the possible scenarios while building the representation. Moreover, in this case, the system is globally studied and the advantages of a component-by-component approach disappear. Why is it so difficult to integrate the evolution of the process variables in a PN? It comes mainly from the fact that continuous variables often constitute strong functional links between the different parts of the system. It is thus impossible to build separately the subnets corresponding to these parts and to simply interconnect them. The building of the PN then requires from the analyst to know a priori the possible scenarios. The potential advantages of PNs on ESDs are then lost.

Let us mention that the hybrid PNs (Chabot et al. 1998) are able to solve this problem. They consist in associating, in a same simulation, two different modellings in interaction: PNs for discrete phenomena and a system of algebraic or integro-differential equations for continuous phenomena. The code providing the dynamic evolution of the system (which depends on its current state) is linked to more basic subnets that give the possible transitions of the system components. This approach is much closer to the theoretical description of dynamic reliability (Devooght & Smidts 1992). It preserves the ability to automatically generate scenarios from the basic data of the problem: component states and transitions, dynamic evolution laws of the process in the different system states, possible transitions between states and dependences. However, it is important to note that the automatic scenario generation is driven by the evolution of the process variables and not thanks to PNs. It is thus possible to replace them by state graphs or other representations. Nevertheless it must be reminded that there is an interest in keeping the PNs for the modelling of

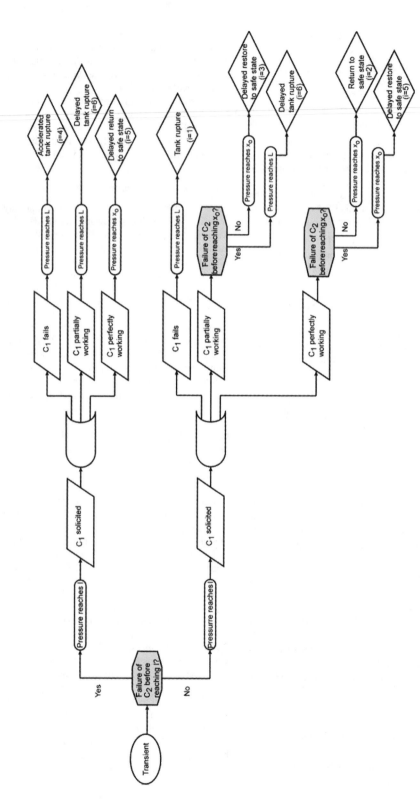

Figure 6.   Example 2 modelled by an ESD.

894

discrete transitions. Indeed, this will allow to keep their advantages on the part of the model that is less affected by the behaviour of the physical variables.

5 CONCLUSION AND FUTURE PROSPECTS

The PN and ESD approaches considered up to now aim to the representation of *data*. However, the a prior knowledge of all the scenarios is needed to build these graphical representations, except when representing systems without process variables with PNs.

As the ability of generating scenarios is more linked to the dynamic approach of the reliability than to the representation technique, other representation methods could be used.

The scenarios obtained by simulation could be grouped and represented by one of the graphical techniques studied above. Indeed, ESDs present a higher graphical potential. Therefore, they should be preferred for the task of representing the *results*.

ACKNOWLEDGEMENTS

This work has been supported by Air Liquide, CEA, EDF, IRSN, PSA Peugeot Citroën and TotalFinaElf.

P.E. Labeau is Research Associate of the National Fund for Scientific Research, Belgium.

REFERENCES

Chabot, J.L., Dutuit, Y. & Rauzy, A. 2001. A Petri net approach to dynamic reliability problems. *Proc. of Esrel 2001* (**2**): 1387–1394. Torino: Politecnico di Torino.

Châtelet, E., Chabot, J.L. & Dutuit, Y. 2000. Events representation in dynamic reliability analysis using stochastic Petri nets. In Smidts, C., Devooght, J. & Labeau, P.E. (eds), *Dynamic reliability: Future directions, International Workshop Series on Advanced Topics in Reliability and Risk Analysis*. College Park, MD: University of Maryland.

Devooght, J. & Smidts, C. 1992. Probabilistic reactor dynamics. I. The theory of continuous event trees. *Nuclear Science and Engineering* **111**: 229–240.

Dutuit, Y., Châtelet, E., Signoret, J.P. & Thomas, P. 1997. Dependability modelling and evaluation by using stochastic Petri nets: application to two test cases. *Reliab. Eng. Syst. Safety* **55**: 117–124.

Labeau, P.E. 2000. A variational principle in probabilistic dynamics. *Ann. Nucl. En.* **27**(**17**): 1543–1575.

Labeau, P.E., Smidts, C. & Swaminathan, S. 2000. Dynamic reliability: towards an integrated platform for probabilistic risk assessment. *Reliab. Eng. Syst. Safety* **68**: 219–254.

Leroy, A. & Signoret, J.P. 1992. *Le risque technologique*. Paris: Presses Universitaires de France.

Swaminathan, S. & Smidts, C. 1999. The event sequence diagram framework for dynamic PRA. *Reliab. Eng. Syst. Safety* **63**: 73–90.

Safety and Reliability – Bedford & van Gelder (eds)
© *2003 Swets & Zeitlinger, Lisse, ISBN 90 5809 551 7*

A study of an expected RAW importance measure

Kilyoo Kim, M.J. Hwang, & D.I. Kang
Integrated Safety Assessment Team, Korea Atomic Energy Research Institute

ABSTRACT: An importance measure called "Expected RAW"(ERAW) is introduced to improve the problem caused by the conventional "Risk Achievement Worth" (RAW) in risk informed regulation and application such as the U.S. Maintenance Rule. ERAW characteristics including the relationship with ERAW, Fussell-Vesely, and event probability P are studied. The threshold value 0.01 of the ERAW for a risk significant is derived, and a SSC having FV < 0.005 and RAW < 2 could be risk significant if its ERAW > 0.01. ERAW importance of a component is considered to be the sum of the ERAWs for the relevant failure modes of the component. ERAW can be used with confidence in the importance categorization in nuclear power plants.

1 INTRODUCTION

Risk Achievement Worth (RAW) as one of the many importance measures is widely used in the risk informed regulations and applications (RIR&A) such as the Maintenance Rule (MRule) (NRC, 1991), the risk categorization processes for special treatment requirements (NRC, 1998), (NEI, 2000), … etc.

RAW is measuring the risk significance if an event failure probability changes from its base value to one, without considering how credible is the assumption that the event probability equals one. Therefore, some structures, systems and components (SSCs) could be determined as risk significant SSCs although their event probabilities are very low. Because RAW does not reflect the event probabilities of SSCs but only shows just the functional redundancy or the defense in depth (DID) of the SSCs (Wall, 1996).

In this paper, "Expected RAW" (ERAW), is introduced to improve the above problem caused by RAW, and its characteristics are described. Also, discussed is how ERAW of the group can be simply expressed by RAWs of the individual members in the group.

2 EXPECTED RAW

2.1 *Definition of ERAW*

The expected importance concept was introduced in Vesely (1985) where Fussell-Vesely (FV) importance was interpreted as the expected importance of Birnbaum importance. Thus, the RAW is a conditional importance assuming or given a failed state. To account for the likelihood that the component actually does fail or is actually down, the RAW importance needs to be multiplied by the appropriate probability. Thus, ERAW of a SSC is defined as follows:

$$ERAW \equiv RAW * P \qquad (1)$$

where P is event probability of the SSC. Thus, ERAW is measuring the risk significance if the event failure probability changes from its base value to one, with considering how credible is the assumption that the event probability equals one. Therefore, ERAW would be useful in the risk significant determination where the use of actual event probability would be better such as in MRule and in the risk categorization processes for special treatment requirements, … etc.

For example, as shown in Table 1, let the unavailability of component A, P(A), be 0.01, and that of component B, P(B), be 0.001. Also, let the RAW of component A and B be 1.5 and 3, respectively. In this situation, component B could be determined as a risk

Table 1. ERAW example.

SSC	P	RAW	ERAW = RAW*P
A	0.01	1.5	0.015
B	0.001	3	0.003

significant component and component A could be determined as a non-risk significant since the RAW of component B is 3 and a widely used criterion for risk significant is RAW >2. However, it is more likely that component A is more important than component B in some RIR&A such as MRule since the ERAW of A is 0.015 and that of B is just 0.003. Because the importance measure to account for the likelihood that the component actually does fail is necessary in an importance categorizing process of SSCs.

2.2 The characteristics of ERAW

Core damage frequency (CDF) can be expressed by minimal cutsets. Each cut indicates a basic event whose probability is usually an unavailability of a SSC. CDF can be expressed by a linear function of the basic event probability P as below:

$$CDF = a * P + b \tag{2}$$

where aP indicates the sum of all minimal cutsets containing P, and b means the sum of all other minimal cutsets which do not have P (Wall et al., 1996). Using Equation 2, FV and RAW can be expressed as below:

$$FV = aP/CDF = aP/(aP + b) \tag{3}$$

$$RAW = (a + b)/(aP + b) \tag{4}$$

The characteristics of RAW are well described in Wall (1996). That is, when RAW is represented by Equation 4, if defense in depth (DID) is poor, a ≫ b, and if DID is good, a ≪ b.

In Equation 1, if DID is poor, i.e., a ≫ b, then ERAW approaches FV value. That is,

$$ERAW = RAW*P$$

$$= (a + b)P/(aP + b)$$

$$\cong aP/(aP + b) = FV \tag{5}$$

In Equation 1, if DID is good, i.e., a ≪ b, ERAW approaches the event probability of the SSC. That is,

$$ERAW = RAW*P$$

$$= (a + b)P/(aP + b) \cong bp/b = P \tag{6}$$

The relationship among ERAW, FV, and P can be further explained as follows.

Risk R (or CDF) can be expressed as below:

$$R = P*R^{+} + (1 - P)*R^{-} \tag{7}$$

Equation 7 can be proved by inserting R = aP + b into Equation 7. From Equation 7, the following relationship can be derived:

$$RAW = 1 + FV*(1 - P)/P \tag{8}$$

Equations 7 and 8 were introduced in Martorell (1996). After multiplying the both sides of Equation 8 by P, we get the following relationship between ERAW, FV, and P.

$$ERAW = P + FV*(1 - P) \tag{9}$$

In Equation 9, if FV = 1, ERAW = 1. Also, if P = 1, ERAW = 1. If P → 0, then ERAW → 0, and FV → 0.

Figure 1 shows the relationship among ERAW, FV, and P (~unavailability).

2.3 The risk significant criterion for ERAW

In the RIR&A such as the U.S. MRule, SSCs are in the risk significant category when RAW > 2 and FV > 0.005 (NUMARC, 1993). Therefore, in Figure 2, if the importance measures of SSCs are located in region I, the SSCs are risk significant, and if they are located in region III, their SSCs are non-risk significant.

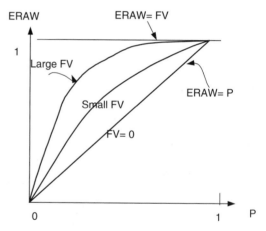

Figure 1. Relationship among ERAW, FV, and P.

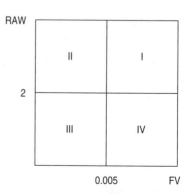

Figure 2. The criteria of FV and RAW to decide risk significant SSCs.

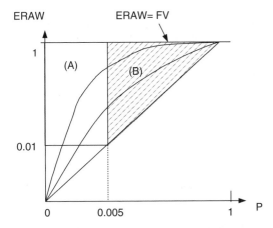

ERAW   ERAW= FV

Figure 3. Relationship among ERAW, FV, and P with ERAW > 0.01 and P < 0.005.

Table 2. Example comparison of FV, RAW and ERAW.

No.	Event	Mean	FV	RAW	ERAW
1	HSMPW00 102	8.90E-05	0.0323	364.362	3.24E-02
2	AFCVW104 849	2.08E-06	0.0313	15069.9	3.13E-02
3	HCOPVCQ HPPB-HD	2.14E-01	0.0051	1.0187	2.18E-01
4	EGDGS01B	1.40E-02	0.0002	1.0164	1.42E-02

However, as a new importance measure, ERAW can be used to find the risk significant SSCs instead of using the combination of FV and RAW values. Thus, an introduction of a threshold value of ERAW for risk significant is necessary. Now, let's derive a threshold value for the ERAW importance measure.

If RAW > 2 and FV > 0.005 are used as criteria for risk significant SSC, from Equation 8,

$$2 < RAW$$

$$2 < 1 + [(1 - P)/P]* FV$$

$$P/(1 - P) < FV$$

In the above equation, since 0.005 < FV,

$$0.005 \geqslant P/(1 - P)$$

$$\sim 0.005 \geqslant P \tag{10}$$

Since FV > 0.005, from Equation 9,

$$ERAW = P + (1-P)*FV$$

$$ERAW > 0.005 + (1-0.005)* 0.005$$

$$ERAW > 0.01 \tag{11}$$

Thus, the SSCs which satisfy Equation 10 and 11 are the risk significant SSCs which satisfy RAW > 2 and FV > 0.005, which belong to region I in Figure 2. The criterion using ERAW > 0.01 and 0.005 ⩾ P for risk significant corresponds to region (A) of Figure 3. (Figure 3 and Figure 1 are identical figures. However, the region of ERAW > 0.01 and P < 0.005 is well depicted in Figure 3.) However, since the SSCs located in region (B) of Figure 3 can be risk significant, the threshold value of the ERAW for risk significant should be 0.01.

In Table 2, the importance measures of several basic events for UCN 3 and 4 units are compared. In Table 2, since No. 1 and 2 events are ERAW > 0.01 and P < 0.005, they belong to region I of Figure 2, and actually, the FV and RAW are larger than 0.005 and 2, respectively. In the No. 3 event, although ERAW is larger than 0.01, since P > 0.005, No. 3 event does not belong to region I of Figure 2. Actually, it belongs to region IV of Figure 2 and region (B) of Figure 3. Also, for the No. 4 event, even though its ERAW is larger than 0.01, since P > 0.005, No. 4 event does not belong to region I of Figure 2. Actually, it belongs to region III of Figure 2 since its FV < 0.005 and RAW < 2. However, No. 3 and 4 events should be regarded as risk significant since ERAW > 0.01 although they are located in region (B) of Figure 3.

2.4 ERAW for groups and individual members of group

As discussed in Cheok (1998) and Wall(1996), for the general case of a group, RAW of the group cannot be simply expressed by RAWs of the individual members in the group. In option 2 (NRC, 1998), (NEI, 2000), for the case of a component consisting of several failure modes, the RAW importance of the component is considered to be the maximum of the RAW values computed for basic events involving the components (NEI, 2000). However, the ERAW importance of a component is considered to be the sum of the ERAW for the relevant failure modes of the component. The following is the explanation of this.

Let component C consist of failure mode C_1, C_2, C_3, C_4. That is, $C = C_1 + C_2 + C_3 + C_4$. Then,

$$RAW (C) = RAW (C_1) = \cdots = RAW (C_4)$$

By multiplying both sides by C,

$$C\ RAW (C) = C\ RAW (C_1) = \cdots = C\ RAW (C_4)$$

Thus,

$$ERAW(C) = (C_1 + C_2 + C_3 + C_4)*RAW (C_1) = \cdots$$
$$= (C_1 + C_2 + C_3 + C_4)*RAW (C_4)$$

899

Therefore,

$$ERAW(C) = C_1\,RAW(C_1) + C_2\,RAW(C_1)$$
$$+ C_3\,RAW(C_1) + C_4\,RAW(C_1)$$

Since $RAW(C_1) = .. = RAW(C_4)$,

$$ERAW(C) = C_1\,RAW(C_1) + C_2\,RAW(C_2)$$
$$+ C_3\,RAW(C_3) + C_4\,RAW(C_4)$$

$$ERAW(C) = ERAW(C_1) + ERAW(C_2)$$
$$+ERAW(C_3) + ERAW(C_4) \qquad (12)$$

In NEI (2000), how FV and RAW of a component can be computed with FVs and RAWs of the component failure modes is well described and it is summarized in Table 3. In Table 3, the ERAW column is newly added. In Table 3, since each event probability of a failure mode can be derived by Equation 8, each ERAW of a failure mode can be computed by Equation 9. The ERAW of the component can be calculated by the sum of the ERAWs for each failure mode of the component like in Equation 12. In Table 3, the interesting thing is that the ERAW of the component can be directly calculated with Equation 8 and 9 by using FV, in which a common cause failure term was subtracted, and the RAW values of the component.

3 CONCLUSIONS

Expected RAW(ERAW) was introduced to improve the problem caused by the conventional RAW in the RIR&A such as the MRule and option 2. The relationship among ERAW, FV, and P were studied. The threshold value 0.01 of the ERAW for risk significant was derived, and a SSC having $FV < 0.005$ and $RAW < 2$ can be risk significant if its $ERAW > 0.01$. ERAW importance of a component is considered to be the sum of the ERAWs for the relevant failure modes of the component.

Table 3. Example importance summary.

Component failure mode	FV	RAW	ERAW
Valve "A" fails to open	0.002	1.7	0.0048
Valve "A" fails to remain closed	0.00002	1.1	0.00022
Valve "A" in maintenance	0.0035	1.7	0.0085
Common cause failure of valves "A" and "B"	0.004	n/a	n/a
Component importance	0.00952	1.7	0.0135

ACKNOWLEDGMENT

This research was supported by "The Mid-&-Long Term Nuclear R&D Program" of MOST (Ministry of Science and Technology), Korea.

REFERENCES

Cheok, M.C. et. al. 1998. Use of importance measures in risk-informed regulatory applications, *Reliability Eng. and System Safety*, Vol. 60, 1998.
Martorell, S. et. al. 1996. Safety-related equipment prioritization for reliability centered maintenance purposes based on a plant specific level 1 PSA, *Reliability Eng. and System Safety*, Vol. 52, 1996.
NEI. 2000. Industry Guideline for Risk-Informed Categorization and Treatment of Structures, Systems, and Components, NEI-00-04, Draft, March, 2000.
NUMARC. 1993. Industry Guideline for Monitoring the Effectiveness of Maintenance at Nuclear Power Plants, NUMARC 93–01, Rev.1, 1993. 5.
NRC. 1991. Requirements for Monitoring the Effectiveness of Maintenance at Nuclear Power Plants, 10 CFR 50.65, July 1991.
NRC. 1998. Options for Risk-Informed Revisions to 10 CFR part 50 – Domestic Licensing of Production and Utilization Facilities, SECY-98-300, December 23, 1998.
Vesely, W.E. & Davis, T.C. 1985. Evaluations and Utilizations of Risk Importances, NUREG/CR-4377, 1985, NRC.
Wall, I.B. & Worledge, D.H. 1996. Some Perspectives on Risk Importance Measure, *PSA '96*, Utah, U.S.A, 1996.

Safety and Reliability – Bedford & van Gelder (eds)
© 2003 Swets & Zeitlinger, Lisse, ISBN 90 5809 551 7

Development of full power risk monitoring system for the UCN 3, 4 nuclear power plants

S.H. Kim, S.C. Jang, K.Y. Kim & S.H. Han
Korea Atomic Energy Research Institute, Daejeon, Korea

ABSTRACT: Recently, risk informed regulations and applications (RIR&A) have been widely used which reduce costs by optimizing resources. One of essential RIR&A is the development of a computerized risk monitoring system which can evaluate risk in real time using plant configuration information such as inoperable equipments and environmental variables. KAERI has developed a full power risk monitoring system called DynaRM for the Korean standard nuclear power plant (NPP). As a risk monitoring system, two main modules have been developed to monitor the current risk of the plant and to assist the decision making for the maintenance schedule. This paper describes the development of DynaRM which is currently being tested in UCN 3, 4 NPPs. We expect that DynaRM will be a good advisory tool for the plant risk monitoring and maintenance scheduling.

1 INTRODUCTION

The need to develop the risk monitor (RM) for plant risk evaluation has been growing rapidly. For example, the US maintenance rule requires evaluating the impact caused by the maintenance activities before performing the maintenance works. So, the risk monitors are widely used for this purpose. In Korea, MOST (Ministry of Science and Technology of Korea) proposes the safety goals to prevent core damage, which require the preparation of a risk monitoring plan and its implementation for all nuclear power plants (NPPs) in Korea. The RM can be applied to the following fields

– Risk Informed Regulation and Application (RIR&A)
– On-Line Maintenance
– Maintenance Schedule Optimization
– Maintenance Rule
– Tech. Spec. Relaxation

Therefore, KAERI has been developing a full power risk monitor, called DynaRM. This paper describes the development of DynaRM for UCN 3,4 NPP.

2 DEVELOPMENT OF DYNARM

In 1990, KAERI developed the first conceptual design of RM called PEPSI, which performs risk evaluation based on the pre-calculated cut sets. Since 1997, KAERI has developed the DynaRM which is the first prototype and it is being installed at the UCN 3&4 NPPs for test purposes.

DynaRM consists of three major components; the RM Model, the RM Software and the Quantification Engine.

The RM Model is a one top model, which is converted from the UCN 3&4 licensed PSA model and it is developed under the PSA QA program. RM Software supports the risk monitoring and maintenance scheduling. We use the KIRAP as a quantification engine which was developed by KAERI and is widely used in the Korean nuclear energy field and other industrial fields.

2.1 *DynaRM model development*

2.1.1 *Status of the PSA for UCN 3&4 NPPs*
The PSA model used for DynaRM has been developed according to following three steps.

– PSA during Construction (1992–97)
  PSA was performed for the Operation License for UCN 3&4 which are the Korean Standard Nuclear Power Plant (KSNPP): 1000MW, PWR type.
– PSA Update (2000.7–2001.3)
  From 2000 to 2001, the previous PSA model was updated. Design changes and plant specific data were incorporated into this new model.
– Conversion of the PSA Model to Risk Monitoring (2001.3–2001.12)

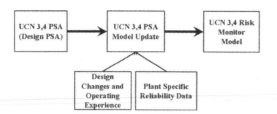

Figure 1. DynaRM model development.

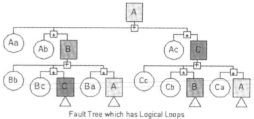

Fault Tree which has Logical Loops

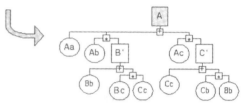

The cut set generation engine rebuilds a new fault tree which has the same logic as the original top gate but logical loops are removed

Figure 2. Solving logical loops in KIRAP.

The Updated PSA model is converted into the DynaRM model. It is developed as a basic tool for configuration risk management and is used for regulatory and non-regulatory risk-informed applications (RIA).

Figure 1 shows the flow diagram of the RM model development process.

2.1.2 Characteristics of UCN 3&4 RM Model
We used the one top model from two units, which is common practice in the rest of the world. The CDF and LERF are used as the surrogates for risk. DynaRM has unique characteristics related to the handling of the initiating events as shown below.

– The first one is the adoption of an impact model for system configuration change of the initiating event frequency, called IE Impact Model.
– The second one is the consideration of plant specific environmental factors (e.g., sea creature, sea surface temperature, typhoon, heavy snowfall, forest fire, etc.).
– The last one is the consideration of initiators affecting two units at the same time (e.g., Double LOOP, etc.).

We modified the PSA modeling assumptions according to plant practice, e.g.

– Realization of location of IE occurrence
– Alternate operating components
– The components shared between units
– Modeling in detail for the previous simplified FT

There are other features in this RM Model:

– the use of the plant-specific reliability data
– solving the circular logic of the fault tree automatically
– rule-based recovery model
– rule-based elimination of nonsense cut sets

It uses the plant-specific reliability data and it solves circular logic automatically. It includes a rule-based recovery model and rule-based elimination of nonsense cut sets.

KIRAP was developed by KAERI and is being used widely in Korea. It has high speed performance, stability and flexibility. It can solve multiple top events and reuse them. The unique characteristics of the KIRAP

quantification engine are that it can solve the fault tree, which has a logical loop, and perform the recovery analysis automatically.

Figure 2 illustrates how the logical loops is solved by the KIRAP.

2.2 DynaRM software development

DynaRM was developed for UCN Units 3&4 Risk Management. It supports real-time risk monitoring and maintenance scheduling so that it can be used as a risk-informed application supporting tool. DynaRM consists of two main modules; the operator module and the scheduler module.

The key roles of the DynaRM are as follows:

– The operator module calculates the current risk with plant configuration data, and indicates the current plant status whether it is in the safety range or not. If a plant configuration change occurs, DynaRM calculates the new risk level automatically.
– The scheduler module assists in maintenance scheduling. Scheduler module can be used to prevent the maintenance activities, which can cause the high peaks in the risk.

Figure 3 is a screen shot of DynaRM. The left one is the screen shot of the operator module and the right one is the screen shot of the scheduler module.

2.2.1 DynaRM operator module
The operator module evaluates current risk in real-time with variable plant configurations. When the plant configuration is changed by some equipment out of service, use of an alternative system, or the change of the environment, this information should

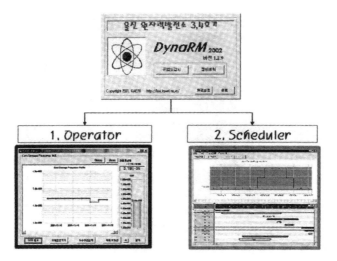

Figure 3. DynaRM operator and scheduler module.

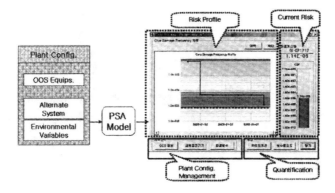

Figure 4. DynaRM operator module.

be input into the operator module to calculate the new core damage frequency.

For out of service equipment management, there are three different ways to update the status of out of service items. The inoperable equipments can be selected from the equipment list, by using the tagging system or by using the P&ID equipment selection system. That is, the selected out of service items change the status of the plant configuration and it brings about the recalculation of the core damage frequency.

The status change of an alternative system has an effect on CDF. We developed the status manager for an alternative system that supports the status change of an alternative system. The status changes of the environmental (external) variables such as sea animals, water temperature, climate, etc, have an effect on the CDF. We also developed an environmental variable management module.

The Figure 5 shows the quantification procedure of CDF. With the given plant configurations, DynaRM generates an IE frequency by IE impact models and after that DynaRM evaluates the core damage frequency with recovery actions.

2.2.2 DynaRM scheduler module

DynaRM scheduler module supports maintenance scheduling. It provides information on which components should be returned to service before particular maintenance activities are performed.

Figure 6 is the screen shot of the scheduler module. At the bottom of the screenshot are the maintenance task activities with Gantt chart and the upper chart shows the planned risk level.

The flow chart (Figure 7) is the scheduling diagram. DynaRM scheduler evaluates the risk with the current maintenance plan. When it is in the safe

903

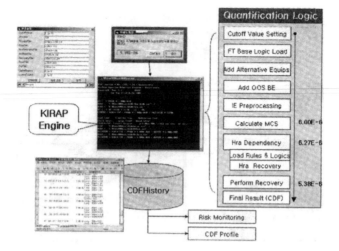

Figure 5. CDF quantification procedure.

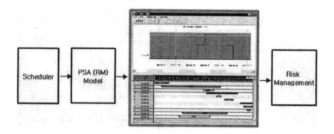

Figure 6. DynaRM scheduler module.

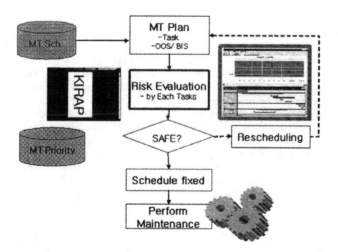

Figure 7. Maintenance scheduling.

range, the schedule is fixed and maintenance will be performed according to the schedule. However, if not, rescheduling is required to meet the safety goal.

3 CONCLUSION

KAERI has successfully developed a risk monitor, called DynaRM. DynaRM evaluates the risk of current plant configuration in real time and it can be used to optimize the maintenance schedule. DynaRM is the first risk monitoring software in Korea, it uses a Korean Standard NPP PSA Model, and it is being tested at the UCN 3&4 NPPs.

At this time, only the UCN 3 & 4 model is developed and applied, but the adaptation for the other NPPs is also easy with a little modification since DynaRM was developed by considering it to be used in the other NPPs. Moreover, we also expect DynaRM will be a good advisory tool for the plant risk monitoring and maintenance scheduling.

REFERENCES

1 Seung-Hwan Kim, et al., Development of Full power risk monitoring system for UCN 3&4 Nuclear power plants – KAERI/TR-2134/2002, 2002.
2 Kil-Yoo Kim, et al., Development of Computerized Risk Management Tool, Proceedings of the 5th Inter. Topical Meeting on Nuclear Thermal Hydraulics, Operations and Safety (NUTHOS), 1997.
3 C. H. Shepherd, Risk Monitor Overview (Draft 3) – A Report on the state of the art, 2002.
4 Seung-Hwan Kim, et al., The Present Status of the Development of Risk Monitor in Korea. OECD/NEA/ WGRISK Workshop on the development and use of Risk monitors, Madrid, 2002.
5 U.S. NRC, "Requirements for Monitoring the Effectiveness of Maintenance at Nuclear Power Plants", 10 CFR 50.65, July 1991.

Safety and Reliability – Bedford & van Gelder (eds)
© 2003 Swets & Zeitlinger, Lisse, ISBN 90 5809 551 7

Analysis of a dynamic system with the stochastics module MCDET and the deterministic code MELCOR

M. Kloos, E. Hofer, B. Krzykacz-Hausmann, J. Peschke & M. Sonnenkalb

Gesellschaft für Anlagen- und Reaktorsicherheit (GRS) mbH, Boltzmannstraße, Germany

ABSTRACT: A method of probabilistic dynamics was developed to achieve a more realistic modeling and analysis of complex dynamic systems. It was realized in the form of the stochastics module MCDET – a combination of Monte Carlo simulation and the Dynamic Event Tree method. MCDET works in tandem with any deterministic code that simulates the process dynamics of interest and is able to account for any stochastic influence expected to have essential effect on the dynamics. In a Level 2 PSA for a nuclear power plant, the combination of MCDET with the code MELCOR (USNRC 1997) was applied to analyze the possible accident consequences after a station black out. A brief description of MCDET was already presented in Peschke et al. (2002). The present paper wants to present more details of the vast amount of results obtained from the analysis.

1 INTRODUCTION

Complex dynamic systems are characterized by the interaction of man, machine, process and environment in the course of time. An analysis that intends to account for the stochastics (aleatory uncertainties) within this frame has to deal with a wide spectrum of possible dynamic situations.

The classical event tree analysis of Level 1 PSA and the accident sequence analysis of Level 2 PSA are not capable of adequately handling the tremendous amount of scenarios. They have to make simplifications and compromises. The trees largely evolve along a so-called effect line rather than along a time axis. Sequences in Level 1 event trees arise from the assumed order of system demands at set points. Sequences in Level 2 trees refer to a rather coarse grid of time (i.e. "early", "late" or "before", "after"), position (i.e. "top", "bottom") and magnitudes (e.g. "small", "medium", "large").

The quality of the classical PSA analysis obviously depends on the selection of scenarios out of the possible spectrum. This involves the risk, that important sequences remain unknown, but also that quite unrealistic sequences are generated. The latter case may happen whenever conditions, that have to be specified for a dynamic sequel, differ from the actual preceding situation. Probabilistic dynamics methods do not have these deficiencies. In their framework, event sequences evolve over time. They account for the interaction between stochastics and dynamics according to

a deterministic model of the process dynamics combined with a probabilistic model of the stochastics.

2 FEATURES OF MCDET

A transition in a dynamic system can be characterized by the time "when" it occurs and by the system states "where to" it may go. The probabilistic analysis considers each of the two characteristics as either a deterministic parameter or as a random variable. The random variable may be discrete or continuous. Most of the probabilistic dynamics methods presented in the literature (Labeau P.E. et al. 2000, Siu N. 1994) are able to adequately handle transitions with deterministic or discrete random "when" and "where to". But they lack a satisfactory treatment of transitions with continuous random "when" and/or "where to". MCDET closes this gap. Discrete random occurrence times or system state changes are generally taken into account by the dynamic event tree analysis. Continuous random occurrence times or system state changes are handled by Monte Carlo simulation.

MCDET operates in tandem with any deterministic code that simulates the process dynamics of interest. The tandem generates a dynamic event tree for each sample obtained by Monte Carlo simulation. It calculates the sequence probabilities of each tree and computes the time histories of the process quantities for each sequence. The calculation is controlled in a way that sequences which have a probability less than a user defined threshold value are ignored, and that sections shared by sequences of a tree are computed only once. With each event tree, a probability distribution is provided for every process quantity and at all points of time. The probability distributions are conditional on the initiating event and on the randomly sampled values from the Monte Carlo simulation. The respective mean distributions over all trees of the Monte Carlo sample are the final outcome of the analysis. They are calculated together with confidence intervals that give an indication of the possible sampling error, with which the mean probability values may estimate the true probabilities. MCDET also permits to perform an approximate epistemic uncertainty and sensitivity analysis (Hofer et al. 2002).

3 ILLUSTRATIVE APPLICATION OF MCDET WITH MELCOR

The current application of MCDET with MELCOR gives a thorough insight into the spectrum of dynamic situations that may follow a total station black out in a reference nuclear power plant (1300 MWe pressurized water reactor of Konvoi type). Due to the total loss of power (also from emergency diesels and other sources), the main coolant pumps and all operational systems fail. Batteries guarantee power supply to all battery supported functions for some period of time. Scram and turbine trip are performed automatically. The pressurizer relief valve and the two safety valves perform the automatic pressure limitation. The primary side pressure release (accident management measure) is initiated by the crew some time after the corresponding system signal. Once the pressure on the primary side has decreased far enough, the accumulators can inject their coolant inventory. In this illustrative application, it is assumed, that external power is restored not before 5700 s, but certainly before 12000 s after the initial event. As soon as the power supply is available, the high and low pressure emergency coolant injection systems can be activated. The four trains are reconnected to the grid one by one, each requiring some preparation time.

Aleatory uncertainties concern the functioning of participating system components (pressurizer relief valve and safety valves, source isolation valves, and additional isolation valves of the accumulator system, high and low pressure safety injection pumps) and some occurrence times (time of the demand cycle with failure of the corresponding pressurizer valve, initiation time of the primary side pressure release, time of external power supply recovery).

Depending on the dynamic situation, the core may experience gradual damage. High system pressure in combination with the high temperatures associated with core melting may lead to the failure of main coolant piping in the hot leg or to the failure of the

pressurizer surge line before the reactor pressure vessel (RPV) integrity is lost. The particular interest of the analysis lies therefore in dynamic quantities like the pressure in the vessel and in the containment, the core exit temperature as well as the total melt mass and the generated hydrogen (H_2) mass as indicators for the degree of core degradation.

Although MCDET is capable of accounting for aleatory and knowledge (epistemic) uncertainties, in this application only aleatory uncertainties were considered. The identified epistemic quantities were represented by their mean values. An aleatory sample of 50 dynamic event trees was generated. Sequences of a tree with a conditional probability below a threshold value of 1.E-04 were not calculated. (The sequence probability is conditional on the total station black out, the power supply recovery between 5700 s and 12000 s, and on the randomly sampled values from the Monte Carlo simulation.)

Each generated tree may be represented as a set of alternative event sequences over time (time-event-plane) and as a set of alternative time histories for each process quantity or state variable (time-state-plane).

4 DYNAMIC EVENT TREE IN THE TIME-EVENT PLANE

Figure 1 shows the time-event-plane for the first 34 (out of a total of 251) sequences of dynamic event tree no. 7 of the sample. In this presentation of a time-event-plane, the vertical lines at branch points are omitted for the purpose of clarity. Instead, each branch is given an identifier at its beginning composed of the sequence (path) number before the slash and the number of its origin (path where it branches off) after the slash. At the end of each path, the path number is repeated together with the conditional path probability. On the time axis, at the bottom of the tree, the randomly chosen demand cycle, when a failure may occur, is indicated for all three pressurizer valves (denoted by AV for the relief valve, by SV1 for the 1st and by SV2 for the 2nd safety valve). Further marks on the time axis (denoted by S1, S2, S3 and S4) are the times of reconnection for the four emergency coolant trains, which are derived from the random time of power restoration (at 9548 s in tree no. 7) plus the preparation times for the reconnection. These preparation times were identified as knowledge uncertainties with a mean value of 750 s for the 1st train and of 450 s for each of the other trains.

Sequence no. 0 – directly above the time axis in Figure 1 – is the base sequence. It is shown from 4500 s onward, since none of the stochastic events in tree no. 7 lead to a branching before this time. Along the base path, the pressurizer valves operate correctly for pressure limitation, and the pressure release of the primary

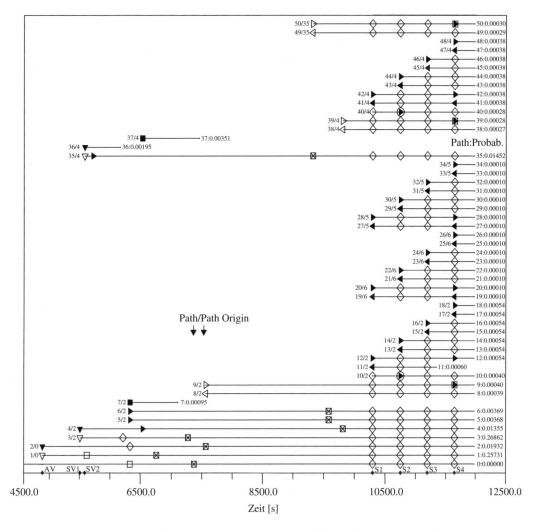

Figure 1.   Time-event plane for the first 34 (out of a total of 251) paths of dynamic event tree no. 7 of the sample.

side (bleeding) is performed after a randomly chosen delay time. All valves open (as is indicated by the open square symbol □) between 6000 s and 6500 s. Once the pressure has decreased far enough, the accumulators can inject coolant. Between 7000 s and 7500 s, the source isolation valves and the additional isolation valves for the hot leg accumulator injection open on demand (pointed out by the crossed square ⊠). The four diamonds ◇ following next on the base path indicate the successful reconnection of each of the four high and low pressure injection pumps and of the opening of the corresponding source isolation valves both in the hot and cold legs.

The paths 1 and 2 of tree no. 7 branch off path 0 at 4886 s – the time of the demand cycle no. 31 of the

pressurizer relief valve. MCDET randomly selected this cycle as the possible failure cycle of the relief valve in tree no. 7. In path 1, the relief valve fails to close (▽), and in path 2, it fails to open (▼). Along paths 1 and 2, everything else functions as intended. Possible failure events lead to branch points on these paths. For instance, in paths 3 and 4, which branch off path 2, the 1st safety valve fails (to close in path 3 (▽), to open in path 4 (▼)) in addition to the stuck closed pressurizer relief valve.

The first diamond ◇ on path 2 points out the initiation of the bleeding operation with reduced total diameter (only the two safety valves open). Paths 5 to 7 represent possible alternatives of the bleeding initiation taking place in path 2. While still one of the safety

909

valves opens in path 5 and 6 (indicated by ▶), both remain closed in path 7 (■). The calculation of path 7 of tree no. 7 stopped automatically before the end, because the critical pressure of 22 MPa was reached in the RPV. It was not intended in the current application to investigate the dynamics following this special situation.

The time of the accumulator injection depends on the extent of pressure decrease. In paths 4 to 6 of tree no. 7, it occurs (⊠) between 2000 s and 3000 s later than in paths 0 to 3. Paths 8 and 9 are possible alternatives of the accumulator injection taking place in path 2. In path 8, the additional isolation valve of the accumulator connected to hot leg no. 4 does not open (◁), in path 9, the source isolation valve in the line to hot leg no. 1 erroneously remains closed (▷).

If a source isolation valve does not open, neither an injection from the corresponding accumulator into the respective (hot or cold) leg nor a high pressure emergency injection by the respective train can take place. In path 9, the missing high pressure injection into leg no. 1 is indicated by the symbol ▶| in the point S4. The emergency coolant train no. 1 is reconnected last because of its failed isolation valve in the hot leg.

Paths 10–34 of tree no. 7 account for possible failure events in the emergency core coolant system. They branch off path 2, 6, or 5, respectively. The filled triangle ▶ at the points S1, S2, S3, or S4 indicates, that the high pressure injection cannot occur, because the corresponding pump does not work (▶ alone), or either a hot (▶ within a square) or a cold leg (▶ within a circle) source isolation valve does not open. The other filled triangle ◀ at the points S1, S2, S3, or S4 indicates, that the low pressure injection cannot occur, if required.

The paths 35–251 not shown in Figure 1 describe further dynamic situations, which evolved on the conditions for tree no. 7 of the sample.

5 DYNAMIC EVENT TREE IN THE TIME-STATE PLANE

Most of the figures in this chapter represent the dynamic event tree shown in Figure 1 in the time-state plane for selected process quantities. The evolution of the quantities is shown nearly from the time onward, when the pressurizer relief valve is demanded the first time (4048 s after the initial event) for automatic pressure limitation. The possible effects of stochastic events occurring in the process of pressure limitation (up to 7000 s) are evidently reflected by the time histories of the pressure in the RPV in Figure 2.

As indicated by the oscillations, the valve automatically opens and closes many times until the requested primary side pressure release is finally performed. Altogether 88 demand cycles are counted for the relief valve in tree no. 7 of the sample. (The number of oscillations in the figure differs from the actual number,

Pressure in the RPV [Pa]

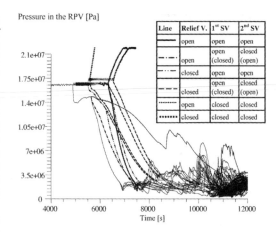

Line	Relief V.	1st SV	2nd SV
——	open	open	open
—·—·	open	open (closed)	closed (open)
—··—	closed	open	open
— —	closed	open (closed)	closed (open)
··········	open	closed	closed
▪▪▪▪▪▪▪	closed	closed	closed

Figure 2. Dynamic event tree no. 7 of the sample in the time-state plane for the pressure in the RPV [Pa].

because the figure does not represent every time step of the process.) The probability, that the valve fails to open or to close, is relatively high at the 1st demand (P1 = 0.06), lower at the 2nd (P2 = 0.1 * P1), and linearly increasing with each further demand (P50 = P1, P100 = 5 * P1). In view of the high number of demands, it is very likely, that the relief valve fails in the course of pressure limitation. From all the possible demand cycles, at which the relief valve might fail, MCDET randomly selected the cycle no. 31 for tree no. 7. It occurred 4886 s after the initial event. The failure in open state at this time leads to an immediate steep descent of the pressure, that is succeeded by a short-time increase.

The following three situations after nearly 5600 s express the possible consequences of aleatory uncertainties in the primary side pressure release. The bleeding operation is induced by the crew some continuous random time after the corresponding signal, and its effect depends – beside the delay time – on the success of the crew to open the two safety valves. Without success, the open relief valve alone would not be sufficient to reduce the pressure fast (gradual decrease). But if the crew can open at least one of the safety valves, the pressure relief is clearly speeded up. The evolution is the same with either only the 1st or the 2nd valve in open state. The steepest descent is reached with the full opening of both safety valves.

As soon as the relief valve erroneously remains closed, the system uses the 1st safety valve for pressure limitation. The aleatory uncertainties with respect to the 1st and also the 2nd safety valve are comparable to those concerning the relief valve except that the probabilities for a failure to open are at a lower level (determined by the probability for a failure to open at the 1st demand, P1 (fail to open) = 0.006. If the 1st

Pressure in the RPV [Pa]

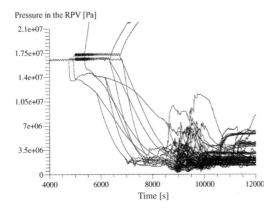

Time [s]

Figure 3. Dynamic event tree no. 2 of the sample in the time-state plane for the pressure in the RPV [Pa].

RPV fluid level [m]

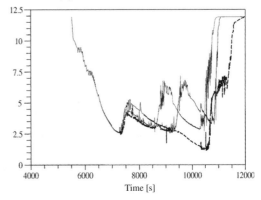

Time [s]

Figure 4. 3 paths of dynamic event tree no. 7 of the sample in the time-state plane for the RPV fluid level [m].

safety valve operates as required, it is able to keep the pressure at about the same level as the relief valve (as indicated by the two overlapping oscillations). If it fails in open state (demand cycle no. 40 at 5508 s), the pressure decreases immediately. The successful initiation of the bleeding (at 6530 s), where the 2nd safety valve can be opened, accelerates the pressure relief.

The 2nd safety valve automatically starts operating for pressure limitation as soon as in addition to the relief valve also the 1st safety valve fails to open. After the pressure slightly goes up, the normally operating 2nd safety valve is able to keep this level. If it fails to open, the pressure rapidly increases and finally passes a critical value. It is expected, that this extreme condition might cause the failure of other components of the reactor system and lead again to a pressure relief. As mentioned above, these alternative cases were not considered in the current application. The calculation terminated as soon as the critical value was reached.

At this point, it should be mentioned that this illustrative application provided some unexpected results, which require detailed investigation by the experts in order to decide, whether they are due to the particular model representation of the plant in MELCOR, or whether they are founded in the physics of the accident. The short-time pressure increase, after the relief valve failed in open state, is an example of a result that cannot be satisfactorily explained so far. Comparisons with other trees showed, that the increase always follows the early failure of the relief valve in open state. Early failures of the safety valves in open state or a later opening of the relief valve for the purpose of bleeding do not have this consequence. The fact, that the opening diameter of a safety valve is twice as large as that of the relief valve, should also be considered in this context. Figure 3 shows, for instance, dynamic event tree no. 2 of the sample, where the

relief valve fails 4757 s and the 1st safety valve 4938 s after the initial event.

As long as the demanded pressurizer valve does not erroneously remain closed, none of the other pressurizer valves is requested for pressure limitation. When the core exit temperature exceeds 400°C and the fluid level passes below 7.5 m, the signal for primary side pressure release is transmitted. Because the core exit temperature is already above 400°C, when the pressurizer relief valve is demanded the first time, the fluid level determines the time of the signal in the current application. The effect of the bleeding initiation is reflected by the extent of pressure decrease. The more valves that can be opened, the faster the pressure relief. It is assumed, that the crew does not succeed in opening a valve that already failed in closed state in the course of automatic pressure limitation.

The accumulators can inject their coolant inventory (34 m^3 each), when the pressure in the reactor coolant loop is below 2.6 MPa. In dynamic event tree no. 7, this is after 6760 s, at the earliest. The injection requires that the source isolation valves and the additional isolation valves of the accumulator system open on demand.

Figure 4 shows, how the RPV fluid level may be affected by the accumulator injection. It represents three time histories of tree no. 7 with the same preceding dynamic situation before 7270 s. The relief valve failed in closed state and the 1st safety valve in open state during the process of pressure limitation, the 2nd safety valve is open since the time of primary pressure release. The solid curve reflects the evolution of the fluid level, if the components of the accumulator system operate normally. It refers to path 3 of tree no. 7 (Fig. 1). The dotted line represents the evolution, if there is no injection into hot leg no. 1, while the broken line indicates the evolution, if there is no injection into hot leg no. 4. The corresponding curve shows,

Core exit temperature [K]

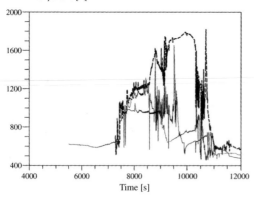

Time [s]

Produced H₂ mass [kg] in core

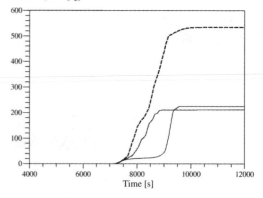

Time [s]

Figure 5.   3 paths of dynamic event tree no. 7 of the sample in the time-state plane for the core exit temperature [K].

Figure 6.   3 paths of dynamic event tree no. 7 of the sample in the time-state plane for the produced H_2 mass [kg] in core.

that, apart from the first injection into the hot legs 1, 2, and 3, which occurs at 7260 s, all following injections are not able to raise the fluid level significantly. The comparison with the dotted curve may lead to the conclusion, that the reason for this behavior is not so much the number of available legs, in which the fluid inventory can be injected, but their location. The fact, that the pressurizer is connected to leg no. 1, might be a possible cause for that. It should also be considered in this context, that the modeling of the reactor coolant loops may have an essential effect. The model is simplified in a way, that the three loops which are not connected with the pressurizer are lumped together.

Figure 5 represents the evolutions of the core exit temperature that correspond to the selected time histories in Figure 4 (in the same line pattern). It is remarkable, that the accumulator injections into the legs 1, 2, and 3 (broken line) are not able to reduce the temperature, but lead – on the contrary – to a significant increase. This might be caused by the larger amount of steam available for exothermal zircaloy oxidation process in the core after evaporation of the injected water. In addition, the amount of water injected in the different phases is relatively small, because the injection is stopped automatically due to the pressure increase caused by the evaporation of the water after contact with hot core structures.

At the end, it is the large amount of coolant from the high and/or low pressure safety injection system, that is capable of reducing the temperature and of stopping the core degradation. The emergency coolant injection system of a train can operate as soon as the power supply is available and the respective train is reconnected to the grid. In the current tree, the first coolant injection occurs 10300 s after station black out.

A high core exit temperature is an indication of the ongoing core damage. Figure 6 shows, that the

produced H_2 mass is more than twice as high in the scenario with accumulator injection into the legs 1, 2, and 3 (broken line) as in the scenarios with a successful injection into all legs or in the legs 2, 3, and 4 (dotted line). Again, there may be some effect of the nodalisation scheme used, which has to be investigated.

6   PROBABILITY DISTRIBUTIONS

The PSA requires the probability distribution of relevant process quantities at specific points of time. The probability distributions calculated in the current analysis are conditional on the station black out, the external power restoration between 5700 s and 12000 s, and on the randomly sampled values from the Monte Carlo simulation. Figures 7–8 show the distribution of the maximum core exit temperature up to 12000 s as obtained from dynamic event trees no. 2 and no. 7 of the sample. The values of the cumulative probability remain below 1.0, because sequences with a probability below a threshold value of 1.0E-04 were not calculated in this application. The difference to 1.0 is the sum of the probabilities of all sequences ignored in the respective trees.

The main stochastic influences that determine the difference between the trees no. 2 and no. 7 are the failure times of the pressurizer valves in the course of pressure limitation (Figs 2–3), the time span between the signal for primary side pressure release and its actual initiation, the time of power supply recovery (it is at 7625 s in tree no. 2, and at 9548 s in tree no. 7), and the source and additional isolation valves, which might fail.

The distribution derived from tree no. 2 (Fig. 7) provides, for instance, the statement, that the maximum core exit temperature does not exceed 1700 K with a conditional probability of 0.82, while the

Probability

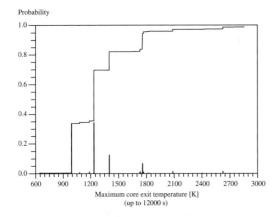

Figure 7. Conditional probability distribution of the maximum core exit temperature up to 12000 s from dynamic event tree no 2 of the sample.

Probability

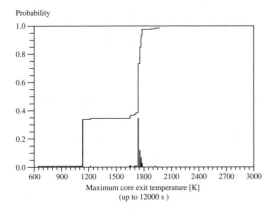

Figure 8. Conditional probability distribution of the maximum core exit temperature up to 12000 s from dynamic event tree no 7 of the sample.

Probability

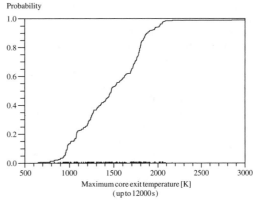

Figure 9. Mean conditional probability distribution of the maximum core exit temperature up to 12000 s from all 50 dynamic event trees of the sample.

Probability

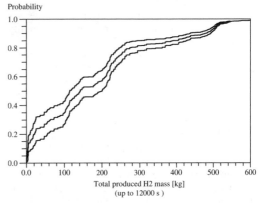

Figure 10. Mean probability distribution of the total produced H_2 mass up to 12000 s from all 50 dynamic event trees of the sample together with 90% confidence intervals.

distribution derived from tree no.7 (Fig. 8) shows, that the corresponding probability is only 0.35.

The actual result of the probabilistic dynamics analysis with MCDET and MELCOR is the mean distribution over all 50 dynamic event trees of the sample. It provides an estimate for the solution of the probabilistic dynamics equations (Devooght J. & Smidts C. 1996). This solution may be defined as the mean cumulative (or complementary cumulative) probability at any (e.g. core exit temperature) value obtained from the infinitely large population of dynamic event trees. Figure 9 represents the mixture distribution of the maximum core exit temperature up to 12000 s derived from all 50 dynamic event trees of the sample. This distribution provides, for instance,

a probability value of 0.65 for the maximum core exit temperature not to exceed 1700 K.

Confidence intervals give an indication of the possible sampling error, with which the resulting mixture distribution may estimate the solution of the probabilistic dynamics equations due to the finite Monte Carlo sample. Figure 10 shows the mixture distribution of the produced H_2 mass up to 12000 s derived from all 50 dynamic event trees of the sample together with 90% confidence intervals.

7 CONCLUSIONS

MCDET is capable of treating discrete random as well as continuous random transitions of a dynamic event

913

tree. Discrete random transitions are generally taken into account by the dynamic event tree analysis, continuous random transitions are handled by Monte Carlo simulation. Each set of sample values from Monte Carlo simulation leads to a new dynamic event tree.

MCDET operates in tandem with any deterministic code that simulates the process dynamics of interest. The tandem generates the sequences of a dynamic event tree on the basis of the underlying dynamics code, the specified discrete random transitions, and the sample values for the specified continuous random transitions. For all trees, each sequence evolves over time and fully accounts for the interaction between stochastics and dynamics. The calculation is controlled in a way, that sections shared by sequences of a tree are computed only once. The tree construction as well as the Monte Carlo simulation can fully exploit the computational capacity of a system of parallel compute nodes.

MCDET provides approximate solutions of the probabilistic dynamics equations. Apart from the probability threshold for the tree construction, the approximation error may be quantified by confidence intervals for cumulative probabilities of any dynamics quantity and at all points of time.

MCDET is also capable of accounting for knowledge uncertainties in the frame of a probabilistic dynamics analysis. An approximate epistemic uncertainty and sensitivity analysis is described in Hofer et al. (2002).

The long term goal of methods development for probabilistic dynamics analysis should be the capability to perform a dynamic PSA (Labeau P.E. et al. 2000). The development of an MCDET module that accounts for the interaction between the operating crew and the dynamics is the next important step in this direction. An example for such a module is published in Chang Y.-H. & Mosleh A. (1998) and in Shukri et al. (1997).

ACKNOWLEDGEMENTS

The development of the stochastics module MCDET was sponsored by the German Ministry of Economics and Technology (BMWi) within the framework of the project RS 1111. The current development of an MCDET module that accounts for the interaction between the operating crew and the dynamics is sponsored by the German Ministry of Economics and Labour (BMWA) within the framework of the project RS 1148.

REFERENCES

Chang Y.-H., Mosleh A. 1998. Dynamic PRA Using ADS with RELAP5 Code as Its Thermal Hydraulic Module, *Proceedings of PSAM-4, A. Mosleh and R.A. Bari (Eds), Springer Verlag*, Berlin.

Devooght J., Smidts C. 1996. Probabilistic dynamics as a tool for dynamic PSA. In *Reliability Engineering and System Safety (52)*: 185–196.

Hofer E., Kloos M., Krzykacz-Hausmann B., Peschke J., Woltereck M. 2002. An Approximate Epistemic Uncertainty Analysis Approach in the Presence of Epistemic and Aleatory Uncertainties. In *Reliability Engineering and System Safety (77)*: 229–238.

Labeau P.E., Smidts C., Swaminathan S. 2000. Dynamic reliability: towards an integrated platform for probabilistic risk assessment, In *Reliability Engineering and System Safety (68)*: 219–254.

U.S. Nuclear Regulatory Commission 1997. "MELCOR 1.8.4 – Reference Manual", USA, Washington.

Peschke J., Hofer E., Kloos M., Krzykacz-Hausmann B., Sonnenkalb M. 2002. Probabilistic dynamics in level 2 PSA: An application of the stochastics module MCDET with the deterministic dynamic code MELCOR. In *Proceedings of PSAM 6, Elsevier Science*. ISBN CD-ROM Edition: 0-0804-4120-3.

Shukri T., Mosleh A., Shen S.H. 1997. Implementation of a Cognitive Human Reliability Model in Dynamic Probabilistic Risk Assessment of a Nuclear Power Plant (ADS-IDA), *UMNE-97*, Center for Technology Risk Studies, University of Maryland, College Park.

Siu N. 1994. Risk assessment for dynamic systems: An overview. In *Reliability Engineering and System Safety (43)*: 43–73.

Safety and Reliability – Bedford & van Gelder (eds)
© 2003 Swets & Zeitlinger, Lisse, ISBN 90 5809 551 7

Derivation of accident sequence conditions for complex systems using global system behavior model

T. Kohda & K. Inoue

Kyoto University, Kyoto, Japan

ABSTRACT: In the safety management and the safety design of technological systems, it is essential to obtain accident sequence conditions. FTA (Fault Tree Analysis) is conventionally used to obtain system failure conditions, but it does not consider event sequence conditions. Further, a FT (Fault Tree) is usually constructed based on the judgment and experience of system analysts, which may yield an erroneous result or omission of serious failure conditions. To obtain objective accident sequence conditions, this paper proposes a systematic method based on the global system behavior model. Utilizing the system bond graph model with information flow as a system model, causal relations among system states can be identified easily. Backward search from the specified system failure state gives a possible disturbance propagation path, which can cause the system failure. By evaluating the effectiveness of a protective action related to the disturbance propagation path, accident sequences can be obtained as the occurrence of disturbance propagation path after the failure of their effective protective actions. An illustrative example shows the details of the proposed method.

1 INTRODUCTION

For the safety management and the safety design of technological systems, it is essential to obtain accident sequence conditions. FTA (Fault Tree Analysis) (U.S. NRC, 1981) is conventionally used for this purpose. However, FTA obtains system failure conditions based on the fault tree (FT), a static logical relation between the system failure and component failure. Further, a FT, which is constructed based on the judgment and experience of system analysts, may yield an erroneous result or omission of serious failure conditions. To obtain objective accident sequence conditions, this paper proposes a systematic method based on the global system behavior model.

The global system model is composed of two types of system behaviors. One is the physical behavior, which must satisfy the natural constraints, or obey physical laws such as Newton's law for the dynamical behaviors. The other is a control behavior, which can be constructed logically as expected. For modeling the physical behavior, bond graphs (Rosenberg, Karnop, & Margollis, 1990) are suitable, because (1) they can model various systems such as mechanical, electrical, and hydraulic systems in a unified way from the viewpoint of energy flows, (2) the intuitive correspondence between bond graph elements (BGE's) and physical behaviors/components makes it easy to construct a system behavior model based on the design assumptions,

(3) their stepwise refinement allows easy model modification and (4) the model can satisfy the physical constraints such as energy conservation.

For the control behavior, functional modeling approach (IFMA), which can model any control action by identifying the designer-defined overall goal it must achieve and the designer/user-defined function it must perform. To obtain causal relations among control signals, the functional model can be simplified into the information flow model such as causal relations between an action and its condition. This information can be summarized as a set of if-then rules. To obtain event sequences using the global system behavior model, causal relations between these two types of behaviors must be specified in the entire system behavior. This is represented as the effect of a control behavior on process variables.

Firstly, the global system model is introduced. Then, how to obtain the accident sequence conditions using the global system model is given. An illustrative example shows the details of the proposed method.

2 GLOBAL SYSTEM MODEL

2.1 *Physical behavior model*

Firstly, the physical system behavior model is constructed using bond graphs. Correspondence between

Table 1. Correspondence between BG and physical systems.

BG	Electrical system	Hydraulic system	Mechanical system
Bond	Electric wire	Pressure pipe	Shaft, Rod
Effort	Electric voltage	Pressure	Force, Torque
Flow	Electric current	Volume flow	Velocity
SE	Voltage source	Pump with pressure regulator	Diesel engine
SF	Current source	Hydrostatic pump	Driving shaft
R	Resistance	Restriction	Friction, Damper
C	Condenser	Hydraulic accumulator	Spring, Compliance
I	Inductor	Mass action, Fluid inertia	Mass, Inertia
TF	Electric transformer	Hydrostatic pump, Cylinder	Gear reducer
GY	Electric DC machine		
0	Parallel circuit	Parallel circuit	Planetary gear
1	Serial circuit	Serial circuit	Lever type joint, Fixed reducer

BGE's and physical phenomena allows the development of a physical system behavior model consistent with design assumptions on normal and abnormal component behaviors. General variables "effort" (e) and "flow" (f) are commonly used whose product is power, common concept connecting these systems. Table 1 summarizes correspondence between basic BGE's and physical phenomena/components. Note that the energy conservation is represented by 0 (zero junction) and 1 (one junction), which correspond to series and parallel circuits, respectively.

The following assumption is made on the system bond graph (SBG).

A-1: The SBG is composed of basic BGE's: source of effort (SE), source of flow (SF), resistance (R), capacitance (C), inertance (I), transformer (TF), gyrator (GY), 0-junction (0), and 1-junction (1). The characteristic function for each BGE is shown in Table 2.

For the SBG under assumption 1, state variables are displacement (integral of flow with respect to time) of a bond connecting to a C-element and momentum (integral of effort with respect to time) of a bond connecting to an I-element, while input variables are effort of a bond connecting to an SE-element and flow of a bond connecting to an SF-element (Karnopp, Margolis & Rosenberg 1990). Based on the causality of the SBG, all efforts and flows can be expressed in functional form of state variables and input variables (Kohda, Nakada, Kimura, & Mitsuoka, 1988). Here, a functional form is defined in such a way that y is represented as f(x) if variable y is given as variable x operated by function f. The functional form shows the causal sequence of physical phenomena. For example, g(f(x)) denotes that variable x is affected by physical phenomena f and g, sequentially. From the definitions of displacement and momentum, the system state

Table 2. BGE and characteristic function

BGE	Symbol	Function	BGE	Symbol	Function
SE	SE—$\vdash^i$	$e_i = E(t)$	TF	$\dashv^i$TF—$\vdash^i$	$e_i = e_j/n$, $f_i = f_j/n$
SF	SF—$\vdash^i$	$f_i = F(t)$		$\vdash^i$TF—$\dashv^i$	$e_j = n\,e_i$, $f_j = n_i f$
R	R—$\dashv^i$	$e_i = R(f_i)$	GY	$\dashv^i$GY—$\vdash^i$	$f_j = e_i/m$, $f_i = e_j/m$
	R—$\vdash^i$	$f_i = R(e_i)$		$\vdash^i$GY—$\dashv^i$	$e_i = m\,f_j$, $e_j = m_i f$
C	C—$\dashv^i$	$e_i = C(\int f_i\, dt)$	0	$\dfrac{i1-0-im}{i2}$	$e_{ij} = e_{ik}$ for any ij, ik; $\Sigma\, f_{ij} = 0$
I	I—$\vdash^i$	$f_i = L(\int e_i\, dt)$	1	$\dfrac{i1-1-im}{i2}$	$f_{ij} = f_{ik}$ for any ij, ik; $\Sigma\, e_{ij} = 0$

equations are obtained as:

$$\frac{dX_i}{dt} = \begin{cases} e_{b(i)}, & \text{if } X_i \text{ represents momentum,} \\ f_{b(i)}, & \text{if } X_i \text{ represents displacement,} \end{cases} \tag{1}$$

where b(i) denotes bond corresponding to state variable Xi. The system state equations obtained from the SBG can give causal relations among component behaviors and system state variables clearly, which makes the system analyst understand the effect of component failure on the system physical behavior easily.

2.2 Information flow model

To maintain the system state as expected, various control systems are installed. For the evaluation of the system behavior under an abnormal condition, the effect of protective actions and control actions must be considered. These actions can be represented by the information flow, which defines input–output relations between conditions and actions. Based on the protection control logic and the decision framework of operation actions, the information flow can be identified as a kind of control signal flow among physical components. Its behavior is not constrained by the physical boundary conditions. Moreover, its action can be made flexible enough to meet their requirement. These actions are represented in terms of input-output logical relation. For example, the behavior of pressure relief valve can be represented as:

If the monitored pressure is larger than the preset pressure, then the relief valve is to be opened.

Otherwise, the relief valve is to be closed.

Since the preset pressure can be specified as desired, the behavior of the relief valve can be modified freely. Similarly, the computerized control system can be constructed as the designer expects. These behaviors do not need to conform to physical constraints. Human operator action based on the operation manual or the emergency procedure can be also represented as the input-output relation. For example, consider a case where the role of an operator is to shutdown the input flow based on the monitored pressure. His normal operation is represented as:

If the monitored pressure is above the threshold value, then the operator closes the valve.

Otherwise, the operator does nothing. Based on this expression, whether the operator behaves normally or not can be determined from the states of the pressure indicator and the valve state.

2.3 Component failure

Component failure/disturbances considered in this analysis are assumed as:

A-2: The effect of a failure is represented as either a deviation of process variable (external disturbance) or a deviation of a component function (component failure).

A hardware component failure can be represented as a deviation of its normal characteristic function. This assumption can be also applied to human errors, which can be represented as a deviation from the normal operation procedure.

For the derivation of the system failure condition, all possible component failure/external disturbances must be assumed beforehand so that the global system state equations can express any failure state condition to be considered. Especially, failures causing additional physical phenomena other than normal behaviors must be added to the SBG or information flow model. For example, consider a leakage failure represented as a deviation of characteristic function of pipeline and tank. If a tank is represented as a C-element corresponding to the storage of flow in a conventional way, its characteristic function must have infinite capacitance to represent its leakage failure. This is not appropriate.

A R-element representing loss of flow must be added to the SBG. Thus, for system failure analysis, the system behavior model must include not only normal behavior, but also all possible failure behaviors.

2.4 Causal relation

Based on this global system behavior model composed of the SBG and the information flow, the causal tree representing cause-effect relations in the entire system can be obtained as follows:

1. Based on the system state equation obtained from the SBG, the relation among system state variables can be obtained as a tree graph, whose nodes and branches represent process variables in the SBG, and cause-effect relations.
2. Considering the effect of a control action/protective action on the controlled variables, its causal relation must be added to the above tree graph representing cause-effect relations in the global system.

3 ACCIDENT SEQUENCE CONDITIONS

The derivation of accident sequence conditions is composed of three steps: the first step is to derive a trigger condition (or disturbance propagation path) leading to system failure. The second step is to derive a complementary condition for the disturbance propagation path to lead to system failure based on the evaluation of its corresponding protective actions. Finally, event sequence conditions are obtained by the examining whether the failure of an effective protective action must occur before the occurrence of the disturbance propagation path.

3.1 Disturbance propagation path

In deriving system failure conditions using the global system behavior model, system failure must be defined as the deviation of a process variable. For example, the explosion of a reservoir is represented as an overpressure in the reservoir. From this system abnormal state, the backward search of its cause is performed using the causal tree. The causal tree obtained from the global system state equations shows cause-effect relations among component behaviors and process variables.

Disturbance propagation paths can be classified into two types:

1. a disturbance caused by an external disturbance such as a deviation of input energy,
2. a disturbance caused by a component failure such as a malfunction of a negative feed-back control system.

In deriving a disturbance propagation path, the backward search can identify the above types of disturbances that can cause a specified deviation at the system state.

3.2 Evaluation of protective actions

Generally speaking, the system has several kinds of protective systems/control systems, which can prevent some abnormal events or maintain the system state as desired. Consider two types of protective or control actions: negative feedback loop and negative feed-forward loop. The former can prevent any external disturbance monitored at the sensing point of the loop, while the latter can prevent nothing but disturbance entering the sensing point of the loop. Human operators can also take an appropriate protective action when they detect an abnormal event. Thus, whether a protective action can prevent a disturbance propagation path can be considered at the first stage by examining whether the protective action can detect the effect of a disturbance propagation path or not. Here, a protective action means either a protective/control system behavior or operator action. If no protective action can detect a deviation caused by the disturbance propagation path, the disturbance propagation path can cause the system failure by itself. Otherwise, its effectiveness must be evaluated.

Depending on the effectiveness of all identified protective actions on the disturbance propagation path, the system failure occurrence condition can be obtained as follows:

If no protective action can, the disturbance propagation path can cause system failure by itself. The system failure occurrence condition in this case is obtained as he condition presented by the disturbance propagation path.

If some protective system can, its protection failure must be accompanied for the disturbance propagation path to lead to system failure.

For the derivation of failure condition of protective actions, Event Tree Analysis (ETA) (Henley & Kumamoto, 1991) is applied, which can consider the dependency among protective actions. Figure 1 shows an example of event tree, where two protective actions are taken after the pressure increases. In this case, the disturbance propagation path (pressure increase due to spurious shout-down of PSD_3) can be protected by either protective action. Thus, the failure of both protective actions is required for the system failure occurrence.

3.3 Accident sequence conditions

When a system failure condition leading to system failure is identified, the system failure condition is represented as AND combination of a disturbance propagation path condition and failure conditions of its effective protective actions which can detect it.

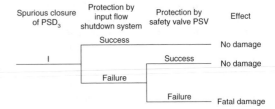

Figure 1. Event tree for spurious shutdown of PSD_3.

To obtain the failure condition of a protective action, the decomposition of a protection action into detection part, diagnosis part and action part is useful, which can analyze the failure conditions of protective/control system behaviors and operator actions in the same manner. At the same time, it must be examined whether failure of each protective action can be detected or not. If it can be detected, some maintenance action, or the overall system maintenance must be taken to nullify its effect before system failure. Thus, if failure of a protective action occurs before the disturbance propagation path, its latency (failure to detect it beforehand) must be considered in derivation of its failure conditions.

Note that the accident sequence is not always equal to the corresponding event sequence in an event tree. For example, if a system failure condition accompanies the failure of a protective system, it must occur before the trigger event occurs. In Figure 1 the failures of input-flow shutdown system and the safety valve must occur before the spurious shutdown of PSD_3.

Thus, accident sequence conditions can be obtained by considering the time sequence constraint in an event sequence of ETA.

4 ILLUSTRATIVE EXAMPLE

Obtain accident sequence conditions for an offshore separator as shown in Figure 2 (Hoyland & Rausand, 1994) using the proposed method.

4.1 System description

Gas from the wellhead manifold is led into the first stage separator. On the inlet pipeline, two process shutdown (PSD) valves, PSD_1 and PSD_2 are installed in series. The valves are held open by hydraulic pressure. When the hydraulic pressure is bled off, the valve will close by the force of a pre-charged actuator. Another PSD valve, PSD_3 of the same type is installed on the gas outlet from the separator. Due to the fail-safe design of the PSD valves, these valves may close spuriously, i.e., without the presence of a hazardous situation. A pressure safety valve (PSV) is installed to

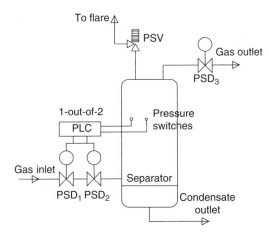

Figure 2. Off-shore separator.

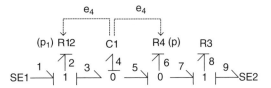

Figure 3. System bond graph for off-shore separator.

Table 3. Functional form expression.

I	Effort i	Flow i
1	U_1	$R_{12}(U_1 - C_1(X_1))$
2	$U_1 - C_1(X_1)$	$R_{12}(U_1 - C_1(X_1))$
3	$C_1(X_1)$	$R_{12}(U_1 - C_1(X_1))$
4	$C_1(X_1)$	$R_{12}(U_1 - C_1(X_1)) - R_4(C_1(X_1))$ $-R_3(C_1(X_1) - U_2)$
5	$C_1(X_1)$	$R_4(C_1(X_1)) + R_3(C_1(X_1) - U_2)$
6	$C_1(X_1)$	$R_4(C_1(X_1))$
7	$C_1(X_1)$	$R_3(C_1(X_1) - U_2)$
8	$C_1(X_1) - U_2$	$R_3(C_1(X_1) - U_2)$
9	U_2	$R_3(C_1(X_1) - U_2)$

relive the pressure in the separator in case the resource increases beyond a specified high pressure (p). The PSV is equipped with a spring-loaded actuator, which may be adjusted to p. The most critical failure mode for the PSV is "fail to open" (FTO); the valve does not open when the pressure in the separator increases beyond p. Two identical pressure switches, PS_1 and PS_2, are installed in the separator. The PS's are on-off devices that are preset to be activated at a pressure (p_1), which is less than p for the PSV. When activated, the PS's will provide signal via the programmable logic controller (PLC) to close PSD_1 and PSD_2. The PLC employs a 1-out-of-2 voting logic.

4.2 *Global system behavior model*

Construct the physical system behavior model based on the above design assumptions using BG. Using analogy between BGE's and physical phenomena in Table 1, the SBG can be obtained as shown in Figure 3. The SBG shows the basic relation between flow and pressure, where other behavior such as thermal effect are neglected for simplicity. The gas inlet is represented as the pressure source SE1, the redundant shutdown valve is simply modeled as an R-element R12 whose characteristic is determined by set-point p1 and pressure in the separator (e_4). Since the separator is simplified as a kind of gas storage, a C-element C1. PSV is represented as an R-element R4 whose characteristic is also determined by set-point p and pressure in the separator (e_4). R3 denotes the shutdown valve of gas outlet.

According to the definition of system and input variables in the SBG, let X_1, U_1, and U_2 denote state variable for C1, input variables for SE1 and SE2, respectively. Using a simple symbolic manipulation method (Kohda, Nakada, Kimura, & Mitsuoka, 1988),

all e's and f's can be obtained as functional forms of X_1, U_1, and U_2 as shown in Table 3, where R_{12}, R_3, R_4, and C_1 represent characteristic functions for R12, R3, R4, and C1, respectively.
The system state equation is obtained as:

$$\frac{dX_1}{dt} = R_{12}(U_1 - C_1(X_1)) - R_4(C_1(X_1)) - R_3(C_1(X_1) - U_2) \qquad (2)$$

Under the steady state or normal condition, the following equality must hold.

$$R_{12}(U_1 - C_1(X_1)) = R_4(C_1(X_1)) + R_3(C_1(X_1) - U_2) \qquad (3)$$

Based on the design assumptions described above, each characteristic function under its normal condition is defined as:

(1) Since PSD_1 and PSD_2 close when the pressure in the separator is greater than set-point p, R_{12} are: If $e_4 \leq p$, $R_{12}(*) > 0$. If $e_4 > p$, $R_{12}(*) = 0$.
(2) Similarly to PSD_1 and PSD_3, R_3 is: If command signal does not issues, $R_3(*) > 0$. Otherwise, $R_3(*) = 0$.
(3) Since PSV open when the monitored pressure in the separator is greater than set-point p_1, R_4 is: If $e_4 \leq p_1$, $R_4(*) = 0$. If $e_4 > p_1$, $R_4(*) > 0$.
(4) Since the pressure in the separator is simplified as an increasing function of gas volume, C_1 is: If $x_1 \leq x_2$, $C_1(x_2) \leq C_1(X_2)$.
(5) Since the manifold determines the output pressure, U_1 is fixed based on its property.

919

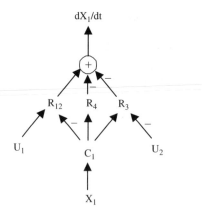

$$dX_1/dt$$

Figure 4.   Causal tree.

Assume the following component failures:

(6) PSD_1, PSD_2:
Spurious shutdown (SC): If $e_4 \leqslant p$, $R_{12}(*) = 0$.
(In case either PSD_1 OR PSD_2 is SC)
Failure to close (FTC): If $e_4 > p$, $R_{12}(*) > 0$.
(In case both PSD_1 AND PSD_2 are FTC)

(7) PSD_3:
Spurious shutdown (SC):
If command signal does not issue, $R_3(*) = 0$.
Failure to close (FTC):
If command signal issues, $R_3(*) = 0$.

(8) PSV:
Failure to open (FTO): If $e_4 > p_1$, $R_4(*) = 0$.

Based on the global system state equation (Eq. (2)), the causal tree representing cause-effect relations among components and process variables is obtained as shown in Figure 4. Here the information flow models are not described explicitly, because they are represented as characteristic functions of BGE's in terms of input variables. Note that each node in the causal tree corresponds to a process variable such as pressure and flow in the off-shore separator.

4.3   Accident sequence conditions

Consider a system failure: "overpressure in the separator", which corresponds to the explosion of the separator. In the SBG, the pressure in the separator corresponds to effort of bond 4 (e_4).

Firstly, obtain potential disturbance paths. Under the normal operation, the PSV is closed, i.e., $R_4(C_1(X_1)) = 0$ because the pressure in the separator is below set-point p. Thus, under the normal condition, the pressure in the separator, $e_4 = C_1(X_1)$, must satisfy

$$R_{12}(U_1 - C_1(X_1)) = R_3(C_1(X_1) - U_2) \text{ and } U_1 > C_1(X_1) > U_2.$$

$$(4)$$

To satisfy this physical constraint, the following relations must hold if a possible deviation of R12 or R3 occurs:

(a) If function R_{12} decreases, $C_1(X_1)$ will decrease.
(b) If function R_3 decreases, $C_1(X_1)$ will increase.

According to the design assumptions on component failure, the spurious shutdown of PSD_1 or PSD_2 corresponds to case (a), while the spurious shutdown of PSD_3 corresponds to case (b). In case (b), the deviation of R_3 can cause the overpressure. Thus, if shutdown valve PSD_3 advertently closes, i.e., $R_3(C_1(X_1) - U_2) = 0$, $R_{12}(U_1 - C_1(X_1))$ must vanish, leading to the increase of the pressure in the separator. Of course, Eq. 4 also shows that the increase of U_1 can cause the increase of $C_1(X_1)$, but this case is not assumed here. Though the expansion of product volume in the separation process may be a more serious phenomenon, the analysis does not consider the reaction in the separator, either. Thus, the spurious shutdown of PSD_3 is considered as the only potential disturbance path in this example. A similar qualitative analysis can be performed using the causal tree in Figure 4. Since node C_1 represents pressure in the separator, the backward search shows that its increase is caused by the increase of volume (X_1). This can be caused by either the increase of input flow (R_{12}) or decrease of output flow (R_3). Considering possible deviations of their input variables, the only possible cause is identified as the decrease of output flow (R_3), which corresponds to the closure of PSD_3.

According to the design assumption, the fail-safe design of PSD_3 cannot prevent the occurrence of its spurious shutdown. In other words, the occurrence probability of a spurious shut-down of PSD_3 cannot be neglected. Thus, the effectiveness of the current protective actions must be evaluated. Consider protective actions for the spurious shutdown of PSD_3. The effect of spurious shutdown, the increase of pressure in the separator, can be reduced by two protective actions. Under the normal condition, the increase of the pressure is monitored by the pressure switches, which activate the shutdown valve of input flow, i.e. $R_{12}(U_1 - C_1(X_1))$ reduces to 0. Thus, the mass balance in the separator can be maintained, i.e., the increase of pressure stops. Further, even if PSD_1 and PSD_2 fail to close, the continuing increase of the pressure in the separator can be prevented by PSV. If the pressure reaches the set-point of PSV, PSV is activated to relieve the pressure. If the PSV fails to relieve the pressure, it leads to the explosion of separator. These scenarios are equivalent to those obtained by ETA as shown in Figure 1. In the proposed method, the effectiveness of protective actions on a disturbance propagation path is evaluated by the system analysts. Further, since these protective systems under their non-response failure condition remain in the same state as their normal state, their failure cannot be detected, in other words, their failure is latent (Kohda and Inoue, 2001). Thus, the failure of PSV is caused by FTO, while the failure of PSD protective system is caused by {FTC of PSD$_1$ AND PSD$_2$} OR {non-response failure of both pressure switches} OR {non-response failure of PLC}.

Thus, the explosion occurs only in case the spurious shutdown of valve PSD$_3$ occurs after both the input flow shutdown system and the safety valve PSV get failed. In other cases where either the input flow shutdown system or the safety valve keeps normal, the explosion due to the occurrence of the spurious shutdown of PSD$_3$ can be prevented.

5 COMPARISON WITH DYNAMIC RELIABILITY ANALYSIS

This paper proposes a framework for the derivation of accident sequence conditions using the global system model composed of physical system behavior model and information flow. The motivation for this study is the inefficiency of FT in representing time dependency of event sequences and inconsistency of a functional ET description with the occurrence order of its constituent events. However, since the object of the proposed method is the development of an interactive safety design support system, the procedure must conform to the conventional procedure of risk-based safety design: identification of hazard, identification of effective safety measures, derivation of accident scenarios and their evaluation. In deriving possible disturbance path, the physical system behavior model is directly utilized instead of a conventional functional model (IFMA) or system functional logic.

For the application of process system models to accident sequence analysis, the extensive studies have been done in the field of "dynamic reliability analysis" such as (Aldemir T. 1987), (Devooght & Smidts 1996) and (Aneziris, Papazoglou & Lygerou 2000). They simulate the detailed model of system dynamics which corresponds to a physical system behavior model in the proposed method, and also utilize Markov modeling approach to represent stochastic behaviors such as component failure. By adjusting the parameters in the system model, various accident conditions can be simulated. More informative analysis results are obtained than the conventional Event Tree/Fault Tree approach. The dynamic reliability analysis is regarded as a kind of simulation-based analysis, while the proposed method is an analytical method based on the qualitative knowledge of system failure mechanism. When the detailed/quantitative information on design assumptions is given, the analysis result obtained by the proposed method must be confirmed using dynamic simulations in the dynamic reliability analysis.

6 CONCLUSIONS

This paper proposes a systematic approach to the derivation of accident sequence conditions using the global system model. The proposed method divides the system failure condition into two parts: (1) a potential disturbance propagation path which can lead to a specific system failure state and (2) failure condition of protective actions which can prevent the potential disturbance propagation path. Using the global system behavior model represented by bond graphs for physical behavior and information flow for computerized actions and human operation actions, all potential disturbance propagation paths to a specific system failure can be obtained. For each potential disturbance propagation path, the system analysts evaluate the effective protective actions, because the analysis of all possible situations in advance is useless. This step can also confirm the completeness of the overall protective actions. Since failure conditions for each protective action can be obtained using the system behavior model, the failure condition of the overall protective actions can be derived easily based on the analysts' judgment. Finally, the accident sequence conditions can be obtained by considering the time sequence of a potential disturbance occurrence and its protective action failure. For the verification of the proposed method, we are now planning to apply the proposed method to a more practical problem.

REFERENCES

Aldemir, T. 1987. Computer-assisted Markov failure modeling of process control systems, *IEEE Trans on Reliability*, R-36,1, 133–144.

Aneziris, O.N., Papazoglou, I. A., & Lygerou, V. 2000. Dynamic safety analysis of process systems with an application to a cryogenic ammonia storage tank, *J. Loss Prevention in Process Industries*, 13, 153–165.

Devooght, J & Smidts, C. 1996. Probabilistic dynamics as a tool for dynamic PSA, *Rel. Eng. and Systems Safety*, 52, 185–196.

Henley, E.J., & Kumamoto, H. 1991. *Probabilistic Risk Assessment, Reliability Engineering, Design, and Analysis*, New York, IEEE Press.

Hoyland, A. & Rausand, M. 1994. *System Reliability Theory, Models and Statistical Methods*, John Wiley & Sons, Inc., IFMA. What is FM?, http://www.enre.umd.edu/ifmaa/ifm02.htm.

Karnopp, D.C., Margolis, D.L., & Rosenberg, R.C. 1990. *System Dynamics: A Unified Approach* (2nd ed.), John Wiley & Sons, Inc., New York.

Kohda, T., & Inoue, K. 2001. Fault tree analysis considering latency of basic events, *Proc. RAMS2001*: 31–35.

Kohda, T., Nakada, T., Kimura, Y., & Mitsuoka, T. 1988. Simulation of bond graphs with nonlinear elements by symbolic manipulation, *Bulletin of Mechanical Engineering Laboratory*, Japan, No. 49.

U.S. NRC. 1981. *Fault Tree Handbook*, NUREG-094.

Safety and Reliability – Bedford & van Gelder (eds)
© 2003 Swets & Zeitlinger, Lisse, ISBN 90 5809 551 7

A simple method to evaluate system failure occurrence probability using minimal cut sets

T. Kohda & K. Inoue
Kyoto University, Kyoto, Japan

ABSTRACT: In conventional FTA (Fault Tree Analysis), system failure conditions are obtained as minimal cut sets (MCS's), minimal combinations of basic events (or component failures) causing system failure. Since a MCS does not express an event sequence condition to system failure explicitly, the inclusion and exclusion method using MCS's does not always mean the system failure occurrence probability. This paper proposes a simple method to evaluate the system failure occurrence probability using MCS's. Based on critical states for a basic event obtained from MCS's, the system failure occurrence probability can be calculated as the sum of occurrence probability of each basic event under its critical conditions. The proposed method can also consider the event sequence dependency by considering the practical meaning of the critical state for a basic event. An illustrative example of a simple system with protective system shows the details and merits of the proposed method.

1 INTRODUCTION

Conventional FTA (Fault Tree Analysis) (U.S. NRC, 1981) obtains system failure conditions as minimal cut sets (MCS's), minimal combinations of basic events (or component failures) leading to system failure. If all basic events at least in a MCS occur, the system failure occurs. A MCS expresses a necessary and sufficient condition for the system failure to occur, but it does not express event sequence conditions to system failure, explicitly. Probability calculation using the inclusion and exclusion method using MCS's does not always give the system failure occurrence probability. For example, consider a system with its safety monitoring system. The system failure condition, or minimal cut set is obtained as {the system failure (event A), the safety monitoring system failure (event B)}. Occurrence of event A after event B leads to an accident, while occurrence of event B after event A does not. Thus, the system failure occurrence probability is not equal to the product of occurrence probabilities of events A and B. This paper proposes a simple method to evaluate the system failure occurrence probability using MCS's.

For the explicit consideration of event sequences in a fault tree (FT), special logic gates such as priority AND gate (Fussell, Aber & Rahl, 1976) must be introduced. However, the introduction of such a special logic gate requires the examination of event sequences in developing each logic gates, which will be more time-consuming. In the proposed method, we assume that FT is coherent (Barlow & Proschan, 1975). In other words, a FT is composed of OR and AND gates and does not include the negative of a basic event. Although we obtain the system failure conditions using a conventional FT, the time dependency (or event sequence) of basic events is considered by examining the practical meaning of a critical state condition, i.e., whether the occurrence of a basic event under its critical state makes sense or not. In the following section, the proposed method is described with a simple example of a redundant system.

2 PROBLEM STATEMENT

2.1 *Assumptions*

The following conventional assumptions in FTA are made:

(1) Basic events (or component failures) occur statistically independently.
(2) The possibility that more than two basic events occur simultaneously is negligible.
(3) Basic events do not occur at time $t = 0$. (In other words, all components are as good as new at time $t = 0$.)
(4) Basic events (or component failures) are not recovered before the system failure occurs.
(5) The FT is coherent (Barlow & Proschan, 1975) and has N MCS's.

2.2 Notations

$X_i(t)$ binary indicator variable for basic event i at time t.

$$X_i(t) \equiv \begin{cases} 1, \text{ basic event } i \text{ is occurring at time } t, \\ 0, \text{ otherwise.} \end{cases} \qquad (1)$$

$\bar{X}_i(t)$ $1 - \bar{X}_i(t)$; negative of $\bar{X}_i(t)$

$F_i(t)$ $\Pr\{\bar{X}_i(t) - 1\}$; cumulative occurrence probability of basic event i at time t

$\bar{F}_i(t)$ $1 - \bar{F}_i(t)$

$f_i(t)$ Occurrence probability density function of basic event i at time t

$r_i(t)$ Occurrence rate of basic event i at time t (occurrence rate at wich basic event i occurs during a unit time interval at time t)

K_j j-th minimal ct set

C set of all minimal cut sets

$C(i)$ saet of all minimal cut sets with basic event i

$\prod_i X_i, \; \coprod_i X_i$ logical AND and OR of binary indicator variables X_i's

2.3 Problem statement

The problem considered here is how to calculate the system failure occurrence probability based on MCS's for a fault tree under assumptions (1)–(5).

3 PROBABILITY EVALUATION OF SYSTEM FAILURE OCCURRENCE

3.1 System failure occurrence condition

Although the examination of occurrence of MCS's is sufficient to confirm whether system failure occurs or not, the cause of system failure cannot always be reduced to the occurrence of a MCS. Consider a FT with MCS's $\{1, 2\}$ and $\{1, 3\}$. If basic events 2, 3 and 1 occur sequentially, the system failure cannot be caused by either $\{1, 2\}$ or $\{1, 3\}$, uniquely. In other words, the system failure occurrence cannot be classified into disjoint cases based on MCS's. However, even in this example, the basic event whose occurrence causes system failure to occur can be identified uniquely under assumption (2). Thus, system failure occurrence conditions can be classified into disjoint cases by considering which basic event causes the system failure. This paper proposes a simple method to obtain the system failure occurrence probability based on basic events leading to system failure.

A basic event which leads to system failure is called an active failure, while a condition where a basic event can be an active failure is called a critical state for the basic event (Barlow & Proschan, 1975(b)). For system failure to occur due to basic event i at time t, two conditions must be satisfied:

(1) a critical state for a basic event exists at time t, AND (2) basic event i occurs at time t. From assumption (2), system failure is caused by some basic event, the system failure occurrence probability at time t is the sum of system failure probabilities caused by all basic events.

3.2 System failure occurrence condition

To obtain the system failure condition, the explicit derivation of critical states for basic event i is necessary. At a critical state for basic event i, {no MCS without basic event i occurs} AND {all basic events except basic event i is occurring at time t at least for a MCS containing basic event i} must be satisfied. Using logical expressions for each condition, critical states for basic event i can be obtained as:

$$\left(\prod_{K_j \in C - C(i)} \coprod_{m \in K_j} \bar{X}_m(t) \right) \left(\coprod_{K_j \in C(i)} \prod_{n \neq i \& n \in K_j} X_n(t) \right) \bar{X}_i(t) \qquad (2)$$

The first bracket term in Eq. (2) represents that no MCS's without basic event i have occurred, the second bracket term indicates that all basic events except basic event i have occurred, and the last term shows that basic event i has not occurred, respectively.

Using the properties of binary indicator variables:

(a) $X_j(t)\bar{X}_j(t) = 0$

(b) $X_j(t)X_j(t) = X_j(t)$

(c) $X_i(t)X_j(t)\coprod X_j(t) = X_j(t)$

Equation (2) can be transformed into logical OR of logical AND combinations of binary indicator variables $X_i(t)$'s. Each logical AND combination represents a condition that a critical state exists for basic event i at time t. For example, consider a FT with MCS's $\{1, 2\}$ and $\{1, 3\}$. Critical state conditions for each basic event is obtained as:

For basic event 1,

$$\left(X_2(t) \coprod X_3(t) \right) \bar{X}_1(t)$$

For basic event 2,

$$\left(\bar{X}_1(t) \coprod \bar{X}_3(t) \right) \left(X_1(t) \right) \bar{X}_2(t) = \bar{X}_3(t) X_1(t) \bar{X}_2(t)$$

For basic event 3,

$$\left(\bar{X}_1(t) \coprod \bar{X}_2(t) \right) \left(X_1(t) \right) \bar{X}_3(t) = \bar{X}_2(t) X_1(t) \bar{X}_3(t)$$

Using the state space vectors $X(t) = (X_1(t), X_2(t), X_3(t))$, critical state are represented as:

For basic event 1, $\{X(t)\} = \{(0,1,0),(0,0,1),(0,1,1)\}$

For basic event 2, $\{X(t)\} = \{(1,0,0)\}$

For basic event 3, $\{X(t)\} = \{(1,0,0)\}$

The above qualitative result shows that basic event 1 causes system failure more frequently than basic events 2 and 3.

3.3 System failure occurrence probability

The system failure occurrence probability caused by basic event i at time t is obtained as the product of {the existence probability of a critical state for basic event i at time t} and {the occurrence probability of basic event i at time t}. The latter probability can be obtained from the occurrence probability of basic event i as follows.

$$\Pr\{X_i(t+dt)=1\,|\,X_i(t)=0\}=r_i(t)dt=\frac{f_i(t)dt}{1-F_i(t)} \quad (3)$$

The former probability can be calculated from the critical state condition represented as Eq. (2). Thus, the system failure occurrence probability caused by basic event i at time t can be obtained as:

$$\Pr\left\{\left(\prod_{K_j\in C-C(i)}\coprod_{m\in K_j}\bar{X}_m(t)\right)\left(\coprod_{K_j\in C(i)}\prod_{n\neq i\&n\in K_j}X_n(t)\right)=1\right\} \quad (4)$$
$$\times f_i(t)dt$$

Integrating the above probability gives the system failure occurrence probability due to basic event i until the current time, and the summation of these probabilities over all basic events gives the system failure occurrence probability. Compared with the probability evaluation method using Markov diagrams (Henley & Kumamoto, 1991), the proposed method does not necessarily assume exponential probability for the occurrence of a basic event. Further, the enumeration of all possible system states is not necessary, which will be a complicated problem for the method using Markov diagrams to solve large complex systems. In the proposed method, only the critical states are necessary, which can be obtained easily by the logical expression approach shown in this paper.

3.4 Event sequence dependency

Since the coherent FT does not represent time dependency or event sequence dependency, the above discussion considers only logical relations. Considering the cold-standby system or system safety monitoring system, the event sequence of component failure must be considered to obtain the meaningful result.

For example, consider a cold-standby system composed of two components. In the coherent FT, the system failure is represented as [{the working component failure (X_1)} AND {the stand-by component failure (X_2)}] OR [{the working component failure (X_1)} AND

{switching failure (X_3)}]. Critical states for each component failure is obtained as:

For X_1, $\left(X_2(t)\coprod X_3(t)\right)\bar{X}_1(t)$

For X_2, $X_1(t)\bar{X}_3(t)\bar{X}_2(t)$

For X_3, $X_1(t)\bar{X}_2(t)\bar{X}_3(t)$

From the above discussion for the coherent FT, the system failure occurs if each basic event occurs under its critical state conditions. However, the system failure does not occur when the switching component fails (X_3) after working component failure (X_1). Because the working component is replaced by the stand-by component before the switching component fails, the switching component failure does not cause any effect on the system state. In this example, the working component failure and the stand-by component failure can become an active failure.

Thus, by considering the system function mechanism under its critical condition, the proposed method can exclude a meaningless basic event as an active failure before the probability calculation. We can also exclude a meaningless critical state condition by examining whether its constituent basic event can be detected or not (Kohda & Inoue, 2001). Since the detection of a component failure can trigger some protective action, the critical state including it cannot be attained.

3.5 Evaluation method

Using the expression in Eq. (4), the system failure occurrence probability caused by basic event i can be obtained. Since the term $\Pr\{*\}$ represents the critical state probability, it can be evaluated using the inclusion–exclusion method by checking the meaning of combined critical state conditions. Using $X_k(t)=1-\bar{X}_k(t)$ and $\Pr\{\bar{X}_k(t)=1\}=\bar{F}_k(t)$, Eq. (4) can be calculated using following product terms.

$$\left\{\prod_{k\neq i\&k\in S}\bar{F}_k(t)\right\}f_i(t)dt \quad (5)$$

where S represents the set of basic events included in a product of critical state conditions. Using occurrence rate $r_k(t)$, the following expressions can be obtained.

$$\bar{F}_k(t)=\exp\left[-\int_0^t r_k(u)du\right] \quad (6)$$

$$f_i(t)dt=r_i(t)\exp\left[-\int_0^t r_i(u)du\right]dt \quad (7)$$

Further, for the exponential distribution of occurrence probability, the probability calculation in Eq. (5) can be simplified as:

$$\left\{\prod_{k\neq i\&k\in S}\bar{F}_k(t)\right\}f_i(t)dt=\lambda_i\exp\left[-\left(\sum_{k\in S}\lambda_k\right)t\right]dt \quad (8)$$

925

For the previous example, failure occurrence probability for each basic event can be obtained as:

For 1, $\left\{1 - \bar{F}_2(t)\bar{F}_3(t)\right\}f_1(t)dt$

$$= \lambda_1\left\{\exp(-\lambda_1 t) - \exp(-(\lambda_1 + \lambda_2 + \lambda_3)t)\right\}dt$$

For 2, $\bar{F}_3(t)(1 - \bar{F}_1(t))f_2(t)dt$

$$=\lambda_2\left\{\exp\left[-(\lambda_2 + \lambda_3)\right] - \exp\left[-(\lambda_1 + \lambda_2 + \lambda_3)\right]\right\}dt$$

For 3, 0.

As shown in the previous section, basic event 3 cannot be an active failure. Integrating the above probability, the cumulative system failure occurrence probability caused by each basic event can be obtained as:

For 1, $1 - \exp[-\lambda_1 t]$

$$-\frac{\lambda_1}{\lambda_1 + \lambda_2 + \lambda_3}\left(1 - \exp[-(\lambda_1 + \lambda_2 + \lambda_3)t]\right)$$

For 2, $\dfrac{\lambda_2}{\lambda_2 + \lambda_3}\left(1 - \exp[-(\lambda_2 + \lambda_3)t]\right)$

$$-\frac{\lambda_2}{\lambda_1 + \lambda_2 + \lambda_3}\left(1 - \exp[-(\lambda_1 + \lambda_2 + \lambda_3)t]\right)$$

For 3, 0.

Thus, the cumulative probability of system failure occurrence is obtained as the sum of the above probabilities:

$$1 - \exp[-\lambda_1 t] + \frac{\lambda_2}{\lambda_2 + \lambda_3}\left(1 - \exp[-(\lambda_2 + \lambda_3)t]\right)$$

$$-\frac{\lambda_1 + \lambda_2}{\lambda_1 + \lambda_2 + \lambda_3}\left(1 - \exp[-(\lambda_1 + \lambda_2 + \lambda_3)t]\right)$$

Apparently, this probability does not equal to the cumulative probability obtained from the corresponding coherent FT without consideration of event sequences.

4 RELATIONS WITH IMPORTANCE

Critical state conditions are conventionally utilized in evaluating several kinds of component importance (Henley & Kumamoto, 1991). For example, Birnbaum basic event importance means the probability that the system is in a state where the occurrence of event i is

critical. The same idea as our proposed method was proposed by Barlow & Proschan (1975(a)) where the expected number of failures caused by basic event i is proposed as a component importance. They considered the system failure probability of the coherent FT structure, but they did not consider constraints on event sequences. We add the consideration of event sequence conditions to MCS's obtained from the coherent FT by considering not only event sequence dependency which is not expressed explicitly in coherent FTs, but also the detectability of basic events which can prevent the evolution of an accident sequence.

5 CONCLUSIONS

This paper proposes a simple novel method to obtain the system failure occurrence probability based on MCS's. Though MCS's are obtained by assuming that the subject system is represented as a coherent FT, event sequence dependency can be considered in the proposed method by examining the practical meaning of the system failure occurrence under a critical state such as the consistency with the system function mechanism. As shown in a simple example of a stand-by system, the cumulative system failure occurrence probability is not always equal to obtained by the conventional FT method based on MCS's. We are now developing a system failure probability evaluation system based on the proposed method.

REFERENCES

Barlow, R.E., & Proschan, F. 1975(a). Importance of System Components and Fault Tree Events, *Stochastic Processes and their Applications*, 3: 153–173.
Barlow, R.E., & Proschan, F. 1975(b). *Statistical Theory of Reliability and Life-Testing, Probability Models*, Holt, Rinehart & Winston, Inc.
Fussell, J.B., Aber, E.F., & Rahl, R.G. 1976. On the Quantitative Analysis of Priority-AND Failure Logic, *IEEE Trans. Reliability*, R-25(5): 324–326.
Henley, E.J., & Kumamoto, H. 1991. *Probabilistic Risk Assessment, Reliability Engineering, Design, and Analysis*, New York, IEEE Press: 418–436.
Kohda, T., & Inoue, K. 2001. Fault Tree Analysis Considering Latency of Basic Events, *Proc. RAMS*: 31–35.
U.S. NRC. 1981. *Fault Tree Handbook*, NUREG-094.

Safety and Reliability – Bedford & van Gelder (eds)
© 2003 Swets & Zeitlinger, Lisse, ISBN 90 5809 551 7

Uncertainty analysis of river flood management in the Netherlands

M. Kok
HKV Consultants, Lelystad, Netherlands
Delft University of Technology, Faculty of Civil Engineering and Geosciences, Delft, Netherlands

J.W. Stijnen
HKV Consultants, Lelystad, Netherlands

W. Silva
Ministry of Transport, Public Works and Water management, RIZA, Arnhem, Netherlands

ABSTRACT: The current flood defense design practice along the major rivers in the Netherlands is to include only the natural variability of water levels (or the discharge) in assessing the exceedance frequency. Other sources of uncertainty which could cause flooding (such as the roughness of the riverbed or the discharge distribution at the bifurcation points) are ignored.

In this paper we will show the influence of other uncertainties on the probability of flooding. Instead of the traditional design method (the exceedance frequency of water levels, using only the river discharge as random variable), we will consider the exceedance probability of (wave) overtopping of the flood defense. We have investigated the failure frequencies of dike sections and not the flood frequency of dike rings, which always consist of a number of failure mechanisms, dike sections and hydraulic structures. Therefore the number of random variables remains small enough so that numerical integration can be used to calculate the frequencies.

It is shown that other sources are a major contribution to the calculated safety against flooding. These uncertainties also influence the efficiency of measures which reduce the risks of flooding, such as the use of retention areas (for example in emergency situations). In the traditional approach this measure seems highly efficient, but if all uncertainties are taken into account this measure is less efficient. However, the attractiveness using retention areas depends on the costs and benefits of this measure, and the approach in this paper is an essential ingredient to assess the benefits. It is recommended to use the approach in this paper in a cost benefit analysis, and to investigate the influence of the assumptions.

1 INTRODUCTION

The Netherlands are situated in the delta of three of Europe's main rivers: the Rhine, the Meuse and the Scheldt. As a result of this, the country has been able to develop into an important, densely populated nation. But living in the Netherlands is not without risks. Large parts of the Netherlands are below mean sea and water levels which may occur on the rivers Rhine and Meuse. High water levels due to storm surges on the North Sea, or due to high discharges of these rivers are a serious threat to the low-lying part of the Netherlands. Proper construction, management and maintenance of flood defences are essential to the population and further development of the country.

Without flood defences much of the Netherlands would be flooded on a regular basis. The influence of the sea would mainly be felt in the West. The influence of the waters of the major rivers has a more (but limited) geographic impact. Along the coast, protection against flooding is predominantly provided by dunes. Where the dunes are absent or too narrow, or where the sea arms have been closed off, flood defences in such as sea dikes or storm surge barriers have been constructed. Along the full length of the Rhine and along parts of the Meuse protection against flooding is provided by dikes. For an overview of the current safety standards along the coast and major rivers, see Brinkhuis-Jak et al, 2003.

2 DESIGN METHODS

The current safety standard has been set after the big 1953 flood disaster in the Netherlands. After this flood the design method of flood protection was improved considerably because of the scientific approach. This approach was invented by the Delta Committee (Delta Committee 1960, Dantzig 1956). The default approach for designing flood protection structures that has been used until then, was based on the highest recorded water level. In relation to this water level a certain safety margin (varying from 0.5 to 1.0 meter) was maintained. The Delta Committee recommended that a certain desired "safe" water level be taken as a starting point. The safety standards should be based on weighting the costs of the construction of flood protection structures against the possible damage caused by floods. An econometric analysis was undertaken by the Delta Commission for Central Holland. Based on information from 1960 this led to an optimum policy of 8×10^{-6} per year. For practical design this was converted into a design water level with a frequency of exceedance of 1/10000 per year. These design frequencies are used for the dike ring areas along the coast. For the major rivers, however, less strict design frequencies are demanded in the Act of Flood Defences, because the consequences of flooding in these riverine areas are less severe than a flood along the coast. The design frequency of flood defences along the major rivers has been set to 1/1250 per year.

The predominantly *deterministic* determination assumes the normative Design Water Level (DWL) that the dike must be able to retain (TAW, 1998). This water level may include wind effects on the local water level. In addition, the wave run-up is subsequently the most important parameter in the determination of the crest height. In the traditional deterministic design method the wave run-up is calculated based on certain wind characteristics at the design water level and the corresponding waves, and taking into account the geometry of the water defence system. Settlement of the soil body over a certain planning period is also taken into account to avoid repair actions, which may be needed if the height of the dike is below the required reference level.

In a *probabilistic* approach the results of the more deterministic method described above are still needed as input to the probabilistic calculations. In a probabilistic approach we are interested in the inundation probability of a dike ring area. Inundation is caused by failure of the flood defence system. Here, the whole range of water levels and waves are included in the analysis (TAW, 1998).

In this paper we will follow a probabilistic approach.

3 ASSUMPTIONS

The following assumptions have been made (see also Stijnen et al, 2002):

a. We investigated only one failure mechanism: overflow and wave overtopping. Other mechanisms (for example sliding of the inner slope, piping and micro instability, see TAW, 1998) are not included. These mechanisms may be important, but in the study TAW, 2001 it is concluded that overflow and wave overtopping is the dominant mechanism in the probability of flooding, assuming that the possible "weak spots" are strengthened;
b. We investigated the following six locations along the major Rhine river branches (see for a map Figure 2):
   - Lobith, Upper Rhine river (km 862)
   - Millingen, Waal river (km 868)
   - Tiel, Waal river (km 915)
   - Opijnen, Waal river (km 929)
   - Amerongen, Lower Rhine river (km 918)
   - Duursche Waarden, IJssel river (km 961)

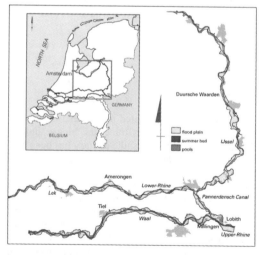

Figure 2. Overview of the Rhine branches in the Netherlands.

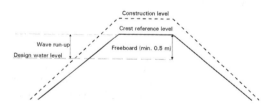

Figure 1. Design of a river dike (TAW, 1998).

c. We did not consider the true dike heights, but the dikes as they should have been designed according to the design rules of the Technical Advisory Committee on Water Defences;

d. In the assessment of flood defences it is useful to distinguish between failure and collapse of a structure. Failure is defined as not fulfilling one or more water defence functions (the crest of a part of the flood defence is too low, for example). Collapse means the loss of cohesion or large deformations in geometry. In this paper we only handle failure of the water defence;

e. The reliability function Z of the failure mechanism wave overtopping is: $Z = q_c - Q(H_s, h)$, where q_c stands for the critical overtopping discharge (which may be stochastic, but in this paper we will assume that it has a deterministic value of $0.001\,\mathrm{m^3/s/m}$, which is equivalent to $1\,\mathrm{l/s/m}$), H_s is the wave height and h is the water level.

4 UNCERTAINTY IN PARAMETERS AND THEIR DISTRIBUTIONS

When designing the height and strength of a water defense section, there are many (stochastic) factors which have to be taken into account. It is important to realize that we are not only dealing with uncertain parameters, but that each of these parameters has a distribution of its own that is unknown. Think about the natural processes such as the river discharge waves (height and shape), the resulting water levels, precipitation, wind (speed and direction), etc. On the other hand, there are also a number of uncertainties in the creation of models: a hydraulic or hydrological model is never perfect, and neither are the required parameters in these models. Finally, we often use measurement data. When we use these measurements to estimate parameters, or distributions, more uncertainties are introduced.

There are basically two categories in which we can place these uncertainties (Noortwijk et al, 2002): natural variability and epistemic uncertainties.

4.1 Natural variability

This is sometimes also called inherent uncertainty, and represents the unpredictability of physical processes. This concerns both uncertainties in time as well as in space. Uncertainties that are a direct consequence of the variability of natural processes fall into this category. Think about the direction or velocity of the wind, but also the local or downstream hydraulic roughness and the discharge distribution near a river bifurcation point. Uncertainties in the discharge itself (both the height of the peak and the shape) are part of this category as well.

4.2 Epistemic uncertainty

The category of epistemic uncertainty (also called knowledge uncertainty) is a large one, and can be further subdivided into statistical uncertainty, model uncertainty and planning uncertainty.

Uncertainties that play a role when determining the water level on the river, or the discharge into a retention area belong in the subcategory of model uncertainties. They arise from prediction models for the river and the retention area. Other examples of uncertainties that play a role within the hydraulic model, are the flow pattern near the inlet construction, and the slope across the inlet construction. These uncertainties maintain a certain amount of subjectivity (Cooke, 1991), because their size and relevance are hard to determine. It is also possible for these results to be influenced by new research results.

Statistical uncertainty arises when there are not enough data to estimate the parameters of a probability distribution of a random variable (Kok et al, 1996, Appendix B). The more data, the smaller becomes the statistical uncertainty. The uncertainties regarding the choice of the type of probability distribution also fall into this subcategory. Examples are the probability distributions for the discharge, and temporal and spatial correlations between the various random variables.

The decision to use a retention area brings with it a number of uncertainties from these different categories. There are, however, uncertainties that fall in yet another subcategory, dealing with the organisational side of a measure, especially in the case of retention. This is closely related to the ability to predict the duration of a discharge wave on a short term (in the order of days to a week). Of special interest in this case are questions that concern the actual use of a retention area. When should the retention area be flooded? Which retention area should be used? There are also social and economic aspects surrounding the decision whether to use a retention area or not, but these will not be discussed in this paper.

5 RESULTS OF DETERMINISTIC AND PROBABILISTIC CALCULATIONS

For each of the six locations mentioned in Section 3 the failure probabilities for the mechanism overflow and wave overtopping have been calculated. We also investigated what happens to these failure probabilities when retention is used as a measure to increase the safety of dike ring areas. In each case we investigated the resulting failure probability *with* and *without* the measure. This in turn enabled us to define the term "efficiency" of the measure retention as follows:

$$\text{Efficiency of retention} = \frac{\text{Failure probability without retention}}{\text{Failure probability with retention}}$$

With the aid of this definition it is possible to obtain insight in the actual safety benefit of a measure. In the computations we used the following random variables and distributions:

- The discharge, with actual exceedance probabilities of the discharge peak according to the working line (Parmet et al, 2002).
- The wind direction, with actual statistics for the measurement station of Schiphol Airport (Geerse et al, 2002).
  The wind speed, with actual exceedance probabilities for the measurement station of Schiphol Airport (Geerse et al, 2002).
- Water level, where a normal distribution is assumed. This is a result from uncertainties around the river bifurcation points, the geometry, hydraulic roughness and lateral inflow (Stijnen et al, 2002).

The results that are presented here are based on a recent study (Stijnen et al, 2002). We made the computations including the entire shape of the discharge wave. Given the peak of the discharge wave, the entire shape is assumed to be known. With the "peak" of the wave we mean in this case the highest discharge within a single wave that has a constant value for a period of 12 hours (an example wave is shown in Figure 3).

Because the number of random variables is relatively small, this enabled us to use numerical integration, instead of other approximation techniques.

The design discharge (the discharge with an annual probability equal to the safety standard of 1/1250) is equal to 16000 m³/s for the Rhine river (Parmet et al, 2002). With the design discharge, the design water levels (DWLs) along the river branches are known. For every location along the Rhine branches, so-called

QH-relations are available that couple the discharge (Q [m³/s]) with the local water levels (H [m]). In order to obtain a consistent set of computations, the height of the dikes at the investigated locations are assumed equal to the design water level plus an additional minimum safety margin of 0.5 meter (see also Section 2).

In this paper we present the results with respect to one flood management measure: the use of retention areas in case of emergency situations. The efficiency of other measures (such as "Room for the River" and dike heightening) is studied in Stijnen et al, 2002, but are not presented in this paper. With regard to retention, we investigated a single area with a volume of 250 Mm³, near the city of Lobith. This volume is inspired by the ideas in a recent advice of the committee Emergency Retention Areas. (Commissie Noodoverloopgebieden, 2002). The inlet construction is considered to be "ideal", meaning that no restrictions are posed on the amount of inflow, etc. The inlet sill is kept at a fixed level of 16000 m³/s (the level it should be according to the current design practice).

In the case of retention the shape and peak of a discharge wave become important, because they determine the volume of water that needs to be withdrawn from the river. The peak of the wave when retention is used, can vary in height per location. It is also possible that the time at which the peak occurs shifts (Figure 4).

Five sets of computations have been performed for each of the six locations. In each of these five sets another random variable has been added, in order to observe the impact that each additional random variable has on the failure probabilities. In the next subsections we will more closely examine these five computations.

5.1 Random variable: discharge (height of dike section: DWL [m + NAP])

In this first calculation, we primarily wanted to establish a base for the rest of the computations. The results

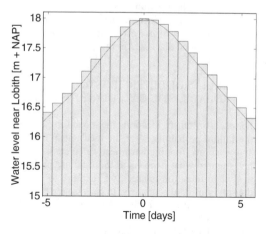

Figure 3. An example of a discretised water level wave for a discharge wave with a peak of 16000 m³/s at Lobith, with independent blocks of 12 hours.

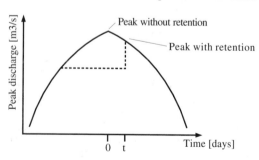

Figure 4. Illustration that shows the shift of the "peak" of a discharge wave in a situation with and without retention. The peak in the situation without retention occurs at time "0" days, while in the situation with retention the peak occurs at time "t" days.

are straightforward, and can be found directly from discharge frequency function (commonly known as the working line). The annual failure probability that is found for each location is equal to the current, designated safety level of the dikes in the upper river branches region: 1/1250. This is indeed equal to the annual exceedance probability of the design discharge of $16000\,\mathrm{m^3/s}$. In this situation, where we have only a single random variable, it is possible to select a single discharge for which the dike section fails for the first time. A discharge that is higher than the $16000\,\mathrm{m^3/s}$ level, will cause a dike section to fail immediately.

When we make use of the measure retention, the failure probability for each of the five locations decreases substantially, and is reduced to 1/4548 per year. In each case we also calculated the corresponding *critical discharge*, which is the lowest discharge that causes failure of the dike section. After retention the critical discharge is no longer equal to $16000\,\mathrm{m^3/s}$ but has increased to $17700\,\mathrm{m^3/s}$ (see also Figure 5). Even though the retention area starts to fill up, there is still a positive effect visible on the water levels downstream. The impact of retention is no longer noticeable for discharges above $18700\,\mathrm{m^3/s}$, where the two working lines are equal again.

Obviously, there is a very large, positive impact of retention for this calculation. The efficiency of the measure retention on the failure probability in this case is equal to $E_{ret} = 4548/1250 = 3.6$.

5.2 Random variable: discharge (dike height: DWL + 0.5 [m + NAP])

Again, we focus only on the discharge as a random variable, but this time the dike sections include an

additional margin of 0.5 meter (which is needed for local wind waves). For each location, the discharge that is required to raise the water level to this additional height is different. Hence, the results are different for each location, because the relations between the discharge and the water level (the *QH*-relations) become important. The failure probabilities are collected in Table 1, and the corresponding critical discharges are shown in Table 2.

The first thing we notice is that the failure probabilities without retention have become smaller (compared to the previous section). From the results it becomes clear that the *QH*-relation at Amerongen is very flat: a huge discharge is required to cause overflowing of the dike when it has been heightened by 0.5 meter. The steepest *QH*-relation is the one for Opijnen, and consequently this location also has one of the lowest critical discharges.

The results for retention vary somewhat. Clearly the effect of retention on the failure probability is positive, although not nearly as large as in the computation of Section 5.1. This can also be noticed in the limited efficiency. The reason for this diminishing effect is largely due to the fact the inlet sill is kept at a fixed level of $16000\,\mathrm{m^3/s}$. For the location of Amerongen, the retention area has already been filled before the peak of the discharge wave arrives, and is therefore useless unless the inlet sill is raised.

Table 1. Failure probabilities, caused by overflow of a dike section with height (DWL + 0.5 [m + NAP]).

| Location | Failure probability [−] | | Efficiency |
	Without retention	With retention	
Lobith	1/3907	1/5712	1.46
Millingen	1/6163	1/6900	1.12
Tiel	1/4215	1/5712	1.36
Opijnen	1/3907	1/5712	1.46
Amerongen	1/28160	1/28160	1.00
D.W.	1/7441	1/8352	1.12

Table 2. Critical discharges in the situation of overflow of a dike section with height (DWL + 0.5 [m + NAP]).

| Location | Critical discharge [m³/s] | |
	Without retention	With retention
Lobith	17500	18000
Millingen	18100	18300
Tiel	17600	18000
Opijnen	17500	18000
Amerongen	20100	20100
D.W.	18400	18500

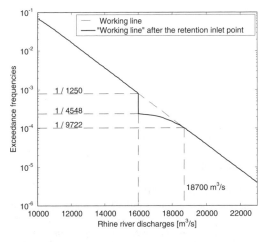

Figure 5. The impact of retention on the exceedance frequencies of the Rhine discharge downstream the inlet construction.

Table 3. The used standard deviations of the water level per river branch.

| River branch | Standard deviation [m] | |
	Without retention	With retention
Upper Rhine	0.11	0.15
Waal	0.12	0.14
Lower Rhine	0.17	0.17
IJssel	0.25	0.25

Table 4. Failure probabilities, caused by overtopping of a dike section, including both discharge and water level as random variables (DWL + 0.5 [m + NAP]).

| | Failure probability [−] | | |
Location	Without retention	With retention	Efficiency
Lobith	1/3265	1/5119	1.57
Millingen	1/4295	1/5965	1.39
Tiel	1/3311	1/5248	1.59
Opijnen	1/3151	1/5107	1.62
Amerongen	1/7387	1/9752	1.32
D.W.	1/2494	1/3547	1.42

Ideally, the discharge level for which water should be drawn-off into the retention area is equal to the critical discharge for that specific location (see Table 2).

5.3 Random variables: discharge and water level (height of dike section: DWL + 0.5 [m + NAP])

In addition to the discharge, we now also include the water level as a random variable. The height of the dike section is again equal to DWL + 0.5 [m + NAP]. The total discharge wave is now included. With two random variables it is possible that failure of a dike section can occur not only at the peak of the discharge waves (as was the case in Sections 5.1 and 5.2), but also at discharge levels *below* the peak.

It is assumed that the uncertainty surrounding the bifurcation points of the river can be modeled by a normal distribution with a standard deviation that is equal to 1% of the inflow at a bifurcation point. The mean of the distribution is assumed to be equal to the local water level.

Because of lateral inflow and differences in roughness and geometry, the uncertainty in the water level is different for specific branches of the river. For each of the river branches, the used standard deviations can be found in Table 3. The branches that start after the two bifurcation points in the Rhine (the Lower Rhine and the IJssel) have the largest standard deviation, which is caused by the fact that the uncertainties surrounding these bifurcation points add up.

When retention is introduced as a measure, an additional uncertainty is incorporated that is related to the uncertainties of retention. It is assumed that this uncertainty is different for each of the branches of the Rhine. The adapted standard deviations per branch in the case of retention are also shown in Table 3.

The results of the calculations can be found in Table 4. The failure probabilities in the situation without and with retention are both shown, as well as the resulting efficiency factors.

We see that the impact of the additional uncertainty in the water level has a distinctly negative impact on the failure probability (compared to previous sections). In particular for the locations with large standard deviations the effects are quite large (such as Amerongen and the Duursche Waarden).

The efficiency of retention has even increased compared to the previous section. The reason for this is the combination of uncertainties in the water level and the use of the shape of the discharge wave. In the case of overflowing of a dike section without uncertainties in the water level, it is possible to select a single discharge for which the dike section will overflow. A location such as Amerongen, for which the retention area has already been completely filled before the peak of the discharge wave passes, will never profit from retention. When the uncertainties in the water level are included there are multiple times within one discharge wave at which overflowing may occur. This increases both the failure probabilities and the efficiency, because other discharges (water levels) besides the peak of the wave are important.

5.4 Random variables: discharge, wind direction and wind speed (height of dike section: DWL + 0.5 [m + NAP])

In this calculation we did not only look at overflowing of a dike section, but at wave run-up due to the effect of the wind as well. So instead of uncertainties in the water level, we now added the speed and direction of the wind as random variables, besides the discharge. Again, the effect of the entire shape of the discharge wave is important, because failure of a dike section may occur not just at the peak. The results for both the failure probabilities and the efficiency can be found in Table 5.

The failure probabilities have increased significantly in comparison to Table 3, even more so than in Table 4. A location that seems particularly vulnerable to effects of wind-induced waves is Opijnen, where the failure probability even drops below the safety standard of 1/1250. The effect of retention is also very poor for this location.

Table 5. Failure probabilities caused by overtopping of a dike section, with discharge, wind direction and wind speed as random variables (DWL + 0.5 [m + NAP]).

Location	Failure probability [−]		
	Without retention	With retention	Efficiency
Lobith	1/1708	1/2398	1.40
Millingen	1/1741	1/2254	1.29
Tiel	1/1856	1/2691	1.45
Opijnen	1/1149	1/1293	1.13
Amerongen	1/5054	1/7357	1.46
D.W.	1/2091	1/2819	1.35

Table 6. Failure probabilities caused by overtopping of a dike section, with discharge, water level, wind direction and wind speed as random variables (DWL + 0.5 [m + NAP]).

Location	Failure probability [−]		
	Without retention	With retention	Efficiency
Lobith	1/1583	1/2055	1.20
Millingen	1/1514	1/1808	1.19
Tiel	1/1678	1/2219	1.32
Opijnen	1/1064	1/1169	1.10
Amerongen	1/2059	1/2458	1.19
D.W.	1/1015	1/1106	1.09

5.5 Random variables: discharge, water level, wind direction and wind speed (height of dike section: DWL + 0.5 [m + NAP]).

The computations in this section are a combination of the random variables in Sections 5.3 and 5.4. Again, we investigated only the failure mechanism of over-topping due to overflowing of a dike and due to wave run-up. The results for both the failure probabilities and the efficiency can be found in Table 6.

We see that besides the location of Opijnen, the location of Duursche Waarden now drops below the annual safety standard of 1/1250 as well. Both locations are sensitive to wind effects and uncertainties in the water level, and these properties translates itself into an unfavorable efficiency for retention. The impacts for the other locations can be less clearly distinguished, although clearly the location of Amerongen benefits from a "flat" QH-relation.

5.6 Random variables: discharge, water level, wind direction, wind speed and the parameters of the frequency-discharge relation (height of dike section: DWL + 0.5 [m + NAP]).

As a further extension to the calculations of Section 5.5, the parameters in the discharge frequency function are

Table 7. Failure probabilities caused by overtopping of a dike section, with discharge, water level, wind direction, wind speed and the parameters of the frequency-discharge relation as random variables (DWL + 0.5 [m + NAP]).

Location	Failure probability [−]		
	Without retention	With retention	Efficiency
Lobith	1/1129	1/1422	1.26
Millingen	1/1101	1/1296	1.18
Tiel	1/1185	1/1515	1.28
Opijnen	1/824	1/903	1.10
Amerongen	1/1469	1/1737	1.18
D.W.	1/807	1/880	1.09

considered to be uncertain. For details we refer to Stijnen et al, 2002). The results are shown in Table 7.

For each of the investigated locations the failure probability increases with approximately 30% to 40%. This is roughly consistent with an increase in the design discharge of 500 m³/s.

Regarding the efficiency, we see that in this case not much has changed compared to the results of Table 6. This can be explained by realizing that the effect of the uncertainties in the frequency discharge relation is present in both the situation with and without the measure retention.

6 DISCUSSION

The results of Section 5 are perhaps somewhat surprising: the retention area seems to have a big influence on the exceedance frequency of water levels, (see section 5.1), but not on the failure probability of wave overtopping. It is important to realize that (many) people and cattle are living in the retention areas, and in order to actually use the areas they have to be evacuated. A flood forecast of 1–2 days is needed for evacuation of the area. The river discharge can be forecasted reasonably accurate, but the other sources of uncertainty cannot. In the calculations it is assumed that the retention area is used when the river discharge exceeds the level of the design discharge, 16000 m³/s. However, since the dikes are higher than the water levels that correspond to the design discharge, the retention area is sometimes used unnecessary. On the other hand, the system may fail for discharges that remain lower than the design discharge (due to uncertainties in the water level).

In order to interpret more detailed the results in Section 5, we distinguish three different classes in which locations can differ: the slope of the QH-relation, the sensitivity to wind-induced waves and the uncertainty in the water levels. For each of the three classes we have made a rough indication (see Table 8).

Table 8. The influence of the QH-relation (1), the uncertainty on the water level (2), and wind-induced wave effects (3). The meaning of the symbols is given in the text.

Location	Impact on exceedance probabilities (small/large)		
	(1)	(2)	(3)
Lobith	ooo	o	o
Millingen	oo	oo	oo
Tiel	ooo	o	o
Opijnen	ooo	o	oo
Amerongen	o	ooo	ooo
D.W.	oo	ooo	ooo

An explanation of the symbols that are used in Table 8 above is given below:

(1) For the QH-relation a "o" indicates a relatively flat gradient, while a "ooo" stands for a relatively steep gradient. It is in fact a comparison of the failure probabilities of Table 1.

(2) The uncertainties in the water level are closely related to the uncertainties regarding the two bifurcation points. The Duursche Waarden and Amerongen are located after the two bifurcation points and the impacts are therefore the largest for these two locations. In this case we have compared Table 3 with Table 5.

(3) The effect of waves is largely influenced by different effective fetches and the orientation of the dike section. A "o" means that a location is not particularly sensitive to the influence of the wind (and waves). In contrast, a "ooo" indicates a location that is sensitive to wind effects. Examples of such locations are the Duursche Waarden en Opijnen. Now we compared the results of Table 3 and Table 6.

In comparison, the weight of the slope of the QH-relation is larger than that of the other two criteria. This follows for example from the results of Amerongen, which is a location that is influenced substantial by uncertainties in the water level as well as by waves, but still comes out as a relatively safe location.

7 CONCLUSIONS & RECOMMENDATIONS

The following conclusions and recommendations can be drawn:

a. The safety standard (as it is given in the Flood Protection Law) is an exceedance frequency of water levels, and along the river Rhine this is equal to 1/1250. In the actual design method only one random variable is included: the discharge at the boundary with Germany (Lobith). If we include uncertainties, the failure probability depends on the properties of the location, but is still in the same order of magnitude of the safety standard (range: 1/800–1/2000).

b. The efficiency of retention is defined by the failure probability without retention divided by the failure probability including retention as an option. This efficiency strongly depends on the inclusion of all the uncertainties which may cause failure or collapse of a dike. If we take only a single random variable into account (the peak of the river discharge) this factor may be equal to 3.6, whereas if we take *all relevant uncertainties* into account this factor falls in the range of 1.1 to 1.3. The efficiency reduces significantly if we take these uncertainties into account, but the retention areas still have a positive impact on the failure probability.

c. It is possible that the conclusions depend on the assumptions that were made in the study. We have used, for example, the heights of dikes as designed by the actual design rules (instead of the real heights of the dikes). Another assumption that has been made is that the retention area is used as soon as the river discharge rises above 16.000 m³/s. It is also very well possible that, instead of such a simple control strategy, more advanced control strategies are desirable. Moreover, the different sources of uncertainty may be reduced with additional measures, which may increase the efficiency of retention. We recommend that the influence of these assumptions on the results of this study are investigated further.

d. In a decision-analysis framework, the efficiency factor as such is not important. However, this factor can be used to assess the benefits of flood management measures with greater accuracy. In a cost-benefit analysis (see for example Vrijling, 1990 or Brinkhuis-Jak et al, 2003) the costs and benefits of measures are optimized. We recommend to use the efficiency factor in a cost-benefit analysis of (different sorts of disaster management) measures to reduce the expected flood damage.

DISCLAIMER

Any opinions expressed in this paper are those of the authors and do not necessarily reflect the position of the Dutch Ministry of Transport, Public Works and Water Management.

REFERENCES

Brinkhuis-Jak, M., Holterman, S.R., Kok, M. & Jonkman, S.N., 2003. Cost benefit analysis and flood damage mitigation in the Netherlands. *ESREL conference*, Maastricht.

Cooke, R.M., 1991. *Experts in Uncertainty. Opinion and subjective probability in Science*. London: Oxford University Press.

Commissie Noodoverloopgebieden, 2002. *Gecontroleerd overstromen (in Dutch)*. The Hague.

Dantzig, D. van, 1956. Economic decision problems for flood protection. *Econometrica*, 276–287, New haven.

Delta Committee, 1960. *Report of the Delta Committee (in Dutch)* The Hague, The Netherlands.

Geerse, C.P.M., Duits, M.T., Kalk, H.J. & Lammers, I.B.M., 2002. *Wind-waterstandstatistiek Hoek van Holland (in Dutch)* HKV CONSULTANTS & RIZA, juli 2002.

Kok, M., Douben, N., Noortwijk J.M. van & Silva, W., 1996. *Integrale Verkenning inrichting Rijntakken. Report 12: Veiligheid (in Dutch)*. WL & RWS Riza, Arnhem.

Noortwijk, J.M. van & Kalk, H.J., 2002. *Onzekerheden in terugkeertijden van rivierafvoeren.(in Dutch)*. Client: RWS-DWW. HKV CONSULTANTS, PR560, September 2002.

Parmet, B.W.A.H., Langemheen, W. van den, Chbab, E.H., Kwadijk, J.C.J., Diermanse, F. & Klopstra, D., 2002. *Analyse van de maatgevende afvoer van de Rijn te Lobith (in Dutch)*. Rijkswaterstaat, RIZA report 2002.012, 2002.

Stijnen, J.W., Kok, M. & Duits, M.T., 2002. *Onzekerheidsanalyse Hoogwaterbescherming Rijntakken, onzekerheidsbronnen en gevolgen van maatregelen (in Dutch)*, HKV CONSULTANTS, November 2002, PR464.

TAW, 1998. *Fundamentals on Water Defenses*. Technical Advisory Committee on Water Defenses, Delft.

TAW, 2000. *Van overschrijdingskans naar overstromingskans (in Dutch)*, Technical Advisory Committee on Water Defenses, Delft.

Vrijling, J.K. & Beurden, I. van, 1990. Sealevel rise: a probabilistic design problem. Coastal Engineering, 1990.

Vrouwenvelder, A.C.W.M., Steenbergen, H.M.G.M & Slijkhuis, K.A.H., 2001. *Theoriehandleiding PC-Ring, Deel A: Mechanismenbeschrijving (in Dutch)*. TNO, Delft.

Vuuren, W. van, 2002. Onzekerheden in de WAQUA – voorspellingen voor de MHW's op de Rijntakken en Maas (in Dutch). Rijkswaterstaat RIZA, Arnhem, Memo WSR 2002–006, February 2002.

Safety and Reliability – Bedford & van Gelder (eds)
© *2003 Swets & Zeitlinger, Lisse, ISBN 90 5809 551 7*

Reliability and availability evaluation of systems related to their operation processes

K. Kolowrocki

Gdynia Maritime University, Poland

ABSTRACT: The semi-markov model of the system operation processes is proposed and its selected parameters are determined. The non-renewable and renewable series-parallel systems are considered and their reliability and availability characteristics are found. Next, the joint model of the system operation process and the system reliability and availability is constructed. Finally, the application of the joint model to the port grain transportation system reliability and availability evaluation is presented.

1 SEMI-MARKOV MODEL OF SYSTEM OPERATION PROCESS

We suppose that the system is composed of n components $E_1, E_2, ..., E_n$ and during its operation process it is performing a repertory of tasks. Namely, the system at each moment $t, t \in <0, \theta>$, where θ is its operation time, is performing at most w tasks. We denote the process of changing of the system task repertory by

$$Z(t) = [Z_1(t), Z_2(t), ..., Z_w(t)],$$

where

$$Z_j(t) = \begin{cases} 1, & \text{if the system is executing} \\ & \text{the } j-\text{th task at the moment } t \\ 0, & \text{if the system is not executing} \\ & \text{the } j-\text{th task at the moment } t, \ j = 1,2,...,w. \end{cases}$$

Thus, $Z(t)$ is the process with the continuous time t, $t \in <0, \theta>$, and discrete states from the set of states $\{0,1\}^w$. We numerate the states of the process $Z(t)$ assuming that it has v different states from the set $Z = \{z^1, z^2, ..., z^v\}$ and they are of the form $z^k = [z_1^k, z_2^k, ... z_w^k]$, $k = 1, 2, ..., v$, where $z_j^k \in \{0, 1\}, j = 1, 2, ..., w$. If the process of changing of the system task repertory $Z(t)$ is semi-markov (Grabski 2002) with its conditional sojourn time θ^{kl} at the state z^k when its next state is z^l, $k, l = 1, 2, ..., v, k \neq l$, then it may be described by the vector of probabilities of the initial states $[p^k(0)]_{1 \times v}$, the matrix of the probabilities of its transitions between the states $[p^{kl}]_{v \times v}$ and the matrix of the conditional distribution functions $[H^{kl}(t)]_{v \times v}$ of the

sojourn times θ^{kl}, $k, l = 1,2, ..., v, k \neq l$, Then, the sojourn time θ^{kl} mean values are given by

$$E[\theta^{kl}] = \int_0^\infty t dH^{kl}(t), \ k,l = 1,2,...,v, \ k \neq l. \quad (1)$$

The unconditional distribution functions of the sojourn times θ^k of the process $Z(t)$ at the states z^k, $k = 1, 2, ..., v$, are given by

$$H^k(t) = \sum_{l=1}^v p^{kl} H^{kl}(t), \ k = 1,2,...,v. \quad (2)$$

The mean values $E[\theta^k]$ of the unconditional sojourn times θ^k are given by

$$E[\theta^k] = \sum_{l=1}^v p^{kl} E[\theta^{kl}], \ k = 1,2,...,v, \quad (3)$$

where $E[\theta^{kl}]$ are defined by (1).

Limit values of the transient probabilities at the states $p^k(t) = P(Z(t) = z^k)$ are given by

$$p^k = \lim_{t \to \infty} p^k(t) = \frac{\pi^k E[\theta^k]}{\sum_{l=1}^v \pi^l E[\theta^l]}, \ k = 1,2,...,v, \quad (4)$$

where the probabilities π^k of the vector $[\pi^k]_{1 \times v}$ satisfy the system of equations

$$\begin{cases} [\pi^k] = [\pi^k][p^{kl}] \\ \sum_{l=1}^v \pi^l = 1. \end{cases}$$

In the case when the sojourn times θ^{kl}, $k, l = 1, 2, \ldots, v$, $k \neq l$, have exponential distributions with the transition rates between the states γ^{kl}, i.e., if for $k, l = 1, 2, \ldots, v, k \neq l$,

$$H^{kl}(t) = P(\theta^{kl} < t) = 1 - \exp[-\gamma^{kl}t], \ t > 0, \quad (5)$$

then their mean values are determined by

$$E[\theta^{kl}] = \frac{1}{\gamma^{kl}}, \quad k, l = 1, 2, \ldots, v, \quad k \neq l, \quad (6)$$

and the probabilities of transitions between the states are given by

$$p^{kl} = \frac{\gamma^{kl}}{\sum_{j \neq k} \gamma^{kj}}, \quad k, l = 1, 2, \ldots, v, \quad k \neq l. \quad (7)$$

The unconditional distribution functions of the process $Z(t)$ sojourn times θ^k at the states z^k, $k = 1, 2, \ldots, v$, according to (2) and (5) are given by

$$H^k(t) = 1 - \sum_{l=1}^{v} p^{kl} \exp[-\gamma^{kl}t], t > 0, \ k = 1, 2, \ldots, v, \quad (8)$$

and their mean values, by (3) and (6), are

$$M^k = E[\theta^k] = \sum_{l=1}^{v} p^{kl} \frac{1}{\gamma^{kl}}, \quad k = 1, 2, \ldots, v. \quad (9)$$

Limit values of the transient probabilities $p^k(t)$ at the states z^k, according to (4), are given by

$$p^k = \lim_{t \to \infty} p^k(t) = \frac{\pi^k \cdot M^k}{\sum_{l=1}^{v} [\pi^l \cdot M^l]}, \quad k = 1, 2, \ldots, v, \quad (10)$$

where the probabilities π^k of the vector $[\pi^k]_{1 \times v}$ satisfy the system of equations

$$\begin{cases} [\pi^k] = [\pi^k][p^{kl}] \\ \sum_{l=1}^{v} \pi^l = 1, \end{cases}$$

with $[p^{kl}]_{v \times v}$ and M^k given by (7) and (9) respectively.

2 SERIES-PARALLEL SYSTEM IN ITS OPERATION PROCESS

We consider a non-homogeneous regular series-parallel system (Kolowrocki 2001) which is composed of a, $1 \leq a \leq k_n, k_n \in N$, different kinds of series subsystems

and the fraction of the i-th kind subsystem in the system is equal to q_i, where $q_i > 0$, $\sum_{i=1}^{a} q_i = 1$. Moreover, the i-th kind series subsystem consists of e_i, $1 \leq e_i \leq l_n$, $l_n \in N$, kinds of components with reliability functions

$$R^{(i,j)}(t) = 1 - F^{(i,j)}(t), j = 1, 2, \ldots, e_i,$$

and the fraction of the j-th kind component in this subsystem is equal to p_{ij}, where $p_{ij} > 0$ and $\sum_{j=1}^{e_i} p_{ij} = 1$.

The scheme of this system is shown in Figure 1.

The reliability function of the non-renewable regular non-homogeneous series-parallel system is given by

$$R_{k_n,l_n}(t) = 1 - \prod_{i=1}^{a}[1 - (R^{(i)}(t))^{l_n}]^{q_i k_n}, t \in (-\infty, \infty), \quad (11)$$

where

$$R^{(i)}(t) = \prod_{j=1}^{e_i}(R^{(i,j)}(t))^{p_{ij}}, \ t \in (-\infty, \infty), \ i = 1, 2, \ldots, a. \quad (12)$$

The lifetimes of the system components E_{ij} depends on the states of the process of changing of repertory of tasks $Z(t)$. The changes of the process $Z(t)$ states have influence on the system components E_{ij} reliability and on the system reliability structure as well. Thus, we denote the conditional reliability function of the system component E_{ij} and the conditional system reliability function, while the system is performing the task z^k, $k = 1, 2, \ldots, v$, respectively by

$$[R^{(i,j)}(t)]^{(k)} = P(T_{ij}^{(k)} \geq t / Z(t) = z^k), \ t \in <0, \infty),$$

$$i = 1, 2, \ldots, a, \ j = 1, 2, \ldots, e_i, \ k = 1, 2, \ldots, v,$$

and

$$R^{(k)}(t)] = P(T^{(k)} \geq t / Z(t) = z^k), \ t \in <0, \infty),$$

$$k = 1, 2, \ldots, v.$$

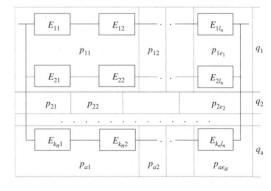

Figure 1. The scheme of a regular non-homogeneous series-parallel system.

The reliability function $[R^{(i,j)}(t)]^{(k)}$ is the conditional probability that the component E_{ij} lifetime $T_{ij}^{(k)}$ is not less than t, while the process $Z(t)$ is at the operation state z^k. Similarly, the reliability function $R^{(k)}(t)$ is the conditional probability that the system lifetime $T^{(k)}$ is not less than t, while the process $Z(t)$ is at the operation state z^k.

Then, the unconditional reliability function of the system $R(t) = P(T > t)$, $t \in <0, \infty)$, where T is the unconditional lifetime of the system, is given by

$$R(t) = \sum_{k=1}^{v} p^k(t) R^{(k)}(t), \ t \in <0, \infty). \tag{13}$$

In the case when the system operation time θ is large enough the transient probability $p^k(t)$ can be replaced by p^k given by (4) or by (10) respectively and the last formula takes form

$$R(t) = \sum_{k=1}^{v} p^k R^{(k)}(t), \ t \in <0, \infty). \tag{14}$$

The formula (14) for the non-renewable regular non-homogeneous series-parallel system takes form

$$R_{k_n, l_n}(t) = \sum_{k=1}^{v} p^k R_{k_n, l_n}^{(k)}(t), \ t \in <0, \infty), \tag{15}$$

where $R_{k_n, l_n}^{(k)}(t)$ is this system conditional reliability function, while it is performing the task $z^k, k = 1, 2, ..., v$.

In the particular exponential case i.e., if the component conditional reliability functions while the system is performing the task z^k, $k = 1, 2, ..., v$, are given by

$$[R^{(i,j)}(t)]^{(k)} = \exp[-\lambda_{ij}^k t] \text{ for } t \geq 0, \ \lambda_{ij}^k > 0,$$
$$i = 1, 2, ..., a, j = 1, 2, ..., e_i, \tag{16}$$

we get

$$[R^{(i)}(t)]^{(k)} = \exp[-\sum_{j=1}^{e_i} p_{ij} \lambda_{ij}^k t] \text{ for } t \geq 0, \ i = 1, 2, ..., a, \tag{17}$$

and next by (12)

$$R_{k_n, l_n}^{(k)}(t) = 1 - \prod_{i=1}^{a} [1 - \exp[-l_n p_{ij} \lambda_{ij}^k t]]^{q_i k_n}, t \geq 0. \tag{18}$$

In this case the mean value and the variance of the regular non-homogeneous series-parallel system lifetime are

$$m = \sum_{k=1}^{v} p^k E[T^k], \tag{19}$$

and

$$\sigma^2 = \sum_{k=1}^{v} p^k \sigma^2[T^k], \tag{20}$$

where for $k = 1, 2, ..., v$,

$$E[T^k] = \int_0^\infty R_{k_n, l_n}^{(k)}(t) dt \ \ t \geq 0, \tag{21}$$

and

$$\sigma^2[T^k] = [2\int_0^\infty t \, R_{k_n, l_n}^{(k)}(t) dt]^2 - [ET^k]^2, \ t \geq 0, \tag{22}$$

and p^k are given by either (4) or by (10).

3   RELIABILITY OF A PORT GRAIN TRANSPORTATION SYSTEM

The grain elevator is the basic object of The Baltic Grain Terminal of the Port of Gdynia assigned to service export and import grain and its clearing (Kolowrocki et al. 2002). One of the basic elevator functions is railway truck loading with grain. The railway trucks loading is performed in the following successive elevator operation steps:

– gravitational passing of grain from the storage placed on the 8-th elevator flour through 45 hall to horizontal conveyers placed in the elevator basement,
– transport of grain through horizontal conveyers to vertical bucket elevators transporting grain to the main distribution station placed at 9-th elevator flour,
– gravitational dumping of grain through main distribution station to the balance placed at 6-th elevator flour,
– dumping weighted grain through the complex of flaps placed on 4-th elevator flour to horizontal conveyers placed at 2-nd elevator flour,
– dumping of grain from horizontal conveyers to worm conveyers,
– dumping of grain from worm conveyers to railway trucks.

In the railway trucks loading with grain the following elevator transportation subsystems take part: S_1 – horizontal conveyers of the first kind, S_2 – vertical bucket elevators, S_3 – horizontal conveyers of the second kind, S_4 – worm conveyers.

The subsystem S_1 is composed of two identical belt conveyers. In each conveyer there are 1 belt with reliability functions

$R^{(1,1)}(t) = \exp[-0.0125t]$ for $t \geq 0$,

2 drums with reliability functions

$R^{(1,2)}(t,1) = \exp[-0.0015t]$ for $t \geq 0$,

117 channelled rollers with reliability functions

$R^{(1,3)}(t,1) = \exp[-0.005t]$ for $t \geq 0$,

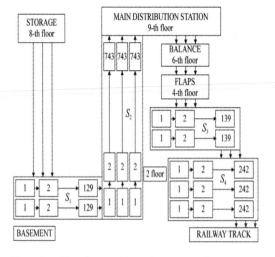

Figure 2. The scheme of the grain transportation system.

and 9 supporting rollers with reliability functions

$$R^{(1,4)}(t,1) = \exp[-0.004t] \text{ for } t \geq 0.$$

The subsystem S_2 is composed of three identical bucket elevators. In each elevator there are 1 belt having reliability functions

$$R^{(1,1)}(t) = \exp[-0.025t] \text{ for } t \geq 0,$$

2 drums having reliability functions

$$R^{(1,2)}(t) = \exp[-0.0015t] \text{ for } t \geq 0,$$

740 buckets having reliability functions

$$R^{(1,3)}(t) = \exp[-0.03t] \text{ for } t \geq 0.$$

The subsystem S_3 is composed of two identical belt conveyers. In each conveyer there are 1 belt having reliability functions

$$R^{(1,1)}(t) = \exp[-0.0125t] \text{ for } t \geq 0,$$

2 drums having reliability functions

$$R^{(1,2)}(t) = \exp[-0.0015t], \text{ for } t \geq 0,$$

117 channelled rollers having reliability functions

$$R^{(1,3)}(t) = \exp[-0.005t] \text{ for } t \geq 0,$$

and 19 supporting rollers having reliability functions

$$R^{(1,4)}(t) = \exp[-0.004t] \text{ for } t \geq 0.$$

The subsystem S_4 is composed of three chain conveyers each of them composed of a wheel driving the belt, a reversible wheel and 160, 160 and 240 links respectively. The subsystem consists of 3 conveyers. Two of them are composed of 162 components and the remaining one is composed of 242 components. Thus, it is a non-homogeneous non-regular multi-state series-parallel system. In order to make it a regular system we conventionally complete two first conveyers having 162 components with 80 components that do not fail. After this supplement the subsystem consists of $k_n = 3$ conveyers each of them composed of $l_n = 242$ components. In two of them there are 2 driving wheels with reliability functions

$$R^{(1,1)}(t) = \exp[-0.005t] \text{ for } t \geq 0,$$

160 links with reliability functions

$$R^{(1,2)}(t) = \exp[-0.012t] \text{ for } t \geq 0,$$

and 80 components with "reliability functions"

$$R^{(1,3)}(t) = \exp[-\beta t] \ t \geq 0, \text{ where } \beta = 0.$$

The third conveyer is composed of 2 driving wheels with reliability functions

$$R^{(2,1)}(t) = \exp[-0.022t] \text{ for } t \geq 0,$$

and 240 links with reliability functions

$$R^{(2,2)}(t) = \exp[-0.034t] \text{ for } t \geq 0.$$

Taking into account the operation process of the considered transportation system we distinguish the following its three tasks: task 1 – the system operation with the largest efficiency when all components of the subsystems S_1, S_2, S_3, and S_4, are used, task 2 – the system operation with less efficiency system when 1 conveyer of the subsystem S_1, 2 elevators of the subsystem S_2, 1 conveyer of the subsystem S_3 and 2 conveyers of the subsystem S_4 are used, task 3 – the system operation with least efficiency when 1 conveyer of the subsystem S_1, 1 elevator of the subsystem S_2, 1 conveyer of the subsystem S_3 and 1 conveyer of the subsystem S_4 are used only.

Since the system tasks are disjoint then its operation states belong to the set $Z = \{z^1, z^2, z^3\}$, where $z^1 = [1, 0, 0]$, $z^2 = [0, 1, 0]$, $z^3 = [0, 0, 1]$. Assuming the following matrix of the conditional distribution functions of the system sojourn times θ^{kl}, $k, l = 1, 2, 3$,

$$[H^{kl}(t)] = \begin{bmatrix} 0 & 1-e^{-5t} & 1-e^{-10t} \\ 1-e^{-40t} & 0 & 1-e^{-50t} \\ 1-e^{-10t} & 1-e^{-20t} & 0 \end{bmatrix}$$

Hence, by (7), the probabilities of transitions between the states are given by

$$[p^{kl}] = \begin{bmatrix} 0 & \dfrac{1}{3} & \dfrac{2}{3} \\ \dfrac{4}{9} & 0 & \dfrac{5}{9} \\ \dfrac{1}{3} & \dfrac{2}{3} & 0 \end{bmatrix}$$

and further, according to (8), the unconditional distribution functions of the process $Z(t)$ sojourn times θ^k in the states z^k, $k = 1, 2, 3$, are given by

$$\begin{cases} H^1(t) = 1 - \dfrac{1}{3}\exp[-5t] - \dfrac{2}{3}\exp[-10t] \\[2mm] H^2(t) = 1 - \dfrac{4}{9}\exp[-40t] - \dfrac{5}{9}\exp[-50t] \\[2mm] H^3(t) = 1 - \dfrac{1}{3}\exp[-10t] - \dfrac{2}{3}\exp[-20t] \end{cases} \quad (23)$$

and their mean values, by (9), are

$$\begin{cases} M^1 = E[\theta^1] = \dfrac{1}{3}\dfrac{1}{5} + \dfrac{2}{3}\dfrac{1}{10} = \dfrac{6}{45} \\[2mm] M^2 = E[\theta^2] = \dfrac{4}{9}\dfrac{1}{40} + \dfrac{5}{9}\dfrac{1}{50} = \dfrac{1}{45} \\[2mm] M^3 = E[\theta^3] = \dfrac{1}{3}\dfrac{1}{10} + \dfrac{2}{3}\dfrac{1}{20} = \dfrac{3}{45} \end{cases} \quad (24)$$

Since from the system of equations

$$\begin{cases} [\pi^1, \pi^2, \pi^3] = [\pi^1, \pi^2, \pi^3]\begin{bmatrix} 0 & \dfrac{1}{3} & \dfrac{2}{3} \\ \dfrac{4}{9} & 0 & \dfrac{5}{9} \\ \dfrac{1}{3} & \dfrac{2}{3} & 0 \end{bmatrix} \\[4mm] \pi^1 + \pi^2 + \pi^3 = 1 \end{cases}$$

we get

$$\pi^1 = \dfrac{1}{3}, \quad \pi^2 = \dfrac{1}{3}, \quad \pi^3 = \dfrac{1}{3},$$

then the limit values of the transient probabilities $p^k(t)$ at the operational states z^k, according to (10), are given by

$$p^1 = 0.6, \quad p^2 = 0.1, \quad p^3 = 0.3. \quad (25)$$

At the operational state 1 the subsystem S_1 becomes a non-homogeneous regular series-parallel system with parameters

$$k_n = k = 2, l_n = 129, a = 1, q_1 = 1, e_1 = 4,$$

$$p_{11} = 1/129, p_{12} = 2/129, p_{13} = 117/129, p_{14} = 9/129.$$

Applying (18), we get the subsystem S_1 reliability function given by

$$R^{(1)}_{2,129}(t) = 1 - [1 - \exp[-0.6365t]]^2 \text{ for } t \geq 0.$$

At the operational state 1 the subsystem S_2 becomes a non-homogeneous regular series-parallel system with parameters

$$k_n = k = 3, l_n = 743, a = 1, q_1 = 1, e_1 = 3,$$

$$p_{11} = 1/743, p_{12} = 2/743, p_{13} = 740/743.$$

Applying (18), we get the subsystem S_2 reliability function given by

$$R^{(1)}_{3,743}(t) = 1 - [1 - \exp[-22.228t]]^3 \text{ for } t \geq 0.$$

At the operational state 1 the sub system S_3 is a non-homogeneous regular series-parallel system with parameters

$$k_n = k = 2, l_n = 139, a = 1, q_1 = 1, e_1 = 4,$$

$$p_{11} = 1/139, p_{12} = 2/139, p_{13} = 117/139, p_{14} = 19/139.$$

Applying (18), we get the subsystem S_3 reliability function given by

$$R^{(1)}_{2,139}(t) = 1 - [1 - \exp[-0.6765t]]^2 \text{ for } t \geq 0.$$

At the operational state 1 the subsystem S_4 becomes a non-homogeneous regular series-parallel system with parameters

$$k_n = k = 3, l_n = 242, a = 2, q_1 = 2/3, q_2 = 1/3,$$

$$e_1 = 3, p_{11} = 2/242, p_{12} = 160/242, p_{13} = 80/242,$$

$$e_2 = 2, p_{21} = 2/242, p_{22} = 240/242,$$

941

Applying (18), we get the subsystem S_4 reliability function given by

$$R_{3,242}^{(1)}(t) = 1 - [1 - \exp[-1.93t]]^2 [1-\exp[-8.204t], \ t \geq 0.$$

Since the considered subsystems create a series structure in a reliability sense, then the reliability function of the whole transportation system, for t $\geq$ 0, is given by

$$\bar{R}^{(1)}(t) = R_{2,129}^{(1)}(t)\, R_{3,743}^{(1)}(t)\, R_{2,139}^{(1)}(t)\, R_{3,242}^{(1)}(t)$$

$$= 24\exp[-25.471t] - 24\exp[-33.675t]$$

$$- 24\exp[-47.699t] + 24\exp[-55.903t]$$

$$- 12\exp[-26.1075t] - 12\exp[-27.401t]$$

$$+ 12\exp[-31.745t] + 12\exp[-34.3115t]$$

$$+ 12\exp[-35.605t] + 12\exp[-48.3355t]$$

$$+ 12\exp[-49.629] - 12\exp[-53.973t]$$

$$- 12\exp[-56.5395t] - 12\exp[-57.833t]$$

$$+ 8\exp[-69.927t] + 8\exp[-78.131t]$$

$$+ 6\exp[-28.0375t] - 6\exp[-32.3815t]$$

$$- 6\exp[-36.241t] - 6\exp[-50.2655t]$$

$$+ 6\exp[-54.6095t] + 6\exp[-58.4695t]$$

$$- 4\exp[-70.5635t] - 4\exp[-71.857t]$$

$$+ 4\exp[-76.201t] + 4\exp[-78.7675t]$$

$$+ 4\exp[-80.061t] + 2\exp[-72.4935t]$$

$$- 2\exp[-76.8375t] - 2\exp[-80.6975t] \tag{26}$$

and according to (21) and (22) the system lifetime mean value and its standard deviation are

$$E[T^{(1)}] = 0.3616, \ \sigma(T^{(1)}) = 0.4204. \tag{27}$$

At the operational state 2 the subsystem S_1 becomes a non-homogeneous regular series-parallel system with parameters

$$k_n = k = 1, l_n = 129, a = 1, q_1 = 1, e_1 = 4,$$

$$p_{11} = 1/129, p_{12} = 2/129, p_{13} = 117/129, p_{14} = 9/129.$$

Applying (18), we get the subsystem S_1 reliability function given by

$$R_{1,129}^{(2)}(t) = \exp[-0.6365t] \text{ for } t \geq 0.$$

At the operational state 2 the sub system S_2 becomes a non-homogeneous regular series-parallel system with parameters

$$k_n = k = 2, l_n = 743, a = 1, q_1 = 1,$$

$$e_1 = 3, p_{11} = 1/743, p_{12} = 2/743, p_{13} = 740/743.$$

Applying (18), we get the S_2 reliability function given by

$$R_{2,743}^{(2)}(t) = 1 - [1-\exp[-22.228t]]^2 \text{ for } t \geq 0.$$

At the operational state 2 the subsystem S_3 becomes a non-homogeneous regular series-parallel system with parameters

$$k_n = k = 1, l_n = 139, a = 1, q_1 = 1, e_1 = 4,$$

$$p_{11} = 1/139, p_{12} = 2/139, p_{13} = 117/139, p_{14} = 19/139.$$

Applying (18), we get the subsystem S_3 reliability function given by

$$R_{1,139}^{(2)}(t) = \exp[-0.6765t] \text{ for } t \geq 0.$$

At the operational state 2 the subsystem S_4 is a non-homogeneous regular series-parallel system with parameters

$$k_n = k = 2, l_n = 242, a = 1, q_1 = 1,$$

$$e_1 = 3, p_{11} = 2/242, p_{12} = 160/242, p_{13} = 80/242,$$

Applying (18), we get the subsystem S_4 reliability function given by

$$R_{2,242}^{(2)}(t) = 1 - [1 - \exp[-1.93t]]^2 \text{ for } t \geq 0.$$

Since the considered subsystems create a series structure in a reliability sense, then the reliability function of the whole transportation system is given by

$$\bar{R}^{(2)}(t) = R_{1,129}^{(2)}(t)\, R_{2,743}^{(2)}(t)\, R_{1,139}^{(2)}(t)\, R_{2,242}^{(2)}(t)$$

$$= 4\exp[-25.471t] - 2\exp[-27.401t] \tag{28}$$

$$- 2\exp[-47.699t] + \exp[-49.629t], \ t \geq 0,$$

and according to (21) and (22) the system lifetime mean value and its standard deviation are

$$E[T^{(2)}] = 0.0623, \ \sigma(T^{(2)}) = 0.0467. \tag{29}$$

942

At the operational state 3 the subsystem S_1 becomes a non-homogeneous regular series-parallel system with parameters

$k_n = k = 1, l_n = 129, a = 1, q_1 = 1, e_1 = 4,$

$p_{11} = 1/129, p_{12} = 2/129, p_{13} = 117/129, p_{14} = 9/129.$

Applying (18), we get the subsystem S_1 reliability function given by

$R^{(3)}_{1,129}(t) = \exp[-0.6365t]$ for $t \geq 0$.

At the operational state 3 the subsystem S_2 becomes a non-homogeneous regular series-parallel system with parameters

$k_n = k = 1, l_n = 743, a = 1, q_1 = 1,$

$e_1 = 3, p_{11} = 1/743, p_{12} = 2/743, p_{13} = 740/743.$

Applying (18), we get the approximate subsystem S_2 reliability function given by

$R^{(3)}_{1,743}(t) = \exp[-22.228t]$ for $t \geq 0$.

At the operational state 3 the subsystem S_3 is a non-homogeneous regular series-parallel system with parameters

$k_n = k = 1, l_n = 139, a = 1, q_1 = 1, e_1 = 4,$

$p_{11} = 1/139, p_{12} = 2/139, p_{13} = 117/139, p_{14} = 19/139.$

Applying (18), we get the approximate subsystem S_3 reliability function given by

$R^{(3)}_{1,139}(t) = \exp[-0.6765t]$ for $t \geq 0$.

At the operational state 3 the subsystem S_4 becomes a non-homogeneous regular series-parallel with parameters

$k_n = k = 1, l_n = 242, a = 1, q_1 = 1,$

$e_1 = 3, p_{11} = 2/242, p_{12} = 160/242, p_{13} = 80/242,$

Applying (18), we get the subsystem S_4 reliability function given by

$R^{(3)}_{1,242}(t) = \exp[-1.93t]$ for $t \geq 0$.

Since the considered subsystems create a series structure in a reliability sense, then the reliability function of the whole transportation system is given by

$$\bar{R}^{(3)}(t) = R^{(3)}_{1,129}(t) R^{(3)}_{1,743}(t) R^{(3)}_{1,139}(t) R^{(3)}_{1,242}(t)$$

$$\tag{30}$$

$$= \exp[-25.471t] \text{ for } t \geq 0$$

and according to (21) and (22) the system lifetime mean value and its standard deviation are

$$E[T^{(3)}] = 0.0393, \ \sigma(T^{(3)}) = 0.0393. \tag{31}$$

Finally, considering (13) and (25), the system unconditional reliability is given by

$$R(t) = 0.6 \ \bar{R}^{(1)}(t) + 0.1 \ \bar{R}^{(2)}(t) + 0.3\bar{R}^{(3)}(t), \tag{32}$$

where $\bar{R}^{(1)}(t)$, $\bar{R}^{(2)}(t)$ and $\bar{R}^{(3)}(t)$ are respectively given by (26), (28) and (30). Hence, applying (19)–(20) and (27), (29) and (31), we get the mean value and the standard deviation of the system unconditional lifetime are

$$m = 0.6 \cdot 0.3616 + 0.1 \cdot 0.0623 + 0.3 \cdot 0.0387$$

$$\cong 0.235 \text{ year}, \tag{33}$$

$$\sigma = \sqrt{0.6(0.4204)^2 + 0.1(0.0467)^2 + 0.3(0.0397)^2}$$

$$\cong 0.327 \text{ year}. \tag{34}$$

In the case when the system is renewable we assume that the time of the system renovation has exponential distribution function, i.e.,

$$G_{k_n,l_n}(t) = 1 - \exp[-200t] \text{ for } t \geq 0,$$

with the mean value and the standard deviations given respectively bay

$$m_0 = 0.005 \text{ year and } \sigma_0 = 0.005 \text{ year} \tag{35}$$

Since by (33) and (35) m_0 is small in comparison to m, then in the system reliability evaluation we may apply characteristics suitable for the renewable systems with the ignored renewal time. Considering the results given in Kolowrocki et al. (2002) and (33)–(34), we obtain the following conclusions:

i) the distribution function $F^{(n)}(t) = P(S_n < t)$ of the time S_n up to the system n-th failure is approximately normal, i.e.,

$$F^{(n)}(t) \cong F_{N(0,1)}(\frac{t - 0.235n}{0.327\sqrt{n}}), \ t \in (-\infty, \infty),$$

$$n = 0,1,2,...,$$

ii) the mean value and the variance of the time S_n up to the system n-th failure respectively are

$$E[S_n] \cong 0.235n, \ D(S_n) \cong 0.107n, \ n = 0,1,2,...,$$

943

iii) the distribution of the number of system failures $N(t)$ up to the moment t, $t \geq 0$, for sufficiently large t, is given by

$$P(N(t) = n) \cong F_{N(0,1)}\left(\frac{t - 0.235n}{0.327\sqrt{n}}\right)$$

$$- F_{N(0,1)}\left(\frac{t - 0.235(n+1)}{0.327\sqrt{n+1}}\right), \quad n = 0,1,2,\ldots,$$

iv) the mean value and the variance of the number of system failures $N(t)$ up to the moment t, t, ≥ 0, for sufficiently large t, are respectively given by

$$H(t) \cong 4.255t, \quad D(t) \cong 4.308t.$$

Considering that the system is renewable and the renewal time is not ignored and applying the results given in Kolowrocki et al. (2002) and in (33)–(35), we obtain the following conclusions:

i) the distribution function $\bar{F}^{(n)}(t) = (\bar{S}_n < t)$ of the time $\bar{S}_n$ up to the system n-th renovation is approximately normal $N(0.24\,n, 0.327\sqrt{n})$, i.e.,

$$\bar{F}^{(n)}(t) \cong F_{N(0,1)}\left(\frac{t - 0.24n}{0.327\sqrt{n}}\right), \quad t \in (-\infty, \infty), \quad n = 1,2,\ldots,$$

ii) the mean value and the variance time $\bar{S}_n$ up to the system n-th renovation respectively are

$$E[\bar{S}_n] \cong 0.24n, \quad D[\bar{S}_n] \cong 0.107n,$$

iii) the distribution function $\bar{F}^{(n)}(t) = P(\bar{S}_n < t)$ of the time $\bar{S}_n$ up to the n-th failure is given by

$$\bar{F}^{(n)}(t) \cong F_{N(0,1)}\left(\frac{t - 0.24n + 0.005}{\sqrt{0.107n - 0.000025}}\right), \quad t \in (-\infty, \infty),$$

$$n = 1,2,\ldots,$$

iv) the mean value and the variance of the time $\bar{S}_n$ up to the system n-th failure respectively are

$$E[\bar{S}_n] \cong 0.24n + 0.05,$$

$$D[\bar{S}_n] \cong 0.107n + 0.000025,$$

v) the distribution of the number of the system renovation $\bar{N}(t)$ up to the moment t, $t \geq 0$, for sufficiently large t, is given by

$$P(\bar{N}(t) = n) \cong F_{N(0,1)}\left(\frac{t - 0.24}{0.327\sqrt{n}}\right)$$

$$- F_{N(0,1)}\left(\frac{t - 0.24(n+1)}{0.327\sqrt{(n+1)}}\right), \quad n = 1,2,\ldots,$$

vi) the mean value and the variance of the system renovation $\bar{N}(t)$ up to the moment t, t, ≥ 0, for sufficiently large t, are respectively given by

$$\bar{H}(t) \cong 4.167t, \quad \bar{D}(t) \cong 7.74t,$$

vii) the distribution function of the number of the system failures $\bar{N}(t)$ up to the moment t, t, ≥ 0, for sufficiently large t, is given by

$$P(\bar{N}(t) = n) \cong F_{N(0,1)}\left(\frac{t - 0.24n + 0.005}{\sqrt{0.107n + 0.000025}}\right)$$

$$- F_{N(0,1)}\left(\frac{t - 0.24(n+1) + 0.005}{\sqrt{0.107(n+1) - 0.000025}}\right), \quad n = 1,2,\ldots,$$

viii) the mean value and the variance of the system failures $\bar{N}(t)$ up to the moment t, t, ≥ 0, for sufficiently large t, are respectively given by

$$\bar{H}(t) \cong 4.167t, \quad \bar{D}(t) \cong 7.74t,$$

ix) the system availability coefficient at the moment t is given by

$$K(t) \cong 0.979, \quad t \geq 0,$$

x) the system availability coefficient at the time interval $(< t, t + \tau)$, $\tau > 0$, is given by

$$K(t, t + \tau) \cong 0.979 \int_{\tau}^{\infty} R(t)dt, \quad t \geq 0, \tau > 0,$$

where $R(t)$ is given by (32).

4 CONCLUSIONS

The paper propose an approach to the solution of practically very important problem of joining the systems reliability and their operation processes. To involve the interactions between the systems operation processes and their changing in time reliability structures a semi-markov model of the system operation processes is applied. This approach gives practically important and not difficult in everyday usage tool for reliability and availability evaluation of the systems with changing reliability structures during their operation processes. Application of the proposed methods is illustrated in the reliability and availability evaluation of the port grain transportation system. The reliability data concerned with the operation process and reliability functions of the port grain transportation system components are not precise. They are coming from experts and are concerned with the mean lifetimes of the system components

944

and with the conditional sojourn times of the system in the operation states under arbitrary assumption that their distributions are exponential. In further developing of the proposed methods it seem to be possible to obtain the results useful in the complex technical systems and their operation processes reliability and availability evaluation.

REFERENCES

Grabski, F. 2002. *Semi-Markov Models of Systems Reliability and Operations.* Warsaw: System Research Institute, Polish Academy of Sciences, (*in Polish*).

Kolowrocki, K. 2001. *Asymptotic Approach to System Reliability Analysis.* Warsaw: System Research Institute, Polish Academy of Sciences, (*in Polish*).

Kolowrocki, K. et al. 2002. Asymptotic approach to reliability analysis and optimisation of complex transport systems. (*in Polish*). Project founded by the Polish Committee for Scientific Research. Gdynia: Maritime University.

BIOGRAPHY

Krzysztof Kolowrocki is a Professor and the Head of Mathematics Department at Faculty of Navigation of Gdynia Maritime University. His field of interest is mathematical modeling of safety and reliability of complex systems and processes. He has published more than 190 books, reports and papers in journals and conference proceedings. He is the Vice-President of Polish Safety and Reliability Association. Website: http://www.wsm.gdynia.pl/~katmatkk/

Safety and Reliability – Bedford & van Gelder (eds)
© 2003 Swets & Zeitlinger, Lisse, ISBN 90 5809 551 7

Modelling NH$_3$-outflow out of sea-going vessels

F. Kootstra

*Netherlands Organisation for Applied Scientific Research, TNO Environment, Energy and
Process Innovation, Department of Industrial Safety, Apeldoorn, The Netherlands*

ABSTRACT: In this paper the validity and uncertainties in the commonly used consequence models for risk assessment of the transport of "liquefied" NH$_3$ in sea-going vessels on the "Westerschelde" are described and discussed. After an accident with a vessel the uncertainties in the commonly used risk consequence methods are mainly due to the assumptions made around the way in which the "refrigerated" NH$_3$ is released. The validity and correctness of the assumptions made in modelling the specific behavior of the liquefied NH$_3$ during release, evaporation, and (finally) dispersion in the atmosphere are investigated. The investigation concentrates on the uncertainties in the physical behavior of the liquefied NH$_3$ in combination with water. A comparison is made between the theoretically possible scenarios and the practical way in which these situations are nowadays modelled. The outline of this paper is as follows: first the uncertainties in the modelling of NH$_3$ are determined. It is shown in some quantitative calculation examples how these uncertainties, and used assumptions, in the NH$_3$ modelling can lead to considerable differences in consequences for external- and environmental safety. After that, several research area are investigated in order to reduce the uncertainties in the physical behavior of NH$_3$ and to improve the currently used models.

1   INTRODUCTION

In the Netherlands large amounts of cooled liquefied ammonia (NH$_3$) are transported by sea-going vessels over the "Westerschelde". Several qualitative risk assessments (QRA) for the transport of NH$_3$ over the Westerschelde have concluded that this transport is coupled with high individual and societal risk. The used techniques in these QRA's are the standard techniques as described in the Yellow [1] and Purple [2] books. The uncertainties in these commonly used risk consequence methods are mainly due to the assumptions made around the way in which the refrigerated NH$_3$ is released, but also the uncertainties around the effect modelling of the liquefied NH$_3$ are of importance. Due to the specific behavior of NH$_3$ after an accident with a sea-going vessel, the release, evaporation, and (finally) dispersion in the atmosphere are in advance unpredictable. As a result of these observed uncertainties in as well the physical behavior of liquefied NH$_3$ in combination with water, and also the used consequence models in QRA's for the transport of liquefied NH$_3$ shipping over the Westerschelde, the department of industrial safety of TNO environment, energy and process innovation was assigned by the department of transport safety from the ministry of

transport, public works and water management to investigate the uncertainties in the modelling of NH$_3$-outflow and dispersion in the atmosphere after an accident with a sea-going vessel. The investigation after the uncertainties and validity of the nowadays used effects models for NH$_3$ will focus upon the determination if the modelling has been done thoroughly and if the modelling has not been done too conservative. A comparison between the theoretically possible scenarios and the practical way in which these situations are nowadays modelled is made to reveal these uncertainties in, and explore the validity of, the models.

2   DEFINITION OF THE PROBLEM, OBJECTIVE AND RESEARCH METHOD

Ammonia is stored and transported in several ways. In general NH$_3$ is transported in large quantities by rail, tube or by road in liquefied form under pressure. In the Netherlands the overland transport is mainly by rail, but also for import and export large quantities are transported by sea-going vessels and inland navigation. The transport of NH$_3$ by sea-going vessels is mainly as a completely cooled (refrigerated) liquid at atmospheric pressure. If a sea-going vessel filled with

NH$_3$ takes part in a serious accident it can lead to a leakage in the cargo tank of the vessel. From this leakage the NH$_3$ will flow out upon the water and will partially dissolve, evaporate from a pool, and eventually disperse as a gas in the atmosphere. A precise and validated way of modelling this scenario from outflow until dispersion is of great importance to determine the eventual risk. The nowadays used standard techniques as described in the colored books [1,2] cannot be followed blindly, while in this case it is seagoing navigation and refrigerated NH$_3$ and in [2] the situation is described for inland navigation and pressurized NH$_3$. Especially the following two properties of NH$_3$ determine the eventual concentration in the neighborhood, and are strongly dependent upon the modelling and assumptions thereby made:

- NH$_3$ dissolves exothermic in water; causing less NH$_3$ to be available for the eventual dispersion in the atmosphere, but also that extra heat becomes available for the evaporation.
- NH$_3$ is lighter than air (also at the boiling point of NH$_3$, -33°C) and will have the tendency to rise after evaporation. When the outflow of the NH$_3$ is quick and the boiling is extreme upon water, there is the possibility of droplet (aërosol) formation in the NH$_3$ cloud. Depending upon the aërosol concentration in the cloud, the cloud can be heavier than air and will shown heavy gas dispersion behavior. Also by mixing in air, which cools down due to the evaporating NH$_3$ aerosols, an NH$_3$/air cloud can develop which is heavier than air.

For being able to give reliable predictions concerning the NH$_3$-concentration in the neighborhood after an accident in which a sea-going vessel filled with NH$_3$ is involved, a precisely validated modelling of such a scenario is an important condition. The careful and precise analysis of the outflow, dissolution, evaporation and dispersion is thereby of great importance. For being able to predict how the dispersion of the NH$_3$ cloud in the neighborhood of the accident will evolve, the above processes cannot be looked at separately but need to be modelled successively, and as accurate as possible. These processes are heavily influenced by:

- the outflow conditions (above or below the waterline).
- the weather conditions (temperature, humidity and wind speed).
- the amount of NH$_3$ that is released, under which conditions, and the outflow rate.
- the percentage of the released NH$_3$ which dissolves in water, and which part will evaporate.
- the spreading of, and evaporation from, the NH$_3$ pool.
- the amount of NH$_3$ "aerosols" in the cloud.

On the basis of the above problem description the following objectives were formulated:

- The determination of the validity, and the uncertainties, concerning the modelling of NH$_3$ releases from a vessel.
- To indicate (if necessary) the effective possibilities to reduce the uncertainties in the NH$_3$ modelling.

To reach these objectives a concrete examination after the validity of the nowadays-common risk analysis methods (and available modelling information) for an accidental release of NH$_3$ from a sea-going vessel will be performed. As a practical example we will examine the performed risk analysis methods used for the NH$_3$ transport on the Westerschelde to evaluate if it has been done correctly, and if it has been performed in an over conservative way. On that basis we will be able to give comments upon the nowadays-used risk analysis modelling methods for NH$_3$.

3 THE BEHAVIOR AND THE MODELLING OF AMMONIA

The above described problems in the NH$_3$ modelling process will be treated here in a theoretical way and the nowadays-used modelling methods will be summarized. By comparing the theory with the used modelling methods, the uncertainties can be deduced. First a short summary of the physical properties of NH$_3$ will be given, after which the different processes of outflow, dissolvation, evaporation and dispersion will be treated.

3.1 Physical properties

- NH$_3$ is gaseous at ambient temperature (20°C).
- NH$_3$ has a boiling point of -33°C (p = p$_0$).
- NH$_3$ will become fluid by: heavy cooling, refrigerated NH$_3$ (T ~ -33°C), or high pressures, pressurized NH$_3$ (p $\sim$ 6 Bar at 10°C), or a combination of both, semi-refrigerated NH$_3$ (T$\downarrow$ and p$\uparrow$).
- liquefied NH$_3$ has a lower density than water (650 kg/m^3 vs. 1000 kg/m^3), and will therefore float on water.
- liquefied NH$_3$ dissolves easily in water. At 20°C dissolves 0.529 kg NH$_3$ in a kg of water (K$_{25°C}$ = 1.81 * 10^{-5}).
- "pure" NH$_3$ vapor (100% NH$_3$) has a density at it's boiling point (-33°C) lower than air (0.86 kg/m^3 vs. 1.2 kg/m^3), and will therefore rise in the air.
- air which "adiabatically" has been completely saturated with NH$_3$, will be (after reaching equilibrium between air and liquefied NH$_3$ (6.1% wt. NH$_3$)) heavier (1.6 kg/m^3, T $= -72$°C) than air. Therefore such a mixture will spread on the ground (or water) surface and not rise into the air.

- the turning point weather being heavier or lighter than air of the NH$_3$/air mixture, is around 45% wt. NH$_3$ [3].

3.2 Outflow

After an accident with a sea-going vessel filled with NH$_3$, it is first of all of great importance to carefully analyse the way in which the outflow on the water surface occurs. This will strongly influence the behavior, the composition, and finally the dispersion of the toxic NH$_3$ cloud in the atmosphere.

3.2.1 Theory

For the exact modelling of the outflow scenario of NH$_3$ out of a sea-going vessel on water, in theory we need to answer the question: *"How does the outflow of the cooled NH$_3$ precisely occur, above or below the water surface?"* For the determination of which outflow scenario should be considered and used, we need to answer more specific questions like: *"Where in the hull of the vessel is the hole situated, and how big is this hole?"* While this influences the outflow rate, and whether liquid or gaseous NH$_3$ escapes from the tank.

3.2.2 Current modelling scenarios

For the modelling of the outflow of NH$_3$ from a vessel were, in all previous safety studies for the Westerschelde [4,5,6], always two outflow scenarios assumed (see Table 1).

Without specifying in [4,5,6] the hole in the hull is assumed to be situated above the water surface of the vessel, and in the liquid part of the tank. In a later study for sea-going and inland navigation [7], a protocol is formulated which includes instructions for how to perform risk analysis (external safety) for the transport of dangerous cargo over (inland) waterways. In [7] the following outflow scenarios for sea-going navigation of cooled tankers were used (see Table 2).

The differences in the used [4,5,6] and proposed [7] outflow scenarios have lead to doubts concerning the validity of the nowadays used scenarios. As described in [8], the area in between a "small" and "big" leakage is not covered, which is especially in the case of NH$_3$ important when considering pool sizes and the behavior of the gas cloud.

3.2.3 Uncertainties

Comparing the theory and current modelling scenarios sections, it is clear that there are still several uncertainties concerning the validity of the outflow scenarios of NH$_3$ from a vessel. First the following uncertainties need to be minimized:

- outflow from a leakage above or below the water line.
- outflow from a hole or a crack in the tank (outflow rate).
- outflow of the total content of the tank (amount of outflow).

Based upon the assumed hole sizes in the protocol [7], outflow rates can be calculated due to the hydrostatic pressure of the NH$_3$ (liquid) column above the hole. In Table 3 a "big" leakage is assumed (hole diameter: 1.1 m) in a tank with a volume of 5.000 m^3 and width and height of 7.2 m.

From Table 3 it can be concluded that the used outflow rates in the protocol [7] are over conservative. The uncertainties around the outflow scenarios in the protocol [7] has also be commented upon by the department MHR-consultancy of "het Gemeentelijk Havenbedrijf Rotterdam" (GHR) [9]. The GHR questioned the "big" outflow scenario. Although imaginable, the probability for this scenario in the protocol is debatable, because it practically means that the collided vessel needs to be sailed into two pieces by the colliding vessel. Therefore the "big" outflow scenarios should be further (experimentally) investigated after their sense of reality.

Table 2. The outflow scenarios in [7].

Outflow	Hole diameter [m]	Outflow rate [m^3/s]	Maximum outflow amount
small leakage	0.25	0.2	0.5 * content tank
big leakage	1.1	15.0	Content tank

Table 3. Calculated outflow rates due to hydrostatic pressure.

Distance hole to liquid surface [m]	Outflow rate [kg/s]	Outflow rate [m^3/s]
7.2	4789	7.37
4.7	3869	5.95
2.2	2646	4.07
0.2	799	1.23

Table 1. The outflow scenarios in [4,5,6].

Outflow	Hole diameter [m]	Outflow rate [m^3/s]
Continuous	0.5	0.3
Quasi-instantaneous	4.0	25.0

3.3 Dissolvation and evaporation

Next to the outflow scenarios, also the dissolvation in the water and the evaporation from the water of NH_3 influences finally the composition and dispersion of the toxic NH_3 cloud.

3.3.1 Theory

To model in theory the scenario of dissolvation, spreading and evaporation of a floating NH_3 pool on a water surface, we initially need to answer the following questions: *"From where in the vessel does the outflow occur? Is the hole/crack in the tank located above or below the waterline of the vessel, and does outflow occur from the liquid or gaseous part of the cargo tank/compartment?"* The above question is directly coupled to: *"Which percentage of the original amount of liquefied NH_3 which has flown out or has been released from the vessel will eventually be able to evaporate from a pool on the water surface?"* This question is important for the determination of the amount which finally can evaporate from an NH_3 pool on the water surface. This amount equals the amount of NH_3 which has flown out minus the amount of NH_3 which has dissolved in the water. The evaporation speed (and thus the duration of the source) is also of relevance here. To model in further detail the evaporation from an NH_3 pool on a water surface, we also need to answer the question: *"How does the spreading of, and the evaporation from, an NH_3 pool on water occur?"*

3.3.2 Current modelling scenarios

After the outflow of the NH_3, in the till now used modelling scenarios for the dissolvation, spreading, and evaporation of the remaining NH_3 pool, in previous safety studies for the Westerschelde [4,5,6] is always assumed:

- a dissolvation:evaporation ratio of 50:50% in [4], 60:40% in [11], and 58:42% in [5,6] (on the basis of [10]).
- a "flash" percentage of 15% at the start of the evaporation [12,13], and
- a boiling NH_3 pool on water with a constant evaporation flux of $0.05\,kg/m^2s$ [12,13].

The above scenarios need verification after their validity because they are outdated. The scale (130 kg of liquefied NH_3) of the so called "large" scale experiments by Raj *et al.* [10] are not in proportion to the amounts of NH_3 which will be released after an accident with a sea-going vessel. Also the influence of a constant evaporation flux from the NH_3 pool in [12,13] needs to be determined, and the validity of such an approximation.

3.3.3 Uncertainties

The uncertainties concerning the modelling of dissolvation, spreading, and evaporation of NH_3 are described by Griffiths and Kaiser in [14]. They particularly mention:

- the partition ratio for dissolvation and evaporation;
- the grow and maximum size of the NH_3 pool;
- the time needed for complete evaporation of the pool.

The highest uncertainty is in the assumptions made for the percentages of originally released amount of NH_3 which will evaporate, because it is not possible to tell in advance if:

- an NH_3 pool will be formed on the water surface, and
- which part of the released NH_3 will dissolve in water, and which part will directly evaporate.

3.4 Dispersion

After the modelling of the outflow, dissolvation, and evaporation of NH_3, finally the modelling of the dispersion of the NH_3 cloud is of great importance.

3.4.1 Theory

For the exact modelling of the dispersion of the formed NH_3 cloud we need in theory answer to the following question: *"Does the released NH_3 gas cloud need to be modelled as a "light" gas (buoyant), as a "neutral" gas (passive), or as a "heavy" gas (dense)?"* The answer to this question is strongly dependent upon the outflow, dissolvation, and evaporation of the released NH_3 prior to the dispersion of the NH_3 gas cloud. For the correct modelling of the dispersion it is of great importance to determine the density of the NH_3 cloud a prior and during the dispersion process. The composition of the NH_3 cloud can exist out of several of the following compounds:

- "pure" ammonia gas [$NH_3(g)$];
- droplets of ammonia, pure [$NH_3(l)$] or dissolved [$NH_3(aq)$];
- air, with or without water vapor [$H_2O(g)$] or condensed water droplets [$H_2O(l)$].

Important conditions which play a role in the final density of the NH_3 gas cloud are:

- the initial temperature of the NH_3 at release.
- the fraction of liquid NH_3 in the NH_3 gas cloud.
- the initial temperature of the ambient air.
- the humidity of the ambient air.

Starting at an initial temperature ($T_{initial}$ [$NH_3(l)$]) of $-33°C$ at the release of the NH_3, Haddock and Williams [15] concluded, according to their calculations:

- if the NH_3 is released on, or above, its boiling point in the purely gaseous composition, the gas cloud will always be lighter (lower density) than the ambient air.

- if a part of the released NH_3 is in the liquid form (as droplets in the gas cloud), the gas cloud can become heavier (higher density) than the ambient air when the gas cloud is diluted by air;
- the critical values for the initial liquid fraction in the NH_3 gas cloud in an ambient air with a temperature of 20°C: at values "lower" than 4–8%, the gas cloud will always be lighter than air; at values "higher" than 16–20% the gas cloud will become heavier than air regardless the humidity of the ambient air; in between these two critical values, the density is dependent upon the humidity and the amount of dilution.

3.4.2 Current modelling scenarios

In the current scenarios used for the modelling of an NH_3 gas cloud, different starting conditions applied:

- dispersion out of a spreading NH_3 pool [4,5];
- dispersion (finally) from a neutral NH_3 gas cloud, which takes into account the initial sagging due to heavy gas dispersion [4,5];
- dispersion of a neutral gas cloud of NH_3 [4,5,6];
- dispersion out of a point source [6].

The use and validation of the above starting conditions in the previous safety studies is not always clearly mentioned. Usually is chosen for an approach which gives the most pessimistic results [4,5], and sometimes for an even more conservative approach [6].

3.4.3 Uncertainties

The currently used modelling scenarios have lead to question concerning: "*To what extent is it allowed, and validated, to model the NH_3 gas cloud after, and due to, the outflow, dissolution, and evaporation of the NH_3 out of a sea-going vessel, as a neutral gas cloud?*" Considering the previously theoretically described scenarios of outflow, dissolution, and evaporation, and the thereby indicated uncertainties, it's clear and obvious that it is not possible to tell already in advance if the to be formed NH_3 cloud can for certain be considered as a neutral gas cloud. This can therefore be considered as the biggest uncertainty in the whole modelling process of NH_3 outflow.

4 DESCRIPTION OF THE POSSIBLE OUTFLOW SCENARIOS

In this section we will give an outline of the specific, and possible, NH_3-outflow scenarios from a sea-going vessel after an accident. We will point out the uncertainties in every possible modelling scenario, and try to indicate the consequences of these uncertainties

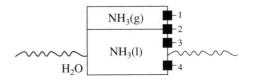

Figure 1. Possible outflow scenarios of NH_3 from a cargo compartment of a sea-going vessel.

upon the final effect distances of the toxic NH_3 gas cloud. In Figure 1, the four possible outflow scenarios from a cargo compartment of a sea-going vessel are indicated:

1. outflow of $NH_3(g)$ from a hole above the waterline.
2. outflow of $NH_3(g$ and $l)$ from a hole above the waterline.
3. outflow of $NH_3(l)$ from a hole above the waterline.
4. outflow of $NH_3(l)$ from a hole below the waterline.

4.1 Case 1: Outflow of $NH_3(g)$ above the waterline

In case of an outflow of NH_3 from a hole in the cargo compartment above the waterline, the NH_3 is released in pure gaseous form. As described in the physical properties of NH_3, such a gas cloud will always be lighter than air and will rise immediately into the air. Therefore little danger is to be expected in the direct surroundings of the accident, and there is no uncertainty in the modelling scenario.

4.2 Case 2: Outflow of $NH_3(g$ and $l)$ above the waterline

In case of an outflow of (at the same time) gaseous (g) and liquid (l) NH_3 from a hole above the waterline, next to a pool of NH_3 on the (sea)water also a gas cloud of NH_3 will be developed in the atmosphere directly after the release. This NH_3 cloud will contain liquid droplets of ammonia, $NH_3(l)$. As described before, such an NH_3 gas cloud can, become heavier than the ambient air. The other part of the NH_3 will be released on the (sea)water. A part will dissolve into the water. The rest will evaporate into the atmosphere, from a spreading NH_3 pool upon the water. The uncertainties in this scenarios are connected with:

1. the amount of liquid NH_3 (droplets) in the NH_3 gas cloud (which develops directly after the release),
2. the amount of NH_3 which dissolves in the (sea)water, and
3. the density of the NH_3 gas cloud which develops from the evaporated NH_3 out of an NH_3 pool on the (sea)water.

These tree uncertainties will have the following consequences upon the calculated effect distances for a dispersing NH_3 gas cloud in the atmosphere:

a. the more ammonia droplets $NH_3(l)$ in the NH_3 gas cloud, the higher the concentration of ammonia $[NH_3]$ in the cloud (higher densities), the more the gas cloud will cool down (the cooling down is due to the fact that heat is withdrawn from the gas cloud to evaporate the droplets in the cloud). Higher NH_3 concentrations in the gas cloud will result in larger effect distances for the toxic NH_3 gas cloud.
b. the more of the liquid part of the NH_3 which dissolves in the water, the less there will evaporate from the NH_3 pool into the atmosphere, the smaller the effect distances for the toxic NH_3 gas cloud.
c. when (and as long as) the density of the NH_3 gas cloud is higher than the ambient air, the dispersing behavior of the toxic NH_3 gas cloud needs to be modelled with a heavy gas model. The modelling of the toxic NH_3 gas cloud with a heavy gas model in comparison with a neutral gas model should in principle, and dependent upon the degree of sagging of the cloud, could result in larger effect distances for the toxic NH_3 gas cloud.

4.3 Case 3: Outflow of $NH_3(l)$ above the waterline

In case of an outflow of $NH_3(l)$ from a hole above the waterline, a spreading NH_3 pool on the (sea)water will be formed. From this NH_3 pool the NH_3 will evaporate into the atmosphere. A part of the released NH_3 on the water will dissolve into the (sea)water. The uncertainties in this scenario are as mentioned in case 2: point (1) and (2). Also the influence of the hole size is an uncertainty. The influence of the hole size upon the complete physical process of the NH_3 outflow, and continuing processes, with finally the influence upon the dispersion behavior of the developed NH_3 gas cloud are still a source of uncertainty.

4.4 Case 4: Outflow of $NH_3(l)$ below the waterline

In case of an outflow of $NH_3(l)$ from a hole below the waterline, there exists the possibility of an NH_3 pool formation on the (sea)water surface from which evaporation of the NH_3 into the atmosphere is possible. Due to the lower density of $NH_3(l)$ in comparison with the density of water, the $NH_3(l)$ will move into the direction of the water surface when released from a hole in the cargo compartment of the sea-going vessel which is situated below the waterline. Dependent upon:

- the counter pressure of the water upon the $NH_3(l)$,
- the size of the hole,
- the location of the hole below the water surface, and
- the percentage (%) of the NH_3 which dissolves,

it is possible that a part of the NH_3 which has flown out from the vessel will also reach the (sea)water surface and manifest itself as a pool on the (sea)water. The percentage of dissolution when released below the waterline will be different than when released above the waterline due to the above mentioned points, and also dependent upon the currents in the (sea)water and the temperature of the (sea)water. The uncertainties with respect to this scenario is mainly the percentage of the released NH_3 which will dissolve into the (sea)water. Till now is assumed that all (below water) release NH_3 will dissolve if:

1. the hole (diameter D) is located at a depth $>10D$, or
2. the outflow rate is in the range of the Reynolds number 10^6–10^8, when the hole is located in between $10D$ and the water surface.

Further experimental research is necessary to reduce these uncertainties.

5 NOWADAYS COMMON RISK-BASED ANALYSIS TECHNIQUES

To determine the validity and uncertainties in the nowadays used risk-based analysis techniques and scenarios, below we will examine the validity of these analysis techniques and scenarios in connection with the information available for modelling. We will explicitly verify if the used modelling techniques in previous safety studies [4,5], for the release of NH_3 along the Westerschelde, have been used correctly or if the used modelling techniques have been used in an over-conservative way at some specific points. Based on the above findings, it is possible to judge if there exist well-founded doubts in the nowadays used risk-based analysis techniques for NH_3, and if it therefore reasonable to deviate from these nowadays used analysis techniques.

5.1 Investigation

In previous safety studies for the transportation of dangerous chemicals over the Westerschelde performed by the safety consultancy office AVIV [4,5] the following assumptions were made. With respect to outflow, two scenarios were used:

- the quasi-instantaneous release of the content of one cargo compartment (outflow rate: $25\,m^3/s$; hole diameter: $4\,m$).
- leakage from a cargo compartment (outflow rate: $0.3\,m^3/s$; hole diameter: $0.5\,m$).

The effect calculations, for the above two scenarios, were performed with the software package HGSYSTEM, in which the model EVAP is used for the spreading and evaporation of the NH_3 pool on

952

(sea)water, and the model HEGADAS for the dispersion of the toxic NH_3 gas cloud in the atmosphere. The pool evaporation model EVAP calculates the spreading and evaporation of boiling and non-boiling liquids. In case of a boiling liquid pool on water, like NH_3, the evaporation flux is assumed constant, in case of NH_3 0.05 kg/m²s [12,13]. While the NH_3 dissolves in the (sea)water, in the calculations is assumed that 50% of the content of the pool will evaporate from the pool into the atmosphere and the other 50% of the pool will dissolves in the (sea)water. The heavy gas dispersion model HEGADAS model calculates the dispersion of non-reactive ideal gasses from a spreading pool. In the for NH_3 used heavy gas dispersion model (initial sagging of the heavy gas cloud) a gradual transition to the conventional passive dispersion model (neutral gas) takes place.

5.2   Validation and qualitative analysis

In a critical examination after the validity of the assumptions in the above safety studies by AVIV the following remarks should be made. In relation to the "big" outflow scenario (quasi-instantaneous release), resulting (finally) in the largest risk contours over the Westerschelde, there exist large uncertainties according to GHR [9], especially concerning the sense of reality of this scenario (outflow rate of 25 m³/s through a hole with a diameter of 4 m). Appropriately is therefore remarked by GHR, and also according to TNO, that this scenario would qualify for further investigation. The foundation of the by AVIV till now used constant evaporation flux (0.05 kg/m²s) in the evporation model is debatable although not inconceivable. The same is true for the ratio dissolvation: evaporation (50:50) of the NH_3 which has been released upon the (sea)water. The dispersion model (neutral- or heavy-gas) to be used for the NH_3 gas cloud will always remain a point of discussion, because it's dependent upon the outflow scenario. To the question, whether or not the till now used heavy gas dispersion model has been chosen to conservative, is difficult to answer. When using the heavy gas dispersion model smaller effect distances are calculated in comparison to the neutral gas dispersion model due to the sagging of the cloud and the initial mixing with air (for the evaporation of the aerosols), however the contours of the cloud are initially broader in the heavy gas dispersion model.

6   QUANTITATIVE INDICATION OF THE UNCERTAINTIES

In this section, some quantitative calculation examples will be performed to show how the uncertainties, and assumptions, in the NH_3-modelling can lead to considerable differences in consequences for the external- and environmental safety. The capacity of the sea-going vessels on the Westerschelde varies. The typically used values are 5.000 m³ per compartment, and a total capacity of 20.000 m³ of NH_3 (4 compartments) in a sea-going vessel. Inquiries after the capacity of the sea-going vessels used by Hydro Agri in Sluiskil (HAS) and BASF in Antwerp showed. (see Table 4)

The assumptions made in the consequence modelling are:

6.1   Outflow

Liquid outflow from a hole in a vessel, which is a cylinder with a length of 15m. Filling degree: 95% of liquefied NH_3 at atmospheric conditions. Hole diameters vary from 4m/1.1m (quasi-instantaneous, big leakage) to 0.5 m/0.25 m (continue, small leakage). The leakage is in the bottom of the tank. These assumptions gave the following results for HAS (maximum volume of a compartment: 2.854 m³ = 1.760.918 kg) (see Table 5) and for BASF (maximum volume of a compartment 6.153m³ = 3.796.401 kg) (see Table 6)

Table 4.   Capacity of the sea-going vessels on the Westerschelde.

Company	Max. volume of a compartment [m³]	Max. total capacity of a sea-going vessel [m³]
HAS	2.854	–
BASF	6.153	16.923

Table 5.   Results for liquid outflow of a HAS vessel.

Hole diameter [m]	Maximum outflow rate [m³/s]	Tank empty after … [s]	Released amount after 1800s [kg]
4.0	132	35	1760918
1.1	10	461	1760918
0.5	2	2233	1584500
0.25	0.5	8933	522460

Table 6.   Results for a liquid outflow of a BASF vessel.

Hole diameter [m]	Maximum outflow rate [m³/s]	Tank empty after … [s]	Released amount after 1800s [kg]
4.0	160	62	3796401
1.1	12	823	3796401
0.5	2.5	3985	2276200
0.25	0.6	15938	658430

Table 7.   Neutral and heavy gas dispersion scenarios.

Scenario	Hole size [m]	Type of release	Pool size [kg]	Release rate [kg/s] for 1800s
HAS-I1	4	Instantaneous	0.88e6	–
HAS-SC2	0.25	Semi-continuous/ Evaporating pool	–	145
BASF-I1	4	Instantaneous	1.9e6	–
BASF-SC2	0.25	Semi-continuous/ Evaporating pool	–	183

Table 8.   Calculated effect distances of the toxic NH_3 cloud for 1% lethality.

Scenario	Dispersion Model	Length of the cloud [m]	Width of the cloud [m]
HAS-I1	neutral	4613	550
HAS-SC2	neutral	2242	244
BASF-I1	neutral	5309	686
BASF-SC2	neutral	2434	270
HAS-I1	heavy	725	1067
HAS-SC2	heavy	153	380
BASF-I1	heavy	974	1481
BASF-SC2	heavy	111	214

Figure 2.   The city Vlissingen along the Westerschelde

6.2   Dissolvation and evaporation

A ratio of 50%:50% is assumed for the amount of released NH_3 between dissolvation and evaporation.

6.3   Dispersion

The dispersion of a boiling evaporating NH_3 pool from an open water surface is calculated in such a way that the formation of the pool is assumed instantaneous and for the size of the pool a thickness of 0.05 m is assumed. The amounts of NH_3 in the pool are according to the calculated outflow amounts after 1800 seconds in the outflow scenarios divided by two due to the dissolvation:evaporation ratio.

The following neutral and heavy gas dispersion calculations were performed (see Table 7)

They gave the following results for the size of the toxic NH_3 cloud for 1% lethality.

As can be seen from Table 8, the shape and size of the toxic NH_3 cloud is different in case of a neutral or heavy gas dispersion. In case of a neutral gas dispersion the cloud is longer and less wide, and the opposite is true in case of heavy gas dispersion due to the initial sagging of the cloud. To give an indication of the effects distances of these toxic NH_3 clouds, in Figure 2 the city Vlissingen along the Westerschelde is depicted together with the calculated neutral gas

dispersion results (the release point of the NH_3 for the dispersion is assumed to be in the middle of the Westerschelde). In Figure 2 the neutral gas dispersion scenarios are depicted for four different wind directions, from S(outh) to W(est)-S(outh)-W(est), and a Pasquill stability class F2. As can be seen from Figure 2, the calculated effect distances of the toxic NH_3 cloud for 1% lethality in case of a neutral gas dispersion overlap considerably with the city of Vlissingen.

7   POSSIBLE MODEL IMPROVEMENTS

7.1   Outflow

The modelling of the outflow of the NH_3 always assumes a specific outflow rate of the NH_3 through a hole or crack in the cargo compartment of a vessel. The outflow is thereby qualified as (quasi-)instantaneous or continuous and/or through a "big" or a "small" leakage. Not taken into account is the way in which the outflow takes place, e.g. the hole size (see also [7]). Already before we described the "large" uncertainties in the sense of reality with respect to the quasi-instantaneous outflow scenario. This scenario should therefore be first further investigated after the sense of reality before considering (or improving) the way in which this scenario has to be modelled in safety studies. In literature a mathematical model is proposed by Wheatley [16] specific for the release (the outflow) of liquefied (cooled or pressurized) NH_3 through a hole in the hull of a cargo compartment. This model is formulated according to experiments for the estimation of the concentration and composition (the starting conditions) of an NH_3 gas cloud prior to the dispersion model calculation. Important quantities are:

- the outflow rate, dependent upon the size of the hole.
- the dilution factor with air (the weather conditions).

- the humidity in the air (interaction of NH_3 with water vapor).

7.2 Dissolvation and evaporation

Based upon the so-called "large scale" (130 kg of NH_3) experiment by Raj [10], the modelling of dissolution and evaporation of the NH_3 is done standard by assuming that 58% of the released amount of NH_3 will dissolve immediately in the (sea)water, and the remaining 42% will evaporate into the air. The amounts of NH_3 which will be released after an accident on the Westerschelde are orders in magnitude higher, and it is therefore highly questionable if extrapolation of the experimental findings by Raj can be justified. After the dissolution, the evaporation from the spreading NH_3 pool on the (sea)water surface is modelled with an constant evaporation flux of $0.05 \, kg/m^2s$ [12,13]. During the modelling of the evaporation of the spreading NH_3 pool, the weather conditions, like wind speed and water temperature, are not taken into account. Therefore it is necessary to develop a general "evaporation" model which takes the solubility of the released chemical into account. Such an evaporation model should at least be dependent upon:

- the amount and outflow rate of the released chemical.
- the temperature of the (sea)water.
- the solubility product (K_T) of the released chemical.
- releases above and below the waterline.

In the Yellow Book [1] there is no integral evaporation model available to correctly model the behavior of a floating, spreading, and also (partially) solvable boiling liquid, like NH_3. The use of the evaporation model GASP (Gas Accumulation over Spreading Pools) would be a large improvement to model the behavior of the boiling NH_3 pool upon water. The evaporation model GASP takes implicitly into account during the evaporation of the spreading pool:

- the weather conditions (wind speed, water temperature),
- the dynamics of the spreading liquid pool,
- the evaporation from the surface of a pool, and
- the transition from boiling to non-boiling.

7.3 Dispersion

As described before, till now in modelling the dispersion of an NH_3 gas cloud, is (finally) always a "neutral" gas cloud of NH_3 assumed. In some cases the initial sagging of the NH_3 cloud is taken into account to discount for possible "heavy" gas behavior. Further is (in most cases) assumed that dispersion takes from a point source instead of a spreading NH_3 pool floating on the (sea)water surface.

Model improvements is desired in at least at two points:

- not to exclude the possible development of a "heavy" gas cloud from an initial "light" or "neutral" gas cloud in the course of the dispersion process. Depending on the composition of the gas cloud during the dispersion process should, if desired, be switched to a "heavy" gas dispersion model.
- not to exclude the possibility of a not rising "light" (buoyant) NH_3 gas cloud due to "bad" weather conditions (high wind velocities).

The last phenomenon, the not rising of a light (buoyant) gas cloud due to high wind velocities, was first described by Briggs [17]. The rising of the gas cloud is reflected in the Richardson number (L) as described in [18] by:

$$L = g \cdot h \cdot \Delta\rho \cdot \left[\ln\left(Z/Z_0\right)\right]^2 / k^2 \cdot \rho_a \cdot \overline{U}^2(Z) \qquad (1)$$

Briggs estimated that if $L > 2$ for a "plume" release, the rising is suppressed, and if $L < 2.5$ for a "puff" release, the rising is suppressed. From Equation (1) it can be seen that that low values of L are connected to high wind velocities ($\overline{U}(Z)$) and high values of the roughness length of the subsoil (Z_0), implying the "higher" the wind velocity and roughness length of the subsoil, the more likely the rising of the gas cloud is suppressed.

8   FURTHER RESEARCH AREA

The various research areas should initially give:

- a better understanding of the physics of an NH_3 outflow from a sea-going vessel, and
- a better determination of the possible danger for the environment and surroundings accompanied by large NH_3 releases on (or below) seawater level.

8.1 Quantitative sensitivity analysis

In this study a quantitative sensitivity analysis will be performed after the influence of several parameters which influence successively the modelling processes of NH_3-outflow. Initially the idea is to vary the important parameters upon the eventual individual and societal risk.

1 outflow: hole sizes, outflow rates, outflow amounts, etc.
2. dissolvation: dissolvation ratio, spreading of an NH_3 pool upon (sea)water, weather conditions (water temperature and wind speed), currents in the (sea) water, etc.
3. evaporation: spreading of an NH_3 pool upon (sea) water, evaporation flux, etc.

4. dispersion: outflow (two-phase flow or pool evaporation), weather conditions (wind speed), type of dispersion (neutral or heavy gas), etc.

8.2 Closer analysis of the NH_3-outflow scenarios

In this study a closer analysis with respect to the outflow scenarios of NH_3 which can happen after accidents with sea-going vessels are examined. Considering the fact that the outflow scenario is the so-called "source term" of the complete modelling process of the NH_3-outflow, it is of great interest to describe (and model) this source term as good as possible.

A closer analysis should be based upon:

1. literature study after the casuistic of accidents with cooled NH_3.
2. information in the navigation area.
3. collision experiments with sea-going vessels.

8.3 Experimental research of NH_3 releases

This research after large releases of NH_3 should be aimed upon:

- the determination of the percentage of "droplets" (aerosols) in the NH_3 gas cloud.
- measurements of the concentration and spreading of the released NH_3 in or upon the (sea)water.
- measurements after the evaporation speed of the NH_3 pool.
- measurements after the dispersion of the NH_3 gas cloud.

8.4 Analytical research of NH_3 releases

The analytical research of NH_3 releases is concerned with the development of analytical methods to model the behavior of large quantities of liquefied cooled NH_3 which are released on water. The following analytical models are necessary to model the consequent stages after the release of the NH_3:

- a model to determine the possible droplet (aerosol) formation in the NH_3 gas cloud.
- a model which describes the concentration in, and the spreading upon the (sea)water of the released NH_3.
- a general evaporation model, which is applicable for the divers outflow scenarios.
- well validated dispersion models for as well a buoyant, neutral as a heavy NH_3 gas cloud for the determination of the concentrations of NH_3 at different heights in the air.

8.5 Study after NH_3 releases below (sea)water level

In this study an analytical model should be developed to describe the behavior of a continuous release of liquefied cooled NH_3 below the (sea)water level. Such an analytical model should be based upon "large scale" experiments. The proposed analytical model should finally give an estimation of:

- the ammonia concentration [NH_3(g)] in the gas cloud, and
- the ammonia concentration [NH_3(l)]in the water.

9 CONCLUSIONS

The central research question in this paper was the determination of the validity, and the uncertainties, in the nowadays used consequence modelling for risk assessment after an accidental release of cooled NH_3 from a sea-going vessel on the "Westerschelde". Indicated are the experimental and/or analytical research areas which could reduce the identified uncertainties in the modelling. The research was performed in two steps:

- Determination of the validity and uncertainties in nowadays used NH_3-consequence modelling for risk assessment.
- Inventory of the possible model improvements to reduce the above uncertainties.

9.1 Uncertainties in the NH_3-consequence modelling

The uncertainties are determined by comparing the processes:

- outflow of NH_3 from a sea-going vessel.
- dissolvation of NH_3 in seawater, and the formation of a NH_3 pool on seawater.
- evaporation of NH_3 from a pool on seawater, and
- dispersion of the formed NH_3 gas cloud in the atmosphere.

In theory with the currently used risk analysis models. It was found that the important question for the eventual individual and societal risk of the formed NH_3 gas cloud was: whether or not formed gas cloud would rise (light gas) or shows a heavy/neutral gas behavior. The rising c.q. neutral/heavy gas behavior is mainly dependent upon (and determined by) the outflow conditions and the way of (heavy) evaporation from the pool. The fluid fraction (F) of NH_3 droplets in the gas cloud will determine the dispersion behavior as follows:

a. low: $F < 4\%$: a buoyant mixture will always be formed.
b. high: $F > 20\%$: with air entrainment will the mixture always be heavy.
c. average: $4\% < F < 20\%$: the mixture can become buoyant, neutral or heavy.

After an outflow of cooled NH_3 on (or below) the seawater surface, the values of F are in the category a (low) or c (average). Therefore it is not possible to tell before-hand, based upon the outflow scenario of the NH$_3$, if the gas cloud will behave as a puff or as a plume. There is still the uncertainty around the mechanism: can an initially rising NH$_3$ gas cloud eventually lead to a heavy gas mixture with an F value in the category b (high). Mainly because of the mixing in of (humid) air in the NH$_3$ gas cloud, and also because of the cooling down of the cloud (heat is extracted from the cloud to evaporate the NH$_3$ droplets (aerosols) in the cloud), could an ini-tially rising (light/neutral) NH$_3$ gas cloud develop into a heavy gas cloud. The uncertainties in the NH$_3$ gas cloud dispersion are mainly based upon the droplet for-mation and fraction F in the gas cloud. They are caused by the flowing uncertainties:

1. outflow: uncertainties wrt. hole sizes in, and out-flow rates from, sea-going vessels in combination with the formed pool sizes and evaporation rates from seawater.
2. evaporation: uncertainties wrt. the division ratio of the total amount of NH$_3$ which is released: which amount will dissolves in the seawater and which amount will evaporate into the air.
3. dispersion: uncertainties wrt. the dispersion behav-ior of the NH$_3$ gas cloud, "light" or "neutral/heavy".

9.2 Validity of the nowadays used NH$_3$-consequence modelling

Next to the determination of the uncertainties was investigated if it is safe to apply the nowadays used risk analysis methods for the transportation of NH$_3$ on the Westerschelde. Found was that there are impor-tant uncertainties in the modelling which will influ-ence the development of the NH$_3$ cloud formed after an accidental NH$_3$ release. Especially the amount of droplets in the gas cloud will determine if the cloud will rise or show neutral/heavy gas behavior. On two main topics in the currently used modelling of NH$_3$ should be made some comments:

- the used outflow scenarios are too conservative. The adjustment of the outflow scenarios in the protocol [7], are too strict and should be further explained. A hydrostatic pressure calculation showed outflow rates which are maximal half of the outflow rates as used in the protocol.
- after the opinion of TNO the physical behavior of NH$_3$ gas cloud formation is insufficiently taken into account. This could lead to also buoyant gas clouds and consequently lower concentrations of NH$_3$ above land.

REFERENCES

TNO Yellow Book, third edition 1997, Methods for the cal-culation of physical effects, resulting from releases of hazardous materials (liquids and gases). Directorate-General of Social Affaires and Employment, CPR 14E third edition, The Hague 1997.

TNO Purple Book, first edition 1999, Guidelines for quanti-tative risk assessment. Directorate-General of Social Affaires and Employment, CPR 18E first edition, The Hague 1999.

J.M. Blanken, Behaviour of ammonia in the event of a spillage, Unie van kunstmestfabrieken (UKF), Holland.

Risicoanalyse Westerschelde vervoer gevaarlijke stoffen, Adviesbureau AVIV, Enschede, april 1994.

H.G. Bos, Risicocontouren Westerschelde 1998, Advies-bureau AVIV, project 98154, Enschede, 29 maart 1999.

J.M. Ham, S.J. Elbers, Overzicht schadeafstanden trans-portroutes gevaarlijke stoffen in de provincie Zeeland, TNO Milieu, Energie en Procesinnovatie, rapport TNO-MEP-R-2000/363, Apeldoorn, oktober 2000.

Risicoanalyse Zee- en Binnenvaart, Het Protocol & Achtergronddocument, Det Norske Veritas (DNV) en Adviesbureau AVIV, 1999.

Gemeentelijk Havenbedrijf Antwerpen (GHA)/Antwerpse Gemeenschap voor de Haven (AGHA), Evaluatiestudie externe risico's en voorgestelde maatregelen transport van gevaarlijke stoffen over Westerschelde, DNV-61201277, augustus 2001.

W. Hoebée, G.C. de Jong, private communication: Model-lering uitstroom NH$_3$ vanuit zeeschepen, Maart 2002.

R.K. Raj, J. Hagopian and A.S. Kalelkar, Production of haz-ards of spills of anhydrous ammonia on water, Arthur D. Little Inc., 1974.

HSC, Major hazard aspects of the transportation of danger-ous substances, 1991.

CPR-14, 1988, Methoden voor het berekenen van Fysische Effecten. Commissie Preventie van Rampen door Gevaarlijke Stoffen.

G. Opschoor, Berekening van de verdamping van tot vloeistof gekoelde en verdichte gassen op water, TNO-77-07236.

R.F. Griffiths, G.D. Kaiser, The accidental release of anhy-drous ammonia to the atmosphere – a systematic study of the factors influencing cloud density and dispersion. UKAEA (SRD Report 154) (1979).

S.R. Haddock, R.J. Williams, The density of an ammonia cloud in the early stages of its atmospheric dispersion. UKAEA Report SRD R103.

C.J. Wheatley, Discharge of liquid ammonia to moist atmos-pheres – survey of experimental data and model for esti-mating initial conditions for dispersion calculations. UKAEA (SRD Report 410) (1987).

G.A. Briggs, Plume rise and buoyant effects, Chapter 8 of Atmospheric Science and Power Production, DOE/T1C-27601, Ed. D. Randerson, U.S. Department of Energy Technical Information Center, Springfield, VA.

G.D. Kaiser, A review of models for predicting the disper-sion of ammonia in the atmosphere, Science Applications International Corporation (SAIC), Denver, Colorado, August 1988.

Safety and Reliability – Bedford & van Gelder (eds)
© 2003 Swets & Zeitlinger, Lisse, ISBN 90 5809 551 7

Coping with uncertainty in sewer system rehabilitation

H. Korving
HKV Consultants, Lelystad, the Netherlands & Department of Civil Engineering, Delft University of Technology, Delft, the Netherlands

J.M. van Noortwijk
HKV Consultants, Lelystad, the Netherlands & Department of Applied Mathematics, Delft University of Technology, Delft, the Netherlands

P.H.A.J.M. van Gelder
Department of Civil Engineering, Delft University of Technology, Delft, the Netherlands

R.S. Parkhi
Management Science Department, University of Strathclyde, Glasgow, UK

ABSTRACT: In decision-making on sewer rehabilitation risk and uncertainty are not taken into account. However, the assessments on which the decisions are based are considerably affected by uncertainties on external inputs, system behaviour and effects. In this paper, a risk-based approach is presented considering uncertainty in sewer system dimensions, natural variability in rainfall and uncertainty in the cost function describing environmental damage. The use of risk based optimisation is illustrated with an example.

1 INTRODUCTION

Decisions on sewer rehabilitation have large, long-lasting consequences and the decisions have to be made under uncertainty. Annually, approximately 1 billion Euro is invested in sewer rehabilitation in the Netherlands. Uncertain information about the structural condition and the hydraulic performance of the sewer system serves as the basis for decision-making. Therefore, the investments involve considerable risks, i.e. sewer rehabilitation that appeared to be dimensioned too large, or too small, or to be even unnecessary later on.

In the past, however, uncertainty analysis with regard to sewer system rehabilitation achieved very limited attention. Only recently, uncertainties influencing decisions on sewer rehabilitation are increasingly examined. For example, when it comes to impacts on receiving waters, such as CSOs (combined sewer overflows) and wwtp (wastewater treatment plant) emissions (Reda & Beck 1997 and Willems 2000), water quality criteria such as dissolved oxygen depletion (Beck 1996 and Hauger et al. 2002.) or the assessment of eco-toxicological risks (Novotny & Witte 1997), risk based approaches are used to some extent.

Decision-making on sewer system rehabilitation requires the use of models to predict compliance of the system with performance criteria, i.e. CSO volumes and flooding. The decisions are usually based on a single computation of CSO volumes using a time series of rainfall as system loads. Consequently, uncertainties in knowledge of sewer system dimensions and natural variability in rainfall are ignored. Besides, statistical uncertainties are not taken into account. Uncertainties in sewer system assessment, however, are not restricted to calculated CSO volumes. The effects of CSOs on natural watercourses are just as much uncertain. Quantification of these effects is problematic because the determinative processed are complex and the knowledge on them is very limited (Harremoës & Madsen 1999). Moreover, measurement data on pollution loads from sewers are lacking and existing sewer models are unable to predict the loads (Ashley et al. 1998).

This paper discusses the sensitivity of optimal storage capacity of a sewer system to uncertainties in model input, i.e. model parameters and rainfall input, using probabilistic cost-benefit analysis. Monte Carlo analysis is applied to systematically study uncertainty propagation in a sewer model. For this purpose, model parameters and rainfall input are varied in each run of the Monte Carlo simulation in order to compute CSO volumes. Statistical uncertainty is treated by means of Bayesian estimation. Environmental damage is translated into a cost function. Based on estimated return periods of CSO volumes the sensitivity of decisions to the input uncertainties is evaluated for a discrete and a continuous damage cost function. The optimal storage capacity is determined by optimally balancing the cost of investment and the damage due to CSOs.

2 SEWER SYSTEM ASSESSMENT FOR REHABILITATION PURPOSES

Sewer systems have been designed to protect society from two important hazards: flooding of urban areas during storms and the endangering of public health due to exposure to faecal contamination. Besides, the environmental effects of CSOs should not exceed the carrying capacity of receiving natural watercourses. Overflow structures serve as emergency outlets to natural watercourses when rainfall volumes exceed the system capacity.

In the Netherlands, the approach to deal with the environmental impacts of sewer systems aims at reducing pollution loads by 50% compared to the 1985 situation. This approach has been translated into practical guidelines for calculations (see e.g. Van Mameren & Clemens 1997). The required pollutant reduction is expressed in terms of maximum allowable CSO discharges from a sewer system of 50 kg COD (chemical oxygen demand) discharged to the receiving waters per ha contributing area and per year (CIW 2001).

Assessing sewer overflows requires the use of a continuous rainfall series of a certain length, taking into account the interdependency of storm events and dry periods, thus enabling the calculation of return periods of the effects of medium and heavy storms. The Dutch guidelines prescribe a rainfall series with an interval of 15 minutes as observed during the years 1955–1979 in De Bilt (the Netherlands). Adding pollutant concentrations to calculated CSO volumes would enable the assessment of environmental impacts. However, these calculated pollutant loads are rather uncertain, since the prevailing pollutant concentrations in overflow volumes are unknown and the knowledge of determinative processes is limited.

In case a sewer system does not comply with the discharge limits several interventions can be planned, such as building additional in-sewer storage, enlarging pumping capacity, cleaning of sewers or improving pumping station performance.

3 UNCERTAINTIES IN SEWER SYSTEM ASSESSMENT

We can conclude from the previous that each assessment contains a certain measure of uncertainty because it is based on calculated CSO volumes and their pollutant loads. Therefore, the question, which arises, is which elements in the assessment of sewer systems should be acknowledged as uncertain and to what extent are decisions sensitive to such uncertainties.

Uncertainties can be part of the external inputs, the system itself or the effects of the functioning of the system (see Figure 1). Input comprises a wide range of relevant driving forces, whereas output reflects the interests of parties that depend on the performance of the sewer system. Uncertainties may cause wrong decisions.

3.1 Input uncertainties

Uncertainties in external inputs may result from rainfall measurement errors (Rauch et al. 1998), spatial and temporal variability in rainfall (Schilling & Fuchs 1986, and Lei & Schilling 1996 and Willems 1999), variation in dry weather flow (dwf) due to varying inputs from households (Butler 1991 and Butler et al. 1995) and leaking groundwater (Clemens 2001).

Besides, uncertainty is introduced because the rainfall runoff process is described in a strongly simplified way in the model. Variability of runoff in time and local differences in runoff parameters (initial losses, infiltration, etc) is not taken into account and knowledge of processes is insufficient (Van de Ven 1989 and Clemens 2001).

Finally, hydraulic performance is assessed assuming perfect technical functioning of all objects in a sewer system leading to uncertainty in model assumptions. For example, risk of technical failure of pumping stations, settling of sewer pipes and clogging of culverts are not taken into consideration.

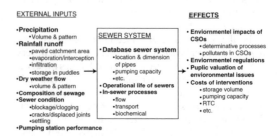

Figure 1. Uncertainties influencing sewer system assessment: external inputs/driving forces, the system itself or effects of the functioning of the system.

3.2 System uncertainties

The data set applied in a sewer model is never entirely perfect. Data errors (geometric structure of the sewer system, catchment area, runoff parameters, etc.) considerably influence calculation results (Price & Osborne 1986 and Clemens 2001).

Sewer models are imperfect because the physical phenomena are not exactly known and some variables of lesser importance are omitted for efficiency reasons. This results in model uncertainty with respect to hydraulics (Beck 1996 and Lei & Schilling 1996) and in-sewer processes determining sewage composition (Ashley et al. 1998). Besides, model uncertainties may stem from estimation (or calibration) of model parameters (Price & Catterson 1997 and Clemens 2001) and numerical calculation errors (Clemens 2001).

In addition, the influence of time dependent sewer deterioration is not accounted for in hydraulic sewer assessments. Except for biogenic sulphuric acid corrosion of sewer pipes there are no reliable models describing sewer deterioration because knowledge of deterioration processed (e.g. clogging, root intrusion, fouling, ingress of soil and longitudinal or radial pipe displacement) is limited. Therefore, assessment of sewer deterioration is performed by means of visual inspection and coding of observations. However, the assumed relationship between observations and actual structural deficiencies is debatable. As a result, prediction of the remaining operational life of sewers highly depends on the limitations of the assessment method.

3.3 Impact uncertainties

It is generally accepted that the quality of natural watercourses deteriorates due to CSOs (see e.g. House et al. 1993). Deterioration comprises water quality changes (dissolved oxygen, polluted sediments, etc.), human health risks and aesthetic contamination (floating waste, algal growth, etc.).

However, the severity is uncertain because CSOs are intermittent loads and their composition strongly varies (Beck 1996). Measurement data of pollution loads from sewers are unavailable and current sewer models are unable to predict them (Ashley et al. 1998). Moreover, translation of uncertain pollutant loads to effects on natural watercourses and their ecology is problematic because the knowledge of water quality processes is rather limited and the resilience of receiving water bodies is uncertain (Shanahan et al. 1998 and Harremoës & Madsen 1999). Therefore, environmental regulations based on available knowledge also incorporate uncertainties.

In addition, the valuation of environmental effects may also give rise to uncertainties in sewer assessments. Some authors claim that environmental effects can be quantitatively expressed in terms of money (see e.g. Crabtree et al. 1999 and Novotny et al. 2001). Others, on the other hand, oppose to this approach and value the effects in a more qualitative way (see e.g. Nijkamp & Van den Berg 1997 and Gilbert & Janssen 1998). An example of the former is the "Contingent Valuation Method" as applied to urban water management by Novotny et al. (2001) which explores the public "willingness to pay" for environmental restoration projects. Authors supporting the more qualitative approach, however, stress that quantitative valuation is unable to take into account uncertain and imprecise information that plays an important role in environmental impact modelling.

3.4 Uncertain future developments

Because of the long operational life of sewers (30–60 years) future developments significantly influence the system performance, not only developments in system input but also in public perception and policy-making. There are a number of examples of infrastructure designs that failed to meet a change in the demand for the goods or services it supplied (Hall 1980). Future developments with respect to sewer assessment include deterioration of sewers, change of regulations, climatic change, change of public perception of the environment and development of receiving water quality.

4 RISK BASED OPTIMISATION OF IN-SEWER STORAGE

As stated before, CSOs may cause serious deterioration of receiving water quality. Therefore, their influence should be reduced. One obvious intervention to reduce effects is to enlarge the in-sewer storage in such a way that CSO emissions diminish.

Currently, however, uncertainty and risk are not taken to account in decision-making on interventions such as enlarging the storage (see e.g. NEN-EN 752-4) As a result wrong decisions are possible because the effectiveness of a proposed intervention (e.g. construction of additional in-sewer storage) depends on the quality of the information supporting the decision-making.

Economic optimisation, as applied for dike design by Van Dantzig (1956), would enable decision-making on additional storage considering uncertainties in sewer system dimensions and natural variability in rainfall. It determines the optimal storage volume by a minimisation of total cost comprising initial investment for construction and cost of environmental damage due to overflows. Expected total costs are discounted over an unbounded horizon assuming that the value of money decreases with time.

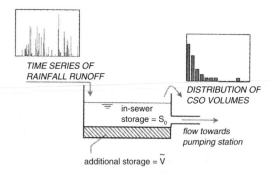

TIME SERIES OF
RAINFALL RUNOFF

DISTRIBUTION OF
CSO VOLUMES

in-sewer
storage = S_0

flow towards
pumping station

additional storage = $\tilde{V}$

Figure 2. Schematic representation of the decision problem: enlarging in-sewer storage capacity in order to prevent CSOs (optimal storage volume = in-sewer storage (S_0) + additional storage ($\tilde{v}$)).

According to Van Dantzig (1956), the decision problem can be formulated as follows (see Figure 2): "Determine the optimal storage volume of the sewer system taking into account the cost of building in-sewer storage, the environmental damage in case of an overflow and the frequency distribution of CSO volumes."

The cost of enlarging the storage capacity is proportional to the volume, i.e. the cost of building an additional m³ diminish with increasing volumes. It can be described as,

$$I = I_0 v^{0.75} \tag{1}$$

where I_0 is investment per m³ storage volume (Euro) and v is storage volume to be built.

The expected cost of damage due to overflows is discounted over an unbounded time horizon,

$$D = D_0 \frac{P_f(v)}{T_{CSO}}\left(\frac{\alpha}{1-\alpha}\right) \tag{2}$$

$$\alpha = 1/(1 + r/100) \tag{3}$$

where D_0 is the cost resulting from an overflow event (Euro), $P_f(v)$ is the probability of failure of sewer system given an overflow event occurs, T_{CSO} is the average return period of overflow events (y), α is the discount factor ($-$) and r is the discount rate (%). Eq. (2) follows from the expected number of overflows exceeding a volume v per year given as,

$$\sum_{k=0}^{\infty} k \cdot \Pr\{\text{number of overflows in period } (0,t] = k\}$$

$$= \sum_{k=0}^{\infty} k \frac{(\lambda \cdot p \cdot t)}{k!} \exp\{-\lambda \cdot p \cdot t\}$$

$$= \lambda \cdot p \cdot t = \lambda \cdot p, \text{ for } t = 1 \tag{4}$$

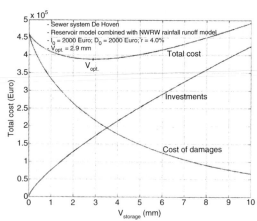

Figure 3. Economic optimisaton of storage volume. (I_0 = 2000 Euro, D_0 = 2000 Euro (constant cost function), r = 4.0%.)

with $\lambda = 1/(T_{CSO})$ being the frequency of overflows, $p = P_f(v)$ being the probability of failure of the sewer system and a Poisson process describing overflow events, the expected costs of failure per year can be discounted over an unbounded time horizon as follows,

$$\frac{P_f(v)}{T_{CSO}} \times [\alpha + \alpha^2 + \alpha^3 + \ldots] = \frac{P_f(v)}{T_{CSO}} \times \frac{\alpha}{1-\alpha} \tag{5}$$

The expected total cost equals,

$$E(TC) = I + D = I_0 v^{0.75} + \frac{D_0 P_f(v)}{T_{CSO}}\left(\frac{\alpha}{1-\alpha}\right). \tag{6}$$

Subsequently, the economic optimum of the storage volume is found by minimising total cost.

Assuming D_0 is constant (see Figure 6), i.e. each overflow event has an immediate effect which is independent of its volume, and CSO volumes are Weibull distributed (Korving 2002), the optimal storage volume can be calculated given the variables I_0, D_0 and r. For example, for the sewer system of "De Hoven", the Netherlands, (see Section 5) with I_0 = 2000 Euro, D_0 = 2000 Euro and r = 4.0% the optimum storage volume becomes 2.9 mm this equals 370 m³ (Figure 3). The cost function appears to be relatively flat around the optimal volume indicating some robustness of the decision to be taken.

5 SENSITIVITY ANALYSIS OPTIMAL STORAGE SEWER SYSTEM "DE HOVEN"

Risk based optimisation is applied to the sewer system of "De Hoven". Storage capacity is optimised taking

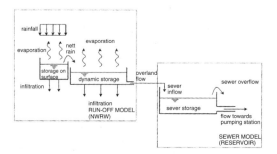

Figure 4. Sewer model comprising a rainfall runoff model (NWRW 4.3) and a reservoir model with external weir and pump.

Table 1. Variations in system parameters (Clemens 2001).

System parameter	μ	σ	CV (%)
S (m^3)	865.00	43.25	5.0
pc (m^3/h)	119.00	5.95	5.0
A (ha)	12.69	0.64	5.0
CC ($m^{0.5}$/s)	1.40	0.35	25.0

into account (1) uncertainties in knowledge of sewer system dimensions, (2) natural variability in rainfall and (3) uncertainties in the cost function describing damage due to CSOs. Uncertainties in both system dimensions and rainfall input are separately modelled by means of Monte Carlo simulation. Subsequently, the expected value of the optimal storage is determined. Finally, the sensitivity of optimal volume to uncertainty in parameters of the cost function describing environmental damage is studied.

The influence of variations in system dimensions (storage capacity, pumping capacity and contributing areas) and natural variability in rainfall on the optimal storage volume of a sewer system is studied by modelling the sewer system of "De Hoven". The catchment "De Hoven" (2200 inhabitants) is situated in the Netherlands on the banks of the river IJssel in the city of Deventer and comprises 12.69 ha paved catchment area. The sewer system (storage 865 m^3 = 6.83 mm) is of the combined type and comprises one pumping station (119 m^3/h = 0.94 mm/h) transporting the sewage to a treatment plant and three CSO structures (external weirs).

5.1 Sewer model

The sewer system is modelled as a reservoir with an external weir and a pump (Figure 4). The rainfall runoff is modelled with the so-called NWRW 4.3 model (Figure 4), the standard rainfall runoff model in the Netherlands. In this model evaporation, infiltration, storage on street surfaces and overland flow are modelled as described in Van Mameren & Clemens (1997). The model input is a 10-year rainfall series (1955–1964) of KNMI (De Bilt, the Netherlands). Dwf is assumed constant (26.4 m^3/h).

5.2 Modelling uncertainties in system dimensions

The influence of variability in four sewer system dimensions is studied: storage volume (S), pumping

capacity (pc), contributing area (A) and overflow coefficient (CC). These dimensions are assumed to be normally distributed and independent. Averages and standard deviations are based on expert judgement (Clemens 2001).

A Monte Carlo simulations of 500 runs is performed. In each run a random value of the model parameters (S, pc, A, CC) is drawn from the probability distribution functions. The parameters values are drawn independently, since their covariances are equal to 0 in the reservoir model. The samples are substituted in the reservoir model.

5.3 Modelling natural variability in rainfall

Natural variability in rainfall is described with a spatial rainfall generator (Willems 2001). The generator has been especially developed for the small spatial scale of urban catchments. Therefore, a detailed description of individual rain cells is required. Spatial distribution of rainfall intensity in an individual rain cell is assumed Gaussian shaped. The generator is based on a model that distinguished rainfall entities at different macroscopic scales, i.e. rain cells, cell clusters, and small and large meso scale areas (rainstorms) (see Figure 5). The model structure is twofold: a physically based part describing individual rain cells and cell clusters and a stochastic part describing the randomness in the sequence of the different rain and storms.

Data from a dense network of rain gauges in Antwerp (Belgium) have been used for generator calibration. The calibrated rain properties comprise moving velocity, moving direction, spatial extent and intensity of rain cells, and inter-arrival times of rain cells and rainstorms.

Based on the above-mentioned properties of the rainfall the spatial rainfall generator has been constructed (see Willems 2001). Subsequently, using a random generator a large number of rain cells can be simulated taking into account their interdependencies. The generated time series of spatial rainfall has the same statistical properties as the data observed with the rain gauge network.

A Monte Carlo simulation of 500 runs is performed. In each run a random time series of rainfall volumes is generated. Although a spatial rainfall field is generated, only the volumes generated at a central

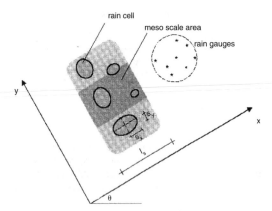

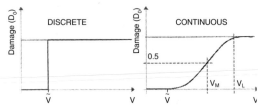

Figure 6. Cost functions describing environmental damage due to CSOs, v is an actual overflow volume, which is a random quantity, and $\tilde{v}$ is the storage volume to be built.

Figure 5. Schematic representation of spatial structure of rainfall including rain cells and meso scale areas or rainstorms (adapted from Willems 2001), where s_x is spatial extent of rain cell in moving direction, s_y is spatial extent of rain cell perpendicular to moving direction, θ is average moving direction of rain cells, l_s is spatial extent of rainstorm (meso scale area) and i is number of rain gauges.

location in Antwerp are used as system loads for the sewer model.

The rainfall generator calibrated for Antwerp can be used in "De Hoven" because the generated rainfall series show considerable agreement with rainfall measurements in Ukkel (Belgium) in terms of IDF (Intensity–Duration–Frequency) relationships (Willems 2001). The Ukkel measurements, for their part, are similar to rainfall data observed in De Bilt (the Netherlands) with respect to IDF relations (Vaes et al. 2002).

5.4 Estimation of distribution type and statistical parameters

The calculated CSO volumes from the Monte Carlo simulations, as described in 5.2 and 5.3, are summed over the individual storm events and analysed statistically. Using Bayes weights the distribution function with the best fit to the CSO data is chosen (see e.g. in Van Noortwijk et al. 2001). Exponential, Rayleigh, normal, lognormal, gamma, Weibull and Gumbel distributions are considered. The Bayesian approach quantifies both inherent and statistical uncertainty.

The Weibull distribution type appears to fit best with the CSO data, has the largest Bayes weight (Korving et al. 2002). Therefore, it is chosen to describe the CSO volumes per storm event statistically. Given the CSO data $v = (v_1, \ldots, v_n)$ the shape parameter a and the scale parameter b of the Weibull distribution,

$$f(v) = \frac{a}{b}\left(\frac{v}{b}\right)^{a-1}\exp\left\{-\left(\frac{v}{b}\right)^a\right\} \tag{7}$$

are estimated using the Maximum Likelihood method. The corresponding survival function of the Weibull distribution is defined as,

$$\bar{F}(v) = \exp\left\{-\left(\frac{v}{b}\right)^a\right\} \tag{8}$$

5.5 Cost functions describing environmental damage

Two types of cost functions are considered to model environmental damage due to overflows: a discrete and a continuous cost function (Figure 6). The discrete case is described in Section 4.

The continuous cost function is a more realistic description of the environmental effects of overflows. If damage is assumed to be a function of the actual overflow volume, the cost function can be described as,

$$D(v) = \begin{cases} D_0\left[1 - \exp\left\{-\left(\dfrac{v-\tilde{v}}{b_1}\right)^{a_1}\right\}\right] & \text{for } v \geq \tilde{v} \\ 0 & \text{for } v < \tilde{v} \end{cases} \tag{9}$$

where v is an actual overflow volume, which is random quantity, $\tilde{v}$ is the storage volume to be built, a_1 and b_1 are parameters on which the shifted Weibull-shaped cost function is dependent. They differ from the parameters a and b of the Weibull distribution describing the inherent uncertainty of overflow volumes.

Since the overflow volume v is a random quantity, in the actual calculation the expectation of Eq. (9) is required,

$$E[D(V)] = \int_{v=\tilde{v}}^{\infty} D_0\left[1 - \exp\left\{-\left(\frac{v-\tilde{v}}{b_1}\right)^{a_1}\right\}\right]f(v)dv$$

$$= \int_{v=\tilde{v}}^{\infty} D_0\left[1 - \exp\left\{-\left(\frac{v-\tilde{v}}{b_1}\right)^{a_1}\right\}\right]$$

$$* \frac{a}{b}\left(\frac{v}{b}\right)^{a-1}\exp\left\{-\left(\frac{v}{b}\right)^a\right\}dv \tag{10}$$

In Eq. (10) the Weibull distribution describing actual CSO volumes has been introduced. Since $v \leqslant \tilde{v}$ for all $\tilde{v}$, the left-truncated Weibull distribution is considered. This distribution can be obtained from the Weibull by conditioning on values larger than $\tilde{v}$,

$$\Pr\{V > v | V > \tilde{v}\} = \frac{\Pr\{V > v \cap V > \tilde{v}\}}{\Pr\{V > \tilde{v}\}} = \frac{\Pr\{V > v\}}{\Pr\{V > \tilde{v}\}}$$

$$= \exp\left\{-\left(\frac{v}{b}\right)^a + \left(\frac{\tilde{v}}{b}\right)^a\right\} \quad (11)$$

Using the left-truncated Weibull distribution, the expected damage (Eq. (10)) can be reformulated as,

$$E[D(V)] = \int_{v=\tilde{v}}^{\infty} D_0 \left[1 - \exp\left\{-\left(\frac{v - \tilde{v}}{b_1}\right)^{a_1}\right\}\right] \frac{a}{b}\left(\frac{v}{b}\right)^{a-1}$$

$$* \exp\left\{-\left(\frac{v}{b}\right)^a + \left(\frac{\tilde{v}}{b}\right)^a\right\} \exp\left\{-\left(\frac{\tilde{v}}{b}\right)^a\right\} dv$$

$$= D_0 P_f(\tilde{v}) \int_{v=\tilde{v}}^{\infty} \left[1 - \exp\left\{-\left(\frac{v - \tilde{v}}{b_1}\right)^{a_1}\right\}\right]$$

$$* \frac{a}{b}\left(\frac{v}{b}\right)^{a-1} \exp\left\{-\left(\frac{v}{b}\right)^a + \left(\frac{\tilde{v}}{b}\right)^a\right\} dv \quad (12)$$

The integral is numerically solved using Monte Carlo simulation. In each Monte Carlo run a value is sampled from the left-truncated Weibull distribution. Subsequently, the expectation of the damage cost can be given by,

$$E[D(V)] \approx D_0 P_f(\tilde{v}) \frac{\sum_{i=1}^{n}\left[1 - \exp\left\{-\left(\frac{v_i - \tilde{v}}{b_1}\right)^{a_1}\right\}\right]}{n} \quad (13)$$

where n is the number of runs in the Monte Carlo simulation and v_i is a sample from left-truncated Weibull distribution.

The parameters a_1 and b_1 in the Weibull-shaped cost function describing environmental damages are estimated as follows. Let v_L be the CSO volume at which the damage cost becomes almost constant, i.e. the damage cost is almost D_0 say, $0.99 * D_0$ (see Figure 6). Let v_M be half of this volume, i.e. $v_M = v_L/2$ (see Figure 6). The cost at v_M is equal to $0.5 * D_0$. The value of v_M determines the steepness of the cost function (Eq. (9)). Chossing v_M completes the description of the Weibull-shaped cost function because it is uniquely described with two percentiles (v_M and $p(v_M)$, v_L and $p(v_L)$). According to Eq. (9) $p(v)$ is defined as,

$$p(v) = 1 - \exp\left\{-\left(\frac{v - \tilde{v}}{b_1}\right)^{a_1}\right\} = \frac{D(v)}{D_0}, \quad \text{for } v \geq \tilde{v} \quad (14)$$

Finally, the expected costs of damage are substituted in the total cost function similar to Eq. (4) and Eq. (5).

5.6 Sensitivity analysis of optimal storage volume

Using both types of cost functions the storage volume is optimised. For each Monte Carlo simulation ($n = 500$) this results in a set of 500 optimal volumes. The set represents the uncertainty in storage volume to be built resulting from either system dimension uncertainty or natural variability in rainfall.

The uncertainty in storage due to input uncertainties can be expressed in the expected value of the optimal volume, which is given by,

$$E(V_{opt}) \approx \frac{1}{n} \sum_{i=1}^{n} V_{opt}(i) \quad (15)$$

where $V_{opt}(i)$ is the optimal storage volume resulting from the ith run and n is the total number of runs in the Monte Carlo simulation ($n = 500$). Besides, the 95% uncertainty interval of optimal volumes (i.e. 95% of calculated values of V_{opt} is within the given boundaries) is calculated to reflect the distribution.

The results based on the discrete damage function are shown in Table 2 (first row), whereas Table 3 (first row) presents the results with respect to the continuous case.

Subsequently, the sensitivity of calculated optimal storage to uncertainty in the assumed cost functions has been tested. For this purpose, the value of D_0 in Eq. (2) is both increased and decreased with 10%. In the continuous damage function the values of D_0 and v_M are changed with 500 € and 2 mm respectively. This implies varying parameters a_1 and b_1 in Eq. (9). The results are presented in the remainder of Table 2 and Table 3.

The results show considerable variation in calculated CSO volumes resulting from uncertainty of sewer dimensions and variability in rainfall. Besides, variation

Table 2. Expected value of optimal volume (in mm) using discrete damage function ($r = 4.0\%$ and $I_0 = 2000$ €). System dimension uncertainty and natural variability in rainfall are separately considered.

	Dimension uncertainty		Variability rainfall	
	$E(V_{opt})$ (mm)	95% interval (mm)	$E(V_{opt})$ (mm)	95% interval (mm)
$D_0 = 2500$ €	4.09	2.25–5.84	1.17	0–2.68
$D_0 = 2250$ €	3.50	1.73–5.23	0.71	0–2.00
$D_0 = 2750$ €	4.63	2.77–6.40	1.67	0–3.32

965

Table 3. Expected value of optimal volume (in mm) using continuous damage function ($r = 4.0\%$ and $I_0 = 2000\,€$). System dimension uncertainty and natural variability in rainfall are separately considered.

	Dimension uncertainty		Variability rainfall	
	$E(V_{opt})$ (mm)	95% interval (mm)	$E(V_{opt})$ (mm)	95% interval (mm)
$D_0 = 7500\,€$ $v_M = 5\,mm$	5.23	3.15–7.01	1.15	0–5.79
$D_0 = 7000\,€$ $v_M = 5\,mm$	4.56	0–6.54	0.67	0–4.81
$D_0 = 8000\,€$ $v_M = 5\,mm$	5.78	3.77–7.46	1.71	0–6.15
$D_0 = 7500\,€$ $v_M = 3\,mm$	7.59	5.86–9.21	4.40	0–8.11
$D_0 = 7500\,€$ $v_M = 7\,mm$	1.76	0–4.80	0.07	0–3.76

increases with increasing return period. Subsequently, variability in CSO volumes is transferred to uncertainty in optimal storage volumes (see 95% uncertainty interval in Table 2 and Table 3).

The cost of damages due to CSOs, however, is difficult to estimate in absence of sufficient data. Therefore, estimation of optimal storage volumes by minimising expected total cost is uncertain. Sensitivity analysis shows that the optimal volume is rather sensitive to changes in the damage cost (D_0) for both cost functions and the steepness of the cost function (v_M) for the continuous case only.

Uncertainty in sewer dimensions results in larger expected optimal volumes but mostly smaller variability within these volumes than rainfall variability. Compared to current practice (CIW 2001) the additional storage should be at least 2.18 mm (given a pumping capacity of 0.7 mm/h plus dwf the storage volume should exceed 7 + 2 mm). With respect to dimension uncertainty the calculated values are slightly larger.

Uncertainty analysis of the cost function is needed to quantify the relative influence of all significant uncertainties in the calculation of the optimal storage volume.

6 CONCLUSIONS

Combining probabilistic optimisation techniques and currently available deterministic sewer models enables assessment of uncertainty in sewer rehabilitation on the basis of calculated CSO volumes.

The paper presents a risk based approach to optimise the required storage volume of a sewer system in order to comply with performance criteria. Sensitivity of optimal storage to uncertainty in model parameters, natural variability in rainfall and uncertainty in type and parameters of the cost function describing environmental damage is taken into account.

The results show that there is considerable variation in optimal storage volume due to the uncertainties in calculated CSO volumes. Besides, uncertainty of sewer dimensions predominantly causes uncertainty. Therefore, investing in more accurate knowledge of the dimensions is worth while. In conclusion, uncertainty analysis of the cost functions is needed in order to allow for the lack of knowledge of environmental damages due to CSOs.

ACKNOWLEDGEMENTS

This paper describes the results of a research, which is financially supported by and carried out in close co-operation with HKV Consultants (Lelystad, the Netherlands) and the RIONED Foundation (Ede, the Netherlands). The authors would like to thank HKV and RIONED for their support. They are also grateful to Patrick Willems (KU Leuven, Belgium) for providing the spatial rainfall generator.

REFERENCES

Ashley, R.M., Hvited-Jacobsen, T. & Bertrand-Krajewski, J.-L. (1998). Quo Vadis sewer process modelling? *Water Science and Technology*, 39(9): 9–22.

Beck, M.B. (1996). Transient polluton events: acute risks to the aquatic environment. *Water Science and Technology*, 33(2): 1–15.

Butler, D., Friedler, E. & Gatt, K. (1995). Characterising the quantity and quality of domestic wastewater inflows. *Water Science and Technology*, 31(7): 13–24.

CIW (2001). *Riooloverstorten Deel 2. Eenduidige basisinspanning. Nadere uitwerking van de definitie van de basisinspanning.* (In Dutch). June 2002. Den Haag: CIW.

Clemens, F.H.L.R. (2001). *Hydrodynamic Models in Urban Drainage: Application and calibration.* PhD thesis. Delft: Technische Universiteit Delft.

Crabtree, B., Hickman, M. & Martin, D. (1999). Integrated water quality and environmental cost-benefit modelling for the management of the River Tame. *Water Science and Technology*, 39(4): 213–220.

Gilbert, A.J. & Janssen, R. (1998). Use of environmental functions to communicate the values of a mangrove ecosystem under different management regimes. *Ecological Economics*, 25: 323–346.

Hall, P. (1980). *Great Planning Disasters.* Weidenfeld & Nicolson: London.

Harremoës, P. & Madsen, H. (1999). Fiction and reality in the modelling world – Balance between simplicity and complexity, calibration and identifiability, verification and falsificationi. *Water Science and Technology*, 39(9): 47–54.

House, M.A., Ellis, J.B., Herricks, E.E., Hvitved-Jacobsen, T., Seager, J., Lijklema, L., Aalderlink, H. & Clifford, I.T. (1993). Urban drainage – Impacts on receiving water quality. *Water Science and Technology*, 27(12): 117–158.

Korving, H., Clemens, F., Van Noortwijk, J. & Van Gelder, P. (2002). Bayesian estimation of return periods of CSO volumes for decision-making in sewer system management. In: *Proc. of 9th Int. Conf. on Urban Drainage. Sept. 2002, Portland, Oregon, USA.*

Lei, J.H. & Schilling, W. (1996). Preliminary uncertainty analysis – A prerequisite for assessing the predictive uncertainty of hydrological models. *Water Science and Technology*, 33(2): 79–90.

Nijkamp, P. & Van den Bergh, J.C.J.M. (1997). New advances in economic modelling and evaluation of environmental issues. *European Journal of Operational Research*, 99: 180–196.

NEN-EN 752-4. *Drian and Sewer Systems Outside Building. Part 4: Hydraulic design and environmental considerations.* March 1998.

Novotny, V., Clark, D., Griffin, R.J. & Booth, D. (2001). Risk based urban watershed management under conflicting objectives. *Water Science and Technology*, 43(5): 69–78.

Novotny, V. & Witte, J.W. (1997). Ascertaining aquatic ecological risks of urban stormwater discharges. *Water Research*, 31(10): 2573–2585.

Parkhi, R. (2002). *Influence of Natural Variability in Rainfall on Decision-Making for Sewer System Management.* MSc thesis. Lelystad: HKV Consultants.

Price, R.K. & Catterson, G.J. (1997). Monitoring and modelling in urban drainage. *Water Science and Technology*, 36(8–9): 283–287.

Price, R.K. & Osborne, M.P. (1986). Verification of sewer simulation models. In: *Proc. of Int. Symp. on Comparison of Urban Drainage Models with Real Catchment Data. UDM'86. Dubrovnik, Yugoslavia*: 99–106.

Rauch, W., Thurner, N. & Harremoës, P. (1998). Required accuracy of rainfall data for integrated urban drainage modeling. *Water Science and Technology*, 37(11): 81–89.

Reda, A.L.L. & Beck, M.B. (1997). Ranking strategies for stormwater management under uncertainty: sensitivity analysis. *Water Science and Technology*, 36(5): 357–371.

Schilling, W. & Fuchs, L. (1986). Errors in stormwater modelling – A quantitative assessment. *Journal of Hydraulic Engineering*, 112(2): 111–123.

Shanahan, P., Henze, M., Koncsos, L., Rauch, W., Reichert, P., Somlyódy, L. & Vanrolleghem, P. (1998) River water quality modelling: II. Problems of the art. *Water Science and Technology*, 38(11): 245–252.

Vaes, G., Clemens, F., Willems, P. & Berlamont, J. (2002). Design rainfall for combined sewer system calculations. Comparison between Fanders and the Netherlands. In: *Proc. of 9th Int. Conf. on Urban Drainage. Sept. 2002, Portland, Oregon*, USA.

Van Dantzig, D. (1956). Economic decision problems for flood prevention. *Econometrica*, 24: 276–287.

Van de Ven, F.H.M. (1989). *Van neerslag tot rioolinloop in vlak gebied.* (In Dutch). Lelystad: Rijkswaterstaat,

Van Mameren, H. & Clemens, F. (1997). Guidelines for hydrodynamic calculations on urban drainage in the Netherlands: overview and principles. *Water Science and Technology*, 36(8–9): 247–252.

Van Noortwijk, J.M., Kalk, H.J., Duits, M.T. & Chbab, E.H. (2001). The use of Bayes factors for model selection in structural reliability. In: *Proc. of 8th Int. Conf. on Structural Safety and Reliability* (ICOSSAR), *June 2001, Newport Beach, California*, USA.

Willems, P. & Berlamont, J. (1999). Probabilistic modelling of sewer system overflow emissions. *Water Science and Technology*, 39(9): 47–54.

Willems, P. (2001). A spatial rainfall generator for small spatial scales. *Journal of Hydrology*, 252: 126–144.

Safety and Reliability – Bedford & van Gelder (eds)
© 2003 Swets & Zeitlinger, Lisse, ISBN 90 5809 551 7

How to approach modelling in a risk analysis

V. Kristensen & T. Aven
Stavanger University College, Stavanger, Norway

ABSTRACT: Models of phenomena like ignition, fire, etc., play an important role in risk analysis. Yet, many risk analysts cannot explain in a satisfactory way the meaning of a model and the various elements of the model. Particular problems are related to the understanding and treatment of uncertainty, and the implications of using assumptions and simplifications in the modelling. Without a proper understanding of the models, it is difficult to select the most appropriate models, and the results of the risk analyses cannot be presented and communicated in a satisfactory manner. In this paper we address these issues, using the phenomenon ignition as an illustrating example. We show that to obtain the necessary insight, it is necessary to define the foundational basis of the analysis, as a model does not mean the same in the classical statistical and the Bayesian framework. The paper has its main focus on a *predictive* Bayesian framework.

1 INTRODUCTION

Models are often used for the assessment of the scenarios that the risk analysts have identified and chosen to be included in a risk analysis. Some models may be used to assess specific phenomena like ignition, fire and explosions, or to assess whether or how often an event like a leak from the process system may occur, while other models may be used to assess, summarize and present an overall picture of the risk.

Within the field of risk analyses, and the phenomena studied within a risk analysis, there is a constant development of new models. The rapid development within the computer technology has made it possible to build and use more and more complex models. This has lead to a situation where the risk analysts spends more and more of their time using complex models on the computer, than on understanding the limitations, strengths and weaknesses of the models and on understanding the theory and assumptions that founds the basis for the models. The time spent on understanding the installation and its systems, identifying new and better technical or operational measures have also been reduced due to the same reason.

It is not our intention to reverse this development. We find the development of new models important and necessary as they simplify and reduce the time spent on complex calculations etc. What we find worrying is that this may lead to a situation where the risk analyst's distance themselves from the understanding of the systems and operations they are set to analyse

and their understanding of the models and tools they are using. When the models are getting more and more complex they may become more and more like a "black box" which the user has no or little knowledge about. This may result in the use of models and tools that are not suitable for the assessment of the systems or operations at interest. Hence, the recommendations and results presented to the decision makers may be more or less useless.

The risk analysts lack of understanding of the models and tools they are using, and also their lack of understanding of what the figures and numbers they are calculating actually mean, is however more influenced or caused by other reasons than the rapid development of computer technology. The most important reason is the foundational basis from which the models and methodology has been developed upon. By the foundational basis we mean the understanding of what a model is, the understanding of risk and uncertainty and the understanding of the various elements that are included in the models. Consider for example the assessment of whether a leak of hydrocarbons may ignite or not. Models used within a "typical" risk analysis, and the analysis itself, are usually based on the classical, statistical and the socalled "probability of frequency" approach, ref. (Apostolakis, G. & Wu, J.S., 1993) and (Kaplan, 1992), which is also referred to as the combined classical and Bayesian approach to risk and uncertainty, ref. (Aven & Kvaløy, 2003). Within these approaches, the risk analysts calculate "best estimates" of fictional quantities, parameters, true probabilities and frequencies. One example is the probability related to whether an ignition might occur in the event of a leak. If sufficient experience data are available, the estimation can be based purely on analysis of the data under the classical, statistical approach. If the data are scarce the probability of frequency approach allows use of engineering judgements to establish subjective uncertainty measures related to what the true value of the probability is. Thus, in both frameworks, uncertainty is related to two levels, i.e. the occurrence of future events and the true value of the probability. In the probability of frequency framework the name probability is used for the subjective probability and frequency for the "objective", relative frequency based probability. Models may also be used to estimate the true ignition probability within the above-mentioned approaches. A Model is viewed as simplified representation of the world, which has uncertainty related to both the true value of the quantities included in the models, and uncertainty related to the models themselves. The model uncertainty may stem from limited understanding of the model as well as deliberate simplification done by the analysts, ref. (Nilsen & Aven, 2003). Although it may be possible to express some of the uncertainty it is difficult to understand the various types of uncertainties and, in practice, difficult to quantify the majority of the uncertainties related to the model and the true value of the quantity assessed. Hence, instead of reducing the uncertainty about the events that may occur, the risk analyses introduce uncertainty.

It seems like this "problem" in a way has become accepted among several risk analysts, their employers, clients and the authorities. The problem may be mentioned in a few sentences in the risk analysis reports, but there seems to be little interest for trying to handle the problem and changing the way risk analysis is done. We will not go into a detailed discussion about the reasons for this in this paper, but only briefly mention three possible reasons that may be of importance. Firstly, the risk analyses have been an important and useful tool that has helped to improve the design and operation of e.g. installations in the North Sea. If the risk analyses are still considered to be good and useful tools by the risk analysts, their employers, clients and the authorities, it may not be so easy to understand why it should be changed. Secondly, or in contrast to the first reason, the risk analysis may not be considered as an important tool by the decision makers, but more as a required documentation or verification of the system or operation at interest. And as they are not important for the decisions, it does not matter whether there are some "problems" related to them. Thirdly, until lately there have been no real alternatives to the classical, statistical and probability of frequency approach. Hence, without having an alternative approach that makes it possible

for the risk analysts to handle and express their uncertainties it is understandable that this issue has not been emphasized.

In this paper we will discuss the problems and issues addressed above, regarding how one should understand models and their various elements, with respect to the way these issues are handled. The framework for our discussion is a predictive, Bayesian approach to risk and uncertainty. This approach will be presented in Section 3.

We will use the phenomenon ignition within a module on an oil installation as an example and basis for our discussion. We will therefore give a brief presentation of the phenomena ignition in Section 2.

In practice, when we are using a model in risk analysis today we are usually using a computer version of the model. However, if not stated otherwise we mean the model itself and not the computer version of a model when we are talking about a model in the remaining part of this paper.

2 IGNITION

The purpose of this section is to give a brief presentation of the phenomena ignition, with respect to the quantities that may affect whether an ignition may occur or not within a module on an oil installation. Readers with knowledge about ignition may therefore skip this section, as it does not include any discussion about the main subject of this paper; the understanding and use of models in risk analysis. The presentation is given in a way that might be considered as inaccurate or imprecise by experts within the subject. However, the purpose is just to present the phenomena to readers with little or no knowledge about it, and has therefore deliberately been given in a simplified way.

In (Drysdale, 1992) ignition is defined as: "the process by which a rapid, exothermic reaction is initiated, which then propagates and causes the material involved undergoing changes, producing temperatures greatly in excess of ambient". It is distinguished between two types of ignition: piloted ignition – in which flaming is initiated in a flammable mixture by a "pilot" such as an electrical spark of an independent flame, and spontaneous ignition – in which flaming develops spontaneously within the mixture (i.e. there are no pilot source causing the ignition).

With respect to risk analyses and the assessment of ignition that may occur on e.g. an oil installation, one are primarily interested in ignition that has an ignition energy that are strong enough to establish a fire or flame or to cause an explosion. A flash or a spark that does not cause a following flame or explosion is therefore here not considered to be an ignition. In order to cause an ignition there has to be a flammable mixture that is exposed to an ignition source. There are several

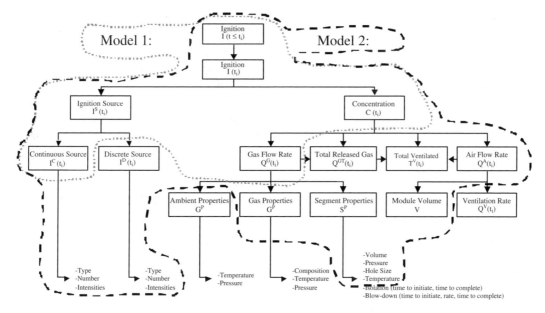

Figure 1. Quantities that may affect whether ignition might occur or not in the event of a leak within a module on an oil platform.

quantities that may affect whether a mixture of vapour/ gas and air is flammable or not. One quantity is the concentration, C, of vapour/gas. On an oil installation there can be several areas or modules that are more or less separated from each other and the surroundings by walls and shielding. Within such an area, which we from now on will refer to as a module, there are several quantities that may affect the concentration, C. Some of these quantities are illustrated on the right side in Figure 1. We have used the sub-notification (t_i) in the figure to illustrate that some of the quantities will vary as a function of time. The figure is however not complete. By looking further into some of the quantities included, similar figures could be made to illustrate the quantities that affect the quantities on the figure on a lower level.

In addition to the concentration of vapour/gas there are several other quantities that may affect the flammability to the mixture that surround the ignition source. Some of these are: other gases in the mixture, direction of flammable propagation, pressure, temperature, turbulence, etc., ref. (SINTEF & Scandpower, 1997).

If the flammable concentration is not ignited spontaneously there has to be an active and strong enough ignition source present. Examples of ignition sources are: Electrical sparks, flames, hot surfaces and mechanical sparks due to violent ruptures. The ignition sources can be divided into discrete and continuous sources. A discrete source can be defined as a source which intensity is affected by time, while a continuous source is not, ref. (SINTEF & Scandpower, 1997).

With respect to the assessment of ignition on an oil installation one also has to consider the size and complexity of such an installation. For instance, there might be several hundred or thousands of potential ignition sources, and there might be local conditions in each cubic meter or decimetre that affects the concentration of vapour/gas in each area.

In Figure 1 we have illustrated the phenomena ignition based on the description given above. We have included two models in the figure; Model 1 and Model 2, which are illustrated by the lines surrounding the quantities included in each of them. The figure and the two models will be used as examples in the discussion throughout the remaining of this paper.

3 THE PREDICTIVE, BAYESIAN APPROACH TO RISK AND UNCERTAINTY

In the predictive, Bayesian approach to risk and risk analysis, focus is on predicting observable quantities, like occurrence, or not, of an accidental event, or the number of fatalities or the magnitude of financial losses in a period of time. Observable quantities express states of the "world", i.e. quantities of the physical reality or the nature; they are unknown at the time of the analysis, but would become (or could become) known in the future. Let Y denote an unknown observable quantity and g the relationship between Y and a vector of unknown observable quantities on a more detailed level, $X = (X_1, X_2, ..., X_n)$, such that:

$$Y = g(X) \tag{1}$$

The function g, which we denote a model, is deterministic. Thus if X were known, Y could be predicted with certainty, given the assumptions underpinning g. However, in most practical cases, such information is not available, and uncertainty related to the predictions has to be taken into account. In the predictive, Bayesian approach uncertainty related to the future values of X is described through an uncertainty distribution $P(X \leq x)$, $x = (x_1, x_2, ..., x_n)$. This uncertainty is epistemic, i.e. a result of lack of knowledge. Then, uncertainty related to Y can be described through a distribution given by:

$$P(Y \leq y) = \int_{\{x:g(x) \leq y\}} dP(X \leq x). \tag{2}$$

The model g is a simplified representation of the world, and it is a tool used, allowing uncertainty of Y to be expressed through the distribution $P(X \leq x)$.

Risk related to Y is described through the entire uncertainty distribution $P(Y \leq y)$. Summarizing measures, such as the mean and the variance, are risk measures that can give more or less information about the risk. The background information used is reported to the decision makers along with the presentation of the risk measures. The models are a part of the background information. All probabilities are conditional on the background information.

An observable quantity represents a state of the world. Thus, it includes also quantities that would have been better described as potentially observable. Consider for example the number of injuries. Provided that a precise definition of an injury has been made, there exists a correct value. The fact that there could be measuring problems in this case – some injuries are not reported – does not change this. The point is that the true number exists and if sufficient resources were made available the number could be found.

The predictive, Bayesian approach is much more than subjective probabilities and Bayesian inference. It relates to fundamental issues like how to express risk and uncertainty, how to understand and use models, which is the main subject of this paper, and how to understand and use parametric distribution classes and parameters in a risk analysis setting. Compared to the more traditional approaches for risk analysis in the engineering community, the predictive, Bayesian approach gives different answers to all these issues.

We define probability as a measure of uncertainty, which means reference to a certain standard such as drawing a ball from an urn. We do not link the

definition to gambling situations with prizes and decision making, as is often done in the literature: When a person says that a probability of occurrence of an event A, is $P(A)$, he implies that he is willing to pay (assuming linear utility for money) $P(A)$ now in exchange for 1\$ later if event A occurs. To us such a definition of probability is not so attractive, it complicates the assignments as it introduces more dimensions; decision making involving money. For a further description of the predictive, Bayesian approach, see (Aven 2003).

4 THE UNDERSTANDING OF MODELS WITHIN THE PREDICTIVE, BAYESIAN APPROACH

4.1 Differences in models representing the same phenomena

There can be several models available for the assessment of one specific phenomenon. With respect to the models themselves, and not the computer versions of the models, the differences can be divided into three main types or categories: different understanding or interpretation of the phenomena, the level of details included and the representation of the quantities included.

As for many phenomena in the world there can be different theories, understanding and/or interpretation of the phenomena included in risk analysis. The understanding and/or the interpretation of the phenomena are decisive for the quantities included in a model and for the functional relationship between the quantities. It may also be decisive with respect to the way the quantities included are represented.

Provided that a set of models is based on the same interpretation of a phenomena, then they may vary with respect to the level of details included. Consider the average concentration of gas/vapour within a module on an oil installation at time t_i, $C(t_i)$, ref. Section 2. This quantity may be (some of) the input given by the user in one model, while a sub-model may be used to represent the quantity in other models. An example of such sub-model is Model 2 in Figure 1, which is given by:

$$C(t_i) = f(Q^G(t_i), Q^{GT}(t_i), T^V(t_i), Q^A(t_i)) \tag{3}$$

where $Q^G(t_i)$ is the gas flow rate at time t_i; $Q^{GT}(t_i)$ is the total amount of released gas at time t_i; $T^V(t_i)$ is the total amount of ventilated air and gas at time t_i and $Q^A(t_i)$ is the air flow rate within the module at time t_i. Further, while the gas flow rate at time t_i, $Q^{GT}(t_i)$, is given as input by the user in Model 1 a sub-model is used to represent this quantity in Model 2.

Whether a model is a main-model or a sub-model depends on the phenomenon being assessed. A model assessing the uncertainty related to the average concentration of vapour/gas, may be a main-model if we explicitly assessed uncertainty related to the concentration, or it may be a sub-model if it is included in a model that assess whether an ignition may occur or not. There are therefore no differences between main-models and sub-models, and the definition of a model given in Section 3 is applicable for them both.

The third category of differences between models is related to the way the observable quantities included are represented. An observable quantity may be represented by either a fixed value or by a probability distribution expressing the uncertainty related to the quantity.

The three categories that we have used to distinguish models assessing the same phenomena in this section are dependent on each other. We will discuss this further in the following sections.

4.2 The relationship between quantities included in a model

One of the quantities that may affect the concentration of gas within a module is the leak rate, ref. Section 2. Let the function h represent a model for the prediction of the release rate (kg/s) at time t_i, $Q(t_i)$. The model, ref. (Det Norske Veritas, 1999), is given by:

$$Q(t_i) = h(Q_R, t) = Q_R \cdot e^{-kt_i} \qquad (4)$$

where Q_R is the initial release at t_0 (kg/s); t is the time after release (s); k is a constant, which is given by:

$$k = \ln(2)/T_{50\%} \qquad (5)$$

where $T_{50\%}$ is the time to reach 50% of the initial release rate (s).

Assume that model h has been chosen for the prediction of the release rate as a function of time for a specific assessment, and that the predicted quantity of Q_R and $T_{50\%}$ are: $q_R^* = 10\,\text{kg/s}$ and $t_{50\%}^* = 120$ seconds. The resulting predicted release rate for this assessment is presented in Figure 2.

Model h is a simplified representation of the amount of gas that may be released in the event of a leak. The amount of gas released at time t_i, is represented by the release rate $Q(t_i)$. One of the assumptions underpinning this model is that the only quantities that affects the $Q(t_i)$ is the initial release rate, Q_R, and the time $T_{50\%}$, at which the release rate has been reduced to 50% of the initial release rate. The assumed relationship between these quantities is given by the model h.

Model h is deterministic. Thus, if Q_0 and $T_{50\%}$ are known, which is the case in the example illustrated in Figure 2, $Q(t_i)$ can be predicted with certainty, given the set of assumptions underpinning the model. It is

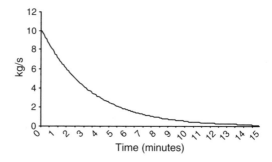

Figure 2.   Prediction of release rate as a function of time.

however in most cases impossible to predict the future values of the quantities with certainty. The uncertainty related to the future values of these quantities must therefore be described. How one should or can describe and handle uncertainty will be discussed in Section 4.3.

A model is a purely deterministic representation of causal mechanisms judge essential by the risk analyst using the model. It provides a framework for mapping uncertainty about the observable quantities of interest, from expression of epistemic uncertainty related to the observable quantities on a lower system level, and does not in itself introduce additional uncertainty. Clearly, a model can be more or less good in describing the world. No model reflect all aspects of the world, per definition it is a model. In this setting the model is merely a tool judge useful for expressing partial knowledge about the phenomenon being assessed.

No mathematical tool or method can be used to calculate whether model h is suitable or appropriate to use in the assessment of the leak rate for a specific case. It is only the (potential) users of the model that can evaluate whether a model is suitable and appropriate to use. By the users we mean the team or group of people that may be involved with the assessment. This may include risk analysts, experts within the phenomena, experts on the oil platform of interest, clients, etc. The evaluation should be based on all the available background at the time of the evaluation. If a model is chosen to be used, then the model itself becomes a part of the background information.

Consider the constant k in model h. Why should the figure $\ln(2) \sim 0.693$ be used in the calculation of the constant k instead of another figure like e.g. 0.5 or 2.0? As there exist no true or correct figure, the answer to this question is the same as it was for the evaluation of whether a model is considered to be suitable or appropriate or not. It can only be done by an evaluation by the potential users of the model. The constant k may therefore be different in different assessments. This does not mean that such figures or quantities are uncertain. What it means is that the users may find different

973

models, or different values to be used to calculate k, as the most suitable or appropriate to be used in their assessment. As we have stated before: The model and the quantities included, is a tool chosen by the users to predict the quantity of interest, by prediction of the observable quantities on a lower level, which are considered to affect the quantity of interest, and by expressing the uncertainty related to these quantities.

This might give the impression that the choice of model, and the value to be used on non-observable quantities included in a model, are more like a lottery or random picking, which certainly would have been the case if the personnel performing the assessment or risk analysis did not have any skills or knowledge about the phenomena assessed. This is of course not the case for the majority of the risk analysis performed. The risk analysts use their knowledge and the available background information in what they consider to be the most appropriate way. Thus, their choices are based on their knowledge and skills and not on lottery or random picking. For several of the phenomena included in a risk analysis there can be a consensus among experts and risk analyst about the understanding and the theories describing the phenomena. Thus, the models used for the assessment of the phenomena may therefore often be the same or similar.

4.3 The representation of uncertainty

The observable quantities included in model h, presented in the section above, are Q_R and $T_{50\%}$. Thus, given the set of assumptions underpinning the model, these two quantities are the only uncertain quantities that affects the release rate at time t_i, $Q(t_i)$. As we stated in the section above, it is in most cases impossible to predict the future values of the quantities with certainty. The uncertainty related to the future values of these quantities must therefore be described.

A probability distribution, for example a known distribution like the Possion distribution or the Normal (Gaussian) distribution, is used to describe the risk analyst's, or risk analysis team's, uncertainty related to the future value of an observable quantity. The probability distribution is not a quantity of the real world, but a description of the risk analyst's uncertainty related to the observable quantity.

The risk analysts should not be forced to express their uncertainty in a specific way, e.g. by only being allowed to use known distributions. Such known distributions should only be used if they are found appropriate and suitable for the representation of the uncertainty. However, some models, and particularly computer versions of models, may force the risk analysts to express their uncertainty in a specific way. If a model is found inappropriate in this respect, it should not be used.

Assume that a risk analysis team has chosen model h for the assessment of the leak rate at time t_i, $Q(t_i)$, for

a specific assessment. Q_R and $T_{50\%}$ are assessed to be independent by the team, and the team's uncertainty related to the quantities Q_R and $T_{50\%}$ are expressed by the distributions $Q_R \sim N(\mu_1, s_1)$ and $T_{50\%} \sim N(\mu_2, s_2)$, with the density functions $f_1(q_R)$ and $f_1(t_{50\%})$. The team's uncertainty related to $Q(t_i)$, could then be expressed by their uncertainty related to Q_R and $T_{50\%}$. This could be done by the use of Monte Carlo simulation or analytically. The analytical expression of the team's uncertainty related to $Q(t_i)$, for any $t_i > 0$, is given by:

$$P(Q(t_i) \leq q) = P(h(Q_R, t) \leq q)$$
$$= \int_{\{q_R, t_{50\%} : h(q_R, t_{50\%}) \leq q\}} f_1(q_R) f_2(t_{50\%}) dq_R dt_{50\%} \qquad (6)$$

If it is chosen to express the uncertainty related to an observable quantity like $Q(t_i)$ by the use of a model there are some issues that have to be addressed. Firstly, one has to decide upon whether one should use Monte Carlo simulation or an analytically approach. The "problem" with analytical approach is that they might get very complex and, in practice, almost impossible to use. Secondly, one has to decide upon whether some of the observable quantities included in the model should be represented by a fixed value. For example: It might be considered appropriate by the risk analysis team in the specific assessment presented above to use a fixed value for $T_{50\%}$ in the representation of their uncertainty related to $Q(t_i)$. The fixed value could be μ_2 or another value found appropriate by the team. Another alternative is to use a set of fixed values, and to represent the uncertainty related to $Q(t_i)$ for each fixed value of $T_{50\%}$. The choice to use a fixed value is a simplification similar to the simplification of choosing a model like model h for the assessment of $Q(t_i)$. They are both choices done in order to simplify the modelling and assessment of $Q(t_i)$. All such simplifications should be described in the set of assumptions underpinning the model.

The risk analysis team's uncertainty related to all observable quantities included in a model should be presented along with the prediction and representation of the uncertainty related to the quantity that the model is used to assess. This applies for both quantities represented by a fixed value and for quantities represented by a probability distribution. Omission of such representation should be described in the assumptions.

4.4 Model assumptions

In order to establish models for the assessment of phenomena like ignition, fire, explosion, etc., there has to be a set of limitations, constrains or assumptions that found the basis for the models. The set of limitations, constrains or assumptions, which we from now on will refer to as the set of assumptions, are used to simplify the complexity of the real world. By simplifying the complexity we refer to the modelling and the assessment of the phenomena and not the complexity of the phenomena itself.

Within the predictive, Bayesian approach there are only two main types of assumptions. The first type is assumptions regarding the relationship between observable quantities, i.e. assumptions describing the models. Assumptions describing the way the quantities on a lower system level are related to the whether an ignition may occur in the event of a leak, ref. Section 2, is an example of such assumptions. Assumptions describing which quantities that are included or excluded in a model are also examples of such assumptions. The second type is assumptions regarding the way the observable quantities included in a model are represented, i.e. whether they are represented by fixed values or probability distributions representing the users uncertainty related to the observable quantities. An example of such assumptions that could be used in a model assessing the phenomena ignition is that the initial leak rate is fixed (e.g. 5 kg/s). Thus, the users uncertainty related to what the initial leak rate may be in the event of a leak (which for some leaks on e.g. an oil installation could be in the range of 0–100 kg/s) is not included in the assessment.

By increasing the amount or extent of the assumptions regarding the phenomena of interest, the complexity may be reduced and the assessment itself may become easier. Consider e.g. the following assumption for the assessment of ignition within a specific module, on a specific oil installation: "in the event of any type of leak there is a constant average concentration of gas within the module in the entire period of interest". How should we understand and use such an assumption? If the purpose of the risk analysis was to assess the uncertainty related to the concentration in the module and to identify the quantities that may have the most effect on the concentration, it could be reasonable to argue that the assumption is unsuitable or inappropriate to use. However, if the purpose of the risk analysis was to assess the toxic effect to the personnel working on the installation in the event of a gas leak, the assumption might be reasonable and appropriate to use.

Whether an assumption or a model is appropriate or reasonable to use cannot be verified or falsified by any tools or methods. It is only through argumentation based on knowledge, skill, and experience about the phenomena, setting (installation, operation, etc.) and the purpose of the analysis that one may justify the choices.

Assumptions may be extremely powerful and decisive for the way phenomena are assessed and the way a risk analysis is performed. For example: the implication of applying the assumption: "No leak will ever occur on the oil installation", would be that there would neither

be any ignition, fire or explosion that could occur in the event of a leak. The assessment of these phenomena would therefore be unnecessary in this case.

Assumptions found the basis for the theories and the models used in risk analyses. They may be decisive for the level of detail included, and the way the quantities included are presented. The applied assumptions should therefore be understood, and argumentation for choosing and using them should be given.

5 DISCUSSION

When selecting models to be used in a risk analysis that is based on a predictive, Bayesian framework to risk and uncertainty, there are several issues or factors that has to be taken into consideration. Although framework conditions, like the available time to perform the analysis and the available resources (personnel, money, etc.), and other factors can be decisive, it should always be the purposes of the risk analysis that is the most important factor when selecting a model. Provided that the main purpose of a risk analysis is to support decision-making, which it always should be, then the decisions that are to be made should be decisive for the choice of models. For example: Some of the decisions to be made during the design of an oil installation are to select between alternative solutions related to e.g. the equipment to be used, the location of equipment or the layout and design of a process module. The models used to assess the phenomena included in the risk analysis, like leak, ignition, fire, explosion, etc. should then be able to reflect the differences, if any, between the alternative solutions being analysed. By reflecting the differences we mean the differences in the risk analysis team's uncertainty related to the future value of the observable quantities included. For a further discussion about selection of models, including a discussion of other issues and factors that should be taken in to consideration, see (Kristensen & Aven, 2002).

The discussion of models and the examples used so far in this paper, has primarily been related to quantities on a detailed level. However, the main concern for the decision-makers when designing an oil installation or planning a complex operation is usually related to whether the personnel working on the installation might be injured or killed, whether there may occur unwanted release of oil or chemicals to the environment or whether there may occur event that may damage the entire installation. Thus, it is the risk analyst's or the risk analysis team's uncertainty related to such quantities that are of most interest and importance to the decision makers when deciding on e.g. the design of an oil installation. To use detailed and complex models may therefore not be necessary in order to represent the risk analyst's or the risk analysis team's uncertainty related to such quantities. It might for example be found appropriate to express the risk analyst's or the risk analysis team's uncertainty related to whether a leak may ignite or not without the use of a model. However, as we stated before, there might be other purposes that justify the use of more or less detailed and complex models.

All models established for the assessment of the phenomena included in a risk analysis should be described and presented in a way that makes it possible to understand their differences, limitations, strengths and weaknesses, etc. A framework for describing models is presented in (Kristensen & Aven, 2002). We find this framework for describing models to be useful in order to select appropriate models to be used in (a specific) risk analysis.

REFERENCES

Apostolakis, G. & Wu, J.S. 1993. In: Barlow RE, Clariotti CA, editors. *Reliability and decision making*. London: Chapman & Hall: p. 311–322.

Aven, T. & Kvaløy, J.T. 2002. *Implementing the Bayesian paradigm in a risk analysis.* Reliability Engineering and System Safety 78(2002): p. 195–201.

Aven, T. 2003. *Foundations of risk analysis.* New York: John Wiley & Sons Ltd, to appear.

Drysdale, D., 1992. *An introduction to fire dynamics.* New York: John Wiley & Sons Ltd.

Kaplan, S. 1992. *Formalisms for handling phenomenological uncertainties: the concepts of probability, frequency, variability, and probability of frequency.* Nucl Technol 1992; 102: p. 137–142.

Kristensen, V. & Aven, T. 2002. *Model description and model selection in the context of risk analysis.* Proceedings of European Safety and Reliability Conference (ESREL) in Lyon.

Nilsen, T. & Aven, T. 2003. Models and model uncertainty in the context of risk analysis. Reliability Engineering and System Safety, to appear.

SINTEF & Scandpower. 1997. *Handbook for fire calculations and fire risk assessment in the process industry.* Lillestrøm: Scandpower A/S.

Det Norske Veritas. 1999. *Papa Explosion 1999, Integrated TDIIM and ProExp prototype.* Technical report No. 99–2046, rev. 01. Oslo: Det Norske Veritas.

Safety and Reliability – Bedford & van Gelder (eds)
© 2003 Swets & Zeitlinger, Lisse, ISBN 90 5809 551 7

Availability modelling of the 3GPP R99 telecommunication networks

Dhananjay Kumar
Nokia Research Center, Farnborough, Hampshire, UK

Akiyoshi Miyabayashi
Nokia Networks, Severo Ochoa, s/n, Edif. de Inst. Universitarios, Parque Tecnológico de Andalucía, Campanillas, Malaga, Spain

Ojala Kari
Nokia Research Center, Nokia Group, Finland

ABSTRACT: In this paper, availability modelling for the 3GPP R99 network architecture is presented. The analytical availability modelling approach has been used. Analytical availability models can be broadly classified into two groups: non-state space (e.g. reliability block diagrams) and state-space models (e.g. Markov models). These models are briefly discussed. The reliability block diagram method is easier to use and has been applied to model the availability of the 3GPP R99 network architecture. An example network with a number of various network elements is considered for availability modelling. The numerical results are presented. The reliability block diagram method is suitable to capture overall availability of a network. However, in order to model features such as software failure, reconfigurations, and fault tolerance, state space modelling approach is needed.

1 INTRODUCTION

During the last decade, rapid technical evolution, market pressures and complexity of telecommunication networks have put a very high demand on performance and availability modelling. The third generation (3G) telecommunication networks are still in development phase. Due to the very high demand on quality of services, very high costs of installation and operations, telecommunication equipment manufacturers have to put special efforts on assuring high reliability and availability of 3G networks. For operators, the compatibility issue, such as inter-workability with legacy networks, is one of the most important factors. Therefore, we have also considered second generation (2G) networks in our modelling.

A variety of measures for network reliability & availability has been proposed. These may be classified broadly into three categories: network survivability, network vulnerability, and network availability. The former two measures are limited to the concept of graph theory, but have penetrated into telecommunication systems. The third one not only concerns the various failure modes of network elements, but also the

degraded performance of a network due to faults in network elements. This paper deals with the network availability models, which can also be extended for network service performance measures.

2 BASIC CONCEPTS OF RELIABILITY AND AVAILABILITY

Recommendation E.800 of the International Telecommunications Union (ITU-T) defines reliability as follows: "The ability of an item to perform a required function under given conditions for a given time interval." In this definition, an item may be a circuit board, a component on a circuit board, a module consisting of several circuit boards, a base transceiver station with several modules, a fiber optic transport system, or a mobile switching center with all of its subtending network elements. The definition includes systems with software. The reliability of an item is a probabilistic measure and is defined mathematically for a time interval "t" by

$$R(t) = P(X > t) = 1 - F(t), \tag{1}$$

where X is time to failure, $F(t)$ is the distribution function of the item's lifetime, and $P(X > t)$ is the probability that the time to failure is greater then time "t".

In practice, Mean Time Between Failure (MTBF) is used as a measure of reliability. The MTBF and reliability are related mathematically as follows:

$$MTBF = \int_0^\infty R(t)dt \qquad (2)$$

Availability is closely related to reliability, and is also defined in the ITU-T Recommendation E.800 as follows: "The ability of an item to be in a state to perform a required function at a given instant of time or at any instant of time within a given time interval, assuming that the external resources, if required, are provided."

The availability at any point t in time, denoted by "$A(t)$", is sometimes called pointwise availability, instantaneous availability, or transient availability. However, in practice, the steady state availability denoted by "A" is often used and is given by

$$A = \frac{MTBF}{MTBF + MTTR}, \qquad (3)$$

where MTTR is "Mean Time To Repair".

An important difference between reliability and availability is that reliability refers to failure-free operation during an interval, while availability refers to failure-free operation at a given instant of time, and usually, at the time when a device or system is first accessed to provide a required function or service. MTBF gives a measure of reliability, while MTBF and MTTR together provide a measure of availability. The availability modelling is more useful as it considers the repair time. More details on these topics can be found in the books by Trivedi, 2002, and by Ross, 1989.

3 AVAILABILITY MODELLING APPROACHES

Approaches to evaluate a system's availability can be broadly categorised as measurement-based and model-based. Measurement-based evaluation is expensive, as it requires building of a real system, taking of measurements and, finally, statistical analysis of the data. Model-based evaluation, on the other hand, is inexpensive and relatively easy to perform. Although easier to perform, model-based availability analysis poses problems such as largeness and complexity of the models, which makes the models difficult to solve.

Model-based availability evaluation can be made through discrete-event simulation, or analytic models, or hybrid models combining simulation and analytic parts. A discrete-event simulation model can depict the detailed system behaviour, as it is essentially a program whose execution simulates the dynamic behaviour of the system and evaluates the required measures. An analytic model consists of a set of equations describing the system's behaviour. The evaluation measures are obtained by solving these equations. In simple cases, closed-form solutions are obtained, but in large real life cases, numerical solutions of the equations are necessary.

The main benefit of discrete-event simulation is the ability to depict detailed system behaviour in the models. The main drawback of discrete-event simulation is the long execution time, particularly when tight confidence bounds are required in the solutions obtained.

Analytical models are more of an abstraction of the real system than a discrete-event simulation model. In general, analytic models tend to be easier to develop and faster to solve than a simulation model. The main drawback is the set of assumptions that are often necessary to make analytic models tractable. Recent advances in model generation and solution techniques as well as computing power make analytic models more attractive. Therefore, availability modelling based only on analytic techniques has been used.

3.1 Analytical models

Analytical models can be broadly classified in two different types of models: non-state space and state space models, depending on the constitutive elements and solution techniques.

3.1.1 Non-state space models
Non-state models do not require the enumeration of the system states. They allow a concise description of the system under study, they can be evaluated efficiently, and a large number of algorithms are available for solving such models (Sun et al., 1999). The main constraint while using these models are the basic assumptions. All failure dependencies must be shown, that is, a component failure leading to a system failure must not make the system operational due to activation of a backup success path. These models cannot represent the system dependency occurring in real systems. Reliability block diagram and fault tree are two non-state space models used for availability prediction.

A reliability block diagram (RBD) is a network diagram of a system that depicts the relationship of the subsystems that are required for successful operation of a system/network. In the RBD, each component/element of the system is represented as a block.

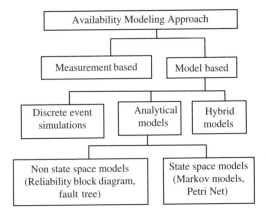

Figure 1. A summary of the availability approaches for telecommunication networks.

The blocks are then connected in series, parallel, or *k* out of *n* configurations based on the operational dependency between the components. If all of the blocks are needed for the system to function, the blocks are connected in series, which means that a failure in any of the blocks leads to a failure in the whole system. If the system can function with at least one block, they are connected in parallel, which means that only simultaneous failures in all of the blocks lead to a system failure.

Fault trees, unlike RBDs, represent the probability of failure approach to availability modelling. It is a pictorial representation of the sequence of events/conditions to be satisfied for a failure to occur. A fault tree uses Boolean gates (such as AND, OR, and *k* of *n* gates) to represent the operational dependency of the system on its components/elements. When a component fails, the corresponding input to the gate becomes TRUE. If any input to an OR gate becomes TRUE, then its output also becomes TRUE. The inputs to an OR gate are those components which are all required to be functioning for the (sub)system to be functioning. The input to an AND gate, on the other hand, are those components, all of which should fail for the (sub)system to fail. Whenever the output of the topmost gate becomes TRUE, the system is considered as failed. To represent situations, where one failure event propagates failure along multiple paths in the fault tree, fault trees can have repeated nodes. Several algorithms for solving fault trees exist (see Luo and Trivedi, 1998; Doyle and Dugan, 1995).

3.1.2 State space models
Many complex systems cannot be represented by non-state space models due to interdependencies and sharing of some of the systems functions, multiple failure modes, fault coverage etc. In reconfigurable systems, the effectiveness of the dynamic reconfiguration process often becomes the critical factor. Normally, Markov modelling or Petri Net modelling techniques are used. The availability modelling approaches are summarised in the following Figure 1.

4 3GPP R99 NETWORK ARCHITECTURE

The 3rd Generation Partnership Project (3GPP) is a collaboration agreement among a number of telecommunications standards bodies. The original scope of the 3GPP was to produce globally applicable Technical Specifications and Technical Reports for a 3rd Generation Mobile System based on evolved GSM core networks and on the radio access technologies that they support. The scope was subsequently amended to include the maintenance and development of the Global System for Mobile communication (GSM) Technical Specifications and Technical Reports, including evolved radio access technologies, such as the General Packet Radio Service (GPRS) and the Enhanced Data rates for GSM Evolution (EDGE). More information can be found at the homepage (www.3GPP.org) of the 3GPP.

The third generation (3G) network architecture contains not only technical evolution, but also expansion to network architecture and services. The Public Land Mobile Network infrastructure is logically divided into a Core Network (CN) and an Access Network (AN) infrastructures. Our availability analysis is based on the 3GPP R99 architecture. The generic 3GPP R99 architecture is shown in Figure 2. The general details of the architecture, mobility, and services can be found in the book by Kaaranen, 2001.

4.1 Access network (AN)

The AN is the radio access network, which is in charge mainly of controlling the use and the integrity of the radio resources and radio channels. Two different types of access are defined: the Base Station Subsystem (BSS) and the Radio Network System (RNS). BSS offers Time Division Multiple Access (TDMA) based radio technology (such as GSM and/or GPRS) whereas RNS offers Wideband Code Division Multiple Access (WCDMA) based radio technology (such as Universal Mobile Telecommunication System, UMTS). BSS and RNS are also called GERAN (GSM/Edge Radio Access Network) and UTRAN (UMTS Terrestrial Radio Access Network), respectively.

Each RNS contains a various number of Nodes B (base stations) and Radio Network Controllers (RNC). In parallel, BSS contains a various number of Base Station Transceivers (BTS) and Base Station Controllers (BSC). The main function of the Node B is to perform the air interface L1 processing (channel coding and interleaving, rate adaptation, spreading etc.).

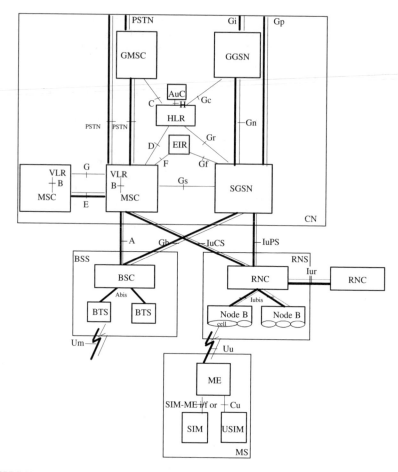

Figure 2. 3GPP R99 architecture. A detailed description can be found in the technical specification document 3GPP TS-TS 23.002, 2002 V3.5.0 (www.3GPP.ORG).

It also performs some basic radio resource management operation. It logically corresponds to the BTS in GERAN. The RNC is the switching and controlling element of the RNS and interfaces with the CN. RNC logically corresponds to the BSC in GERAN. The 3GPP R99 does not define interconnection of Radio Access Network nodes (RAN) to multiple CN nodes. This means that any particular RNC/BSC is connected to a predefined CN node.

4.2 Core network (CN)

The core network is constituted of a Circuit Switched (CS) domain and a Packet Switched (PS) domain. These two domains differ by the way that they support user traffic. The entities specific to the CS domain are Mobile Switching Centre (MSC), Gateway MSC (GMSC) and Visitor Location Register (VLR). On the other hand, the entities specific to the PS domain are Serving GPRS Support Node (SGSN) and Gateway GPRS Support Node (GGSN). The rest of the CN network elements (NEs), e.g. Home Location Register (HLR), Equipment Identify Register (EIR), and Authentication Center (AuC), are common to both CS and PS domains.

In 3GPP R99, the BSC is connected to the MSC (CS domain) via an A interface (as the basic 2G GSM network). In case of PS domain, the BSC is connected to the 2G-SGSN via Gb interface. The RNC is connected to the MSC via Iu-CS interface and to the 3G-SGSN via Iu-PS interface. The HLR is connected to the SGSN, GGSN, MSC and GMSC.

The MSC is responsible for CS connection management, paging and securities activities. It also performs the call control and mobility management. The GMSC is the MSC acting as a bridge between the

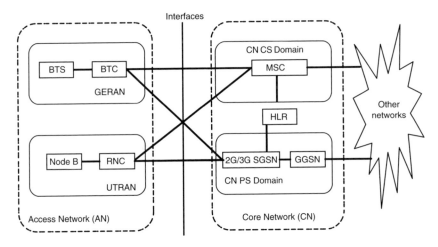

Interfaces

Figure 3. A simplified version of the CS and PS Network scenarios for 3GPP R99 networks.

mobile network and the fixed network. The VLR is a database, which stores information regarding the subscribers under the MSC area (temporarily).

In PS domain, the SGSN functions in 2G and 3G are different. In 2G, protocol conversion, ciphering, compression and mobility management are the major tasks. In 3G, packet processing is the major task. SGSN is in charge of the mobility management, session management, packet transfer, charging, and admission control. GGSN is the interface to external data networks. GGSN has router functionality and charging functions.

The HLR is located in the users home network and it contains subscription data and routing information. The EIR handles security functions related to the verification and identification of the mobile equipment. The AuC handles security functions related to the verification of the identification of the user.

The simplified network architecture to be used in our availability analysis is shown in Figure 3. In order to simplify our availability analysis, the following assumptions related to the network architecture are made:

- A simplified 3GPP R99 network architecture is considered for modelling (see Figure 3). No transport elements or transmission failures have been considered
- SGSN and GGSN belong to the same network.
- VLR is integrated in the MSC
- AuC and EIR are integrated in the HLR
- HLR is connected to only SGSN
- MSC and GMSC are part of the same physical NE called "MSC". In reality the GMSC functions may be supported by another available MSC in the network.

5   AVAILABILITY OF THE 3GPP R99 NETWORK

Availability analysis based on reliability block diagrams (RBD) is the simplest approach. However, it masks the maintainability, reconfiguration, process delays, and resilience aspects of a modern system. Therefore, one should adopt hierarchical modelling approach, where the top level is analyzed using RBD, and all blocks in the RBD are analyzed using state-space models if needed. The following assumptions have been made for availability modelling:

- The 3G R99 architecture is completely represented by the RBD
- There is no dependency between the NEs represented by blocks
- All network elements exist in one of the two states: failed or operational
- No reconfiguration, processing delay or resilience in the network
- No call blockages, drops or handovers.

Similar approaches for the CS and PS networks have been adopted. The availability modelling of a PS is discussed in detail. The generic reliability block diagram for the PS network is shown in Figure 4. This RBD diagram is generic in order to represent any number of NEs. The RBD of the CS network will be similar except that the SGSN is replaced by MSC and there will be no GGSN. The number of network elements in "Block B" may be the same as shown in the "Block A". The symbols and number of elements in the "Block A" are explained below.

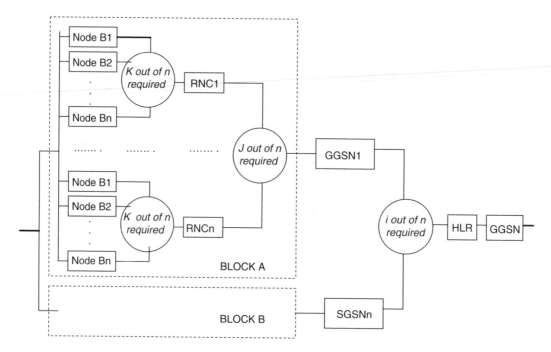

Figure 4. Reliability Block Diagram showing series and parallel operational relationship between different network elements of the 3GPP PS R99 network. A similar diagram will represented the CS network except that Node B is replaced by BTS, RNC by BSC, SGSNs by MSC and there will be no GGSN.

The Node Bs are shown in parallel combination, where k out of n are required for the system to be available ($k = 1, 2, \ldots , n$). If all the "n" Node Bs are required for system to be assumed to be operational, the corresponding block diagram will convert to a series system. Similarly, if j out of n RNCs are required, the block diagram shows parallel combination ($j = 1, 2, \ldots , n$). On the other hand, if all of the RNCs are required for the system to be available, the corresponding block diagram will convert to a series system.

Similarly, if i out of n 2G/3G SGSN are required, the block diagram shows a parallel combination ($i = 1, 2, \ldots , n$). On the other hand, if all of the 2G/3G SGSN are required for the system to be available, the corresponding block diagram will convert to a series system. The mathematical equations for the availability of the subnetworks were derived on the basis of series, parallel or k out of n combinations of the blocks (See Trivedi, 2002; Ross, 1989).

The availability of the GERAN, where k out of n BTS are needed for the GERAN to be considered available, is given by

$$A_{GERAN}=\prod_{j=1}^{n}\left(A_{BSCj}\times\sum_{k=l}^{n}\frac{n!}{l!(n-l)!}A^{l}{}_{BTS_{jk}}(1-A_{BTS_{jk}})^{n-l}\right) \quad (4)$$

The availability of the UTRAN, where k out of n Node Bs are needed for the UTRAN to be considered available, is given by

$$A_{UTRAN}=\prod_{j=1}^{n}\left(A_{RNCj}\times\sum_{k=l}^{n}\frac{n!}{l!(n-l)!}A^{l}{}_{Node B_{jk}}(1-A_{Node B_{jk}})^{n-l}\right) \quad (5)$$

The availability equations for GERAN and UTRAN will be same as the above equations (4) and (5) for CS network.

The availability of the CN for the PS network, where i out of n SGSN are needed for the CN to be considered available, is given by

$$A_{CNPS} = A_{HLR}\times A_{GGSN}\sum_{i=l}^{n}\frac{n!}{l!(n-l)!}A^{l}{}_{SGSN}(1-A_{SGSN})^{n-l} \quad (6)$$

The 2G-service availability for the PS is given by

$$A_{2Gservice} = A_{HLR}\prod_{i=1}^{n}\left[A_{MSC_i}\times A_{GERAN_i}\right] \quad (7)$$

The 3G-service availability for the PS is given by

$$A_{3Gservice} = A_{GGSN}\times A_{HLR}\prod_{i=1}^{n}\left[A_{SGSN_i}\times A_{UTRAN_i}\right] \quad (8)$$

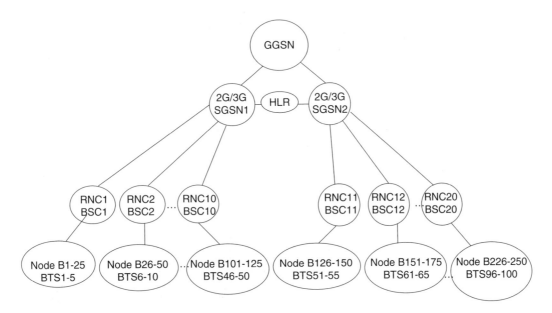

Figure 5. An example of a 3G R99 packet switch network. A group of Node B/BTSs are dedicated to a particular RNC/BSC which itself is dedicated to a particular SGSN. In case of the CS network, the network element SGSN is replaced by an MSC and there is no GGSN.

Table 1. The availability values of the network shown in Figure 5 and its corresponding network for the CS. The values in the parenthesis are down time per year in minutes. It was calculated using the relationship down-time = 8760 × 60 (1-availability) min/year.

Failure criteria	All NEs needed	One Node B/BTS from each group needed and all other NEs needed	80% of BTS/Node B from each group needed and all other NEs needed
Service Availability for the 2G PS network	99.987601% (65.2 min)	99.997600% (12.6 min)	99.997600% (12.6 min)
Service Availability for the 3G PS network	99.947614% (275.3 min)	99.997600% (12.6 min)	99.997600% (12.6 min)
Availability for Maintenance of the PS network	99.935421% (339.4 min)	99.995400% (24.2 min)	99.995400% (24.2 min)
Service Availability for the 2G CS network	99.987701% (64.64 min)	99.997700% (12.09 min)	99.997700% (12.09 min)
Service Availability for the 3G CS network	99.947714% (274.82 min))	99.997700% (12.09 min)	99.997700% (12.09 min)
Availability for Maintenance of the CS network	99.935721% (337.85 min)	99.995700% (22.60 min)	99.995700% (22.60 min)

The availability for maintenance may also be of importance to operators, for their resource planning. The availability for maintenance for the PS network is

$$A_{PS\,maintenancee} = A_{GGSN} \times A_{HLR} \prod_{i=1}^{n}\left[A_{SGSN_i} \times A_{UTRAN_i}\right]$$
$$\prod_{i=1}^{n}\left[A_{MSC_i} \times A_{GERAN_i}\right]$$

(9)

6 AN EXAMPLE OF THE 3GPP R99 NETWORK

Let us consider a network in which the GERAN consists of 100 BTSs and 20 BSCs, and the UTRAN consists of 250 Node Bs and 20 RNCs. In case of the PS CN, 2 SGSNs (2G/3G), one GGSN and one HLR are considered. Similarly, in case of the CS CN, 2 MSCs (2G/3G) and one HLR are considered. The 2G and 3G SGSNs are physically two different NEs while 2G and

3G MSCs are physically in one NE. For the sake of simplicity, it is assumed that each of these NEs has the availability of 99.9999%. The hypothetical network for a PS is represented in the Figure 5. Note that $i = 1, 2$, $j = 1, 2, \ldots, 10$, $k = 1, 2, \ldots, 25$ for the Node Bs, and $K = 1, 2, \ldots, 5$ for the BTSs. The 3G and 2G service availability using the equations in Section 5, are listed in Table 1. In case of the CS network, the availability of the CN was calculated by replacing the SGSN with MSC and eliminating the term for GGSN in Equation (6).

The numerical results show that by allowing at least one of the Node Bs/BTSs to fail, the down time per year decreases many times compared to when no failure is allowed. However, the down time does not decrease if a larger number of the Node Bs/BTSs are allowed to fail. The operator can plan their network based on the subscribers for their 2G and 3G services. The operator can plan maintenance resources based on the maintenance availability.

7 CONCLUSIONS

At high levels, the availability of telecommunication networks can be modelled by using a Reliability Block Diagram (RBD). Other components of the network can be easily modelled as long as their operational relationships do not violate the assumptions of the RBD. In order to include the network features such as resilience, reconfigurations etc., state space models are necessary. It is suggested that the initial availability modelling is carried out by using such simple methods such as RBD. In order to capture the complexity of the network, state space models may be used for each block in the RBD.

Operators may use the results of availability modelling to plan their resources for maintenance and also for revenue purposes from each type of service provided to their subscribers.

ACKNOWLEDGEMENTS

We are thankful to Mr. Veikko Juusola, Mr. Erik Salo, and Mr. Heikki Almay from Nokia/Networks, Dr. Jukka Rantala from Nokia Research Center, and Ms. Raquel Sanchez from University of Malaga for their support during the research work.

REFERENCES

3GPP TS-TS 23.002, 2002, 3rd Generation Partnership Project: Technical specification group services and systems aspects: Network architecture V3.5.0 (Release 1999), www.3GPP.ORG.

Doyle, S.A., and Dugan, J.B., 1995, Dependability assessment using binary decisions diagram, Proceedings of 5th international symposium on fault tolerant computing, pp. 249–258.

ITU-T, 1994, International Telecommunications Union Telecommunication Standardization Sector Recommendations E.800.

Kaaranen, H. et al., 2001, UMTS Networks, UK, Wiley & Sons.

Luo, T., and Trivedi, K.S., 1998, An improved algorithm for coherent-system reliability", IEEE Transactions on Reliability, Vol. 47, No. 1, pp. 73–78, 1998.

Ross, S.M., 1989, Introduction to Probability Models, Academic Press Inc., New York.

Sun, H.R., Cao, Y., Han, J.J., and Trivedi, K.S., 1999, Availability and performance evaluation of automatic protection switching in TDMA wireless systems, Pacific Rim dependence Conference.

Trivedi, K.S., 2002, Probability and Statistics with reliability, queuing, and computer science applications, John Wiley, New York.

Author index

Offshore Reliability Data Handbook 2002

for the offshore engineering business

is now for sale

The 4th edition of the OREDA handbook will give you a unique data source on failure rates, failure mode distribution and repair times for equipment used in the offshore industry. The data can also be used for other applications via a quantification process. Such data are necessary for reliability as well as risk analysis. The reliability, availability, maintenance and safety (RAMS) of offshore exploration and production (E & P) facilities are of considerable concern to employees, companies and authorities. RAMS analyses are carried out to provide a basis for decisions in offshore engineering, fabrication and operations. In order to allow these analyses to be conducted, a source of reliability data is required.

Ordering:
The book costs NOK 4000, (EUR 545,00) plus handling/shipping charges and is sold by
DET NORSKE VERITAS, N-1322 Høvik, Norway
You may order the book by:
http://www.dnv.com/, by e-mail: oreda@dnv.com or by fax No +47 67 57 99 11.

TNO contributes to Public Safety

Do you want to know more?
You are welcome during ESREL 2003 at our stand no. 1.

TNO is a large research organisation whose expertise and research make a substantial contribution to the competitiveness of businesses and organisations, to the economy and to the quality of our society as a whole. One of our aims is to develop our resources into a centre of expertise in the field of Public Safety at the service of the relevant departments and administering services, such as the police. Think about expertise area's such as infrastructural safety, training, risk and accident analysis.

For more information:
www.mv.tno.nl

TNO Public Safety